LIST OF EXAMPLES IN TEXT (*continued*)

Manufacturing Processes for Engineering Materials

SECOND EDITION

Manufacturing Processes for Engineering Materials

SECOND EDITION

SEROPE KALPAKJIAN
Illinois Institute of Technology

ADDISON-WESLEY PUBLISHING COMPANY
Reading, Massachusetts • Menlo Park, California • New York
Don Mills, Ontario • Wokingham, England • Amsterdam • Bonn
Sydney • Singapore • Tokyo • Madrid • San Juan

Sponsoring Editor	*Don Fowley*
Production Supervisor	*Loren Hilgenhurst Stevens*
Design, Editorial, and Production Services	*Quadrata, Inc.*
Illustrator	*Capricorn Design*
Cover Design	*T. A. Philbrook*
Manufacturing Supervisor	*Hugh Crawford*

Library of Congress Cataloging-in-Publication Data

Kalpakjian, Serope
Manufacturing processes for engineering materials / Serope Kalpakjian. — 2nd ed.
p. cm.
Includes bibliographical references and index.
ISBN 0-201-50806-0
1. Manufacturing processes. I. Title.
TS183.K34 1991
670—dc20 90-39438
CIP

Reprinted with corrections April, 1991

2 3 4 5 6 7 8 9 10 HA 9594939291

To **Kuku,**
for all you have done

About the author

Professor Serope Kalpakjian has been teaching and conducting research in manufacturing at the Illinois Institute of Technology since 1963. After graduating (with High Honors) from Robert College, Harvard University, and the Massachusetts Institute of Technology, he joined Cincinnati Milacron, Inc., where he was a research supervisor in charge of metal-forming processes and machine tools. He has published numerous papers in technical journals and is the author of several articles in the *Encyclopedia of Materials Science and Engineering*, the *McGraw-Hill Encyclopedia of Science and Technology*, and the *Marks' Standard Handbook for Mechanical Engineers*. He is a co-founder and former co-editor of the *Journal of Applied Metalworking* (now the *Journal of Materials Shaping Technology*) and an associate editor of the *Journal of Tribology*. He presently serves on the editorial boards of *Manufacturing Review*, the *Journal of Manufacturing Systems*, and the *Journal of Engineering Manufacture*.

Professor Kalpakjian is also the author of four books: *Mechanical Processing of Materials* (Van Nostrand, 1967); *Manufacturing Processes for Engineering Materials* (first edition, Addison-Wesley, 1984), which received the M. Eugene Merchant Manufacturing Textbook Award in 1985; *Lubricants and Lubrication in Metalworking Operations* (with E. S. Nachtman; Marcel Dekker, 1985); and *Manufacturing Engineering and Technology* (Addison-Wesley, 1989), which received the M. Eugene Merchant Manufacturing Textbook Award in 1990. He has also edited *Tool and Die Failures: A Source Book* (ASM, 1982). He is a Fellow of the American Society of Mechanical Engineers and American Society for Metals International, and is a full member of the International Institution for Production Engineering Research (CIRP). He is a founding member and a past president of the North American Manufacturing Research Institution (NAMRI/SME).

Professor Kalpakjian has received a number of awards, including citations by the Forging Industry Educational and Research Foundation for best paper (1966) and the Society of Carbide and Tool Engineers (1977), the "Excellence in Teaching Award" from the Illinois Institute of Technology (1970), the "Centennial Medallion" from the American Society of Mechanical Engineers (1980), and the "Education Award" from the Society of Manufacturing Engineers (1989).

Preface

Manufacturing is a broad activity comprising many subjects; among these are the mechanical and physical behavior and properties of materials, mechanical metallurgy, stress analysis, tribology, primary and secondary processing of these materials, computer controls, machine tools, and economics. In view of these diverse topics, the proper method of teaching manufacturing processes in colleges and universities has been a subject of considerable discussion and evaluation. After many years of neglect, this subject has finally acquired the academic status and significance that it so rightly deserves. Students, as well as the public at large, have come to the realization that without a sound manufacturing base, no nation can hope for economic survival in an increasingly competitive international marketplace.

I firmly believe that, in addition to the applied aspects, manufacturing engineering should be taught with a sound analytical basis, as are many courses in engineering curricula. Manufacturing processes and their control for optimal production have become an increasingly complex subject, and only through analytical approaches can these topics and their interrelationship be truly understood and appreciated. The first edition of this textbook, which appeared in 1984, covered this subject matter from a largely analytical viewpoint, although it also described practical applications with various examples. In a subsequent textbook, *Manufacturing Engineering and Technology*, published in 1989, I presented a less analytical but more comprehensive approach to the subject matter for those who preferred less emphasis on analysis but a broader coverage of this topic touching on all relevant applied and industrial aspects.

The favorable reception and wide adoption of the first edition of *Manufacturing Processes for Engineering Materials*, the many valuable suggestions received from colleagues, and the more recent advances and trends in the analysis of manufacturing processes all led me to prepare this revised edition. Although the book has basically the same format, balance, and introductory nature, it has certain additional features:

- The emphasis on analysis of manufacturing processes has been increased.
- The number of chapters has been increased from thirteen to fifteen to include and describe more broadly the fundamentals of computer-integrated manufacturing and the economics and competitive aspects of manufacturing processes.
- The number of numerical example problems has been increased to help the student better understand the significance of the analytical approach to the subject matter.
- The number of questions and problems at the end of each chapter has been greatly increased to better reflect the broad range of topics covered in the text and allow instructors to have a wider selection of questions to assign as homework.
- Every attempt is made to present the subject matter with a balanced coverage of the relevant fundamentals, analytical approaches, and applications to enable the student to properly assess up-to-date capabilities and potentials as well as limitations of manufacturing processes.

The text has been written for students in mechanical, manufacturing, industrial, chemical, and metallurgical and materials engineering programs. It is hoped that in reading and studying this text, students will appreciate the importance and relevance of manufacturing engineering as a subject that is as exciting and challenging as any other engineering discipline.

Acknowledgments

It is a pleasure to acknowledge the help of many people who assisted me in the preparation and publication of this second edition.

As always, I am very grateful to Donald A. Fowley, Jr., Senior Editor at Addison-Wesley Publishing Company, for his enthusiastic and very knowledgeable help. No author could possibly hope to work with a better and more dedicated editor. Many thanks to Laurie McGuire, Assistant Editor, and Loren Hilgenhurst Stevens, Production Supervisor, for all their help and diligence in the preparation of the manuscript. As usual, Martha Morong and Geri Davis of Quadrata, Inc., were most helpful and cooperative in producing this book.

Many thanks to my colleagues at the Illinois Institute of Technology and at other institutions for their help and constructive suggestions: My chairman, H. M.

Nagib of the Mechanical and Aerospace Engineering Department, for his constant encouragement and support, and K. J. Weinmann (Michigan Technological University), D. L. Bourell (University of Texas at Austin), S. A. Spiewak (University of Wisconsin–Madison), D. A. Lucca (Oklahoma State University), M. C. Shaw and J. Shah (Arizona State University), A. Henkin (Iowa State University), J. S. Colton (Georgia Institute of Technology), J. S. Kallend (IIT), and D. Durham (University of Vermont). The assistance of my present and former students at IIT is also acknowledged, especially R. J. Rogalla, X. Z. Li, and S. R. Schmid. In addition, I appreciate the help of many organizations who supplied me with numerous illustrations and photographs.

Finally, many thanks to my family, Jean, Claire, and Kent, for their patience during the writing and production of this book.

Chicago, Illinois S. K.

Contents

2 FUNDAMENTALS OF THE MECHANICAL BEHAVIOR OF MATERIALS 29

3 STRUCTURE AND MANUFACTURING PROPERTIES OF METALS 109

4 SURFACES, DIMENSIONAL CHARACTERISTICS, INSPECTION, AND QUALITY ASSURANCE 159

5 CASTING PROCESSES 231

6 BULK DEFORMATION PROCESSES 313

7 SHEET-METAL FORMING PROCESSES 403

8 MATERIAL-REMOVAL PROCESSES: CUTTING 473

9 MATERIAL-REMOVAL PROCESSES: ABRASIVE, CHEMICAL, ELECTRICAL, AND HIGH-ENERGY BEAMS 583

Manufacturing Processes for Engineering Materials

SECOND EDITION

1

General Introduction

1.1 What Is Manufacturing?

As you begin to read this Introduction, take a few moments and inspect the different objects around you: your watch, chair, stapler, pencil, calculator, telephone, and light fixtures. You will soon realize that all these objects had a different shape at one time. You could not find them in nature as they appear in your room. They have been transformed from various raw materials and assembled into the shapes that you now see. A paper clip, for example, was once a piece of wire. The wire was once a piece of metal obtained from ores.

Some objects are made of one part, such as nails, bolts, wire or plastic coat hangers, metal brackets, and forks. However, most objects—aircraft engines (Fig. 1.1), ballpoint pens, toasters, bicycles, computers and thousands more—consist of several parts made from a variety of materials. A typical automobile, for example, consists of about 15,000 parts, and a C-5A transport plane is made of more than

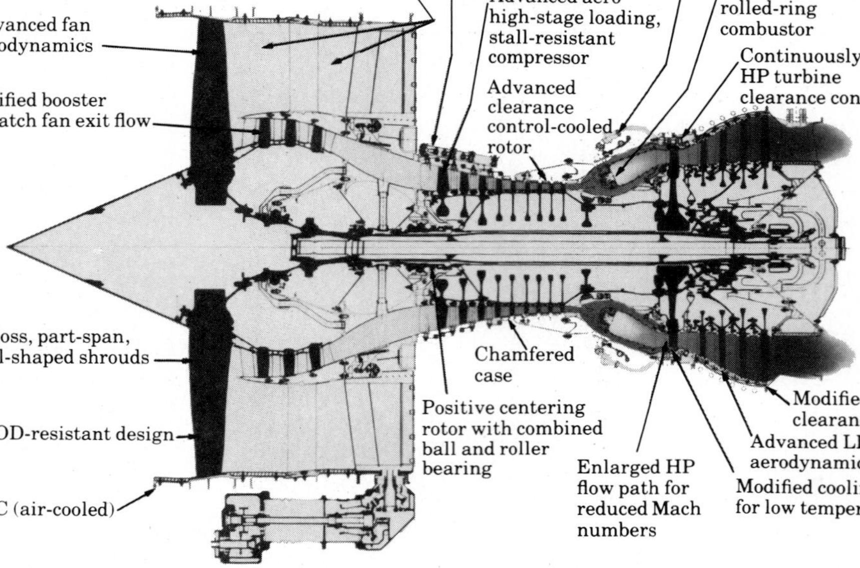

FADEC—Full authority digital electronic control, FOD—Foreign object damage, HP—High pressure, LP—Low pressure, VSV—Variable stator vane

FIGURE 1.1
Cross-sectional view of a jet engine showing various components. Many of the materials used in this engine must maintain their strength and resist oxidation at high temperatures. *Source:* Courtesy of General Electric Company.

4,000,000 parts. All are made by various processes that we call manufacturing. *Manufacturing*, in its broadest sense, is the process of converting raw materials into products. It encompasses the design and production of goods, using various production methods and techniques.

Manufacturing is the backbone of any industrialized nation. Its importance is emphasized by the fact that, as an economic activity, it comprises approximately one-third of the value of all goods and services produced in industrialized nations. The level of manufacturing activity is directly related to the economic health of a country. Generally, the higher the level of manufacturing activity in a country, the higher is the standard of living of its people.

Manufacturing also involves activities in which the manufactured product is itself used to make other products. Examples are large presses to form sheet metal for car bodies, metalworking machinery used to make parts for other products, and sewing machines for making clothing. An equally important aspect of manufacturing activities is servicing and maintaining this machinery during its useful life.

The word *manufacturing* is derived from the Latin *manu factus*, meaning made by hand. The word manufacture first appeared in 1567, and the word manufacturing appeared in 1683. In the modern sense, manufacturing involves making products from raw materials by various processes, machinery, and operations, following a well-organized plan for each activity required. The word *product* means something that is produced, and the words product and production first appeared sometime during the fifteenth century. The word production is often used interchangeably with the word manufacturing. Whereas *manufacturing engineering* is the term used widely in the United States to describe this area of industrial activity, the equivalent term in Europe and Japan is *production engineering*.

Because a manufactured item has undergone a number of changes in which a piece of raw material has become a useful product, it has a *value*—defined as monetary worth or marketable price. For example, as the raw material for ceramics, clay has a certain value as mined. When the clay is used to make a ceramic dinner plate, cutting tool, or electrical insulator, value is added to the clay. Similarly, a wire coathanger or a nail has a value over and above the cost of a piece of wire. Thus manufacturing has the important function of adding value.

Manufacturing is generally a complex activity (Fig. 1.2), involving people who have a broad range of disciplines and skills and a wide variety of machinery,

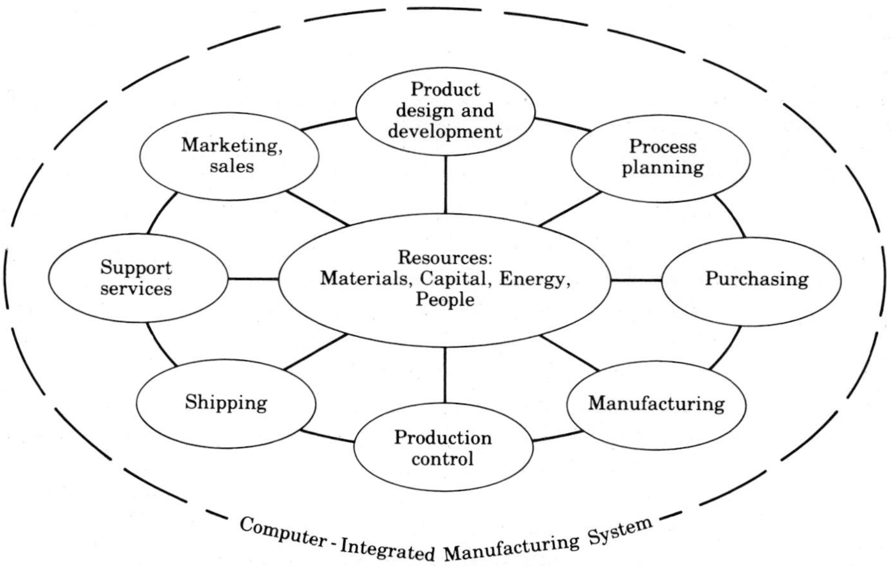

FIGURE 1.2
Chart showing relationships among many activities in manufacturing, involving materials, processes, machinery, and people.

equipment, and tooling with various levels of automation, including computers, robots, and material-handling equipment. Manufacturing activities must be responsive to several demands and trends:

- A product must fully meet design requirements and specifications.
- A product must be manufactured by the most economical methods in order to minimize costs.
- Quality must be built into the product at each stage, from design to assembly, rather than relying on quality testing after the product is made.
- In a highly competitive environment, production methods must be sufficiently flexible so as to respond to changing market demands, types of products, production rates, production quantities, and on-time delivery to the customer.
- New developments in materials, production methods, and computer integration of both technological and managerial activities in a manufacturing organization must constantly be evaluated with a view to their timely and economic implementation.
- Manufacturing activities must be viewed as a large system, each part of which is interrelated to others. Such systems can be modeled in order to study the effect of factors such as changes in market demands, product design, material and various other costs, and production methods on product quality and cost.
- The manufacturing organization must constantly strive for higher *productivity*, defined as the optimum use of all its resources: materials, machines, energy, capital, labor, and technology. Output per employee per hour in all phases must be maximized.

1.1.1 Brief history of manufacturing

Manufacturing dates back to 5000–4000 B.C. and began with the production of various articles made of wood, ceramic, stone, and metal (Table 1.1). The materials and processes first used to shape those products by casting and hammering have been gradually developed over the centuries, using new materials and more complex operations, at increasing rates of production and higher levels of quality.

The first materials used for making household utensils and ornamental objects included metals such as gold, copper, and iron. These were followed by silver, lead, tin, bronze, and brass. The production of steel in about A.D. 600–800 was a major development. Since then a wide variety of ferrous and nonferrous metals have been developed. Today, the materials used in advanced products such as computers and supersonic aircraft include engineered or tailor-made materials with unique properties, such as ceramics, reinforced plastics, composite materials, and specially alloyed metals.

Until the Industrial Revolution, which began in England in the 1750s, goods had been produced in batches, with heavy reliance on manual labor in all aspects of production. Modern mechanization began in England and Europe with the development of textile machinery and machine tools for cutting metals. This technology soon moved to the United States where it was developed further, including the important advance of designing, making, and using *interchangeable* parts. Prior to the introduction of interchangeable parts, a great deal of hand fitting was necessary because no two parts were made exactly alike. We now take for granted that we can replace a broken bolt of a certain size with an identical one purchased years later from a local hardware store.

Further developments soon followed, resulting in numerous products that we can't imagine being without because they are so common, such as typewriters, sewing machines, and engines. Since the early 1940s, major milestones have been established in all aspects of manufacturing. Note from Table 1.1 the progress made during the past 100 years, and especially the last two decades with the advent of the computer age, as compared to the period of 4000 to 1 B.C..

Although the Romans had factories for mass producing glassware, manufacturing methods were at first very primitive and generally very slow, with much manpower involved in handling parts and running the machinery. Today, with the help of *computer-controlled* manufacturing systems, production methods have been advanced to such an extent that, for example, aluminum beverage cans are manufactured at a rate of 10 cans per second. Screws and bolts are made at rates of thousands per minute.

Manufacturing may produce *discrete products*, meaning individual parts or part pieces, or *continuous products*. Nails, gears, steel balls, beverage cans, and engine blocks are examples of discrete parts, even though they are mass produced at high rates. On the other hand, a spool of wire, metal or plastic sheet, tubes, hose, and pipe are continuous products, which may be cut into individual pieces and thus become discrete parts.

1.1.2 Manufacturing a product

Let us now examine briefly the thought processes and procedures involved in designing and manufacturing common products. We identify the important factors involved and show how intimately designing and manufacturing are interrelated through two simple examples and by brief reference to a third, more complex example.

• Example 1.1: Paper clip.

Assume that you are asked to design and produce ordinary paper clips. What type of material would you choose to make this product? Does it have to be metallic or can it be nonmetallic, such as plastic? If you choose metal, what kind of metal? If the material that you start with is wire, what should be its diameter? Should it even be round or should it have some other cross-section? Is the wire's surface finish important and, if so, what should be its roughness? How would you shape a piece of wire into a paper clip? Would you shape it by hand on a simple fixture and if not,

TABLE 1.1
HISTORICAL DEVELOPMENT OF MATERIALS AND MANUFACTURING PROCESSES (DATES ARE APPROXIMATE)

	PERIOD	METALS AND CASTING	FORMING PROCESSES
Egypt: ~3100 B.C. to ~300 B.C.; Greece: ~1100 B.C. to ~146 B.C.; Roman empire: ~500 B.C. to A.D. 476; Middle ages: ~476 to 1492; Renaissance: 14th to 16th centuries	Before 4000 B.C.	Gold, copper, meteoritic iron	Hammering
	4000–3000 B.C.	Copper casting, stone and metal molds, lost wax process, silver, lead, tin, bronze	Stamping, jewelry
	3000–2000 B.C.	Bronze casting	Wire by cutting sheet and drawing; gold leaf
	2000–1000 B.C.	Wrought iron, brass	
	1000–1 B.C.	Cast iron, cast steel	Stamping of coins
	A.D. 1–1000	Zinc, steel	Armor, coining, forging, steel swords
	1000–1500	Blast furnace, type metals, casting of bells, pewter	Wire drawing, gold and silver smith work
	1500–1600	Cast iron cannon, tinplate	Water power for metalworking, rolling mill for coinage strips
Industrial revolution: ~1750 to 1850	1600–1700	Permanent mold casting, brass from copper and metallic zinc	Rolling (lead, gold, silver), shape rolling (lead)
	1700–1800	Malleable cast iron, crucible steel	Extrusion (lead pipe), deep drawing, rolling (iron bars and rods)
	1800–1900	Centrifugal casting, Bessemer process, electrolytic aluminum, nickel steels, babbitt, galvanized steel, powder metallurgy, tungsten steel, open-hearth steel	Steam hammer, steel rolling, seamless tube piercing, steel rail rolling, continuous rolling, electroplating
WWI	1900–1920		Tube rolling, hot extrusion
WWII	1920–1940	Die casting	Tungsten wire from powder
	1940–1950	Lost wax for engineering parts	Extrusion (steel), swaging, powder metals for engineering parts
Space age	1950–1960	Ceramic mold, nodular iron, semiconductors, continuous casting	Cold extrusion (steel), explosive forming, thermomechanical treatment
	1960–1970	Squeeze casting, single crystal turbine blades	Hydrostatic extrusion; electroforming
	1970–1980s	Compacted graphite, vacuum casting, organically bonded sand, automation of molding and pouring, large aluminum castings for aircraft structures, rapid solidification technology	Precision forging, isothermal forging, superplastic forming, die design by analytical methods, net-shape forming

Source: After J. A. Schey, C. S. Smith, R. F. Tylecote, T. K. Derry, T. I. Williams, and S. Kalpakjian.

(*continued*)

TABLE 1.1 (*continued*)

JOINING PROCESSES	TOOLS, TOOL MATERIALS, AND MACHINING	NONMETALLIC MATERIALS
	Tools of stone, flint, wood, bone, ivory, composite tools	Earthenware, glazing, natural fibers
Soldering (Cu-Au, Cu-Pb, Pb-Sn)	Corundum	
Riveting, brazing	Hoe making, hammered axes, tools for ironmaking and carpentry	Glass beads, potter's wheel, glass vessels
Forge welding of iron and steel, gluing	Improved chisels, saws, files, woodworking lathes	Glass pressing and blowing
	Etching of armor	Venetian glass
	Sandpaper, windmill driven saw	Crystal glass
	Boring, turning, screw cutting lathe, drill press	Porcelain
	Hand lathe (wood)	Cast plate glass, flint glass
	Shaping, milling, copying lathe for gunstocks; turret lathe, universal milling machine, vitrified grinding wheel	Window glass from slit cylinder, light bulb, vulcanization, rubber processing, polyester, styrene, celluloid, rubber extrusion, molding
Oxyacetylene; arc, electrical resistance, and thermit welding	Geared lathe, automatic screw machine, hobbing, high-speed steel tools, aluminum oxide and silicon carbide (synthetic)	Automatic bottle making, Bakelite, borosilicate glass
Coated electrodes	Tungsten carbide, mass production, transfer machines	Development of plastics, casting, molding, PVC, cellulose acetate, polyethylene, glass fibers
Submerged arc welding		Acrylics, synthetic, rubber, epoxies, photosensitive glass
Gas metal–arc, gas tungsten–arc, and electroslag welding, explosive welding	Electrical and chemical machining, automatic control	ABS, silicones, fluorocarbons, polyurethane, float glass, tempered glass, glass ceramics
Plasma arc and electron beam, adhesive bonding	Titanium carbide, synthetic diamond, numerical control	Acetals, polycarbonates, cold forming of plastics, reinforced plastics, filament winding
Laser beam, diffusion bonding (also combined with superplastic forming)	Cubic boron nitride, coated tools, computer integrated manufacturing, adaptive control, industrial robots, flexible manufacturing systems, unmanned factory	Adhesives, composite materials, optical fibers, structural ceramics, ceramic components for automotive and aerospace engines

what kind of machine would you design or purchase to make paper clips? If, as the owner of a company, you were given an order of 100 clips versus a million clips, would your approach to this manufacturing problem be different?

The paper clip must meet its basic functional requirement: to hold pieces of paper together with sufficient clamping force so that the papers do not slip away from each other. It must be designed properly, including its shape and size. The design process is based partly on our knowledge of strength of materials and mechanics of solids, dealing with the stresses and strains involved in the manufacturing and normal use of the clip.

The material selected for a paper clip must have certain stiffness and strength. For example, if the stiffness (a measure of how much it deflects under a given force) is too high, a great deal of force may be required to open the clip, just as a stiff spring requires a greater force to stretch or compress it than does a softer spring. If the material is not sufficiently stiff, the clip will not exert enough clamping force on the papers. Also, if the yield stress of the material (the stress required to cause permanent deformation) is too low, the clip will bend permanently during its normal use and will be difficult to reuse. These factors also depend on the diameter of the wire and the design of the clip.

Included in the design process are considerations such as style, appearance, and surface finish or texture of the clip. Note, for example, that some clips have serrated surfaces for better clamping. After finalizing the design, a suitable material has to be selected. Material selection requires a knowledge of the *function* and *service requirements* of the product and the materials that, preferably, are available commercially to fulfill these requirements at the lowest possible cost. The selection of the material also involves considerations of its corrosion resistance, since the clip is handled often and is subjected to moisture and other environmental attack. Note, for example, the rust marks left by paper clips on documents stored in files for a long period of time.

Many questions concerning production of the clips must be asked. Will the material selected be able to undergo bending during manufacturing without cracking or breaking? Can the wire be easily cut from a long piece without causing excessive wear on the tooling? Will the cutting process produce a smooth edge on the wire, or will it leave a burr (a sharp edge)? A burr is undesirable in the use of paper clips since it may tear the paper or even cut the user's finger. Finally, what is the most economical method of manufacturing this part at the desired production rate, so that it can be competitive in the national and international marketplace and the manufacturer can make a profit? A suitable manufacturing method, tools, machinery, and related equipment must then be selected to shape the wire into a paper clip.

• Example 1.2: Bicycle.

Consider the design, material selection, and processing methods of the major components of a bicycle (Fig. E1.1). Some of the important factors are safety, strength, corrosion resistance, weight, and appearance. Note that a number of

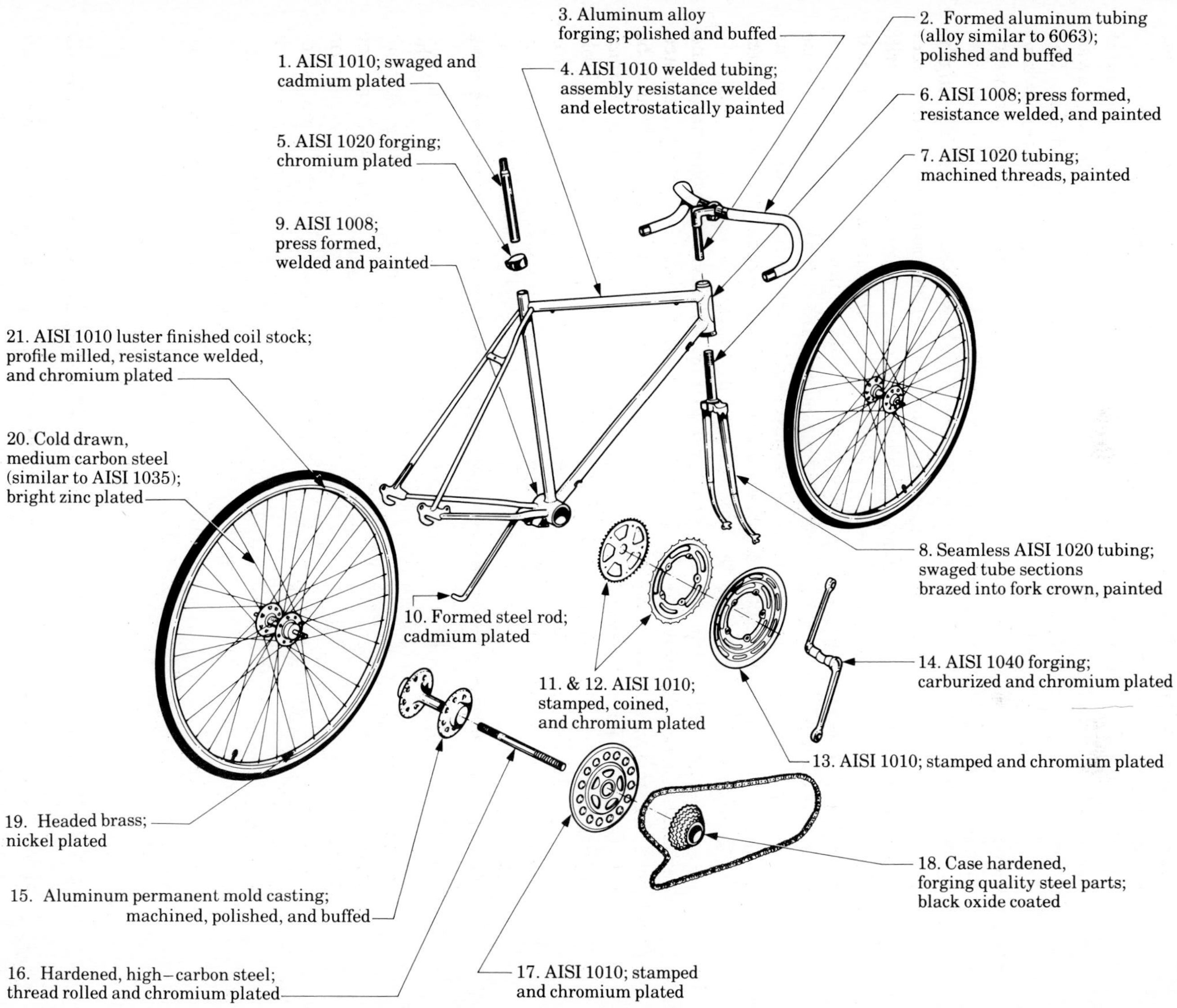

FIGURE E1.1
Components of a bicycle (Schwinn Continental) showing materials and processes used in manufacturing this bicycle. Note that there are about a dozen materials and more than a dozen individual manufacturing processes involved. *Source: Metal Progress,* July 1973, p. 64, © ASM International.

materials have been selected for the various components, although these are mostly plain carbon steels because of their relatively low cost and sufficient strength. There are about a dozen manufacturing processes involved in producing this bicycle, including forging, machining, casting, welding, plating, and polishing. Although not shown in Fig. E1.1, each part is also specified by the design engineer with a range of dimensions, tolerances, and surface finish.

The handlebar (item 2 in Fig. E1.1) is made of aluminum tubing, which is shaped and then polished and buffed for appearance. From the many types of aluminum available, the manufacturer selected the 6063 alloy as the optimal material for this application. The stem of the handlebar (item 3) is made of an aluminum forging. Malleable iron and aluminum castings were considered for this part, but these materials were unacceptable because of weight and safety reasons. Handlebars tend to be misused; the bicycle may be dropped on a hard surface or the handlebars may have to support a passenger.

The stem and the front fork (item 8) on a bicycle are regarded by the designers as the most important components of a bicycle from a safety standpoint. These highly stressed parts must be designed and the materials selected carefully without sacrificing the bicycle's weight. For light racing bicycles, just as for aircraft and aerospace applications, strength-to-weight and stiffness-to-weight ratios are important factors in material selection. The spokes of the wheels of the bicycle are made of steel wire with a specified minimum tensile strength of about 1000 MPa (150,000 psi). They are bright zinc electroplated for corrosion resistance and appearance.

• Example 1.3: Jet engines.

Compared to the preceding two examples, designing and manufacturing a jet engine is a much more challenging and demanding task (Fig. 1.1). We presented this figure merely to point out the complexity of this important product, which—depending on its size and capacity—costs up to a few million dollars. Selection of materials and processes, inspection and testing, and quality control are particularly critical for this engine because of its application. Failure of any of the major components in this engine can be catastrophic in terms of loss of life and property.

These diverse examples show that each of the manufacturing operations requires thought processes common to making all products. In the following sections we present an overview of the important interrelationships among product design, material selection, and manufacturing processes. The end result should be a product that meets high quality standards and expected service requirements—and that is also economical to produce.

1.2 Design for Manufacture

Because a particular design is eventually made into a product, design and manufacturing must be intimately interrelated. Design and manufacturing should never be viewed as separate disciplines or activities. Each part or component of a product must be designed so that it not only meets design requirements and specifications, but also can be manufactured economically and with relative ease. This approach improves productivity and allows a manufacturer to remain competitive.

This broad view has now become recognized as the area of *design for manufacture*. Also known as *simultaneous* or *concurrent engineering*, it is a comprehensive approach to production of goods and integrates the design process with materials, manufacturing methods, process planning, assembly, testing, and quality control. Effectively implementing design for manufacture requires that designers have a fundamental understanding of the characteristics, capabilities, and limitations of materials, manufacturing processes, and related operations, machinery, and equipment. This knowledge includes characteristics such as variability in machine performance, surface finish and dimensional accuracy of the workpiece, processing time, and the effect of processing method on part quality.

Designers must be able to assess the impact of design modifications on manufacturing process selection, assembly, inspection, tools and dies, and product cost. Establishing quantitative relationships is essential in order to optimize the design for ease of manufacturing and assembly at minimum product cost (also called *producibility*). Computer-aided design, manufacturing, and process planning techniques, using powerful computer programs, have become indispensable to such analysis. New developments include expert systems (see the section, "Automation and the Impact of Computers on Manufacturing"), which have optimization capabilities, thus expediting the traditional iterative process in design optimization.

1.3 The Design Process

The design process for a product first requires a clear understanding of the functions and the performance expected of that product (Fig. 1.3). The product may be new, or it may be a revised version of an existing product. We all have observed, for example, how the design and style of radios, toasters, watches, automobiles, and washing machines have changed. The market for a product and its anticipated uses

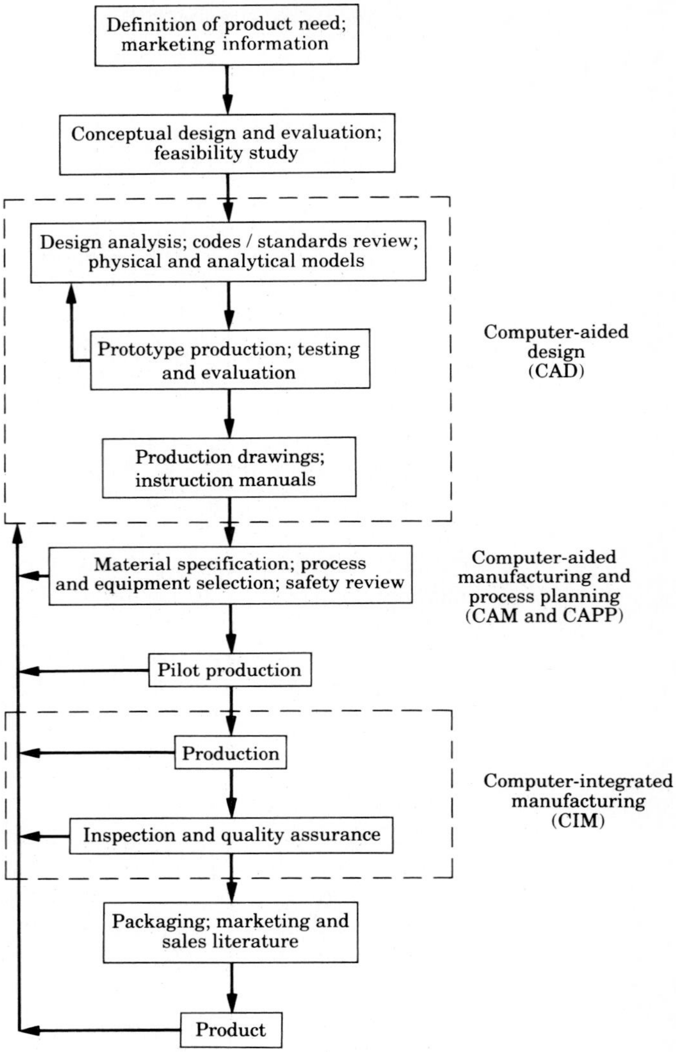

FIGURE 1.3 Chart showing various steps involved in designing and manufacturing a product. Depending on the complexity of the product and the type of materials used, the time span between the original concept and marketing a product may range from a few months to many years. *Simultaneous engineering* combines these stages to reduce the time span.

must be defined clearly, with the assistance of sales personnel, market analysts, and others in the organization.

The design process begins with the development of an original product concept. An innovative approach to design is highly desirable—and even essential—at this stage for the product to be successful in the marketplace. Innovative approaches can also lead to major savings in material and production costs. The design engineer or designer in charge of the product must be knowledgeable of the interrelationships among materials, design, and manufacturing, as well as the overall economics of the operation. In most cases, the majority of the decisions for selecting materials and manufacturing processes for a product are made by the designer, with input from others in the organization.

Product design often involves preparing analytical and physical models of the product, as an aid to analyzing factors such as forces, stresses, deflections, and optimal part shape. The necessity for such models depends on product complexity. Today, constructing and studying analytical models is simplified through the use of *computer-aided design and manufacturing* techniques. On the basis of these models, the designer selects and specifies the final shape and dimensions of the product, its surface finish and dimensional accuracy, and the materials to be used. The selection of materials is often made with the advice and cooperation of materials engineers, unless the design engineer is also experienced and qualified in this area.

An important design consideration is how a particular component is to be assembled into the final product. Take apart a ballpoint pen or a toaster, or lift the hood of your car and observe how hundreds of components are put together in a limited space. Note also how difficult it is on some cars to remove a spark plug or an oil filter, much less to make repairs or perform maintenance on the engine. The components of a product may be assembled by a variety of means, such as with bolts, screws, and rivets or by welding, soldering, or adhesive bonding. The method of assembly should be reliable and economical and require as little time as possible to perform, particularly for mass-produced items such as hair dryers, calculators, and automobiles.

The next step in the production process is to make and test a prototype, that is, an original working model of the product. Testing, either at this stage or periodically during production, is an important aspect of product manufacturing. Testing is now done statistically, and the proper interpretation of test results is crucial to maintaining the quality of a product (*statistical process control*). Total quality control of a product throughout the manufacturing process is one of the most important considerations in manufacturing engineering.

Tests must be designed to simulate as closely as possible the conditions under which the product is to be used. These include environmental conditions such as temperature and humidity, as well as the effects of vibration and repeated use and misuse of the product. Computer-aided design techniques are now capable of comprehensively and rapidly performing such simulations. During this stage, modifications in the original design, materials selected, or production methods may be necessary. Difficulties are often encountered in making the product function properly while fulfilling design, quality, and service requirements—or producing it economically.

After this phase has been completed, appropriate manufacturing methods, equipment, and tooling should be selected with the cooperation of manufacturing engineers, process planners, and all others that are to be directly involved in production. As Table 1.2 shows, various manufacturing processes are available for making parts with a variety of shapes, depending on the material and various other factors that we discuss later in this Introduction.

TABLE 1.2
SHAPES AND SOME COMMON METHODS OF PRODUCTION

SHAPE	PRODUCTION METHOD
Flat surfaces	Rolling, planing, broaching, milling, shaping, grinding
Parts with cavities	End milling, electrical-discharge machining, electrochemical machining, ultrasonic machining, cast-in cavity
Parts with sharp features	Permanent-mold casting, machining, grinding, fabricating
Thin hollow shapes	Slush casting, electroforming, fabricating
Tubular shapes	Extrusion, drawing, roll forming, spinning, centrifugal casting
Shaping of tubular parts	Rubber forming, expanding with hydraulic pressure, explosive forming, spinning
Curvature on thin sheets	Stretch forming, peen forming, fabricating
Openings in thin sheets	Blanking, chemical blanking, photochemical blanking
Reducing cross-sections	Drawing, extruding, shaving, turning, centerless grinding
Producing square edges	Fine blanking, machining, shaving, belt grinding
Producing small holes	Laser, electrical-discharge machining, electrochemical machining
Producing surface textures	Knurling, wire brushing, grinding, belt grinding, shot blasting, etching
Detailed surface features	Coining, investment casting, permanent-mold casting
Threaded parts	Thread cutting, thread rolling, thread grinding, chasing
Very large parts	Casting, forging, fabricating
Very small parts	Investment casting, machining

1.4 Selecting Materials

An ever-increasing variety of materials is available, each having its own characteristics, applications, advantages, and limitations. The following are the general types of materials used in manufacturing today:

- Irons and steels (carbon, alloy, stainless, and tool and die steels).
- Nonferrous metals and alloys (aluminum, magnesium, copper, nickel, titanium, superalloys, refractory metals, beryllium, zirconium, low-melting alloys, and precious metals).
- Plastics (thermoplastics, thermosets, and elastomers).
- Ceramics, glass ceramics, glasses, graphite, and diamond.
- Composite materials (reinforced plastics, metal-matrix and ceramic-matrix composites, and honeycomb structures).

1.4.1 Properties of materials

When selecting materials for products, we first consider their mechanical properties: strength, toughness, ductility, hardness, elasticity, fatigue, and creep. The strength-to-weight and stiffness-to-weight ratios of material are also important, particularly for aerospace and automotive applications. Aluminum, titanium, and reinforced plastics, for example, have higher ratios than steels and cast irons. The mechanical properties specified for a product and its components should of course be for the conditions under which the product is expected to function. We then consider the physical properties of density, specific heat, thermal expansion and conductivity, melting point, and electrical and magnetic properties.

Chemical properties also play a significant role in hostile as well as normal environments. Oxidation, corrosion, general degradation of properties, toxicity, and flammability of materials are among the important factors to be considered. In some commercial airline disasters, for example, many deaths have been caused by toxic fumes from burning nonmetallic materials in the aircraft cabin.

Manufacturing properties of materials determine whether they can be cast formed, machined, welded, and heat treated with relative ease. The method(s) used to process materials to the desired shapes can adversely affect the product's final properties and service life.

1.4.2 Availability and cost

Availability and cost of raw and processed materials and manufactured components are major concerns in manufacturing. Competitively, the economic aspects of material selection are as important as the technological considerations of properties and characteristics of materials.

If raw or processed materials or manufactured components are not available in the desired quantities, shapes, and dimensions, substitutes and/or additional processing will be required, which can contribute significantly to product cost. For example, if we need a round bar of a certain diameter and it is not available in standard form, then we have to purchase a larger rod and reduce its diameter by some means, such as machining, drawing through a die, or grinding.

Reliability of supply, as well as demand, affects cost. Most countries import numerous raw materials that are essential for production. The broad political implications of such reliance on other countries is self-evident.

Different costs are involved in processing materials by different methods. Some methods require expensive machinery, others require extensive labor, and still others require personnel with special skills or a high level of education or specialized training.

1.4.3 Appearance, service life, and disposal

The appearance of materials after they have been manufactured into products influences their appeal to the consumer. Color, feel, and surface texture are characteristics that we all consider when making a decision about purchasing a product. Time- and service-dependent phenomena such as wear, fatigue, creep, and dimensional stability are important. These phenomena can significantly affect a product's performance and, if not controlled, can lead to total failure of the product. Similarly, compatibility of materials used in a product is important. Friction and wear, corrosion, and other phenomena can shorten a product's life or cause it to fail. An example is galvanic action between mating parts made of dissimilar metals, which corrodes the parts.

Recycling or proper disposal of materials at the end of their useful service lives has become increasingly important in an age conscious of maintaining a clean and healthy environment. Note, for example, the use of biodegradable packaging materials or recyclable glass bottles and aluminum beverage cans. The proper disposal of toxic wastes and materials is also a crucial consideration.

1.5 Selecting Manufacturing Processes

Many processes are used to produce parts and shapes. As Table 1.2 shows, there is usually more than one method of manufacturing a part from a given material. The broad categories of processing methods for materials are:

- Casting (expendable mold and permanent mold).
- Forming and shaping (rolling, forging, extrusion, drawing, sheet forming, powder metallurgy, and molding).
- Machining (turning, boring, drilling, milling, planing, shaping, broaching, grinding, ultrasonic machining; and chemical, electrical, and electrochemical machining; and high-energy beam machining).
- Joining (welding, brazing, soldering, diffusion bonding, adhesive bonding, and mechanical joining).
- Finishing operations (honing, lapping, polishing, burnishing, deburring, surface treating, coating, and plating).

Selection of a particular manufacturing process depends not only on the shape to be produced but also on a large number of other factors. The type of material and its properties are basic considerations. Brittle and hard materials, for example, cannot be formed easily, whereas they can be cast or machined by several methods. The manufacturing process usually alters the properties of materials. Metals that are formed at room temperature, for example, become stronger, harder, and less ductile than they were before processing.

1.5.1 Dimensional and surface finish requirements

Size, thickness, and shape complexity of the part have a major bearing on the process selected to produce it. Flat parts with thin cross-sections, for example, cannot be cast properly. Complex parts cannot be formed easily and economically, whereas they may be cast or fabricated from individual pieces.

Tolerances and surface finish obtained in hot-working operations cannot be as good as those obtained in cold-working operations. Dimensional changes, warpage, and surface oxidation occur during processing at elevated temperatures. Some casting processes produce a better surface finish than others because of the different types of mold materials used and their surface finish.

Since not all manufacturing operations produce finished products, additional operations such as grinding or polishing may be necessary. These operations can add significantly to product cost. Consequently, there is now a trend, called *near-net* or *net-shape processing*, in which the part is made as close to the final dimensions and specifications as possible.

1.5.2 Operational and cost considerations

The design and cost of tooling, the lead time required to begin production, and the effect of workpiece material on tool and die life are major considerations. Depending on its size, design, and expected life, the cost of tooling can be substantial. For example, a set of steel dies for stamping sheet-metal fenders for automobiles may cost about $2 million.

For parts made from expensive materials, the lower the scrap rate, the more economical the production process will be. Because it generates chips, machining may not be more economical than forming operations, all other factors being the same.

Availability of machines and equipment and operating experience within the manufacturing facility are also important. If they are not available, some parts may have to be manufactured by outside firms. Automakers, for example, purchase many parts from outside vendors, or have them made by outside firms according to their specifications.

The number of parts or products required and the desired production rate (pieces per hour) help determine the processes to be used and the economics of production. Beverage cans or transistors, for example, are consumed in numbers and at rates much higher than telescopes and propellers for ships.

The operation of machinery has significant environmental and safety implications. Depending on the type of operation and the machinery involved, some processes adversely affect the environment. Unless properly controlled, such processes can cause air, water, and noise pollution. The safe use of machinery is another important consideration, requiring precautions to eliminate hazards in the workplace.

1.5.3 Consequences of improperly selecting materials and processes

Numerous examples of product failure can be traced to improper selection of material or manufacturing processes or improper control of process variables. A component or a product is generally considered to have failed when:

- It stops functioning (broken shaft, gear, bolt, cable, or turbine blade).
- It does not function properly or perform within required specification limits (worn bearings, gears, tools, and dies).
- It becomes unreliable or unsafe for further use (frayed cable in a winch, crack in a shaft, poor connection in a printed-circuit board, or delamination of a reinforced plastic component).

We describe in detail in this text the types of failure of a component or product that result from design deficiencies, improper material selection, material defects, manufacturing-induced defects, improper component assembly, and improper product use.

• Example 1.4: Manufacturing a part by various methods.

Often there is more than one method of manufacturing a part. The selection of a particular process over others depends on a large number of factors. Consider, for instance, the following example of forming a simple dish-shaped part from sheet metal. Such a part can be formed by placing a flat piece of sheet metal between a pair of male and female dies and closing the dies through proper application of closure force (Fig. E1.2a). A large number of parts can be formed at high production

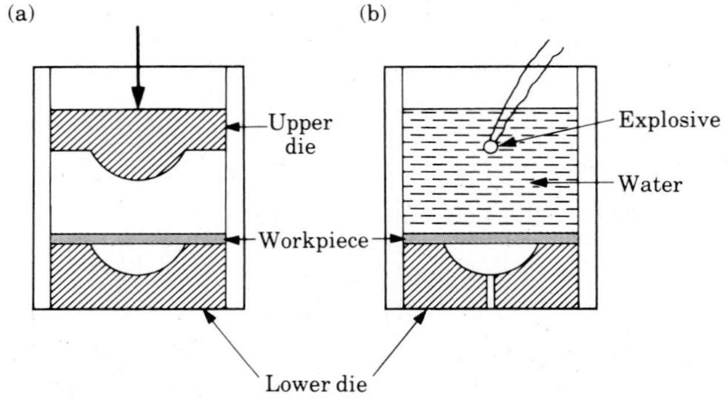

FIGURE E1.2

Two methods of forming a dish-shaped part from sheet metal: (a) conventional mechanical or hydraulic press, using a pair of male and female dies; and (b) explosive forming method using only one die. In manufacturing operations, choice of the process for a particular application depends on a wide variety of technical and economic considerations.

rates by this method, known as stamping or pressworking. The dies are activated by a machine, which can be either a mechanical or a hydraulic press.

Assume now that the size of the part is very large, e.g., 2 m (80 in.) in diameter. Further assume that only 50 of these parts are required. In response to these constraints, we have to reconsider the total operation and ask a number of questions. Is it economical to manufacture a set of dies 2 m in diameter (which are very costly) when the production run (quantity of production) is so low? Are machines available with sufficient capacity to accommodate such large dies? What are the alternative methods of manufacturing this part? Does it have to be made in one piece?

This part can also be made by welding together smaller pieces of metal formed by other methods. (Ships and water tanks are made by this method.) Would a part manufactured by welding be acceptable for its intended purpose? Will it have the required properties and the correct shape after welding, or will it require additional processing? The same part can be made by a spinning process similar to making ceramic dishes and pottery on a potter's wheel. It can also be made by the explosive forming process shown in Fig. E1.2(b). In this method, the male die is replaced by water, and the shock wave from the explosion generates sufficiently high pressure to form the part.

In comparing the two processes in Fig. E1.2, we note that, in the explosive forming process, the deformation of the material takes place at a very high rate. In selecting this process over conventional pressworking, we have to consider various factors. For instance, is the material capable of undergoing deformation at high rates without fracture? Does the high rate have any detrimental effect on the final properties of the formed part? Can tolerances be held within acceptable limits? Is the life of the die sufficiently long under the high transient pressures generated in this process? Can this operation be performed in a manufacturing plant in a city or should it be carried out in open country? Although having only one die is an advantage, is the overall operation economical?

1.6 Assembly

After individual parts have been manufactured, they are assembled into a product. Assembly is an important phase of the overall manufacturing operation and requires consideration of the ease, speed, and cost of putting parts together. Many products are also designed so that they can be taken apart for maintenance and servicing.

There are several methods of assembly, each with its own characteristics and requiring different operations. The use of a bolt and nut, for example, requires preparation of holes that must match in location and size. Hole generation requires operations such as drilling or punching, which take additional time, require separate operations, and produce scrap. Furthermore, the presence of holes could reduce the useful life of components because of stress concentration, leading to fatigue failure of the part. On the other hand, products assembled with bolts and nuts can be taken apart and reassembled with relative ease.

Parts can also be assembled with adhesives. This method, which is being used extensively in aircraft and automobile production, does not require holes. However, surfaces to be assembled must match properly and be clean because joint strength is affected by the presence of contaminants, such as dirt, dust, oil, and moisture. Production rates in assembly with adhesives may be low because of the time required for the adhesive to be applied and to develop full strength. Unlike mechanical fastening, adhesively joined components, as well as those that are welded, are not usually designed to be taken apart and reassembled.

Parts may be assembled by hand or by automatic equipment. The choice depends on factors such as the complexity of the product and the number of parts to be assembled, the protection required to prevent damage or scratching of finished surfaces of the parts, and the relative costs of labor and machinery required for automated assembly. Variations in the dimensions of parts may be tolerated in hand assembly since the operator can make minor adjustments. In automated assembly, however, parts must have consistently uniform dimensions. Otherwise, their movement through the assembly equipment can be impeded and parts not assembled properly.

1.6.1 Design for assembly

Because assembly operations can contribute significantly to product cost, design for assembly is now recognized as an important area of manufacturing. It involves the following basic considerations:

a) Design of the product to permit assembly with relative ease. Note in Fig. 1.4, for example, how a design can be simplified by reducing the number of pieces required, thus making assembly of the product easier and faster—and resulting in lower product cost. Also modifying a part's shape can allow easier assembly or make it adaptable for robotic assembly.
b) Possibility of designing multipurpose parts so that they can be used in different lines of products, thus reducing the number of different parts to be made.
c) Capabilities and limitations of each manufacturing process, particularly in regard to consistency in part shape and dimensions. The greater the variations in these characteristics, the more difficult proper assembly becomes.
d) Methods by which individual parts are brought together for assembly and are packaged, including the use of conveyors, feeders, transfer equipment, and programmable industrial robots.

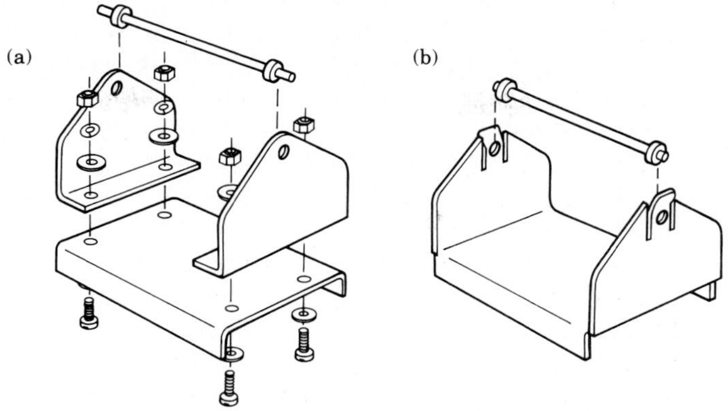

FIGURE 1.4
An example of design for assembly. (a) As designed, this product consists of many components and takes considerable time to assemble. (b) Redesigned product consists of only two parts and is easy to assemble, either by hand or in automated machinery. *Source:* Courtesy of G. Boothroyd.

Design for assembly is an integral part of design for manufacture. Thus it should always be given due consideration as a way to improve productivity in manufacturing.

1.7 Product Quality

We all have used terms like poor quality and good quality in describing a product. What do we mean by quality? In a broad sense, *quality* is a characteristic or property consisting of several well-defined technical and aesthetic, hence subjective, considerations. The general public's perception is that a high-quality product functions reliably and as expected over a long period of time.

Product quality has always been one of the most important aspects of manufacturing, as it directly influences the marketability of a product and customer satisfaction. Traditionally, quality assurance has been obtained by inspecting parts after they have been manufactured. Parts are inspected to ensure that they conform to a detailed set of specifications and standards, such as dimensions, surface finish, and mechanical and physical properties.

However, quality cannot be inspected into a product after it is made. The practice of inspecting products after they are made is now being replaced rapidly by the broader view that *quality must be built into a product*—from the design stage through all subsequent stages of manufacture and assembly. Producing defective products can be very costly to the manufacturer, creating difficulties in assembly operations, necessitating repairs in the field, and resulting in customer dissatisfaction. Contrary to general public opinion, high-quality products do not necessarily cost more than poor-quality products.

Total quality control and quality assurance have become the responsibility of everyone involved in designing and manufacturing a product. Our awareness of the technological and economic importance of built-in product quality has been heightened further by recent pioneers in quality control, primarily Deming and Taguchi. They pointed out the importance of the management's commitment to product quality, pride of workmanship at all levels of production, and the use of modern statistical techniques to identify the causes of quality control problems.

Important developments in quality control include the implementation of *experimental design*, a technique in which the factors involved in a manufacturing process and their interactions are studied simultaneously. Thus, for example, variables affecting dimensional accuracy or surface finish in a machining operation can be readily identified and appropriate actions taken. The use of computers has greatly enhanced our capability to utilize such techniques rapidly and effectively.

1.8 Automation and the Impact of Computers on Manufacturing

The major goals of automation in manufacturing facilities are to integrate various operations to improve productivity, increase product quality and uniformity, minimize cycle times and effort, and reduce labor costs. Beginning in the 1940s, automation has accelerated because of rapid advances in control systems for machines and in computer technology.

Few developments in the history of manufacturing have had a more significant impact than computers. Beginning with computer graphics, which are rapidly replacing drafting boards, the use of computers has been extended to computer-aided design and manufacturing and, ultimately, to *computer-integrated manufacturing systems*. The use of computers covers a broad range of applications, including control and optimization of manufacturing processes, material handling, assembly, and automated inspection and testing of products.

1.8.1 Machine control systems

Numerical control (NC) of machines is a method of controlling the movements of machine components by direct insertion of coded instructions in the form of numerical data. Numerical control was first implemented in the early 1950s and was a major advance in automation of machines.

In *adaptive control* (AC), process parameters are adjusted automatically to optimize production rate and product quality and minimize cost. Parameters such as forces, temperatures, surface finish, and dimensions of the part are monitored constantly. If they move outside the acceptable range, the system adjusts the process variables until the parameters again fall within the acceptable range.

Major advances have been made in *automated handling of materials* in various stages of completion, such as from storage to machines, from machine to machine, inspection, inventory, and shipment. Introduced in the early 1960s, *industrial robots* are replacing humans in operations that are repetitive, boring, and dangerous, thus reducing the possibility of human error, decreasing variability in product quality, and improving productivity. Robots with sensory perception capabilities are being developed, with movements that are beginning to simulate those of humans. *Automated assembly systems* are replacing costly assembly by operators. Products are being designed or redesigned so that they can be assembled more easily by machine.

1.8.2 Computer technology

Computers allow us to *integrate* virtually all phases of manufacturing operations, which consist of various technical as well as managerial activities. With sophisticated software and hardware, manufacturers are now able to minimize manufacturing costs, improve product quality, reduce product development time, and maintain a competitive edge in the domestic and international marketplace.

Computer-integrated manufacturing (CIM) is particularly effective in responding to recent trends of short product life cycles, market demand for product types and quantities, and customer requirements for high-quality, low-cost products. The benefits of CIM are:

- Responsiveness to rapid changes in market demand, product modification, and shorter product life cycles.
- High-quality products at low cost.
- Better use of materials, machinery, and personnel and reduced inventory.
- Better control of production and management of the *total* manufacturing operation.

Computer-aided design (CAD) allows the designer to conceptualize objects more easily without having to make costly illustrations, models, or prototypes. These systems are now capable of rapidly and completely analyzing designs, from a simple bracket to complex structures, such as aircraft wings. The performance of structures subjected to static or fluctuating loads and temperatures can now be simulated, analyzed, and tested efficiently, accurately, and more quickly than ever on the computer. The information developed can be stored, retrieved, displayed, printed, and transferred anywhere in the organization. Designs can be optimized, and modifications can be made directly and easily at any time in the life of a product.

Computer-aided manufacturing (CAM) involves all phases of manufacturing by utilizing and processing further the large amount of information on materials and processes collected and stored in the organization's database. Computers now assist manufacturing engineers and others in organizing tasks such as programming numerical control of machines; programming robots for material handling; designing tools, dies, and fixtures; and quality control and inspection.

Computer-aided process planning (CAPP) is capable of improving productivity in a plant by optimizing process plans, reducing planning costs, and improving the consistency of product quality and reliability. Functions such as cost estimating and work standards (time required to perform a certain operation) can also be incorporated into the system.

The high level of automation and flexibility achieved with numerically controlled machines has led to the development of *group technology* and *cellular manufacturing*. The concept of group technology (GT) is that parts can be grouped and produced by classifying them according to similarities in design and manufacturing processes. In this way, part designs and process plans can be standardized, and families of parts can be produced efficiently and economically.

Flexible manufacturing systems (FMS) integrate manufacturing cells into a large unit, containing industrial robots serving several machines, all interfaced with a central computer. Flexible manufacturing systems have the highest level of efficiency, sophistication, and productivity in manufacturing. They are capable of producing parts randomly and changing manufacturing sequences on different parts quickly. Thus they can meet rapid changes in market demand for various types of products.

Artificial intelligence (AI) involves the use of machines and computers to replace human intelligence. Computer-controlled systems are becoming capable of learning from experience and making decisions so as to optimize operations and minimize costs. *Expert systems*, defined as intelligent computer programs that have the capability to solve difficult problems with knowledge and inference, are now used for guiding missiles, prescribing treatments for medical ailments, and diagnosing problems in machine performance.

We can now envisage the *factory of the future* in which production takes place with little or no direct human intervention. The human role will be confined to supervision, maintenance, and upgrading machines, computers, and software. The impact of such advances on the workforce remains to be fully assessed.

1.9 Economics of Manufacture

The cost of a product is often the overriding consideration in its marketability and general customer satisfaction. Consequently, all the costs involved in manufacturing a high-quality product must be thoroughly analyzed. Typically, manufacturing costs represent about 40 percent of a product's selling price.

The concepts of design for manufacture and assembly that we described earlier also include design principles for economic production:

- The design should be as simple as possible to manufacture and assemble.
- Materials should be chosen for their appropriate manufacturing characteristics.
- Dimensional accuracy and surface finish specified should be as broad as permissible.
- Because they can add significantly to cost, secondary and additional processing of parts should be avoided.

The total cost of manufacturing a product consists of costs of materials, tooling, and labor, as well as fixed costs and capital costs. Several factors are involved in each cost category. The cost of materials is usually the largest percentage of total manufacturing costs. This cost can be minimized by analyzing the product design to determine whether part size and shape are optimal and the materials selected are the least costly, while possessing the desired properties and characteristics. The possibility of substituting materials is an important consideration in minimizing costs.

Tooling costs depend on the complexity of part shape, the materials involved, the manufacturing process, and the number of parts to be made. Complex shapes, difficult-to-machine materials, and stringent dimensional accuracy requirements all add to the cost of tooling. Direct labor costs are usually a small proportion of the total, typically ranging from 10 percent to 15 percent. The trend toward increased automation and computer control of all aspects of manufacturing helps minimize labor involvement and direct labor costs, which continue to decline steadily. Automation also has other benefits, such as higher production rates and more consistent product quality.

Fixed costs and capital costs depend on the particular manufacturer and plant facilities. Computer-controlled machinery, which constitutes a capital cost, can be very expensive. However, economic analysis indicates that, more often than not, such an expenditure is warranted in view of the long-range benefits, such as improved productivity. In mass production of products, such as automobile engines, specialized machines are arranged in various product flow lines. These machines, too, require a major capital investment, yet the high production rate makes the cost per part competitive.

SUMMARY

Manufacturing a product, whether a light bulb, a typewriter, or an automobile, requires a comprehensive approach involving design, materials, and processing.

Moreover, close cooperation among the product designer, materials engineer, and manufacturing engineer is essential for manufacturing a product that functions well and is competitive in the marketplace. The more complicated the product, the greater is the need for coordination and communication among the parties involved. In view of the economic significance and competitive aspects of manufacturing today, the need for thorough assessment of the advantages and limitations of each material and each process is more important than ever. The rest of this book describes in detail the properties and behavior of materials and principles of all manufacturing methods, including the economics of manufacturing.

BIBLIOGRAPHY

Selected General References on Manufacturing

Alexander, J.M., R.C. Brewer, and G.W. Rowe, *Manufacturing Technology*, 2 vols. Chichester, England: Ellis Horwood, 1987.

Alting, L., *Manufacturing Engineering Processes*. New York: Marcel Dekker, 1982.

Amstead, B.H., P.F. Ostwald, and M.L. Begeman, *Manufacturing Processes*, 8th ed. New York: Wiley, 1987.

DeGarmo, E.P., J.T. Black, and R.A. Kohser, *Materials and Processes in Manufacturing*, 7th ed. New York: Macmillan, 1988.

Dieter, G.E., *Engineering Design: A Materials and Processing Approach*. New York: McGraw-Hill, 1983.

Doyle, L.E., C.A. Keyser, J.L. Leach, G.F. Schrader, and M.S. Singer, *Manufacturing Processes and Materials for Engineers*, 3d ed. Englewood Cliffs, N.J.: Prentice-Hall, 1985.

Kalpakjian, S., *Manufacturing Engineering and Technology*. Reading, Mass.: Addison-Wesley, 1989.

Lascoe, O.D., *Handbook of Fabrication Processes*. Metals Park, Ohio: ASM International, 1988.

Lindberg, R.A., *Processes and Materials of Manufacture*, 4th ed. Boston: Allyn & Bacon, 1990.

Ludema, K.C., R.M. Caddell, and A.G. Atkins, *Manufacturing Engineering: Economics and Processes*. Englewood Cliffs, N.J.: Prentice-Hall, 1987.

Niebel, B.W., A.B. Draper, and R.A. Wysk, *Modern Manufacturing Process Engineering*. New York: McGraw-Hill, 1989.

Schey, J.A., *Introduction to Manufacturing Processes*, 2d ed. New York: McGraw-Hill, 1987.

Tijunelis, D., and K.E. McKee (eds.), *Manufacturing High Technology Handbook*. New York: Marcel Dekker, 1987.

Wright, R.T., *Processes of Manufacturing*. South Holland, Ill.: Goodheart-Willcox Co., 1988.

Young, J.F., and R.S. Shane (eds.), *Materials and Processes*, 3d ed., 2 vols. Schenectady, N.Y.: Genium, 1990.

General References on Materials and Processes

Bralla, J.G. (ed.), *Handbook of Product Design and Manufacturing*. New York: McGraw-Hill, 1986.
Encyclopedia of Chemical Technology, Kirk-Othmer, 3d ed. New York: Wiley, 1978.
The Encyclopedia of Materials Science and Engineering. Oxford: Pergamon, 1986.
Machinery's Handbook, 22nd ed. New York: Industrial Press, 1984 (revised periodically).
Metals Handbook, Desk Edition. Metals Park, Ohio: American Society for Metals, 1985.
Metals Handbook, 10th ed. Materials Park, Ohio: ASM International (various volumes and dates).
Smithells Metals Reference Book. Woburn, Mass.: Butterworths, 1986.
Standard Handbook for Mechanical Engineers, 9th ed. New York: McGraw-Hill, 1987.
Tool and Manufacturing Engineers Handbook, 4th ed., 5 vols. Dearborn, Mich.: Society of Manufacturing Engineers.

Conference Proceedings

Advances in Machine Tool Design and Research
Annals of the CIRP (International Institution for Production Engineering Research)
Proceedings of the North American Manufacturing Research Conference

2

Fundamentals of the Mechanical Behavior of Materials

2.1 Introduction

The manufacturing methods and techniques by which materials can be shaped into useful products were outlined in Chapter 1. In manufacturing, one of the most important groups of processes is *plastic deformation*, namely, shaping materials by applying forces in various ways (also known as deformation processing). It includes bulk deformation (forging, rolling, extrusion, and rod and wire drawing) and sheet-forming processes (bending, drawing, spinning, and general pressworking).

This chapter deals with the fundamental aspects of the mechanical behavior of materials during deformation. Individual topics are deformation modes, stresses, forces, effects of rate of deformation and temperature, hardness, residual stresses, and yield criteria.

In stretching a piece of metal to make an object such as a fender of an automobile or a length of wire, the material is subjected to tension. A solid cylindrical piece of metal is forged in the making of a turbine disk, thus subjecting the material to compression. Sheet metal undergoes shearing stresses when, for instance, a hole is punched through its cross-section. A piece of plastic tubing is expanded by internal pressure to make a bottle, thus subjecting the material to tension in various directions.

In all these processes, the material is subjected to one or more of the basic modes of deformation shown in Fig. 2.1, namely, tension, compression, and shear. The degree of deformation to which the material is subjected is defined as *strain*. For tension or compression the *engineering strain*, or *nominal strain*, is defined as

$$e = \frac{\ell - \ell_o}{\ell_o}. \tag{2.1}$$

Note that in tension the strain is positive and in compression it is negative. The *shear strain* is defined as

$$\gamma = \frac{a}{b}. \tag{2.2}$$

In order to change the geometry of the bodies or elements in Fig. 2.1, forces must be applied to them as shown by the arrows. The determination of these forces as a function of strain is very important in manufacturing processes. We have to know the forces in order to design the proper equipment, select the tool and die materials for proper strength, and determine whether a specific metalworking operation can be accomplished on certain equipment. Thus the relation between a force and the deformation it produces is an essential parameter in manufacturing.

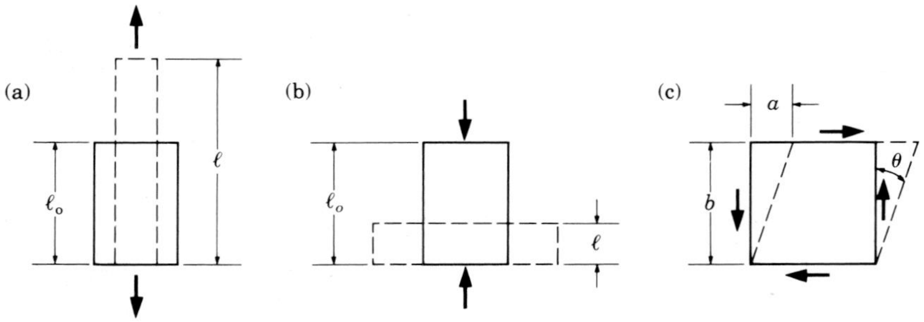

FIGURE 2.1

Types of strain. (a) Tensile. (b) Compressive. (c) Shear. All deformation processes in manufacturing involve strains of these types. Tensile strains are involved in stretching sheet metal to make car bodies, compressive strains in forging metals to make turbine disks, and shear strains in making holes by punching.

2.2 Tension

Because of its relative simplicity, the tension test is the most common test for determining the *strength-deformation characteristics* of materials. It involves the preparation of a test specimen (according to ASTM specifications) and testing it under tension on any of a variety of available testing machines.

The specimen has an original length ℓ_o and an original cross-sectional area A_o. Although most specimens are solid and round, flat sheet or tubular specimens are also tested under tension. The original length is the distance between *gage marks* on the specimen and is generally 50 mm (2 in.). Longer lengths may be used for larger specimens, such as structural members.

Typical results from a tension test are shown in Fig. 2.2. The *engineering stress*, or *nominal stress*, is defined as the ratio of the applied load to the original area,

$$\sigma = \frac{P}{A_o}, \tag{2.3}$$

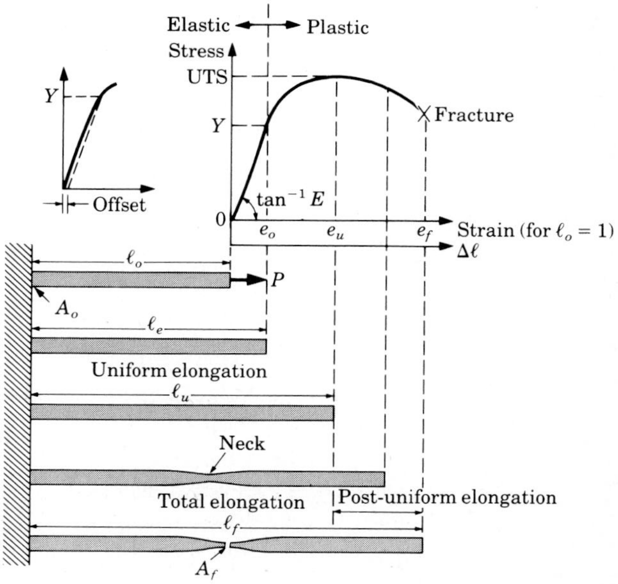

FIGURE 2.2
Outline of a tensile test sequence showing different stages in the elongation of the specimen.

and the engineering strain is

$$e = \frac{\ell - \ell_o}{\ell_o}.$$

When the load is applied, the specimen elongates proportionately to load up to the *proportional limit*. This is the range of *linear elastic behavior*. The material will continue to deform elastically, although not strictly linearly, up to the *yield point*, Y. If the load is removed before the yield point is exceeded, the specimen will return to its original length. The *modulus of elasticity*, or *Young's modulus*, E, is defined as

$$E = \frac{\sigma}{e}. \tag{2.4}$$

This linear relationship between stress and strain is known as *Hooke's law*, the more generalized forms of which are given in Section 2.13.3.

The elongation of the specimen is accompanied by a contraction of its lateral

TABLE 2.1
MODULUS OF ELASTICITY *E* AND POISSON'S RATIO ν FOR VARIOUS MATERIALS AT ROOM TEMPERATURE

METALS	E (GPa)	ν
Aluminum and its alloys	69–79	0.31–0.34
Cast irons	105–150	0.21–0.30
Copper and its alloys	105–150	0.33–0.35
Lead and its alloys	14	0.43
Magnesium and its alloys	41–45	0.29–0.35
Molybdenum	325	0.32
Nickel and its alloys	180–214	0.31
Steel (plain carbon)	200	0.33
Steel (austenitic stainless)	190–200	0.28
Titanium and its alloys	80–130	0.31–0.34
Tungsten	400	0.27
NONMETALLIC		
Acrylics	1.4–3.4	0.35–0.40
Epoxies	3.5–17	0.34
Nylons	1.4–2.8	0.32–0.40
Rubbers	0.01–0.1	0.5
Plastics, reinforced	2–50	—
Glass and porcelain	70–80	0.24
Diamond	820–1050	—
Graphite	240–390	—

Note: Multiply GPa by 145,000 to obtain psi.

dimensions. The absolute value of the ratio of the lateral strain to longitudinal strain is known as *Poisson's ratio*, ν. Typical values for E and ν for various materials are given in Table 2.1.

The area under the stress–strain curve up to the yield point Y of a material (Table 2.2) is known as the *modulus of resilience*:

$$\text{Modulus of resilience} = \frac{Ye_o}{2} = \frac{Y^2}{2E}. \tag{2.5}$$

TABLE 2.2
MECHANICAL PROPERTIES OF VARIOUS MATERIALS AT ROOM TEMPERATURE

METAL	Y (MPa)	UTS (MPa)	ELONGATION IN 50 mm (%)
Aluminum	35	90	45
Aluminum alloys	35–550	90–600	45–4
Beryllium	185–260	230–350	3.5–1
Columbium (niobium)	205	275	30
Copper	70	220	45
Copper alloys	76–1100	140–1310	65–3
Iron	40–200	185–285	60–3
Steels	205–1725	415–1750	65–2
Lead	7–14	17	50
Lead alloys	14	20–55	50–9
Magnesium	90–105	160–195	15–3
Magnesium alloys	130–305	240–380	21–5
Molybdenum alloys	80–2070	90–2340	40–30
Nickel	58	320	30
Nickel alloys	105–1200	345–1450	60–5
Tantalum alloys	480–1550	550–1550	40–20
Titanium	140–550	275–690	30–17
Titanium alloys	344–1380	415–1450	25–7
Tungsten	550–690	620–760	0
NONMETALLIC			
Ceramics	—	140–2600	0
Glass	—	140	0
Glass, fibers	—	3500–4500	0
Graphite, fibers	—	2100–2500	0
Thermoplastics	—	7–80	5–1000
Thermoplastics, reinforced	—	20–120	1–10
Thermosets	—	35–170	0
Thermosets, reinforced	—	200–520	0

Note: Multiply MPa by 145 to obtain psi.

This area has the units of *energy per unit volume* and indicates the *specific energy* that the material can store elastically.

With increasing load, the specimen begins to yield; that is, it begins to undergo *plastic (permanent) deformation* and the relationship between stress and strain is no longer linear. The rate of change in the slope of the stress–strain curve beyond the yield point is rather small for most materials, so the determination of Y may be difficult. The usual practice is to define the yield stress as the point on the curve that is offset by a strain of (usually) 0.2%, or 0.002 (Fig. 2.2). Other offset strains may also be used and should be specified in reporting the yield stress.

As the specimen continues to elongate under increasing load, its cross-sectional area decreases uniformly throughout its length. The load (hence the engineering stress) reaches a maximum and then begins to decrease. The maximum stress is known as the *tensile strength* or *ultimate tensile strength (UTS)* of the material (Table 2.2).

When the specimen is loaded beyond its UTS, it begins to *neck* (Fig. 2.2) and the elongation is no longer uniform. That is, the change in the cross-sectional area of the specimen is no longer uniform along the length of the specimen but is concentrated locally in a "neck" formed in the specimen (known as *necking*, or *necking down*). As the test progresses, the engineering stress drops further and the specimen finally fractures in the necked region. The final stress level (marked by an × in Fig. 2.2) is known as *breaking* or *fracture stress*.

2.2.1 Ductility

Ultimate tensile strength is a practical measure of the overall strength of a material. Likewise, the strain at fracture is a measure of its *ductility*, that is, how large a strain the material withstands before failure. Note from Fig. 2.2 that until the UTS is reached, elongation is uniform. The strain up to the UTS is known as *uniform strain*. The elongation at fracture is known as the *total elongation*. It is measured between the original gage marks after the two pieces of the broken specimen are placed together.

Two quantities commonly used to define ductility in a tension test are percent elongation and reduction of area. Percent *elongation* is defined as

$$\% \text{ elongation} = \frac{\ell_f - \ell_o}{\ell_o} \times 100 \tag{2.6}$$

and is based on the total elongation (Table 2.2).

Necking is a *local* phenomenon. If we put a series of gage marks at different points on the specimen, pull, and break it under tension and then calculate the percent elongation for each pair of gage marks, we find that with decreasing gage length the percent elongation increases. The closest pair of gage marks undergo the largest elongation because they are closest to the necked region. However, the curves do not approach zero elongation with increasing gage length. The reason is that the specimens have all undergone a finite uniform and permanent elongation before fracture. It is thus important to report gage length in conjunction with elongation data, whereas other tensile properties are generally independent of gage length.

A second measure of ductility is *reduction of area*, defined as

$$\% \text{ reduction of area} = \frac{A_o - A_f}{A_o} \times 100. \tag{2.7}$$

Thus a material that necks down to a point at fracture, such as a glass rod at elevated temperature, has 100% reduction of area.

The elongation and reduction of area are generally related to each other for many common engineering metals and alloys. Elongation ranges approximately between 10 and 60%, and values between 20 and 90% are typical for reduction of area for most materials. *Thermoplastics* (Chapter 10) and *superplastic* materials (Section 2.2.7) exhibit much higher ductility. Brittle materials, by definition, have little or no ductility. Examples are glass at room temperature and a piece of chalk.

• Example 2.1: Calculation of percent elongation and reduction of area.

A tension test specimen has an original gage length of 50 mm and a diameter of 4 mm. The maximum load during the test is 10 kN. The final gage length is 80 mm and the diameter of the necked region is 3 mm. Calculate the UTS, percent elongation, and reduction of area.

SOLUTION.

$$\text{UTS} = \frac{P}{A_o} = \frac{10{,}000}{\pi \frac{4^2}{4}} = 796 \text{ N/mm}^2 = 796 \text{ MPa}.$$

$$\text{Elongation} = \frac{80 - 50}{50} \times 100 = 60\%.$$

$$A_o = \pi \frac{4^2}{4} = 12.57 \text{ mm}^2.$$

$$A_f = \pi \frac{3^2}{4} = 7.07 \text{ mm}^2.$$

$$\text{Reduction of area} = \frac{12.57 - 7.07}{12.57} \times 100 = 43.8\%.$$

•

2.2.2 True stress and true strain

Because stress is defined as the ratio of force to area, *true stress* should be defined as

$$\sigma = \frac{P}{A}, \tag{2.8}$$

where A is the actual (hence true) or instantaneous area supporting the load.

TABLE 2.3
COMPARISON OF ENGINEERING AND TRUE STRAINS IN TENSION

e	0.01	0.05	0.1	0.2	0.5	1	2	5	10
ε	0.01	0.049	0.095	0.18	0.4	0.69	1.1	1.8	2.4

Likewise, the complete tension test may be regarded as a series of incremental tension tests where, for each succeeding increment, the original specimen is a little longer than the previous one. We can now define *true strain* (or *natural* or *logarithmic strain*) ε as

$$\varepsilon = \int_{\ell_o}^{\ell} \frac{d\ell}{\ell} = \ln\left(\frac{\ell}{\ell_o}\right). \tag{2.9}$$

Note that, for small values of engineering strain, $e = \varepsilon$ since $\ln(1 + e) = \varepsilon$. For larger strains, however, the values diverge rapidly, as shown in Table 2.3.

The volume of a metal specimen in the plastic region of the test remains constant (Section 2.3.5). Hence the true strain within the uniform elongation range can be expressed as

$$\varepsilon = \ln\left(\frac{\ell}{\ell_o}\right) = \ln\left(\frac{A_o}{A}\right) = \ln\left(\frac{D_o}{D}\right)^2 = 2\ln\left(\frac{D_o}{D}\right). \tag{2.10}$$

Once necking begins, the true strain at any point in the specimen can be calculated from the change in cross-sectional area at that point. Thus, by definition, the largest strain is at the narrowest region of the neck.

We have shown that, at small strains, the engineering and true strains are very close and hence either one can be used in calculations. However, for the large strains encountered in metalworking, the true strain should be used. This is the true measure of the strain and can be illustrated by the following two examples.

Assume that a tension specimen is elongated to twice its original length. This deformation is equivalent to compressing a specimen to one half its original height. Using the subscripts t and c for tension and compression, respectively, we find that $\varepsilon_t = 0.69$ and $\varepsilon_c = -0.69$, whereas $e_t = 1$ and $e_c = -0.5$. Thus true strain is a correct measure of strain. For the second example, assume that a specimen 10 mm in height is compressed to a final thickness of zero. Thus $\varepsilon_c = -\infty$, whereas $e_c = -1$. Thus we have deformed the specimen infinitely, which is exactly what the value of the true strain indicates. From these examples, true strains obviously are consistent with the actual physical phenomenon and engineering strains are not.

2.2.3 True stress–true strain curves

The relation between engineering and true values for stress and strain can now be used to construct true stress–true strain curves from a curve such as that in Fig. 2.2. The procedure for this construction is given in Example 2.2.

A typical true stress–true strain curve is shown in Fig. 2.3. For convenience, such a curve can be approximated by the equation

$$\sigma = K\varepsilon^n. \tag{2.11}$$

Note that Eq. (2.11) indicates neither the elastic region nor the yield point Y of the material. These quantities, however, are readily available from the engineering stress–strain curve. (Since the strains at the yield point are very small, the difference between the true and engineering yield stress is negligible for metals. The reason is that the difference in the cross-sectional areas A_o and A at yielding is very small.)

We can rewrite Eq. (2.11) as

$$\log \sigma = \log K + n \log \varepsilon.$$

Thus, if we now plot the true stress–true strain curve on a log-log scale, we obtain Fig. 2.3(b). The slope n is known as the *strain-hardening exponent*, and K is known as the *strength coefficient*. Note that K is the true stress at a true strain of unity.

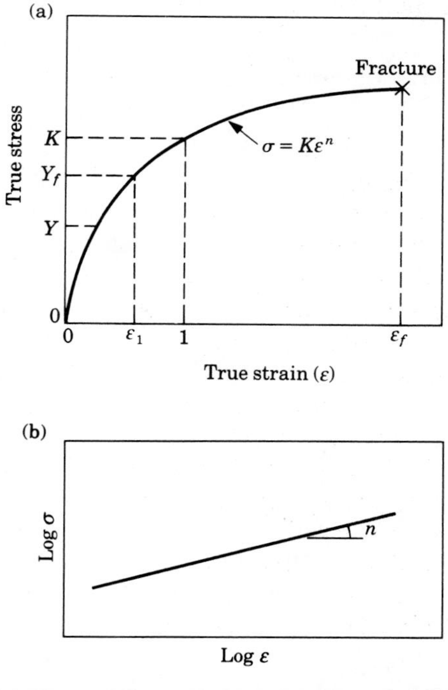

FIGURE 2.3

(a) True stress–true strain curve in tension. Note that, unlike in an engineering stress–strain curve, the slope is always positive and that the slope decreases with increasing strain. Although in the elastic range stress and strain are proportional, the total curve can be approximated by the power expression shown. On this curve, Y is the yield stress and Y_f is the flow stress. (b) True stress–true strain curve plotted on a log–log scale.

Values of K and n for a variety of engineering materials are given in Table 2.4. The true stress–true strain curves for some materials are given in Fig. 2.4. Some differences between Table 2.4 and these curves may exist because of different sources of data and test conditions.

In Fig. 2.3 Y_f is the *flow stress*, defined as the true stress required to continue plastic deformation at a particular true strain, ε_1. Thus for strain-hardening materials, the flow stress increases with increasing strain.

The area under the true stress–true strain curve is known as *toughness* and can be expressed as

$$\text{Toughness} = \int_o^{\varepsilon_f} \sigma \, d\varepsilon, \tag{2.12}$$

where ε_f is the true strain at fracture. Toughness is defined as the energy per unit volume (specific energy) that has been dissipated up to fracture. It is important to remember that this specific energy pertains only to that volume of material at the narrowest region of the neck. Any volume of material away from the neck has undergone less strain and hence has dissipated less energy.

As used here, *toughness* is different from the concept of fracture toughness as treated in textbooks on fracture mechanics. Fracture mechanics (the study of the

TABLE 2.4
TYPICAL VALUES FOR *K* AND *n* IN EQ. (2.11) AT ROOM TEMPERATURE

Material	K (MPa)	n
Aluminum, 1100-O	180	0.20
2024-T4	690	0.16
5052-O	210	0.13
6061-O	205	0.20
6061-T6	410	0.05
7075-O	400	0.17
Brass, 60-39-1 Pb, annealed	800	0.33
70-30, annealed	895	0.49
85-15, cold-rolled	580	0.34
Bronze (phosphor), annealed	720	0.46
Cobalt-base alloy, heat-treated	2070	0.50
Copper, annealed	315	0.54
Molybdenum, annealed	725	0.13
Steel, low-carbon annealed	530	0.26
1045 hot-rolled	965	0.14
1112 annealed	760	0.19
1112 cold-rolled	760	0.08
4135 annealed	1015	0.17
4135 cold-rolled	1100	0.14
4340 annealed	640	0.15
17-4 P-H annealed	1200	0.05
52100 annealed	1450	0.07
302 stainless, annealed	1300	0.30
304 stainless, annealed	1275	0.45
410 stainless, annealed	960	0.10

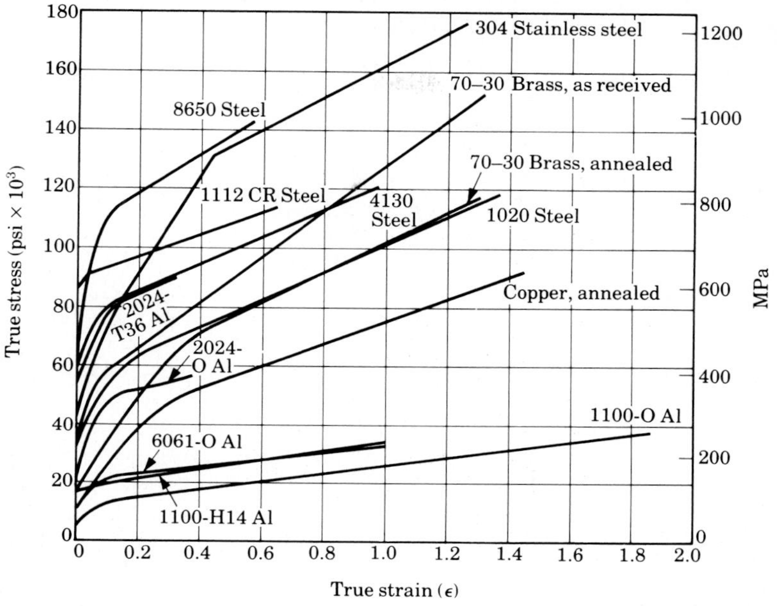

FIGURE 2.4 True stress–true strain curves in tension at room temperature for various metals. The point of intersection of each curve at the ordinate is the yield stress Y; thus the elastic portions of the curves are not indicated. When the K and n values are determined from these curves, they may not agree with those given in Table 2.4 because of the different sources from which they were collected. *Source*: S. Kalpakjian.

initiation and propagation of cracks in a solid medium) is beyond the scope of this text. It is of limited relevance to metalworking processes, except in die design and die life.

• Example 2.2: Construction of true stress–true strain curve.

The following data are taken from a stainless steel tension test specimen:

$A_o = 0.056 \text{ in}^2$
$A_f = 0.016 \text{ in}^2$
$\ell_o = 2$ in.

Elongation in 2 in. = 49%

LOAD, P, (lb)	EXTENSION, $\Delta\ell$ (in.)
1600	0
2500	0.02
3000	0.08
3600	0.20
4200	0.40
4500	0.60
4600 (max.)	0.86
3300 (fracture)	0.98

Draw the true stress–true strain curve for the material.

SOLUTION. In order to determine the true stress and true strain, we use the following relationships:

$$\text{True stress } \sigma = \frac{P}{A}$$

$$\text{True strain } \varepsilon = \ln\left(\frac{\ell}{\ell_o}\right) \text{ up to necking}$$

$$\varepsilon_f = \ln\left(\frac{A_o}{A_f}\right) \text{ at fracture}$$

$$\ell = \ell_o + \Delta\ell \text{ up to necking.}$$

Assuming that the volume of the specimen remains constant, we have

$$A_o\ell_o = A\ell.$$

Thus

$$A = \frac{A_o\ell_o}{\ell} = \frac{(0.056)(2)}{\ell} = \frac{0.112}{\ell}\ \text{in}^2.$$

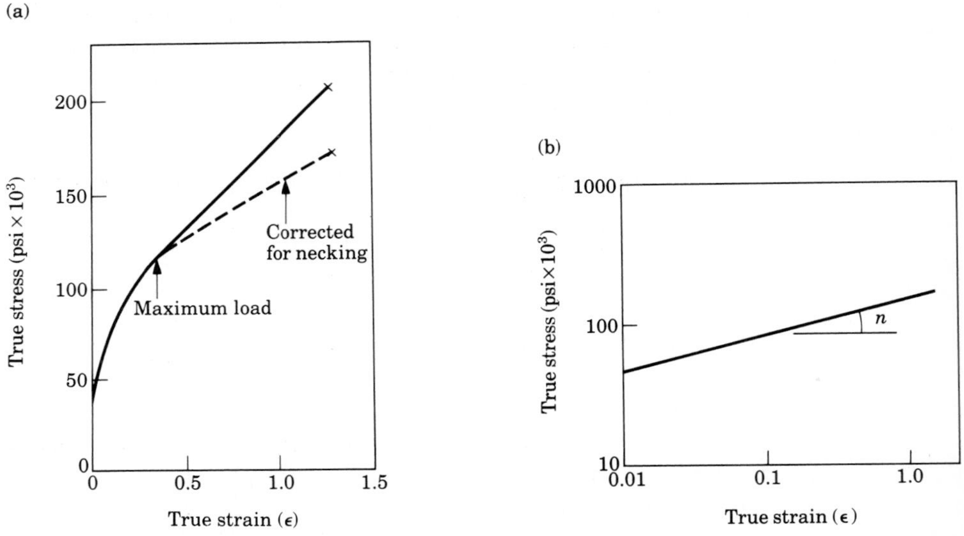

FIGURE E2.1
(a) True stress–true strain curve developed from the data given. Note that this curve has a positive slope, indicating that the material is becoming stronger as it is strained. (b) True stress–true strain curve plotted on log–log paper and based on the corrected curve. The steeper the slope n, the higher will be the capacity of the material to become stronger and harder as it is worked.

Using these relationships, we obtain the following data:

$\Delta \ell$ (in.)	ℓ (in.)	ε	A (in^2)	TRUE STRESS, (psi)
0	2.00	0	0.056	28,600
0.02	2.02	0.01	0.0555	45,000
0.08	2.08	0.039	0.054	55,800
0.20	2.20	0.095	0.051	70,800
0.40	2.40	0.182	0.047	90,000
0.60	2.60	0.262	0.043	104,500
0.86	2.86	0.357	0.039	117,300
0.98	2.98	1.253	0.016	206,000

The true stress and true strain are plotted (solid line) in Fig. E2.1. The point at necking is connected to the point at fracture by a straight line, because there are no data on the instantaneous areas after necking begins. The correction on this curve (broken line) is explained in Section 2.13.2.

2.2.4 Instability in tension

As we noted previously, the onset of necking in a tension test corresponds to the ultimate tensile strength (UTS) of the material. The slope of the load–elongation curve at this point is zero (or $dP = 0$), and it is here that instability begins. That is, the specimen begins to neck and cannot support the load because the neck is becoming smaller in cross-sectional area.

Using the relationships

$$\varepsilon = \ln\left(\frac{A_o}{A}\right), \qquad A = A_o e^{-\varepsilon}, \quad \text{and} \quad P = \sigma A = \sigma A_o e^{-\varepsilon},$$

we determine $\frac{dP}{d\varepsilon}$. Thus

$$\frac{dP}{d\varepsilon} = A_o\left(\frac{d\sigma}{d\varepsilon} e^{-\varepsilon} - \sigma e^{-\varepsilon}\right).$$

Because $dP = 0$ at the UTS where necking begins, we set this expression equal to zero and obtain

$$\frac{d\sigma}{d\varepsilon} = \sigma.$$

However,

$$\sigma = K\varepsilon^n,$$

and, consequently,

$$nK\varepsilon^{n-1} = K\varepsilon^{n}$$

$$\varepsilon = n. \tag{2.13}$$

Thus the true strain at the onset of necking (termination of uniform elongation) is numerically equal to the strain-hardening exponent, n. Hence the higher the value of n, the greater will be the strain to which a piece of material can be stretched before necking begins. Table 2.4 shows that metals such as annealed copper, brass, and stainless steel can be stretched uniformly to a greater extent than the other materials listed. These observations are covered in greater detail in Chapters 7 and 10 because of their relevance to sheet-forming processes.

Instability in a tension test can be viewed as a phenomenon where two competing processes are simultaneously taking place. As the load on the specimen is increased, its cross-sectional area becomes smaller, which is more pronounced in the region where necking begins. With increasing strain, however, the material is becoming stronger due to strain hardening. Since the load on the specimen is the product of area and strength, instability sets in when the rate of decrease in area is greater than the rate of increase in strength. This condition is also known as *geometric softening*.

• Example 2.3: Calculation of an ultimate tensile strength.

A material has a true stress–true strain curve given by

$$\sigma = 100{,}000\varepsilon^{0.5} \text{ psi}.$$

Calculate the true ultimate tensile strength and the engineering UTS of this material.

SOLUTION. Since the necking strain corresponds to the maximum load and the necking strain for this material is

$$\varepsilon = n = 0.5,$$

we have as the true ultimate tensile strength

$$\begin{aligned}\sigma &= Kn^{n}\\ &= 100{,}000(0.5)^{0.5} = 70{,}710 \text{ psi}.\end{aligned}$$

The true area at the onset of necking is obtained from

$$\ln\left(\frac{A_o}{A_{\text{neck}}}\right) = n = 0.5.$$

Thus

$$A_{\text{neck}} = A_o e^{-0.5}$$

and the maximum load P is

$$P = \sigma A = \sigma A_0 e^{-0.5},$$

where σ is the true ultimate tensile strength. Hence

$$P = (70{,}710)(0.606)(A_o) = 42{,}850 A_o \text{ lb.}$$

Since UTS $= P/A_o$,

$$\text{UTS} = 42{,}850 \text{ psi.}$$

2.2.5 Types of stress–strain curves

Every material has a differently shaped stress–strain curve. Its shape depends on its composition and many other factors to be treated in detail later. Some of the major types of curves are shown in Fig. 2.5 and have the following characteristics:

1. A perfectly elastic material behaves like a spring with stiffness E. The behavior of brittle materials, such as glass, ceramics, and some cast irons, may be represented by such a curve (Fig. 2.5a). There is a limit to the stress the material can sustain, after which it breaks. Permanent deformation, if any, is negligible.
2. A rigid, perfectly plastic material has, by definition, an infinite value of E. Once the stress reaches the yield stress Y, it continues to undergo deformation at the same stress level. When the load is released, the material has undergone permanent deformation, with no elastic recovery (Fig. 2.5b).
3. An elastic, perfectly plastic material is a combination of the first two: It has a finite elastic modulus and it undergoes elastic recovery when the load is released (Fig. 2.5c).
4. A rigid, linearly strain-hardening, material requires an increasing stress level to undergo further strain. Thus its *flow stress* (magnitude of the stress required to maintain plastic deformation at a given strain) increases with

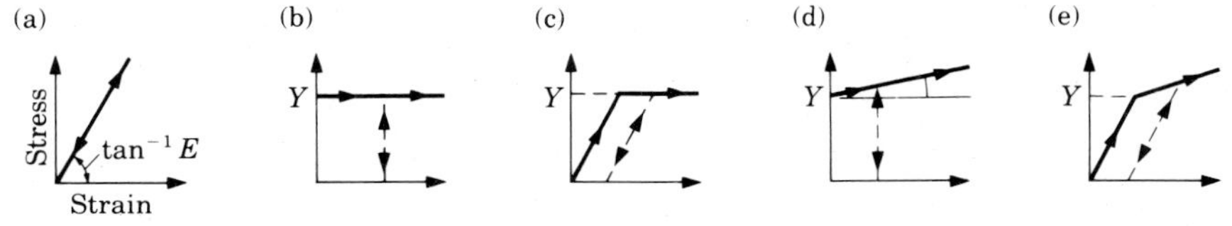

FIGURE 2.5 Schematic illustration of various types of idealized stress–strain curves. (a) Perfectly elastic. (b) Rigid, perfectly plastic. (c) Elastic, perfectly plastic. (d) Rigid, linearly strain hardening. (e) Elastic, linearly strain hardening. The broken lines and arrows indicate unloading and reloading during the test. Most engineering metals exhibit a behavior similar to curve (e).

increasing strain. It has no elastic recovery upon unloading (Fig. 2.5d). See Fig. 2.51 for flow stress.

5. An elastic, linearly strain-hardening curve (Fig. 2.5e) is an approximation of the behavior of most engineering materials, with the modification that the plastic portion of the curve has a decreasing slope with increasing strain (Fig. 2.3).

Note that some of these curves can be expressed by Eq. (2.11) by changing the value of n (Fig. 2.6) or by other equations of a similar nature.

2.2.6 Effects of temperature

In this and subsequent sections, the various factors that have an influence on the shape of stress–strain curves will be discussed. The first factor is temperature. Although somewhat difficult to generalize, increasing temperature usually increases ductility and toughness and lowers the modulus of elasticity, yield stress, and ultimate tensile strength. These effects are shown in Figs. 2.7 and 2.8.

Temperature also affects the strain-hardening exponent n of most metals, in that n decreases with increasing temperature. Depending on the type of material and its composition and level of impurities, elevated temperatures can have other significant effects, as detailed in Section 2.10 and Chapter 3. However, the influence of temperature is best discussed in conjunction with strain rate for the reasons explained in the next section.

2.2.7 Effects of strain rate

Depending on the particular manufacturing operation and equipment, a piece of material may be formed at low or at high speeds. In performing a tension test, the

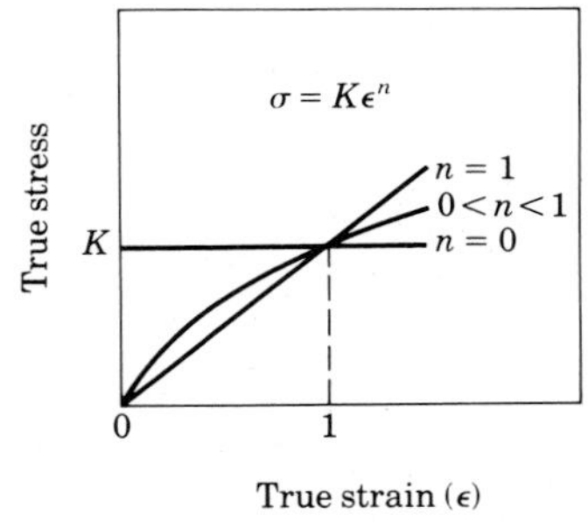

FIGURE 2.6
The effect of strain-hardening exponent n on the shape of true stress–true strain curves. When $n = 1$, the material is elastic, and when $n = 0$ it is rigid, perfectly plastic.

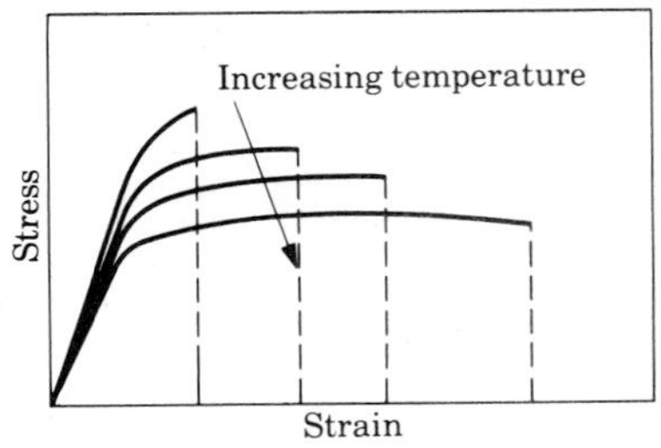

FIGURE 2.7
Typical effects of temperature on engineering stress–strain curves. Temperature affects the modulus of elasticity, yield stress, ultimate tensile strength, and toughness of materials.

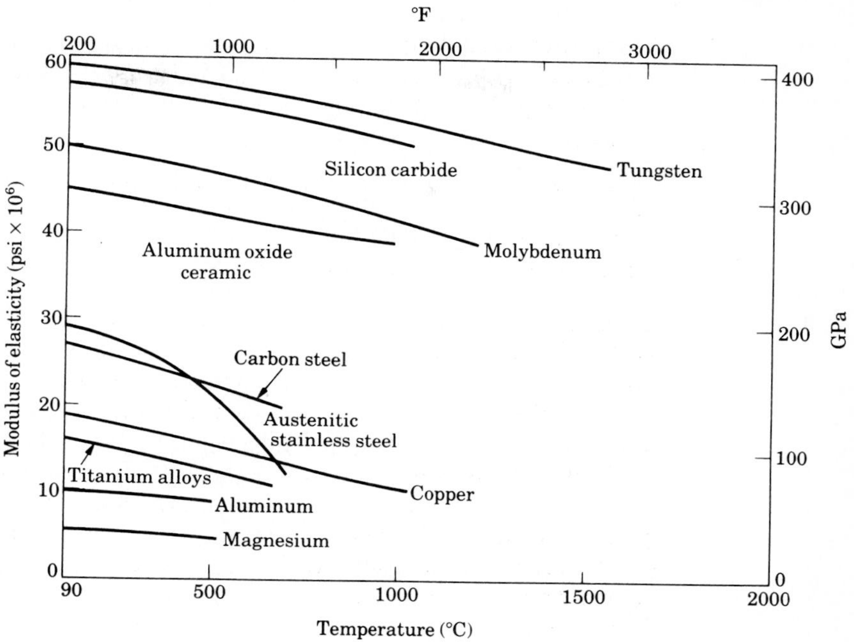

FIGURE 2.8 The effect of temperature on the modulus of elasticity for various materials.

specimen can be strained at different rates to simulate the actual deformation process.

Whereas the *deformation rate* may be defined as the speed (in meters per second or feet per minute, for instance) at which a tension test is being carried out (e.g., the rate at which a rubber band is stretched), the *strain rate* is a function of the geometry of the specimen.

The *engineering strain rate* $\dot{e}$ is defined as

$$\dot{e} = \frac{de}{dt} = \frac{d\left(\frac{\ell - \ell_o}{\ell_o}\right)}{dt} = \frac{1}{\ell_o} \cdot \frac{d\ell}{dt} = \frac{v}{\ell_o}, \tag{2.14}$$

and the *true strain rate* $\dot{\varepsilon}$ as

$$\dot{\varepsilon} = \frac{d\varepsilon}{dt} = \frac{d\left[\ln\left(\frac{\ell}{\ell_o}\right)\right]}{dt} = \frac{1}{\ell} \cdot \frac{d\ell}{dt} = \frac{v}{\ell}, \tag{2.15}$$

where v is the speed of deformation, e.g., the speed of the jaws of the testing machine in which the specimen is clamped.

Although the deformation rate v and engineering strain rate $\dot{e}$ are equivalent, the true strain rate $\dot{\varepsilon}$ is not identical to v. Thus in a tension test with v constant, the true strain rate decreases as the specimen becomes longer. In order to maintain a constant $\dot{\varepsilon}$, the speed must be increased accordingly. However, for small changes in length of the specimen during a test, this difference is not significant.

Typical deformation speeds employed in various metalworking processes, and the strain rates involved, are shown in Table 2.5. Note that there are considerable differences in the magnitudes. Because of this wide range, strain rates are generally quoted in orders of magnitude, such as 10^2 per second, 10^4 per second, etc.

The typical effects of temperature and strain rate on the strength of metals are shown in Fig. 2.9. It clearly indicates that increasing strain rate increases strength and that the sensitivity of strength to the strain rate increases with temperature. Note, however, that this effect is relatively small at room temperature.

Figure 2.9 shows that the same strength can be obtained either at low temperature, low strain rate or at high temperature, high strain rate. These relationships are important in estimating the resistance of materials to deformation when processing them at various strain rates and temperatures.

The effect of strain rate on strength also depends on the particular level of strain: It increases with strain. The strain rate also affects the strain-hardening exponent n, because it decreases with increasing strain rate. The effect of strain rate on the strength of materials is generally expressed by

$$\sigma = C\dot{\varepsilon}^m, \tag{2.16}$$

where C is the *strength coefficient*, similar to K in Eq. (2.11), and m is the *strain-rate sensitivity exponent* of the material.

A general range of values for m is: up to 0.05 for cold working, 0.05 to 0.4 for hot working, and 0.3 to 0.85 for superplastic materials. For a *Newtonian* fluid, where the shear stress increases linearly with rate of shear, the value of m is 1. Some specific values for C and m are given in Table 2.6.

TABLE 2.5
TYPICAL RANGES OF STRAIN, DEFORMATION SPEED, AND STRAIN RATES IN METALWORKING PROCESSES

PROCESS	TRUE STRAIN	DEFORMATION SPEED (m/s)	STRAIN RATE (s^{-1})
Cold working			
Forging, rolling	0.1–0.5	0.1–100	1–10^3
Wire and tube drawing	0.05–0.5	0.1–100	1–10^4
Explosive forming	0.05–0.2	10–100	10–10^5
Hot working and warm working			
Forging, rolling	0.1–0.5	0.1–30	1–10^3
Extrusion	2–5	0.1–1	10^{-1}–10^2
Machining	1–10	0.1–100	10^3–10^6
Sheet-metal forming	0.1–0.5	0.05–2	1–10^2
Superplastic forming	0.2–3	10^{-4}–10^{-2}	10^{-4}–10^{-2}

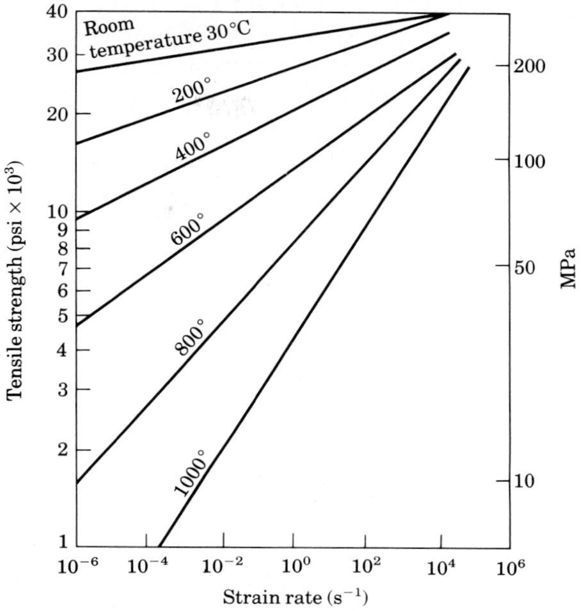

FIGURE 2.9
The effect of strain rate on the ultimate tensile strength of aluminum. Note that as temperature increases, the slope increases. Thus tensile strength becomes more and more sensitive to strain rate as temperature increases. *Source*: After J. H. Hollomon.

TABLE 2.6
APPROXIMATE RANGE OF VALUES FOR *C* AND *m* IN EQ. (2.16) FOR VARIOUS ANNEALED METALS AT TRUE STRAINS RANGING FROM 0.2 TO 1.0. (After T. Altan and F. W. Boulger, *Jr. Eng. Ind.* 95:1009–1019, 1973.)

		C		
MATERIAL	TEMPERATURE, °C	psi × 10³	MPa	*m*
Aluminum	200–500	12–2	82–14	0.07–0.23
Aluminum alloys	200–500	45–5	310–35	0–0.20
Copper	300–900	35–3	240–20	0.06–0.17
Copper alloys (brasses)	200–800	60–2	415–14	0.02–0.3
Lead	100–300	1.6–0.3	11–2	0.1–0.2
Magnesium	200–400	20–2	140–14	0.07–0.43
Steel				
Low-carbon	900–1200	24–7	165–48	0.08–0.22
Medium-carbon	900–1200	23–7	160–48	0.07–0.24
Stainless	600–1200	60–5	415–35	0.02–0.4
Titanium	200–1000	135–2	930–14	0.04–0.3
Titanium alloys	200–1000	130–5	900–35	0.02–0.3
Ti-6Al-4V*	815–930	9.5–1.6	65–11	0.50–0.80
Zirconium	200–1000	120–4	830–27	0.04–0.4

*At a strain rate of 2×10^{-4}/s.
Note: As temperature increases, *C* decreases and *m* increases. As strain increases, *C* increases and *m* may increase or decrease, or it may become negative within certain ranges of temperature and strain.

The term *superplastic* refers to the capability of some materials to undergo large uniform elongation prior to failure. The elongation may be on the order of a few hundred to as much as 2000%. An extreme example is hot glass. Among other materials exhibiting superplastic behavior are polymers at elevated temperatures, very fine-grain alloys of zinc–aluminum and titanium alloys.

The magnitude of m has a significant effect on necking in a tension test. Experimental observations show that, with high m values, the material stretches to a greater length before it fails. This is an indication that necking is delayed with increasing m. When necking is about to begin, its strength with respect to the rest of the specimen increases due to strain hardening. However, the strain rate in the neck region is also higher than in the rest of the specimen, because the material is elongating faster there. Since the material in the necked region is becoming stronger as it is strained at a higher rate, this region exhibits a greater resistance to necking. This increased resistance to necking thus depends on the magnitude of m.

As the test progresses, necking becomes more diffuse and the specimen becomes longer before fracture. Thus total elongation increases with increasing m value. As expected, the elongation after necking (postuniform elongation) also increases with increasing m. Moreover, the value of m decreases with metals of increasing strength.

The effect of strain rate on ductility is difficult to generalize, as Fig. 2.10 shows. Since the formability of materials depends largely on their ductility, it is important

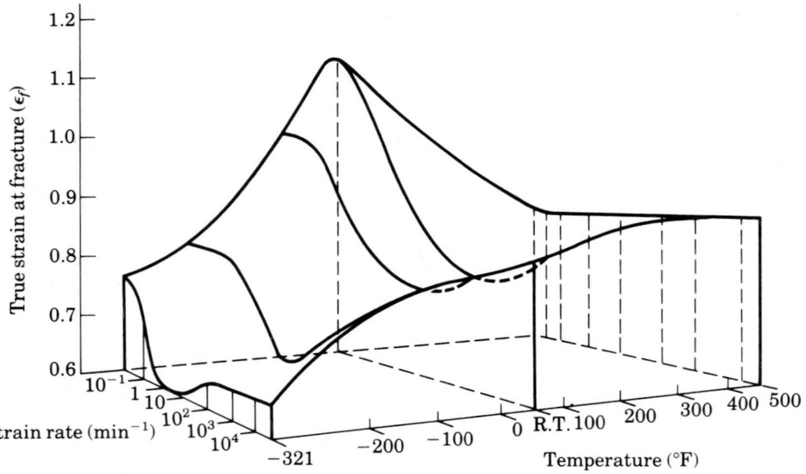

FIGURE 2.10
True strain at fracture for 303 stainless steel as a function of strain rate and temperature. This curve shows that it is difficult to generalize the combined effects of temperature and strain rate on the ductility of metals. *Source*: After G. W. Form and W. M. Baldwin, Jr., *Trans. ASM*, vol. 48, 1956, pp. 474–485.

to recognize the effect of temperature and strain rate on ductility. Generally, higher strain rates have an adverse effect on the ductility of materials. (The increase in ductility due to the strain-rate sensitivity of materials has been exploited in superplastic forming of metals, as described in Section 7.10.)

2.2.8 Effects of hydrostatic pressure

Note that the tests described thus far have been carried out at ambient pressure. Tests have been performed under hydrostatic conditions, with pressures ranging up to 10^3 MPa (10^5 psi). Three important observations have been made concerning the effects of high hydrostatic pressure on the behavior of materials: (1) it substantially increases the strain at fracture (Fig. 2.11); (2) it has little or no effect on the shape of the true stress–true strain curve, but only extends it; and (3) it has no effect on the strain or the maximum load at which necking begins.

The increase in ductility due to hydrostatic pressure has also been observed in other tests, such as compression and torsion tests. This increase in ductility has been observed not only with ductile metals, but also with brittle metals and nonmetallic materials. Materials such as cast iron, marble, and various rocks acquire some ductility (or increase in ductility) and thus deform plastically when subjected to hydrostatic pressure. The level of the pressure required to enhance ductility depends

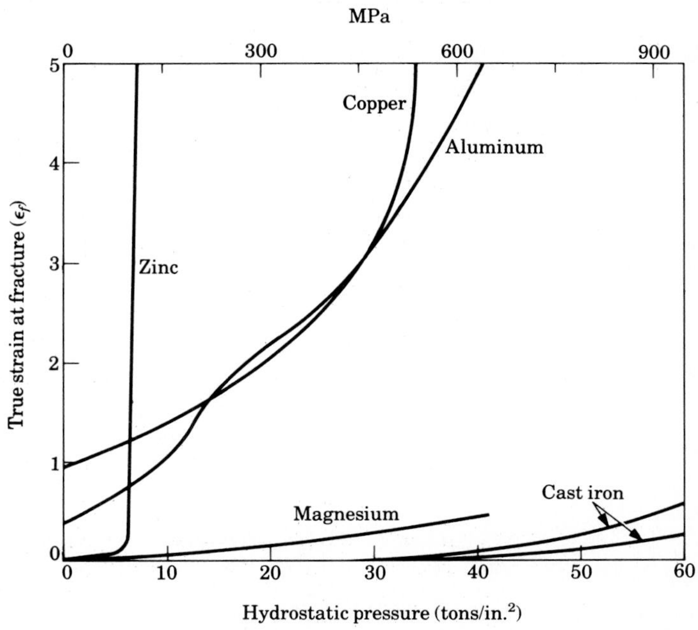

FIGURE 2.11 The effect of hydrostatic pressure on true strain at fracture in tension for various metals. Even cast iron becomes ductile under high pressure. *Source*: After H. Ll. D. Pugh and D. Green, *Proc. Inst. Mech. Eng.*, vol. 179, Part 1, No. 12, 1964–1965.

on the material. Experiments have shown that, generally, the mechanical properties of metals are not altered after being subjected to hydrostatic pressure.

2.2.9 Effects of radiation

In view of the nuclear applications of many metals and alloys, studies have been conducted on the effects of radiation on material properties. Typical changes in the mechanical properties of steels and other metals exposed to high-energy radiation are increased yield stress, increased tensile strength and hardness, and decreased ductility and toughness. The magnitudes of these changes depend on the material and its condition, temperature, and level of radiation.

2.3 Compression

Many operations in metalworking, such as forging, rolling, and extrusion, are performed with the workpieces under compressive loads. The compression test, where the specimen is subjected to a compressive load as shown in Fig. 2.1(b), can give useful information for these processes, such as stresses required and the behavior of the material under compression. However, the deformation shown in Fig. 2.1(b) is ideal. The compression test is usually carried out by compressing a solid cylindrical specimen between two flat platens; the friction between the specimen and the dies is an important factor. Friction causes *barreling* (Fig. 2.12). In other words, friction prevents the top and bottom surfaces from expanding freely.

This situation makes it difficult to obtain relevant data and to construct properly the compressive stress–strain curve because (1) the cross-sectional area of the specimen changes along its height; and (2) friction dissipates energy, and this energy is supplied through an increased compressive force. Thus it is difficult to obtain results that are truly indicative of the properties of the material. With effective lubrication or other means, it is possible, of course, to minimize friction—and hence barreling—to obtain a reasonably constant cross-sectional area during this test.

The engineering strain rate $\dot{e}$ in compression is given by

$$\dot{e} = -\frac{v}{h_o}, \tag{2.17}$$

where v is the speed of the die and h_o is the original height of the specimen. The true strain rate $\dot{\varepsilon}$ is given by

$$\dot{\varepsilon} = -\frac{v}{h}, \tag{2.18}$$

where h is the instantaneous height of the specimen. If v is constant, the true strain rate increases as the test progresses. In order to conduct this test at a constant true

FIGURE 2.12 Barreling in compressing a round solid cylindrical specimen (7075–O aluminum) between flat dies. Barreling is caused by friction at the die–specimen interfaces, which retards the free flow of the material. See also Figs. 6.1 and 6.2. *Source*: K. M. Kulkarni and S. Kalpakjian.

strain rate, a *cam plastometer* has been designed that, through a cam action, reduces v proportionately as h decreases during the test.

The compression test can also be used to determine the ductility of a metal by observing the cracks that form on the barreled cylindrical surfaces of the specimen. Incidentally, hydrostatic pressure has a beneficial effect in delaying the formation of these cracks. With a sufficiently ductile material and effective lubrication, compression tests can be carried out uniformly to large strains. This result is unlike that of the tension test, where, even for very ductile materials, necking sets in after relatively little elongation.

2.3.1 Plane-strain compression test

Another test is the *plane-strain compression test* (Fig. 2.13), which simulates processes such as rolling. In this test, the die and workpiece geometries are such that

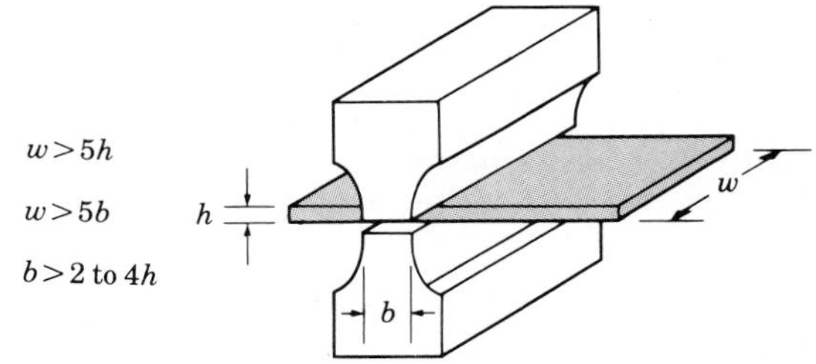

FIGURE 2.13 Schematic illustration of the plane-strain compression test. The dimensional relationships shown should be satisfied for this test to be useful and reproducible. This test gives the yield stress of the material in plane strain, Y'. *Source*: After A. Nadai and H. Ford.

the width of the specimen does not undergo any significant change during compression; that is, the material under the dies is in the condition of *plane strain* (see Section 2.13.3). The yield stress of a material in plane strain Y' is

$$Y' = \frac{2}{\sqrt{3}} Y = 1.15Y, \qquad \textbf{(2.19)}$$

according to the distortion–energy yield criterion (see Section 2.13.2).

As the geometric relationships in Fig. 2.13 show, the test parameters must be chosen properly to make the results meaningful. Furthermore, caution should be exercised in test procedures, such as preparing the die surfaces and aligning the dies, lubricating the surfaces, and accurately measuring the load.

When the results of tension and compression tests on the same material are compared, for *ductile* metals the true stress–true stain curves for both tests coincide (Fig. 2.14). However, this is not true for brittle materials, particularly in regard to ductility.

• Example 2.4: Calculation of strain rates.

A 40-mm-high cylindrical specimen is being compressed between flat platens at a speed of 0.1 m/s. Calculate the engineering and true strain rates to which the material is being subjected when its height is 10 mm.

SOLUTION. The engineering strain rate is

$$\dot{e} = -\frac{v}{h_o} = -\frac{0.1}{0.040} = -2.5 \text{ s}^{-1}.$$

The true strain rate is

$$\dot{\varepsilon} = -\frac{v}{h} = -\frac{0.1}{0.010} = -10 \text{ s}^{-1}.$$

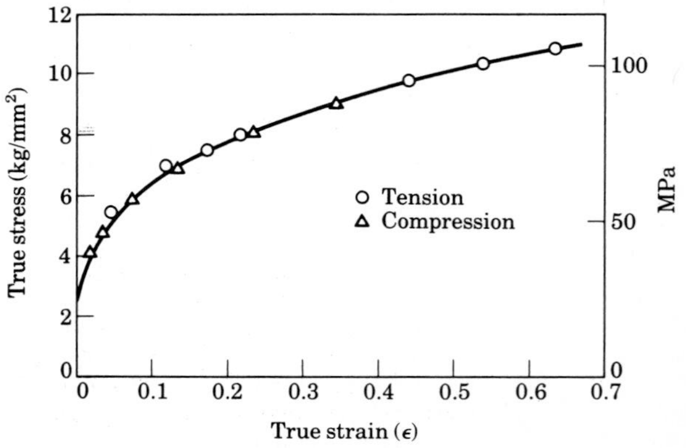

FIGURE 2.14 True stress–true strain curve in tension and compression for aluminum. For ductile metals the curves for tension and compression are identical. *Source*: After A. H. Cottrell, *The Mechanical Properties of Matter*, New York: Wiley, 1964, p. 289.

2.3.2 Bauschinger effect

Sometimes in deformation processing of materials a workpiece is first subjected to tension and then to compression, or vice versa. Examples are bending and unbending, roller leveling (see Section 6.3.3), and reverse drawing (see Section 7.12.2). When a metal with a tensile yield stress Y is subjected to tension into the plastic range and then the load is released and applied in compression, the yield stress in compression is lower than that in tension (Fig. 2.15).

This phenomenon is known as the *Bauschinger effect* and is exhibited in varying degrees by all metals and alloys. This effect is also observed when the loading path is reversed, i.e., compression followed by tension. Because of the lowered yield stress in the reverse direction of load application, this phenomenon is also called *strain softening* or *work softening.*

2.3.3 The disk test

For brittle materials such as ceramics and glasses, a *disk test* has been developed in which the disk is subjected to compression between two hardened flat platens (Fig. 2.16). When loaded as shown, tensile stresses develop perpendicular to the vertical centerline along the disk, fracture begins, and the disk splits vertically in half.

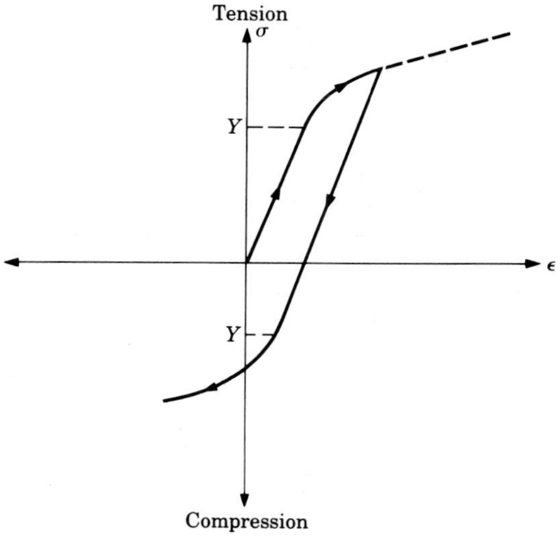

FIGURE 2.15 Schematic illustration of the Bauschinger effect. Arrows show loading and unloading paths. Note the decrease in the yield stress in compression after the specimen has been subjected to tension. The same result is obtained if compression is applied first, followed by tension, whereby the yield stress in tension decreases.

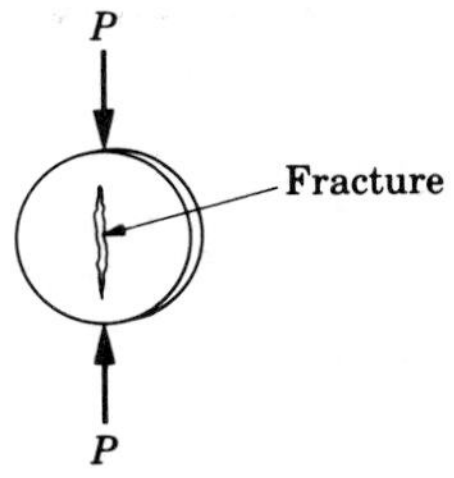

FIGURE 2.16 Disk test on a brittle material, showing the direction of loading and the fracture path. This test is useful for brittle materials, such as ceramics and carbides.

The tensile stress σ in the disk is uniform along the centerline and can be calculated from the formula

$$\sigma = \frac{2P}{\pi d t}, \tag{2.20}$$

where P is the load at fracture, d is the diameter of the disk, and t is its thickness. In order to avoid premature failure at the contact points, thin strips of soft metal are placed between the disk and the platens. These also protect the platens from being damaged during the test.

2.4 Torsion

Another method of determining material properties is the torsion test. In order to obtain an approximately uniform stress and strain distribution along the cross-section, this test is generally carried out on a tubular specimen (Fig. 2.17). The shear stress τ can be determined from the equation

$$\tau = \frac{T}{2\pi r^2 t}, \tag{2.21}$$

where T is the torque, r is the mean radius, and t is the thickness of the reduced section of the tube. The shear strain γ is determined from the equation

$$\gamma = \frac{r\phi}{\ell}, \tag{2.22}$$

where ℓ is the length of the reduced section and ϕ is the angle of twist, in radians.

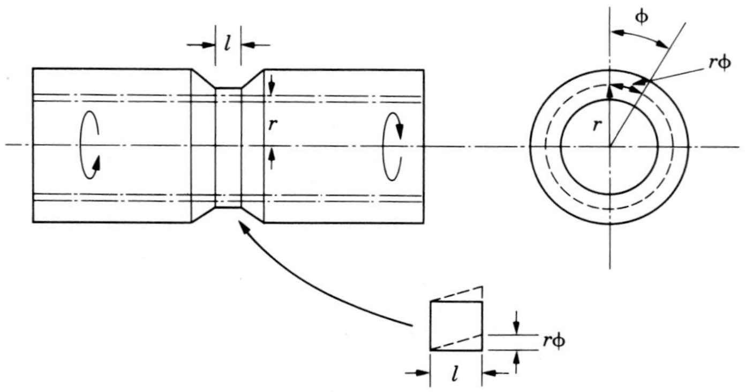

FIGURE 2.17

A typical torsion-test specimen. It is mounted between the two heads of a machine and is twisted. Note the shear deformation of an element in the reduced section.

With the shear stress and shear strain thus obtained from torsion tests, we can construct the shear stress–shear strain curve of the material (see Section 2.13.7).

In the elastic range, the ratio of the shear stress to shear strain is known as the *shear modulus* or the *modulus of rigidity*, G:

$$G = \frac{\tau}{\gamma}. \quad (2.23)$$

The shear modulus and the modulus of elasticity E are related by the formula

$$G = \frac{E}{2(1 + \nu)}. \quad (2.24)$$

Thus for most metals, E is about 2.6 times G.

Note that unlike tension and compression tests, we do not have to be concerned with changes in the cross-sectional area of the specimen in torsion testing. The shear stress–shear strain curves in torsion increase monotonically, hence they are analogous to true stress–true strain curves.

Torsion tests are performed on solid round bars at elevated temperatures in order to estimate the formability of a metal in forging. The greater the number of twists prior to failure, the greater the forgeability of the material. Tests have also been conducted on round bars that are compressed axially. The maximum shear strain at fracture is measured as a function of the compressive stress. The shear strain at fracture increases substantially as the compressive stress increases (Fig. 2.18). This observation again indicates the beneficial effect of a compressive

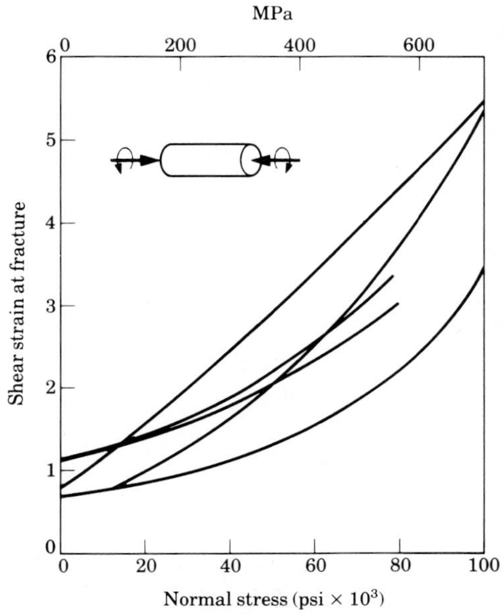

FIGURE 2.18 The effect of axial compressive stress on the shear strain at fracture in torsion for various steels. Note that the effect on ductility is similar to that of hydrostatic pressure, shown in Fig. 2.11. *Source*: Based on data in P. W. Bridgman, *Large Plastic Flow and Fracture*, New York: McGraw-Hill, 1952.

environment on the ductility of materials. Other experiments show that, with tensile normal stresses, the curves in Fig. 2.18 follow a trend downward and to the left, signifying reduced ductility.

The effect of compressive stresses in increasing the maximum shear strain at fracture has also been observed in metal cutting. The normal compressive stress has no effect on the magnitude of shear stresses required to cause yielding or to continue the deformation, just as hydrostatic pressure has no effect on the general shape of the stress–strain curve.

2.5 Bending (Flexure)

Preparing specimens from brittle materials, such as ceramics and carbides, is difficult because of the problems involved in shaping and machining them to proper dimensions. Furthermore, because of their sensitivity to surface defects and notches, clamping brittle test specimens for testing is difficult. Improper alignment of the test specimen may result in nonuniform stress distribution along the cross-section of the specimen.

A commonly used test method for brittle materials is the *bend* or *flexure test*, usually involving a specimen with a rectangular cross-section and supported at its ends (Fig. 2.19). The load is applied vertically—either at one or two points; hence these tests are referred to as three-point or four-point bending, respectively. The stresses in these specimens are tensile at their lower surfaces and compressive at their upper surfaces. We calculate these stresses using simple beam equations described in texts on mechanics of solids. The stress at fracture in bending is known as the *modulus of rupture*, or *transverse rupture strength*, and is obtained from the formula

$$\sigma = \frac{Mc}{I}, \tag{2.25}$$

where M is the bending moment, c is one half of the specimen depth, and I is the moment of inertia of the cross-section.

Note that there is a basic difference between the two loading conditions in Fig. 2.19. In 3-point bending, the maximum stress is at the center of the beam, whereas in the 4-point test the maximum stress is constant between the two loading points. The stress is the same in both cases when all other parameters are maintained. However, there is a greater probability for defects and imperfections to be present in the volume of material between the loading points in the 4-point test than in the

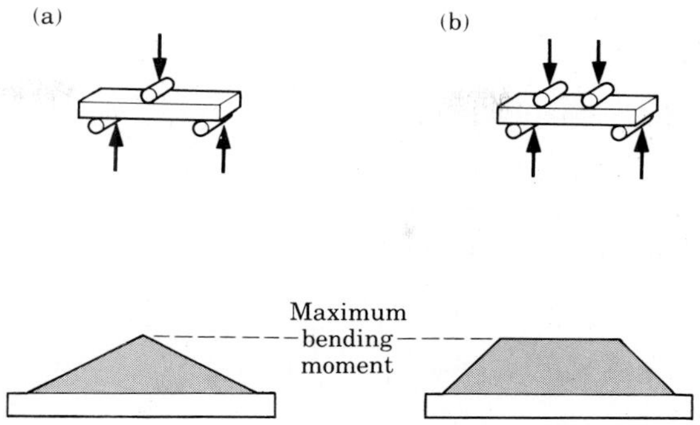

FIGURE 2.19

Two bend-test methods for brittle materials: (a) three-point bending; (b) four-point bending. The shaded areas on the beams represent the bending-moment diagrams, described in texts on mechanics of solids. Note the region of constant maximum bending moment in (b), whereas the maximum bending moment occurs only at the center of the specimen in (a).

much smaller volume under the single load in the 3-point test. This means that the 4-point test is likely to result in a lower modulus of rupture than the 3-point test, which has been verified by experiments. Furthermore, the results of the 4-point test show less scatter than the 3-point test. The reason is that the strength of a small volume of brittle material (such as under the single load in the 3-point test) is subject to greater variations than is the strength of the weakest point in a larger volume (such as in the 4-point test).

2.6 Hardness

One of the most common tests for assessing the mechanical properties of materials is the hardness test. *Hardness* of a material is generally defined as its resistance to permanent indentation. Less commonly, hardness may also be defined as resistance to scratching or to wear. Various techniques have been developed to measure the hardness of materials using different indenter materials and geometries. Because resistance to indentation depends on the shape of the indenter and the load applied, hardness is not a fundamental property. Among the most common standardized hardness tests are the Brinell, Rockwell, Vickers, Knoop, and Scleroscope tests (Fig. 2.20).

2.6.1 Brinell test

In the Brinell test a steel or tungsten carbide ball 10 mm in diameter is pressed against a surface with a load of 500, 1500, or 3000 kg. The Brinell hardness number (HB) is defined as the ratio of the load P to the curved area of indentation, or

$$HB = \frac{2P}{(\pi D)(D - \sqrt{D^2 - d^2})} \text{ kg/mm}^2, \tag{2.26}$$

where D is the diameter of the ball and d is the diameter of the impression in millimeters.

Test	Indenter	Shape of indentation (Side view / Top view)	Load	Hardness number
Brinell	10-mm steel or tungsten carbide ball		500 kg 1500 kg 3000 kg	
Vickers	Diamond pyramid	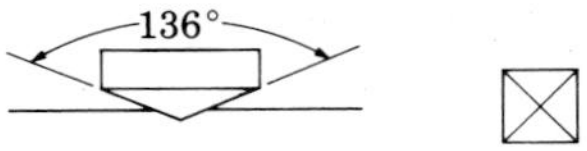	1–120 kg	
Knoop	Diamond pyramid	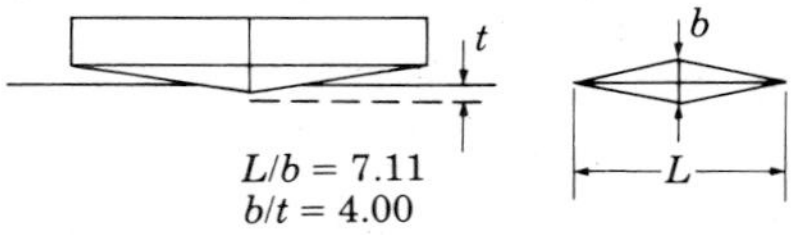	25 g–5 kg	
Rockwell			kg	Hardness number
A C D	Diamond cone	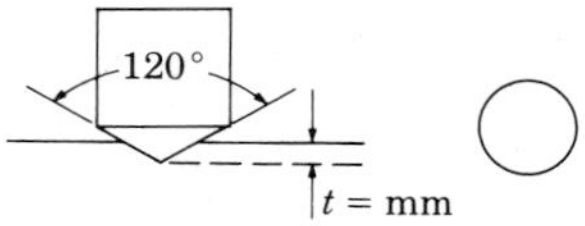	60 150 100	HRA HRC HRD } $= 100 - 500t$
B F G	$\frac{1}{16}$-in. diameter steel ball	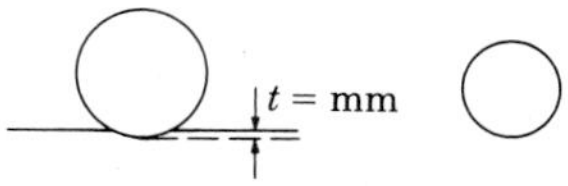	100 60 150	HRB HRF HRG
E	$\frac{1}{8}$-in. diameter steel ball		100	HRE } (B, F, G, E) $= 130 - 500t$

FIGURE 2.20
General characteristics of hardness testing methods. The Knoop test is known as a microhardness test because of the light load and small impressions. *Source*: After H. W. Hayden, W. G. Moffatt, and V. Wulff.

Depending on the condition of the material, different types of impressions are obtained on the surface after a Brinell hardness test has been performed. Annealed materials generally have a rounded profile, whereas cold-worked (strain-hardened) materials have a sharp profile (Fig. 2.21). The correct method of measuring the indentation diameter d for both cases is shown.

Because the indenter (with a finite elastic modulus) also undergoes elastic deformation under the applied load P, hardness measurements may not be as accurate as expected. One method of minimizing this effect is to use tungsten carbide balls, which, because of their high modulus of elasticity, deform less than steel balls. Also, since harder workpiece materials produce very small impressions, a 1500-kg or 3000-kg load is recommended in order to obtain impressions that are sufficiently large for accurate measurement. Tungsten carbide balls are generally recommended for Brinell hardness numbers higher than 500. In reporting the test results for these high hardnesses the type of ball used should be cited.

Because the impressions made by the same indenter at different loads are not geometrically similar, the Brinell hardness number depends on the load used. Thus the load employed should also be cited with the test results. The Brinell test is generally suitable for materials of low to medium hardness.

Brinelling is a term used to describe permanent indentations on a surface between contacting bodies, for example, a component with a hemispherical protrusion or a ball bearing resting on a flat surface. Under fluctuating loads or vibrations, such as during transportation or vibrating foundations, a permanent indentation may be produced on the flat surface by dynamic loading.

2.6.2 Rockwell test

In the Rockwell test the depth of penetration is measured. The indenter is pressed on the surface, first with a minor load and then a major one. The difference in the depth of penetration is a measure of the hardness. There are several Rockwell hardness test scales that use different loads, indenter materials, and indenter geometries. Some of the more common hardness scales and the indenters used are listed in Fig. 2.20. The Rockwell hardness number, which is read directly from a dial on the testing machine, is expressed as follows: If the hardness number is 55 using

FIGURE 2.21
Indentation geometry for Brinell hardness testing: (a) annealed metal; (b) work-hardened metal. Note the difference in metal flow at the periphery of the impressions.

the C scale, then it is written as 55 HRC. *Rockwell superficial hardness* tests have also been developed using lighter loads and the same type of indenters.

2.6.3 Vickers test

The *Vickers* test, also known as the *diamond pyramid hardness* test, uses a pyramid-shaped diamond indenter (see Fig. 2.20) with loads ranging from 1 to 120 kg. The Vickers hardness number (HV) is given by the formula

$$\mathrm{HV} = \frac{1.854P}{L^2}. \tag{2.27}$$

The impressions are typically less than 0.5 mm on the diagonal. The Vickers test gives essentially the same hardness number regardless of the load. It is suitable for testing materials with a wide range of hardness, including very hard steels.

2.6.4 Knoop test

The Knoop test uses a diamond indenter in the shape of an elongated pyramid (see Fig. 2.20) with loads ranging generally from 25 g to 5 kg. The Knoop hardness number (HK) is given by the formula

$$\mathrm{HK} = \frac{14.2P}{L^2}. \tag{2.28}$$

The Knoop test is a *microhardness* test because of the light loads utilized and hence is suitable for very small or thin specimens and for brittle materials, such as gemstones, carbides, and glass. This test is also used in measuring the hardness of individual grains in a metal. The size of the indentation is generally in the range of 0.01 to 0.10 mm; thus surface preparation is very important. Because the hardness number obtained depends on the applied load, test results should always cite the load applied.

2.6.5 Scleroscope

The Scleroscope is an instrument in which a diamond-tipped indenter (hammer), enclosed in a glass tube, is dropped on the specimen from a certain height. The hardness is determined by the rebound of the indenter; the higher the rebound, the harder the specimen. Indentation is slight. Since the instrument is portable it is useful for measuring the hardness of large objects.

2.6.6 Mohs test

The Mohs test is based on the capability of one material to scratch another. Mohs hardness is a scale of 10, with 1 for talc and 10 for diamond (hardest substance known). Thus a material with a higher Mohs hardness can scratch ones with a lower hardness. The Mohs scale is used generally by mineralogists and geologists. However, some of the materials tested are of interest to manufacturing engineers, as discussed in Chapters 8, 9, and 11. Although the Mohs scale is qualitative, good correlation is obtained with Knoop hardness. Soft metals have a Mohs hardness of 2 to 3, hardened steels about 6, and aluminum oxide (used in grinding wheels) a 9.

2.6.7 Durometer

The hardness of rubbers, plastics, and similar soft and elastic materials is generally measured with an instrument called a *durometer*. This is an empirical test in which an indenter is pressed against the surface, with a constant load applied rapidly. The depth of penetration is measured after one second, and the hardness is inversely related to the penetration. There are two different scales for this test. Type A has a blunt indenter, a load of 1 kg, and is used for softer materials. Type D has a sharper indenter, a load of 5 kg, and is used for harder materials. The hardness numbers in these tests range from 0 to 100.

2.6.8 Hardness and strength

Since hardness is the resistance to permanent indentation, testing for it is equivalent to performing a compression test on a small portion of a material's surface. Thus we would expect some correlation between hardness and yield stress Y in the form of

$$\text{Hardness} = cY, \tag{2.29}$$

where c is a proportionality constant.

Theoretical studies, based on plane strain slip-line analysis with a smooth flat punch indenting the surface of a semi-infinite body, show that for a perfectly plastic material of yield stress Y the value of c is about 3. This value is in reasonably good agreement with experimental data (Fig. 2.22). Note that cold-worked materials

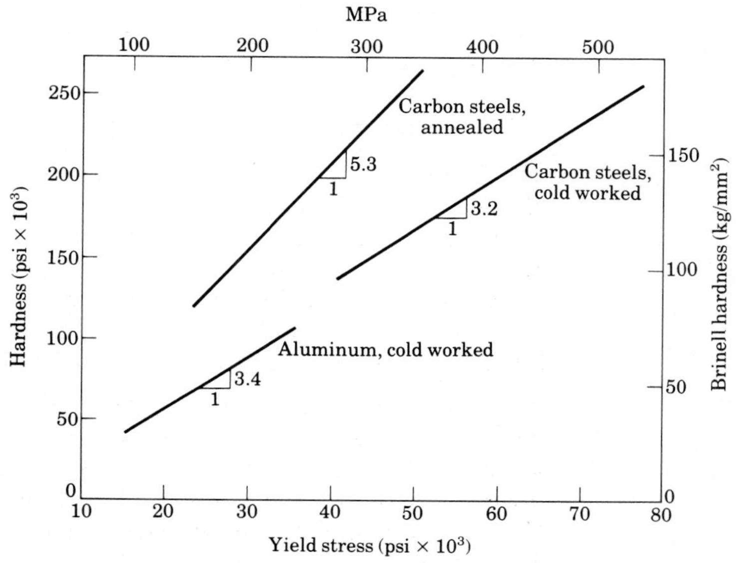

FIGURE 2.22 Relation between Brinell hardness and yield stress for aluminum and steels. For comparison, the Brinell hardness (which is always measured in kg/mm²) is converted to psi units on the left scale.

(which are close to being perfectly plastic in their behavior) show better agreement than annealed ones. The higher c value for the annealed materials is explained by the fact that, due to strain hardening, the average yield stress they exhibit during indentation is higher than their initial yield stress.

The reason that hardness, as a compression test, gives higher values than the uniaxial yield stress Y of the material can be seen in the following analysis. If we assume that the volume under the indenter is a column of material, it would exhibit a uniaxial compressive yield stress Y. However, the volume being deformed under the indenter is, in reality, surrounded by a rigid mass (Fig. 2.23). The surrounding mass prevents this volume of material from deforming freely. In fact, this volume is under *triaxial compression.* As we show in Section 2.13 on yield criteria, this material requires a normal compressive yield stress that is higher than the uniaxial yield stress of the material.

More practically, a relationship also has been observed between the ultimate tensile strength (UTS) and Brinell hardness number (HB) for steels:

$$\text{UTS} = 500(\text{HB}), \tag{2.30}$$

where UTS is in psi and HB in kg/mm^2 as measured with a load of 3000 kg. In SI units, the relationship is given by

$$\text{UTS} = 3.5(\text{HB}), \tag{2.31}$$

where UTS is in MPa.

Hot hardness tests can also be carried out using conventional testers with certain modifications, such as surrounding the specimen and indenter with a small electric furnace. The hot hardness of materials is important in applications where

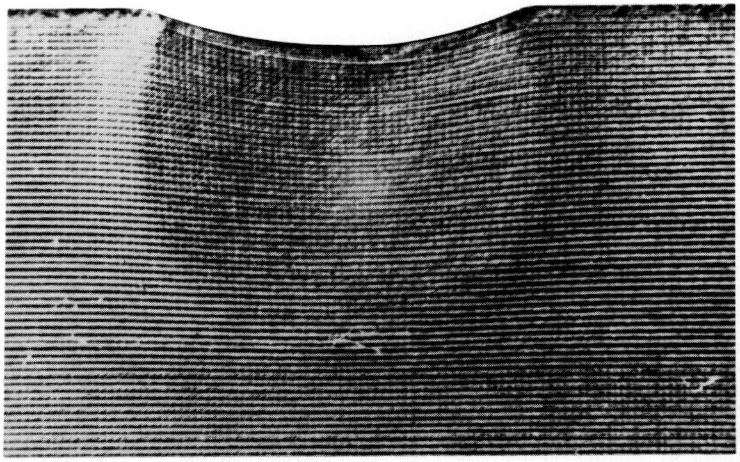

FIGURE 2.23 Bulk deformation in mild steel under a spherical indenter. Note that the depth of the deformed zone is about one order of magnitude larger than the depth of indentation. For a hardness test to be valid, the material should be allowed to fully develop this zone. This is why thinner specimens require smaller impressions. *Source*: Courtesy of M. C. Shaw and C. T. Yang.

the materials are subjected to elevated temperatures, such as in cutting tools in machining and dies for metalworking.

Example 2.5: Calculation of modulus of resilience.

A piece of steel is highly deformed at room temperature. Its hardness is found to be 300 HB. Estimate the modulus of resilience for this material in lb/in^2.

SOLUTION. Since the steel has been subjected to large strains at room temperature, we may assume that its stress–strain curve has flattened considerably, thus approaching the shape of a perfectly plastic curve. According to Eq. (2.29) and using a value of $c = 3$, we obtain

$$Y = \frac{300}{3} = 100 \text{ kg/mm}^2 = 142{,}250 \text{ psi}.$$

The modulus of resilience is defined as in Eq. (2.5),

$$\text{Modulus of resilience} = \frac{Y^2}{2E}.$$

For steel, $E = 30 \times 10^6$ psi. Hence,

$$\text{Modulus of resilience} = \frac{(142{,}250)^2}{2 \times 30 \times 10^6} = 337 \text{ in.}\cdot\text{lb/in}^3.$$

2.7 Fatigue

Various structures and components in manufacturing operations, such as tools, dies, gears, cams, shafts, and springs, are subjected to rapidly fluctuating (cyclic or periodic) loads, as well as static loads. Cyclic stresses may be caused by fluctuating mechanical loads, such as gear teeth, or by thermal stresses, such as a cool die coming into repeated contact with hot workpieces. Under these conditions, the part fails at a stress level below which failure would occur under static loading. This phenomenon is known as *fatigue failure*, and is responsible for the majority of failures in mechanical components.

Fatigue test methods involve testing specimens under various states of stress, usually in a combination of tension and compression, or torsion. The test is carried out at various stress amplitudes (S), and the number of cycles (N) to cause total failure of the specimen or part is recorded. Stress amplitude is the maximum stress, in tension and compression, to which the specimen is subjected.

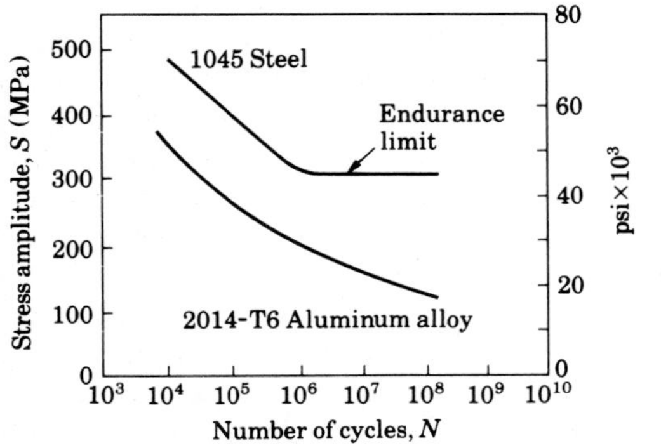

FIGURE 2.24
Typical *S–N* curves for two metals. Note that, unlike steel, aluminum does not have an endurance limit.

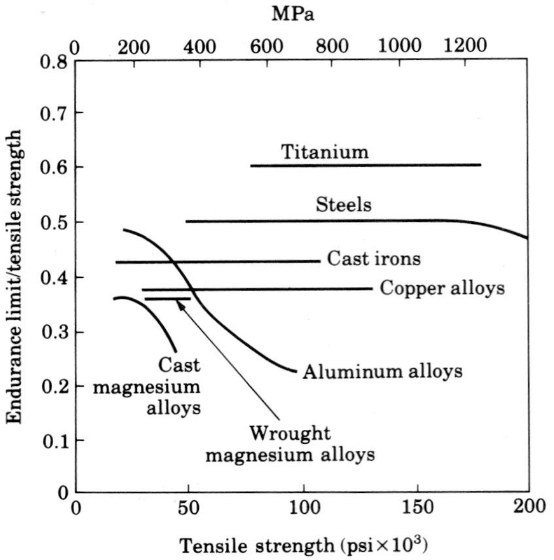

FIGURE 2.25
Ratio of endurance limit to tensile strength for various metals, as a function of tensile strength. Fatigue strength of a metal can thus be estimated from its tensile strength.

A typical plot, known as *S–N curves*, is shown in Fig. 2.24. These curves are based on complete reversal of the stress, that is, maximum tension, maximum compression, maximum tension, and so on, such as that obtained by bending an eraser or piece of wire alternately in one direction, then the other. The test can also be performed on a rotating shaft with a constant downward load. The maximum stress to which the material can be subjected without fatigue failure, regardless of the number of cycles, is known as the *endurance limit* or *fatigue limit*.

The endurance limit for metals is related to their ultimate tensile strength (Fig. 2.25). For steels, the endurance limit is about one-half their tensile strength. Although many metals, especially steels, have a definite endurance limit, aluminum alloys do not have one, and the *S–N* curve continues its downward trend. For metals exhibiting such behavior, the fatigue strength is specified at a certain number of cycles, such as 10^7. In this way the useful service life of the component can be specified.

2.8 Creep

Creep is the permanent elongation of a component under a static load maintained for a period of time. It is a phenomenon of metals and certain nonmetallic materials, such as thermoplastics and rubbers, and can occur at any temperature. Lead, for example, creeps under a constant tensile load at room temperature. The thickness of window glass in old houses has been found to be greater at the bottom than at the top of windows, the glass having undergone creep by its own weight over many years.

For metals and their alloys, creep of any significance occurs at elevated temperatures, beginning at about 200 °C (400 °F) for aluminum alloys, and up to about 1500 °C (2800 °F) for refractory alloys. The mechanism of creep at elevated temperature in metals is generally attributed to grain-boundary sliding. Creep is especially important in high-temperature applications, such as gas-turbine blades and similar components in jet engines and rocket motors. High-pressure steam lines and nuclear-fuel elements also are subject to creep. Creep deformation also can occur in tools and dies that are subjected to high stresses at elevated temperatures during hot-working operations, such as forging and extrusion.

A creep test typically consists of subjecting a specimen to a constant tensile load, hence constant engineering stress, at a certain temperature and measuring the change in length over a period of time. A typical creep curve (Fig. 2.26) usually consists of primary, secondary, and tertiary stages. The specimen eventually fails by necking and fracture, as in a tension test, which is called *rupture* or *creep rupture*. As expected, the creep rate increases with temperature and the applied load.

Design against creep usually involves a knowledge of the secondary (linear) range and its slope, since the creep rate can be determined reliably when the curve has a constant slope. Generally, resistance to creep increases with the melting temperature of a material, which serves as a general guideline for design purposes. Stainless steels, superalloys, and refractory metals and alloys are commonly used in applications where creep resistance is required.

Stress relaxation is closely related to creep. In stress relaxation the stresses resulting from external loading of a structural component decrease in magnitude over a period of time, even though the dimensions of the component remain constant. Examples are rivets, bolts, guy wires, and similar parts under tension,

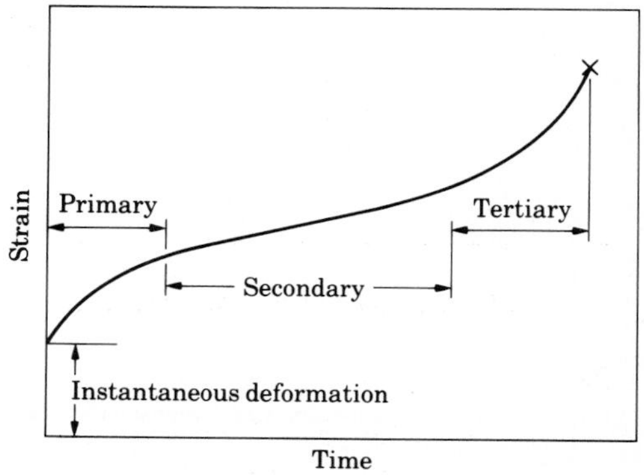

FIGURE 2.26 Schematic illustration of a typical creep curve. The linear segment of the curve (constant slope) is useful in designing components for a specific creep life.

compression, or bending. This phenomenon is particularly common and important in thermoplastics (see Section 10.3).

2.9 Impact

In many manufacturing operations, as well as during the service life of components, materials are subjected to *impact* (or *dynamic*) *loading*, as in high-speed metalworking operations such as drop forging. Tests have been developed to determine the impact toughness of metallic and nonmetallic materials under impact loading. A typical impact test consists of placing a notched specimen in an impact tester and breaking it with a swinging pendulum. In the Charpy test the specimen is supported at both ends (Fig. 2.27a), whereas in the Izod test it is supported at one end like a cantilever beam (Fig. 2.27b). From the amount of swing of the pendulum, the energy dissipated in breaking the specimen is obtained. This energy is the *impact toughness* of the material.

Impact tests are particularly useful in determining the ductile–brittle transition temperature of materials (see Section 2.10.1). Materials that have impact resistance are generally those that have high strength and high ductility, hence high toughness. Sensitivity to surface defects (*notch sensitivity*) is important, as it lowers impact toughness.

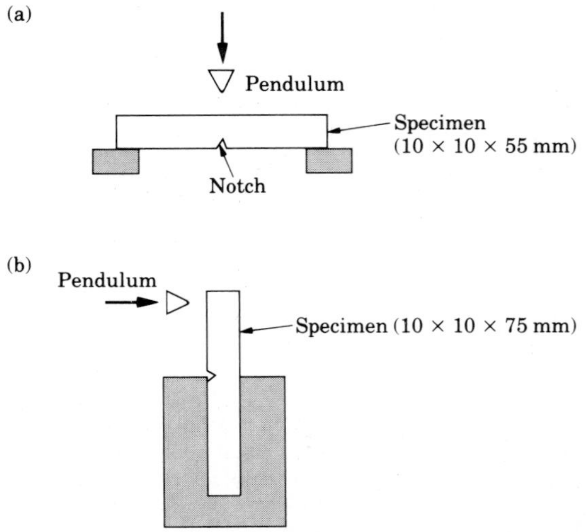

FIGURE 2.27
Impact test specimens: (a) Charpy; (b) Izod.

2.10 Failure and Fracture

Failure is one of the most important aspects of a material's behavior because it directly influences the selection of a material for a certain application, the methods of manufacturing, and the service life of the component. Because of the many factors involved, failure and fracture of materials is a complex area of study. In this section, we consider only those aspects of failure that are of particular significance to selecting and processing materials. There are two general types of failure: (1) fracture and separation of the material, either through internal or external cracking; and (2) buckling (Fig. 2.28). Fracture is further divided into two general categories: ductile and brittle (Fig. 2.29).

Although failure of materials is generally regarded as undesirable, certain products are designed in such a way that failure is essential for their function. Typical examples are food and beverage containers with tabs or entire tops, which are removed by tearing the sheet metal along a prescribed path, and screw caps for bottles.

2.10.1 Ductile fracture

Ductile fracture is characterized by plastic deformation, which precedes failure of the part. In a tension test, highly ductile materials such as gold and lead may neck down to a point and then fail (Fig. 2.29d). Most metals and alloys, however, neck down to a finite area and then fail. Ductile fracture generally takes place along

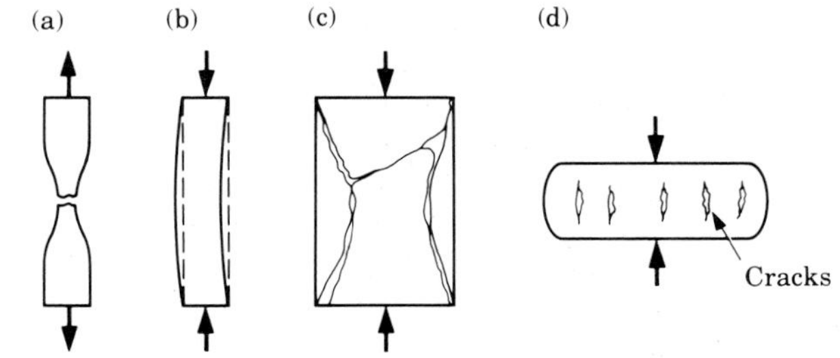

FIGURE 2.28 Schematic illustration of types of failures in materials: (a) necking and fracture of ductile materials; (b) buckling of ductile materials under a compressive load; (c) fracture of brittle materials in compression; (d) cracking on the barreled surface of ductile materials in compression.

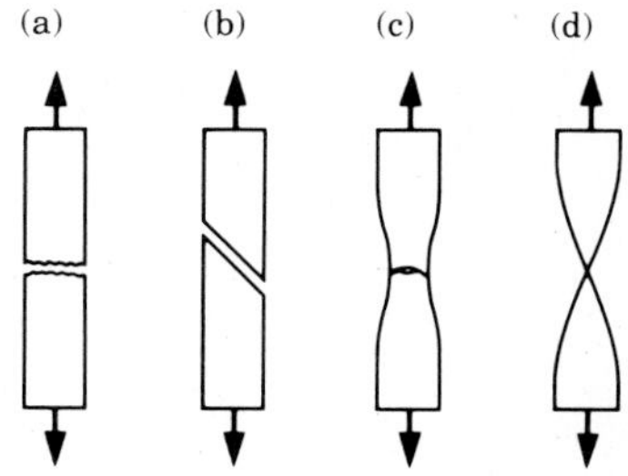

FIGURE 2.29 Schematic illustration of the types of fracture in tension: (a) brittle fracture in polycrystalline metals; (b) shear fracture in ductile single crystals; (c) ductile cup-and-cone fracture in polycrystalline metals; (d) complete ductile fracture in polycrystalline metals, with 100 percent reduction of area.

planes on which the *shear stress is a maximum*. In torsion, for example, a ductile metal fractures along a plane perpendicular to the axis of twist, that is, the plane on which the shear stress is a maximum. Fracture in shear is a result of extensive slip along slip planes within the grains.

Upon close examination of the surface of a ductile fracture (Fig. 2.30) we see a *fibrous* pattern with *dimples*, as if a number of very small tension tests have been carried out over the fracture surface. Failure is initiated with the formation of tiny voids, usually around small inclusions or preexisting *voids*, which then *grow* and *coalesce*, developing cracks which grow in size and lead to fracture.

In a tension-test specimen, fracture begins at the center of the necked region from the growth and coalescence of cavities (Fig. 2.31). The central region becomes one large crack and propagates to the periphery of this necked region. Because of its appearance, the fracture surface of a tension-test specimen is called a *cup-and-cone* fracture.

Effects of inclusions. Because they are nucleation sites for voids, *inclusions* have an important influence on ductile fracture and thus on the formability of materials. Inclusions may consist of impurities of various kinds and second-phase particles, such as oxides, carbides, and sulfides. The extent of their influence depends on factors such as their shape, hardness, distribution, and volume fraction. The greater the volume fraction of inclusions, the lower will be the ductility of the material. Voids and porosity developed during processing, such as from casting, reduce the ductility of a material.

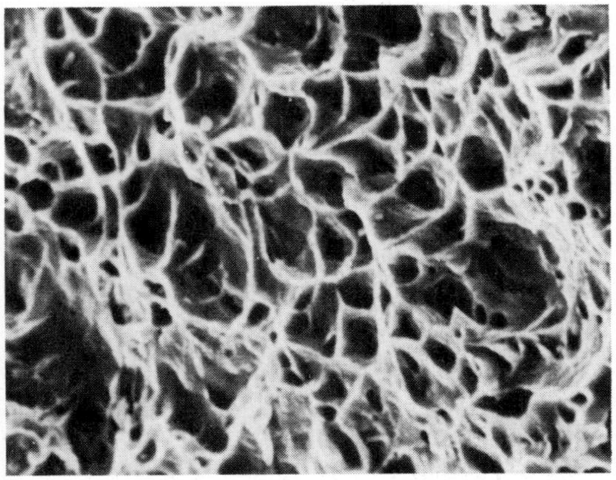

FIGURE 2.30
Surface of ductile fracture in low-carbon steel showing dimples. Fracture is usually initiated at impurities, inclusions, or preexisting voids in the metal. *Source*: K.-H. Habig and D. Klaffke. Photo by BAM Berlin/Germany.

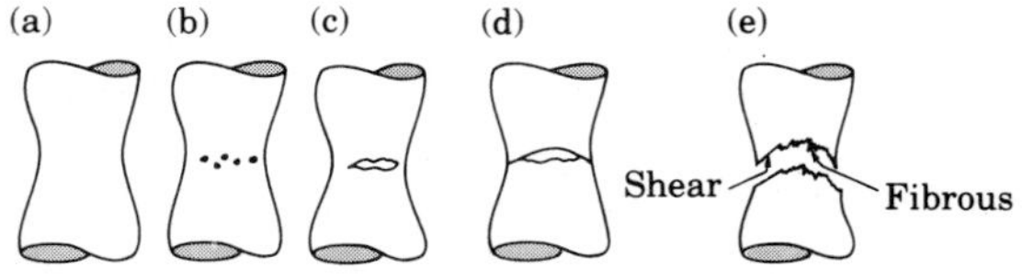

FIGURE 2.31
Sequence of events in necking and fracture of a tensile-test specimen: (a) early stage of necking; (b) small voids begin to form within the necked region; (c) voids coalesce, producing an internal crack; (d) rest of cross-section begins to fail at the periphery by shearing; (e) final fracture surfaces, known as cup- (top fracture surface) and-cone (bottom surface) fracture.

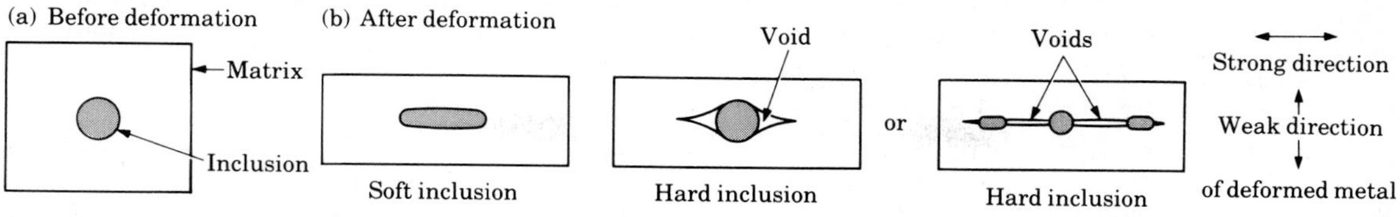

FIGURE 2.32
Schematic illustration of the deformation of soft and hard inclusions and their effect on void formation in plastic deformation. Note that hard inclusions, because they do not comply with the overall deformation of the ductile matrix, can cause voids.

Two factors affect void formation. One is the strength of the bond at the interface of an inclusion and the matrix. If the bond is strong, there is less tendency for void formation during plastic deformation. The second factor is the hardness of the inclusion. If it is soft, such as manganese sulfide, it will conform to the overall change in shape of the specimen or workpiece during plastic deformation. If it is hard, such as carbides and oxides, it could lead to void formation (Fig. 2.32). Hard inclusions may also break up into smaller particles during deformation because of their brittle nature. The alignment of inclusions during plastic deformation leads to *mechanical fibering*. Subsequent processing of such a material must therefore involve considerations of the proper direction of working for maximum ductility and strength.

Transition temperature. Many metals undergo a sharp change in ductility and toughness across a narrow temperature range, called the *transition temperature* (Fig. 2.33). This phenomenon occurs in body-centered cubic and some hexagonal close-packed metals; it is rarely exhibited by face-centered cubic metals. The transition temperature depends on factors such as composition, microstructure, grain size, surface finish and shape of the specimen, and rate of deformation. High rates, abrupt changes in shape, and surface notches raise the transition temperature;

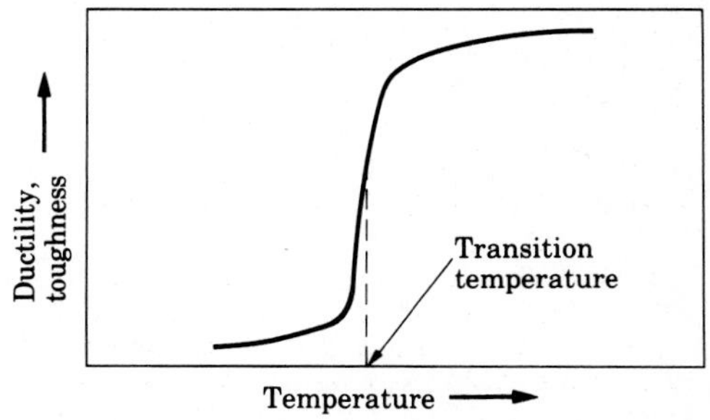

FIGURE 2.33
Schematic illustration of transition temperature. Note the narrow temperature range across which the behavior of the metal undergoes a major transition.

that is, higher temperatures are required to make materials become less ductile and less tough.

Strain aging. *Strain aging* is a phenomenon in which carbon atoms in steels segregate to dislocations, thereby pinning them and thus increasing the resistance to dislocation movement. The result is increased strength and reduced ductility. The effects of strain aging on the shape of the stress–strain curve for low-carbon steel at room temperature are shown in Fig. 2.34. Curve *abc* shows the original curve, with upper and lower yield points typical of these steels. If the tension test is stopped at point *e* and the specimen is unloaded and tested again, curve *dec* is obtained. Note that the upper and lower yield points have disappeared. However, if 4 hours pass before the specimen is strained again, curve *dfg* is obtained. If 126 hours pass, curve *dhi* is obtained. Instead of taking place over several days at room temperature, this same phenomenon can occur in a few hours at a higher temperature, and is called *accelerated strain aging*.

An example of accelerated strain aging in steels is *blue brittleness*, so named because it occurs in the blue-heat range where the steel develops a bluish oxide film. This phenomenon causes a marked decrease in ductility and toughness and an increase in strength of plain carbon and some alloy steels.

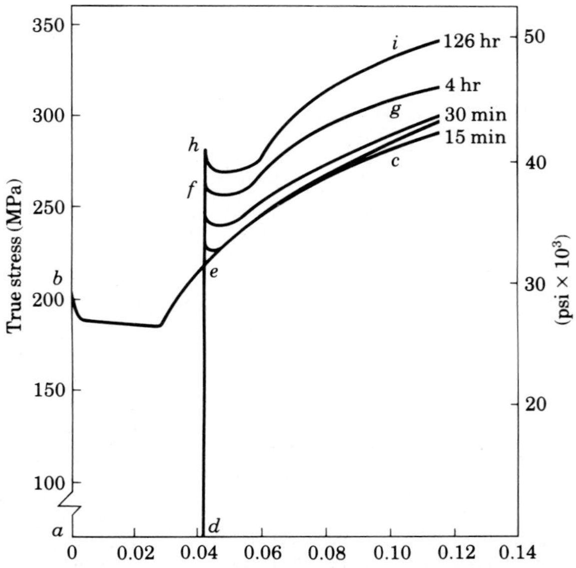

FIGURE 2.34 Strain aging and its effect on the shape of the true stress–true strain curve for 0.03% C rimmed steel at 60 °C (140 °F). *Source*: A. S. Keh and W. C. Leslie.

2.10.2 Brittle fracture

Brittle fracture occurs with little or no gross plastic deformation preceding the separation of the material into two or more pieces. In tension, fracture takes place along a crystallographic plane, called a *cleavage plane*, on which the normal tensile stress is a maximum. Face-centered cubic metals usually do not fail in brittle fracture, whereas body-centered cubic and some hexagonal close-packed metals fail by cleavage. In general, low temperature and high rates of deformation promote brittle fracture.

In a polycrystalline metal under tension, the fracture surface has a bright granular appearance because of the changes in the direction of the cleavage planes as the crack propagates from one grain to another. An example of the surface of brittle fracture is shown in Fig. 2.35. Brittle fracture of a specimen in compression is more complex and may follow a path that is theoretically 45° to the applied force direction.

Examples of fracture along a cleavage plane are the splitting of rock salt or peeling of layers of mica. Tensile stresses normal to the cleavage plane, caused by pulling, initiate and control the propagation of fracture. Another example is the behavior of brittle materials, such as chalk, gray cast iron, and concrete. In tension, they fail in the manner shown in Fig. 2.29(a). In torsion, they fail along a plane at 45° to the axis of twist, that is, along a plane on which the tensile stress is a maximum.

Defects. An important factor in fracture is the presence of defects such as scratches, flaws, and external or internal cracks. Under tension, the tip of the crack is subjected to high tensile stresses, which propagate the crack rapidly because the material has little capacity to dissipate energy. The tensile strength of a specimen,

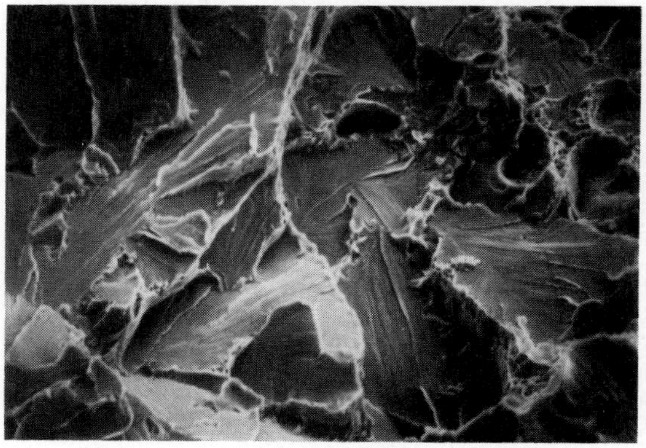

FIGURE 2.35 Typical surface of steel that has failed in a brittle manner. The fracture path is transgranular (through the grains). Compare with the ductile fracture surface in Fig. 2.30. Magnification: 200×. *Source*: Courtesy of B. J. Schulze and S. L. Meiley and Packer Engineering Associates, Inc.

with a crack perpendicular to the direction of pulling, is related to the length of the crack as follows:

$$\sigma \alpha \frac{1}{\sqrt{\text{crack length}}}. \tag{2.32}$$

The presence of defects is essential in explaining why brittle materials are so weak in tension compared to their strength in compression. The difference is on the order of 10 for rocks and similar materials, about 5 for glass, and about 3 for gray cast iron. Under tensile stresses cracks propagate rapidly, causing what is known as catastrophic failure.

With polycrystalline metals, fracture paths most commonly observed are *transgranular* (transcrystalline or intragranular), meaning that the crack propagates through the grain. *Intergranular* fracture, where the crack propagates along the grain boundaries (Fig. 2.36), generally occurs when the grain boundaries are soft, contain a brittle phase, or have been weakened by liquid- or solid-metal embrittlement.

The maximum crack velocity in a brittle material is about 38% of the elastic wave-propagation (or acoustic) velocity of the material. (This velocity is given by the formula $\sqrt{E/\rho}$, where E is the elastic modulus and ρ is the mass density.) Thus, for steel, the maximum crack velocity is 2000 m/s (6600 ft/s).

As Fig. 2.37 shows, cracks may be subjected to stresses in different directions. Mode I is tensile stress applied perpendicular to the crack. Modes II and III are shear

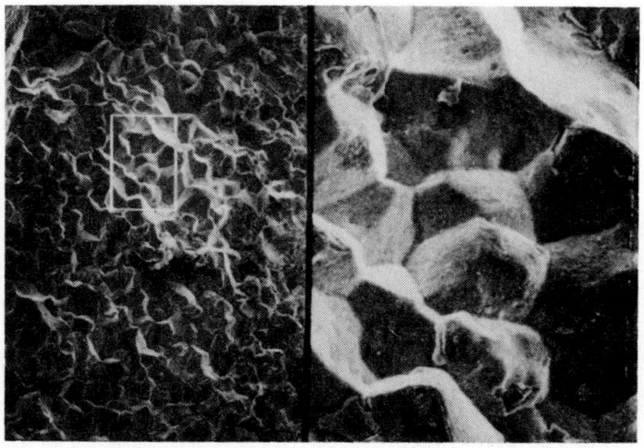

FIGURE 2.36
Intergranular fracture, at two different magnifications. Grains and grain boundaries are clearly visible in this micrograph. The fracture path is along the grain boundaries. Magnification: left, 100×; right, 500×. *Source:* Courtesy of B. J. Schulze and S. L. Meiley and Packer Engineering Associates, Inc.

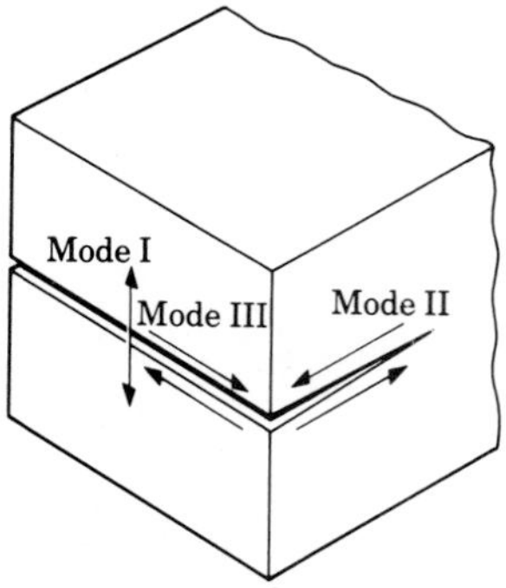

FIGURE 2.37
Three modes of fracture. Mode I has been studied extensively because it is the most commonly observed situation in engineering structures and components. Mode II is rare. Mode III is the tearing process, similar to opening a flip-top can, tearing a piece of paper or cutting materials with a pair of scissors.

stresses applied in two different directions. Tearing paper, cutting sheet metal with shears, or opening flip-top cans are examples of Mode III fracture.

• Example 2.6: Brittle fracture.

A 5-lb block is dropped from a height of 12 in. on a 1-in.-cube-shaped specimen. The specimen is made of a perfectly brittle material with a surface energy of 0.01 in.·lb/in^2. Estimate the number of pieces into which the specimen will break.

SOLUTION. Assume that each broken piece is a cube of dimension x on each side and that all broken pieces are of equal size. Thus, let

x = lateral dimension of each broken piece;
n = number of broken pieces.

Then,

surface area of each broken piece = $6x^2$,
volume of each broken piece = x^3, and
total volume of broken pieces = 1 in^3.

Hence

$$n = \frac{1}{x^3}.$$

The potential energy of the falling block is dissipated as surface energy, so

$$(5)(12) = (n)(6x^2)(0.01) = \frac{0.06}{x}.$$

Thus

$$x = 0.001 \text{ in.}$$

and

$$n = 10^9 \text{ pieces.}$$

Note that, because it represents a very small percentage of the energy involved, the original surface of the specimen (6 in^2) has been ignored in these calculations. Although n appears to be a large number (because of the data in this problem), a piece of chalk, for example, indeed breaks into a large number of small pieces when stepped upon.

• Example 2.7: Comparison of strengths.

Two pieces of glass rod, one tubular and the other solid and round, are subjected to tension. The rods are of equal length and net cross-sectional area. Which rod is likely to carry the higher tensile load before fracture occurs?

SOLUTION. Glass, like other brittle materials, is very notch sensitive and its strength depends on surface flaws, cracks and scratches. Hence the tubular specimen, with its larger surface area, has a greater probability of flaws and a greater number of them. Therefore its strength is likely to be lower than that of the solid, round rod. As discussed in Section 11.12.2, residual stresses on the surface of glass (due to the nature of manufacturing or preparation of the glass specimens) also play a significant role in the strength of glass.

Fatigue fracture. Fatigue fracture is basically of a brittle nature. Minute external or internal cracks develop at flaws or defects in the material, which then propagate and eventually lead to total failure of the part. The fracture surface in fatigue is generally characterized by the term *beach marks*, because of its appearance (Fig. 2.38). With large magnification, such as higher than 1000 ×, a series of *striations* can be seen on fracture surfaces, each beach mark consisting of a number of striations.

Fatigue life is greatly influenced by the method of preparation of the surfaces of the part or specimen (Fig. 2.39). The fatigue strength of manufactured products can be improved generally by the following methods:

a) Inducing compressive residual stresses on surfaces, such as by shot peening or roller burnishing.
b) Surface (case) hardening by various means.

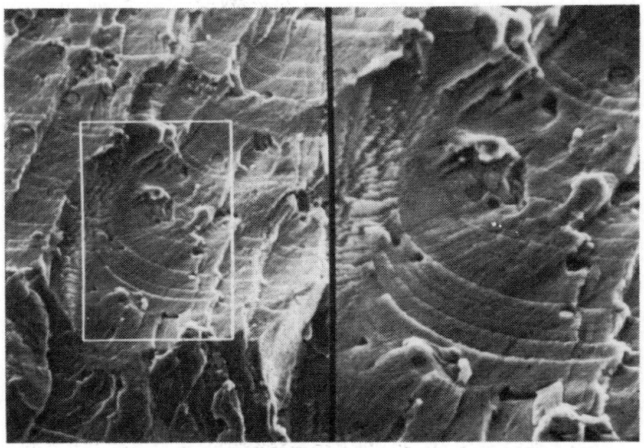

FIGURE 2.38
Typical fatigue fracture surface on metals, showing beach marks. Most components in machines and engines fail by fatigue and not by excessive static loading. Magnification: left, 500 × ; right, 1000 ×. *Source:* Courtesy of B. J. Schulze and S. L. Meiley and Packer Engineering Associates, Inc.

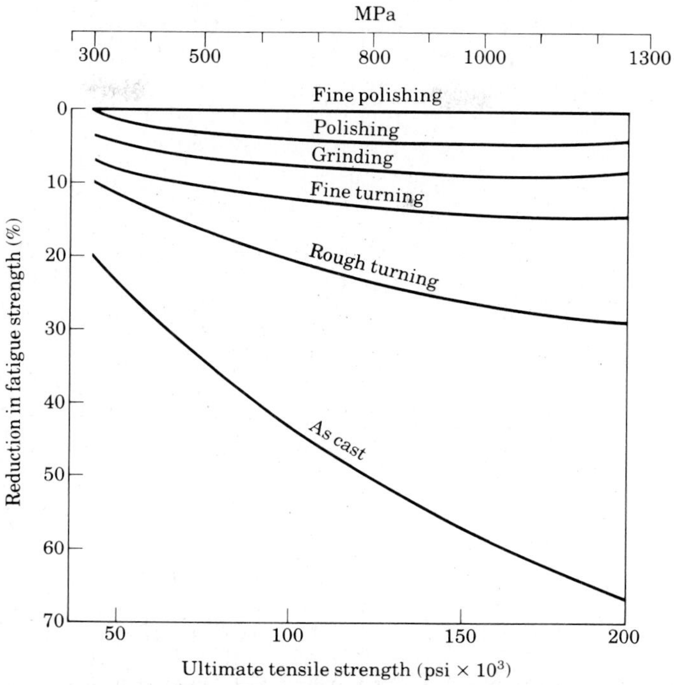

FIGURE 2.39

Reduction in fatigue strength of cast steels subjected to various surface-finishing operations. Note that the reduction is greater as surface roughness and strength of the steel increase. *Source*: M. R. Mitchell.

c) Providing fine surface finish, thus reducing effects of notches and other surface imperfections.
d) Selecting appropriate materials and ensuring that they are free from significant amounts of inclusions, voids, and impurities.

Conversely, the following factors and processes can reduce fatigue strength: decarburization, surface pits due to corrosion that act as stress raisers, hydrogen embrittlement, galvanizing, and electroplating.

Stress–corrosion cracking. An otherwise ductile metal can fail in a brittle manner by *stress–corrosion cracking* (*stress cracking* or *season cracking*). Parts free from defects after forming may, either over a period of time or soon after being made into a product, develop cracks. Crack propagation may be intergranular or transgranular.

The susceptibility of metals to stress–corrosion cracking depends mainly on the material, the presence and magnitude of tensile residual stresses, and the environment. Brass and austenitic stainless steels are among metals that are highly susceptible to stress cracking. The environment could be corrosive media such as salt water or other chemicals. The usual procedure to avoid stress–corrosion cracking is to stress relieve the part just after it is formed. Full annealing may also be done, but this treatment reduces the strength of cold-worked parts.

Hydrogen embrittlement. The presence of hydrogen can reduce ductility and cause severe embrittlement in many metals, alloys, and nonmetallic materials and cause premature failure. This phenomenon is known as *hydrogen embrittlement* and is especially severe in high-strength steels. Possible sources of hydrogen are during melting of the metal, during pickling (removing of surface oxides by chemical or electrochemical reaction), through electrolysis in electroplating, from water vapor in the atmosphere, or from moist electrodes and fluxes used during welding.

2.10.3 Size effect

Dependence of the properties of a material on its size is known as *size effect*. Note from the foregoing discussions and from Eq. (2.32) that as the size decreases, defects, cracks, imperfections, and the like are less likely to occur. Hence the strength and ductility of a specimen increase with decreasing size. Strength is related to diameter, but the length of the specimen is also important. The greater the length, the greater is the probability of defects. As an analogy, a long chain is likely to be weaker than a shorter one, because the probability of one of its links being weak increases with the number of links (length of chain). Size effect is also demonstrated by *whiskers*, which because of their small size, are either free from imperfections or do not contain the type of imperfections that affect their strength. The actual strength in such cases approaches the theoretical strength.

2.11 Deformation-Zone Geometry

The observations made concerning Fig. 2.23 are important in estimating and calculating forces in metalworking operations. As shown previously for hardness testing, the compressive stress required for indentation is, ideally, about 3 times the yield stress Y required for uniaxial compression. Moreover, (a) the deformation under the indenter is localized, making the overall deformation highly nonuniform;

and (b) the deformation zone is relatively small compared to the dimensions of the specimen.

However, in a simple frictionless compression test with flat dies, the top and bottom surfaces of the specimen are always in contact with the dies and the specimen deforms uniformly. Various situations occur between these two extreme examples of specimen-deformation geometry. The *deformation zones* and the pressures required are shown in Fig. 2.40 for the frictionless condition as obtained from *slip-line analysis*. Note that the ratio h/L is the important parameter in determining the inhomogeneity of deformation. The frictionless nature of these examples is important, because friction significantly affects forces, particularly at small values of h/L. Deformation-zone geometry depends on the particular metalworking process and such parameters as die geometry and percent reduction of the

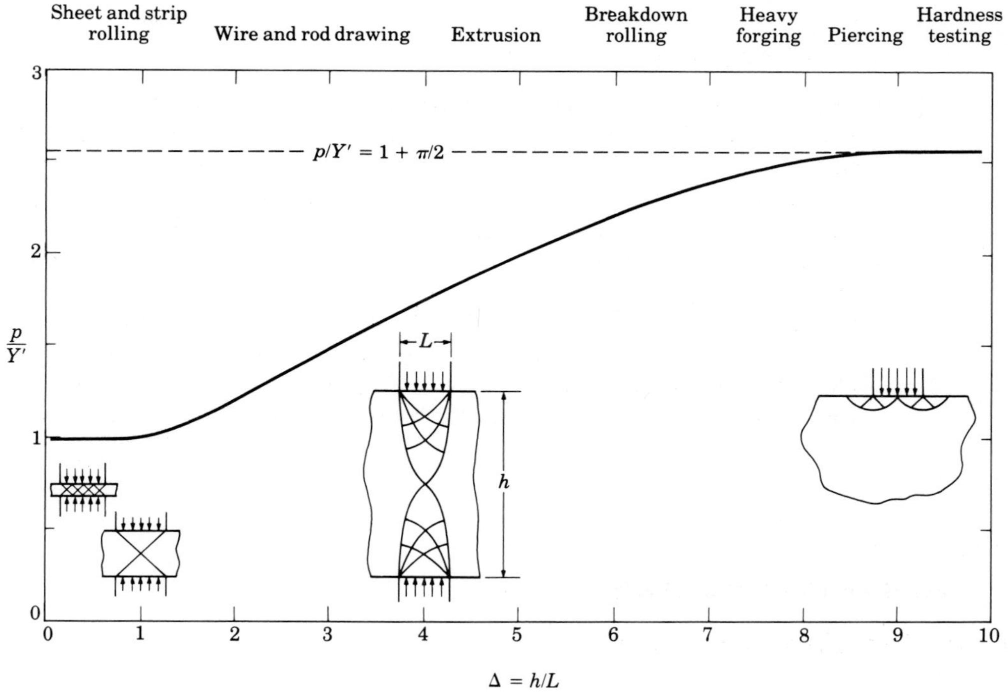

FIGURE 2.40

Die pressures required in *frictionless* plane-strain conditions for a variety of metalworking operations. The geometric relationship between contact area of the dies and workpiece dimensions is an important factor in predicting forces in plastic deformation of materials. *Source*: After W. A. Backofen, *Deformation Processing*, Reading, Mass.: Addison-Wesley, 1972, p. 135.

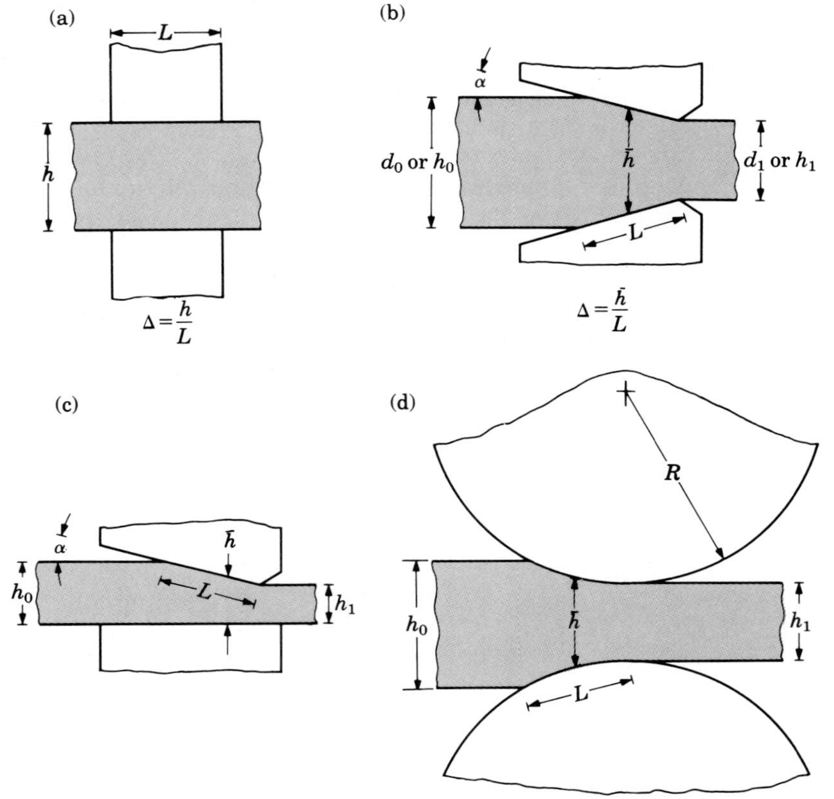

FIGURE 2.41
Examples of plastic deformation processes in plane strain showing the h/L ratio. (a) Indenting with flat dies. This operation is similar to cogging, shown in Fig. 6.18. (b) Drawing or extrusion of strip with a wedge-shaped die, described in Chapter 6. (c) Ironing. See also Fig. 7.56. (d) Rolling, described in Section 6.3. As shown in Fig. 2.40, the larger the h/L ratio, the higher the die pressure becomes. In actual processing, however, the smaller this ratio, the greater is the effect of friction at the die–workpiece interfaces. The reason is that contact area, hence friction, increases with decreasing h/L ratio.

material (Fig. 2.41). Details of these processes and the role of the deformation geometry are covered in Chapter 6.

2.12 Residual Stresses

In this section we show that inhomogeneous deformation leads to *residual stresses*—stresses that remain within a part after it has been deformed and all external forces have been removed. A typical example of inhomogeneous deformation is the bending of a beam (Fig. 2.42). The bending moment first produces a linear elastic stress distribution. As the moment is increased, the outer fibers begin to yield and, for a typical strain-hardening material, the stress distribution shown in Fig. 2.42(b) is eventually obtained. After the part is bent (permanently, since it has undergone plastic deformation), the moment is removed by unloading. This unloading is equivalent to applying an equal and opposite moment to the beam.

As previously shown in Fig. 2.5, all recovery is elastic. Thus in Fig. 2.42(c) the moments of the areas *oab* and *oac* about the neutral axis must be equal. (For purposes of this treatment we assume that the neutral axis does not shift.)

The difference between the two stress distributions produces the residual stress pattern within the beam. Note that there are compressive residual stresses in layers *ad* and *oe*, and tensile residual stresses in layers *do* and *ef*. With no external forces, the residual stress system in the beam must be in static equilibrium. Although this example involves stresses in one direction only, in most situations in deformation processing the residual stresses are three dimensional.

In this example, the equilibrium of residual stresses may be disturbed by altering the geometry of the beam, such as by removing a layer of material. The beam will then acquire a new radius of curvature in order to ensure the balance of internal forces. Another example of this effect is the drilling of round holes on surfaces with residual stresses. It may be found that, as a result of removing this material, the equilibrium of the residual stresses is disturbed and the hole becomes oval.

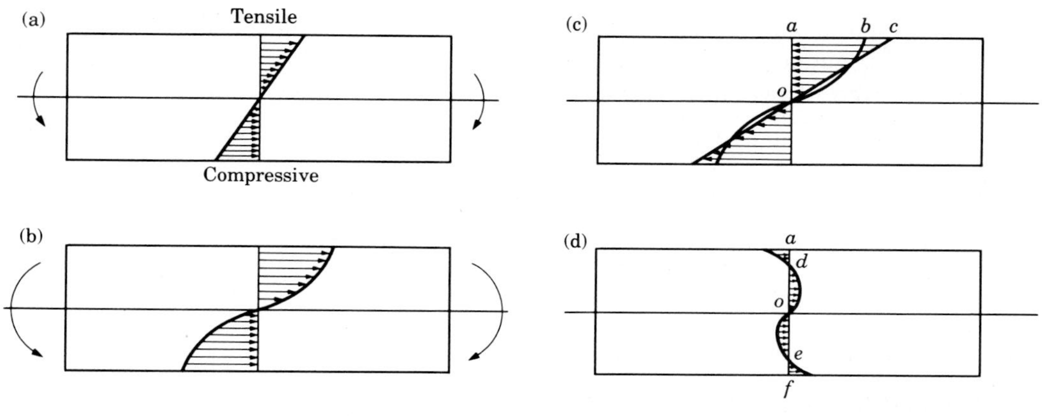

FIGURE 2.42
Residual stresses developed in bending a beam made of an elastic, strain-hardening material. Note that unloading is equivalent to applying an equal and opposite moment to the part, as shown in (b). Because of nonuniform deformation, most parts made by plastic deformation processes contain residual stresses. Note that the forces and moments due to residual stresses must be internally balanced.

Such disturbances of residual stresses lead to *warping*, some simple examples of which are shown in Fig. 2.43. The equilibrium of residual stresses may also be disturbed by *relaxation* of these stresses over a period of time, which results in instability of the dimensions and shape of the component. These dimensional changes can be an important consideration for precision machinery and measuring equipment.

Residual stresses are also caused by *phase changes* in metals during or after processing because of density differences between phases (such as between ferrite and martensite in steels). Phase changes cause microscopic volumetric changes and result in residual stresses. This phenomenon is important in warm and hot working and in heat treatment following cold working (such as deformation at room temperature).

Residual stresses can also be caused by *temperature gradients* within a body, as during the cooling cycle of a casting, when applying brakes to a railroad wheel, or in a grinding operation. The expansion and contraction due to temperature gradients are analogous to nonuniform plastic deformation.

2.12.1 Effects of residual stresses

Tensile residual stresses on the surface of a part are generally considered to be undesirable, because they lower the fatigue life and fracture strength of the part. A surface with tensile residual stresses can sustain lower additional tensile stresses (due to external loading) than can a surface that is free from residual stresses. This is particularly true for relatively brittle materials, where fracture takes place with little or no plastic deformation.

Tensile residual stresses in manufactured products can also lead to *stress cracking* or *stress–corrosion cracking* over a period of time. Conversely, compressive residual stresses on a surface are generally desirable. In fact, in order to increase the fatigue life of components, compressive residual stresses are imparted on surfaces by techniques such as *shot peening* and *surface rolling*.

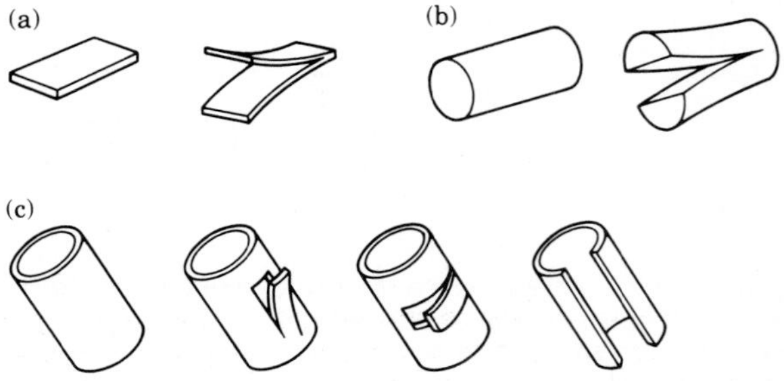

FIGURE 2.43 Distortion of parts with residual stresses after cutting or slitting: (a) rolled sheet or plate; (b) drawn rod; (c) thin-walled tubing. Because of the presence of residual stresses on the surfaces of parts, a round drill may produce an oval-shaped hole because of relaxation of stresses when a portion is removed.

2.12.2 Reduction of residual stresses

Residual stresses may be reduced or eliminated either by *stress-relief annealing* or by further *plastic deformation*. Given sufficient time, residual stresses may also be diminished at room temperature by *relaxation*. The time required can be greatly reduced by raising the temperature of the component. Relaxation of residual stresses by stress-relief annealing is generally accompanied by warpage of the part. Hence a "machining allowance" is commonly provided to compensate for dimensional changes during stress relieving.

The mechanism of reduction or elimination of residual stresses by plastic deformation is as follows: Assume that a piece of metal has the residual stresses shown in Fig. 2.44, namely, tensile on the outside and compressive on the inside. The part with these stresses, which are in the elastic range, is in equilibrium. Also assume that the material is elastic–perfectly plastic, as shown in Fig. 2.44(d).

The levels of residual stress are shown on the stress–strain diagram, both being below the yield stress Y. If a uniformly distributed tension is applied to this specimen, points σ_c and σ_t in the diagram move up on the stress–strain curve, as shown by the arrows. The maximum level that these stresses can reach is the tensile yield stress Y. With sufficiently high loading, the stress distribution becomes uniform throughout the part, as shown in Fig. 2.44(c). If the load is then removed, the stresses recover elastically and the part has no residual stresses. Note that very little stretching is required to relieve these residual stresses. The reason is that the elastic portions of the stress–strain curves for metals are very steep; hence the elastic stresses can be raised to the yield stress with very little strain.

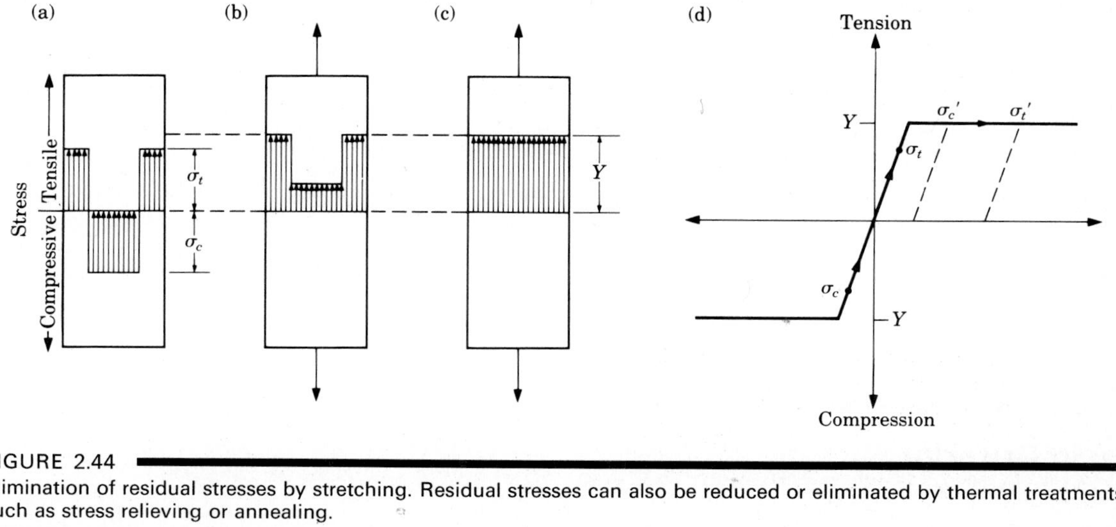

FIGURE 2.44 Elimination of residual stresses by stretching. Residual stresses can also be reduced or eliminated by thermal treatments, such as stress relieving or annealing.

The technique for reducing or relieving residual stresses by plastic deformation, such as by stretching as described, requires sufficient straining to establish a uniformly distributed stress in the part. A material such as the elastic, linearly strain-hardening type, Fig. 2.5(e), therefore can never reach this condition since the compressive stress $\sigma_{c'}$ will always lag behind $\sigma_{t'}$. If the slope of the stress–strain curve in the plastic region is small, the difference between $\sigma_{c'}$ and $\sigma_{t'}$ will be rather small and little residual stress will be left in the part after unloading.

• Example 2.8: Elimination of residual stresses.

Refer to Fig. 2.44 and assume that $\sigma_t = 140$ MPa and $\sigma_c = -140$ MPa. The material is aluminum and the length of the specimen is 0.25 m. Calculate the length to which this specimen should be stretched so that, when unloaded, it will be free from residual stresses. Assume that the yield stress of the material is 150 MPa.

SOLUTION. Stretching should be to the extent that σ_c reaches the yield stress in tension, Y. Thus the total strain should be equal to the sum of the strain required to bring the compressive residual stress to zero and the strain required to bring it to the tensile yield stress. Hence

$$\varepsilon_{\text{total}} = \frac{\sigma_c}{E} + \frac{Y}{E}.$$

For aluminum, let $E = 70$ GPa. Thus

$$\varepsilon_{\text{total}} = \frac{140}{70 \times 10^3} + \frac{150}{70 \times 10^3} = 0.00414.$$

Hence the stretched length should be

$$\ln\left(\frac{\ell_f}{0.25}\right) = 0.00414 \quad \text{or} \quad \ell_f = 0.2510 \text{ m}.$$

As the strains are very small, we may use engineering strains in these calculations. Thus

$$\frac{\ell_f - 0.25}{0.25} = 0.00414 \quad \text{or} \quad \ell_f = 0.2510 \text{ m}.$$

2.13 Yield Criteria

In most manufacturing operations involving deformation processing, the material, unlike that in a simple tension or compression test specimen, is generally subjected to *triaxial* stresses. For example, in the expansion of a thin-walled spherical shell under internal pressure, an element in the shell is subjected to equal biaxial tensile stresses (Fig. 2.45a). In drawing a rod or wire through a conical die, an element in the deformation zone is subjected to tension in its length direction and to compression on its conical surface (Fig. 2.45b). An element in the flange in deep drawing of sheet metal is subjected to a tensile radial stress and compressive stresses on its surface and in the circumferential direction (Fig. 2.45c). As we show in subsequent chapters, many similar examples can be given where the material is subjected to various normal and shear stresses during processing.

In a simple tension or compression test, when the applied stress reaches the uniaxial yield stress Y, the material will deform plastically. However, if the material is subjected to a more complex state of stress, there are relationships between these stresses that will predict yielding. These relationships are known as *yield criteria*, the most common ones being the maximum-shear-stress criterion and the distortion-energy criterion.

2.13.1 Maximum-shear-stress criterion

This criterion, also known as the *Tresca criterion*, states that yielding occurs when the *maximum shear stress* within an element is equal to or exceeds a critical value. As we show in Section 3.3, this critical value of the shear stress is a material property called *shear yield stress*, *k*. Thus for yielding,

$$\tau_{max} \geq k. \tag{2.33}$$

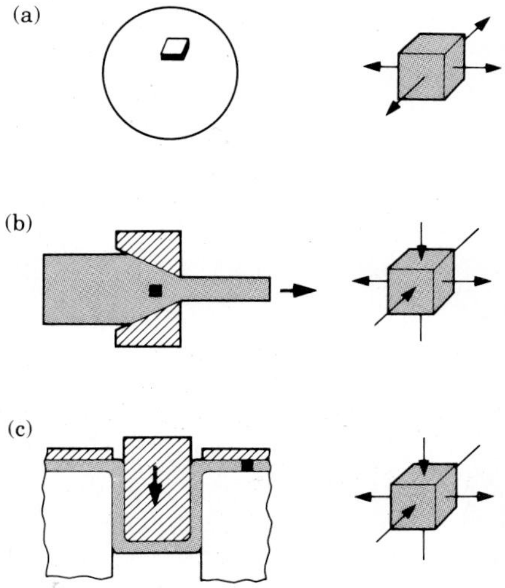

FIGURE 2.45 The state of stress in various metalworking operations. (a) Expansion of a thin-walled spherical shell under internal pressure. (b) Drawing of round rod or wire through a conical die to reduce its diameter. (c) Deep drawing of sheet metal with a punch and die to make a cup.

The most convenient way of determining the stresses on an element is by the use of *Mohr circles*. Some typical examples are shown in Fig. 2.46. Note that in these examples the normal stresses are *principal stresses*; in other words, they act on planes on which there are no shear stresses.

If the maximum shear stress, as determined from Fig. 2.46 or from appropriate equations, is equal to or exceeds k, then yielding will occur. This condition can be best visualized by the construction in Fig. 2.47. In order to cause yielding, the largest circle must touch the "ceiling" represented by the shear yield stress k. Note the many combinations of stresses (or states of stress) that can give the same maximum shear stress.

If, for some reason, we are unable to increase the stresses on the element in order to cause yielding, the solution is simply to lower the ceiling by raising the temperature of the material. This solution is the basis and one major reason for hot working of materials.

From the simple tension test, we find that

$$k = \frac{Y}{2}. \tag{2.34}$$

We assumed that the material is *continuous*, *homogeneous*, and *isotropic*; that is, it has the same properties in all directions. Also, we assumed that tensile stresses are positive, compressive stresses negative, and the yield stress in tension and in compression are equal. All are important assumptions.

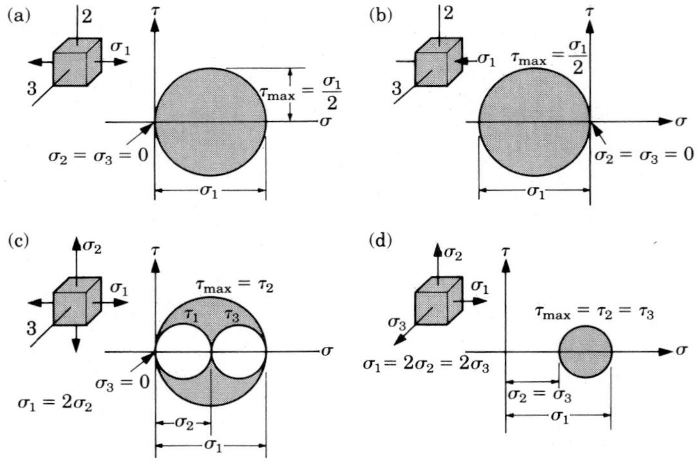

FIGURE 2.46
Mohr's circles for various states of stress: (a) uniaxial tension; (b) uniaxial compression; (c) biaxial tension (plane stress); (d) triaxial tension. These states of stress are encountered in most metalworking processes.

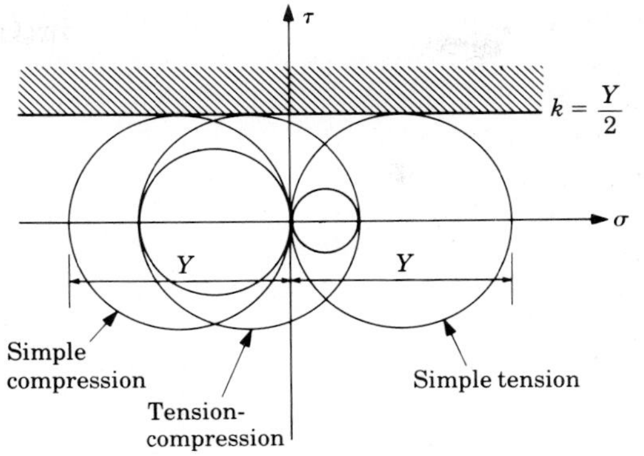

FIGURE 2.47
Three sets of Mohr's circles representing stresses that are large enough to cause yielding. Note that there is an infinite number of states of stress that produce circles of the same maximum diameter, hence yielding. For any state of stress, the radius of the largest circle must be equal to the shear yield stress k to cause yielding. The height of the ceiling depends on the particular material and its condition, temperature, and strain rate.

We can now write the maximum-shear-stress criterion as

$$\sigma_{max} - \sigma_{min} = Y = 2k. \tag{2.35}$$

This means that the maximum and minimum normal stresses produce the largest circle and hence the *largest* shear stress. Consequently, *the intermediate stress has no effect on yielding*. It should be emphasized that the left-hand side of Eq. (2.35) represents the *applied stresses* and that the right-hand side is a *material property*.

2.13.2 Distortion-energy criterion

The distortion-energy criterion, also called the *von Mises criterion*, states that yielding occurs when the relationship between the principal applied stresses and the uniaxial yield stress Y of the material is

$$(\sigma_1 - \sigma_2)^2 + (\sigma_2 - \sigma_3)^2 + (\sigma_3 - \sigma_1)^2 = 2Y^2. \tag{2.36}$$

Note that, unlike the maximum-shear-stress criterion, the intermediate principal stress is included in this equation. Here again, the left-hand side of the equation represents the applied stresses and the right-hand side a material property.

- **Example 2.9: Yielding of thin-walled shell.**

A thin-walled spherical shell is under internal pressure p. The shell is 20 in. in diameter and 0.1 in. thick. It is made of a perfectly plastic material with a yield stress of 20,000 psi. Calculate the pressure required to cause yielding of the shell according to both yield criteria discussed above.

SOLUTION. For this shell under internal pressure, the membrane stresses are given by

$$\sigma_1 = \sigma_2 = \frac{pr}{2t}, \tag{2.37}$$

where $r = 10$ in. and $t = 0.1$ in. The stress in the thickness direction, σ_3, is negligible because of the high r/t ratio of the shell. Thus, according to the maximum-shear-stress criterion,

$$\sigma_{max} - \sigma_{min} = Y$$

or

$$\sigma_1 - 0 = Y$$

and

$$\sigma_2 - 0 = Y.$$

Hence $\sigma_1 = \sigma_2 = 20{,}000$ psi.

The pressure required is then

$$p = \frac{2tY}{r} = \frac{(2)(0.1)(20{,}000)}{10} = 400 \text{ psi}.$$

According to the distortion-energy criterion,

$$(\sigma_1 - \sigma_2)^2 + (\sigma_2 - \sigma_3)^2 + (\sigma_3 - \sigma_1)^2 = 2Y^2$$

or

$$0 \quad + \quad \sigma_2^2 \quad + \quad \sigma_1^2 \quad = 2Y^2.$$

Hence $\sigma_1 = \sigma_2 = Y$.

Thus the answer is the same, or

$$p = 400 \text{ psi}.$$

• Example 2.10: Correction factor for true stress–true strain curves.

Explain why a correction factor has to be applied in the construction of a true stress–true strain curve, shown in Fig. E2.1, from tensile-test data.

SOLUTION. The stress distribution at the neck of a specimen, where there is a triaxial state of stress, is shown in Fig. E2.2. The reason for this state is that each element in the region has a different cross-sectional area; the smaller the area, the

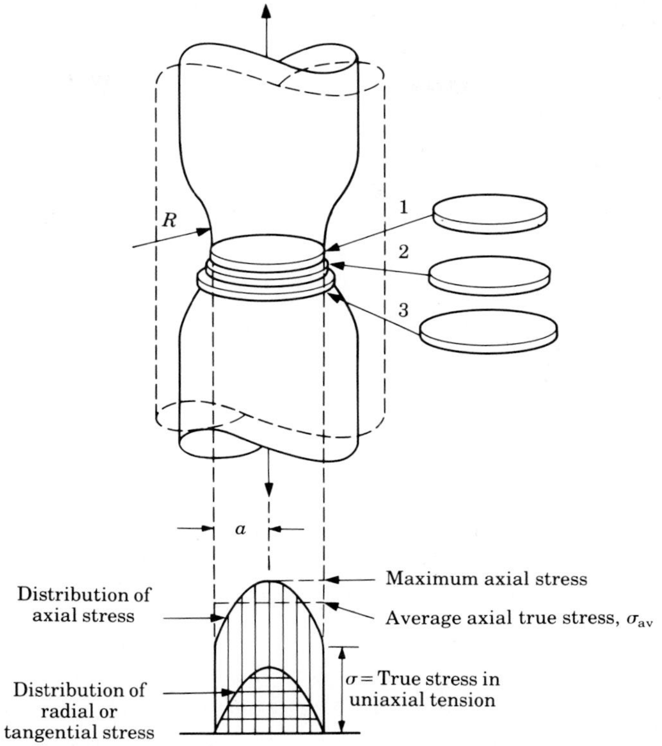

FIGURE E2.2

greater is the tensile stress on the element. Hence element 1 will contract more than element 2, and so on. However, element 1 is restrained from contracting freely by element 2, and element 2 is restrained by element 3, and so on. This restraint causes radial and circumferential tensile stresses in the necked region. This situation results in an axial tensile stress distribution as shown in Fig. E2.2.

The true uniaxial stress in tension is σ, whereas the calculated value of true stress at fracture is the average stress. Hence a correction has to be made. A mathematical analysis by P. W. Bridgman gives the ratio of true to average stress as

$$\frac{\sigma}{\sigma_{av}} = \frac{1}{(1 + 2R/a)[\ln(1 + a/2R)]}, \tag{2.38}$$

where R is the radius of curvature of the neck and a is the radius of the specimen at the neck. Since R is not easy to measure during a test, an empirical relation has been

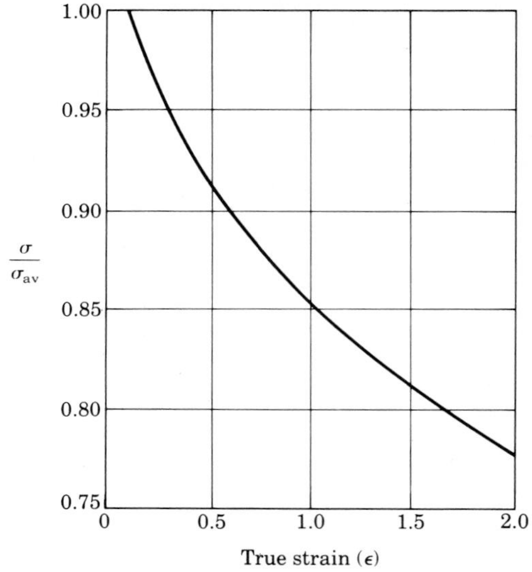

FIGURE E2.3

established between a/R and the true strain at the neck, Fig. E2.3. The corrected true stress–true strain curve is shown in Fig. E2.1(a), in Example 2.2.

2.13.3 Plane stress and plane strain

Plane stress and plane strain are important in the application of yield criteria. *Plane stress* is the state of stress in which one or two of the pairs of faces on an elemental cube are free from stress. An example is the torsion of a thin-walled tube. There are no stresses normal to the inside or outside surface of the tube; hence the state of the stress of the tube is one of plane stress. Other examples are shown in Figs. 2.45(a) and 2.48(a).

The state of stress where one of the pairs of faces on an element undergoes zero strain is known as *plane strain*. An example is shown in Fig. 2.48(b), which depicts a piece of material being compressed in a die; note that one pair of faces is touching the walls (groove) of the die and cannot expand. Another example is the *plane-strain compression test* shown in Fig. 2.13. There, by proper choice of specimen dimensions, the width of the specimen is kept essentially constant. (Note that an element does not have to be physically constrained on the pair of faces for plane-strain conditions to exist.) A third example is the torsion of a thin-walled tube in which the wall thickness remains constant (see Section 2.13.7).

FIGURE 2.48
Two examples of states of stress: (a) plane stress; (b) plane strain.

A review of the two yield criteria previously outlined indicates that the plane-stress condition (where $\sigma_2 = 0$) can be represented by the diagram in Fig. 2.49. The maximum-shear-stress criterion gives an envelope of straight lines. In the first quadrant, where $\sigma_1 > 0$, $\sigma_3 > 0$, and σ_2 is always zero for plane stress, Eq. (2.35) reduces to $\sigma_{max} = Y$. Hence the maximum value that either σ_1 or σ_3 can acquire is Y; thus the straight lines in the diagram. In the third quadrant the same situation exists because σ_1 and σ_3 are both compressive. In the second and fourth quadrants, σ_2 (which is zero for the plane-stress condition) is the intermediate stress. Thus for the second quadrant, Eq. (2.35) reduces to

$$\sigma_3 - \sigma_1 = Y, \tag{2.39}$$

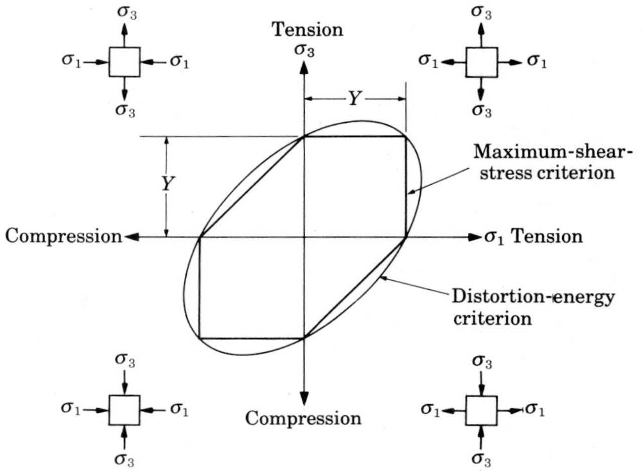

FIGURE 2.49
Plane-stress diagrams for maximum-shear-stress and distortion-energy criteria. Note that $\sigma_2 = 0$.

and for the fourth quadrant, reduces to

$$\sigma_1 - \sigma_3 = Y. \tag{2.40}$$

Equations (2.39) and (2.40) represent the 45° lines in Fig. 2.49.

The distortion-energy criterion for plane stress reduces to

$$\sigma_1^2 + \sigma_3^2 - \sigma_1\sigma_3 = Y^2 \tag{2.41}$$

and is shown graphically in the figure. Whenever a point (with its coordinates representing the two principal stresses) falls on these boundaries, the element will yield.

Yielding under plane-strain conditions requires determining the stress level, if any, on the faces of the element undergoing plane strain (Fig. 2.48b). We do so by using *generalized Hooke's law* equations:

$$\varepsilon_1 = \frac{1}{E}[\sigma_1 - \nu(\sigma_2 + \sigma_3)]; \tag{2.42a}$$

$$\varepsilon_2 = \frac{1}{E}[\sigma_2 - \nu(\sigma_1 + \sigma_3)]; \tag{2.42b}$$

$$\varepsilon_3 = \frac{1}{E}[\sigma_3 - \nu(\sigma_1 + \sigma_2)]. \tag{2.42c}$$

For plane strain, one of the strains in Eqs. (2.42a)–(2.42c) is zero. Thus a particular relationship exists between the principal stresses. At yielding, the Poisson's ratio ν is 0.5 (as the volume of an element undergoing plastic deformation remains constant) and all three stresses therefore can be determined. Thus for the case in Fig. 2.48(b),

$$\sigma_2 = \frac{\sigma_1 + \sigma_3}{2}. \tag{2.43}$$

Note that σ_2 is now an intermediate stress.

For the plane-strain compression of Fig. 2.48, the distortion-energy criterion (which includes the intermediate stress) reduces to

$$\sigma_1 - \sigma_3 = \frac{2}{\sqrt{3}}Y = 1.15Y = Y'. \tag{2.44}$$

Note that, whereas for the maximum-shear-stress criterion $k = Y/2$, for the distortion-energy criterion $k = Y/\sqrt{3}$ for the plane-strain condition.

• Example 2.11: Calculation of compressive force.

A specimen in the shape of a cube 10 mm on each side is being compressed without friction in a die cavity, as shown in Fig. 2.48(b), where the width of the groove is 15 mm. Assume that the material is made of an elastic, linearly strain-hardening material given by

$$\sigma = 100 + 20\varepsilon \text{ MPa}.$$

Calculate the compressive force required when the height of the specimen is 3 mm, according to both yield criteria.

SOLUTION. When the height of this specimen is reduced to 3 mm, the surface area A will be

$$(10)(10)(10) = (3)(A).$$

Hence

$$A = 333.3 \text{ mm}^2.$$

The groove is only 15 mm wide, so the specimen will touch the walls of the groove, because $(15)(15) = 225 \text{ mm}^2$, which is smaller than the final surface area required. Thus this problem involves both a plane strain and a plane stress.

The true strain after deformation is (in absolute value)

$$|\varepsilon| = \ln\left(\frac{10}{3}\right) = 1.204.$$

Hence the strength level (*flow stress*, see Fig. 2.3) that the material will exhibit at this strain is

$$Y_f = 100 + (20)(1.204) = 124.08 \text{ MPa}.$$

According to the maximum-shear-stress criterion, the force required is

$$P = Y_f A = (124.08)(333.3) = 41{,}350 \text{ N} = 41.35 \text{ kN}.$$

According to the distortion-energy criterion,

$$Y_f = (1.15)(124.08) = 142.69 \text{ MPa}.$$

Hence

$$P = (142.69)(333.3) = 47{,}560 \text{ N} = 47.56 \text{ kN}.$$

• Example 2.12: Force–distance curve in compression.

A cylindrical slug 1 in. in diameter and 1 in. high is placed at the center of a 2-in. diameter cavity in a rigid die (Fig. E2.4). The slug is surrounded by a compressible matrix, the pressure of which is given by the relation

$$p_m = 100{,}000 \frac{\Delta V}{V_{om}} \quad \text{psi},$$

where the subscript m denotes matrix and V_{om} is the original volume of the matrix. Both the slug and the matrix are being compressed by a piston without friction between the piston and the cavity or the slug. The initial pressure of the matrix is zero and the slug material has the curve $\sigma = 30{,}000\varepsilon^{0.2}$. Plot the force F versus piston travel d up to $d = 0.5$ in.

SOLUTION. The total force F on the piston will be

$$F = F_w + F_m,$$

where the subscript w denotes the workpiece and m the matrix. As d increases, the matrix pressure increases, subjecting the slug to transverse compressive stresses on

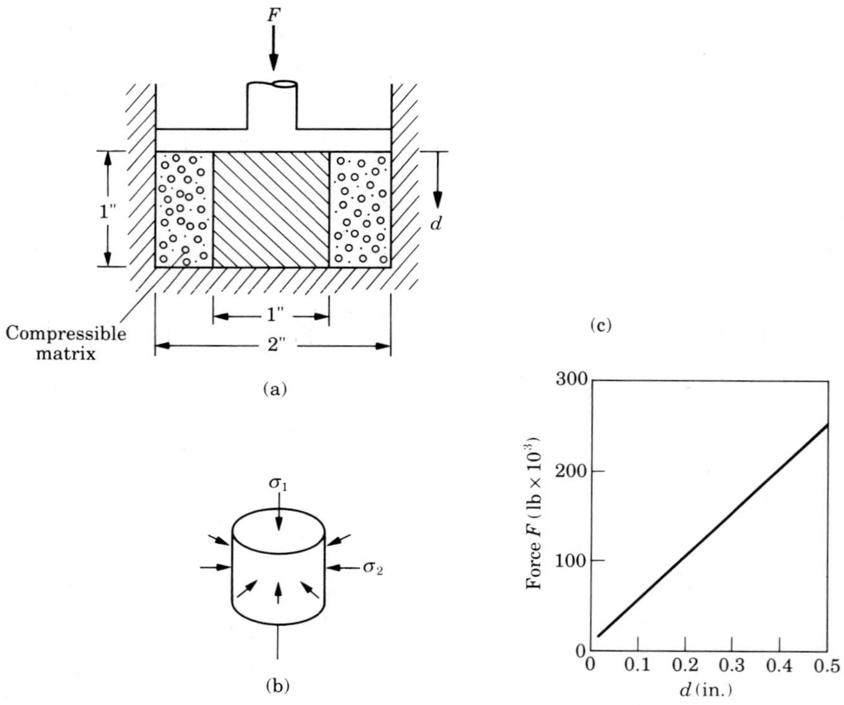

FIGURE E2.4

its circumference. Hence the slug will be subjected to triaxial compressive stresses (Fig. E2.4b). Using the maximum-shear-stress criterion for simplicity, we have

$$\sigma_1 = Y_f + \sigma_2,$$

where σ_1 is the required compressive stress on the slug, Y_f is the flow stress of the slug material corresponding to a particular strain, that is, reduction in height, and σ_2 is the compressive stress due to matrix pressure. Let us now determine the matrix pressure in terms of d.

The volume of the slug is equal to $\pi/4$ and the volume of the cavity when $d = 0$ is π. Hence the original volume of the matrix is $V_{om} = \frac{3}{4}\pi$. The volume of the matrix at any value of d then is

$$V_m = \pi(1 - d) - \frac{\pi}{4} = \pi\left(\frac{3}{4} - d\right) \quad \text{in}^3,$$

from which we obtain

$$\frac{\Delta V}{V_{om}} = \frac{V_{om} - V_m}{V_{om}} = \frac{4}{3}d.$$

Note that when $d = \frac{3}{4}$ in., the volume of the matrix becomes zero. The matrix pressure, hence σ_2, is now given by

$$\sigma_2 = \frac{400{,}000}{3}d \quad \text{psi}.$$

The absolute value of the true strain in the slug is given by

$$\varepsilon = \ln\frac{1}{1 - d},$$

with which we can determine the value of Y_f for any d.

The cross-sectional area of the workpiece at any d is

$$A_w = \frac{\pi}{4(1 - d)} \quad \text{in}^2$$

and that of the matrix is

$$A_m = \pi - \frac{\pi}{4(1 - d)} \quad \text{in}^2.$$

The required compressive stress on the slug is

$$\sigma_1 = Y_f + \sigma_2 = Y_f + \frac{400{,}000}{3}d.$$

We may now write the total force on the piston as

$$F = A_w\left(Y_f + \frac{400{,}000}{3}d\right) + A_m\frac{400{,}000}{3}d \quad \text{lb}.$$

The following table gives the complete data up to $d = 0.5$. The plot of F versus d is shown in Fig. E2.4(c).

d (in.)	A_w (in²)	A_m (in²)	ε	Y_f (psi)	F (lb)
0.1	0.872	2.270	0.105	19,100	58,400
0.2	0.98	2.162	0.223	22,200	105,700
0.3	1.121	2.021	0.357	24,600	153,300
0.4	1.31	1.832	0.510	26,200	201,800
0.5	1.571	1.571	0.692	27,900	252,200

2.13.4 Experimental verification of yield criteria

The yield criteria described have been tested experimentally. A suitable specimen commonly used is a thin-walled tube under internal pressure and/or torsion. Under such loading, it is possible to generate different states of plane stress. Various experiments, with a variety of ductile materials, have shown that the distortion-energy criterion agrees better with the experimental data than does the maximum-shear-stress criterion.

These findings suggest use of the distortion-energy criterion for the analysis of metalworking processes, which generally make use of ductile materials. The simpler maximum-shear-stress criterion, however, can also be used, particularly by designers. The difference between the two criteria is negligible for most practical applications.

2.13.5 Volume strain

By summing the three equations of the generalized Hooke's law (Eq. 2.42), we obtain

$$\varepsilon_1 + \varepsilon_2 + \varepsilon_3 = \frac{1 - 2\nu}{E}(\sigma_1 + \sigma_2 + \sigma_3), \tag{2.45}$$

where the left-hand side of the equation can be shown to be the *volume strain* or *dilatation*, Δ. Thus

$$\Delta = \frac{\text{volume change}}{\text{original volume}} = \frac{1 - 2\nu}{E}(\sigma_1 + \sigma_2 + \sigma_3). \tag{2.46}$$

In the plastic range, where $\nu = 0.5$, the volume change is zero. Hence in plastic working,

$$\varepsilon_1 + \varepsilon_2 + \varepsilon_3 = 0, \tag{2.47}$$

which is a convenient means of determining a third strain if two strains are known.

The *bulk modulus* is defined as

$$\text{Bulk modulus} = \frac{\sigma_{av}}{\Delta} = \frac{E}{3(1-2\nu)}, \tag{2.48}$$

where

$$\sigma_{av} = \tfrac{1}{3}(\sigma_1 + \sigma_2 + \sigma_3). \tag{2.49}$$

From Eq. (2.46), in the elastic range, where $0 < \nu < 0.5$, the volume of a tension-test specimen increases and that of a compression-test specimen decreases.

2.13.6 Effective stress and effective strain

A convenient means of expressing the state of stress on an element is the *effective* (*equivalent* or *representative*) *stress* $\bar{\sigma}$ and *effective strain* $\bar{\varepsilon}$. For the maximum-shear-stress criterion,

$$\bar{\sigma} = \sigma_1 - \sigma_3, \tag{2.50}$$

and for the distortion-energy criterion,

$$\bar{\sigma} = \frac{1}{\sqrt{2}}[(\sigma_1 - \sigma_2)^2 + (\sigma_2 - \sigma_3)^2 + (\sigma_3 - \sigma_1)^2]^{1/2}. \tag{2.51}$$

The factor $1/\sqrt{2}$ is chosen so that, for simple tension, the effective stress is equal to the uniaxial yield stress Y.

The strains are likewise related to the effective strain. For the maximum-shear-stress criterion,

$$\bar{\varepsilon} = \tfrac{2}{3}(\varepsilon_1 - \varepsilon_3), \tag{2.52}$$

and for the distortion-energy criterion,

$$\bar{\varepsilon} = \frac{\sqrt{2}}{3}[(\varepsilon_1 - \varepsilon_2)^2 + (\varepsilon_2 - \varepsilon_3)^2 + (\varepsilon_3 - \varepsilon_1)^2]^{1/2}. \tag{2.53}$$

Again, the factors 2/3 and $\sqrt{2}/3$ are chosen so that for simple tension the effective strain is equal to the uniaxial tensile strain. It is apparent that stress–strain curves may also be called effective stress–effective strain curves.

2.13.7 Comparison of normal stress–strain and shear stress–strain

Stress–strain curves in tension and torsion for the same material are, of course, comparable. Also, it is possible to construct one curve from the other since the material is the same, the procedure for which is outlined.

The following observations are made with regard to Fig. 2.50, which shows the tension and torsional states of stress:

a) In the tension test, the uniaxial stress σ_1 is also the effective stress and the principal stress.
b) In the torsion test, the principal stresses occur on planes whose normals are at 45° to the longitudinal axis; the principal stresses σ_1 and σ_3 are equal in magnitude but opposite in sign.
c) The magnitude of the principal stress in torsion is the same as the maximum shear stress.

We now have the following relationships:

$$\sigma_1 = -\sigma_3, \qquad \sigma_2 = 0, \quad \text{and} \quad \sigma_1 = \tau_1.$$

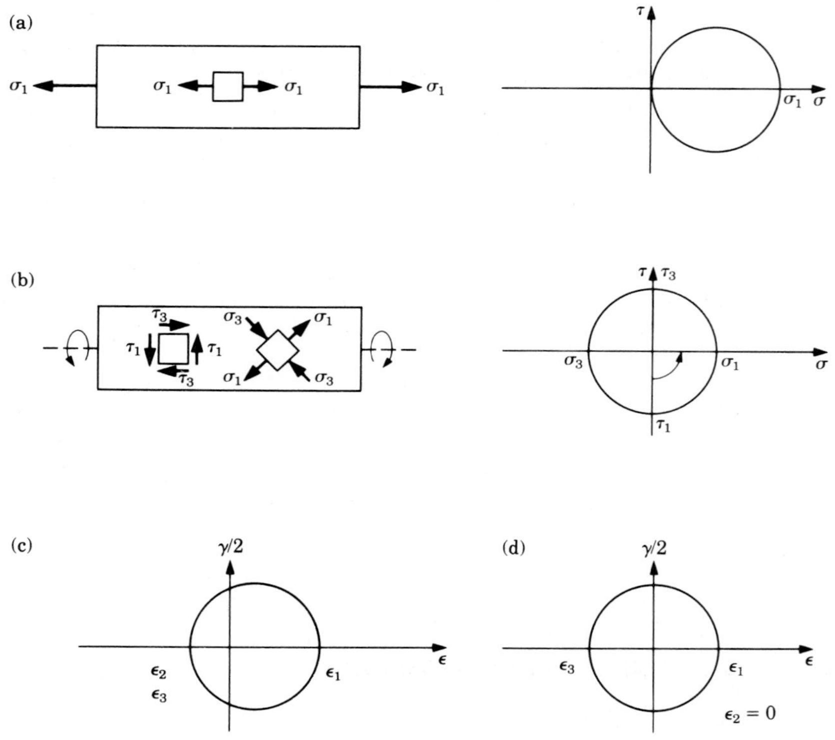

FIGURE 2.50 Mohr's circle diagrams for stress and strain in uniaxial tension, (a) and (c), and torsion, (b) and (d). Guy wires for antennas, or the spokes of a bicycle, are subjected to uniaxial tension (because they are thin and long), whereas the drive shaft of an automobile is subjected to the stresses shown in (b).

Substituting these stresses into Eqs. (2.50) and (2.51) for effective stress, we obtain the following relationships: For the maximum-shear-stress criterion,

$$\bar{\sigma} = \sigma_1 - \sigma_3 = \sigma_1 + \sigma_1 = 2\sigma_1 = 2\tau_1, \tag{2.54}$$

and for the distortion-energy criterion,

$$\bar{\sigma} = \frac{1}{\sqrt{2}}[(\sigma_1 - 0)^2 + (0 + \sigma_1)^2 + (-\sigma_1 - \sigma_1)^2]^{1/2} = \sqrt{3}\sigma_1 = \sqrt{3}\tau_1. \tag{2.55}$$

With regard to strains, the following observations about Fig. 2.50 can be made:

a) In the tension test, $\varepsilon_2 = \varepsilon_3 = -\dfrac{\varepsilon_1}{2}$.

b) In the torsion test, $\varepsilon_1 = -\varepsilon_3 = \dfrac{\gamma}{2}$.

c) The strain in the thickness direction of the tube is zero, i.e., $\varepsilon_2 = 0$.

Observation (c) is true because the thinning caused by the principal tensile stress is countered by the thickening under the principal compressive stress of the same magnitude. Hence $\varepsilon_2 = 0$. Since σ_2 is also zero, a thin-walled tube under torsion is both a plane stress and a plane strain situation.

Substituting these strains into Eqs. (2.52) and (2.53) for effective strain, we obtain the following relationships: For the maximum-shear-stress criterion,

$$\bar{\varepsilon} = \tfrac{2}{3}(\varepsilon_1 - \varepsilon_3) = \tfrac{2}{3}(\varepsilon_1 + \varepsilon_1) = \tfrac{4}{3}\varepsilon_1 = \tfrac{2}{3}\gamma, \tag{2.56}$$

and for the distortion-energy criterion,

$$\bar{\varepsilon} = \frac{\sqrt{2}}{3}[(\varepsilon_1 - 0)^2 + (0 + \varepsilon_1)^2 + (-\varepsilon_1 - \varepsilon_1)^2]^{1/2} = \frac{2}{\sqrt{3}}\varepsilon_1 = \frac{1}{\sqrt{3}}\gamma. \tag{2.57}$$

This set of equations provides a means by which tensile-test data can be converted to torsion-test data, and vice versa.

2.14 Work of Deformation

In this section we discuss the work required for plastic deformation of materials. Work is defined as the product of collinear force and distance, so a quantity equivalent to work per unit volume is the product of stress and strain. Because the relation between stress and strain in the plastic range depends on the particular stress–strain curve, this work is best calculated by referring to Fig. 2.51.

Note that the area under the true stress–true strain curve for any strain ε_1 is the *energy per unit volume u* (*specific energy*) of the material deformed. It is expressed as

$$u = \int_0^{\varepsilon_1} \sigma \, d\varepsilon. \tag{2.58}$$

As seen in Section 2.2.3, true stress–true strain curves can be represented by the expression

$$\sigma = K\varepsilon^n.$$

Hence Eq. (2.58) can be written as

$$u = K\int_0^{\varepsilon_1} \varepsilon^n d\varepsilon$$

or

$$u = \frac{K\varepsilon_1^{n+1}}{n+1} = \bar{Y}\varepsilon_1, \tag{2.59}$$

where $\bar{Y}$ is the *average flow stress* of the material.

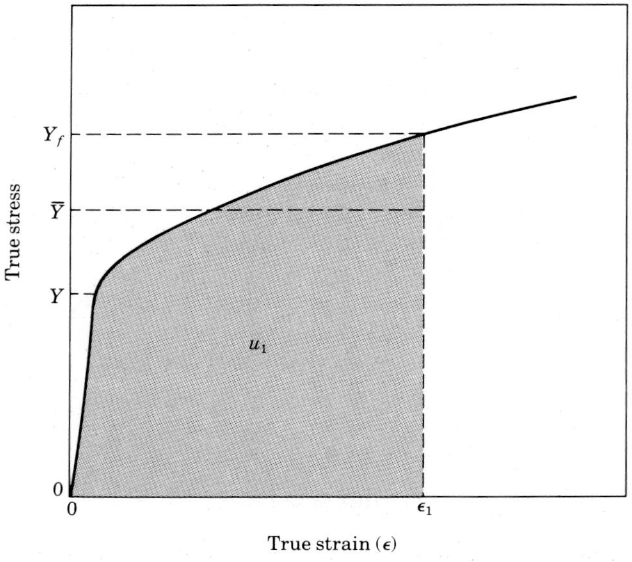

FIGURE 2.51 Schematic illustration of true stress–true strain curve showing yield stress Y, average flow stress $\bar{Y}$, specific energy u_1, and flow stress Y_f.

This energy represents the work dissipated in uniaxial deformation. For a more general condition, where the workpiece is subjected to triaxial stresses and strains, the *effective stress* and *effective strain* can be used. The energy per unit volume is then

$$u = \int_0^{\bar{\varepsilon}} \bar{\sigma}\, d\bar{\varepsilon}. \tag{2.60}$$

To obtain the work expended, we multiply u by the volume of the material deformed. Thus

$$\text{Work} = (u)(\text{volume}). \tag{2.61}$$

The energy represented by Eq. (2.61) is the minimum energy or the *ideal* energy required for uniform (homogeneous) deformation. The energy required for actual deformation involves two additional factors. One is the energy required to overcome friction at the die–workpiece interfaces. The other is the *redundant work* of deformation, which is described as follows:

In Fig. 2.52(a), a block of material is being deformed into shape by forging, extrusion, or drawing through a die, as described in Chapter 6. As shown in sketch (b) this deformation is uniform, or homogeneous. In reality, however, the material more often than not deforms as in sketch (c) from the effects of friction and die geometry. The difference between (b) and (c) is that (c) has undergone additional shearing along horizontal planes.

This shearing requires expenditure of energy, because additional plastic work has to be done in subjecting the various layers to shear strains. This is known as *redundant work*; the word redundant reflects the fact that this work does not contribute to the shape change of the material. [Note that (b) and (c) have the same overall shape and dimensions.]

The total specific energy required can now be written as

$$u_{\text{total}} = u_{\text{ideal}} + u_{\text{friction}} + u_{\text{redundant}}. \tag{2.62}$$

The efficiency of a process is defined as

$$\eta = \frac{u_{\text{ideal}}}{u_{\text{total}}}. \tag{2.63}$$

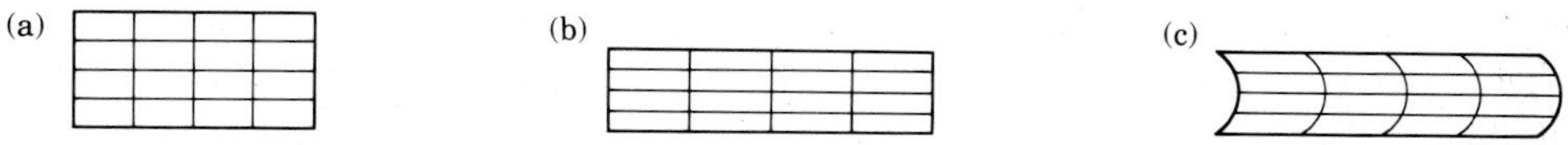

FIGURE 2.52
Deformation of grid patterns in a workpiece: (a) original pattern; (b) after ideal deformation; (c) after inhomogeneous deformation, requiring redundant work of deformation. Note that (c) is basically (b) with additional shearing, especially at the outer layers. Thus (c) requires greater work of deformation than (b).

The magnitude of this efficiency varies widely, depending on the particular process, frictional conditions, die geometry, and other process parameters. Typical values are estimated to be 30 to 60 % for extrusion and 75 to 95 % for rolling.

• Example 2.13: Expansion of a thin-walled shell.

A thin-walled spherical shell made of a perfectly plastic material of yield stress Y, original radius r_o, and thickness t_o is being expanded by internal pressure. Calculate the work done in expanding this shell to a radius of r_f. If the diameter expands at a constant rate, what changes take place in the power consumed as the radius increases?

SOLUTION. The membrane stresses are given by

$$\sigma_1 = \sigma_2 = Y$$

(from Example 2.9), where r and t are instantaneous dimensions. The true strains in the membrane are given by

$$\varepsilon_1 = \varepsilon_2 = \ln\left(\frac{2\pi r_f}{2\pi r_o}\right) = \ln\left(\frac{r_f}{r_o}\right).$$

Because an element in this shell is subjected to equal biaxial stretching, the specific energy is

$$u = \int_o^{\varepsilon_1} \sigma_1\, d\varepsilon_1 + \int_o^{\varepsilon_2} \sigma_2\, d\varepsilon_2 = 2\sigma_1\varepsilon_1 = 2Y \ln\left(\frac{r_f}{r_o}\right).$$

Since the volume of the shell material is $4\pi r_o^2 t_o$, the work done is

$$W = (u)(\text{volume}) = 8\pi Y r_o^2 t_o \ln\left(\frac{r_f}{r_o}\right).$$

The specific energy can also be calculated from the effective stresses and strains. Thus, according to the distortion-energy criterion,

$$\bar{\sigma} = \frac{1}{\sqrt{2}}\left[(o)^2 + (\sigma_2)^2 + (-\sigma_1)^2\right]^{1/2} = \sigma_1 = \sigma_2$$

and

$$\bar{\varepsilon} = \frac{\sqrt{2}}{3}\left[(o)^2 + (\varepsilon_2 + 2\varepsilon_2)^2 + (-2\varepsilon_2 - \varepsilon_2)^2\right]^{1/2} = 2\varepsilon_2 = 2\varepsilon_1.$$

(The thickness strain $\varepsilon_3 = -2\varepsilon_2 = -2\varepsilon_1$ because of volume constancy in plastic deformation, where $\varepsilon_1 + \varepsilon_2 + \varepsilon_3 = 0$.) Hence

$$u = \int_o^{\bar{\varepsilon}} \bar{\sigma}\, d\bar{\varepsilon} = \int_o^{2\varepsilon_1} \sigma_1\, d\varepsilon_1 = 2\sigma_1\varepsilon_1.$$

Thus the answer is the same.

Power is defined as work per unit time; thus

$$\text{Power} = \frac{dW}{dt}.$$

The expression for work can be written as

$$W = k \ln\left(\frac{r}{r_o}\right) = k(\ln r - \ln r_o),$$

since all other factors in the expression are constant. Thus

$$\text{Power} = \frac{k}{r}\frac{dr}{dt}.$$

Because the shell is expanding at a constant rate, dr/dt = constant. Hence the power is related to the instantaneous radius r by

$$\text{Power} \propto \frac{1}{r}.$$

2.14.1 Work and heat

Almost all the mechanical work of deformation in plastic working is converted into *heat*. This conversion is not 100% because a small portion of this energy is stored within the deformed material as elastic energy. This is known as *stored energy* (see Section 3.6). Stored energy is generally 5% to 10% of the total energy input. However, it may be as high as 30% in some alloys.

In a simple frictionless process—assuming that work is completely converted into heat—the temperature rise is given by

$$\Delta T = \frac{u_{\text{total}}}{\rho c}, \tag{2.64}$$

where u_{total} is the specific energy from Eq. (2.62), ρ is the density, and c is the specific heat of the material. Higher temperatures are associated with large areas under the stress–strain curve and smaller values of specific heat.

The theoretical temperature rise for a true strain of 1 (such as a 27-mm-high specimen compressed down to 10 mm) has been calculated to be as follows: aluminum, 75 °C (165 °F); copper, 140 °C (285 °F); low-carbon steel, 280 °C (535 °F); and titanium, 570 °C (1060 °F).

The temperature rise given by Eq. (2.64) is for an ideal situation, where there is no heat loss. In actual operations heat is lost to the environment, to tools and dies, and to any lubricants or coolants used. If the process is performed very rapidly, these losses are relatively small.

Under extreme conditions, an adiabatic process is approached, with very high temperature rise, leading to *incipient melting*. This rise in temperature can be calculated provided the stress–strain curve used is at the appropriate strain rate level. However, if the process is carried out slowly, the actual temperature rise will be a small portion of the calculated value. Properties such as specific heat and thermal conductivity also depends on temperature and they should be taken into account in the calculations.

• Example 2.14: Calculation of temperature rise.

Calculate the total work done for the specimen in Example 2.2. Calculate the specific energy for an element in the necked area and the theoretical temperature rise.

SOLUTION. The total work done on the specimen can be obtained from the area under the load-extension curve. This area is determined graphically to be

$$\text{Work} = 3950 \text{ in.} \cdot \text{lb.}$$

The specific energy of an element in the necked area, that is, at fracture, is the area under the true stress–true strain curve (corrected) in Fig. E2.1(a). Thus

$$u = \int_o^{1.253} \sigma \, d\varepsilon = 155{,}000 \text{ in.} \cdot \text{lb/in}^3.$$

The theoretical temperature rise, that is, adiabatic, is obtained from Eq. (2.64), where for stainless steel, $\rho = 0.29$ lb/in^3 and $c = 0.12$ Btu/lb · °F. Thus

$$\Delta T = \frac{155{,}000}{(0.29)(0.12)(778)(12)} = 477\ {}^\circ\text{F} = 247\ {}^\circ\text{C}.$$

Note that in these calculations only the work of plastic deformation is considered, because no friction or redundant work is involved in a simple tension test. Also, the actual temperature rise will be lower because of the heat loss from the necked zone to the rest of the specimen and to the environment.

SUMMARY

The mechanical behavior of materials in manufacturing processes is related to their strength, ductility, elasticity, and hardness and the energy required for plastic deformation. The behavior of materials depends on the particular material and other variables, such as temperature, strain rate, and state of stress.

Tests have been developed to obtain the relation between stress and strain as a function of the variables. These relationships are important in assessing the behavior of a particular material in a manufacturing process, especially with regard to forces required and the capability of the material to undergo the desired deformation without failure.

Two important parameters are the strain-hardening capability (indicated by the strain-hardening exponent n) and the strain-rate sensitivity of the material (indicated by the exponent m). An important factor is fracture of the material. Basic types of fracture are brittle and ductile and intergranular and transgranular. Temperature, strain rate, surface conditions, and state of stress are significant factors in fracture.

How a material is subjected to a shape change is important also and requires the study of deformation-zone geometry. The manner in which a material is subjected to plastic deformation is important in determining the nature and extent of residual stresses in the part after all external forces are removed. These stresses are important in subsequent processing of the part or during its service life.

In actual metalworking operations, material is generally subjected to three-dimensional stresses. Thus yield criteria have been developed to establish relationships between the uniaxial yield stress of the material (generally obtained from a tension test) and the stresses applied.

BIBLIOGRAPHY

General Texts

Alexander J.M., and R.C. Brewer, *Manufacturing Properties of Materials*. London: Van Nostrand, London, 1963.

Ashby, M.F., and D.R.H. Jones, *Engineering Materials*, Vol. 1, *An Introduction to Their Properties and Applications*, 1980; Vol. 2, *An Introduction to Microstructures. Processing and Design*, 1986. New York: Pergamon.

Boyer, H.E. (ed.), *Atlas of Stress–Strain Curves*. Metals Park, Ohio: ASM International, 1986.

Case Histories in Failure Analysis. Metals Park, Ohio: American Society for Metals, 1979.

Colangelo, V.J., and F.A. Heiser, *Analysis of Metallurgical Failures*, 2d ed. New York: John Wiley, 1987.

Courtney, T.H., *Mechanical Behavior of Materials*. New York: McGraw-Hill, 1989.

Dieter, G.E., *Mechanical Metallurgy*, 3d ed. New York: McGraw-Hill, 1986.

Dieter, G.E. (ed.), *Workability Testing Techniques*. Metals Park, Ohio: American Society for Metals, 1984.

Hardness Testing. Metals Park, Ohio: ASM International, 1987.

Hsu, T.H., *Stress and Strain Data Handbook*. Houston, Texas: Gulf Publishing Co., 1986.

Metals Handbook, 9th ed., Vol. 8: *Mechanical Testing*. Metals Park, Ohio: American Society for Metals, 1985.

O'Neill, H., *Hardness Measurement of Metals and Alloys*. London: Chapman and Hall, 1967.

Parkins, R.N., *Mechanical Treatment of Metals*. London: Allen and Unwin, 1968.

Polakowski, N.H., and E.J. Ripling, *Strength and Structure of Engineering Materials*. Englewood Cliffs, N.J.: Prentice-Hall, 1964.

Pugh, H.L.D., *Mechanical Behaviour of Materials under Pressure*. New York: Elsevier, 1971.
Suh, N.P., and A.P.L. Turner, *Elements of the Mechanical Behavior of Solids*. New York: McGraw-Hill, 1975.
Tabor, D., *The Hardness of Metals*. New York: Oxford, 1951.

Advanced Texts

Avitzur, B., *Handbook of Metal-Forming Processes*. New York: Wiley, 1983.
Avitzur, B., *Metal Forming: Processes and Analysis*. New York: McGraw-Hill, 1968.
Backofen, W.A., *Deformation Processing*. Reading, Mass.: Addison-Wesley, 1972.
Ford, H., and J.M. Alexander, *Advanced Mechanics of Materials*, 2d ed. New York: Halsted, 1977.
Hill, R., *The Mathematical Theory of Plasticity*. New York: Oxford, 1950.
Hoffman, O., and G. Sachs, *Introduction to the Theory of Plasticity for Engineers*. New York: McGraw-Hill, 1953.
Johnson, W., and P.B. Mellor, *Engineering Plasticity*. New York: Van Nostrand, 1973.
Nadai, A., *Theory of Flow and Fracture of Solids*, 2d ed. New York: McGraw-Hill, Vol. I, 1950, Vol. II, 1963.
Prager, W., and P.G. Hodge, *Theory of Perfectly Plastic Solids*. New York: Wiley, 1951.
Slater, R.A., *Engineering Plasticity: Theory and its Application to Metal Forming Processes*. New York: Halsted, 1974.
Thomsen, E.G., C.T. Yang, and S. Kobayashi, *Mechanics of Plastic Deformation in Metal Processing*. New York: Macmillan, 1964.

QUESTIONS

2.1 Explain why you cannot calculate the percent elongation of materials based on the information given in Fig. 2.4 only.

2.2 With a simple sketch, explain whether it is necessary to use the offset method to determine the yield stress Y of a material that has been highly cold worked.

2.3 We know that the percent elongation of a tensile-test specimen (at fracture) decreases as the gage length increases. Explain whether it is possible for the elongation to approach zero if the gage length is great.

2.4 If we use the same scale for stress, the tensile true stress–true strain curve is higher than the engineering stress–strain curve. Explain whether this condition holds for a compression test.

2.5 Which of the two tests, tension or compression, requires the higher capacity testing machine? Why?

2.6 Explain how the modulus of resilience of a strain-hardening material changes, if any, as it is cold worked.

2.7 Explain the strain-rate sensitivity of materials, giving some examples showing their engineering relevance.

2.8 If you pull and break a tensile-test specimen rapidly, where would the temperature be highest? Why?

2.9 Explain why the difference between engineering strain and true strain becomes larger as strain increases. Is this condition true for both tensile and compressive strains? Explain.

2.10 In a tension test the area under the true stress–true strain curve is the work done per unit volume (specific work). We also know that the area under the load-elongation curve represents the work done on the specimen. If we divide this work by the volume of the specimen between the gage marks, we can determine the work done per unit volume (assuming that all deformation is confined between the gage marks). Is this specific work the same as the area under the true stress–true strain curve? Explain. Is your answer the same for any value of strain? Explain.

2.11 Briefly describe the advantages and limitations of using engineering stresses and strains rather than true stresses and true strains in engineering calculations.

2.12 You are given the K and n values of two different materials, respectively. Is this information sufficient to determine which material is the tougher? If not, what additional information do you need? Why?

2.13 Modify the curves in Fig. 2.5 to indicate the effects of temperature.

2.14 Explain why deformation rate and true strain rate are not the same.

2.15 In Example 2.4 we noted the large differences between the two strain rates obtained. Explain the significance of this observation.

2.16 It has been stated that the higher the value of m, the more diffuse the neck is; likewise, the lower the m value, the more localized the neck is. Explain the reason for this behavior.

2.17 Explain why materials with high m values, such as hot glass and silly putty, when stretched slowly undergo large elongations before failure. Consider events taking place in the necked region of the specimen.

2.18 You are running 4-point bending tests on a number of identical specimens of the same length and cross-section but with increasing distance between the upper points of loading (see Fig. 2.19b). What changes, if any, would you expect in the test results? Explain.

2.19 Explain the significance of the modulus of elasticity of the ball (indenter) on hardness-test results. What is the trend as the modulus decreases? Explain.

2.20 Which hardness tests and scales would you use for very thin strips of material, such as aluminum foil? Why?

2.21 If a material such as aluminum does not have an endurance limit, how would you estimate its fatigue life?

2.22 Describe the difference between creep and stress relaxation, giving two examples for each as they relate to engineering applications.

2.23 If you remove the layer of material *ad* from the part shown in Fig. 2.42 by machining or grinding, which way will the specimen curve? *Hint*: Assume that the part in (d) is composed of four horizontal springs held at the ends. Thus from the top down you have compression, tension, compression, and tension springs.

2.24 Is it possible to completely remove residual stresses in a piece of material by the technique described in Fig. 2.44 if the material is elastic and linearly strain hardening? Explain.

2.25 Using the generalized Hooke's law Eq. (2.42), show whether a thin-walled tube undergoes any thickness change when subjected to torsion in the elastic range.

2.26 Make a sketch showing the nature and distribution of residual stresses in Figs. 2.43(a) and (b) before they were cut. Assume that the split parts are free from any stresses. *Hint*: Force these parts back to the shape they were in before they were cut.

2.27 It is possible to calculate the work of plastic deformation by measuring the temperature rise in a workpiece, assuming that there is no heat loss and that the temperature distribution is uniform throughout. If the specific heat of the material decreases with increasing temperature, will the work of deformation calculated using the specific heat at room temperature be higher or lower than the actual work done? Explain.

2.28 Explain whether the volume of a homogeneous, isotropic specimen changes when subjected to a uniaxial compressive stress in the elastic range.

2.29 The diagram for the distortion-energy criterion in Fig. 2.49 depicts Eq. (2.36) for the condition when the principal stress σ_2 is zero (plane-stress condition). What geometric figure does this equation represent? *Hint*: Draw the ellipse for conditions when σ_2 is greater than zero and when it is less than zero. Thus all three principal stresses have finite values, and the state of stress is not one of plane strain any more.

2.30 Subjecting a specimen to hydrostatic compression is relatively easy. Devise a means whereby the specimen (say in the shape of a cube or a round disk) can be subjected to hydrostatic tension or one approaching this state of stress. (A thin-walled, internally pressurized spherical shell is not a correct answer, because it is subjected only to a state of plane stress.)

2.31 It has been stated that the magnitude of residual stresses in a body in any of the principal directions cannot be higher than the uniaxial yield stress of the material. Assuming that there is no strain hardening and considering yield criteria, is this statement strictly correct? Explain.

2.32 Comment on the statement: "In plane stress, an element can have principal normal stresses without any shear stresses. However, principal shear stresses may or may not be associated with any normal stresses." Give an appropriate example to support your answer.

2.33 Is an element in the deformation zone in the plane-strain compression test subjected to simple shear or pure shear? Explain.

2.34 A circular disk of soft metal is being compressed between two flat circular steel punches having the same diameter as the disk. Assume that the disk material is perfectly plastic and that there are no frictional or any temperature effects. Explain the change, if any, in the magnitude of the punch force as the disk is being compressed.

PROBLEMS

2.1 A strip of metal is originally 1.5 m long. It is stretched in three steps, first to a length of 2.0 m, then to 2.5 m, and finally to 3.0 m. Show that the total true strain is the sum of the true strains in each step; that is, the strains can be added. Show that, using engineering strains, the strains for each step cannot be added to obtain the total strain.

2.2 A paper clip is made of wire 1 mm in diameter. If the original material from which the wire is made is a rod 10 mm in diameter, calculate the longitudinal engineering and true strains that the wire has undergone during processing.

2.3 A material has the following properties: UTS = 60,000 psi and $n = 0.4$. Calculate its strength coefficient K.

2.4 A cable is made of two parallel strands of different materials, A and B, having cross-sections as follows:

Material A: $K = 100{,}000$ psi; $n = 0.5$; $A_0 = 0.2$ in^2.

Material B: $K = 50{,}000$ psi; $n = 0.5$; $A_0 = 0.1$ in^2.

(a) Calculate the maximum tensile force that this cable can withstand prior to necking. (b) How would you arrive at an answer if the n values of the two strands were different?

2.5 Using only Fig. 2.4, calculate the maximum load in tension testing of an 8650 steel specimen with an original diameter of 0.3 in.

2.6 Derive an expression for the toughness of a material represented by $\sigma = K(\varepsilon + 0.2)^n$ and having a fracture strain denoted as ε_f.

2.7 In a disk test performed on a specimen that is 25 mm in diameter and 3 mm thick, the specimen fractures at a stress of 350 MPa. What was the load on the specimen at fracture?

2.8 On the basis of information given in Fig. 2.4, calculate the ultimate tensile strength of annealed copper.

2.9 Identify two materials in Fig. 2.4 that have the lowest and highest uniform elongations, respectively. Calculate these quantities as percentages of the original gage lengths.

2.10 Show that necking cannot take place in the torsion of a thin-walled tube made of a material having a true stress–true strain curve represented by Eq. (2.11). *Hint*: Use the expression for torque and an approach similar to that shown in Section 2.2.4.

2.11 Assume that a material with a uniaxial yield stress Y yields under a stress system of principal stresses σ_1, σ_2, and σ_3, where $\sigma_1 > \sigma_2 > \sigma_3$. Show that the superposition of a hydrostatic stress p on this system (such as placing the specimen in a pressurized chamber) does not affect yielding. In other words, the material will still yield according to yield criteria.

2.12 A cylindrical specimen made of a brittle material 1 in. high and 1 in. in diameter is subjected to a compressive force. Fracture occurs at an angle of 45° under a load of 30,000 lb. Calculate the shear stress and the normal stress on the fracture surface.

2.13 Calculate the work done in frictionless compression of a solid cylinder 30 mm high and 20 mm in diameter to a reduction in height of 50% for the following materials: (a) 6061-T6 aluminum, (b) annealed copper, (c) annealed 4340 steel, and (d) annealed 304 stainless steel.

2.14 For the material in Example 2.2, calculate the toughness by the area under the stress–strain curve. Compare your answer with that obtained by first determining the K and n values for this material and then obtaining the toughness by integration.

2.15 A material has a strength coefficient $K = 100{,}000$ psi and $n = 0.2$. Assuming that a tensile-test specimen made from this material begins to neck at a true strain of 0.2, show that the ultimate tensile strength of this material is 59,340 psi.

2.16 A tensile-test specimen is made of a material represented by $\sigma = K(\varepsilon + n)^n$. (a) Determine the true strain at which necking will begin. (b) Show that it is possible for an engineering material to exhibit this behavior.

2.17 A cylindrical specimen 1 in. in diameter and 1 in. high is being compressed by dropping a weight of 100 lb on it from a certain height. After deformation, the temperature rise in the specimen is 100 °F. Assuming no heat loss and no friction, calculate the final height of the specimen, using the following data for the material: $K = 15{,}000$ psi, $n = 0.5$, density = 0.1 lb/in^3, and specific heat = 0.3 Btu/lb °F.

2.18 A thin-walled spherical shell with a yield stress Y is subjected to an internal pressure p. With appropriate equations, show whether the pressure required to yield this shell depends on the particular yield criterion used.

2.19 Show that you can take a bent bar made of an elastic, perfectly plastic material and straighten it by stretching it into the plastic range. *Hint*: Consider the events shown in Fig. 2.44.

2.20 In Fig. 2.44(a) let the tensile and compressive residual stresses both be 16,000 psi and the modulus of elasticity of the material be 15×10^6 psi, with a modulus of resilience of 18 in.-lb/in^3. If the original length in (a) is 40 in., what should be the stretched length in (b) so that when unloaded the strip will be free of residual stresses?

2.21 A 0.5 m long bar is bent and then stress relieved. The radius of curvature to the neutral axis is 0.25 m. The bar is 20 mm thick and is made of an elastic, perfectly plastic material with $Y = 400$ MPa and $E = 200$ GPa. Calculate the length to which this bar should be stretched so that, after unloading, it will become and remain straight.

2.22 Show that, according to the distortion-energy criterion, the yield stress in plane strain is $1.15Y$, where Y is the uniaxial yield stress of the material.

2.23 A closed-end, thin-walled cylinder of original length l, thickness t, and internal radius r is subjected to an internal pressure p. Using the generalized Hooke's law equations, show the change, if any, that occurs in the length of this cylinder when it is pressurized.

2.24 A 1 in. square cube made of a material with $K = 80{,}000$ psi and $n = 0.5$ is placed in the center of a rigid die with a groove that is 1.5 in. wide (see Fig. 2.48b). It is being compressed without friction. (a) Calculate the true strain at which the workpiece touches the die walls. (b) Calculate the work done up to that point of deformation. (c) Calculate the compressive force required when the height of the workpiece is down to 0.2 in. Give your answers using both of the yield criteria described in this chapter.

2.25 Take a cubic piece of metal with length of the sides of l_0. Deform it plastically to the shape of a rectangular parallelpiped of dimensions l_1, l_2, and l_3. Assuming that the material is rigid, perfectly plastic, show that volume constancy requires that the expression $\varepsilon_1 + \varepsilon_2 + \varepsilon_3 = 0$ be satisfied.

2.26 What is the diameter of an originally 50 mm diameter solid steel sphere when subjected to a hydrostatic pressure of 2 GPa?

2.27 Determine the effective stress and effective strain in plane-strain compression according to the distortion-energy criterion.

2.28 A thin-walled tube under longitudinal tension (in the elastic range) decreases in diameter. Of course, the tube can be pressurized internally so that its diameter remains constant while it is being stretched. Assuming that the tube is open ended, derive an expression for the pressure required to keep the diameter constant as the tube is stretched.

2.29 What is the magnitude of compressive stress for the plane-strain compression test shown in Fig. 2.13, according to the maximum shear-stress criterion?

2.30 Calculate the work done in expanding a thin-walled spherical shell 0.1 in. thick from a diameter of 10 in. to 12 in. Let the shell material have the characteristics:

$$\sigma = 6000 + 2000\varepsilon^{0.5} \text{ psi}.$$

Solve this problem first by the plastic work of deformation method and then compare your answer with that obtained by the pressure–volume method.

3

Structure and Manufacturing Properties of Metals

3.1 Introduction

The structure of metals greatly influences their behavior and properties. A knowledge of structures guides us in controlling and predicting the behavior and performance of metals in various manufacturing processes. Understanding the structure of metals also allows us to predict and evaluate their *properties*. This helps us make appropriate selections for specific applications under particular force, temperature, and environmental conditions. For example, you will come to appreciate the reasons for developing single-crystal turbine blades (Fig. 3.1) for use in jet engines; these blades have better properties than those made conventionally by casting or forging.

In addition to atomic structure, various other factors also influence the properties and behavior of metals. Among these are the composition of the metal,

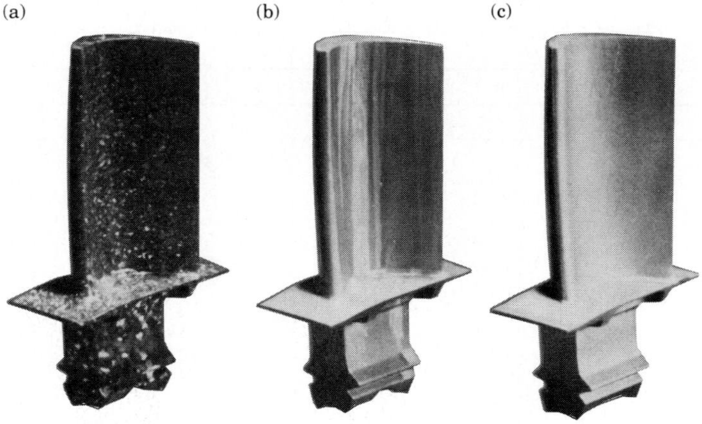

FIGURE 3.1
Turbine blades for jet engines, manufactured by three different methods: (a) conventionally cast; (b) directionally solidified, with columnar grains, as can be seen from the vertical streaks; and (c) single crystal. Although more expensive, single-crystal blades have properties at high temperatures that are superior to other blades. *Source:* Courtesy of United Technologies' Pratt and Whitney.

impurities and vacancies in the atomic structure, grain size, grain boundaries, environment, and size and surface condition of the metal, as well as the methods by which metals and alloys are made into useful products.

3.2 The Crystal Structure of Metals

When metals solidify from a molten state, the atoms arrange themselves into various orderly configurations, called *crystals*. This arrangement of the atoms in the crystal is called *crystalline structure*. A crystalline structure is like metal scaffolding in front of a building under construction, with uniformly repetitive horizontal and vertical metal pipes and braces.

The smallest group of atoms showing the characteristic *lattice structure* of a particular metal is known as a *unit cell*. It is the building block of a crystal, and a single crystal can have many unit cells. Think of each brick in a wall as a unit cell. The wall has a crystalline structure; that is, it consists of an orderly arrangement of bricks. Thus the wall is like a single crystal, consisting of many unit cells.

The three basic patterns of atomic arrangement found in most metals are (a) body-centered cubic (bcc); (b) face-centered cubic (fcc); and(c) hexagonal close-packed (hcp) (Fig. 3.2).

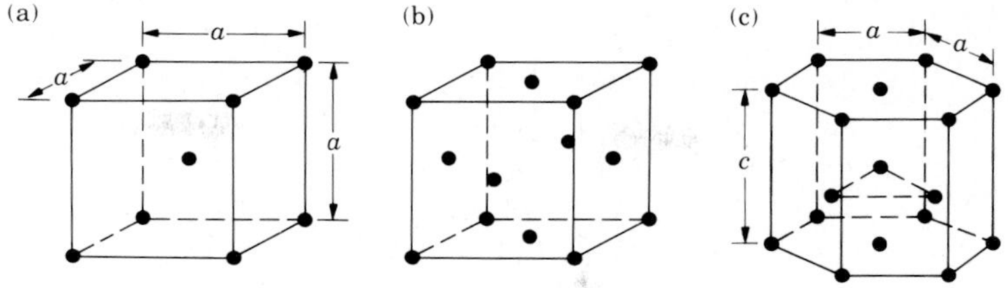

FIGURE 3.2
Crystal structures in metals. (a) Body-centered cubic unit cell, bcc. (b) Face-centered cubic unit cell, fcc. (c) Hexagonal close-packed, hcp.

The order of magnitude of the distance between the atoms in these crystal structures is 0.1 nm (10^{-8} in.). The way in which these atoms are arranged determines the properties of a particular metal. We can modify these arrangements by adding atoms of some other metal or metals. This is known as *alloying* and often improves the properties of the metal.

Of the three structures illustrated, the fcc and hcp crystals have the most densely packed configurations. In the hcp structure, the top and bottom planes are called *basal planes*. The reason that different crystal structures form is a matter of minimizing the energy required to form these structures. Thus tungsten forms a bcc structure because it requires less energy than other structures. Similarly, aluminum forms an fcc structure. However, at different temperatures the same metal may form different structures because of the lower energy required at that temperature. For example, iron forms a bcc structure below 912 °C (1674 °F), and above 1394 °C (2541 °F), but an fcc structure between 912 °C and 1394 °C. As we describe in Section 5.10, these structural changes are an important aspect of heat treatment of metals and alloys.

3.3 Deformation and Strength of Single Crystals

When crystals are subjected to an external force, they first undergo *elastic deformation*; that is, they return to their original shape when the force is removed. An analogy to this type of behavior is a helical spring that stretches when loaded and returns to its original shape when the load is removed. However, if the force on the crystal structure is increased sufficiently, the crystal undergoes *plastic deformation* (or permanent deformation); that is, it does not return to its original shape when the force is removed.

There are two basic mechanisms by which plastic deformation may take place in crystal structures. One is the slipping of one plane of atoms over an adjacent plane—the *slip plane*—under a shearing force (Fig. 3.3). The deformation of a single-crystal specimen by slip is shown schematically in Fig. 3.4. This situation is much like sliding playing cards against each other.

Just as it takes a certain amount of force to slide playing cards against each other, so a crystal requires a certain amount of shear stress (*critical shear stress*) to undergo permanent deformation. Thus there must be a shear stress of sufficient magnitude within a crystal for plastic deformation to take place.

The maximum *theoretical shear stress* to cause permanent deformation in a perfect crystal is obtained as follows: When there is no stress, the atoms in the crystal are in equilibrium (Fig. 3.5). Under a shear stress the upper row of atoms is moved to the right, the position of the atom being denoted as x. Thus when $x = 0$ or $x = b$ the shear stress is zero. Each atom of the upper row is attracted to the nearest atom of the lower row, resulting in nonequilibrium at positions 2 and 4, where the stresses are maximum but opposite in sign. At position 3 the shear stress is again zero, since this is a symmetric position. In Fig. 3.5 we assumed that, as a first approximation, the shear stress varies sinusoidally. Hence the shear stress at a displacement x is

$$\tau = \tau_{\max} \sin \frac{2\pi x}{b}, \tag{3.1}$$

which for small values of x/b can be written as

$$\tau = \tau_{\max} \frac{2\pi x}{b}.$$

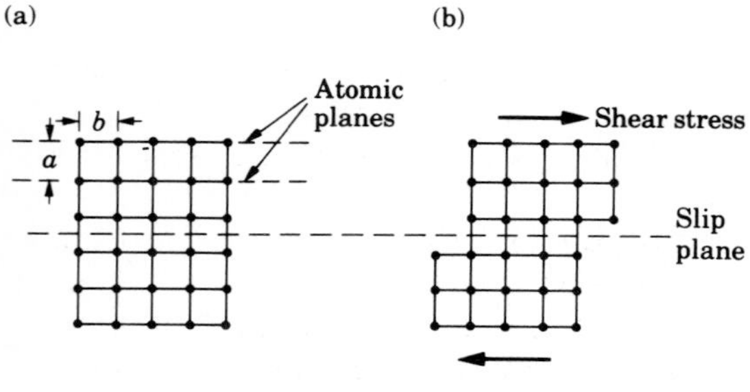

FIGURE 3.3
Permanent deformation, also called plastic deformation, of a single crystal subjected to a shear stress: (a) structure before deformation; and (b) deformation by slip. The b/a ratio influences the magnitude of shear stress required to cause slip.

FIGURE 3.4
Permanent deformation of a single crystal under a tensile load. Note that the slip planes tend to align themselves in the direction of pulling. This behavior can be simulated using a deck of cards with a rubber band around them.

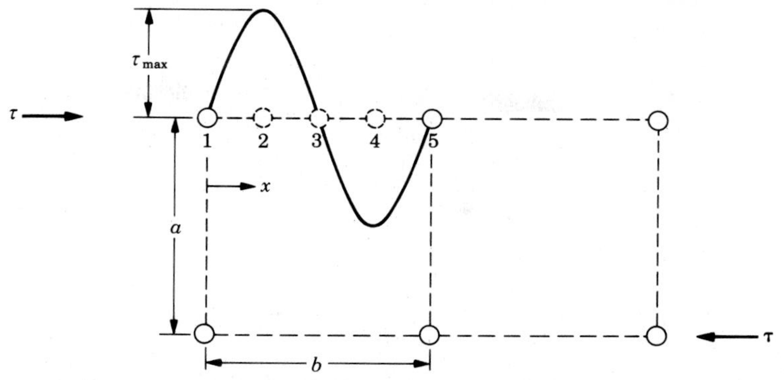

FIGURE 3.5
Variation of shear stress in moving a plane of atoms over another plane.

From Hooke's law we have

$$\tau = G\gamma = G\left(\frac{x}{a}\right).$$

Hence

$$\tau_{max} = \frac{G}{2\pi} \cdot \frac{b}{a}. \tag{3.2}$$

If we assume that b is approximately equal to a, we obtain

$$\tau_{max} = \frac{G}{2\pi}. \tag{3.3}$$

More refined calculations have set the value of the maximum theoretical shear strength as between $G/10$ and $G/30$.

The shear stress required to cause slip in single crystals is directly proportional to the ratio b/a in Fig. 3.5. We can therefore state that slip in a crystal takes place along planes of maximum atomic density, or that slip takes place in closely packed planes and in closely packed directions.

Because the b/a ratio is different for different directions within the crystal, a single crystal has different properties when tested in different directions. Thus a crystal is anisotropic. A common example of *anisotropy* is woven cloth, which stretches differently when we pull it in different directions, or plywood, which is much stronger in the planar direction than along its thickness direction (it splits easily).

The second mechanism of plastic deformation is *twinning*, in which a portion of the crystal forms a mirror image of itself across the plane of twinning. Twins form abruptly and are the cause of the creaking sound (tin cry) when a tin or zinc rod is bent at room temperature. Twinning usually occurs in hcp metals.

3.3.1 Slip systems

The combination of a slip plane and its direction of slip is known as a *slip system*. In general, metals with five or more slip systems are ductile, whereas those with less than five are not ductile.

a) In body-centered cubic crystals there are 48 possible slip systems. Thus the probability is high that an externally applied shear stress will operate on one of these systems and cause slip. However, because of the relatively high b/a ratio, the required shear stress is high. Metals with bcc structures, such as titanium, molybdenum, and tungsten, have good strength and moderate ductility.
b) In face-centered cubic crystals there are 12 slip systems. The probability of slip is moderate, and the required shear stress is low. These metals, such as aluminum, copper, gold, and silver, have moderate strength and good ductility.
c) The hexagonal close-packed crystal has three slip systems, thus having a low probability of slip. However, more systems become active at elevated temperatures. Metals with hcp structures, such as beryllium, magnesium, and zinc, are generally brittle at room temperature.

Note in Fig. 3.4 that the portions of the single crystal that have slipped have rotated from their original angular position toward the direction of the tensile force. Note also that slip has taken place along certain planes only. With the use of electron microscopy, it has been shown that what appears to be a single slip plane is actually a *slip band*, consisting of a number of slip planes (Fig. 3.6).

3.3.2 Ideal tensile strength of metals

In addition to obtaining the maximum theoretical shear strength, we can also calculate the theoretical or ideal *tensile* strength of materials. In the bar shown in Fig. 3.7, the interatomic distance is a when no stress is applied. In order to increase this distance we have to apply a force to overcome the cohesive force between the atoms, the cohesive force being zero in the unstrained equilibrium condition. When the tensile stress reaches σ_{max}, the atomic bonds between two neighboring atomic planes break. This σ_{max} is known as the *ideal tensile strength*.

We may approximate the curve for cohesive force by a sine curve of wavelength λ to obtain

$$\sigma = \sigma_{max} \sin \frac{2\pi x}{\lambda}, \tag{3.4}$$

where x is the displacement from a. For small values of x we write

$$\sigma = \sigma_{max} \frac{2\pi x}{\lambda}.$$

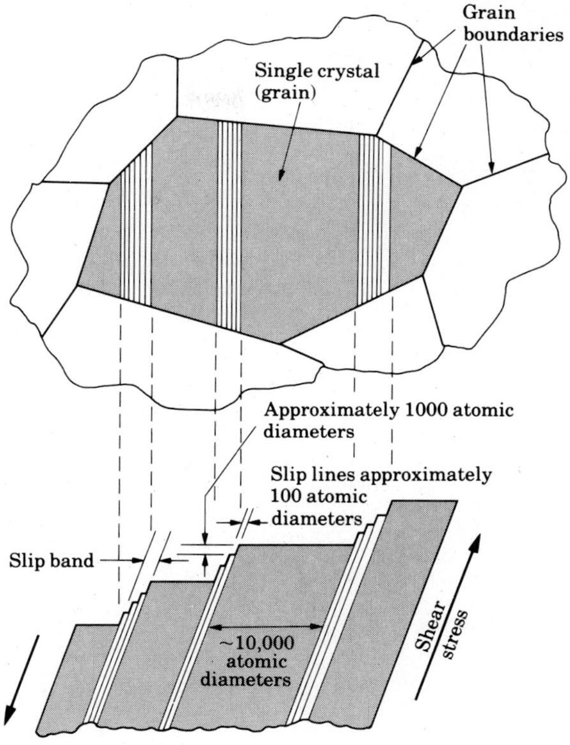

FIGURE 3.6 Schematic illustration of slip lines and slip bands in a single crystal subjected to a shear stress. A slip band consists of a number of slip planes. The crystal at the center of the upper part is an individual grain surrounded by other grains.

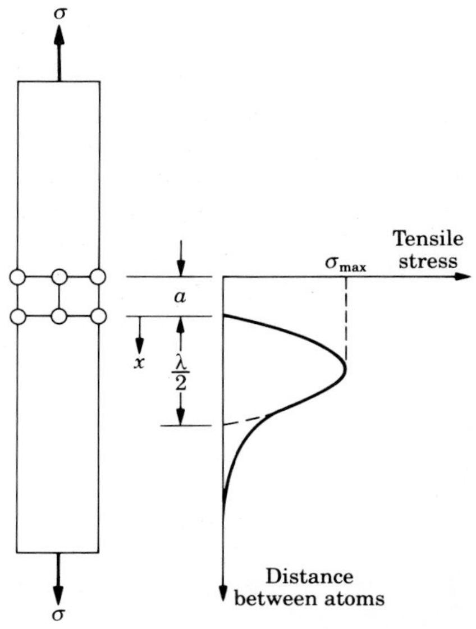

FIGURE 3.7 Variation of cohesive stress as a function of distance between atoms.

Also, for small values of x we have, from Hooke's law,

$$\sigma = \frac{E \cdot x}{a}$$

and hence

$$\sigma_{\max} = \frac{E \cdot \lambda}{2\pi a}. \qquad (3.5)$$

The work done per unit area of the tensile specimen is the area under the cohesive force curve. Thus

$$\text{Work} = \int_0^{\lambda/2} \sigma_{\max} \sin \frac{2\pi x}{\lambda}\, dx = \frac{\sigma_{\max} \cdot \lambda}{\pi}. \qquad (3.6)$$

The work consumed may be assumed to go into generating *two new surfaces* on the specimen where separation takes place. This involves surface energy. The atoms within the bulk of a solid or liquid have bonds to their neighbors in all directions; hence they are subjected to equal forces in all directions. The atoms at or near the surface, however, have about one half the number of bonds. To bring an atom from within the bulk to the surface, work is required to overcome the unbalanced attractive forces. Thus the energy of an atom on the surface is higher than an atom in the bulk. The energy required to create a new surface is known as *surface energy* γ.

The surface energy involved in pulling a tensile specimen apart is

$$\text{Surface energy} = 2\gamma. \qquad (3.7)$$

Combining Eqs. (3.5)–(3.7), we obtain

$$\sigma_{\max} = \sqrt{E\gamma/a}. \qquad (3.8)$$

For most solids, when we substitute appropriate values into this equation, we have roughly

$$\sigma_{\max} \simeq E/10. \qquad (3.9)$$

3.3.3 Imperfections

The actual strength of metals is approximately one to two orders of magnitude lower than the strength levels obtained from theoretical calculations. This discrepancy has been explained in terms of imperfections in the crystal structure. Unlike the idealized models we have described, actual metal crystals contain a large number of defects and imperfections, which we may categorize as:

a) Line defects, called *dislocations* (Fig. 3.8).

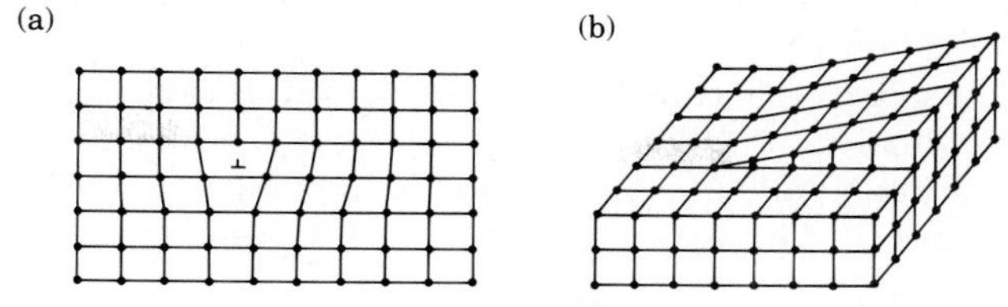

FIGURE 3.8
Types of dislocations in a single crystal: (a) edge dislocation; and (b) screw dislocation.

b) Point defects, such as a *vacancy* (missing atom), an *interstitial atom* (extra atom in the lattice), or an *impurity* (foreign atom) that has replaced the atom of the pure metal (Fig. 3.9).
c) Volume or bulk imperfections, such as *voids* or *inclusions* (nonmetallic elements such as oxides, sulfides, and silicates).
d) Planar imperfections, such as *grain boundaries*.

Dislocations. First observed in the 1930s, *dislocations* are defects in the orderly arrangement of a metal's atomic structure. They are the most significant defects that help explain the discrepancy between the actual and theoretical strength of metals. There are two types of dislocations: *edge* and *screw* (Fig. 3.8).

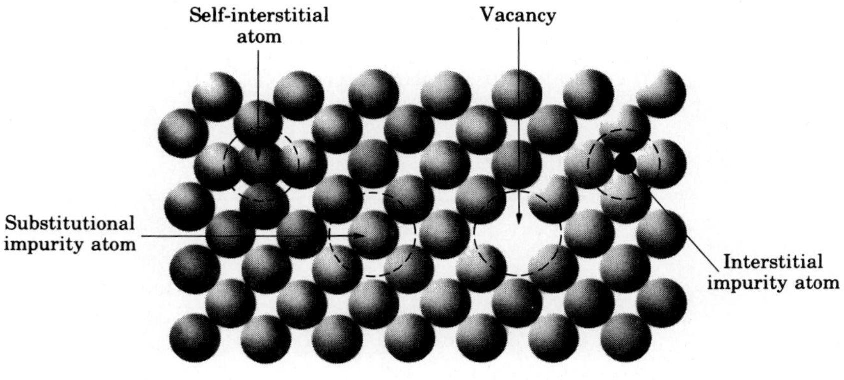

FIGURE 3.9
Defects in a single-crystal lattice. Self-interstitial, vacancy, interstitial, and substitutional. *Source:* After Moffatt et al.

A slip plane containing a dislocation requires less shear stress to cause slip than a plane in a perfect lattice (Fig. 3.10). One analogy used to describe the movement of an edge dislocation is the earthworm, which moves forward through a hump that starts at the tail and moves toward the head. A second is moving a large carpet by forming a hump at one end and moving it to the other end. The force required to move a carpet in this way is much less than that required to slide the carpet along the floor. Screw dislocations are so named because the atomic planes form a spiral ramp.

The *density* of dislocations (total length of dislocation line per unit volume) increases with increasing plastic deformation by as much as 10^6 at room temperature. The dislocation density for some conditions are:

Very pure single crystals, 0 to 10^3
Annealed single crystals, 10^5 to 10^6
Annealed polycrystals, 10^7 to 10^8
Highly cold-worked metals, 10^{11} to 10^{12}

The mechanical properties of metals such as yield and fracture strength, as well as electrical conductivity, are affected by lattice defects. These are known as *structure-sensitive* properties. On the other hand, melting point, specific heat, coefficient of thermal expansion, and elastic constants are not sensitive to these defects, and are known as *structure-insensitive* properties.

3.3.4 Work hardening (strain hardening)

Although the presence of a dislocation lowers the shear stress required to cause slip, dislocations can (a) become entangled and interfere with each other and (b) be impeded by barriers, such as grain boundaries and impurities and inclusions in the material. Entanglement and impediments increase the shear stress required for slip. This entanglement is like moving two humps at different angles across a carpet, with the two humps interfering with each other's movement, thus making it more difficult to move the carpet.

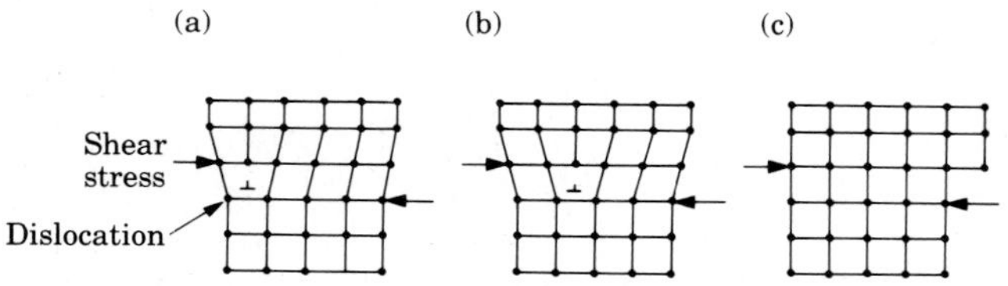

FIGURE 3.10
Movement of an edge dislocation across the crystal lattice under a shear stress. Dislocations help explain why the actual strength of metals is much lower than that predicted by theory.

The increase in shear stress, and hence the increase in the overall strength of the metal, is known as *work hardening* or *strain hardening*. The greater the deformation, the more the entanglements, which increases the metal's strength. Work hardening is used extensively in strengthening metals in metalworking processes at ambient temperature. Typical examples are strengthening wire by reducing its cross-section by drawing it through a die, producing the head on a bolt by forging it, and producing sheet metal for automobile bodies and aircraft fuselages by rolling.

3.4 Grains and Grain Boundaries

Metals commonly used for manufacturing various products are composed of many individual, randomly oriented crystals (*grains*). We are thus dealing with metal structures that are not single crystals but *polycrystals*, that is, many crystals.

When a mass of molten metal begins to solidify, crystals begin to form independently of each other, with random orientations and at various locations within the liquid mass (Fig. 3.11). Each of these crystals grows into a crystalline

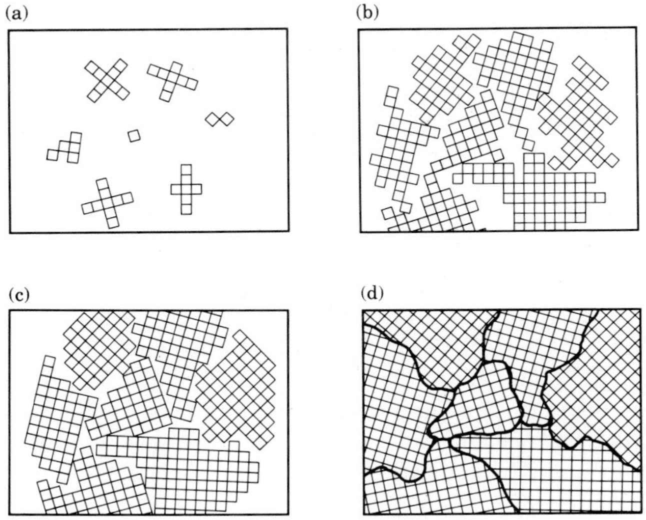

FIGURE 3.11 Schematic illustration of the various stages during solidification of molten metal. Each small square represents a unit cell. (a) Nucleation of crystals at random sites in the molten metal. Note that the crystallographic orientation of each site is different. (b) and (c) Growth of crystals as solidification continues. (d) Solidified metal, showing individual grains and grain boundaries. Note the different angles at which neighboring grains meet each other. *Source:* W. Rosenhain.

structure or grain. The number and size of the grains developed in a unit volume of the metal depends on the rate at which *nucleation* (initial stage of formation of crystals) takes place. The number of different sites in which individual crystals begin to form (seven in Fig. 3.11a) and the rate at which these crystals grow are important in the size of grains developed. Generally, rapid cooling produces smaller grains, whereas slow cooling produces larger grains.

If the crystal nucleation rate is high, the number of grains in a unit volume of metal will be greater, and consequently grain size will be small. Conversely, if the rate of growth of the crystals is high, compared to their nucleation rate, there will be fewer—but larger—grains per unit volume.

Note in Fig. 3.11 how the grains eventually interfere with and impinge upon one another. The surfaces that separate these individual grains are called *grain boundaries*. Each grain consists of either a single crystal, for pure metals, or a polycrystalline aggregate, for alloys. Note that the crystallographic orientation changes abruptly from one grain to the next across the grain boundaries. Recall from Section 3.3 that the behavior of a single crystal or single grain is anisotropic. The ideal behavior of a piece of polycrystalline metal (Fig. 3.12) is *isotropic* because the grains have random crystallographic orientations. Thus its properties do not vary with the direction of testing. In practice, however, this situation rarely exists.

3.4.1 Grain size

Grain size significantly influences the mechanical properties of metals. Large grain size is generally associated with low strength, hardness, and ductility. Large grains, particularly in sheet metals, also result in a rough surface appearance after being stretched. The yield strength Y is the most sensitive property and is related to grain size by the empirical formula (Hall–Petch equation):

$$Y = Y_i + kd^{-1/2}, \tag{3.10}$$

where Y_i is a basic yield stress that can be regarded as the stress opposing the motion of dislocations, k is a constant indicating the extent to which dislocations are piled up at barriers, such as grain boundaries, and d is the grain diameter. Equation (3.10) is valid below the recrystallization temperature of the material.

We usually measure grain size by counting the number of grains in a given area or the number of grains that intersect a given length of a line randomly drawn on an enlarged photograph of the grains taken under a microscope on a polished and etched specimen. Grain size may also be determined by comparing it to a standard chart. The ASTM (The American Society for Testing and Materials) grain size number, n, is related to the number of grains, N, per square inch at a magnification of 100× (equal to 0.0645 mm^2 of actual area) by

$$N = 2^{n-1}. \tag{3.11}$$

Because grains are generally extremely small, many grains occupy a unit volume of metal. Grain sizes between 5 and 8 are generally considered fine grains. A grain

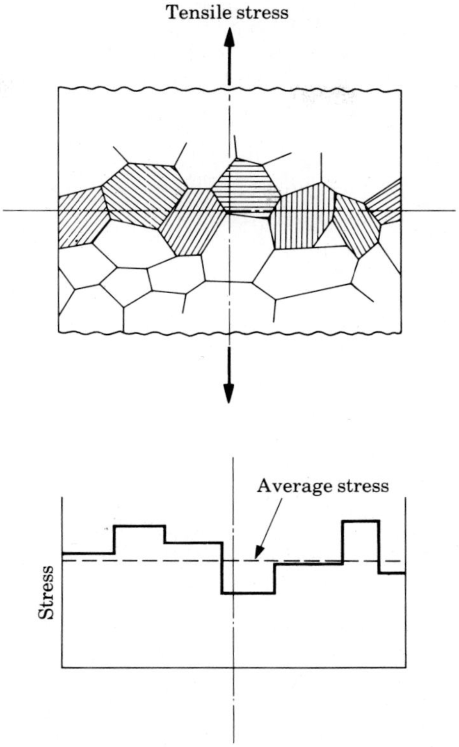

FIGURE 3.12

Variation of tensile stress across a plane of polycrystalline metal specimen subjected to tension. Note that the strength exhibited by each grain depends on its orientation. *Source:* L. H. Van Vlack, *Materials for Engineering.* Addison-Wesley Publishing Company, Inc., Reading, Massachusetts, 1982.

size of 7 is generally acceptable for sheet metals used to make car bodies, appliances, and kitchen utensils. Grains can be large enough to be visible by the naked eye, such as those of zinc on the surface of galvanized sheet steels.

3.4.2 Influence of grain boundaries

Grain boundaries have an important influence on the strength and ductility of metals. Because they interfere with the movement of dislocations, grain boundaries also influence strain hardening. These effects depend on temperature, rate of deformation, and the type and amount of impurities present along the grain boundaries. Grain boundaries are more reactive than the grains themselves. This is because the atoms along the grain boundaries are packed less efficiently and are more disordered and hence have a higher energy than the atoms in the orderly arrangement within the grains.

At elevated temperatures and in materials whose properties depend on the rate of deformation, plastic deformation also takes place by means of *grain boundary sliding*. The *creep mechanism* (elongation under stress over a period of time, usually at elevated temperatures) results from grain-boundary sliding.

Grain boundary embrittlement. When brought into close atomic contact with certain low-melting-point metals, a normally ductile and strong metal can crack under very low stresses. Examples are aluminum wetted with a mercury–zinc amalgam or liquid gallium and copper at elevated temperature wetted with lead or bismuth. These elements weaken the grain boundaries of the metal by *embrittlement*. We use the term *liquid-metal embrittlement* to describe such phenomena because the embrittling element is in a liquid state. However, embrittlement can also occur at temperatures well below the melting point of the embrittling element. This phenomenon is known as *solid-metal embrittlement*.

Hot shortness is caused by local melting of a constituent or an impurity in the grain boundary at a temperature below the melting point of the metal itself. When such a metal is subjected to plastic deformation at elevated temperatures (hot working), the piece of metal crumbles and disintegrates along the grain boundaries. Examples are antimony in copper and leaded steels and brass. To avoid hot shortness, the metal is usually worked at a lower temperature in order to prevent softening and melting along the grain boundaries. Another form of embrittlement is *temper embrittlement* in alloy steels, which is caused by segregation (movement) of impurities to the grain boundaries.

3.5 Plastic Deformation of Polycrystalline Metals

If a piece of polycrystalline metal with uniform equiaxed grains (having equal dimensions in all directions, as shown in the model in Fig. 3.13) is subjected to plastic deformation at room temperature (cold working), the grains become deformed and elongated. The deformation process may be carried out either by compressing the metal, as is done in forging, or by subjecting it to tension, as is done in stretching sheet metal. The deformation within each grain takes place by the mechanisms we described in Section 3.3 for a single crystal.

During plastic deformation, the grain boundaries remain intact and mass continuity is maintained. The deformed metal exhibits greater strength because of the entanglement of dislocations with grain boundaries. The increase in strength depends on the amount of deformation (strain) to which the metal is subjected; the greater the deformation, the stronger the metal becomes. Furthermore, the increase

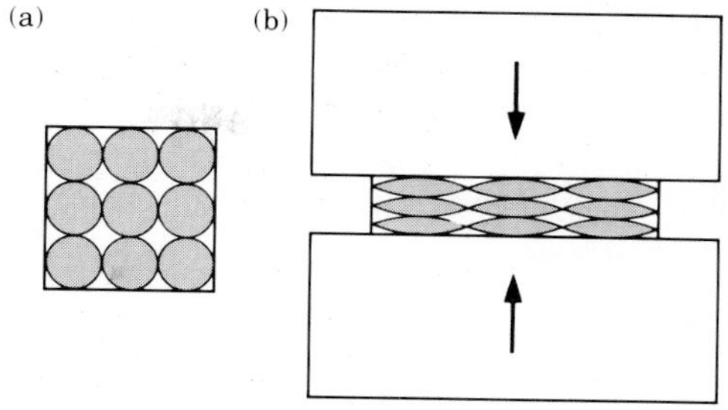

FIGURE 3.13 Plastic deformation of idealized (equiaxed) grains in a specimen subjected to compression, such as is done in rolling or forging of metals: (a) before deformation; and (b) after deformation. Note the alignment of grain boundaries along a horizontal direction.

in strength is greater for metals with smaller grains because they have a larger grain-boundary surface area per unit volume of metal.

3.5.1 Anisotropy (texture)

Figure 3.13 shows that, as a result of plastic deformation, the grains have elongated in one direction and contracted in the other. Consequently, this piece of metal has become *anisotropic*, and its properties in the vertical direction are different from those in the horizontal direction.

Many products develop anisotropy of mechanical properties after they have been processed by metalworking techniques. The degree of anisotropy depends on how uniformly the metal is deformed. Note from the direction of the crack in Fig. 3.14, for example, that the ductility of the cold-rolled sheet in the vertical (transverse) direction is lower than in its longitudinal direction.

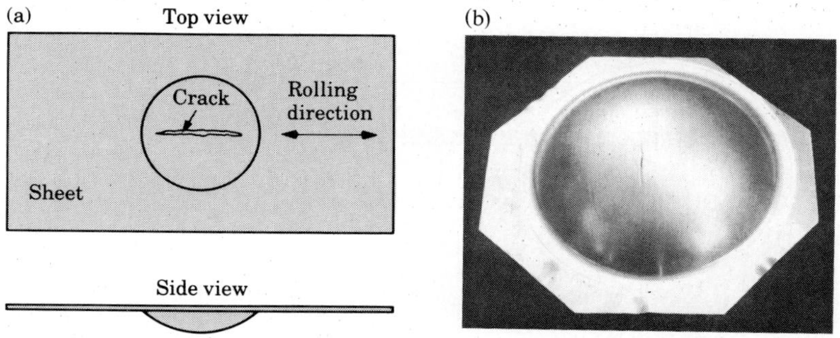

FIGURE 3.14 (a) Schematic illustration of a crack in sheet metal subjected to bulging, such as caused by pushing a steel ball against the sheet. Note the orientation of the crack with respect to the rolling direction of the sheet. This material is anisotropic. (b) Aluminum sheet with a crack (vertical dark line at the center) developed in a bulge test. *Source:* J. S. Kallend.

Anisotropy influences both mechanical and physical properties of metals. For example, sheet steel for electrical transformers is rolled in such a way that the resulting deformation imparts anisotropic magnetic properties to the sheet, thus reducing magnetic-hysteresis losses and improving the efficiency of transformers (see also "Amorphous Alloys," Section 3.10.8). There are two general types of anisotropy in metals: *preferred orientation* and *mechanical fibering*.

Preferred orientation. Also called *crystallographic anisotropy*, preferred orientation can be best described by reference to Fig. 3.4. When a metal crystal is subjected to tension, the sliding blocks rotate toward the direction of pulling. Thus slip planes and slip bands tend to align themselves with the direction of deformation. Similarly, for a polycrystalline aggregate with grains in various orientations (Fig. 3.12), all slip directions tend to align themselves with the direction of pulling. Conversely, under compression the slip planes tend to align themselves in a direction perpendicular to the direction of compression.

Mechanical fibering. Mechanical fibering results from the alignment of impurities, inclusions (stringers), and voids in the metal during deformation. Note that if the spherical grains in Fig. 3.13 were coated with impurities, these impurities would align themselves generally in a horizontal direction after deformation. Since impurities weaken the grain boundaries, this piece of metal would be weak and less ductile when tested in the vertical direction. An analogy would be plywood, which is strong in tension along its planar direction, but peels off easily when tested in tension in its thickness direction.

3.6 Recovery, Recrystallization, and Grain Growth

We have shown that plastic deformation at room temperature results in the deformation of grains and grain boundaries, a general increase in strength, and a decrease in ductility, as well as causing anisotropic behavior. These effects can be reversed and the properties of the metal brought back to their original levels by heating it in a specific temperature range for a period of time. The temperature range and amount of time depend on the material and several other factors. Three events take place consecutively during the heating process:

- *Recovery.* During recovery, which occurs at a certain temperature range below the *recrystallization temperature* of the metal, the stresses in the highly deformed regions are relieved. Subgrain boundaries begin to form—called *polygonization*—with no appreciable change in mechanical properties, such as hardness and strength (Fig. 3.15).

- *Recrystallization.* The process in which, at a certain temperature range, new equiaxed and strain-free grains are formed, replacing the older grains, is called *recrystallization.* The temperature for recrystallization ranges approximately between $0.3T_m$ and $0.5T_m$, where T_m is the melting point of the

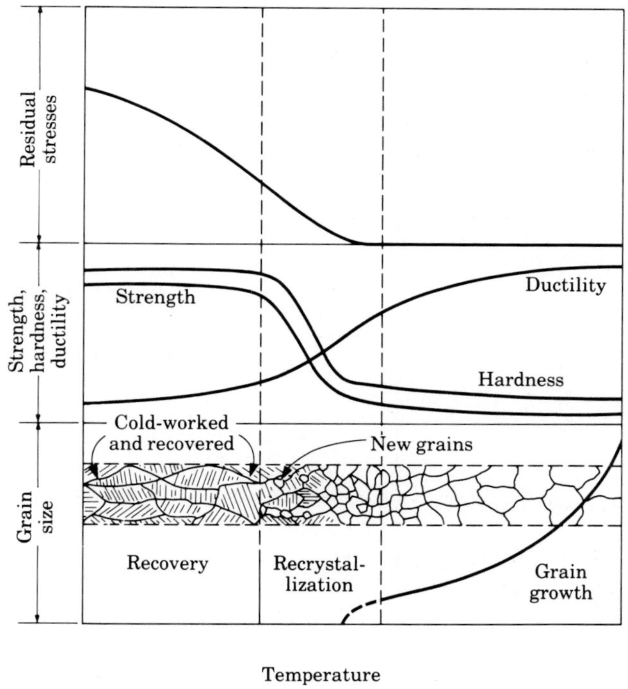

FIGURE 3.15 Schematic illustration of the effects of recovery, recrystallization, and grain growth on mechanical properties and shape and size of grains. Note the formation of small new grains during recrystallization. *Source:* G. Sachs.

metal on the absolute scale (Fig. 3.16). The recrystallization temperature is generally defined as the temperature at which complete recrystallization occurs within approximately one hour. Recrystallization decreases the density of dislocations and lowers the strength and raises the ductility of the metal (Fig. 3.15). Figure 3.16 shows that metals such as lead, tin, cadmium, and zinc recrystallize at about room temperature. Thus when these metals are cold worked (Section 3.7), they do not work-harden.

Recrystallization depends on the degree of prior cold work (work hardening); the higher the cold work, the lower the temperature required for recrystallization to occur. The reason is that as the amount of cold work increases, the number of dislocations and the amount of energy stored in dislocations (*stored energy*) also increase. This energy supplies the work required for recrystallization. Recrystallization is a function of time because it involves diffusion, that is, movement and exchange of atoms across grain boundaries.

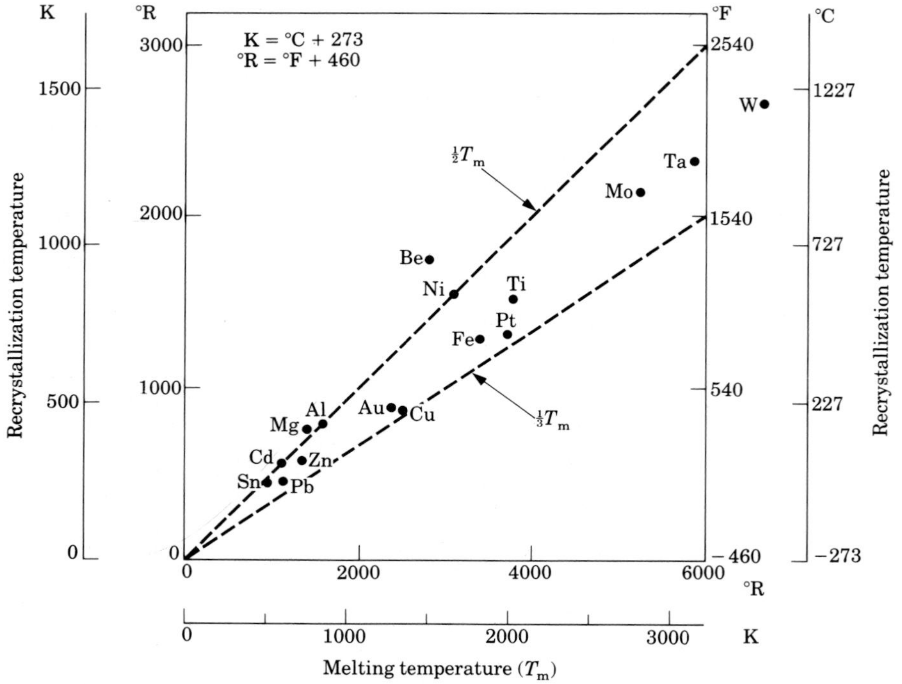

FIGURE 3.16

Recrystallization temperature for various metals. This temperature is approximately between one-half and one-third of the absolute melting point of metals. Note that metals such as lead (Pb) and tin (Sn) recrystallize at room temperature. *Source:* L. H. Van Vlack, *Materials for Engineering*. Addison-Wesley Publishing Company, Inc., Reading, Massachusetts, 1982.

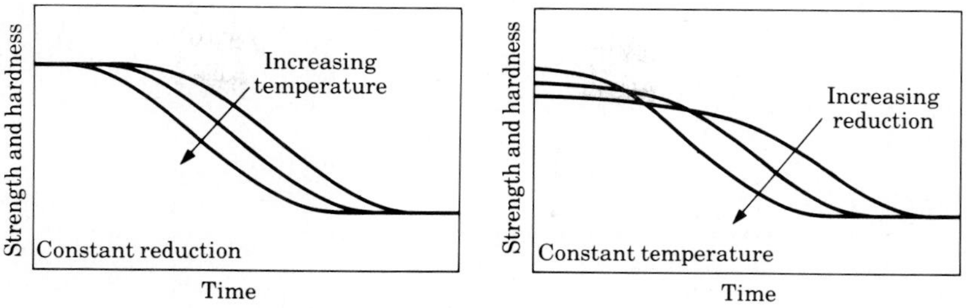

FIGURE 3.17
Variation of strength and hardness with recrystallization temperature, time, and prior cold work. Note that the more a metal is cold worked, the less time it takes to recrystallize. This is because of the higher stored energy from cold working due to increased dislocation density.

The effects on recrystallization of temperature, time, and reduction in the thickness or height of the workpiece by cold working are as follows (Fig. 3.17):

a) For a constant amount of deformation by cold working, the time required for recrystallization decreases with increasing temperature.
b) The greater the prior cold work done, the lower is the temperature required for recrystallization.
c) The higher the amount of deformation, the smaller the grain size becomes during recrystallization (Fig. 3.18). This is a common method of converting a coarse-grained structure to one of fine grain, with improved properties.
d) Anisotropy due to preferred orientation usually persists after recrystallization. To restore isotropy, a temperature higher than that required for recrystallization may be necessary.

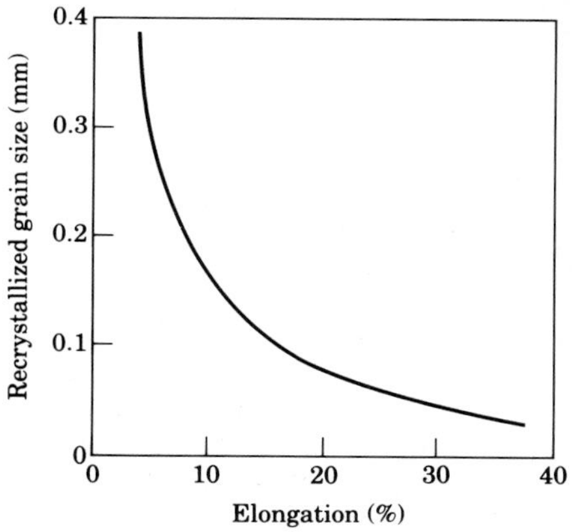

FIGURE 3.18
The effect of prior cold work on the recrystallized grain size of alpha brass. Below a critical elongation (strain), typically 5%, no recrystallization occurs.

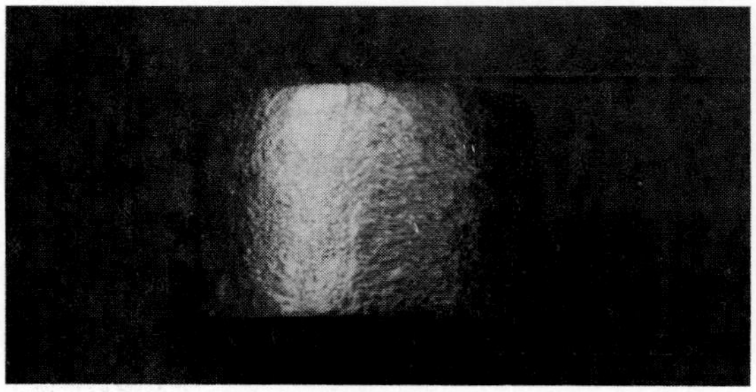

FIGURE 3.19
Surface roughness on the cylindrical surface of an aluminum specimen subjected to compression. *Source:* A. Mulc and S. Kalpakjian.

- *Grain growth.* If we continue to raise the temperature of the metal, the grains begin to grow and their size may eventually exceed the original grain size. This phenomenon is known as *grain growth*, and it adversely affects mechanical properties slightly (Fig. 3.15). However, large grains produce a rough surface appearance on sheet metals—called *orange peel*—when they are stretched to form a part or when a piece of metal is subjected to compression (Fig. 3.19), such as in forging operations.

3.7 Cold, Warm, and Hot Working

When plastic deformation is carried out at room temperature, it is called *cold working*, and when carried out at above the recrystallization temperature, it is called *hot working*. As the name implies, *warm working* is carried out at higher than room temperature but lower than hot-working temperature. Thus warm working is a compromise between cold and hot working. The temperature range for these three categories of plastic deformation are given in Table 3.1 in terms of a ratio, where T is the working temperature and T_m is the melting point of the metal,

TABLE 3.1
HOMOLOGOUS TEMPERATURE RANGES FOR VARIOUS PROCESSES

PROCESS	T/T_m
Cold working	<0.3
Warm working	0.3 to 0.5
Hot working	>0.6

both on the absolute scale. Although it is a dimensionless quantity, this ratio is known as the *homologous temperature*.

There are important technological differences in products that are processed by cold, warm, or hot working. For example, compared to cold-worked products, hot-worked products generally have less dimensional accuracy, because of uneven thermal expansion and contraction during processing, and rougher surface appearance and finish, because of the oxide layer that usually develops during heating. Other important manufacturing characteristics, such as formability, machinability, and weldability, are also affected by cold, warm, and hot working to various degrees.

3.8 Physical Properties

In addition to their mechanical properties, the *physical properties* of materials should also be considered in selecting and processing materials. Properties of particular interest are density, melting point, specific heat, thermal conductivity and expansion, electrical and magnetic properties, and resistance to oxidation and corrosion.

3.8.1 Density

Density of a material is its weight per unit volume. If expressed relative to the density of water, it is known as *specific gravity* and thus has no units. The range of densities for a variety of materials at room temperature is given in Table 3.2. The density of materials depends on atomic weight, atomic radius, and the packing of the atoms. Alloying elements generally have a minor effect on the density of metals, the effect depending on the density of the alloying elements.

The most significant role that density plays is in the *specific strength* (strength-to-weight ratio) and *specific stiffness* (stiffness-to-weight ratio) of materials and structures. Saving weight is particularly important for aircraft and aerospace structures, automotive bodies and components, and other products where energy consumption and power limitations are major concerns. Substitution of materials for weight savings and economy is a major factor in advanced equipment and machinery, as well as in consumer products such as automobiles.

Density is an important factor in the selection of materials for high-speed equipment, such as the use of magnesium in printing and textile machinery, many components of which usually operate at very high speeds. To obtain exposure times of 1/4000 s in cameras without sacrificing accuracy, the shutters of some high-quality 35-mm cameras are made of titanium. The resulting light weight of the components in these high-speed operations reduces inertial forces that otherwise could lead to vibrations, inaccuracies, and even part failure over a period of time.

TABLE 3.2
PHYSICAL PROPERTIES OF SELECTED MATERIALS AT ROOM TEMPERATURE

METAL	DENSITY (kg/m^3)	MELTING POINT (°C)	SPECIFIC HEAT (J/kg K)	THERMAL CONDUCTIVITY (W/m K)
Aluminum	2700	660	900	222
Aluminum alloys	2630–2820	476–654	880–920	121–239
Beryllium	1854	1278	1884	146
Columbium (niobium)	8580	2468	272	52
Copper	8970	1082	385	393
Copper alloys	7470–8940	885–1260	377–435	29–234
Iron	7860	1537	460	74
Steels	6920–9130	1371–1532	448–502	15–52
Lead	11350	327	130	35
Lead alloys	8850–11350	182–326	126–188	24–46
Magnesium	1745	650	1025	154
Magnesium alloys	1770–1780	610–621	1046	75–138
Molybdenum alloys	10210	2610	276	142
Nickel	8910	1453	440	92
Nickel alloys	7750–8850	1110–1454	381–544	12–63
Tantalum alloys	16600	2996	142	54
Titanium	4510	1668	519	17
Titanium alloys	4430–4700	1549–1649	502–544	8–12
Tungsten	19290	3410	138	166
NONMETALLIC				
Ceramics	2300–5500	—	750–950	10–17
Glasses	2400–2700	580–1540	500–850	0.6–1.7
Graphite	1900–2200	—	840	5–10
Plastics	900–2000	110–330	1000–2000	0.1–0.4
Wood	400–700	—	2400–2800	0.1–0.4

Because of their low density, ceramics are being considered for use in components of automated machine tools in high-speed operations. However, there are applications where weight is desirable. Examples are counterweights for various mechanisms (using lead and steel) and components for self-winding watches (using high-density materials such as tungsten).

3.8.2 Melting point

The *melting point* of a material depends on the energy required to separate its atoms. As Table 3.2 shows, the melting point of an alloy can have a wide range, unlike pure metals which have a definite melting point. The melting points of alloys depend on their particular composition.

The melting point of a material has a number of indirect effects on manufacturing operations. The choice of a material for high-temperature applications is the most obvious consideration, such as in jet engines and furnaces, or where high frictional heat is generated, such as with high sliding speeds of machine components. Sliding down a rope or rubbing your hands together fast is a simple demonstration of how high temperatures can become with increasing sliding speed.

Since the recrystallization temperature of a metal is related to its melting point, operations such as annealing, heat treating, and hot working require a knowledge of the melting points of the materials involved. These considerations, in turn, influence the selection of tool and die materials in manufacturing operations. Another major influence of the melting point is in the selection of the equipment and melting practice in casting operations. The higher the melting point of the material, the more difficult the operation becomes. Also, the selection of die materials, as in die casting, depends on the melting point of the workpiece material. In the electrical-discharge machining process, the melting points of metals are related to the rate of material removal and tool wear.

3.8.3 Specific heat

Specific heat is the energy required to raise the temperature of a unit mass of material by one degree. Alloying elements have a relatively minor effect on the specific heat of metals.

The temperature rise in a workpiece, resulting from forming or machining operations, is a function of the work done and the specific heat of the workpiece material. Thus the lower the specific heat, the higher the temperature will rise in the material. If excessive, temperature can have detrimental effects on product quality by adversely affecting its surface finish and dimensional accuracy, cause excessive tool and die wear, and result in adverse metallurgical changes in the material.

3.8.4 Thermal conductivity

Whereas specific heat indicates the temperature rise in the material as a result of energy input, *thermal conductivity* indicates the ease with which heat flows within and through the material. Metallically bonded materials (metals) generally have high thermal conductivity, whereas ionically or covalently bonded materials (ceramics and plastics) have poor conductivity. Because of the large difference in their thermal conductivities, alloying elements can have a significant effect on the thermal conductivity of alloys, as we can see in Table 3.2 by comparing the metals with their alloys.

When heat is generated by plastic deformation or friction, the heat should be conducted away at a high enough rate to prevent a severe rise in temperature. The main difficulty experienced in machining titanium, for example, is caused by its very low thermal conductivity. Low thermal conductivity can also result in high thermal gradients and thus cause inhomogeneous deformation in metalworking processes.

3.8.5 Thermal expansion

Thermal expansion of materials can have several significant effects. Generally, the coefficient of thermal expansion is inversely proportional to the melting point of the material. Alloying elements have a relatively minor effect on the thermal expansion of metals.

Shrink fits utilize thermal expansion. Other examples where relative expansion or contraction is important are electronic and computer components, glass-to-metal seals, struts on jet engines, and moving parts in machinery that require certain clearances for proper functioning. Improper selection of materials and assembly can cause *thermal stresses*, cracking, warping, or loosening of components in the structure during their service life.

Thermal conductivity, in conjunction with thermal expansion, plays the most significant role in causing thermal stresses, both in manufactured components and in tools and dies. This is particularly important in a forging operation, for example, when hot workpieces are placed over relatively cool dies, making the die surfaces undergo thermal cycling. To reduce thermal stresses, a combination of high thermal conductivity and low thermal expansion is desirable.

Thermal stresses can lead to cracks in ceramic parts and in tools and dies made of relatively brittle materials. *Thermal fatigue* results from thermal cycling and causes a number of surface cracks. *Thermal shock* is the term generally used to describe development of cracks after a single thermal cycle. Thermal stresses may be caused both by temperature gradients and by *anisotropy of thermal expansion*, which we generally observe in hexagonal close-packed metals and in ceramics.

Because of their resistance to high temperature and wear, ceramics have become attractive materials for components of heat engines. One of the most attractive ceramics is partially stabilized zirconia, which has thermal expansion characteristics close to those of cast iron. Thus components such as cylinder liners do not undergo significant dimensional changes with respect to the engine block as engine temperature changes.

The influence of temperature on dimensions and modulus of elasticity can be a significant problem in precision instruments and equipment. A spring, for example, will have a lower stiffness as its temperature increases because of reduced elastic modulus. Similarly, a tuning fork or a pendulum will have different frequencies at different temperatures. Dimensional changes resulting from thermal expansion can be significant in measurements for quality control and in equipment such as measuring tapes for geodetic applications.

To alleviate some of the problems with thermal expansion, a family of iron–nickel alloys that have very low thermal-expansion coefficients have been developed, called *low-expansion alloys*. A typical composition is 64 percent iron–36 percent nickel. This alloy is called Invar, because the length of a component made of this alloy is almost invariable with respect to temperature, although over a narrow range. Consequently, these alloys also have good thermal-fatigue resistance.

3.8.6 Electrical and magnetic properties

Electrical conductivity and dielectric properties of materials are of great importance not only in electrical equipment and machinery, but also in manufacturing processes such as magnetic-pulse forming of sheet metals, and electrical-discharge machining and electrochemical grinding of hard and brittle materials.

The *electrical conductivity* of a material can be defined as a measure of the ease with which a material conducts electric current. Materials with high conductivity, such as metals, are generally referred to as *conductors*. The units of electrical conductivity are mho/m or mho/ft, where mho is the inverse of ohm, the unit for electrical resistance. *Electrical resistivity* is the inverse of conductivity. Materials with high resistivity are referred to as *dielectrics* or *insulators*. The influence of the type of atomic bonding on the electrical conductivity of materials is the same as that for thermal conductivity. Alloying elements have a major effect on the electrical conductivity of metals: the higher the conductivity of the alloying element, the higher the conductivity of the alloy. Dielectric strength of materials is the resistivity to direct electric current, and is defined as the voltage required per unit distance for electrical breakdown, and has the units of V/m or V/ft.

Superconductivity is the phenomenon of almost zero electrical resistivity that occurs in some metals and alloys below a critical temperature. The highest temperature at which superconductivity has to date been exhibited is with an alloy of lanthanum, strontium, copper, and oxygen. Developments in superconductivity indicate that the efficiency of electrical components such as large high-power magnets, high-voltage power lines, and various other electronic and computer components can be markedly improved.

The electrical properties of materials such as single-crystal silicon, germanium, and gallium arsenide are extremely sensitive to the presence and type of minute impurities, as well as to temperature. Thus by controlling the concentration and type of impurities (*dopants*), such as phosphorus and boron in silicon, electrical conductivity can be controlled. This property is utilized in *semiconductor* (solid state) devices used extensively in miniaturized electronic circuitry. They are very compact, efficient, and relatively inexpensive, consume little power, and require no warmup time for operation.

Ferromagnetism is the large and permanent magnetization resulting from alignment of iron, nickel, and cobalt atoms into domains. It has important applications in electric motors, generators, transformers, and microwave devices.

The *piezoelectric* effect (*piezo* from Greek, meaning to press) is exhibited by some materials, such as certain ceramics and quartz crystals, in which there is a reversible interaction between an elastic strain and an electric field. This property is utilized in making transducers, which are devices that convert the strain from an external force to electrical energy. Typical applications are force or pressure transducers, strain gages, sonar detectors, and microphones.

The phenomenon of expansion or contraction of a material when subjected to a magnetic field is called *magnetostriction*. Some materials, such as pure nickel and some iron–nickel alloys, exhibit this behavior. Magnetostriction is the principle behind ultrasonic machining equipment.

3.8.7 Corrosion resistance

Metals, ceramics, and plastics are all subject to corrosion. We apply the word *corrosion* to deterioration of metals and ceramics, while calling similar phenomena in plastics *degradation*. Corrosion resistance is an important aspect of material selection for applications in the chemical, food, and petroleum industries, as well as in manufacturing operations. In addition to various possible chemical reactions from the elements and compounds present, environmental oxidation and corrosion of components and structures is a major concern, particularly at elevated temperatures and in automobiles and other transportation vehicles.

Resistance to corrosion depends on the particular environment, as well as the composition of the material. Corrosive media may be chemicals (acids, alkali, and salts), the environment (oxygen, pollution, and acid rain), and water (fresh or salt). Nonferrous metals, stainless steels, and nonmetallic materials generally have high corrosion resistance. Steels and cast irons generally have poor resistance and must be protected by various coatings and surface treatments.

Corrosion can occur over an entire surface, or it can be localized, which we call *pitting*. It can occur along grain boundaries of metals as *intergranular corrosion* and at the interface of bolted or riveted joints as *crevice corrosion*. Two dissimilar metals may form a galvanic cell—that is, two electrodes in an electrolyte in a corrosive environment (including moisture)—and cause *galvanic corrosion*. Two-phase alloys are more susceptible to galvanic corrosion, because of the two different metals involved, than single-phase alloys or pure metals. Thus heat treatment can have an influence on corrosion resistance.

Moreover, corrosion may act in indirect ways. *Stress-corrosion cracking* is an example of the effect of a corrosive environment on the integrity of a product that, as manufactured, had residual stresses in it. Likewise, metals that are cold worked are likely to contain residual stresses and hence are more susceptible to corrosion, compared to hot-worked or annealed metals.

Tool and die materials also can be susceptible to chemical attack by lubricants and coolants. The chemical reaction alters their surface finish and adversely influences the metalworking operation. An example is carbide tools and dies having cobalt as a binder, in which the cobalt is attacked by elements in the cutting fluid (*selective leaching*). Thus compatibility of the tool, die, and workpiece materials and the metalworking fluid under actual operating conditions is an important consideration in the selection of materials.

Chemical reactions should not be regarded as having adverse effects only.

Nontraditional machining processes such as chemical machining and electrochemical machining are indeed based on controlled reactions. These processes remove material by chemical reactions, similar to the etching of metallurgical specimens.

The usefulness of some level of *oxidation* is exhibited by the corrosion resistance of aluminum, titanium, and stainless steel. Aluminum develops a thin (a few atomic layers), strong, and adherent hard oxide film (Al_2O_3) that protects the surface from further environmental corrosion. Titanium develops a film of titanium oxide (TiO_2). A similar phenomenon occurs in stainless steels, which because of the chromium present in the alloy, develop a similar protective film on their surfaces. These processes are known as *passivation*. When the protective film is scratched, thus exposing the metal underneath, a new oxide film forms in due time.

3.9 Properties of Ferrous Alloys

By virtue of their wide range of mechanical, physical, and chemical properties, *ferrous alloys* are among the most useful of all metals. Ferrous metals and alloys contain iron as their base metal; these metals are carbon and alloy steels, stainless steels, tool and die steels, cast irons, and cast steels. Ferrous alloys are produced as sheet steel for automobiles, appliances, and containers; as plates for ships, boilers, and bridges; as structural members, such as I-beams, bar products for leaf springs, gears, axles, crankshafts, and railroad rails; as stock for tools and dies; as music wire; and as fasteners such as bolts, rivets, and nuts.

A typical U.S. passenger car contains about 800 kg (1750 lb) of steel, accounting for about 55–60 percent of its weight. As an example of their widespread use, ferrous materials comprise 70–85 percent by weight of virtually all structural members and mechanical components. Carbon steels are the least expensive of all metals, but stainless steels can be costly.

3.9.1 Carbon and alloy steels

Carbon and alloy steels are among the most commonly used metals and have a wide variety of applications. The composition and processing of steels are controlled in a manner that makes them suitable for numerous applications. They are available in various basic product shapes: plate, sheet, strip, bar, wire, tube, castings, and forgings.

Various elements are added to steels to impart properties of hardenability, strength, hardness, toughness, wear resistance, workability, weldability, and machinability. Generally, the higher the percentages of these elements in steels, the higher are the particular properties that they impart. Thus, for example, the higher the carbon content, the higher is the hardenability of the steel and the higher are its strength, hardness, and wear resistance. Conversely, ductility, weldability, and toughness are reduced with increasing carbon content.

Carbon steels. *Carbon steels* are generally classified as:

- *Low-carbon steel*, also called *mild steel*, has less than 0.30 percent carbon. It is generally used for common industrial products, such as bolts, nuts, sheet, plate, tubes, and machine components that do not require high strength.
- *Medium-carbon steel* has 0.30 percent to 0.60 percent carbon. It is generally used in applications requiring higher strength than low-carbon steels, such as machinery, automotive and agricultural equipment parts (gears, axles, connecting rods, crankshafts), railroad equipment, and parts for metal-working machinery.
- *High-carbon steel* has more than 0.60 percent carbon. It is generally used for parts requiring strength, hardness, and wear resistance, such as cutting tools, cable, music wire, springs, and cutlery. After being manufactured into shapes, the parts are usually heat treated and tempered. The higher the carbon content of the steel, the higher are its hardness, strength, and wear resistance after heat treatment.

Alloy steels. Steels containing significant amounts of alloying elements are called *alloy steels*. Alloy steels are usually made with more care than are carbon steels. *Structural-grade* alloy steels, as identified by ASTM specifications, are used mainly in the construction and transportation industries because of their high strength. Other alloy steels are used in applications where strength, hardness, creep and fatigue resistance, and toughness are required. These steels may also be heat treated to obtain the desired properties.

High-strength low-alloy steels. In order to improve the strength-to-weight ratio of steels, a number of high-strength low-alloy (HSLA) steels have been developed. These steels have a low carbon content, usually less than 0.30 percent, and are characterized by a microstructure consisting of fine-grain ferrite and a hard second phase of martensite and austenite. First developed in the 1930s, HSLA steels are usually produced in sheet form by microalloying and controlled hot rolling. Plates, bars, and structural shapes are made from these steels. However, the ductility, formability, and weldability of HSLA steels are generally inferior to conventional low-alloy steels. To improve these properties, dual-phase steels have been developed.

Sheet products of HSLA steels typically are used in certain parts of automobile bodies to reduce weight, and hence fuel consumption, and in transportation equipment, mining, agricultural, and various other industrial applications. Plates are used in ships, bridges, building construction, and shapes such as I-beams, channels, and angles are used in buildings and various other structures.

Dual-phase steels. *Dual-phase steels*, designated with the letter "D," are processed specially and have a mixed ferrite and martensite structure. Developed in the late 1960s, these steels have high work-hardening characteristics (high n value), which improve their ductility and formability.

3.9.2 Stainless steels

Stainless steels are characterized primarily by their corrosion resistance, high strength and ductility, and high chromium content. They are called *stainless* because, in the presence of oxygen (air), they develop a thin, hard adherent film of chromium oxide that protects the metal from corrosion (passivation). This protective film builds up again in the event the surface is scratched. For passivation to occur, the minimum chromium content of the steel should be 10–12 percent by weight.

In addition to chromium, other alloying elements in stainless steels typically are nickel, molybdenum, copper, titanium, silicon, manganese, columbium, aluminum, nitrogen, and sulfur. The higher the carbon content, the lower is the corrosion resistance of stainless steels. The reason is that the carbon combines with chromium in the steel and forms chromium carbide, which lowers the passivity of the steel. The chromium carbides introduce a second phase in the metal, which promotes galvanic corrosion.

Developed in the early 1900s, stainless steels are made by techniques similar to those used in other types of steelmaking, using electric furnaces or the basic-oxygen process. The level of purity is controlled by various refining techniques. Stainless steels are available in a wide variety of shapes. Typical applications are for cutlery, kitchen equipment, health care and surgical equipment, and automotive trim and in the chemical, food processing, and petroleum industries.

Types of stainless steels. Stainless steels are generally divided into five types (Table 3.3): austenitic, ferritic, martensitic, precipitation hardening, and duplex structure steels.

The *austenitic* (200 and 300 series) *steels* are generally composed of chromium, nickel, and manganese in iron. They are nonmagnetic and have excellent corrosion resistance, but are susceptible to stress-corrosion cracking. Austenitic stainless steels are hardened by cold working. They are the most ductile of all stainless steels, hence can be formed easily. However, with increasing cold work their formability is reduced. These steels are used in a wide variety of applications, such as kitchenware, fittings, welded construction, lightweight transportation equipment, furnace and heat-exchanger parts, and components for severe chemical environments.

The *ferritic* (400 series) *steels* have a high chromium content: up to 27 percent. They are magnetic and have good corrosion resistance, but have lower ductility than austenitic stainless steels. Ferritic stainless steels are hardened by cold working and are not heat treatable. They are generally used for nonstructural applications, such as kitchen equipment and automotive trim.

Most of the *martensitic* (400 and 500 series) *stainless steels* do not contain nickel and are hardenable by heat treatment. Their chromium content may be as

TABLE 3.3
ROOM-TEMPERATURE MECHANICAL PROPERTIES AND TYPICAL APPLICATIONS OF ANNEALED STAINLESS STEELS

AISI (UNS)	ULTIMATE TENSILE STRENGTH (MPa)	YIELD STRENGTH (MPa)	ELONGATION (%)	CHARACTERISTICS AND TYPICAL APPLICATIONS
303 (S30300)	550–620	240–260	50–53	Screw machine products, shafts, valves, bolts, bushings, and nuts; aircraft fittings; bolts; nuts; rivets; screws; studs.
304 (S30400)	565–620	240–290	55–60	Chemical and food processing equipment, brewing equipment, cryogenic vessels, gutters, downspouts, and flashings.
316 (S31600)	550–590	210–290	55–60	High corrosion resistance and high creep strength. Chemical and pulp handling equipment, photographic equipment, brandy vats, fertilizer parts, ketchup cooking kettles, and yeast tubs.
410 (S41000)	480–520	240–310	25–35	Machine parts, pump shafts, bolts, bushings, coal chutes, cutlery, finishing, tackle, hardware, jet engine parts, mining machinery, rifle barrels, screws, and valves.
416 (S41600)	480–520	275	20–30	Aircraft fittings, bolts, nuts, fire extinguisher inserts, rivets, and screws.

much as 18 percent. These steels are magnetic and have high strength, hardness and fatigue resistance, and good ductility, but moderate corrosion resistance. Martensitic stainless steels are used for cutlery, surgical tools, instruments, valves, and springs.

The *precipitation hardening* (PH) *steels* contain chromium and nickel, along with copper, aluminum, titanium, or molybdenum. They have good corrosion resistance, ductility, and high strength at elevated temperatures. Their main application is in aircraft and aerospace structural components.

The *duplex structure steels* have a mixture of austenite and ferrite. They have good strength and also have higher resistance to corrosion in most environments and stress-corrosion cracking than the 300 series of austenitic steels. Typical applications are in water-treatment plants and heat-exchanger components.

3.9.3 Tool and die steels

Tool and die steels are specially alloyed steels (Tables 3.4 and 3.5) designed for high strength, impact toughness, and wear resistance at room and elevated temperatures. They are commonly used in forming and machining of metals.

TABLE 3.4
BASIC TYPES OF TOOL AND DIE STEELS

TYPE	AISI
High speed	M (molybdenum base) T (tungsten base)
Hot work	H1 to H19 (chromium base) H20 to H39 (tungsten base) H40 to H59 (molybdenum base)
Cold work	D (high carbon, high chromium) A (medium alloy, air hardening) O (oil hardening)
Shock resisting	S
Mold steels	P1 to P19 (low carbon) P20 to P39 (others)
Special purpose	L (low alloy) F (carbon–tungsten)
Water hardening	W

High-speed steels. *High-speed steels* (HSS) are the most highly alloyed tool and die steels and maintain their hardness and strength at elevated operating temperatures. First developed in the early 1900s, there are two basic types of high-speed steels: the molybdenum type (M series) and the tungsten type (T series).

The M-series steels contain up to about 10 percent molybdenum, with chromium, vanadium, tungsten, and cobalt as other alloying elements. The T-series steels contain 12–18 percent tungsten, with chromium, vanadium, and cobalt as other alloying elements. The M-series steels generally have higher abrasion resistance than the T-series steels, have less distortion in heat treatment, and are less expensive. The M series constitutes about 95 percent of all high-speed steels produced in the United States. High-speed steel tools can be coated with titanium nitride and titanium carbide for better wear resistance.

Hot-work, cold-work, and shock-resisting die steels. *Hot-work steels* (H series) are designed for use at elevated temperatures and have high toughness and high resistance to wear and cracking. The alloying elements are generally tungsten, molybdenum, chromium, and vanadium. *Cold-work steels* (A, D, and O series) are used for cold-working operations. They generally have high resistance to wear and cracking. These steels are available as oil-hardening or air-hardening types. *Shock-resisting steels* (S series) are designed for impact toughness and are used in applications such as header dies, punches, and chisels. Other properties of these steels depend on the particular composition. Various tool and die materials for a variety of manufacturing applications are presented in Table 3.6.

TABLE 3.5
PROCESSING AND SERVICE CHARACTERISTICS OF COMMON TOOL AND DIE STEELS

AISI DESIGNATION	RESISTANCE TO DECARBURIZATION	RESISTANCE TO CRACKING	APPROXIMATE HARDNESS (*HRC*)	MACHINABILITY	TOUGHNESS	RESISTANCE TO SOFTENING	RESISTANCE TO WEAR
M2	Medium	Medium	60–65	Medium	Low	Very high	Very high
T1	High	High	60–65	Medium	Low	Very high	Very high
T5	Low	Medium	60–65	Medium	Low	Highest	Very high
H11, 12, 13	Medium	Highest	38–55	Medium to high	Very high	High	Medium
A2	Medium	Highest	57–62	Medium	Medium	High	High
A9	Medium	Highest	35–56	Medium	High	High	Medium to high
D2	Medium	Highest	54–61	Low	Low	High	High to very high
D3	Medium	High	54–61	Low	Low	High	Very high
H21	Medium	High	36–54	Medium	High	High	Medium to high
H26	Medium	High	43–58	Medium	Medium	Very high	High
P20	High	High	28–37	Medium to high	High	Low	Low to medium
P21	High	Highest	30–40	Medium	Medium	Medium	Medium
W1, W2	Highest	Medium	50–64	Highest	High	Low	Low to medium

Source: Adapted from *Tool Steels*, American Iron and Steel Institute, 1978.

TABLE 3.6
TYPICAL TOOL AND DIE MATERIALS

PROCESS	MATERIAL
Die casting	H13, P20
Powder metallurgy	
Punches	A2, S7, D2, D3, M2
Dies	WC, D2, M2
Molds for plastics and rubber	S1, O1, A2, D2, 6F5, 6F6, P6, P20, P21, H13
Hot forging	6F2, 6G, H11, H12
Hot extrusion	H11, H12, H13
Cold heading	W1, W2, M1, M2, D2, WC
Cold extrusion	
Punches	A2, D2, M2, M4
Dies	O1, W1, A2, D2
Coining	52100, W1, O1, A2, D2, D3, D4, H11, H12, H13
Drawing	
Wire	WC, diamond
Shapes	WC, D2, M2
Bar and tubing	WC, W1, D2
Rolls	
Rolling	Cast iron, cast steel, forged steel, WC
Thread rolling	A2, D2, M2
Shear spinning	A2, D2, D3
Sheet metals	
Shearing	
Cold	D2, A2, A9, S2, S5, S7
Hot	H11, H12, H13
Pressworking	Zinc alloys, 4140 steel, cast iron, epoxy composites, A2, D2, O1
Deep drawing	W1, O1, cast iron, A2, D2
Machining	Carbides, high-speed steels, ceramics

Notes: Tool and die materials are usually hardened to 55 to 65 *HRC* for cold working, and 30 to 55 for hot working.

Tool and die steels contain one or more of the following major alloying elements: chromium, molybdenum, tungsten, and vanadium. For further details see the bibliography at the end of this chapter.

3.10 Properties of Nonferrous Metals and Alloys

Nonferrous metals and alloys cover a wide range of materials, from the more common metals, such as aluminum, copper, and magnesium, to high-strength high-temperature alloys, such as tungsten, tantalum, and molybdenum. Although more expensive than ferrous metals, nonferrous metals and alloys have important applications because of their numerous mechanical, physical, and chemical properties.

A turbofan jet engine for the Boeing 757 aircraft typically contains the following nonferrous metals and alloys: 38 percent titanium, 37 percent nickel, 12 percent chromium, 6 percent cobalt, 5 percent aluminum, 1 percent columbium (niobium), and 0.02 percent tantalum. Without these materials, a jet engine (Fig. 3.20) could not be designed, manufactured, and operated at the energy and efficiency levels required.

Typical examples of the applications of nonferrous metals and alloys are aluminum for cooking utensils and aircraft bodies, copper wire for electricity and copper tubing for water in residences, zinc for carburetors, titanium for jet-engine turbine blades and prosthetic devices (such as artificial limbs), and tantalum for rocket engines. In this section we discuss the general properties, production methods, and important engineering applications of nonferrous metals and alloys.

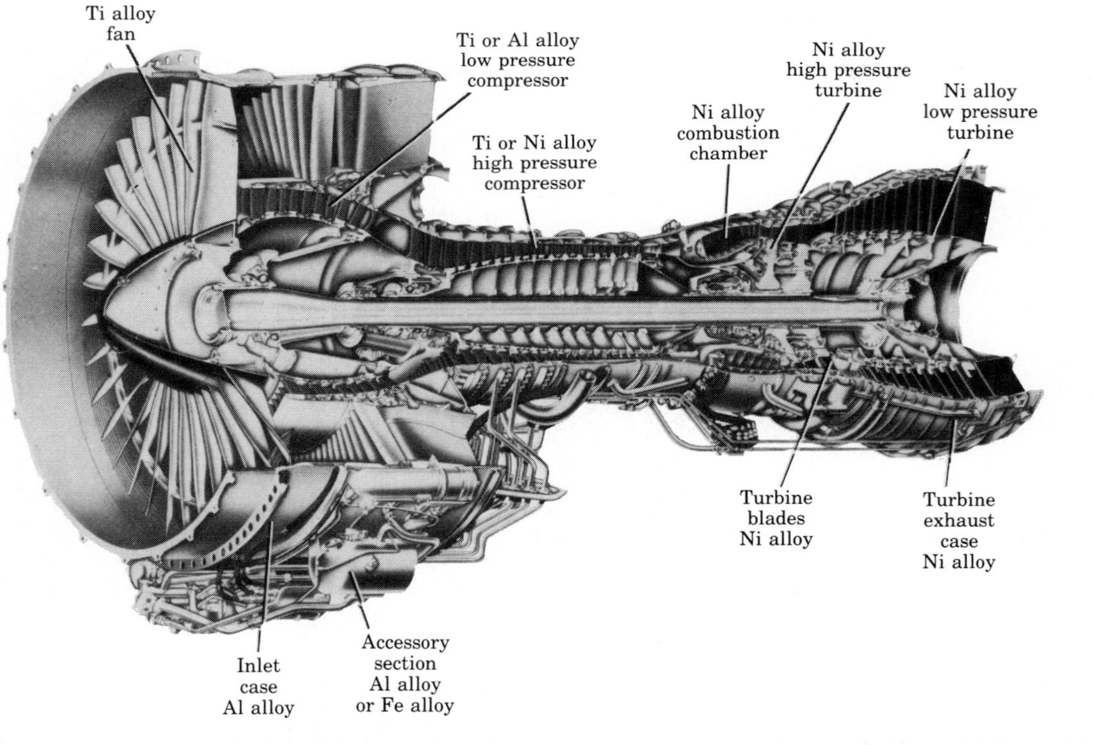

FIGURE 3.20

Cross section of a jet engine (PW2037) showing various components and the alloys used in making them. *Source:* Courtesy of United Aircraft Pratt & Whitney.

We describe the manufacturing properties of these materials, such as formability, machinability, and weldability, in various chapters throughout this text.

3.10.1 Aluminum

The important factors in selecting aluminum (Al) and its alloys are their high strength-to-weight ratio, resistance to corrosion by many chemicals, high thermal and electrical conductivity, nontoxicity, reflectivity, appearance, and ease of formability and machinability; they are also nonmagnetic.

Principal uses of aluminum and its alloys, in decreasing order of consumption, are containers and packaging (aluminum cans and foil), buildings and other types of construction, transportation (aircraft and aerospace applications, buses, automobiles, railroad cars, and marine craft), electrical (economical and nonmagnetic electrical conductor), consumer durables (appliances, cooking utensils, and furniture), and portable tools (Tables 3.7 and 3.8). Nearly all high-voltage transmission wiring is made of aluminum. Of their structural (load bearing) components, 82 percent of a Boeing 747 aircraft and 79 percent of a Boeing 757 aircraft are made of aluminum.

Aluminum alloys are available as mill products, that is, wrought products made into various shapes by rolling, extrusion, drawing, and forging. Aluminum ingots are available for casting, as are powder metals for powder metallurgy applications. There are two types of wrought alloys of aluminum: (1) alloys that can be hardened by cold working and are not heat treatable, and (2) alloys that are hardenable by heat treatment. Techniques have been developed whereby most aluminum alloys can be machined, formed, and welded with relative ease.

TABLE 3.7
PROPERTIES OF SELECTED ALUMINUM ALLOYS AT ROOM TEMPERATURE

ALLOY (UNS)	TEMPER	ULTIMATE TENSILE STRENGTH (MPa)	YIELD STRENGTH (MPa)	ELONGATION IN 50 mm (%)
1100 (A91100)	O	90	35	35–45
1100	H14	125	120	9–20
2024 (A92024)	O	190	75	20–22
2024	T4	470	325	20–19
3003 (A93003)	O	110	40	30–40
3003	H14	150	145	8–16
5052 (A95052)	O	190	90	25–30
5052	H34	260	215	10–14
6061 (A96061)	O	125	55	25–30
6061	T6	310	275	12–17
7075 (A97075)	O	230	105	17–16
7075	T6	570	500	11

TABLE 3.8
MANUFACTURING PROPERTIES AND TYPICAL APPLICATIONS OF WROUGHT ALUMINUM ALLOYS

	CHARACTERISTICS*			
ALLOY	*CORROSION RESISTANCE*	*MACHINABILITY*	*WELDABILITY*	TYPICAL APPLICATIONS
1100	A	D–C	A	Sheet metal work, spun hollow ware, tin stock
2024	C	C–B	C–B	Truck wheels, screw machine products, aircraft structures
3003	A	D–C	A	Cooking utensils, chemical equipment, pressure vessels, sheet metal work, builders' hardware, storage tanks
5052	A	D–C	A	Sheet metal work, hydraulic tubes, and appliances; bus, truck and marine uses
6061	B	D–C	A	Heavy-duty structures where corrosion resistance is needed, truck and marine structures, railroad cars, furniture, pipelines, bridge railings, hydraulic tubing
7075	C	B–D	D	Aircraft and other structures, keys, hydraulic fittings

* A, excellent; D, poor.

3.10.2 Magnesium

Magnesium (Mg) is the lightest engineering metal available and also has good vibration-damping characteristics. Its alloys are used in structural and nonstructural applications where weight is of primary importance. Magnesium is also an alloying element in various nonferrous metals.

Typical uses of magnesium alloys are for aircraft and missile components, material-handling equipment, portable power tools (such as drills and sanders), ladders, luggage, bicycles, sporting goods, and general lightweight components. These alloys are available either as castings or wrought products, such as extruded bars and shapes, forgings, and rolled plate and sheet. Magnesium alloys are also used in printing and textile machinery to minimize inertial forces in high-speed components.

Because it is not sufficiently strong in its pure form, magnesium is alloyed with various elements (Table 3.9) to impart certain specific properties, particularly high strength-to-weight ratio. A variety of magnesium alloys have good casting, forming, and machining characteristics. Because they oxidize rapidly—that is, they are *pyrophoric*—a fire hazard exists and precautions must be taken when machining,

TABLE 3.9
PROPERTIES AND TYPICAL FORMS OF WROUGHT MAGNESIUM ALLOYS

ALLOY	COMPOSITION (%) Al	Zn	Mn	Zr	CONDITION	ULTIMATE TENSILE STRENGTH (MPa)	YIELD STRENGTH (MPa)	ELONGATION IN 50 mm (%)	TYPICAL FORMS
AZ31B	3.0	1.0	0.2		F	260	200	15	Extrusions
					H24	290	220	15	Sheet and plates
AZ80A	8.5	0.5	0.2		T5	380	275	7	Extrusions and forgings
HK31A			3Th	0.7	H24	255	200	8	Sheet and plates
ZK60A		5.7		0.55	T5	365	300	11	Extrusions and forgings

grinding, or sand casting magnesium alloys. However, products made of magnesium and its alloys are not a fire hazard.

3.10.3 Copper

First produced in about 4000 B.C., copper (Cu) and its alloys have properties somewhat similar to those of aluminum alloys. In addition, they are among the best conductors of electricity and heat and have good corrosion resistance. Because of these properties, copper and its alloys are among the most important metals. They can be processed easily by various forming, machining, casting and joining techniques.

Copper alloys often are attractive for applications where combined properties, such as electrical, mechanical, corrosion resistance, nonmagnetic, thermal conductivity, and wear resistance, are required. Applications include electrical and electronic components, springs, cartridges for small arms, plumbing, heat exchangers, and marine hardware, as well as consumer goods, such as cooking utensils, jewelry, and other decorative objects.

Copper alloys can acquire a wide variety of properties by the addition of alloying elements and by heat treatment to improve their manufacturing characteristics. The most common copper alloys are brasses and bronzes. *Brass*, which is an alloy of copper and zinc, is one of the earliest alloys developed and has numerous applications, including decorative objects (Table 3.10). *Bronze* is an alloy of copper and tin (Table 3.11). There are also other bronzes, such as aluminum bronze, which is an alloy of copper and aluminum, and tin bronzes. Beryllium–copper, or *beryllium bronze*, and *phosphor bronze* have good strength and hardness for applications such as springs and bearings. Other major copper alloys are copper nickels and nickel silvers.

3.10.4 Nickel

Nickel (Ni), a silver-white metal discovered in 1751, is a major alloying element that imparts strength, toughness, and corrosion resistance. It is used extensively in

TABLE 3.10
PROPERTIES AND TYPICAL APPLICATIONS OF WROUGHT COPPER AND BRASSES

TYPE AND UNS NUMBER	NOMINAL COMPOSITION (%)	ULTIMATE TENSILE STRENGTH (MPa)	YIELD STRENGTH (MPa)	ELONGATION IN 50 mm (%)	TYPICAL APPLICATIONS
Electrolytic tough pitch copper (C11000)	99.90 Cu, 0.04 O	220–450	70–365	55–4	Downspouts, gutters, roofing, gaskets, auto radiators, busbars, nails, printing rolls, rivets
Red brass, 85% (C23000)	85.0 Cu, 15.0 Zn	270–725	70–435	55–3	Weather-stripping, conduit, sockets, fasteners, fire extinguishers, condenser and heat exchanger tubing
Cartridge brass, 70% (C26000)	70.0 Cu, 30.0 Zn	300–900	75–450	66–3	Radiator cores and tanks, flashlight shells, lamp fixtures, fasteners, locks, hinges, ammunition components, plumbing accessories
Free-cutting brass (C36000)	61.5 Cu, 3.0 Pb, 35.5 Zn	340–470	125–310	53–18	Gears, pinions, automatic high-speed screw machine parts
Naval brass (C46400 to C46700)	60.0 Cu, 39.25 Zn, 0.75 Sn	380–610	170–455	50–17	Aircraft turnbuckle barrels, balls, bolts, marine hardware, propeller shafts, rivets, valve stems, condenser plates

TABLE 3.11
PROPERTIES AND TYPICAL APPLICATIONS OF WROUGHT BRONZES

TYPE AND UNS NUMBER	NOMINAL COMPOSITION (%)	ULTIMATE TENSILE STRENGTH (MPa)	YIELD STRENGTH (MPa)	ELONGATION IN 50 mm (%)	TYPICAL APPLICATIONS
Architectural bronze (C38500)	57.0 Cu, 3.0 Pb, 40.0 Zn	415 (As extruded)	140	30	Architectural extrusions, store fronts, thresholds, trim, butts, hinges
Phosphor bronze, 5% A (C51000)	95.0 Cu, 5.0 Sn, trace P	325–960	130–550	64–2	Bellows, clutch disks, cotter pins, diaphragms, fasteners, wire brushes, chemical hardware, textile machinery
Free-cutting phosphor bronze (C54400)	88.0 Cu, 4.0 Pb, 4.0 Zn, 4.0 Sn	300–520	130–435	50–15	Bearings, bushings, gears, pinions, shafts, thrust washers, valve parts
Low silicon bronze, B (C65100)	98.5 Cu, 1.5 Si	275–655	100–475	55–11	Hydraulic pressure lines, bolts, marine hardware, electrical conduits, heat exchanger tubing
Nickel silver, 65–10 (C74500)	65.0 Cu, 25.0 Zn, 10.0 Ni	340–900	125–525	50–1	Rivets, screws, slide fasteners, hollow ware, nameplates

stainless steels and nickel-base alloys, also called *superalloys*. These alloys are used for high-temperature applications, such as jet engine components, rockets, and nuclear power plants, as well as in food-handling and chemical processing equipment, coins, and marine applications. Because nickel is magnetic, nickel alloys are also used in electromagnetic applications, such as solenoids. The principal use of nickel as a metal is in electroplating for appearance and for improving corrosion and wear resistance.

Nickel alloys have high strength and corrosion resistance at elevated temperatures. Alloying elements in nickel are chromium, cobalt, and molybdenum. The behavior of nickel alloys in machining, forming, casting, and welding can be modified by various other alloying elements.

A variety of nickel alloys having a range of strength at different temperatures have been developed (Table 3.12). *Monel* is a nickel–copper alloy, and *Inconel* is a nickel–chromium alloy. A nickel–molybdenum–chromium alloy (*Hastelloy*) has good corrosion resistance and high strength at elevated temperatures. *Nichrome*, an alloy of nickel, chromium, and iron, has high oxidation and electrical resistance, and is used for electrical heating elements. *Invar*, an alloy of iron and nickel, has relatively low sensitivity to temperature (see Section 3.8.5).

TABLE 3.12
PROPERTIES AND TYPICAL APPLICATIONS OF NICKEL ALLOYS (ALL ARE TRADE NAMES)

ALLOY [CONDITION]	PRINCIPAL ALLOYING ELEMENTS (%)	ULTIMATE TENSILE STRENGTH (MPa)	YIELD STRENGTH (MPa)	ELONGATION IN 50 mm (%)	TYPICAL APPLICATIONS
Nickel 200 [annealed]	None	380–550	100–275	60–40	Chemical and food processing industry, aerospace equipment, electronic parts
Duranickel 301 [age hardened]	4.4 Al, 0.6 Ti	1300	900	28	Springs, plastics extrusion equipment, molds for glass, diaphragms
Monel R-405 [hot-rolled]	30 Cu	525	230	35	Screw-machine products, water meter parts
Monel K-500 [age hardened]	29 Cu, 3 Al	1050	750	30	Pump shafts, valve stems, springs
Inconel 600 [annealed]	15 Cr, 8 Fe	640	210	48	Gas turbine parts, heat-treating equipment, electronic parts, nuclear reactors
Hastelloy C-4 [solution-treated and quenched]	16 Cr, 15 Mo	785	400	54	High temperature stability, resistance to stress-corrosion cracking

3.10.5 Superalloys

Superalloys are important in high-temperature applications, hence they are also known as *heat-resistant*, or *high-temperature, alloys*. Major applications of superalloys are in jet engines and gas turbines, with other applications in reciprocating engines, rocket-engines, tools and dies for hot working of metals, and in the nuclear, chemical, and petrochemical industries.

These alloys are referred to as *iron-base*, *cobalt-base*, or *nickel-base superalloys*. They contain nickel, chromium, cobalt, and molybdenum as major alloying elements. Other alloying elements are aluminum, tungsten, and titanium. Superalloys are generally identified by trade names or by special numbering systems and are available in a variety of shapes. Most superalloys have a maximum service temperature of about 1000 °C (1800 °F) for structural applications. The temperatures can be as high as 1200 °C (2200 °F) for nonload-bearing components. Superalloys generally have good resistance to corrosion, mechanical and thermal fatigue, mechanical and thermal shock, creep, and erosion at elevated temperatures.

Iron-base superalloys generally contain 32–67 percent iron, 15–22 percent chromium, and 9–38 percent nickel. Common alloys in this group are the *Incoloy* series. Cobalt-base superalloys generally contain 35–65 percent cobalt, 19–30 percent chromium, and up to 35 percent nickel. Cobalt (Co) is a white-colored metal that resembles nickel. These superalloys are not as strong as nickel-base superalloys, but they retain their strength at higher temperatures. Nickel-base superalloys are the most common of the superalloys and are available in a wide variety of compositions (Table 3.13). The range of nickel is from 38–76 percent. They also contain up to 27 percent chromium and 20 percent cobalt. Common alloys in this group are the *Hastelloy*, *Inconel*, *Nimonic*, *René*, *Udimet*, *Astroloy*, and *Waspaloy* series.

3.10.6 Titanium

Titanium (Ti) was discovered in 1791 but was not commercially produced until the 1950s. Although expensive, the high strength-to-weight ratio of titanium and its corrosion resistance at room and elevated temperatures make it attractive for applications such as aircraft, jet engine, racing-car, chemical, petrochemical, and marine components, submarine hulls, and biomaterials, such as prosthetic devices (Table 3.14). Titanium alloys have been developed for service at 550 °C (1000 °F) for long periods of time and up to 750 °C (1400 °F) for shorter periods.

Unalloyed titanium, known as commercially pure titanium, has excellent corrosion resistance for applications where strength considerations are secondary. Aluminum, vanadium, molybdenum, manganese, and other alloying elements are added to titanium alloys to impart properties such as improved workability, strength, and hardenability.

The properties and manufacturing characteristics of titanium alloys are extremely sensitive to small variations in both alloying and residual elements. Thus control of composition and processing are important, including prevention of

TABLE 3.13
PROPERTIES AND TYPICAL APPLICATIONS OF SELECTED NICKEL-BASE SUPERALLOYS AT 870 °C (1600 °F) (ALL ARE TRADE NAMES)

ALLOY	CONDITION	ULTIMATE TENSILE STRENGTH (MPa)	YIELD STRENGTH (MPa)	ELONGATION IN 50 mm (%)	TYPICAL APPLICATIONS
Astroloy	Wrought	770	690	25	Forgings for high temperature
Hastelloy X	Wrought	255	180	50	Jet engine sheet parts
IN-100	Cast	885	695	6	Jet engine blades and wheels
IN-102	Wrought	215	200	110	Superheater and jet engine parts
Inconel 625	Wrought	285	275	125	Aircraft engines and structures, chemical processing equipment
Inconel 718	Wrought	340	330	88	Jet engine and rocket parts
MAR-M 200	Cast	840	760	4	Jet engine blades
MAR-M 432	Cast	730	605	8	Integrally cast turbine wheels
René 41	Wrought	620	550	19	Jet engine parts
Udimet 700	Wrought	690	635	27	Jet engine parts
Waspaloy	Wrought	525	515	35	Jet engine parts

surface contamination by hydrogen, oxygen, or nitrogen during processing. These elements cause embrittlement of titanium, resulting in reduced toughness and ductility.

The body-centered cubic structure of titanium (beta-titanium, above 880 °C, 1600 °F) is ductile, whereas its hexagonal close-packed structure (alpha-titanium) is somewhat brittle and is very sensitive to stress corrosion. A variety of other structures (alpha, near alpha, alpha–beta, and beta) can be obtained by alloying and heat treating, such that the properties can be optimized for specific applications. New developments include the so-called *titanium aluminide intermetallics*, TiAl and Ti_3Al. They have higher stiffness and lower density and can withstand temperatures higher than conventional titanium alloys.

3.10.7 Refractory metals

Refractory metals are molybdenum, columbium, tungsten, and tantalum. They are called *refractory* because of their high melting point. Although refractory metal elements were discovered about 200 years ago—and have been used as important alloying elements in steels and superalloys—their use as engineering metals and alloys did not begin until about the 1940s.

More than most other metals and alloys, these metals maintain their strength at elevated temperatures. Thus they are of great importance and use in rocket engines, gas turbines, and various other aerospace applications, in the electronics, nuclear power, and chemical industries, and as tool and die materials. The temperature

TABLE 3.14
PROPERTIES AND TYPICAL APPLICATIONS OF WROUGHT TITANIUM ALLOYS

			ROOM TEMPERATURE				VARIOUS TEMPERATURES					
NOMINAL COMPOSITION (%)	UNS	CONDITION	*ULTIMATE TENSILE STRENGTH* (MPa)	*YIELD STRENGTH* (MPa)	*ELONGATION* (%)	*REDUCTION OF AREA* (%)	*TEMP.* (°C)	*ULTIMATE TENSILE STRENGTH* (MPa)	*YIELD STRENGTH* (MPa)	*ELONGATION* (%)	*REDUCTION OF AREA* (%)	TYPICAL APPLICATIONS
99.5 Ti	R50250	Annealed	330	240	30	55	300	150	95	32	80	Airframes; chemical, desalination, and marine parts; plate type heat exchangers
5 Al, 2.5 Sn	R54520	Annealed	860	810	16	40	300	565	450	18	45	Aircraft engine compressor blades and ducting; steam turbine blades
6 Al, 4 V	R56400	Annealed	1000	925	14	30	300 425 550	725 670 530	650 570 430	14 18 35	35 40 50	Rocket motor cases; blades and disks for aircraft turbines and compressors; structural forgings and fasteners
		Solution + age	1175	1100	10	20	300	980	900	10 12 22	28 35 45	
13 V, 11 Cr, 3 Al	R58010	Solution + age	1275	1210	8	—	425	1100	830	12	—	High strength fasteners; aerospace components; honeycomb panels

range for some of these applications is on the order of 1100 °C to 2200 °C (2000 °F to 4000 °F), where strength and oxidation are of major concern.

Molybdenum. *Molybdenum* (Mo), a silvery white metal, was discovered in the 18th century. It has a high melting point, a high modulus of elasticity, good resistance to thermal shock, and good electrical and thermal conductivity. Typical applications of molybdenum are in solid-propellent rockets, jet engines, honeycomb structures, electronic components, heating elements, and dies for die casting.

Principal alloying elements for molybdenum are titanium and zirconium. Molybdenum is used in greater amounts than any other refractory metal. It is an important alloying element in casting and wrought alloys, such as steels and heat-resistant alloys, and imparts strength, toughness, and corrosion resistance. A major disadvantage of molybdenum alloys is their low resistance to oxidation at temperatures above 500 °C (950 °F), thus necessitating the use of protective coatings.

Columbium (niobium). *Columbium* (after the mineral columbite and Nb, for niobium) possesses good ductility and formability and has greater oxidation resistance than other refractory metals. With various alloying elements, columbium alloys can be produced with moderate strength and good fabrication characteristics. These alloys are used in rockets and missiles and nuclear, chemical, and superconductor applications. Columbium, first identified in 1801, is also an alloying element in various alloys and superalloys.

Tungsten. *Tungsten* (W, from wolframite) was first identified in 1781 and is the most plentiful of all refractory metals. Tungsten has the highest melting point of any metal (3410 °C, 6170 °F), and thus it is characterized by high strength at elevated temperatures. On the other hand, it has high density, brittleness at low temperatures, and poor resistance to oxidation.

Tungsten and its alloys are used for applications involving temperatures above 1650 °C (3000 °F), such as nozzle throat liners in missiles and in the hottest parts of jet and rocket engines, circuit breakers, welding electrodes, and spark-plug electrodes. The filament wire in incandescent light bulbs is made of pure tungsten, using powder metallurgy and wire drawing techniques. Because of its high density, tungsten is used as balancing weights and counterbalances in mechanical systems, including self-winding watches. Tungsten is an important element in tool and die steels, imparting strength and hardness at elevated temperatures. Tungsten carbide, with cobalt as a binder for the carbide particles, is one of the most important tool and die materials.

Tantalum. Identified in 1802, *tantalum* (Ta) is characterized by a high melting point (3000 °C, 5425 °F), good ductility, and resistance to corrosion. However, it has high density and poor resistance to chemicals at temperatures above 150 °C (300 °F). Tantalum is also used as an alloying element.

Tantalum is used extensively in electrolytic capacitors and various components in the electrical, electronic, and chemical industries, as well as for thermal applications, such as in furnaces and acid-resistant heat exchangers. A variety of tantalum-base alloys is available in many forms for use in missiles and aircraft.

3.10.8 Various other nonferrous metals

Beryllium. Steel gray in color, *beryllium* (Be) has a high strength-to-weight ratio. Unalloyed beryllium is used in nuclear and x-ray applications, because of its low neutron absorption characteristics, and in rocket nozzles, space and missile structures, aircraft disc brakes, and precision instruments and mirrors. Beryllium is also an alloying element, and its alloys of copper and nickel are used in applications such as springs (beryllium–copper), electrical contacts, and nonsparking tools for use in explosive environments, such as mines and metal powder production. Beryllium and its oxide are toxic and should not be inhaled.

Zirconium. *Zirconium* (Zr) is silvery in appearance, has good strength and ductility at elevated temperatures, and has good corrosion resistance because of an adherent oxide film. The element is used in electronic components and nuclear power reactor applications because of its low neutron absorption.

Low-melting metals. The major metals in this category are lead, zinc, and tin. *Lead* (Pb, after plumbium, the root for the word plumber) has properties of high density, resistance to corrosion (by virtue of the stable lead oxide layer that forms and protects the surface), softness, low strength, ductility, and good workability. Alloying with various elements, such as antimony and tin, enhances lead's properties, making it suitable for piping, collapsible tubing, bearing alloys, cable sheathing, roofing, and lead–acid storage batteries.

Lead is also used for damping sound and vibrations, radiation shielding against x-rays, printing (type metals), ammunition, and weights and in the chemical and paint industries. The oldest lead artifacts were made in about 3000 B.C. Lead pipes made by the Romans and installed in the Roman baths in Bath, England, two millenia ago are still in use. Lead is also an alloying element in solders, steels, and copper alloys and promotes corrosion resistance and machinability. Because of its toxicity, environmental contamination by lead is a significant concern.

Industrially, *zinc* (Zn), bluish-white in color, is the fourth most utilized metal, after iron, aluminum, and copper. Although known for many centuries, zinc was not studied and developed until the 18th century. It has two major uses: one is for galvanizing iron and steel sheet and wire, and the other is as an alloy base for casting. In *galvanizing*, zinc serves as the anode and protects the steel (cathode) from corrosive attack should the coating be scratched or punctured. Zinc is also used as an alloying element. Brass, for example, is an alloy of copper and zinc.

The major use of zinc is structural, but pure zinc is rarely used for this purpose. Major alloying elements in zinc are aluminum, copper, and magnesium. They impart strength and provide dimensional control during casting of the metal. Zinc-base alloys are used extensively in die casting for making products such as carburetors, fuel pumps, and grills for automobiles, components for household appliances (such as vacuum cleaners, washing machines, and kitchen equipment), various other machine parts, and photoengraving plates. Another use for zinc is in superplastic alloys, which have good formability characteristics by virtue of their capacity to undergo large deformation without failure. Very fine-grained 78% Zn–22% Al sheet is a common example of a superplastic zinc alloy that can be formed by methods used for forming plastics or metals.

Although used in small amounts, *tin* (Sn, after stannum) is an important metal. The most extensive use of tin, a silvery white, lustrous metal, is as a protective coating on steel sheet (tin plate), which is used in making containers—*tin cans*—for food and various other products. The low shear strength of the tin coatings on steel sheet improves its performance in deep drawing and general pressworking operations. However, unlike galvanized steels, if this coating is punctured, the steel corrodes because it is anodic to tin (cathode).

Unalloyed tin is used in applications such as lining material for water distillation plants, and as a molten layer of metal over which plate glass is made. Tin-base alloys (also called *white metals*) generally contain copper, antimony, and lead. The alloying elements impart hardness, strength, and corrosion resistance. Because of their low friction coefficients, which result from low shear strength and low adhesion, tin alloys are used as journal bearing materials. These alloys are known as *babbitts* and contain tin, copper, and antimony. *Pewter* is an alloy of tin, copper, and antimony. It was developed in the 15th century and is used for tableware, hollowware, and decorative artifacts. Organ pipes are made of tin alloys. Tin is an alloying element for type metals, dental alloys, and bronze (copper–tin alloy), titanium, and zirconium alloys. Tin–lead alloys are common *soldering* materials, with a wide range of compositions and melting points.

Precious metals. Gold, silver, and platinum are the most important *precious* (that is, costly) metals and are also called *noble metals*. *Gold* (Au, after aurum) is soft and ductile, and has good corrosion resistance at any temperature. Typical applications include jewelry, coinage, reflectors, gold leaf for decorative purposes, dental work, and electroplating, as well as important applications as electrical contacts and terminals.

Silver (Ag, after argentum) is a ductile metal and has the highest electrical and thermal conductivity of any metal. It does, however, develop an oxide film that affects its surface properties and appearance. Typical applications for silver include tableware, jewelry, coinage, electroplating, and photographic film, as well as electrical contacts, solders, bearings, and food and chemical equipment. Sterling silver is an alloy of silver and 7.5 percent copper.

Platinum (Pt) is a soft, ductile, grayish-white metal that has good corrosion resistance, even at elevated temperatures. Platinum alloys are used as electrical contacts, spark-plug electrodes, catalysts for automobile pollution-control devices, filaments, nozzles, dies for extruding glass fibers, and thermocouples, and in the electrochemical industry. Other applications include jewelry and dental work.

Shape-memory alloys. *Shape-memory alloys*, after being plastically deformed at room temperature into various shapes, return to their original shapes upon heating. For example, a piece of straight wire made of this material can be wound into a helical spring. When heated with a match, the spring uncoils and returns to the original straight shape. A typical shape-memory alloy is 55% Ni–45% Ti. One potential application for these alloys is in structures, such as an antenna, for space travel. They would be folded at room temperature into a small volume to save space and for ease of transportation. After reaching their destination, the structures would be heated by some suitable means and return to their original shapes.

Amorphous alloys. A class of metal alloys that, unlike ordinary metals, does not have a long-range crystalline structure is called *amorphous alloys* (see also Section 5.9.8). These alloys have no grain boundaries, and the atoms are randomly and tightly packed. Because their structure resembles that of glasses, these alloys are also called *metallic glasses*. Amorphous alloys typically contain iron, nickel, and chromium, alloyed with carbon, phosphorus, boron, aluminum, and silicon. They are available in the form of wire, ribbon, strip, and powder.

These alloys exhibit excellent corrosion resistance, good ductility, high strength, and very low loss from magnetic hysteresis. The latter property is utilized in making magnetic steel cores for transformers, generators, motors, lamp ballasts, magnetic amplifiers, and linear accelerators, with greatly improved efficiency.

The amorphous structure was first obtained in the late 1960s by extremely rapid cooling of the molten alloy. One method is called *splat cooling*, in which the alloy is propelled at very high speed against a rotating metal surface. Since the rate of cooling is on the order of 10^6 to 10^8 K/s, the molten alloy does not have sufficient time to crystallize. If an amorphous alloy is raised in temperature and then cooled, it develops a crystalline structure.

SUMMARY

The manufacturing properties of metals and alloys depend largely on their mechanical and physical properties. These, in turn, are governed by the crystal structure of metals, imperfections, grain boundaries, grain size, and texture.

Metals and alloys can be worked at room, warm, or high temperatures. Their overall behavior, force requirements, and workability depend largely on whether the working temperature is below or above the recrystallization temperature.

A variety of metals and alloys is available with a wide range of properties, such as strength, toughness, hardness, ductility, and resistance to oxidation. They can be classified generally as ferrous alloys, nonferrous metals and alloys, refractory metals and their alloys, and superalloys.

The selection of a material for a particular application requires careful consideration of many factors. Among these are design considerations, service requirements, long-term effects, compatibility with other materials, environmental attack, and economic factors.

BIBLIOGRAPHY

Materials Science

Ashby, M.F., and D.R.H. Jones, *Engineering Materials*, Vol. 1, *An Introduction to Their Properties and Applications*, 1980; Vol. 2, *An Introduction to Microstructures, Processing and Design*, 1986. Oxford: Pergamon.

Askeland, D.R., *The Science of Engineering Materials*. Monterey, Calif.: Brooks/Cole, 1984.

Callister, W.D., Jr., *Materials Science and Engineering*. New York: Wiley, 1985.

Dieter, G.E., *Mechanical Metallurgy*, 3d ed. New York: McGraw-Hill, 1986.

Flinn, R.A., and P.K. Trojan, *Engineering Materials and Their Applications*, 3d ed. Boston: Houghton Mifflin, 1986.

Hertzberg, R.W., *Deformation and Fracture Mechanics of Engineering Materials*, 2d ed. New York: Wiley, 1983.

LeMay, I., *Principles of Mechanical Metallurgy*. New York: Elsevier, 1981.

Shackelford, J.F., *Introduction to Materials Science for Engineers*, 2d ed. New York: Macmillan, 1988.

Smith, W.F., *Principles of Materials Science and Engineering*, 2d ed. New York: McGraw-Hill, 1990.

Thornton, P.A., and V.J. Colangelo, *Fundamentals of Engineering Materials*. Englewood Cliffs, N.J.: Prentice-Hall, 1985.

Van Vlack, L.H., *Materials for Engineering*. Reading, Mass.: Addison-Wesley, 1982.

Van Vlack, L.H., *Elements of Materials Science and Engineering*, 6th ed. Reading, Mass.: Addison-Wesley, 1989.

Ferrous Metals

Aerospace Structural Metals Handbook, 5 vols. Columbus, Ohio: Metals and Ceramics Information Center, Battelle, 1987.

ASM Metals Reference Book, 2d ed. Metals Park, Ohio: American Society for Metals, 1983.

Brady, G.S., and H.R. Clauser, *Materials Handbook*, 12th ed. New York: McGraw-Hill, 1985.

Design Guidelines for the Selection and Use of Stainless Steels. Washington, D.C.: American Iron and Steel Institute, 1977.

Harvey, P.D. (ed.), *Engineering Properties of Steel*. Metals Park, Ohio: American Society for Metals, 1982.

Kern, R.F., and M.E. Suess, *Steel Selection: A Guide for Improving Performance and Profits*. New York: Wiley-Interscience, 1979.

Lankford, W.T., Jr., N.L. Samways, R.F. Craven, and H.E. McGannon, *The Making, Shaping and Treating of Steel*, 10th ed. Pittsburgh: United States Steel Co., 1985.

Lula, R.A., *Stainless Steels*. Metals Park, Ohio: American Society for Metals, 1985.

Metals Handbook, Desk Edition. Metals Park, Ohio: American Society for Metals, 1985.

Metals Handbook, 9th ed., Vol. 1: *Properties and Selection: Irons and Steels*, 1978; Vol. 3, *Properties and Selection: Stainless Steels, Tool Materials and Special Purpose Metals*, 1980; 10th ed., Vol. 1: *Properties and Selection: Irons, Steels, and High-Performance Alloys*, 1990. Materials Park, Ohio: ASM International.

Peckner, D., and I.M. Bernstein (eds.), *Handbook of Stainless Steels.* New York: McGraw-Hill, 1977.

Roberts, G.A., and R.A. Cary, *Tool Steels*, 4th ed. Metals Park, Ohio: American Society for Metals, 1980.

Tool and Manufacturing Engineers Handbook, 4th ed., Vol. 3: *Materials, Finishing and Coating.* Dearborn, Mich.: Society of Manufacturing Engineers, 1985.

Nonferrous Metals

Aluminum Standards and Data, revised periodically. Washington, D.C.: The Aluminum Association.

Barry, B.T.K., and C.J. Thwaites, *Tin and Its Alloys and Components.* New York: John Wiley, 1983.

Betteridge, W., *Nickel and Its Alloys.* New York: John Wiley, 1984.

Boyer, H.E. (ed.), *Selection of Materials for Component Design: Source Book.* Metals Park, Ohio: American Society for Metals, 1986.

Brady, G.S., and H.R. Clauser, *Materials Handbook*, 12th ed. New York: McGraw-Hill, 1985.

Donachie, M.J. (ed.), *Superalloys—Source Book.* Metals Park, Ohio: American Society for Metals, 1983.

Donachie, M.J. (ed.), *Titanium and Titanium Alloys—Source Book.* Metals Park, Ohio: American Society for Metals, 1982.

Hatch, J.E. (ed.), *Aluminum: Properties and Physical Metallurgy.* Metals Park, Ohio: American Society for Metals, 1984.

Luborsky, F.E., *Amorphous Metallic Alloys.* Woburn, Mass.: Butterworth, 1983.

Material Selector, annual publication of *Materials Engineering Magazine*, Penton/IPC, Cleveland, Ohio.

Metals Handbook, 9th ed., Vol. 2: *Properties and Selection: Nonferrous Alloys and Pure Metals.* Metals Park, Ohio: American Society for Metals, 1979.

Metals Reference Book, 2d ed. Metals Park, Ohio: American Society for Metals, 1983.

Sedlacek, V., *Nonferrous Metals and Alloys.* New York: Elsevier, 1986.

Source Book on Copper and Copper Alloys. Metals Park, Ohio: American Society for Metals, 1979.

Tool and Manufacturing Engineers Handbook, 4th ed., Vol. 3: *Materials, Finishing and Coating.* Dearborn, Mich.: Society of Manufacturing Engineers, 1985.

West, E.G., *Copper and Its Alloys.* New York: John Wiley, 1982.

QUESTIONS

3.1 Explain the difference between a unit cell and a single crystal.

3.2 What effects does recrystallization have on the properties of metals?

3.3 What is the significance of a slip system?
3.4 Explain what is meant by structure-sensitive and structure-insensitive properties of metals.
3.5 What is the relationship between nucleation rate and the number of grains per unit volume of a metal?
3.6 Explain the difference between recovery and recrystallization.
3.7 Describe the significance of some metals undergoing allotropism as far as manufacturing is concerned.
3.8 Is it possible for two pieces of the same metal to have different recrystallization temperatures? Explain.
3.9 Is it possible for recrystallization to take place in some regions of a workpiece before other regions in the same workpiece? Explain.
3.10 Explain why different crystal structures exhibit different strengths and ductilities.
3.11 Describe the difference between preferred orientation and mechanical fibering.
3.12 A cold-worked piece of metal has been recrystallized. When tested, it is found to be anisotropic. Explain the probable reason.
3.13 Draw some analogies to the phenomenon of hot shortness in metals.
3.14 Explain why the orange peel effect on metal surfaces may be of concern.
3.15 How can you tell the difference between two parts made of the same metal, one of which was formed by cold working and the other by hot working? Explain the differences you might observe.
3.16 Explain why the strength of a polycrystalline metal at room temperature decreases as its grain size increases.
3.17 Describe the technique by which you would reduce the orange peel effect.
3.18 What is the significance of metals such as lead and tin having recrystallization temperatures at about room temperature?
3.19 Explain why the heat content of a cold-worked piece of metal is not equal to the amount of work done in plastic deformation.
3.20 Using the information in Chapters 2 and 3, describe the conditions that induce brittle fracture in an otherwise ductile piece of metal.
3.21 A part fractures while being subjected to plastic deformation at an elevated temperature. List the factors that could have played a role in the fracture.
3.22 Make a list of metals that would be suitable for: (a) paper clip, (b) buckle for safety belt, (c) nail, (d) battery cable, and (e) gas turbine blade.
3.23 What might be the adverse effects of using a material with low specific heat in manufacturing?
3.24 Describe the significance of structures and machine components made of two materials with different coefficients of thermal expansion.
3.25 Explain why parts may crack when suddenly subjected to extremes of temperature.
3.26 From your own experience and observations, list three applications each for the following metals and their alloys: (a) steel, (b) aluminum, (c) copper, (d) lead, (e) gold, and (f) titanium.
3.27 Name products that would not have been developed to their advanced stages, as we find them today, if alloys with high strength, corrosion resistance, and creep resistance at elevated temperatures had not been developed.
3.28 Inspect several metal products and components and make an educated guess as to what materials they are made from. Give reasons for your guess. If you list two or more possibilities, explain your reasoning.

3.29 List three engineering applications each where the following physical properties would be desirable: (a) high density, (b) low melting point, and (c) high thermal conductivity.

3.30 Two physical properties that have a major influence on the cracking of workpieces or dies during thermal cycling are thermal conductivity and thermal expansion. Explain why.

PROBLEMS

3.1 The unit cells shown in Fig. 3.2 can be represented by tennis balls arranged in various configurations in a box. In such an arrangement, *atomic packing factor* (APF) is defined as the ratio of the volume of atoms to the volume of the unit cell. Show that the packing factor for the bcc structure is 0.68 and for the fcc structure 0.74.

3.2 Show that the lattice constant a in Fig. 3.2(b) is related to the atomic radius by $a = 2\sqrt{2}R$, where R is the radius of the atom as depicted by the tennis-ball model.

3.3 Show that for the fcc unit cell, the radius r of the largest hole is given by $r = 0.414R$. Determine the size of the largest hole for the iron atom in the fcc structure.

3.4 Calculate the theoretical (a) shear strength and (b) tensile strength for aluminum, plain-carbon steel, and tungsten. Estimate the ratios of their theoretical to actual strengths.

3.5 A technician determines that the grain size of a certain etched specimen is 6. Upon further checking, he or she finds that the magnification used was 150 ×, instead of 100 × that is required by ASTM standards. Determine the correct grain size.

3.6 Estimate the number of grains in the spherical head of a pin 1 mm in diameter. Assume that the ASTM grain size is 8.

3.7 Estimate the number of grains in a regular paper clip if the ASTM grain size is 5.

3.8 Determine the maximum crack velocity in the three materials listed in Problem 3.4 (see Section 2.10.2 and Tables 2.1 and 3.2).

3.9 Calculate the maximum theoretical shear stress τ_{max}, assuming that in Fig. 3.5 the shear stress τ varies linearly with x.

3.10 The natural frequency f of a cantilever beam is given by $f = 0.56\sqrt{EIg/wL^4}$, where E is the modulus of elasticity, I the moment of inertia, g the gravitational constant, w the weight of the beam per unit length, and L the length of the beam. How does the natural frequency of the beam change, if any, as its temperature is increased?

4

Surfaces, Dimensional Characteristics, Inspection, and Quality Assurance

4.1 Introduction

Because of the various mechanical, physical, thermal, and chemical effects induced by its processing history, the *surface* of a manufactured part generally has properties and behavior that are considerably different from those of its bulk. Although the *bulk* material generally determines the component's overall mechanical properties, the component's surface directly influences several important properties and characteristics of the manufactured part:

- Friction and wear properties of the part during subsequent processing when it comes directly into contact with tools and dies and when it is placed in service.
- Effectiveness of lubricants during the manufacturing processes, as well as throughout the part's service life.

- Appearance and geometric features of the part and their role in subsequent operations such as painting, coating, welding, soldering, and adhesive bonding, as well as the resistance of the part to corrosion.
- Initiation of cracks because of surface defects, such as roughness, scratches, seams, and heat-affected zones, which could lead to weakening and premature failure of the part by fatigue or other fracture mechanisms.
- Thermal and electrical conductivity of contacting bodies. For example, rough surfaces have higher thermal and electrical resistance than smooth surfaces.

Friction, wear, and lubrication, now called *tribology*, are surface phenomena. Friction influences forces, power requirements, and surface quality of parts. Wear alters the surface geometry of tools and dies, which in turn, adversely affects the quality of manufactured products and the economics of production.

Lubrication, with few exceptions, is an integral aspect of all manufacturing operations, as well as in the proper functioning of machinery and equipment. We describe those aspects of lubrication relevant to manufacturing operations, such as the types of wear encountered, how to reduce wear, and the proper selection and application of lubricants and coolants. Another important aspect in surface technology is the treatment of surfaces to modify their properties and characteristics. We describe several mechanical, thermal, electrical, and chemical methods that can be used to modify surfaces of parts to improve frictional behavior, effectiveness of lubricants, resistance to wear and corrosion, and surface finish and appearance.

Measurement of the relevant dimensions and features of parts is an integral aspect of interchangeable manufacture—the basic concept of standardization and mass production. We describe the principles involved and the various instruments used in measurement. Another important aspect is testing and inspection of manufactured products, using both destructive and nondestructive testing methods. One of the most critical aspects of manufacturing is product quality. We discuss the technologic and economic importance of building quality into a product, rather than inspecting the product after it is made. Finally, we describe modern trends in quality engineering using statistical techniques and control charts.

4.2 Surface Structure and Properties

Upon close examination of the surface of a piece of metal you will find that it generally consists of several layers (Fig. 4.1). Starting at the interior of the metal and moving outward to the surface is the bulk metal. Also known as the metal *substrate*,

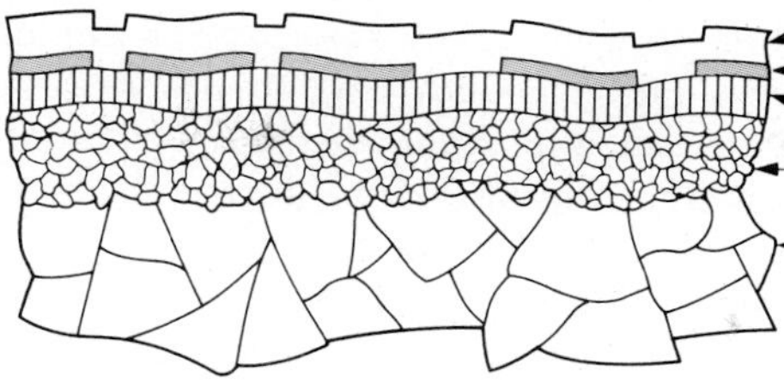

FIGURE 4.1
Schematic illustration of the cross-section of the surface structure of metals. The thickness of the individual layers depends on processing conditions and the environment. *Source:* After E. Rabinowicz.

its structure depends on the composition and processing history of the metal. Above this bulk metal is a layer that usually has been plastically deformed and work hardened during the manufacturing process.

The depth and properties of the work-hardened layer—the *surface structure*—depend on factors such as the processing method used and frictional sliding to which the surface was subjected. Sharp tools and selection of proper process parameters produce surfaces with little or no disturbance. For example, if the surface is produced by machining with a dull tool or under poor cutting conditions—or is ground with a dull grinding wheel—this layer will be relatively thick. Also, nonuniform surface deformation or severe temperature gradients during manufacturing operations usually cause residual stresses in this work-hardened layer.

Unless the metal is processed and kept in an inert (oxygen-free) environment—or it is a noble metal, such as gold or platinum—an *oxide layer* usually lies on top of the work-hardened layer. Some examples are:

a) Iron has an oxide structure with FeO adjacent to the bulk metal, followed by a layer of Fe_3O_4 and then a layer of Fe_2O_3, which is exposed to the environment.
b) Aluminum has a dense amorphous (without crystalline structure) layer of Al_2O_3 with a thick, porous hydrated aluminum-oxide layer over it.
c) Copper has a bright shiny surface when freshly scratched or machined. Soon after, however, it develops a Cu_2O layer, which is then covered with a layer of CuO. This gives copper its somewhat dull color, such as we see in kitchen utensils.
d) Stainless steels are "stainless" because they develop a protective layer of chromium oxide, CrO (passivation).

Under normal environmental conditions, surface oxide layers are generally covered with adsorbed layers of gas and moisture. Finally, the outermost surface of the metal may be covered with contaminants, such as dirt, dust, grease, lubricant residues, cleaning-compound residues, and pollutants from the environment.

Thus surfaces generally have properties that are very different from those of the substrate. The oxide on a metal's surface, for example, is generally much harder than the base metal. Hence oxides tend to be brittle and abrasive. This surface characteristic has several important effects on friction, wear, and lubrication in materials processing and coatings on products. The factors involved in the surface structure of metals are also relevant to a great extent to the surface structure of plastics and ceramics. The surface texture of these materials depends, as with metals, on the method of production. Environmental conditions also influence the surface characteristics of these materials.

4.2.1 Surface integrity

Surface integrity describes not only the topological (geometric) aspects of surfaces, but also their mechanical and metallurgical properties and characteristics. Surface integrity is an important consideration in manufacturing operations because it influences the properties of the product, such as its fatigue strength and resistance to corrosion, and its service life.

Several defects caused by and produced during component manufacturing can be responsible for lack of surface integrity. These defects are usually caused by a combination of factors, such as defects in the original material, the method by which the surface is produced, and lack of proper control of process parameters that can result in excessive stresses and temperatures.

4.3 Surface Texture

Regardless of the method of production, all surfaces have their own characteristics, which we refer to as *surface texture*, roughness, and finish. The description of surface texture as a geometrical property is complex. However, certain guidelines have been established for identifying surface texture in terms of well-defined and measurable quantities (Fig. 4.2).

Flaws, or *defects*, are random irregularities, such as scratches, cracks, holes, depressions, seams, tears, and inclusions. *Lay*, or *directionality*, is the direction of the predominant surface pattern and is usually visible to the naked eye.

Roughness consists of closely spaced, irregular deviations on a scale smaller than that for waviness. Roughness may be superimposed on waviness. Roughness is expressed in terms of its height, its width, and the distance on the surface along which it is measured.

Waviness is a recurrent deviation from a flat surface, much like waves on the surface of water. It is measured and described in terms of the space between adjacent crests of the waves (*waviness width*) and height between the crests and

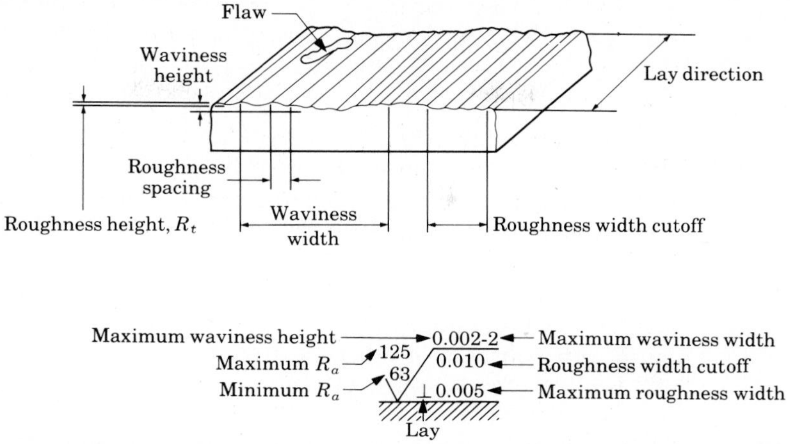

FIGURE 4.2
Standard terminology and symbols to describe surface finish. The quantities are given in μin.

valleys of the waves (*waviness height*). Waviness may be caused by deflections of tools, dies, and the workpiece, warping from forces or temperature, uneven lubrication, and vibration or any periodic mechanical or thermal variations in the system during the manufacturing process.

4.3.1 Surface roughness

Surface roughness is generally described by two methods: arithmetic mean value (R_a); and root-mean-square average (R_q). The *arithmetic mean value* (R_a; formerly identified as AA for *arithmetic average* or CLA for *center-line average*) is based on the schematic illustration of a rough surface shown in Fig. 4.3. The arithmetic mean value R_a is defined as

$$R_a = \frac{a + b + c + d + \cdots}{n}, \tag{4.1}$$

where all ordinates, a, b, c, ..., are absolute values.

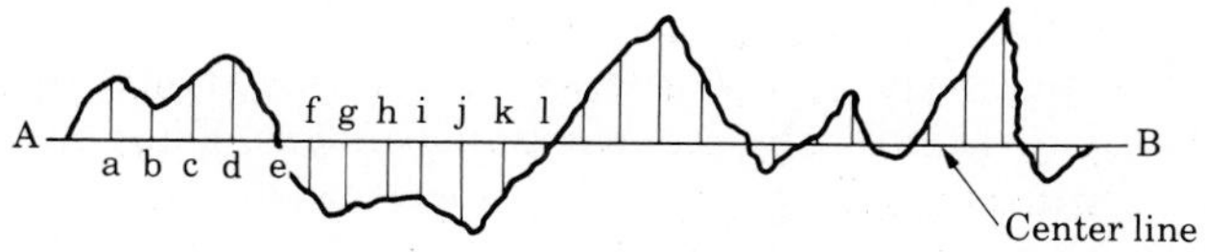

FIGURE 4.3
Coordinates used for surface-roughness measurement, using Eqs. (4.1) and (4.2).

The *root-mean-square average* (R_q; formerly identified as RMS) is defined as

$$R_q = \sqrt{\frac{a^2 + b^2 + c^2 + d^2 + \cdots}{n}}. \tag{4.2}$$

The datum line AB in Fig. 4.3 is located so that the sum of the areas above the line is equal to the sum of the areas below the line. The units generally used for surface roughness are μm (micrometer, or micron) or μin. (microinch), where 1 μm = 40 μin. and 1 μin. = 0.025 μm.

Additionally, we may also use the *maximum roughness height* (R_t) as a measure of roughness. It is defined as the height from the deepest trough to the highest peak. It indicates the amount of material that has to be removed to obtain a smooth surface by polishing or other means.

Because of its simplicity, the arithmetic mean value R_a was adopted internationally in the mid-1950s and is widely used in engineering practice. Equations (4.1) and (4.2) show that there is a relationship between R_q and R_a. For a surface roughness in the shape of a sine curve, R_q is larger than R_a by a factor of 1.11. This factor is 1.1 for most machining processes by cutting, 1.2 for grinding, and 1.4 for lapping and honing.

In general, we cannot adequately describe a surface by its R_a or R_q value alone, since these values are averages. Two surfaces may have the same roughness value but their actual topography may be quite different. A few deep troughs, for example, affect the roughness values insignificantly. However, differences in the surface profile can be significant in terms of fatigue, friction, and the wear characteristics of a manufactured product.

Symbols for surface roughness. Acceptable limits for surface roughness are specified on technical drawings by the symbols shown around the check mark in the lower portion of Fig. 4.2, and their values are placed to the left of the check mark. Symbols used to describe a surface only specify the roughness, waviness, and lay; they do not include flaws. Whenever important, a special note is included in technical drawings to describe the method to be used to inspect for surface flaws.

Measuring surface roughness. Various commercially available instruments, called *surface profilometers*, are used to measure and record surface roughness. The most commonly used instruments feature a diamond stylus traveling along a straight line over the surface (Fig. 4.4). The distance that the stylus travels, which can be varied, is called the *cut off* (see Fig. 4.2).

To highlight the roughness, profilometer traces are recorded on an exaggerated vertical scale (a few orders of magnitude greater than the horizontal scale; Fig. 4.4), called *gain* on the recording instrument. Thus the recorded profile is significantly distorted, and the surface appears to be much rougher than it actually is. The recording instrument compensates for any surface waviness and indicates only roughness. A record of the surface profile is made by mechanical and electronic means.

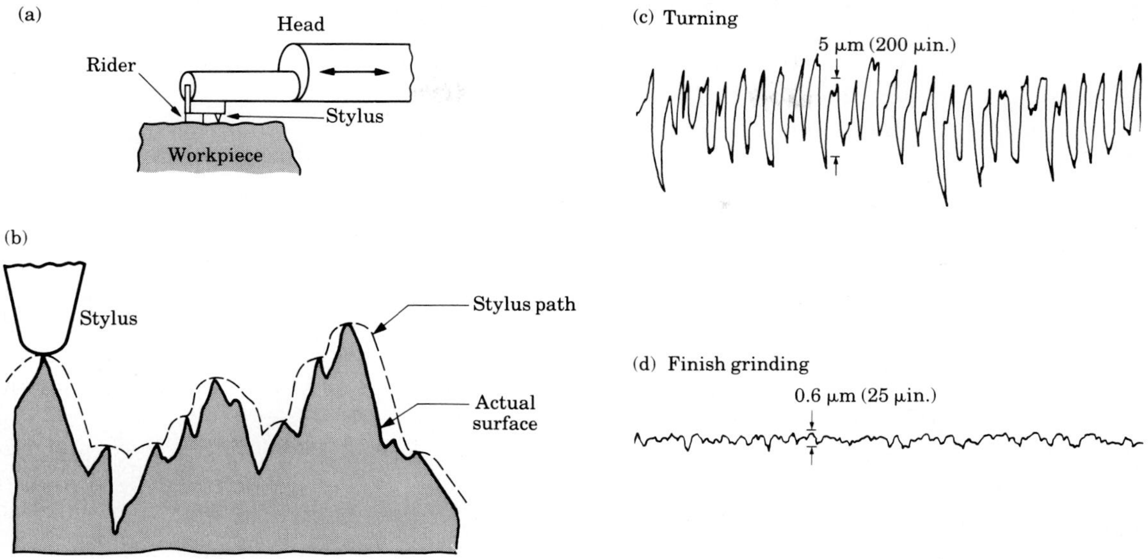

FIGURE 4.4
(a) Measuring surface roughness with a stylus. The rider supports the stylus and guards against damage. (b) Path of stylus in surface roughness measurements (broken line) compared to actual roughness profile. Note that the profile of the stylus path is smoother than the actual surface profile. *Source:* D. H. Buckley. Typical surface profiles produced by (c) turning and (d) finish grinding processes. Note the difference between the vertical and horizontal scales. *Source:* D. B. Dallas (ed.), *Tool and Manufacturing Engineers Handbook,* 3d ed. Copyright © 1976, McGraw-Hill Publishing Company. Used with permission.

We can observe surface roughness directly through an optical or scanning electron microscope. Stereoscopic photographs are particularly useful for three-dimensional views of surfaces, and we can use them to measure surface roughness.

Surface roughness in practice. Surface roughness design requirements for typical engineering applications can vary by as much as two orders of magnitude for different parts. The reasons and considerations for this wide range include:

a) *Accuracy required on mating surfaces*, such as seals, fittings, gaskets, tools, and dies. For example, ball bearings and gages require very smooth surfaces, whereas surfaces for gaskets and brake drums can be quite rough.
b) *Frictional considerations*, that is, the effect of roughness on friction, wear, and lubrication.
c) *Fatigue and notch sensitivity*, because the rougher the surface is, the shorter the fatigue life will be.
d) *Electrical and thermal contact resistance*, because the rougher the surface is, the higher the resistance will be.

e) *Corrosion resistance*, because the rougher the surface is, the greater will be the possibility of entrapped corrosive media.
f) *Subsequent processing*, such as painting and coating, in which a certain roughness can result in better bonding.
g) *Appearance*.
h) *Cost considerations*, because the finer the finish is, the higher the cost will be.

All these factors should be carefully considered before a decision is made as to the recommendation about surface roughness for a certain product. As in all manufacturing processes, the cost involved in the selection should also be a major consideration.

4.4 Tribology

Tribology is the science and technology of interacting surfaces, thus involving friction, wear, and lubrication. All can significantly influence the technology and economics of manufacturing processes and operations.

4.4.1 Friction

Friction is defined as the resistance to relative sliding between two bodies in contact under a normal load. It plays an important role in all metalworking and manufacturing processes because of the relative motion and forces always present between tools, dies, and workpieces. Friction is an energy-dissipating process that results in the generation of heat. The subsequent rise in temperature can have a major effect on the overall operation. Furthermore, because it impedes free movement at interfaces, friction can significantly affect the flow and deformation of the material in many metalworking processes.

However, friction should not always be regarded as undesirable. The presence of friction is necessary for the success or optimization of many operations, the best example of which is the rolling of metals to make sheet and plate. Without friction it would be impossible to roll metals, just as it would be impossible to drive a car on the road. Various theories have been proposed to explain the phenomenon of friction. An early Coulomb model states that friction results from the mechanical interlocking of rough surfaces; this interlocking will require some force to slide the two bodies against each other.

For any model of friction to be valid, it must explain the frictional behavior of two bodies under different loads, speed of relative sliding, temperature, surface conditions, environment, etc., as observed in practice. Consequently, many models have been proposed with varying degrees of success. The most commonly used theory of friction is based on adhesion, because it agrees reasonably well with experimental observations.

Adhesion theory of friction. The *adhesion theory of friction* is based on the observation that two clean, dry (unlubricated) surfaces, regardless of how smooth they are, contact each other at only a fraction of their apparent area of contact (Fig. 4.5). The static load at the interface is thus supported by the minute asperities in contact with each other. The sum total of the contacting areas is known as the *real* area of contact, A_r. For most engineering surfaces, the average angle that the asperity hill makes with a horizontal line is found to be between 5° and 15°. (Surface roughness illustrations are highly exaggerated because of the scales involved.)

Under light loads and with a large real area of contact, the normal stress at the asperity contact is *elastic*. As the load increases, the stresses increase and eventually *plastic* deformation takes place at the *junctions*. Furthermore, with increasing load, the asperities in contact increase in area; also, new junctions are formed with other asperities on the two surfaces in contact with each other. Because of the random height of the asperities, some junctions are in elastic and others in plastic contact.

The intimate contact of asperities creates an *adhesive bond*. The nature of this bond is complex, involving interactions on an atomic scale, and may include mutual solubility and diffusion. The strength of the bond depends on the physical and mechanical properties of the materials in contact, temperature, and the nature and thickness of any oxide film or other contaminants present on the surfaces.

In metalworking processes, the load at the interface is generally high so that the normal stress on the asperity reaches the yield stress. Plastic deformation of the asperity and adhesion then take place at the junctions. In other words, the asperities form *microwelds*. The cleaner the interface, the stronger are the adhesive bonds. To

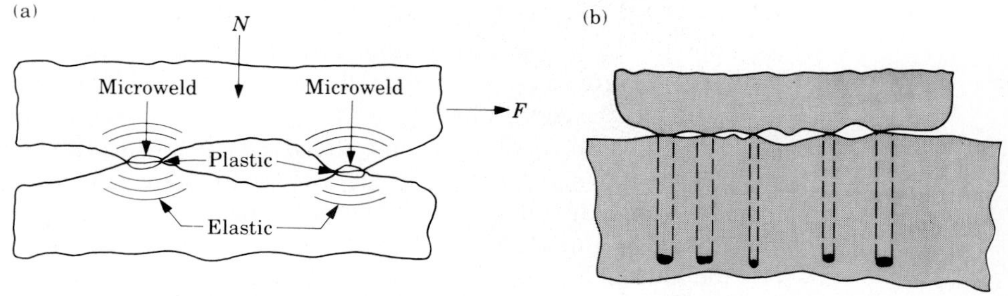

FIGURE 4.5
(a) Schematic illustration of the interface of two contacting surfaces, showing the real areas of contact. (b) Sketch illustrating the proportion of the apparent area to real area of contact. The ratio of the areas can be as high as four to five orders of magnitude.

pull the two surfaces apart requires a certain force, much like pulling apart a welded joint. The coefficient of adhesion is defined as

$$\text{Coefficient of adhesion} = \frac{\text{Tensile force}}{\text{Compressive force applied}}. \tag{4.3}$$

Experimental observations indicate that this coefficient can be quite high with softer metals and that, with few exceptions, it requires a certain finite force to pull the surfaces apart. The more ductile the materials, the greater is the plastic deformation and the stronger is the adhesion of the junctions.

Coefficient of friction. Sliding between the two bodies with such an interface under a normal load N is possible only by the application of a tangential force F. According to the adhesion theory, this is the force required to *shear* the junctions (friction force). The *coefficient of friction* μ at the interface is defined as

$$\mu = \frac{F}{N} = \frac{\tau A_r}{\sigma A_r} = \frac{\tau}{\sigma}, \tag{4.4}$$

where τ is the shear strength of the junction and σ is the normal stress, which, for a plastically deformed asperity, is the yield stress of the asperity. Thus the numerator in Eq. (4.4) is a *surface* property, whereas the denominator is a *bulk* property. Because an asperity is surrounded by a large mass of material, the normal stress on an asperity is equivalent to the hardness of the material (see Section 2.6). Thus, the coefficient of friction μ can now be defined as

$$\mu = \frac{\tau}{\text{hardness}}. \tag{4.5}$$

Two additional phenomena take place in asperity interactions under plastic stresses: For a strain-hardening material, the peak of the asperity is stronger than the bulk material because of plastic deformation at the junction. Thus, under ideal conditions with clean surfaces, the breaking of the bonds under a tensile load will likely follow a path below or above the geometric interface.

Second, a tangential motion at the junction under load will cause *junction growth*; that is, the contact area increases. The reason is that, according to the yield criteria described in Section 2.13, the effective yield stress of the material decreases when subjected to a shear stress. Hence the junction has to grow in area to support the normal load. In other words, the two surfaces tend to approach each other under a tangential force.

If the normal load is increased further, the real area of contact A_r in this model ideally continues to increase, especially with soft and ductile metals. In the absence of contaminants or fluids trapped at the interface, the real area of contact eventually reaches the apparent area of contact. This is the maximum contact area that can be obtained. Because the shear strength at the interface is constant, the friction force (shearing force) reaches a maximum and stabilizes (Fig. 4.6).

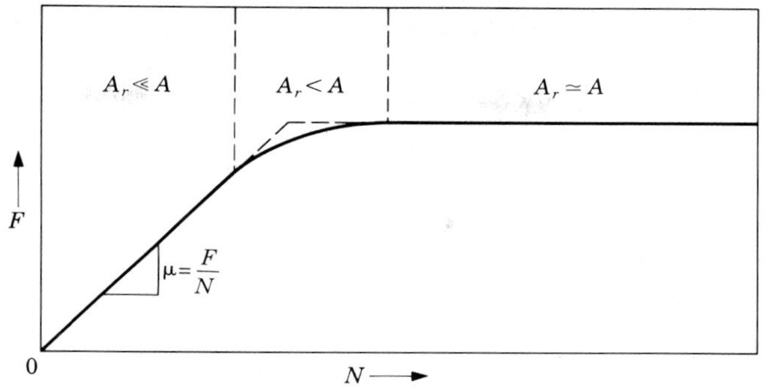

FIGURE 4.6
Schematic illustration of the relation between friction force (*F*) and normal force (*N*). Note that as the real area of contact approaches the apparent area, the friction force reaches a maximum and stabilizes. Most machine components operate in the first region. The second and third regions are encountered in metalworking operations because of the high contact pressures involved between sliding surfaces, i.e., die and workpiece.

This condition, known as *sticking*, creates difficulty in defining friction. Sticking in a sliding interface does not necessarily mean complete adhesion at the interface, as in welding. Rather, it means that the frictional stress at the surface has reached the shear yield stress k of the material. However, when two clean surfaces are pressed together with sufficiently high forces, cold pressure welding can indeed take place.

As the normal load N is increased further, the friction force F remains constant and hence, by definition, the coefficient of friction decreases. This situation, which is an anomaly, indicates that there might be a different and more realistic means of expressing the frictional condition at an interface.

A more recent trend is to define a *friction factor* or *shear factor* m as

$$m = \frac{\tau_i}{k}, \tag{4.6}$$

where τ_i is the shear strength of the interface and k is the *shear yield stress* of the softer material in a sliding pair. [The quantity k is equal to $Y/2$ according to Eq. (2.34) and to $Y/\sqrt{3}$ according to the distortion-energy criterion, where Y is the uniaxial yield stress of the material.]

The value of τ_i is difficult to measure because we have to know the real area of contact; average values will not be meaningful. However, note that in this definition when $m = 0$, there is no friction and when $m = 1$ complete sticking takes place at the interface. The magnitude of m is independent of the normal force or stress. The reason is that the shear yield stress of a thin layer of material is unaffected by the magnitude of the normal stress.

If the upper body in the model shown in Fig. 4.5 is harder than the lower one, or if its surface contains protruding hard particles, then as it slides over the softer body it can scratch and produce grooves on the lower surface. This is known as *ploughing* (or *plowing*) and is an important aspect in frictional behavior; in fact, it can be a dominant mechanism for situations where adhesion is not strong.

Ploughing may involve two different mechanisms. One is the generation of a *groove*, whereby the surface is deformed plastically. The other is the formation of a groove because the cutting action of the upper body generates a *chip* or *sliver*. Either of these processes involves work supplied by a force that manifests itself as friction force. The ploughing force can contribute significantly to friction and to the measured coefficient of friction at the interface.

The nature and strength of the interface is the most significant factor in friction. A strong interface requires a high friction force for relative sliding and one with little strength requires a low friction force. Equation (4.5) indicates that the coefficient of friction can be reduced not only by decreasing the numerator but also by increasing the denominator.

This observation suggests that thin films of low shear strength over a substrate with high hardness is the ideal method for reducing friction. In fact, this is exactly what is achieved with a surface lubricated either with a fluid or solid lubricant.

With no low-shear-strength layer or contaminant present, and under ideal clean conditions with a rigid, perfectly plastic material pair, the coefficient of friction at an asperity can be calculated theoretically from Eq. (4.5) as

$$\mu = \frac{\tau}{\text{hardness}} = \frac{Y/\sqrt{3}}{3Y} \simeq 0.2. \tag{4.7}$$

If the number of asperities increases, such as with increasing load, the combined shear strength of the junctions may exceed the bulk shear strength of the softer metal. In this case, plastic flow takes place in a sublayer (sticking). The value of the normal stress in Eq. (4.4) then becomes that of the uniaxial compressive yield stress of the softer metal. The coefficient of friction is then

$$\mu = \frac{\tau}{\sigma} = \frac{Y/\sqrt{3}}{Y} = 0.577. \tag{4.8}$$

Equation (4.7) can be modified to include a term for ploughing. This term can be estimated roughly to be on the order of 0.1, which should be added algebraically to the value obtained from Eq. (4.7).

The sensitivity of friction to different factors can be illustrated by the following examples. The coefficient of friction of steel sliding on lead is 1.0, steel on copper is 0.9. However, for steel sliding on copper coated with a thin layer of lead, it is 0.2. The coefficient of friction of pure nickel sliding on nickel in a hydrogen or nitrogen atmosphere is 5, in air or oxygen it is 3, and in the presence of water vapor it is 1.6.

Coefficients of friction in sliding contact, as measured experimentally, vary from as low as 0.02 to 100, or even higher. This range is not surprising in view of the many variables involved in the friction process. In metalworking operations that use various lubricants, the range for μ is much narrower, as shown in Table 4.1.

The effects of load, temperature, speed, and environment on the coefficient of friction are difficult to generalize, as each situation has to be studied individually.

Temperature rise caused by friction. Almost all the energy dissipated in overcoming friction is converted into heat. A small portion remains in the bodies as stored energy, and some of it goes into surface energy during generation of new surfaces and wear particles.

Frictional heat raises the interface temperature. The magnitude of this temperature rise and its distribution depends not only on the friction force but also on speed, surface roughness, and the physical properties of the materials, such as thermal conductivity and specific heat. The interface temperature increases with increasing friction, speed, and with low thermal conductivity and specific heat of the materials. The interface temperature can be high enough to soften and melt the surface. (The temperature cannot, of course, exceed the melting point of the material.) A number of analytical expressions have been obtained to calculate the temperature rise in sliding.

Friction in plastics and ceramics. Although their strength is low compared to metals, plastics generally possess low frictional characteristics. This property makes polymers attractive for bearings, gears, seals, prosthetic joints, and general friction-reducing applications. In fact, polymers are sometimes called self-lubricating. The factors involved in friction and metal wear are also generally applicable to polymers. In sliding, the ploughing component of friction in thermoplastics and elastomers is a significant factor because of their viscoelastic behavior (exhibiting both viscous and elastic behavior) and subsequent hysteresis loss.

An important factor in plastics applications is the effect of temperature rise at the sliding interfaces caused by friction. Thermoplastics lose their strength and become soft as temperature increases. Their low thermal conductivity and low melting points are significant in terms of heat generation by friction. If the

TABLE 4.1
COEFFICIENT OF FRICTION IN METALWORKING PROCESSES

	COEFFICIENT OF FRICTION (μ)	
PROCESS	*COLD*	*HOT*
Rolling	0.05–0.1	0.2–0.7
Forging	0.05–0.1	0.1–0.2
Drawing	0.03–0.1	—
Sheet-metal forming	0.05–0.1	0.1–0.2
Machining	0.5–2	—

temperature rise is not controlled, sliding surfaces can undergo deformation and thermal degradation. In view of the importance of ceramics, their frictional behavior is now being studied intensively. Initial investigations indicate that the origin of friction in ceramics is similar to that in metals. Thus adhesion and ploughing at interfaces contribute to the friction force in ceramics.

Reducing friction. Friction can be reduced by selecting materials that have low adhesion, such as carbides and ceramics, and by using surface films and coatings. Lubricants, such as oils, or solid films, such as graphite, interpose an adherent film between tool, die, and workpiece. The film minimizes adhesion and interactions of one surface to the other, thus reducing friction.

Friction can also be reduced significantly by subjecting the die–workpiece interface to *ultrasonic vibrations*, generally at 20 kHz. These vibrations momentarily separate die and workpiece, thus allowing the lubricant to flow more freely into the interface. An additional factor is the high-frequency variation of the relative velocity between die and workpiece, thus reducing the friction at the interface.

Measuring friction. The coefficient of friction is usually determined experimentally, either during actual manufacturing processes or in simulated tests using small-scale specimens of various shapes. The techniques used generally involve measurement of either forces or dimensional changes in the specimen.

One test that has gained wide acceptance, particularly for bulk deformation processes such as forging, is the *ring compression test*. A flat ring is compressed plastically between two flat platens (Fig. 4.7). As its height is reduced, the ring expands radially outward. If friction at the interfaces is zero, both the inner and outer diameters of the ring expand as if it were a solid disk. With increasing friction, the inner diameter becomes smaller.

For a particular reduction in height, there is a critical friction value at which the internal diameter increases (from the original) if μ is low and decreases if μ is high.

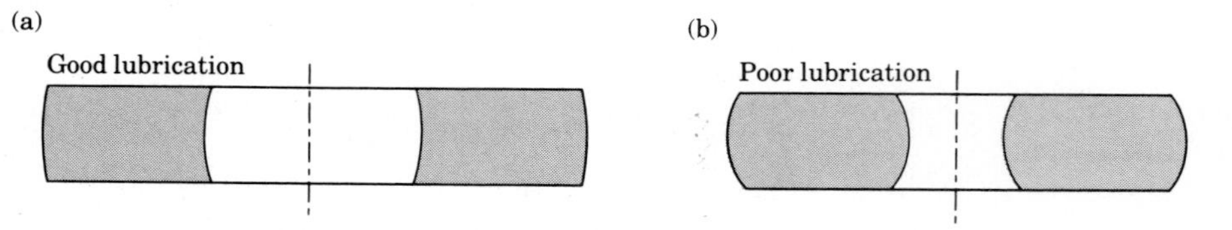

FIGURE 4.7

Effect of lubrication on barreling in ring compression test: (a) with good lubrication, both the inner and outer diameters increase as the specimen is compressed; and (b) with poor or no lubrication, friction is high and the inner diameter decreases. The direction of barreling depends on the relative motion of the cylindrical surfaces with respect to the flat dies.

By measuring the change in the specimen's internal diameter, and using the curves shown in Fig. 4.8, which are obtained through theoretical analyses, we can determine the coefficient of friction.

Each ring geometry has its own specific set of curves. The most common geometry has outer diameter to inner diameter to height proportions of the specimen of 6:3:2. The actual size of the specimen usually is not relevant in these tests. Thus, once you know the percentage reductions in internal diameter and height, you can determine μ from the appropriate chart.

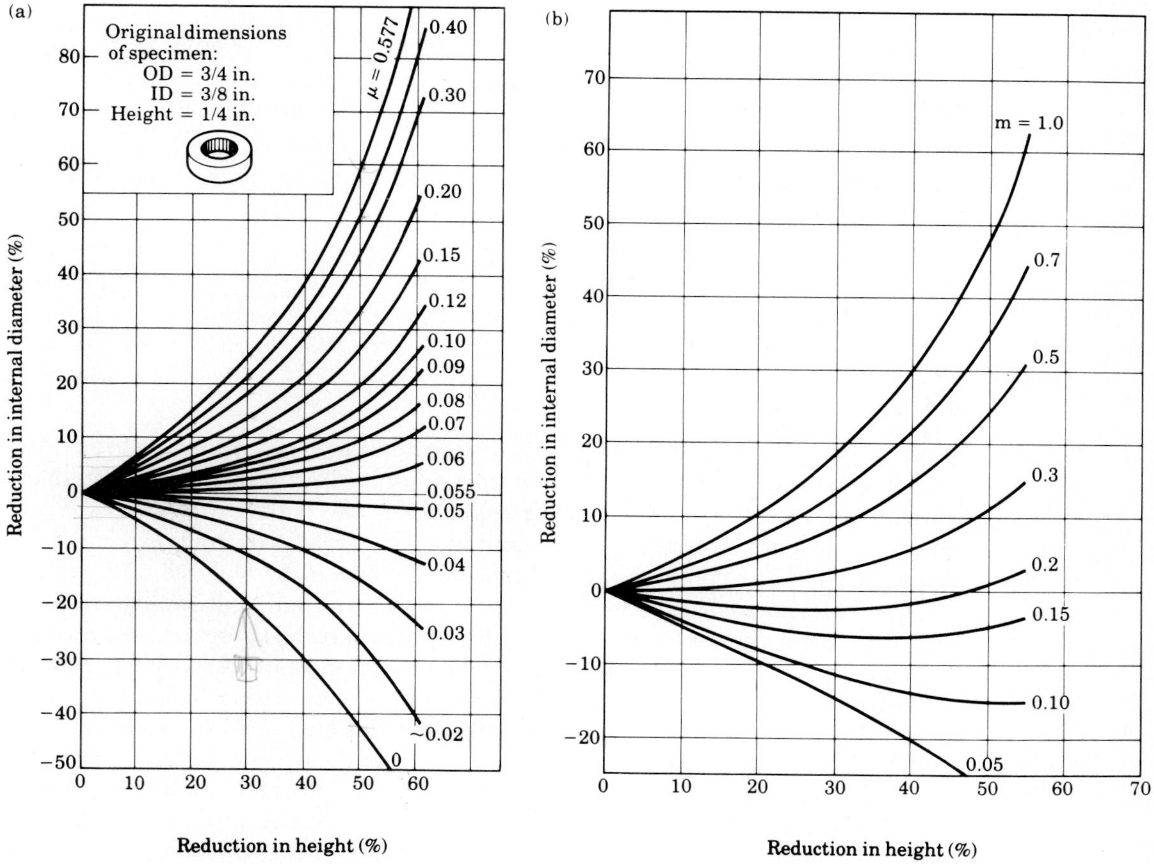

FIGURE 4.8

Charts to determine friction in ring compression tests: (a) coefficient of friction μ; (b) friction factor m. Friction is determined from these charts, from the percent reduction in height and by measuring the percent change in the internal diameter of the specimen after compression.

• Example 4.1: Determining coefficient of friction.

In a ring compression test, a specimen 10 mm in height with outside diameter (OD) = 30 mm and inside diameter (ID) = 15 mm is reduced in thickness by 50%. Determine the coefficient of friction μ and the friction factor m if the OD after deformation is 38 mm.

SOLUTION. We first have to determine the new ID. We do so from volume constancy, as follows:

$$\text{Volume} = \frac{\pi}{4}(30^2 - 15^2)10 = \frac{\pi}{4}(38^2 - \text{ID}^2)5.$$

From this, the new ID = 9.7 mm. Thus the change in internal diameter is

$$\text{Change in ID} = \frac{15 - 9.7}{15} \times 100 = 35\% \text{ (decrease)}.$$

For a 50% reduction in height and a reduction in internal diameter of 35%, the following values are obtained from Fig. 4.8:

$$\mu = 0.21 \quad \text{and} \quad m = 0.72.$$

4.4.2 Wear

Wear is defined as the progressive loss or removal of material from a surface. Wear has important technologic and economic significance because it changes the shape of the workpiece, tool, and die interfaces. Hence it affects the process and size and quality of the parts produced. The magnitude of the wear problem is evident in the countless parts and components that continually have to be replaced or repaired.

Examples of wear in manufacturing processes are dull drills that have to be reground, worn cutting tools that have to be indexed or resharpened, and forming tools and dies that have to be repaired or replaced. Important components in some metalworking machinery are *wear plates*, which are subjected to high loads. These plates, also known as *wear parts* because they are expected to wear, can be replaced easily.

Although wear generally alters the surface topography and may result in severe surface damage, it also has a beneficial effect: It can reduce surface roughness by removing the peaks from asperities (Fig. 4.9). Thus, under controlled conditions, wear may be regarded as a kind of smoothening or polishing process. The *running-in* period for various machines and engines produces this type of wear.

Wear is usually classified as adhesive, abrasive, corrosive, fatigue, erosion, fretting, and impact. We describe below the major types of wear relevant to manufacturing operations.

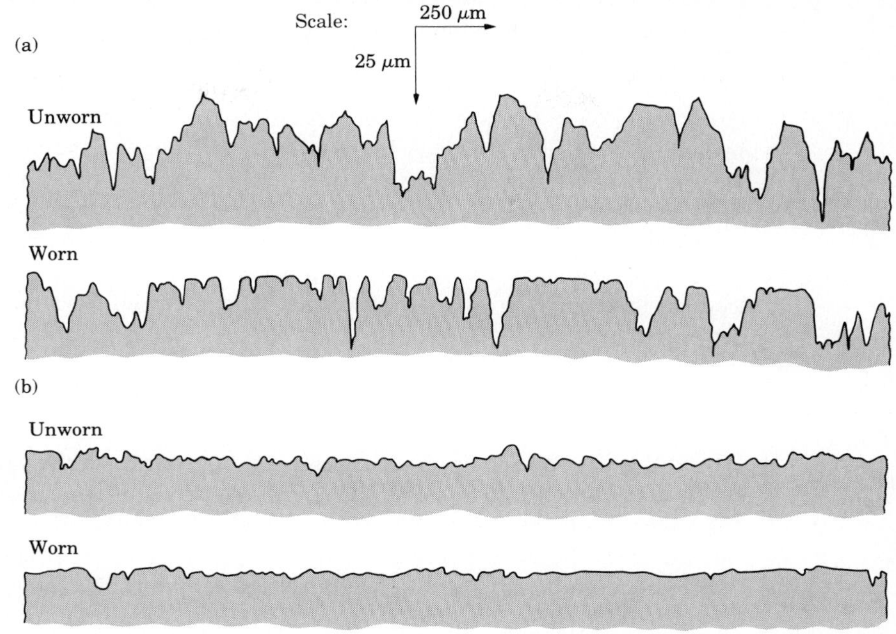

FIGURE 4.9
Changes in originally (a) wire-brushed and (b) ground-surface profiles after wear. *Source:* E. Wild and K. J. Mack.

Adhesive wear. If a tangential force is applied to the model shown in Fig. 4.5, shearing can take place either at the original interface or along a path below or above it (Fig. 4.10), causing *adhesive wear*. The fracture path depends on whether or not the strength of the adhesive bond of the asperities is greater than the cohesive strength of either of the two sliding bodies.

Because of factors such as strain hardening at the asperity contact, diffusion, and mutual solid solubility, the adhesive bonds are often stronger than the base

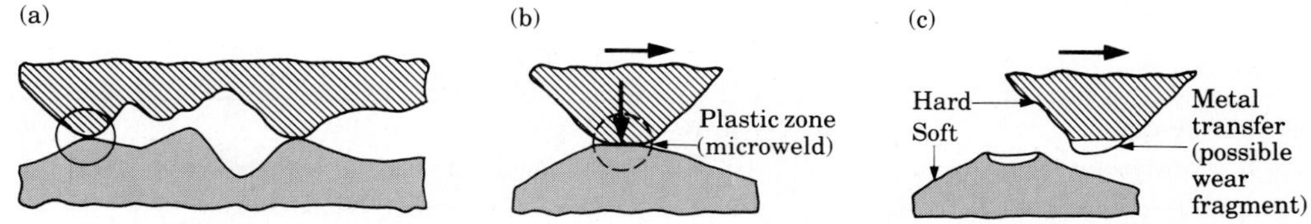

FIGURE 4.10
Schematic illustration of (a) two asperities contacting, (b) adhesion between two asperities, and (c) the formation of a wear particle.

metals. Thus during sliding, fracture at the asperity usually follows a path in the weaker or softer component. A wear fragment is then generated. Although this fragment is attached to the harder component (upper member in Fig. 4.10), it eventually becomes detached during further rubbing at the interface and develops into a loose wear particle. This process is known as adhesive wear or sliding wear. In more severe cases, such as high loads and strongly bonded asperities, adhesive wear is described as scuffing, smearing, tearing, galling, or seizure. Oxide layers on surfaces greatly influence adhesive wear. They can act as a protective film, resulting in what is known as *mild wear*, consisting of small wear particles.

Based on the probability that a junction between two sliding surfaces would lead to the formation of a wear particle, an expression for adhesive wear has been developed as

$$V = k\frac{LW}{3p}, \tag{4.9}$$

where V is the volume of material removed by wear from the surface, k is the wear coefficient (dimensionless), L is the length of travel, W is the normal load, and p is the indentation hardness of the softer body. In some references, the factor 3 is deleted from the denominator and incorporated in the wear coefficient k.

Typical values of k for a combination of materials sliding in air are given in Table 4.2. The wear coefficient for the same pair of materials can vary by a factor of 3, depending on whether wear is measured as a loose particle or as a transferred particle (to the other body). Loose particles have a lower k value. Likewise, when mutual solubility is a significant parameter in adhesion, similar metal pairs have higher k values than dissimilar pairs.

Thus far we have based treatment of adhesive wear on the assumption that the surface layers of the two contacting bodies are clean and free from contaminants. Under these conditions, the adhesive wear rate can be very high, which is known as *severe wear*. As described in Section 4.2, however, metal surfaces are almost always covered with oxide layers, with thicknesses generally ranging between 20 and 100 Å. Although this thickness may at first be regarded as insignificant, the oxide layer has a profound effect on wear behavior.

TABLE 4.2
APPROXIMATE ORDER OF MAGNITUDE FOR WEAR COEFFICIENT *k* IN AIR

UNLUBRICATED	k	LUBRICATED	k
Mild steel on mild steel	10^{-2} to 10^{-3}	52100 steel on 52100 steel	10^{-7} to 10^{-10}
60–40 brass on hardened tool steel	10^{-3}	Aluminum bronze on hardened steel	10^{-8}
Hardened tool steel on hardened tool steel	10^{-4}	Hardened steel on hardened steel	10^{-9}
Polytetrafluoroethylene (PTFE) on tool steel	10^{-5}		
Tungsten carbide on mild steel	10^{-6}		

Oxide layers are usually hard and brittle. When such a surface is subjected to rubbing, this layer can have the following effects. If the load is light and the oxide layer is strongly adhering to the bulk metal, the strength of the junctions between the asperities is weak and wear is low. In this situation, the oxide layer acts as a protective film and the type of wear is known as *mild wear*.

The oxide layer can be broken up under high normal loads if the oxide layer is brittle and is not strongly adhering to the bulk metal, or if the asperities are rubbing against each other repeatedly, so that the oxide layer breaks up by fatigue. However, if the surfaces are smooth, the oxide layers are more difficult to break up. When this layer is eventually broken up, a wear particle is formed. An upper asperity can then form a strong junction with the lower asperity, which is now unprotected by the oxide layer. The wear rate will now be higher until a fresh oxide layer is formed.

In addition, a surface exposed to the environment is covered with adsorbed layers of gas and usually with other contaminants. Such films, even if they are very thin (5 Å), generally weaken the interfacial bond strength of contacting asperities by separating the metal surfaces. The net effect of these films is to provide a protective layer by reducing adhesion and thus reducing wear. The magnitude of the effect of adsorbed gases and contaminants depends on many factors. For instance, it has been shown repeatedly that even small differences in humidity can have a profound influence on the wear rate.

• Example 4.2: Adhesive wear in sliding.

The end of a rod made of 60–40 brass is sliding over the unlubricated surface of hardened tool steel with a load of 200 lb. The hardness of brass is 120 HB. What is the distance traveled to produce a wear volume of 0.001 in^3 by adhesive wear of the brass rod?

SOLUTION. The parameters in Eq. (4.9) for adhesive wear are as follows:

$$V = 0.001 \text{ in}^3$$
$$k = 10^{-3} \text{ (from Table 4.2)}$$
$$W = 200 \text{ lb}$$
$$p = 120 \text{ kg/mm}^2 = 170{,}700 \text{ lb/in}^2.$$

Thus the distance traveled is

$$L = \frac{3Vp}{kW} = \frac{(3)(0.001)(170{,}700)}{(10^{-3})(200)} = 2560 \text{ in.} = 213 \text{ ft.}$$

•

Abrasive wear. *Abrasive wear* is caused by a hard and rough surface—or a surface containing hard, protruding particles—sliding across a surface. This type of wear removes particles by forming microchips or slivers, thereby producing grooves or scratches on the softer surface (Fig. 4.11). In fact, the abrasive processes that we describe in Chapter 9, such as grinding, ultrasonic machining, and abrasive-jet machining, act in this manner. The difference is that in those operations we control the process parameters to produce desired shapes and surfaces, whereas abrasive wear is unintended and unwanted.

The abrasive wear resistance of pure metals and ceramics is directly proportional to their hardness (Fig. 4.12). Abrasive wear can thus be reduced by increasing the hardness of materials (such as by heat treating and microstructural changes) or by reducing the normal load. Other materials that resist abrasive wear are elastomers and rubbers, because they deform elastically and then recover when abrasive particles cross over their surfaces. The best example is automobile tires, which have long lives even though they are operated on abrasive road surfaces. Even hardened steels would not last long under such conditions.

There are two basic types of abrasive wear: 2-body and 3-body wear. The first type is the basis of *erosive* wear. The abrasive particles (such as sand) are usually carried in a jet of fluid or air, and they remove material from the surface by erosion. The 3-body wear is important in metalworking processes because a lubricant between the die and workpiece may carry with it wear particles (which are generated over a period of time) and cause abrasive wear. Thus the necessity for proper inspection and filtering of a metalworking fluid is apparent. This type of wear is also significant in the maintenance of various machinery and equipment. There is a possibility that particles (from machining, grinding, the environment, etc.) may contaminate the lubricating system of the machine components and thus scratch and damage the surfaces.

Corrosive wear. *Corrosive wear*, also known as oxidation, or chemical, wear, is caused by chemical or electrochemical reactions between the surface and the environment. The fine corrosive products on the surface constitute the wear particles. When the corrosive layer is destroyed or removed, as by sliding or

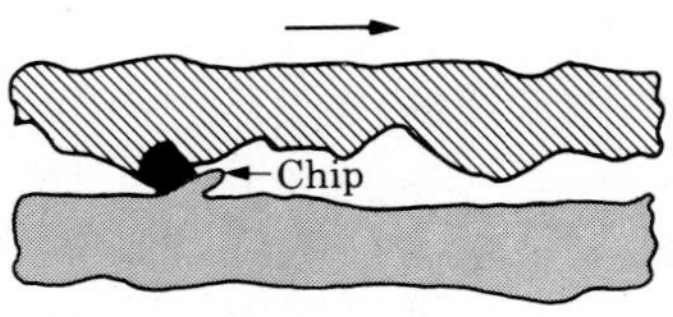

FIGURE 4.11 Schematic illustration of abrasive wear in sliding. Longitudinal scratches on a surface usually indicate abrasive wear.

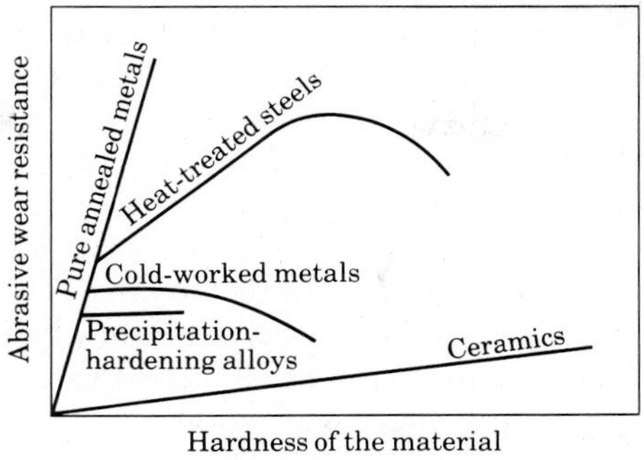

FIGURE 4.12
Abrasive wear resistance as a function of the hardness of several classes of materials.

abrasion, another layer begins to form, and the process of removal and corrosive-layer formation is repeated. Among corrosive media are water, seawater, oxygen, acids and chemicals, and atmospheric hydrogen sulfide and sulfur dioxide. We can reduce corrosive wear by selecting materials that will resist environmental attack, controlling the environment, and reducing operating temperatures to lower the rate of chemical reaction.

Fatigue wear. *Fatigue wear*, also called surface fatigue or surface-fracture wear, is caused when the surface of a material is subjected to cyclic loading, such as in rolling contact in bearings. The wear particles are usually formed by spalling or pitting. Another type of fatigue wear is *thermal fatigue*. Cracks on the surface are generated by thermal stresses from thermal cycling, such as a cool die repeatedly contacting hot workpieces (heat checking). These cracks then join, and the surface begins to spall, producing fatigue wear. This type of wear usually occurs in hot-working and die-casting dies. We can reduce fatigue wear by lowering contact stresses, reducing thermal cycling, and improving the quality of materials by removing impurities, inclusions, and various other flaws that may act as local points for crack initiation.

Other types of wear. Several other types of wear are important in manufacturing processes. *Erosion* is caused by loose abrasive particles abrading a surface. *Fretting corrosion* occurs at interfaces that are subjected to very small movements, such as in machinery. *Impact wear* is the removal of small amounts of materials from a surface by impacting particles. Deburring by vibratory finishing and tumbling (Chapter 9) is an example of this type of wear. In many cases component wear is the result of a combination of different types of wear. Note in Fig. 4.13, for example, that even in the same forging die, various types of wear take place in different locations. A similar situation exists in cutting tools.

Wear of plastics and ceramics. The wear behavior of plastics is similar to that of metals. Thus wear may occur in ways similar to those described earlier. Abrasive wear behavior depends partly on the ability of the polymer to deform and recover elastically, similar to the behavior of elastomers. There is evidence that the parameter describing this behavior may be the ratio of hardness to elastic modulus. Thus the abrasive wear resistance of the polymer increases as this ratio increases.

Typical polymers with good wear resistance are polyimides, nylons, polycarbonate, polypropylene, acetals, and high-density polyethylene. These polymers are molded or machined to make gears, pulleys, sprockets, and similar mechanical components. Because plastics can be made with a wide variety of compositions, they can also be blended with internal lubricants, such as PTFE, silicon, graphite, and molybdenum disulfide, and with rubber particles that are interspersed within the polymer matrix.

Wear resistance of reinforced plastics depends on the type, amount, and direction of reinforcement in the polymer matrix. Carbon, glass, and aramid fibers all improve wear resistance. Wear takes place when fibers are pulled out of the matrix (fiber pullout). Wear is greatest when the sliding direction is parallel to the fibers because they can be pulled out more easily. Long fibers increase the wear

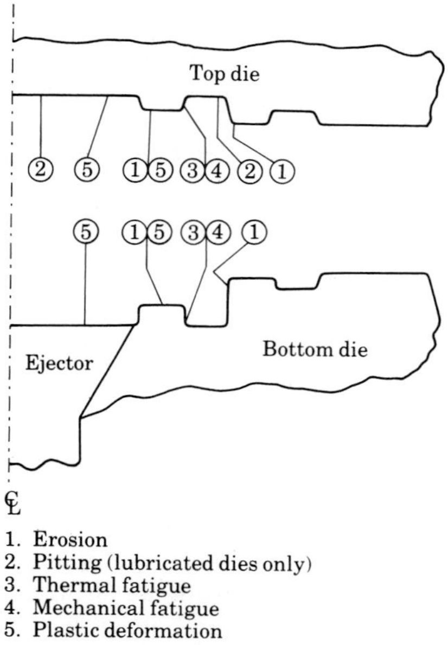

FIGURE 4.13

Types of wear observed in a single die used for hot forging. *Source:* T. A. Dean.

resistance of composites because they are more difficult to pull out, and cracks in the matrix cannot propagate to the surface as easily.

When ceramics slide against metals, wear is caused by small-scale plastic deformation and brittle surface fracture, surface chemical reactions, ploughing, and possibly some fatigue. Metals can be transferred to the oxide-type ceramic surfaces, forming metal oxides. Thus sliding actually takes place between the metal and metal-oxide surface. Conventional lubricants do not appear to influence ceramics wear significantly.

General observations. It is apparent from the foregoing that wear is a complex phenomenon and that each type of wear is influenced by many factors. Numerous theories have been advanced to explain and interpret the experimental data and observations made regarding wear. Because of the many factors involved, quantifying wear data is difficult. In fact, it is well known that wear data are difficult to reproduce.

The most significant aspects of wear are the nature and geometry of the contacting asperities. The strength of the bond at the asperity junction relative to the cohesive strength of the weaker body is an important factor. When a polymer such as PTFE, that is, *Teflon*, is rubbed against steel, the surface of the polymer (which is much weaker than the steel) becomes embedded with steel wear particles. Also, when you rub gold with your fingers, wear particles of gold remain on the fingers. The mechanisms for this phenomenon are not clear.

The question also arises as to whether friction and wear for a pair of metals are directly related. Although at first we would expect such a relation, experimental evidence for unlubricated metal pairs indicates that this is not always the case. These results are not surprising in view of the many factors involved in friction and wear. For instance, ball bearings and races have very little friction, yet they can undergo wear after a large number of cycles.

Measuring wear. Several methods can be used to observe and measure wear. The choice of a particular method depends on the accuracy desired and the physical constraints of the system, such as specimen or part size and difficulty of disassembly.

Although not quantitative, the simplest method is visual and tactile (touching) inspection. Measuring dimensional changes, gaging the worn component, profilometry, and weighing are more accurate methods. However, for large workpieces or tools and dies, the weighing method is not accurate because the amount of wear is usually very small compared to the overall weight of the components involved. Performance and noise level can be monitored. Worn machinery components emit more noise than new parts.

Radiography is a method in which wear particles from an irradiated surface are transferred to the mating surface, which is then measured for the amount of radiation. An example is the transfer of wear particles from irradiated cutting tools to the back side of chips. In other situations, the lubricant can be analyzed for wear particles (spectroscopy). This is a precision method and is used widely for applications such as checking jet-engine component wear.

4.4.3 Lubrication

In manufacturing processes the interface between tools, dies, and workpieces is usually subjected to a wide range of variables. Among the major ones are:

a) Contact pressures, ranging from low values of elastic stresses to multiples of the yield stress of the workpiece material.
b) Relative speeds, ranging from very low (such as in some superplastic metal forming operations) to very high speeds (such as in explosive forming, thin wire drawing, grinding, and high-speed metal cutting).
c) Temperatures, ranging from ambient to almost melting, such as in hot extrusion and squeeze casting.

If two surfaces slide against each other under these conditions, with no protective layers at the interface, friction and wear would be high. To reduce friction and wear, surfaces should be held as far apart as possible. This generally is done by metalworking lubricants, which may be solid, semisolid, or liquid in nature. Metalworking lubricants are not only fluids or solids with certain desirable physical properties, such as *viscosity*, but also are *chemicals* that can react with the surfaces of the tools, dies, and workpieces and alter their physical properties.

Lubrication regimes. Basically, four regimes of lubrication are relevant to metalworking processes.

a) *Thick Film*: The surfaces are completely separated by a fluid film having a thickness of about one order of magnitude greater than the surface roughness; thus there is no metal-to-metal contact. The normal load is light and is supported by the *hydrodynamic fluid film* by the wedge effect caused by the relative velocity of the two bodies and the viscosity of the fluid. The coefficient of friction is very low, usually ranging between 0.001 and 0.02.
b) *Thin Film*: As the normal load increases, or as the speed and viscosity of the fluid decrease (caused by, say, a rise in temperature), the film thickness is reduced to about three to five times the surface roughness. There may be some metal-to-metal contact at the higher asperities; this contact increases friction and leads to slightly higher wear.
c) *Mixed Lubrication*: In this situation, a significant portion of the load is carried by the metal-to-metal contact of the asperities, and the rest of the load is carried by the pressurized fluid film present in hydrodynamic pockets, such as in the valleys of asperities. The film thickness is less than three times the surface roughness. With proper selection of lubricants, a strongly adhering *boundary* film a few molecules thick can be formed on the surfaces. This film prevents direct metal-to-metal contact and thus reduces wear. Depending on the strength of the boundary film and other parameters, the friction coefficient in mixed lubrication may range up to about 0.4.
d) *Boundary Lubrication*: The load here is supported by the contacting surfaces, which are covered with a *boundary layer*. Depending on the boundary film thickness and its strength, the friction coefficient ranges from about 0.1 to 0.4. Wear can be relatively high if the boundary layer is

destroyed. Typical *boundary lubricants* are natural oils, fats, fatty acids, and soaps. Boundary films form rapidly on metal surfaces. As the film thickness decreases and metal-to-metal contact takes place, the chemical aspects of surfaces and their roughness become significant. (In thick-film lubrication the viscosity of the lubricant is the important parameter in controlling friction and wear; chemical aspects are not particularly significant, except as they affect corrosion and staining of metal surfaces.)

A boundary film can *break down* or it can be removed by being disturbed or rubbed off during sliding, or because of desorption due to high temperatures at the interface. The metal surfaces may thus be deprived of this protective layer. The clean metal surfaces then contact each other and, as a consequence, severe wear and scoring can occur. Thus the adherence of boundary films is an important aspect in lubrication.

The role of surface roughness, particularly in mixed lubrication, should be recognized. Roughness can serve to create local reservoirs or pockets for lubricants. The lubricant can be trapped in the valleys of the surface and, because these fluids are incompressible, they can support a substantial portion of the normal load. These pockets also supply lubricant to regions where the boundary layer has been destroyed. There is an optimal roughness for lubricant entrapment. In metalworking operations, it is generally desirable for the workpiece, not the die, to have the rougher surface. Otherwise the workpiece surface will be damaged by the rougher and harder die surface. The recommended surface roughness on most dies is about 0.40 μm (15 μin).

Geometric effects. In addition to surface roughness, the overall geometry of the interacting bodies is an important consideration in lubrication. Figure 4.14

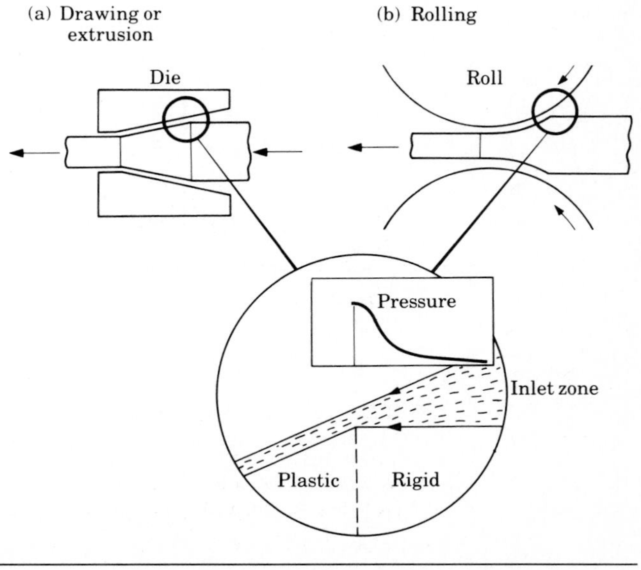

FIGURE 4.14 Inlet zones for entrainment of lubricants in drawing, extrusion, and rolling. Entrainment of lubricants is important in reducing friction and wear in metalworking operations. *Source:* After W. R. D. Wilson.

shows the typical metalworking processes of drawing, extrusion, and rolling. The movement of the workpiece into the deformation zone must allow a supply of the lubricant to be carried into the die–workpiece interfaces.

Analysis of hydrodynamic lubrication shows that the inlet angle is an important parameter. As this angle decreases, more lubricant is entrained; thus lubrication is improved and friction is reduced. With proper selection of parameters, a relatively thick film of lubricant can be maintained at the die–workpiece interface. A similar situation exists in the compression of a straight solid cylinder between flat platens (Fig. 4.15). As the approach speed of the platen increases, the lubricant is trapped more and more because of its viscosity; a thick film is thus developed. The pressure distribution at the interface is given by the parabolic expression

$$p = \frac{3\eta V(R - r)^2}{t^3}, \tag{4.10}$$

where η is the viscosity of the lubricant and t is the film thickness. Thus the pressure increases with increasing viscosity, approach speed, and workpiece diameter; the pressure decreases with increasing film thickness.

The reduction of friction and wear by entrainment or entrapment of lubricants may not always be desirable. The reason is that with a thick fluid film the metal surface cannot come into full contact with the die surfaces and hence does not acquire the shiny appearance that may be required on the product. (Note, for instance, the shiny appearance of a piece of copper wire.) A thick film of lubricant generates a grainy dull surface on the workpiece; the degree of roughness depends on the grain size of the material (Fig. 4.15b). In operations such as coining and forging, trapped lubricants are undesirable because they interfere with the forming process and prevent precise shape generation.

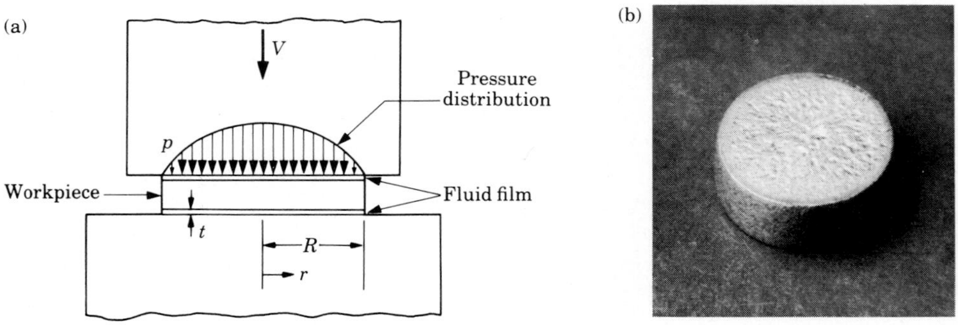

FIGURE 4.15

(a) Pressure distribution at the lubricated interface between a flat die (platen) and a solid cylindrical workpiece. (b) Rough surface produced in compressing an aluminum specimen, due to the high viscosity of the lubricant and high speed of compression. The lubricant is trapped at the die–specimen interface. The surface roughness produced depends on the grain size.

Metalworking fluids. On the basis of our previous discussions, we can summarize the functions of metalworking fluids as follows:

- Reduce friction, thus reducing force and energy requirements and temperature rise.
- Reduce wear, seizure, and galling.
- Improve material flow in tools, dies, and molds.
- Act as a thermal barrier between the workpiece and tool and die surfaces, thus preventing workpiece cooling in hot-working processes.
- Act as a release or parting agent to help in the removal or ejection of parts from dies and molds.

Many types of metalworking fluids are now available to fulfill these requirements. Because of their diverse chemistries, properties, and characteristics, the behavior and performance of lubricants can be complex. In this section we describe the general properties of only the most commonly used lubricants.

Oils have high film strength on the surface of a metal, as you well know if you have ever tried to clean an oily surface. Although they are very effective in reducing friction and wear, oils have low thermal conductivity and specific heat. Thus they are not effective in conducting away the heat generated by friction and plastic deformation in metalworking operations. Oils also are difficult to remove from component surfaces that subsequently are to be painted or welded, and they are difficult to dispose of. The sources of oils are mineral (petroleum), animal, vegetable, and fish. Oils may be *compounded* with a variety of additives or with other oils to impart special properties.

Metalworking fluids are usually blended with several *additives*. These include oxidation inhibitors, rust preventatives, odor control agents, antiseptics, and foam inhibitors. Important additives in oils are sulfur, chlorine, and phosphorus. Known as *extreme-pressure (EP) additives* and used singly or in combination, they react chemically with metal surfaces and form adherent surface films of metallic sulfides and chlorides. These films have low shear strength and good antiweld properties and thus effectively reduce friction and wear. While EP additives are important in boundary lubrication, these lubricants may preferentially attack the cobalt binder in tungsten carbide tools and dies, causing changes in their surface roughness and integrity.

An *emulsion* is a mixture of two immiscible liquids, usually mixtures of oil and water in various proportions, along with additives. Emulsions, also known as *water-soluble fluids*, are of two types: direct and indirect. In a direct emulsion, mineral oil is dispersed in water as very small droplets. In an indirect emulsion, water droplets are dispersed in the oil. Direct emulsions are important fluids because the presence of water gives them high cooling capacity. They are particularly effective in high-speed metal machining where temperature rise has detrimental effects on tool life, workpiece surface integrity, and dimensional accuracy.

Synthetic solutions are chemical fluids containing inorganic and other chemicals dissolved in water. Various chemical agents are added to impart different properties. *Semisynthetic* solutions are basically synthetic solutions to which small amounts of emulsifiable oils have been added.

Soaps are generally reaction products of sodium or potassium salts with fatty acids. Alkali soaps are soluble in water, but other metal soaps are generally insoluble. They are effective boundary lubricants and can also form thick film layers at die–workpiece interfaces, particularly when applied on conversion coatings for cold metalworking applications.

Greases are solid or semisolid lubricants and generally consist of soaps, mineral oil, and various additives. They are highly viscous and adhere well to metal surfaces. Although used extensively in machinery, greases have limited use in manufacturing processes.

Waxes may be of animal or plant (paraffin) origin and have complex structures. Compared to greases, waxes are less "greasy" and are more brittle. They have limited use in metalworking operations, except for copper and, as chlorinated paraffin, for stainless steels and high-temperature alloys.

Solid lubricants. Because of their unique properties and characteristics, several solid materials are used as lubricants in manufacturing operations. In this section we describe four of the most commonly used *solid lubricants*.

The properties of *graphite* are described in Section 11.14. Graphite is weak in shear along its layers and has a low coefficient of friction in that direction. Thus it can be a good solid lubricant, particularly at elevated temperatures. However, the graphite friction is low only in the presence of air or moisture. In a vacuum or an inert gas atmosphere, friction is very high; in fact, graphite can be quite abrasive in these environments. Graphite may be applied either by rubbing it on surfaces or as a *colloidal* (dispersion of small particles) suspension in liquid carriers, such as water, oil, or alcohols.

Another widely used lamellar solid lubricant, *molybdenum disulfide* (MoS_2) is somewhat similar in appearance to graphite. However, unlike graphite it has a high friction coefficient in an ambient environment. Oils are commonly used as carriers for molybdenum disulfide and are used as a lubricant at room temperature. Molybdenum disulfide can also be rubbed onto the surfaces of a workpiece.

Because of their low strength, thin layers of *soft metals* and *polymer coatings* are used as solid lubricants. Suitable metals are lead, indium, cadmium, tin, silver, and polymers such as PTFE, polyethylene, and methacrylates. However, these coatings have limited applications because of their lack of strength under high stresses and at elevated temperatures. Soft metals are used to coat high-strength metals such as steels, stainless steels, and high-temperature alloys. Copper or tin, for example, is chemically deposited on the surface of the metal before it is processed. If the oxide of a particular metal has low friction and is sufficiently thin, the oxide layer can serve as a solid lubricant, particularly at elevated temperatures.

Although a solid material, *glass* becomes viscous at high temperatures and hence can serve as a liquid lubricant. Viscosity is a function of temperature, but not of pressure, and depends on the type of glass. Poor thermal conductivity also makes glass attractive, since it acts as a thermal barrier between hot workpieces and relatively cool dies. Typical glass lubrication applications are in hot extrusion and forging.

Conversion coatings. Lubricants may not always adhere properly to workpiece surfaces, particularly under high normal and shearing stresses. This condition is a special problem in forging, extrusion, and wire drawing of steels, stainless steels, and high-temperature alloys. For these applications, acids transform the workpiece surface by chemically reacting with it, leaving a somewhat rough and spongy surface that acts as a carrier for the lubricant. After treatment, borax or lime is used to remove any excess acid from the surfaces. A liquid lubricant, such as a soap, is then applied to the coated surface. The lubricant film adheres to the surface and cannot be scraped off easily. *Zinc phosphate* conversion coatings are often used on carbon and low-alloy steels. *Oxalate* coatings are used for stainless steels and high-temperature alloys.

Lubricant selection. Selecting a lubricant for a particular process and workpiece material involves consideration of several factors:

a) The particular manufacturing process.
b) Compatibility of the lubricant with the workpiece and tool and die materials.
c) Surface preparation required.
d) Method of lubricant application.
e) Removal of lubricant after processing.
f) Contamination of the lubricant by other lubricants, such as those used to lubricate the machinery.
g) Treatment of waste lubricant.
h) Storage and maintenance of lubricants.
i) Biological and ecological considerations.

In selecting an oil as a lubricant, we should recognize the importance of its viscosity–temperature–pressure characteristics. Low viscosity can have a significant detrimental effect on friction and wear. The different functions of a metalworking fluid, whether primarily a lubricant or a coolant, must also be taken into account. Water-base fluids are very effective coolants but as lubricants are not as effective as oils.

Metalworking fluids should not leave any harmful residues that could interfere with machinery operations. The fluids should not stain or corrode workpiece or equipment. The fluids should be checked periodically for deterioration caused by bacterial growth, accumulation of oxides, chips, and wear debris—and also general degradation and breakdown because of temperature and time. A lubricant may carry with it wear particles and cause damage to the system, so proper inspection and filtering of metalworking fluids are important. These precautions are necessary for all types of machinery, including internal combustion engines and jet engines.

After completion of manufacturing operations, metal surfaces are usually covered with lubricant residues, which should be removed prior to further workpiece processing, such as welding or painting. Various cleaning solutions and techniques can be used for this purpose (Section 4.5.2). Biological and ecological considerations, with their accompanying legal ramifications, are also important considerations. Potential health hazards may be involved in contacting or inhaling some metalworking fluids. Recycling of waste fluids and their disposal are additional important factors to be considered.

4.5 Surface Treatments and Cleaning

After a component is manufactured, all or parts of its surfaces may have to be processed further in order to impart certain properties and characteristics. Surface treatment may be necessary to:

- Improve resistance to wear, erosion, and indentation (slideways in machine tools, wear surfaces of machinery, and shafts, rolls, cams, and gears).
- Control friction (sliding surfaces on tools, dies, bearings, and machine ways).
- Reduce adhesion (electrical contacts).
- Improve lubrication (surface modification to retain lubricants).

4.5.1 Processes

In *shot peening*, the workpiece surface is hit repeatedly with a large number of cast steel, glass, or ceramic shot (small balls), making overlapping indentations on the surface. This action causes plastic deformation of surfaces, to depths up to 1.25 mm (0.05 in.), using shot sizes ranging from 0.125 mm to 5 mm (0.005 in. to 0.2 in.) in diameter. Because plastic deformation is not uniform throughout a part's thickness, shot peening imparts compressive residual stresses on the surface, thus improving the fatigue life of the component. This process is used extensively on shafts, gears, springs, oil-well drilling equipment, and jet-engine parts (such as turbine and compressor blades).

In *roller burnishing*, also called *surface rolling*, the surface of the component is cold worked by a hard and highly polished roller or series of rollers. This process is used on various flat, cylindrical, or conical surfaces (Fig. 4.16). Roller burnishing improves surface finish by removing scratches, tool marks, and pits. Consequently, corrosion resistance is also improved since corrosive products and residues cannot

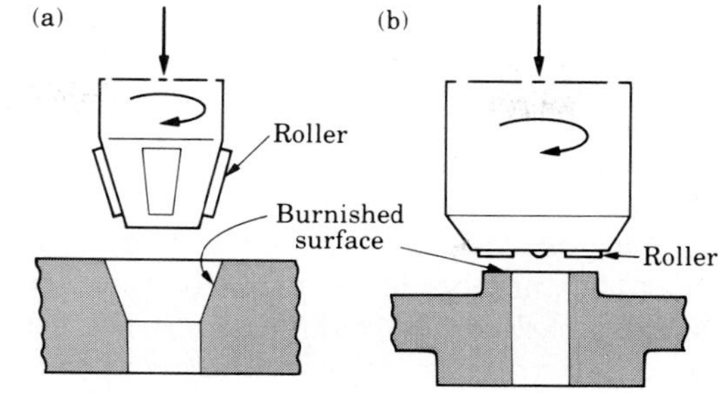

FIGURE 4.16
Examples of roller burnishing of (a) a conical surface and (b) a flat surface and the burnishing tools used. *Source:* Sandvik, Inc.

be entrapped. Internal cylindrical surfaces are burnished by a similar process, called *ballizing* or *ball burnishing*. A smooth ball, slightly larger than the bore diameter, is pushed through the length of the hole.

Roller burnishing is used to improve the mechanical properties of surfaces, as well as the shape and surface finish of components. It can be used either singly or in combination with other finishing processes, such as grinding, honing, and lapping. Soft and ductile, as well as very hard metals, can be roller burnished. Typical applications include hydraulic-system components, seals, valves, spindles, and fillets on shafts.

In *explosive hardening*, surfaces are subjected to high transient pressures by placing a layer of explosive sheet directly on the workpiece surface and detonating it. Large increases in surface hardness can be obtained by this method, with very little change (less than 5 percent) in the shape of the component. Railroad rail surfaces can be hardened by this method.

In *cladding*, metals are bonded with a thin layer of corrosion-resistant metal by applying pressure with rolls or other means. A typical application is cladding of aluminum (*Alclad*) in which a corrosion-resistant layer of aluminum alloy is clad over pure aluminum. Other applications are steels clad with stainless steel or nickel alloys. The cladding material may also be applied through dies, as in cladding steel wire with copper, or by explosives.

In *mechanical plating* (also called mechanical coating, impact plating, or peen plating), fine metal particles are compacted over the workpiece surfaces by impacting them with spherical glass, ceramic, or porcelain beads. The beads are propelled by rotary means. The process is used typically for hardened-steel parts for automobiles, with plating thickness usually less than 0.025 mm (0.001 in.).

We describe traditional methods of *case hardening* (carburizing, carbonitriding, cyaniding, nitriding, flame hardening, and induction hardening) in Section 5.10.3 and summarize them in Table 5.6. In addition to the common heat sources of gas and electricity, laser beams are also used as a heat source in surface hardening of both metals and ceramics. Case hardening, as well as some of the other surface-treatment processes, induce residual stresses on surfaces. The formation of martensite in case hardening causes compressive residual stresses on surfaces. Such stresses are desirable because they improve the fatigue life of components by delaying the initiation of fatigue cracks.

In *hard facing*, a relatively thick layer, edge, or point of wear-resistant hard metal is deposited on the surface by any of the welding techniques we describe in Chapter 12. A number of layers are usually deposited (*weld overlay*). Hard coatings of tungsten carbide, chromium, and molybdenum carbide can also be deposited using an electric arc (*spark hardening*). Hard-facing alloys are available as electrodes, rod, wire, and powder. Typical applications for hard facing are valve seats, oil-well drilling tools, and dies for hot metalworking. Worn parts are also hard faced for extended use.

In *thermal spraying*, metal in the form of rod, wire, or powder is melted in a stream of oxyacetylene flame, electric arc, or plasma arc, and the droplets are sprayed on the preheated surface, at speeds up to 100 m/s (20,000 ft/min) with a compressed-air spray gun. All of the surfaces to be sprayed should be cleaned and roughened to improve bond strength. Typical applications for thermal spraying, also called *metallizing*, are steel structures, storage tanks, and tank cars, sprayed with zinc or aluminum up to 0.25 mm (0.010 in.) in thickness.

Techniques have also been developed for spraying ceramic coatings for high-temperature and electrical-resistance applications, such as to withstand repeated arcing. Powders of hard metals and ceramics are also used as spraying materials. Plasma-arc temperatures may reach 15,000 °C (27,000 °F), which is much higher than that obtained with flames. Typical applications are nozzles for rocket motors and wear-resistant parts.

Vapor deposition. *Vapor deposition* is a process in which the substrate (workpiece surface) is subjected to chemical reactions by gases that contain chemical compounds of the materials to be deposited. The coating thickness is usually a few μm. The deposited materials may consist of metals, alloys, carbides, nitrides, borides, ceramics, or various oxides. The substrate may be metal, plastic, glass, or paper. Typical applications are coating cutting tools, drills, reamers, milling cutters, punches, dies, and wear surfaces.

There are two major deposition processes: physical vapor deposition and chemical vapor deposition. These techniques allow effective control of coating composition, thickness, and porosity. The three basic types of *physical vapor deposition* (PVD) processes are vacuum or arc evaporation (PV/ARC), sputtering, and ion plating. These processes are carried out in a high vacuum at temperatures in the range of 200–500 °C (400–900 °F). In physical vapor deposition, the particles to be deposited are carried physically to the workpiece, rather than by chemical reactions as in chemical vapor deposition.

In *vacuum evaporation*, the metal to be deposited is evaporated at high temperatures in a vacuum and is deposited on the substrate, which is usually at room temperature or slightly higher. Uniform coatings can be obtained on complex shapes. In PV/ARC, which was developed recently, the coating material (cathode) is evaporated by a number of arc evaporators, using highly localized electric arcs. The arcs produce a highly reactive plasma consisting of ionized vapor of the coating material. The vapor condenses on the substrate (anode) and coats it. Applications for this process may be functional (oxidation-resistant coatings for high temperature applications, electronics, and optics) or decorative (hardware, appliances, and jewelry).

In *sputtering*, an electric field ionizes an inert gas (usually argon). The positive ions bombard the coating material (cathode) and cause sputtering (ejecting) of its atoms. These atoms then condense on the workpiece, which is heated to improve bonding. In *reactive sputtering*, the inert gas is replaced by a reactive gas, such as oxygen, in which case the atoms are oxidized and the oxides are deposited. *Radio-frequency* (RF) *sputtering* is used for nonconductive materials such as electrical insulators and semiconductor devices.

Ion plating is a generic term describing the combined processes of sputtering and vacuum evaporation. An electric field causes a glow discharge, generating a plasma. The vaporized atoms in this process are only partially ionized.

Chemical vapor deposition (CVD) is a thermochemical process. In a typical application, such as for coating cutting tools with titanium nitride (TiN), the tools are placed on a graphite tray and heated to 950–1050 °C (1740–1920 °F) at atmospheric pressure in an inert atmosphere. Titanium tetrachloride (a vapor), hydrogen, and nitrogen are then introduced into the chamber. The chemical reactions form titanium nitride on the tool surfaces. For coating with titanium carbide, methane is substituted for the gases. Chemical vapor deposition coatings are usually thicker than those obtained from PVD.

Ion implantation. Here ions are introduced into the surface of the workpiece material. The ions are accelerated in a vacuum to such an extent that they penetrate the substrate to a depth of a few μm. Ion implantation (not to be confused with ion plating) modifies surface properties by increasing surface hardness and improving friction, wear, and corrosion resistance. This process can be controlled accurately, and the surface can be masked to prevent ion implantation in unwanted places. When used in specific applications, such as semiconductors, this process is called *doping* (meaning alloying with small amounts of various elements).

Diffusion coating. This is a process in which an alloying element is diffused into the surface of the substrate, thus altering its properties. Such elements can be supplied in solid, liquid, or gaseous states. This process acquires different names, depending on the diffused element, which describes diffusion processes such as carburizing, nitriding, and boronizing (see Table 5.6).

Electroplating. In *electroplating*, the workpiece (cathode) is plated with a different metal (anode), while both are suspended in a bath containing a water-base electrolyte solution. Although the plating process involves a number of reactions, basically the metal ions from the anode are discharged under the potential from the external source of electricity, combine with the ions in the solution, and are deposited on the cathode.

All metals can be electroplated, with thicknesses ranging from a few atomic layers to a maximum of about 0.05 mm (0.002 in.). Complex shapes may have varying plating thicknesses. Common plating materials are chromium, nickel, cadmium, copper, zinc, and tin. Chromium plating is carried out by first plating the metal with copper, then with nickel, and finally with chromium. *Hard chromium plating* is done directly on the base metal and has a hardness up to 70 *HRC*.

Typical electroplating applications are copper plating aluminum wire and phenolic boards for printed circuits, chrome plating hardware, tin plating copper electrical terminals for ease of soldering, and components requiring resistance to wear and corrosion and good appearance. Because they do not develop oxide films, noble metals (such as gold, silver, and platinum) are important electroplating materials for the electronics and jewelry industries. Plastics such as ABS, polypropylene, polysulfone, polycarbonate, polyester, and nylon also can be electroplated. Because they are not electrically conductive, plastics must be preplated by such processes as electroless nickel plating. Parts to be coated may be simple or complex, and size is not a limitation.

A variation of electroplating, *electroforming* is actually a metal fabricating process. Metal is electrodeposited on a mandrel (also called mold or matrix), which is then removed. Thus the coating itself becomes the product. Simple and complex shapes can be produced by electroforming, with wall thicknesses as small as 0.025 mm (0.001 in.). Parts may weigh from a few grams to as much as 270 kg (600 lb). The electroforming process is particularly suited to low production quantities or intricate parts (such as molds, dies, waveguides, nozzles, and bellows) made of nickel, copper, gold, and silver. It is also suitable for aerospace, electronics, and electrooptics applications. Production rates can be increased with multiple mandrels.

Electroless plating. This process is carried out by chemical reactions, without the use of an external source of electricity. The most common application utilizes nickel, although copper is also used. In electroless nickel plating, nickel chloride (a metallic salt) is reduced—with sodium hypophosphite as the reducing agent—to nickel metal, which is then deposited on the workpiece. The hardness of nickel plating ranges between 425 *HV* and 575 *HV*, and can be heat treated to 1000 *HV*. The coating has excellent wear and corrosion resistance.

Anodizing. This is an oxidation process (*anodic oxidation*) in which the workpiece surfaces are converted to a hard and porous oxide layer that provides corrosion resistance and a decorative finish. The workpiece is the anode in an electrolytic cell immersed in an acid bath, resulting in chemical adsorption of oxygen from the bath. Organic dyes of various colors (typically black, red, bronze, gold, gray) can be used to produce stable, durable surface films. Typical applications for anodizing are aluminum furniture and utensils, architectural shapes, automobile trim, picture frames, keys, and sporting goods. Anodized surfaces also serve as a good base for painting, especially for aluminum, which otherwise is difficult to paint.

Conversion coating. Also called *chemical reaction priming*, in this process a coating forms on metal surfaces as a result of chemical or electrochemical reactions. Various metals, particularly steel, aluminum, and zinc, can be conversion coated. Oxides that naturally form on their surfaces are a form of conversion coating. Phosphates, chromates, and oxalates are used to produce conversion coatings. These coatings are for purposes such as corrosion protection, prepainting, and decorative finish. An important application is in conversion coating of workpieces as a lubricant carrier in cold forming operations. The two common methods of coating are immersion and spraying. The equipment involved depends on the method of application, the type of product, and considerations of quality.

As the name implies, *coloring* involves processes that alter the color of metals, alloys, and ceramics. It is caused by the conversion of surfaces (by chemical, electrochemical, or thermal processes) into chemical compounds, such as oxides, chromates, and phosphates.

Hot dipping. In this process the workpiece, usually steel or iron, is dipped into a bath of molten metal, such as zinc (for galvanized-steel sheet and plumbing supplies), tin (for tinplate and tin cans for food containers), aluminum (aluminizing), and terne (lead alloyed with 10–20 percent tin). Hot-dipped coatings on discrete parts or sheet metal provide long-term corrosion resistance to galvanized pipe, plumbing supplies, and many other products. The coating thickness is usually given in terms of coating weight per unit surface area of the sheet, typically 150–900 g/m^2 (0.5–3 oz/ft^2). Service life depends on the thickness of the zinc coating and the environment to which it is exposed. Various precoated sheet steels are used extensively in automobile bodies. Proper draining to remove excess coating materials is an important consideration.

Porcelain enameling and ceramic coating. Metals may be coated with a variety of glassy (vitreous) coatings to provide corrosion and electrical resistance and for service at elevated temperatures. These coatings are usually classified as porcelain enamels and generally include enamels and ceramics. The word *enamel* is also used for glossy paints, indicating a smooth, hard coating.

Porcelain enamels are glassy inorganic coatings consisting of various metal oxides. *Enameling* involves fusing the coating material on the substrate by heating them both to 425–1000 °C (800–1800 °F) to liquefy the oxides. Depending on their composition, enamels have varying resistances to alkali, acids, detergents, cleansers, and water—and come in different colors.

Typical applications for porcelain enameling are household appliances, plumbing fixtures, chemical processing equipment, signs, cookware, and jewelry. Porcelain enamels are also used as protective coatings on jet-engine components. The coating may be applied by dipping, spraying, or electrodeposition, and thicknesses are usually 0.05–0.6 mm (0.002–0.025 in.). Metals that are coated are typically steels, cast iron, and aluminum. Glasses are used as lining for chemical resistance, and the thickness is much greater than in enameling. *Glazing* is the application of glassy coatings on ceramic wares to give them decorative finishes and to make them impervious to moisture.

Ceramic coatings such as aluminum oxide or zirconium oxide are applied, with the use of binders, to the substrate at room temperature. Such coatings have been used in hot extrusion dies to extend their life.

4.5.2 Cleaning surfaces

We have stressed the importance of surfaces and the influence of deposited or adsorbed layers of various elements and contaminants on surfaces. A clean surface can have both beneficial and detrimental effects. Although an unclean surface may reduce the tendency for adhesion and galling, in general cleanliness is essential for more effective application of metalworking fluids, coating and painting, adhesive bonding, welding, brazing, soldering, reliable functioning of manufactured parts in machinery, food and beverage containers, storage, and in assembly operations.

Cleaning involves removal of solid, semisolid, or liquid contaminants from a surface, and it is an important part of manufacturing operations and the economics of production. The word *clean*, or the degree of cleanliness of a surface, is somewhat difficult to define. How, for example, would you test the cleanliness of a dinner plate? Two simple and common tests are based on:

1. Wiping with a clean cloth and observing any residues on the cloth, as we all have done at one time or another.
2. Observing whether water continuously coats the surface. If water collects as individual droplets, the surface is not clean (*waterbreak test*). Test this phenomenon yourself by wetting dinner plates that have been cleaned to varying degrees.

The type of cleaning process required depends on the type of contaminants to be removed. Contaminants, also called *soils*, may consist of rust, scale, chips and other metallic and nonmetallic debris, metalworking fluids, solid lubricants, pigments, polishing and lapping compounds, and general environmental elements.

Basically there are two types of cleaning methods: mechanical and chemical. *Mechanical* methods consist of physically disturbing the contaminants, as with wire or fiber brushing, dry or wet abrasive blasting, tumbling, and steam jets. Many of these processes are particularly effective in removing rust, scale, and other solid contaminants. Ultrasonic cleaning may also be placed in this category.

Chemical cleaning usually involves the removal of oil and grease from surfaces. It consists of one or more of the following processes:

- *Solution.* The soil dissolves in the cleaning solution.
- *Saponification.* A chemical reaction that converts animal or vegetable oils into a soap that is soluble in water.
- *Emulsification.* The cleaning solution reacts with the soil or lubricant deposits and forms an emulsion. The soil and the emulsifier then become suspended in the emulsion.
- *Dispersion.* The concentration of soil on the surface is decreased by surface-active materials in the cleaning solution.
- *Aggregation.* Lubricants are removed from the surface by various agents in the cleaner and collect as large dirt particles.

Some common *cleaning fluids* are used in conjunction with electrochemical processes for more effective cleaning. These fluids include: alkaline solutions, emulsions, solvents, hot vapors, acids, salts, and organic compound mixtures.

Cleaning discrete parts having complex shapes can be difficult. Design engineers should be aware of this difficulty and provide alternative designs, such as avoiding deep blind holes, making several smaller components instead of one large component that may be difficult to clean, and provide appropriate drain holes in the part to be cleaned.

4.6 Engineering Metrology

Engineering metrology is defined as the measurement of dimensions: length, thickness, diameter, taper, angle, flatness, profiles, and others. Traditionally, measurements have been made after the part has been produced, which is known as postprocess inspection. The current trend in manufacturing is to make measurements while the part is being produced on the machine, which is known as in-process, on-line, or real-time inspection. Here the term *inspection* means to check the dimensions of what we have produced or are producing and to see whether it complies with the specified dimensional accuracy. Much progress has been made in developing new and automated instruments that are highly precise and sensitive.

An important aspect of metrology in manufacturing processes is *dimensional tolerances*, that is, the permissible variation in the dimensions of a part. Tolerances are important not only for proper functioning of products, but they also have a major economic impact on manufacturing costs. The smaller we make the tolerance, the higher the production costs.

4.6.1 Instruments

Line-graduated instruments. Line-graduated instruments are used for measuring length (linear measurements) or angles (angular measurements). *Graduated* means marked to indicate a certain quantity.

The simplest and most commonly used instrument for making linear measurements is a steel rule (*machinist's rule*), bar, or tape, with fractional or decimal graduations. Lengths are measured directly, to an accuracy that is limited to the nearest division, usually 1 mm or $\frac{1}{64}$ in. Rules may be rigid or flexible and may be equipped with a hook at one end for ease of measuring from an edge. Rule depth gages are similar to rules and slide along a special head.

Vernier calipers have a graduated beam and a sliding jaw with a *vernier*. These instruments are also called *caliper gages*. The two jaws of the caliper contact the part being measured, and the dimension is read at the matching graduated lines. Vernier calipers can be used to measure inside or outside lengths. The vernier improves the sensitivity of a simple rule by indicating fractions of the smallest division on the graduated beam, usually to 25 μm (0.001 in.). Vernier calipers are also equipped with digital readouts, which are easy to read and less subject to human error than reading verniers. Vernier height gages are vernier calipers with setups similar to a depth gage and have similar sensitivity.

Micrometers. Commonly used for measuring the thickness and inside or outside diameters of parts, the *micrometer* has a graduated, threaded spindle. Circumferential vernier readings to a sensitivity of 2.5 μm (0.0001 in.) can be obtained. Micrometers are also available for measuring depths (*micrometer depth gage*) and internal diameters (*inside micrometer*) with the same sensitivity. Micrometers are also equipped with digital readout to reduce errors in reading. The anvils on micrometers can be equipped with conical or ball contacts. They are used to measure inside recesses, threaded rod diameters, and wall thicknesses of tubes and curved sheets.

Diffraction gratings consist of two flat optical glasses with closely spaced parallel lines scribed on their surfaces (Fig. 4.17). The grating on the shorter glass is inclined slightly. As a result, interference fringes develop when it is viewed over the longer glass. The position of these fringes depends on the relative position of the two sets of glasses. With modern equipment, using electronic counters and photoelectric sensors, sensitivities of 2.5 μm (0.0001 in.) can be obtained with gratings having 40 lines/mm (1000 lines/in.).

Indirect-reading instruments are typically *calipers* and *dividers* without any graduated scales. They are used to transfer the size measured to a direct-reading instrument, such as a rule. After adjusting the legs to contact the part at the desired location, the instrument is held against a graduated rule, and the dimension is read. Because of the experience required in using them and their dependence on graduated scales, the accuracy of this type of indirect measurement is limited. *Telescoping gages* are available for indirect measurement of holes or cavities.

Angles are measured in degrees, radians, or minutes and seconds of arc. Because of the geometry involved, angles are usually more difficult to measure than are

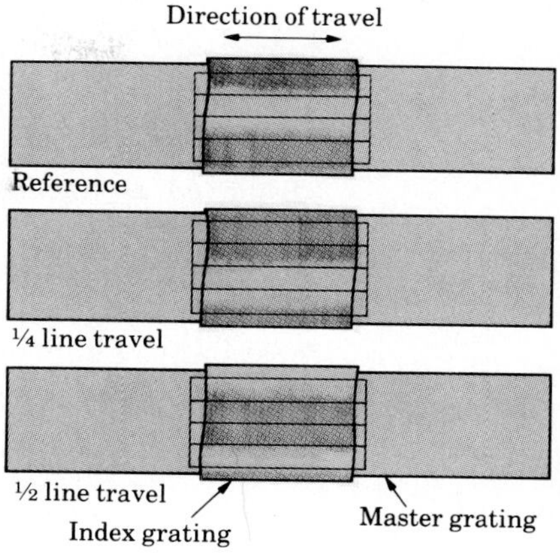

FIGURE 4.17 Measurement of very small linear displacements by fringes (moiré fringes). Note the shift in the fringe patterns. Gratings can typically be 40 lines/mm (1000 lines/in.).

linear dimensions. A *bevel protractor* is a direct-reading instrument similar to a common protractor, except that it has a movable member. The two blades of the protractor are placed in contact with the part being measured, and the angle is read directly on the vernier scale. The sensitivity of the instrument depends on the graduations of the vernier. Another type of bevel protractor is the *combination square*, which is a steel rule equipped with devices for measuring 45° and 90° angles.

Measuring with a *sine bar* involves placing the part on an inclined bar or plate and adjusting the angle by placing gage blocks on a surface plate. After the part is placed on the sine bar, a dial indicator is used to scan the top surface of the part. Gage blocks are added or removed as necessary until the top surface is parallel to the surface plate. The angle on the part is then calculated from geometric relationships. Angles can also be measured using *angle gage blocks*. These are blocks with different tapers that can be assembled in various combinations and used in a manner similar to sine bars. Angles on small parts can be measured through microscopes, with graduated eyepieces, or with optical projectors.

Comparative length-measuring instruments. Unlike the instruments we have just described, instruments used for measuring comparative lengths, also called *deviation-type* instruments, amplify and measure variations or deviations in distance between two or more surfaces. These instruments compare dimensions, hence the word *comparative*. We describe below common types of instruments used for making comparative measurements.

Dial indicators are simple mechanical devices that convert linear displacements of a pointer to rotation of an indicator on a circular dial. The indicator is set to zero at a certain reference surface, and the instrument or the surface to be measured —either external or internal—is brought into contact with the pointer. The movement of the indicator is read directly on the circular dial—either plus or minus—to accuracies as high as 1 μm (40 μin.).

Unlike mechanical systems, *electronic gages* sense the movement of the contacting pointer through changes in the electrical resistance of a strain gage or through inductance or capacitance. The electrical signals are then converted and displayed as linear dimensions. A commonly used electronic gage is the *linear variable differential transformer* (LVDT), used extensively for measuring small displacements. Although they are more expensive than other types, electronic gages have advantages such as ease of operation, rapid response, digital readout, less possibility of human error, versatility, flexibility, and the capability to be integrated into automated systems through microprocessors and computers.

Measuring straightness, flatness, roundness, and profile. The geometric features of straightness, flatness, roundness, and profile are important aspects of engineering design and manufacturing. For example, piston rods, instrument components, and machine-tool slideways should all meet certain requirements with regard to these characteristics in order to function properly. Consequently, their accurate measurement is essential.

Straightness can be checked with straight edges or with dial indicators. *Autocollimators*, resembling a telescope with a light beam that bounces back from the object, are used for accurately measuring small angular deviations on a flat surface. Optical means such as *transits* and *laser beams* are used for aligning individual machine elements in the assembly of machine components.

Flatness can be measured by mechanical means, using a surface plate and a dial indicator. This method can be used for measuring perpendicularity, which can also be measured with the use of precision steel squares.

Another method for measuring flatness is by *interferometry*, using an *optical flat*. The device—a glass or fused quartz disk with parallel flat surfaces—is placed on the surface of the workpiece (Fig. 4.18a). When a monochromatic (one wavelength) light beam is aimed at the surface at an angle, the optical flat splits it into two beams, appearing as light and dark bands to the naked eye (Fig. 4.18b).

The number of fringes that appear is related to the distance between the surface of the part and the bottom surface of the optical flat (Fig. 4.18c). Consequently, a truly flat workpiece surface (that is, when the angle between the two surfaces in Fig. 4.18a is zero) will not split the light beam and fringes will not appear. When surfaces are not flat, fringes are curved (Fig. 4.18d). The interferometry method is also used for observing surface textures and scratches (Fig. 4.18e) through microscopes for better visibility.

Roundness is usually described as deviations from true roundness (mathematically, a circle). The term *out of roundness* is actually more descriptive of the shape of the part. Roundness is very important to the proper functioning of rotating shafts, bearing races, pistons and cylinders, and steel balls in bearings.

The various methods of measuring roundness basically fall into two categories. In the first, the round part is placed on a V-block or between centers and is rotated,

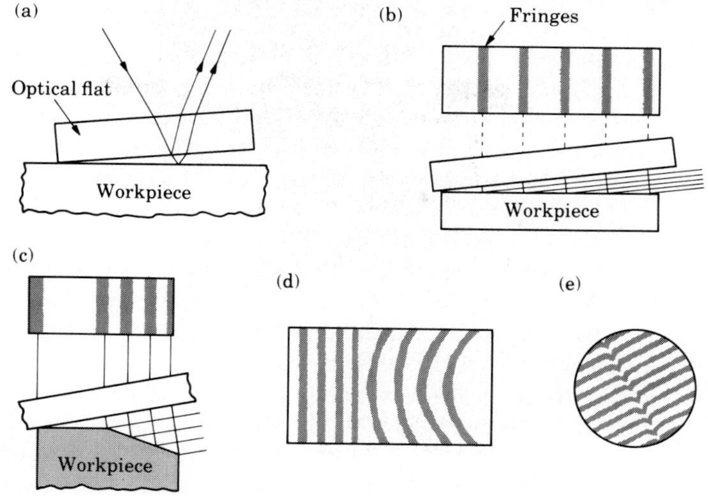

FIGURE 4.18
(a) Interferometry method for measuring flatness using an optical flat. (b) Fringes on a flat inclined surface. An optical flat resting on a perfectly flat workpiece surface will not split the light beam, and no fringes will be present. (c) Fringes on a surface with two inclinations. Note: the greater the incline, the closer the fringes. (d) Curved fringe patterns indicate curvatures on the workpiece surface. (e) Fringe pattern indicating a scratch on the surface.

with the point of a dial indicator in contact with the surface. After a full rotation of the workpiece, the difference between the maximum and minimum readings on the dial is noted. This difference is called the *total indicator reading* (TIR) or full indicator movement. This method is also used for measuring the straightness (squareness) of shaft end faces. In the second method, called *circular tracing*, the part is placed on a platform, and its roundness is measured by rotating the platform. Conversely, the probe can be rotated around a stationary part to make the measurement.

Profile may be measured by several methods. In one method, a surface is compared with a template or profile gage to check shape conformity. Radii or fillets can be measured by this method. Profile may also be measured with a number of dial indicators or similar instruments. Profile-tracing instruments are the latest development.

Threads and *gear teeth* have several features with specific dimensions and tolerances. These dimensions must be produced accurately for smooth operation of gears, reducing wear and noise level, and part interchangeability. These features are measured basically by means of thread gages of various designs that compare the thread produced against a standard thread. Some of the gages used are threaded plug gages, screw-pitch gages (similar to radius gages), micrometers with cone-shaped points, and snap gages with anvils in the shape of threads. Gear teeth are measured with instruments that are similar to dial indicators, with calipers and with micrometers using pins or balls of various diameters. Special profile-measuring equipment is also available, including optical projectors.

Optical projectors, also called *optical comparators*, were first developed in the 1940s to check the geometry of cutting tools for machining screw threads but are now used for checking all profiles. The part is mounted on a table, or between centers, and the image is projected on a screen at magnifications up to 100× or higher. Linear and angular measurements are made directly on the screen, which is equipped with reference lines and circles. The screen can be rotated to allow angular measurements as small as 1 min, using verniers.

Coordinate measuring and layout machines. Coordinate measuring and layout machines are the latest development in measurement technology. Basically, they consist of a platform on which the workpiece being measured is placed and moved linearly or rotated (Fig. 4.19). A stylus, attached to a head capable of lateral and vertical movements, records all measurements. *Coordinate measuring machines* (CMM), also called *measuring machines*, are versatile in their capability to record measurements of complex profiles with high sensitivity (0.25 μm; 10 μin.).

These machines are built rigidly and are very precise. They are equipped with

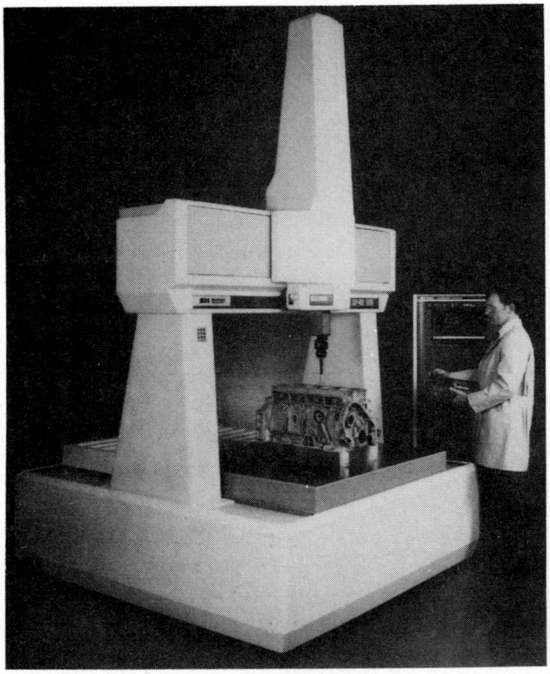

FIGURE 4.19
A coordinate measuring machine, measuring dimensions on an engine block. *Source:* Courtesy of Sheffield Measurement Division, Warner & Swasey Company.

digital readout or can be linked to computers for on-line inspection of parts. They can be placed close to machine tools for efficient inspection and rapid feedback for correction of processing parameters before the next part is made. They are also being made more rugged to resist environmental effects in manufacturing plants, such as temperature variations, vibration, and dirt. Dimensions of large parts are measured by *layout machines*, which are equipped with digital readout. These machines are equipped with scribing tools for marking dimensions on large parts, with an accuracy of ± 0.04 mm (0.0016 in.).

Gages. Thus far we have used the word *gage* to describe some types of measuring instruments, such as a caliper gage, depth gage, telescoping gage, electronic gage, strain gage, and radius gage. The reason is that the words instrument and gage (also spelled gauge) have traditionally been used interchangeably. However, gage has a variety of meanings, such as pressure gage, gage length of a tension-test specimen, and gages for sheet metal, wire, railroad rail, and the bore of shotguns.

Gage blocks are individual square, rectangular, or round metal blocks of various sizes, made very precisely from heat-treated and stress-relieved alloy steels or from carbides. Their surfaces are lapped and are flat and parallel within a range of 0.02–0.12 μm (1–5 μin.). Gage blocks are available in sets of various sizes, some sets containing almost a hundred gage blocks. The blocks can be assembled in many combinations to obtain desired lengths. Dimensional accuracy can be as high as 0.05 μm (2 μin.). Environmental temperature control is important in using gages for high-precision measurements. Although their use requires some skill, gage-block assemblies are commonly utilized in industry as an accurate reference length. Angle blocks are made similarly and are available for angular gaging.

Fixed gages are replicas of the shapes of the parts to be measured. *Plug gages* are commonly used for holes. The *GO gage* is smaller than the *NOT GO* (or *NO GO*) *gage* and slides into any hole whose smallest dimension is less than the diameter of the gage. The *NOT GO* gage must not go into the hole. Two gages are required for such measurements, although both may be on the same device, either at opposite ends or in two steps at one end (step-type gage). Plug gages are also available for measuring internal tapers (in which deviations between the gage and the part are indicated by the looseness of the gage), splines, and threads (in which the *GO* gage must screw into the threaded hole).

Ring gages are used to measure shafts and similar round parts. Ring thread gages are used to measure external threads. The *GO* and *NOT GO* features on these gages are identified by the type of knurling on the outside diameters of the rings. *Snap gages* are commonly used to measure external dimensions. They are made with adjustable gaging surfaces for use with parts having different dimensions. One of the gaging surfaces may be set at a different gap from the other, thus making a one-unit *GO–NOT GO* gage.

Although fixed gages are easy to use and inexpensive, they only indicate whether a part is too small or too large, compared to an established standard. They do not measure actual dimensions.

There are several types of *pneumatic gages*, also called *air gages*. The gage head has holes through which pressurized air, supplied by a constant-pressure line, escapes. The smaller the gap between the gage and the hole, the more difficult it is for the air to escape, and hence the back pressure is higher. The back pressure, sensed and indicated by a pressure gage, is calibrated to read dimensional variations of holes.

Microscopes are optical instruments used to view and measure very fine details, shapes, and dimensions on small and medium-sized tools, dies, and workpieces. The most common and versatile microscope used in tool rooms is the *toolmaker's microscope*. It is equipped with a stage that is movable in two principal directions and can be read to 2.5 μm (0.0001 in.). Several models of microscopes are available with various features for specialized inspection, including models with digital readout.

The *light section microscope* is used to measure small surface details, such as scratches, and the thickness of deposited films and coatings. A thin light band is applied obliquely to the surface and the reflection is viewed at 90°, showing surface roughness, contours, and other features. Unlike ordinary optical microscopes, the *scanning electron microscope* (SEM) has excellent depth of field. As a result, all regions of a complex part are in focus and can be viewed in and photographed to show extremely fine detail. This type of microscope is particularly useful for studying surface textures and fracture patterns. Although expensive, such microscopes are capable of magnifications greater than 100,000 ×.

4.6.2 General characteristics of measuring instruments

The characteristics and quality of measuring instruments are generally described by certain specific terms. These terms, in alphabetical order, are defined as follows:

a) *Accuracy*. The degree of agreement of the measured dimension with its true magnitude.
b) *Amplification*. See Magnification.
c) *Calibration*. Adjusting or setting an instrument to give readings that are accurate within a reference standard.
d) *Drift*. See Stability.
e) *Linearity*. The accuracy of the readings of an instrument over its full working range.
f) *Magnification*. The ratio of instrument output to the input dimension.
g) *Precision*. Degree to which an instrument gives repeated measurement of the same standard.
h) *Repeat accuracy*. Same as accuracy, but repeated many times.
i) *Resolution*. Smallest dimension that can be read on an instrument.
j) *Rule of 10*. An instrument or gage should be 10 times more accurate than the dimensional tolerances of the part being measured.

k) *Sensitivity.* Smallest difference in dimension that an instrument can distinguish or detect.
l) *Speed of response.* How rapidly an instrument indicates the measurement, particularly when a number of parts are measured in rapid succession.
m) *Stability.* An instrument's capability to maintain its calibration over a period of time (also called *drift*).

Selection of an appropriate measuring instrument for a particular application depends on the foregoing factors. In addition, the size and type of parts to be measured, the environment (temperature, humidity, dust, and so on), operator skills required, and costs have to be considered in the purchase of such equipment.

4.6.3 Automated measurement

With increasing automation in all aspects of manufacturing processes and operations, the need for automated measurement (also called *automated inspection*) has become much more apparent. Flexible manufacturing systems and manufacturing cells have led to the adoption of advanced measurement techniques and systems. In fact, installation and utilization of these systems is now a necessary—not an optional—manufacturing technology.

Traditionally, a batch of parts was manufactured and sent for measurement in a separate quality-control room, and if they passed measurement inspection, they were put into inventory. Automated inspection, however, is based on various on-line sensor systems that monitor the dimensions of parts being made and use these measurements to correct the process.

To appreciate the importance of on-line monitoring of dimensions, let's find the answer to the following question: If a machine has been producing a certain part with acceptable dimensions, what factors contribute to subsequent deviation in the dimensions of the same part produced by the same machine? The major factors are:

- Static and dynamic deflections of the machine because of vibrations and fluctuating forces, caused by variations such as in the properties and dimensions of the incoming material.
- Deformation of the machine because of thermal effects. These effects include changes in temperatures of the environment, metalworking fluids, and machine bearings and components.
- Wear of tools and dies, which, in turn, affects the dimensions of the parts produced.

As a result of these factors, the dimensions of parts produced will vary, necessitating monitoring of dimensions during production. *In-process* workpiece control is accomplished by special gaging and is used in a variety of applications, such as high-quantity machining and grinding.

4.7 Dimensional Tolerances

Dimensional tolerance is defined as the permissible or acceptable variation in the dimensions (height, width, depth, diameter, angles) of a part. Tolerances are unavoidable because it is virtually impossible (and unnecessary) to manufacture two parts that have precisely the same dimensions. Furthermore, because close tolerances substantially increase the product cost, a narrow tolerance range is undesirable economically.

Tolerances become important only when a part is to be assembled or mated with another part. Surfaces that are free and not functional do not need close tolerance control. Thus, for example, the accuracies of the holes and the distance between the holes for a connecting rod are far more critical than the rod's width and thickness at various locations along its length. By reviewing the figures throughout this text, you can determine which dimensions and features of the parts illustrated are more critical than others.

To illustrate the importance of dimensional tolerances, let's assemble a simple shaft (axle) and a wheel with a hole, assuming that we want the axle's diameter to be 1 in. (Fig. 4.20). We go to the hardware store and purchase a 1-in. round rod and a wheel with a 1-in. hole. Will the rod fit into the hole without forcing it, or will it be loose in the hole?

The 1-in. dimension is the *nominal* size of the shaft. If we purchase such a rod from different stores or at different times—or select one randomly from a lot of, say, 50 shafts—the chances are that each rod will have a slightly different diameter. Machines may, with the same setup, produce rods of slightly different diameters, depending on a number of factors, such as speed of operation, temperature, lubrication, variations in the incoming material, and similar variables.

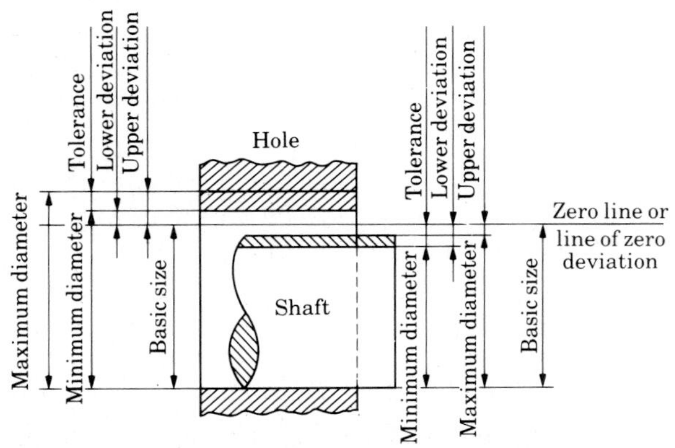

FIGURE 4.20
Basic size, deviation, and tolerance on a shaft, according to the ISO system.

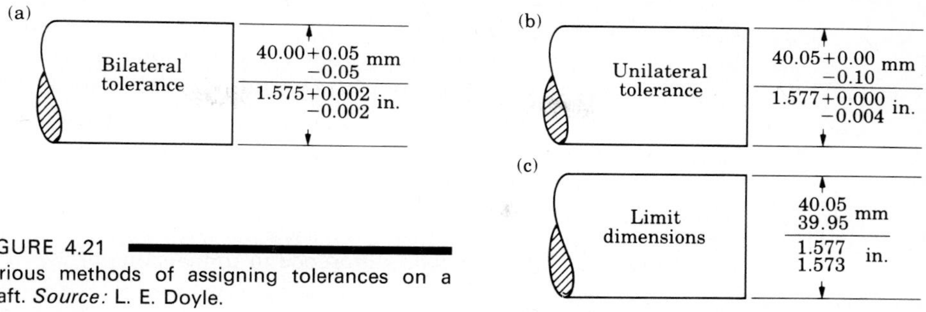

FIGURE 4.21

Various methods of assigning tolerances on a shaft. *Source:* L. E. Doyle.

If we now specify a range of diameters for both the rod and the hole of the wheel, we can predict correctly the type of fit that we will have after assembly. Certain terminology has been established to clearly define these geometric quantities, such as the ISO system shown in Fig. 4.20. Note that both the shaft and the hole have minimum and maximum diameters, respectively, the difference being the tolerance for each member. A proper engineering drawing should specify these parameters with numerical values, as shown in Fig. 4.21.

The range of tolerances obtained in various manufacturing processes is given in Fig. 4.22. There is a general relationship between tolerances and surface finish of

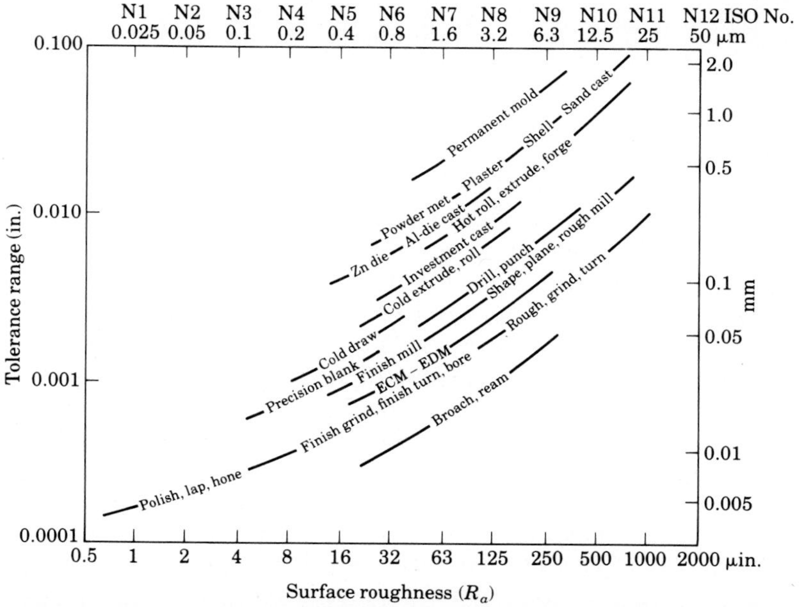

FIGURE 4.22

Tolerances and surface roughness obtained in various manufacturing processes. These tolerances apply to a 25-mm (1-in.) workpiece dimension. *Source:* J. A. Schey.

parts manufactured by various processes. Note the wide range of tolerances and surface finishes obtained. Also, the larger the part, the greater the obtainable tolerance range becomes. Experience has shown that dimensional inaccuracies of manufactured parts are approximately proportional to the cube root of the size of the part. Thus doubling the size of a part increases the inaccuracies by $\sqrt[3]{2} = 1.26$ times, or 26 percent.

Several terms are used to describe features of dimensional relationships between mating parts. These terms, in alphabetical order, are defined as follows:

a) *Allowance.* The specified difference in dimensions between mating parts; also called *functional dimension* or *sum dimension.*
b) *Basic size.* Dimension from which limits of size are derived, using tolerances and allowances.
c) *Bilateral tolerance.* Deviation—plus or minus—from the basic size.
d) *Clearance.* The space between mating parts.
e) *Clearance fit.* Fit that allows for rotation or sliding between mating parts.
f) *Datum.* A theoretically exact axis, point, line, or plane.
g) *Feature.* Physically identifiable portion of a part, such as hole, slot, pin, or chamfer.
h) *Fit.* The amount of clearance or interference between mating parts.
i) *Geometric tolerancing.* Tolerances that involve shape features of the part.
j) *Hole-basis system.* Tolerances based on a zero line on the hole; also called *standard hole practice* or *basic hole system.*
k) *Interference.* Negative clearance.
l) *Interference fit.* Fit that allows for alignment and stiffness of the mating parts.
m) *Limit dimensions.* The maximum and minimum dimensions of a part; also called *limits.*
n) *Maximum material condition.* A part whose dimensions are at the maximum limit dimensions.
o) *Nominal size.* Dimension that is used for the purpose of general identification.
p) *Positional tolerancing.* A system of specifying the true position, size, and form of the features of a part, including allowable variations.
q) *Shaft-basis system.* Tolerances based on a zero line on the shaft; also called *standard shaft practice* or *basic shaft system.*
r) *Standard size.* Nominal size in integers and common subdivisions of length.
s) *Transition fit.* Fit with small clearance or interference that allows for accurate location of mating parts.
t) *Unilateral tolerancing.* Deviation in one direction only from the nominal dimension.
u) *Zero line.* Reference line along the basic size from which a range of tolerances and deviations are specified.

Because the dimensions of holes are more difficult to control than those of shafts, the hole-basis system is commonly used for specifying tolerances in shaft and hole assemblies.

4.8 Testing and Inspection

Before they are marketed, manufactured parts and products are inspected for several characteristics. This inspection routine is particularly important for products or components whose failure or malfunction has potentially serious implications, such as bodily injury or fatality. Typical examples are cables breaking, switches malfunctioning, brakes failing, grinding wheels breaking, railroad wheels fracturing, turbine blades failing, pressure vessels bursting, and weld or joints failing. In this section we identify and describe the various methods that are commonly used to inspect manufactured products.

Product quality has always been one of the most important elements in manufacturing operations, and with increasing domestic and international competition, it has become even more important. Prevention of defects in products and on-line inspection are now major goals in all manufacturing activities. We again emphasize that quality must be built into a product and not merely checked after the product has been made. Thus close cooperation and communication between design and manufacturing engineers are essential.

4.8.1 Nondestructive testing

Nondestructive testing (NDT) is carried out in such a way that product integrity and surface texture remain unchanged (Table 4.3). These techniques generally require considerable operator skill. Interpreting test results accurately may be difficult because test results can be quite subjective. However, the use of computer graphics and other enhancement techniques have reduced the likelihood of human error in nondestructive testing. We describe below the basic principles of the more commonly used nondestructive testing techniques.

In the *liquid-penetrants technique*, fluids are applied to the surfaces of the part and allowed to penetrate into surface openings, cracks, seams, and porosity. The penetrant can seep into cracks as small as 0.1 μm (4 μin.) in width. Two common types of liquids are (1) fluorescent penetrants with various sensitivities, which fluoresce under ultraviolet light; and (2) visible penetrants, using dyes usually red in color, which appear as bright outlines on the surface.

The surface to be inspected is first thoroughly cleaned and dried. The liquid is brushed or sprayed on the surface to be inspected and allowed to remain long enough to seep into surface openings. Excess penetrant is then wiped off or washed away with water or solvent. A developing agent is then added to allow the penetrant

TABLE 4.3
ADVANTAGES AND LIMITATIONS OF NONDESTRUCTIVE AND DESTRUCTIVE TESTING

NONDESTRUCTIVE TESTING	DESTRUCTIVE TESTING
ADVANTAGES	*ADVANTAGES*
1. Can be done directly on production items without regard to part cost or quantity available, and no scrap losses are incurred except for bad parts	1. Can often directly and reliably measure response to service conditions
2. Can be done on 100% of production or on representative samples	2. Measurements are quantitative, and usually valuable for design or standardization
3. Can be used when variability is wide and unpredictable	3. Interpretation of results by a skilled technician usually not required
4. Different tests can be applied to the same item simultaneously or sequentially	4. Correlation between tests and service usually direct, leaving little margin for disagreement among observers as to meaning and significance of test results
5. The same test can be repeated on the same item	
6. May be performed on parts in service	
7. Cumulative effect of service usage can be measured directly	
8. May reveal failure mechanism	
9. Little or no specimen preparation is required	
10. Equipment is often portable for use in field	
11. Labor costs are usually low, especially for repetitive testing of similar parts	
LIMITATIONS	*LIMITATIONS*
1. Results often must be interpreted by a skilled, experienced technician	1. Can be applied only to a sample, and separate proof that the sample represents the population is required
2. In absence of proven correlation, different observers may disagree on meaning and significance of test results	2. Tested parts cannot be placed in service
3. Properties are measured indirectly, and often only qualitative or comparative measurements can be made	3. Repeated tests of same item are often impossible, and different types of tests may require different samples
4. Some nondestructive tests require large capital investments	4. Extensive testing usually cannot be justified, because of large scrap losses
	5. May be prohibited on parts with high material or fabrication costs, or on parts of limited availability
	6. Cumulative effect of service usage cannot be measured directly, but only inferred from tests on parts used for different lengths of time
	7. Difficult to apply to parts in service, and usually terminates their useful life
	8. Extensive machining or other preparation of test specimens is often required
	9. Capital investment and manpower costs are often high

Source: ASM Metals Reference Book, 2d ed., 1983.

to seep back to the surface and spread to the edges of openings, thus magnifying the size of defects. The surface is then inspected for defects, either visually in the case of dye penetrants or with fluorescent lighting. This method is capable of detecting a variety of surface defects and is used extensively. The equipment is simple and easy to use, can be portable, and is less costly to operate than other methods. However, this method can only detect defects that are open to the surface, not internal defects.

The *magnetic-particle inspection technique* consists of placing fine ferromagnetic particles on the surface. The particles can be applied either dry or in a liquid carrier such as water or oil. When the part is magnetized with a magnetic field, a discontinuity (defect) on the surface causes the particles to gather visibly around it.

The collected particles generally take the shape and size of the defect. Subsurface defects can also be detected by this method, provided they are not deep. The ferromagnetic particles may be colored with pigments for better visibility on metal surfaces. Wet particles are used for detecting fine discontinuities, such as fatigue cracks.

In *ultrasonic inspection*, an ultrasonic beam travels through the part. An internal defect, such as a crack, interrupts the beam and reflects back a portion of the ultrasonic energy. The amplitude of the energy reflected and the time required for return indicates the presence and location of any flaws in the workpiece. The ultrasonic waves are generated by transducers, called search units or probes, of various types and shapes. They operate on the principle of piezoelectricity, using materials such as quartz, lithium sulfate, and various ceramics. Most inspections are carried out at a frequency range of 1–25 MHz. Couplants are used to transmit the ultrasonic waves from the transducer to the test piece. Typical couplants are water, oil, glycerin, and grease.

The ultrasonic inspection method has high penetrating power and sensitivity. It can be used to inspect flaws in large volumes of material, such as railroad wheels, pressure vessels, and die blocks, from various directions. Accuracy is higher than that of other nondestructive inspection methods. However, this method requires experienced personnel to carry out the inspection and interpret the results correctly.

The *acoustic-emission technique* detects signals (high-frequency stress waves) generated by the workpiece itself during plastic deformation, crack initiation and propagation, phase transformation, and sudden reorientation of grain boundaries. Bubble formation during boiling and friction and wear of sliding interfaces are other sources of acoustic signals. Acoustic-emission inspection is typically performed by stressing elastically the part or structure, such as bending a beam, applying torque to a shaft, and pressurizing a vessel. Acoustic emissions are detected by sensors consisting of piezoelectric ceramic elements. This method is particularly effective for continuous surveillance of load-bearing structures.

The *acoustic-impact technique* consists of tapping the surface of an object and listening to and analyzing the signals to detect discontinuities and flaws. The principle is basically the same as tapping walls, desktops, or countertops in various locations with your fingers or a hammer and listening to the sound emitted. Vitrified grinding wheels are tested in a similar manner (ring test) to detect cracks in the wheel that may not be visible to the naked eye. The acoustic-impact technique can be instrumented and automated and is easy to perform. However, the results depend on part geometry and mass, thus requiring a reference standard to identify flaws.

Radiography involves x-ray inspection to detect internal flaws or density and thickness variations in the part. The radiation source is typically an x-ray tube, and a visible permanent image is made on an x-ray film or radiographic paper.

The *eddy-current inspection method* is based on the principle of electromagnetic induction. The part is placed in or adjacent to an electric coil through which alternating current (exciting current) flows at frequencies ranging from 60 Hz to

6 MHz. This current causes eddy currents to flow in the part. Defects in the part impede and change the direction of eddy currents, causing changes in the electromagnetic field. These changes affect the exciting coil (inspection coil) whose voltage is monitored to determine the presence of flaws.

Thermal inspection involves observing temperature changes by contact- or noncontact-type heat-sensing devices. Defects in the workpiece, such as cracks, debonded regions in laminated structures, and poor joints, cause a change in temperature distribution. In *thermographic inspection*, materials such as heat-sensitive paints and papers, liquid crystals, and other coatings are applied to the surface. Any changes in their color or appearance indicate defects.

The *holography technique* creates a three-dimensional image of the workpiece, utilizing an optical system. This technique is generally used on simple shapes and highly polished surfaces, and the image is recorded on a photographic film. Its use has been extended to inspection of parts (*holographic interferometry*) having various shapes and surface conditions. Using double- and multiple-exposure techniques while the part is being subjected to external forces or time-dependent variations, changes in the images reveal defects in the part.

In *acoustic holography*, information on internal defects is obtained directly from the image of the interior of the part. In *liquid-surface acoustical holography*, the workpiece and two ultrasonic transducers (one for the object beam and the other for the reference beam) are immersed in a water-filled tank. The holographic image is then obtained from the ripples in the tank. In *scanning acoustical holography*, only one transducer is used and the hologram is produced by electronic-phase detection. This system is more sensitive, the equipment is usually portable, and very large workpieces can be accommodated by using a water column instead of a tank.

4.8.2 Destructive testing

As the name suggests, the part or product tested using destructive testing methods no longer maintains its integrity, original shape, or surface texture. Mechanical test methods are all destructive, in that a sample or specimen has to be removed from the product in order to test it. In addition to mechanical testing, other destructive tests include speed testing of grinding wheels to determine their bursting speed and high-pressure testing of pressure vessels to determine their bursting pressure. Destructive testing has several advantages and limitations, compared to nondestructive testing, which are outlined in Table 4.3.

Hardness tests leaving large impressions may be regarded as destructive testing. However, microhardness tests may be regarded as nondestructive because of the very small permanent indentations. This distinction is based on the assumption that the material is not notch sensitive. However, most glasses, highly heat-treated metals, and ceramics are notch sensitive; that is, the small indentation produced by the indenter may lower their strength and toughness.

4.8.3 Automated inspection

Note that in all the preceding examples we discussed testing parts or products that had already been manufactured. Traditionally, individual parts and assemblies of parts have been manufactured in batches, sent to inspection in quality-control rooms, and if approved, put in inventory. If products do not pass the quality inspection, they are either scrapped or kept on the basis of a certain acceptable deviation from the standard. Obviously, such a system lacks flexibility, requires maintaining an inventory, and inevitably results in some defective parts going through the system. The traditional method actually counts the defects after they occur (*postprocess inspection*), and in no way attempts to prevent defects.

In contrast, one of the important trends in modern manufacturing is *automated inspection.* This method uses a variety of sensor systems that monitor the relevant parameters *during* the manufacturing process (*on-line inspection*). Then, using these measurements, the process automatically corrects itself to produce acceptable parts. Thus further inspection of the part at another location in the plant is unnecessary. Parts may also be inspected immediately after they are produced (*in-process inspection*).

The use of accurate sensors and computer-control systems has integrated automated inspection into manufacturing operations. Such a system ensures that no part is moved from one manufacturing process to another (for example, a turning operation on a lathe followed by cylindrical grinding) unless the part is made correctly and meets the standards of the first operation. Automated inspection is flexible and responsive to product design changes. Furthermore, because of automated equipment, less operator skill is required, productivity is increased, and parts have higher quality, reliability, and dimensional accuracy.

Sensors for automated inspection. Recent advances in sensor technology are making on-line or real-time monitoring of manufacturing processes feasible. Directly or indirectly, and with the use of various probes, sensors can detect dimensional accuracy, surface roughness, temperature, force, power, vibration, tool wear, and the presence of external or internal defects. Sensors operate on the principles of strain gages, inductance, capacitance, ultrasonics, acoustics, pneumatics, infrared radiation, optics, lasers, and various electronic gages. Sensors may be tactile (touching) or nontactile.

Sensors, in turn, are linked to microprocessors and computers for graphic data display. This capability allows rapid on-line adjustment of any processing parameters in order to produce parts that are consistently within specified standards of tolerance and quality. Such systems have already been implemented as standard equipment on many metal-cutting machine tools and grinding machines.

4.9 Quality Assurance

Quality assurance is the total effort by a manufacturer to ensure that its products conform to a detailed set of specifications and standards. These standards cover several parameters, such as dimensions, surface finish, tolerances, composition, color, and mechanical, physical, and chemical properties. In addition, standards are usually written to ensure proper assembly using interchangeable, defect-free components and a product that performs as intended by its designers.

Quality assurance is the responsibility of everyone involved with design and manufacturing. The often-repeated statement that quality must be built into a product reflects this important concept. Quality cannot be inspected into a finished product. Although product quality has always been important, increased domestic and international competition has caused quality assurance to become even more important. Every aspect of design and manufacturing operation, such as material selection, production, and assembly, is now being analyzed in detail to ensure that quality is truly built into the final product.

If you were in charge of product quality for a manufacturing plant, how would you make sure that the final product is of acceptable quality? The best method is to control materials and processes in such a manner that the products are made correctly in the first place. However, 100 percent inspection is usually too costly to maintain. Therefore several methods of inspecting smaller, statistically relevant sample lots have been devised. These methods all use statistics to determine the probability of defects occurring in the total production batch.

Inspection involves a series of steps:

1. Inspecting incoming materials to make sure that they meet certain property, dimension, and surface finish and integrity requirements.
2. Inspecting individual product components to make sure that they meet specifications.
3. Inspecting the product to make sure that individual parts have been assembled properly.
4. Testing the product to make sure that it functions as designed and intended.

Why must inspections be continued once a product of acceptable quality has been produced? Because there will always be variations in the dimensions and properties of incoming materials, variations in the performance of tools, dies, and machines used in various stages of manufacturing, possibilities of human error, and errors made during assembly of the product. As a result, no two products are ever made exactly alike.

Another important aspect of quality control is the capability to analyze defects and promptly eliminate them or reduce them to acceptable levels. In an even broader sense, quality control involves evaluating the product design and customer satisfaction. The sum total of all these activities is referred to as *total quality control*.

Thus in order to control quality we have to be able to (1) measure quantitatively the level of quality, and (2) identify all the material and process variables that can be controlled. The level of quality obtained during production can then be established by inspecting the product to determine whether it meets the specifications for tolerances, surface finish, defects, and other characteristics. The identification of material and process variables and their effect on product quality is now possible through the extensive knowledge gained from research and development activities in all aspects of manufacturing.

4.9.1 Statistical methods of quality control

Statistics deals with the collection, analysis, interpretation, and presentation of large amounts of numerical data. The use of statistical techniques in modern manufacturing operations is necessary because of the large number of material and process variables involved. Those events that occur randomly, that is, without any particular trend or pattern, are called *chance variations*. Those that can be traced to specific causes are called *assignable variations*. The existence of *variability* in production operations has been recognized for centuries, but Eli Whitney (1765–1825) first grasped its full significance when he found that interchangeable parts were indispensible to the mass production of firearms. Modern statistical concepts relevant to manufacturing engineering were first developed in the early 1900s.

To understand *statistical quality control* (SQC), we first need to define some of the terms that are commonly used in this field.

- *Sample size*: The number of parts to be inspected in a sample, whose properties are studied to gain information about the whole population.
- *Random sampling*: Taking a sample from a population or lot in which each item has an equal chance of being included in the sample. Thus when taking samples from a large bin, the inspector does not take only those that happen to be within reach.
- *Population*: The totality of individual parts of the same design from which samples are taken (also called the *universe*).
- *Lot size*: A subset of population. A lot or several lots can be considered subsets of the population and may be treated as representative of the population.

The sample is inspected for certain characteristics and features, such as tolerances, surface finish, and defects, with the instruments and techniques that we described in this chapter. These characteristics fall into two categories: those that can be measured quantitatively (method of variables) and those that are qualitative (method of attributes).

The *method of variables* is the quantitative measurement of characteristics such as dimensions, tolerances, surface finish, or physical or mechanical properties. Such measurements are made for each of the units in the group under consideration, and the results are compared against specifications.

The *method of attributes* involves observing the presence or absence of qualitative characteristics, such as external or internal defects in machined, formed, or welded parts or dents in sheet-metal products, for each of the units in the group under consideration. Sample size for attributes-type data is generally larger than for variables-type data.

During the inspection process, measurement results will vary. For example, assume that you are measuring the diameter of turned shafts as they are produced on a lathe, using a micrometer. You soon note that their diameters vary, even though ideally you want all the shafts to be exactly the same size. Let's now turn to consideration of statistical quality-control techniques, which allow us to evaluate these variations and set limits for the acceptance of parts.

If we list the measured diameters of the turned shafts in a given population, we note that one or more parts have the smallest diameter, and one or more have the largest diameter. The majority of the turned shafts have diameters that lie between these extremes. If we group these diameters and plot them, the plot consists of a bar graph representing the number of parts in each diameter group (Fig. 4.23). The bars show a *distribution*, also called a *spread* or *dispersion*, of the shaft-diameter measurements. The bell-shaped curve in Fig. 4.23 is called *frequency distribution*, or the frequency with which parts within each diameter group are being produced.

Data from manufacturing processes often fit curves represented by a mathematically derived *normal distribution curve* (Fig. 4.24). These curves are also called *Gaussian*, developed on the basis of probability. The bell-shaped normal distribution curve fitted to the data in Fig. 4.23 has two important features.

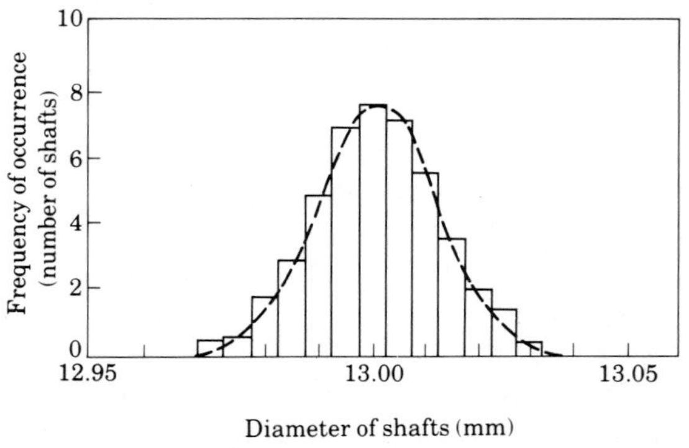

FIGURE 4.23
A plot of the number of shafts measured and their respective diameters. This type of curve is called a frequency distribution.

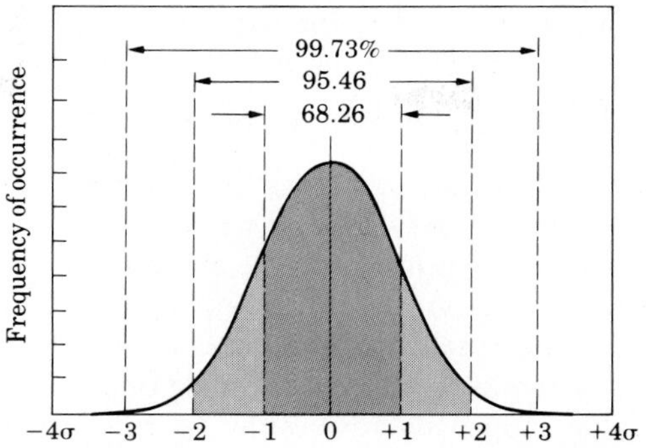

FIGURE 4.24
A normal distribution curve indicating areas within each range of standard deviation. Note: The greater the range, the higher the percentage of parts that fall within it.

First, it shows that most part diameters tend to cluster around an *average* value (*arithmetic mean*). This average is usually designated as $\bar{x}$ and is calculated from the expression

$$\bar{x} = \frac{x_1 + x_2 + x_3 + \cdots + x_n}{n}, \tag{4.11}$$

where the numerator is the sum of all measured values (diameters), and n is the number of measurements (number of shafts).

The second feature of this curve is its width, indicating the *dispersion* of the diameters measured. The wider the curve, the greater the dispersion. The difference between the largest value and smallest value is called the *range*, R:

$$R = x_{max} - x_{min}. \tag{4.12}$$

Dispersion is estimated by the *standard deviation*, which is generally denoted as σ and is obtained from the expression

$$\sigma = \sqrt{\frac{(x_1 - \bar{x})^2 + (x_2 - \bar{x})^2 + (x_3 - \bar{x})^2 + \cdots + (x_n - \bar{x})^2}{n - 1}}, \tag{4.13}$$

where x is the measured value for each part. Note from the numerator in Eq. (4.13) that as the curve widens, the standard deviation becomes greater. Also note that σ has the unit of linear dimension. In comparing Eqs. (4.12) and (4.13), we note that the range R is a simpler and more convenient measure of dispersion.

Since we know the number of turned parts that fall within each group, we can calculate the percentage of the total population represented by each group. Thus Fig. 4.24 shows that the diameters of 99.73 percent of the turned shafts fall within the range of $\pm 3\sigma$, 95.46 percent within $\pm 2\sigma$, and 68.26 percent within $\pm 1\sigma$. Thus only 0.27 percent fall outside the $\pm 3\sigma$ range.

4.9.2 Statistical process control

If the number of parts that do not meet set standards (defective parts) increases during a production run, we must be able to determine the cause (incoming materials, machine controls, degradation of metalworking fluids, operator boredom, or others) and take appropriate action. Although this statement at first appears to be self-evident, it was only in the early 1950s that a systematic statistical approach was developed to guide operators in manufacturing plants.

This approach advises the operator to take certain measures and when to take them in order to avoid producing further defective parts. Known as *statistical process control* (SPC), this technique consists of several elements: control charts and setting control limits, capabilities of the particular manufacturing process, and characteristics of the machinery involved.

Control charts. The frequency distribution curve in Fig. 4.23 shows a range of shaft diameters being produced that may fall beyond the predetermined design tolerance range. Figure 4.25 shows the same bell-shaped curve, which now includes the specified tolerances for the diameter of turned shafts.

Control charts graphically represent the variations of a process over a period of time. They consist of data plotted during production and, typically, there are two plots. The quantity $\bar{x}$ (Fig. 4.26a) is the average for each subset of samples taken and inspected, say, a subset consisting of 5 parts. A sample size of between 2 and 10 parts is sufficiently accurate, provided that sample size is held constant throughout the inspection.

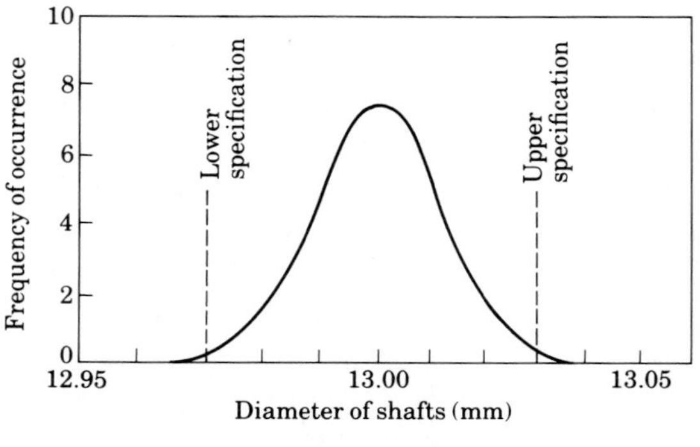

FIGURE 4.25
Frequency distribution curve, showing lower and upper specification limits.

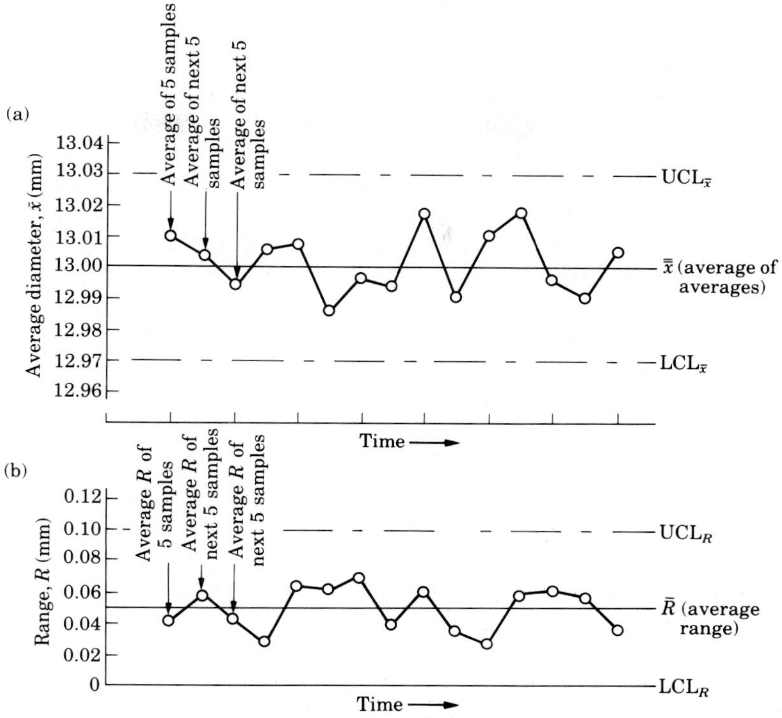

FIGURE 4.26

Control charts used in statistical quality control. The process shown is in statistical control because all points fall within the lower and upper control limits. In this illustration sample size is 5 and the number of samples is 15.

The frequency of sampling depends on the nature of the process. Some processes may require continuous sampling, whereas others may require only one sample per day. Quality-control analysts are best qualified to determine this frequency for a particular situation. Since the measurements in Fig. 4.26(a) are made consecutively, the abscissa of these control charts also represents time. The solid horizontal line in this figure is the *average of averages* (*grand average*), denoted as $\bar{\bar{x}}$, and represents the population mean. The upper and lower horizontal broken lines in these control charts indicate the control limits for the process.

The *control limits* are set on these charts according to statistical-control formulas designed to keep actual production within the usually acceptable $\pm 3\sigma$ range. Thus for $\bar{x}$

$$\text{Upper control limit } (\text{UCL}_{\bar{x}}) = \bar{x} + 3\sigma = \bar{\bar{x}} + A_2\bar{R} \qquad \textbf{(4.14)}$$

and

$$\text{Lower control limit } (\text{LCL}_{\bar{x}}) = \bar{x} - 3\sigma = \bar{\bar{x}} - A_2\bar{R}, \tag{4.15}$$

where A_2 is read from Table 4.4 and $\bar{R}$ is the average of R values.

These limits are calculated based on the historical production capability of the equipment itself. They are not generally associated with either design tolerance specifications or dimensions. They indicate the limits within which a certain percentage of measured values are normally expected to fall because of the inherent variations of the process itself, upon which the limits are based. The major goal of statistical process control is to improve the manufacturing process with the aid of control charts to eliminate assignable causes. The control chart continually indicates progress in this area.

The second control chart (Fig. 4.26b) shows the range R in each subset of samples. The solid horizontal line represents the average of R values in the lot, denoted as $\bar{R}$, and is a measure of the variability of the samples. The upper and lower control limits for R are obtained from the equations

$$\text{UCL}_R = D_4\bar{R} \tag{4.16}$$

and

$$\text{LCL}_R = D_3\bar{R}, \tag{4.17}$$

where the constants D_4 and D_3 are obtained from Table 4.4. This table also includes the constant d_2, which is used in estimating the standard deviation from the equation

$$\sigma = \frac{\bar{R}}{d_2}. \tag{4.18}$$

When the curve of a control chart is like that shown in Fig. 4.26(b), we say that the process is "in good statistical control." In other words, there is no clear

TABLE 4.4
CONSTANTS FOR CONTROL CHARTS

SAMPLE SIZE	A_2	D_4	D_3	d_2
2	1.880	3.267	0	1.128
3	1.023	2.575	0	1.693
4	0.729	2.282	0	2.059
5	0.577	2.115	0	2.326
6	0.483	2.004	0	2.534
7	0.419	1.924	0.078	2.704
8	0.373	1.864	0.136	2.847
9	0.337	1.816	0.184	2.970
10	0.308	1.777	0.223	3.078
12	0.266	1.716	0.284	3.258
15	0.223	1.652	0.348	3.472
20	0.180	1.586	0.414	3.735

discernible trend in the pattern of the curve, the points (measured values) are random with time, and they do not exceed the control limits. However, you can see that curves such as those shown in Fig. 4.27(a), (b), and (c) indicate certain trends. For example, note in the middle of curve (a) that the diameter of the shafts increases with time. The reason for this increase may be a change in one of the process variables, such as wear of the cutting tool. If, as in curve (b), the trend is toward consistently larger diameters, hovering around the upper control limit, it could mean that the tool settings on the lathe may not be correct and, as a result, the parts being turned are consistently too large. Curve (c) shows two distinct trends that may

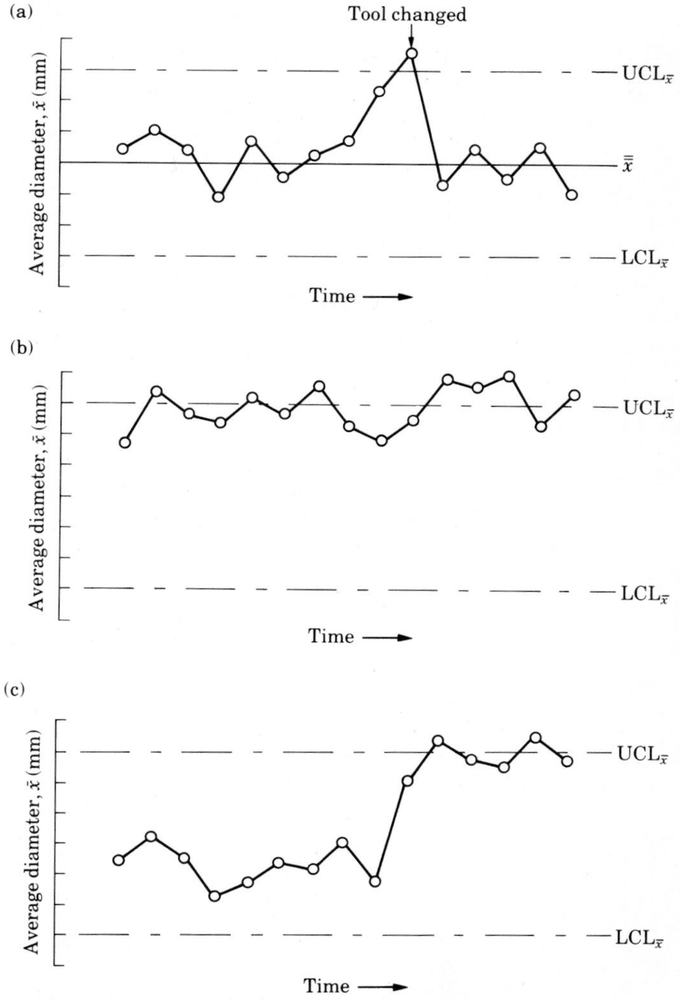

FIGURE 4.27 Control charts. (a) Process begins to become out of control because of factors such as tool wear. The tool is changed and the process is then in statistical control. (b) Process parameters are not set properly; thus all parts are around the upper control limit. (c) Process becomes out of control because of factors such as a sudden change in the properties of the incoming material.

be caused by factors such as a change in the properties of the incoming material or a change in the performance of the cutting fluid (for example, its degradation). These situations place the process "out of control."

Analyzing control-chart patterns and trends requires considerable experience in order to identify the specific cause(s) of an out-of-control situation. Overcontrol of the manufacturing process—that is, setting upper and lower control limits too close to each other (smaller standard deviation range)—is a further reason for out-of-control situations. This is the reason for calculating control limits on process capability rather than on a potentially inapplicable statistic.

Process capability. *Process capability* is defined as the limits within which individual measurement values resulting from a particular manufacturing process would normally be expected to fall for random variations only. Thus process capability tells us that the process can produce parts within certain limits of precision. Since a manufacturing process involves materials, machinery, and operators, each can be analyzed individually to identify a problem when process capabilities do not meet part specifications.

• Example 4.3: Calculation of control limits and standard deviation.

The data in the accompanying table show length measurements (in.) taken on a machined workpiece. Sample size is 5 and the sample number is 10; thus the total number of parts measured is 50. The quantity $\bar{x}$ is the average of five measurements in each sample. We first calculate the average of averages $\bar{\bar{x}}$,

$$\bar{\bar{x}} = \frac{44.296}{10} = 4.430 \text{ in.},$$

and the average of R values,

$$\bar{R} = \frac{1.03}{10} = 0.103 \text{ in.}$$

Since the sample size is 5, we determine from Table 4.4 the following constants: $A_2 = 0.577$, $D_4 = 2.115$, and $D_3 = 0$. We can now calculate the control limits using Eqs. (4.14)–(4.17). Thus, for averages we have

$$\text{UCL}_{\bar{x}} = 4.430 + (0.577)(0.103) = 4.489 \text{ in.},$$

$$\text{LCL}_{\bar{x}} = 4.430 - (0.577)(0.103) = 4.371 \text{ in.},$$

and for ranges we have

$$\text{UCL}_R = (2.115)(0.103) = 0.218 \text{ in.},$$

$$\text{LCL}_R = (0)(0.103) = 0 \text{ in.}$$

We can calculate the standard deviation using Eq. (4.18) and a value of $d_2 = 2.326$. Thus,

$$\sigma = \frac{0.103}{2.326} = 0.044 \text{ in.}$$

SAMPLE NUMBER	x_1	x_2	x_3	x_4	x_5	$\bar{x}$	R
1	4.46	4.40	4.44	4.46	4.43	4.438	0.06
2	4.45	4.43	4.47	4.39	4.40	4.428	0.08
3	4.38	4.48	4.42	4.42	4.35	4.410	0.13
4	4.42	4.44	4.53	4.49	4.35	4.446	0.18
5	4.42	4.45	4.43	4.44	4.41	4.430	0.04
6	4.44	4.45	4.44	4.39	4.40	4.424	0.06
7	4.39	4.41	4.42	4.46	4.47	4.430	0.08
8	4.45	4.41	4.43	4.41	4.50	4.440	0.09
9	4.44	4.46	4.30	4.38	4.49	4.414	0.19
10	4.42	4.43	4.37	4.47	4.49	4.436	0.12

4.9.3 Acceptance sampling and control

Acceptance sampling consists of taking only a few random samples from a lot and inspecting them for the purpose of judging whether the entire lot is acceptable or should be rejected or reworked. Developed in the 1920s and used extensively during World War II for military hardware (MIL STD 105), this statistical technique is widely used and valuable. It is particularly useful for inspecting high production rate parts, where 100 percent inspection would be too costly.

A variety of acceptance sampling plans have been prepared for both military and national standards, based on an acceptable, predetermined, and limiting percentage of nonconforming parts in the sample. If this percentage is exceeded, the entire lot is rejected, or it is reworked if economically feasible. Note that the actual number of samples (not percentages of the lot that are in the sample) can be significant in acceptance sampling. The greater the number of samples taken from a lot, the greater will be the percentage of nonconforming parts and the lower the probability of acceptance of the lot. *Probability* is defined as the relative occurrence of an event. The probability of acceptance is obtained from various operating characteristic curves, one example of which is shown in Fig. 4.28.

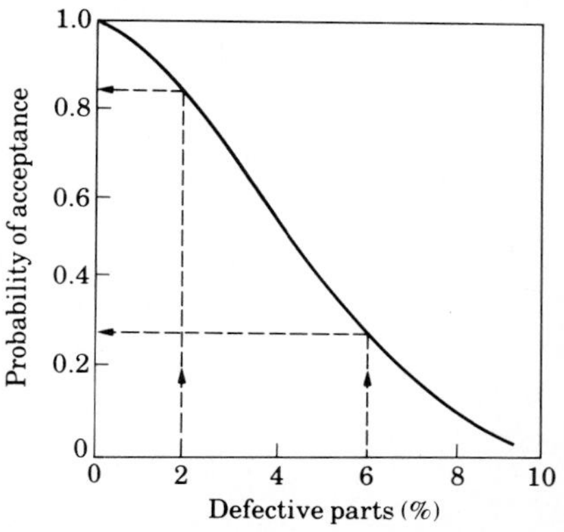

FIGURE 4.28
A typical operating-characteristic curve used in acceptance sampling. The higher the percentage of defective parts, the lower is the probability of their being accepted by the consumer. There are several methods of obtaining these curves.

The *acceptance quality level* (AQL) is commonly defined as the level at which there is a 95 percent probability of acceptance of the lot. This percentage would indicate to the manufacturer that 5 percent of the parts in the lot may be rejected by the consumer (called *the producer's risk*). Likewise, the consumer knows that 95 percent of the parts are acceptable (called *the consumer's risk*).

Lots that do not meet the desired quality standards can be salvaged by the manufacturer by a secondary rectifying inspection. In this method, a 100 percent inspection is made of a rejected lot, and the defective parts are removed. This time-consuming and costly process is an incentive for the manufacturer to control the production process.

Acceptance sampling requires less time and fewer inspections than do other sampling methods. Consequently, inspection of the parts can be more detailed. Keep in mind, however, that automated inspection techniques are being developed rapidly so that 100 percent inspection of all parts is indeed feasible and can be economical.

4.9.4 Total quality control and the quality circle

The *total quality control* (TQC) concept is a management system emphasizing the fact that quality must be designed and built into a product. Defect *prevention*, rather than defect detection, is the major goal. Total quality control is a systems approach in that both management and workers make an integrated effort to manufacture high-quality products consistently. All tasks concerning quality improvements and responsibilities in the organization are clearly identified. The TQC concept also requires 100 percent inspection of parts, usually by automated inspection systems. No defective parts are allowed to continue through the production line.

A related concept that has gained wide acceptance is *the quality circle*. This activity consists of regular meetings by groups of workers who discuss how to improve and maintain product quality at all stages of the manufacturing process. Worker involvement and responsibility are emphasized. Comprehensive training is provided so that the worker can become capable of analyzing statistical data, identifying causes of poor quality, and taking immediate action to correct the situation. Putting this concept into practice recognizes the importance of quality assurance as a major company-wide management policy, affecting all personnel and all aspects of production.

Quality engineering as a philosophy. Many of the quality-control concepts and methods that we have described have been put into larger perspective by certain experts in quality control. Notable among these experts are W. E. Deming and G. Taguchi, whose philosophies of quality and product cost have had a major impact on modern manufacturing.

During World War II, *Deming* and several others developed new methods of statistical process control in manufacturing plants for wartime industry. The need for statistical control arose from the recognition that there were variations in the performance of machines and people and the quality and dimensions of raw materials. Their efforts, however, involved not only statistical methods of analysis, but a new way of looking at manufacturing operations to improve quality and lower costs. Recognizing the fact that manufacturing organizations are systems of management, workers, machines, and products, Deming emphasizes communication, direct worker involvement, and education in statistics, as well as in modern manufacturing technology. His ideas have been widely accepted in Japan but only in some segments of the U.S. industry.

In the *Taguchi* methods, high quality and low costs are achieved by combining engineering and statistical methods to optimize product design and manufacturing processes. Loss of quality is defined as the financial loss to society after the product is shipped, with the following results: (a) poor quality leads to customer dissatisfaction; (b) costs are involved in servicing and repairing defective products, some in the field; (c) the manufacturer's credibility is diminished in the marketplace; and (d) the manufacturer eventually loses its share of the market.

The Taguchi methods of quality engineering emphasize the importance of:

- Enhancing cross-functional team interaction. In this interaction, design engineers and process or manufacturing engineers communicate with each other in a common language. They quantify the relationships between design requirements and the manufacturing process.
- Implementing experimental design, in which the factors involved in a process or operation and their interactions are studied simultaneously.

In *experimental design*, the effects of controllable and uncontrollable variables on the product are identified. This approach minimizes variations in product dimensions and properties, bringing the mean to the desired level. The methods used for experimental design are complex, involving the use of fractional factorial design and orthogonal arrays, which reduce the number of experiments required. These methods are also capable of identifying the effect on the product of variables that cannot be controlled (called *noise*), such as changes in environmental conditions.

The use of these methods allows rapid identification of the controlling variables and determination of the best method of process control. These variables are then controlled, without the need for costly new equipment or major modifications to existing equipment. For example, variables affecting tolerances in machining a particular component can be readily identified, and the correct cutting speed, feed, cutting tool, and cutting fluids can be specified.

SUMMARY

In manufacturing processes, surfaces and their properties are as important as the bulk properties of the materials. Surfaces involve not only a particular geometry and appearance, but also are composed of a layer with properties that are generally different from those of the bulk material. Measurements and descriptions of surfaces and their roughness are a complex problem. Certain standards have been developed and are specified along with other design requirements in manufacturing.

Friction and wear are among the most significant factors in the processing of materials. In spite of the difficulties involved, considerable progress has been made in understanding these phenomena and in identifying the factors that govern them. Among these are the affinity and reactivity of the two materials in contact, the nature of surface films, the presence of contaminants, and process parameters such as load, speed, and temperature.

Surface treatments are used to impart certain physical and mechanical properties, such as resistance to environmental attack and wear and fatigue resistance. Various techniques employed are heat treatment, coatings, surface treatments, and mechanical working of surfaces.

In modern manufacturing technology, many parts are made with a high degree of precision, thus requiring measuring instrumentation with several features and characteristics. The selection of a particular measuring instrument depends on factors such as the type of measurement, the environment, and the accuracy of measurement. Great advances have been made in automated measurement, linking them to microprocessors and computers for accurate in-process control of manufacturing operations.

Dimensional tolerances and their selection are important factors in manufacturing. Tolerances not only affect the accuracy and operation of all types of machinery and equipment, but can also significantly influence product cost. The smaller the range of tolerances specified, the greater becomes the cost of achieving it.

Several nondestructive and destructive testing techniques, each having its own applications, advantages, and limitations, are available for inspection of completed parts and products. The traditional approach has been to inspect the part or product after it is manufactured and to accept a certain number of defective parts. The trend now is toward on-line, 100 percent inspection of all parts and products being manufactured.

Quality must be built into products. Quality assurance is concerned with various aspects of production, such as design, manufacturing, and assembly, and inspection at each step of production for conformance to specifications. Statistical quality control and process control have become indispensable to modern manufacturing. They are particularly important for interchangeable parts and in reducing manufacturing costs.

BIBLIOGRAPHY

Surfaces, Friction, and Wear

Bowden, F.P., and D. Tabor, *The Friction and Lubrication of Solids*. New York: Oxford, Vol. I, 1950, Vol. II, 1964.

Czichos, H., *Tribology—A Systems Approach to the Science and Technology of Friction, Lubrication and Wear*. New York: Elsevier, 1978.

Dowson, D., *The History of Tribology*. New York: Longman, 1979.

Engel, P.A., *Impact Wear of Materials*. New York: Elsevier, 1976.

Friction and Wear Devices, 2d ed. Park Ridge, Ill.: American Society of Lubrication Engineers, 1976.

Glaeser, W.A., et al. (eds.), *Wear of Materials*. New York: American Society of Mechanical Engineers, biannual, since 1977.

Halling, J., *Principles of Tribology*. New York: Macmillan, 1975.

Kragelskii, I.V., *Friction and Wear*. Washington, DC: Butterworth, 1965.

Lansdown, A.R., and A.L. Price, *Materials to Resist Wear: A Guide to Their Selection and Use*. Oxford: Pergamon Press, 1986.

Lipson, C., *Wear Considerations in Design*. Englewood Cliffs, N.J.: Prentice-Hall, 1967.

Moore, D.F., *Principles and Applications of Tribology*. New York: Pergamon, 1975.

Moore, D.F., *The Friction and Lubrication of Elastomers*. New York: Pergamon, 1972.

Neale, M.J. (ed.), *Tribology Handbook*. New York: Butterworth, 1973.

Peterson, M.B., and W.O. Winer (eds.), *Wear Control Handbook*. New York: American Society of Mechanical Engineers, 1980.

Rabinowicz, E., *Friction and Wear of Materials*. New York: Wiley, 1965.

Rigney, D.V. (ed.), *Fundamentals of Friction and Wear of Materials*. Metals Park, Ohio: American Society for Metals, 1981.

Sarkar, A.D., *Wear of Metals*. New York: Pergamon, 1976.

Scott, D. (ed.), *Wear*. Volume 13 in *Treatise on Materials Science and Technology*. New York: Academic Press, 1979.

Source Book on Wear Control Technology. Metals Park, Ohio: American Society for Metals, 1978.

Tribology in Metalworking

Billet, M., *Industrial Lubrication*. New York: Pergamon, 1979.

Braithwaite, E.R. (ed.), *Lubrication and Lubricants*. New York: Elsevier, 1967.

Braithwaite, E.R., *Solid Lubricants and Surfaces*. New York: Macmillan, 1964.

Clauss, F.J., *Solid Lubricants and Lubricating Solids*. New York: Academic Press, 1971.

Cutting and Grinding Fluids. Dearborn, Mich.: Society of Manufacturing Engineers, 1967.

Friction and Lubrication in Metal Processing. New York: American Society of Mechanical Engineers, 1966.

Kalpakjian, S., and S. Jain (eds.), *Metalworking Lubrication*. New York: American Society of Mechanical Engineers, 1980.

Nachtman, E.S., and S. Kalpakjian, *Lubricants and Lubrication in Metalworking Operations*. New York: Marcel Dekker, 1985.

Olds, N.J., *Lubricants, Cutting Fluids and Coolants*. Boston: Cahners, 1973.

Schey, J.A. (ed.), *Metal Deformation Processes: Friction and Lubrication*. New York: Marcel Dekker, 1970.

Schey, J.A., *Tribology in Metalworking—Friction, Lubricating and Wear*. Metals Park, Ohio: American Society for Metals, 1983.

Surface Treatments

Auciello, O., and R. Kelly (eds.), *Ion Bombardment Modification of Surfaces*. Amsterdam: Elsevier, 1984.

Banov, A. (ed.), *Paints and Coatings Handbook.* New York: McGraw-Hill, 1982.

Budinski, K.G., *Surface Engineering for Wear Resistance.* Englewood Cliffs, N.J.: Prentice-Hall, 1988.

Bunshah, R.F., et al., *Deposition Technologies for Films and Coatings.* Park Ridge, N.J.: Noyes Publications, 1982.

Gabe, D.R., *Principles of Metal Surface Treatment and Protection.* New York: Pergamon, 1972.

Metals Handbook, 9th ed., *Vol. 5: Surface Cleaning, Finishing, and Coating.* Metals Park, Ohio: American Society for Metals, 1982.

Peterson, M.B., and W.O. Winer (eds.), *Wear Control Handbook.* New York: ASME, 1980.

Rudzki, G.J., *Surface Finishing Systems: Metal and Non-Metal Finishing Handbook-Guide.* Middlesex, UK: Finishing Publications Ltd., 1983.

Schey, J.A., *Tribology in Metalworking—Friction, Lubrication and Wear.* Metals Park, Ohio: American Society for Metals, 1983.

Source Book on Wear Control Technology. Metals Park, Ohio: American Society for Metals, 1978.

Strafford, K.N., P.K. Datta, and C.G. Googan (eds.), *Coatings and Surface Treatments for Corrosive and Wear Resistance.* New York: Wiley, 1984.

Stuart, R.V., *Vacuum Technology, Thin Films, and Sputtering.* New York: Academic Press, 1983.

Tool and Manufacturing Engineers Handbook, 4th ed., *Vol. 3: Materials, Finishing and Coating.* Dearborn, Mich.: Society of Manufacturing Engineers, 1985.

Metrology

Bentley, J.P., *Principles of Measurement Systems,* 2d ed. New York: Wiley, 1988.

Farago, F.T., *Handbook of Dimensional Measurement,* 2d ed. New York: Industrial Press, 1982.

Kennedy, C.W., and E.G. Hoffman, *Inspection and Gaging,* 6th ed. New York: Industrial Press, 1987.

Lange, J.C., *Design Dimensioning with Computer Graphics Applications.* New York: Marcel Dekker, 1984.

Lenk, J.D., *Handbook of Controls and Instrumentation.* Englewood Cliffs, N.J.: Prentice-Hall, 1980.

Lowell, W.F., *Modern Geometric Dimensioning and Tolerancing,* 2d ed. Fort Washington, Md.: National Tooling and Machining Association, 1982.

Machinery's Handbook. New York: Industrial Press (revised periodically).

Metals Handbook, 8th ed., *Vol. 11: Nondestructive Testing and Quality Control.* Metals Park, Ohio: American Society for Metals, 1976.

Spotts, M.F., *Dimensioning and Tolerancing for Quantity Production.* Englewood Cliffs, N.J.: Prentice-Hall, 1983.

Tool and Manufacturing Engineers Handbook, 4th ed., *Vol. 4: Quality Control and Assembly.* Dearborn, Mich.: Society of Manufacturing Engineers, 1987.

Warnecke, H.J., and W. Dutschke (eds.), *Metrology in Manufacturing Technology.* Berlin: Springer, 1984.

Testing and Inspection

Buck, O., and S.M. Wolf, *Nondestructive Evaluation: Application to Materials Processing.* Metals Park, Ohio: American Society for Metals, 1984.

Metals Handbook, 9th ed., *Vol. 17: Nondestructive Evaluation and Quality Control*. Metals Park, Ohio: ASM International, 1989.

Nondestructive Testing Handbook, Vol. 1: Leak Testing. R.C. McMaster (ed.), 1982; *Vol. 2: Liquid Penetrant Tests*. R.C. McMaster (ed.), 1982; *Vol. 3: Radiography and Radiation Testing*. L.E. Bryant (ed.), 1985. Metals Park, Ohio: American Society for Metals.

Robinson, S.L., and R.K. Miller, *Automated Inspection and Quality Assurance*. New York: Marcel Dekker, 1989.

Tool and Manufacturing Engineering Handbook, Vol. 4: Assembly, Testing, and Quality Control. Dearborn, Mich.: Society of Manufacturing Engineers, 1986.

Quality Assurance

Aft, L.S., *Fundamentals of Industrial Quality Control*. Reading, Mass.: Addison-Wesley, 1986.

Besterfield, D.H., *Quality Control*, 2d ed. Englewood Cliffs, N.J.: Prentice-Hall, 1986.

Deming, W.E., *Out of the Crisis*. Cambridge, Mass.: MIT Press, 1986.

Dhillon, B.S., *Quality Control, Reliability, and Engineering Design*. New York: Marcel Dekker, 1985.

Enrick, N.L., *Quality, Reliability, and Process Improvement*, 8th ed. New York: Industrial Press, 1985.

Feigenbaum, A.V., *Total Quality Control*, 3d ed. New York: McGraw-Hill, 1983.

Grant, E.I., and R.S. Leavenworth, *Statistical Quality Control*, 6th ed. New York: McGraw-Hill, 1987.

Halpern, S., *The Assurance Sciences: An Introduction to Quality Control and Reliability*. Englewood Cliffs, N.J.: Prentice-Hall, 1978.

Hansen, B.L., and P.M. Ghare, *Quality Control and Application*. Englewood Cliffs, N.J.: Prentice-Hall, 1987.

Ishikawa, K., *Guide to Quality Control*, rev. ed. New York: UNIPUB, 1984.

Juran, J., *Quality Control Handbook*. New York: McGraw-Hill, 1979.

Juran, J., and F.M. Gryna, Jr., *Quality Planning and Analysis*, 2d ed. New York: McGraw-Hill, 1980.

Lester, R.H., N.L. Enrick, and H.E. Mottley, Jr., *Quality Control for Profit*, 2d ed. New York: Marcel Dekker, 1985.

Lewis, E.E., *Introduction to Reliability Engineering*. New York: Wiley, 1987.

Metals Handbook, 8th ed., *Vol. 11: Nondestructive Inspection and Quality Control*. Metals Park, Ohio: American Society for Metals, 1976.

Montgomery, D.C., *Introduction to Statistical Quality Control*. New York: Wiley, 1985.

Nondestructive Testing Handbook, Vol. 1: Leak Testing. R.C. McMaster (ed.), 1982; *Vol. 2: Liquid Penetrant Tests*. R.C. McMaster (ed.), 1982; *Vol. 3: Radiography and Radiation Testing*. L.E. Bryant (ed.), 1985. Metals Park, Ohio: American Society for Metals.

Oakland, J.S., *Statistical Process Control: A Practical Guide*. New York: Wiley, 1986.

Simmons, D.A., *Practical Quality Control*. Reading, Mass.: Addison-Wesley, 1979.

Taguchi, G., *Introduction to Quality Engineering*. Lanham, Maryland: UNIPUB/Kraus International, 1986.

QUESTIONS

4.1 Explain what is meant by surface integrity. Why is it of interest?

4.2 Why are surface roughness design requirements in engineering so broad?

4.3 A surface has various layers. Describe the factors that influence the thickness of these layers.

4.4 What factors would you consider in specifying the lay of a surface?

4.5 Explain why the same surface roughness values do not necessarily represent the same type of surface.

4.6 In using a surface roughness measuring instrument, how would you go about determining the cutoff value?

4.7 What is the significance of the fact that the stylus path and the actual surface profile are not generally the same?

4.8 Explain the significance of surface temperature rise caused by friction.

4.9 To what factors would you attribute the fact that the coefficient of friction in hot working is higher than in cold working, as shown in Table 4.1?

4.10 Describe the tribological differences between ordinary machine elements (such as gears, cams, and bearings) and metalworking processes involving tools and dies.

4.11 Explain why an originally round specimen in a ring compression test may become oval after deformation.

4.12 Describe the applicability of simulated friction and wear tests to actual manufacturing operations.

4.13 Can the temperature rise at a sliding interface exceed the melting point of the metals? Explain.

4.14 It has been stated that, as the normal load decreases, abrasive wear is reduced. Explain why this is so.

4.15 Why is the abrasive-wear resistance of a material a function of its hardness?

4.16 Explain why surface treatment of manufactured products may be necessary.

4.17 Which surface treatments are functional and which are decorative? Give examples.

4.18 Give some typical applications of mechanical surface treatment.

4.19 Explain the difference between case hardening and hard facing.

4.20 List several applications for coated sheet metal.

4.21 Explain how roller burnishing processes induce residual stresses on the surface of workpieces.

4.22 How would you go about estimating forces required for roller burnishing?

4.23 Explain the difference between direct- and indirect-reading linear measurements.

4.24 Explain what is meant by comparative length measurement.

4.25 Why have coordinate measuring machines become important instruments in modern manufacturing?

4.26 Why is the control of dimensional tolerances important?

4.27 Why do you think that the words *accuracy* and *precision* are so often incorrectly interchanged?

4.28 Explain why a measuring instrument may not have sufficient precision.

4.29 It has been stated that tolerances for nonmetallic stock are usually wider than for metals. Explain why.

4.30 Describe the basic features of nondestructive testing techniques that use electrical energy.

4.31 Identify the nondestructive techniques that are capable of detecting internal flaws and those that detect external flaws only.

4.32 Which of the nondestructive inspection techniques are suitable for nonmetallic materials? Why?

4.33 Why is automated inspection becoming an important part of manufacturing engineering?

4.34 Describe situations in which nondestructive testing techniques are unavoidable.

4.35 Should products be designed and built for a certain expected life? Explain.

4.36 What are the consequences of setting lower and upper specifications closer to the peak of the curve in Fig. 4.25?

4.37 Identify factors that can cause a process to become out of control.

PROBLEMS

4.1 Using Fig. 4.8(a), make a plot of the coefficient of friction versus change in internal diameter for a reduction in height of (a) 10 percent, (b) 30 percent, and (c) 50 percent.

4.2 Refer to Example 4.1 and assume that the coefficient of friction is 0.15. If all other parameters remain the same, what is the new internal diameter of the ring specimen?

4.3 Assume that a steel rule expands by 2 percent because of an increase in environmental temperature. What is the indicated diameter of a shaft whose actual diameter is 2.000 in.?

4.4 Assume that in Example 4.3 the number of samples was 5 instead of 10. Using the top half of the data given in the table, recalculate control limits and the standard deviation. Compare your results with the results obtained by using 10 samples.

4.5 Calculate the control limits for averages and ranges for the following: number of samples = 4, $\bar{\bar{x}} = 70$, $\bar{R} = 7$.

4.6 Calculate the control limits for the following: number of samples = 3, $\bar{\bar{x}} = 36.5$, $UCL_R = 4.75$.

4.7 In an inspection with sample size 7 and a sample number of 50, it was found that the average range was 12 and the average of averages was 75. Calculate the control limits for averages and ranges.

4.8 Determine the control limits for the data shown in the table below.

x_1	x_2	x_3	x_4
0.55	0.60	0.57	0.55
0.59	0.55	0.60	0.58
0.55	0.50	0.55	0.51
0.54	0.57	0.50	0.50
0.58	0.58	0.60	0.56
0.60	0.61	0.55	0.61

4.9 The average of averages of a number of samples of size 9 was determined to be 125. The average range was 17.82 and the standard deviation was 6. The following measurements were taken in a sample: 120, 132, 124, 130, 118, 132, 135, 121, and 127. Is the process in control?

5

Casting Processes

5.1 Introduction

We can use several methods to shape materials into useful products. Making parts by casting molten metal into a mold and letting it solidify is an obvious choice. Indeed, casting is among the oldest methods of manufacturing and was first used in about 4000 B.C. to make ornaments, copper arrowheads, and various other objects.

Basically, metal-casting processes involve the introduction of molten metal into a mold cavity where, upon solidification, the metal conforms to the shape of the cavity. The casting process is thus capable of producing intricate shapes in a single piece, including those with internal cavities. Very large, very small, and hollow parts can be produced economically by casting techniques. Typical cast products are engine blocks, crankshafts, pistons, valves, railroad wheels, and ornamental artifacts.

Almost all metals can be cast in (or nearly in) the final shape desired, often with only minor finishing required. With appropriate control of material and process parameters, parts can be cast with uniform properties throughout. As with all other manufacturing processes, certain fundamental relationships are essential to the production of good quality and economical castings. Knowledge of these relationships helps us establish proper techniques for mold design and casting practice. The objective is to produce castings that are free from defects and meet requirements for strength, dimensional accuracy, and surface finish.

Fundamentally, the important factors in casting operations are

- solidification of the metal from its molten state;
- flow of the molten metal into the mold cavity;
- heat transfer during solidification and cooling of the metal in the mold; and
- influence of the type of mold material.

We first discuss these aspects of casting, then continue with the description of melting practice and the characteristics of casting alloys. The remainder of the chapter covers ingot and continuous casting, various expendable-mold and permanent-mold casting processes, including their advantages and limitations, processing of castings, casting design, and the economics of casting.

5.2 Solidification of Metals

In this section we present an overview of the solidification of metals and alloys. We cover topics that are particularly relevant to casting and essential for our understanding of the structures developed in casting and the structure–property relationships obtained in the casting processes described in this chapter.

Pure metals have clearly defined melting or freezing points, and solidification takes place at a constant temperature (Fig. 5.1). When the temperature of the molten metal is reduced to the freezing point, the latent heat of solidification is given off while the temperature remains constant. At the end of this thermal cycle, solidification is complete and the solid metal cools to room temperature.

Unlike pure metals, *alloys* solidify over a range of temperatures. Solidification begins when the temperature of the molten metal drops below the *liquidus*; it is completed when the temperature reaches the *solidus*. Within this temperature range the alloy is in a mushy or pasty state. Its composition and state is described by the particular alloy's phase diagram.

5.2.1 Solid solutions

Two terms are essential in describing alloys: solute and solvent. *Solute* is the minor element (such as salt or sugar) that is added to the *solvent*, which is the major

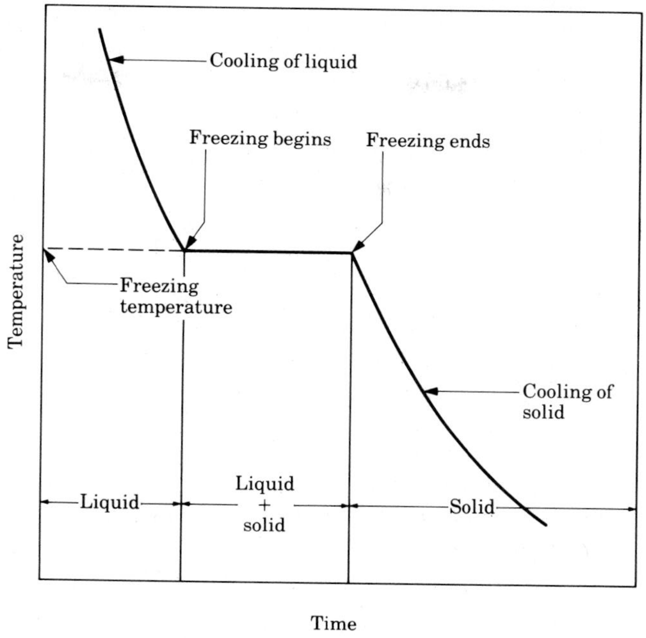

FIGURE 5.1

Cooling curve for the solidification of pure metals. Note that freezing takes place at a constant temperature.

element (such as water). In terms of the elements involved in a metal's crystal structure, the solute is the element (solute atoms) added to the solvent (host atoms). When the particular crystal structure of the solvent is maintained during alloying, the alloy is called a *solid solution.*

Substitutional solid solutions. If the size of the solute atom is similar to that of the solvent atom, the solute atoms can replace solvent atoms and form a *substitutional solid solution* (see Fig. 3.9). An example of this phenomenon is brass, an alloy of zinc and copper, in which zinc (solute atom) is introduced into the lattice of copper (solvent atoms). The properties of brasses can thus be altered over a certain range by controlling the amount of zinc in copper.

Interstitial solid solutions. If the size of the solute atom is much smaller than that of the solvent atom, the solute atom occupies an interstitial position and forms an *interstitial solid solution.* An important example of interstitial solution is steel, an alloy of iron and carbon, in which carbon atoms are present in an interstitial position between iron atoms. As we show in Section 5.10, we can vary the properties of steel over a wide range by controlling the amount of carbon in iron. This is one reason that, in addition to being inexpensive, steel is such a versatile and useful material with a wide variety of properties and applications.

5.2.2 Intermetallic compounds

Intermetallic compounds are complex structures in which solute atoms are present among solvent atoms in certain specific proportions. Thus some intermetallic compounds have solid solubility. The type of atomic bonds may range from metallic to ionic. Intermetallic compounds are strong, hard, and brittle.

5.2.3 Two-phase systems

A solid solution is one in which two or more elements are soluble in a solid state, forming a single homogeneous solid phase in which the alloying elements are uniformly distributed throughout the solid mass. However, there is a limit to the concentration of solute atoms in a solvent-atom lattice, just as there is a solubility limit to sugar in water. Most alloys consist of two or more solid phases and may be regarded as mechanical mixtures. Thus we call a system with two solid phases a *two-phase system*. Each phase is a homogeneous part of the total mass and has its own characteristics and properties.

A typical example of a two-phase system in metals is lead added to copper in the molten state. After the mixture solidifies, the structure consists of two phases: One phase has a small amount of lead in solid solution in copper; in the other phase, lead particles, roughly spherical in shape, are *dispersed* throughout the structure (Fig. 5.2a). This copper–lead alloy has properties that are different from those of either copper or lead alone. Alloying with finely dispersed particles (*second-phase particles*) is an important method of strengthening alloys and controlling their properties. In two-phase alloys the second-phase particles present obstacles to dislocation movement, which increases the alloy's strength.

Another example of a two-phase alloy is the aggregate structure shown in Fig. 5.2(b). This alloy system contains two sets of grains, each with its own composition

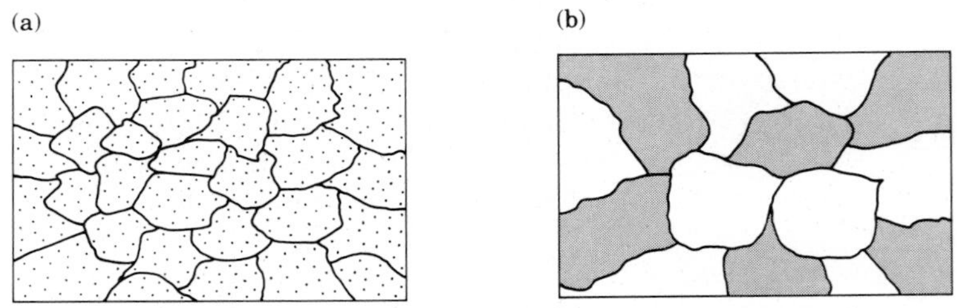

FIGURE 5.2
(a) Schematic illustration of grains, grain boundaries, and particles dispersed throughout the structure of a two-phase system, such as lead–copper alloy. The grains represent lead in solid solution of copper, and the particles are lead as a second phase. (b) Schematic illustration of a two-phase system consisting of two sets of grains: dark and light. Dark and light grains have their own compositions and properties, respectively.

and properties. The darker grains may, for example, have a different structure than the lighter grains and be brittle, whereas the lighter grains may be ductile.

5.2.4 Phase diagrams

A *phase diagram*, also called an *equilibrium diagram* or a *constitutional diagram*, shows the relationships among temperature, composition, and the phases present in a particular alloy system. *Equilibrium* means that the state of a system remains constant over an indefinite period of time. *Constitutional* indicates the relationships among structure, composition, and physical makeup of the alloy.

One example of a phase diagram is shown in Fig. 5.3 for the nickel–copper alloy. It is called a *binary phase diagram* because of the two elements (nickel and

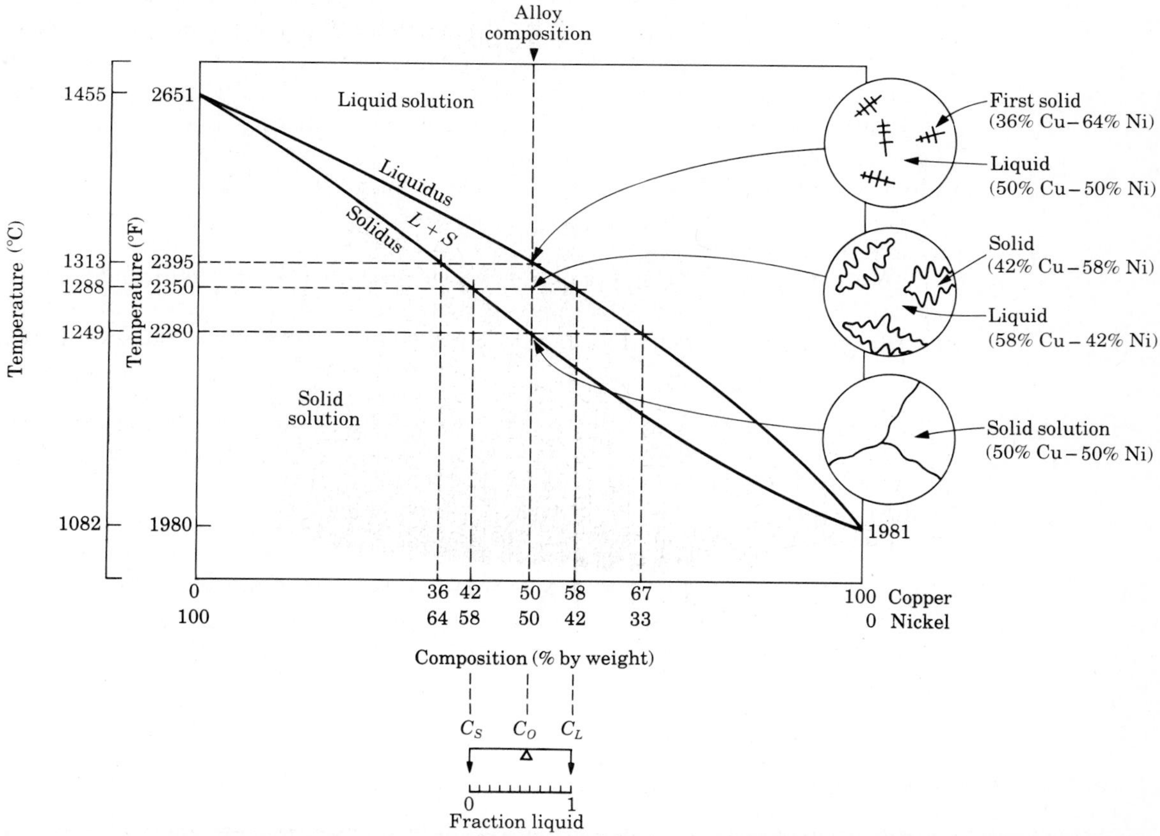

FIGURE 5.3 Phase diagram for nickel–copper alloy system obtained by slow rate of solidification. Note that pure nickel and pure copper each have one freezing or melting temperature. The top circle on the right depicts the nucleation of crystals. The second circle shows the formation of dendrites. The bottom circle shows the solidified alloy with grain boundaries. See the text for an explanation of the lever rule illustrated in the bottom portion of the figure.

copper) in the system. The left boundary of this phase diagram (100% Ni) indicates the melting point of nickel, and the right boundary (100% Cu) the melting point of copper. (All percentages in this discussion are by weight.)

Note that for a composition of, say, 50% Cu–50% Ni, the alloy begins to solidify at a temperature of 1313 °C (2395 °F), and solidification is complete at 1249 °C (2280 °F). Above 1313 °C, a homogeneous liquid of 50% Cu–50% Ni exists. When cooled slowly to 1249 °C, a homogeneous solid solution of 50% Cu–50% Ni results. However, between the liquidus and solidus curves, say, at a temperature of 1288 °C (2350 °F), is a two-phase region: a *solid phase* composed of 42% Cu–58% Ni, and a *liquid phase* of 58% Cu–42% Ni. To determine the solid composition, go left horizontally to the solidus curve and read down to obtain 42% Cu–58% Ni. Obtain the liquid composition similarly by going to the right to the liquidus curve (58% Cu–42% Ni).

The completely solidified alloy in the phase diagram shown in Fig. 5.3 is a solid solution, because the alloying element (Cu, solute atom) is completely dissolved in the host metal (Ni, solvent atom), and each grain has the same composition. We can show that copper can become the host metal and nickel the solute by rotating Fig. 5.3. The mechanical properties of solid solutions of Cu–Ni depend on their composition. By increasing the nickel content, the properties of pure copper are improved. The improvements in properties result from pinning (blocking) of dislocations at solute atoms of nickel, which may also be regarded as impurity atoms. As a result, dislocations cannot move as freely, and consequently the strength of the alloy increases.

Another example of a binary phase diagram is shown in Fig. 5.4 for the tin–lead system. The single phases alpha and beta are solid solutions. Note that the single-phase regions are separated from the liquid phase by two two-phase regions:

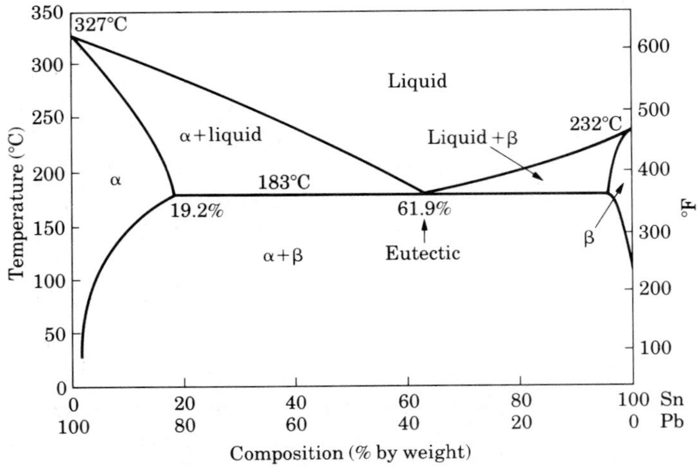

FIGURE 5.4 The lead–tin phase diagram. Note that the composition of the eutectic point for this alloy is 61.9% Sn–28.1% Pb. A composition either lower or higher than this ratio will have a higher liquidus temperature.

alpha + liquid on the left and beta + liquid on the right. Figure 5.4 shows the composition of the alloy (61.9% Sn–38.1% Pb) that has the *lowest* temperature at which the alloy is still completely liquid, namely, 183 °C (361 °F). This point is known as the *eutectic point* (*eutectic* from the Greek *eutektos*, meaning easily melted).

Lever rule. The composition of various phases in a phase diagram can be determined by a procedure called the *lever rule*. As shown in the lower portion of Fig. 5.3, we first construct a lever between the solidus and liquidus lines (*tie line*), which is balanced (on the triangular support) at the nominal weight composition C_O of the alloy. The left end of the lever represents the composition C_S of the solid phase and the right end the composition C_L of the liquid phase. Note from the graduated scale in the figure that we also have indicated the liquid fraction along this tie line, ranging from 0 at the left (fully solid) to 1 at the right (fully liquid).

The lever rule states that the weight fraction of solid is proportional to the distance between C_O and C_L:

$$\frac{S}{S+L} = \frac{C_O - C_L}{C_S - C_L}. \tag{5.1}$$

Likewise, the weight fraction of liquid is proportional to the distance between C_S and C_O, hence

$$\frac{L}{S+L} = \frac{C_S - C_O}{C_S - C_L}. \tag{5.2}$$

Note that these quantities are fractions and must be multiplied by 100 to obtain percentages.

From inspection of the tie line in Fig. 5.3—and for a nominal alloy composition of C_O = 50% Cu–50% Ni—we see that, because C_O is closer to C_L than it is to C_S, the solid phase contains less copper than the liquid phase. Measuring on the phase diagram and using the lever-rule equations, we find that the composition of the solid phase is 42% Cu and of the liquid phase is 58% Cu, as stated in the middle circle at the right in Fig. 5.3.

Note that these calculations refer to copper. If we now reverse the phase diagram in the figure, so that the left boundary is 0% nickel (whereby nickel becomes the alloying element in copper), these calculations give us the compositions of the solid and liquid phases in terms of nickel. The lever rule is also known as the *inverse lever rule* because, as the foregoing equations show, the amount of each phase is proportional to the length of the opposite end of the lever.

5.2.5 The iron–carbon system

Steel and cast iron are represented by the iron–carbon binary system. Commercially pure iron contains up to 0.008 percent carbon, steels up to 2.11 percent carbon, and cast irons up to 6.67 percent carbon, although most cast irons contain less than

4.5 percent carbon. In this section we discuss the iron–carbon system in order to evaluate and modify the properties of these important materials for specific applications.

The *iron–iron carbide phase diagram* is shown in Fig. 5.5. Although this diagram can be extended to the right—to 100 percent carbon (pure graphite)—the range that is significant to engineering applications is up to 6.67 percent carbon, where cementite forms. Pure iron melts at a temperature of 1538 °C (2800 °F), as shown at the left in Fig. 5.5. As iron cools, it first forms delta ferrite, then austenite, and finally alpha ferrite.

Ferrite. Delta ferrite is stable only at very high temperatures and has no practical engineering significance. Alpha ferrite, or simply *ferrite*, is a solid solution of body-centered cubic iron and has a maximum solid solubility of 0.022 percent carbon at a temperature of 727 °C (1341 °F). Just as there is a solubility limit to salt in water—with any extra amount precipitating as solid salt at the bottom of the container—so there also is a solid solubility limit to carbon in iron.

Ferrite is relatively soft and ductile and is magnetic from room temperature to 768 °C (1414 °F). Although very little carbon can dissolve interstitially in bcc iron,

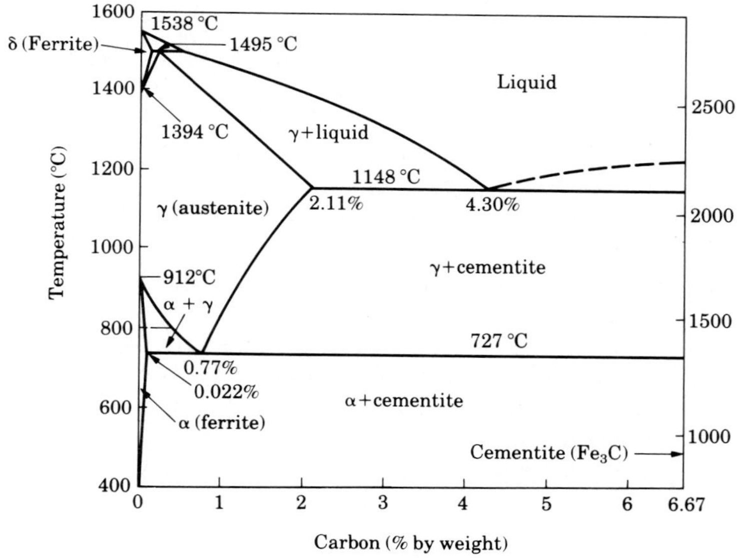

FIGURE 5.5 The iron–iron carbide phase diagram. Because of the importance of steel as an engineering material, this diagram is one of the most important phase diagrams.

the amount of carbon can significantly affect the mechanical properties of ferrite. Also, significant amounts of chromium, manganese, nickel, molybdenum, tungsten, and silicon can be contained in iron in solid solution, imparting certain desirable properties.

Austenite. Between 1394 °C (2541 °F) and 912 °C (1674 °F) iron undergoes a *polymorphic transformation* from the bcc to fcc structure, becoming what is known as gamma iron or, more commonly, *austenite*. This structure has a solid solubility of up to 2.11 percent carbon at 1148 °C (2098 °F). Thus the solid solubility of austenite is about two orders of magnitude higher than that of ferrite, with the carbon occupying interstitial positions (Fig. 5.6).

Austenite is an important phase in the heat treatment of steels (Section 5.10). It is denser than ferrite, and its single-phase fcc structure is ductile at elevated temperatures; thus it possesses good formability. Large amounts of nickel and manganese can also be dissolved in fcc iron to impart various properties. Austenitic steel is nonmagnetic at high temperatures; austenitic stainless steels are nonmagnetic at room temperature.

Cementite. The right boundary of Fig. 5.5 represents *cementite*, also called *carbide*, which is 100 percent iron carbide (Fe_3C), with a carbon content of 6.67 percent. (This carbide should not be confused with carbides used for tool and die materials.) Cementite is a very hard and brittle intermetallic compound and significantly influences the properties of steels. It can include other alloying elements, such as chromium, molybdenum, and manganese.

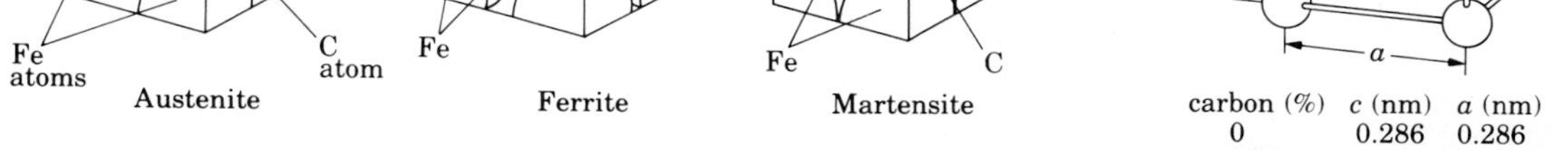

carbon (%)	c (nm)	a (nm)
0	0.286	0.286
0.20	0.288	0.2858
0.40	0.291	0.2856

FIGURE 5.6
The unit cell for (a) austenite, (b) ferrite, and (c) martensite. The effect of percentage of carbon (by weight) on the lattice dimensions for martensite is shown in (d). Note the interstitial position of the carbon atoms (see Fig. 3.9) and the increase in dimension c with increasing carbon content. Thus the unit cell of martensite is in the shape of a rectangular prism.

5.2.6 The iron–iron carbide phase diagram

The region of the iron–iron carbide phase diagram that is significant for steels is shown in Fig. 5.7, an enlargement of the lower left portion of Fig. 5.5. Various microstructures can be developed in steels, depending on the carbon content and the method of heat treatment. For example, let's consider iron with a 0.77 percent carbon content being cooled very slowly from a temperature of, say, 1100 °C (2012 °F) in the austenite phase. The reason for the *slow* cooling rate is to maintain equilibrium; higher rates of cooling are used in heat treating. At 727 °C (1341 °F) a reaction takes place in which austenite is transformed into alpha ferrite (bcc) and cementite. As the solid solubility of carbon in ferrite is only 0.022 percent, the extra carbon forms cementite.

This reaction is called a *eutectoid* (meaning *eutecticlike*) *reaction*, indicating that at a certain temperature a single solid phase (austenite) is transformed into two other solid phases (ferrite and cementite). The structure of eutectoid steel is called *pearlite*, because it resembles mother of pearl at low magnifications. The microstructure of pearlite consists of alternate layers (*lamellae*) of ferrite and cementite

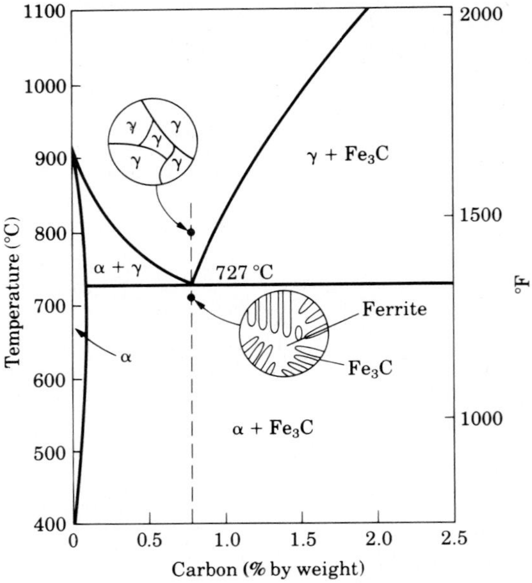

FIGURE 5.7
Schematic illustration of the microstructures for an iron–carbon alloy of eutectoid composition (0.77 percent carbon) above and below the eutectoid temperature of 727 °C (1341 °F).

(Fig. 5.7). Consequently, the mechanical properties of pearlite are intermediate between ferrite (soft and ductile) and cementite (hard and brittle).

In iron with less than 0.77 percent carbon, the microstructure formed consists of a pearlite phase (ferrite and cementite) and a ferrite phase. The ferrite in the pearlite is called *eutectoid ferrite*. The ferrite phase is called *proeutectoid ferrite*, *pro* meaning before, because it forms at a temperature higher than the eutectoid temperature of 727 °C (1341 °F). If the carbon content is greater than 0.77 percent, the austenite transforms into pearlite and cementite. The cementite in the pearlite is called *eutectoid cementite*, and the cementite phase is called *proeutectoid cementite*, because it forms at a temperature higher than the eutectoid temperature.

Effects of alloying elements in iron. Although carbon is the basic element that transforms iron into steel, other elements are also added to impart various desirable properties. The effect of these alloying elements on the iron–iron carbide phase diagram is to shift the eutectoid temperature and eutectoid composition (the percentage of carbon in steel at the eutectoid point).

The eutectoid temperature may be raised or lowered from 727 °C (1341 °F), depending on the particular alloying element. However, alloying elements always lower the eutectoid composition; that is, the carbon content becomes less than 0.77 percent. Lowering the eutectoid temperature means increasing the austenite range. Thus an alloying element such as nickel is known as an *austenite former*. Because nickel has an fcc structure, it probably favors the fcc structure of austenite. Conversely, chromium and molybdenum have the bcc structure, causing these elements to favor the bcc structure of ferrite; these elements are known as *ferrite formers*.

5.2.7 Cast irons

The term *cast iron* refers to a family of ferrous alloys composed of iron, carbon (ranging from 2.11 percent to about 4.5 percent), and silicon (up to about 3.5 percent). Cast irons are usually classified according to their solidification morphology:

a) Gray cast iron or gray iron.
b) Nodular cast iron, ductile cast iron, or spheroidal graphite cast iron.
c) White cast iron.
d) Malleable iron.
e) Compacted graphite iron.

Cast irons are also classified by their structure: ferrite, pearlite, quenched and tempered, or austempered.

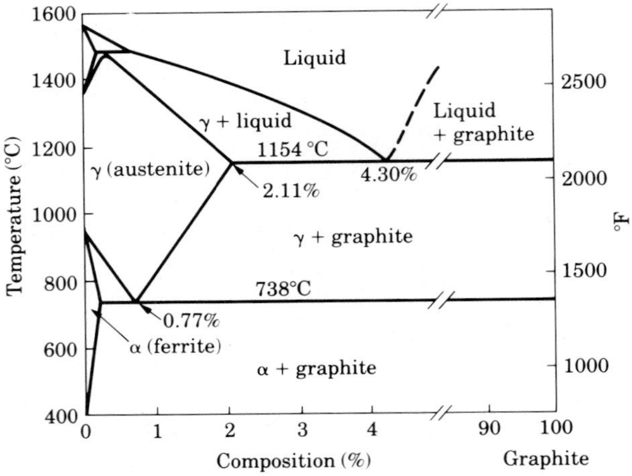

FIGURE 5.8
Phase diagram for the iron–carbon system with graphite, instead of cementite, as the stable phase. Note that this figure is an extended version of Fig. 5.5.

The equilibrium diagram relevant to cast irons is shown in Fig. 5.8, in which the right boundary is 100% C, that is, pure graphite. The horizontal liquid + graphite line is at 1154 °C (2109 °F). Thus cast irons are completely liquid at temperatures lower than those required for liquid steels. Consequently, cast irons have lower melting temperatures, which is why the casting process is so suitable for iron with high carbon content.

Although cementite exists in steels almost indefinitely, it is not completely stable. That is, it is *metastable*, with an extremely low rate of decomposition. However, cementite can be made to decompose into alpha ferrite and graphite. The formation of graphite (*graphitization*) can be controlled, promoted, and accelerated by modifying the composition and the rate of cooling, and by the addition of silicon.

• Example 5.1: Determining the amount of phases.

Determine the amount of gamma and alpha phases in a 10-kg, 1040 steel casting as it is being cooled slowly to the following temperatures: (a) 900 °C, (b) 728 °C, and (c) 726 °C.

SOLUTION: (a) Referring to Fig. 5.7, we draw a vertical line at 0.40% C at 900 °C. We are in the single-phase austenite region, so percent gamma is 100 (10 kg) and percent alpha is 0. (b) At 728 °C, the alloy is in the two-phase gamma–alpha field. When we draw the phase diagram in greater detail, we can find the weight percentages of each phase by the lever rule:

$$\text{Percent alpha} = \left(\frac{C_\gamma - C_O}{C_\gamma - C_\alpha}\right)100 = \left(\frac{0.77 - 0.40}{0.77 - 0.022}\right)100 = 50\%, \text{ or } 5 \text{ kg};$$

$$\text{Percent gamma} = \left(\frac{C_O - C_\alpha}{C_\gamma - C_\alpha}\right)100 = \left(\frac{0.40 - 0.022}{0.77 - 0.022}\right)100 = 50\%, \text{ or } 5 \text{ kg}.$$

(c) At 726 °C, the alloy will be in the two-phase alpha and Fe_3C region. No gamma phase will be present. Again, we use the lever rule to find the amount of alpha present:

$$\text{Percent alpha} = \left(\frac{6.67 - 0.40}{6.67 - 0.022}\right)100 = 94\%, \text{ or } 9.4 \text{ kg.}$$

5.3 Cast Structures

The type of cast structure developed during solidification of metals and alloys depends not only on the composition of the particular metal but also on the rate of heat transfer and the flow of the liquid metal during its solidification. As we describe in other sections in this chapter, the structures developed, in turn, affect the properties of castings.

5.3.1 Pure metals

The grain structure of a pure metal solidified in a square mold is shown in Fig. 5.9(a). At the mold walls the metal cools rapidly, because the walls are at ambient

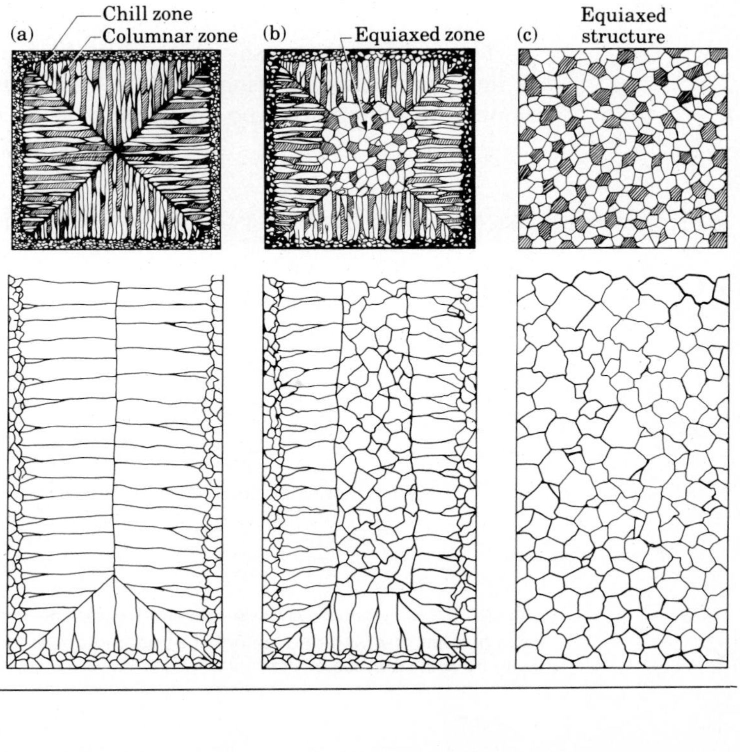

FIGURE 5.9 Schematic illustration of three cast structures of metals solidified in a square mold: (a) pure metals; (b) solid-solution alloys; and (c) structure obtained by heterogeneous nucleation of grains using nucleating agents. *Source:* G. W. Form, J. F. Wallace, J. L. Walker, and A. Cibula.

temperature. As a result of cooling first at the mold walls, the casting develops a solidified *skin*, or *shell*, of fine equiaxed grains. The grains grow in the direction opposite to the heat transfer from the mold. Grains that have a favorable orientation, called *columnar grains*, grow preferentially (Fig. 5.10). Grains that have substantially different orientations are blocked from further growth. Grain development is known as *homogeneous nucleation*, meaning that the grains (crystals) grow upon themselves, starting at the mold wall.

5.3.2 Alloys

Pure metals have limited mechanical properties. However, their properties can be enhanced and modified by *alloying*. The vast majority of metals used in engineering applications are some form of an *alloy*, consisting of two or more chemical elements, at least one of which is a metal. Because many products consist of cast alloys, a study of their solidification is essential.

Solidification in alloys begins when the temperature drops below the liquidus, T_L, and is complete when it reaches the solidus, T_S (Fig. 5.11). Within this temperature range the alloy is in a mushy or pasty state with *columnar dendrites* (from the Greek *dendron* meaning akin to, and *drys* meaning tree). Note the presence of liquid metal between the dendrite arms. Dendrites have three-dimensional arms and branches (secondary arms) and they eventually interlock, as shown in Fig. 5.12. The study of dendritic structures, although complex, is important because of detrimental factors such as compositional variations, segregation, and microporosity that generally exist within a casting.

The width of the mushy zone, where both liquid and solid phases are present, is an important factor during solidification. We describe this zone in terms of a temperature difference, known as the *freezing range*, as follows:

$$\text{Freezing range} = T_L - T_S. \tag{5.3}$$

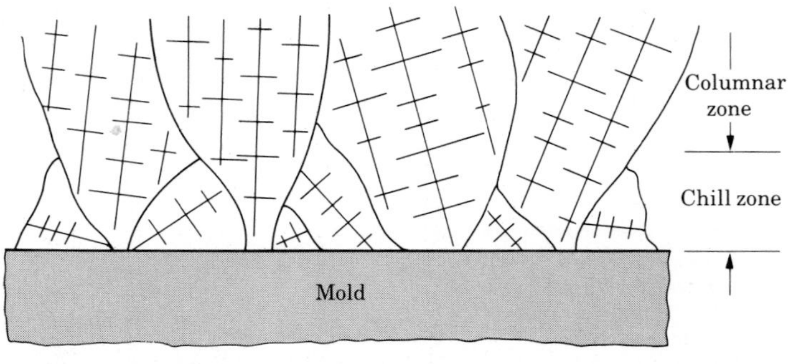

FIGURE 5.10 Development of a preferred texture at a cool mold wall. Note that only favorably oriented grains grow away from the surface of the mold.

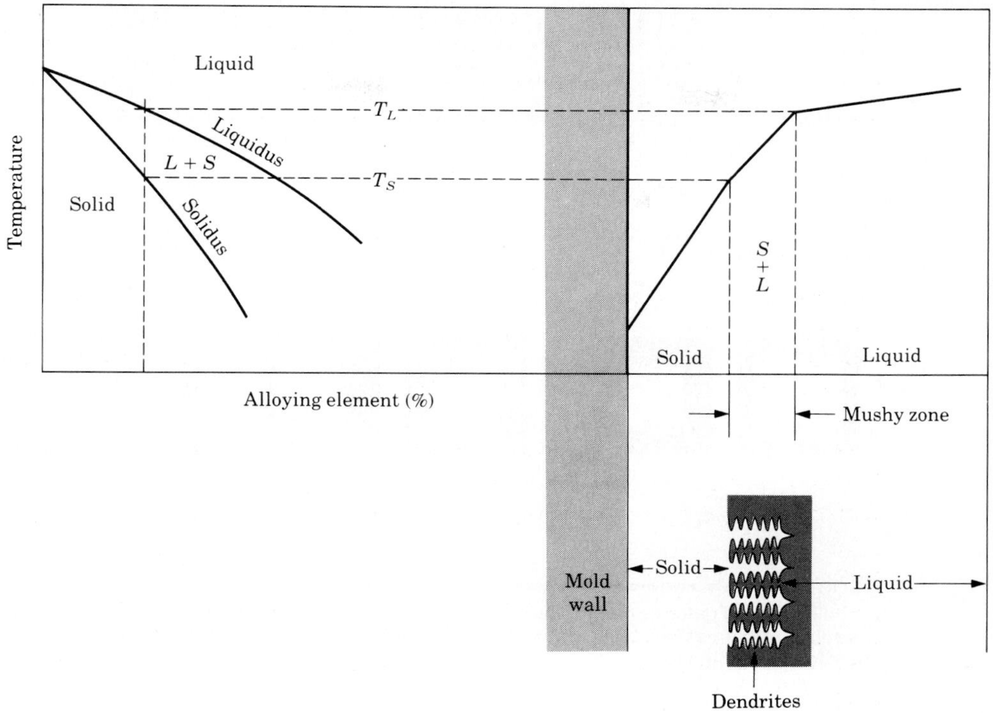

FIGURE 5.11 Schematic illustration of alloy solidification and temperature distribution in the solidifying metal. Note the formation of dendrites in the mushy zone.

Figure 5.11 shows that pure metals have a freezing range that approaches zero and that the solidification front moves as a plane front, without forming a mushy zone. Eutectics solidify in a similar manner with an approximately plane front. The type of solidification structure developed depends on the composition of the eutectic. For alloys with a nearly symmetrical phase diagram, the structure is generally lamellar with two or more solid phases present, depending on the alloy system. When the volume fraction of the minor phase of the alloy is less than about 25 percent, the structure generally becomes fibrous. These conditions are particularly important for cast irons.

For alloys, although it is not precise, a short freezing range generally involves a temperature difference of less than 50 °C (90 °F), and a long freezing range greater than 110 °C (200 °F). Ferrous castings generally have narrow mushy zones, whereas aluminum and magnesium alloys have wide mushy zones. Consequently, these alloys are in a mushy state throughout most of the solidification process.

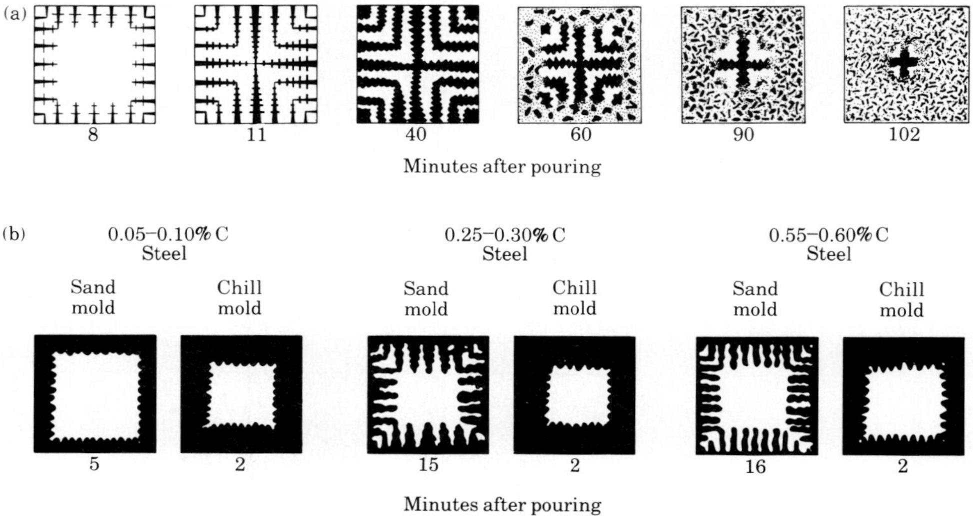

FIGURE 5.12 (a) Solidification patterns for gray cast iron in a 180-mm (7-in.) square casting. Note that after 11 min of cooling, dendrites reach each other, but the casting is still mushy throughout. It takes about two hours for this casting to solidify completely. (b) Solidification of carbon steels in sand and chill (metal) molds. Note the difference in solidification patterns as the carbon content increases. *Source:* H. F. Bishop and W. S. Pellini.

Effects of cooling rates. Slow cooling rates—on the order of 10^2 K/s—or long local solidification times result in coarse dendritic structures with large spacing between the dendrite arms. For faster cooling rates—on the order of 10^4 K/s—or short local solidification times, the structure becomes finer with smaller dendrite arm spacing. For still faster cooling rates—on the order of 10^6 to 10^8 K/s—the structures developed are amorphous, as described in Section 5.9.7. The structures developed and the resulting grain size, in turn, influence the properties of the casting. As grain size decreases (a) the strength and ductility of the cast alloy increase (see Hall–Petch equation, Section 3.4.1), (b) microporosity (interdendritic shrinkage voids) in the casting decreases, and (c) the tendency for the casting to crack (*hot tearing*) during solidification decreases. Lack of uniformity in grain size and distribution results in castings with anisotropic properties.

A criterion describing the kinetics of the liquid–solid interface is the ratio G/R, where G is the thermal gradient and R is the rate at which the liquid–solid interface moves. Typical values for G range from 10^2 to 10^3 K/m and for R from 10^{-3} to 10^{-4} m/s. Dendritic type structures, shown in Fig. 5.13(a) and (b), typically have a G/R ratio in the range of 10^5 to 10^7, whereas ratios of 10^{10} to 10^{12} produce a plane front nondendritic liquid–solid interface (Fig. 5.14).

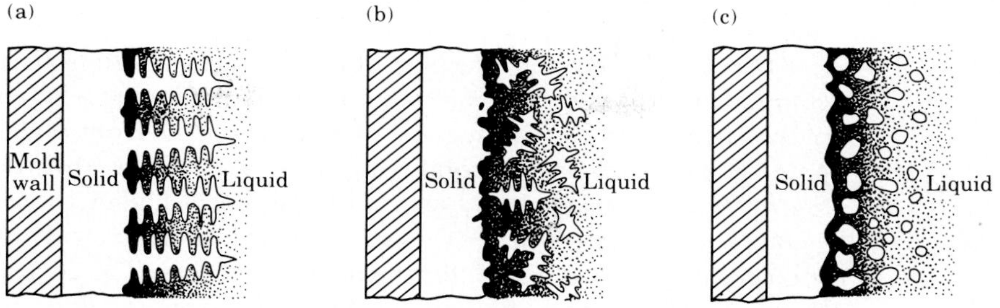

FIGURE 5.13
Schematic illustration of three basic types of cast structures: (a) columnar dendritic; (b) equiaxed dendritic; and (c) equiaxed nondendritic. *Source:* After D. Apelian.

5.3.3 Structure–property relationships

Because all castings are expected to possess certain properties to meet design and service requirements, the relationships between these properties and the structures developed during solidification are important aspects of casting. In this section, we describe these relationships in terms of dendrite morphology and the concentration of alloying elements in various regions.

The compositions of dendrites and the liquid metal are given by the phase diagram of the particular alloy. When the alloy is cooled very slowly, each dendrite develops a uniform composition. Under normal cooling encountered in practice, however, *cored dendrites* are formed and have a surface composition different from that at their centers (concentration gradient). The surface has a higher concentration of alloying elements than at the core of the dendrite owing to solute rejection from the core toward the surface during solidification of the dendrite, which is called *microsegregation.* The darker shading in the interdendritic liquid near the dendrite roots in Fig. 5.13 indicates that these regions have a higher solute

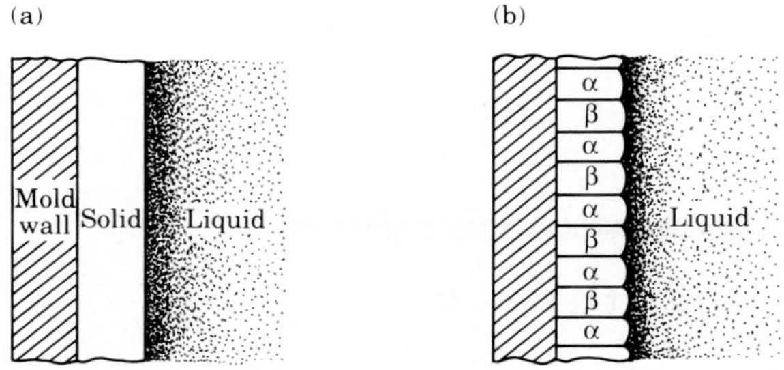

FIGURE 5.14
Schematic illustration of cast structures in (a) plane front, single phase, and (b) plane front, two phase. *Source:* After D. Apelian.

concentration. Thus microsegregation in these regions is much more pronounced than in others.

In contrast to microsegregation, *macrosegregation* involves differences in composition throughout the casting itself. In situations where the solidifying front moves away from the surface of a casting as a plane front (Fig. 5.14), lower melting point constituents in the solidifying alloy are driven toward the center (*normal segregation*). Consequently, such a casting has a higher concentration of alloying elements at its center than at its surfaces. In dendritic structures such as those for solid-solution alloys (Fig. 5.9b) the opposite occurs: The center of the casting has a lower concentration of alloying elements (*inverse segregation*). The reason is that liquid metal (having a higher concentration of alloying elements) enters the cavities developed from solidification shrinkage in the dendrite arms, which have solidified sooner. Another form of segregation is the result of gravity (*gravity segregation*), whereby higher density inclusions or compounds sink, and lighter elements (such as antimony in an antimony–lead alloy) float to the surface.

A typical cast structure of a solid-solution alloy with an inner zone of equiaxed grains is shown in Fig. 5.9(b). This inner zone can be extended throughout the casting, as shown in Fig. 5.9(c), by adding an *inoculant* (nucleating agent) to the alloy. The inoculant induces nucleation of grains throughout the liquid metal (*heterogeneous nucleation*).

Because of the presence of thermal gradients in a solidifying mass of liquid metal and because of gravity (hence density differences), convection has a strong influence on the structures developed. Convection promotes the formation of an outer chill zone, refines grain size, and accelerates the transition from columnar to equiaxed grains. The structure shown in Fig. 5.13(b) can also be obtained by increasing convection within the liquid metal, whereby dendrite arms separate (*dendrite multiplication*). Conversely, reducing or eliminating convection results in coarser and longer columnar dendritic grains. The dendrite arms are not particularly strong, and they can be broken up by agitation or mechanical vibration in the early stages of solidification (*rheocasting*). This results in finer grain size, with equiaxed nondendritic grains distributed more uniformly throughout the casting (Fig. 5.13c). Convection can be enhanced by the use of mechanical or electromagnetic methods. Experiments are now being conducted during space flights concerning the effects of gravity on the microstructure of castings.

5.4 Fluid Flow and Heat Transfer

5.4.1 Fluid flow

To emphasize the importance of fluid flow, let's briefly describe the basic casting system, as shown in Fig. 5.15. The molten metal is poured through a *pouring basin*

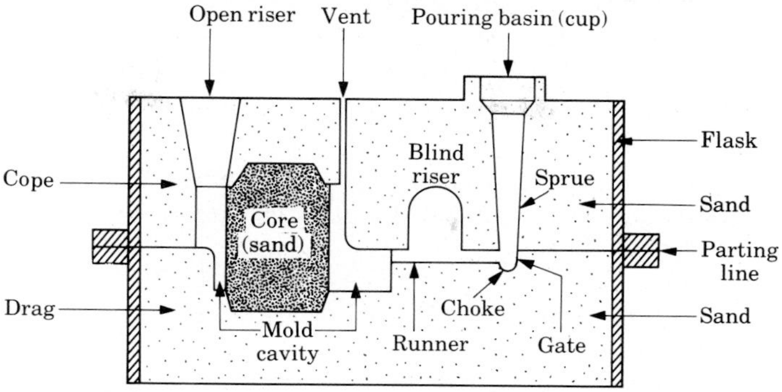

FIGURE 5.15
Schematic illustration of a sand mold showing various features.

or *cup*. It then flows through the *sprue* and *runners* into the mold cavity. *Risers* serve as reservoirs of molten metal to supply the metal necessary to prevent shrinkage. Although such a *gating system* appears to be relatively simple, successful casting requires careful design and control of the solidification process to ensure adequate fluid flow in the system. For example, one of the most important functions of the gating system is to trap contaminants (such as oxides and other inclusions) in the molten metal by having the contaminants adhere to the walls of the gating system, thereby preventing their reaching the actual mold cavity. Furthermore, the proper design of the gating system avoids or minimizes problems such as premature cooling, turbulence, and gas entrapment. Even before it reaches the mold cavity, the molten metal must be handled carefully to avoid forming oxides on molten metal surfaces from exposure to the environment or introduction of impurities into the molten metal.

Two basic principles of fluid flow are relevant to gating design: Bernoulli's theorem and the law of mass continuity.

Bernoulli's theorem. Bernoulli's theorem is based on the principle of conservation of energy and relates pressure, velocity, elevation of the fluid at any location in the system, and the frictional losses in a system that is full of liquid:

$$h + \frac{p}{\rho g} + \frac{v^2}{2g} = \text{constant}, \tag{5.4}$$

where h is the elevation above a certain reference plane, p is the pressure at that elevation, v is the velocity of the liquid at that elevation, ρ is the density of the fluid (assuming that it is incompressible), and g is the gravitational constant. Conservation of energy requires that, at a particular location in the system, the following relationship be satisfied:

$$h_1 + \frac{p_1}{\rho g} + \frac{v_1^2}{2g} = h_2 + \frac{p_2}{\rho g} + \frac{v_2^2}{2g} + f, \tag{5.5}$$

where the subscripts 1 and 2 represent two different elevations, respectively, and f represents the frictional loss in the liquid as it travels downward through the system. The frictional loss includes such factors as energy loss at the liquid–mold wall interfaces and turbulence in the liquid.

Continuity. The continuity law states that, for incompressible liquids and in a system with impermeable walls, the rate of flow is constant:

$$Q = A_1 v_1 = A_2 v_2, \tag{5.6}$$

where Q is the rate of flow, such as m^3/s, A is the cross-sectional area of the liquid stream, and v is the velocity of the liquid in that cross-sectional location. The subscripts 1 and 2 pertain to two different locations in the system. Thus the flow rate must be maintained anywhere in the system. The permeability of the walls of the system is important, because otherwise some liquid will permeate through (for example, in sand molds) and the flow rate will decrease as the liquid moves through the system.

Application. An application of the two principles stated is the traditional tapered design of sprues (Fig. 5.15). We can determine the shape of the sprue by using Eqs. (5.5) and (5.6). Assuming that the pressure at the top of the sprue is equal to the pressure at the bottom and that there are no frictional losses, at any point in the sprue the relationship between height and cross-sectional area is given by the parabolic relationship:

$$\frac{A_1}{A_2} = \sqrt{\frac{h_2}{h_1}}, \tag{5.7}$$

where, for example, the subscript 1 denotes the top of the sprue and 2 the bottom. Moving downward from the top, the cross-sectional area of the sprue must decrease. Recall that in a free-falling liquid, such as water from a faucet, the cross-sectional area of the stream decreases as it gains velocity downward. If we design a sprue with a constant cross-sectional area and pour the molten metal, regions may develop where the liquid loses contact with the sprue walls. As a result, *aspiration* may take place whereby air will be sucked in or entrapped in the liquid. On the other hand, in many systems tapered sprues are now replaced by straight-sided sprues with a *choking* mechanism at the bottom, consisting of either a choke core or a runner choke (as shown in Fig. 5.15).

Flow characteristics. An important consideration in fluid flow in gating systems is the presence of *turbulence*, as opposed to *laminar flow* of fluids. We use the *Reynolds number* to quantify this aspect of fluid flow. The Reynold's number, Re, represents the ratio of the inertia to the viscous forces in fluid flow, and is defined as

$$\text{Re} = \frac{vD\rho}{\eta}, \tag{5.8}$$

where v is the velocity of the liquid, D is the diameter of the channel, and ρ and η are the density and viscosity, respectively, of the liquid. The higher this number, the greater is the tendency for turbulent flow. In ordinary gating systems Re ranges from 2000 to 20,000. An Re value of up to 2000 represents laminar flow; between 2000 and 20,000 it is a mixture of laminar and turbulent flow and is generally regarded as harmless in gating systems. However, Re values in excess of 20,000 represent severe turbulence, resulting in air entrainment and *dross* formation (the scum that forms on the surface of molten metal) from the reaction of the liquid metal with air and other gases. Techniques for minimizing turbulence generally involve avoidance of sudden changes in flow direction and in the geometry of channel cross-sections in the design of the gating system.

The elimination of dross or slag is another important consideration in fluid flow. It can be achieved by skimming, properly designing pouring basins and gating systems, or using filters. Filters are usually made of ceramics, mica, or fiberglass and their proper location and placement is important for effective filtering of dross and slag.

5.4.2 Fluidity of molten metal

A term commonly used to describe the capability of the molten metal to fill mold cavities is *fluidity*. This term consists of two basic factors: (1) characteristics of the molten metal and (2) casting parameters. The following characteristics of molten metal influence fluidity.

a) *Viscosity*. As viscosity and its sensitivity to temperature (*viscosity index*) increase, fluidity decreases.
b) *Surface tension*. A high surface tension of the liquid metal reduces fluidity. Oxide films developed on the surface of the molten metal thus have a significant adverse effect on fluidity. For example, the oxide film on the surface of pure molten aluminum triples the surface tension.
c) *Inclusions*. As insoluble particles, inclusions can have a significant adverse effect on fluidity. This effect can be verified by observing the viscosity of a liquid such as oil with and without sand particles in it; the latter has higher viscosity.
d) *Solidification pattern of the alloy*. The manner in which solidification occurs, as described in Section 5.3, can influence fluidity. Moreover, fluidity is inversely proportional to the freezing range (see Eq. 5.3). Thus the shorter the range (as in pure metals and eutectics), the higher the fluidity becomes. Conversely, alloys with long freezing ranges (such as solid-solution alloys) have lower fluidity.

The following casting parameters influence fluidity and also influence the fluid flow and thermal characteristics of the system.

a) *Mold design*. The design and dimensions of components such as the sprue, runners, and risers all influence fluidity.

b) *Mold material and its surface characteristics.* The higher the thermal conductivity of the mold and the rougher its surfaces, the lower the fluidity of the molten metal becomes. Heating the mold improves fluidity, even though it slows down solidification of the metal and the casting develops coarse grains; hence it has less strength.
c) *Degree of superheat.* Defined as the increment of temperature above the melting point of an alloy, superheat improves fluidity by delaying solidification.
d) *Rate of pouring.* The slower the rate of pouring the molten metal into the mold, the lower the fluidity becomes because of the faster rate of cooling.
e) *Heat transfer.* This factor directly affects the viscosity of the liquid metal.

Although the interrelationships are complex, we use the term *castability* generally to describe the ease with which a metal can be cast to obtain a part with good quality. Obviously, this term includes not only fluidity but casting practices as well.

Tests for fluidity. Although none is accepted universally, several tests have been developed to quantify fluidity. One such test is shown in Fig. 5.16, where the molten metal is made to flow along a channel at room temperature. The distance of metal flow before it solidifies and stops is a measure of its fluidity. Obviously this length is a function of the thermal properties of the metal and the mold, as well as the design of the channel. Such tests are useful and simulate casting situations to a reasonable degree.

5.4.3 Heat transfer

An important consideration in casting is the heat transfer during the complete cycle from pouring to solidification and cooling to room temperature. Heat flow at different locations in the system is a complex phenomenon; it depends on many

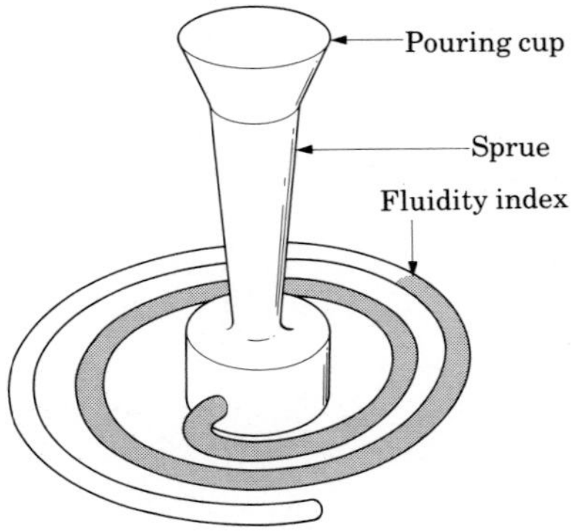

FIGURE 5.16

A test for fluidity using a spiral mold. The fluidity index is the length of the solidified metal in the spiral passage. The greater the length of the solidified metal, the greater is its fluidity.

factors relating to the casting material and the mold and process parameters. For instance, in casting thin sections the metal flow rates must be high enough to avoid premature chilling and solidification. But, the flow rate must not be so fast as to cause excessive turbulence, with its detrimental effects on the casting process.

A typical temperature distribution in the mold–liquid metal interface is shown in Fig. 5.17. Heat from the liquid metal is given off through the mold wall and the surrounding air. The temperature drop at the air–mold and mold–metal interfaces is caused by the presence of boundary layers and imperfect contact at these interfaces. The shape of the curve depends on the thermal properties of the molten metal and the mold.

5.4.4 Solidification time

During the early stages of solidification, a thin solidified skin begins to form at the cool mold walls and, as time passes, the skin thickens. With flat mold walls, this thickness is proportional to the square root of time. Thus doubling the time will make the skin $\sqrt{2} = 1.41$ times, or 41 percent, thicker.

The *solidification time* is a function of the volume of a casting and its surface area (*Chvorinov's rule*):

$$\text{Solidification time} = C\left(\frac{\text{volume}}{\text{surface area}}\right)^2, \tag{5.9}$$

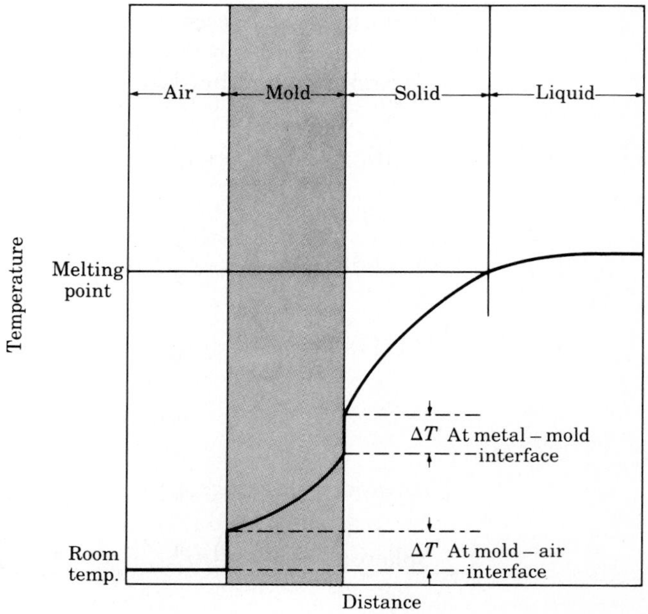

FIGURE 5.17 Temperature distribution at the mold wall and liquid–metal interface during solidification of metals in casting.

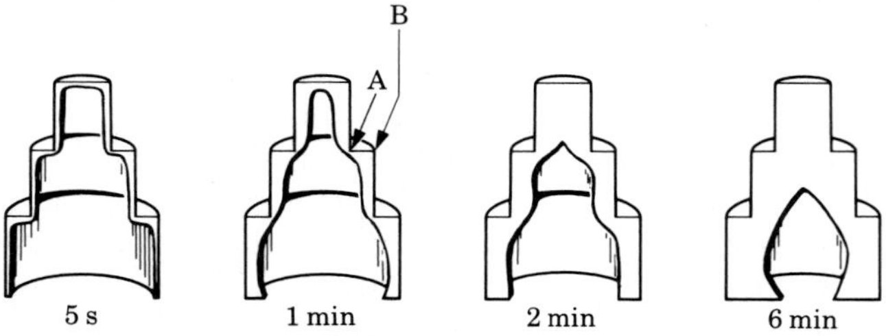

FIGURE 5.18 Solidified skin on a steel casting. The remaining molten metal is poured out at the times indicated in the figure. Hollow ornamental and decorative objects are made by a process called slush casting, which is based on this principle. *Source:* H. F. Taylor, J. Wulff, and M. C. Flemings.

where C is a constant that reflects mold material, metal properties (including latent heat), and temperature. Thus a large sphere solidifies and cools to ambient temperature at a much slower rate than does a smaller sphere. The reason is that the volume of a sphere is proportional to the cube of its diameter, and the surface area is proportional to the square of its diameter. Similarly, we can show that molten metal in a cube-shaped mold will solidify faster than in a spherical mold of the same volume.

The effects of mold geometry and elapsed time on skin thickness and shape are shown in Fig. 5.18. As illustrated, the unsolidified molten metal has been poured from the mold at different time intervals, ranging from 5 s to 6 min. Note that the skin thickness increases with elapsed time but that the skin is thinner at internal angles (location A in the figure) than at external angles (location B). This latter condition is caused by slower cooling at internal angles than at external angles.

• Example 5.2: Solidification times for various shapes.

Three pieces being cast have the same volume but different shapes. One is a sphere, one a cube, and the other a cylinder with a height equal to its diameter. Which piece will solidify the fastest and which one the slowest?

SOLUTION. The volume is unity, so we have from Eq. (5.9):

$$\text{Solidification time} \propto \frac{1}{(\text{surface area})^2}.$$

The respective surface areas are

$$\text{Sphere:}\quad V = \left(\frac{4}{3}\right)\pi r^3, \qquad r = \left(\frac{3}{4\pi}\right)^{1/3}, \quad \text{and} \quad A = 4\pi r^2 = 4\pi\left(\frac{3}{4\pi}\right)^{2/3} = 4.84$$

$$\text{Cube:}\quad V = a^3, \qquad a = 1, \quad \text{and} \quad A = 6a^2 = 6.$$

$$\text{Cylinder:} \quad V = \pi r^2 h = 2\pi r^3, \qquad r = \left(\frac{1}{2\pi}\right)^{1/3}, \quad \text{and}$$

$$A = 2\pi r^2 + 2\pi r h = 6\pi r^2 = 6\pi\left(\frac{1}{2\pi}\right)^{2/3} = 5.54.$$

Thus the respective solidification times t are

$$t_{\text{sphere}} = 0.043C, \qquad t_{\text{cube}} = 0.028C, \quad \text{and} \quad t_{\text{cylinder}} = 0.033C.$$

Hence the cube-shaped casting will solidify the fastest and the sphere-shaped casting will solidify the slowest.

5.4.5 Shrinkage

Because of their thermal expansion characteristics, metals shrink (contract) during solidification and cooling. Shrinkage, which causes dimensional changes—and, sometimes, cracking—is the result of:

1. contraction of the molten metal as it cools prior to its solidification;
2. contraction of the metal during phase change from liquid to solid (latent heat of fusion); and
3. contraction of the solidified metal (the casting) as its temperature drops to ambient temperature.

The largest amount of shrinkage occurs during cooling of the casting. The amount of contraction for various metals during solidification is shown in Table 5.1. Note that gray cast iron expands. The reason is that graphite has a relatively

TABLE 5.1
SOLIDIFICATION CONTRACTION FOR VARIOUS CAST METALS

METAL OR ALLOY	VOLUMETRIC SOLIDIFICATION CONTRACTION (%)	METAL OR ALLOY	VOLUMETRIC SOLIDIFICATION CONTRACTION (%)
Aluminum	6.6	70% Cu–30% Zn	4.5
Al–4.5% Cu	6.3	90% Cu–10% Al	4
Al–12% Si	3.8	Gray iron	Expansion to 2.5
Carbon steel	2.5–3	Magnesium	4.2
1% carbon steel	4	White iron	4–5.5
Copper	4.9	Zinc	6.5

Note: After R. A. Flinn.

high specific volume, and when it precipitates as graphite flakes during solidification, it causes a net expansion of the metal.

5.4.6 Defects

Depending on casting design and practice, several defects can develop in castings. Because different names have been used to describe the same defect, the International Committee of Foundry Technical Associations has developed standardized nomenclature consisting of seven basic categories of casting defects:

A. *Metallic projections*, consisting of fins, flash, or massive projections such as swells and rough surfaces.
B. *Cavities*, consisting of rounded or rough internal or exposed cavities, including blowholes, pinholes, and shrinkage cavities (see porosity below).
C. *Discontinuities*, such as cracks, cold or hot tearing, and cold shuts. If the solidifying metal is constrained from shrinking freely, cracking and tearing can occur. Although many factors are involved in tearing, coarse grain size and the presence of low-melting segregates along the grain boundaries (intergranular) increase the tendency for hot tearing. Incomplete castings result from the molten metal being at too low a temperature or pouring the metal too slowly. Cold shut is an interface in a casting that lacks complete fusion because of the meeting of two streams of liquid metal from different gates.
D. *Defective surface*, such as surface folds, laps, scars, adhering sand layers, and oxide scale.
E. *Incomplete casting*, such as misruns (due to premature solidification), insufficient volume of metal poured, and runout (due to loss of metal from mold after pouring).
F. *Incorrect dimensions or shape*, owing to factors such as improper shrinkage allowance, pattern mounting error, irregular contraction, deformed pattern, or warped casting.
G. *Inclusions*, which form during melting, solidification, and molding. Generally nonmetallic, they are regarded as harmful because they act like stress raisers and reduce the strength of the casting. They can be filtered out during processing of the molten metal. Inclusions may form during melting because of reaction of the molten metal with the environment (usually oxygen) or the crucible material. Chemical reactions among components in the molten metal may produce inclusions; slags and other foreign material entrapped in the molten metal also become inclusions. Reactions between the metal and the mold material may produce inclusions. Spalling of the mold and core surfaces also produces inclusions, indicating the importance of the quality and maintenance of molds.

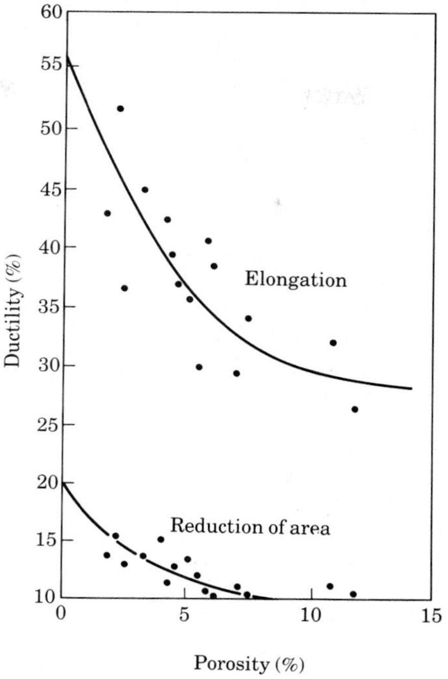

FIGURE 5.19
The effect of microporosity on the ductility of quenched and tempered 1% Cr–0.25% Mo cast steel. *Source:* T. Grousatier.

Porosity. Porosity in a casting may be caused by shrinkage or gases, or both. Porosity is detrimental to the ductility of a casting (Fig. 5.19) and its surface finish, making it permeable and thus affecting pressure tightness of a cast pressure vessel.

Porous regions can develop in castings because of shrinkage of the solidified metal. Thin sections in a casting solidify sooner than thicker regions. As a result, molten metal cannot be fed into the thicker regions that have not yet solidified. Because of contraction, as the surfaces of the thicker regions begin to solidify, porous regions develop at their centers. Microporosity can also develop when the liquid metal solidifies and shrinks between dendrites and between dendrite branches.

Porosity caused by shrinkage can be reduced or eliminated by various means. Basically, adequate liquid metal should be provided to avoid cavities caused by shrinkage. External or internal chills, used in sand casting (Fig. 5.20), also are an effective means of reducing shrinkage porosity. The function of chills is to increase the rate of solidification in critical regions. Internal chills are usually made of the same material as the castings. External chills may be made of the same material or may be iron, copper, or graphite. With alloys, porosity can be reduced or eliminated by making the temperature gradient steep, using, for example, mold materials that have high thermal conductivity. Subjecting the casting to hot isostatic pressing is another method of reducing porosity.

Liquid metals have much greater solubility for gases than do solids. When a metal begins to solidify, the dissolved gases are expelled from the solution. Gases may also result from reactions of the molten metal with the mold materials. Gases either accumulate in regions of existing porosity, such as in interdendritic areas, or they cause microporosity in the casting, particularly in cast iron, aluminum, and copper. Dissolved gases may be removed from the molten metal by flushing or purging with an inert gas or by melting and pouring the metal in a vacuum. If the dissolved gas is oxygen, the molten metal can be *deoxidized*. Steel is usually deoxidized with aluminum or silicon and copper-base alloys with phosphorus copper.

Whether microporosity is a result of shrinkage or is caused by gases may be difficult to determine. If the porosity is spherical and has smooth walls, much like the shiny surfaces of holes in Swiss cheese, it is generally from gases. If the walls are

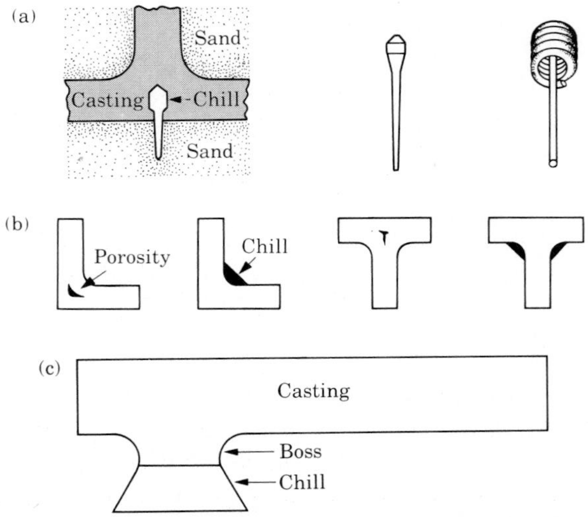

FIGURE 5.20 Various types of (a) internal and (b) external chills (dark areas at corners), used in castings to eliminate porosity caused by shrinkage. Chills are placed in regions where there is a larger volume of metal, as shown in (c).

rough and angular, porosity is likely from shrinkage between dendrites. Gross porosity is from shrinkage and is usually called *shrinkage cavities*.

5.5 Melting Practice

Melting practice is an important aspect of casting operations, because it has a direct bearing on the quality of castings. Furnaces are charged with melting stock consisting of metal, alloying elements, and various other materials such as flux. *Fluxes* are inorganic compounds, such as limestone and dolomite, and may include secondary fluxes, such as sodium carbonate and calcium fluoride (for cast iron) and borax–silica mixtures (for copper alloys). These compounds refine the molten metal by removing dissolved gases and various impurities. However, fluxes can have a detrimental effect on the lining of some furnaces, particularly induction furnaces.

To protect the surface of the molten metal against atmospheric reaction and contamination—and to refine the melt—the pour must be insulated against heat loss. Insulation is usually provided by covering the surface or mixing the melt with compounds that form a *slag*. In casting steels, the composition of the slag includes CaO, SiO_2, MnO, and FeO. A small quantity of liquid metal is usually tapped and its composition analyzed. Necessary additions or inoculations are then made prior to pouring the metal into the molds.

The metal charge may be composed of commercially pure *primary* metals, which are remelted scrap. Clean scrapped castings, gates, and risers may also be included in the charge. If the melting points of the alloying elements are sufficiently low, pure alloying elements are added to obtain the desired composition in the melt. If the melting points are too high, the alloying elements do not mix readily with the low-melting-point metals. In this case, *master alloys*, or *hardeners*, are often used. They usually consist of lower melting-point alloys with higher concentrations of one or two of the needed alloying elements. Differences in specific gravity of master alloys should not be great enough to cause segregation in the melt.

5.5.1 Foundries

Casting operations are usually carried out in *foundries*. Although these operations have traditionally involved much manual labor, modern foundries are equipped with automated and computer-integrated facilities in all aspects of their operations. They produce a variety of shapes and sizes of castings at high production rates, at low cost, and with good quality control.

Foundry operations initially involve two separate activities. The first is pattern and mold making, which now utilize computer-aided design and manufacturing techniques (see Chapter 14), minimizing trial and error and improving efficiency.

The second activity in a foundry is melting the metals, controlling their composition and impurities, and pouring them into appropriate molds. These operations, as well as shakeout, cleaning, heat treatment, and inspection, are also automated. Automation minimizes labor costs, reduces the possibility of human error, increases production, and attains higher quality.

Furnace selection requires careful consideration of several factors that can significantly influence the quality of castings, as well as the economics of casting operations. A variety of furnaces meet requirements for melting and casting metals and alloys in foundries, so proper selection of a furnace generally depends on:

a) economic considerations, such as initial cost and operating and maintenance costs;
b) the composition and melting point of the alloy to be cast and ease of controlling metal chemistry;
c) control of the atmosphere to avoid contamination of the metal;
d) capacity and the rate of melting required;
e) environmental considerations, such as air pollution and noise;
f) power supply and its availability and cost of fuels;
g) ease of superheating the metal; and
h) type of charge material that can be used.

The most commonly used furnaces in ferrous foundries today are *cupolas* and *electric-arc* and *induction* furnaces. Gas-fired furnaces are very common for nonferrous metals, particularly aluminum. Cupolas are basically refractory-lined vertical steel vessels that are charged with alternating layers of metal, coke, and flux. Although they require major investments, cupolas produce large amounts of molten metal (as much as 120 tons per hour). They operate continuously and have high melting rates.

5.6 Casting Alloys

The general properties of various casting alloys are summarized in Fig. 5.21 and Tables 5.2–5.4.

5.6.1 Ferrous alloys

Cast irons represent the largest amount of all metals cast. They generally possess several desirable properties, such as wear resistance, hardness, and good machinability. These alloys can easily be cast into intricate shapes. The term *cast iron* refers to a family of alloys: gray cast iron (gray iron), nodular (ductile or spheroidal) iron, white cast iron, malleable iron, and compacted-graphite iron.

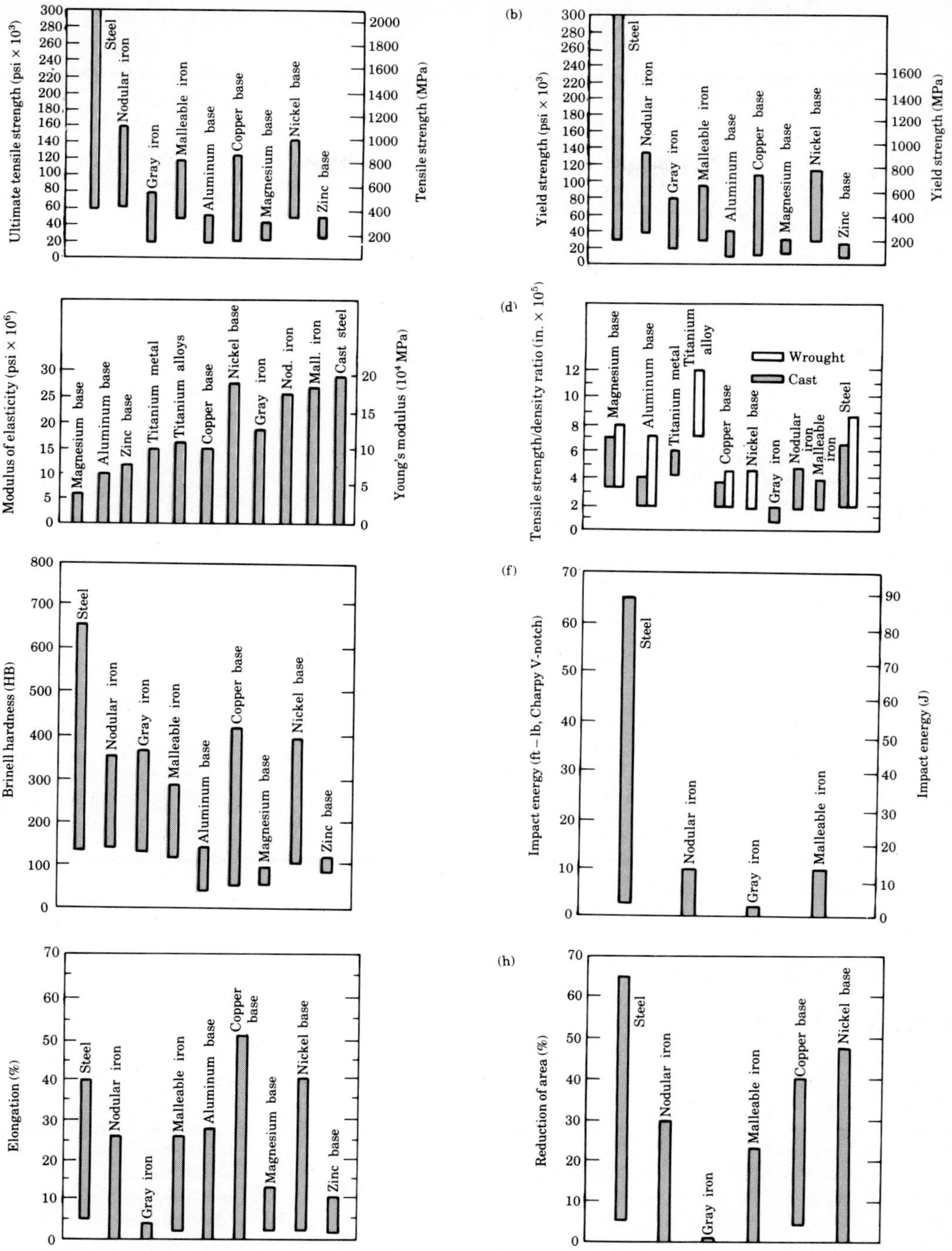

FIGURE 5.21

Mechanical properties for various groups of cast alloys. Compare with various tables of properties in Chapter 3. *Source:* Courtesy of Steel Founders' Society of America.

TABLE 5.2
TYPICAL APPLICATIONS FOR CASTINGS AND CASTING CHARACTERISTICS

TYPE OF ALLOY	APPLICATION	CASTABILITY*	WELDABILITY*	MACHINABILITY*
Aluminum	Pistons, clutch housings, exhaust manifolds	G–E	F	G–E
Copper	Pumps, valves, gear blanks, marine propellers	F–G	F	G–E
Gray iron	Engine blocks, gears, brake disks and drums, machine bases	E	D	G
Magnesium	Crankcase, transmission housings	G–E	G	E
Malleable iron	Farm and construction machinery, heavy-duty bearings, railroad rolling stock	G	D	G
Nickel	Gas turbine blades, pump and valve components for chemical plants	F	F	F
Nodular iron	Crankshafts, heavy-duty gears	G	D	G
Steel (carbon and low alloy)	Die blocks, heavy-duty gear blanks, aircraft undercarriage members, railroad wheels	F	E	F–G
Steel (high alloy)	Gas turbine housings, pump and valve components, rock crusher jaws	F	E	F
White iron	Mill liners, shot blasting nozzles, railroad brake shoes, crushers and pulverizers	G	VP	VP
Zinc	Door handles, radiator grills, carburetor bodies	E	D	E

* E, excellent; G, good; F, fair; VP, very poor; D, difficult.

TABLE 5.3
PROPERTIES AND TYPICAL APPLICATIONS OF CAST IRONS

CAST IRON	TYPE	ULTIMATE TENSILE STRENGTH (MPa)	YIELD STRENGTH (MPa)	ELONGATION IN 50 mm (%)	TYPICAL APPLICATIONS
Gray	Ferritic	170	140	0.4	Pipe, sanitary ware
	Pearlitic	275	240	0.4	Engine blocks, machine tools
	Martensitic	550	550	0	Wearing surfaces
Nodular (Ductile)	Ferritic	415	275	18	Pipe, general service
	Pearlitic	550	380	6	Crankshafts, highly stressed parts
	Tempered martensite	825	620	2	High-strength machine parts, wear resistance
Malleable	Ferritic	365	240	18	Hardware, pipe fittings, general engineering service
	Pearlitic	450	310	10	Couplings
	Tempered martensite	700	550	2	Gears, connecting rods
White	Pearlitic	275	275	0	Wear-resistance, mill rolls

TABLE 5.4
TYPICAL PROPERTIES OF NONFERROUS CASTING ALLOYS

ALLOY	CONDITION	CASTING METHOD*	UTS (MPa)	YIELD STRESS (MPa)	ELONGATION IN 50 mm (%)	HARDNESS (HB)
Aluminum						
357	T6	S	345	296	2.0	90
380	F	D	331	165	3.0	80
390	F	D	279	241	1.0	120
Magnesium						
AZ63A	T4	S, P	275	95	12	—
AZ91A	F	D	230	150	3	—
QE22A	T6	S	275	205	4	—
Copper						
Brass C83600	—	S	255	117	30	60
Bronze C86500	—	S	490	193	30	98
Bronze C93700	—	S	240	124	20	60
Zinc						
No. 3	—	D	283	—	10	82
No. 5	—	D	331	—	7	91
ZA27	—	P	425	365	1	115

* S, sand; D, die; P, permanent mold.

Gray cast iron. In this structure, graphite exists largely in the form of flakes (Fig. 5.22a). It is called *gray cast iron*, or *gray iron*, because when broken, the fracture path is along the graphite flakes and thus it has a gray, sooty appearance. These flakes act as stress raisers. Consequently, gray iron has negligible ductility

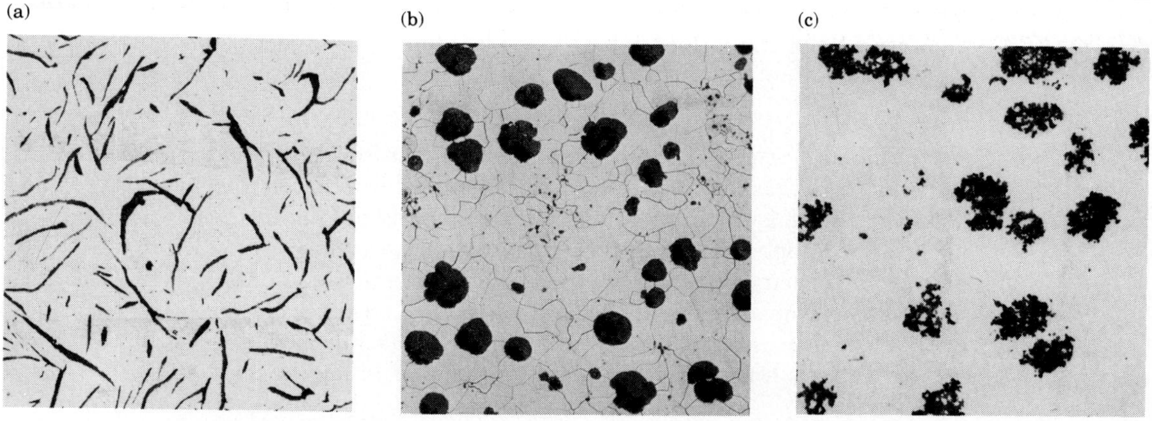

FIGURE 5.22
Microstructure for cast irons. Magnification: 100×. (a) Ferritic gray iron with graphite flakes. (b) Ferritic nodular iron, (ductile iron) with graphite in nodular form. (c) Ferritic malleable iron. This cast iron solidified as white cast iron, with the carbon present as cementite, and was heat treated to graphitize the carbon. *Source:* ASM International.

and is weak in tension, although strong in compression, as are other brittle materials. The graphite flakes give this material the capacity to dampen vibrations by the internal friction (thus energy dissipation) caused by these flakes. This capacity makes gray cast iron a suitable and commonly used material for constructing machine tool bases and structures.

Various types of gray cast iron are called ferritic, pearlitic, and martensitic (see Section 5.10). Because of the different structures, each has different properties and applications. In *ferritic gray iron*, also known as fully gray iron, the structure consists of graphite flakes in an alpha ferrite matrix. *Pearlitic gray iron* has a structure of graphite in a matrix of pearlite. Although still brittle, it is stronger than gray iron. *Martensitic gray iron* is obtained by austenitizing a pearlitic gray iron, followed by rapid quenching to produce a structure of graphite in a martensite matrix. As a result, this cast iron is very hard.

Castings of gray iron have relatively few shrinkage cavities and little porosity. Typical uses of gray cast iron are for engine blocks, machine bases, electric-motor housings, pipes, and machine wear surfaces. Gray cast irons are specified by a two-digit ASTM designation. Class 20, for example, specifies that the material must have a minimum tensile strength of 140 MPa (20 ksi).

Nodular iron (ductile iron). In the structure of nodular iron, graphite is in *nodular* or *spheroid* form (Fig. 5.22b). This shape permits the material to be somewhat ductile and shock resistant. The shape of graphite flakes is changed into nodules (spheres) by small additions of magnesium and/or cerium to the molten metal prior to pouring. Nodular iron can be made ferritic or pearlitic by heat treatment. It can also be heat treated to obtain a structure of tempered martensite.

Typically used for machine parts, pipe, and crankshafts, nodular cast irons are specified by a set of two-digit numbers. Thus class or grade 80-55-06, for example, indicates that the material has a minimum tensile strength of 550 MPa (80 ksi), a minimum yield strength of 380 MPa (55 ksi), and a 6 percent elongation in 50 mm (2 in.).

White cast iron. The structure of white cast iron is very hard, wear resistant, and brittle because it contains large amounts of iron carbide instead of graphite. White cast iron is obtained either by rapid cooling of gray iron or by adjusting the material's composition by keeping the carbon and silicon content low. This cast iron is also called white iron, because the lack of graphite gives the fracture surface a white crystalline appearance. Because of its extreme hardness and wear resistance, white cast iron is used mainly for liners for machinery that processes abrasive materials, rolls for rolling mills, and railroad-car brake shoes.

Malleable iron. Malleable iron is obtained by annealing white cast iron in an atmosphere of carbon monoxide and carbon dioxide, between 800 °C and 900 °C (1470 °F and 1650 °F) for up to several hours, depending on the size of the part.

During this process the cementite decomposes (*dissociates*) into iron and graphite. The graphite exists as *clusters* (Fig. 5.22c) in a ferrite or pearlite matrix and thus has a structure similar to nodular iron. This structure promotes ductility, strength, and shock resistance—hence the term malleable. Typical uses are for railroad equipment and various hardware. Malleable irons are specified by a five-digit designation. Thus 35018, for example, indicates that the yield strength of the material is 240 MPa (35 ksi), and its elongation is 18 percent in 50 mm (2 in.).

Compacted-graphite iron. The compacted-graphite structure has short, thick, and interconnected flakes with undulating surfaces and rounded extremities. The mechanical and physical properties of this cast iron are intermediate between those of flake graphite and nodular graphite cast irons. Its machinability is better than nodular iron. Typical applications are crankcases, ingot molds, cylinder heads, and brake disks. It is particularly suitable for components at elevated temperatures and resists thermal fatigue.

Steels. Because of the high temperatures involved, casting of steels requires considerable knowledge and experience. Steels shrink considerably during solidification and hence require extensive risering and proper mold design to allow for shrinkage. The high temperatures required also present problems in the selection of mold materials and melting and pouring, particularly in view of the high reactivity of steels with oxygen.

However, steel castings have more uniform (isotropic) properties than shapes made by mechanical working. Also, they can be welded without losing properties to the heat generated during welding. Cast weldments have gained great favor where complex configurations, or the size of the casting, may prevent casting in one piece.

5.6.2 Nonferrous alloys

Aluminum-base alloys. Alloys with an aluminum base have a wide range of mechanical properties, mainly because of various hardening mechanisms and heat treatments that can be used with them. Their fluidity depends on oxides and alloying elements in the metal. These alloys have high electrical conductivity and generally good corrosion resistance to most elements (except alkali). They are nontoxic and lightweight and have good machinability. However, except for alloys with silicon, they generally have low resistance to wear and abrasion. Aluminum-base alloys have many applications, including architectural, decorative, aerospace, and electrical. Engine blocks of some automobiles are made of aluminum-alloy castings.

Magnesium-base alloys. Alloys with a magnesium base have good corrosion resistance and moderate strength. These characteristics depend on the particular heat treatment used and on the proper use of protective coating systems.

Copper-base alloys. Alloys with a copper base have the advantages of good electrical and thermal conductivity, corrosion resistance, nontoxicity (unless they contain lead), and wear properties suitable for bearing materials. Mechanical properties and fluidity are influenced by the alloying elements. These alloys also possess good machinability characteristics.

Zinc-base alloys. Alloys with a zinc base have good fluidity and sufficient strength for structural applications. These alloys are commonly used in die casting.

High-temperature alloys. Having a wide range of properties, high-temperature alloys typically require temperatures of up to 1650 °C (3000 °F) for casting titanium and superalloys and higher for refractory alloys. Special techniques are used to cast these alloys into parts for jet and rocket engine components. Some of these alloys are more suitable and economical for casting than for shaping by other manufacturing methods such as forging.

5.7 Ingot Casting and Continuous Casting

Traditionally, the first step in metal processing is the shaping of the molten metal into a solid form—an *ingot*—for further processing by rolling it into shapes, casting it into semifinished forms, or forging. The shaping process is being rapidly replaced by *continuous casting*, thus improving efficiency and eliminating the need for ingots (see Section 5.7.2). The molten metal is poured (*teemed*) from the ladle into ingot molds in which the metal solidifies.

The cooled ingots are removed (*stripped*) from the molds and lowered into soaking pits, where they are reheated to a uniform temperature of about 1200 °C (2200 °F) for subsequent processing by rolling. Ingots may be square, rectangular, or round in cross-section, and their weight ranges from a few hundred pounds to 40 tons.

5.7.1 Ferrous alloy ingots

Certain reactions take place during solidification of an ingot, which in turn have an important influence on the quality of the steel produced. For example, significant amounts of oxygen and other gases can dissolve in the molten metal during steelmaking. However, much of these gases are rejected during solidification of the metal, because the solubility limit of gases in the metal decreases sharply as its temperature decreases. The rejected oxygen combines with carbon, forming carbon monoxide, which causes porosity in the solidified ingot. Depending on the amount of gas evolved during solidification, three types of steel ingots can be produced: killed, semi-killed, and rimmed.

Killed steel is a fully deoxidized steel; that is, oxygen is removed and thus porosity is eliminated. In the deoxidation process the dissolved oxygen in the molten metal is made to react with elements such as aluminum, silicon, manganese, and vanadium that are added to the melt. These elements have an affinity for oxygen and form metallic oxides. If aluminum is used, the product is called aluminum-killed steel. The term *killed* comes from the fact that the steel lies quietly after being poured into the mold.

If sufficiently large, the oxide inclusions in the molten bath float out and adhere to, or are dissolved in, the slag. A fully killed steel is thus free from porosity caused by gases. It is also free from *blowholes* (large spherical holes near the surfaces of the ingot). Consequently, the chemical and mechanical properties of killed steels are relatively uniform throughout the ingot. However, because of metal shrinkage during solidification, an ingot of this type develops a *pipe* at its top. Also called a *shrinkage cavity*, it has a funnel-like appearance. This pipe can comprise a substantial portion of the ingot and has to be cut off and scrapped.

Semi-killed steel is a partially deoxidized steel. It contains some porosity, generally in the upper central section of the ingot but has little or no pipe; thus scrap is reduced. Piping in semi-killed steels is less because it is compensated for by the presence of porosity in that region. Semi-killed steels are economical to produce.

In a *rimmed steel*, which generally has a low carbon content (less than 0.15 percent), the evolved gases are only partially killed or controlled by the addition of elements such as aluminum. The gases form blowholes along the outer rim of the ingot—hence the term *rimmed*. Blowholes are generally not objectionable unless they break through the outer skin of the ingot. Rimmed steels have little or no piping and have a ductile skin with good surface finish. However, blowholes may break through the skin if they are not controlled properly. Furthermore, impurities and inclusions tend to segregate toward the center of the ingot. Products made from this steel may thus be defective and should be inspected.

5.7.2 Continuous casting

The traditional method of casting ingots is a batch process; that is, each ingot has to be stripped from its mold after solidification and processed individually. Piping and microstructural and chemical variations also are present throughout the ingot. These problems are alleviated by continuous casting processes, which produce higher quality steels for less cost.

Conceived in the 1860s, *continuous*, or *strand*, *casting* was first developed for casting nonferrous metal strip. Now used for steel production, the process yields major improvements in efficiency and productivity and significant reductions in cost. One system for continuous casting is shown schematically in Fig. 5.23. The molten metal in the ladle is cleaned and equalized in temperature by blowing nitrogen gas through it for 5 to 10 minutes. The metal is then poured into a refractory-lined intermediate pouring vessel (*tundish*) where impurities are skimmed off. The tundish holds as much as three tons of metal. The molten metal travels through water-cooled copper molds and begins to solidify as it travels downward along a path supported by rollers (*pinch rolls*).

Before starting the casting process, a solid *starter*, or *dummy bar*, is inserted into the bottom of the mold. The molten metal is then poured and solidifies on the starter bar (Fig. 5.24). The bar is withdrawn at the same rate the metal is poured. The cooling rate is such that the metal develops a solidified skin (shell) to support itself during its travel downward, at speeds typically 25 mm/s (1 in./s). The shell thickness at the exit end of the mold is about 12–18 mm (0.5–0.75 in.) Additional cooling is provided by water sprays along the travel path of the solidifying metal. The molds are generally coated with graphite or similar solid lubricants to reduce friction and adhesion at the mold–metal interfaces. The molds are vibrated to further reduce friction and sticking.

The continuously cast metal may be cut into desired lengths by shearing or torch cutting, or it may be fed directly into a rolling mill for further reductions in thickness and for shape rolling of products such as channels and I-beams. In addition to costing less, continuously cast metals have more uniform compositions

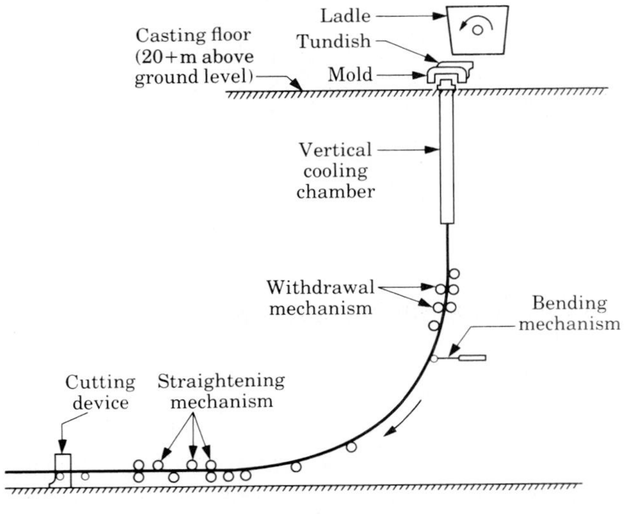

FIGURE 5.23
Schematic illustration of continuous casting of steel. Because of high efficiency and favorable economics, continuous casting is rapidly replacing traditional steelmaking processes, as well as for nonferrous metals. *Source:* USX Corporation.

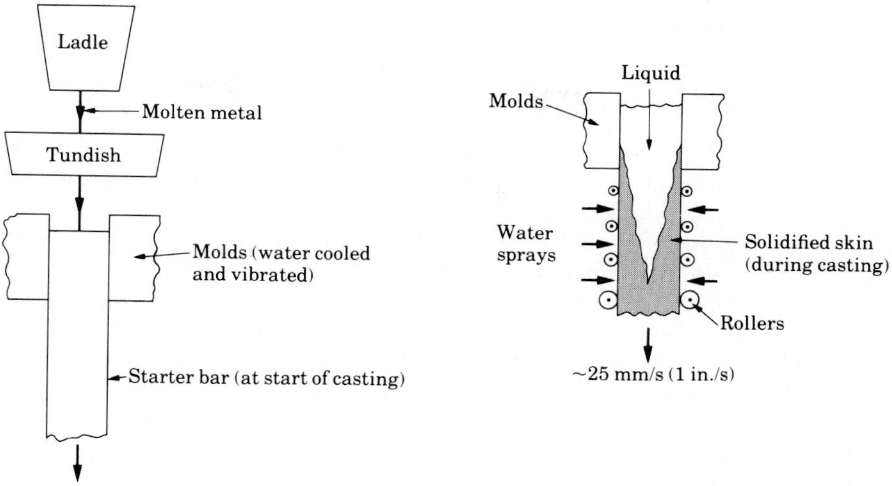

FIGURE 5.24
Method of starting a continuous casting process with a starter bar, and the cross-section of a continuously cast metal as it begins to solidify.

and properties than those obtained by ingot casting. Although the thickness of steel strand is usually about 250 mm (10 in.), new developments have reduced this thickness to about 25 mm (1 in.) or less. The thinner strand reduces the number of rolling operations required and improves the economy of the overall operation.

After they are hot rolled, steel plates or shapes undergo one or more further processes, such as cleaning and pickling by chemicals to remove surface oxides, cold rolling to improve strength and surface finish, annealing, and coating (galvanizing or aluminizing) for resistance to corrosion.

5.8 Casting Processes: Expendable Mold

Casting processes are generally classified according to mold materials, molding processes, and methods of feeding the mold with the molten metal. The two major categories are expendable-mold and permanent-mold casting (Section 5.9). *Expendable molds* are made of sand, plaster, ceramics, and similar materials, which generally are mixed with various binders or bonding agents. These materials are refractories; that is, they have the capability to withstand the high temperatures of molten metals. After the casting has solidified, the molds in these processes are broken up to remove the casting.

5.8.1 Sand casting

The traditional method of casting metals is in sand molds and has been used for millenia. Simply stated, *sand casting* consists of placing a pattern (having the shape of the desired casting) in sand to make an imprint, incorporating a gating system, filling the resulting cavity with molten metal, allowing the metal to cool until it solidifies, breaking away the sand mold, and removing the casting. While the origins of sand casting date to ancient times, it is still the most prevalent form of casting. In the United States alone, about 15 million tons of metal are cast by this method each year. Typical parts made by sand casting are machine-tool bases, engine blocks, cylinder heads, and pump housings.

Sands. Most sand casting operations use silica sands (SiO_2). Sand is the product of the disintegration of rocks over extremely long periods of time. It is inexpensive and is suitable as mold material because of its resistance to high temperatures. There are two general types of sand: *naturally bonded* (*bank* sands) and *synthetic* (*lake* sands). Because its composition can be controlled more accurately, synthetic sand is preferred by most foundries. Several factors are important in the selection of sand for molds. Sand having fine, round grains can be closely packed and forms a smooth mold surface. Good permeability of molds and cores allows gases and steam evolved during casting to escape easily. The mold should have good collapsibility (the casting shrinks while cooling) to avoid defects in the casting, such as hot tearing and cracking. The selection of sand involves certain tradeoffs with respect to properties. For example, fine-grained sand enhances mold strength, but the fine grains also lower mold permeability. Sand is typically conditioned before use. *Mulling* machines are used to uniformly mull (mix thoroughly) sand with additives. Clay is used as a cohesive agent to bond sand particles, giving the sand strength.

Types of sand molds. Sand molds are characterized by the types of sand that comprise them and by the methods used to produce them. There are three basic types of sand molds: green-sand, cold-box, and no-bake molds.

The most common mold material is *green molding sand*. The term *green* refers to the fact that the sand in the mold is moist or damp while the metal is being poured into it. Green molding sand is a mixture of sand, clay, and water. Green-sand molding is the least expensive method of making molds.

In the *skin-dried* method the mold surfaces are dried, either by storing the mold in air or drying it with torches. Skin-dried molds are generally used for large castings because of their higher strength. Sand molds are also oven dried (*baked*) prior to receiving the molten metal. They are stronger than green-sand molds and impart better dimensional accuracy and surface finish to the casting. However, distortion of the mold is greater, the castings are more susceptible to hot tearing because of the lower collapsibility of the mold, and the production rate is slower because of the drying time required.

In the *cold-box mold* process, various organic and inorganic binders are blended into the sand to bond the grains chemically for greater strength. These molds are dimensionally more accurate than green-sand molds but are more expensive.

In the *no-bake mold* process, a synthetic liquid resin is mixed with the sand, and the mixture hardens at room temperature. Because bonding of the mold in this and the cold-box process takes place without heat, they are called *cold-setting processes*.

Major components of sand molds, as depicted in Fig. 5.15, are:

- The mold itself, which is supported by a *flask*. Two-piece molds consist of a *cope* on top and a *drag* on the bottom. The seam between them is the parting line. When more than two pieces are used, the additional parts are called *cheeks*.
- A *pouring basin* or *pouring cup*, into which the molten metal is poured.
- A *sprue*, through which the molten metal flows downward.
- A *gate*, which is located at the base of the sprue. Molds typically contain a system of gates constructed to minimize turbulence in the molten metal and control flow so that metal is supplied at a rate to adequately supply the critical section thickness of the casting. Gating systems often include passageways called *runners*.
- *Risers*, which supply additional metal to the casting as it shrinks during solidification. Figure 5.15 shows two different types of risers: a *blind riser* and an *open riser*.
- *Cores*, which are inserts made from sand. They are placed in the mold to form hollow regions or otherwise define the interior surface of the casting.
- *Vents*, which are placed in molds to carry off gases produced when the molten metal comes into contact with the sand in the molds and core.

Patterns. Patterns are used to mold the sand mixture into the shape of the casting. They may be made of wood, plastic, or metal. The selection of a pattern material depends on the size and shape of the casting, the dimensional accuracy and the quantity of castings required, and the molding process to be used. Because patterns are used repeatedly to make molds, the strength and durability of the material selected for patterns must reflect the number of castings the mold will produce. Patterns may be made of a combination of materials to reduce wear in critical regions. Patterns are usually coated with a *parting agent* to facilitate their removal from the molds.

Patterns can be designed with a variety of features to fit application and economic requirements. *One-piece patterns* are generally used for simpler shapes and low-quantity production. They are generally made of wood and are inexpensive. *Split patterns* are two-piece patterns made so that each part forms a portion of the cavity for the casting. In this way castings having complicated shapes can be produced. *Match-plate patterns* are a popular type of mounted pattern in which two-piece patterns are constructed by securing each half of one or more split patterns to the opposite sides of a single plate (see Fig. 5.25). In such constructions, the gating system can be mounted on the drag side of the pattern. This type of pattern most often is used in conjunction with molding machines and large production runs.

Cores. For castings with internal cavities or passages, such as in an automotive engine block or a valve body, cores are utilized. Cores are placed in the mold cavity before casting to form the interior surfaces of the casting and are removed from the finished part during shakeout and further processing. Like molds, cores must possess strength, permeability, ability to withstand heat, and collapsibility. Therefore cores are made of sand aggregates.

The core is anchored by *core prints*. These are recesses that are added to the pattern to support the core and to provide vents for the escape of gases. A common problem with cores is that for certain casting requirements, as in the case where a recess is required, they may lack sufficient structural support in the cavity. To keep the core from shifting, metal supports, known as *chaplets*, may be used to anchor the core in place.

Cores are generally made in a manner similar to that used in making molds, and the majority are made with shell, no-bake, or cold-box processes. Cores are formed in *core boxes*, which are used much like patterns are used to form sand molds. The sand can be packed into the boxes with sweeps or blown into the box by compressed air from *core blowers*. Core blowers have the advantages of producing uniform cores and operating at a very high production rate.

Sand-molding machines. The oldest known method of molding, which is still used for simple castings, is to compact the sand by hand hammering or ramming it around the pattern. For most operations, however, the sand mixture is compacted around the pattern by molding machines. These machines eliminate arduous labor,

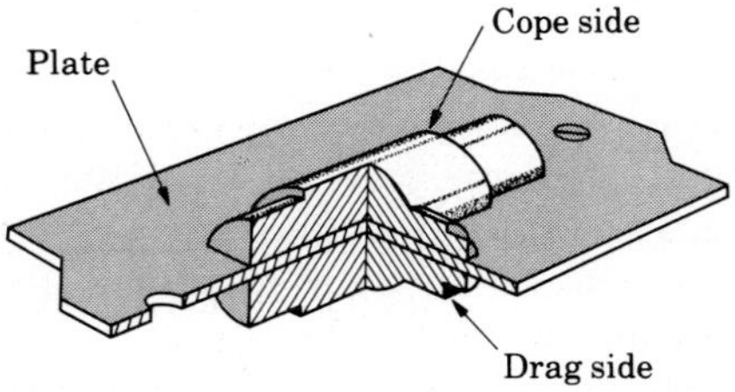

FIGURE 5.25
A typical metal match-plate pattern used in sand casting.

offer higher quality casting by improving the application and distribution of forces, manipulate the mold in a carefully controlled fashion, and increase the rate of production.

Mechanization of the molding process can be further assisted by *jolting* the assembly. The flask, molding sand, and pattern are placed on a pattern plate mounted on an anvil, and jolted upward by air pressure at rapid intervals. The inertial forces compact the sand around the pattern.

In *vertical flaskless molding* the halves of the pattern form a vertical chamber wall against which sand is blown and compacted. Then the mold halves are packed horizontally, with the parting line oriented vertically and moved along a pouring conveyor. This operation is simple and eliminates the need to handle flasks, making potential production rates very high, particularly when other aspects of the operation, such as coring and pouring, are automated.

Sandslingers fill the flask uniformly with sand under a stream of high pressure. Sandthrowers are used to fill large flasks and are typically operated by machine. An impeller in the machine throws sand from its blades or cups at such high speeds that the machine not only places the sand but also rams it appropriately.

In *impact molding* the sand is compacted by controlled explosion or instantaneous release of compressed gases. This method produces molds with uniform strength and good permeability.

In *vacuum molding*, also known as the "*V*" *process*, the pattern is covered tightly by a thin sheet of plastic. A flask is placed over the coated pattern and is filled with sand. A second sheet of plastic is placed on top of the sand, and a vacuum action hardens the sand so that the pattern can be withdrawn. Both halves of the mold are made this way and then assembled. During pouring, the mold remains under a vacuum but the casting cavity does not. When the metal has solidified, the vacuum is turned off and the sand falls away, releasing the casting. Vacuum molding produces castings having very good detail and accuracy. It is especially well suited for large, relatively flat castings.

The sand casting operation. After the mold has been shaped and the cores have been placed in position, the two halves (*cope* and *drag*) are closed, clamped, and weighted down. They are weighted to prevent the separation of the mold sections under the pressure exerted when the molten metal is poured into the mold cavity.

The design of the gating system is important for proper delivery of the molten metal into the mold cavity. Turbulence must be minimized, air and gases must be allowed to escape by vents or other means, and proper temperature gradients must be established and maintained to eliminate shrinkage and porosity. The design of risers is also important for supplying the necessary molten metal during solidification of the casting; the pouring basin may also serve as a riser. After solidification, the casting is shaken out of its mold, and the sand and oxide layers adhering to the casting are removed by vibration (using a shaker) or by sand blasting. The risers and gates are cut off by oxyfuel-gas cutting, sawing, shearing, and abrasive wheels, or they are trimmed in dies.

Almost all metals can be sand cast. The surface finish obtained is largely a function of the materials used in making the mold. Dimensional accuracy is not as good as that of other casting processes. However, intricate shapes such as cast-iron engine blocks and very large propellers for ocean liners and impellers can be cast by this process. Sand casting can be economical for relatively small production runs as well as large production runs; equipment costs are generally low. The characteristics of sand casting and other casting processes are given in Table 5.5.

5.8.2 Shell molding

The use of *shell molding* has grown significantly, because it can produce many types of castings with close tolerances and good surface finishes at a low cost. In this process, a mounted pattern, made of a ferrous metal or aluminum, is heated to 175–370 °C (350–700 °F), coated with a parting agent such as silicone, and clamped to a box or chamber containing a fine sand containing a 2.5–4.0 percent thermosetting resin binder, such as phenol-formaldehyde, which coats the sand particles. The sand mixture is blown over the heated pattern, coating it evenly. The assembly is often—but not always—placed in an oven for a short period of time to complete the curing of the resin. The shell hardens around the pattern and is removed from the pattern using built-in ejector pins. Two half-shells are made in this manner and are bonded or clamped together in preparation for pouring.

The shells are light and thin, usually 5–10 mm (0.2–0.4 in.), and consequently their thermal characteristics are different from those for thicker molds. The thin shells allow gases to escape during solidification of the metal. The mold is generally used vertically and is supported by surrounding it with steel shot in a cart. The walls of the mold are relatively smooth, offering low resistance to flow of the molten metal and producing castings with sharper corners, thinner sections, and smaller projections than are possible in green-sand molds. With the use of multiple gating systems, several castings can be made in a single mold.

Shell molding may be more economical than other casting processes, depending on various production factors, particularly energy cost. The relatively high cost of metal patterns becomes a smaller factor as the size of production run increases. The high quality of the finished casting can significantly reduce cleaning, machining, and other finishing costs. Complex shapes can be produced with less labor, and the process can be automated fairly easily. Shell-molding applications include small mechanical parts requiring high precision, such as gear housings, cylinder heads, and connecting rods. Shell molding is also widely used in producing high-precision molding cores.

Composite molds. Composite molds are made of two or more different materials and are used in shell molding and other casting processes. They are generally utilized in casting complex shapes, such as impellers for turbines. An

TABLE 5.5
GENERAL CHARACTERISTICS OF CASTING PROCESSES

PROCESS	TYPICAL MATERIALS CAST	WEIGHT (kg)		TYPICAL SURFACE FINISH (μm, R_a)	POROSITY*	SHAPE COMPLEXITY*	DIMENSIONAL ACCURACY*	SECTION THICKNESS (mm)	
		MINIMUM	*MAXIMUM*					*MINIMUM*	*MAXIMUM*
Sand	All	0.05	No limit	5–25	4	1–2	3	3	No limit
Shell	All	0.05	100+	1–3	4	2–3	2	2	—
Plaster	Nonferrous (Al, Mg, Zn, Cu)	0.05	50+	1–2	3	1–2	2	1	—
Investment	All (High-melting pt.)	0.005	100+	1–3	3	1	1	1	75
Permanent mold	All	0.5	300	2–3	2–3	3–4	1	2	50
Die	Nonferrous (Al, Mg, Zn, Cu)	<0.05	50	1–2	1–2	3–4	1	0.5	12
Centrifugal	All	—	5000+	2–10	1–2	3–4	3	2	100

* Relative rating: 1 best, 5 worst. *Note:* These ratings are only general; significant variations can occur, depending on the methods used.

example of composite molds is shown in Fig. 5.26. Molding materials commonly used are shells (made as previously described), plaster, sand with binder, metal, and graphite. Composite molds may also include cores and chills to control the rate of solidification in critical areas of castings. Composite molds increase the strength of the mold, improve the dimensional accuracy and surface finish of castings, and may help reduce overall costs and processing time.

5.8.3 Evaporative pattern casting

The *evaporative pattern casting* process uses a polystyrene pattern, which evaporates upon contact with molten metal to form a cavity for the casting. The process is also known as *lost-foam* or *lost-pattern* casting and under the trade name *Full-Mold* process. First, raw polystyrene beads, containing 5–8 percent pentane (a volatile hydrocarbon) are placed in a preheated die, usually made of aluminum. The polystyrene expands and takes the shape of the die cavity. Additional heat is applied to fuse and bond the beads together. The die is then cooled, opened, and the polystyrene pattern is removed.

The pattern is then coated with a refractory slurry (a watery mixture), dried, and placed in a flask. The flask is filled with sand, which surrounds and supports the pattern. The sand may be dried or mixed with bonding agents to give it additional strength. Then, without removing the polystyrene pattern, the molten metal is poured into the mold. This action immediately vaporizes the pattern and fills the mold cavity, completely replacing the space previously occupied by the polystyrene.

This process is relatively simple because there are no parting lines or cores. Inexpensive flasks are sufficient for the process, and the need for cleaning and

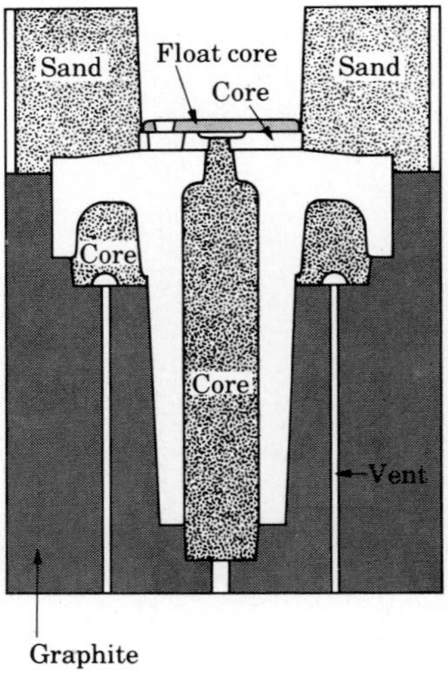

FIGURE 5.26 Schematic illustration of a semipermanent composite mold. *Source: Steel Castings Handbook,* 5th ed. Steel Founders' Society of America, 1980.

machining are relatively low. Polystyrene is inexpensive, can be easily processed into a wide variety of shapes, and thus can be used to make complex castings of various sizes economically. Typical applications are cylinder heads, crankshafts, brake components, and manifolds for automobiles and machine bases.

5.8.4 Plaster-mold casting

In the *plaster-mold casting* process the mold is made of plaster of paris (gypsum, or calcium sulfate), with the addition of talc and silica flour to improve strength and control the time required for the plaster to set. These components are mixed with water, and the resulting slurry is poured over the pattern. After the plaster sets, usually within 15 minutes, the pattern is removed and the mold is dried at 120–260 °C (250–500 °F) to remove the moisture. Higher drying temperatures may be used, depending on the type of plaster. The mold halves are then assembled to form the mold cavity and preheated to about 120 °C (250 °F) for 16 hours. The molten metal is then poured into the mold.

Because plaster molds have very low permeability, gases evolved during solidification of the metal cannot escape. Consequently, the molten metal is poured either in a vacuum or under pressure. Plaster-mold permeability can be increased substantially by the *Antioch* process. The molds are dehydrated in an autoclave (pressurized oven) for 6–12 hours, then rehydrated in air for 14 hours. Another method of increasing permeability is to use foamed plaster, containing trapped air bubbles.

Patterns for plaster molding are generally made of aluminum alloys, thermosetting plastics, brass, or zinc alloys. Wood patterns are not suitable for making a large number of molds, because the patterns are repeatedly subjected to the water-based plaster slurry which causes the wood to swell. Because there is a limit to the maximum temperature that the plaster mold can withstand, generally about 1200 °C (2200 °F), plaster-mold casting is used only for aluminum, magnesium, zinc, and some copper-base alloys. The castings have fine detail and good surface finish. Because plaster molds have lower thermal conductivity than other types of molds, the castings cool slowly, yielding more uniform grain structure with less warpage.

This process and the ceramic-mold and investment casting processes are known as *precision casting* because of the high dimensional accuracy and good surface finish obtained. Typical parts made are lock components, gears, valves, fittings, tooling, and ornaments.

5.8.5 Ceramic-mold casting

The *ceramic-mold casting* process is similar to the plaster-mold process, with the exception that it uses refractory mold materials suitable for high-temperature applications. The process is also called *cope-and-drag investment casting*. The slurry is a mixture of fine-grained zircon ($ZrSiO_4$), aluminum oxide, and fused silica, which are mixed with bonding agents and poured over the pattern (Fig. 5.27), which has been placed in a flask.

The pattern may be made of wood or metal. After setting, the molds (ceramic facings) are removed, dried, burned off to remove volatile matter, and baked. The molds are clamped firmly and used as all-ceramic molds. In the *Shaw* process, the ceramic facings are backed by fireclay (clay used in making firebricks that resist high temperatures) to give the molds strength. The facings are then assembled into a complete mold, ready to be poured.

The high-temperature resistance of the refractory molding materials allows these molds to be used in casting ferrous and other high-temperature alloys, stainless steels, and tool steels. The castings have good dimensional accuracy and surface finish over a wide range of sizes and intricate shapes, but the process is somewhat expensive. Typical parts made are impellers, cutters for machining, dies for metalworking, and molds for making plastic or rubber components.

5.8.6 Investment casting

The *investment-casting* process, also called the *lost-wax* process, was first used during the period 4000–3000 B.C. The pattern is made of wax or a plastic such as polystyrene. The sequences involved in investment casting are shown in Fig. 5.28.

Solid mold. The pattern is made by injecting *semisolid* or *liquid* wax or plastic into a metal die in the shape of the pattern. The pattern is then removed and dipped into a slurry of refractory material, such as very fine silica and binders, ethyl silicate, and acids. After this initial coating has dried, the pattern is coated repeatedly to increase its thickness. The term *investment* comes from investing the

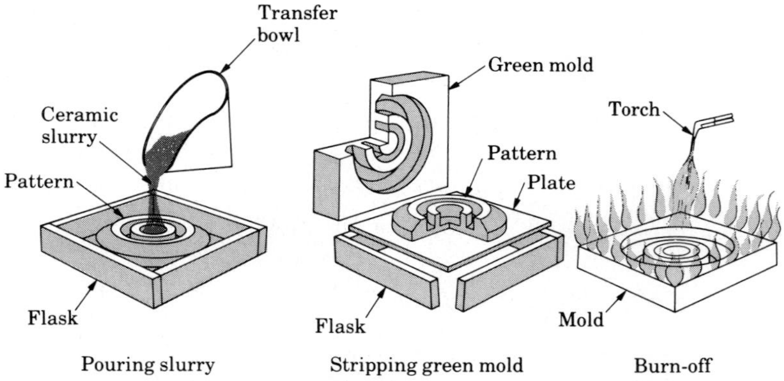

FIGURE 5.27
Sequence of operations in making a ceramic mold. *Source: Metals Handbook,* vol. 5, 8th ed.

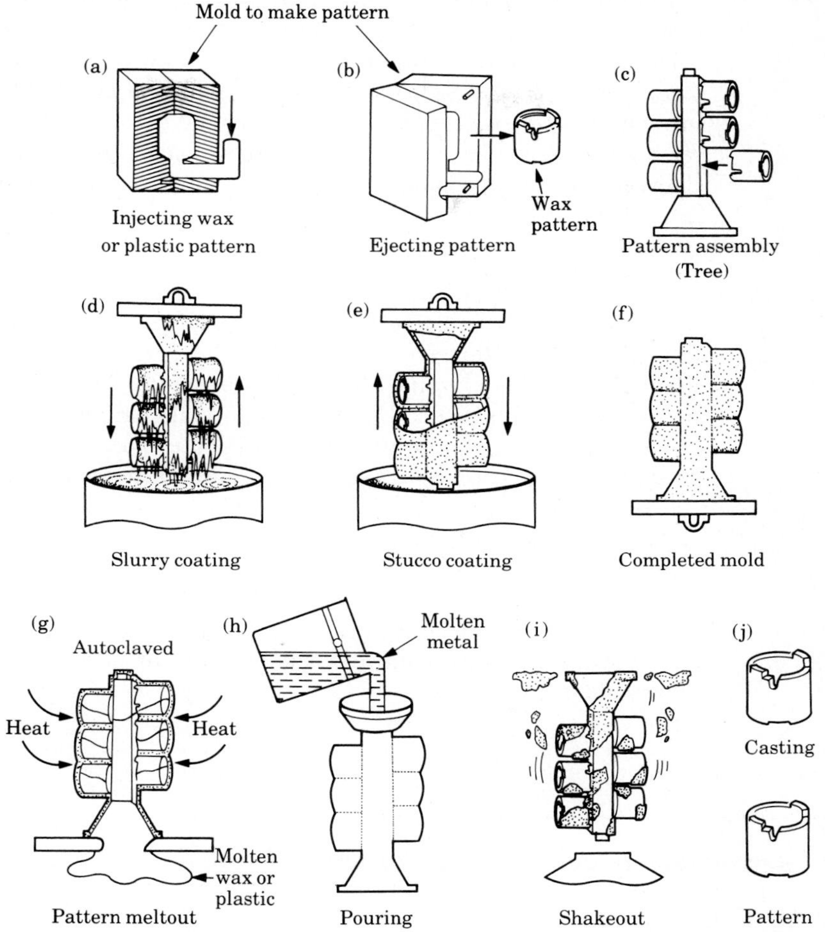

FIGURE 5.28
Schematic illustration of investment casting (lost-wax process). Castings by this method can be made with very fine detail and from a variety of metals. *Source:* Steel Founders' Society of America.

pattern with the refractory material. Wax patterns require careful handling because they are not strong enough to withstand the forces involved during mold making.

The one-piece mold is dried in air and heated to a temperature of 90–175 °C (200–350 °F) for about 4 hours, depending on the metal to be cast, to drive off the water of crystallization (chemically combined water). After the metal has been poured and has solidified, the mold is broken up and the casting is removed. A number of patterns can be joined to make one mold called a *tree*, thus increasing the production rate.

Although the labor and materials involved make the lost-wax process costly, it is suitable for casting high-melting-point alloys with a good surface finish and close tolerances. Thus little or no finishing is required, which otherwise would add significantly to the total cost of the casting. This process is capable of producing intricate shapes, with parts weighing from 5 g to 100 kg, from a wide variety of ferrous and nonferrous metals and alloys. Typical parts made are components for office equipment; mechanical components such as gears, cams, valves, and ratchets; and jewelry.

Ceramic-shell investment casting. A variation of the investment-casting process is *ceramic-shell casting*. It uses the same type of wax or plastic pattern, which is dipped first in a slurry with colloidal silica or ethyl silicate binder, then into a fluidized bed of fine-grained fused silica or zircon flour. The pattern is then dipped into coarse-grain silica to build up additional coatings and thickness to withstand the thermal shock of pouring. This process is economical and is used extensively for precision casting of steels, aluminum, and high-temperature alloys.

If ceramic cores are used in the casting, they are removed by leaching with caustic solutions under high pressure and temperature. The molten metal may be poured in a vacuum to extract evolved gases and reduce oxidation, thus improving the quality of the casting. To further reduce microporosity, the castings made by this and other processes are subjected to hot isostatic pressing. Aluminum castings, for example, are subjected to a gas pressure of up to about 100 MPa (15 ksi) at 500 °C (900 °F).

5.9 Casting Processes: Permanent Mold

Permanent molds, as the name implies, are used repeatedly and are designed so that the casting can be easily removed and the mold reused. These molds are made of metals that maintain their strength at high temperatures and thus can be used repeatedly. Because metal molds are better heat conductors than expendable molds, the solidifying casting is subjected to a higher rate of cooling, which in turn affects the microstructure and grain size within the casting, as described in Section 5.3.

In permanent-mold casting, two halves of a mold are made from materials such as cast iron, steel, bronze, refractory metal alloys, or graphite (semipermanent mold). The mold cavity and gating system are machined into the mold and thus become an integral part of it. To produce castings with internal cavities, cores made of metal or sand aggregate are placed in the mold prior to casting. Typical core materials are shell or no-bake cores, gray iron, low-carbon steel, and hot-work die steel. Gray iron is the most commonly used core material, particularly for large molds for aluminum and magnesium castings. Inserts are also used for various parts of the mold.

In order to increase the life of permanent molds, the surfaces of the mold cavity are coated with a refractory slurry or sprayed with graphite every few castings. These coatings also serve as parting agents and thermal barriers, controlling the rate of cooling of the casting. Mechanical ejectors, such as pins located in various parts of the mold, may be needed for removal of complex castings. Ejectors usually leave small round impressions on castings and gating systems.

The molds are clamped together by mechanical means and heated to facilitate metal flow and reduce thermal damage to the dies. The molten metal is then poured through the gating system. After solidification, the molds are opened and the casting is removed. Special methods of cooling the mold include water or the use of fins, similar to those found on motorcycle or lawnmower engines that cool the engine block.

Although the permanent-mold casting operation can be performed manually, the process can be automated for large production runs. This process is used mostly for aluminum, magnesium, and copper alloys because of their generally lower melting points. Steels can also be cast in graphite or heat-resistant metal molds.

This process produces castings with good surface finish, close tolerances, uniform and good mechanical properties and at high production rates. Typical parts made by permanent-mold castings are automobile pistons, cylinder heads, and connecting rods, gear blanks for appliances, and kitchenware. Although equipment costs can be high because of die costs, the process can be mechanized, thus keeping labor costs low. Permanent-mold casting is not economical for small production runs. Furthermore, because of the difficulty in removing the casting from the mold, intricate shapes cannot be cast by this process. However, easily collapsed sand cores can be used and removed from castings to leave intricate internal cavities.

5.9.1 Slush casting

We noted in Fig. 5.18 that a solidified skin develops first in a casting and that this skin becomes thicker with time. Hollow castings with thin walls can be made by permanent-mold casting using this principle—a process called *slush casting*. The molten metal is poured into the metal mold, and after the desired thickness of solidified skin is obtained, the mold is inverted or slung, and the remaining liquid metal is poured out. The mold halves are then opened and the casting is removed. The process is suitable for small production runs and is generally used for making ornamental and decorative objects and toys from low-melting-point metals, such as zinc, tin, and lead alloys.

5.9.2 Pressure casting

In the two permanent-mold processes that we just described, the molten metal flows into the mold cavity by gravity. In the *pressure-casting* process, also called *pressure pouring* or *low-pressure casting* (Fig. 5.29), the molten metal is forced upward by gas pressure into a graphite or metal mold. The pressure is maintained until the metal has completely solidified in the mold. The molten metal may also be forced upward by a vacuum, which also removes dissolved gases and produces a casting with lower porosity.

5.9.3 Die casting

The *die-casting* process, developed in the early 1900s, is a further example of permanent-mold casting. The molten metal is forced into the die cavity at pressures ranging from 0.7 to 700 MPa (0.1 to 100 ksi). The European term *pressure die casting*, or simply die casting, that we describe in this section, is not to be confused with the term *pressure casting* that we described in Section 5.9.2. Typical parts made by die casting are carburetors, motors, business machine and appliance components, hand tools, and toys. The weight of most castings ranges from less than 90 g (3 oz) to about 25 kg (55 lb).

Hot-chamber process. The *hot-chamber* process (Fig. 5.30) involves the use of a piston, which traps a certain volume of molten metal and forces it into the die cavity through a gooseneck and nozzle. The pressures range up to 35 MPa (5000 psi), with an average of about 15 MPa (2000 psi). The metal is held under pressure until it solidifies in the die. To improve die life and to aid in rapid metal cooling, thus reducing cycle time, dies are usually cooled by circulating water or oil through various passageways in the die block. Cycle times usually range up to 900 shots (individual injections) per hour for zinc, although very small components such as zipper teeth can be cast at 18,000 shots per hour. Low-melting-point alloys such as zinc, tin, and lead are commonly cast by this process.

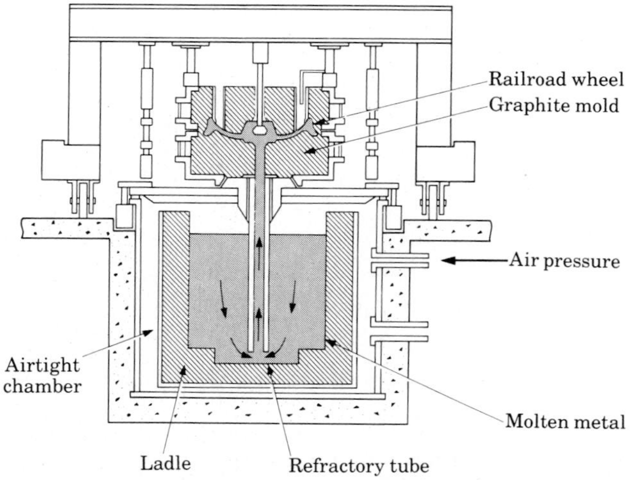

FIGURE 5.29 The bottom-pressure casting process utilizes graphite molds for the production of steel railroad wheels. *Source:* Griffin Wheel Division of Amsted Industries Incorporated.

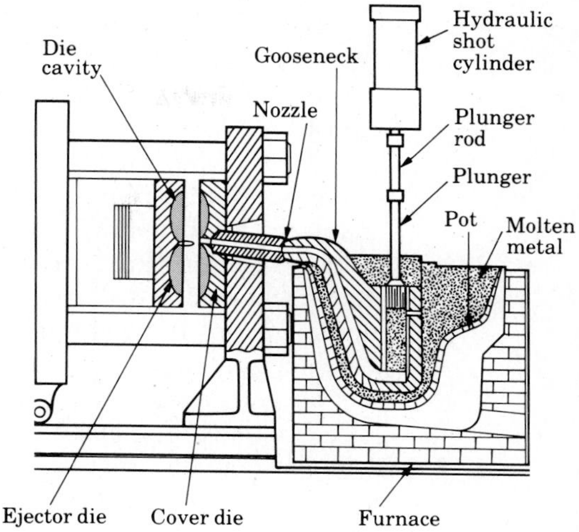

FIGURE 5.30 Sequence of steps in die casting of a part in the hot-chamber process. *Source:* Courtesy of *Foundry Management and Technology.*

Cold-chamber process. In the *cold-chamber* process (Fig. 5.31) molten metal is poured into the injection cylinder (*shot chamber*) with a ladle. The shot chamber is not heated—hence the term *cold* chamber. The metal is forced into the die cavity at pressures usually ranging from 20 MPa to 70 MPa (3 ksi to 10 ksi), although they may be as high as 150 MPa (20 ksi). The machines may be horizontal or vertical; in the latter the shot chamber is vertical and the machine is similar to a vertical press.

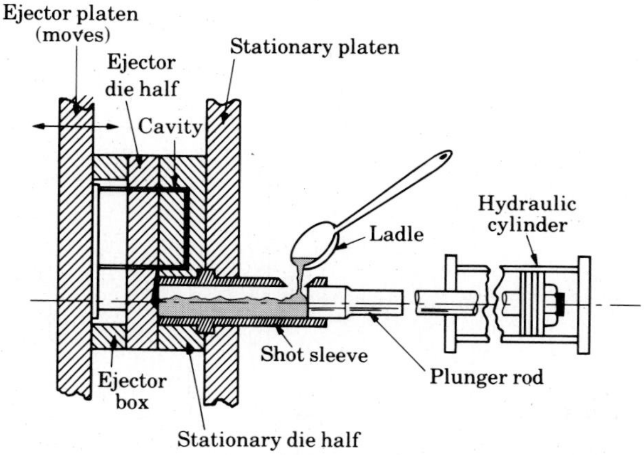

FIGURE 5.31 Sequence of operations in die casting of a part in the cold-chamber process. *Source:* Courtesy of *Foundry Management and Technology.*

High-melting-point alloys of aluminum, magnesium, and copper are normally cast by this method, although other metals (including ferrous metals) can also be cast in this manner. Molten metal temperatures start at about 600 °C (1150 °F) for aluminum and magnesium alloys and increase considerably for copper-base and iron-base alloys.

Process capabilities and machine selection. Because of the high pressures involved, the dies have a tendency to part unless clamped together tightly. Die-casting machines are rated according to the clamping force that can be exerted to keep the dies closed. The capacities of commercially available machines range from about 25 tons to 3000 tons. Other factors involved in the selection of die-casting machines are die size, piston stroke, shot pressure, and cost.

Die-casting dies (Fig. 5.32) may be made single cavity, multiple cavity (several identical cavities), combination cavity (several different cavities), or unit dies (simple small dies that can be combined in two or more units in a master holding die). Dies are usually made of hot-work die steels or mold steels. Die wear increases with the temperature of the molten metal. *Heat checking* of dies (surface cracking from repeated heating and cooling of the die) can be a problem. When die materials are selected and maintained properly, dies may last more than half a million shots before die wear becomes significant.

Die design includes taper (draft) to allow the removal of the casting. The sprues and runners may be removed either manually or by using trim dies in a press. The entire die-casting and finishing process can be highly automated. Lubricants (parting agents) are usually applied as thin coatings on die surfaces. Alloys, except magnesium alloys, generally require lubricants. They usually have a water base, with graphite or other compounds in suspension. Because of the high cooling capacity of water, these lubricants also are effective in keeping die temperatures low.

Die casting has the capability for high production rates with good strength, high-quality parts with complex shapes, and good dimensional accuracy and surface detail, thus requiring little or no subsequent machining or finishing operations. Components such as pins, shafts, and fasteners can be cast integrally

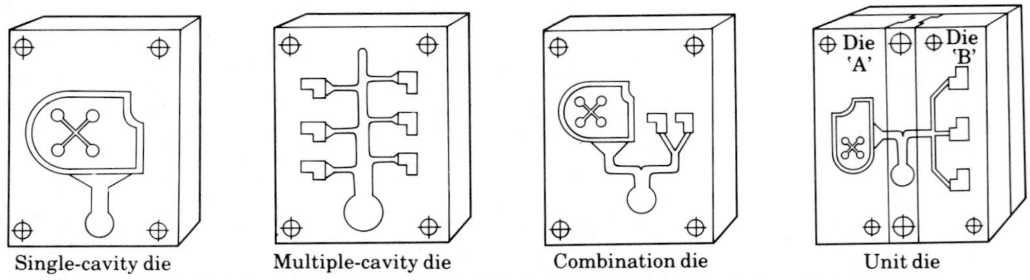

FIGURE 5.32
Various types of cavities in die-casting dies. *Source:* Courtesy of American Die Casting Institute.

(*insert molding*), a process similar to putting wooden sticks (the pin) in popsicles (the casting). Ejector marks remain, as do small amounts of flash (thin material squeezed out between the dies) at the die parting line.

In the fabrication of certain parts, die casting can compete favorably with other manufacturing methods, such as sheet-metal stamping and forging, or other casting processes. Additionally, because the molten metal chills rapidly at the die walls, the casting has a fine-grain, hard skin with higher strength than in the center. Consequently, the strength-to-weight ratio of die-cast parts increases with decreasing wall thickness. With good surface finish and dimensional accuracy, die casting can produce bearing surfaces that would normally be machined. Equipment costs, particularly the cost of dies, are somewhat high, but labor costs are generally low because the process is semi- or fully automated. Die casting is economical for large production runs.

5.9.4 Centrifugal casting

As its name implies, the *centrifugal casting* process utilizes the inertial forces caused by rotation to distribute the molten metal into the mold cavities. This method was first suggested in the early 1800s. There are three types of centrifugal casting: true centrifugal casting, semicentrifugal casting, and centrifuging.

True centrifugal casting. In *true centrifugal casting*, hollow cylindrical parts, such as pipes, gun barrels, and lampposts, are produced by the technique shown in Fig. 5.33, in which molten metal is poured into a rotating mold. The axis of rotation is usually horizontal but can be vertical for short workpieces. Molds are made of steel, iron, or graphite and may be coated with a refractory lining to increase mold life. Mold surfaces can be shaped so that pipes with various outer shapes, including square or polygonal, can be cast. The inner surface of the casting remains cylindrical because the molten metal is uniformly distributed by centrifugal forces.

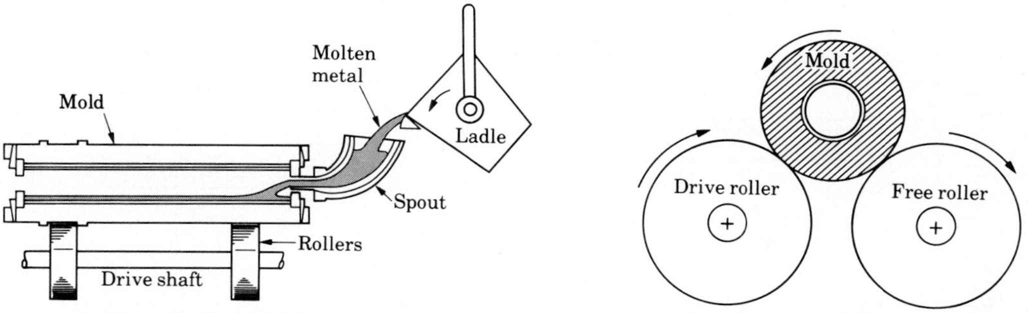

FIGURE 5.33
Schematic illustration of the centrifugal casting process. Pipes, cylinder liners, and similarly shaped parts can be cast by this process.

However, because of density differences, lighter elements such as dross, impurities, and pieces of the refractory lining tend to collect on the inner surface of the casting.

Cylindrical parts ranging from 13 mm (0.5 in.) to 3 m (10 ft) in diameter and 16 m (50 ft) long can be cast centrifugally, with wall thicknesses ranging from 6 mm to 125 mm (0.25 in. to 5 in.). The pressure generated by the centrifugal force is high, as much as 150 gs, and is necessary for casting thick-walled parts. Castings of good quality, dimensional accuracy, and external surface detail are obtained by this process. In addition to pipes, typical parts made are bushings, engine cylinder liners, and bearing rings with or without flanges.

Semicentrifugal casting. An example of *semicentrifugal casting* is shown in Fig. 5.34(a). This method is used to cast parts with rotational symmetry, such as a wheel with spokes.

Centrifuging. In *centrifuging*, also called *centrifuge casting*, mold cavities of any shape are placed at a certain distance from the axis of rotation. The molten metal is poured at the center and is forced into the mold by centrifugal forces (Fig. 5.34b). The properties within the castings vary by the distance from the axis of rotation.

5.9.5 Squeeze casting

The *squeeze-casting* process, developed in the 1960s, involves solidification of the molten metal under high pressure. Thus it is a combination of casting and forging (Fig. 5.35). The machinery includes a die, punch, and ejector pin. The pressure applied by the punch keeps the entrapped gases in solution, and the high-pressure contact at the die–metal interface promotes rapid heat transfer, resulting in a fine microstructure with good mechanical properties.

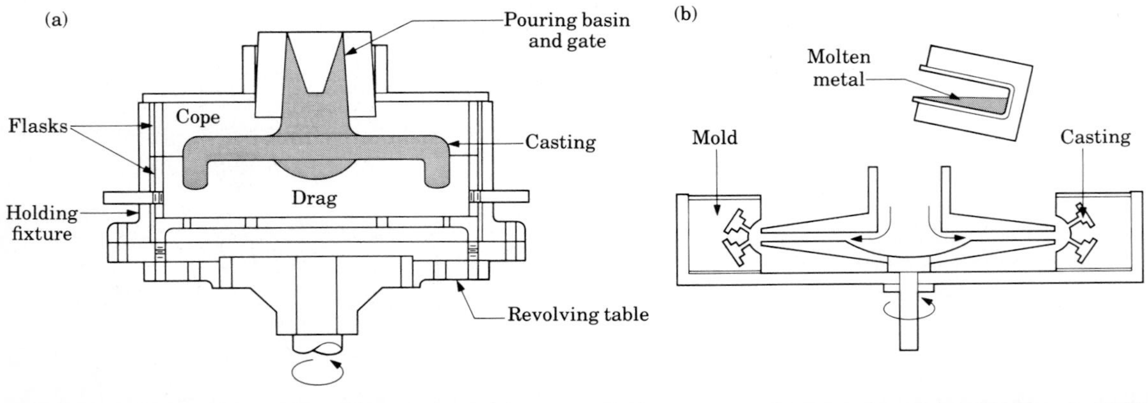

FIGURE 5.34
(a) Schematic illustration of the semicentrifugal casting process. Wheels with spokes can be cast by this process. (b) Schematic illustration of casting by centrifuging. The molds are placed at the periphery of the machine, and the molten metal is forced into the molds by centrifugal forces.

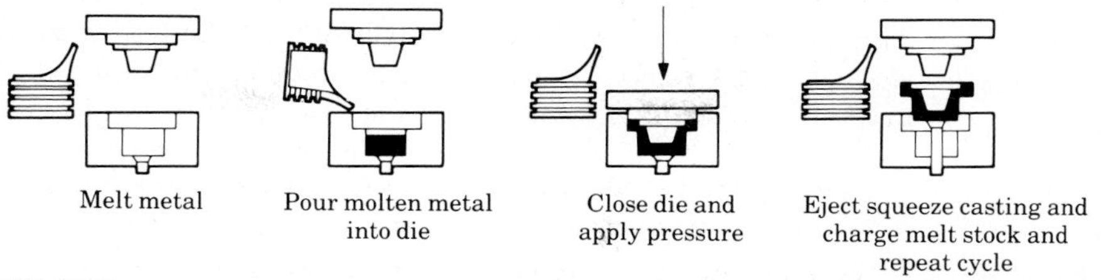

FIGURE 5.35
Sequence of operations in the squeeze-casting process. This process combines the advantages of casting and forging.

Parts can be made to near-net shape, with complex shapes and fine surface detail, from both nonferrous and ferrous alloys. Typical products made are automotive wheels and mortar bodies (a short-barreled cannon). The pressures required in squeeze casting are lower than those for hot or cold forging.

5.9.6 Casting techniques for single-crystal components

We illustrate these techniques by describing the casting of gas turbine blades, which are generally made of nickel-base superalloys. The procedures involved can also be used for other alloys and components.

Conventional casting of turbine blades. The *conventional casting* process involves investment casting using a ceramic mold as shown in Fig. 5.36(a). The

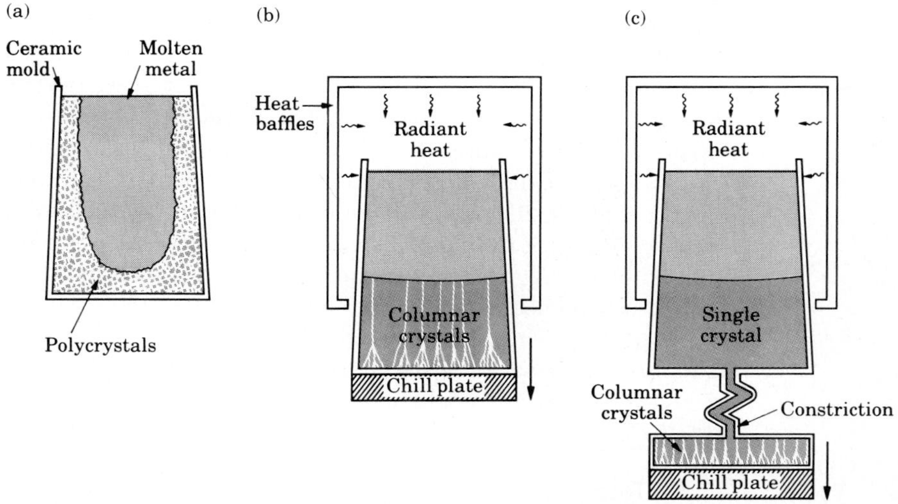

FIGURE 5.36
Three methods of casting turbine blades: (a) conventional casting with ceramic mold; (b) directional solidification; and (c) method to produce single-crystal blade. (See Fig. 3.1 for photographs of these blades.) *Source:* B. H. Kear, *Scientific American,* October 1986.

molten metal is poured into the mold and begins to solidify at the ceramic walls. The grain structure developed is polycrystalline, and the presence of grain boundaries makes this structure susceptible to creep and cracking along those boundaries under the centrifugal forces at elevated temperatures.

Directionally solidified blades. In the *directional solidification* process (Fig. 5.36b), first developed in 1960, the ceramic mold is prepared basically by investment casting techniques and is preheated by radiant heating. The mold is supported by a water-cooled chill plate. After the metal is poured into the mold, the assembly is lowered slowly. Crystals begin to grow at the chill-plate surface and upward. The blade is thus directionally solidified, with longitudinal but no transverse grain boundaries. Consequently, the blade is stronger in the direction of centrifugal forces developed in the gas turbine.

Single-crystal blades. In the growing process for single-crystal blades, developed in 1967, the mold is prepared basically by investment casting techniques and has a constriction in the shape of a corkscrew (Fig. 5.36c), the cross-section of which allows only one crystal through. As the assembly is lowered slowly, a single crystal grows upward through the constriction and begins to grow in the mold. Strict control of the rate of movement is necessary. The solidified mass in the mold is a single-crystal blade. Although more expensive than other blades, the lack of grain boundaries makes these blades resistant to creep and thermal shock. Thus they have a longer and more reliable service life.

Single-crystal growing. With the advent of the semiconductor industry, *single-crystal growing* has become a major activity in the manufacture of microelectronic devices. There are basically two methods of crystal growing. In the *crystal pulling* method, known as the *Czochralski* process (Fig. 5.37a), a seed crystal is dipped into the molten metal and then pulled out slowly, at a rate of about 10 μm/s, while being rotated at about 1 rev/s. The liquid metal begins to solidify on the seed, and the crystal structure of the seed is continued throughout. *Dopants* (alloying elements) may be added to the liquid metal to impart special electrical properties. Single crystals of silicon, germanium, and various other elements are grown by this process. Single-crystal ingots typically 50–150 mm (2–6 in.) in diameter and over 1 m (40 in.) in length have been produced by this technique.

The second technique for crystal growing is the *floating-zone* method (Fig. 5.37b). Starting with a rod of polycrystalline silicon resting on a single crystal, an induction coil heats these two pieces while moving slowly upward. The single crystal grows upward while maintaining its orientation. Thin wafers are then cut from the rod, cleaned, and polished for use in microelectronic device fabrication.

5.9.7 Rapid solidification

First developed in the 1960s, *rapid solidification* involves cooling of molten metal at rates as high as 10^6 K/s, whereby the metal does not have sufficient time to crystallize. These alloys are called *amorphous alloys* or *metallic glasses* because they do not have a long-range crystalline structure. They typically contain iron, nickel,

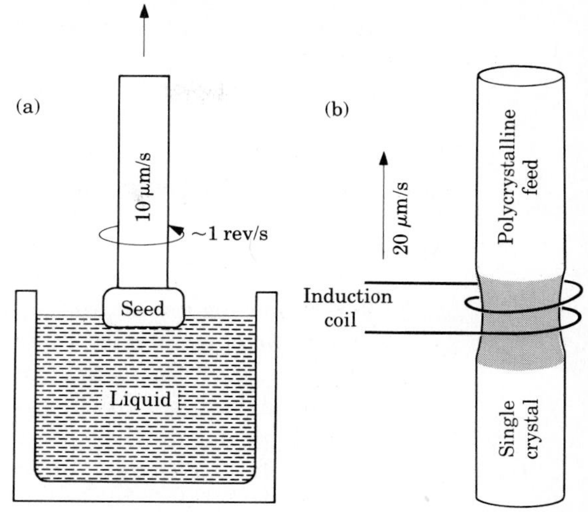

FIGURE 5.37 Two methods of crystal growing: (a) crystal pulling (Czochralski process) and (b) floating-zone method. Crystal growing is especially important in the semiconductor industry. *Source:* L. H. Van Vlack, *Materials for Engineering.* Addison-Wesley Publishing Co., Inc., 1982.

and chromium, which are alloyed with carbon, phosphorus, boron, aluminum, and silicon. Among other effects, rapid solidification results in a significant extension of solid solubility, grain refinement, and reduced microsegregation.

Amorphous alloys exhibit excellent corrosion resistance, good ductility, and high strength. Furthermore, they exhibit very little loss from magnetic hysteresis, high resistance to eddy currents, and high permeability. The latter properties are utilized in making magnetic steel cores for transformers, generators, motors, lamp ballasts, magnetic amplifiers, and linear accelerators, with greatly improved efficiency. Another major application is superalloys of rapidly solidified powders consolidated into near-net shapes for use in aerospace engines.

These alloys are produced in the form of wire, ribbon, strip, and powder. In one method, called *melt spinning* (Fig. 5.38), the alloy is melted by induction in a

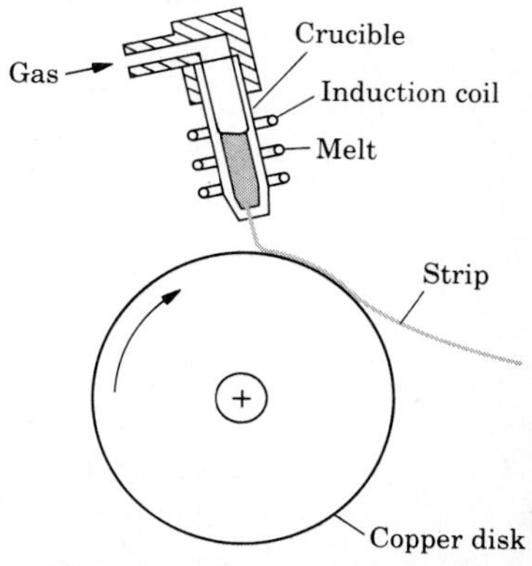

FIGURE 5.38 Schematic illustration of the melt-spinning process to produce thin strips of amorphous metal.

ceramic crucible and propelled under high gas pressure at very high speed against a rotating copper disk (chill block) and chills rapidly (splat cooling).

5.10 Processing of Castings

The various microstructures developed during metal processing can be modified by *heat treatment* techniques, that is, by controlled heating and cooling of the alloys at various rates. These treatments induce phase transformations that greatly influence mechanical properties such as strength, hardness, ductility, toughness, and wear resistance of the alloys.

The effects of thermal treatment depend primarily on the alloy, its composition and microstructure, the degree of prior cold work, and the rates of heating and cooling during heat treatment. The processes of recovery, recrystallization, and grain growth (see Section 3.6) are examples of thermal treatment, involving changes in the grain structure of the alloy.

5.10.1 Heat treating ferrous alloys

In this section we describe the microstructural changes that occur in the iron–carbon system discussed in Section 5.2.5.

Pearlite. If the ferrite and cementite lamellae in the pearlite structure (see Section 5.2.6) of the eutectoid steel are thin and closely packed, the microstructure is called *fine pearlite*. If the lamellae are thick and widely spaced, it is called *coarse pearlite*. The difference between the two depends on the rate of cooling through the eutectoid temperature, a reaction in which austenite is transformed into pearlite. If the rate of cooling is relatively high, as in air, fine pearlite is produced; if slow, as in a furnace, coarse pearlite is produced.

The transformation from austenite to pearlite (and for other structures) is best illustrated by Fig. 5.39(b) and (c). These diagrams are called *isothermal transformation* (IT) *diagrams* or *time–temperature–transformation* (TTT) diagrams. They are constructed from the data in Fig. 5.39(a), which shows the percentage of austenite transformed into pearlite as a function of temperature and time. The higher the temperature and/or the longer the time, the greater is the percentage of austenite transformed to pearlite. Note that for each temperature a minimum time is required for the transformation to begin and that some time later all the austenite is transformed to pearlite.

Spheroidite. When pearlite is heated to just below the eutectoid temperature and held at that temperature for a period of time, say, for a day at 700 °C (1300 °F),

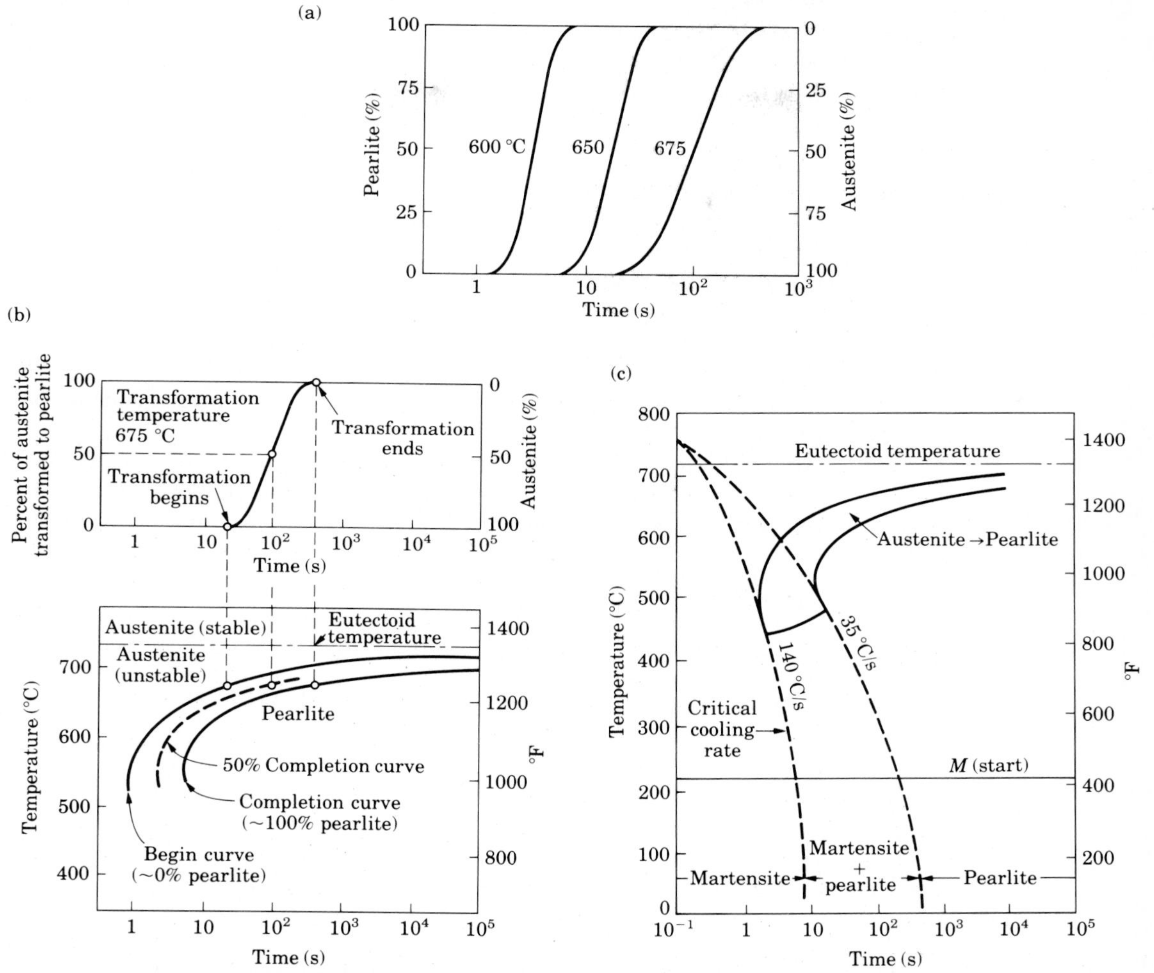

FIGURE 5.39

(a) Austenite to pearlite transformation of iron–carbon alloy as a function of time and temperature. (b) Isothermal transformation diagram obtained from (a) for a transformation temperature of 675 °C (1247 °F). (c) Microstructures obtained for a eutectoid iron–carbon alloy as a function of cooling rate. *Source:* ASM International.

the cementite lamellae (see Fig. 5.7) transform to *spherical* shapes. Unlike the lamellar shape of cementite, which acts as stress raisers, *spheroidites* (spherical particles) are less conducive to stress concentration because of their rounded shapes. Consequently, this structure has higher toughness and lower hardness than the pearlite structure. In this form it can be cold worked since the ductile ferrite has high toughness, and the spheroidal carbide particles prevent the propagation of cracks within the material.

Bainite. Visible only using electron microscopy, *bainite* has a very fine microstructure, consisting of ferrite and cementite. It can be produced in steels with alloying elements and at cooling rates that are higher than those required for transformation to pearlite. This structure, called *bainitic steel*, is generally stronger and more ductile than pearlitic steel at the same hardness level.

Martensite. When austenite is cooled rapidly, as by quenching in water, its fcc structure is transformed to a tetragonal body-centered structure (see Fig. 5.6d). It is a body-centered rectangular prism, which is slightly elongated along one of its principal axes called *martensite*. Because it does not have as many slip systems as a bcc structure, and the carbon is in interstitial positions, martensite is extremely hard and brittle, lacks toughness, and thus has limited use. Martensite transformation takes place almost instantaneously (Fig. 5.39c), because it does not involve the diffusion process, a time-dependent phenomenon that is the mechanism in other transformations.

Transformations involve volume changes because of the different densities of the various phases in the structure. For example, when austenite transforms to martensite, its volume increases (hence its density decreases) by as much as 4 percent. A similar but smaller volume expansion also occurs when austenite transforms to pearlite. These expansions, and thus the thermal gradients present in a quenched part, can cause internal stresses within the body. These stresses may cause parts to crack during heat treatment, as in *quench cracking* of steels caused by rapid cooling during quenching.

Retained austenite. If the temperature to which the alloy is quenched is not sufficiently low, only a portion of the structure is transformed to martensite. The rest is *retained austenite*, which is visible as white areas in the structure along with dark needlelike martensite. Retained austenite can cause dimensional instability and cracking and lowers alloy hardness and strength.

Tempered martensite. *Tempering* is a heating process that reduces martensite's hardness and improves its toughness. The body-centered tetragonal martensite is heated to an intermediate temperature, where it transforms to a two-phase microstructure, consisting of body-centered cubic alpha ferrite and small particles of cementite. Longer tempering time and higher temperature decrease martensite's hardness. The reason is that the cementite particles coalesce and grow, and the distance between the particles in the soft ferrite matrix increases as the less stable, smaller carbide particles dissolve.

Hardenability of ferrous alloys. The capability of an alloy to be hardened by heat treatment is called its *hardenability*. It is a measure of the depth of hardness that can be obtained by heating and subsequent quenching. The term hardenability should not be confused with hardness, which is the resistance of a material to indentation or scratching. Hardenability of ferrous alloys depends on the carbon content, the grain size of the austenite, and the alloying elements present in the material. A test (the *Jominy* test) has been developed in order to determine alloy hardenability.

Quenching media. The fluid used for quenching the heated alloy also affects hardenability. Quenching may be carried out in water, brine (saltwater), oils, molten salts, or air. Caustic solutions, polymer solutions, and gases are also used. Because of the differences in the thermal conductivity, specific heat, and heat of vaporization of these media, the rate of cooling of the alloy (*severity of quench*) will also be different.

In relative terms and in decreasing order, the cooling capacity of several quenching media is: agitated brine, 5; still water, 1; still oil, 0.3; cold gas, 0.1; and still air, 0.02. Agitation is also a significant factor in the rate of cooling. In tool steels the quenching medium is specified by a letter (see Table 3.4), such as W for water hardening, O for oil hardening, and A for air hardening. The cooling rate also depends on the surface area-to-volume ratio of the part (see Eq. 5.9). The higher this ratio, the higher is the cooling rate. Thus, for example, a thick plate cools more slowly than a thin plate with the same surface area.

Water is a common medium for rapid cooling. However, the heated metal may form a *vapor blanket* along its surfaces from water-vapor bubbles that form when water boils at the metal–water interface. This blanket creates a barrier to heat conduction because of the lower thermal conductivity of the vapor. Agitating the fluid or the part helps to reduce or eliminate the blanket. Also, water may be sprayed on the part under high pressure. Brine is an effective quenching medium because salt helps to nucleate bubbles at the interfaces, thus improving agitation. However, brine can corrode the part. *Die quenching* is a term used to describe the process of clamping the part to be heat treated to a die, which chills selected regions of the part. In this way cooling rates and warpage can be controlled.

5.10.2 Heat treating nonferrous alloys and stainless steels

Nonferrous alloys and some stainless steels generally cannot be heat treated by the techniques used with ferrous alloys. The reason is that nonferrous alloys do not undergo phase transformations as steels do. The hardening and strengthening mechanisms for these alloys are fundamentally different.

Heat-treatable aluminum alloys, copper alloys, and martensitic and precipitation-hardening stainless steels are hardened and strengthened by a process called *precipitation hardening*. This is a technique in which small particles—of a different phase and called *precipitates*—are uniformly dispersed in the matrix of the original phase (see Fig. 5.2a). In this process precipitate forms, because the solid solubility of one element (one component of the alloy) in the other is exceeded.

Three stages are involved in precipitation hardening. We can best describe them by referring to Fig. 5.40, the phase diagram for the aluminum–copper system. For an alloy with the composition of 95.5% Al–4.5% Cu, a single-phase (kappa) substitutional solid-solution of copper (solute) in aluminum (solvent) exists between 500 °C and 570 °C (930 °F and 1060 °F). The kappa phase is aluminum rich, with an fcc structure, and is ductile. Below the lower temperature, that is, below the lower solubility curve, two phases are present: kappa and theta (a hard intermetallic compound of $CuAl_2$). This alloy can be heat treated and its properties modified by solution treatment or precipitation.

Solution treatment. In *solution treatment* the alloy is heated to within the solid-solution kappa phase, say, 540 °C (1000 °F), and cooled rapidly, such as by quenching in water. The structure obtained soon after quenching (*A* in Fig. 5.40b), consists only of the single phase kappa. This alloy has moderate strength and considerable ductility.

Precipitation hardening. The structure obtained in *A* in Fig. 5.40(b) can be strengthened by precipitation hardening. The alloy is reheated to an intermediate temperature and held there for a period of time, during which precipitation takes place. The copper atoms diffuse to nucleation sites and combine with aluminum

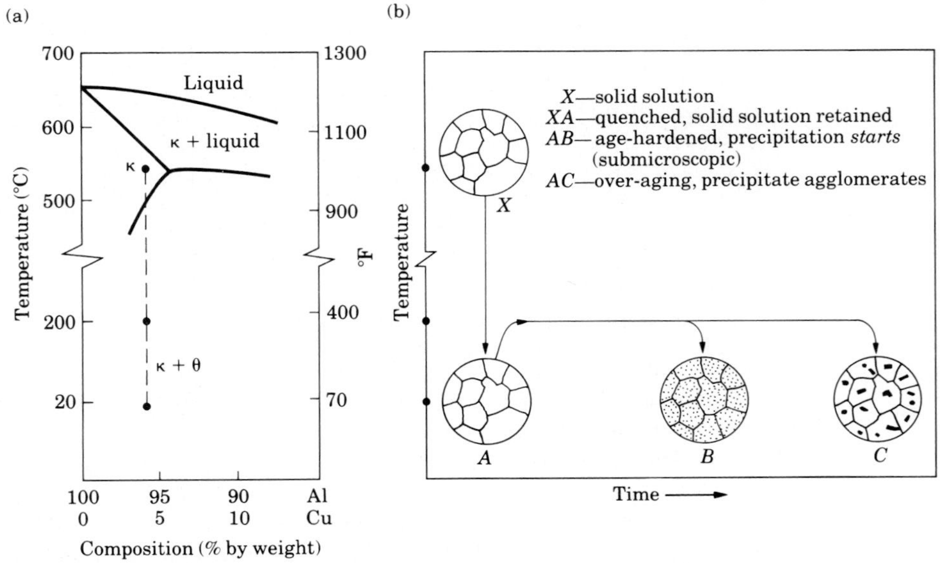

FIGURE 5.40 (a) Phase diagram for the aluminum–copper alloy system. (b) Various microstructures obtained during the age-hardening process. *Source:* L. H. Van Vlack, *Materials for Engineering.* Addison-Wesley Publishing Co., Inc., 1982. Reprinted with permission.

atoms, producing the theta phase, which form as submicroscopic precipitates, shown in *B* by the small dots within the grains of the kappa phase. This structure is stronger than that in *A*, although it is less ductile. The increase in strength is attributed to increased resistance to dislocation movement in the region of the precipitates.

Aging. Because the precipitation process is one of time and temperature, it is also called *aging*, and the property improvement is known as *age hardening*. If carried out above room temperature, the process is called *artificial aging*. However, several aluminum alloys harden and become stronger over a period of time at room temperature, by a process known as *natural aging*. Such alloys are first quenched and then, if desired, are formed at room temperature into various shapes and allowed to gain strength and hardness by natural aging. Natural aging can be slowed by refrigerating the quenched alloy.

In the precipitation process, if the reheated alloy is held at that temperature for an extended period of time, the precipitates begin to coalesce and grow. They become larger but fewer, as shown by the larger dots in *C* in Fig. 5.40(b). This process is called *overaging*, which makes the alloy softer and weaker. Thus there is an optimal time–temperature relationship in the aging process for obtaining desired properties. Obviously, an aged alloy can be used only up to a certain maximum temperature in service; otherwise it will overage and lose its strength and hardness. Although weaker, an overaged part has better dimensional stability.

Maraging. Maraging is a precipitation-hardening treatment for a special group of high-strength iron-base alloys. The word *maraging* is derived from the words martensite and aging. In this process one or more intermetallic compounds are precipitated in a matrix of low-carbon martensite. A typical maraging steel may contain 18 percent nickel, in addition to other elements, and aging is done at 480 °C (900 °F). Hardening by maraging does not depend on the cooling rate. Thus full uniform hardness can be obtained throughout large parts with minimal distortion. Typical uses of maraging steels are for dies and tooling for casting, molding, forging, and extrusion.

5.10.3 Case hardening

The heat treatment processes we have described so far involve microstructural alterations and property changes in the *bulk* of the material or component by *through hardening*. In many situations, however, alteration of only the *surface* properties of a part—hence the term *case hardening*—is desirable. This method is particularly useful for improving resistance to surface indentation, fatigue, and wear. Typical applications for case hardening are gear teeth, cams, shafts, bearings, fasteners, pins, automotive clutch plates, and tools and dies. Through hardening of these parts would not be desirable, because a hard part lacks the necessary toughness for these applications. A small surface crack can propagate rapidly through the part and cause total failure.

Various surface-hardening processes are available (Table 5.6): *carburizing* (gas, liquid, and pack carburizing), *carbonitriding*, *cyaniding*, *nitriding*, *boronizing*, and *flame* and *induction hardening*. Basically, these are heat-treating operations in which the component is heated in an atmosphere containing elements (such as carbon, nitrogen, or boron) that alter the composition, microstructure, and properties of surfaces.

For steels with sufficiently high carbon content, surface hardening takes place without using any of these additional elements. Only the heat-treatment processes described in Section 5.10.1 are needed to alter the microstructures, usually by flame hardening or induction hardening. Laser beams and electron beams are also used effectively to harden both small and large surfaces and also for through hardening of relatively small parts.

Because case hardening is a localized heat treatment, case-hardened parts have a hardness gradient. Typically, the hardness is greatest at the surface and decreases below the surface, the rate of decrease depending on the composition of the metal and the process variables. Surface-hardening techniques can also be used for tempering, thus modifying the properties of surfaces that have been subjected to heat treatment. Various other processes and techniques for surface hardening, such as shot peening and surface rolling, improve wear resistance and various other characteristics (see Section 4.5.1).

Decarburization is the phenomenon in which alloys containing carbon lose carbon from their surfaces as a result of heat treatment or hot working in a medium, usually oxygen, that reacts with the carbon. Decarburization is undesirable because it affects the hardenability of the surfaces of the part by lowering the carbon content. It also adversely affects the hardness, strength, and fatigue life of steels by significantly lowering their endurance limits. Decarburization is best avoided by processing the alloy in an inert atmosphere or a vacuum or using neutral salt baths during heat treatment.

5.10.4 Annealing

Annealing is a general term used to describe the restoration of a cold-worked or heat-treated metal or alloy to its original properties, so as to increase ductility (hence formability), reduce hardness and strength, or modify the microstructure. Annealing is also used to relieve residual stresses in a manufactured part for improved machinability and dimensional stability. (The term annealing also applies to thermal treatment of glasses and similar products and weldments.)

The annealing process involves (a) heating the workpiece to a specific range of temperature, (b) holding it at that temperature for a period of time (soaking), and (c) cooling it slowly. The process may be carried out in an inert or controlled atmosphere or performed at low temperatures to prevent or minimize surface oxidation. Annealing temperatures may be higher than the recrystallization temperature, depending on the degree of cold work (hence stored energy). For example, the recrystallization temperature for copper ranges between 200 °C and 300 °C (400 °F and 600 °F), whereas the annealing temperature needed to fully recover the original

TABLE 5.6
OUTLINE OF HEAT TREATMENT PROCESSES FOR SURFACE HARDENING

PROCESS	METALS HARDENED	ELEMENT ADDED TO SURFACE	PROCEDURE	GENERAL CHARACTERISTICS	TYPICAL APPLICATIONS
Carburizing	Low-carbon steel (0.2%C), alloy steels (0.08–0.2% C)	C	Heat steel at 870–950 °C (1600–1750 °F) in an atmosphere of carbonaceous gases (gas carburizing) or carbon-containing solids (pack carburizing). Then quench.	A hard, high-carbon surface is produced. Hardness 55 to 65 *HRC*. Case depth $<$0.5–1.5 mm ($<$0.020 to 0.060 in.) Some distortion of part during heat treatment.	Gears, cams, shafts, bearings, piston pins, sprockets, clutch plates
Carbonitriding	Low-carbon steel	C and N	Heat steel at 700–800 °C (1300–1600 °F) in an atmosphere of carbonaceous gas and ammonia. Then quench in oil.	Surface hardness 55 to 62 *HRC*. Case depth 0.07 to 0.5 mm (0.003 to 0.020 in.) Less distortion than in carburizing.	Bolts, nuts, gears
Cyaniding	Low-carbon steel (0.2% C), alloy steels (0.08–0.2% C)	C and N	Heat steel at 760–845 °C (1400–1550 °F) in a molten bath of solutions of cyanide (e.g., 30% sodium cyanide) and other salts.	Surface hardness up to 65 *HRC*. Case depth 0.025 to 0.25 mm (0.001 to 0.010 in.) Some distortion.	Bolts, nuts, screws, small gears
Nitriding	Steels (1% Al, 1.5% Cr, 0.3% Mo), alloy steels (Cr, Mo), stainless steels, high-speed tool steels	N	Heat steel at 500–600 °C (925–1100 °F) in an atmosphere of ammonia gas or mixtures of molten cyanide salts. No further treatment.	Surface hardness up to 1100 *HV*. Case depth 0.1 to 0.6 mm (0.005 to 0.030 in.) and 0.02 to 0.07 mm (0.001 to 0.003 in.) for high-speed steel.	Gears, shafts, sprockets, valves, cutters, boring bars, fuel-injection pump parts
Boronizing	Steels	B	Part is heated using boron-containing gas or solid in contact with part.	Extremely hard and wear resistant surface. Case depth 0.025–0.075 mm (0.001–0.003 in.)	Tool and die steels
Flame hardening	Medium-carbon steels, cast irons	None	Surface is heated with an oxyacetylene torch, then quenched with water spray or other quenching methods.	Surface hardness 50 to 60 *HRC*. Case depth 0.7 to 6 mm (0.030 to 0.25 in.) Little distortion.	Gear and sprocket teeth, axles, crankshaft, piston rod, lathe beds and centers
Induction hardening	Same as above	None	Metal part is placed in copper induction coils and is heated by high-frequency current, then quenched.	Same as above.	Same as above.

properties ranges from 260 °C to 650 °C (500 °F to 1200 °F), depending on the degree of prior cold work.

Full annealing is a term applied to annealing ferrous alloys, generally low- and medium-carbon steels. The steel is heated to above A_1 or A_3 (Fig. 5.41), and cooling takes place slowly, say, 10 °C (20 °F) per hour, in a furnace after it is turned off. The structure obtained in full annealing is coarse pearlite, which is soft and ductile and has small uniform grains. Excessive softness in the annealing of steels can be avoided if the entire cooling cycle is carried out in still air. This process is called *normalizing*, in which the part is heated to a temperature of A_3 or A_{cm} to transform the structure to austenite. It results in somewhat higher strength and hardness and lower ductility than in full annealing. The structure obtained is fine pearlite with small uniform grains. Normalizing is generally done to refine the grain structure, obtain uniform structure (*homogenization*), decrease residual stresses, and improve machinability.

Process annealing. In *process annealing* (also called *intermediate annealing*, *subcritical annealing*, or *in-process annealing*) the workpiece is annealed to restore its ductility, part or all of which may have been exhausted by work hardening during cold working. In this way, the part can be worked further into the final desired shape. If the temperature is high and/or the time of annealing is long, grain growth may result, with adverse effects on formability of annealed parts.

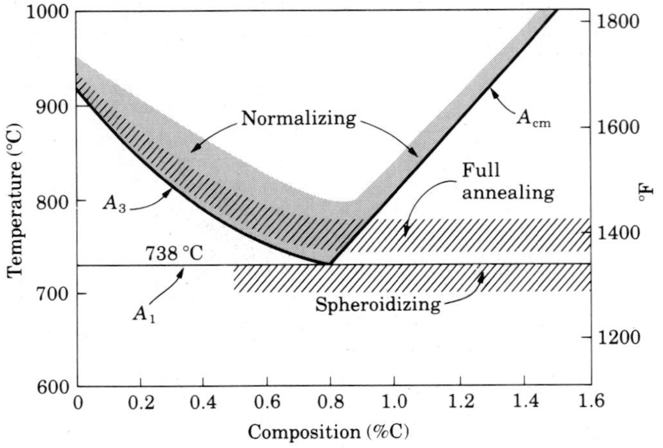

FIGURE 5.41 Heat-treating temperature ranges for plain-carbon steels, as indicated on the iron–iron carbide phase diagram. *Source:* ASM International.

Stress-relief annealing. To reduce or eliminate residual stresses, a workpiece is generally subjected to *stress-relief annealing*, or simply *stress relieving*. The temperature and time required for this process depend on the material and the magnitude of residual stresses present. The residual stresses may have been induced during forming, machining, or other shaping processes or caused by volume changes during phase transformations. For steels, the part is heated to below A_1, thus avoiding phase transformations. Slow cooling rates, such as in still air, are generally employed. Stress relieving promotes dimensional stability in situations where subsequent relaxing of residual stresses present may cause distortion of the part over a period of time when it is in service. It also reduces the tendency for stress-corrosion cracking.

Tempering. If steels are hardened by heat treatment, *tempering* (or *drawing*, not to be confused with wire drawing or deep drawing, described in Chapters 6 and 7, respectively), is used in order to reduce brittleness, increase ductility and toughness, and reduce residual stresses. The term tempering is also used for glasses. In tempering, the steel is heated to a specific temperature, depending on composition, and cooled at a prescribed rate. Alloy steels may undergo *temper enbrittlement*, caused by the segregation of impurities along the grain boundaries at temperatures between 480 °C and 590 °C (900 °F and 1100 °F).

Austempering. In *austempering*, the heated steel is quenched from the austenitizing temperature rapidly enough to avoid formation of ferrite or pearlite. It is then held at a certain temperature until isothermal transformation from austenite to bainite is complete. It is then cooled to room temperature, usually in still air, at a moderate rate to avoid thermal gradients within the part. The quenching medium most commonly used is molten salt, at temperatures ranging from 160 °C to 750 °C (320 °F to 1380 °F).

Austempering is often substituted for conventional quenching and tempering, either to reduce the tendency for cracking and distortion during quenching or to improve ductility and toughness while maintaining hardness. Because of the relatively short cycle time, this process is economical for many applications. In *modified austempering*, a mixed structure of pearlite and bainite is obtained. The best example of this practice is *patenting*, which provides high ductility and moderately high strength, such as the patented wire used in the wire industry.

Martempering (marquenching). In *martempering*, the steel or cast iron is quenched from the austenitizing temperature into a hot fluid medium, such as hot oil or molten salt. It is held at that temperature until the temperature is uniform throughout the part and then cooled at a moderate rate, such as in air, to avoid temperature gradients within the part. The part is then tempered, because the structure thus obtained is primarily untempered martensite and is not suitable for most applications. Martempered steels have less tendency to crack, distort, and develop residual stresses during heat treatment. In modified martempering the quenching temperature is lower and thus the cooling rate is higher. The process is suitable for steels with lower hardenability.

Ausforming. In *ausforming*, also called *thermomechanical processing*, the steel is formed into desired shapes within controlled ranges of temperature and time to avoid formation of nonmartensitic transformation products. The part is then cooled at various rates to obtain the desired microstructures. Ausformed parts have superior mechanical properties.

5.10.5 Design considerations for heat treating

In addition to the metallurgical factors we described, successful heat treating involves design considerations of avoiding problems such as cracking, warping, and nonuniform properties throughout the heat-treated part. The rate of cooling during quenching may not be uniform, particularly with complex shapes of varying cross-sections and thicknesses, thus producing severe temperature gradients. These gradients lead to variations in contraction, which induce thermal stresses and may cause the part to crack. Furthermore, nonuniform cooling causes residual stresses in the part, which can lead to stress-corrosion cracking. Thus the method selected and care taken in quenching, as well as the proper choice of quenching media and temperatures, are important considerations.

As a general guideline for part design for heat treating, internal or external sharp corners should be avoided. Otherwise stress concentrations at these corners raise stress levels high enough to cause cracking. Parts should have as nearly uniform thicknesses as possible, or the transition between regions of different thicknesses should be smooth. Parts with holes, grooves, keyways, splines, and unsymmetrical shapes may also be difficult to heat treat because they may crack during quenching. Large surfaces with thin cross-sections are likely to warp. Hot forgings and hot steel-mill products may have a decarburized skin and thus may not successfully respond to heat treatment.

5.10.6 Cleaning, finishing, and inspection

After solidification and removal from the mold or die, castings may be subjected to a variety of additional processes. In sand casting, the casting is shaken out of its mold and the sand and oxide layers adhering to the casting are removed by vibration (using a shaker) or by sand blasting. In sand casting and other casting processes—and depending on size—the risers and gates are cut off by oxyfuel-gas cutting, sawing, shearing, and abrasive wheels, or they are trimmed in dies. Castings may be cleaned electrochemically or by pickling with chemicals to remove surface oxides. The surface of castings is important in subsequent machining operations, because machinability can be adversely affected if the castings are not cleaned properly. If regions of the casting have not formed properly or have formed incompletely, the defects may be repaired by welding, thus filling them with weld metal. Sand castings generally have rough, grainy surfaces, although the extent depends on the quality of the mold and the materials used. Finishing operations for castings may involve straightening or forging with dies and machining to obtain final dimensions.

Inspection is an important final step and is done to ensure that the casting meets all design and quality-control requirements. Several methods are available for inspection of castings to determine quality and the presence of any defects. Castings can be inspected visually or optically for surface defects. Subsurface and internal defects are investigated using various nondestructive techniques described in Section 4.8. In destructive testing, test specimens are removed from various sections of a casting and tested for strength, ductility, and other mechanical properties and to determine the presence and location of any defects.

Pressure tightness of cast components (valves, pumps, pipes) is usually determined by sealing the openings in the casting and pressurizing it with water, oil, or air. The casting is then inspected for leaks while the pressure is maintained. Unacceptable or defective castings are melted down for reprocessing. Because of the major economic impact, the types of defects present in castings and their causes must be investigated. Control of all stages during casting, from mold preparation to removal of castings from molds or dies, is important in maintaining good quality.

5.11 Casting Design

As in all engineering practice and manufacturing operations, certain guidelines and design principles pertaining to casting have been developed over many years. Although these principles were established primarily through practical experience, analytical methods and computer-aided design and manufacturing techniques are now coming into wider use, improving productivity and the quality of castings. Moreover, careful design can result in significant cost savings.

5.11.1 Designing for expendable-mold casting

The following guidelines generally apply to all types of castings. The most significant design considerations are identified and addressed.

Corners, angles, and section thickness. Sharp corners, angles, and fillets should be avoided (Fig. 5.42a), because they may cause cracking and tearing during solidification of the metal. Fillet radii should be selected to reduce stress concentrations and to ensure proper liquid-metal flow during the pouring process. If the fillet radii are too large, the volume of the material in those regions is also large and, consequently, the rate of cooling is less.

Section changes in castings should smoothly blend into each other. The location of the largest circle that can be inscribed in a particular region is critical as far as shrinkage cavities are concerned (Fig. 5.42b and c). Because the cooling rate in regions with the larger circles is less, they are called *hot spots*. These regions could develop *shrinkage cavities* and *porosity*. Although they increase the cost of production, metal paddings in the mold can eliminate or minimize hot spots. These paddings act as external chills.

Flat areas. Large flat areas (plain surfaces) should be avoided, as they may warp because of temperature gradients during cooling or develop poor surface finish because of uneven flow of metal during pouring. Flat surfaces can be broken up with ribs and serrations.

Shrinkage. Allowance for shrinkage during solidification should be provided for, so as to avoid cracking of the casting, particularly in permanent molds. Pattern dimensions should also provide for shrinkage of the metal during solidification and cooling. Allowances for shrinkage, also known as *patternmaker's shrinkage allowances*, usually range from about 10 mm/m to 20 mm/m (1/8 in./ft to 1/4 in./ft). Table 5.7 gives the normal shrinkage allowance for metals commonly sand cast.

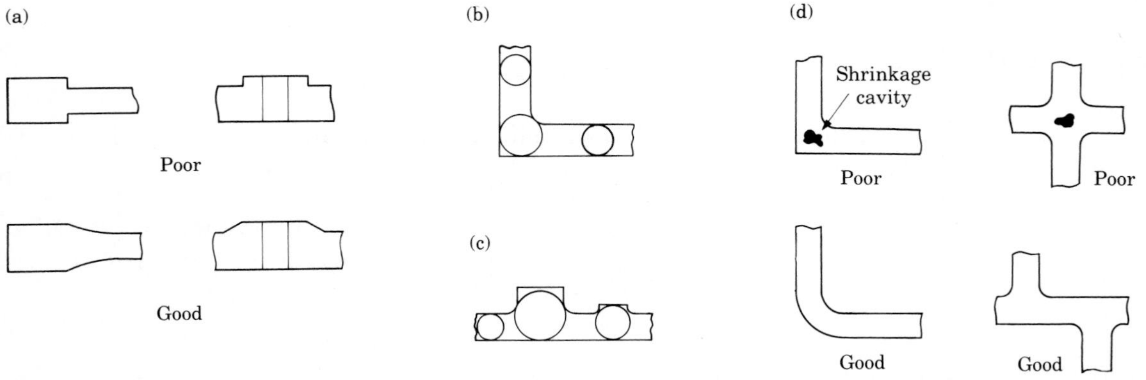

FIGURE 5.42
(a) Suggested design modifications to avoid defects in castings. Note that sharp corners are avoided to reduce stress concentrations. (b, c, d) Examples of designs showing the importance of maintaining uniform cross-sections in castings to avoid hot spots and shrinkage cavities.

TABLE 5.7
NORMAL SHRINKAGE ALLOWANCE FOR METALS CAST IN SAND MOLDS

METAL	PERCENT
Gray cast iron	0.83–1.3
White cast iron	2.1
Malleable cast iron	0.78–1.0
Aluminum alloys	1.3
Magnesium alloys	1.3
Yellow brass	1.3–1.6
Phosphor bronze	1.0–1.6
Aluminum bronze	2.1
High-manganese steel	2.6

Parting lines. In general, the parting line should be along a flat plane, rather than contoured. Whenever possible, the parting line should be at the corners or edges of castings, rather than on flat surfaces in the middle of the casting. In this way, the *flash* at the parting line (material running out between the two halves of the mold) will not be as visible. The location of the parting line is important because it influences mold design, ease of molding, number and shape of cores, method of support, and the gating system. Preparation of dry-sand cores requires additional time and cost, so they should be avoided or minimized. We can usually do so by reviewing the design of castings and simplifying them. Two examples of casting design modifications are shown in Fig. 5.43.

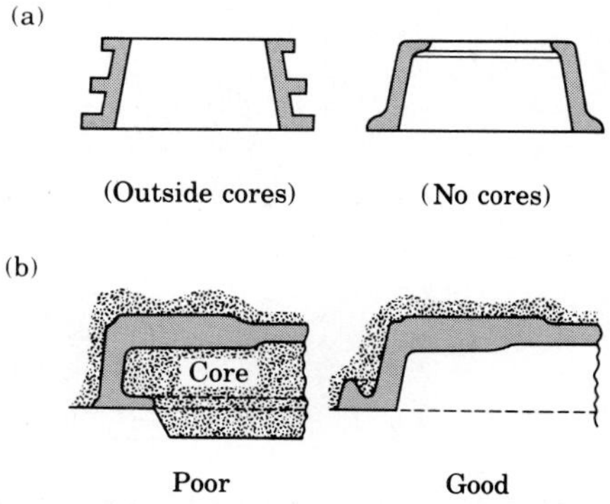

FIGURE 5.43
Examples of casting design modifications. *Source: Steel Castings Handbook,* 5th ed. Steel Founders' Society of America, 1980. Used with permission.

Draft. A small *draft* (taper) is provided in sand-mold patterns to enable removal of the pattern without damaging the mold. Depending on the quality of the pattern, draft angles usually range from 0.5° to 2°. The angles on inside surfaces of molds are typically twice this range. They have to be higher than those for outer surfaces, because the casting shrinks inward toward the core.

Tolerances. Tolerances depend on the particular casting process, size of the casting, and type of pattern used. Tolerances are smallest within one part of the mold and, because they are cumulative, increase between different parts of the mold. Tolerances should be as wide as possible, within the limits of good part performance; otherwise the cost of the casting increases. In commercial practice, tolerances usually are about ± 0.8 mm (1/32 in.) for small castings and increase with the size of castings, say, to 6 mm (1/4 in.) for large castings.

Allowances. Because most expendable-mold castings require some additional finishing operations, such as machining, allowances should be made in casting design for these operations. *Machining allowances* are included in pattern dimensions and depend on the type of casting. They increase with the size and section thickness of castings, usually ranging from about 2 mm to 5 mm (0.1 in. to 0.2 in.) for small castings to more than 25 mm (1 in.) for large castings.

5.11.2 Designing for permanent-mold casting

The design principles for permanent-mold casting are similar to those for expendable-mold casting. Typical design guidelines and examples for permanent-mold casting are shown schematically in Fig. 5.44 for die casting. Note that the cross-

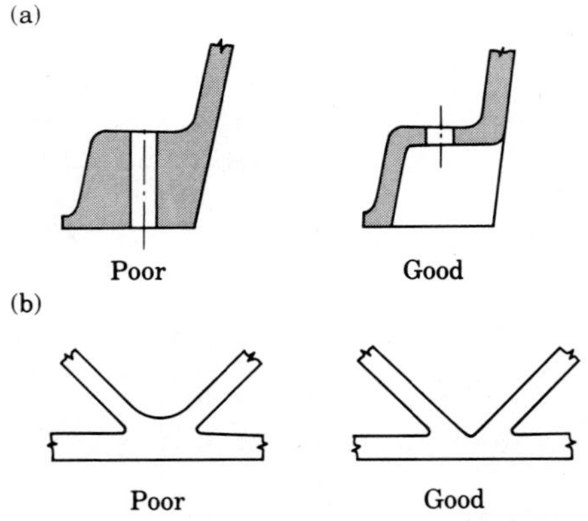

FIGURE 5.44
Examples of undesirable and desirable design practices for die-cast parts. Note that section-thickness uniformity is maintained throughout the part. *Source:* American Die Casting Institute.

sections have been reduced in order to decrease solidification time and save material. Special considerations are involved in designing and tooling for die casting. Designs may be modified to eliminate the draft for better dimensional accuracy.

5.11.3 Modeling of casting processes

Because casting processes involve complex interactions among material and process variables, a quantitative study of these interactions is essential to proper design of castings and the production of high quality castings. In the past such studies have presented major difficulties. However, rapid advances in computers and modeling techniques have led to important innovations in modeling various aspects of casting, including fluid flow, heat transfer, and microstructures developed during solidification under various conditions.

Modeling of *fluid flow* is based on Bernoulli's and the continuity equations (see Section 5.4). It predicts the behavior of the metal during pouring into the gating system and its travel into the mold cavity, as well as velocity and pressure distributions in the system. Progress has also been made in modeling of *heat transfer* in casting. These studies include investigation of the coupling of fluid flow and heat transfer and the effects of surface conditions, thermal properties of the materials involved, and natural and forced convection on cooling. The capability to calculate isotherms, for example, gives us insight into possible hot spots and subsequent development of shrinkage cavities.

Similar studies are being conducted on modeling the development of *micro-structures* in casting. These studies encompass heat flow, temperature gradients, nucleation and growth of crystals, formation of dendritic and equiaxed structures, impingement of grains on each other, and movement of the liquid–solid interface during solidification. Such models are now capable of predicting, for example, the width of the mushy zone (see Fig. 5.11) during solidification and grain size in castings. With the availability of user-friendly computers and advances in computer-aided design and manufacturing (see Chapter 14), modeling techniques are becoming easier to implement. The benefits are increased productivity, improved quality, easier planning and cost estimating, and quicker response to design changes.

5.12 Economics of Casting

When looking at various casting processes, we noted that some require more labor than others, some require expensive dies and machinery, and some take a great deal of time to complete. Outlined in Table 5.8, each of these important factors affects to a greater or lesser degree the overall cost of a casting operation.

As we describe in greater detail in Chapter 15, the total cost of a product involves the costs of materials, labor, tooling, and equipment. Preparations for casting a product include making molds and dies that require raw materials, time, and effort, which we can translate into costs. As Table 5.8 shows, relatively little cost is involved in molds for sand casting. At the other extreme, die-casting dies require expensive materials and a great deal of machining and preparation. In addition to molds and dies, facilities are required for melting and pouring the molten metal into the molds or dies. These facilities include furnaces and related machinery, their costs depending on the level of automation desired. Finally, costs are involved in cleaning and inspecting castings.

The amount of labor required for these operations can vary considerably, depending on the particular process and level of automation. Investment casting, for example, requires a great deal of labor because of the large number of steps involved in this operation. Conversely, operations such as highly automated die casting can maintain high production rates with little labor required.

The cost of equipment per casting (*unit cost*), however, decreases as the number of parts cast increases (Fig. 5.45). Thus sustained high production rates can justify the high cost of dies and machinery. However, if demand is relatively small, the cost per casting increases rapidly. It then becomes more economical to manufacture the parts by sand casting or by other manufacturing processes. Note that Fig. 5.45 can be expanded to include other casting processes suitable for making the same part.

The two processes (sand and die casting) we compared produce castings with significantly different dimensional and surface-finish characteristics. Thus not all manufacturing decisions are based purely on economic considerations. In fact, parts can usually be made by more than one or two processes. The final decision depends

TABLE 5.8
GENERAL COST CHARACTERISTICS OF CASTING PROCESSES

PROCESS	COST*			PRODUCTION RATE† (Pieces/hr)
	DIE	*EQUIPMENT*	*LABOR*	
Sand	L	L	L–M	<20
Shell	L–M	M–H	L–M	<50
Plaster	L–M	M	M–H	<10
Investment	M–H	L–M	H	<1000
Permanent mold	M	M	L–M	<60
Die	H	H	L–M	<200
Centrifugal	M	H	L–M	<50

* L, low; M, medium; H, high.
† Depends greatly on casting size.

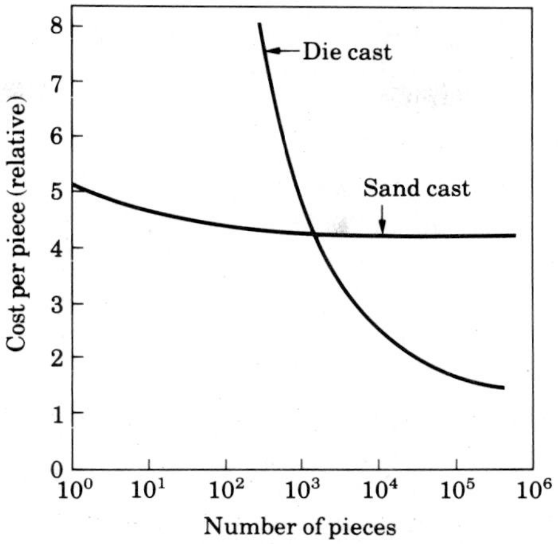

FIGURE 5.45
Economic comparison of making a part by two different casting processes. Note that because of the high cost of equipment, die casting is economical for large production runs. *Source:* American Die Casting Institute.

on both economic and technical considerations. We describe competitive aspects of manufacturing processes further in Chapter 15.

SUMMARY

Casting is a solidification process in which molten metal is poured into a mold and allowed to cool to ambient temperature. Fluid flow and heat transfer are important considerations, as is the design of the mold and gating system to ensure proper flow of the metal into the mold cavities.

Because metals contract during solidification and cooling, cavities can form in the casting. Porosity caused by gases evolving during solidification is a significant problem, particularly because of its adverse effect on the mechanical properties of castings. The grain structure of castings can be controlled by various means to obtain the desired properties.

In this chapter we described expendable-mold and permanent-mold casting processes. The most common processes in the first group are sand, shell, plaster, and ceramic mold and investment casting. Compared to permanent-mold casting, castings made by these processes usually involve relatively low mold and equipment costs. However, they generally tend to produce castings having high porosity and low dimensional accuracy at a low production rate.

The molds used in permanent-mold casting are made of metal or graphite and are used repeatedly to produce many parts. Because metals are good heat conductors but do not allow gases to escape, permanent molds perform in fundamentally different ways from sand and other aggregate mold materials. Processes that use permanent molds are pressure, slush, die, centrifugal, and squeeze casting. Die and equipment costs are relatively high, but the processes are economical for large production runs. Scrap loss is low and dimensional accuracy is relatively high, with good surface details.

Castings are subjected to further heat treatment processes. Phase diagrams show the relationships among temperature, composition, and the phases present in a particular alloy system. Of the binary systems, the most important is the iron–carbon system, which includes steels and cast irons. Various transformations take place in microstructures that have widely varying characteristics and properties, as temperature is decreased at different rates.

Important mechanisms of hardening and strengthening involve thermal treatments, including heat treatment by quenching and precipitation hardening. Also important are the methods and techniques of heat treating, quenching, annealing, control of the furnace atmosphere, and the characteristics of the equipment used, as well as the shape of the parts to be heat treated.

General principles have been derived to aid designers in producing castings that are free from defects and meet tolerances and service requirements. These principles concern shape of casting and various techniques to minimize hot spots that could lead to shrinkage cavities. Because of the large number of variables involved, close control of all parameters is essential, particularly those related to the nature of liquid metal flow into the molds and dies and the rate of cooling in different regions of the mold or die.

BIBLIOGRAPHY

Allsop, D.F., and D. Kennedy, *Pressure Die Casting—Part II: The Technology of the Casting and the Die*. Oxford: Pergamon, 1983.

An Introduction to Die Casting. Des Plaines, Ill.: American Die Casting Institute, 1981.

Analysis of Casting Defects. Des Plaines, Ill.: American Foundrymen's Society, 1974.

Angus, H.T., *Cast Iron: Physical and Engineering Properties*. New York: Butterworths, 1976.

Beeley, P.R., *Foundry Technology*. London: Butterworths, 1988.

Bradley, E.F., *High Performance Castings: A Technical Guide*. Metals Park, Ohio: ASM International, 1989.

Brody, H.D., and D. Apelian (eds.), *Modeling of Casting and Welding Processes*. Warrendale, Pa.: The Metallurgical Society of AIME, 1981.

Casting Design Handbook. Metals Park, Ohio: American Society for Metals, 1962.

Dantzig, J.A., and J.T. Berry (eds.), *Modeling of Casting and Welding Processes*. Warrendale, Pa.: The Metallurgical Society of AIME, 1984.

Davies, G.J., *Solidification and Casting*. Essex, England: Applied Science Publishers, 1973.

Flemings, M.C., *Solidification Processing*. New York: McGraw-Hill, 1974.

Flinn, R.A., *Fundamentals of Metal Casting*. Reading, Mass.: Addison-Wesley, 1963.
Heine, R.W., C.R. Loper, Jr., and C. Rosenthal, *Principles of Metal Casting*, 2d ed. New York: McGraw-Hill, 1967.
Investment Casting Handbook. Chicago, Ill.: Investment Casting Institute, 1979.
Kaye, A., and A.C. Street, *Die Casting Metallurgy*. London: Butterworths, 1982.
Kondic, V., *Metallurgical Principles of Founding*. London: Arnold, 1968.
Metal Casting and Molding Processes. Des Plaines, Ill.: American Foundrymen's Society, 1981.
Metals Handbook, 9th ed., Vol. 15: *Casting*. Metals Park, Ohio: American Society for Metals, 1988.
Mikelonis, P.J. (ed.), *Foundry Technology: A Source Book*. Metals Park, Ohio: American Society for Metals, 1982.
Minkoff, I., *Solidification and Cast Structure*. New York: Wiley, 1986.
Rauch, A.H. (ed.), *Source Book on Ductile Iron*. Metals Park, Ohio: American Society for Metals, 1977.
Romanoff, R., *Centrifugal Casting*. Blue Ridge Summit, Pa.: TAB Books, 1981.
Rowley, M.T. (ed.), *International Atlas of Casting Defects*. Des Plaines, Ill.: American Foundrymen's Society, 1974.
Smith, T.J. (ed.), *Modelling the Flow and Solidification of Metals*. Boston: Martinus Nijhoff, 1987.
Street, A.C., *The Diecasting Book*. Surrey, England: Portcullis Press, 1977.
Sylvia, J.G., *Cast Metals Technology*. Reading, Mass.: Addison-Wesley, 1972.
Szekely, J., *Fluid Flow Phenomena in Metal Processing*. New York: Academic Press, 1979.
Upton, B., *Pressure Die Casting—Part I: Metals-Machines-Furnaces*. Oxford: Pergamon, 1982.
Walton, C.F., and T.J. Opar (eds.), *Iron Castings Handbook*, 3d ed. Des Plaines, Ill.: Iron Castings Society, 1981.
Wieser, P.F. (ed.), *Steel Casting Handbook*, 5th ed. Des Plaines, Ill.: Steel Founders' Society of America, 1980.

QUESTIONS

5.1 Explain why carbon is so effective in imparting strength to iron in the form of steel.

5.2 Describe the engineering significance of the existence of a eutectic point in phase diagrams.

5.3 Explain the difference between hardness and hardenability.

5.4 Describe the characteristics of (a) an alloy, (b) pearlite, (c) austenite, (d) martensite, and (e) cementite.

5.5 How does the shape of graphite in cast iron affect its properties?

5.6 Explain the difference between short and long freezing ranges. Why are they important?

5.7 We know that pouring molten metal very fast into a mold has certain disadvantages. Are there any disadvantages to pouring it very slowly? Explain.

5.8 Explain why subjecting castings to various heat treatments may be desirable.

5.9 Why does porosity adversely affect the mechanical properties of castings? Which physical properties are also affected adversely by porosity?

5.10 A spoked wheel is to be cast in gray iron. To prevent hot tearing of the spokes, would you insulate the spokes or chill them? Explain.

5.11 Which of the following considerations are important for a riser to function properly? (a) Have a surface area larger than the part being cast. (b) Be kept open to atmospheric pressure. (c) Solidify first. Why?

5.12 Explain why the constant C in Eq. (5.9) depends on mold material, metal properties, and temperature.

5.13 Explain why gray iron expands rather than contracts during solidification.

5.14 What differences, if any, would you expect in the properties of items made by permanent-mold casting and sand casting?

5.15 Would you recommend preheating the molds in permanent-mold casting? Would you remove the casting soon after it has solidified? Explain.

5.16 In sand casting, what factors determine the time at which you would remove the casting from the mold?

5.17 Explain why the strength-to-weight ratio of die-cast parts increases with decreasing wall thickness.

5.18 Explain how ribs and serrations are helpful in casting flat surfaces that otherwise may warp. Give an illustration.

5.19 The ductility of some cast alloys is nil (see Fig. 5.21). Should this be a significant concern in engineering applications of castings? Explain.

5.20 The modulus of elasticity E of gray irons varies significantly with its type, such as the ASTM class. Explain why.

5.21 What is the purpose of heat treating? Annealing?

5.22 Describe the differences between case hardening and through hardening insofar as engineering applications are concerned.

PROBLEMS

5.1 Using Fig. 5.3, estimate the following quantities for a 20% Cu–80% Ni alloy: (a) liquidus temperature; (b) solidus temperature; (c) percentage of nickel in the liquid at 1400 °C (2550 °F); (d) the major phase at 1400 °C; and (e) the ratio of solid to liquid at 1400 °C.

5.2 Determine the amount of gamma and alpha phases (Fig. 5.7) in a 10-kg, 1050 steel casting as it is being cooled to the following temperatures: (a) 900 °C, (b) 728 °C, and (c) 726 °C. Compare with Example 5.1.

5.3 A round casting is 0.1 m in diameter and 0.5 m in length. Another casting of the same metal is elliptical in cross-section, with a major-to-minor axis ratio of 2, and has the same length and cross-section area as the round casting. Both pieces are cast under the same conditions. What is the difference in the solidification times of the two castings?

5.4 Prove Eq. (5.7).

5.5 Estimate the clamping force for a die casting machine in which the casting is rectangular, with projected dimensions of 100 mm × 200 mm.

5.6 Would your answer to Problem 5.5 depend on whether it is a hot-chamber or cold-chamber process? Explain.

5.7 We stated in Section 5.8.1 that the two halves of the mold (cope and drag) are weighted down to keep them from separating under the pressure exerted by the molten metal (buoyancy). Consider a solid, spherical steel casting, 10 in. in diameter, that is being produced by sand casting. Each flask (see Fig. 5.15) is 30 in. × 30 in. and 15 in. deep. The parting line is at the middle of the part. Estimate the clamping force required.

5.8 Would the position of the parting line in Problem 5.7 influence your answer? Explain.

5.9 Plot the clamping force in Problem 5.7 as a function of increasing diameter of the casting, from 10 in. to 20 in.

5.10 Figure P5.1 shows casting designs that may result in defects, such as hot tears and shrinkage cavities. Make specific recommendations for design changes or use of various techniques to avoid or minimize these defects.

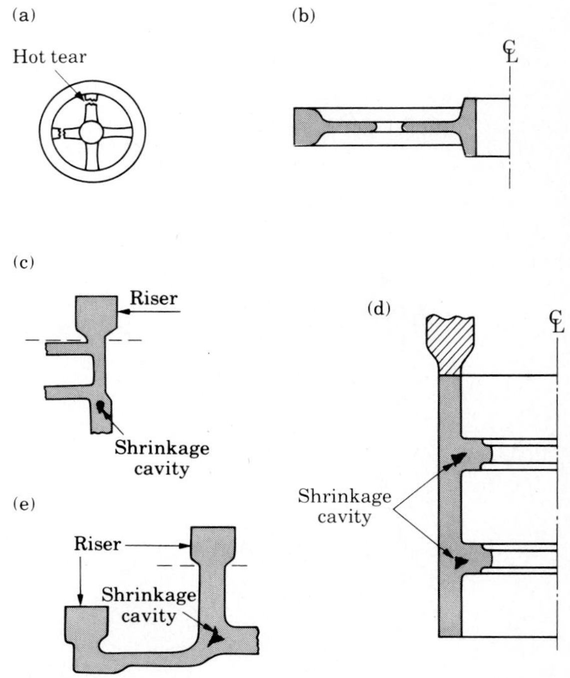

FIGURE P5.1

6

Bulk Deformation Processes

6.1 Introduction

Deformation processes in manufacturing are operations that induce shape changes on the workpiece by plastic deformation under forces applied by various tools and dies. Deformation processes can be grouped by temperature, size, and shape of the workpiece and type of operation. For example, using temperature as a criterion, deformation processes may be divided into three basic categories of cold (room temperature), warm, and hot working. Also, deformation processes may be classified by type of operation as primary working or secondary working.

Primary working operations are those that take a solid piece of metal (generally from a cast state, such as an ingot) and break it down successively into shapes such as slabs, plates, and billets. Traditionally, primary working processes are forging, rolling, and extrusion. *Secondary working* involves further processing of the products from primary working into final or semifinal products such as bolts, sheet metal parts, and wire.

These classifications are not rigid because some operations belong to both primary and secondary working. For instance, some forgings result in a product as final as those that result from secondary working. Likewise, in extrusion some products are ready for use without any further processing.

The recent trend has been to classify deformation processes according to the size and shape of the workpiece. In these terms, all deformation processes are classified as either bulk deformation or sheet forming.

Bulk deformation is the processing of workpieces having a relatively small surface area-to-volume (or surface area-to-thickness) ratio—hence the term *bulk*. In all bulk deformation processing, the thickness or cross-section of the workpiece changes. In *sheet-forming* operations the surface area-to-thickness ratio is relatively large. In general, the material is subjected to shape changes by various dies. Thickness changes are usually undesirable and, in fact, can lead to failure.

In this chapter we describe four basic bulk deformation processes for metal: forging, rolling, extrusion, and drawing of rod and wire (Table 6.1). We cover bulk deformation processing of plastics and nonmetallic materials in Chapters 10 and 11, respectively.

TABLE 6.1
GENERAL CHARACTERISTICS OF BULK DEFORMATION PROCESSES

PROCESS	CHARACTERISTICS
Forging	Production of discrete parts with a set of dies; some finishing operations usually required; similar parts can be made by casting and powder-metallurgy techniques; usually performed at elevated temperatures; die and equipment costs are high; moderate to high labor cost; moderate to high operator skill.
Rolling	
Flat	Production of flat plate, sheet, and foil in long lengths, at high speeds, and with good surface finish, especially in cold rolling; requires high capital investment; low to moderate labor cost.
Shape	Production of various structural shapes, such as I-beams, at high speeds; includes thread rolling; requires shaped rolls and expensive equipment; low to moderate labor cost; moderate operator skill.
Extrusion	Production of long lengths of solid or hollow products with constant cross-section; usually performed at elevated temperatures; product is then cut into desired lengths; can be competitive with roll forming; cold extrusion has similarities to forging and is used to make discrete products; moderate to high die and equipment cost; low to moderate labor cost; low to moderate operator skill.
Drawing	Production of long rod and wire, with round or various cross-sections; smaller cross-sections than extrusions; good surface finish; low to moderate die, equipment, and labor cost; low to moderate operator skill.

6.2 Forging

Forging denotes a family of processes by which plastic deformation of the workpiece is carried out by *compressive forces.* Forging is one of the oldest metalworking operations known, dating back to 5000 B.C., and is used in making parts of widely varying sizes and shapes from a variety of metals. Typical parts made by forging today are crankshafts and connecting rods for engines, turbine disks, gears, wheels, bolt heads, hand tools, and many types of structural components for machinery and transportation equipment.

Forging can be carried out at room temperature (cold working), or at elevated temperatures, called warm and hot forging, depending on the temperature. The temperature range for these categories is given in Table 3.1 in terms of the homologous temperature, T/T_m, where T_m is the melting point of the workpiece material on the absolute scale. Note that the homologous recrystallization temperature for metals is about 0.5 (see also Fig. 3.16).

Very simple forgings can be made with a heavy hammer and an anvil by techniques used by blacksmiths for centuries. Usually, though, a set of dies and a press are required. The three basic categories of forging are open die, impression die, and closed die.

6.2.1 Open-die forging

In its simplest form, *open-die forging* generally involves placing a solid cylindrical workpiece between two flat dies (platens) and reducing its height by compressing it (Fig. 6.1a). This operation is also known as *upsetting.* The die surfaces may be

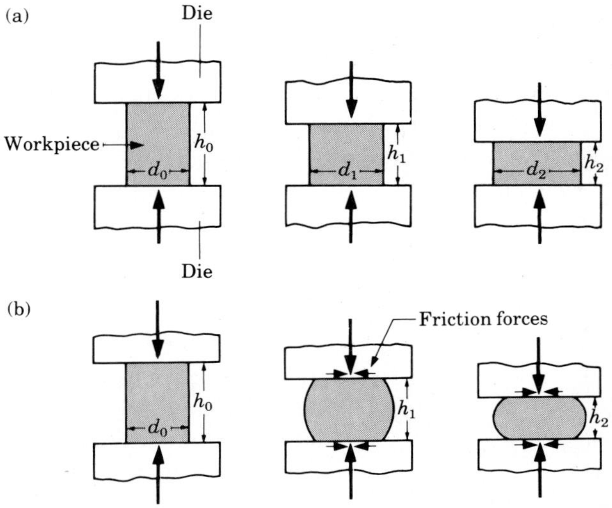

FIGURE 6.1
(a) Ideal deformation of a solid cylindrical specimen compressed between flat frictionless dies. This process is known as upsetting. (b) Deformation in upsetting with friction at the die–workpiece interfaces.

shaped, such as a conical or curved cavity, thereby forming the ends of the cylindrical workpiece during upsetting.

Under ideal conditions, a solid cylinder deforms as shown in Fig. 6.1(a). This is known as *homogeneous deformation*. Because volume is constant, any reduction in height increases the diameter of the cylinder. For a specimen that has been reduced in height from h_0 to h_1, we have

$$\text{Reduction in height} = \frac{h_0 - h_1}{h_0} \times 100\,\%. \tag{6.1}$$

From Eqs. (2.1) and (2.9) and using absolute values (as is generally the case with bulk deformation processes), we have

$$e_1 = \frac{h_0 - h_1}{h_0} \tag{6.2}$$

and

$$\varepsilon_1 = \ln\left(\frac{h_0}{h_1}\right). \tag{6.3}$$

With a relative velocity v between the platens, the specimen is subjected to a strain rate, according to Eqs. (2.17) and (2.18), of

$$\dot{e}_1 = -\frac{v}{h_0} \tag{6.4}$$

and

$$\dot{\varepsilon}_1 = -\frac{v}{h_1}. \tag{6.5}$$

If the specimen is reduced in height from h_0 to h_2, the subscript 1 in the foregoing equations is replaced by subscript 2. Note that the true strain rate $\dot{\varepsilon}$ increases rapidly as the height of the specimen approaches zero. Positions 1 and 2 in Fig. 6.1(a) may be regarded as instantaneous positions during a continuous upsetting operation.

Actually, the specimen develops a *barrel* shape, as shown in Fig. 6.1(b). Barreling is caused primarily by *frictional* forces at the die–workpiece interfaces that oppose the outward flow of the material at these interfaces. Barreling also occurs in upsetting *hot* workpieces between *cool* dies. The material at and near the interfaces cools rapidly, while the rest of the specimen is relatively hot. Because the strength of the material decreases with temperature, the ends of the specimen show a greater resistance to deformation than does the center. Thus the central portion of the cylinder deforms to a greater extent than do its ends.

In barreling, the material flow within the specimen becomes *nonuniform*, or *inhomogeneous*, as can be seen from the polished and etched section of a barreled

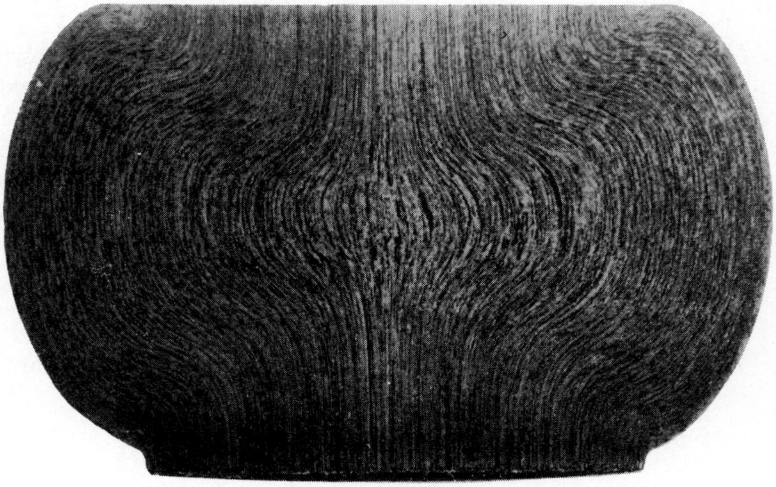

FIGURE 6.2
Grain flow lines in upsetting a solid steel cylinder at elevated temperatures. Note the highly inhomogeneous deformation and barreling. The different shape of the bottom section of the specimen (as compared to the top) results from the hot specimen resting on the lower cool die before deformation proceeded. The bottom surface was chilled; thus it exhibits greater strength and hence deforms less than the top surface. *Source:* J. A. Schey, et al., IIT Research Institute.

cylindrical specimen in Fig. 6.2. Note the roughly triangular stagnant (*dead*) zones at the top and bottom surfaces, which have a wedge effect on the rest of the material, thus causing barreling. This inhomogeneous deformation can also be observed in grid patterns, as shown in Fig. 6.3, which are obtained experimentally using various techniques (Section 6.4.1).

In addition to the single barreling shown in Fig. 6.2, *double barreling* can also be observed. It occurs at large ratios of height-to-cross-sectional area and when friction at the die–workpiece interfaces is high. Under these conditions, the stagnant zone under the platen is sufficiently far from the midsection of the workpiece to

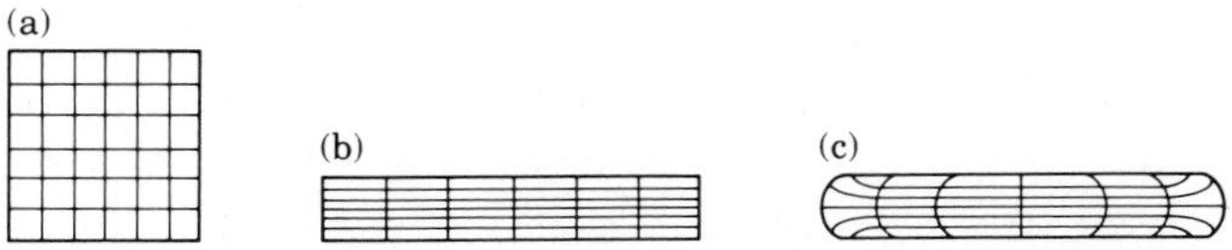

FIGURE 6.3
Schematic illustration of grid deformation in upsetting: (a) original grid pattern; (b) after deformation, without friction; (c) after deformation, with friction. Such deformation patterns can be used to calculate the strains within a deforming body.

allow the midsection to deform uniformly while the top and bottom portions barrel out—hence the term double barreling.

Barreling caused by friction can be minimized by applying an effective lubricant or ultrasonically vibrating the platens. Also, the use of heated platens or a thermal barrier at interfaces will reduce barreling in hot working.

Forces and work of deformation under ideal conditions. If friction at the interfaces is zero and the material is perfectly plastic with a yield stress of Y, then the normal compressive stress on the cylindrical specimen is uniform at a level Y. The force at any height h_1 is then

$$F = YA_1, \tag{6.6}$$

where A_1 is the cross-sectional area and is obtained from volume constancy. Thus

$$A_1 = \frac{A_0 h_0}{h_1}.$$

The work of deformation is the product of the volume of the specimen and the specific energy u (Eq. 2.58):

$$\text{Work} = \text{Volume} \int_0^{\varepsilon_1} \sigma \, d\varepsilon, \tag{6.7}$$

where ε_1 is obtained from Eq. (6.3).

If the material is strain hardening, with a true stress–true strain curve given by

$$\sigma = K\varepsilon^n,$$

then the expression for force at any stage during deformation becomes

$$F = Y_f A_1, \tag{6.8}$$

where Y_f is the *flow stress* of the material, corresponding to the true strain given by Eq. (6.3) and Fig. 2.51.

The expression for the work done is

$$\text{Work} = (\text{Volume})(\bar{Y})(\varepsilon_1), \tag{6.9}$$

where $\bar{Y}$ is the *average flow stress* and is given by

$$\bar{Y} = \frac{K \int_0^{\varepsilon_1} \varepsilon^n \, d\varepsilon}{\varepsilon_1} = \frac{K\varepsilon_1^n}{n+1}. \tag{6.10}$$

6.2.2 Methods of analysis

Slab method. The *slab method* is one of the simpler methods of analyzing the stresses and loads in forging and other bulk deformation processes. It requires the selection of an element in the workpiece and identifying all the normal and frictional forces acting on that element. Let's take the case of simple compression with friction (Fig. 6.4), which is the basic deformation in forging. The purpose of this analysis is to determine the die pressure distribution from which we can then calculate the forging load required.

As the flat dies compress the part, it is reduced in thickness and, as the volume remains constant, the part expands laterally. This relative movement at the die–workpiece interfaces causes frictional forces acting in opposition to the movement of the piece. These frictional forces are shown by the horizontal arrows in Fig. 6.4. For simplicity, let's also assume that the deformation is in *plane strain*; that is, the workpiece is not free to flow in the direction perpendicular to this page.

Let's now take an element and indicate all the forces acting on it (Fig. 6.4b). Note the correct direction of the frictional forces. Also note the difference in the horizontal forces acting on the sides of the element; this difference is caused by the presence of frictional forces on the element. We assume that the lateral stress distribution σ_x is uniform along the height h.

The next step in this analysis is to balance the horizontal forces on this element, because it must be in static equilibrium. Thus

$$(\sigma_x + d\sigma_x)h + 2\mu\sigma_y\,dx - \sigma_x h = 0 \qquad \textbf{(6.11)}$$

or

$$d\sigma_x + \frac{2\mu\sigma_y}{h}\,dx = 0.$$

Note that we have one equation but two unknowns: σ_x and σ_y. We obtain the necessary second equation from the yield criteria as follows. As shown in Fig. 6.4(c), this element is subjected to triaxial compression (compare this element with that in Fig. 2.48b). Using the distortion-energy criterion for plane strain, we obtain

$$\sigma_y - \sigma_x = \frac{2}{\sqrt{3}}\,Y = Y'. \qquad \textbf{(6.12)}$$

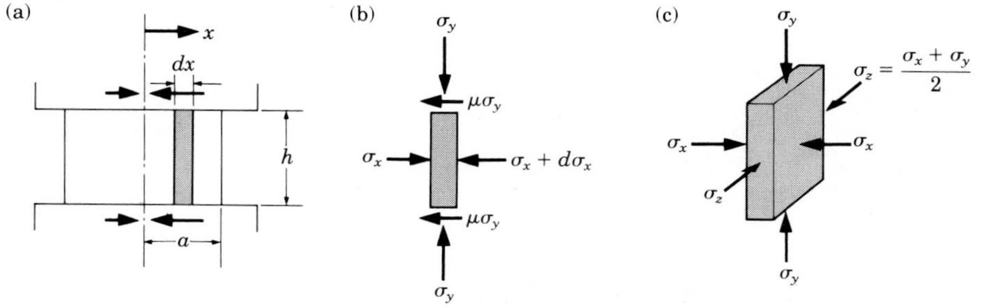

FIGURE 6.4

Stresses on an element in plane-strain compression (forging) between flat dies. The stress σ_x is assumed to be uniformly distributed along the height h of the element. Identifying the stresses on an element (slab) is the first step in the slab method of analysis for metalworking processes.

Thus

$$d\sigma_y = d\sigma_x.$$

Note that we assumed σ_y and σ_x to be *principal stresses*. In the strictest sense, σ_y cannot be a principal stress because a shear stress is also acting on the same plane. However, this assumption is acceptable for low values of the coefficient of friction μ and is the standard practice in this method of analysis. Note also that σ_z in Fig. 6.4(c) is derived in a manner similar to that in Eq. (2.43).

We now have two equations that can be solved by noting that

$$\frac{d\sigma_y}{\sigma_y} = -\frac{2\mu}{h}\,dx \quad \text{or} \quad \sigma_y = Ce^{-2\mu x/h}.$$

The boundary conditions are such that at $x = a$, $\sigma_x = 0$, and thus $\sigma_y = Y'$ at the edges of the specimen. (All stresses are compressive, so we may ignore negative signs for stresses, which is traditional in such analyses.) Hence the value of C becomes

$$C = Y'e^{2\mu a/h},$$

and thus

$$p = \sigma_y = Y'e^{2\mu(a-x)/h}. \qquad \textbf{(6.13)}$$

Also,

$$\sigma_x = \sigma_y - Y' = Y'[e^{2\mu(a-x)/h} - 1]. \qquad \textbf{(6.14)}$$

Equation (6.13) is plotted qualitatively in Fig. 6.5 in dimensionless form. Note that the pressure increases exponentially toward the center of the part and also that it increases with the a/h ratio and increasing friction.

For a strain-hardening material, Y' in Eqs. (6.13) and (6.14) is replaced by Y'_f. Because of its shape, the pressure-distribution curve in Fig. 6.5 is referred to as the *friction hill*. Note that the pressure with friction is higher than it is without friction. This result is expected, because the work required to overcome friction must be supplied by the upsetting force.

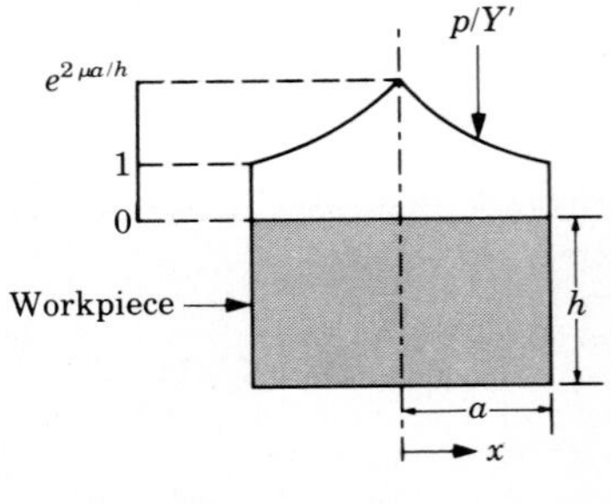

FIGURE 6.5 Distribution of die pressure p in plane-strain compression with sliding friction. Note that the pressure at the left and right boundaries is equal to the yield stress in plane strain, Y'. Sliding friction means that the frictional stress is directly proportional to the normal stress.

FIGURE 6.6
Normal stress (pressure) distribution in the compression of a rectangular workpiece with sliding friction under conditions of plane stress using the distortion-energy criterion. Note that the stress at the corners is equal to the uniaxial yield stress of the material, Y.

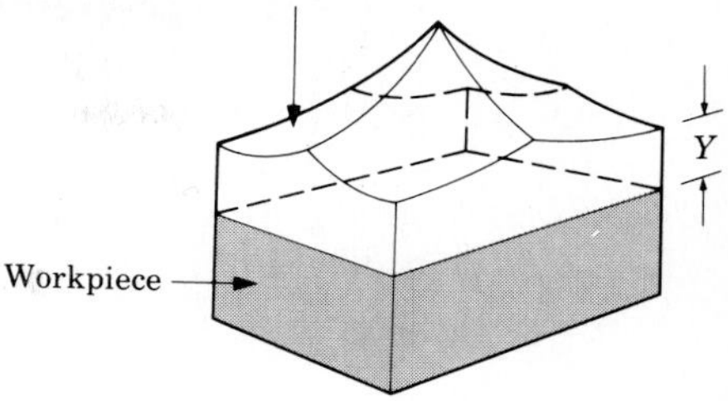

The area under the pressure curve in Fig. 6.5 is the upsetting *force per unit width* of the specimen. This area can be obtained by integration, but an approximate expression for the average pressure p_{av} is

$$p_{av} \simeq Y'\left(1 + \frac{\mu a}{h}\right). \tag{6.15}$$

Again note the significant roles of a/h and friction on the pressure required, especially at high a/h ratios. The upsetting force F is the product of the average pressure and the contact area; that is,

$$F = (p_{av})(2a)(\text{width}). \tag{6.16}$$

Note that the expressions for the pressure p are in terms of an instantaneous height h. Thus the force at any h during a continuous upsetting operation must be calculated individually.

A *rectangular specimen* can be upset without being constrained on its sides (plane stress). According to the distortion-energy criterion, the normal stress distribution can be given qualitatively by Fig. 6.6. The pressure is Y at the corners because the elements at the corners are in uniaxial compression. A friction hill remains along the edges of the specimen because of friction along the edges.

The lateral expansion (top view) of the edges of a rectangular specimen in actual plane stress upsetting is shown in Fig. 6.7. Whereas the increase in the length of the specimen is 40 percent, the increase in width is 230 percent. The reason is

FIGURE 6.7
Increase in contact area of a rectangular specimen (top view) compressed between flat dies with friction. Note that the length of the specimen has increased proportionately less than its width. Likewise, a specimen in the shape of a cube acquires the shape of a pancake after deformation with friction.

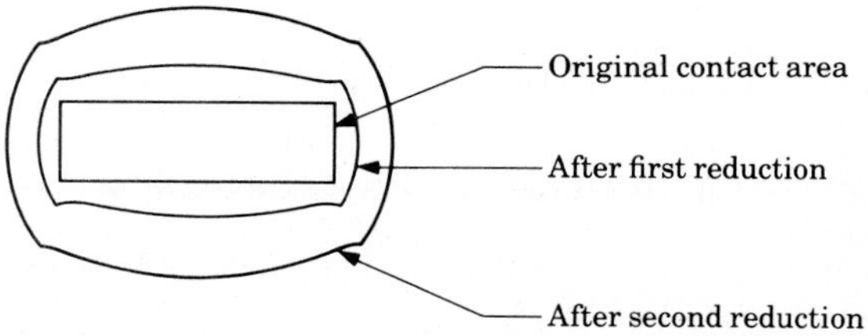

that, as expected, *the material flows in the direction of least resistance.* The width has less frictional resistance than the length. Likewise, after upsetting, a specimen in the shape of an equilateral cube acquires a pancake shape. The diagonal direction expands at a slower rate than the other directions.

For a *solid cylindrical specimen* of radius r, the normal stress distribution is similar to that shown in Fig. 6.5, where the expression for the normal stress p at any radius x is

$$p = Ye^{2\mu(r-x)/h}. \tag{6.17}$$

If we use the slab method of analysis, the average pressure p_{av} can be given approximately as

$$p_{av} \simeq Y\left(1 + \frac{2\mu r}{3h}\right). \tag{6.18}$$

The upsetting force is thus

$$F = (p_{av})(\pi r^2). \tag{6.19}$$

For strain-hardening materials, Y in Eqs. (6.17) and (6.18) is replaced by the flow stress Y_f. In forging operations, the value of the coefficient of friction μ in Eqs. (6.17) and (6.18) can be estimated to be 0.05 to 0.1 for cold forging, and 0.1 to 0.2 for hot forging. The exact value depends on the effectiveness of the lubricant and can be higher than stated, especially in regions where the workpiece surface is devoid of lubricant.

The effects of friction and the aspect ratio of the specimen (a/h or r/h) on the average pressure p_{av} in upsetting are shown in Fig. 6.8. The pressure is given in a dimensionless form and can be regarded as a pressure-multiplying factor. These

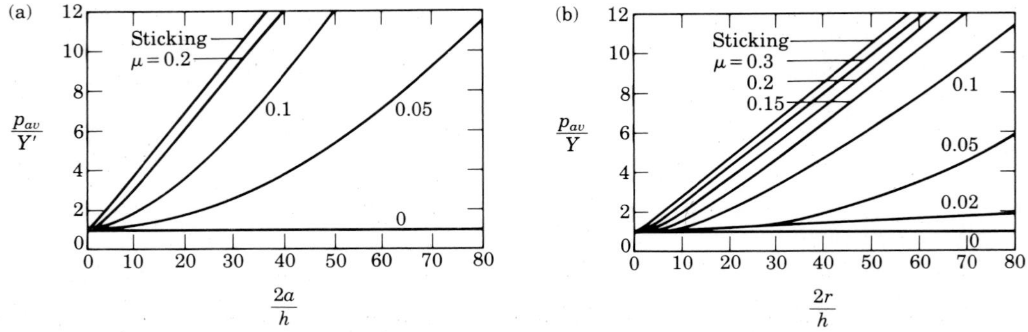

FIGURE 6.8

Ratio of average die pressure to yield stress as a function of friction and aspect ratio of the specimen: (a) plane-strain compression; and (b) compression of a solid cylindrical specimen. Note that the yield stress in (b) is Y, and not Y' as in plane-strain compression in (a). *Source:* (a) After J. F. W. Bishop, *J. Mech. Phys. Solids*, vol. 6, 1958, pp. 132–144. (b) Adapted from W. Schroeder and D. A. Webster, *Trans. ASME*, vol. 71, 1949, pp. 289–294.

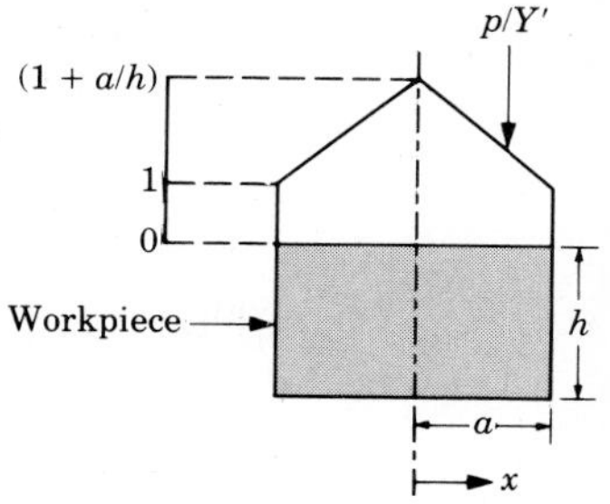

FIGURE 6.9 Normal stress (pressure) distribution in the compression of a rectangular specimen in plane strain and under sticking condition. The pressure at the edges is the uniaxial yield stress of the material in plane strain, Y'.

curves are convenient to use and again show the significance of friction and the aspect ratio of the specimen on upsetting pressure.

The product of μ and p is the frictional stress (surface shear stress) at the interface at any location x from the center of the specimen. As p increases toward the center, μp also increases. However, the value of μp cannot be greater than the shear yield stress k of the material. When $\mu p = k$, *sticking* takes place. (In plane strain the value of k is $Y'/2$.) Sticking does not necessarily mean adhesion at the interface; it reflects the fact that, relative to the platen surfaces, the material does not move.

For the sticking condition, the normal stress distribution in plane strain is

$$p = Y'\left(1 + \frac{a - x}{h}\right). \tag{6.20a}$$

The pressure varies linearly with x, as shown in Fig. 6.9. The normal stress distribution for a cylindrical specimen under sticking conditions is

$$p = Y\left(1 + \frac{r - x}{h}\right). \tag{6.20b}$$

The stress distribution again is linear, as shown in Fig. 6.9 for plane strain.

• Example 6.1: Calculating upsetting force.

A cylindrical specimen made of annealed 4135 steel is 6 in. in diameter and 4 in. high. It is upset by open-die forging with flat dies to a height of 2 in. at room temperature. Assuming that the coefficient of friction is 0.2, calculate the force required at the end of the stroke. Use the average-pressure formula.

SOLUTION. The average-pressure formula, from Eq. (6.18), is

$$p_{av} \simeq Y\left(1 + \frac{2\mu r}{3h}\right),$$

where Y is replaced by Y_f because the workpiece material is strain hardening. From Table 2.4, $K = 1015$ MPa (147,000 psi) and $n = 0.17$.

The absolute value of the true strain is

$$\varepsilon_1 = \ln\left(\frac{4}{2}\right) = 0.693.$$

Thus

$$Y_f = K\varepsilon_1^n = (147{,}000)(0.693)^{0.17} = 138{,}000 \text{ psi}.$$

The final height h_1 is 2 in. The radius r at the end of the stroke is found from volume constancy:

$$\frac{\pi 6^2}{4} \cdot 4 = \pi r_1^2 \cdot 2 \quad \text{and} \quad r_1 = 4.24 \text{ in}.$$

Therefore

$$p_{av} \simeq 138{,}000\left[1 + \frac{(2)(0.2)(4.24)}{(3)(2)}\right] = 177{,}000 \text{ psi}.$$

The upsetting force then is

$$F = (177{,}000)\pi \cdot (4.24)^2 = 10^7 \text{ lb}.$$

[Note: Figure 6.8(b) can also be used to solve this problem.]

• Example 6.2: Calculating peak pressure.

A block made of a perfectly plastic material with a yield stress Y of 150 MPa has the dimensions of $2a = 0.2$ m, $h = 0.1$ m, and a width of 0.15 m at one instant during plane-strain forging. Assuming there is sticking friction, calculate the peak pressure.

SOLUTION. For this case the applicable formula is Eq. (6.20a):

$$p = Y'\left(1 + \frac{a - x}{h}\right),$$

where $Y' = 1.15Y = 172.5$ MPa, $a = 0.1$ m, $h = 0.1$ m, and $x = 0$ (because the peak pressure is at $x = 0$). Thus

$$p = 172.5\left(1 + \frac{0.1}{0.1}\right) = 345 \text{ MPa}.$$

Note that knowing the width of the block is not necessary for this problem. •

Slip-line analysis. The *slip-line analysis* method is applied generally to plane-strain conditions. The deforming body is assumed to be rigid, perfectly plastic, and isotropic. The technique consists of the construction of a family of straight or curvilinear lines that intersect each other orthogonally. These lines, known as a *slip-line field*, correspond to the directions of yield stress of the material in shear, k.

The network of slip-lines, the construction of which depends largely on intuition and experience and are postulated a priori, must satisfy certain conditions: static equilibrium of forces, yield criteria, and boundary conditions. The slip-line fields must be checked for equilibrium conditions so that, for instance, the slip-lines meet at a free surface at a 45° angle. Also, they must be compatible with the velocity field. This means that the movement of the material must maintain mass continuity.

Upper-bound technique. In the *upper-bound technique*, the overall deformation zone is divided into a number of smaller zones within which the velocity of a particle is continuous. However, the particle velocities in adjacent zones may be different. As in slip-line analysis, at the boundaries between the zones or between a zone and the die surface, all movement must be such that discontinuities in velocity occur only in the tangential direction.

This technique, which has been used successfully for axisymmetric cases, such as extrusion and rod drawing, is lengthy and gives results similar to those in slip-line analysis. In this analysis, the total power consumed in an operation is the sum of the (a) *ideal* power of deformation, (b) power consumed in *shearing* the material along velocity discontinuities, and (c) power required in overcoming *friction* at the die–workpiece interfaces.

Finally, a velocity field that minimizes the total calculated power is taken as the actual one and is subsequently compared with experimental data.

Theorems have been developed to obtain lower- and upper-bound solutions for the required loads. The lower-bound solution underestimates the load and hence is of no practical value. Conversely, the upper-bound solutions overestimate the load and are therefore of practical interest in metalworking operations.

Finite-element method. In the *finite-element method*, the deformation zone in an elastic–plastic body is divided into a number of elements interconnected at a finite number of nodal points. The actual velocity distribution is then approximated for each element. A set of simultaneous equations is then developed representing unknown velocity vectors. From the solution of these equations actual velocity distributions and the stresses are calculated.

This technique can incorporate friction conditions at the die–workpiece interfaces and actual properties of the workpiece material. It has been applied to relatively complex geometries in bulk deformation and in sheet-forming problems. Accuracy is influenced by the number and shape of the finite elements, the deformation increment, and the methods of calculation. To ensure accuracy, complex problems require extensive computations. The finite-element method gives a detailed picture of the actual stresses and strain distributions throughout the workpiece.

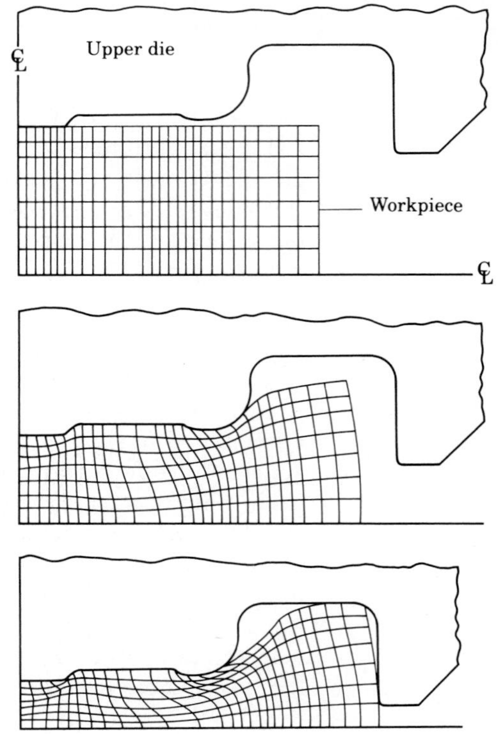

FIGURE 6.10
Plastic deformation in forging as predicted by the finite-element method of analysis. *Source:* After T. Altan, et al.

Figure 6.10 illustrates the results of applying the finite-element method of analysis on a solid cylindrical workpiece in impression-die forging. Only one quarter of the workpiece is shown, with the vertical and horizontal lines being the axes of symmetry. The grid pattern is known as a *mesh*, and its distortion is predicted from theory, aided by the use of computers with the results displayed on the video output screen. Note how the workpiece deforms as it is being forged.

This technique is also capable of predicting microstructural changes in the material during hot-working operations and the onset of defects in workpieces without actually running experiments. This knowledge is then utilized in modifying die design to avoid making defective parts. Application of this technique requires inputs such as the stress–strain characteristics of the material, as a function of strain rate and temperature, and frictional and heat transfer characteristics of the die and workpiece.

6.2.3 Impression-die forging

In *impression-die forging*, the workpiece acquires the shape of the die cavities (impressions) while it is being upset between the closing dies. A typical example is

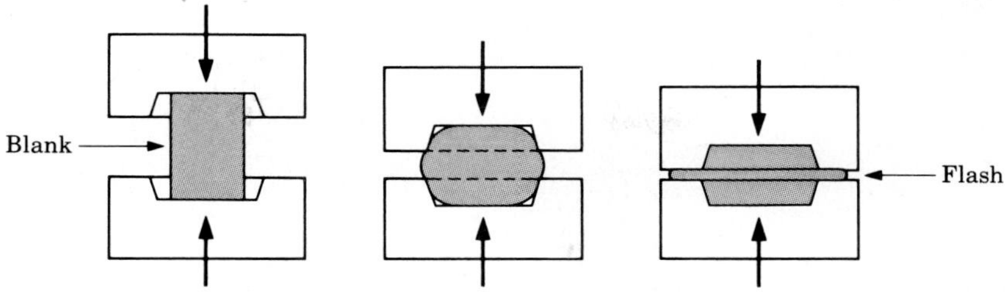

FIGURE 6.11
Schematic illustration of stages in impression-die forging. Note the formation of flash, or excess material that is subsequently trimmed off.

shown in Fig. 6.11. Some of the material flows radially outward and forms a *flash*. Because of its high length-to-thickness ratio (equivalent to a high a/h ratio), the flash is subjected to high pressure. These pressures in turn mean high frictional resistance to material flow in the radial direction in the flash gap. Because high friction encourages the filling of the die cavities, the flash has a significant role in the flow of material in impression-die forging.

Furthermore, if the operation is carried out at elevated temperatures, the flash, because of its high surface-to-thickness ratio, cools faster than the bulk of the workpiece. Thus the flash resists deformation more than the bulk does and helps fill the die cavities.

Because of the more complex shapes involved, accurate calculation of forces in impression-die forging is difficult. Depending on its position, each element within the workpiece is generally subject to different strains and strain rates. Consequently, the level of strength that the material exhibits at each location depends not only on the strain and strain rate, but also on the exponents n and m of the workpiece material in Eqs. (2.11) and (2.16).

Because of the many difficulties involved in calculating forces in impression-die forging, certain *pressure-multiplying factors* K_p have been recommended, as shown in Table 6.2. These factors are to be used with the expression

$$F = (K_p)(Y_f)(A), \tag{6.21}$$

TABLE 6.2
RANGE OF K_p VALUES FOR EQ. (6.21) IN IMPRESSION-DIE FORGING

Simple shapes, without flash	3–5
Simple shapes, with flash	5–8
Complex shapes, with flash	8–12

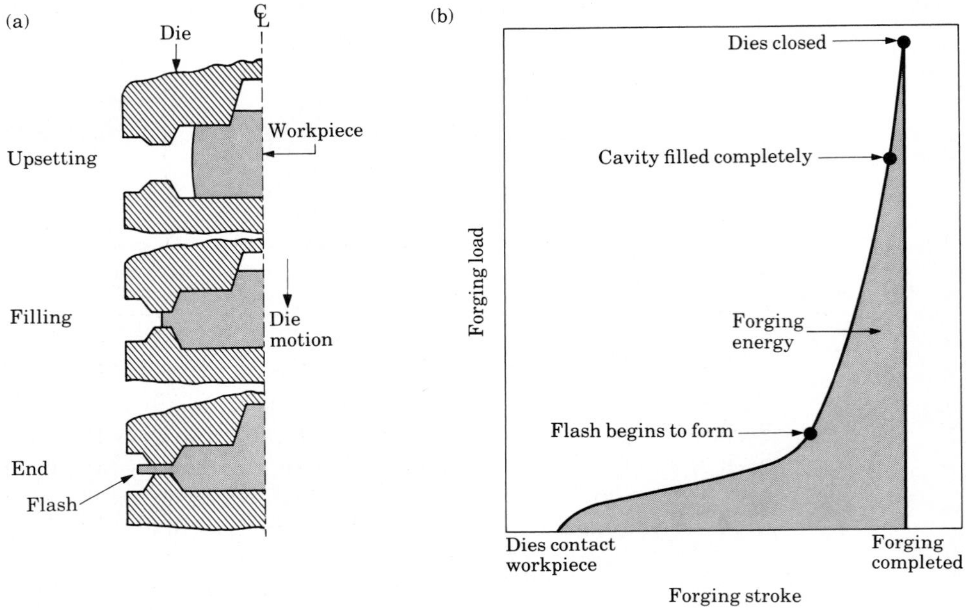

FIGURE 6.12
Typical load–stroke curve for closed-die forging. Note the sharp increase in load after the flash begins to form. In hot forging operations, the flash requires high levels of stress because it is thin, that is, small *h*, and cooler than the bulk of the forging. *Source:* After T. Altan.

where F is the forging load, A is the projected area of the forging (including the flash), and Y_f is the flow stress of the material at the strain and strain rate to which the material is subjected.

A typical impression-die forging load as a function of the stroke of the die is shown in Fig. 6.12. For this axisymmetric workpiece, the force increases gradually as the cavity is filled (Fig. 6.12b). The force then increases rapidly as flash forms. In order to obtain the final dimensions and details on the forged part, the dies must close further, with an even steeper rise in the forging load.

Note that the flash has a finite contact length with the die called *land* (Fig. 6.12). The land ensures that the flash generates sufficient resistance of the outward flow of the material to aid in die filling without contributing excessively to the forging load.

• Example 6.3: Calculating flow stress.

Assume that a forging with a very complex shape has a projected cross-sectional area of 30 in^2. If the tonnage of the forging press is 10,000, what is the maximum flow stress that the material can have? What type of material could this be?

SOLUTION. Because this is a complex forging, the value for K_p in Eq. (6.21) is taken as 12 (from Table 6.2). Thus

$$Y_f = \frac{F}{K_p A} = \frac{(10{,}000)(2{,}000)}{(12)(30)} = 55{,}500 \text{ psi}.$$

By referring to Fig. 2.4, we find that a number of materials could exhibit this level of flow stress. Drawing a horizontal line at a true stress of 55,500 psi, we note that the material could be annealed copper at a true strain of about 0.5, or it could be 1020 steel at a true strain of a little over 0.1. Thus copper will have to be compressed more than steel to reach the same flow stress level.

Closed-die forging. Although not quite correct, the example shown in Fig. 6.11 is also referred to as *closed-die forging*. In true closed-die forging no flash is formed and the workpiece is completely surrounded by the dies. In impression-die forging any excess metal in the die cavity is formed into a flash, but this is not the case in closed-die forging. Thus proper control of the volume of material is essential to obtain a forging of desired dimensions. Undersized blanks in closed-die forging prevent the complete filling of the die; oversized blanks may cause premature die failure or jamming of the dies.

Precision forging, or *flashless forging*, and similar operations where the part formed is close to the final dimensions of the desired component, are also known as *near–net-shape* production (Fig. 6.13). Any excess material is subsequently removed by various machining processes.

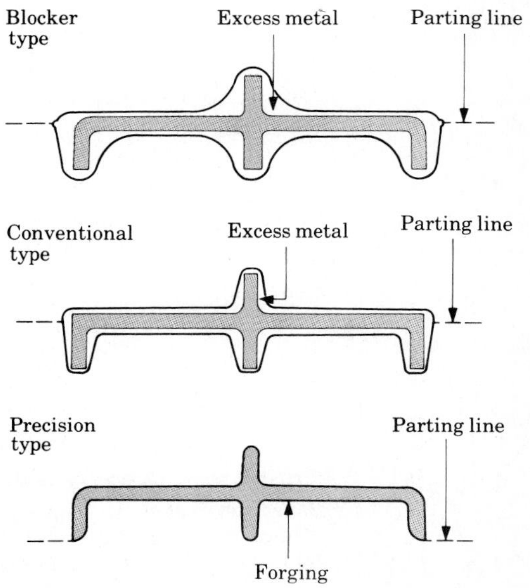

FIGURE 6.13 Three basic types of forging operations. Note that precision forging requires little or no machining to produce the final product. Blocker types require the removal of the largest amount of material. Production of parts that are at or near the final desired dimensions is known as *near–net-shape production.* Forging, squeeze casting, and powder metallurgy techniques are the most commonly used processes in near–net-shape forming. The parting line is the location where the two dies meet.

In precision forging, special dies are made and machined to greater accuracy than in ordinary impression-die forging. Precision forging requires higher capacity forging equipment than do other forging processes. Aluminum and magnesium alloys are particularly suitable for precision forging because of the low forging loads and temperatures required. Also, they result in little die wear and produce a good surface finish. Steels and other alloys are more difficult to precision forge. The choice between conventional forging and precision forging requires an economic analysis. Precision forging requires special dies. However, much less machining is involved, because the part is closer to the desired final shape.

In the *isothermal forging* process, also known as *hot-die forging*, the dies are heated to the same temperature as the hot blank. In this way, cooling of the workpiece is eliminated, the low flow stress of the material is maintained, and material flow within the die cavities is improved. The dies are generally made of nickel alloys, and complex parts with good dimensional accuracy can be forged in one stroke in hydraulic presses. Isothermal forging generally is expensive; however, it can be economical for intricate forgings of expensive materials, provided that the quantity required is large enough to justify die costs.

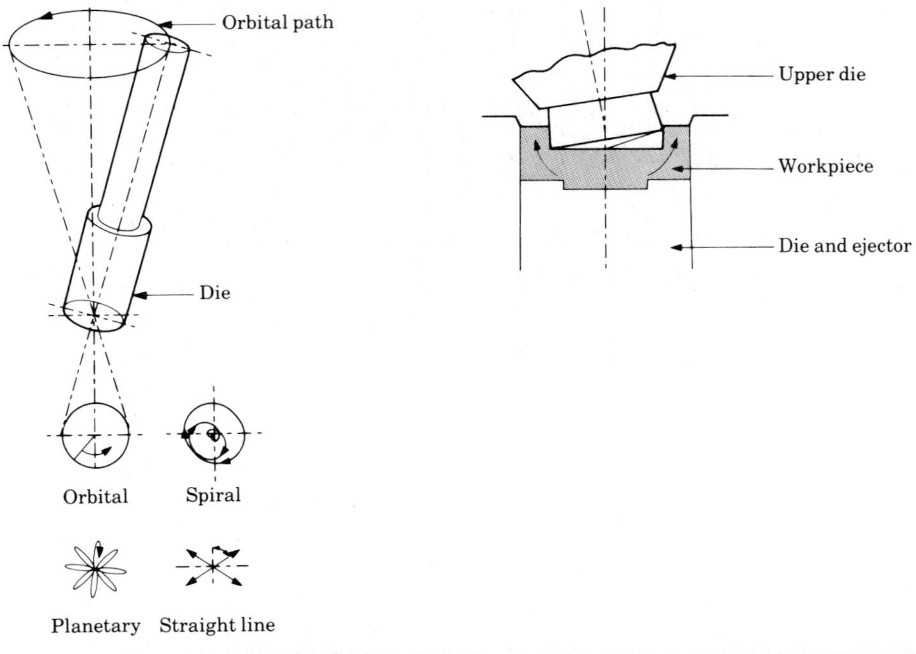

FIGURE 6.14

Schematic illustration of the orbital forging process. Note that the die is in contact only with a portion of the workpiece surface. This process is also called rotary forging, swing forging, or rocking-die forging and can be used for forming bevel gears, wheels, and bearing rings.

Orbital forging is a relatively new process in which the die moves along an orbital path (Fig. 6.14) and forges or forms the part incrementally.

6.2.4 Miscellaneous forging operations

Coining. Another example of closed-die forging is the minting of coins (Fig. 6.15), where the slug is shaped in a completely closed cavity. The pressures required can be as great as five to six times the flow stress of the material in order to produce the fine details of a coin or a medallion. Several coining operations may be necessary in order to obtain full detail on some parts.

The coining process is also used with forgings and other products to improve surface finish and impart the desired dimensional accuracy (*sizing*) of the products. The pressures are large, and usually very little change in shape takes place. Lubricants cannot be tolerated in coining, because they can be trapped in die cavities and prevent reproduction of fine die surface details.

Heading. Basically an upsetting operation, *heading* is performed at the end of a rod (usually round) to form a shape with a larger cross-section. Typical examples

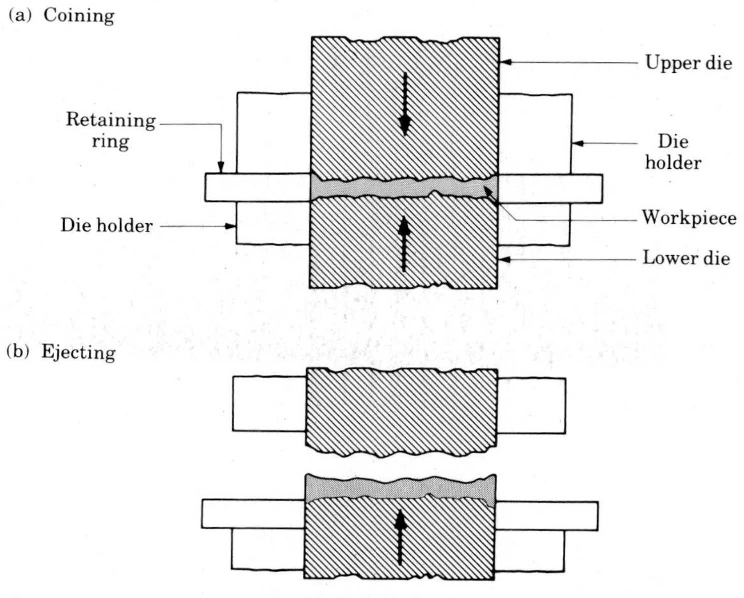

FIGURE 6.15
Schematic illustration of coining. The earliest coins were made by the Romans using simple tools. This process is also used for decorative items, medallions, patterned tableware, and in sizing parts.

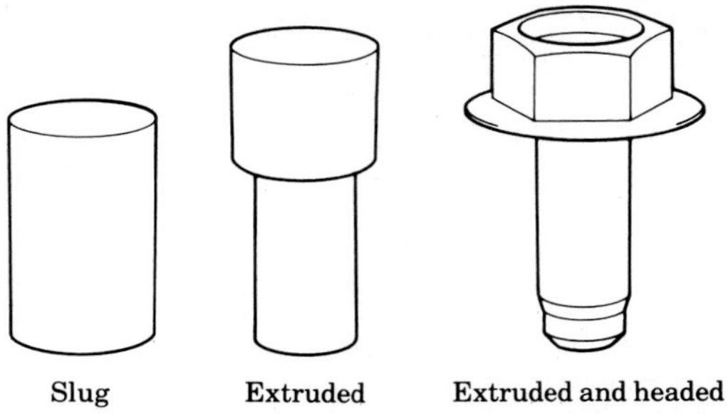

FIGURE 6.16 Stages in heading operation for a bolt blank. Note that the head undergoes basically an upsetting operation. The threads on the bolt are produced either by cutting or forming operations.

are the heads of bolts, screws, and nails (Fig. 6.16). An important aspect of the upsetting process during heading is the tendency for buckling if the length-to-diameter ratio is too high. Heading is done on machines called *headers*, which are usually highly automated horizontal machines with high production rates. The operation can be carried out cold, warm, or hot. Heading forces can be estimated reliably using the slab method of analysis described in Section 6.2.2.

Piercing. In *piercing* a punch indents the surface of a workpiece to produce a cavity or an impression (Fig. 6.17). The workpiece may be confined in a die cavity or it may be unconstrained. The piercing force depends on the punch cross-sectional area and its tip geometry, the flow stress of the material, and the friction at the interfaces. Punch pressures may range from three to five times the flow stress of the material. The term piercing is also used for the process of cutting of holes with a punch and die.

Hubbing. In *hubbing* a hardened punch with a particular tip geometry is pressed into the surface of a block. This process is used to produce a die cavity, which is then used for subsequent forming operations. The cavity is usually shallow; for deeper cavities some material may be removed from the surface by machining prior to hubbing. The pressure required to generate a cavity by hubbing

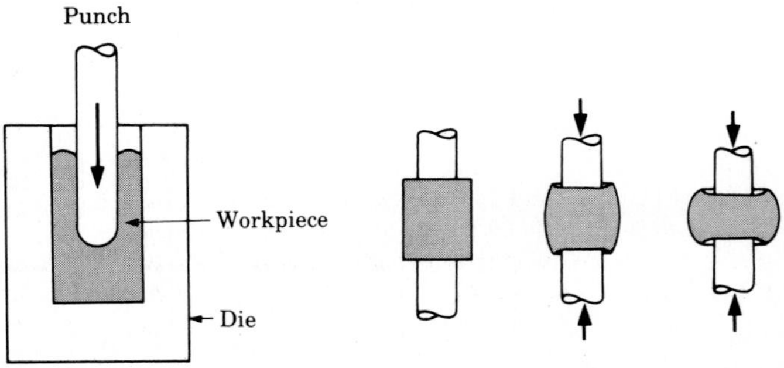

FIGURE 6.17 Examples of piercing operations.

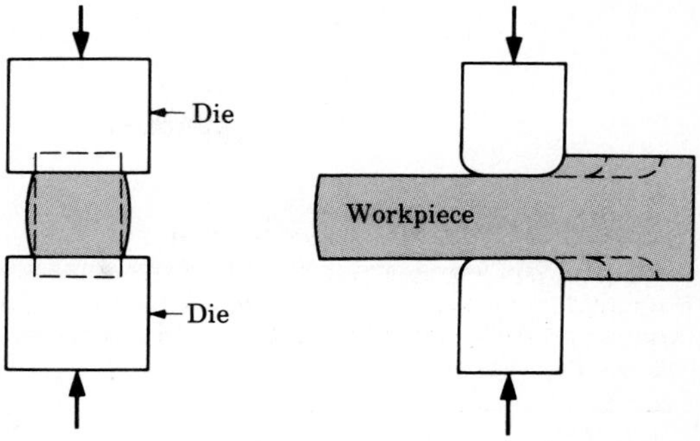

FIGURE 6.18 Schematic illustration of a cogging operation on a rectangular bar. With simple tools, the thickness and cross-section of a bar can be reduced by multiple cogging operations. Note the barreling after cogging. Blacksmiths use a similar procedure to reduce the thickness of parts in small increments by heating the workpiece and hammering it numerous times.

is approximately three times the ultimate tensile strength (UTS) of the material of the block. Thus the force required is

$$\text{Hubbing force} = 3(\text{UTS})(A), \tag{6.22}$$

where A is the projected area of the impression. Note that the factor 3 is in agreement with the observations made with regard to hardness of materials in Section 2.6.8.

Cogging. In *cogging*, also called *drawing out*, the thickness of a bar is reduced by successive steps at certain intervals (Fig. 6.18). A long section of a bar can be reduced in thickness without large forces, because the contact area is small.

Fullering and edging. The operations of *fullering* and *edging* are performed —usually on bar stock—to distribute the material in certain regions of a forging. The parts are then formed into final shapes by other forging processes.

Roll forging. In *roll forging* the cross-sectional area of a bar is reduced and altered in shape by passing it through a pair of rolls with grooves of various shapes (Fig. 6.19). This operation may also be used to produce a part that is basically the

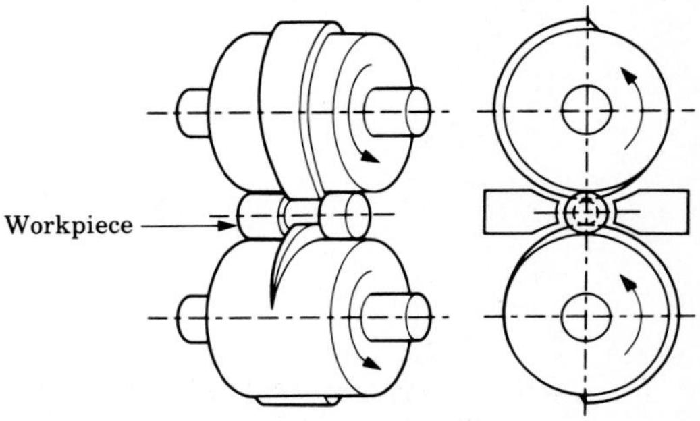

FIGURE 6.19 Schematic illustration of a roll forging (cross rolling) operation. Tapered leaf springs and knives can also be made by this process with specially designed rolls. *Source:* After J. Holub, *Machinery,* vol. 102, Jan. 16, 1963.

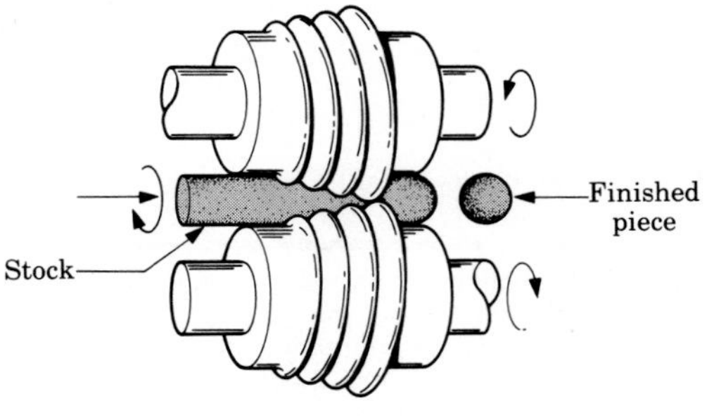

FIGURE 6.20
Production of steel balls for bearings by the skew rolling process. Balls for bearings can also be made by the forging process shown in Fig. 6.21.

final product, such as tapered shafts, tapered leaf springs, table knives, and numerous tools. Roll forging is also used as a preliminary forming operation, followed by other forging processes. Typical products are crankshafts and other automotive components.

Skew rolling. A process similar to roll forging is *skew rolling* (Fig. 6.20), which is used for making ball bearings. Round wire or rod stock is fed into the roll gap and spherical blanks are formed continuously by the rotating rolls. (Another method of forming blanks for ball bearings is by cutting pieces from a round bar and upsetting them, as shown in Fig. 6.21. The balls are then ground and polished in special machinery.)

6.2.5 Defects

Besides surface cracking, other defects in forgings can be caused by the material flow patterns in the die cavity. As shown in Fig. 6.22, excess material in the web of a forging can buckle during forging and develop *laps*. If the web is thick, the excess material flows past the already forged portions and develops internal cracks (Fig. 6.23). These examples indicate the importance of properly distributing material and controlling the flow in the die cavity.

The various radii in the die cavity can significantly affect formation of defects. In Fig. 6.24, the material follows a large corner radius better than a small radius.

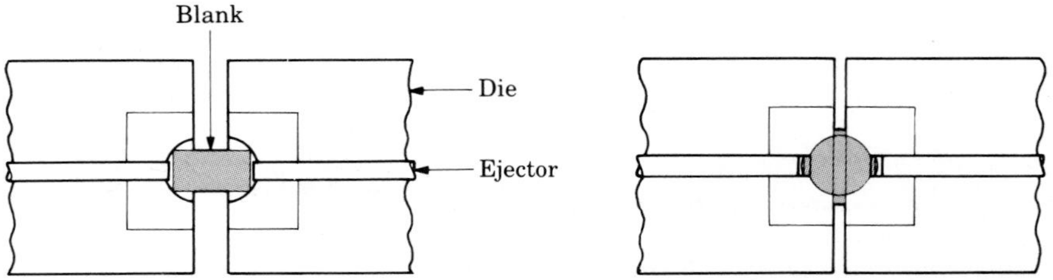

FIGURE 6.21
Production of steel balls by upsetting of a cylindrical blank. Note the formation of flash. The balls are subsequently ground and polished for use in ball bearings and other mechanical components.

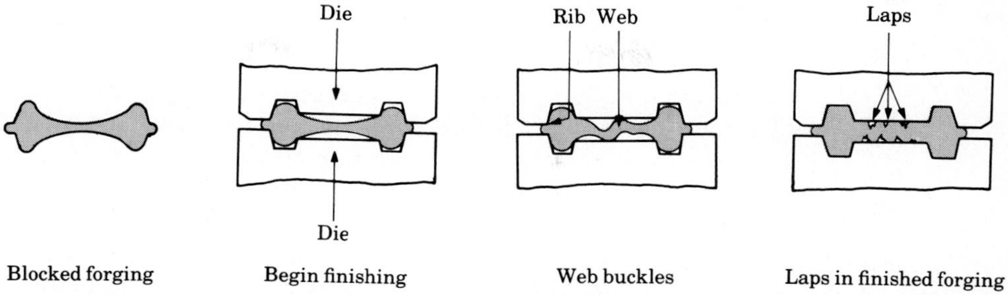

FIGURE 6.22
Laps formed by buckling of the web during forging. The solution to this problem is to increase the initial web thickness to avoid buckling.

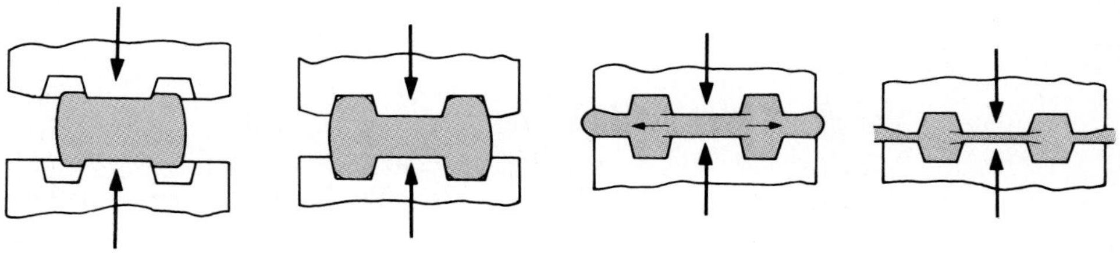

FIGURE 6.23
Internal defects produced in a forging because of an oversized billet. The die cavities are filled prematurely and the material at the center of the part flows past the filled regions as deformation continues.

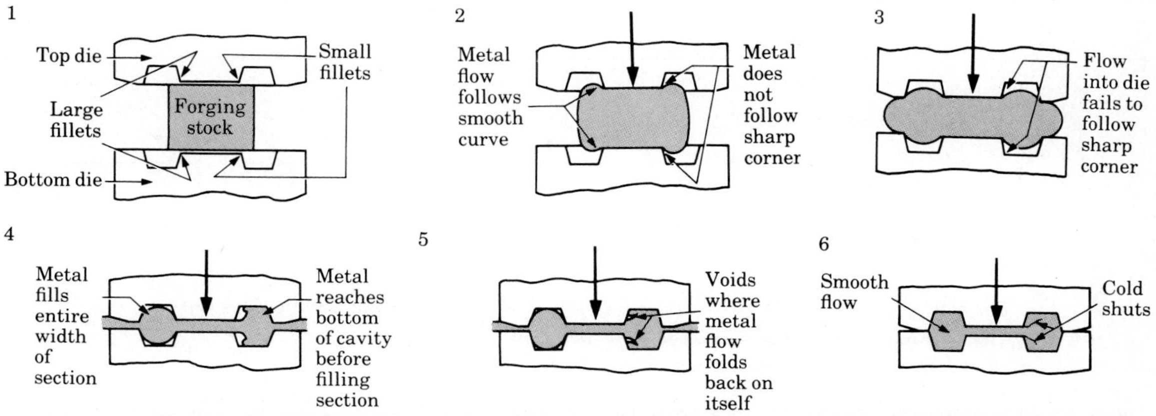

FIGURE 6.24
Effect of fillet radius on defect formation in forging. Small fillets (right side of drawings) cause the defects. *Source:* ALCOA.

With small radii, the material can fold over itself and produce a lap, called *cold shut*.

During the service life of the forged component, such defects can lead to fatigue failure and other problems, such as excessive corrosion and wear. The importance of inspecting forgings before they are put into service, particularly for critical applications, is obvious. (For inspection techniques, see Section 4.8.)

Although it may not be considered a flaw, another important aspect of quality in a forging is the *grain flow pattern*. At times the grain flow lines reach a surface perpendicularly, exposing the grain boundaries directly to the environment. They are known as *end grains*. In service they can be attacked by the environment and develop a rough surface and act as stress raisers. For critical components, end grains in forgings can be avoided by proper selection of the original workpiece orientation in the die cavity and by control of material flow.

Because the metal flows in various directions in a forging—and because of temperature variations—the properties of a forging are generally anisotropic. As Fig. 6.25 shows, strength and ductility vary significantly in test specimens taken from different locations and orientations in a forged part.

6.2.6 Forgeability

The *forgeability* of a metal can be defined as its capability to undergo deformation by forging without cracking. This definition can be expanded to include the flow strength of the metal. Thus a material with good forgeability is one that can be shaped with low forces without cracking. A number of tests have been developed to measure forgeability, although none is universally accepted. We describe the more commonly used tests.

Upsetting test. A standard test is to upset a solid cylindrical specimen and observe any cracking on the barreled surfaces (Fig. 2.28d). The greater the reduction

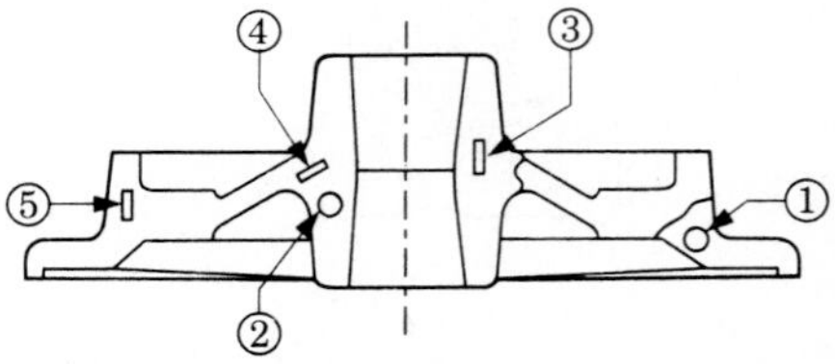

Specimen location	Yield strength, psi	Ultimate tensile strength, psi	Elongation, %
1.	26,500	41,200	20.0
2.	23,000	39,400	17.7
3.	13,800	38,000	18.6
4.	25,400	40,400	16.2
5.	17,100	37,500	14.0

FIGURE 6.25 Mechanical properties of five tensile test specimens taken at various locations and directions in an AZ61 magnesium alloy forging. Note the anisotropy of properties caused by inhomogeneous deformation during forging. *Source:* After S. M. Jablonski, *Modern Metals,* vol. 16, 1963, pp. 62–70.

in height prior to cracking, the greater is the forgeability of the metal. These cracks are caused by *secondary* tensile stresses on the barreled surfaces. They are called secondary because no external tensile stress is applied to the material, and the stresses develop because of changes in the material's shape.

Because barreling is a result of friction at the die–workpiece interfaces, friction has a marked effect on cracking in upsetting. Figure 6.26 shows that, as friction increases the specimen cracks at a lower reduction in height. Of interest also is whether the crack is at 45 degrees or vertical; the angle depends on the sign of the axial stress σ_z on the barreled surface. If this stress is negative (compressive) the crack is at 45 degrees; if positive (tensile) it is longitudinal. Depending on the notch sensitivity of the material, any surface defects will also affect the results by causing premature cracking. A typical surface defect is a *seam*: a longitudinal scratch, a string of inclusions, or a fold from prior working of the material.

Upsetting tests can be performed at various temperatures and strain rates. An optimal range for these parameters can then be specified for forging a particular material. Such tests can serve only as guidelines, since in actual forging the metal is subjected to a different state of stress as compared with a simple upsetting operation.

Hot-twist test. A round specimen is twisted continuously until it fails. The test is performed at various temperatures, and the number of turns that each specimen undergoes before failure is observed. The optimal forging temperature is then determined. The hot-twist test is particularly useful in determining the forgeability of steels, although upsetting tests can also be used for that purpose. Small changes in the composition of or impurities in the metal can have a significant effect on forgeability.

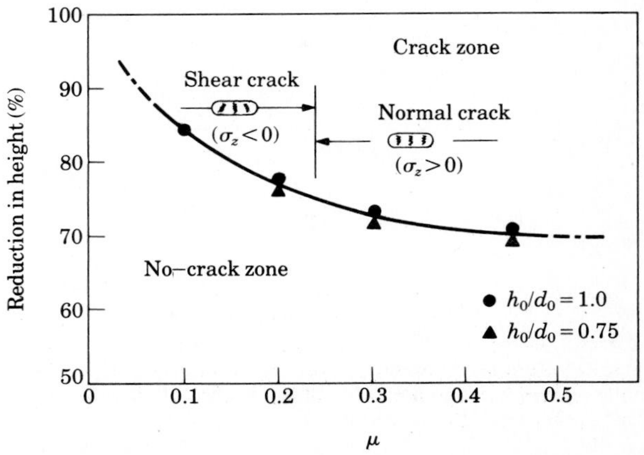

FIGURE 6.26 Cracking in upsetting solid cylindrical AISI 1045 steel specimens, as a function of reduction in height and friction. No cracks are observed below the curve. *Source:* After S. Kobayashi.

Forgeability of various metals. Based on the results of various tests and observations, the forgeability of several metals and alloys has been determined. It is based on such considerations as the ductility and strength of the metals, forging temperature required, frictional behavior, and quality of the forgings obtained.

Effect of hydrostatic pressure on forgeability. As described in Section 2.2.8, hydrostatic pressure has a significant beneficial effect on the ductility of metals and nonmetallic materials. Experiments have indicated that the results of room-temperature forgeability tests, outlined above, are improved (that is, cracking takes place at higher strain levels) if the tests are conducted in an environment of high hydrostatic pressure. Although such a test would be difficult to perform at elevated temperatures, techniques have been developed to forge metals in a high compressive environment in order to take advantage of this phenomenon. The pressure-transmitting medium is usually a low-strength ductile metal.

6.2.7 Dies

Proper design of forging dies and selection of die materials require considerable experience and knowledge of the strength and ductility of the workpiece material, its sensitivity to strain rate and temperature, and its frictional characteristics. Die distortion under high forging loads can also be an important consideration in die design, particularly if close tolerances are required.

The most important rule in die design is that the workpiece material flow in the direction of least resistance. Thus common practice is to distribute the material so that it can properly fill die cavities. An example is the forging of a connecting rod, as shown in Fig. 6.27. Starting with a round bar stock, the bar is first preformed (*intermediate shape*) by some of the techniques previously described and formed into the final shape in two additional forging operations, and then trimmed. Note how proper distribution of the material fills the die cavities to forge the part successfully.

Another example of intermediate shapes is shown in Fig. 6.28. Selection of these shapes requires experience and involves calculating cross-sectional areas at each location in the forging. The dies used for intermediate forging operations are called *blocker* dies. Care in calculating the volume of the material required to properly fill

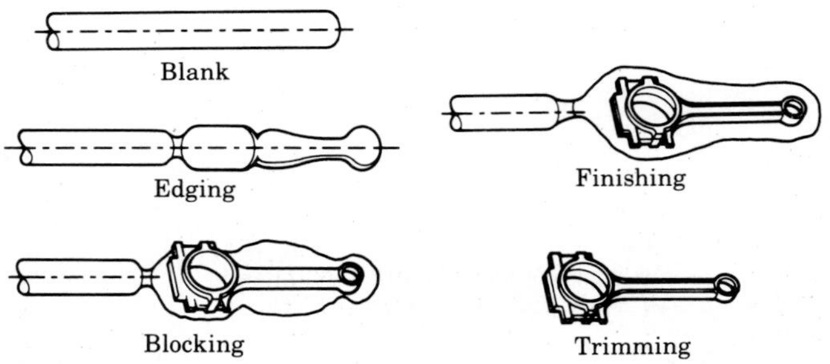

FIGURE 6.27
Stages in forging a connecting rod for an internal combustion engine. Note the amount of flash that is necessary to fill the die cavities properly.

the die cavities is essential. Computer techniques are now being developed to expedite these calculations. The sequence of operations involved in making a railroad wheel is shown in Fig. 6.29. The terminology used in die design is shown in Fig. 6.30. The significance of various parameters are as follows.

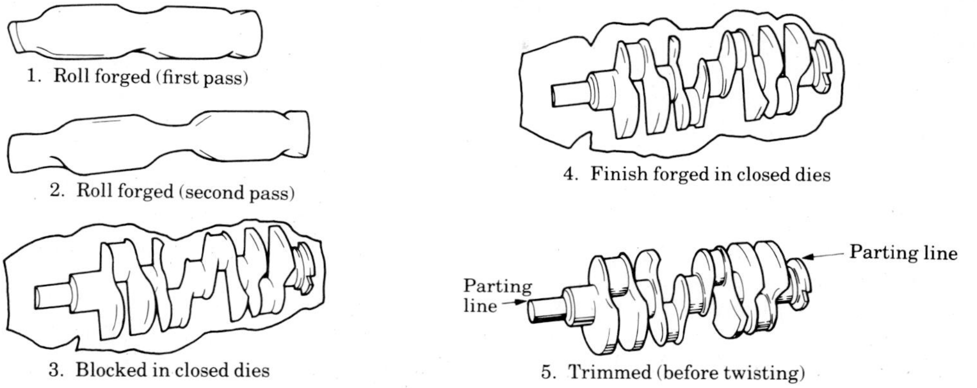

FIGURE 6.28
Intermediate stages in forging a crankshaft. These intermediate stages are important for distributing the material and filling the die cavities properly.

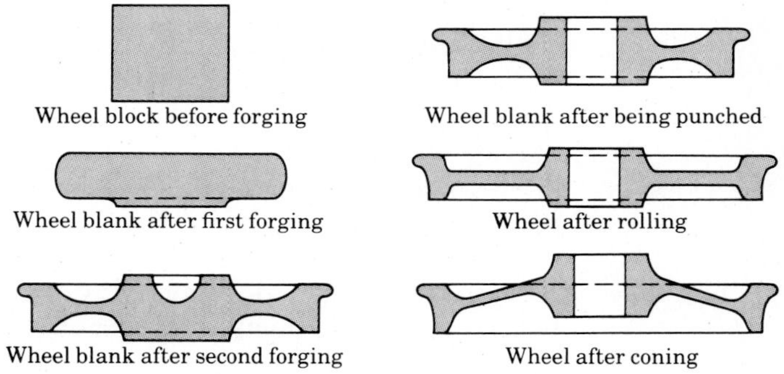

FIGURE 6.29
Stages in the manufacture of a railroad wheel. *Source:* Copyright 1964 by United States Steel Corporation.

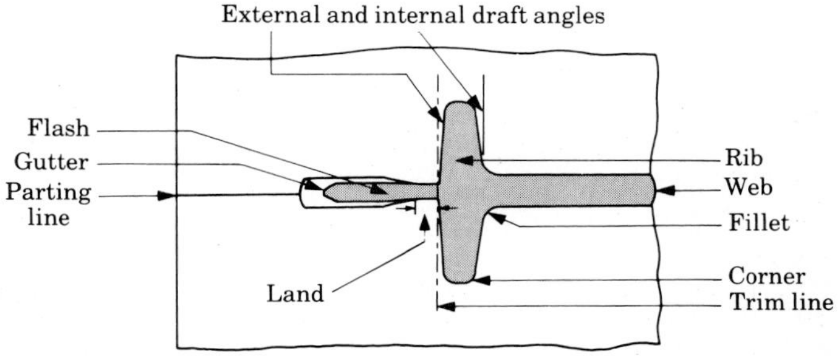

FIGURE 6.30
Standard terminology for various features of a typical forging die.

The *parting line* is where the two dies meet. For simple symmetrical shapes the parting line is a straight line at the center of the forging; for more complex shapes the line may be offset and may not be in a single plane. Selection of the proper location for the parting line is based on the shape of the part, metal flow, balance of forces, and the flash.

The significance of *flash* is described in Section 6.2.3. After sufficiently constraining lateral flow (by the length of the land), the flash is allowed to flow into a *gutter*. Thus the extra flash does not increase the forging load unnecessarily. A general guideline for flash clearance (between the dies) is 3 percent of the maximum thickness of the forging. The length of the land is usually five times the flash clearance.

Draft angles are necessary in almost all forgings to facilitate the removal of the part from the die. Draft angles usually range between 3° and 10°. Because the forging shrinks in its radial direction (as well as in other directions) as it cools, internal draft angles are made larger than external ones. Thus internal angles are about 7° to 10° and external angles about 3° to 5°.

Proper selection of the radii for corners and fillets ensures smooth flow of the metal in the die cavity and improves die life. Small radii are generally not desirable because of their adverse effect on metal flow and their tendency to wear rapidly from stress concentration and thermal cycling. Small radii in fillets can cause fatigue cracking in dies.

Die materials. Because most forgings, particularly large ones, are performed at elevated temperatures, die materials generally must have strength and toughness at elevated temperatures, hardenability, mechanical and thermal shock resistance, and wear resistance—particularly, abrasive wear resistance because of the presence of scale on heated forgings.

Selection of die materials depends on die size, composition and properties of the workpiece, complexity of its shape, forging temperature, type of operation, cost of die material, the number of forgings required, and the heat transfer and distortion characteristics of the die material. Common die materials are tool and die steels, which contain chromium, nickel, molybdenum, and vanadium.

Forging temperatures and lubrication. The range of temperatures for hot forging are given in Table 6.3.

TABLE 6.3
FORGING TEMPERATURE RANGES FOR VARIOUS METALS

	°C	°F		°C	F
Aluminum alloys	400–450	750–850	Titanium alloys	750–975	1400–1800
Copper alloys	625–950	1150–1750	Refractory alloys	975–1650	1800–3000
Alloy steels	925–1250	1700–2300			

Lubrication in forging plays an important role. It affects friction and wear and, consequently, the flow of metal into the die cavities. Lubricants can also serve as a thermal barrier between the hot forging and the relatively cool dies, thus slowing the rate of cooling of the workpiece. Another important role of a lubricant is to serve as a parting agent, that is, to prevent the forging from sticking to the dies. Lubricants can significantly affect the wear pattern in forging dies. A wide variety of lubricants can be used in forging. For hot forging, graphite, molybdenum disulfide, and (sometimes) glass are commonly used as lubricants. For cold forging, mineral oils and soaps are common lubricants (see Table 6.4).

6.2.8 Equipment

Forging equipment of various designs, capacities, speeds, and speed–stroke characteristics is available (Fig. 6.31).

Hydraulic presses. Hydraulic presses have a constant low speed and are *load limited*. Large amounts of energy can be transmitted to the workpiece by a constant load available throughout the stroke. Ram speed can be varied during the stroke. These presses are used for both open-die and closed-die forging operations. The largest hydraulic press in existence has a capacity of 670 MN (75,000 tons).

Mechanical presses. Mechanical presses are *stroke limited*. They are basically crank or eccentric types, with speeds varying from a maximum at the center of the stroke to zero at the bottom. The force available depends on the stroke position and becomes extremely large at the bottom dead center. Thus proper setup is essential to avoid breaking the dies or other equipment. The largest mechanical press has a capacity of 107 MN (12,000 tons).

Screw presses. Screw presses derive their energy from a flywheel. The forging load is transmitted through a vertical screw. These presses are *energy limited* and can be used for many forging operations. They are particularly suitable for producing small quantities, for parts requiring precision (such as turbine blades), and for control of ram speed. The largest screw press has a capacity of 280 MN (31,500 tons).

Hammers. Hammers derive their energy from the potential energy of the ram, which is then converted to kinetic energy; thus they are *energy limited*. The speeds are high; therefore the low forming times minimize cooling of the hot forging, allowing the forging of complex shapes, particularly with thin and deep recesses. To complete the forging, several blows may have to be made on the part. In *power hammers* the ram is accelerated in the downstroke by steam or air, in addition to gravity. The highest energy available in power hammers is 1150 kJ (850,000 ft-lb).

TABLE 6.4
LUBRICANTS COMMONLY USED IN METALWORKING OPERATIONS (SEE ALSO CHAPTER 4)

MATERIAL AND ALLOY	TEMPERATURE	FORGING	ROLLING	EXTRUSION	ROD AND WIRE DRAWING	SHEET METALWORKING
Aluminum	Cold	FA + MO, S	FA + MO, MO	D, G, MS, L, S	MO, E, Wa, FA + MO	FO + MO, E, L, Wa, S
	Hot	MO + G, MS	FA + CO, FA + E	D, G, PI	—	G
Beryllium	Hot	MO + G, J	G	MS, G, J	G	—
Copper	Cold	S, E, T	E, MO	S, T, L, Wa, G, MS	FO + E + S, MO, Wa	FO + E, FO + MO, S + Wa
	Hot	G	D, E	D, MO + G	—	G
Lead	Cold	FO + MO	FA + MO, EM	D, S, T, E, FO	FO	FO + MO
	Hot	—	—	D	—	—
Magnesium	Cold	FA + MO, CC + S	D, FA + MO	D, T	—	FO + MO
	Hot	W + G, MO + G	MO + FA + EM, D, G	D, PI	—	G + MO, S + Wa, T + G
Nickel	Cold	—	MO + CL	CC + S	CC + S, MO + CL	E, MO + EP, CL, CC + S
	Hot	MO + G, W + G, GI	W, E	GI, J	—	G, MS
Refractory	Hot	GI, G, MS	G + MS	J + GI	G + MS, PI	MS, G
Steels (carbon and low alloy)	Cold	EP + MO, CC + S	E, CO, MO	CC + S, T, W + MS	S, CC + S	E, S, Wa, FO + MO, PI, CSN
	Hot	MO + G, Sa, GI	W, MO, CO, G + E	GI, G	—	G
Steels (stainless)	Cold	CL + MO, CC + S	CL + EM, CL + MO	CC + S, CP, MO	CC + S, CL + MO	FO + MO, Wa, PI, EP + MO, S
	Hot	MO + G, GI	D, W, E	GI	—	G
Titanium	Cold	S, MO	MO, CC + FO, G	CC + G, CC + S	CC + PI	CL, S, PI, Wa
	Hot	W + G, GI, MS	PI, CC, G, MS	J + GI, GI	—	G, MS

CC	Conversion coatings	CSN	Chemicals and synthetics	EP	Extreme pressure	G	Graphite	L	Lanolin	PI	Polymer	T	Tallow
CL	Chlorinated paraffin	D	Dry	FA	Fatty acid	GI	Glass	MO	Mineral oil	S	Soap	W	Water
CO	Compounded oil	E	Emulsion	FO	Fatty oil	J	Canned or jacketed	MS	Molybdenum disulfide	Sa	Salts	Wa	Wax
CP	Copper plate												

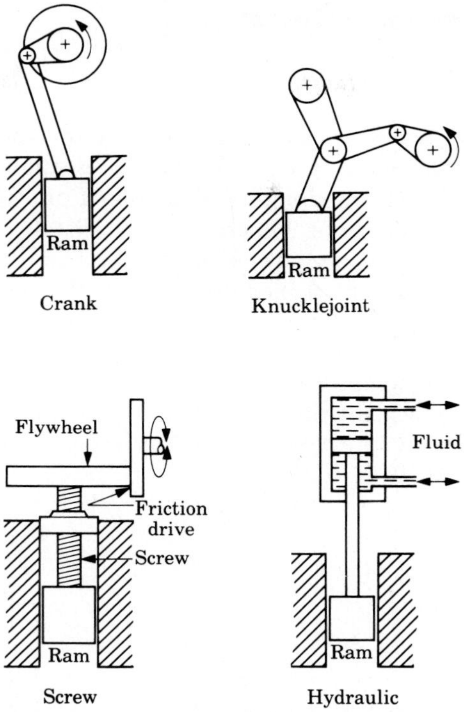

FIGURE 6.31 Schematic illustration of various types of presses used in metalworking. The choice of the press is an important factor in the overall operation.

Counterblow hammers. Counterblow hammers have two rams that simultaneously approach each other to forge the part. They are generally of the mechanical–pneumatic or mechanical–hydraulic type. These machines transmit less vibration to the foundation. The largest counterblow hammer has a capacity of 1200 kJ (900,000 ft-lb).

Equipment selection. Selection of forging equipment depends on the size and complexity of the forging, strength of the material and its sensitivity to strain rate, and degree of deformation required. Generally, presses are usually preferred for aluminum, magnesium, beryllium, bronze, and brass; hammers are preferred for copper, steels, titanium, and refractory alloys. Production rate is also a consideration in equipment selection. The number of strokes per minute ranges from a few for hydraulic presses to as many as 300 for power hammers.

6.3 Rolling

Rolling is the process of reducing the thickness or changing the cross-section of a long workpiece by compressive forces applied through a set of *rolls* (Fig. 6.32) similar to rolling dough with a rolling pin to reduce its thickness. Rolling, which accounts for about 90 percent of all metals produced by metalworking processes, was first developed in the late 1500s. The basic operation is flat rolling, or simply rolling, where the rolled products are flat plate and sheet.

Plates, which are generally regarded as having a thickness greater than 6 mm (1/4 in.), are used for structural applications such as ship hulls, boilers, bridges, girders, machine structures, and nuclear vessels. Plates can be as much as 0.3 m (12 in.) thick for the supports for large boilers, 150 mm (6 in.) for reactor vessels, and 100–125 mm (4–5 in.) for battleships and tanks.

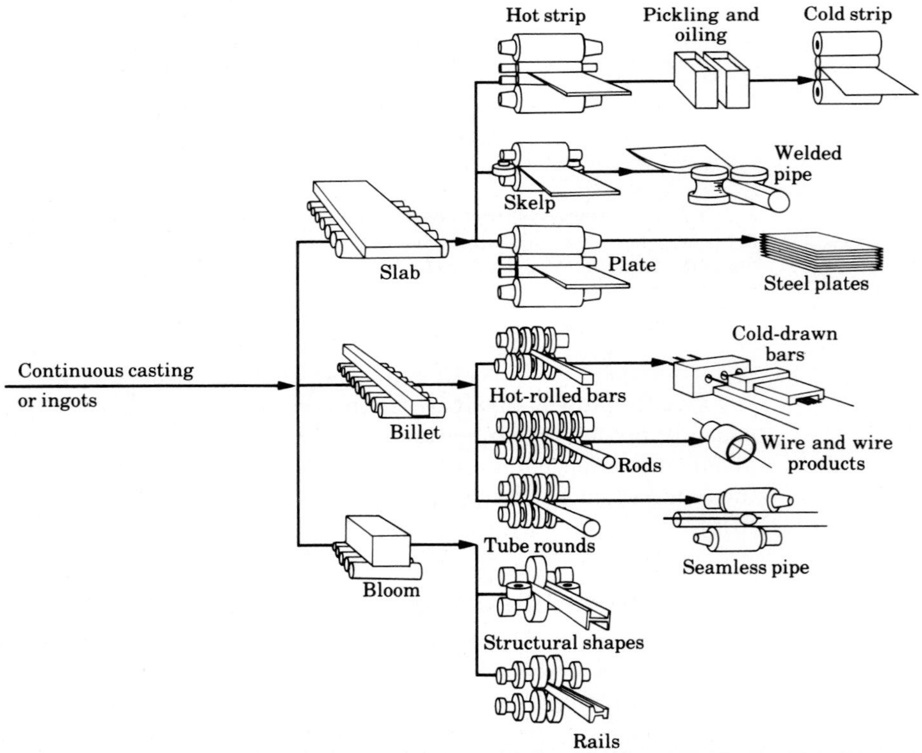

FIGURE 6.32 Schematic outline of various flat- and shape-rolling processes. *Source:* American Iron and Steel Institute.

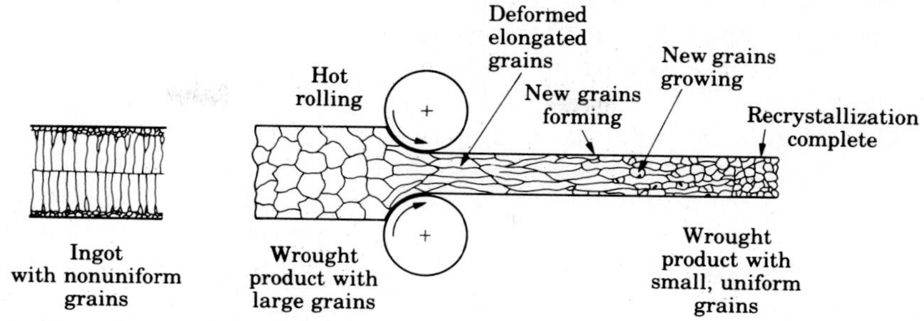

FIGURE 6.33
Changes in the grain structure of cast or large-grain wrought metals during hot rolling. Hot rolling is an effective way to reduce grain size in metals for improved strength and ductility. Cast structures of ingots or continuous castings are converted to a wrought structure by hot working.

Sheets are generally less than 6 mm thick and are used for automobile bodies, appliances, containers for food and beverages, and kitchen and office equipment. Commercial aircraft fuselages are usually made of about 1 mm (0.040 in.) thick aluminum-alloy sheet, while beverage cans are made of 0.15 mm (0.006 in.) aluminum sheet. Aluminum *foil* used to wrap candy and cigarettes has a thickness of 0.008 mm (0.0003 in.) Sheets are provided as flat pieces or as strip in coils to manufacturing facilities for further processing into products.

Traditionally, the initial material form for rolling is an ingot. However, this practice is now being rapidly replaced by continuous casting and rolling, with much higher efficiency and lower cost. Rolling is first carried out at elevated temperatures (hot rolling), wherein the coarse-grained, brittle, and porous structure of the ingot or continuously cast metal is broken down into a *wrought structure*, with finer grain size (Fig. 6.33).

6.3.1 Mechanics of flat rolling

The flat rolling process is shown schematically in Fig. 6.34. A strip of thickness h_0 enters the roll gap and is reduced to h_f by the rotating (powered) rolls. The surface

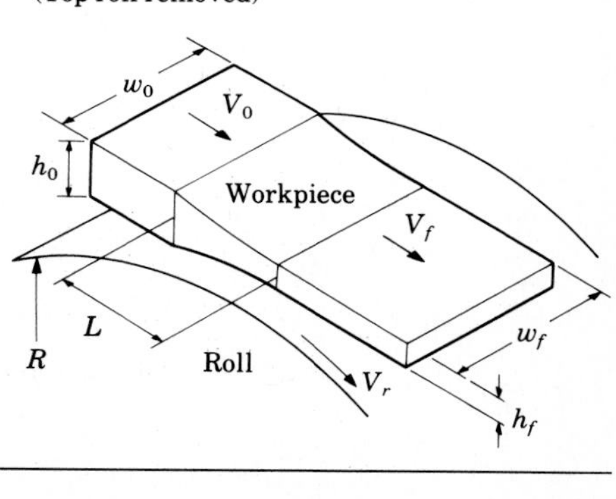

FIGURE 6.34
Schematic illustration of the flat rolling process. A greater volume of metal is formed by rolling than by any other metalworking process.

speed of the roll is V_r. To keep the volume rate of metal flow constant, the velocity of the strip must increase as it moves through the roll gap (similar to fluid flow through a converging channel). At the exit of the roll gap, the velocity of the strip is V_f (Fig. 6.35). Because V_r is constant along the roll gap, sliding occurs between the roll and the strip.

At one point along the arc of contact the two velocities are the same. It is known as the *neutral point* or *no-slip point.* To the left of this point, the roll moves faster than the workpiece, and to the right the workpiece moves faster than the roll. Because of friction at the interfaces, the frictional forces—which oppose motion—act on the strip surfaces, as shown in Fig. 6.35.

Note the similarity between Figs. 6.35 and 6.1(b). The friction forces oppose each other at the neutral point. In upsetting, these forces are equal to each other because of the symmetry of the operation. In rolling, the frictional force on the left of the neutral point must be greater than the force on the right. This difference yields a net frictional force to the right, which makes the rolling operation possible by pulling the strip into the roll gap. Furthermore, the net frictional force and the surface velocity of the roll must be in the same direction in order to supply work to the system. Thus the neutral point should be toward the exit in order to satisfy these requirements.

Forward slip in rolling is defined in terms of the exit velocity of the strip V_f and the surface speed of the roll V_r as

$$\text{Forward slip} = \frac{V_f - V_r}{V_r} \tag{6.23}$$

and is a measure of the relative velocities involved.

Roll pressure. Although the deformation zone in the roll gap is subjected to a state of stress similar to that in upsetting, the calculation of forces and stress distribution in flat rolling is more involved because of the curved surface of contact.

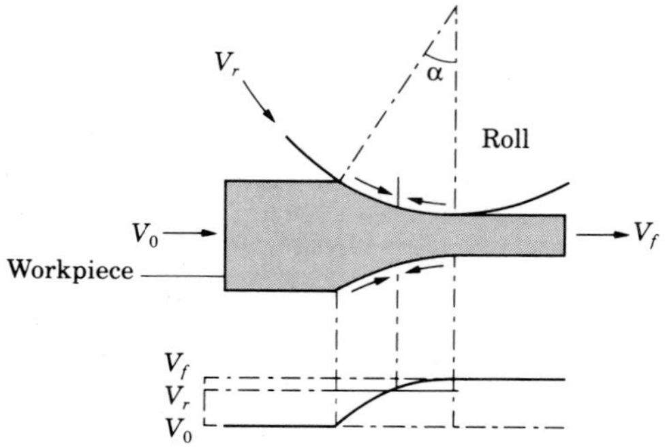

FIGURE 6.35 Relative velocity distribution between roll and strip surfaces. Note the difference in the direction of frictional forces. The arrows represent the frictional forces acting on the strip.

In addition, the material at the exit is strain hardened, so the flow stress at the exit is higher than that at the entry.

The stresses on an element in the entry and exit zones, respectively, are shown in Fig. 6.36. Note that the only difference between the two elements is the direction of the friction force. Using the *slab method* of analysis for *plane strain* (described in Section 6.2.2), we may analyze the stresses in rolling as follows.

From the equilibrium of the horizontal forces on the element shown in Fig. 6.36, we have

$$(\sigma_x + d\sigma_x)(h + dh) - 2pR\,d\phi \sin\phi - \sigma_x h \pm 2\mu pR\,d\phi \cos\phi = 0.$$

When we simplify, neglecting second-order terms, this expression reduces to

$$\frac{d(\sigma_x h)}{d\phi} = 2pR(\sin\phi \mp \mu\cos\phi).$$

In rolling, the angle α is only a few degrees; hence we may assume that $\sin\phi = \phi$ and $\cos\phi = 1$. Thus

$$\frac{d(\sigma_x h)}{d\phi} = 2pR(\phi \mp \mu). \tag{6.24}$$

Because the angles involved are small, we assume that p is a principal stress, σ_x being the other principal stress. The relationship between these two stresses and the flow stress Y_f of the material is given by Eq. (2.44) for plane strain, or

$$p - \sigma_x = \frac{2}{\sqrt{3}} Y_f = Y_f'. \tag{6.25}$$

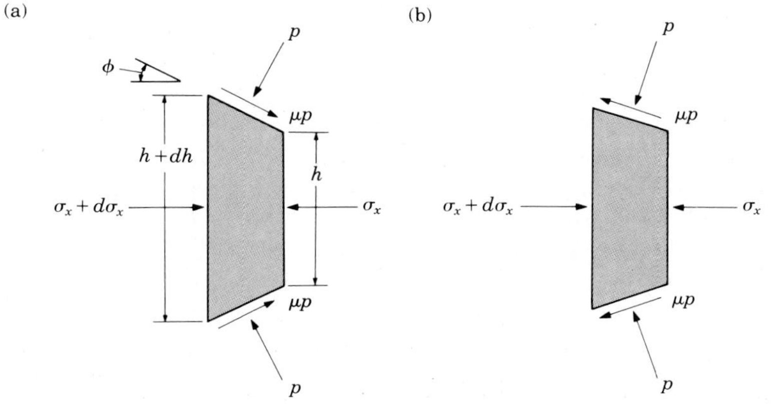

FIGURE 6.36

Stresses on an element in rolling: (a) entry zone and (b) exit zone.

For a strain-hardening material, the flow stress Y_f in these expressions corresponds to the strain that the material has undergone at that particular location in the roll gap.

We now have two equations and two unknowns (p and σ_x). We define h as

$$h = h_f + 2R(1 - \cos\phi)$$

or, approximately,

$$h = h_f + R\phi^2. \tag{6.26}$$

From these relationships we obtain the pressure at the entry and exit zones, respectively:

$$p = Y'_f \frac{h}{h_o} ed^{\mu(H_0 - H)} \tag{6.27}$$

and

$$p = Y'_f \frac{h}{h_f} e^{\mu H}, \tag{6.28}$$

where the quantity H is defined as

$$H = 2\sqrt{\frac{R}{h_f}} \tan^{-1}\left(\sqrt{\frac{R}{h_f}}\,\phi\right). \tag{6.29}$$

At entry, $\phi = \alpha$; hence $H = H_0$ with ϕ replaced by α. At exit, $\phi = 0$; hence $H = H_f = 0$.

Note that the pressure p is a function of h and the angular position ϕ along the arc of contact. These expressions also indicate that the pressure increases with increasing strength of the material, increasing coefficient of friction, and increasing R/h_f ratio. The R/h_f ratio in rolling is equivalent to the a/h ratio in upsetting.

The dimensionless theoretical pressure distribution in the roll gap is shown in Fig. 6.37. Note the similarity of this curve to Fig. 6.5 (*friction hill*). Also note that the neutral point shifts toward the exit as friction decreases, which is to be expected. When friction approaches zero, the rolls begin to slip instead of pulling the strip in, so the neutral point must approach the exit. (This situation is similar to the spinning of automobile wheels when accelerating on wet pavement.)

The effect of reduction in thickness of the strip on the pressure distribution is shown in Fig. 6.38. As reduction increases, the length of contact in the roll gap increases, which in turn increases the peak pressure. The curves shown are theoretical; actual pressure distributions, as determined experimentally, have smoother curves with their peaks rounded off.

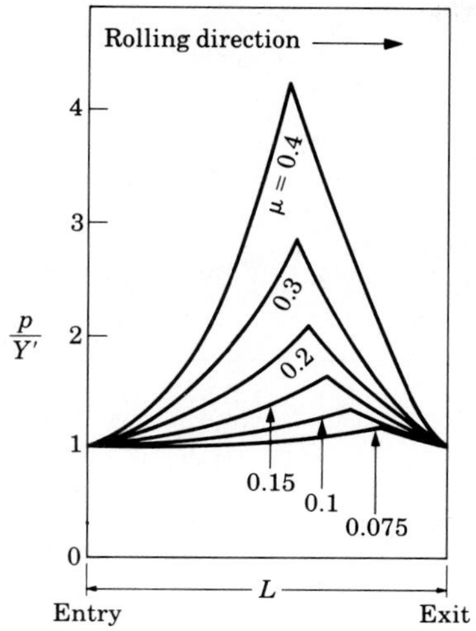

FIGURE 6.37 Pressure distribution in the roll gap as a function of the coefficient of friction. Note that, as friction increases, the neutral point shifts toward the entry. Without friction, the rolls slip and the neutral point shifts completely to the exit.

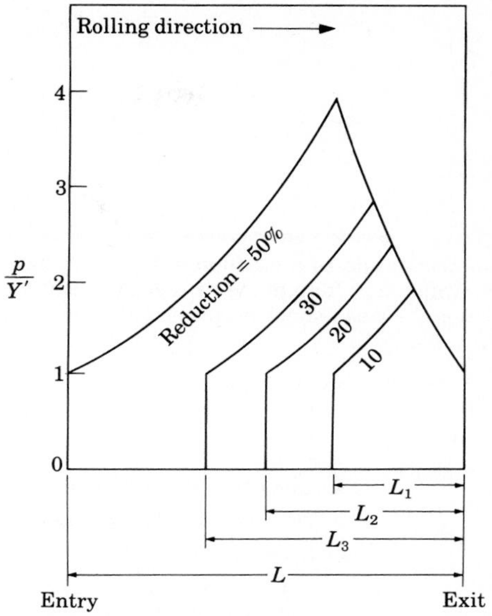

FIGURE 6.38 Pressure distribution in the roll gap as a function of reduction in thickness. Note the increase in the area under the curves with increasing reduction in thickness, thus increasing the roll-separating force.

Determination of the neutral point. We determine the neutral point simply by equating Eqs. (6.27) and (6.28). Thus at the neutral point,

$$\frac{h_0}{h_f} = \frac{e^{\mu H_0}}{e^{2\mu H_n}} = e^{\mu(H_0 - 2H_n)}$$

or

$$H_n = \frac{1}{2}\left(H_0 - \frac{1}{\mu}\ln\frac{h_0}{h_f}\right). \tag{6.30}$$

Substituting Eq. (6.30) into Eq. (6.29), we obtain

$$\phi_n = \sqrt{\frac{h_f}{R}}\tan\left(\sqrt{\frac{h_f}{R}}\cdot\frac{H_n}{2}\right). \tag{6.31}$$

Front and back tension. The roll force F can be reduced by various means, such as lower friction, smaller roll radii, smaller reductions, and higher workpiece temperatures. A particularly effective method is to reduce the apparent compressive yield stress of the material by applying longitudinal tension. Recall from our

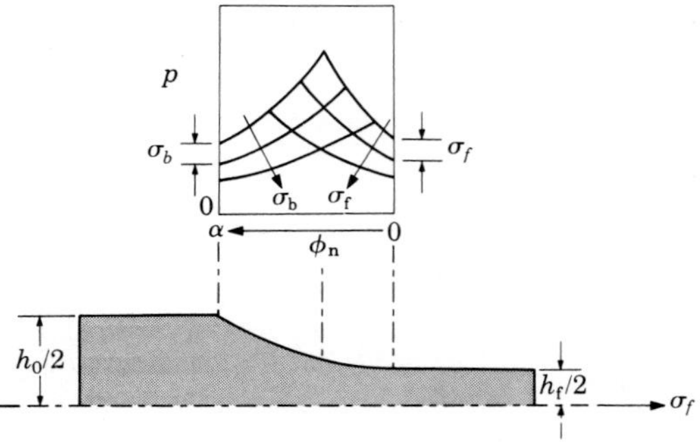

FIGURE 6.39 Pressure distribution as a function of front and back tension. Note the shifting of the neutral point and the reduction in the area under the curves with increasing tension.

discussion of yield criteria (Section 2.13) that if tension is applied to a strip (Fig. 6.39), the yield stress normal to the strip surface drops and hence the roll pressure will decrease.

Tensions in rolling can be applied either at the entry (*back tension* σ_b) or at the exit of the strip (*front tension* σ_f) or both. We then modify Eqs. (6.27) and (6.28) to include the effect of tension for the entry and exit zones, respectively, as follows:

$$p = (Y'_f - \sigma_b)\,\frac{h}{h_0}\,e^{\mu(H_0 - H)} \tag{6.32}$$

and

$$p = (Y'_f - \sigma_f)\,\frac{h}{h_f}\,e^{\mu H}. \tag{6.33}$$

Depending on the relative magnitudes of the tensions applied, the neutral point may shift, as shown in Fig. 6.39. This shift affects the pressure distribution, torque, and power requirements in rolling. Front tension is controlled by the torque on the coiler (*delivery* reel) around which the rolled sheet is coiled. The back tension is controlled by a braking system in the uncoiler (*payoff* reel). Special instrumentation is available for these controls. Tensions are particularly important in rolling thin, high-strength materials because they require high roll forces.

Roll forces. The area under these pressure–contact length curves multiplied by the strip width, w, is the *roll force*, F, on the strip. This area can be obtained graphically after the points are plotted. This force can also be calculated from the expression

$$F = \int_0^{\phi_n} wpR\,d\phi + \int_{\phi_n}^{\alpha} wpR\,d\phi. \tag{6.34}$$

A simpler method of calculating the roll force is to multiply the contact area with an average contact stress:

$$F = Lwp_{\text{av}}, \tag{6.35}$$

where L is the arc of contact. We can approximate the dimension L as the projected area:

$$L = \sqrt{R\,\Delta h}, \tag{6.36}$$

where R is the roll radius and Δh is the difference between the original and final thicknesses of the strip (*draft*).

The estimation of p_{av} depends on the h/L ratio, where h is now the *average* thickness of the strip in the roll gap (see Figs. 2.40 and 2.41). For large h/L ratios (small reductions and/or small roll diameters), the rolls act as indenters in a hardness test. Friction is not significant, and p_{av} is obtained from Fig. 2.40, where $a = L/2$. Small h/L ratios (large reductions and/or large roll diameters) are equivalent to high a/h ratios. Thus friction is predominant, and p_{av} is obtained from Eq. (6.15). For strain-hardening materials the appropriate flow stresses must be calculated.

As a rough approximation and for low frictional conditions, we can simplify Eq. (6.35) as follows:

$$F = Lw\bar{Y}', \tag{6.37}$$

where $\bar{Y}'$ is the *average flow stress* in plane strain (see Fig. 2.51) of the material in the roll gap.

For higher frictional conditions, we may write an expression similar to Eq. (6.15), or

$$F = Lw\bar{Y}'\left(1 + \frac{\mu L}{2h_{\text{av}}}\right). \tag{6.38}$$

Roll torque and power. The *roll torque* T for each roll can be calculated analytically from the expression

$$T = \underset{\text{(entry)}}{\int_{\phi_n}^{\alpha} w\mu p R^2\, d\phi} - \underset{\text{(exit)}}{\int_{0}^{\phi_n} w\mu p R^2\, d\phi}. \tag{6.39}$$

Note that the negative sign indicates the change in direction of the friction force at the neutral point. Thus, if the frictional forces are equal to each other, the torque is zero.

The torque in rolling can also be estimated by assuming that the roll-separating force F acts in the *middle* of the arc of contact, that is, a moment arm of $0.5L$, and that F is perpendicular to the plane of the strip. (Whereas $0.5L$ is a good estimate for hot rolling, $0.4L$ is a better estimate for cold rolling.)

The *torque per roll* then is

$$T = \frac{FL}{2}.$$

The *power required per roll* is

$$\text{Power} = T\omega, \tag{6.40}$$

where $\omega = 2\pi N$ and N is the revolutions per minute of the roll. Consequently, the power per roll is

$$\text{Power} = \frac{\pi FLN}{60{,}000} \quad \text{kW}, \tag{6.41}$$

where F is in newtons, L is in meters, and N is the rpm of the roll. We can also express the power as

$$\text{Power} = \frac{\pi FLN}{33{,}000} \quad \text{hp}, \tag{6.42}$$

where F is in lb and L is in ft.

• Example 6.4: Calculating power required in rolling.

A 9 in. wide 6061-O aluminum strip is rolled from a thickness of 1.00 in. to 0.80 in. If the roll radius is 12 in. and the roll rpm is 100, estimate the horsepower required for this operation.

SOLUTION. The power needed for a set of *two* rolls is given by Eq. (6.42):

$$\text{Power} = \frac{2\pi FLN}{33{,}000} \quad hp,$$

where F is given by Eq. (6.37) and L by Eq. (6.36). Thus

$$F = Lw\bar{Y}' \quad \text{and} \quad L = \sqrt{R\Delta h}.$$

We now have

$$L = \sqrt{(12)(1.0 - 0.8)} = 1.55 \text{ in.} = 0.13 \text{ ft} \quad \text{and} \quad w = 9 \text{ in.}$$

For 6061-O aluminum, $K = 30{,}000$ psi and $n = 0.2$ (from Table 2.4). The true strain in this operation is

$$\varepsilon_1 = \ln\left(\frac{1.0}{0.8}\right) = 0.223.$$

Thus, from Eq. (6.10),

$$\bar{Y} = \frac{(30{,}000)(0.223)^{0.2}}{1.2} = 18{,}500 \text{ psi}, \quad \text{and} \quad \bar{Y}' = (1.15)(18{,}500) = 21{,}275 \text{ psi}.$$

Therefore

$$F = (1.55)(9)(21{,}275) = 297{,}000 \text{ lb},$$

and

$$\text{Power} = \frac{(2\pi)(297{,}000)(0.13)(100)}{33{,}000} = 735 \text{ hp}.$$

Forces in hot rolling. Because ingots and slabs are usually hot rolled, the calculation of forces and torque in hot rolling is important. However, there are two major difficulties. One is the proper estimation of the coefficient of friction μ at elevated temperatures. The other is the strain-rate sensitivity of metals at high temperatures.

The average strain rate in flat rolling can be obtained by dividing the strain by the time required for an element to undergo this strain in the roll gap. The time can be approximated as L/V_r, so

$$\dot{\bar{\varepsilon}} = \frac{V_r}{L} \ln\left(\frac{h_0}{h_f}\right). \tag{6.43}$$

The flow stress Y_f of the material corresponding to this strain rate must first be obtained and then substituted into the proper equations. These calculations are approximate. Variations in μ and the temperature within the strip in hot rolling further contribute to the difficulties in calculating forces accurately.

Friction. In rolling, although the rolls cannot pull the strip into the roll gap without some friction, forces and power requirements rise with increasing friction. In cold rolling, the coefficient of friction μ usually ranges between 0.02 and 0.3, depending on the materials and lubricants used. The low ranges for the coefficient of friction are obtained with effective lubricants and regimes approaching hydrodynamic lubrication, such as in cold rolling of aluminum at high speeds. In hot rolling, the coefficient of friction may range from about 0.2, with effective lubrication, to as high as 0.7, indicating sticking, which usually occurs with steels, stainless steels, and high-temperature alloys.

The maximum possible draft, that is, $h_0 - h_f$, in flat rolling is a function of friction and roll radius:

$$\Delta h_{max} = \mu^2 R. \tag{6.44}$$

Hence the higher the friction and the larger the roll radius, the greater is the maximum draft. As expected, the draft is zero when there is no friction. The maximum value of the angle α (*angle of acceptance*) in Fig. 6.35 is geometrically related to Eq. (6.44). From the simple model of a block sliding down an inclined plane, we can see that

$$\alpha_{max} = \tan^{-1}\mu. \tag{6.45}$$

If α_{max} is larger than this value, the rolls begin to slip, because the friction is not great enough to pull the material through the roll gap.

Roll deflections. Roll forces tend to bend the rolls, as shown in Fig. 6.40(a), with the result that the strip is thicker at its center than at its edges (*crown*). The usual method of avoiding this problem is to grind the rolls so that their diameter at the center is slightly larger than at the edges. This is known as *camber*. In rolling sheet metals the camber is generally less than 0.25 mm (0.01 in.) on the radius.

When properly designed, such rolls produce flat strips as shown in Fig. 6.40(b). However, a particular camber is correct only for a certain load and width of strip. In hot rolling, uneven temperature distribution in the roll can also cause its diameter to vary along the length of the roll. In fact, camber can be controlled by varying the location of the coolant (lubricant) on the rolls in hot rolling.

Forces also tend to flatten the rolls elastically, much like the flattening of tires on automobiles. Flattening of the rolls increases roll radius and hence yields a larger contact area for the same reduction in thickness. Thus the roll force F increases.

The new (distorted) roll radius R' is

$$R' = R\left(1 + \frac{CF'}{h_0 - h_f}\right). \tag{6.46}$$

In Eq. (6.46), C is 2.3×10^{-2} mm²/kN (1.6×10^{-4} in²/klb) for steel rolls and 4.57×10^{-2} (3.15×10^{-4}) for cast iron rolls. F' is the *roll force per unit width* of strip, kN/mm (klb/in.). The higher the elastic modulus of the roll material, the less the roll distorts. Note from Eq. (6.46) that reducing the roll force, such as by using an effective lubricant or taking smaller reductions, also reduces roll flattening.

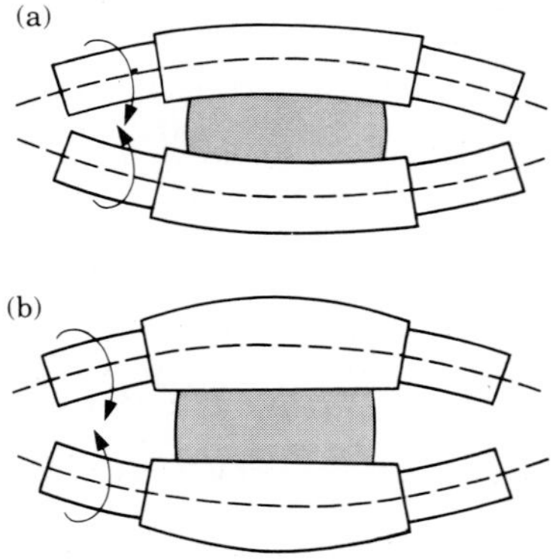

FIGURE 6.40
(a) Bending of straight cylindrical rolls because of the roll-separating force. (b) Bending of rolls, ground with camber, that produce a sheet of uniform thickness during rolling.

From Eq. (6.46) we cannot determine R' directly, because it is a function of the roll force F, which itself is a function of the roll radius. The solution is obtained by trial and error. (Note that R in all previous equations should be replaced by R' when significant roll flattening occurs.)

Spreading. Although rolling plates and sheets with high width-to-thickness ratios is essentially a process of plane strain, with smaller ratios, such as a square cross-section, the width increases considerably during rolling. This increase in width is known as *spreading* (Fig. 6.41). Spreading decreases with increasing width-to-thickness ratios of the entering material, increasing friction, and increasing ratios of roll radius-to-strip thickness.

6.3.2 Defects

Successful rolling practice requires balancing many factors, including material properties, process variables, and lubrication. Defects may be on the surfaces of the rolled plates and sheets, or they may be structural defects within the material. *Surface* defects may result from inclusions and impurities in the material, scale, rust, dirt, roll marks, and other causes related to the prior treatment and working of the material. In hot rolling blooms, billets, and slabs, the surface is usually preconditioned by various means, such as by torch (*scarfing*), to remove scale.

Structural defects are those that distort or affect the integrity of the rolled product. Some typical defects are shown in Fig. 6.42. Wavy edges are caused by

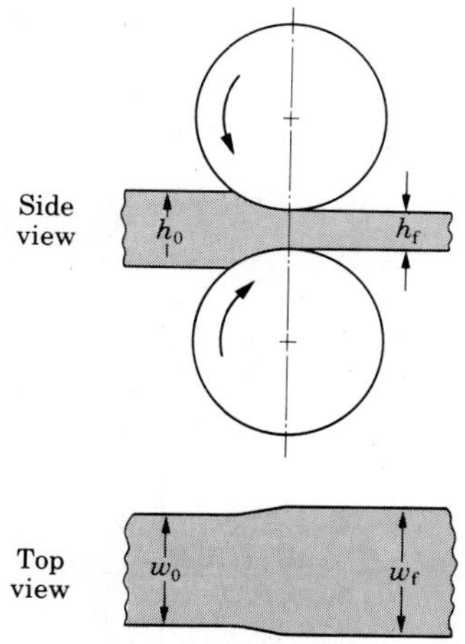

FIGURE 6.41 Increase in the width (spreading) of a strip in flat rolling. Spreading can be similarly observed when dough is rolled with a rolling pin.

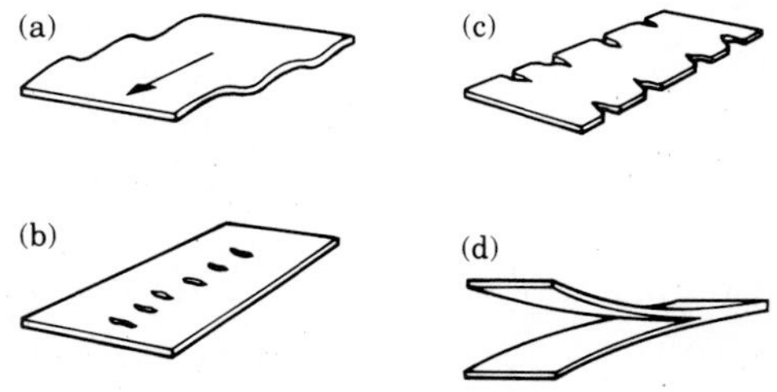

FIGURE 6.42 Schematic illustration of typical defects in flat rolling: (a) wavy edges; (b) zipper cracks in center of strip; (c) edge cracks; (d) alligatoring.

bending of the rolls; the edges of the strip are thinner than the center. Because the edges elongate more than the center and are restrained from expanding freely, they buckle. The cracks shown in Fig. 6.42(b) and (c) are usually caused by low ductility and barreling. *Alligatoring* is a complex phenomenon resulting from inhomogeneous deformation of the material during rolling or defects in the original cast ingot, such as piping.

Residual stresses. Residual stresses can be generated in rolled sheets and plates due to inhomogeneous plastic deformation in the roll gap. Small-diameter rolls or small reductions tend to work the metal plastically at its surfaces (similarly to shot peening or roller burnishing). This working generates compressive residual stresses on the surfaces and tensile stresses in the bulk (Fig. 6.43a). Large-diameter rolls and high reductions, however, tend to deform the bulk to a greater extent than the surfaces. The reason is the frictional constraint at the surfaces along the arc of contact. This situation generates residual stresses that are opposite to the previous case, as shown in Fig. 6.43(b).

6.3.3 Flat-rolling practice

The initial breaking down of an ingot (or continuous casting) by hot rolling converts the coarse-grained, brittle, and porous cast structure to a wrought structure (see Fig. 6.33). This structure has finer grains and enhanced ductility resulting from breaking up brittle grain boundaries and closing up internal defects, such as porosity. Traditional methods of rolling ingots are now being rapidly replaced by continuous casting. Temperature ranges for hot rolling are similar to those for forging (see Table 6.3).

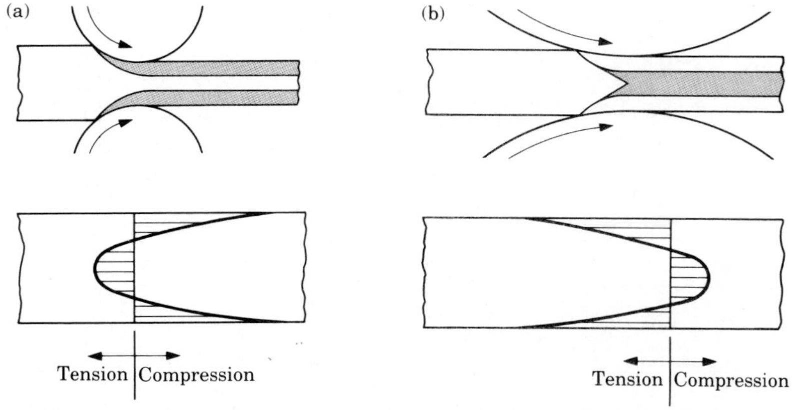

FIGURE 6.43
The effect of roll radius on the type of residual stresses developed in flat rolling: (a) small rolls, or small reduction; and (b) large rolls, or large reduction in thickness.

The product of the first hot-rolling operation is called a *bloom* or *slab* (see Fig. 6.32). A bloom usually has a square cross-section, at least 150 mm (6 in.) on the side; a slab is usually rectangular in cross-section. Blooms are processed further by shape rolling into structural shapes, such as I-beams and railroad rails. Slabs are rolled into plates and sheet. *Billets* are usually square, with a cross-sectional area smaller than blooms, and are rolled into various shapes, such as round rods and bars, using shaped rolls. (Hot-rolled round rods are used as the starting material for rod and wire drawing and are called *wire rods*.)

In hot rolling blooms, billets, and slabs, the surface of the material is usually *conditioned* (prepared for a subsequent operation) prior to rolling. Conditioning is done by various means, such as with a torch (scarfing) to remove heavy scale, or by rough grinding to smoothen surfaces. Prior to cold rolling, the scale developed during hot rolling or other defects may be removed by pickling with acids or by mechanical means, such as blasting with water, or by grinding.

Pack rolling is a flat-rolling operation in which two or more layers of metal are rolled together, thus improving productivity. Aluminum foil, for example, is pack rolled in two layers. You probably have noticed that one side of aluminum foil is matte and the other side shiny. The foil-to-foil side has a matte and satiny finish, and the foil-to-roll side is shiny and bright because it has been in contact with the polished roll.

Mild steel, when stretched during sheet-forming operations, undergoes *yield-point elongation*. This phenomenon causes surface irregularities called *stretcher strains or Lueder's bands* (see also Section 7.2). To avoid this situation, the sheet metal is subjected to a light pass of 0.5–1.5 percent reduction, known as *temper rolling*.

A rolled sheet may not be sufficiently flat as it leaves the roll gap because of variations in the material or in the processing parameters during rolling. To improve flatness, the strip is then passed through a series of *leveling rolls*. Each roll is usually driven separately with individual electric motors. The strip is flexed in opposite directions as it passes through the sets of rollers. Several different roller arrangements are used (Fig. 6.44).

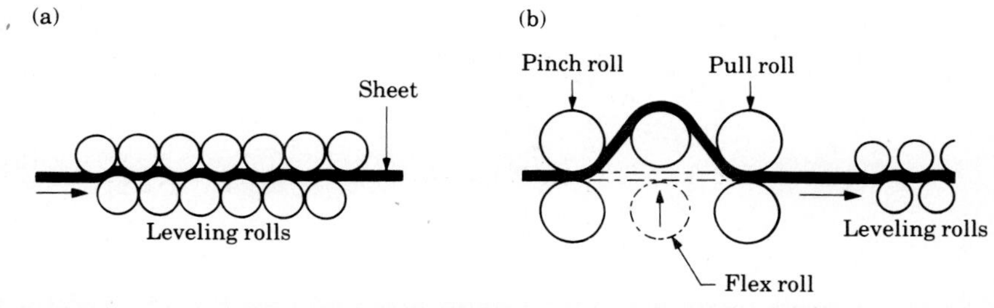

FIGURE 6.44
Schematic illustration of methods of roller leveling. These processes are used to flatten rolled sheets.

Sheet thickness is identified by a *gage* number; the smaller the number, the thicker the sheet. There are different numbering systems for different sheet metals. Rolled sheets of copper and brass are also identified by thickness changes.

Lubrication. Ferrous alloys are usually hot rolled without a lubricant, although graphite may be used. Aqueous solutions are used to cool the rolls and break up the scale on the workpiece. Nonferrous alloys are hot rolled with a variety of compounded oils, emulsions, and fatty acids. Cold rolling is done with low-viscosity lubricants, including mineral oils, emulsions, paraffin, and fatty oils (see Table 6.4).

Equipment. A wide variety of rolling equipment is available with a number of roll arrangements, the major types of which are shown in Fig. 6.45. *Two-high* or *three-high* (developed in the mid 1800s) rolling mills are used for initial breakdown passes on cast ingots with roll diameters ranging from 600 to 1400 mm (24 to 55 in.). Small-diameter rolls are preferable, because the smaller the roll radius the lower the roll force. However, small rolls deflect under roll forces and have to be supported by other rolls (Fig. 6.45c and d).

The range of dimensions in rolling is extremely wide. The rolled product can be as wide as 5000 mm (200 in.) and as thin as 0.0025 mm (0.0001 in.) obtained in a cluster mill. The *cluster mill* (*Sendzimir*) is particularly suitable for cold rolling thin strips of high-strength metals. One arrangement for the rolls is shown in Fig. 6.46. The *work roll* (smallest roll) can be as small as 6 mm (1/4 in.) in diameter and is usually made of tungsten carbide for rigidity, strength, and wear resistance.

Stiffness in rolling mills is important for controlling dimensions. Mills can be highly automated with rolling speeds as high as 25 m/s (5000 ft/min). Flat rolling can also be carried out with front tension only, using idling rolls (*Steckel rolling*). For this case, the torque T on the roll is zero, assuming frictionless bearings.

Requirements for roll materials are mainly strength and resistance to wear. Three common roll materials are cast iron, cast steel, and forged steel. For hot rolling, roll surfaces are generally rough and may even have notches or grooves in order to pull the metal through the roll gap at high reductions. Rolls for cold rolling are ground to a fine finish and, for special applications, are also polished.

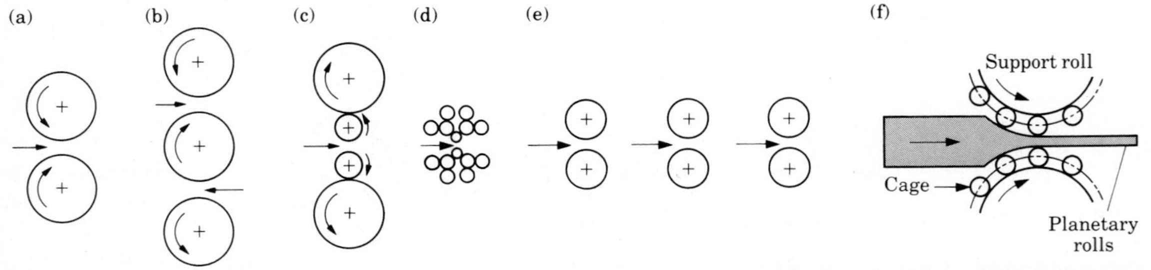

FIGURE 6.45 Schematic illustration of various roll arrangements: (a) 2 high; (b) 3 high; (c) 4 high; (d) cluster; (e) tandem rolling, with three stands; and (f) planetary.

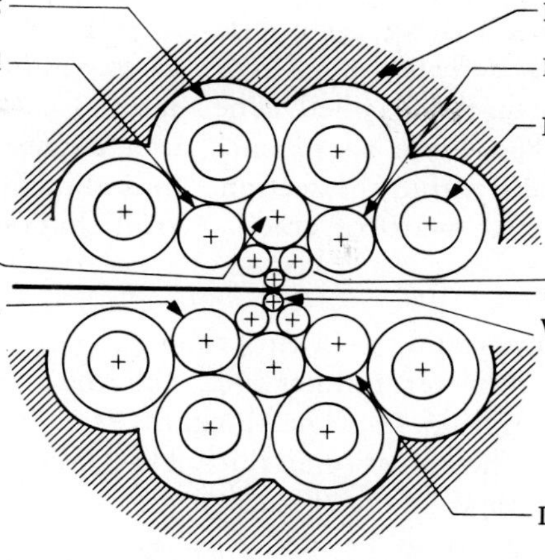

FIGURE 6.46

Schematic illustration of a cluster (*Sendzimir*) mill. These mills are very rigid and are used in rolling thin sheets of high-strength materials, with good control of dimensions.

6.3.4 Miscellaneous rolling operations

Shape rolling. Straight structural shapes such as bars of various cross-sections, channel sections, I-beams, and railroad rails are rolled by passing the stock through a number of pairs of specially designed rollers (Fig. 6.47). These processes

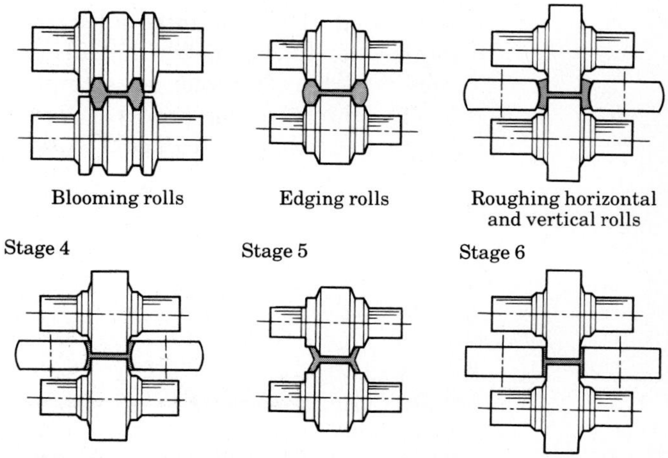

FIGURE 6.47

Stages in shape rolling of an H-section part. Various other structural sections, such as channels and I-beams, are also rolled by this process.

were first developed in the late 1700s. The original material is usually a *bloom* (Fig. 6.32). Designing a series of rolls (*roll-pass design*) requires experience in order to avoid defects and hold tolerances, although some of these defects may also be due to the material rolled. The material elongates as it is reduced in cross-section. However, for a shape such as a channel the reduction is different in different locations within the section. Thus elongation is not uniform, which can cause the product to warp or crack. Airfoil shapes can also be produced by shape-rolling techniques.

Ring rolling. In ring rolling, a small-diameter, thick ring is expanded into a larger-diameter, thinner ring. The ring is placed between two rolls, one of which is driven (Fig. 6.48). The thickness is reduced by bringing the rolls closer as they rotate. The reduction in thickness is compensated for by an increase in the diameter of the ring. A great variety of cross-sections can be ring rolled with shaped rolls. This process can be carried out at room or elevated temperatures, depending on the size and strength of the product. The advantages of ring rolling, compared with other processes for making the same part, are short production runs, material savings, close tolerances, and favorable grain flow direction. Typical applications of ring rolling are large rings for rockets and turbines, gearwheel rims, ball and roller bearing races, flanges and reinforcing rings for pipes, and pressure vessels.

Thread and gear rolling. In the cold-forming thread and gear rolling process, threads are formed on round rods or workpieces by passing them between reciprocating or rotating dies (Fig. 6.49a). Typical products are screws, bolts, and similar threaded parts. With flat dies the threads are rolled on the rod or wire with each stroke of the reciprocating die. Production rates are very high but depend on the diameter of the product. With small diameters the rates can be as high as eight pieces per second and with larger diameters (as much as 25 mm [1 in.]) about one per second. Two- or three-roller thread-rolling machines are also available. In another design (Fig. 6.49b) threads are formed with a rotary die with production rates as high as 80 pieces per second.

The thread-rolling process generates threads without any metal loss and with greater strength because of cold working. The surface finish is very smooth, and the

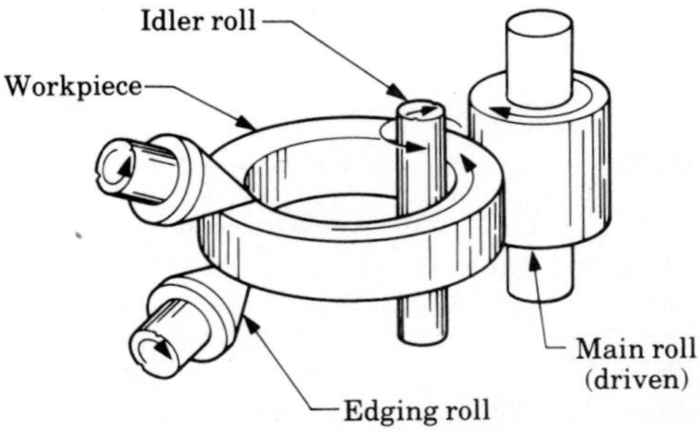

FIGURE 6.48
A horizontal ring-rolling operation. The reduction in the thickness of the ring is compensated for by the increase in ring diameter. Large bearing races and turbine components are made by this process.

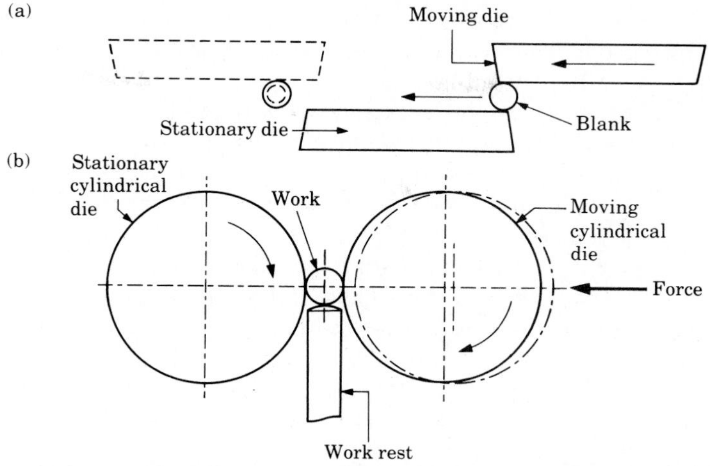

FIGURE 6.49
Thread rolling processes: (a) flat dies and (b) two-roller dies. These processes are used extensively in making threaded fasteners at high rates of production.

process induces compressive residual stresses on the surfaces, which improves fatigue life. The product is superior to that made by thread cutting and is used in the production of almost all externally threaded fasteners.

Because of volume constancy in plastic deformation, a rolled thread requires a smaller diameter round stock to produce the same major diameter as a machined thread (Fig. 6.50). Also, whereas machining removes material by cutting through

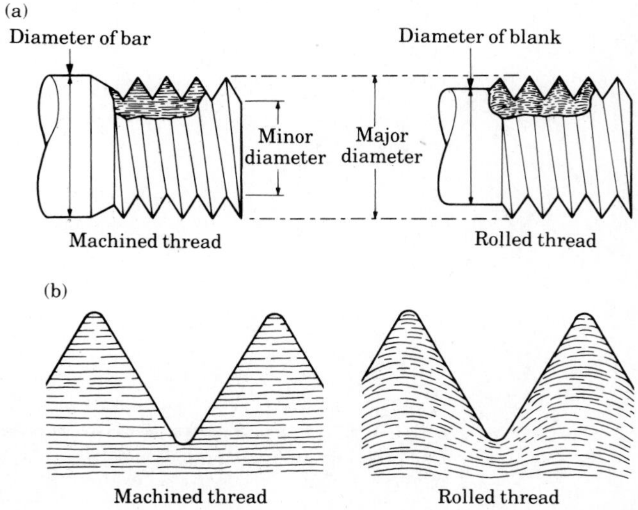

FIGURE 6.50
(a) Schematic illustration of machined and rolled threads. Note the increase in the blank diameter for the rolled thread. (b) Grain flow lines in machined and rolled threads. Unlike machined threads which are cut through the grains of the metal, rolled threads follow the grains and are stronger because of cold working.

the grain flow lines of the material, rolled threads have a grain flow pattern that improves the strength of the thread because of cold working. Thread rolling can also be carried out internally with a fluteless forming tap. The process is similar to external thread rolling and produces accurate threads with good strength. In all thread-rolling processes it is essential that the material have sufficient ductility and that the rod or wire be of proper size. Lubrication is also important for good surface finish and to minimize defects.

Spur and helical gears are also produced by cold-rolling processes similar to thread rolling. The process may be carried out on solid cylindrical blanks or on precut gears. Helical gears can also be made by a direct extrusion process, using specially shaped dies. Cold rolling of gears has many applications in automatic transmissions and power tools.

Rotary tube piercing. A hot-working process, rotary tube piercing is used to make long, thick-walled seamless tubing, as shown in Fig. 6.51. The process is based on the principle that when a round bar is subjected to radial compression in the manner shown in Fig. 6.51(a), tensile stresses develop at the center of the rod. When subjected to cycling compressive stresses, as shown in Fig. 6.51(b), a cavity begins to form at the center of the rod.

The *rotary tube piercing* (*Mannesmann* process, developed in the 1880s) process is carried out by an arrangement of rotating rolls, as shown in Fig. 6.51(c). The axes of the rolls are skewed in order to pull the round bar through the rolls by their rotary action. A mandrel assists the operation by expanding the hole and sizing the inside diameter of the tube. Because of the severe deformation that the metal undergoes in this process, high-quality, defect-free bars must be used.

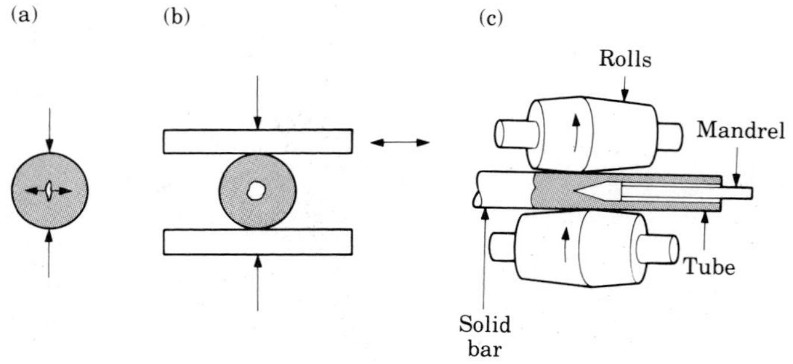

FIGURE 6.51 Cavity formation by secondary tensile stresses in a solid round bar and its utilization in the rotary tube piercing process. This is the principle of the Mannesmann mill for seamless tubemaking.

6.4 Extrusion

In the basic extrusion process, developed in the late 1700s for lead pipe, a round billet is placed in a chamber and forced through a die opening by a ram. The die may be round or of various other shapes. There are four basic types of extrusion: direct, indirect, hydrostatic, and impact (Fig. 6.52).

Direct extrusion (*forward* extrusion) is similar to forcing the paste through the opening of a toothpaste tube. The billet slides relative to the container wall; the wall friction increases the ram force considerably.

In *indirect extrusion* (*reverse*, *inverted*, or *backward* extrusion), the die moves toward the billet; thus, except at the die, there is no relative motion at the billet–container interface.

In *hydrostatic* extrusion, the chamber is filled with a fluid that transmits the pressure to the billet, which is then extruded through the die. There is no friction along the container walls.

Impact extrusion is a form of indirect extrusion and is particularly suitable for hollow shapes.

Extrusion processes can be carried out hot or cold. Because a chamber is involved, each billet is extruded individually and hence extrusion is basically a *batch* process. *Cladding* by extrusion can also be carried out with coaxial billets (such as copper clad with silver), provided that the flow stresses of the two metals are similar.

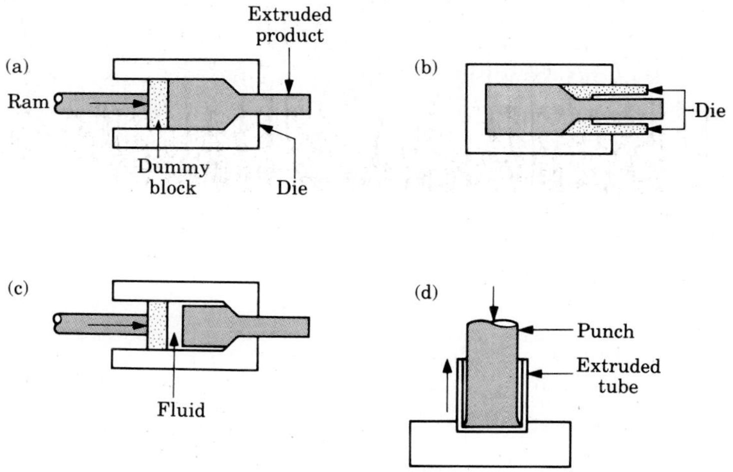

FIGURE 6.52
Types of extrusion: (a) direct; (b) indirect; (c) hydrostatic; (d) impact.

6.4.1 Metal flow in extrusion

Because the billet is forced through a die, with a substantial reduction in its cross-section, the metal flow pattern in extrusion is an important factor in the overall process. A common technique for investigating the flow pattern is to halve the round billet lengthwise and mark one face with a grid pattern. The two halves are then placed together in the container (they may also be fastened together or *brazed* to keep the two halves intact) and extruded. They are then taken apart and inspected.

Figure 6.53 shows three typical results in direct extrusion with square dies. The conditions under which these different flow patterns are obtained are as follows.

a) The most homogeneous flow pattern is obtained when there is no friction at the billet–container–die interfaces (Fig. 6.53a). This type of flow occurs when the lubricant is very effective or with indirect extrusion.
b) When friction along all interfaces is high, a *dead-metal zone* develops (Fig. 6.53b). Note the high-shear area as the material flows into the die exit, somewhat like a funnel. This configuration may indicate that the billet surfaces (with their oxide layer and lubricant) could enter this high-shear zone and be extruded, causing defects in the extruded product.
c) The high-shear zone extends farther back (Fig. 6.53c). This extension can result from high container-wall friction, which retards the flow of the billet, or materials in which the flow stress drops rapidly with increasing temperature. In hot working, the material near the container walls cools rapidly and hence increases in strength. Thus the material in the central regions flows toward the die more easily than that at the outer regions. As a result, a large dead-metal zone forms and the flow is inhomogeneous. This flow pattern leads to a defect known as a *pipe* or *extrusion defect.*

Thus the two factors that greatly influence metal flow in extrusion are the frictional conditions at billet–container–die interfaces and thermal gradients in the billet.

6.4.2 Mechanics of extrusion

The ram force in direct extrusion can be calculated in the following ways for different situations.

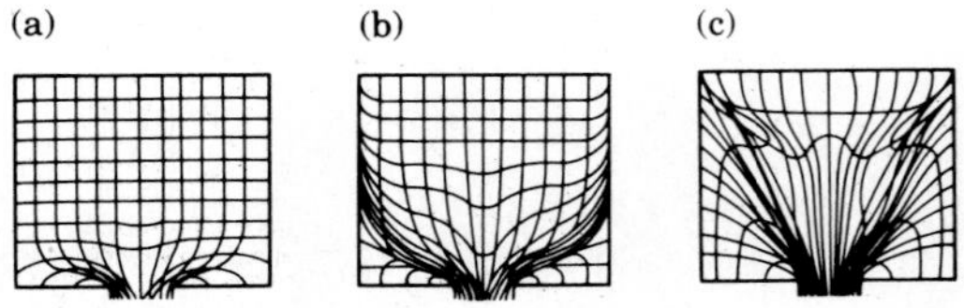

FIGURE 6.53 Schematic illustration of three different types of metal flow in direct extrusion.

Ideal deformation. The *extrusion ratio* R is defined as

$$R = \frac{A_0}{A_f}, \tag{6.47}$$

where A_0 is the billet cross-sectional area and A_f is the area of the extruded product (Fig. 6.54). The absolute value of the true strain then is

$$\varepsilon_1 = \ln\left(\frac{A_0}{A_f}\right) = \ln\left(\frac{L_f}{L_0}\right) = \ln R, \tag{6.48}$$

where L_0 and L_f are the lengths of the billet and the extruded product, respectively.

For a perfectly plastic material with a yield stress Y, the energy dissipated in plastic deformation per unit volume, u, is

$$u = Y\varepsilon_1. \tag{6.49}$$

Hence the work done on the billet is

$$\text{Work} = (A_0)(L_0)(u). \tag{6.50}$$

This work is supplied by the ram force F, which travels a distance L_0. Thus

$$\text{Work} = FL_0 = pA_0L_0, \tag{6.51}$$

where p is the *extrusion pressure* at the ram. Equating the work of plastic deformation to the external work done, we find that

$$p = u = Y \ln\left(\frac{A_0}{A_f}\right) = (Y)(\ln R). \tag{6.52}$$

For strain-hardening materials, Y should be replaced by the *average flow stress* $\bar{Y}$. Note that Eq. (6.52) is the same as the area under the true stress–true strain curve for the material.

Ideal deformation and friction. Equation (6.52) pertains to ideal deformation without any friction. Based on the slab method of analysis, when friction at the die–billet interface is included (but not the container-wall friction) and for small die angles, the pressure p is

$$p = Y\left(1 + \frac{\tan\alpha}{\mu}\right)[(R)^{\mu\cot\alpha} - 1]. \tag{6.53}$$

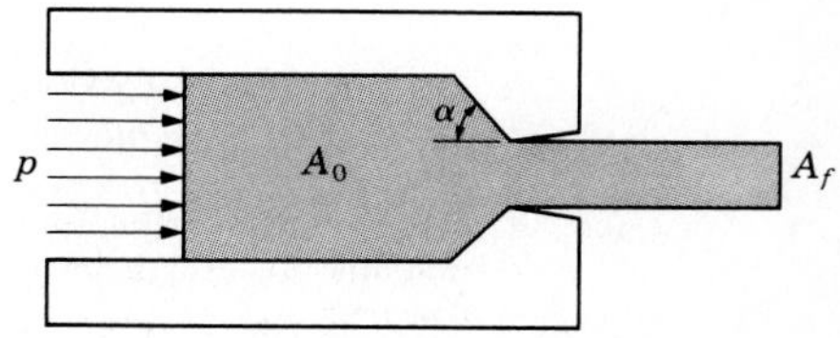

FIGURE 6.54
Variables in the direct extrusion process.

An estimate of p can also be obtained by assuming that, because of the dead zone, material flow in the container takes place along a 45° angle. If the frictional stress along this 45° "die" is assumed to be the yield stress in shear k of the material (where $k = Y/2$), then

$$p = 1.7Y \ln R. \tag{6.54}$$

In this analysis, we neglected the force required to overcome friction at the billet–container interface. If we assume that the frictional stress is equal to the shear yield stress of the material k, we can obtain the additional ram pressure required due to friction, p_f:

$$(p_f)\left(\frac{\pi D_0^2}{4}\right) = \pi D_0 kL$$

or

$$p_f = k\frac{4L}{D_0} = Y\frac{2L}{D_0}, \tag{6.55}$$

where L is the length of the billet remaining in the container. Thus Eq. (6.54) becomes

$$p = Y\left(1.7 \ln R + \frac{2L}{D_0}\right). \tag{6.56}$$

For strain-hardening materials, Y in these expressions should be replaced by $\bar{Y}$.

Note that as the billet is extruded farther, L decreases and thus the ram force decreases (Fig. 6.55), whereas in indirect extrusion the ram force is not a function of billet length.

Actual forces. The derivation of analytical expressions, including friction, die angle, and redundant work due to inhomogeneous deformation of the material, can be difficult. Furthermore, there are difficulties in estimating the coefficient of friction, the flow stress of the material, and the redundant work in a particular operation. Consequently, a convenient *empirical* formula has been developed:

$$p = Y(a + b \ln R), \tag{6.57}$$

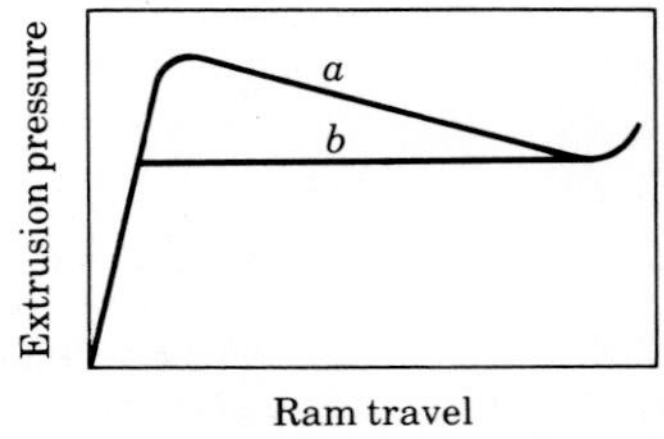

FIGURE 6.55 Schematic illustration of typical extrusion pressure as a function of ram travel: (a) direct extrusion and (b) indirect extrusion. The pressure in direct extrusion is higher because of frictional resistance in the chamber as the billet moves toward the die.

where a and b are constants determined experimentally. Approximate values for a and b are 0.8 and 1.2 to 1.5, respectively. Again, note that for strain-hardening materials Y is replaced by $\bar{Y}$.

Optimum die angle. The die angle has an important effect on forces in extrusion. Its relationship to work is as follows.

a) The *ideal* work of deformation is independent of the die angle (Fig. 6.56), because it is a function only of the extrusion ratio.
b) The *frictional* work increases with decreasing die angle because the length of contact at the billet–die increases, thus requiring more work.
c) The *redundant* work caused by inhomogeneous deformation increases with die angle.

Because the total ram force is the sum of these three components, there is an angle where this force is a *minimum* (Fig. 6.56). Unless the behavior of each component as a function of the die angle is known, determination of this *optimum angle* is difficult.

Forces in hot extrusion. Because of the strain-rate sensitivity of metals at elevated temperatures, forces in hot extrusion are difficult to calculate. The average true strain rate $\dot{\bar{\varepsilon}}$ is

$$\dot{\bar{\varepsilon}} = \frac{6V_0 D_0^2 \tan\alpha}{D_0^3 - D_f^3} \ln R, \tag{6.58}$$

where V_0 is the ram velocity. Note from this equation that for high extrusion ratios ($D_0 \gg D_f$) and for $\alpha = 45^\circ$, as may be the case with a square die (thus developing a dead zone) and poor lubrication, the strain rate reduces to

$$\dot{\bar{\varepsilon}} = \frac{6V_0}{D_0} \ln R. \tag{6.59}$$

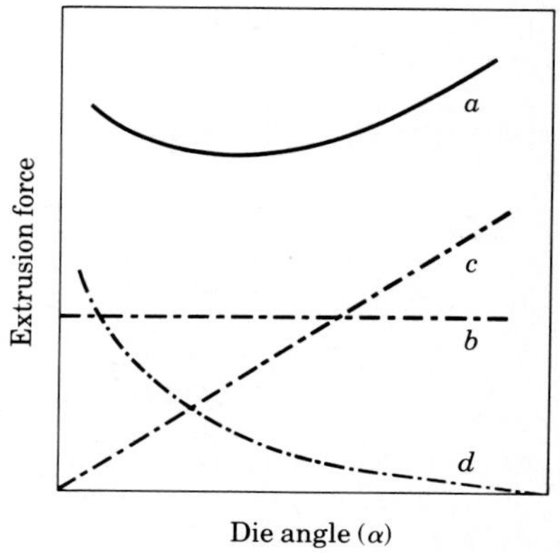

FIGURE 6.56 Schematic illustration of extrusion force as a function of die angle: (a) total force; (b) ideal force; (c) force required for redundant deformation; and (d) force required to overcome friction. Note that there is an optimum die angle where the total extrusion force is a minimum.

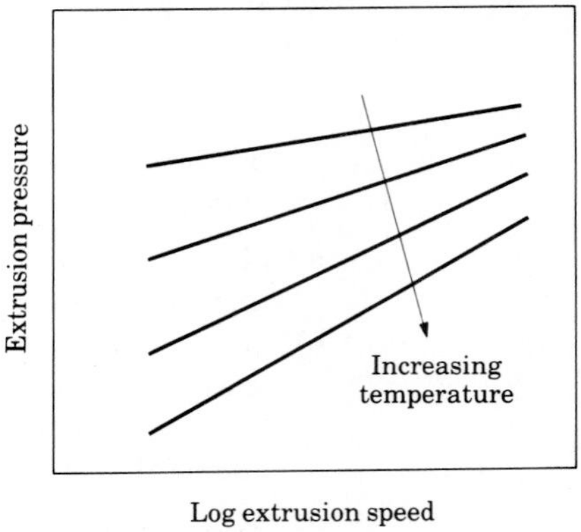

FIGURE 6.57 Schematic illustration of the effect of temperature and ram speed on extrusion pressure. Compare with Fig. 2.9.

The effect of ram speed and temperature on extrusion pressure is shown in Fig. 6.57. As expected, pressure increases rapidly with ram speed, especially at elevated temperatures. As extrusion speed increases, the rate of work done per unit time also increases. Because work is converted into heat, the heat generated at high speeds may not be dissipated fast enough. The subsequent rise in temperature can cause *incipient melting* of the workpiece material and cause defects. Circumferential surface cracks caused by *hot shortness* may also develop; in extrusion this is known as *speed cracking*. These problems can be eliminated by reducing the extrusion speed.

A convenient parameter that is used to estimate forces in extrusion is an experimentally determined *extrusion constant* K_e, which includes various factors and can be determined from

$$p = K_e \ln R. \qquad (6.60)$$

Figure 6.58 gives some typical values of K_e for various materials.

• Example 6.5: Estimating force in hot extrusion.

A copper billet 5 in. in diameter and 10 in. long is extruded at 1500 °F at a speed of 10 in./s. Using square dies and assuming poor lubrication, estimate the force required in this operation if the final diameter is 2 in.

SOLUTION. The extrusion ratio is

$$R = \frac{5^2}{2^2} = 6.25.$$

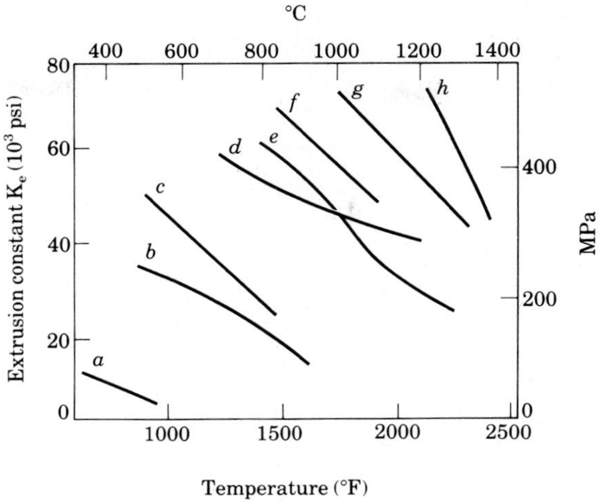

FIGURE 6.58
Extrusion constant K_e for various materials as a function of temperature. Note the ranges of temperature for various materials. (a) 1100 aluminum; (b) copper; (c) 70–30 brass; (d) beryllium; (e) cold-rolled steel; (f) stainless steel; (g) molybdenum; and (h) chromium. *Source*: After P. Loewenstein, ASTME Paper SP63-89.

The average true strain rate, from Eq. (6.51), is

$$\dot{\bar{\varepsilon}} = \frac{(6)(10)}{(5)} \ln 6.25 = 22/\text{s}.$$

From Table 2.6 let's assume that $C = 19{,}000$ psi (an average value) and $m = 0.06$. Then

$$\sigma = C\dot{\varepsilon}^m = (19{,}000)(22)^{0.06} = 22{,}870 \text{ psi}.$$

Assuming that $\bar{Y} = \sigma$, then from Eq. (6.56),

$$p = \bar{Y}\left(1.7 \ln R + \frac{2L}{D_0}\right) = (22{,}870)\left[(1.7)(1.83) + \frac{(2)(10)}{(5)}\right] = 162{,}630 \text{ psi}.$$

Hence

$$F = (p)(A_0) = (162{,}630)\,\frac{(\pi)(5)^2}{4} = 3.2 \times 10^6 \text{ lb}.$$

6.4.3 Miscellaneous extrusion processes

Cold extrusion. *Cold extrusion* is a general term often used to denote a combination of processes, such as direct and indirect extrusion and forging (Fig. 6.59). Many materials can be extruded into various configurations, with the billet either at room temperature or at a temperature of a few hundred degrees.

Cold extrusion has gained wide acceptance in industry because of the following advantages over hot extrusion:

a) Improved mechanical properties resulting from strain hardening, provided that the heat generated by plastic deformation and friction does not recrystallize the extruded metal.
b) Good control of tolerances, thus requiring a minimum of machining operations.
c) Improved surface finish, provided that lubrication is effective.
d) Lack of oxide layers.
e) High production rates and relatively low cost.

However, the stresses on tooling in cold extrusion are very high, especially with steel workpieces. The stress levels on tooling are on the order of the hardness of the material, that is, at least three times its flow stress. The design of tooling and selection of appropriate tool materials are therefore crucial to success in cold extrusion. The hardness of tooling usually ranges between 60 and 65 HRC for the punch and 58 to 62 HRC for the die. Punches are a critical component; they must have sufficient strength, toughness, wear, and fatigue resistance.

Lubrication also is crucial, especially with steels, because of the generation of new surfaces and the possibility of seizure between the metal and the tooling caused by the breakdown of lubrication. The most effective lubrication is phosphate conversion coatings on the workpiece and soap (or wax in some cases) as the lubricant (see Table 6.4). Temperature rise in cold extrusion is an important factor, especially at high extrusion ratios. The temperature may be sufficiently high to initiate and complete the recrystallization process of the cold-worked metal, thus reducing the advantages of cold working.

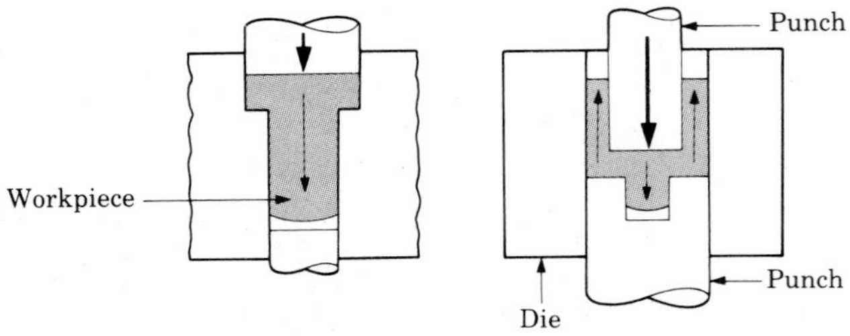

FIGURE 6.59 Examples of cold extrusion. Arrows indicate the direction of material flow. These parts may also be considered as forgings.

Impact extrusion. A process often included in the category of cold extrusion, *impact extrusion* is similar to indirect extrusion (Fig. 6.52d). The punch descends at a high speed and strikes the *blank* (*slug*), extruding it upward. The thickness of the extruded tubular section is a function of the clearance between the punch and the die cavity.

The impact-extrusion process usually produces tubular sections having wall thicknesses that are small in relation to their diameters. This ratio can be as small as 0.005. The concentricity between the punch and the blank is important for uniform wall thickness. A typical example of impact extrusion is the production of collapsible tubes, such as for toothpaste (Fig. 6.60a). The punch travel is determined by the setting of the press. A variety of nonferrous metals are impact extruded in this manner into various shapes (Fig. 6.60b), using vertical presses at production rates as high as two parts per second.

Hydrostatic extrusion. In *hydrostatic extrusion* the pressure required for extrusion is supplied through a fluid medium surrounding the billet (Fig. 6.52c). Consequently, there is no container-wall friction. Pressures are usually about 1400 MPa (200 ksi). The high pressure in the chamber transmits some of the fluid to the die surfaces, thus significantly reducing friction and forces. Hydrostatic extrusion, which was developed in the early 1950s, has been improved by extruding the part into a second pressurized chamber, which is under lower pressure (*fluid-to-fluid extrusion*). This operation reduces the defects in the extruded product.

Because the hydrostatic pressure increases the ductility of the material, brittle materials can be extruded successfully by this method. However, the main reasons for this success appear to be low friction and use of low die angles and high

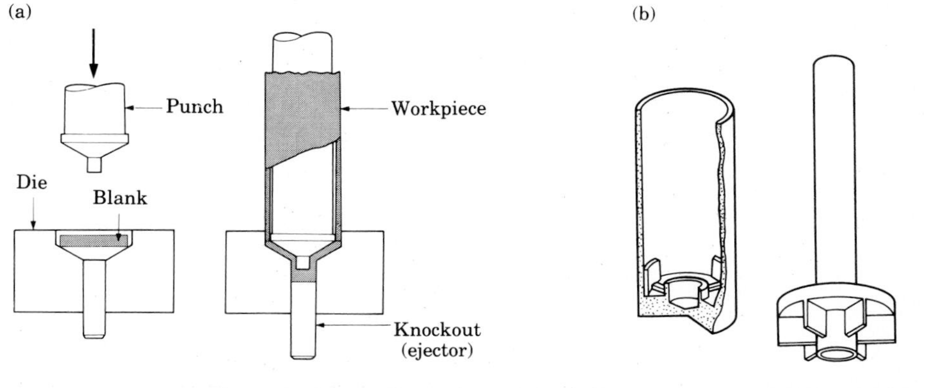

FIGURE 6.60
(a) Impact extrusion of a collapsible tube (Hooker process). (b) Two examples of products made by impact extrusion. These parts may also be made by casting, forging, and machining, depending on the dimensions and materials involved and the properties desired. Economic considerations are also important in final process selection.

extrusion ratios. Most commercial hydrostatic-extrusion operations use ductile materials. However, a variety of metals and polymers, solid shapes, tubes and other hollow shapes, as well as honeycomb and clad profiles, have been extruded successfully.

Hydrostatic extrusion is usually carried out at room temperature, typically using vegetable oils as the fluid, particularly castor oil because it is a good lubricant and its viscosity is not influenced significantly by pressure. For elevated-temperature extrusion, waxes, polymers, and glass are used as the fluid. These materials also serve as thermal insulators and help maintain the billet temperature during extrusion. In spite of the success obtained, hydrostatic extrusion has had limited industrial applications, largely because of the somewhat complex nature of tooling, the experience required with high pressures and design of specialized equipment, and the long cycle times required.

6.4.4 Defects

There are three principal defects in extrusion. They are surface cracking, extrusion defect, and internal cracking.

Surface cracking. If the extrusion temperature, friction, or extrusion speed are too high, surface temperatures rise significantly and can lead to surface cracking and tearing (*fir-tree cracking* or *speed cracking*). These cracks are intergranular, are usually the result of hot shortness, and occur especially with aluminum, magnesium, and zinc alloys. This defect is also observed with other metals, such as molybdenum alloys. This situation can be avoided by using lower temperatures and speeds.

However, surface cracking may also occur at low temperatures and has been attributed to periodic sticking of the extruded product along the die land. When the product being extruded sticks to the die land, the extrusion pressure increases rapidly. Shortly thereafter, the product moves forward again and the pressure is released. The cycle is then repeated.

Extrusion defect. The type of metal flow observed in Fig. 6.53(c) tends to draw surface oxides and impurities toward the center of the billet, much like a funnel. This defect is known as *extrusion defect*, *pipe*, *tailpipe*, or *fishtailing*. A considerable portion of the material can be rendered useless as an extruded product because of it—as much as one third the length of the extrusion.

This defect can be reduced by modifying the flow pattern to a less inhomogeneous one, such as by controlling friction and minimizing temperature gradients. Another method is to machine the surface of the billet prior to extrusion to eliminate scale and impurities. The extrusion defect can also be avoided by using a dummy block (Fig. 6.52a) that is smaller in diameter than the container, thus leaving a thin shell along the container wall as extrusion progresses.

Internal cracking. The center of an extruded product can develop cracks (variously known as *centerburst*, *center cracking*, *arrowhead fracture*, or *chevron*

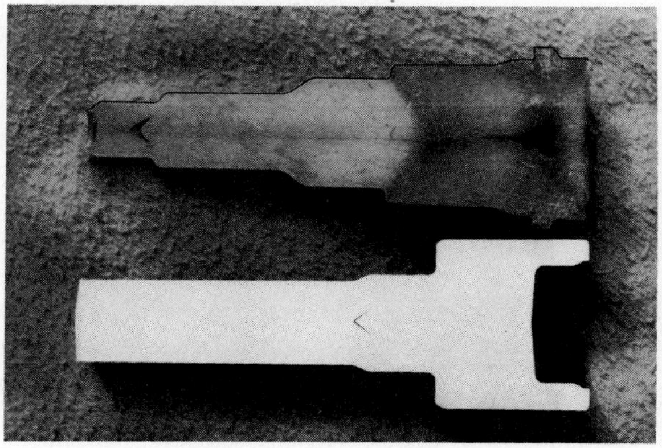

FIGURE 6.61 Chevron cracking in round steel bars during extrusion. Unless the part is inspected properly, such internal defects may remain undetected and possibly cause failure of the part in service.

cracking), as shown in Fig. 6.61. These cracks are attributed to a state of hydrostatic tensile stress (also called *secondary tensile stresses*) at the centerline of the deformation zone in the die. This situation is similar to the necked region in a uniaxial tensile-test specimen.

The major variables affecting hydrostatic tension are the die angle, extrusion ratio (reduction in cross-sectional area), and friction. We can understand them best by observing the extent of inhomogeneous deformation in extrusion. Experimental results indicate that, for the same reduction, as the die angle becomes larger, the deformation across the part becomes more inhomogeneous.

In addition to the die angle, another factor in internal cracking is the die contact length. The smaller the die angle, the longer is the contact length. This situation is similar to a hardness test with a flat indenter. The size and depth of the deformation zone increases with increasing contact length. This condition is illustrated in Fig. 6.62.

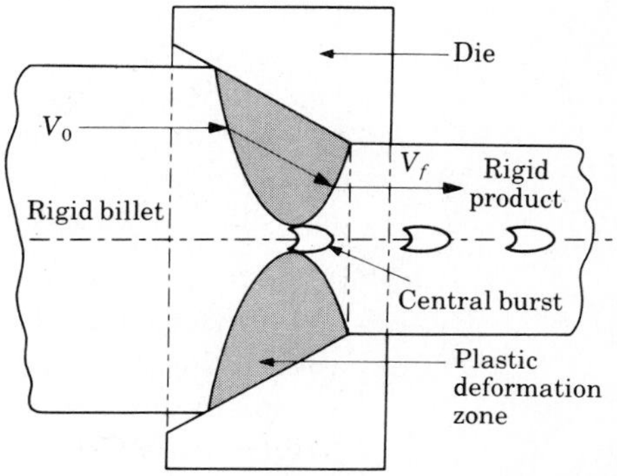

FIGURE 6.62 Deformation zone in extrusion showing rigid and plastic zones. Note that the plastic zones do not meet, leading to chevron cracking. The same observations are also made in drawing round bars through conical dies, and drawing flat sheet plate through wedge-shaped dies. *Source*: After B. Avitzur.

As shown in Fig. 2.40, an important parameter is the h/L ratio: the higher this ratio, the more inhomogeneous the deformation. High ratios mean small reductions and large die angles. Inhomogeneous deformation indicates that the center of the billet is not in a fully plastic state (it is rigid). The reason is that the plastic deformation zones under the die contact lengths do not reach each other (Fig. 6.62). Likewise, small reductions and high die angles retard the flow of the material at the surfaces, while the central portions are freer to move through the die.

The high h/L ratios described generate hydrostatic tensile stresses in the center of the billet, causing the type of defects shown in Fig. 6.61. These defects form more readily in materials with impurities, inclusions, and voids, because they serve as nucleation sites for defect formation. As for the role of friction, high friction in extrusion apparently delays formation of these cracks.

Such cracks have also been observed in tube extrusion and in spinning of tubes. The cracks appear on the inside surfaces for the reasons already given. In summary, the tendency for center cracking increases with increasing die angles and levels of impurities, and decreases with increasing extrusion ratio. These observations are also valid for drawing of rod and wire (see Section 6.5).

6.4.5 Extrusion practice

Numerous materials can be extruded to a wide variety of cross-sectional shapes and dimensions. Extrusion ratios can range from about 10 to 100 or over. Ram speeds may be up to 0.5 m/s (100 ft/min). Generally, slower speeds are preferable for aluminum, magnesium, and copper and higher speeds for steels, titanium, and refractory alloys. Presses for hot extrusion are generally hydraulic and horizontal and are usually vertical for cold extrusion.

Hot extrusion. In addition to the strain-rate sensitivity of the material at elevated temperatures, hot extrusion requires other special considerations. Cooling of the billet in the container can result in highly inhomogeneous deformation. Furthermore, because the billet is heated prior to extrusion, it is covered with an oxide layer (unless heated in an inert atmosphere). The resulting different frictional properties can affect the flow of the material and may produce an extrusion covered with an oxide layer. In order to avoid this problem, the diameter of the dummy block ahead of the ram (Fig. 6.52a) is made a little smaller than that of the container. A thin cylindrical shell (*skull*), consisting mainly of the oxidized layer, is thus left in the container, and the extruded product is free of oxides. Temperature ranges for hot extrusion are similar to those for forging (see Table 6.3).

Lubrication. Lubrication is important in hot extrusion. For steels, stainless steels, and high-temperature materials, glass is an excellent lubricant. It maintains its viscosity at elevated temperatures, has good wetting characteristics, and acts as a thermal barrier between the billet and the container and the die, thus minimizing cooling. A circular glass pad is usually placed at the die entrance. This pad softens and melts away slowly as extrusion progresses and forms an optimal die geometry. The viscosity–temperature index of the glass is an important factor in this application. Solid lubricants such as graphite and molybdenum disulfide are also used in hot extrusion. Nonferrous metals are usually extruded without a lubricant, although graphite may be used (see Table 6.4).

For materials that have a tendency to stick to the container and the die, the billet can be enclosed in a *jacket* of a softer metal such as copper or mild steel (*canning*). Besides acting as a low-friction interface, canning prevents contamination of the billet by the environment or the billet material from contaminating the environment if the material is toxic or radioactive. The canning technique is also used for processing metal powders.

Dies. Die materials for hot extrusion are usually hot-work die steels. To extend die life, coatings may be applied to the dies. Designing dies requires considerable experience because of the variety of products extruded. Dies with angles of 90° (square dies or *shear dies*) can also be used in extrusion of nonferrous metals, especially aluminum. Tubing is also extruded with a ram fitted with a mandrel (Fig. 6.63). For billets with a pierced hole, the mandrel may be attached to the ram. If the billet is solid, it must first be pierced in the container by the mandrel.

The complexity of an extrusion is a function of the ratio of the perimeter to the cross-sectional area of the part, known as the *shape factor*. Thus a solid, round extrusion is the simplest shape. Hollow shapes (Fig. 6.64) can also be extruded by *welding-chamber* methods using various special dies known as *spider*, *porthole*, and *bridge* dies. The metal flows around the arms of the die into strands, which are then rewelded under the high die pressures at the exit. This process is suitable only for

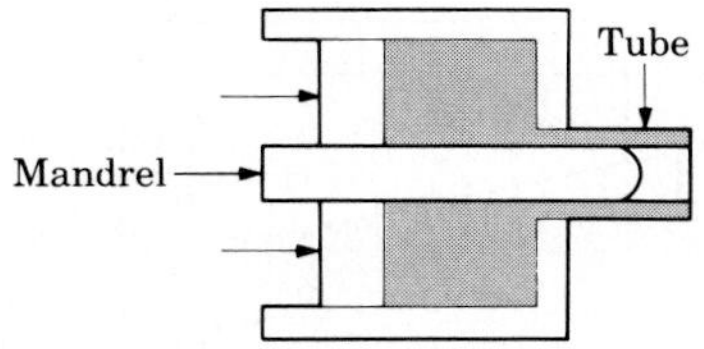

FIGURE 6.63 Extrusion of a seamless tube. The hole in the billet may be prepunched or pierced, or it may be generated during extrusion.

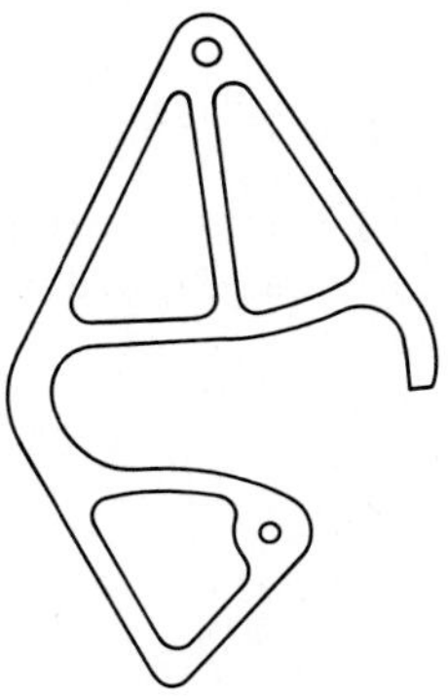

FIGURE 6.64 An example of extrusion that requires special dies to produce the five hollow sections in the part. This component, made of 6063-T6 aluminum, is a ladder lock for aluminum extension ladders.

aluminum and some of its alloys because of their capacity for pressure welding. Lubricants cannot be used as they prevent rewelding during extrusion.

Equipment. The basic equipment for extrusion is a hydraulic press, usually horizontal. These presses are designed for a variety of extrusion operations. Crank-type mechanical presses are also used for cold extrusion and piercing and for mass production of steel tubing. The largest hydraulic press for extrusion has a capacity of 160 MN (16,000 tons).

6.5 Rod, Wire, and Tube Drawing

Drawing is an operation in which the cross-sectional area of a bar or tube is reduced by pulling it through a converging die (Fig. 6.65). The die opening may be any shape. Wire drawing involves smaller diameter materials than rod drawing, with sizes as small as 0.025 mm (0.001 in.). The drawing process, which was an established art by the 11th century, is somewhat similar to extrusion, except that in drawing the bar is under tension, whereas in extrusion it is under compression.

Rod and wire drawing are usually finishing processes. The product is either used as produced or is further processed into other shapes, usually by bending or machining. Rods are used for various applications, such as small pistons, tension-carrying structural members, shafts, and spindles, and as the raw material for fasteners such as bolts and screws. Wire and wire products have a wide range of applications, such as electrical and electronic equipment and wiring, cables, springs, musical instruments, fencing, bailing, wire baskets, and shopping carts.

6.5.1 Mechanics of rod and wire drawing

The major variables in the drawing process are reduction in cross-sectional area, die angle, and friction. These variables are illustrated in Fig. 6.65.

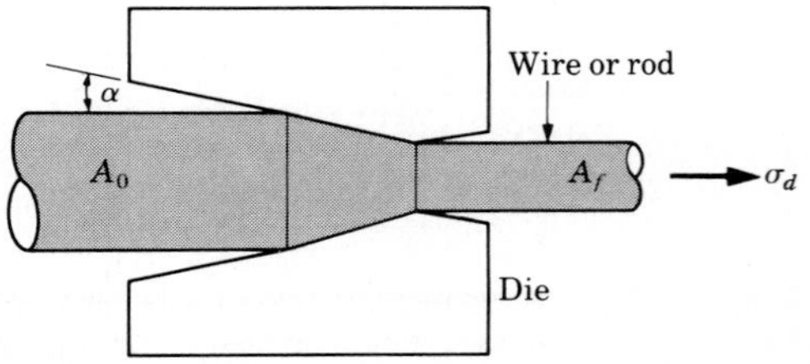

FIGURE 6.65
Variables in drawing round rod or wire.

Ideal deformation. For a round rod or wire, the *drawing stress* σ_d for the simplest case of ideal deformation (no friction or redundant work) can be obtained by the same approach as that in extrusion. Thus

$$\sigma_d = Y \ln\left(\frac{A_0}{A_f}\right). \tag{6.61}$$

Note that this expression is the same as Eq. (6.52) and that it also represents the energy per unit volume u. For strain-hardening materials Y is replaced by an average flow stress $\bar{Y}$ in the deformation zone (Fig. 6.66). Thus for a material that exhibits the true stress–true strain behavior of

$$\sigma = K\varepsilon^n,$$

the average flow stress $\bar{Y}$ is

$$\bar{Y} = \frac{K\varepsilon_1^n}{n+1}.$$

The drawing force F then is

$$F = \bar{Y}A_f \ln\left(\frac{A_0}{A_f}\right). \tag{6.62}$$

Note that the greater the reduction in cross-sectional area and the stronger the material are, the higher the drawing force is.

Ideal deformation and friction. In drawing, friction increases the drawing force, because work has to be supplied to overcome that friction. Using the slab

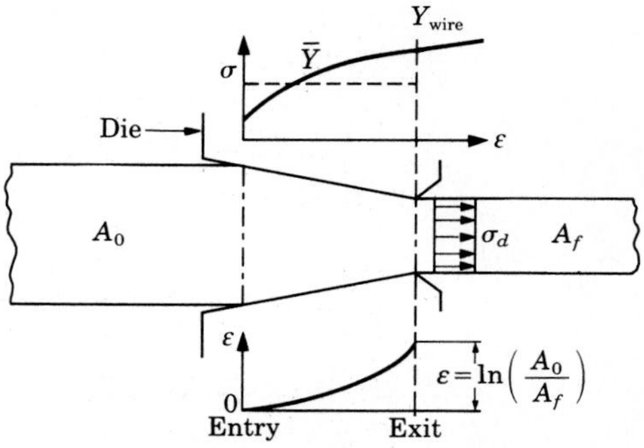

FIGURE 6.66 Variation in strain and flow stress in the deformation zone in drawing. Note that the strain increases rapidly toward the exit. The reason is that when the exit diameter is zero, the true strain reaches infinity. The point Y_{wire} represents the yield stress of the wire.

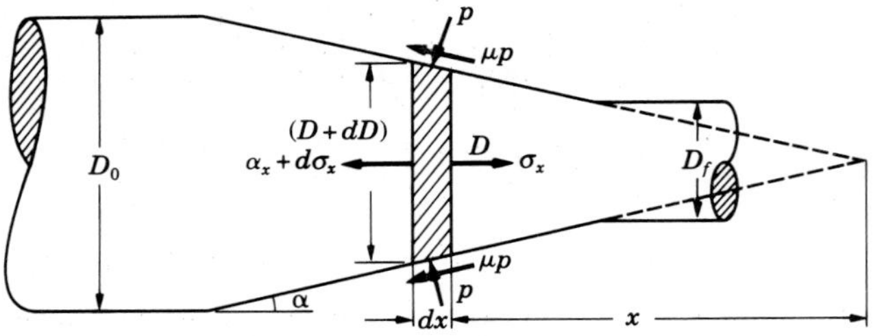

FIGURE 6.67
Stresses acting on an element in drawing of a solid cylindrical rod or wire through a conical converging die.

method of analysis and on the basis of Fig. 6.67, we equate the forces in the horizontal direction as follows:

$$(\sigma_x + d\sigma_x)\frac{\pi}{4}(D + dD)^2 - \sigma_x\frac{\pi}{4}D^2 + p\frac{\pi D\,dx}{\cos\alpha}\sin\alpha + \mu p\frac{\pi D\,dx}{\cos\alpha}\cos\alpha = 0.$$

Simplifying and ignoring the second-order terms, we obtain

$$Dd\sigma_x + 2\sigma_x dD + 2p\left(1 + \frac{\mu}{\tan\alpha}\right)dD = 0.$$

Note that we have two unknowns (σ_x and p) but only one equation. We obtain the second equation from yield criteria, recognizing that this is an axisymmetric case and that both the maximum shear-stress and the distortion-energy criterion give the relationship

$$\sigma_x + p = Y. \tag{6.63}$$

Letting $\mu/\tan\alpha = B$ and using the above relationships, we now obtain

$$\frac{dD}{D} = \frac{d\sigma_x}{2B\sigma_x - 2Y(1 + B)}.$$

Integrating this equation between the limits D_f and D_0—and by noting that at $D = D_0$, $\sigma_x = 0$, and at $D = D_f$, $\sigma_x = \sigma_d$—yields

$$\sigma_d = Y\frac{1 + B}{B}\left[1 - \left(\frac{D_f}{D_0}\right)^{2B}\right]$$

or

$$\sigma_d = Y\frac{1 + B}{B}\left[1 - \left(\frac{A_f}{A_0}\right)^{B}\right]$$

$$= Y\left(1 + \frac{\tan\alpha}{\mu}\right)\left[1 - \left(\frac{A_f}{A_0}\right)^{\mu\cot\alpha}\right]. \tag{6.64}$$

For a strain-hardening material, the yield stress Y is replaced by the average flow stress $\bar{Y}$, as shown in Fig. 6.66. Investigations have shown that, even though Eq. (6.64) does not include the redundant work, it is in good agreement with experimental data for small die angles and for a wide range of reductions.

Redundant work of deformation. Depending on the die angle and reduction, the material in drawing undergoes inhomogeneous deformation, much as it does in extrusion. Thus the redundant work of deformation also has to be included in the expression for the drawing stress. One such expression is as follows:

$$\sigma_d = \bar{Y}\left\{\left(1 + \frac{\tan\alpha}{\mu}\right)\left[1 - \left(\frac{A_f}{A_o}\right)^{\mu\cot\alpha}\right] + \frac{4}{3\sqrt{3}}\alpha^2\left(\frac{1-r}{r}\right)\right\}, \tag{6.65}$$

where r is the fractional reduction of area and α is the die angle in radians.

The first term in Eq. (6.65) represents the ideal and frictional work components. The second term represents the redundant-work component, which, as expected, is a function of the die angle. The larger the angle is, the greater the inhomogeneous deformation and hence the greater the redundant work are.

Another expression for the drawing stress for small die angles (including all three components of work) is

$$\sigma_d = \bar{Y}\left[\left(1 + \frac{\mu}{\alpha}\right)\ln\left(\frac{A_o}{A_f}\right) + \frac{2}{3}\alpha\right]. \tag{6.66}$$

The last term in this expression is the redundant-work component, whereby this work increases linearly with the die angle, as shown in Fig. 6.56.

Because redundant deformation is a function of the h/L ratio (see Fig. 2.40), an inhomogeneity factor Φ can replace the last term in Eq. (6.66). For drawing of round sections this factor is, approximately,

$$\Phi = 1 + 0.12\left(\frac{h}{L}\right). \tag{6.67}$$

A simple expression for the drawing stress then is

$$\sigma_d = \Phi\bar{Y}\left(1 + \frac{\mu}{\alpha}\right)\ln\left(\frac{A_o}{A_f}\right). \tag{6.68}$$

Equations (6.64), (6.65), (6.66), and (6.68), while not always agreeing with experimental data, approximate reasonably well the stresses required for wire drawing. More important, they identify the effects of the various parameters involved. With good lubrication, the value of μ in these equations ranges from about 0.03 to 0.1.

Die pressure. The die pressure p at any diameter along the die contact length can be conveniently obtained from

$$p = Y_f - \sigma, \tag{6.69}$$

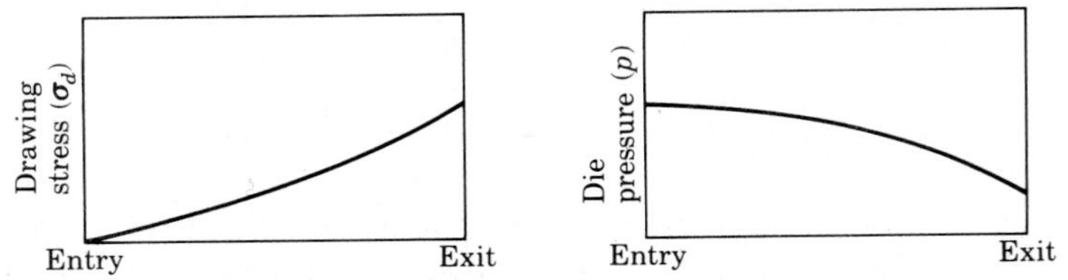

FIGURE 6.68
Variation in the drawing stress and die contact pressure along the deformation zone. Note that as the drawing stress increases, the die pressure decreases. This condition can be observed from the yield criteria, described in Section 2.13.

where σ is the tensile stress in the deformation zone at any diameter. (Thus σ is equal to σ_d at the exit and zero at entry). And Y_f is the flow stress of the material at any diameter corresponding to that strain.

Equation (6.69) is based on yield criteria for an element subjected to the stress system shown in Fig. 2.45(b), where the compressive stresses in the two principal directions are both equal to p. Note that as the tensile stress increases toward the exit, the die pressure drops toward the exit. Tensile stress and die pressure distributions are shown qualitatively in Fig. 6.68.

• Example 6.6: Calculation of power and die pressure.

A round rod of annealed 302 stainless steel is being drawn from a diameter of 10 mm to one of 8 mm at a speed of 0.5 m/s. Assume that the frictional and redundant work together constitute 40% of the ideal work of deformation. (a) Calculate the power required in this operation. (b) Calculate the die pressure at the die exit.

SOLUTION.

a) The true strain in this operation is

$$\varepsilon_1 = \ln\left(\frac{10^2}{8^2}\right) = 0.446.$$

From Table 2.4, $K = 1300$ MPa and $n = 0.30$. Thus

$$\bar{Y} = \frac{K\varepsilon_1^n}{n+1} = \frac{(1300)(0.446)^{0.30}}{1.30} = 785 \text{ MPa}.$$

From Eq. (6.62)

$$F = \bar{Y} A_f \ln\left(\frac{A_0}{A_f}\right),$$

where

$$A_f = \frac{(\pi)(0.008)^2}{4} = 5 \times 10^{-5}\ \text{m}^2.$$

Hence

$$F = (785)(5 \times 10^{-5})(0.446) = 0.0175\ \text{MN},$$

$$\begin{aligned}\text{Power} &= (F)(V_f) = (0.0175)(0.5)\\ &= 0.00875\ \text{MN}\cdot\text{m/s}\\ &= 0.00875\ \text{MW} = 8.75\ \text{kW},\end{aligned}$$

and

$$\text{Actual power} = (1.4)(8.75) = 12.25\ \text{kW}.$$

b) From Eq. (6.69)

$$p = Y_f - \sigma,$$

where Y_f represents the flow stress of the material at the exit. Thus

$$Y_f = K\varepsilon_1^n = (1300)(0.446)^{0.30} = 1020\ \text{MPa}.$$

In this equation σ is the drawing stress σ_d. Hence, using the actual force,

$$\sigma_d = \frac{F}{A_f} = \frac{(1.4)(0.0175)}{0.00005} = 490\ \text{MPa}.$$

Therefore the die pressure at the exit is

$$p = 1020 - 490 = 530\ \text{MPa}.$$

Drawing at elevated temperatures. At elevated temperatures, the flow stress of metals is a function of the strain rate. The average true strain rate $\dot{\bar{\varepsilon}}$ in the deformation zone in drawing is

$$\dot{\bar{\varepsilon}} = \frac{6V_0}{D_0} \ln\left(\frac{A_0}{A_f}\right). \tag{6.70}$$

[Note that this expression is the same as Eq. (6.59)]. For a particular drawing operation at an elevated temperature, we first calculate the average strain rate. From that we determine the flow stress and the average flow stress $\bar{Y}$ of the material.

Optimum die angle. Because of the various effects of the die angle on the three work components (ideal, friction, and redundant work), there is an optimum die angle where the extrusion force is a minimum. Drawing involves a similar type of deformation, so there is an optimum die angle in rod and wire drawing also, as shown in Fig. 6.56. Figure 6.69 shows a typical example from an experimental study. The optimum die angle for the minimum force increases with reduction. Note that the optimum angles in drawing are rather small.

Maximum reduction per pass. With greater reduction the drawing stress increases, but obviously the magnitude of the drawing stress has a limit. If it reaches the yield stress of the material, the material will simply continue to yield further as it leaves the die. This result is unacceptable because the product will undergo further deformation. Thus the maximum possible drawing stress can equal only the yield stress of the exiting material.

In the ideal case of a perfectly plastic material with a yield stress Y, the limiting condition is

$$\sigma_d = Y \ln\left(\frac{A_0}{A_f}\right) = Y, \tag{6.71}$$

or

$$\ln\left(\frac{A_0}{A_f}\right) = 1.$$

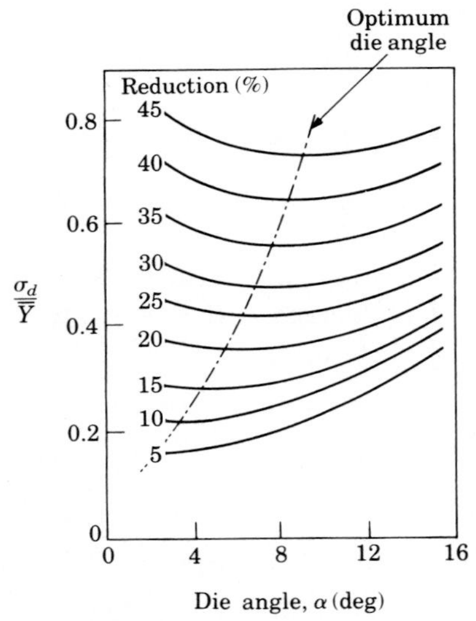

FIGURE 6.69 The effect of reduction in cross-sectional area on the optimum die angle in drawing copper wire. Note that the optimum angle increases with reduction. *Source*: After J. G. Wistreich.

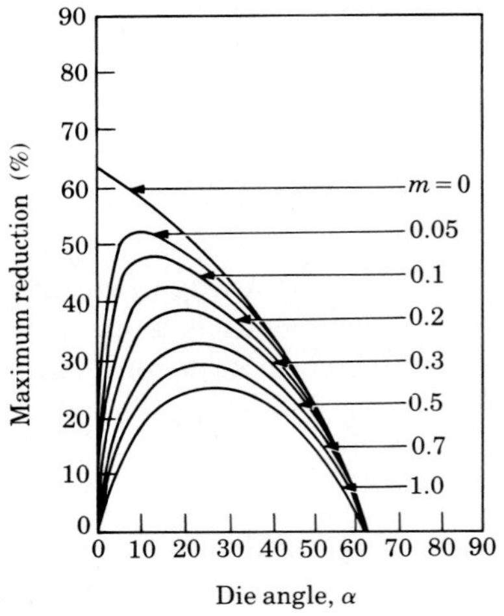

FIGURE 6.70 Effect of friction factor m and die angle on maximum possible reduction in wire drawing. A value of $m = 1$ indicates complete sticking (see Section 4.4). Note that the maximum possible reduction is 63%, as derived in Eq. (6.72). *Source*: After B. Avitzur.

Hence

$$\frac{A_0}{A_f} = e$$

and therefore

$$\text{Maximum reduction per pass} = \frac{A_0 - A_f}{A_0} = 1 - \frac{1}{e} = 0.63 = 63\%. \qquad (6.72)$$

The effects of friction and die angle on maximum reduction per pass are similar to those shown in Fig. 6.56. Because both friction and redundant work increase the drawing stress, the maximum reduction per pass will be lower than the ideal. In other words, the exiting material must have a larger cross-sectional area to sustain the higher drawing forces. These effects are shown in Fig. 6.70 for the friction factor *m*. With strain hardening, the exiting material is stronger than the rest, and the maximum reduction per pass increases.

• Example 6.7: Maximum reduction per pass for a strain-hardening material.

Obtain an expression for the maximum reduction per pass for a material with a true stress–true strain curve of $\sigma = K\varepsilon^n$. Ignore friction and redundant work.

SOLUTION. From Eq. (6.61) for this material,

$$\sigma_d = \bar{Y} \ln\left(\frac{A_0}{A_f}\right) = \bar{Y}\varepsilon_1,$$

where

$$\bar{Y} = \frac{K\varepsilon_1^n}{n+1}$$

and σ_d, for this problem, can have a maximum value equal to the flow stress at ε_1, or

$$\sigma_d = Y_f = K\varepsilon_1^n.$$

Hence we can write Eq. (6.71) as

$$K\varepsilon_1^n = \frac{K\varepsilon_1^n}{n+1}\varepsilon_1, \quad \text{or} \quad \varepsilon_1 = n + 1.$$

With $\varepsilon_1 = \ln(A_0/A_f)$ and maximum reduction $= (A_0 - A_f)/A_0$, these expressions reduce to

$$\text{Maximum reduction per pass} = 1 - e^{-(n+1)}. \qquad (6.73)$$

Note that when $n = 0$ (perfectly plastic material) this expression reduces to Eq. (6.72). As n increases, the maximum reduction per pass increases. •

Drawing of flat strip. Whereas drawing of round sections is axisymmetric, flat-strip drawing with high width-to-thickness ratios can be regarded as a *plane-strain* problem. The dies are wedge-shaped and thus the process is somewhat similar to rolling of wide strips. There is little or no change in the width of the strip during drawing. This process, although not of any industrial significance, is the fundamental deformation mechanism in ironing, as described in Section 7.12.

The treatment of this subject, as far as forces and maximum reductions are concerned, is similar to that for round sections. The drawing stress for the ideal condition is

$$\sigma_d = Y' \ln\left(\frac{t_0}{t_f}\right), \tag{6.74}$$

where Y' is the yield stress of the material in plane strain and t_0 and t_f are the original and final thicknesses of the strip, respectively. The effects of friction and redundant deformation in strip drawing are similar to those for round sections.

We obtain the maximum reduction per pass by equating the drawing stress (Eq. 6.74) to the *uniaxial* yield stress of the material, because the drawn strip is subjected only to simple tension. Thus

$$\sigma_d = Y' \ln\left(\frac{t_0}{t_f}\right) = Y, \qquad \ln\left(\frac{t_0}{t_f}\right) = \frac{Y}{Y'} = \frac{\sqrt{3}}{2} \quad \text{and} \quad \frac{t_0}{t_f} = e^{\sqrt{3}/2}.$$

We can then show that

$$\text{Maximum reduction per pass} = 1 - \frac{1}{e^{\sqrt{3}/2}} = 0.58 = 58\%. \tag{6.75}$$

Drawing of tubes. Tubes produced by extrusion or other processes can be reduced in thickness or diameter (*tube sinking*) by the tube-drawing processes illustrated in Fig. 6.71. A variety of mandrels are used to produce different shapes.

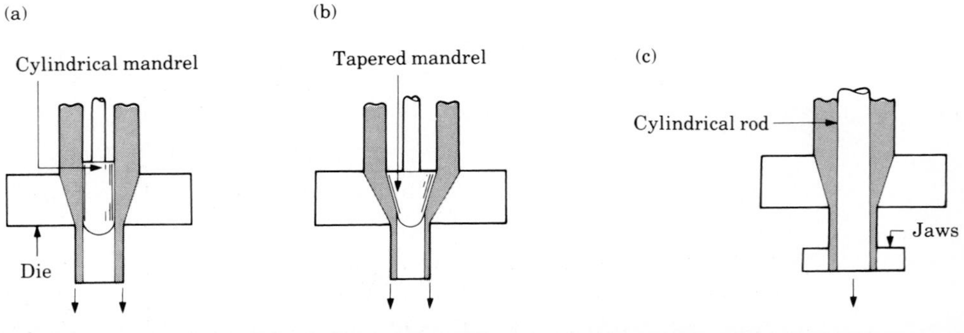

FIGURE 6.71
Various methods of tube drawing.

Shape changes can also be imparted by using dies and mandrels with various profiles. Drawing forces, die pressures, and maximum reduction per pass in tube drawing can be calculated by methods similar to those described for round rods.

6.5.2 Defects

Defects in drawing are similar to those observed in extrusion, especially *center cracking*. The factors influencing these internal cracks are the same, namely, the tendency for cracking increases with increasing die angle and decreasing reduction per pass, and with friction and the presence of inclusions in the material.

Another type of defect is the formation of *seams*, which are longitudinal scratches or folds in the material. Such defects can open up during subsequent forming operations by upsetting, heading, thread rolling, or bending of the rod or wire. Various surface defects can also result from improper selection of process parameters and lubrication.

Because of inhomogeneous deformation, a cold-drawn rod, wire, or tube usually contains residual stresses. Typically, as shown in Fig. 6.72, a wide range of residual stresses are present within the rod in three principal directions. However, for very light reductions, the surface residual stresses are compressive. Light reductions are equivalent to shot peening or surface rolling, thus improving fatigue

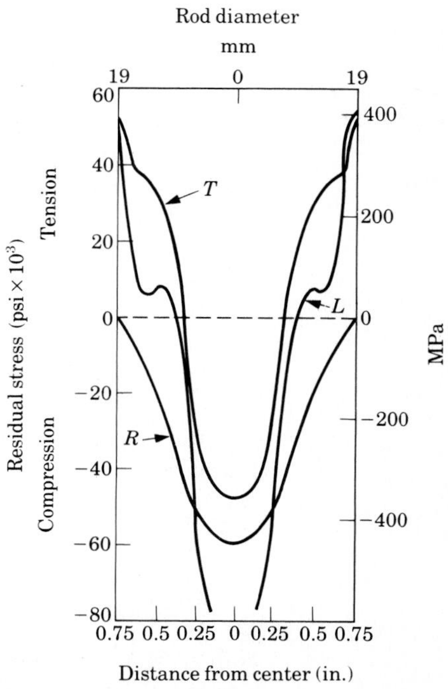

FIGURE 6.72 Residual stresses in cold-drawn AISI 1045 carbon steel round rod: T = transverse, L = longitudinal, and R = radial directions. *Source*: After E. S. Nachtman.

life. In addition to fatigue, residual stresses can also be significant in stress–corrosion cracking over a period of time, and in warping of the component when a layer is subsequently removed, as by machining or grinding.

6.5.3 Drawing practice

Successful drawing operations require careful selection of process parameters and consideration of many factors. A typical die design for drawing, with its characteristic features, is shown in Fig. 6.73. Die angles usually range from 6° to 15°. The purpose of the land is to size and set the final diameter of the product. Also, when the die is reground after use, the land maintains the exit dimension of the die opening. Reductions in cross-sectional area per pass range from about 10 percent to 45 percent; usually, the smaller the cross-section the smaller is the reduction per pass. Reductions per pass greater than 45 percent may result in breakdown of lubrication and deterioration of surface finish. Light reductions may also be made (*sizing pass*) on rods to improve surface finish and dimensional accuracy.

A rod or wire is fed into the die by first *pointing* it by swaging (forming the tip of the rod into a conical shape, Section 6.6). After the rod or wire is placed in the die, the tip is clamped into the jaws of the wire-drawing machine and the rod or wire is drawn continuously through the die. In most wire drawing operations the wire passes through a series of dies (*tandem drawing*). In order to avoid excessive tension in the exiting wire, it is wound one or two turns around a capstan between each pair of dies. The speed of the capstan is adjusted so that it supplies not only tension but also a small back tension to the wire entering the next die. Back tension reduces the die pressure and extends die life.

Drawing speeds depend on the material and cross-sectional area. They may be as low as 0.15 m/s (30 ft/min) for heavy sections, to as high as 50 m/s (10,000 ft/min) for very fine wire. Temperature can rise substantially at high drawing speeds.

Rods and tubes that are not sufficiently straight (or if supplied in a coiled form) can be straightened by passing them through pairs of rolls placed at different axes. The rolls subject the material to a series of bending and unbending operations, similar to the method shown in Fig. 6.44 for rolling.

Large sections can be drawn at elevated temperatures. In cold drawing, because of strain hardening, intermediate annealing between passes may be necessary to maintain sufficient ductility. Steel wires for springs and musical instruments are made by a heat-treatment process that precedes or follows the drawing operation (*patenting*). These wires have ultimate tensile strengths as high as 4800 MPa (700,000 psi), with tensile reduction of area of about 20 percent.

Dies. Die materials are usually alloy tool steels, carbides, or diamond. A diamond die (used for drawing of fine wires) may be a single crystal or a polycrystalline diamond in a metal matrix. Carbide and diamond dies are made as inserts or nibs, which are then supported in a steel casing.

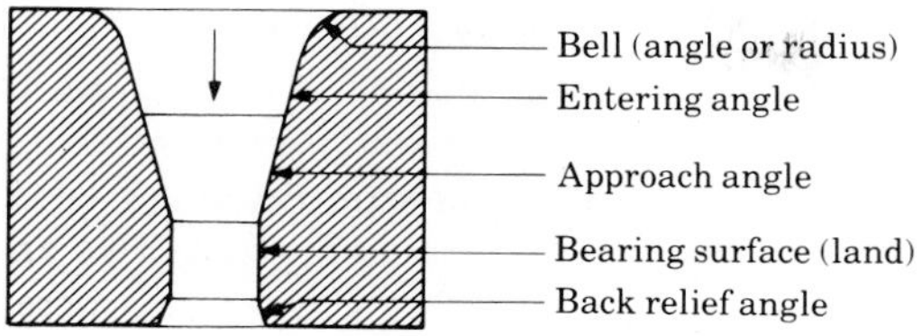

FIGURE 6.73
Terminology for a typical die for drawing round rod or wire.

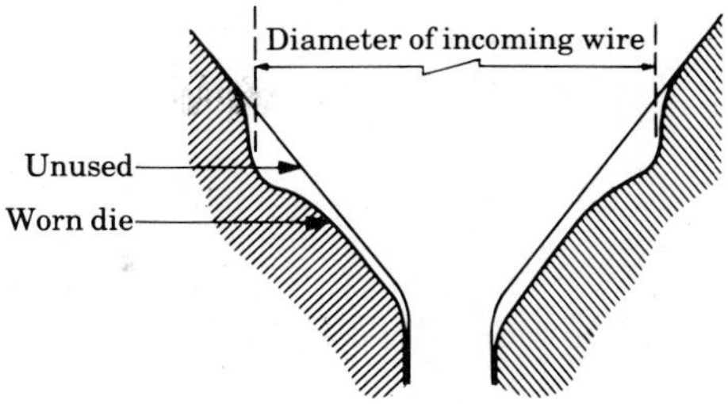

FIGURE 6.74
Schematic illustration of typical wear pattern in a wire drawing die.

In addition to rigid dies, a set of idling rolls is also used in drawing of rods or bars of various shapes. This arrangement is known as *Turk's head* and is more versatile than ordinary dies since the rolls can be adjusted to various positions for different products.

A typical wear pattern on a drawing die is shown in Fig. 6.74. Note that die wear is highest at the entry. Although the die pressure is highest in this region and may be partially responsible for wear, other factors are involved. They are variation in the diameter of the entering wire and vibration (thus subjecting the entry contact zone to fluctuating stresses) and the presence of abrasive scale on the surface of the entering wire.

Lubrication. Proper lubrication is essential in rod, tube, and wire drawing, regardless of whether the process is dry or wet drawing. In *dry drawing* the surface of the wire is coated with various lubricants, depending on its strength and frictional characteristics. A common lubricant used is soap. The rod to be drawn is first surface treated by *pickling*. This treatment removes the surface scale that could lead to surface defects and, being quite abrasive, would considerably reduce die life. The soap is picked up by the wire as it goes through a box filled with soap powder (stuffing box).

With high-strength materials, such as steels, stainless steels, and high-temperature alloys, the surface of the rod or wire may be coated either with a softer metal or with a *conversion coating*. Copper or tin can be chemically deposited on the surface of the metal. This thin layer of softer metal acts as a solid lubricant during drawing. Conversion coatings may consist of sulfate or oxalate coatings on the rod, which are then typically coated with soap, as a lubricant. Polymers are also used as solid lubricants, such as in drawing of titanium.

In *wet drawing*, the dies and rod are completely immersed in a lubricant. Typical lubricants are oils and emulsions containing fatty or chlorinated additives, and various chemical compounds.

Equipment. Two types of equipment are used in drawing. A *draw bench* is similar to a long horizontal tensile testing machine. It is used for single draws of straight, large cross-section rods and tubes for lengths up to 30 m (100 ft) with a hydraulic or chain-drive mechanism. Rod and wire of smaller cross-section are drawn by a *bull block* (*capstan*), which is a rotating drum around which the wire is wrapped. The tension in this setup provides the force required to draw the wire.

6.6 Swaging

In swaging, also known as *rotary swaging* or *radial forging*, a solid rod or a tube is reduced in diameter by the reciprocating radial movement of two or four dies (Fig. 6.75). The die movements are generally obtained by means of a set of rollers in a cage. The internal diameter and the thickness of the tube can be controlled with or

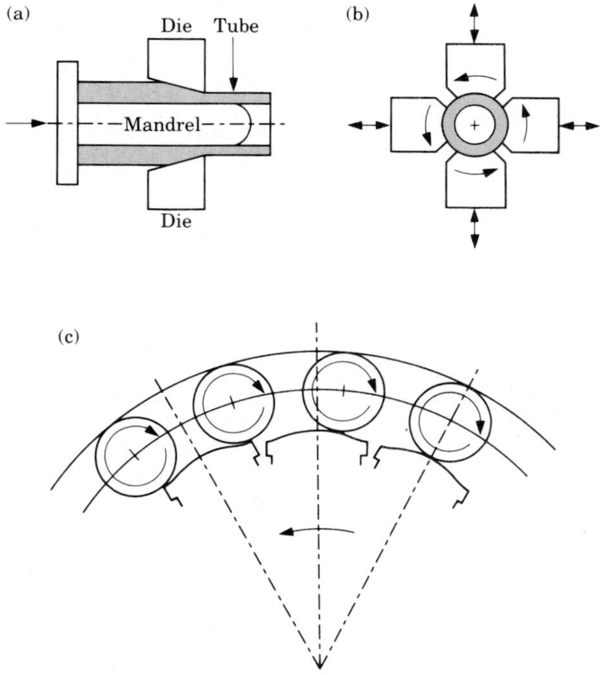

FIGURE 6.75 Schematic illustration of the swaging process: (a) side view and (b) front view. (c) Schematic illustration of roller arrangement, curvature on the four radial hammers (that give motion to the dies), and the radial movement of a hammer as it rotates over the rolls.

without mandrels as shown in Fig. 6.76. Mandrels can also be made with longitudinal grooves (similar in appearance to a splined shaft); thus internally shaped tubes can be swaged (Fig. 6.77). The rifling in gun barrels is made by swaging a tube over a mandrel with spiral grooves.

The swaging process is usually limited to a diameter of about 50 mm (2 in.), although special machinery has been built to swage gun barrels of larger diameter. The length of the product in this process is limited only by the length of the mandrel (if needed). Die angles are usually only a few degrees and may be compound—the die may have more than one angle, for more favorable material flow during swaging. Lubricants are used for improved surface finish and longer die life. The process is generally carried out at room temperature. Parts produced by swaging have improved mechanical properties and good dimensional accuracy.

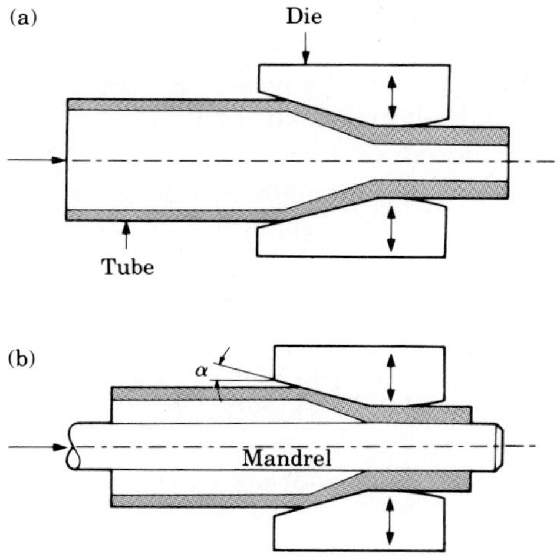

FIGURE 6.76

Reduction of outer and inner diameters of tubes by swaging. (a) Free sinking without a mandrel. The ends of solid bars and wire are tapered (pointing) by this process in order to feed the material into the conical die. (b) Sinking on a mandrel. Coaxial tubes of different materials can also be swaged in one operation.

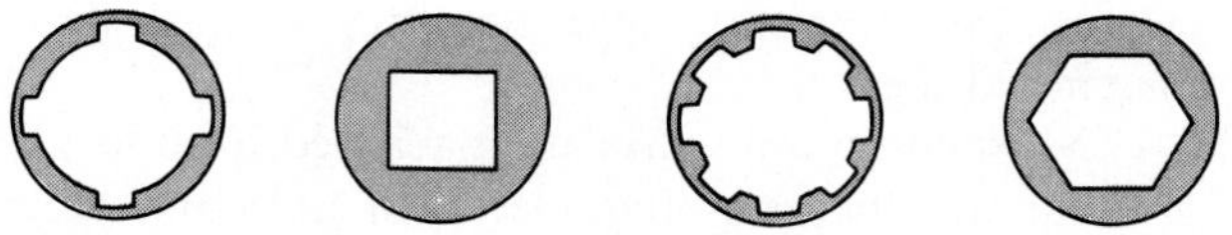

FIGURE 6.77

Typical cross-sections produced by swaging constant-wall-thickness tube blanks on shaped mandrels. Rifling of small gun barrels can also be made by swaging, using a specially shaped mandrel. The formed tube is then removed by slipping it out of the mandrel.

6.7 Die Manufacturing Methods

Various manufacturing methods, either singly or in combination, are used in making dies. These processes include casting, forging, machining, grinding, electrical, and electrochemical methods of die sinking and finishing operations, such as honing and polishing. The process of hubbing, either cold or hot, may also be used to make small dies with shallow cavities. Dies are usually heat treated for greater strength and wear resistance. If necessary, their surface profile and finish are improved by finish grinding and polishing, either by hand or with programmable industrial robots.

The choice of a die manufacturing method depends on the particular manufacturing process in which the die is to be used and the size and shape of the die. Cost often dictates the process selected because, as we have shown, tool and die costs can be significant in manufacturing operations. For example, the cost of a set of dies for automotive body panels may run $2 million. Even small and relatively simple dies can cost hundreds of dollars. On the other hand, because a large number of parts are usually made from the same die, die cost per piece is generally only a small portion of a part's manufacturing cost.

Dies may be classified as male and female. Male dies are generally machined externally, whereas female dies (with recesses and die cavities) are produced by internal machining. Because removing material from deep recesses is hard to do, internal machining is generally more difficult than external machining.

Dies may also be classified by their size. Small dies generally are those that have a surface area of 10^3–10^4 mm^2 (2–15 in^2), whereas large dies have surface areas of 1 m^2 (9 ft^2) and larger, such as those used for pressworking automotive body panels.

Dies of various sizes and shapes can be cast from steels, cast irons, and nonferrous alloys. The processes used may be sand casting for large dies weighing many tons to shell molding for small dies. Cast steels are generally preferred as die materials because of their strength and toughness and the ease with which their composition, grain size, and properties can be controlled and modified.

Depending on how they solidify, cast dies, unlike dies made of wrought metal dies made by forging or rolling, may not have directional properties. Thus they exhibit the same properties on all working surfaces. However, because of shrinkage, control of dimension and accuracy in cast dies can be difficult, compared to machined dies.

Most commonly, dies are machined from forged die blocks by processes such as milling, turning, grinding, electrical and electrochemical machining, and polishing. Typically, a die for hot-working operations is machined by milling on an automatic copy machine that traces a master model (pattern), or on computer-controlled machine tools that use various software. The patterns may be made of wood, epoxy resins, aluminum, or gypsum. Conventional machining can be difficult for high-strength and wear-resistant die materials that are hard or are heat treated. These operations can be very time consuming even with copy milling machines. As a result, nontraditional machining processes are used extensively, particularly for small- or medium-sized dies. These processes are generally faster and more economical, and the dies usually do not require additional finishing. Diamond dies for drawing fine wire are manufactured by producing holes with a thin rotating needle, coated with diamond dust suspended in oil. The complete procedure for producing a diamond die can take many days, and repolishing of worn diamond dies can take as much as 12 hours.

To obtain improved hardness, wear resistance, and strength, die steels are usually heat treated. Improper heat treatment is one of the most common causes of die failure. Particularly important are the condition and composition of the die surfaces. Thus the proper selection of temperatures and atmospheres for heat treatment, quenching media, quenching practice, tempering procedures, and handling is important. Dies may be subjected to various surface treatments for improved frictional and wear characteristics (see Chapter 4).

After heat treatment, tools and dies are subjected to grinding, polishing, or chemical and electrical machining processes to obtain the desired surface finish and dimensional accuracy. Recall that heat treatment may distort dies because of microstructural changes and uneven thermal cycling. The grinding process, if not controlled properly, can cause surface damage by excessive heat and induce harmful tensile residual stresses on the surface of the die, thus reducing its fatigue life. Scratches on a die's structure can act as stress raisers. Likewise, commonly used die-making processes, such as electrical-discharge machining, can cause surface damage and cracks, unless the process parameters are carefully controlled.

6.8 Die Failures

Failure of dies in manufacturing operations generally result from one or more of the following causes: improper design, defective material, improper heat treatment and finishing operations, overheating and heat checking (cracking caused by temperature cycling), excessive wear, overloading, misuse, and improper handling. We describe in the following paragraphs some of the important factors leading to die failure. These factors apply to dies made of die steels, but many are also applicable to other die materials. The proper design of dies is as important as the proper selection of die materials. In order to withstand the forces in manufacturing processes, a die must have proper cross sections and clearances. Sharp corners, radii, and fillets, as well as sudden changes in cross-section, act as stress raisers and can have detrimental effects on die life. Dies may be made in segments and prestressed during assembly for improved strength.

The proper handling, installation, assembly, and aligning of dies are important. Overloading of tools and dies can cause premature failure. A common cause of failure of cold-extrusion dies, for example, is the failure of the operator to remove a formed part from the die before loading it with another blank.

In spite of their hardness and resistance to abrasion, die materials such as carbides and diamond are susceptible to cracking and chipping from impact forces or thermal stresses caused by temperature gradients within the die. Thus surface preparation and finishing are important. Even metalworking fluids can adversely affect tool and die materials. Sulfur and chlorine additives in lubricants and coolants, for example, can leach away the cobalt binder in tungsten carbide and lower its strength and toughness.

Even if they are manufactured properly, dies are subjected to stresses and temperature during their use, which cause wear and hence shape changes. Die wear is important because when die shape changes, the parts, in turn, have improper dimensions. Thus the economics of the manufacturing operation is adversely affected.

During use, dies may also undergo heat checking from thermal cycling. To reduce heat checking (which has the appearance of parched land) and eventual die breakage in hot-working operations, dies are usually preheated to temperatures of about 150 °C to 250 °C (300 °F to 500 °F). Cracked or worn dies may be repaired by welding and metal-deposition techniques, including lasers. Dies may be designed and constructed with inserts that can be replaced when worn or cracked. The proper design and placement of these inserts is important, because otherwise they can crack.

Die failure and fracture in manufacturing plants can be hazardous to employees. It is not unusual for a set of dies resting on the floor or a shelf to suddenly disintegrate because of the highly stressed condition of its components. The broken pieces are propelled at high speed and can cause serious injury or death. Therefore highly stressed dies and tooling should always be surrounded by metal shielding. These shields should be properly designed and sufficiently strong to contain the fractured pieces in the event of die failure.

SUMMARY

In bulk deformation processes the workpiece using dies and various other tooling is subjected to major changes in its dimensions. The products of bulk deformation processing, such as plates, sheets, forgings, rod, wire, and extruded shapes, are then processed by various other methods described in subsequent chapters.

The bulk properties, as well as the surface properties, of the workpiece material are important, because the material is in contact with dies under pressures sufficiently high to cause yielding. The strain-hardening capability, as well as its strain-rate sensitivity, of the material is significant, particularly if high temperatures are involved.

Various methods of analysis are available to calculate the stresses, forces, and energies required in deformation processing of materials. These analyses are important not only for the selection of appropriate equipment for metalworking, but also in their design.

Forging is an ancient and important metalworking process. It is capable of producing a wide variety of parts with favorable characteristics of strength, toughness, dimensional accuracy, and reliability in service. Material behavior during deformation, as well as friction, heat transfer, and material-flow characteristics in a die cavity, are important considerations. Also important is the proper selection of die materials, lubricants, temperatures, speeds, and equipment. Defects can develop if the process is not controlled properly.

A variety of forging machines are available, each with its own characteristics and capabilities. Forging processes have been highly automated with industrial robots and computer controls. Computer-aided design and manufacturing techniques are now being used extensively in die design and manufacturing, as well as in preform design and predicting material flow and possibility of defects during forging.

Rolled plates, sheets, pipe and tubing, and foil are used in a wide variety of products, ranging from beverage cans and packaging to car bodies, boilers, and ship hulls. In addition to flat rolling, shape rolling is used to make products with various cross-sections, such as bars and rods and structural shapes for buildings and transportation equipment. An important development is continuous casting of billets, which are rolled directly into semifinished products.

As in all metalworking processes, rolling involves a number of process and material variables that should be controlled in order to roll products having proper quality, properties, surface finish, and dimensional accuracy. These variables include rolling temperature and speed, lubrication, and the condition of the rolls and the characteristics of the equipment.

The extrusion process is capable of producing lengths of solid and hollow sections with constant cross-sectional area. Important factors in successful extrusion are die design, extrusion ratio, lubrication, billet temperature, and extrusion speed. Cold extrusion, which is a combination of various extrusion and forging operations, is capable of producing parts economically and with good mechanical properties.

Rod, wire, and tube have numerous important applications. These products are made basically by a drawing process in which the material is pulled through one or more dies. Although the cross-sections of most drawn products are round, rectangular and other shapes can be drawn. Drawing tubular products usually requires internal mandrels. Proper die design and selection of materials and lubricants are essential to obtaining a product with good quality and surface finish, dimensional accuracy, and strength.

BIBLIOGRAPHY

General Introductory Texts with Some Analytical Treatment

Alexander, J. M., and R. C. Brewer, *Manufacturing Properties of Materials*. New York: Van Nostrand Reinhold, 1963.

Altan, T., S.-I. Oh, and H. Gegel, *Metal Forming—Fundamentals and Applications*. Metals Park, Ohio: American Society for Metals, 1983.

Cook, N. H., *Manufacturing Analysis*. Reading, Mass.: Addison-Wesley, 1966.

Crane, F. A. A., *Mechanical Working of Metals*. New York: Macmillan, 1964.

Dieter, G. E., *Mechanical Metallurgy*, 3d ed. New York: McGraw-Hill, 1986.

Harris, J. N., *Mechanical Working of Metals*. New York: Pergamon, 1983.

Parkins, R. N., *Mechanical Treatment of Metals*. London: Allen and Unwin, 1968.

Rowe, G. W., *Elements of Metalworking Theory*. London: Edward Arnold, 1979.

Rowe, G. W., *Principles of Industrial Metalworking Processes*. London: Edward Arnold, 1977.

Schey, J. A., *Introduction to Manufacturing Processes*, 2d ed. New York: McGraw-Hill, 1987.

Advanced Texts

Avitzur, B., *Handbook of Metal-Forming Processes*. New York: Wiley, 1983.

Avitzur, B., *Metal Forming: Processes and Analysis*. New York: McGraw-Hill, 1968.

Backofen, W. A., *Deformation Processing*. Reading, Mass.: Addison-Wesley, 1972.

Blazynski, T. Z., *Metal Forming: Tool Profiles and Flow*. New York: Halsted Press, 1976.

Boer, C. R., N. Rebelo, H. Rydstad, and G. Schroder, *Process Modelling of Metal Forming and Thermomechanical Treatment*. New York: Springer-Verlag, 1986.

Ford, H., and J. M. Alexander, *Advanced Mechanics of Materials*, 2d ed. New York: Halsted Press, 1977.

Hoffman, O., and G. Sachs, *Introduction to the Theory of Plasticity for Engineers*. New York: McGraw-Hill, 1953.

Hosford, W. F., and R. M. Caddell, *Metal Forming, Mechanics and Metallurgy*. Englewood Cliffs, N.J.: Prentice-Hall, 1983.

Johnson, W., and P. B. Mellor, *Engineering Plasticity*. New York: Van Nostrand Reinhold, 1973.

Kobayashi, S., S.-I. Oh, and T. Altan, *Metal Forming and the Finite-Element Method*. New York: Oxford, 1989.

Lippmann, H. (ed.) *Engineering Plasticity: Theory of Metal Forming Processes* (2 vols.). New York: Springer, 1977.

Lippmann, H., *Metal Forming Plasticity*. New York: Springer, 1979.

Slater, R. A., *Engineering Plasticity: Theory and its Application to Metal Forming Processes*. New York: Halsted Press, 1974.

Thomsen, E. G., C. T. Yang, and S. Kobayashi, *Mechanics of Plastic Deformation in Metal Processing*. New York: Macmillan, 1964.

Forging

Altan, T., S.-I. Oh, and H. C. Gegel, *Metal Forming—Fundamentals and Applications*. Metals Park, Ohio: American Society for Metals, 1983.

Altan, T., S.-I. Oh, and H. C. Gegel, *Forging: Equipment, Materials and Practices*. Columbus, Ohio: Battelle Memorial Institute, 1973.

Avitzur, B., and C. J. van Tyne (eds.), *Production to Near Net Shape: Source Book*. Metals Park, Ohio: American Society for Metals, 1983.

Byrer, T. G. (ed.), *Forging Handbook*. Metals Park, Ohio: American Society for Metals, 1985.

Forging Design Handbook. Metals Park, Ohio: American Society for Metals, 1972.

Hosford, W. F., and R. M. Caddell, *Metal Forming: Mechanics and Metallurgy*. Englewood Cliffs, N.J.: Prentice-Hall, 1983.

Jenson, J. E. (ed.), *Forging Industry Handbook*. Cleveland: Forging Industry Association, 1970.

Lange, K. (ed.), *Handbook of Metal Forming*. New York: McGraw-Hill, 1985.

Metals Handbook, 9th ed., *Vol. 14: Forming and Forging*. Metals Park, Ohio: ASM International, 1988.

Open Die Forging Manual, 3d ed. Cleveland: Forging Industry Association, 1982.

Sabroff, A. M., F. W. Boulger, and H. J. Henning, *Forging Materials and Practices*. New York: Reinhold, 1968.

Thomas, A., *DFRA Forging Handbook: Die Design*. Sheffield, England: Drop Forging Research Association, 1980.

Watkins, M. T., *Metal Forming I: Forging and Related Processes*. Oxford: Oxford University Press, 1975.

Rolling

Lange, K. (ed.), *Handbook of Metal Forming*. New York: McGraw-Hill, 1985.

Larke, E. C., *The Rolling of Strip, Sheet, and Plate*, 2d ed. London: Chapman and Hall, 1963.

Roberts, W. L., *Cold Rolling of Steel*. New York: Marcel Dekker, 1978.

Roberts, W. L., *Hot Rolling of Steel*. New York: Marcel Dekker, 1983.

Starling, C. W., *The Theory and Practice of Flat Rolling*. London: The University of London Press, 1962.

Underwood, L. R., *The Rolling of Metals*, Vol. 1. New York: Wiley, 1950.

Wusatowski, Z., *Fundamentals of Rolling*. New York: Pergamon, 1969.

Extrusion

Alexander, J. M., and B. Lengyel, *Hydrostatic Extrusion*. London: Mills and Boon, 1971.

Everhart, J. E., *Impact and Cold Extrusion of Metals*. New York: Chemical Publishing Company, 1964.

Inoue, N., and M. Nishihara (eds.) *Hydrostatic Extrusion: Theory and Applications*. New York: Elsevier, 1985.

Lange, K. (ed.), *Handbook of Metal Forming*. New York: McGraw-Hill, 1985.

Laue, K., and H. Stenger, *Extrusion—Processes, Machinery, Tooling*. Metals Park, Ohio: American Society for Metals, 1981.

Metals Handbook, 9th ed., *Vol. 14: Forming and Forging*, Metals Park, Ohio: ASM International, 1988.

Michaeli, W., *Extrusion Dies*. New York: Macmillan, 1984.

Pearson, C. E., and R. N. Parkins, *The Extrusion of Metals*, 2d ed. New York: Wiley, 1961.

Source Book on Cold Forming. Metals Park, Ohio: American Society for Metals, 1975.

Drawing

Bernhoeft, C. P., *The Fundamentals of Wire Drawing*. London: The Wire Industry Ltd., 1962.

Developments in the Drawing of Metals, Book no. 301. London: The Metals Society, 1983.

Lange, K. (ed.), *Handbook of Metal Forming*. New York: McGraw-Hill, 1985.

Nonferrous Wire Handbook, 2 vols. Branford, Conn.: The Wire Association International, Inc., 1977 and 1981.

Pomp, A., *The Manufacture and Properties of Steel Wire*. London: The Wire Industry Ltd., 1954.

Steel Wire Handbook, vol. 1, 1968; vol. 2, 1969; vol. 3, 1972; vol. 4, 1980. Guilford, Conn.: Wire Association International.

Tool and Die Failures

Kalpakjian, S. (ed.), *Source Book on Tool and Die Failures*. Metals Park, Ohio: American Society for Metals, 1982.

Lubrication

See Bibliography on Tribology in Metalworking, Chapter 4.

QUESTIONS

Forging

6.1 How can you tell whether a certain part is forged or cast? Describe the features that you would investigate to arrive at a conclusion.

6.2 Why is the control of blank volume important in closed-die forging?

6.3 What are the advantages and limitations of a cogging operation? Die inserts in forging?

6.4 Explain why there are so many different kinds of forging machines.

6.5 Devise an experimental method whereby you can measure the force required for forging only the flash in impression-die forging.

6.6 A manufacturer is successfully hot forging a certain part using material supplied by Company A. A new supply of material is obtained from Company B, with the same nominal composition of the major alloying elements as the previous material. However, the new forgings are cracking even though the same procedure is followed as before. What is the probable reason?

6.7 Explain why there might be a change in the density of a forged product compared to that of the blank.

6.8 Describe the role of surface oxide layers on a blank as it is being deformed by impression-die forging.

6.9 Glass is a good lubricant for hot extrusion, but would you use glass for impression-die forging also? Explain.

6.10 Describe the factors that influence spread in cogging operations on square billets.

Rolling

6.11 Rolling reduces the thickness of plates and sheets. However, it is possible to reduce the thickness by simply stretching the material. Would this be a feasible process? Explain.
6.12 Explain why the neutral point moves toward the roll-gap entry as friction increases in Fig. 6.37.
6.13 Explain how a cast structure is converted into a wrought structure by hot rolling.
6.14 Three factors influence spreading in rolling: strip width-to-thickness ratio, friction, and ratio of roll radius to strip thickness. Explain how these factors affect spreading.
6.15 Explain how the residual stress patterns in Fig. 6.43 become reversed when roll radius or reduction per pass is changed.
6.16 Explain how to apply front and back tensions to sheet metals during rolling.
6.17 Rolls tend to flatten under roll forces. Which property(ies) of the roll material can be increased to reduce flattening? Describe the methods by which flattening can be reduced.
6.18 Explain the technical and economic reasons for taking larger rather than smaller reductions per pass in flat rolling.
6.19 Surface roughness in hot-rolled products is higher than in cold-rolled products. Explain why.
6.20 List and explain the methods that can be used to reduce the roll-separating force.
6.21 Explain the advantages and limitations of using small diameter rolls on flat rolling.

Extrusion

6.22 Explain the different ways in which die geometry affects the extrusion process.
6.23 The extrusion ratio, die geometry, extrusion speed, and billet temperature all affect the extrusion pressure. Explain why.
6.24 How would you avoid centerburst defects in extrusion? Explain why your methods would be effective.
6.25 Assume that you are reducing the diameter of two round rods, one by simple tension and the other by indirect extrusion. Which method requires more force? Why?
6.26 How would you make a stepped extrusion that has increasingly larger cross-sections along its length. Would your process be economical and suitable for large production runs? Explain.
6.27 The temperature ranges for extruding various metals are similar to those for forging (see Table 6.3). Describe the consequences of extruding at a temperature (a) below and (b) above these ranges.
6.28 From Eq. (6.52), for low values of extrusion ratio such as $R = 2$, the ideal extrusion pressure p can be lower than the yield stress Y of the material. Is this result logical? Explain.
6.29 Refer to Eq. (6.59) and explain why the three variables in that equation should influence the strain rate in extrusion.

Drawing

6.30 What changes would you expect in the strength, hardness, ductility, and anisotropy of annealed metals after they have been cold drawn through dies? Why?
6.31 In rod and wire drawing the maximum die pressure is at the die entry. Why?

6.32 Describe the conditions under which wet drawing and dry drawing, respectively, are desirable.
6.33 Name the important process variables in drawing and explain how they affect the drawing process.
6.34 Assume that a rod drawing operation can be carried out either in one pass or in two passes in tandem. If all die angles and the total reduction are the same, will the drawing forces be different? Explain.
6.35 Assume that the reduction in the cross-section in Fig. 6.65 is taking place by pushing the rod through the die instead of pulling it. Sketch the die pressure distribution for this case and explain your reasoning.
6.36 In deriving Eq. (6.72) we did not mention the ductility of the original material being drawn. Explain why that was unnecessary.
6.37 Explain why the die pressure in drawing decreases toward the exit.
6.38 What is the value of the die pressure at the exit when an ideal drawing operation is being carried out at the maximum theoretical reduction per pass?
6.39 Explain why the maximum reduction per pass in drawing should increase with increasing strain hardening exponent n.
6.40 Explain why the maximum reduction per pass is higher in drawing round bars (63 percent) compared to plane-strain drawing of flat sheet or plate (58 percent).
6.41 If in deriving Eq. (6.72) we include friction, is the maximum reduction per pass the same, higher, or lower than 63 percent? Explain.

Swaging

6.42 State the reasons for development of the swaging process.
6.43 How would you go about estimating the swaging force acting on each die?
6.44 In swaging a tube its thickness increases. How would you go about calculating the change in thickness without measuring it? You may make any other measurements.

General

6.45 Take any three topics from Chapter 2 and with a specific example for each, show their relevance to the topics covered in this chapter.
6.46 Same as Question 6.45, but for Chapter 3.
6.47 From the topics covered in this chapter, list and explain specifically two examples where (a) friction is desirable and (b) friction is not desirable.
6.48 List and explain the reasons why there are so many different types of die materials available for the processes described in this chapter.

PROBLEMS

Forging

6.1 Plot the force versus reduction-in-height curve in open-die forging a cylindrical, annealed copper specimen 1 in. high and 1 in. in diameter, up to a reduction of 75 percent for the case of (a) no friction between the flat dies and the specimen, (b) $\mu = 0.2$, and (c) $\mu = 0.4$. Ignore barreling.

6.2 Determine the temperature rise in the specimen for each case in Problem 6.1, assuming that the process is adiabatic and the temperature is uniform throughout the specimen.

6.3 Calculate the work done for each case in Problem 6.1.

6.4 To determine forgeability, a hot twist test is performed on a round bar 25 mm in diameter and 300 mm long. It undergoes 200 turns before it fractures. Calculate the shear strain at the outer surface of the bar at fracture.

6.5 Based on Eq. (6.20a), derive an expression for the average pressure in plane-strain compression with sticking friction.

6.6 For plane-strain compression, what is the value of μ when the forging load with sliding friction is equal to the load with sticking friction? Use average pressure formulas.

6.7 In plane-strain upsetting the frictional stress cannot be greater than the shear yield stress k of the material. Thus there may be a distance x in Fig. 6.4 where a transition occurs from sliding to sticking friction. Derive an expression for x in terms of a, h, and μ only.

6.8 Assume that the workpiece in Fig. 6.5 is being pushed to the right by a lateral force F while being compressed between flat dies. (a) Make a sketch of the die pressure distribution for the condition where F is not large enough to slide the workpiece to the right. (b) Make a similar sketch for an F that is large enough that the workpiece is sliding to the right while being compressed.

6.9 For the sticking example in Fig. 6.9, derive an expression for the lateral force F required to slide the workpiece to the right while it is being compressed between flat dies.

6.10 Two solid cylindrical specimens A and B made of a perfectly plastic material are being forged with friction and isothermally at room temperature to a reduction in height of 50 percent. Specimen A is 2 in. high and 1 in^2 in cross-sectional area, and specimen B is 1 in. high and 2 in^2 in cross-section. Will the work done be the same for the two specimens? Explain.

6.11 In Fig. 6.6 does the pressure distribution along the four edges of the workpiece depend on the particular yield criterion used? Explain.

6.12 Derive Eq. (6.17) for the pressure in upsetting a solid cylindrical specimen, using the same approach as that for plane-strain upsetting given in Section 6.2.2 but for a polar coordinate system.

6.13 Derive the average die pressure formula given by Eq. (6.15). *Hint*: Obtain the volume under the friction hill over the surface (by integration) and divide it by the cross-sectional area of the workpiece.

6.14 Take two solid cylindrical specimens of equal diameter but different heights and compress them (frictionless) to the same percent reduction in height. Show that the final diameters will be the same.

6.15 A rectangular workpiece has the following original dimensions: $a = 50$ mm, $h = 25$ mm, and $w = 20$ mm. The metal has a strength coefficient of 400 MPa and a strain hardening exponent of 0.5. It is being forged in plane strain with $\mu = 0.3$. Calculate the force required when the height is reduced by 20 percent. Do not use average pressure formulas.

Rolling

6.16 In Example 6.5 what is the velocity of the strip leaving the rolls?

6.17 With appropriate sketches explain the changes that occur in the roll pressure distribution if one of the rolls is idling; that is, power is shut off to that roll.

6.18 It is possible to determine μ in flat rolling without measuring torque or forces. By inspecting rolling equations, describe an experimental procedure to do so. You may measure any quantity other than torque or forces.

6.19 Derive a relationship between back tension σ_b and front tension σ_f in rolling so that when both are increased, the neutral point remains in the same position.

6.20 Take an element at the center of the deformation zone in flat rolling. Assuming that all the stresses acting on this element are principal stresses, place these stresses qualitatively and state whether they are tension or compression. Explain why. Is it possible for these three principal stresses to be equal in magnitude? Explain.

6.21 In Steckel rolling the rolls are idling and thus there is no net torque, assuming frictionless bearings. Where then is the energy coming from to supply the work of deformation in rolling? Explain with appropriate sketches and state the conditions that have to be satisfied.

6.22 In rolling a flat strip, the roll-separating force is reduced about twice as effectively by back tension than it is by front tension. With appropriate sketches explain this result. *Hint*: Note the position of the neutral point.

6.23 In rolling a strip the rolls will begin to slip if the back tension σ_b is too high. Derive an analytical expression for the magnitude of the back tension in order to make the powered rolls begin to slip. Use the same terminology as in the text.

6.24 Prove Eq. (6.44).

6.25 Derive an expression for the tension required in Steckel rolling of a flat sheet, without friction, for a workpiece whose true stress–true strain curve is given by $\sigma = a + b\varepsilon$.

6.26 Make a neat sketch of the pressure distribution in the roll gap in Steckel rolling without roll bearing friction. Superimpose on this diagram the pressure distribution when the roll bearings exhibit frictional resistance. Is there a difference in these two diagrams? If so, how and why?

6.27 Using Eqs. (6.27)–(6.29), derive an expression for the neutral angle ϕ_n in terms of h_f, H_n, and R.

6.28 In Fig. 6.38 asume that $L = 2L_2$. Is the roll force F for L twice or more than twice the force for L_2? Explain.

6.29 A flat rolling operation is being carried out where $h_0 = 0.125$ in., $h_f = 0.100$ in., $w_0 = 10$ in., $R = 10$ in., $\mu = 0.1$, and the average flow stress of the material is 40,000 psi. Estimate the roll force and the torque. Include roll flattening.

Extrusion

6.30 Derive an expression for the true strain rate $\dot{\varepsilon}$ in the deformation zone in extrusion of a round billet, in terms of the entering velocity v_0, die angle α, initial billet radius r_0, and distance x from entry.

6.31 Calculate the force required in direct extrusion of 1100-O aluminum from a diameter of 6 in. to 2 in. Assume that the redundant work is 40 percent of the ideal work of deformation and that the friction work is 25 percent of the total work of deformation.

6.32 Prove Eq. (6.54).

Drawing

6.33 Calculate the power required in Example 6.7 if the workpiece material is annealed 70-30 brass.

6.34 Using Eq. (6.64), make a plot similar to Fig. 6.69 for $K = 70$ MPa, $n = 0.25$, and $\mu = 0.04$.

6.35 Derive an analytical expression for the die pressure in wire drawing, without friction or redundant work, as a function of the diameter in the deformation zone.

6.36 A material with a true stress–true strain curve $\sigma = 5{,}000 + 20{,}000\varepsilon$ psi is being drawn into a wire. If the original diameter of the wire is 0.2 in., what is the minimum possible diameter at the die exit? Assume no redundant work and that the frictional work is 25 percent of the ideal work of deformation. *Hint*: The yield stress of the exiting wire is the point on the true stress–true strain curve that corresponds to the strain that the material has undergone.

6.37 In Fig. 6.72 assume that the longitudinal residual stress at the center is $-80{,}000$ psi. Using the distortion-energy criterion, calculate the minimum yield stress that this particular steel must have in order to sustain these levels of residual stresses.

6.38 Derive an expression for the die separating force in frictionless wire drawing of a perfectly plastic material. Use the same terminology as in the text.

6.39 A material with a true stress–true strain curve $\sigma = 10{,}000\varepsilon^{0.5}$ is used in wire drawing. Assuming that friction and redundant work comprise a total of 50 percent of the ideal work of deformation, calculate the maximum reduction in cross-sectional area per pass that is possible.

6.40 Derive an expression for the maximum reduction per pass for a material of $\sigma = K\varepsilon^n$, assuming that the friction and redundant work contribute a total of 20 percent to the ideal work of deformation.

6.41 Prove that the true strain rate $\dot{\varepsilon}$ in drawing or extrusion in plane strain with wedge-shaped dies is given by the expression

$$\dot{\varepsilon} = \frac{-2 \tan \alpha v_0 t_0}{(t_0 - 2x \tan \alpha)^2},$$

where α is the die angle, t_0 is the original thickness, and x is the distance from die entry. *Hint*: Note that $d\varepsilon = dA/A$.

7

Sheet-Metal Forming Processes

7.1 Introduction

In Chapter 6 we noted that workpieces could be characterized by the ratio of their surface area to their volume or thickness. Sheet forming, unlike bulk deformation processes, involves workpieces with a high ratio of surface area to thickness. A sheet thicker than 6 mm (1/4 in.) is generally called a *plate*. Although relatively thick plates, such as those used in boilers, bridges, ships, and nuclear power plants, may have smaller ratios of surface area to thickness, sheet forming usually involves relatively thin materials.

The products made by sheet-forming processes include a large variety of shapes and sizes, ranging from simple bends to double curvatures with shallow or deep recesses. Typical examples are metal desks, appliance bodies, hubcaps, aircraft panels, beverage cans, car bodies, and kitchen utensils. Thus sheet forming, which is also called *pressworking*, is among the most important of metalworking processes.

It dates back to as early as 5000 B.C. when household utensils, jewelry, and other objects were made by hammering and stamping metals such as gold, silver, and copper.

In this chapter, we first review the characteristic features of sheet metals and their formability. We then describe the major sheet-forming processes, products made, and equipment used.

Sheet metal is produced by a rolling process, as described in Section 6.3. If the sheet is thin, it is generally coiled; if thick, it is available as flat sheets or plates, which may have been decoiled and flattened.

Before a sheet-metal part is formed, a blank of suitable dimensions is first removed from a large sheet. Removal is usually done by a shearing process, as described in Section 7.3. However, there are several other methods for cutting sheet, particularly plates. The sheet or plate may be cut with a band saw, which is a chip removal process. Flame cutting is another common method, particularly for thick steel plates, and is widely used in shipbuilding and heavy structural components. Laser cutting has also become an important process and is used with computer-controlled equipment to cut a variety of shapes consistently. In friction sawing, a disk or blade rubs against the sheet or plate at high surface speeds. Because shearing is the most commonly used method, we focus on it as a method of cutting sheet metal.

7.2 Sheet-Metal Characteristics

Sheet metals are generally characterized by a high ratio of surface area to thickness. Forming of sheet metals generally is carried out by tensile forces in the plane of the sheet; otherwise, the application of compressive forces could lead to buckling, folding, and wrinkling of the sheet.

In bulk deformation processes such as forging, rolling, extrusion, and wire drawing (Chapter 6), the thickness or the lateral dimensions of the workpiece are intentionally changed. However, in most sheet-forming processes any thickness change is caused by stretching of the sheet under tensile stresses (Poisson's ratio). Thickness decreases generally should be avoided, as they could lead to necking and failure.

Because the basic mechanisms of all sheet-forming processes are stretching and bending, certain factors significantly influence the overall operation. The major ones are: elongation, yield-point elongation, anisotropy, grain size, residual stresses, springback, and wrinkling.

7.2.1 Elongation

Although sheet-forming operations rarely involve simple uniaxial stretching, the observations made in regard to simple tensile testing can be useful in understanding the behavior of sheet metals. Recall from Fig. 2.2 that a specimen subjected to tension first undergoes uniform elongation (which corresponds to the UTS, after which necking begins). This elongation is then followed by additional nonuniform elongation until fracture takes place (postuniform elongation). The material is being stretched in sheet forming, so high uniform elongation is desirable for good formability.

In Section 2.2.4 we showed that, for a material having a true stress–true strain curve that can be represented by the equation

$$\sigma = K\varepsilon^n, \tag{7.1}$$

the strain at which necking begins (*instability*) is given by

$$\varepsilon = n. \tag{7.2}$$

Hence the true uniform strain in a simple stretching operation (uniaxial tension) is numerically equal to the strain-hardening exponent n. A large n indicates large uniform elongation and thus it is desirable for sheet forming.

Necking of a sheet specimen generally takes place at an angle ϕ to the direction of tension, as shown in Fig. 7.1(a) (*localized necking*). For an isotropic sheet

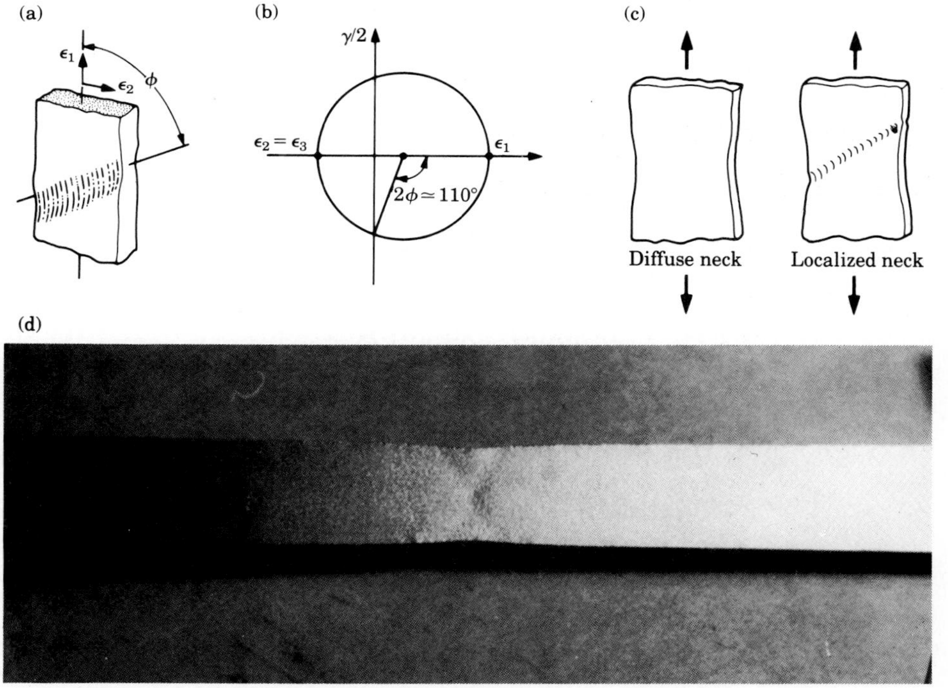

FIGURE 7.1
(a) Localized necking in a sheet specimen under tension. (b) Determination of the angle of neck from Mohr's circle for strain. (c) Schematic illustrations for diffuse and localized necking, respectively. (d) Localized necking in an aluminum strip stretched in tension. Note the double neck.

specimen in simple tension, the Mohr circle for this situation is constructed as follows (Fig. 7.1b). The strain ε_1 is the longitudinal strain, and ε_2 and ε_3 are the lateral strains. Poisson's ratio in the plastic range is 0.5, so the lateral strains have the value $-\varepsilon_1/2$. The narrow neck band in Fig. 7.1(a) is in *plane strain* along its length because it is constrained by the material above and below the neck.

The angle ϕ can now be determined from the Mohr circle by a rotation (either clockwise or counterclockwise) of 2ϕ from the ε_1 position (Fig. 7.1b). This angle is about 110°; thus the angle ϕ is about 55°. Note that although the length of the neck band remains essentially constant during the test, its thickness decreases (because of volume constancy), and the specimen eventually fractures. The angle ϕ will be different for materials that are anisotropic in the plane of the sheet.

Whether necking is *localized* or *diffuse* (Fig. 7.1c) depends on the strain rate sensitivity m of the material, as given by the equation

$$\sigma = C\dot{\varepsilon}^m. \tag{7.3}$$

As described in Section 2.2.7, the higher the value of m, the more diffuse the neck becomes. An example of localized necking on an aluminum strip in tension is shown in Fig. 7.1(d). Note the double localized neck, that is, ϕ can be in the clockwise and counterclockwise position in Fig. 7.1(a).

In addition to uniform elongation, the *total* elongation [such as 50 mm (2 in.)] gage length of a tension-test specimen is also a significant factor in formability of sheet metals. Note that total elongation is the sum of uniform elongation and postuniform elongation. Uniform elongation is governed by the strain-hardening exponent n, whereas postuniform elongation is governed by the strain-rate–sensitivity index m. The higher the m value, the more diffuse the neck and hence the greater the postuniform elongation before fracture become. Thus the total elongation of the material increases with increasing values of both n and m.

7.2.2 Yield-point elongation

Low-carbon steels exhibit a behavior called *yield-point elongation*, with upper and lower yield points, as shown in Fig. 7.2(a). This behavior indicates that, after the material yields, it stretches farther in certain regions in the specimen with no increase in the lower yield point, while other regions have not yet yielded. When the overall elongation reaches the yield-point elongation, the entire specimen has been deformed uniformly. The magnitude of the yield-point elongation depends on the strain rate (with higher rates, the elongation generally increases) and the grain size of the sheet metal. As grain size decreases, yield-point elongation increases. Yield-point elongation is usually on the order of a few percent.

This behavior of low-carbon steels produces *Lueder's bands* (also called *stretcher strain marks* or *worms*) on the sheet, as shown in Fig. 7.2(b). These are elongated depressions on the surface of the sheet and can be objectionable in the final product because of surface appearance. They can also cause difficulty in subsequent coating and painting operations. Stretcher strain marks on the curved

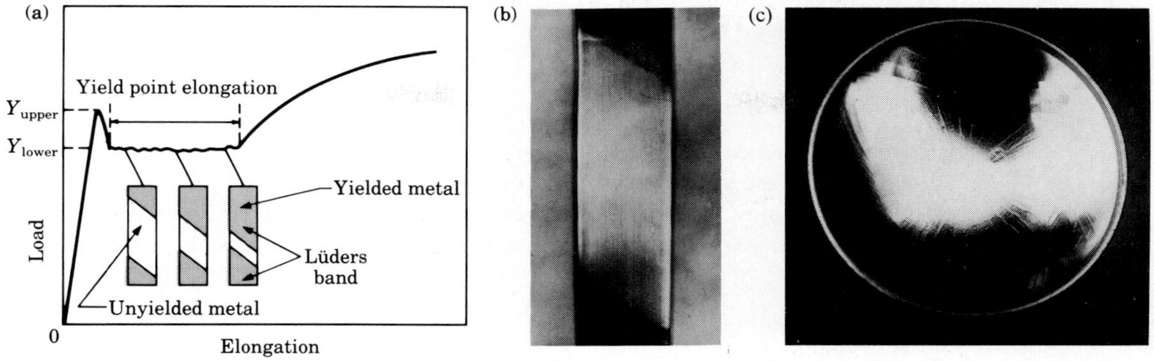

FIGURE 7.2 (a) Yield point elongation and Lueder's bands in tension testing. (b) Lueder's bands in annealed low-carbon steel sheet. (c) Stretcher strains at the bottom of a steel can for household products. *Source:* (b) Courtesy of R. B. Liss, Caterpillar Tractor Co.

bottom of steel cans for common household products are shown in Fig. 7.2(c); aluminum cans do not exhibit this behavior.

The usual method of avoiding this problem is to eliminate or to reduce yield-point elongation by reducing the thickness of the sheet 0.5 to 1.5 percent by cold rolling, known as *temper rolling* or *skin rolling*. Because of strain aging, yield-point elongation reappears after a few days at room temperature or after a few hours at higher temperatures. Thus the sheet metal should be formed within a certain period of time (from one to three weeks for rimmed steel, Section 5.7) to avoid reappearance of stretcher strains.

7.2.3 Anisotropy

Another important factor influencing sheet-metal forming is *anisotropy*, or *directionality*, of the sheet metal. Anisotropy is acquired during the thermo-mechanical processing history of the sheet. Recall from Section 3.5 that there are two types of anisotropy: *crystallographic anisotropy* (from preferred grain orientation) and *mechanical fibering* (from alignment of impurities, inclusions, voids, and the like, throughout the thickness of the sheet during processing).

Anisotropy may be present not only in the plane of the sheet, but also in its thickness direction. The former is called *planar* anisotropy and the latter *normal* or *plastic* anisotropy.

7.2.4 Grain size

Grain size of sheet metal is important for two reasons: first, because of its effect on the mechanical properties of the material, and second, because of its effect on the surface appearance of the formed part. The coarser the grain, the rougher the surface appears (*orange peel*). An ASTM grain size of No. 7 or finer is preferred for general sheet-metal forming.

7.2.5 Residual stresses

Residual stresses can be present in sheet metal parts because of the nonuniform deformation of the sheet during forming. When disturbed, as by removing a portion of it, the part may distort. Tensile residual stresses on surfaces can also lead to *stress–corrosion cracking* of sheet-metal parts (Fig. 7.3) unless they are properly stress relieved.

7.2.6 Springback

Sheet-metal parts, because they are generally thin and are subjected to relatively small strains, are likely to experience considerable *springback*. This is particularly significant in bending and other sheet-forming operations where the bend radius-to-thickness ratio is high, such as in automotive panels.

7.2.7 Wrinkling

Although in sheet forming the metal is generally subjected to tensile stresses, the method of forming may be such that compressive stresses are developed in the plane of the sheet. An example is the wrinkling of the flange in deep drawing because of circumferential *compression*. Other terms used to describe similar phenomena are *buckling*, *folding*, and *collapsing*. The tendency for wrinkling increases with the

FIGURE 7.3
Stress–corrosion cracking in a deep-drawn brass part for a light fixture. The cracks developed over a period of time. Brass and austenitic (300 series) stainless steels are among metals that are susceptible to stress–corrosion cracking.

unsupported or unconstrained length or surface area of the sheet metal, decreasing thickness, and nonuniformity of the thickness of the sheet. Such defects may also be initiated by lubricants that are trapped or distributed nonuniformly between the surface of the sheet and the dies.

7.2.8 Coated sheet metal

Sheet metals, especially steel, precoated with a variety of organic coatings, films, and laminates are available. Coated sheet metals are used primarily for appearance and eye appeal and also offer corrosion resistance. Coatings are applied to the coil stock on continuous lines. Coating thicknesses generally range from 0.0025 to 0.2 mm (0.0001 to 0.008 in.) on flat surfaces. Coatings are available with a wide range of properties, such as flexibility, durability, hardness, resistance to abrasion and chemicals, color, texture, and gloss. Coated sheet metal is subsequently formed into products such as TV cabinets, appliance housings, paneling, shelving, residential siding, and metal furniture.

7.3 Shearing

The shearing process involves cutting sheet metal by subjecting it to shear stresses, usually between a *punch* and a *die* much like a paper punch (Fig. 7.4). The punch and die may be any shape; for example, they may be circular or straight blades similar to a pair of scissors. The major variables in the shearing process are

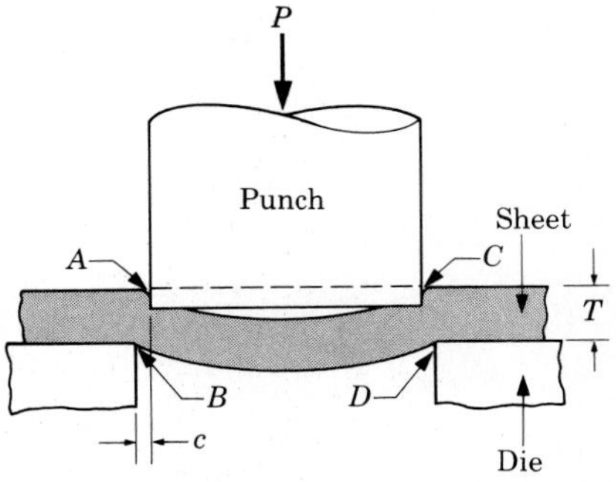

FIGURE 7.4 Schematic illustration of the shearing process with a punch and die. This is a common method of producing various openings in sheet metals.

the punch force P, the speed of the punch, lubrication, surface condition and materials of the punch and die, their corner radii, and the clearance between punch and die.

The overall features of typical sheared edges for the two sheared surfaces (the slug and the sheet) are shown in Fig. 7.5. Note that the edges are neither smooth nor perpendicular to the plane of the sheet. The *clearance* c (Fig. 7.4) is the major factor determining the shape and quality of the sheared edge. As shown in Fig. 7.6(a), as clearance increases, the edges become rougher and the zone of deformation becomes larger. The material is pulled into the clearance area, and the edges of the sheared zone become more and more rounded. In fact, if the clearance is too large, the sheet

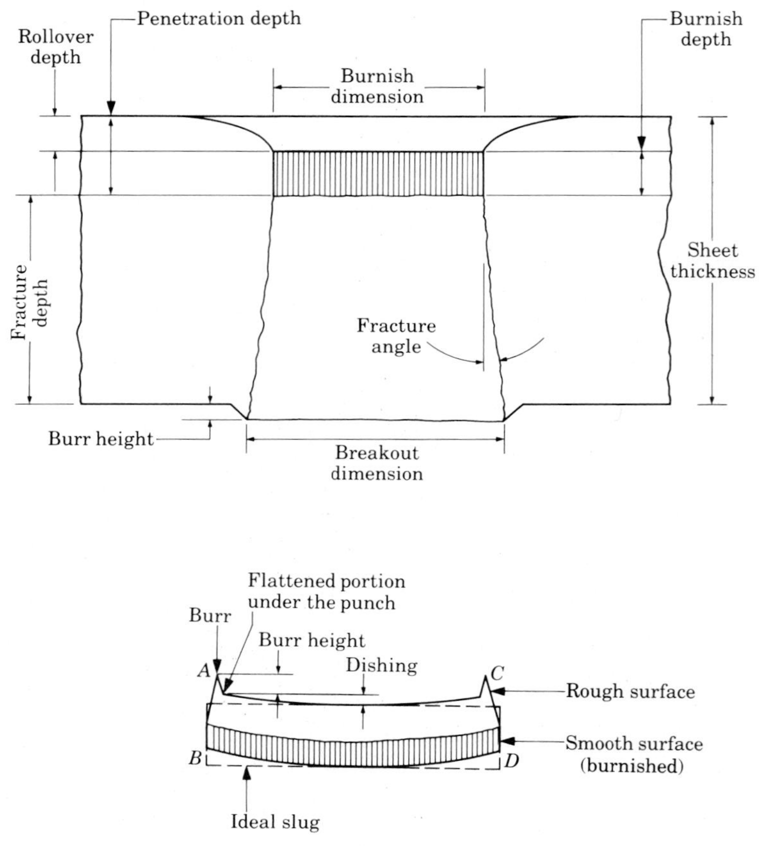

FIGURE 7.5
Characteristic features of a punched hole and the punched slug.

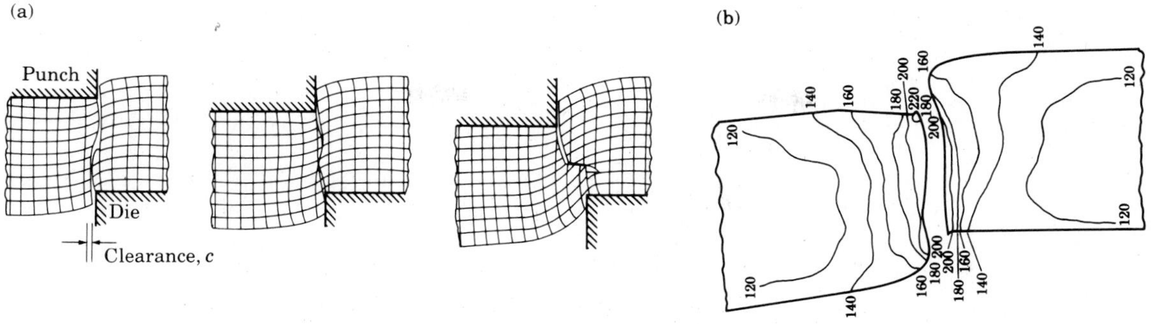

FIGURE 7.6
(a) Effect of clearance *c* between the punch and die on the deformation zone in shearing. As clearance increases, the material tends to be pulled into the die, rather than being sheared. In practice, clearances usually range between 2 and 10 percent of the thickness of the sheet. (b) Microhardness (HV) contours for a 6.4-mm (0.25-in.) thick AISI 1020 hot-rolled steel in the sheared region. *Source:* After H. P. Weaver and K. J. Weinmann.

metal is bent and subjected to tensile stresses, instead of undergoing a shearing deformation. As Fig. 7.6(b) shows, sheared edges can undergo severe cold working because of the high strains involved. This result, in turn, can adversely affect the formability of the sheet during subsequent operations.

Note the formation of a *burr* in Fig. 7.5. Burr height increases with increasing clearance and increasing ductility of the metal. Tools with dull edges are also a major factor in burr formation. The height, shape, and size of the burr can significantly affect many subsequent forming operations, such as in flanging, where a burr could lead to cracks.

Observing the shearing mechanism reveals that shearing usually starts with the formation of cracks on both the top and bottom edges of the workpiece (*A* and *B* in Fig. 7.4). These cracks eventually meet, and complete separation takes place. The rough fracture surface in the slug in Fig. 7.5 is caused by these cracks. The smooth, shiny, and burnished surface is from the contact and rubbing of the sheared edge against the walls of the die. In the slug shown, the burnished surface is in the lower region, because this is the section that rubs against the die wall. On the other hand, inspection of the sheared surface on the sheet itself reveals that the burnished surface is on the upper region in the sheared edge and results from rubbing against the punch.

The ratio of the burnished to rough areas on the sheared edge increases with increasing ductility of the sheet metal; it decreases with increasing material thickness and clearance. The punch travel required to complete the shearing process depends on the maximum shear strain that the material can undergo before fracture. Thus a brittle metal, or one that is highly cold worked, requires little travel of the punch to complete shearing.

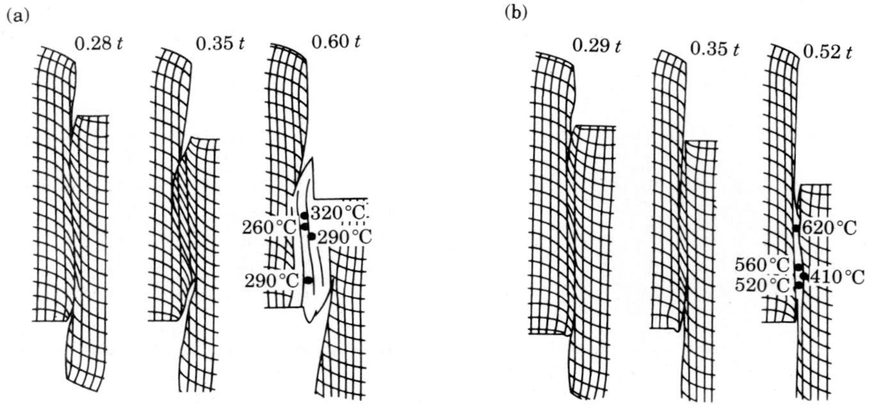

FIGURE 7.7

Deformation and temperature rise in the shearing zone. The temperature was measured by thermocouples. Punching at (a) slow speed and (b) high speed. Note that the deformation is confined to a narrow zone in high-speed shearing and that the temperature is higher than in slow-speed shearing. *Source:* After N. Yanagihara, H. Saito, and T. Nakagawa. Numbers above the figures indicate punch penetration.

Note from Fig. 7.7 that the deformation zone is subjected to high shear strains. The width of this zone depends on the rate of shearing, that is, punch speed. With increasing punch speed, the heat generated by plastic deformation is confined to a smaller zone (approaching a narrow adiabatic zone), and the sheared surface is smoother.

The punch force P is basically the product of the shear strength of the sheet metal and the cross-sectional area being sheared. However, friction between the punch and the workpiece can increase this force substantially. Because the sheared zone is subjected to cracks, plastic deformation, and friction, the punch force–stroke curves can have various shapes. One typical curve for a ductile material is shown in Fig. 7.8. The area under this curve is the *total work* done in shearing.

An approximate empirical formula for estimating the *maximum punch force* P is

$$P = 0.7(\text{UTS})(t)(L), \tag{7.4}$$

FIGURE 7.8

Typical punch-penetration curve in shearing. The area under the curve is the work done in shearing. The shape of the curve depends on process parameters and material properties.

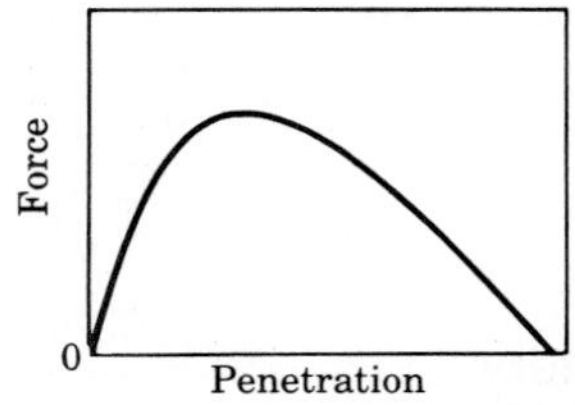

where UTS is the *ultimate tensile strength* of the sheet metal, t is its thickness, and L is the total length of the sheared edge. Thus for a round hole of diameter D, we have $L = \pi D$.

• Example 7.1: Calculation of punch force.

Estimate the force required in punching a 1-in. (25-mm) diameter hole through a $\frac{1}{8}$-in. (3.2-mm) thick annealed titanium-alloy Ti-6Al-4V sheet at room temperature.

SOLUTION. The force is estimated from Eq. (7.4), and the UTS for this alloy is found from Table 3.15 to be 1000 MPa, or 140,000 psi. Thus

$$F = 0.7(\tfrac{1}{8})(\pi)(1)(140{,}000) = 38{,}500 \text{ lb}$$
$$= 19.25 \text{ tons} = 0.17 \text{ MN}.$$

•

7.3.1 Shearing operations

Various operations based on the shearing process are performed. We first define two terms. In *punching*, the sheared slug is discarded (Fig. 7.9a). In *blanking*, the slug is the part and the rest is scrap.

Die cutting. *Die cutting* consists of the following operations (Fig. 7.9b): (a) *perforating*, or punching a number of holes in a sheet; (b) *parting*, or shearing the sheet into two or more pieces; (c) *notching*, or removing pieces or various shapes from the edges; and (d) *lancing*, or leaving a tab without removing any material.

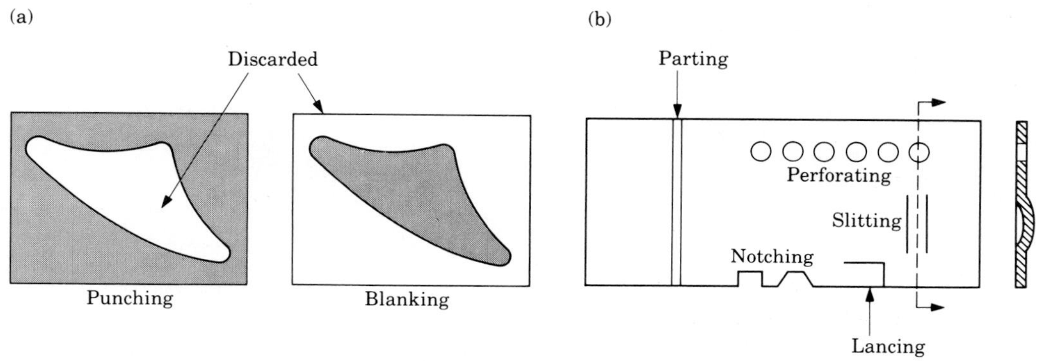

FIGURE 7.9
(a) Punching (piercing) and blanking. (b) Examples of various shearing operations on sheet metal.

Parts produced by these processes have various uses, particularly in assembly with other components.

Fine blanking. Very smooth and square edges can be produced by *fine blanking* (Fig. 7.10a). One basic die design is shown in Fig. 7.10(b). A V-shaped stinger, or impingement, locks the sheet metal tightly in place and prevents the type of distortion of the material shown in Fig. 7.6. The fine-blanking process involves clearances on the order of 1 percent of the sheet thickness, which may range from 0.13 mm to 13 mm (0.005 in. to 0.5 in.). The operation is usually carried out on triple-action hydraulic presses where the movements of the punch, pressure pad, and die are controlled individually. Fine blanking usually involves parts having holes, which are punched simultaneously with blanking.

Slitting. Shearing operations can be carried out with a pair of circular blades (*slitting*), similar to those in a can opener (Fig. 7.11). The blades follow either a straight line or a circular or curved path. A slit edge normally has a burr, which may be removed by rolling. There are two types of slitting equipment. In the *driven* type, the blades are powered. In the *pull-through* type, the strip is pulled through idling blades. Slitting operations, if not performed properly, may cause various distortions of the sheared part.

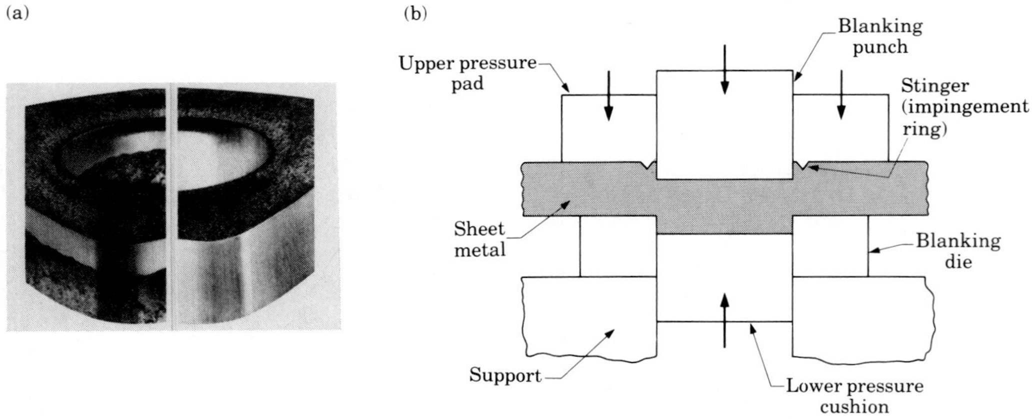

FIGURE 7.10
(a) Comparison of sheared edges by conventional (left) and fine-blanking (right) techniques. (b) Schematic illustration of setup for fine blanking. *Source:* Feintool U.S. Operations.

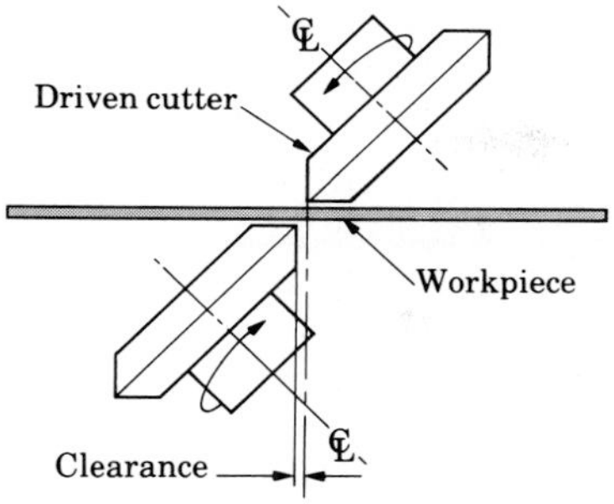

FIGURE 7.11
Slitting with rotary knives. This process is similar to opening cans.

Steel rules. Soft metals, paper, leather, and rubber can be blanked with *steel-rule dies*. Such a die consists of a thin strip of hardened steel bent to the shape to be sheared (similar to a cookie cutter) and held on its edge on a flat wooden base. The die is pressed against the sheet, which rests on a flat surface, and cuts the sheet to the shape of the steel rule.

Nibbling. In *nibbling*, a machine called a nibbler moves a straight punch up and down rapidly into a die. The sheet metal is fed through the gap, and a number of overlapping holes are made. This operation is similar to making a large elongated hole by successively punching holes with a paper punch. Sheets can be cut along any desired path by manual control. The process is economical for small production runs since no special dies are required.

Scrap in shearing. The amount of scrap produced, or *trim loss*, in shearing operations can be significant. On large stampings it can be as high as 30 percent. A significant factor in manufacturing cost, scrap can be reduced significantly by proper arrangement of the shapes on the sheet to be cut. Computer-aided design techniques have been developed to minimize scrap in shearing operations.

7.3.2 Shearing dies

Because the formability of the sheared part can be influenced by the quality of its sheared edges, clearance control is important. In practice, clearances usually range between 2 percent and 8 percent of the sheet's thickness. The thicker the sheet is, the larger the clearance (as much as 10 percent). However, the smaller the clearance, the

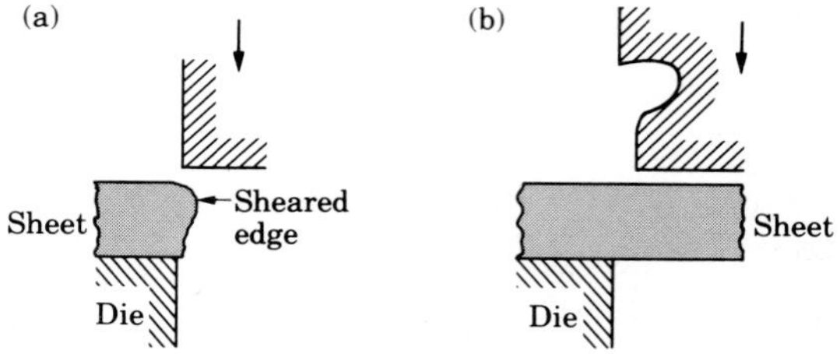

FIGURE 7.12
Schematic illustrations of shaving on a sheared edge. (a) Shaving a sheared edge. (b) Shearing and shaving, combined in one stroke.

better is the quality of the edge. In a process called *shaving* (Fig. 7.12), the extra material from a rough sheared edge is trimmed by cutting.

Punch and die shapes. Note in Fig. 7.4 that the surfaces of the punch and die are flat. Thus the punch force builds up rapidly during shearing because the entire thickness is sheared at the same time. The area being sheared at any moment can be controlled by beveling the punch and die surfaces (Fig. 7.13). The geometry is similar to that of a paper punch, which you can see by looking closely at the tip. This geometry is particularly suitable for shearing thick blanks because it reduces the force at the beginning of the stroke. It also reduces the operation's noise level.

Compound dies. Several operations on the same strip may be performed in one stroke with a *compound die* in one station. These operations are usually limited to relatively simple shearing because they are somewhat slow, and the dies are more expensive than those for individual shearing operations.

Progressive dies. Parts requiring multiple operations, such as punching, blanking, and notching, are made at high production rates in *progressive dies*. The sheet metal is fed through a coil strip, and a different operation is performed at the

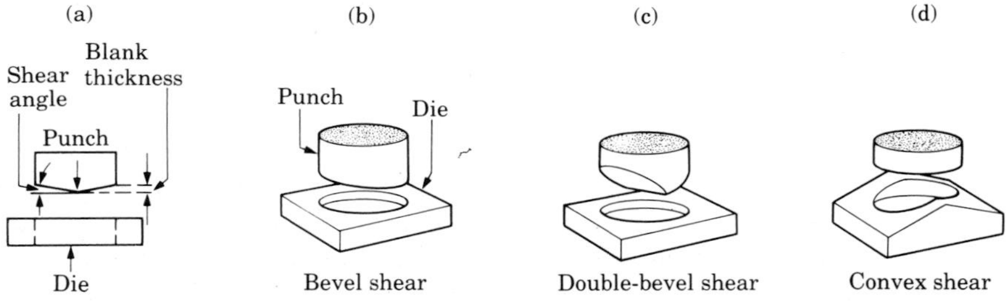

FIGURE 7.13
Examples of the use of shear angles on punches and dies.

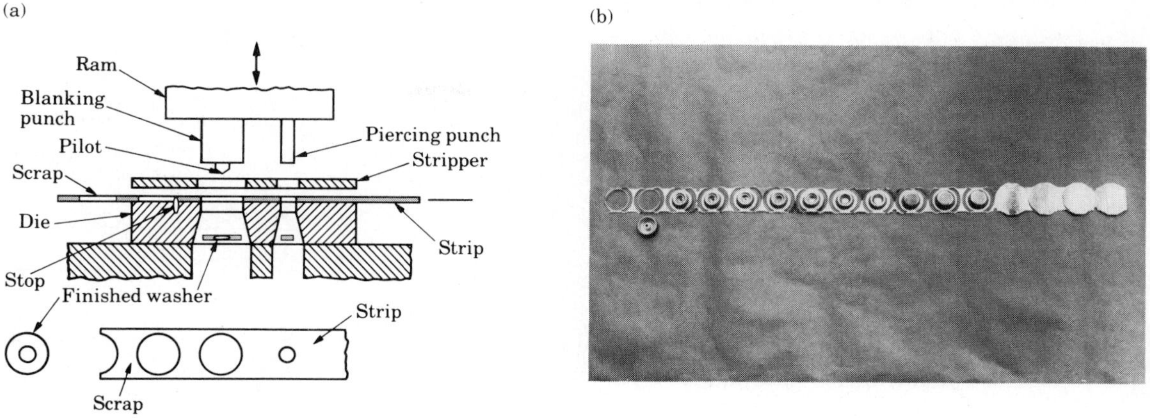

FIGURE 7.14
(a) Schematic illustration of making a washer in a progressive die. (b) Forming of the top piece of an aerosol spray can in a progressive die. Note that the part is attached to the strip until the last operation is completed.

same station with each stroke of a series of punches (Fig. 7.14a). An example of a part made in progressive dies is shown in Fig. 7.14(b).

Transfer dies. In a *transfer die* setup, the sheet metal undergoes different operations at different stations, which are arranged along a straight line or a circular path. After each operation, the part is transferred to the next station for additional operations.

Tool and die materials. Tool and die materials for shearing are generally tool steels and, for high production rates, carbides (see Table 3.6). Lubrication is important for reducing tool and die wear and improving edge quality.

7.4 Bending of Sheet and Plate

One of the most common metalworking operations is *bending*. This process is used not only to form parts such as flanges, curls, seams, and corrugations, but also to impart stiffness to the part by increasing its moment of inertia (note, for instance, that a long strip of metal is much less rigid when flat than when it is formed into the shape of a gutter). In this section, we describe the mechanics and practices of bending operations.

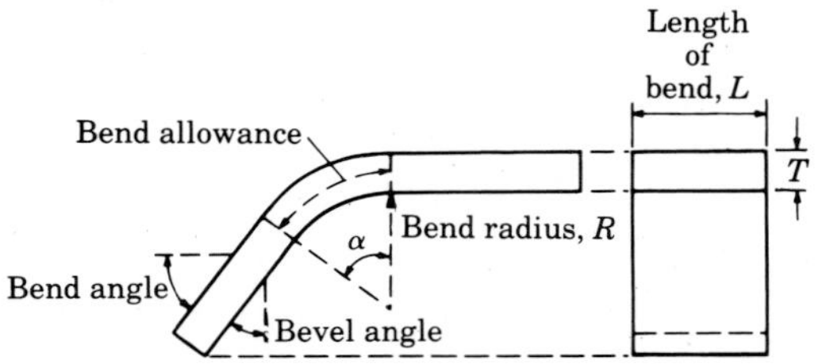

FIGURE 7.15
Bending terminology. The bend radius is measured to the inner surface of the bend. Note that the length of the bend is the width of the sheet. Also note that the bend angle and the bend radius (sharpness of the bend) are two different variables.

The terminology used in bending is shown in Fig. 7.15. In bending, the outer fibers of the material are in tension and the inner fibers are in compression. Theoretically, the strains at the outer and inner fibers are equal in magnitude and are given by the equation

$$e_o = e_i = \frac{1}{(2R/T) + 1}, \tag{7.5}$$

where R is the bend radius and T is the sheet thickness.

Experimental evidence indicates that, although Eq. (7.5) holds reasonably well for the inner fiber strain e_i, actual values of e_o are considerably higher than e_i. The reason is the shifting of the neutral axis toward the inner surface. The length of bend (Fig. 7.15) is smaller in the outer region than in the inner region (as can easily be observed when you bend a rectangular eraser). The difference between the outer and inner strains increases with decreasing R/T ratio, that is, sharper bends.

7.4.1 Minimum bend radius

It is apparent from Eq. (7.5) that, as the R/T ratio decreases, the tensile strain at the outer fiber increases and that the material may crack after a certain strain is reached. The radius R at which a crack appears on the outer surface of the bend is called the *minimum bend radius*. The minimum radius to which a part can be bent safely is normally expressed in terms of its thickness, such as $2T$, $3T$, $4T$, and so on. Thus a $3T$ bend radius indicates that the smallest radius to which the sheet can be bent without cracking is 3 times its thickness. The minimum bend radius for various materials has been determined experimentally and is available in various handbooks. Some typical results are given in Table 7.1.

Studies have also been conducted to establish a relationship between the minimum R/T ratio and a given mechanical property of the material. One such analysis is based on the following assumptions.

a) The true strain at cracking on the outer fiber in bending is equal to the true strain at fracture ε_f of that material in a simple tension test.
b) The material is homogeneous and isotropic.
c) The sheet is bent in a state of plane stress, that is, its L/T ratio is small.

TABLE 7.1
MINIMUM BEND RADIUS FOR VARIOUS MATERIALS AT ROOM TEMPERATURE

	CONDITION	
MATERIAL	*SOFT*	*HARD*
Aluminum alloys	0	6T
Beryllium copper	0	4T
Brass, low-leaded	0	2T
Magnesium	5T	13T
Steels		
austenitic stainless	0.5T	6T
low-carbon, low-alloy and HSLA	0.5T	4T
Titanium	0.7T	3T
Titanium alloys	2.6T	4T

The true strain at fracture in tension is

$$\varepsilon_f = \ln\left(\frac{A_0}{A_f}\right) = \ln\left(\frac{100}{100 - r}\right),$$

where r is the percent reduction of area in a tension test. From Section 2.2.2,

$$\varepsilon_0 = \ln(1 + e_o) = \ln\left(1 + \frac{1}{(2R/T) + 1}\right) = \ln\left(\frac{R + T}{R + (T/2)}\right).$$

Equating the two expressions and simplifying, we obtain

$$\text{Minimum } \frac{R}{T} = \frac{50}{r} - 1. \tag{7.6}$$

For the experimental data shown in Fig. 7.16, the curve that best fits the data is

$$\text{Minimum } \frac{R}{T} = \frac{60}{r} - 1. \tag{7.7}$$

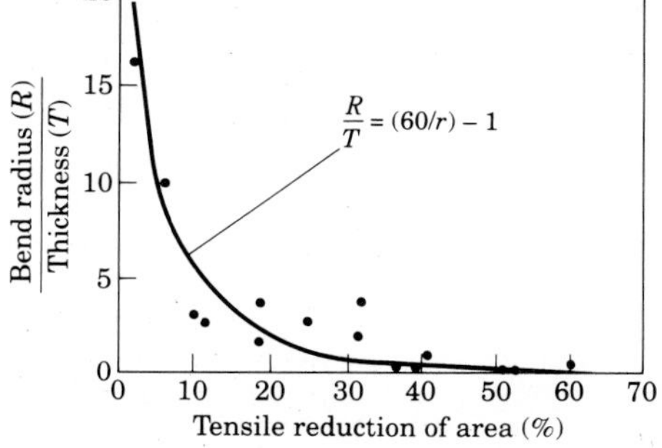

FIGURE 7.16 Relationship between the ratio of bend radius-to-sheet thickness and tensile reduction of area for various materials. Note that sheet metal with a reduction of area of about 50 percent can be bent and flattened over itself without cracking. *Source:* After J. Datsko and C. T. Yang, *J. Eng. Ind.*, vol. 82, 1960, pp. 309–314.

The R/T ratio approaches zero (complete bendability, that is, the material can be folded over itself) at a tensile reduction of area of 50 percent. Interestingly, this is the same value obtained in spinnability of metals (described in Section 7.8), where a material with 50 percent reduction of area is found to be completely spinnable. A curve similar to that in Fig. 7.16 also is obtained when the data are plotted against the total elongation (percent) of the material.

The bendability of a metal may be increased by increasing its tensile reduction of area, either by *heating* or by the application of *hydrostatic pressure*. Other techniques may also be employed to increase the compressive environment in bending, such as applying compressive forces in the plane of the sheet during bending to minimize tensile stresses in the outer fibers of the bend area.

As the length of the bend increases, the state of stress at the outer fibers changes from uniaxial stress to a biaxial stress. The reason is that L tends to become smaller due to stretching of the outer fibers, but it is constrained by the material around the bend area. Biaxial stretching tends to reduce ductility, that is, strain to fracture. Thus as the length L increases, the minimum bend radius increases (Fig. 7.17). However, at a length of about $10T$, the minimum bend radius increases no further, and a plane-strain condition is fully developed.

As the R/T ratio decreases, narrow sheets (smaller length of bend) crack at the edges and wider sheets crack at the center, where the biaxial stress is the highest. Bendability also depends on the edge condition of the sheet being bent. Because rough edges are points of stress concentration, bendability decreases as edge roughness increases.

Another significant factor in edge cracking is the amount and shape of inclusions in the sheet metal and the amount of cold working that the edges undergo

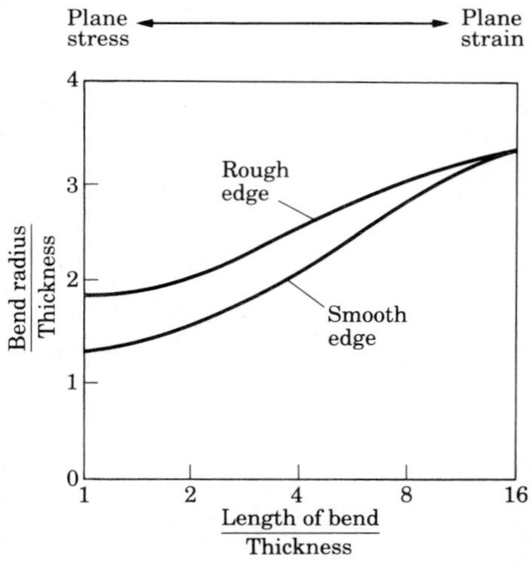

FIGURE 7.17 The effect of length of bend and edge condition on bend radius–thickness ratio of 7075-T aluminum. *Source:* After G. Sachs and G. Espey.

during shearing (see Fig. 7.6b). Inclusions in the form of *stringers* are more detrimental than globular-shaped inclusions. Removal of the cold-worked regions (by machining or heat treating) greatly improves resistance to edge cracking. Thus anisotropy of the sheet is also important in bendability. As depicted in Fig. 7.18, cold rolling of sheets produces anisotropy because of alignment of impurities, inclusions, and voids (mechanical fibering). Thus transverse ductility is reduced, as shown in Fig. 7.18(c). In bending such a sheet, caution should be used in cutting the blank in the proper direction from the rolled sheet, although this may not always be possible in practice.

7.4.2 Springback

Because all materials have a finite modulus of elasticity, plastic deformation is followed by *elastic recovery* upon removal of the load. In bending, this recovery is

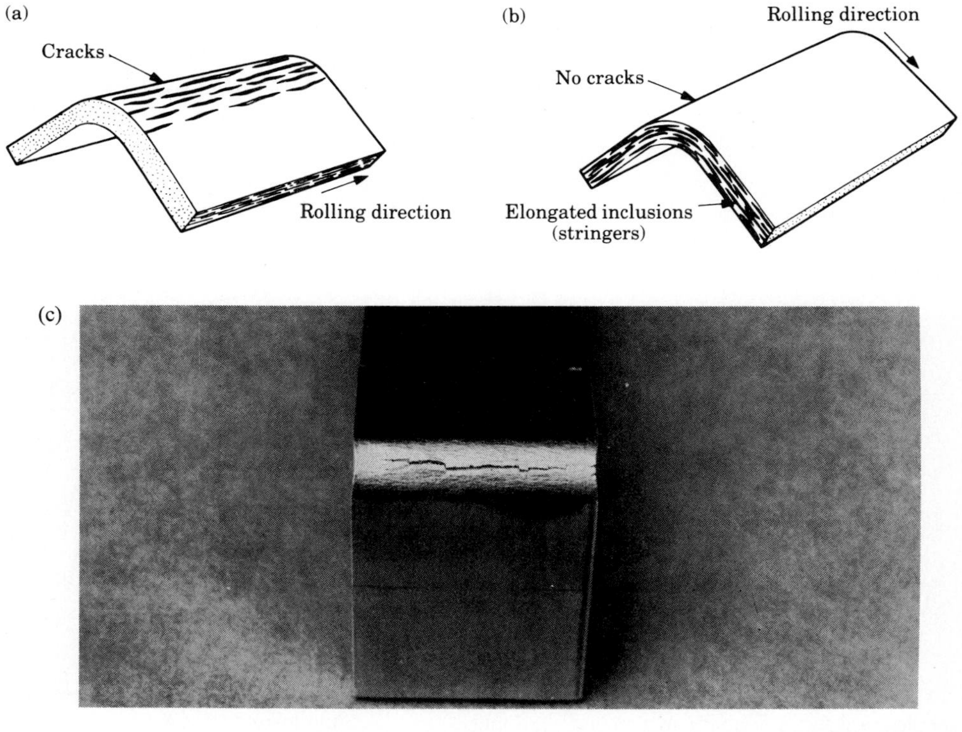

FIGURE 7.18
(a) and (b) The effect of elongated inclusions (stringers) on cracking as a function of the direction of bending with respect to the original rolling direction of the sheet. This example shows the importance of the direction of cutting from large sheets workpieces that are subsequently bent to make a product. (c) Cracks on the outer radius of an aluminum strip bent to an angle of 90°.

known as *springback*. As shown in Fig. 7.19, the final bend angle after springback is smaller and the final bend radius is larger than before. (This can be easily observed by bending a piece of wire.) Springback occurs not only in flat sheets or plate, but also in bars, rod, and wire of any cross-section.

A quantity characterizing springback is the *springback factor* K_s, which is determined as follows. As the bend allowance (Fig. 7.15) is the same before and after bending, the relationship obtained for pure bending is

$$\text{Bend allowance} = \left(R_i + \frac{T}{2}\right)\alpha_i = \left(R_f + \frac{T}{2}\right)\alpha_f.$$

From this relationship, K_s is defined as

$$K_s = \frac{\alpha_f}{\alpha_i} = \frac{(2R_i/T) + 1}{(2R_f/T) + 1}, \tag{7.8}$$

where R_i and R_f are the initial and final bend radii, respectively.

We note from this expression that the springback factor K_s depends only on the R/T ratio. A springback factor of $K_s = 1$ indicates no springback, and $K_s = 0$ indicates complete elastic recovery (Fig. 7.20), as in the leaf spring of an automobile.

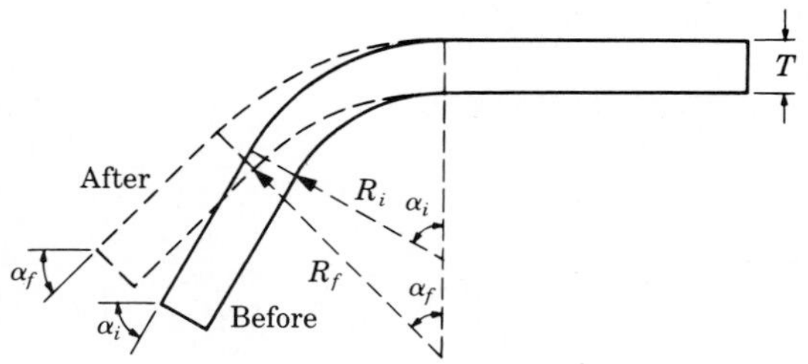

FIGURE 7.19 Terminology for springback in bending. Springback is caused by the elastic recovery of the material upon unloading. In this example, the material tends to recover toward its originally flat shape. However, there are situations where the material bends farther upon unloading (negative springback), as shown in Fig. 7.21.

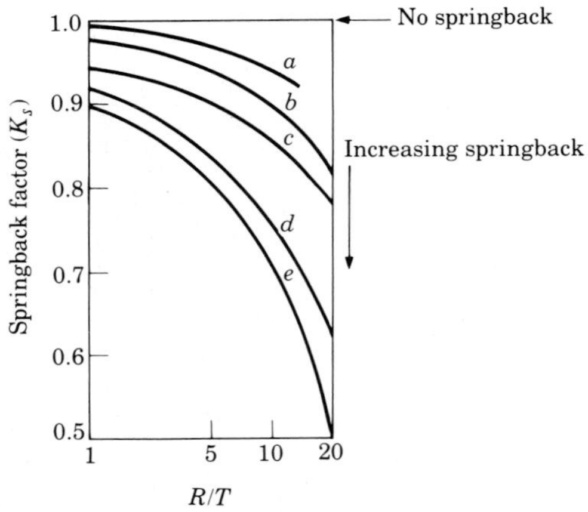

FIGURE 7.20 Springback factor K_s for various materials: (a) 2024-0 and 7075-0 aluminum; (b) austenitic stainless steels; (c) 2024-T aluminum; (d) 1/4 hard austenitic stainless steels; (e) 1/2 hard to full-hard austenitic stainless steels. *Source:* After G. Sachs, *Principles and Methods of Sheet-Metal Fabricating.* Reinhold, 1951, p. 100.

Recall that the amount of elastic recovery depends on the stress level and the modulus of elasticity E of the material. Thus elastic recovery increases with the stress level and with decreasing elastic modulus. Based on this observation, an approximate formula has been developed to estimate springback:

$$\frac{R_i}{R_f} = 4\left(\frac{R_i Y}{ET}\right)^3 - 3\left(\frac{R_i Y}{ET}\right) + 1, \tag{7.9}$$

where Y is the yield stress of the material at 0.2 percent offset.

Negative springback. The springback observed in Fig. 7.19 is called positive springback. Under certain conditions negative springback is also possible. In other words, the bend angle becomes larger after the bend has been completed and the load is removed. This phenomenon is generally associated with V-die bending (Fig. 7.21).

The development of negative springback can be explained by observing the sequence of deformation in Fig. 7.21. If we remove the piece at stage (b), it will undergo positive springback. At stage (c) the ends of the part are touching the male punch. Note that between stages (c) and (d) the part is actually being bent in the direction opposite to that between (a) and (b). Note also the lack of conformity of the punch radius and the inner radius of the part in (b) and (c). In (d), however, the radii are the same.

Upon unloading, the part in stage (d) will spring back inwardly because it is being *unbent* from stage (c) both at the tip of the punch and in the arms of the part. This inward (negative) springback can be greater than the positive springback that would result upon unloading from stage (c) because of the large strains that the material is undergoing in the bend area in stage (b).

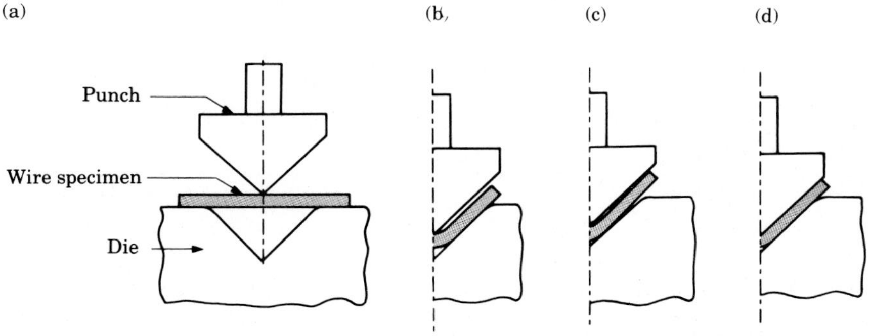

FIGURE 7.21

Schematic illustration of the stages in bending round wire in a V-die. This type of bending can lead to negative springback, which does not occur in air bending (shown in Fig. 7.26a). *Source:* After K. S. Turke and S. Kalpakjian, *Proc. NAMRC III*, 1975, pp. 246–262.

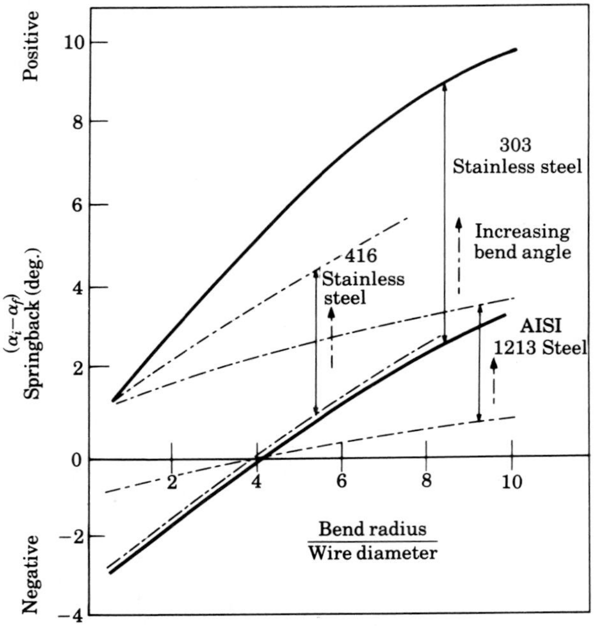

FIGURE 7.22 Range of positive and negative springback for various materials (with the same modulus of elasticity) as a function of the ratio of bend radius to wire diameter. *Source:* After K. S. Turke and S. Kalpakjian, *Proc. NAMRC III,* 1975, pp. 246–262.

Some experimental results with round wires using three different materials are shown in Fig. 7.22. Note that positive springback increases with increasing R/D ratio, where D is the wire diameter. (This result is similar to the R/T effect in Fig. 7.20). However, springback becomes negative at low R/D ratios because low R/D ratios indicate less conformity at the punch radius–wire interface in Fig. 7.21.

Compensation for springback. In forming practice, springback is usually compensated for by *overbending* the part (Figs. 7.23a and b). Several trials may be necessary to obtain the desired results. Another method is to coin the bend area by subjecting it to high localized compressive stresses between the tip of the punch and the die surface (Figs. 7.23c and d)—known as *bottoming* the punch. Another method is *stretch bending*, in which the part is subjected to tension while being bent.

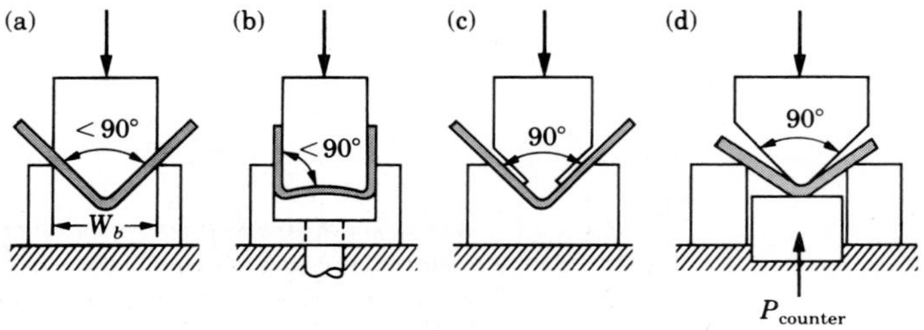

FIGURE 7.23 Methods of reducing or eliminating springback in bending operations. *Source:* V. Cupka, T. Nakagawa, and H. Tyamoto.

Because springback decreases as yield stress decreases—all other parameters being the same—bending may also be carried out at elevated temperatures to reduce springback.

• Example 7.2: Estimating springback.

A 20-gage steel sheet is bent to a radius of 0.5 in. Assuming that its yield stress is 40,000 psi, calculate the radius of the part after it is bent.

SOLUTION. The appropriate formula is Eq. (7.9), where

$R_i = 0.5$ in., $Y = 40{,}000$ psi, $E = 29 \times 10^6$ psi, and $T = 0.0359$ in.,

for 20-gage steel, as found in handbooks. Thus

$$\frac{R_i Y}{ET} = \frac{(0.5)(40{,}000)}{(29 \times 10^6)(0.0359)} = 0.0192,$$

and

$$\frac{R_i}{R_f} = 4(0.0192)^3 - 3(0.0192) + 1 = 0.942.$$

Hence

$$R_f = \frac{0.5}{0.942} = 0.531 \text{ in.}$$

•

7.4.3 Forces

Bending forces can be estimated by assuming that the process is one of simple bending of a rectangular beam. Thus the bending force is a function of the strength of the material, the length and thickness of the part, and the die opening W, as shown in Fig. 7.24. Excluding friction, the general expression for the *maximum bending force* P is

$$P = k\frac{YLT^2}{W}, \tag{7.10}$$

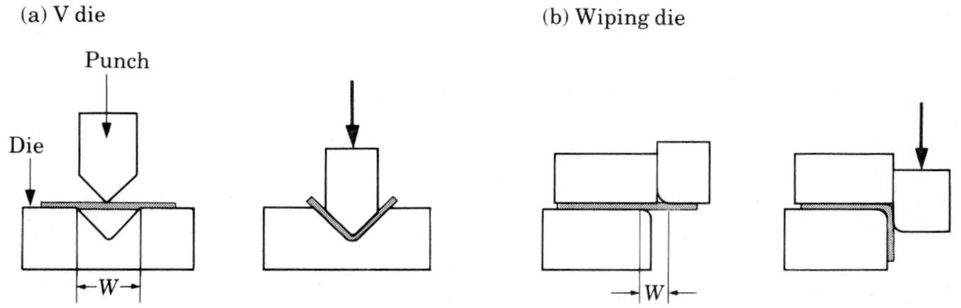

FIGURE 7.24

Common die-bending operations, showing the die-opening dimension W used in calculating bending forces [see Eq. (7.11)].

where the factor k ranges from about 0.3 for a wiping die to 0.7 for a U-die to 1.3 for a V die. We can simplify Eq. (7.10) to

$$P = \frac{(\mathrm{UTS})LT^2}{W} \qquad (7.11)$$

for a V die, which can be used as a general guide for other die geometries. Equation (7.11) applies well to situations in which the punch radius and sheet thickness are small compared to die opening W.

The bending force is also a function of punch travel. It increases from zero to a maximum and may decrease as the bend is completed. The force then increases sharply as the punch bottoms in the case of die bending. In *air bending*, or *free bending*, the force does not increase again after it begins to decrease.

7.4.4 Common bending operations

In this section we describe common bending operations. Some of these processes are performed on discrete sheet-metal parts; others are done continuously, as in roll forming of coiled sheet stock.

Press brake forming. Sheet metal or plate can be bent easily with simple fixtures using a press. Long—7 m (20 ft) or more—and relatively narrow pieces are

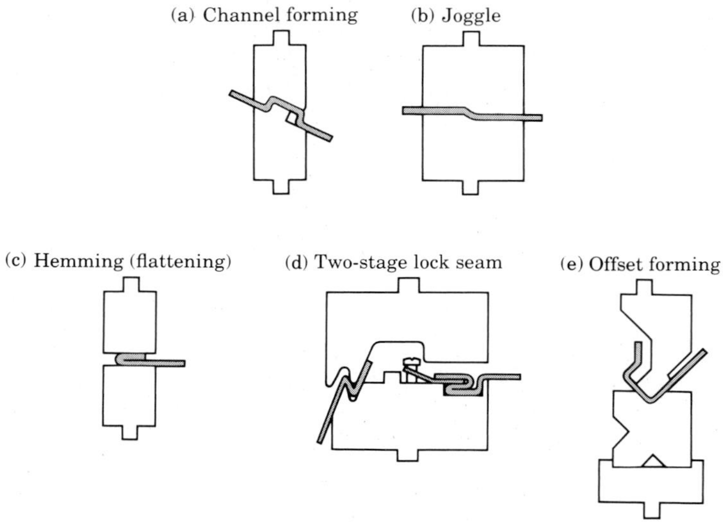

FIGURE 7.25
Schematic illustrations of various bending operations in a press brake.

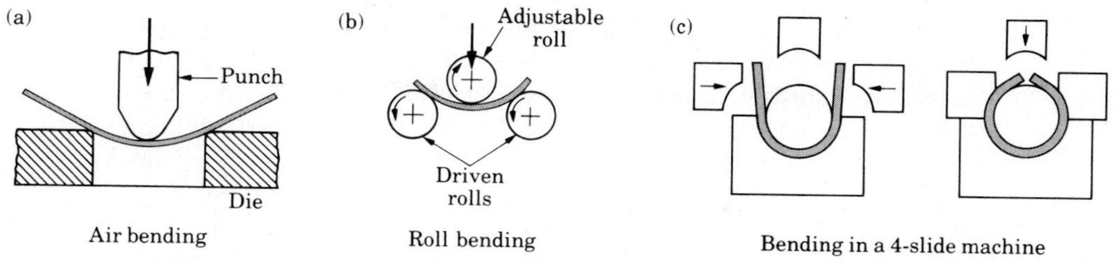

FIGURE 7.26
Examples of various bending operations.

usually bent in a *press brake*. This machine utilizes long dies in a mechanical or hydraulic press and is suitable for small production runs. The tooling is simple and adaptable to a wide variety of shapes (Fig. 7.25). Die materials may range from hardwood, for low-strength materials and small production runs, to carbides. For most applications, carbon-steel or gray-iron dies are generally used.

Other bending operations. Sheet metal may be bent by a variety of processes (Fig. 7.26). Plates are bent by the *roll-bending* process shown in Fig. 7.26(b). Adjusting the distance between the three rolls produces various curvatures.

Beading. In *beading*, the edge of the sheet metal is bent into the cavity of a die (Fig. 7.27). The bead gives stiffness to the part by increasing the moment of inertia of the edges. Also, it improves the appearance of the part and eliminates exposed sharp edges.

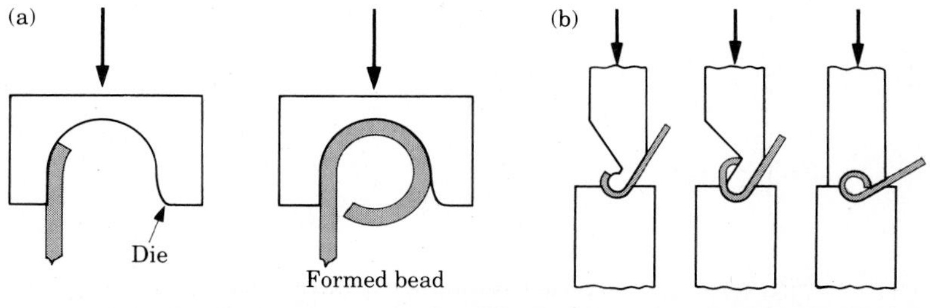

FIGURE 7.27
(a) Bead forming with a single die. (b) Bead forming with two dies in a press brake.

Flanging. Flanging is a process of bending the edges of sheet metals. In *shrink flanging* (Fig. 7.28a), the flange is subjected to compressive hoop stresses which, if excessive, cause the flange edges to wrinkle. The wrinkling tendency increases with decreasing radius of curvature of the flange. In *stretch flanging*, the flange edges are subjected to tensile stresses and, if excessive, can lead to cracking at the edges.

In the *dimpling* operation (Fig. 7.28b), a hole is first punched and then expanded into a flange, or a shaped punch pierces the sheet metal and expands the hole. Flanges may be produced by *piercing* with a shaped punch (Fig. 7.28c). The ends of tubes are flanged by a similar process (Fig. 7.28d). When the angle of bend is less than 90°, as in fittings with conical ends, the process is called *flaring*.

The condition of the edges is important in these operations. Stretching the material causes high tensile stresses at the edges, which could lead to cracking and tearing of the flange. As the ratio of flange to hole diameters increases, the strains

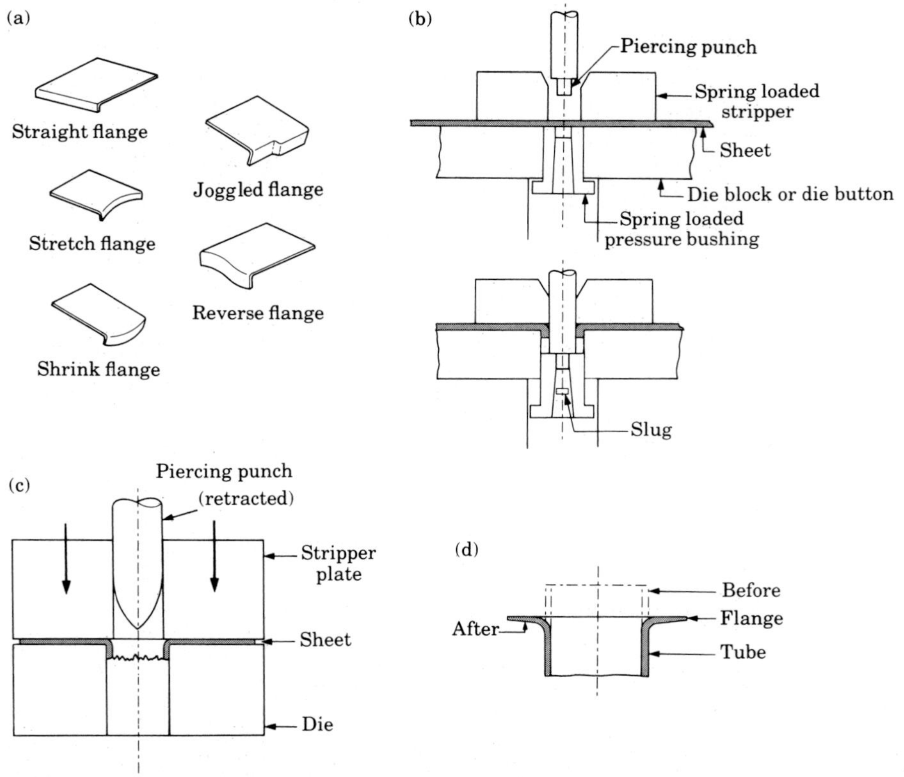

FIGURE 7.28
Various flanging operations. (a) Flanges on flat sheet. (b) Dimpling. (c) Piercing sheet metal to form a flange. In this operation a hole does not have to be prepunched before the punch descends. Note, however, the rough edges along the circumference of the flange. (d) Flanging of a tube. Note thinning of the edges of the flange.

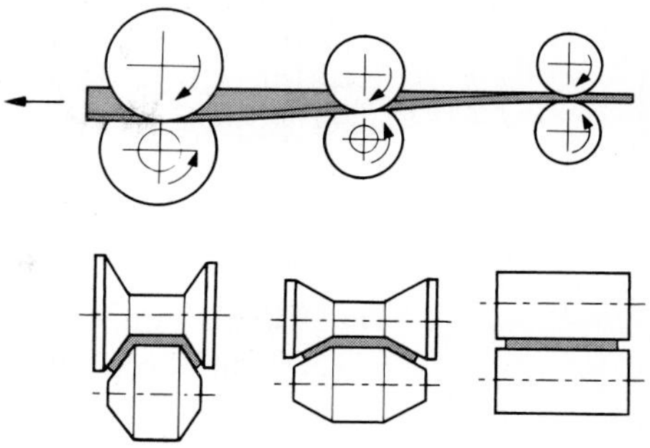

FIGURE 7.29
The roll-forming process.

increase proportionately. The rougher the edge, the greater will be the tendency for cracking. Sheared or punched edges may be shaved with a sharp tool (see Fig. 7.12) to improve the surface finish of the edge and reduce the tendency for cracking.

Hemming. In the *hemming* process (also called *flattening*), the edge of the sheet is folded over itself (Fig. 7.25c). Hemming increases the stiffness of the part, improves its appearance, and eliminates sharp edges. *Seaming* involves joining two edges of sheet metal by hemming (Fig. 7.25d). Double seams are made by a similar process, using specially shaped rollers, for watertight and airtight joints, such as in food and beverage containers.

Roll forming. For bending continuous lengths of sheet metal and for large production runs, *roll forming* is used (contour roll forming or cold roll forming). The metal strip is bent in stages by passing it through a series of rolls (Fig. 7.29). Typical products are channels, gutters, siding, panels, frames (Fig. 7.30), and pipes and tubing with lock seams. The length of the part is limited only by the amount of material supplied from the coiled stock. The parts are usually sheared and stacked continuously. The sheet thickness typically ranges from about 0.125 mm to 20 mm (0.005 in. to 0.75 in.). Forming speeds are generally below 1.5 m/s (300 ft/min), although they can be much higher for special applications.

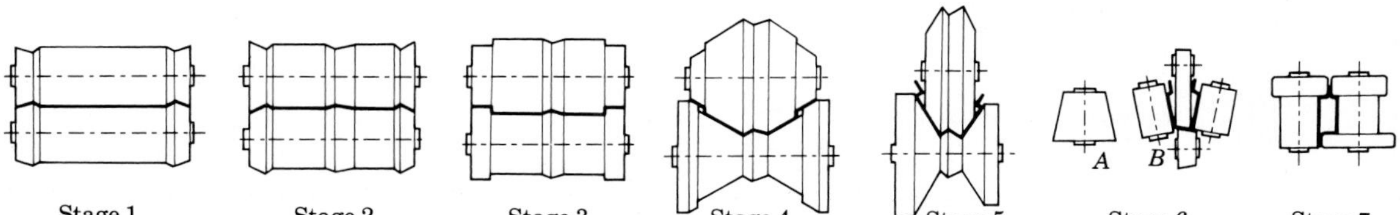

FIGURE 7.30
Stages in roll forming of sheet-metal door frame. In stage 6, the rolls may be shaped as in *A* or *B*. *Source:* G. Oehler.

Proper design and sequencing of the rolls, which usually are mechanically driven, requires considerable experience. Tolerances, springback, and tearing and buckling of the strip have to be considered. The rolls are generally made of carbon steel or gray iron and may be chromium plated—for better surface finish of the product and wear resistance of the rolls. Lubricants may be used to improve roll life and surface finish and to cool the rolls and the workpiece.

Tube bending. Bending and forming tubes and other hollow sections require special tooling to avoid buckling and folding. The oldest and simplest method of bending a tube or pipe is to pack the inside with loose particles, commonly sand, and bend the part in a suitable fixture. This technique prevents the tube from buckling. After the tube has been bent, the sand is shaken out. Tubes can also be plugged with various flexible internal mandrels, as shown in Fig. 7.31, which also illustrates various bending techniques and fixtures for tubes and sections. A relatively thick tube having a large bend radius can be bent without filling it with particulates or using plugs.

The beneficial effect of forming metals with highly compressive forces is demonstrated in the bending of a tube with relatively sharp corners (Fig. 7.32). In this operation, the tube is subjected to longitudinal compressive stresses, which reduce the stresses in the outer fibers in the bend area, thus improving the bendability of the material.

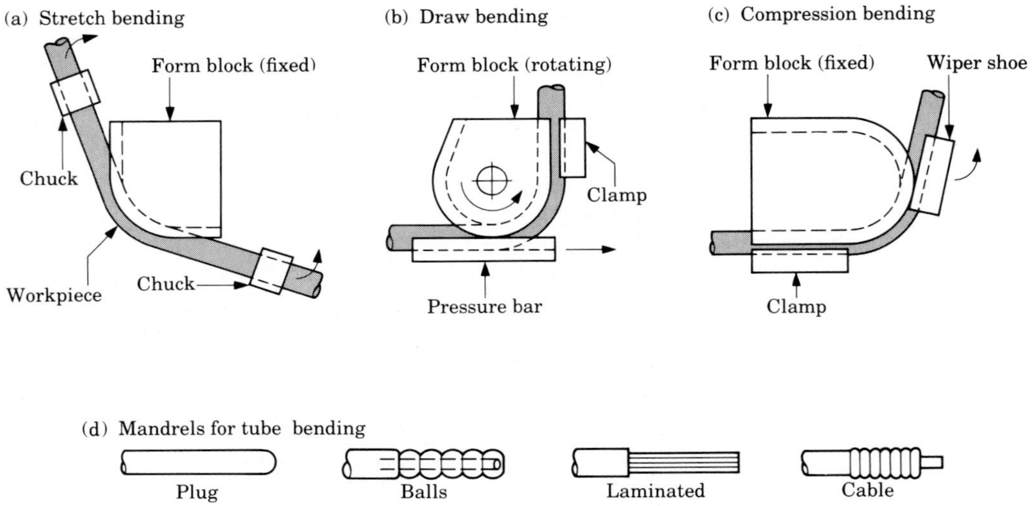

FIGURE 7.31
Methods of bending tubes. Internal mandrels, or filling tubes with particulate materials such as sand, are often necessary to prevent collapsing of the tubes during bending. Solid rods and structural shapes are also bent by these techniques.

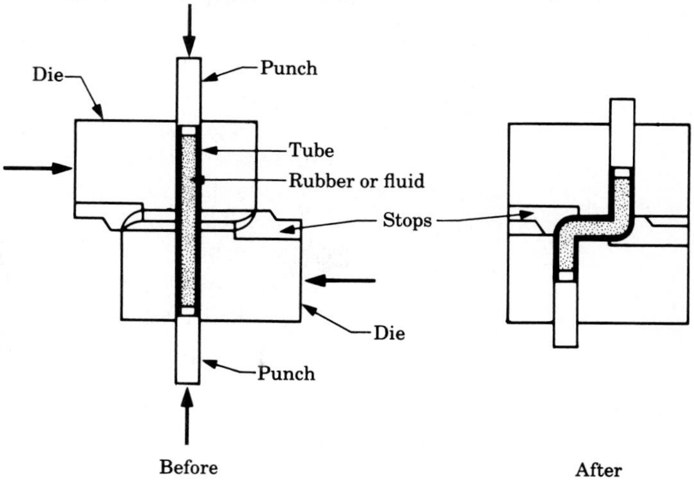

FIGURE 7.32
A method of forming a tube with sharp angles using axial compressive forces. Compressive stresses are beneficial in forming operations because they delay fracture. Note that the tube is supported internally with rubber or fluid to avoid collapsing during forming. *Source:* After J. L. Remmerswaal and A. Verkaik, *Int. Conf. Manufacturing Technology,* ASME, 1967, pp. 1171–1181.

7.5 Stretch Forming

In *stretch forming*, the sheet metal is clamped around its edges and stretched over a die or form block, which moves upward, downward, or sideways, depending on the particular machine (Fig. 7.33). Stretch forming is used primarily to make aircraft wing-skin panels, automobile door panels, and window frames. Although this process is generally used for low-volume production, it is versatile and economical.

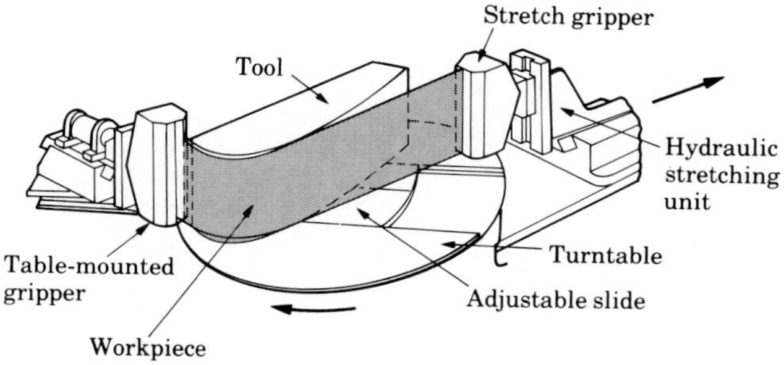

FIGURE 7.33
Schematic illustration of a stretch-forming process. Aluminum skins for aircraft can be made by this process. *Source:* Cyril Bath Co.

Aluminum skins for the Boeing 767 and 757 aircraft are made by stretch forming, with a tensile force of 9 MN (2 million lb). The rectangular sheets are 12 m × 2.5 m × 6.4 mm (40 ft × 8.3 ft × 0.25 in.).

In most operations, the blank is a rectangular sheet, clamped along its narrower edges and stretched lengthwise, thus allowing the material to shrink in width. Controlling the amount of stretching is important to avoid tearing. Stretch forming cannot produce parts with sharp contours or re-entrant corners (depressions on the surface of the die).

Dies for stretch forming are generally made of zinc alloys, steel, plastics, or wood. Most applications require little or no lubrication. Various accessory equipment can be used in conjunction with stretch forming, including additional forming with both male and female dies while the part is in tension.

• Example 7.3: Work done in stretch forming.

A 15 in. long workpiece (Fig. E7.1) is stretched by a force F until $\alpha = 20°$. The original cross-sectional area is 0.5 in^2. The material has a curve $\sigma = 100{,}000\varepsilon^{0.3}$. Find the total work done, ignoring end effects or bending. What is α_{max} before necking begins?

SOLUTION. This situation is equivalent to stretching a workpiece from 15 in. to a length of $a + b$. For $\alpha = 20°$, we find that the final length $L_f = 16.8$ in. The true strain then is

$$\varepsilon = \ln\left(\frac{L_f}{L_0}\right) = \ln\left(\frac{16.8}{15}\right) = 0.114.$$

The work done per unit volume is

$$u = \int_0^{0.114} \sigma \cdot d\varepsilon = 10^5 \int_0^{0.114} \varepsilon^{0.3}\, d\varepsilon = 10^5 \left.\frac{\varepsilon^{1.3}}{1.3}\right]_0^{0.114} = 4570 \text{ in} \cdot \text{lb/in}^3.$$

The volume of the workpiece is

$$V = (15)(0.5) = 7.5 \text{ in}^3.$$

Hence

$$\text{Work} = (u)(V) = 34{,}275 \text{ in.} \cdot \text{lb}.$$

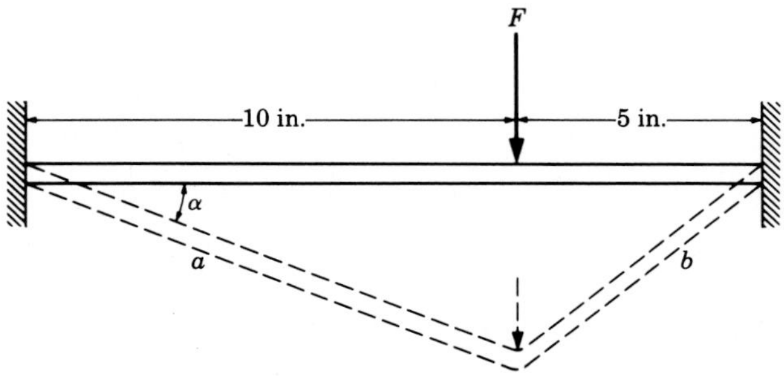

FIGURE E7.1

The necking limit for uniaxial tension is given by Eq. (7.2), the problem being similar to a stretch-forming operation. Thus

$$L_{max} = L_0 e^n = 15e^{0.3} = 20.2 \text{ in.}$$

Hence $a + b = 20.2$ in., and from similarities in triangles, we obtain

$$a^2 - 10^2 = b^2 - 5^2, \quad \text{or} \quad a^2 = b^2 + 75.$$

Hence $a = 12$ in. and $b = 8.2$ in. Therefore

$$\cos \alpha = \frac{10}{12} = 0.833 \quad \text{or} \quad \alpha_{max} = 33.6°.$$

7.6 Bulging

The basic forming process of *bulging* involves placing a tubular, conical, or curvilinear part in a split-female die and expanding it with, say, a polyurethane plug (Fig. 7.34a). The punch is then retracted, the plug returns to its original shape, and

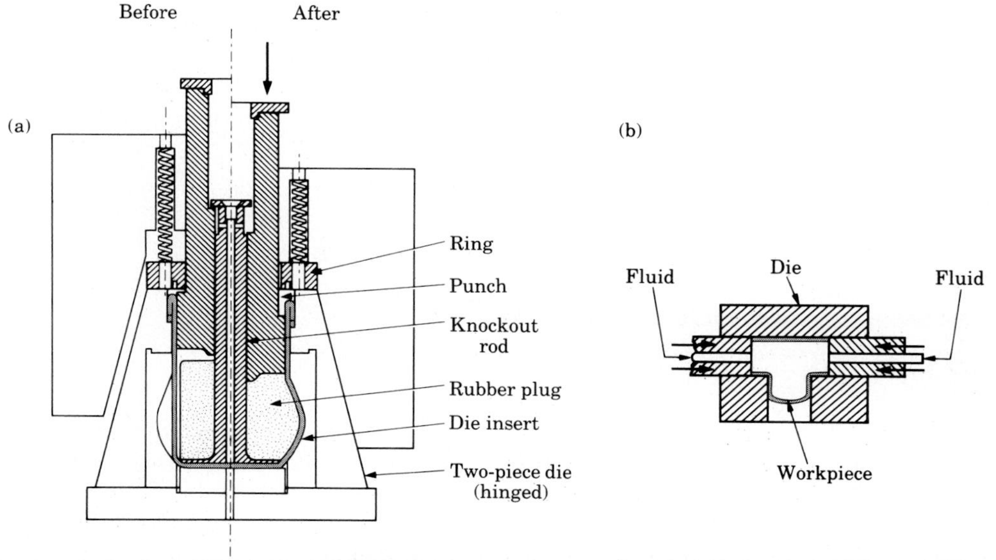

FIGURE 7.34 (a) Bulging of a tubular part with a flexible plug. Water pitchers can be made by this method. (b) Production of fittings for plumbing by expanding tubular blanks with internal pressure. The bottom of the piece is then punched out to produce a "T." *Source:* J. A. Schey, *Introduction to Manufacturing Processes,* 2d ed. New York: McGraw-Hill Publishing Company, 1987.

the part is removed by opening the dies. Typical products are coffee or water pitchers, barrels, and beads on drums. For parts with complex shapes, the plug, instead of being cylindrical, may be shaped in order to apply greater pressure at critical points. The major advantage of using polyurethane plugs is that they are very resistant to abrasion, wear, and lubricants, and they do not damage the surface finish of the part being formed.

Hydraulic pressure may be used in bulging operations, although they require sealing and need hydraulic controls (Fig. 7.34b). *Segmented dies* that are expanded and retracted mechanically may also be used. These dies are relatively inexpensive and can be used for large production runs. Formability in bulging processes can be enhanced by the application of longitudinal compressive stresses.

Example 7.4: Manufacturing bellows.

Bellows are manufactured by a bulging process, as shown in the accompanying figure. After the tube is bulged at several equidistant locations, it is compressed axially to collapse the bulged regions, thus forming bellows. The tube material must be able to undergo the large strains involved during the collapsing process.

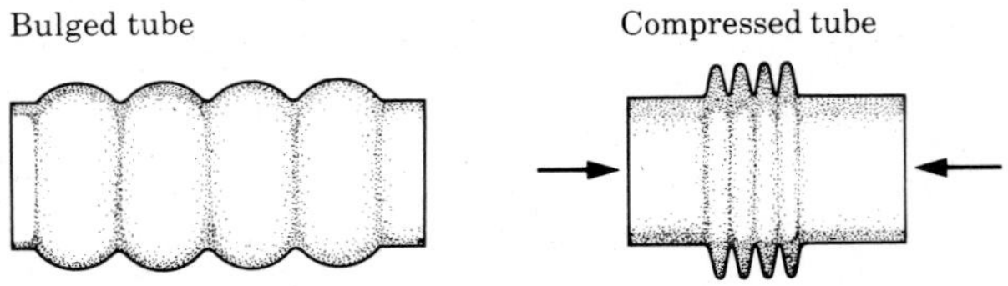

7.7 Rubber Forming

In the processes described in the preceding sections, we noted that the dies are made of solid materials. However, in *rubber forming* one of the dies in a set can be made of flexible material, such as a rubber or polyurethane membrane. Polyurethanes are used widely because of their resistance to abrasion, resistance to cutting by burrs or sharp edges of the sheet metal, and long fatigue life.

In bending and embossing sheet metal, as shown in Fig. 7.35, the female die has been replaced with a rubber pad. Note that the outer surface of the sheet is protected from damage or scratches because it is not in contact with a hard metal surface during forming. Pressures are usually on the order of 10 MPa (1500 psi).

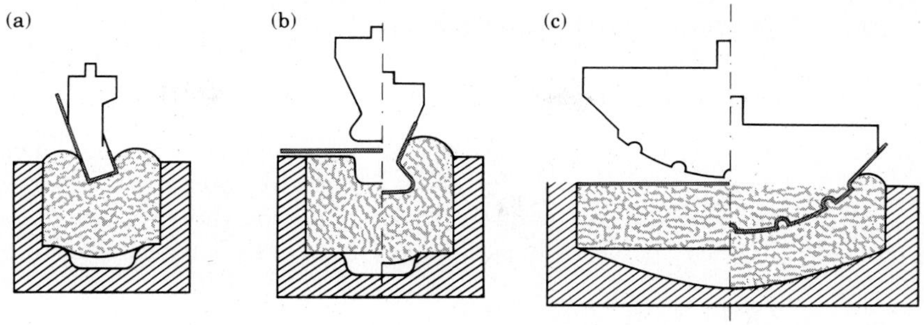

FIGURE 7.35 Examples of bending and embossing sheet metal with a metal punch and flexible pad serving as the female die. *Source:* Polyurethane Products Corporation.

In the *hydroform* or *fluid-forming* process (Fig. 7.36), the pressure over the rubber membrane is controlled throughout the forming cycle, with maximum pressures of up to 100 MPa (15,000 psi). This procedure allows close control of the part during forming to prevent wrinkling or tearing. Deeper draws are obtained than in conventional deep drawing because the pressure around the rubber membrane forces the cup against the punch. The friction at the punch–cup interface reduces the longitudinal tensile stresses in the cup and thus delays fracture. The control of frictional conditions in rubber forming as well as other sheet-forming operations can be a critical factor in making parts successfully. Selection of proper lubricants and application methods is important.

When selected properly, rubber forming processes have the advantages of low tooling cost, flexibility and ease of operation, low die wear, no damage to the surface of the sheet, and capability to form complex shapes. Parts can also be formed with laminated sheets of various nonmetallic materials or coatings.

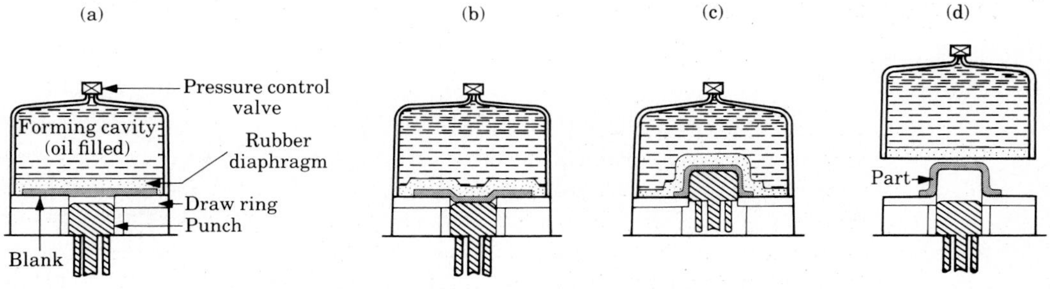

FIGURE 7.36 The hydroform, or fluid forming, process. Note that unlike the ordinary deep-drawing process, the dome pressure forces the cup walls against the punch. The cup travels with the punch and thus deep drawability is improved.

7.8 Spinning

Spinning involves the forming of axisymmetric parts over a rotating mandrel with the use of rigid tools or rollers. There are three basic types of spinning processes: conventional (or manual), shear, and tube spinning. The equipment for all these processes is basically similar to a lathe with various special features.

7.8.1 Conventional spinning

In the *conventional spinning* process, a circular blank of flat or preformed sheet metal is held against a rotating mandrel while a rigid tool deforms and shapes the material over the mandrel (Fig. 7.37a). The tools may be actuated either manually or by a hydraulic mechanism. The operation involves a sequence of passes (Fig. 7.38) and requires considerable skill. This process has been found to be quite difficult to study analytically.

Typical shapes made by conventional spinning processes are shown in Fig. 7.39. This process is particularly suitable for conical and curvilinear shapes, which would otherwise be difficult or uneconomical to form by other methods. Part diameters may range up to 6 m (20 ft). Although most spinning is performed at room temperature, thick parts or metals with low ductility or high strength require spinning at elevated temperatures. Tooling costs in spinning are relatively low. However, because the operation requires multiple passes to form the final part, it is economical for small production runs only.

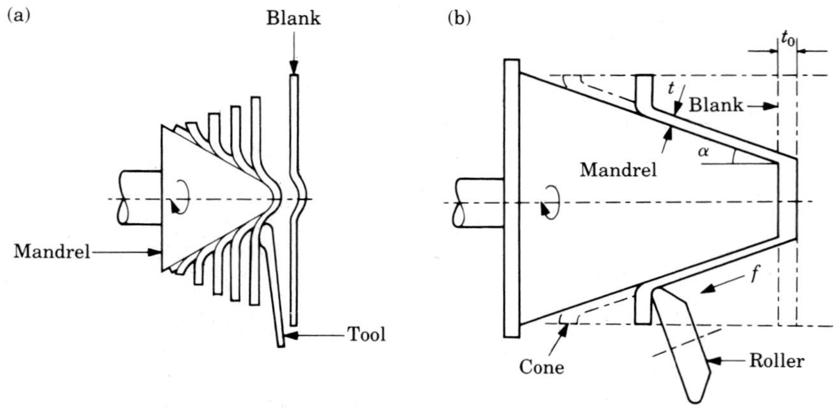

FIGURE 7.37 Schematic illustration of spinning processes: (a) conventional spinning and (b) shear spinning. Note that in shear spinning the diameter of the spun part, unlike conventional spinning, is the same as that of the blank. The quantity f is the feed (mm/rev or in./rev).

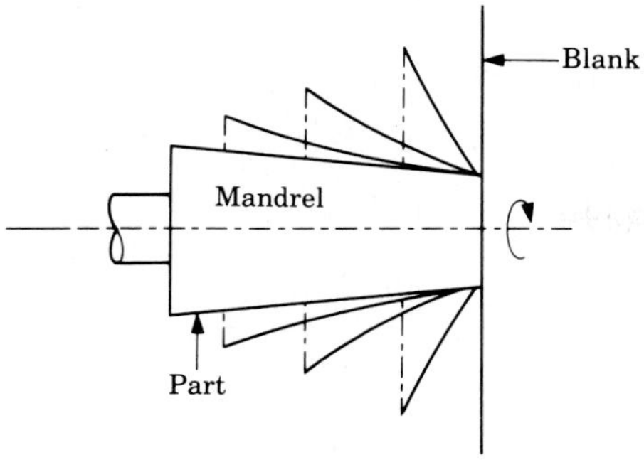

FIGURE 7.38 Stages in conventional spinning of a tubular component from a flat, circular metal disk. This operation requires considerable skill to prevent the part from collapsing or buckling during spinning.

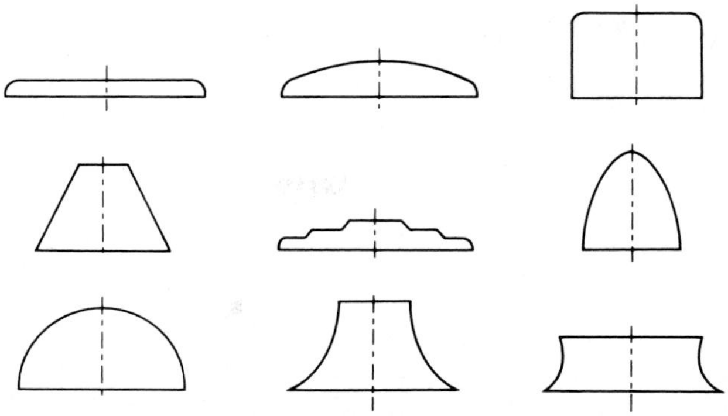

FIGURE 7.39 Typical shapes produced by the conventional spinning process. Circular marks on the external surfaces of components usually indicate that the parts have been made by spinning. Examples are aluminum kitchen utensils and light reflectors.

7.8.2 Shear spinning

In the *shear spinning* process (also known as *power spinning*, *flow turning*, *hydrospinning*, or *spin forging*), an axisymmetrical conical or curvilinear shape is generated by holding the diameter of the part constant (see Fig. 7.37b). Parts typically made by this process are rocket-motor casings and missile nose cones. Although a single roller can be used, two rollers are desirable to balance the radial forces acting on the mandrel. Large parts, up to about 3 m (10 ft), can be spun to close tolerances by shear spinning. Little material is wasted, and the operation is completed in a relatively short time. If carried out at room temperature, the part has a higher yield strength than the original material, but lower ductility and toughness. A great variety of shapes can be spun with relatively simple tooling, generally made of tool steel. Because of the large plastic deformation involved, the process generates considerable heat. It is usually carried away by a coolant-type fluid applied during spinning.

In shear spinning over a conical mandrel, the thickness t of the spun part is simply

$$t = t_0 \sin \alpha. \tag{7.12}$$

The force primarily responsible for supplying energy is the tangential force, F_t. For an ideal situation in the shear spinning of a cone, this force is

$$F_t = ut_0 f \sin \alpha, \tag{7.13}$$

where u is the specific energy of deformation, Eq. (2.58). It is the area under the true stress–true strain curve corresponding to a true strain related to the shear strain by

$$\varepsilon = \frac{\gamma}{\sqrt{3}} = \frac{\cot \alpha}{\sqrt{3}} \tag{7.14}$$

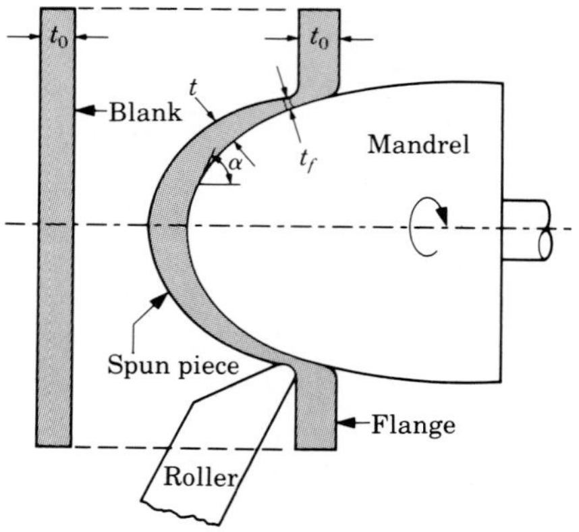

FIGURE 7.40 Schematic illustration of a shear-spinnability test. As the roller advances, the part thickness is reduced. The reduction in thickness at fracture is called the maximum spinning reduction per pass. *Source:* After R. L. Kegg, *J. Eng. Ind.*, vol. 83, 1961, pp. 119–124.

(see Section 2.13.7). Actual forces can be as much as 50 percent higher than those given by Eq. (7.13) because of factors such as redundant work and friction.

An important factor in shear spinning is the *spinnability* of the metal, that is, the smallest thickness to which a part can be spun without fracture. To determine spinnability, a simple test method has been developed (Fig. 7.40), in which a circular blank is spun over an ellipsoid mandrel, according to Eq. (7.12). As the thickness is eventually reduced to zero, all materials will fail at some thickness t_f.

Data for various materials are given in Fig. 7.41. The maximum spinning reduction in thickness is defined as

$$\text{Maximum reduction} = \frac{t_0 - t_f}{t_0} \times 100\% \qquad (7.15)$$

and is plotted against the tensile reduction of area of the material. Note that if a metal has a tensile reduction of area of about 50 percent or greater, it can be reduced in thickness by 80 percent by spinning in one pass. For less ductile materials the spinnability is lower. Process variables such as feed and speed do not appear to have a significant influence on spinnability.

Figures 7.16 and 7.41 show that both maximum bendability and spinnability coincide with a tensile reduction of area of about 50 percent. Any further increase in the ductility of the original material does not improve its formability. For materials with low ductility, spinnability can usually be improved by forming at elevated temperatures.

7.8.3 Tube spinning

In the *tube spinning* process, tubes are reduced in thickness by spinning them on a mandrel using rollers. The operation may be carried out externally or internally

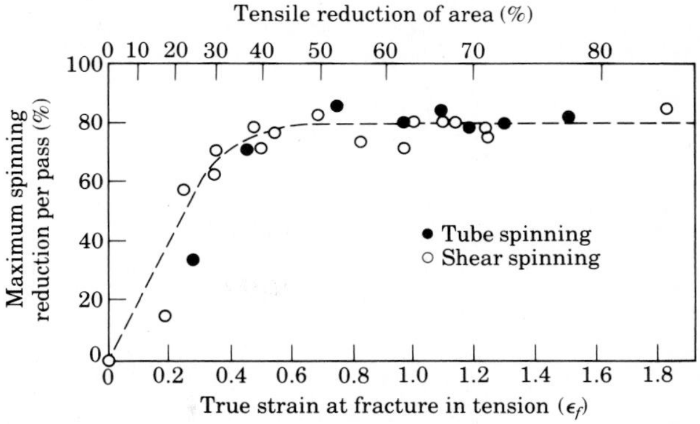

FIGURE 7.41

Experimental data showing the relationship between maximum spinning reduction per pass and the tensile reduction of area of the original material. Note that once a material has about 50 percent reduction of area in a tension test, further increase in the ductility of the original material does not improve its spinnability. *Source:* S. Kalpakjian, *J. Eng. Ind.,* vol. 86, 1964, pp. 49–54.

(Fig. 7.42). Also, the part may be spun forward or backward, similar to a drawing or a backward extrusion process. In either case, the reduction in wall thickness results in a longer tube. Various internal and external profiles can be generated by changing the path of the roller during its travel along the mandrel. Tube spinning can be used for making pressure vessels, automotive components, and rocket and missile parts.

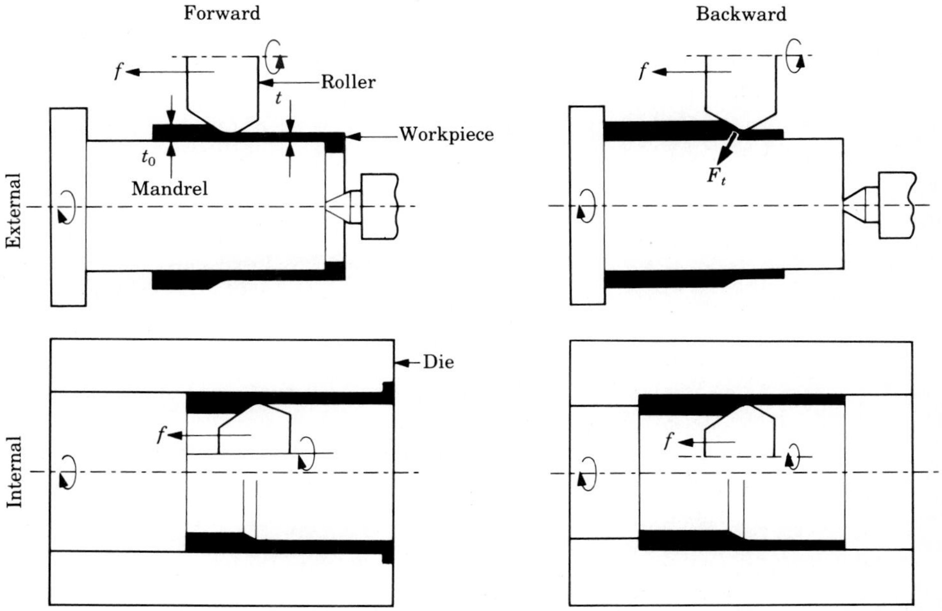

FIGURE 7.42

Examples of external and internal tube spinning and the variables involved.

It can also be combined with shear spinning, as producing the compressor shaft for an aircraft engine shown in Fig. 7.43.

Spinnability in this process is determined by a test method similar to that for shear spinning (Fig. 7.44). Maximum reduction per pass in tube spinning is related to the tensile reduction of area of the material (see Fig. 7.41), with results very similar to shear spinning. As in shear spinning, ductile metals fail in tension after the reduction in thickness has taken place, whereas the less ductile metals fail in the deformation zone under the roller.

The ideal tangential force F_t in forward tube spinning can be expressed as

$$F_t = \bar{Y} \Delta t f, \tag{7.16}$$

where Δt is $(t_0 - t)$ (see Fig. 7.42) and $\bar{Y}$ is the average flow stress of the material. For backward spinning, the ideal tangential force is roughly twice that given by Eq. (7.16). Friction and redundant work of deformation cause the actual forces to be about twice those obtained by Eq. (7.16).

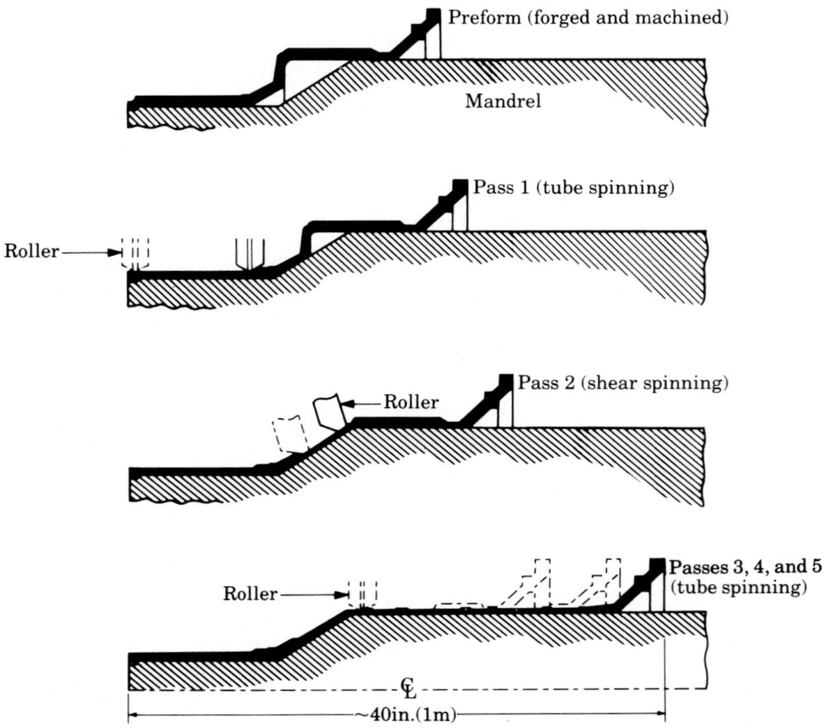

FIGURE 7.43

Stages in tube and shear spinning of a compressor shaft for the jet engine of the supersonic Concorde aircraft. Economic analysis indicated that the best method of manufacturing this part was to spin a preformed (forged and machined) tubular blank.

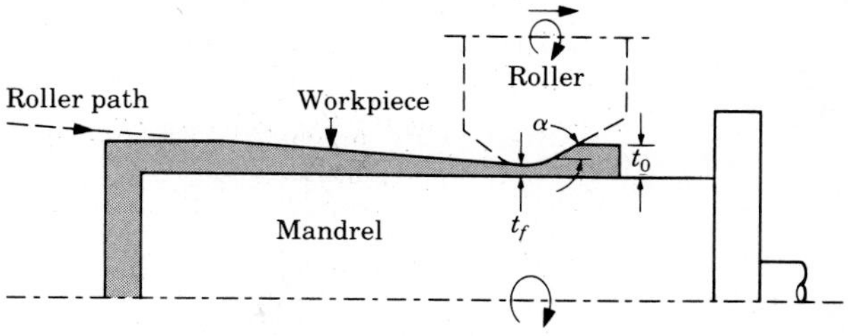

FIGURE 7.44 Schematic illustration of a tube-spinnability test. As the roller advances, the wall thickness is reduced and the part eventually fractures. *Source:* S. Kalpakjian, *J. Eng. Ind.*, vol 86, 1964, pp. 49–54.

7.9. High-Energy-Rate Forming

Sheet-metal–forming processes that use chemical, electrical, or magnetic sources of energy are described in this section. They are called *high-energy-rate* processes because the energy is released in a very short time. The basic operations are explosive forming, electrohydraulic forming, and magnetic-pulse forming.

7.9.1 Explosive forming

The most common *explosive forming* process (first proposed in the early 1900s) is shown in Fig. 7.45. The workpiece is clamped over a die, the air in the die cavity is evacuated, and the workpiece is lowered into a tank filled with water. An explosive charge is then placed at a certain height and detonated. The rapid conversion of the explosive into gas generates a shock wave. The pressure of this wave is sufficiently high to form the metal.

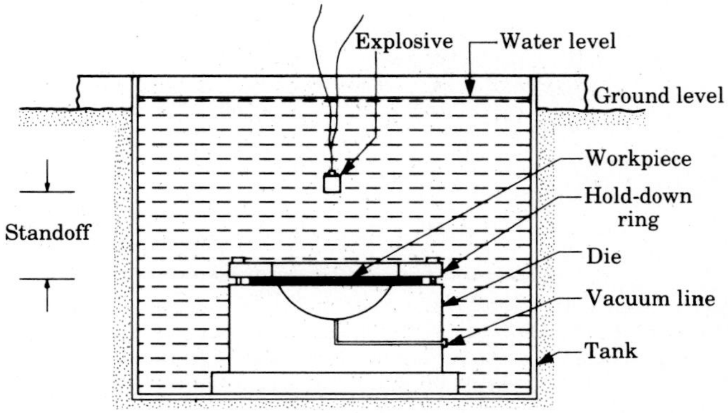

FIGURE 7.45 Schematic illustration of the explosive forming process. Although explosives are generally used for destructive purposes, their energy can be controlled and utilized in forming large parts that would otherwise be difficult or expensive to produce by other methods.

The peak pressure p, generated in water, is given by the expression

$$p = K\left(\frac{\sqrt[3]{W}}{R}\right)^a, \tag{7.17}$$

where p is the peak pressure in pounds per square inch; K is a constant that depends on the explosive, e.g., 21,600 for TNT; W is the weight of the explosive in pounds; R is the distance of the explosive from the workpiece (standoff) in feet; and a is a constant, generally taken as 1.15.

An important factor in determining peak pressure is the *compressibility* of the energy-transmitting medium (say, water) and its *acoustic impedance*, that is, the product of mass density and sound velocity in the medium. The lower the compressibility and the higher the density of the medium, the higher the peak pressure rises (Fig. 7.46). The distance between the water level and the explosive should not be too small; otherwise, the energy is dissipated rapidly to the environment. Detonation speeds are typically 6700 m/s (22,000 ft/s), and the speed at which the metal is formed is estimated to be on the order of 30 to 200 m/s (100 to 600 ft/s).

A great variety of shapes can be formed explosively, provided that the material is sufficiently ductile at high strain rates. Depending on the number of parts to be formed, dies may be made of aluminum alloys, steel, ductile iron, zinc alloys, reinforced concrete, wood, plastics, and composite materials. The final properties of parts made by this process are basically the same as those made by conventional methods. Safety is an important aspect in these operations.

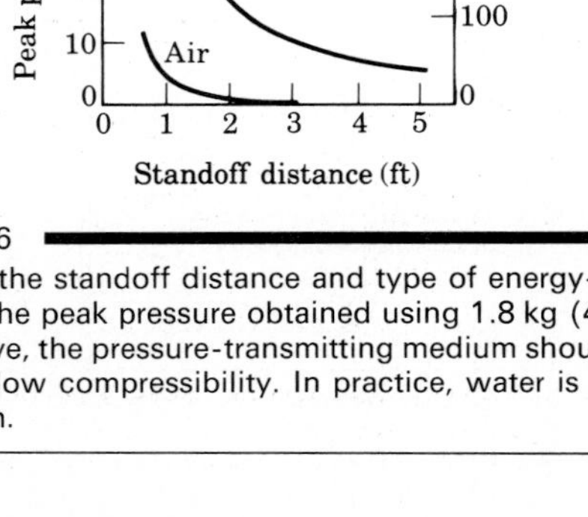

FIGURE 7.46 Influence of the standoff distance and type of energy-transmitting medium on the peak pressure obtained using 1.8 kg (4 lb) of TNT. To be effective, the pressure-transmitting medium should have high density and low compressibility. In practice, water is a commonly used medium.

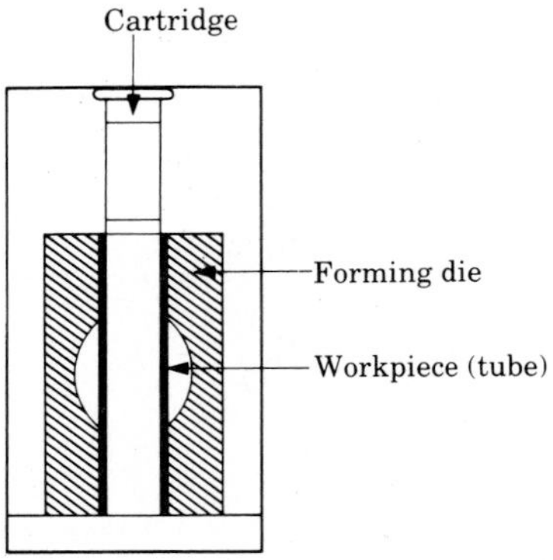

FIGURE 7.47 Schematic illustration of the confined method of explosive bulging of tubes. Thin-walled tubes of nonferrous metals can be formed to close tolerances by this process.

Needing only one die and being versatile, with no limit to the size of the workpiece, explosive forming is particularly suitable for small-quantity production runs of large parts. Steel plates 25 mm (1 in.) thick and 3.6 m (12 ft) in diameter have been formed by this method. Tubes of 25 mm (1 in.) wall thickness have also been bulged by explosive forming techniques. Another explosive forming method, which uses a cartridge as the source of energy, is shown in Fig. 7.47; no other energy-transmitting medium is used. The process can be used for bulging and expanding of thin-walled tubes. Die wear and die failure can be a significant problem in this operation.

• Example 7.5: Peak pressure in explosive forming.

Calculate the peak pressure in water for 0.1 lb of TNT at a standoff of 1 ft. Is this pressure sufficiently high for forming sheet metals?

SOLUTION. Using Eq. (7.17), we find that

$$p = (21{,}600)\left(\frac{\sqrt[3]{0.1}}{1}\right)^{1.15} \simeq 9000 \text{ psi}.$$

This pressure would be sufficiently high to form sheet metals. Note in Example 2.9, for instance, that the pressure required to expand a thin-walled spherical shell of a material similar to soft aluminum alloys is only 400 psi. Also note that a process such as hydroform (Section 7.7) has a maximum hydraulic pressure in the dome of about 100 MPa (15,000 psi). Other rubber-forming processes use pressures from about 1500 to 7500 psi. Thus the pressure obtained in this problem is sufficient for most sheet-forming processes.

7.9.2 Electrohydraulic forming

In the *electrohydraulic forming* process, also called *underwater-spark* or *electric-discharge forming*, the source of energy is a spark from electrodes connected with a thin wire (Fig. 7.48). The energy is stored in a bank of condensers charged with direct current. The rapid discharge of this energy through the electrodes generates a

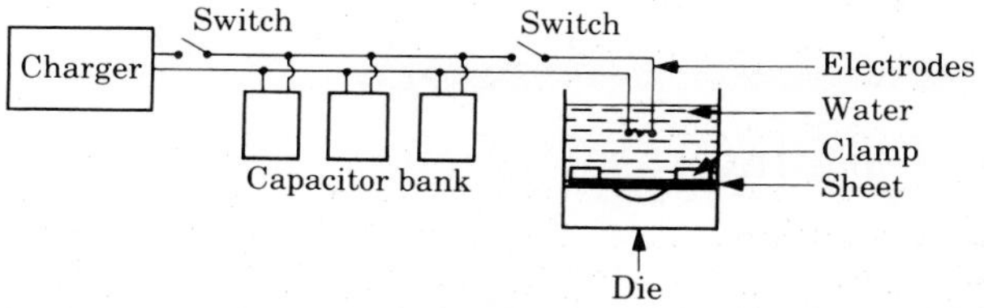

FIGURE 7.48
Schematic illustration of the electrohydraulic forming process.

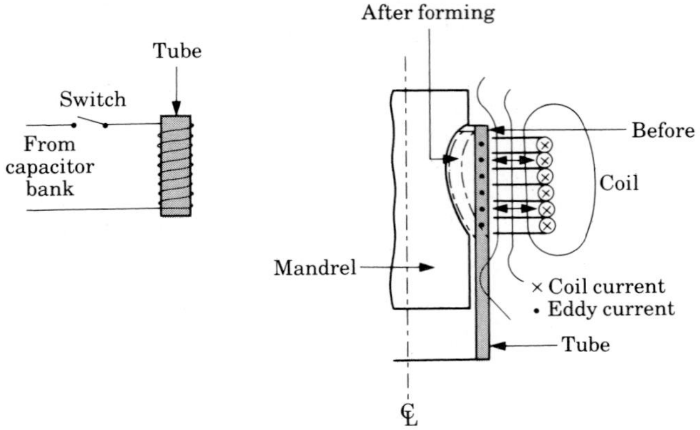

FIGURE 7.49
Schematic illustration of the magnetic pulse forming process. The part is formed without physical contact with any object.

shock wave, which then forms the part. This process is essentially similar to explosive forming, except that it utilizes a lower level of energy and is used with smaller workpieces. It is also a safer operation.

7.9.3 Magnetic-pulse forming

In the *magnetic-pulse forming* process, the energy stored in a capacitor bank is discharged rapidly through a magnetic coil. In a typical example, a ring-shaped coil is placed over a tubular workpiece to be formed (collapsed) over another solid piece to make an integral part (Fig. 7.49). The magnetic field produced by the coil crosses the metal tube, generating eddy currents in the tube. This current, in turn, produces its own magnetic field. The forces produced by the two magnetic fields oppose each other; thus there is a repelling force between the coil and the tube. The high forces generated collapse the tube over the inner piece.

The higher the electrical conductivity of the workpiece, the higher is the magnetic force. The metal does not have to have any special magnetic properties. Magnetic-pulse forming is used for a variety of operations, such as swaging of thin-walled tubes over rods, cables, plugs, etc., and for bulging and flaring. Flat coils are also made for forming of flat sheet metal, as for embossing and shallow drawing operations.

7.10 Superplastic Forming

The superplastic behavior of certain very fine-grained alloys (normally less than 10 to 15 μm), where very large elongations (up to 2000 percent) are obtained at certain temperatures and low strain rates, was described in Section 2.2.7. These alloys, such as Zn–Al and titanium, can be formed into complex shapes with common metalworking or polymer processing techniques. The high ductility and relatively low strength of superplastic alloys present the following advantages:

a) Lower strength of tooling, because of the low strength of the material at forming temperatures, hence, lower tooling costs.
b) Forming of complex shapes in one piece, with fine detail and close tolerances, and elimination of secondary operations.
c) Weight and material savings because of formability of the materials.
d) Little or no residual stresses in the formed parts.

The limitations of superplastic forming are:

a) The material must not be superplastic at service temperatures.
b) Because of the extreme strain-rate sensitivity of the superplastic material, it must be formed at sufficiently low rates (typically at strain rates of 10^{-4} to 10^{-2}/s). Forming times range anywhere from a few seconds to several hours. Thus cycle times are much longer than conventional forming processes. Superplastic forming is therefore a batch-forming process.

Superplastic alloys (particularly Zn-22Al and Ti-6Al-4V) can be formed by bulk deformation processes, such as compression molding, closed-die forging, coining, hubbing, and extrusion. Sheet forming of these materials can also be done using operations such as thermoforming, vacuum forming, and blow molding.

An important development is the ability to fabricate sheet-metal structures by combining diffusion bonding with superplastic forming. Typical structures in which flat sheets are diffusion bonded and formed are shown in Fig. 7.50. After diffusion

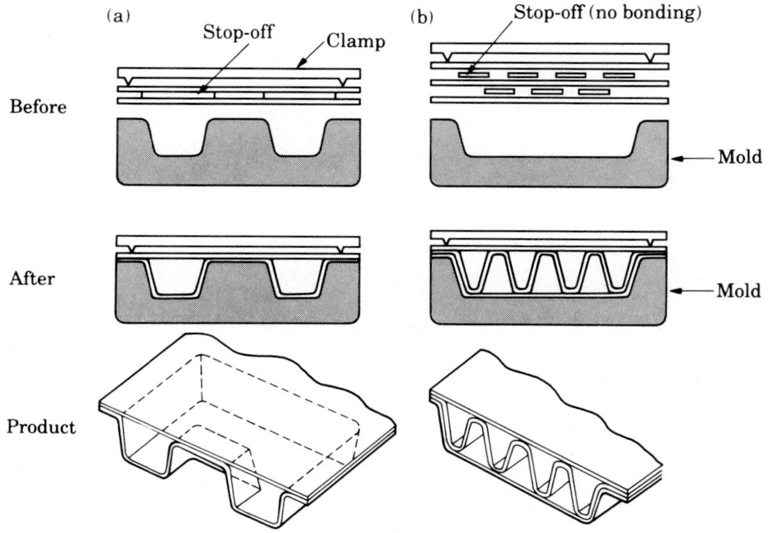

FIGURE 7.50 Two types of structures made by diffusion bonding and superplastic forming of sheet metal. Such structures have a high stiffness-to-weight ratio. *Source:* Rockwell International Corp.

bonding of selected locations of the sheets, the unbonded regions (*stop-off*) are expanded into a mold by air pressure. These structures are thin and have high stiffness-to-weight ratios. Hence they are particularly important in aircraft and aerospace applications.

This process improves productivity by eliminating mechanical fasteners and produces parts with good dimensional accuracy and low residual stresses. The technology is now well advanced for titanium structures (typically Ti-6Al-4V alloy) for aerospace applications. Structures made of 7475-T6 aluminum alloy are also being developed using this technique. Other metals for superplastic forming are Inconel 100 and Incoloy 718 nickel alloys and iron-base, high-carbon alloys. Commonly used die materials in superplastic forming are low-alloy steels, cast tool steels, ceramics, graphite, and plaster of paris. Selection depends on the forming temperature and strength of the superplastic alloy.

7.11 Various Forming Methods

Peen forming is used to produce curvatures on thin sheet metals by *shot peening* one surface of the sheet. Peening is done with cast-iron or steel shot, discharged either from a rotating wheel or with an air blast from a nozzle. Peen forming is used by the aircraft industry to generate smooth and complex curvatures on aircraft wing skins. Cast-steel shot about 2.5 mm (0.1 in.) in diameter at speeds of 60 m/s (200 ft/s) has been used to form wing panels 25 m (80 ft) long. For heavy sections, shot diameters as large as 6 mm (1/4 in.) may be used.

In peen forming, the surface of the sheet is subjected to compressive stresses, which tend to expand the surface layer. Since the material below the peened surface remains rigid, the surface expansion causes the sheet to develop a curvature. The process also induces compressive surface residual stresses, with improved fatigue strength of the sheet. The peen forming process is also used for straightening twisted or bent parts. Out-of-round rings, for example, can be straightened by this method.

Gas mixtures in a closed container have been utilized as an energy source. When ignited, the pressures generated are sufficient to form parts. The principle is similar to the generation of pressure in an internal combustion engine.

Liquefied gases, such as liquid nitrogen, may be used to develop pressures high enough to form sheet metals. When allowed to reach room temperature in a closed container, liquefied nitrogen becomes gaseous and expands, developing the necessary pressure to form the part.

7.11.1 Manufacturing honeycomb materials

There are two principal methods of manufacturing honeycomb materials. In the *expansion* process (Fig. 7.51a)—the most common method—sheets are cut from a

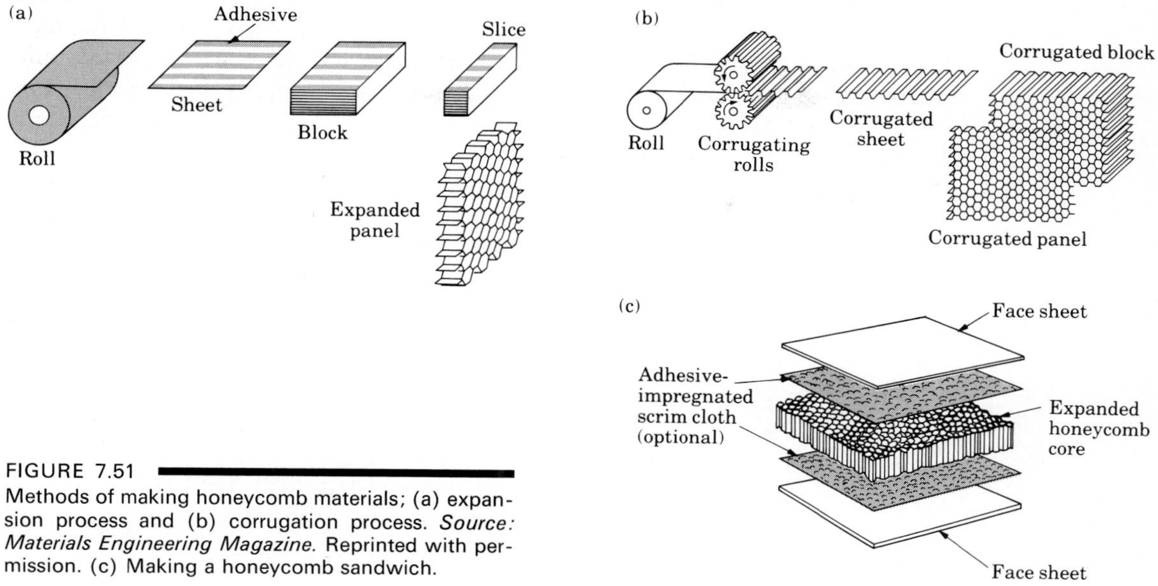

FIGURE 7.51
Methods of making honeycomb materials; (a) expansion process and (b) corrugation process. *Source: Materials Engineering Magazine.* Reprinted with permission. (c) Making a honeycomb sandwich.

coil and an adhesive is applied at intervals (node lines). The sheets are stacked and cured in an oven, whereby strong bonds develop at the adhesive joints. The block is then cut into slices of desired dimension and stretched to produce a honeycomb structure. This procedure is similar to expanding folded paper structures into the shape of decorative objects.

In the *corrugation* process (Fig. 7.51b), the sheet passes through a pair of specially designed rolls, which produce corrugated sheets that are then cut into desired lengths. Again, adhesive is applied to the node lines, and the block is cured. Note that no expansion process is involved. The honeycomb material is then made into a sandwich structure (Fig. 7.51c). Face sheets are subsequently joined with adhesives to the top and bottom surfaces.

7.12 Deep Drawing

In *deep drawing*, a flat sheet-metal blank is formed into a cylindrical or box-shaped part by means of a punch that presses the blank into the die cavity (Fig.

7.52a). Although the process is generally called deep drawing, meaning deep parts, the basic operation also produces parts that are shallow or have moderate depth. Deep drawing, first developed in the 1700s, has been studied extensively and has become an important metalworking process. Typical parts produced are beverage cans, pots and pans, containers of all shapes and sizes, sinks, and automobile panels.

The basic parameters in deep drawing a cylindrical cup are shown in Fig. 7.52(b). A circular sheet blank, with a diameter D_0 and thickness t_0, is placed over a die opening with a corner radius of R_d. The blank is held in place with a blankholder, or hold-down ring, with a certain force. A punch, with a diameter of D_p and a corner radius R_p, moves downward and pushes the blank into the die

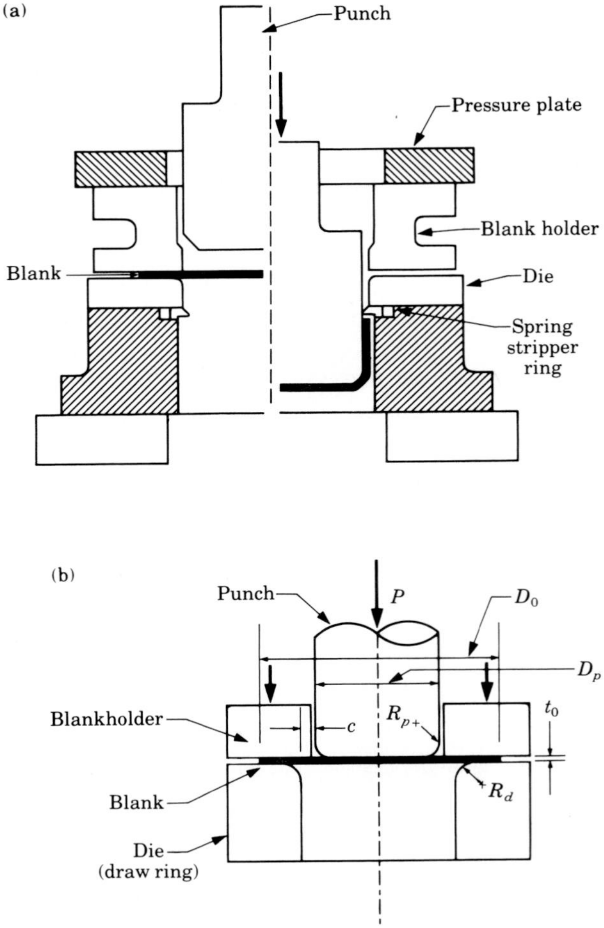

FIGURE 7.52 (a) Schematic illustration of the deep drawing process. This is the first step in the basic process by which aluminum beverage cans are produced today. The stripper ring facilitates the removal of the formed cup from the punch. (b) Variables in deep drawing of a cylindrical cup. Only the punch force in this illustration is a dependent variable; all others are independent variables, including the blankholder force.

cavity, thus forming a cup. The significant independent variables in deep drawing are

a) properties of the sheet metal,
b) the ratio of blank diameter to punch diameter,
c) the clearance between the punch and the die,
d) punch and die corner radii,
e) blankholder force,
f) friction and lubrication at the punch, die, and workpiece interfaces, and
g) speed of the punch.

At an intermediate stage during the deep drawing operation, the workpiece is subjected to the states of stress shown in Fig. 7.53. On element *A* in the blank, the radial tensile stress is due to the blank being pulled into the cavity, and the compressive stress normal to the element is due to the blankholder pressure. With a free-body diagram of the blank along its diameter, the radial tensile stresses lead to compressive hoop stresses on element *A*. (It is these hoop stresses that tend to cause the flange to wrinkle during drawing, thus requiring a blankholder under a certain force.) Under this state of stress, element *A* contracts in the hoop direction and elongates in the radial direction.

The cup wall, which is already formed, is subjected principally to a longitudinal tensile stress, as shown in element *B*. The punch transmits the drawing force *P* (see Fig. 7.52b) through the walls of the cup and to the flange that is being drawn into the cavity. The tensile hoop stress on element *B* is caused by the cup being held tightly on the punch because of its contraction under tensile stresses in the cup wall. (A thin-walled tube, when subjected to tension, becomes smaller in diameter, as can be shown from the generalized Hooke's law equations.) Thus element *B* tends to

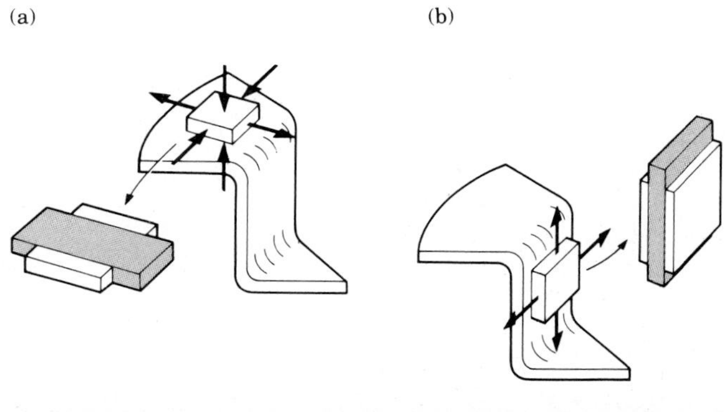

FIGURE 7.53
Deformation of elements in the flange (a) and the cup wall (b) in deep drawing of a cylindrical cup.

elongate in the longitudinal (axial) direction with no change in its width, because it is constrained by the rigid punch.

An important aspect of this operation is determining how much *stretching* and how much *pure drawing* is taking place (Fig. 7.54). Note that either with a high blankholder force or with the use of *draw beads* (Figs. 7.54b and 7.55) the blank can be prevented from flowing freely into the die cavity. The deformation of the sheet metal takes place mainly under the punch and the sheet begins to stretch, eventually resulting in necking and tearing. Whether necking is localized or diffused depends on (a) the strain rate sensitivity index m of the sheet metal (the higher the m value the more diffuse the neck), (b) geometry of the punch, and (c) lubrication.

Conversely, a low blankholder force will allow the blank to flow freely into the die cavity, whereby the blank diameter is reduced as drawing progresses. This process is referred to as *pure drawing*; the deformation of the sheet is mainly in the flange, and the cup wall is subjected only to elastic stresses. However, these stresses increase with increasing D_0/D_p ratio and can eventually lead to failure when the cup wall cannot support the load required to draw in the flange (Fig. 7.54a). Also note that, in pure drawing, element A in Fig. 7.53 tends to increase in thickness as it moves toward the die cavity because it is being reduced in diameter.

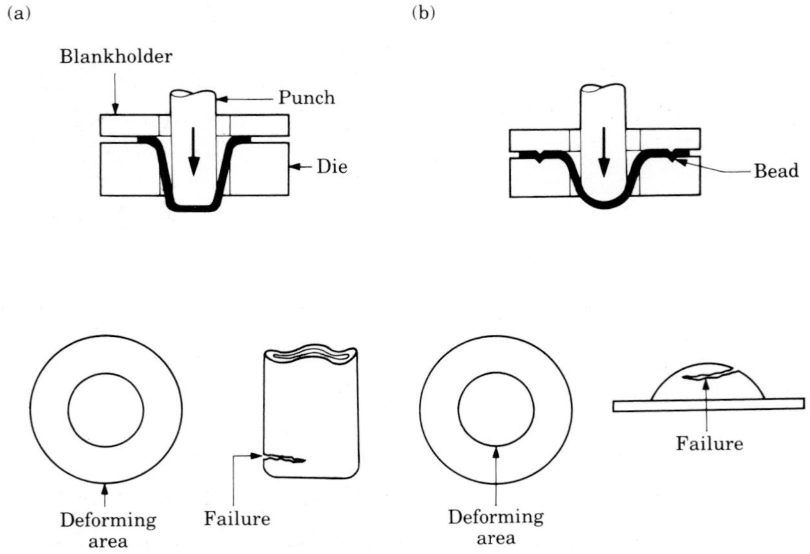

FIGURE 7.54
Examples of drawing operations: (a) pure drawing and (b) pure stretching. The bead prevents the sheet metal from flowing freely into the die cavity.

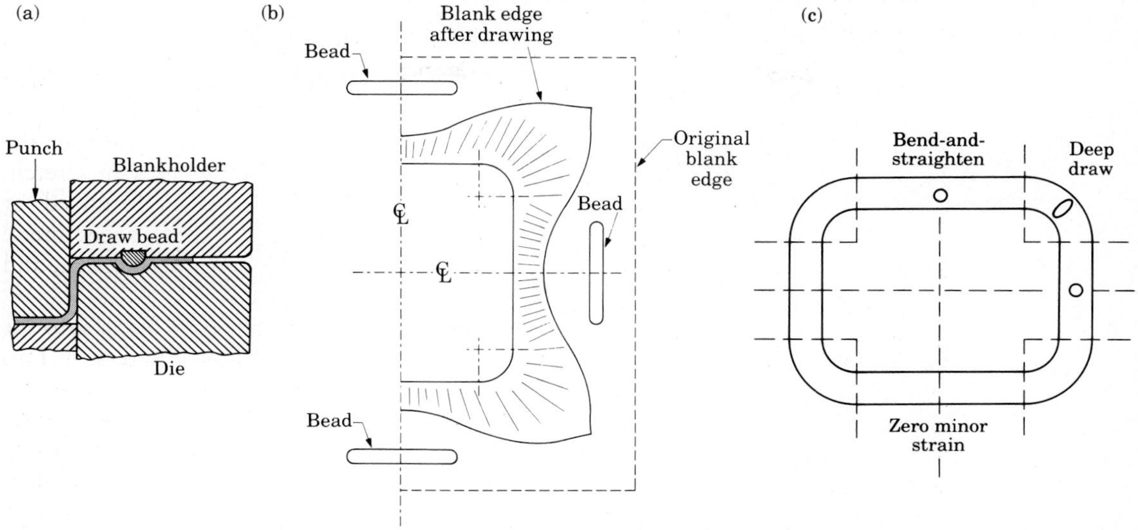

FIGURE 7.55 (a) Schematic illustration of a draw bead. (b) Metal flow during drawing of a box-shaped part, using beads to control the movement of the material. (c) Deformation of circular grids in drawing (see Section 7.13). *Source:* After S. Keeler.

Ironing. If the thickness of the sheet as it enters the die cavity is greater than the clearance between the punch and the die, the thickness will be reduced. This effect is known as *ironing*; it produces a cup with constant wall thickness (Fig. 7.56). Thus the smaller the clearance, the greater is the ironing. Obviously, because of volume constancy, an ironed cup will be longer than one produced with a large clearance.

FIGURE 7.56 Schematic illustration of the ironing process. Note that the cup wall is thinner than its bottom. All beverage cans without seams (known as 2-piece cans) are ironed, generally in three steps, after being drawn into a cup by deep drawing. (Cans with separate tops and bottoms are known as 3-piece cans.)

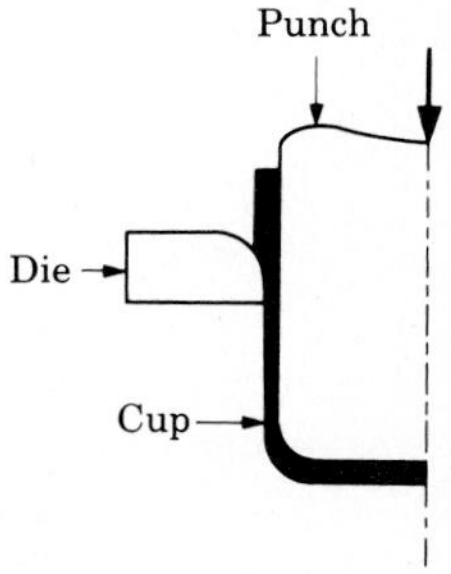

7.12.1 Deep drawability (limiting drawing ratio)

The *limiting drawing ratio* (LDR) is defined as the maximum ratio of blank diameter to punch diameter that can be drawn without failure, or D_0/D_p. Many attempts have been made to correlate this ratio with various mechanical properties of the sheet metal.

In an ordinary deep drawing process, failure generally occurs by thinning in the cup wall under high longitudinal tensile stresses. By observing the movement of the material into the die cavity (Fig. 7.53), we note that the material should be capable of undergoing a reduction in width (by being reduced in diameter), yet it should resist thinning under the longitudinal tensile stresses in the cup wall.

The ratio of width (w) to thickness (t) strain (Fig. 7.57) is

$$R = \frac{\varepsilon_w}{\varepsilon_t} = \frac{\ln\left(\dfrac{w_0}{w_f}\right)}{\ln\left(\dfrac{t_0}{t_f}\right)}, \tag{7.18}$$

where R is known as the *normal anisotropy* of the sheet metal (also known as *plastic anisotropy* or *strain ratio*). The subscripts o and f refer to the original and

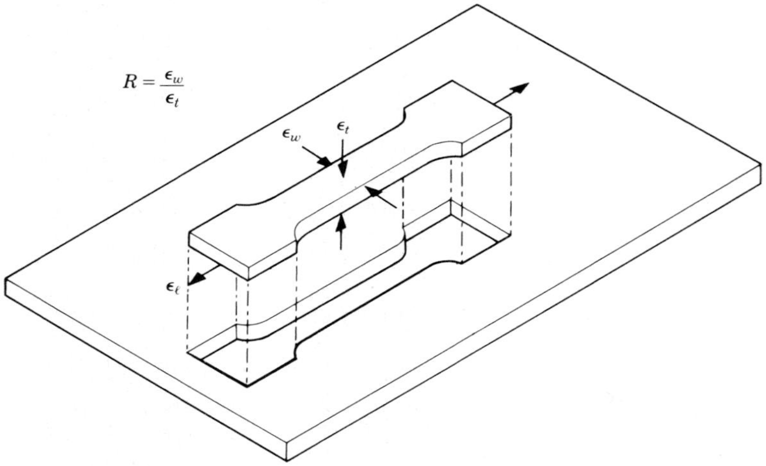

FIGURE 7.57 Definition of normal anisotropy ratio R in terms of width and thickness strains in a tensile-test specimen cut from a rolled sheet. Note that the specimen can be cut in different directions with respect to the length, or rolling direction, of the sheet.

TABLE 7.2
TYPICAL RANGE OF AVERAGE NORMAL ANISOTROPY RATIO $\bar{R}$ FOR VARIOUS SHEET METALS

Zinc	0.2
Hot-rolled steel	0.8–1.0
Cold-rolled rimmed steel	1.0–1.35
Cold-rolled aluminum-killed steel	1.35–1.8
Aluminum	0.6–0.8
Copper and brass	0.8–1.0
Titanium	4–6

final dimensions, respectively. An R value of one indicates that the width and thickness strains are equal to each other; that is, the material is isotropic.

As sheet metals are generally thin compared to their surface area, error in the measurement of small thickness is possible. Equation (7.18) can be modified, based on volume constancy, to

$$R = \frac{\ln\left(\dfrac{w_o}{w_f}\right)}{\ln\left(\dfrac{w_f \ell_f}{w_o \ell_o}\right)}, \tag{7.19}$$

where ℓ refers to the gage length of the sheet specimen. To calculate R, the final length and width in a test specimen are generally measured at an elongation of 15 percent to 20 percent, or for materials with lower ductility, below the elongation where necking begins.

Rolled sheets generally have *planar anisotropy*. Thus the R value of a specimen cut from a rolled sheet (Fig. 7.57) will depend on its orientation with respect to the rolling direction of the sheet. In this case, an average R value, $\bar{R}$, is calculated as follows:

$$\bar{R} = \frac{R_0 + 2R_{45} + R_{90}}{4}, \tag{7.20}$$

where the subscripts 0, 45, and 90 refer to angular orientation (in degrees) of the test specimen with respect to the rolling direction of the sheet. Thus an isotropic material has an $\bar{R}$ value of 1. Some typical $\bar{R}$ values are given in Table 7.2. Although hexagonal close-packed metals usually have high $\bar{R}$ values, the low value for zinc is the result of its high c/a ratio in the crystal lattice (see Fig. 3.2). The $\bar{R}$ value also depends on the grain size of the sheet metal. For cold-rolled steels, for example, $\bar{R}$ increases as grain size increases. For hot-rolled sheet steels, $\bar{R}$ is approximately 1, because the texture developed has a random orientation.

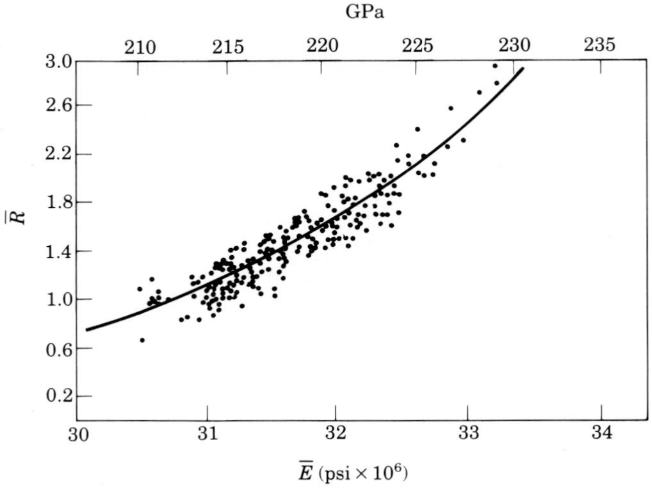

FIGURE 7.58 Relationship between average normal anisotropy $\bar{R}$ and the average modulus of elasticity $\bar{E}$ for steel sheet. *Source:* After P. R. Mould and T. R. Johnson, Jr., *Sheet Met. Ind.,* vol. 50, 1973, p. 328.

Experiments with sheet steels have shown that $\bar{R}$ is related to the average modulus of elasticity, $\bar{E}$, as shown in Fig. 7.58. Thus a high value of $\bar{R}$ is associated with a high value of $\bar{E}$. The average modulus, $\bar{E}$, is determined in the same manner as $\bar{R}$. Techniques and equipment are now available to measure the modulus of elasticity of sheet specimens by observing their natural frequencies.

The direct relationship between $\bar{R}$ and LDR, as determined experimentally, is shown in Fig. 7.59. In spite of its scatter, no other mechanical property of sheet metal indicates as consistent a relationship to LDR as $\bar{R}$ does. For an isotropic material and based on ideal deformation, the maximum LDR is equal to $e = 2.718$.

The *planar anisotropy* of a sheet, ΔR, can also be defined in terms of directional R values as follows:

$$\Delta R = \frac{R_0 - 2R_{45} + R_{90}}{2}, \tag{7.21}$$

which is the difference between the average of the R values in the 0° and 90° directions to rolling and the R value at 45°. And ΔR is related to ΔE, where ΔE is determined in the same manner as ΔR is in Eq. (7.21), with R replaced by E.

Earing. Planar anisotropy causes *ears* to form in drawn cups (Fig. 7.60), producing a wavy edge. The number of ears produced may be two, four, or six. The

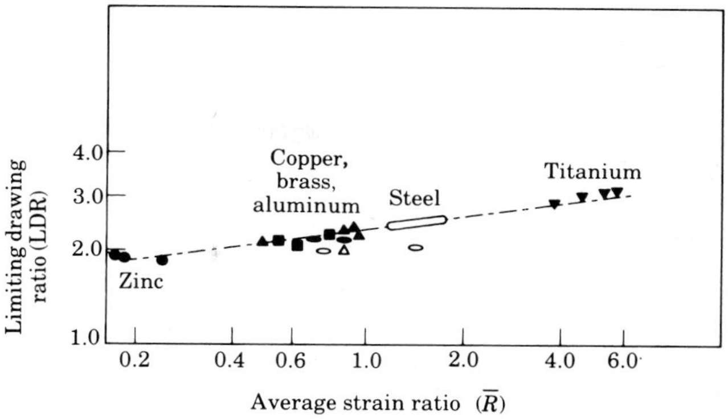

FIGURE 7.59 Effect of average normal anisotropy $\bar{R}$ on limiting drawing ratio (LDR) for a variety of sheet metals. Zinc has a high *c/a* ratio (see Fig. 3.2), whereas titanium has a low ratio. *Source:* After M. Atkinson, *Sheet Met. Ind.*, vol. 44, 1967, p. 167.

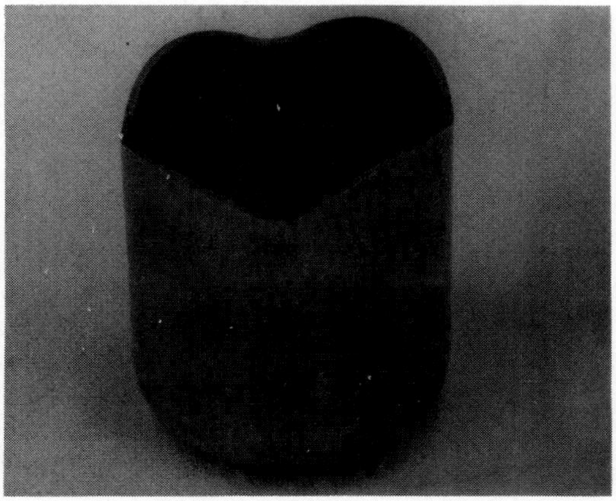

FIGURE 7.60 Earing in a drawn steel cup, caused by the planar anisotropy of the sheet metal.

height of the ears increases with increasing ΔR. When $\Delta R = 0$, no ears form. Ears are objectionable because they have to be trimmed off, wasting material.

Deep drawability is thus enhanced with a high $\bar{R}$ and a low ΔR. Generally, however, sheet metals with a high $\bar{R}$ also have a high ΔR. Attempts are being made to develop textures in sheet metals to improve drawability. The controlling parameters in processing metals have been found to be: alloying elements, additives, processing temperatures, annealing cycles after processing, thickness reduction in rolling, and cross rolling (biaxial) of plates in processing them into sheets.

• Example 7.6: Estimating the limiting drawing ratio.

Estimate the limiting drawing ratio (LDR) that you would expect from a sheet metal, which, when stretched by 23% in length, decreases in thickness by 10%.

SOLUTION. From volume constancy of the test specimen, we have

$$w_0 t_0 \ell_0 = w_f t_f \ell_f \quad \text{or} \quad \frac{w_f t_f \ell_f}{w_0 t_0 \ell_0} = 1.$$

From the information given,

$$\frac{\ell_f - \ell_0}{\ell_0} = 0.23 \quad \text{or} \quad \frac{\ell_f}{\ell_0} = 1.23$$

and

$$\frac{t_f - t_0}{t_0} = -0.10 \quad \text{or} \quad \frac{t_f}{t_0} = 0.90.$$

Hence

$$\frac{w_f}{w_0} = 0.903.$$

From Eq. (7.18), we obtain

$$R = \frac{\ln\left(\dfrac{w_0}{w_f}\right)}{\ln\left(\dfrac{t_0}{t_f}\right)} = \frac{\ln 1.107}{\ln 1.111} = 0.965.$$

If the sheet has planar isotropy, then $R = \bar{R}$ and from Fig. 7.59 we estimate that

$$\text{LDR} = 2.4.$$

The *punch force P* supplies the work required in deep drawing. The work, as in other deformation processes, consists of ideal work of deformation, redundant work, friction work and, when present, the work required for ironing (Fig. 7.61). Because of the many variables involved in this operation and because deep drawing is not a steady-state process, calculating the punch force is difficult. Of the various expressions that have been developed, one simple and very approximate formula for the punch force P is

$$P = \pi D_p t_0 (\text{UTS}) \left(\frac{D_0}{D_p} - 0.7 \right). \tag{7.22}$$

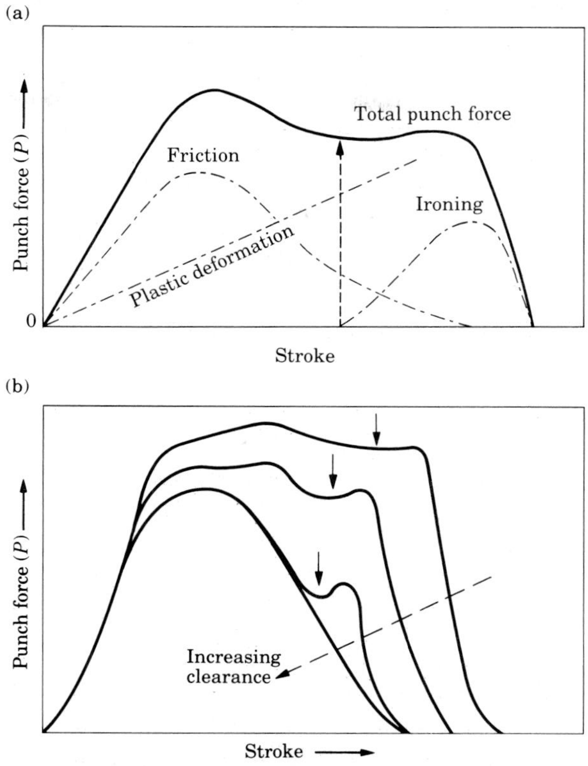

FIGURE 7.61
(a) Schematic illustrations of the variation of punch force with stroke in deep drawing, showing different components of the punch force. (b) Note that ironing does not begin until after the punch has traveled a certain distance and the cup is formed partially. Arrows indicate the start of ironing.

Note that Eq. (7.22) does not include friction, the punch and die corner radii, or the blankholder force. However, this empirical equation makes rough provision for these factors. The punch force is supported basically by the cup wall. If this force is excessive, *tearing* occurs, as shown in Fig. 7.54(a). Note that the cup was drawn to a considerable depth before failure occurred. This result can be expected from Fig. 7.61, where we observe that the punch force does not reach a maximum until after the punch has traveled a certain distance.

The punch corner radius and die radius (if they are greater than 10 times the sheet thickness) do not affect the maximum punch force significantly.

7.12.2 Drawing practice

The *blankholder pressure* is generally 0.7 percent to 1.0 percent of the sum of the yield and ultimate tensile strength of the sheet metal. Too high a blankholder force

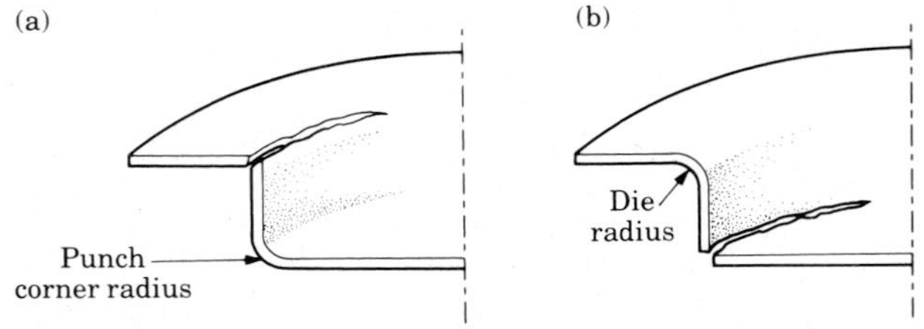

FIGURE 7.62
Effect of die and punch radii in deep drawing on fracture of a cylindrical cup. (a) Die radius too small. The die radius should generally be 5 to 10 times the sheet thickness. (b) Punch corner radius too small. Because friction between the cup and the punch aids in the drawing operation, excessive lubrication of the punch is detrimental to drawability.

increases the punch load (because of friction) and leads to tearing of the cup wall. However, if the force is too low, wrinkling occurs in the flange.

Clearances generally are 7–14 percent greater than the original sheet thickness. The choice of clearance depends on the thickening of the cup wall (which is a function of the drawing ratio). As the clearance decreases, ironing increases. If the clearance is too small, the blank may simply be pierced and sheared by the punch.

The *corner radius* of the punch and the die are important. If they are too small they can cause fracture at the corners (Fig. 7.62). If they are too large, the unsupported area wrinkles. Wrinkling in this region (and from the flange area) causes a defect on the cup wall called *puckering*.

Draw beads (see Fig. 7.55) are useful in controlling the flow of the blank into the die cavity. They are essential in drawing box-shaped and nonsymmetric parts. Draw beads also help in reducing the blankholder forces required because of the stiffness imparted to the flange by the bent regions at the beads. Proper design and location of draw beads requires considerable experience.

Redrawing. Containers or shells that are too difficult to draw in one operation are generally *redrawn*, as shown in Fig. 7.63(a). Another process is *reverse* drawing, shown in Fig. 7.63(b), where the metal is subjected to bending in the direction opposite to its original bending configuration. This reversal in bending results in *strain softening* and is another example of the Bauschinger effect (described in Section 2.3.2). This operation requires lower forces, and the material behaves in a more ductile manner.

Drawing without a blankholder. Deep drawing may be carried out without a blankholder, provided that the sheet metal is thick enough to prevent wrinkling. The dies are specially contoured for this operation, an example of which is shown in Fig. 7.64. An approximate limit for drawing without a blankholder is given by

$$D_0 - D_p < 5t_0. \tag{7.23}$$

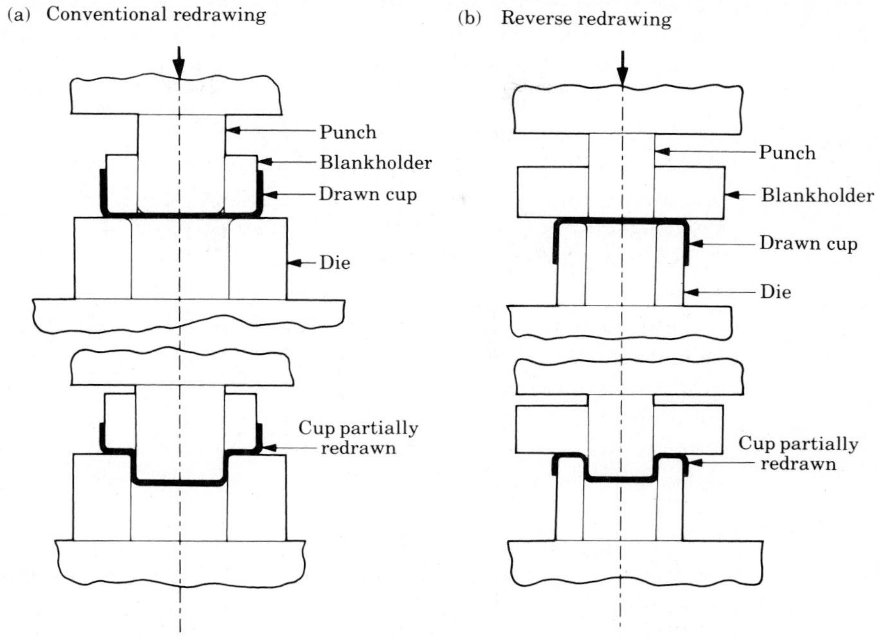

FIGURE 7.63

Reducing the diameter of drawn cups by redrawing operations: (a) conventional redrawing and (b) reverse redrawing. Small-diameter deep containers undergo many drawing and redrawing operations.

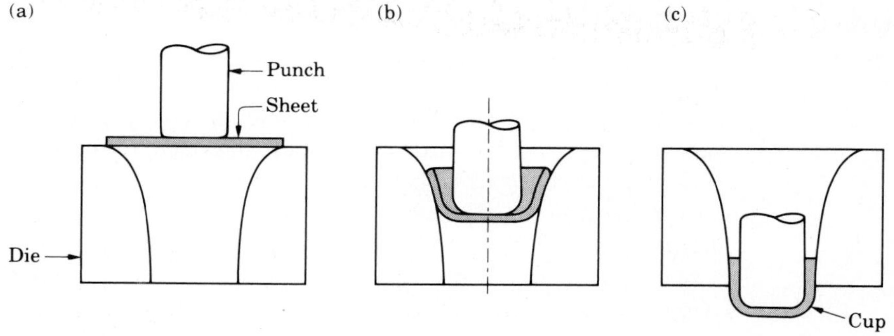

FIGURE 7.64

Deep drawing without a blankholder, using a *tractrix* die profile. The tractrix is a special curve, the construction for which can be found in texts on analytical geometry or in handbooks.

It is apparent from Eq. (7.23) that the process can be used with thin materials for shallow draws. Although the punch stroke is greater, a major advantage of this process is the reduced cost of tooling and equipment.

Various operations. Numerous types of parts are also made with shallow or moderate depths. Some involve drawing or stretching, or a combination of these operations. Parts may be embossed with male and female dies or by other means. *Embossing* involves a number of very shallow draws (made with matching dies) on a sheet. The process is used principally for decorative purposes.

Lubrication. Lubrication in deep drawing is important in lowering forces, increasing drawability, reducing tooling wear, and reducing part defects. In general, lubrication of the punch should be minimized, as friction between the punch and the cup improves drawability. For general applications, commonly used lubricants are mineral oils, soap solutions, and heavy-duty emulsions. For more difficult applications, coatings, wax, and solid lubricants are used (see Table 6.4).

Tooling and equipment. The most commonly used tool materials for deep drawing are tool steels and alloyed cast iron. Other materials, including carbides and plastics, may also be used, depending on the particular application (see Table 3.7). A double-action hydraulic press is generally used for deep drawing, although mechanical presses are also used. The double-action press controls the punch and blankholder independently and forms the part at a constant speed. Punch speeds generally range between 0.1 and 0.3 m/s (20 and 60 ft/min). Speed is generally not important in drawability, although lower speeds are used for high-strength metals.

7.13 Formability of Sheet Metals

Sheet-metal *formability* is generally defined as the ability of the metal to undergo the desired shape change without failure, such as by necking or tearing. Although not generally done, the definition can include the strength of the metal: the stronger it is, the lower is its formability. The formability of sheet metals has been of great interest because of its technologic significance. In this section we describe the different methods that are used to predict formability.

7.13.1 Testing for formability

Cupping tests. Because sheet forming is basically a biaxial stretching process, the earliest tests developed to determine or to predict formability were cupping tests, such as the *Erichsen* and *Olsen* tests (*stretching*) and the *Swift* and *Fukui* tests

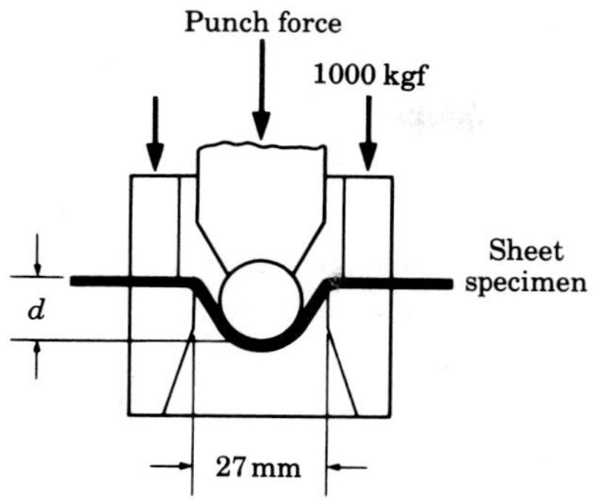

FIGURE 7.65 The Erichsen cupping test to determine formability of sheet metal. The sheet metal under the steel ball is subjected to biaxial stretching. The greater the distance *d* before failure, the greater is the formability of the material. Ball diameter = 20 mm.

(*drawing*). A typical cupping test is the Erichsen test, shown in Fig. 7.65. The sheet-metal specimen is clamped over a circular flat die with a load of 1000 kg. A 20-mm–diameter steel ball is then hydraulically pushed into the sheet metal until a crack appears on the stretched specimen or the punch force reaches a maximum. The distance d, in millimeters, is known as the Erichsen number. The greater the value of d, the greater is the formability of the sheet. Cupping tests measure the capability of the material to be stretched before fracturing and are relatively easy to perform. However, they do not simulate the exact conditions of actual forming operations, because the stretching under the ball is axisymmetric.

Bulge test. Equal (balanced) biaxial stretching of sheet metals is also performed in the *bulge test*, which has been used extensively to simulate sheet-forming operations. In this test, a circular blank is clamped at its periphery and is bulged by *hydraulic* pressure, thus replacing the punch. The process is one of pure stretch forming, and no friction is involved, as in using a punch. This test can be used to provide effective stress–effective strain curves for biaxial loading under frictionless conditions.

Forming-limit diagrams. An important development in testing formability of sheet metals is the construction of *forming-limit diagrams* (FLD). In these tests, the sheet blank is marked with a grid pattern of circles or similar patterns, using chemical etching or photoprinting techniques. The blank is then stretched over an unlubricated punch and the deformation of the circles is observed and measured in regions where necking and tearing occur. For improved accuracy of measurement, the circles are made as small as practical.

In order to develop unequal biaxial stretching, the specimens are prepared with varying widths (Fig. 7.66). Thus a square specimen produces equal (balanced) biaxial stretching under the punch, whereas a specimen with a small width approaches uniaxial stretching. After a series of such tests is performed on a particular sheet metal (Fig. 7.67), the boundaries between failed and safe regions are plotted on the forming-limit diagram (Fig. 7.68). Another example of a failed part with round and square grid patterns is shown in Fig. 7.69.

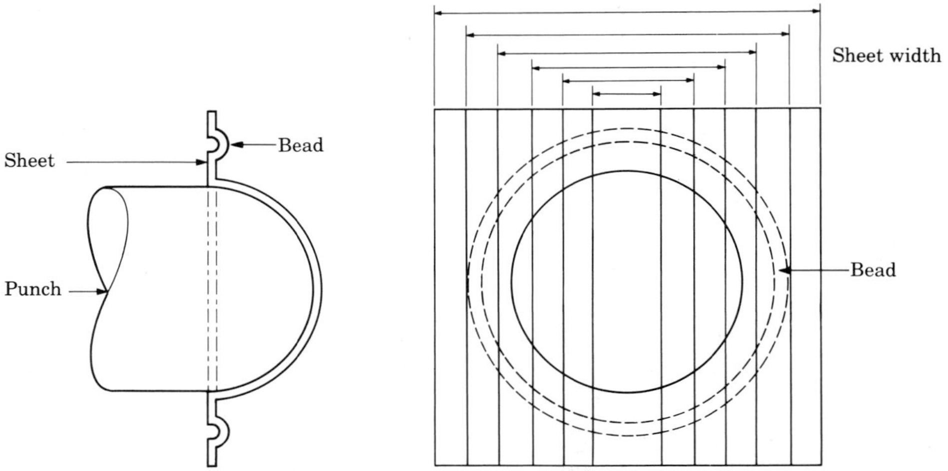

FIGURE 7.66 Schematic illustration of the punch–stretch test on sheet specimens with different widths and clamped at the edges. The narrower the specimen, the more uniaxial is the stretching. A large square specimen stretches biaxially under the hemispherical punch (see also Fig. 7.67).

FIGURE 7.67 Bulge test results on steel sheets of various widths. The first specimen (left) stretched farther before cracking than the last specimen. From left to right, the state of stress changes from uniaxial to biaxial stretching. *Source:* Courtesy of R. W. Thompson, Inland Steel Research Laboratories.

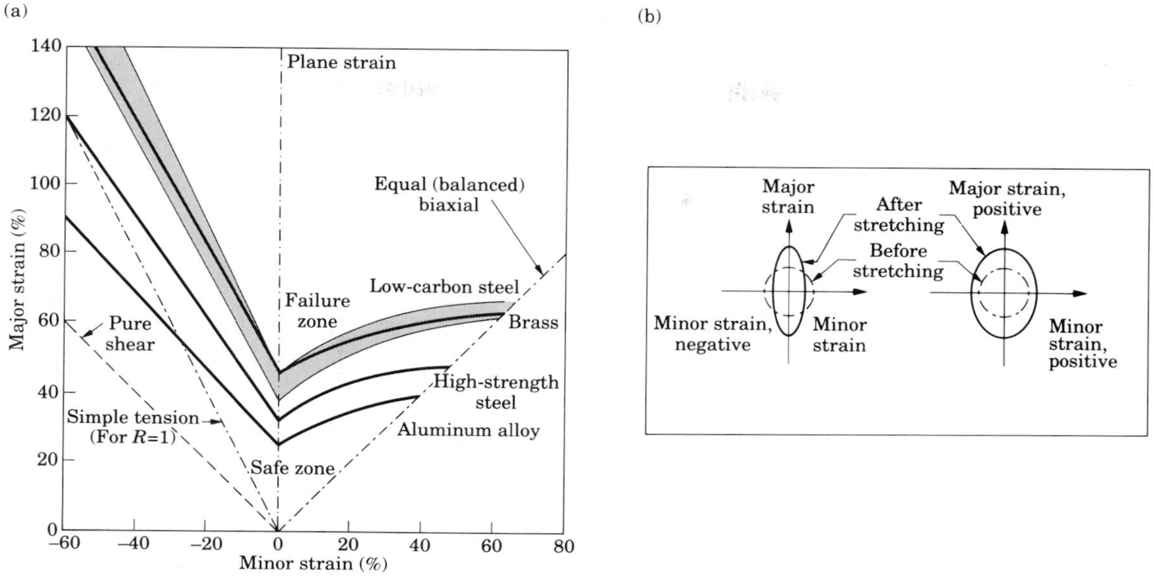

FIGURE 7.68

(a) Forming-limit diagram (FLD) for various sheet metals. The major strain is always positive. The region above the curves is the failure zone; hence the state of strain in forming must be such that it falls below the curve for a particular material; R is the normal anisotropy. (b) Note the definition of positive and negative minor strains. If the area of the deformed circle is larger than the area of the original circle, the sheet is thinner than the original, because the volume remains constant during plastic deformation. *Source:* After S. S. Hecker and A. K. Ghosh.

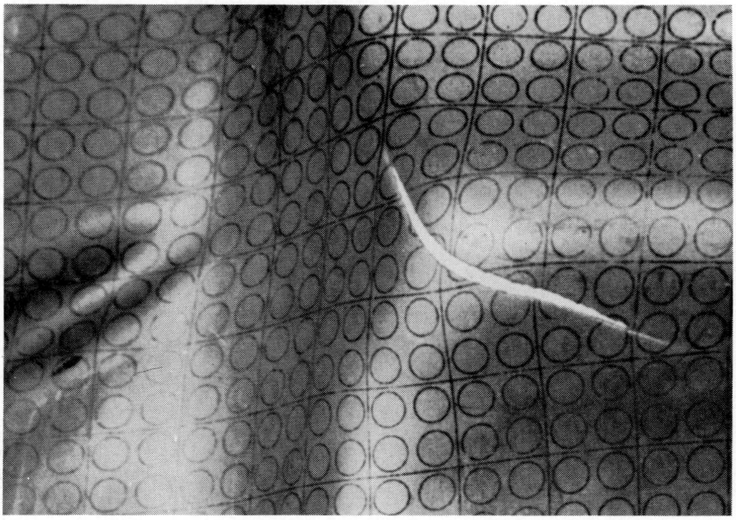

FIGURE 7.69

An example of using grid marks (circular and square) to determine the magnitude and direction of surface strains in sheet-metal forming. Note that the crack (tear) is generally perpendicular to the major (positive) strain. *Source:* After S. P. Keeler.

In the forming-limit diagram the *major* and *minor strains* (engineering) are obtained as follows. Note in Fig. 7.68(b) that after stretching the original circle has deformed into an ellipse. The major axis of the ellipse represents the major direction and magnitude of stretching. The major strain plotted in Fig. 7.68(a) is the *engineering strain* (percent) in this direction. Likewise, the minor axis of the ellipse represents the magnitude of the stretching *or* shrinking in the transverse direction. Thus the minor strain can be negative or positive. (The major strain is always positive, because forming sheet metal takes place by stretching.)

For example, if we draw a circle in the center of a sheet-metal tensile-test specimen and then stretch it, the minor strain will be negative. The reason is Poisson's ratio (that is, the specimen becomes narrower as it is stretched), which can be easily observed by experimenting with a wide rubber band. However, if we place a circle on a spherical rubber balloon and inflate it, the minor strain is positive and equal in magnitude to the major strain. The circle simply becomes a larger circle.

By observing the difference in surface area between the original circle and the ellipse, we can also determine whether the thickness of the sheet has changed. If the area of the ellipse is larger than the original circle, the sheet has become thinner (because volume remains constant in plastic deformation).

Friction at the punch–metal interface can also be important to test results. With well-lubricated interfaces, the strains are more uniformly distributed over the punch. Depending on the notch sensitivity of the sheet metal, surface scratches, deep gouges, and blemishes can reduce formability and cause premature tearing and failure during testing and in actual forming operations. With effective lubrication the coefficient of friction in sheet-metal forming generally ranges from about 0.05 to 0.1 for cold forming, and from 0.1 to 0.2 for elevated temperatures.

From Fig. 7.68(a), we observe, as expected, that different materials have different forming-limit diagrams. The higher the curve, the better is the formability of the material. Also note that for the same minor strain, say 20 percent, a compressive minor strain is associated with a higher major strain before failure than is a tensile (positive) minor strain. In other words, it is desirable for the minor strain to be negative, that is, shrinking in the minor direction. Special tools have been designed for forming sheet metals that take advantage of the beneficial effect of negative minor strains on extending formability (see Fig. 7.32).

The effect of sheet-metal thickness on forming-limit diagrams is to raise the curves in Fig. 7.68(a). Thus the thicker the sheet, the higher its formability curve, hence the more formable it is. In actual forming operations, however, a thick blank may not bend as easily around small radii. The possible effect of the rate of deformation on forming-limit diagrams should also be assessed for each material. Extensive studies are being carried out to develop new test methods to predict the behavior of metals in sheet-forming operations.

• Example 7.7: Estimating diameter of expansion.

A thin-walled spherical shell made of the aluminum alloy shown in Fig. 7.68(a) is being expanded by internal pressure. If the original diameter is 200 mm, what is the maximum diameter to which it can be expanded safely?

SOLUTION. Because the material is being stretched in a state of equal (balanced) biaxial tension, we find from Fig. 7.68(a) that the maximum allowable engineering strain is about 40%. Thus

$$e = \frac{\pi D_f - \pi D_0}{\pi D_0} = \frac{D_f - 200}{200} = 0.40.$$

Hence

$$D_f = 280 \text{ mm}.$$

• Example 7.8: Strains in the sheet metal body of an automobile.

Computer programs have been developed that compute the major and minor strains and their orientations from the measured distortions of grid patterns placed on the surface of sheet metal. One application for an automobile body is shown in Fig. E7.2 for various panels. Note that the trunk lid and the roof are subjected mainly to

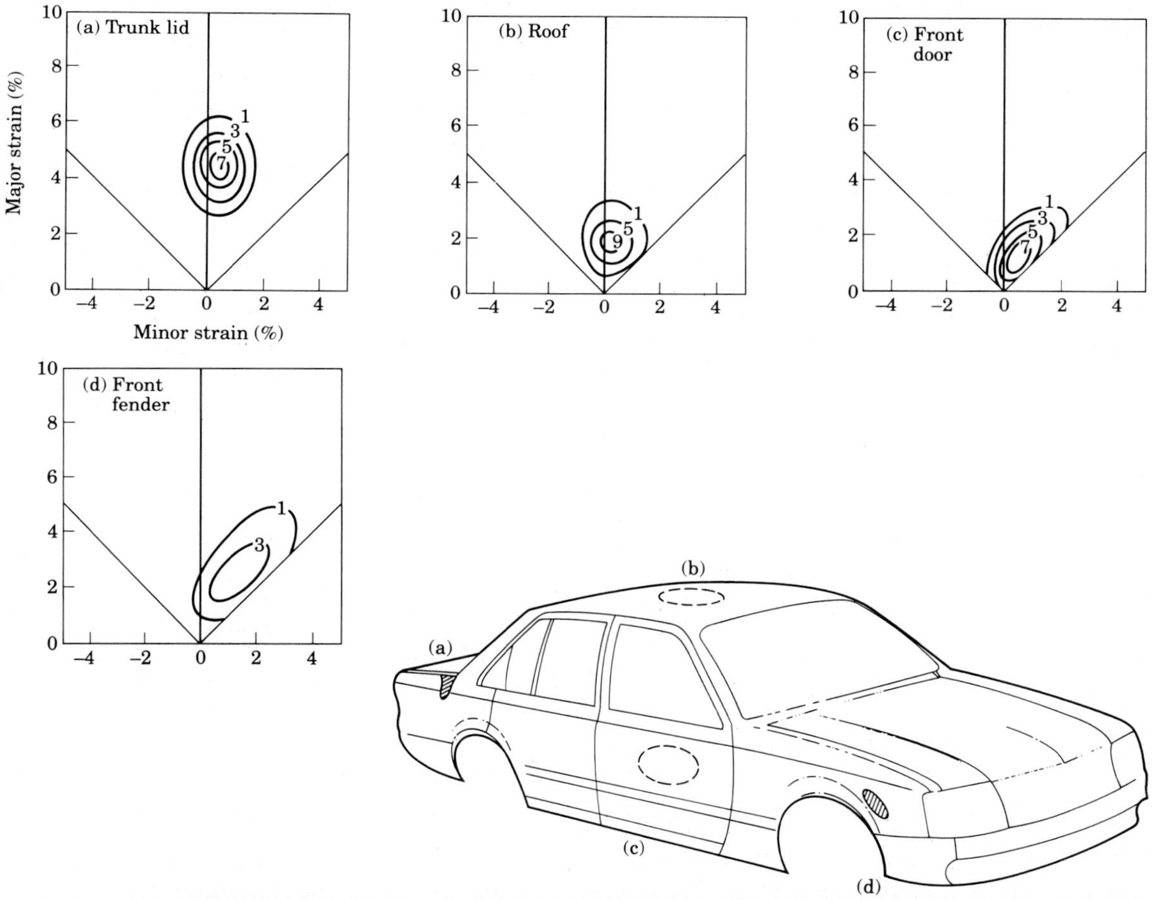

FIGURE E7.2

plane strain (see Fig. 7.68a), whereas the front door and front fender are subjected to biaxial strains. The numbers in the strain paths indicate the frequency of occurrence. *Source*: After T. J. Nihill and W. R. Thorpe.

Limiting-dome height test. Recall that in the test illustrated in Fig. 7.66, the major and minor strains were measured and plotted on the forming-limit diagram. In the *limiting-dome height* (LDH) test, performed with similarly prepared specimens, the *height* of the dome at failure of the sheet—or when the punch force reaches a maximum—is measured, as is done in the test shown in Fig. 7.65. Because the specimens are clamped, the LDH test indicates the capability of the material to stretch without failure. It has been shown that high LDH values are related to sheet-metal properties such as high n and m values, as well as high total elongation (percent) of the sheet metal.

7.13.2 Dent resistance of sheet-metal parts

In certain applications involving sheet-metal parts, such as automotive body panels, appliances, and office furniture, an important consideration is the dent resistance of the sheet-metal panel. The factors significant in dent resistance are the yield stress Y, thickness t, and the shape of the panel. *Dent resistance* can be defined as

$$\text{Dent resistance} \propto \frac{Y^2 t^4}{S}, \tag{7.24}$$

where S is the panel stiffness, which in turn is defined as

$$S = (E)(t^a)(\text{shape}), \tag{7.25}$$

where the value of a ranges from 1 to 2 for most panels. As for shape, the smaller the curvature (hence the flatter the panel), the greater the dent resistance is because of its flexibility. Thus dent resistance increases with increasing strength and thickness and decreases with increasing elastic modulus E and stiffness and decreasing curvature.

Dents are usually caused by *dynamic* forces, such as those of falling objects or other objects that hit the sheet-metal panel. In typical automotive panels, for instance, impact velocities range up to 45 m/s (150 ft/s). Thus the dynamic yield stress (high strain rate), rather than the static yield stress, is the significant strength parameter. Denting under quasistatic forces could also be important.

For materials in which yield stress increases with strain rate, denting requires higher energy levels than under static conditions. Furthermore, dynamic forces tend to cause more *localized* dents than static forces, which tend to spread the dented area. Because a portion of the energy goes into elastic deformation, the modulus of resilience of the sheet metal is an additional factor to be considered.

7.13.3 Modeling of sheet-forming processes

In Section 6.2.2 we outlined the techniques for studying bulk deformation processes and described modeling of impression-die forging using the finite-element method (see Fig. 6.10). Such mathematical modeling is also being applied to sheet-forming processes. Deformation of an originally square mesh on the surface of a sheet-metal part is shown in Fig. 7.70, as developed by computer-aided modeling. The ultimate goal of such simulation techniques is rapid analysis of stresses, strains, flow patterns, and springback as functions of parameters such as material characteristics, friction, anisotropy, deformation speed, and temperature. Through such interactive analysis, we can determine the optimum tool and die geometry to make a certain part, thus reducing or eliminating costly die tryouts. In addition, these simulation techniques can be used to determine blank size and shape (including intermediate shapes), press characteristics, and process parameters to optimize the forming operation. The application of computer modeling, although requiring powerful computers and extensive software, has already proven to be a cost-effective tool in sheet-metal forming, particularly in the automotive industry.

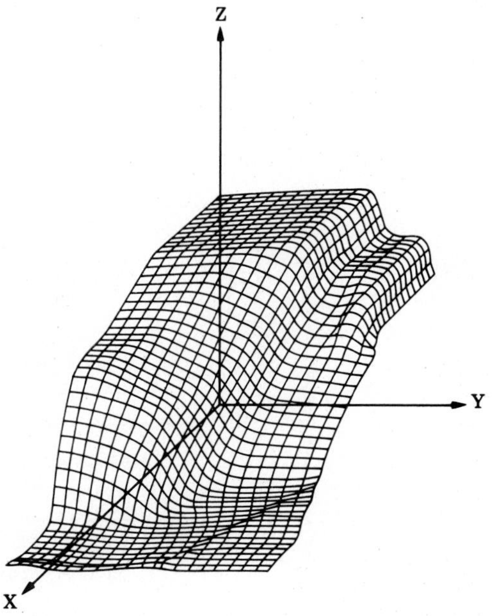

FIGURE 7.70

Deformation of a square mesh in computer simulation of forming a sheet-metal part. *Source:* J. L. Duncan, R. Sowerby, and E. Chu.

7.14 Economics of Sheet-Metal Forming

Sheet-metal forming involves economic considerations similar to those for the other processes that we have discussed. Sheet-forming operations compete with each other, as well as with other processes, more than other processes do. We have noted that sheet-forming operations are versatile and that a number of different processes can be used to produce the same part. For example, a cup-shaped part can be formed by deep drawing, spinning, rubber forming, or explosive forming. Similarly, a cup can be formed by impact extrusion or casting, or by fabricating it from different pieces.

The part shown in Fig. 7.71 can be made either by deep drawing or by conventional spinning. However, the die costs for the two processes are significantly different. Deep-drawing dies have many components and cost much more than the relatively simple mandrels and tools employed in spinning. Consequently, the die cost per part in drawing will be high if few parts are needed. On the other hand, this part can be formed by deep drawing in a much shorter time (seconds) than by spinning (minutes), even if the latter operation is automated. Furthermore, spinning requires more skilled labor. Considering all these factors, we find that the breakeven point is at about 700 parts. Thus deep drawing is more economical for quantities greater than that.

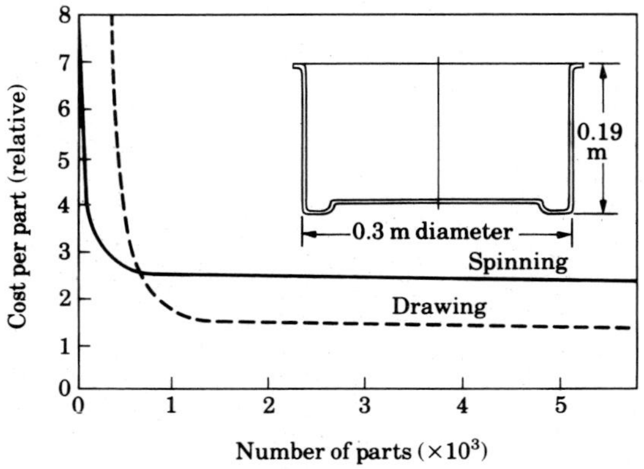

FIGURE 7.71
Cost comparison for manufacturing a round sheet-metal container by conventional spinning and deep drawing. Note that for small quantities spinning is more economical.

SUMMARY

Sheet-metal forming processes generally involve workpieces that have a high ratio of surface area to thickness. Unlike in bulk deformation processes, in sheet forming, the material is generally prevented from being reduced in its cross-sectional area in order to avoid necking and fracture.

Sheet forming generally involves the application of tensile stresses in the plane of the sheet by the use of various tools and dies. Because of the large surfaces involved, friction can be a significant factor in the overall operation as far as forces and formability are concerned.

The surfaces of the parts formed by these processes are generally not subjected to further processing, except surface coatings, painting, and joining. Consequently, surface finish can be a significant factor. Lubrication is also an important parameter, as it directly affects surface finish. Surface conditions are important, because many sheet-metal parts are subsequently joined or welded.

Sheet metals are generally thin compared to the shapes to which they are formed. Thus springback is a significant problem, especially in bending.

The forces and energy required in sheet forming can be transmitted to the workpiece not only through solid tools and dies, but also by means such as flexible rubber or polyurethane members, and through electrical, chemical, and magnetic means. Unlike in bulk deformation processes, forces in sheet forming are not particularly significant as far as energy requirements are concerned. A wide variety of processes are available for forming sheet metals into complex shapes.

The important material parameters are the capacity of the sheet metal to stretch uniformly (hence the desirability for a high strain-hardening exponent n) and its resistance to thinning (hence high normal anisotropic properties). Various tests have been developed to predict the formability of sheet metals in actual processing. Because of the low thickness-to-length ratios in most sheet-forming operations, buckling and wrinkling are significant problems. They can be avoided by properly designing tools and dies and minimizing the unsupported length of the material during processing.

BIBLIOGRAPHY

Benjamin, W.P., *Plastic Tooling*. New York: McGraw-Hill, 1972.

Blickwede, D., *New Knowledge about Sheet Steel*. Metals Park, Ohio: American Society for Metals, 1970.

Bowman, H.B., *Handbook of Precision Sheet, Strip and Foil*. Metals Park, Ohio: American Society for Metals, 1980.

Bruno, E.J. (ed.), *High-Velocity Forming of Metals*, rev. ed. Dearborn, Mich.: Society of Manufacturing Engineers, 1968.

Crane, E.V., *Plastic Working of Metals and Non-metallic Materials in Presses*, 3d ed. New York: Wiley, 1964.

Davies, R.S. and E.R. Austin, *Developments in High Speed Metal Forming*. New York: Industrial Press, 1970.

Eary, D.F. and E.A. Reed, *Techniques of Pressworking Sheet Metal*, 2d ed. Englewood Cliffs, N.J.: Prentice-Hall, 1974.

Ezra, A.A., *Principles and Practice of Explosive Metalworking*. London: Industrial Newspapers Ltd, 1973.

Grainger, J.A., *Flow Turning of Metals*. Brighton, England: The Machinery Publishing Co., 1969.

Hoffmann, E.G., *Fundamentals of Tool Design*, 2d ed. Dearborn, Mich.: Society of Manufacturing Engineers, 1984.

Koistinen, D.P., and N.M. Wang (eds.), *Mechanics of Sheet Metal Forming*. New York: Plenum, 1978.

Metals Handbook, 8th ed., Vol. 4: *Forming*. Metals Park, Ohio: American Society for Metals, 1969.

Metals Handbook, 9th ed., Vol. 14: *Forming and Forging*. Metals Park, Ohio: ASM International, 1988.

Morgan, E., *Tinplate and Modern Canmaking Technology*. Oxford, England: Pergamon, 1985.

Pacquin, J.R., and R.E. Crowley, *Die Design Fundamentals*. New York: Industrial Press, 1987.

Sachs, G., *Principles and Methods of Sheet Metal Fabricating*, 2d ed. New York: Reinhold, 1966.

Source Book on Forming of Steel Sheet. Metals Park, Ohio: American Society for Metals, 1976.

Strasser, F., *Metal Stamping Plant Productivity Handbook*. New York: Industrial Press, 1983.

Tool and Manufacturing Engineers Handbook, 4th ed., Vol. 2: *Forming*. Dearborn, Mich.: Society of Manufacturing Engineers, 1984.

Watkins, M.T., *Metal Forming II: Pressing and Related Processes*. New York: Oxford, 1975.

QUESTIONS

7.1 Take any three topics from Chapter 2 and, with specific examples for each, show their relevance to the topics covered in this chapter.

7.2 Repeat Problem 7.1, but for Chapter 3.

7.3 Describe the (a) similarities and (b) dissimilarities between the bulk deformation processes described in Chapter 6 and the sheet-forming processes covered in this chapter.

7.4 Describe the material and process variables that influence the shape of the punch force versus stroke curve for shearing, such as that shown in Fig. 7.8, including its height and width.

7.5 In preparing a large number of blanks for sheet-forming operations, the blanks are spaced (nesting) to minimize material waste. Describe the considerations involved in this procedure, with particular emphasis on subsequent forming operations.

7.6 Describe your observations concerning Figs. 7.6 and 7.7.

7.7 Inspect a common paper punch and comment on the shape of the punch end, compared to those shown in Fig. 7.13.

7.8 Can the presence of burrs be beneficial in certain applications? If so, give specific examples.

7.9 Refer to Eq. (7.11) and explain why the maximum bending force P depends on (a) L, (b) W, and (c) T^2.

7.10 Describe the difference between compound, progressive, and transfer dies.

7.11 It has been stated that the quality of the sheared edges can influence the formability of sheet metals. Explain why.

7.12 Explain why and how various factors influence springback in bending of sheet metals.

7.13 We note in Fig. 7.17 that the state of stress shifts from plane stress to plane strain as the ratio of length of bend to sheet thickness increases. Explain why.

7.14 Inspect Fig. 7.20 and describe those material properties that have an effect on the relative position of the curves.

7.15 In Table 7.1 we note that the hard material conditions have higher R/T ratios than do soft ones. Explain why.

7.16 Why do tubes buckle when bent?

7.17 Based on Fig. 7.24, sketch the shape of a U-die to produce channel-shaped bends.

7.18 Explain why negative springback does not occur in air bending of sheet metals.

7.19 Give examples for which the presence of beads on parts is beneficial.

7.20 Explain why cupping tests do not always predict the behavior of the sheet metal in actual forming operations.

7.21 Assume that you are carrying out a sheet-forming operation and find that the material is not sufficiently ductile. What would you do to improve ductility?

7.22 Refer to Fig. 7.41 and explain why a further increase in the ductility of the material (beyond about 50 percent tensile reduction of area) does not improve the maximum spinning reduction per pass.

7.23 Many missile components are made by spinning. What methods would you use if spinning processes were not available?

7.24 In deep drawing of a cylindrical cup is it always necessary for tensile circumferential stresses to be on the element in the cup wall? (See Fig. 7.53b.)

7.25 When comparing the hydroform process with the deep drawing process, we stated that deeper draws are obtained in the former method. With appropriate sketches explain why.

7.26 We note in Fig. 7.53(a) that element A in the flange is subjected to compressive circumferential (hoop) stresses. Using a free-body diagram, explain why.

7.27 From the topics covered in this chapter, list and explain specifically two examples for which (a) friction is desirable and (b) friction is not desirable.

7.28 Explain why increasing the value of normal anisotropy R improves the deep drawability of sheet metals.

7.29 Make a list of the independent variables that influence the punch force in deep drawing of a cylindrical cup and explain why and how they influence the force.

7.30 Explain why the simple tension line in the forming-limit diagram in Fig. 7.68 states that it is for $R = 1$, where R is the normal anisotropy.

7.31 Explain the reasons for developing forming-limit diagrams. Do you have any criticisms of such diagrams? Explain.

7.32 Explain the reasoning behind Eq. (7.20) for normal anisotropy and Eq. (7.21) for planar anisotropy, respectively.

7.33 Describe why earing occurs. How would you avoid it?

7.34 What is the significance of the size of the grid patterns used in studying sheet-metal formability?
7.35 We stated that the thicker the sheet metal is, the higher the curve in the forming-limit diagram is. Explain why.
7.36 Inspect the earing shown in Fig. 7.60 and identify the direction in which the blank was cut.
7.37 Describe the factors that influence the size and length of beads in sheet-forming operations.
7.38 We showed that the strength of metals depends on their grain size. Would you then expect that strength influences the *R* value of sheet metals? Explain.
7.39 A general rule for dimensional relationships for successful drawing without a blank-holder is given in Eq. (7.23). Explain what happens if this limit is exceeded.
7.40 Explain why the three broken lines (simple tension, plane strain, and equal biaxial stretching) in Fig. 7.68(a) have those particular slopes.
7.41 Think of a specific part and explain which of the processes from Chapters 6 and 7, respectively, can be used to make that part. Then make a final choice, explaining your reasoning.
7.42 Discuss the process or processes by which you think some of the sheet-metal products around you were made.

PROBLEMS

7.1 Derive Eq. (7.5).
7.2 Regarding Eq. (7.5), we stated that actual values of e_o are higher than e_i because of the shifting of the neutral axis during bending. With an appropriate sketch explain this phenomenon.
7.3 Estimate the maximum bending force required for a $\frac{1}{8}$-in. thick and 10-in. wide titanium alloy Ti-5Al-2.5Sn in a V die with a width of 8 in.
7.4 In Example 7.3, calculate the work done by the force–distance method.
7.5 What would be the answer to Example 7.3 if the tip of the force *F* is fixed to the strip by some means, thus maintaining its lateral position. *Hint*: Note that the left portion of the strip will now be strained more than the right portion.
7.6 Calculate the magnitude of force *F* in Example 7.3 for $\alpha = 15°$.
7.7 Calculate the press force needed in punching 5052-O aluminum foil, 0.1 mm thick, in the shape of a square hole 30 mm on each side.
7.8 In Example 7.5 calculate the amount of TNT required to develop a pressure of 15,000 psi on the workpiece surface.
7.9 Estimate the limiting drawing ratio (LDR) for the materials listed in Table 7.2.
7.10 For the same material and thickness as in Problem 7.3, calculate the force required for deep drawing, with a blank diameter of 12 in. and punch diameter of 9 in.
7.11 A steel sheet has *R* values of 1.3, 0.9, and 1.9 for the 0°, 45°, and 90° directions to rolling, respectively. If a round blank is 100 mm in diameter, estimate the cup diameter to which it can be drawn. Also state whether ears will form and, if so, why.

8

Material-Removal Processes: Cutting

8.1 Introduction

Parts manufactured by casting, forming, and various shaping processes often require further operations before the product is ready for use. Also, in many engineering applications, parts have to be interchangeable in order to function properly and reliably during their expected service lives, thus requiring control of dimensional accuracy. Often, critical choices have to be made about the extent of shaping and forming versus the extent of machining to be done on a workpiece to produce an acceptable part.

Although *machining* is the broad term used to describe *removal* of material from a workpiece, it covers several processes, which we usually divide into the following categories:

- Cutting, generally involving single-point or multipoint cutting tools, each with a clearly defined geometry.

- Abrasive processes, such as grinding.
- Nontraditional machining processes, utilizing electrical, chemical, and optical sources of energy.

We can summarize why material-removal processes are desirable or even necessary in manufacturing operations as follows:

a) Closer dimensional accuracy may be required than is available from casting, forming, or shaping processes alone. In a forged crankshaft, for example, the bearing surfaces and the holes cannot be produced with good dimensional accuracy and surface finish by forming and shaping processes alone.
b) Parts may have external and internal profiles, as well as sharp corners and flatness, that cannot be produced by forming and shaping processes.
c) Some parts are heat treated for improved hardness and wear resistance. Since heat-treated parts may undergo distortion and surface discoloration, they generally require additional finishing operations, such as grinding, to obtain the desired final dimensions and surface finish.
d) Special surface characteristics or texture that cannot be produced by other means may be required on all or part of the surfaces of the product. Copper mirrors with very high reflectivity, for example, are made by machining with a diamond cutting tool.
e) Machining the part may be more economical than manufacturing it by other processes, particularly if the number of parts desired is relatively small.

Against these advantages, material-removal processes have certain limitations:

a) Removal processes inevitably waste material and generally require more energy, capital, and labor than forming and shaping operations. Thus they should be avoided whenever possible.
b) Removing a volume of material from a workpiece generally takes longer than to shape it by other processes.
c) Unless carried out properly, material-removal processes can have adverse effects on the surface quality and properties of the product.

In spite of these limitations, material-removal processes and machines are indispensable to manufacturing technology. Ever since lathes were introduced in the 1700s, these processes have developed continuously. We now have available a variety of computer controlled machines, as well as new techniques using electrical, chemical, and optical energy sources. The machines on which material-removal operations are performed are generally called *machine tools*. Their construction and characteristics influence greatly these operations and product quality. It is important to view machining, as well as all manufacturing operations, as a *system* consisting of the workpiece, the tool, and the machine. Cutting operations cannot be carried out efficiently and economically without a knowledge of the interactions among these elements.

8.2 Mechanics of Chip Formation

Cutting processes, such as turning on a lathe, drilling, milling, or thread cutting remove material from the surface of the workpiece by producing *chips*. The basic mechanics of chip formation is essentially the same for all these operations, which we represent by the two-dimensional model in Fig. 8.1. In this model a tool moves along the workpiece at a certain velocity V and a depth of cut t_0. A chip is produced ahead of the tool by shearing the material continuously along the shear plane.

The major *independent variables* (those that we can change directly) in the cutting process are:

- Tool material and its condition.
- Tool shape, surface finish, and sharpness.
- Workpiece material, condition, and temperature.
- Cutting conditions, such as speed and depth of cut.
- Use of a cutting fluid.
- The characteristics of the machine tool, such as its stiffness and damping.

Dependent variables are those that are influenced by changes in the independent variables and are:

- Type of chip produced.
- Force and energy dissipated in the cutting process.
- Temperature rise in the workpiece, the chip, and the tool.
- Wear and failure of the tool.
- The surface finish produced on the workpiece after machining.

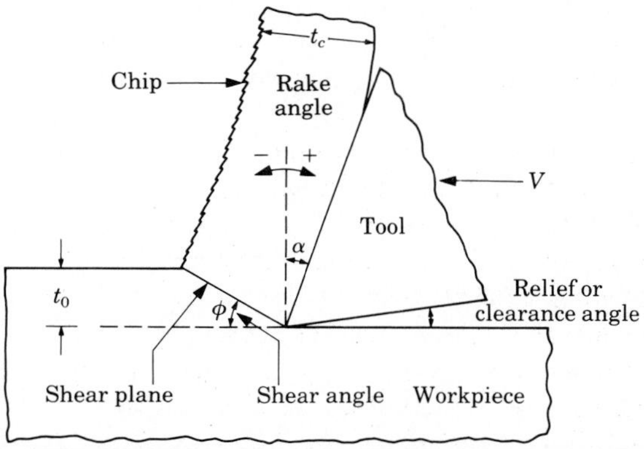

FIGURE 8.1
Schematic illustration of a two-dimensional cutting process (also called orthogonal cutting).

Ask yourself the following questions. If, for example, the surface finish of the workpiece being cut is poor and unacceptable, which of the independent variables do you change first? The angle of the tool? If so, do you increase it or decrease it? If the tool wears and becomes dull rapidly, do you change the cutting speed, the depth of cut, or the tool material? If the cutting tool begins to vibrate, what should be done to eliminate this vibration? In this section we will describe the mechanics of chip formation in order to establish the effects of various parameters on the overall cutting process.

Although almost all cutting processes are three-dimensional in nature, the model shown in Fig. 8.1 is very useful in studying the basic mechanics of cutting. In this model, known as *orthogonal cutting*, the tool has a *rake angle* of α (positive, as shown in the figure) and a *relief*, or *clearance*, *angle*. Note that the sum of the rake, relief, and included angles of the tool is 90°.

Microscopic examinations have revealed that chips are produced by the shearing process shown in Fig. 8.2(a), and that shearing takes place along a *shear plane* making an angle ϕ, called the *shear angle*, with the surface of the workpiece. Below the shear plane the workpiece is undeformed, and above it is the chip, already formed and climbing up the face of the tool as cutting progresses. Because of the relative velocity, there is friction between the chip and the rake face of the tool. Note also that this shearing process is like cards in a deck sliding against each other.

The thickness of the chip, t_c, can be determined by knowing t_o, α, and ϕ. The ratio of t_o to t_c is known as the *cutting ratio*, r, which we can express as

$$r = \frac{t_o}{t_c} = \frac{\sin \phi}{\cos (\phi - \alpha)}. \quad (8.1)$$

The chip thickness is always greater than the depth of cut; hence the value of r is less than unity. The reciprocal of r is known as the *chip compression ratio* and is a

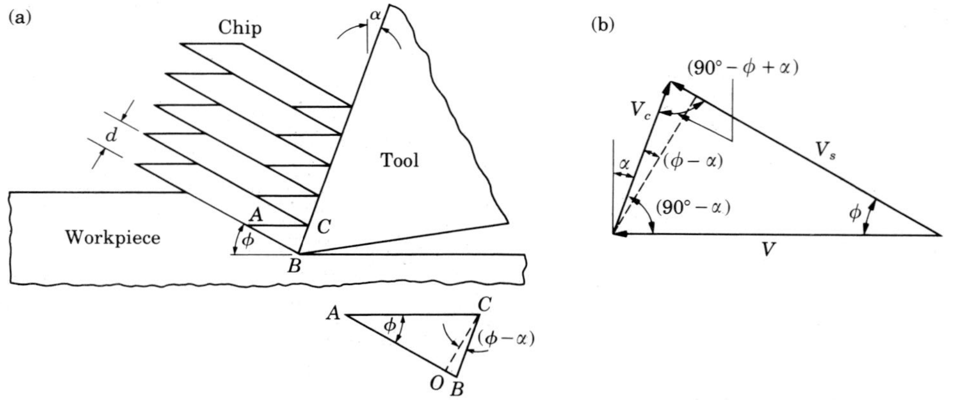

FIGURE 8.2
(a) Schematic illustration of the basic mechanism of chip formation in cutting. (b) Velocity diagram in the cutting zone.

measure of how thick the chip has become compared to the depth of cut. Thus the chip compression ratio is always greater than unity.

On the basis of Fig. 8.2(a), we can express the *shear strain*, γ, that the material undergoes as

$$\gamma = \frac{AB}{OC} = \frac{AO}{OC} + \frac{OB}{OC}, \tag{8.2}$$

or

$$\gamma = \cot \phi + \tan (\phi - \alpha). \tag{8.3}$$

Thus large shear strains are associated with low shear angles and low or negative rake angles (Fig. 8.3). Shear strains of 5 or larger have been observed in actual cutting operations. Compared to forming and shaping processes, therefore, the material undergoes greater deformation during cutting. Also, deformation in cutting takes place within a very narrow deformation zone; that is, the dimension $d = OC$ is very small.

From Fig. 8.1 we note that since chip thickness t_c is greater than the depth of cut t_o, the velocity of the chip, V_c, has to be lower than the cutting speed, V. Since mass continuity has to be maintained, we have

$$Vt_o = V_c t_c \qquad \text{or} \qquad V_c = Vr. \tag{8.4}$$

Hence

$$V_c = V \frac{\sin \phi}{\cos (\phi - \alpha)}. \tag{8.5}$$

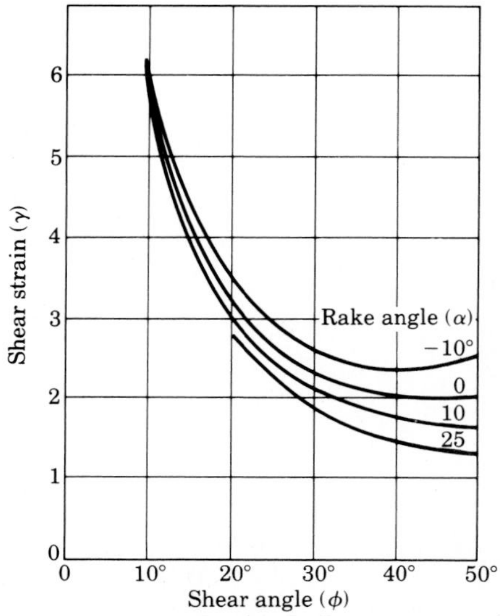

FIGURE 8.3 Shear strain in cutting as a function of shear angle and rake angle. These curves are plots of Eq. (8.3). Note that the shear strains can be on the order of 5 or higher.

We can construct a velocity diagram (Fig. 8.2b) and from trigonometric relationships obtain the following relationships:

$$\frac{V}{\cos(\phi - \alpha)} = \frac{V_s}{\cos \alpha} = \frac{V_c}{\sin \phi}, \tag{8.6}$$

where V_s is the velocity at which shearing takes place in the shear plane. The *shear strain rate* is the ratio of V_s to the thickness d of the sheared element (shear zone), or

$$\dot{\gamma} = \frac{V_s}{d}. \tag{8.7}$$

Experimental evidence indicates that d is on the order of 10^{-2} to 10^{-3} mm (10^{-3}

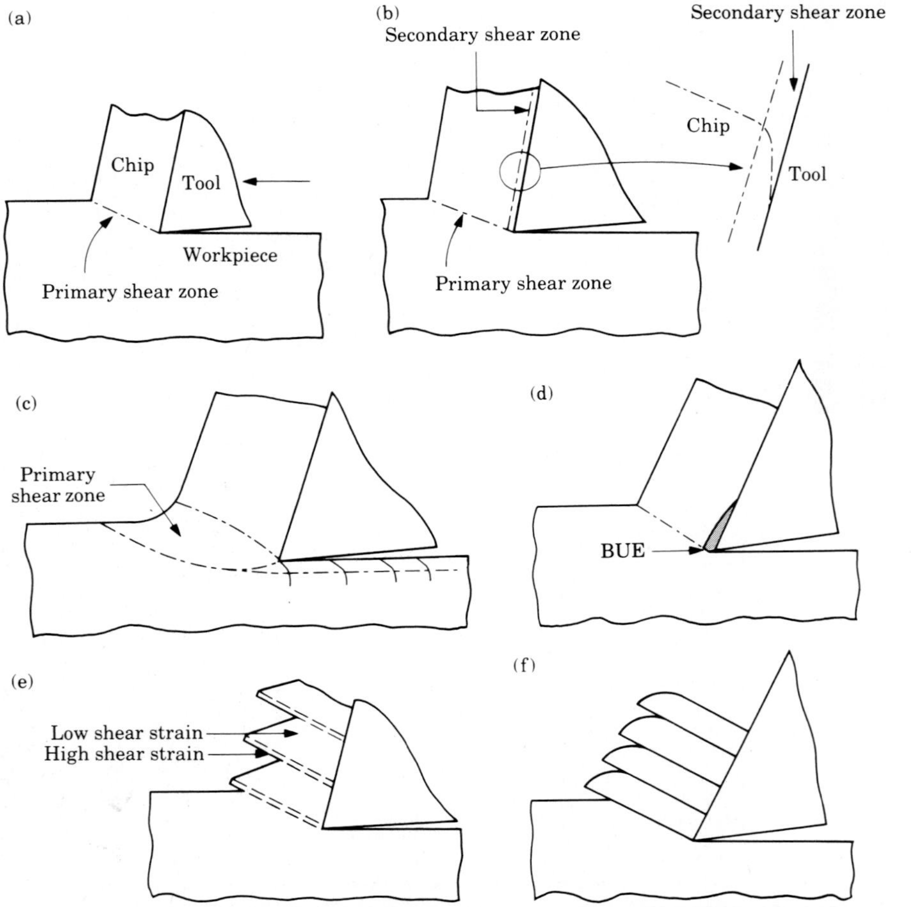

FIGURE 8.4
Basic types of chips produced in metal cutting: (a) continuous chip with narrow, straight primary shear zone; (b) secondary shear zone at the chip–tool interface; (c) continuous chip with large primary shear zone; (d) continuous chip with built-up edge; (e) segmented or nonhomogeneous chip; and (f) discontinuous chip. *Source*: After M. C. Shaw.

to 10^{-4} in). This means that, even at low cutting speeds, the shear strain rate is very high, on the order of 10^3 to 10^6/s. A knowledge of the shear strain rate is essential because of its effects on the strength and ductility of the material, and the chip morphology (chip form).

8.2.1 Types of chips

When we observe actual chip formation under different metal-cutting conditions, we find significant deviations from the ideal model shown in Figs. 8.1 and 8.2(a). Some types of metal chips commonly observed in practice are shown schematically in Fig. 8.4, with micrographs in Fig. 8.5. Because the type of chips produced

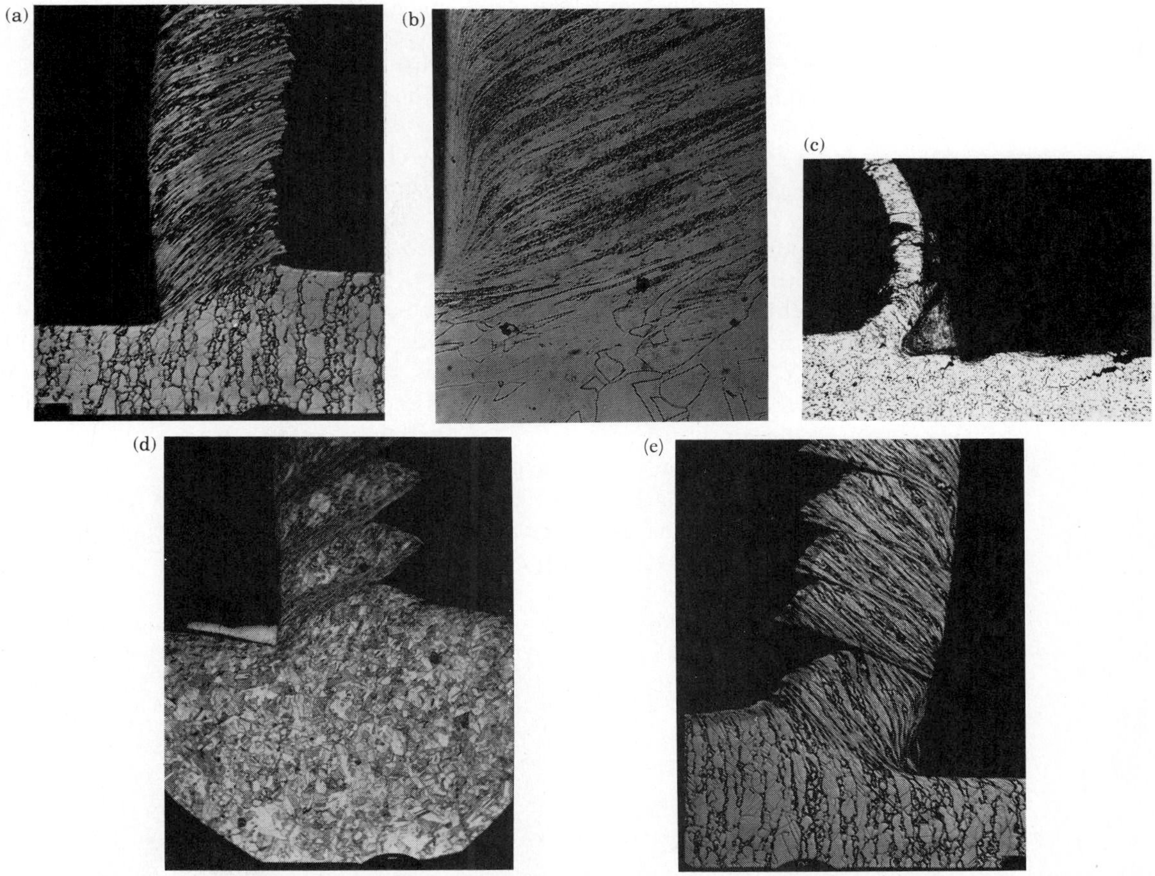

FIGURE 8.5
Photomicrographs of basic types of chips obtained in cutting metals. (a) Continuous chip in cutting 60–40 brass at 3.4 m/s (670 ft/min) at a depth of cut of 0.2 mm (0.008 in). (b) Secondary shear zone in cutting copper at 1.7 m/s (330 ft/min). (c) Built-up edge in cutting sintered tungsten. (d) Inhomogeneous (serrated) chip in cutting 321 stainless steel at 1.25 m/s (250 ft/min) at a depth of cut of 0.2 mm (0.008 in). (e) Discontinuous chip in cutting 60–40 brass at 0.25 m/s (50 ft/min). Note the small built-up edge at the root of the chip. *Source*: Parts (a), (b), (d), (e) courtesy of P. K. Wright, Carnegie-Mellon University; part (c), A. J. Moser and S. Kalpakjian.

significantly influences the surface finish produced and the overall cutting operation, we will discuss the type of chips in the following order:

1. Continuous
2. Built-up edge
3. Serrated
4. Discontinuous

Let's first note that a chip has two surfaces: one that is in contact with the tool face (rake face), and the other from the original surface of the workpiece. The tool side of the chip surface is shiny, or *burnished* (Fig. 8.6), which is caused by rubbing of the chip as it climbs up the tool face. The other surface of the chip does not come into contact with any solid body. This surface has a jagged, steplike appearance (Fig. 8.5a), which is caused by the shearing mechanism of chip formation.

Continuous chips. *Continuous chips* are usually formed at high cutting speeds and/or high rake angles (Figs. 8.4a and 8.5a). The deformation of the material takes place along a narrow shear zone, called the *primary shear zone*. Continuous chips may develop a *secondary shear zone* at the tool–chip interface (Figs. 8.4b and 8.5b), caused by friction. The secondary zone becomes deeper as tool–chip friction increases.

In continuous chips, deformation may also take place along a wide primary shear zone with curved boundaries (Fig. 8.4c). Note that the lower boundary is below the machined surface, which subjects the machined surface to distortion, as depicted by the distorted vertical lines. This situation occurs particularly in machining soft metals at low speeds and low rake angles. It can produce poor surface finish and induce residual surface stresses, which may be detrimental to the properties of the machined part.

Although they generally produce good surface finish, continuous chips are not always desirable, particularly in automated machine tools. They tend to get tangled around the tool holder, and the operation has to be stopped to clear away the chips. This problem can be alleviated with chip breakers (see p. 484).

FIGURE 8.6
Shiny (burnished) surface on the tool-side of a continuous chip produced in turning.

As a result of strain hardening (caused by the shear strains to which it is subjected), the chip usually becomes harder, less ductile, and stronger than the original workpiece material. The increase in hardness and strength of the chip depends on the shear strain (Table 8.1). As the rake angle decreases, the shear strain increases and the chip becomes stronger and harder. With increasing strain, the material tends to behave like a rigid, perfectly plastic body.

Built-up edge chips. A *built-up edge* (BUE) may form at the tip of the tool during cutting (Figs. 8.4d and 8.5c). This edge consists of layers of material from the workpiece that are gradually deposited on the tool (hence the term *built-up*). As it becomes larger, the BUE becomes unstable and eventually breaks up. Part of the BUE material is carried away by the tool side of the chip; the rest is deposited randomly on the workpiece surface. The process of BUE formation and destruction is repeated continuously during the cutting operation.

The built-up edge is commonly observed in practice. It is one of the factors that most adversely affects surface finish in cutting, as Figs. 8.5(c) and 8.7 show. A built-up edge, in effect, changes the geometry of cutting. Note, for example, the large tip radius of the BUE and the rough surface finish produced. Because of work hardening and deposition of successive layers of material, BUE hardness increases significantly (Fig. 8.7a). Although BUE is generally undesirable, a thin, stable BUE is usually regarded as desirable because it protects the tool's surface.

The exact mechanism of formation of the BUE is not yet clearly understood. However, investigations have identified two distinct mechanisms that contribute to formation. One is the adhesion of the workpiece material to the rake face of the tool. The strength of this bond depends on the affinity of the workpiece and tool materials. The other mechanism is the growth of the adhered metal layers to form a BUE.

One of the important factors in forming the BUE is the propensity for strain hardening of the workpiece material; the higher the strain-hardening exponent, the greater the tendency is for BUE formation. Experimental evidence indicates that as the cutting speed increases, the BUE decreases or is eliminated. It is not clear

TABLE 8.1

STRENGTH AND HARDNESS OF CHIPS IN TURNING MILD STEEL (After K. Nakayama).

	Original material						
Rake angle, α		45°	35°	27°	10°	10°	10°
Feed (f), mm/rev		0.30	0.30	0.20	0.20	0.20	0.20
Cutting speed (V), m/min		50	50	168	168	168	76
Cutting fluid		Soluble oil	None	None	Soluble oil	None	None
Shear strain, γ		1.1	1.7	2.1	2.9	3.1	4.0
Tensile strength (UTS), kg/mm^2	46	75	84	91	92	93	95
Vickers hardness number (HV)	209	272	289	302	320	314	325
HV/UTS	4.5	3.6	3.4	3.3	3.5	3.4	3.4

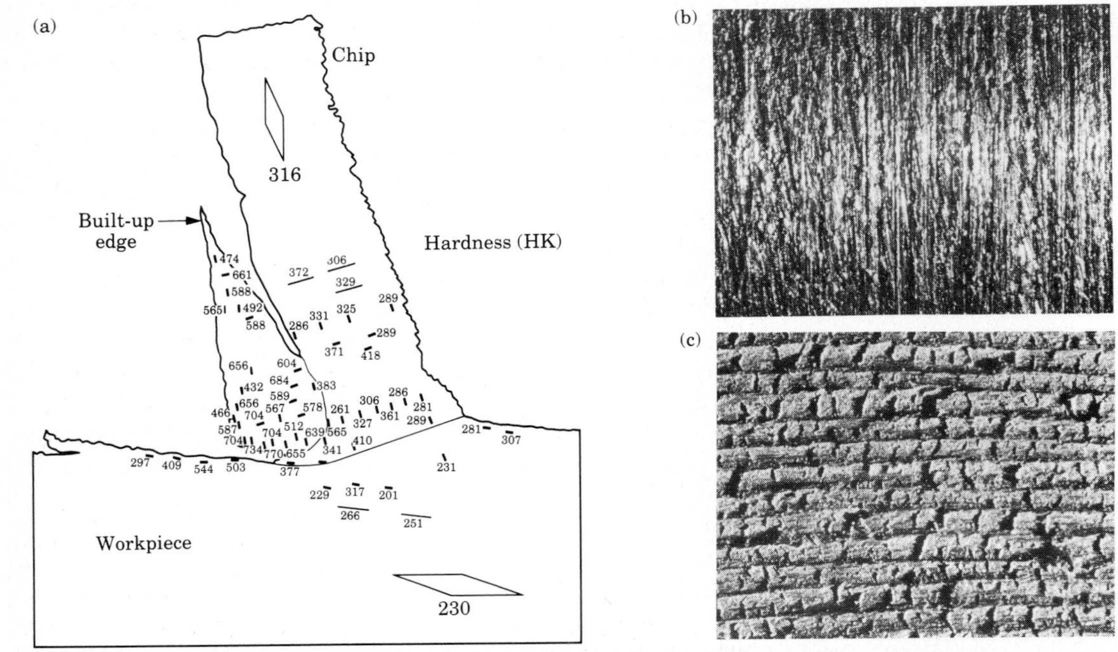

FIGURE 8.7
(a) Hardness distribution in the cutting zone for 3115 steel. Note that some regions in the built-up edge are as much as three times harder than the bulk metal. (b) Surface finish in turning 5130 steel with a built-up edge. (c) Surface finish on 1018 steel in face milling. Magnifications: 15 ×. *Source:* Courtesy of Metcut Research Associates, Inc.

whether the speed, as such, plays a significant role (such as by increasing the strain rate), or whether the effect results from the increase in temperature with speed. The latter appears to contradict the observation that higher temperatures improve adhesion, because as speed increases, temperature increases but BUE decreases. In addition to the factors outlined, the tendency for BUE formation can be reduced by decreasing the depth of cut, increasing the rake angle, using a tool with a small tip radius, and using an effective cutting fluid.

Serrated chips. *Serrated chips* (also called *segmented* or *nonhomogeneous* chips) are semicontinuous chips, with zones of low and high shear strain (Figs. 8.4e and 8.5d). Metals with low thermal conductivity and strength that decreases sharply with temperature, such as titanium, exhibit this behavior. The chips have a sawtoothlike appearance.

Discontinuous chips. *Discontinuous chips* consist of segments that may be firmly or loosely attached to each other (Figs. 8.4f and 8.5e). Discontinuous chips usually form under the following conditions: (a) brittle workpiece materials, because they do not have the capacity to undergo the high shear strains developed in cutting; (b) materials that contain hard inclusions and impurities; (c) very low or very high cutting speeds; (d) large depths of cut and low rake angles; (e) low stiffness of the machine tool; and (f) lack of an effective cutting fluid.

Impurities and hard particles act as nucleation sites for cracks, thereby producing discontinuous chips. A large depth of cut increases the probability of such defects being present in the cutting zone. Higher cutting speeds mean higher temperatures; hence the material is more ductile and has less tendency to form discontinuous chips. Another factor in discontinuous-chip formation is the magnitude of the compressive stresses on the shear plane. Recall from Fig. 2.18 that the maximum shear strain at fracture increases with increasing compressive stress. If the normal stress is not sufficiently high, the material is unable to undergo the shear strain required to form a continuous chip.

Because of the discontinuous nature of chip formation, forces continually vary during cutting. Consequently, the stiffness of the cutting-tool holder and the machine tool is important in cutting with discontinuous-chip as well as serrated-chip formation. If not stiff enough the machine tool may begin to vibrate and chatter. This, in turn, adversely affects the surface finish and dimensional accuracy of the machined component and may damage or cause excessive wear of the cutting tool.

Chip formation in nonmetallic materials. Much of the discussion thus far for metals is also generally applicable to nonmetallic materials. A variety of chips are obtained in cutting thermoplastics, depending on the type of polymer and process parameters such as depth of cut, tool geometry, and cutting speed. Because they are brittle, thermosetting plastics and ceramics generally produce discontinuous chips.

Chip curl. *Chip curl* (Fig. 8.8a) is common to all cutting operations with metals, as well as with nonmetallic materials, such as plastics and wood. The

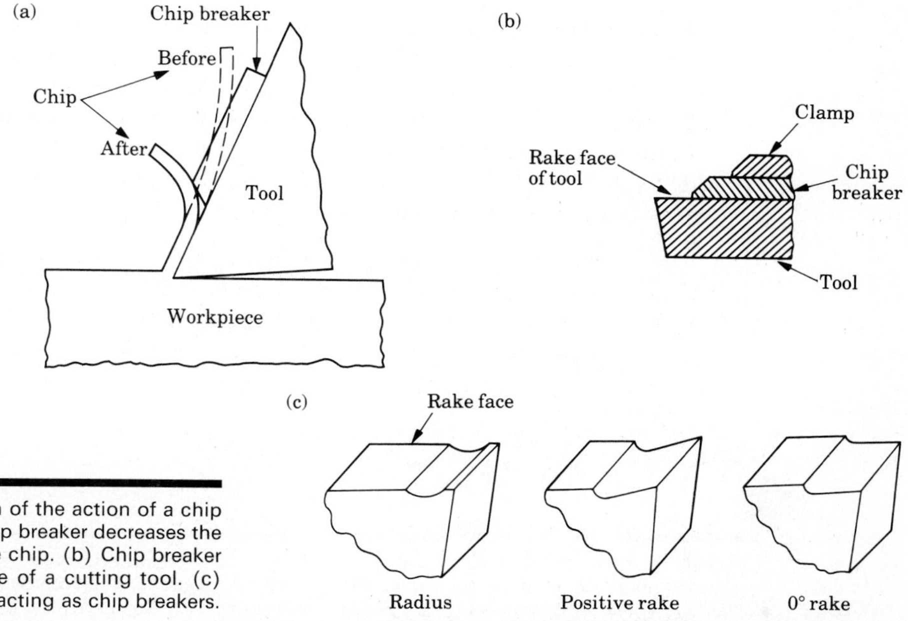

FIGURE 8.8 (a) Schematic illustration of the action of a chip breaker. Note that the chip breaker decreases the radius of curvature of the chip. (b) Chip breaker clamped on the rake face of a cutting tool. (c) Grooves in cutting tools acting as chip breakers.

reasons for chip curl are still not clearly understood. Among the possible factors contributing to it are the distribution of stresses in the primary and secondary shear zones, thermal effects, and the work-hardening characteristics of the workpiece material. Process variables also affect chip curl. Generally, the radius of curvature decreases (the chip becomes curlier) with decreasing depth of cut, increasing rake angle, and decreasing friction at the tool–chip interface. The use of cutting fluids and various additives in the workpiece material also influence chip curl.

Chip breakers. As we stated earlier, long, continuous chips are undesirable. They tend to become entangled, and interfere with cutting operations, and can be a safety hazard. This situation is especially troublesome in high-speed automated machinery. The usual procedure to avoid it is to break the chip intermittently with a *chip breaker*. The chip breaker can be a piece of metal clamped to the rake face of the tool (Fig. 8.8b), or it can be an integral part of the tool (Fig. 8.8c).

Chips can also be broken by changing the tool geometry, thus controlling chip flow, as in the turning operations shown in Fig. 8.9. Various cutting tools with chip-breaker features are available. In interrupted cutting operations, such as milling, chip breakers are generally not necessary, since the chips already have finite lengths resulting from the intermittent nature of the operation.

8.2.2 Oblique cutting

Thus far we have described the cutting process two-dimensionally. However, the majority of cutting operations involve tool shapes that are three-dimensional (*oblique*). The basic difference between two-dimensional and oblique cutting is shown in Fig. 8.10(a). As we have shown, in orthogonal cutting the tool edge is perpendicular to the movement of the tool, and the chip slides directly up the face of the tool. In oblique cutting, the cutting edge is at an angle i, the *inclination angle* (Fig. 8.10b). Note the lateral direction of chip movement in oblique cutting. This situation is similar to an angled snow-plow blade, which throws the snow sideways.

Note that the chip in Fig. 8.10(a) flows up the rake face of the tool at angle α_c (*chip flow angle*), measured in the plane of the tool face. Angle α_n is known as the *normal rake angle*, which is a basic geometric property of the tool. This is the angle between the normal o_z to the workpiece surface and the line o_a on the tool face.

The workpiece material approaches the tool at a velocity V, and leaves the surface (as a chip) with a velocity V_c. We calculate the effective rake angle α_e in

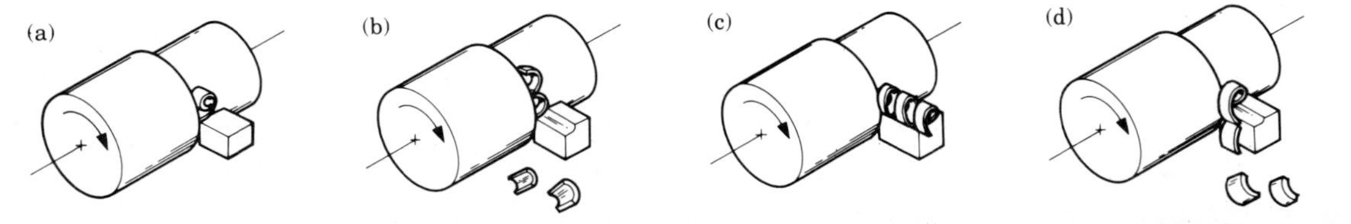

FIGURE 8.9
Various chips produced in turning: (a) tightly curled chip; (b) chip hits workpiece and breaks; (c) continuous chip moving away from workpiece; and (d) chip hits tool shank and breaks off. *Source:* G. Boothroyd, *Fundamentals of Metal Machining and Machine Tools.* Copyright © 1975; McGraw-Hill Publishing Company. Used with permission.

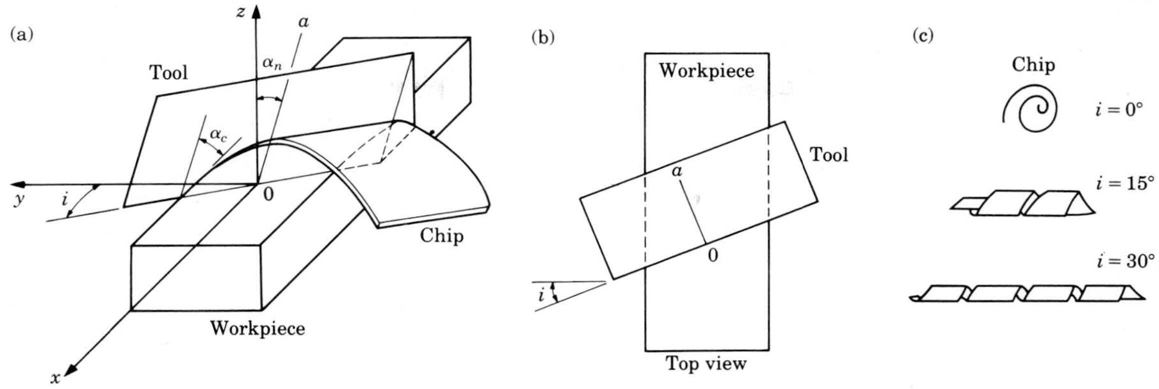

FIGURE 8.10
(a) Schematic illustration of cutting with an oblique tool. (b) Top view showing the inclination angle *i*. (c) Types of chip produced with different inclination angles.

the plane of these two velocities. Assuming that the chip flow angle α_c is equal to the inclination angle i (which is experimentally found to be approximately correct), the effective rake angle α_e is

$$\alpha_e = \sin^{-1}(\sin^2 i + \cos^2 i \sin \alpha_n). \tag{8.8}$$

Since we can measure both i and α_n directly, we can calculate the effective rake angle. As i increases, the effective rake angle increases, and the chip becomes thinner and longer. The effect of the inclination angle on chip shape is shown in Fig. 8.10(c).

A typical single-point turning tool used on a lathe is shown in Fig. 8.11. Note the various angles involved, each of which has to be selected properly for efficient

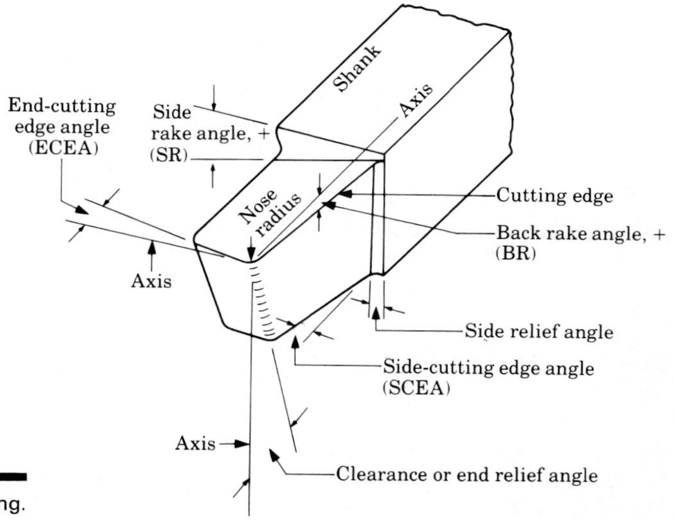

FIGURE 8.11
Terminology for a right-hand cutting tool for turning.

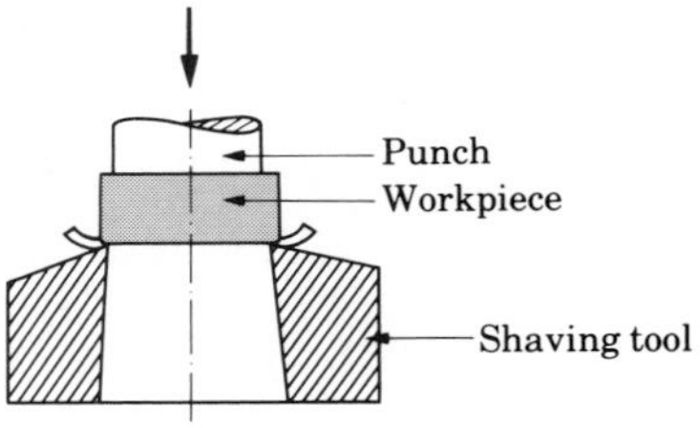

FIGURE 8.12 Schematic illustration of the shaving process for improving the accuracy of a slug. Note that the tool has a small positive rake angle (see also Fig. 7.12).

cutting. We discuss various three-dimensional cutting tools in greater detail in Sections 8.8 and 8.9. These include tools for drilling, tapping, milling, planing, shaping, broaching, sawing, and filing.

Shaving and skiving. Thin layers of material can be removed from straight or curved surfaces by a process similar to the use of a plane to shave wood. *Shaving* is particularly useful in improving the surface finish and dimensional accuracy of sheared parts and punched slugs (Fig. 8.12). Another application of shaving is in finishing gears with a cutter that has the shape of the gear tooth. Parts that are long or have a combination of shapes are shaved by *skiving*. The skiving tool moves tangentially across the length of the workpiece.

8.2.3 Cutting forces

Knowledge of the forces and power involved in cutting operations is important for the following reasons:

- Power requirements have to be determined so that a motor of suitable capacity can be installed in the machine tool.
- Data on forces are necessary for the proper design of machine tools for cutting operations that avoid excessive distortion of the machine elements and maintain desired tolerances for the machined part.
- Whether the workpiece can withstand the cutting forces without excessive distortion has to be determined in advance.

The forces acting on the tool in orthogonal cutting are shown in Fig. 8.13. The *cutting force*, F_c, acts in the direction of the cutting speed V and supplies the energy required for cutting. The *thrust force*, F_t, acts in the direction normal to the cutting velocity, that is, perpendicular to the workpiece. These two forces produce the *resultant force*, R. Note that the resultant force can be resolved into two components on the tool face: a *friction force*, F, along the tool–chip interface, and a *normal force*, N, perpendicular to it. From Fig. 8.13, we can show the force F to be

$$F = R \sin \beta, \tag{8.9}$$

and the force N as

$$N = R \cos \beta. \tag{8.10}$$

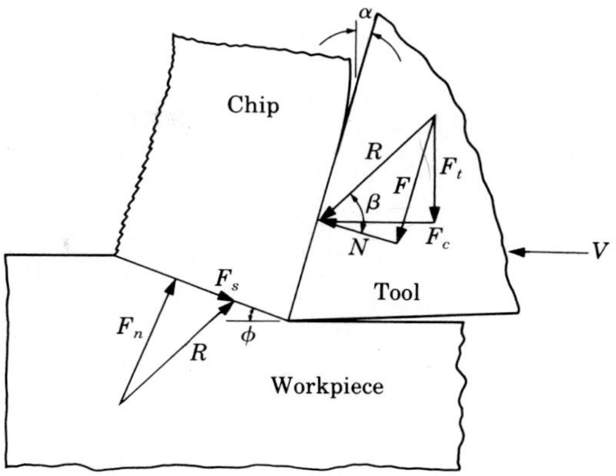

FIGURE 8.13 Forces acting on a cutting tool in two-dimensional cutting.

Note that the resultant force is balanced by an equal and opposite force along the shear plane and is resolved into a *shear force*, F_s, and a *normal force*, F_n.

The ratio of F to N is the *coefficient of friction*, μ, at the tool–chip interface, and the angle β is the *friction angle*. We can express μ as

$$\mu = \tan\beta = \frac{F_t + F_c \tan\alpha}{F_c - F_t \tan\alpha}. \tag{8.11}$$

The coefficient of friction in metal cutting generally ranges from about 0.5 to 2.0, thus indicating that the chip encounters considerable frictional resistance while climbing up the face of the tool.

Although the magnitude of forces in actual cutting operations is generally on the order of a few hundred newtons, the local stresses in the cutting zone and the pressures on the tool are very high because the contact areas are very small. The chip–tool contact length (Fig. 8.1), for example, is typically on the order of 1 mm (0.04 in.). Thus the tool is subjected to very high stresses, which lead to wear and sometimes chipping and fracture of the tool.

Thrust force. Knowing the thrust force in cutting is important. The tool holder and the machine tool must be stiff enough to minimize deflections caused by this force. For example, if the thrust force is too high or if the machine tool is not sufficiently stiff, the tool will be pushed away from the surface being machined. This movement will, in turn, reduce the depth of cut, causing lack of dimensional accuracy in the machined part.

Referring to Fig. 8.13, note that the cutting force must always be in the direction of cutting in order to supply energy to the system. Although the thrust force does not contribute to the total work done, knowing its magnitude is important. For instance, if the tool holder is not sufficiently stiff, the thrust force will simply deflect the tool and reduce the depth of cut. Thus, tolerances cannot be held.

Also note from Fig. 8.13 that the thrust force is downward. We will now show that this force can act upward. First, we express the thrust force as

$$F_t = R \sin(\beta - \alpha) \tag{8.12}$$

or

$$F_t = F_c \tan(\beta - \alpha). \tag{8.13}$$

Because the magnitude of F_c is always positive (say, as shown in Fig. 8.13), the sign of F_t can be either positive or negative, depending on the relative magnitudes of β and α. When $\beta > \alpha$, the sign of F_t is positive (downward); when $\beta < \alpha$, it is negative (upward). Thus an upward thrust force at high rake angles and/or with low friction at the tool–chip interface is possible. We can visualize the situation in Fig. 8.13: When $\mu = 0$, then $\beta = 0$ and the resultant force R coincides with the force N. In this case, R will have an upward thrust force component. Also note that for $\alpha = 0$ and $\beta = 0$, the thrust force is zero.

These observations have been verified experimentally, as shown in Fig. 8.14. The influence of the depth of cut is obvious: As t_o increases, R must also increase so that F_c will also increase. This action supplies the additional energy required to remove the extra material produced by the increased depth of cut. The change in direction and magnitude of the thrust force can play a significant role. Within a certain range of operating conditions, it leads to instability problems in machining,

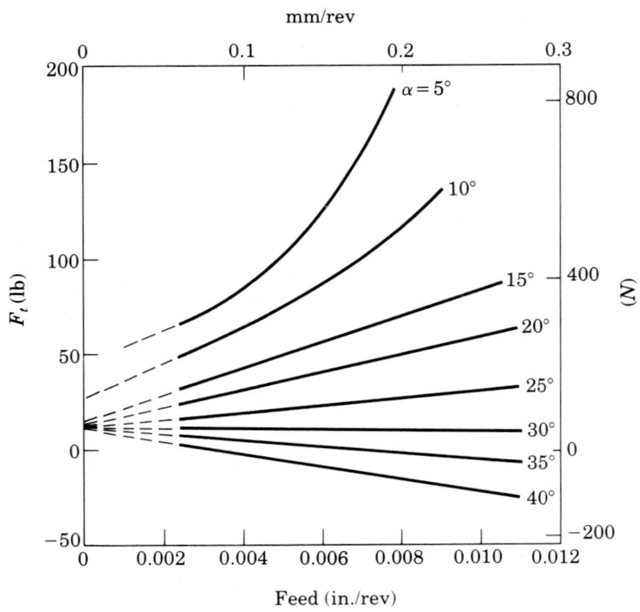

FIGURE 8.14 Thrust force as a function of rake angle and feed in orthogonal cutting of AISI 1112 cold-rolled steel. Note that at high rake angles the thrust force is negative. A negative thrust force has important implications in the design of machine tools and in controlling the stability of the cutting process. *Source:* S. Kobayashi and E. G. Thomsen, *J. Eng. Ind.*, vol. 81, 1959, pp. 251–262.

particularly if the machine tool is not stiff enough. However, such high rake angles are rarely used in machining most metals.

Observations on forces in the cutting zone. In addition to being a function of the strength of the workpiece material, forces in cutting are also influenced by other variables. Extensive data are available, and one set, Table 8.2, shows that the cutting force F_c increases with increasing depth of cut, decreasing rake angle, and decreasing speed. From analysis of the data in Table 8.3, the effect of cutting speed can be attributed to the fact that as speed decreases, the shear angle decreases and the friction coefficient increases. Both effects increase the cutting force.

Another factor that can significantly influence the cutting force is the tip radius of the tool. The larger the radius (the duller the tool), the greater is the force. Experimental evidence indicates that, for depths of cut on the order of five times the tip radius or higher, the effect of dullness on the cutting forces is negligible. Note

TABLE 8.2
ORTHOGONAL CUTTING DATA FOR 4130 STEEL*

α	ϕ	γ	μ	β	F_c (lb)	F_t (lb)	$u_t \left(\frac{\text{in.-lb}}{\text{in}^3} \times 10^3\right)$	u_s	u_f	$\frac{u_f}{u_t}$ (%)
25°	20.9°	2.55	1.46	56	380	224	320	209	111	35
35	31.6	1.56	1.53	57	254	102	214	112	102	48
40	35.7	1.32	1.54	57	232	71	195	94	101	52
45	41.9	1.06	1.83	62	232	68	195	75	120	62

* t_o = 0.0025 in.; w = 0.475 in.; V = 90 ft/min; tool: high-speed steel.
Source: After E. G. Thomsen.

TABLE 8.3
ORTHOGONAL CUTTING DATA FOR 9445 STEEL*

α	V (ft/min)	ϕ	γ	μ	β	F_c	F_t	u_t	u_s	u_f	$\frac{u_f}{u_t}$ (%)
+10	197	17	3.4	1.05	46	370	273	400	292	108	27
	400	19	3.1	1.11	48	360	283	390	266	124	32
	642	21.5	2.7	0.95	44	329	217	356	249	107	30
	1186	25	2.4	0.81	39	303	168	328	225	103	31
−10	400	16.5	3.9	0.64	33	416	385	450	342	108	24
	637	19	3.5	0.58	30	384	326	415	312	103	25
	1160	22	3.1	0.51	27	356	263	385	289	96	25

* t_0 = 0.037 in.; w = 0.25 in.; tool: cemented carbide.
Source: After M. E. Merchant.

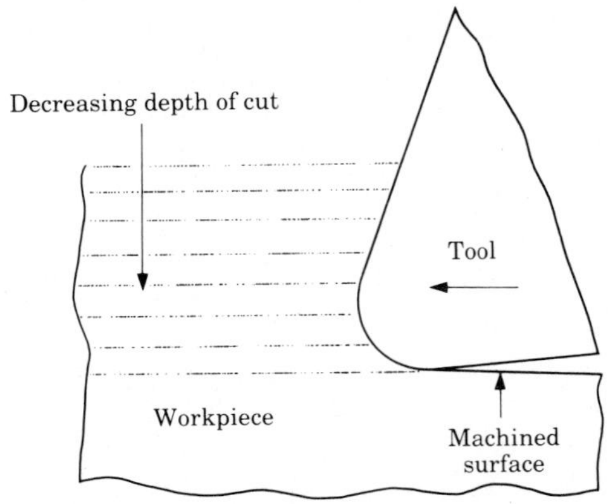

FIGURE 8.15 Schematic illustration of a dull tool (large tool-tip radius) in orthogonal cutting and dependence of the effective rake angle on depth of cut. Note that at small depths of cut the rake angle is effectively negative. Chips may not be formed, because the tool tends to ride over the surface of the workpiece, producing a burnished surface, causing temperature rise, and possibly damaging surface.

from Fig. 8.15 that, if the depth of cut is very small compared to the tip radius, the tip radius in effect determines the rake angle. Even though the tool may have a positive rake angle, the effective angle is actually highly negative. This is why dull tools cannot remove thin layers of material. Such tools may actually rub against the surface without removing any material. This action smears and burnishes the workpiece surface and may lead to damage. However, dull tools could also impart compressive residual stresses to the surface.

Stresses. We can analyze the stresses in the shear plane and at the tool–chip interface by assuming that they are uniformly distributed. The forces in the shear plane can be resolved into shear and normal forces and stresses. The average *shear stress* in the shear plane is

$$\tau = \frac{F_s}{A_s} \tag{8.14}$$

and the average *normal stress* is

$$\sigma = \frac{F_n}{A_s}, \tag{8.15}$$

where A_s is the area of the shear plane, namely,

$$A_s = \frac{wt_o}{\sin \phi}.$$

Figure 8.16 gives some data pertaining to these average stresses. The rake angle is a parameter, and the shear plane area is increased by increasing the depth of cut. We can draw the following conclusions from these curves.

a) The shear stress on the shear plane is independent of the rake angle.
b) The normal stress on the shear plane decreases with increasing rake angle.

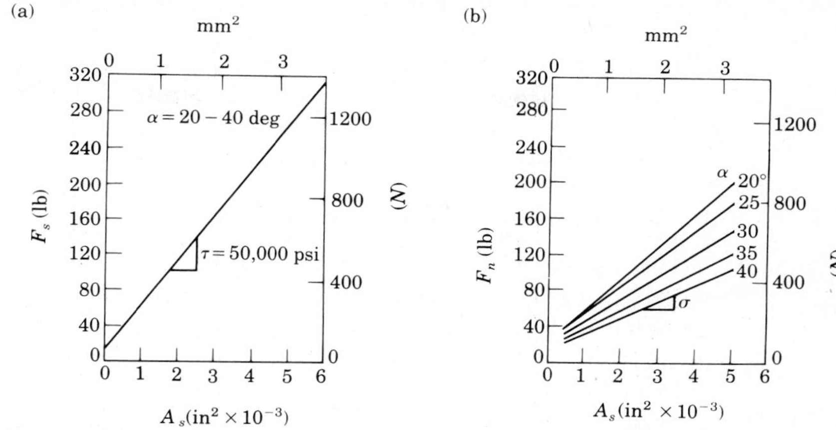

FIGURE 8.16
Shear force and normal force as a function of shear plane area and rake angle for 85–15 brass. Note that the shear stress in the shear plane is constant, regardless of the magnitude of the normal stress. Thus normal stress has no effect on the shear flow stress of the material. *Source:* After S. Kobayashi and E. G. Thomsen, *J. Eng. Ind.*, vol. 81, pp. 251–262.

c) Consequently, the normal stress in the shear plane has no effect on the magnitude of the shear stress. This phenomenon has also been verified by other mechanical tests. (However, normal stress strongly influences the magnitude of the shear strain in the shear zone. The maximum shear strain to fracture increases with the normal compressive stress.)

Determining the stresses on the rake face of the tool presents considerable difficulties. One problem is accurately determining the length of contact at the tool–chip interface. This length increases with decreasing shear angle, indicating that the contact length is a function of rake angle, cutting speed, and friction at the tool–chip interface. Another problem is that the stresses are not uniformly distributed on the rake face.

Photoelastic studies have determined that the actual stress distribution is qualitatively as shown in Fig. 8.17. Note that the stress normal to the tool face is

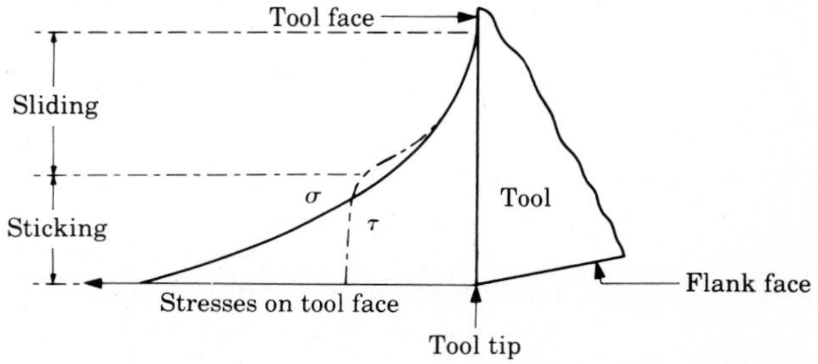

FIGURE 8.17
Schematic illustration of the distribution of normal and shear stresses at the tool–chip interface (rake face). Note that, whereas the normal stress increases continuously toward the tip of the tool, the shear stress reaches a maximum and remains at that value (known as sticking).

maximal at its tip and decreases rapidly toward the end of the contact length. The shear stress has a similar trend, except that it levels off at approximately the center of the contact length. This behavior indicates that *sticking* is taking place, whereby the shear stress has reached the shear yield strength of the material. Such a sticking region has been observed on some chips. (Sticking regions also can be observed in various other metal-forming processes.)

8.2.4 Shear-angle relationships

Because the shear angle and the shear zone have great significance in the mechanics of cutting, a great deal of effort has been expended to determine the relationship of the shear angle to material properties and process variables. One of the earliest analyses (by M. E. Merchant) is based on the assumption that the shear angle adjusts itself so that the cutting force is a minimum, or so that the maximum shear stress occurs in the shear plane. From the force diagram in Fig. 8.13, the following relationships can be obtained:

$$F_c = R \cos(\beta - \alpha) \tag{8.16}$$

and

$$F_s = R \cos(\phi + \beta - \alpha). \tag{8.17}$$

The area of the shear plane is

$$A_s = \frac{wt_o}{\sin\phi}, \tag{8.18}$$

where w is the width of the cut. Thus the shear stress in the shear plane is

$$\tau = \frac{F_s}{A_s} = \frac{F_c \sec(\beta - \alpha)\cos(\phi + \beta - \alpha)\sin\phi}{wt_o}. \tag{8.19}$$

If we assume that β is independent of ϕ, we can find the shear angle corresponding to the maximum shear stress by differentiating Eq. (8.19) with respect to ϕ and equating it to zero:

$$\frac{d\tau}{d\phi} = \cos(\phi + \beta - \alpha)\cos\phi - \sin(\phi + \beta - \alpha)\sin\phi = 0.$$

Thus

$$\tan(\phi + \beta - \alpha) = \cot\phi = \tan(90^\circ - \phi)$$

or

$$\phi = 45^\circ + \frac{\alpha}{2} - \frac{\beta}{2}. \tag{8.20}$$

Note that Eq. (8.20) indicates that, as the rake angle decreases and/or as the friction at the tool–chip interface increases, the shear angle decreases and the chip is thus thicker. This result is to be expected, because decreasing α and increasing β tend to lower the shear angle.

A second method of determining ϕ is based on slip-line analysis (of E. H. Lee and B. W. Shaffer) and on the following assumptions.

a) The shear plane AB in Fig. 8.18(a) is a plane of maximum shear stress.
b) There are no shear or normal stresses along AC, as there should not be, as the chip beyond AC is unconstrained.
c) The material is rigid, perfectly plastic.

The appropriate Mohr's circle construction for this shear zone is shown in Fig. 8.18(b). The angle η is given by

$$\eta = 45^\circ - \beta.$$

From geometry,

$$\eta = \phi - \alpha,$$

so the expression for the shear angle is

$$\phi = 45^\circ + \alpha - \beta. \tag{8.21}$$

Note that this expression is similar to Eq. (8.20) and indicates the same trends, although numerically it gives different values.

In another study (by T. Sata and M. Mizuno), the following simple relationships were proposed and could serve as a guide for estimating the shear angle for practical purposes:

$$\phi = \alpha \quad \text{for} \quad \alpha > 15^\circ, \tag{8.22}$$

$$\phi = 15^\circ \quad \text{for} \quad \alpha < 15^\circ. \tag{8.23}$$

A number of other expressions based on various models and different assumptions have been obtained for the shear angle. However, many of these expressions

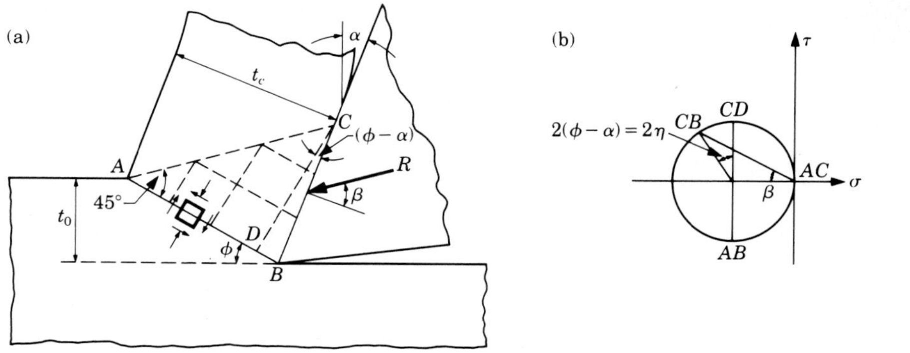

FIGURE 8.18

(a) Slip-line field in orthogonal cutting. Line AB is the shear plane along which the shear stress is assumed to be a maximum. (b) Mohr's circle for stresses in the cutting zone. Point CB on the circle represents the normal and shear stresses at the tool–chip interface, respectively.

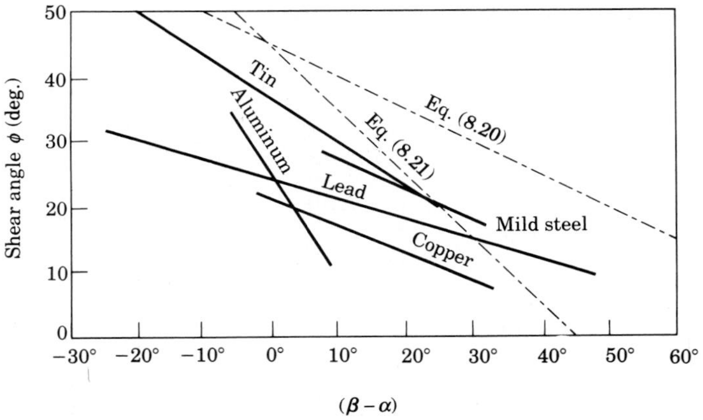

FIGURE 8.19 Comparison of experimental and theoretical shear angle relationships. More recent analytical studies have resulted in better agreement with experimental data.

do not agree well with experimental data over a wide range of conditions (Fig. 8.19), largely because shear rarely occurs in a thin plane. However, the shear angle always decreases with increasing $(\beta - \alpha)$ as shown in Fig. 8.20. Recent and more comprehensive studies appear to accurately predict the shear angle analytically, especially for continuous chips.

8.2.5 Specific energy

Refer to Fig. 8.13 and note that the *total power* input in cutting is

$$\text{Power} = F_c V.$$

If we let the width of the cut be w, then the *total energy per unit volume* of material

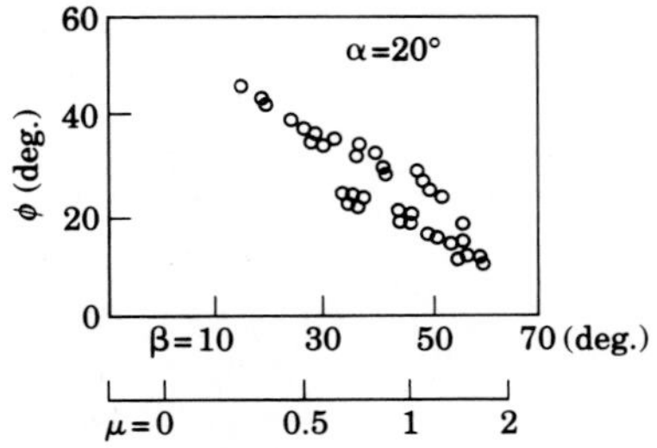

FIGURE 8.20 Relation between the shear angle and friction angle for various alloys and cutting speeds. (*Source:* S. Kobayashi, et al., *Trans. ASME, J. Eng. Ind.*, vol. 82, 1960, pp. 333–347.

removed (*specific energy*), u_t, is

$$u_t = \frac{F_c V}{w t_o V} = \frac{F_c}{w t_o}. \tag{8.24}$$

In other words, u_t is simply the ratio of the cutting force to the projected area of the cut. Figures 8.2 and 8.13 also show that the power required to overcome friction at the tool–chip interface is the product of F and V_c or in terms of frictional specific energy, u_f,

$$u_f = \frac{F V_c}{w t_o V} = \frac{F r}{w t_o} = \frac{(F_c \sin \alpha + F_t \cos \alpha) r}{w t_o}. \tag{8.25}$$

Likewise, the power required for shearing along the shear plane is the product of F_s and V_s. Hence the specific energy for shear, u_s, is

$$u_s = \frac{F_s V_s}{w t_o V}. \tag{8.26}$$

The total specific energy, u_t, is the sum of the two, or

$$u_t = u_f + u_s. \tag{8.27}$$

There are two other sources of energy in cutting. One is the *surface energy* resulting from the formation of new surfaces when a layer of material is removed by cutting. However, this source represents a very small amount of energy compared to the shear and frictional energies involved. The other source is the energy associated with the *momentum change* as the metal crosses the shear plane. (This source is similar to the forces involved in a turbine blade from momentum changes of the fluid or gas.) Although in ordinary metal-cutting operations the momentum energy is negligible, this energy can be significant at very high cutting speeds [above 125 m/s (25,000 ft/min) or so].

Experimental data on specific energies are given in Tables 8.2 and 8.3. Note that, as the rake angle increases, the frictional specific energy remains more or less constant, whereas the shear specific energy is reduced rapidly. Thus the ratio u_f/u_t increases considerably as α increases. This trend can also be predicted by obtaining an expression for the ratio as follows:

$$\frac{u_f}{u_t} = \frac{F V_c}{F_c V} = \frac{R \sin \beta}{R \cos (\beta - \alpha)} \cdot \frac{V r}{V} = \frac{\sin \beta}{\cos (\beta - \alpha)} \cdot \frac{\sin \phi}{\cos (\phi - \alpha)}. \tag{8.28}$$

Experimental observations have revealed that as α increases, both β and ϕ increase. Thus inspection of Eq. (8.28) indicates that the ratio u_f/u_t should also increase with α. Obviously, u_f and u_s are related. Although u_f is not required for the cutting action to take place, it affects the magnitude of u_s. The reason is that, as friction increases, the shear angle decreases. A decreasing shear angle, in turn, increases the magnitude of u_s.

The calculation of all the parameters involved in these specific energies in cutting presents considerable difficulties. Good theoretical computations are avail-

TABLE 8.4
APPROXIMATE POWER REQUIREMENTS IN VARIOUS CUTTING OPERATIONS (at spindle motor, corrected for 80% efficiency)

MATERIAL	UNIT POWER ($W \cdot s/mm^3$)*
Aluminum alloys	0.4–1.1
Cast irons	1.6–5.5
Copper alloys	1.4–3.3
High-temperature alloys	3.3–8.5
Magnesium alloys	0.4–0.6
Nickel alloys	4.9–6.8
Refractory alloys	3.8–9.6
Stainless steels	3.0–5.2
Steels	2.7–9.3
Titanium alloys	3.0–4.1

* Divide by 2.73 to obtain $hp \cdot min/in^3$.

able, but they are difficult to perform. The reliable prediction of cutting forces and energies is therefore still based largely on experimental data (Table 8.4). These data should serve as a useful guide. The wide range of values in Table 8.4 can be attributed to differences in strength within each material group and other variables, such as friction and operating conditions.

• Example 8.1: Percentage of frictional energy.

We are carrying out an orthogonal cutting process in which $t_o = 0.005$ in., $V = 400$ ft/min, $\alpha = 10^\circ$, and the width of cut $= 0.25$ in. We observe that $t_c = 0.009$ in., $F_c = 125$ lb, and $F_t = 50$ lb. We want to calculate the percentage of the total energy that goes into overcoming friction at the tool–chip interface.

SOLUTION. We can express the percentage as:

$$\frac{\text{Friction energy}}{\text{Total energy}} = \frac{FV_c}{F_cV} = \frac{Fr}{F_c},$$

where

$$r = \frac{t_o}{t_c} = \frac{5}{9} = 0.555, \qquad F = R \sin \beta, \qquad F_c = R \cos(\beta - \alpha),$$

and

$$R = \sqrt{F_t^2 + F_c^2} = \sqrt{50^2 + 125^2} = 135 \text{ lb}.$$

Thus

$$125 = 135 \cos (\beta - 10),$$

from which

$$\beta = 32^\circ \quad \text{and} \quad F = 135 \sin 32^\circ = 71.5 \text{ lb.}$$

Hence

$$\text{Percentage} = \frac{(71.5)(0.555)}{125} = 0.32 = 32\%.$$

• Example 8.2: Comparison of forming and machining energies.

You are given two pieces of annealed 304 stainless steel rods, each 0.500 in. in diameter and 6 in. long. You are asked to reduce the diameters to 0.480 in., one piece by pulling it in tension and the other by machining it on a lathe in one pass. Calculate the respective amounts of work involved and explain the difference in the energies dissipated.

SOLUTION. The work done in pulling the rod is

$$W_{\text{tension}} = (u)(\text{volume}),$$

where

$$u = \int_0^{\varepsilon_1} \sigma \, d\varepsilon.$$

The strain is found from

$$\varepsilon_1 = \ln \left(\frac{0.500}{0.480}\right)^2 = 0.0816.$$

From Table 2.4, the following values for K and n are obtained for this material:

$$K = 1275 \text{ MPa} = 185{,}000 \text{ psi} \quad \text{and} \quad n = 0.45.$$

Thus

$$u = \frac{K\varepsilon_1^{n+1}}{n+1} = \frac{(185{,}000)(0.0816)^{1.45}}{1.45} = 3370 \text{ in.} \cdot \text{lb/in}^3$$

and

$$W_{\text{tension}} = (3370)(\pi)(0.25)^2(6) = 3970 \text{ in.} \cdot \text{lb.}$$

From Table 8.4 an average value for the specific energy in machining stainless steels is obtained as 1.5 hp · min/in^3. The volume of material machined is

$$\frac{\pi}{4}[(0.5)^2 - (0.480)^2](6) = 0.092 \text{ in}^3.$$

The specific energy, in appropriate units, is

$$\text{Specific energy} = (1.5)(33{,}000)(12) = 594{,}000 \text{ in.} \cdot \text{lb/in}^3.$$

Thus the machining work is

$$W_{\text{mach}} = (594{,}000)(0.092) = 54{,}650 \text{ in.} \cdot \text{lb}.$$

Note that the work in machining is about 14 times that for tension. The difference between the energies is that tension requires very little straining and involves no friction. Machining involves friction, and the material removed—even though small in volume—has undergone much higher strains than the bulk material in tension. Assuming an average shear strain of 3 (from Tables 8.2 and 8.3), which is equivalent to an effective strain of 1.7 (Eq. 2.57), we note that in machining the material is subjected to a strain 21 times that in tension. These differences explain why machining consumes much more energy than reducing the diameter of this rod by stretching. However, as the diameter of the rod decreases, the difference between the two energies becomes smaller, assuming that the same depth of material is to be removed.

• Example 8.3: Calculating maximum depth of cut.

A workpiece of width w is attached to a round tensile specimen with an original cross-sectional area of A_o (Fig. E8.1). The workpiece is resting on a plane foundation, and the coefficient of friction between the workpiece and the foundation is μ_1. A wide tool with a rake angle of α is removing a layer of the workpiece material at a depth of cut d. The total energy required to machine is u in.-lb per cubic inch of material removed. The coefficient of friction at the tool–chip interface is μ_2. Ignoring the weight of the workpiece, derive an expression for the maximum value of d at which the tensile specimen begins to neck.

SOLUTION. The true strain at necking was shown in Eq. (2.13) to be

$$\varepsilon = n.$$

Thus the true stress at necking is

$$\sigma = Kn^n \quad \text{and} \quad A = A_o e^{-\varepsilon}.$$

The true area at necking is

$$A = A_0 e^{-n}.$$

Hence the tensile force on the specimen at necking is

$$F = \sigma A = KA_o n^n e^{-n}.$$

The power consumed in cutting is the product of the cutting force and cutting speed V. As the volume of material removed per unit time is wdV, the power consumed is $uwdV$. Hence the cutting force is

$$F_c = uwd.$$

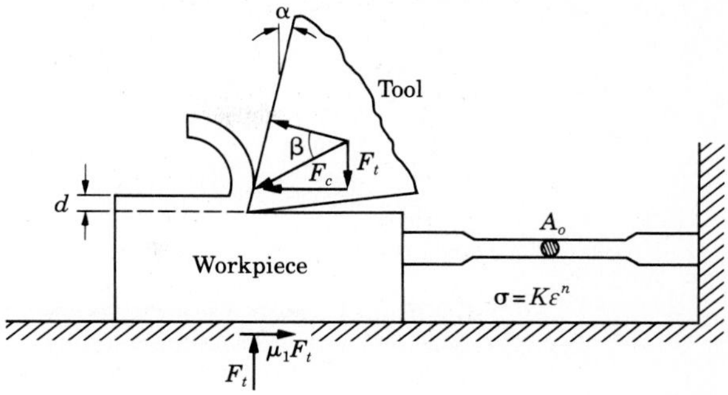

FIGURE E8.1

From Eq. (8.13) we have

$$F_t = F_c \tan(\beta - \alpha).$$

Hence the friction force between the workpiece and the foundation is $\mu_1 F_t$. The force balance in the horizontal direction therefore is

$$F_c = F + \mu_1 F_t \quad \text{or} \quad uwd = KA_o n^n e^{-n} + \mu_1 uwd \tan(\beta - \alpha).$$

Hence

$$d = \frac{KA_o n^n e^{-n}}{uw[1 - \mu_1 \tan(\beta - \alpha)]}$$

where $\beta = \tan^{-1} \mu_2$.

8.2.6 Temperature

As in all metalworking operations, the energy dissipated in cutting operations is converted into heat, which in turn, raises the temperature in the cutting zone. Knowledge of the temperature rise in cutting is important because it

- adversely affects the strength, hardness, and wear resistance of the cutting tool;
- causes dimensional changes in the part being machined, making control of dimensional accuracy difficult; and
- can induce thermal damage to the machined surface, adversely affecting its properties.

Because of the work done in shearing and in overcoming friction on the rake face of the tool, the main sources of heat generation are the primary shear zone and the tool–chip interface. Additionally, if the tool is dull or worn, heat is also generated by the tool tip rubbing against the machined surface.

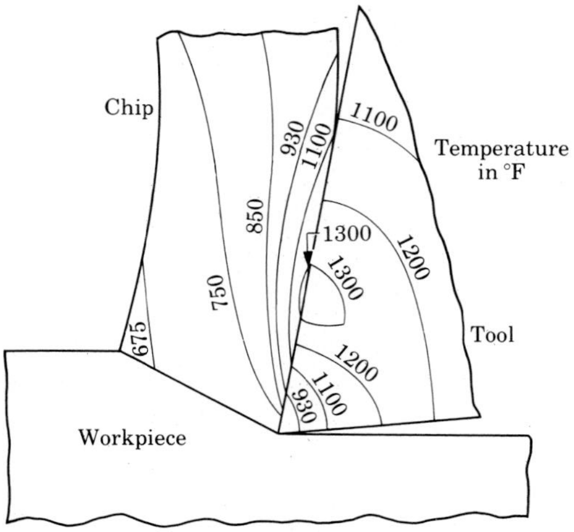

FIGURE 8.21 Typical temperature distribution in the cutting zone. Note that the maximum temperature is about halfway up the face of the tool, and that there is a steep temperature gradient across the thickness of the chip. Some chips may become red hot, causing safety hazards to the operator and thus requiring safety guards. *Source:* After G. Vieregge, *Werkstatt und Betrieb*, No. 11, 1953, p. 696.

Various studies have been made of temperatures in cutting, based on heat transfer and dimensional analysis using experimental data. Although Fig. 8.21 shows that there are considerable temperature gradients in the cutting zone, a simple but approximate expression for the mean temperature for orthogonal cutting is

$$T = \frac{1.2Y_f}{\rho c}\sqrt[3]{\frac{Vt_o}{K}}, \tag{8.29}$$

where T is the mean temperature of the tool–chip interface (°F); Y_f is the flow stress of the material (psi); V is the cutting speed (in./s); t_o is the depth of cut (in.); ρc is the volumetric specific heat of the workpiece (in. · lb/in^3 · °F); and K is the thermal diffusivity of the workpiece (in^2/s). Thermal diffusivity is the ratio of thermal conductivity to volumetric specific heat.

Equation (8.29) indicates that temperature increases with the strength of the workpiece material, cutting speed, and depth of cut. Also, the lower the thermal conductivity of the workpiece material, the higher the temperature. Thermal properties of the tool are relatively unimportant compared to those of the workpiece. Because some of the parameters in Eq. (8.29) depend on temperature, using appropriate values that are compatible with the predicted temperature range is important.

Based on Eq. (8.29), an expression for the mean temperature T in turning (on a lathe) is

$$T \propto V^a f^b, \tag{8.30}$$

where a and b are constants, V is the cutting speed, and f is the feed of the tool (Fig. 8.22).

FIGURE 8.22
Terminology in a turning operation on a lathe, where f is the feed (in./rev or mm/rev) and d is the depth of cut. Note that feed in turning is equivalent to the depth of cut in orthogonal cutting (Fig. 8.1), and depth of cut in turning is equivalent to width of cut in orthogonal cutting.

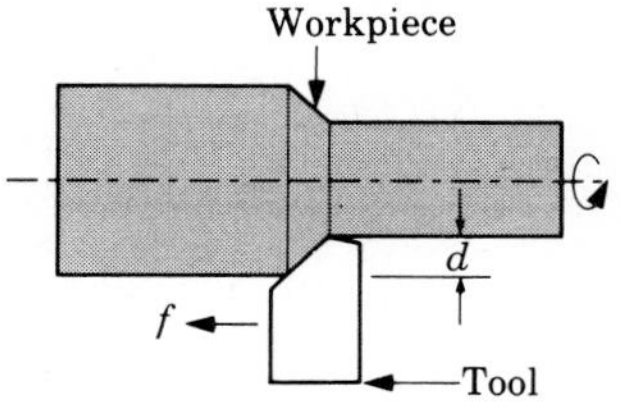

Approximate values for a and b are

TOOL	a	b
Carbide	0.2	0.125
High-speed steel	0.5	0.375

The depth of cut has negligible influence on the mean temperature for depths exceeding twice the tip radius of the tool. Results from one experimental measurement of temperature (using thermocouples) are shown in Fig. 8.23. Note that the maximum temperature is at a point away from the tip of the tool and that it increases with cutting speed.

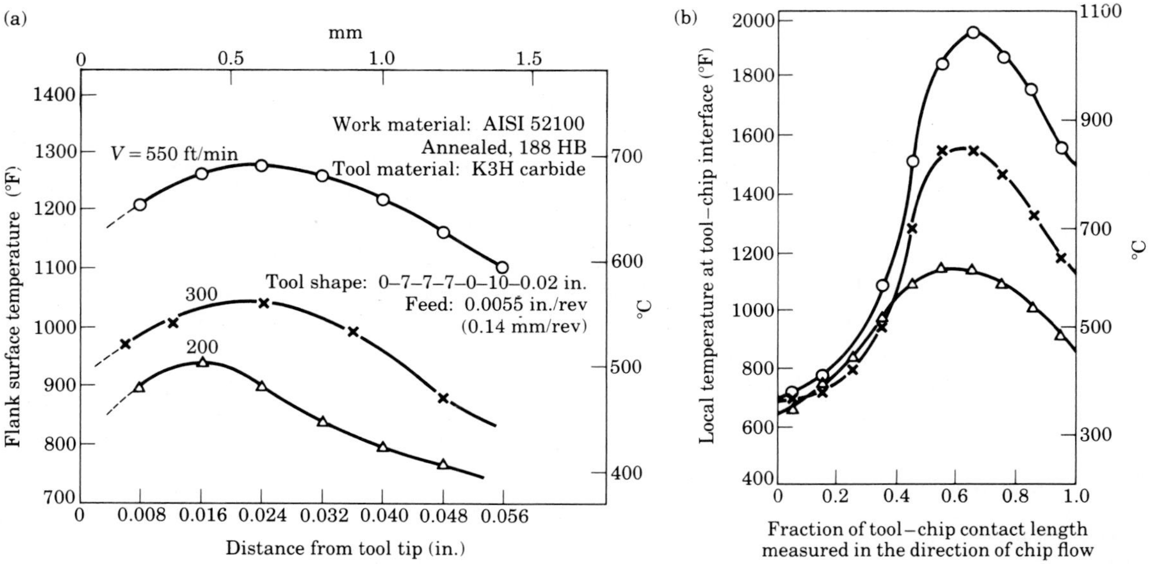

FIGURE 8.23
Temperature distribution in turning: (a) flank temperature for tool shape (see Fig. 8.54); (b) tool–chip interface temperature. Note that the rake face temperature is higher than at the flank surface. *Source:* After B. T. Chao and K. J. Trigger, *J. Eng. Ind.*, vol. 83, 1961, pp. 496–504.

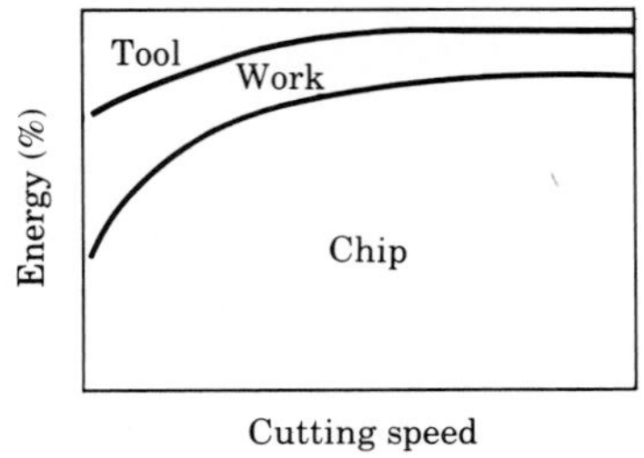

FIGURE 8.24 Typical energy distribution as a function of cutting speed. Note that most of the cutting energy is carried away by the chip (in the form of heat), particularly as speed increases. For dimensional accuracy during cutting, it is important not to allow the workpiece temperature to rise significantly.

The temperature generated in the shear plane is a function of the specific energy for shear, u_s, and the specific heat of the material. Hence temperature rise is highest in cutting materials with high strength and low specific heat, as Eq. (2.64) shows. The temperature rise at the tool–chip interface is a function of the coefficient of friction. Flank wear (see Section 8.3) is also a source of heat because of rubbing of the tool on the machined surface.

Cutting speed has a major influence on temperature. As the speed increases, there is little time for the heat to be dissipated and hence temperature rises. The chip is a good heat sink, in that it carries away most of the heat generated. (This action of the chip is similar to *ablation*, where layers of metal melt away from a surface, thus carrying away the heat.) As cutting speed increases, a larger proportion of the heat is carried away by the chip, as shown in Fig. 8.24.

Techniques for measuring temperature. Temperatures and their distribution in the cutting zone may be determined from thermocouples embedded in the tool and/or the workpiece. This technique has been used successfully, although it involves considerable effort. A simpler technique for determining the average temperature is by the thermal emf (electromotive force) at the tool–chip interface, which acts as a hot junction between two different (tool and chip) materials. Infrared radiation from the cutting zone may also be monitored with a radiation pyrometer. However, this technique indicates only surface temperatures, and the accuracy of the results depends on the emissivity of the surfaces, which is difficult to determine accurately.

8.3 Tool Wear

We have shown that tool surfaces are subjected to forces, temperature, and sliding—all conditions that induce wear. Because of its effects on the quality of the machined surface and the economics of machining, tool wear is one of the most

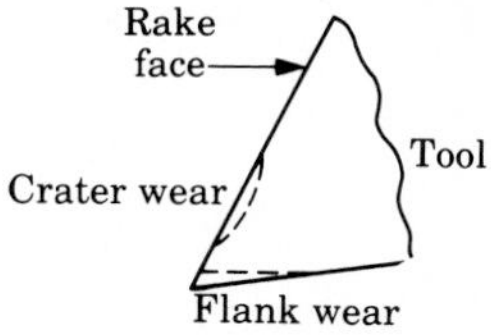

FIGURE 8.25 Schematic illustration of crater and flank wear on a cutting tool. Wear of tools is one of the most important aspects in cutting and has a major influence on the economics of the overall machining operation.

important and complex aspects of machining operations. Although cutting speed is an independent variable, the forces and temperatures generated are dependent variables and are functions of numerous parameters. Similarly, wear depends on tool and workpiece materials (their physical, mechanical, and chemical properties), tool geometry, cutting fluid properties, and various other operating parameters. The types of wear on a tool depend on the relative roles of these variables. Analytical studies of tool wear present considerable difficulties, and therefore our knowledge of wear is based largely on experimental data.

The basic wear behavior of a tool is shown in Fig. 8.25 (for a two-dimensional cut) and Figs. 8.26 and 8.27 (for a three-dimensional cut, where the position of the tool is shown in Fig. 8.22). The various regions of wear are identified as *flank wear*, *crater wear*, *nose wear*, and *chipping* of the cutting edge.

The tool profile can be altered by these various wear and fracture processes, and this change is bound to influence the cutting operation. In addition to these wear processes, plastic deformation of the tool can also take place to some extent, such as in softer tools at elevated temperatures. Gross chipping of the tool is called *catastrophic failure*, whereas wear is generally a gradual process. Because of the complex interactions between material and process variables, the interpretation of these wear patterns and study of the wear mechanisms must be carried out cautiously.

8.3.1 Flank wear

Flank wear has been studied extensively and is generally attributed to

a) sliding of the tool along the machined surface causing adhesive and/or abrasive wear, depending on the materials involved; and
b) temperature, because of its influence on tool material properties.

Following a classic and extensive study by F. W. Taylor (published in 1907), the following relationship was established for cutting various steels:

$$VT^n = C, \tag{8.31}$$

where V is the cutting speed, T is the time that it takes to develop a flank wear land (VB in Fig. 8.26c) of certain dimensions, n is an exponent that depends on cutting conditions, and C is a constant. Equation (8.31) is a simple version of the many relationships developed by Taylor among the variables involved. Each combination

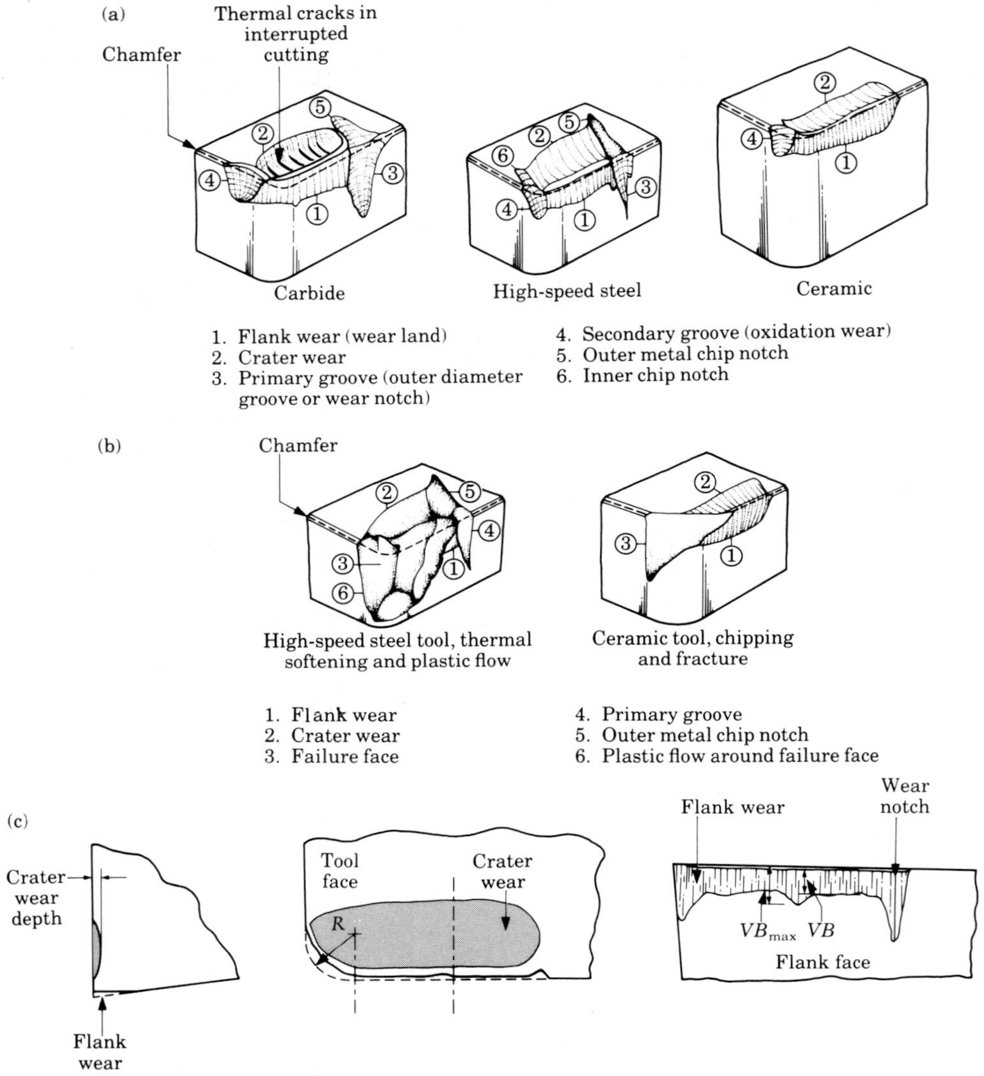

FIGURE 8.26 (a) Types of wear observed in cutting tools. The thermal cracks shown are usually observed in interrupted cutting operations, such as in milling. (b) Catastrophic failure of tools. (c) Features of tool wear in a turning operation. The *VB* indicates average flank wear. *Source:* (a) and (b) After V. C. Venkatesh. (c) International Standards Organization (ISO).

of workpiece and tool material and each cutting condition has its own n value and a different constant C.

Tool-life curves. *Tool-life curves* are plots of experimental data obtained in cutting tests (Fig. 8.28). Note the rapid decrease in tool life as cutting speed increases and the strong influence of the condition of the workpiece material on tool

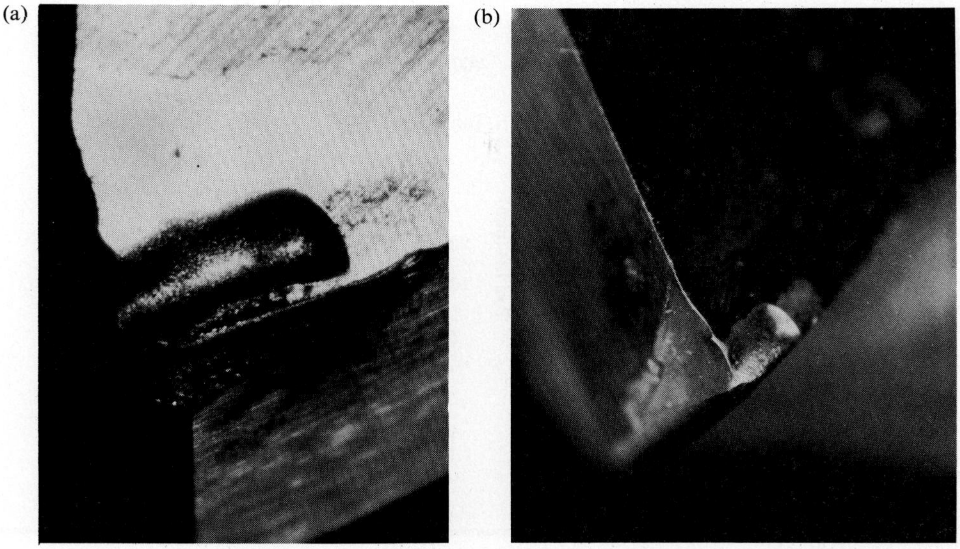

FIGURE 8.27
(a) Crater wear on rake face (top surface) and flank wear (side front view) on a carbide insert used in cutting sintered tungsten on a lathe. (b) Crater wear and chipping (lower left side) on a carbide insert used in machining tungsten. *Source:* A. J. Moser and S. Kalpakjian.

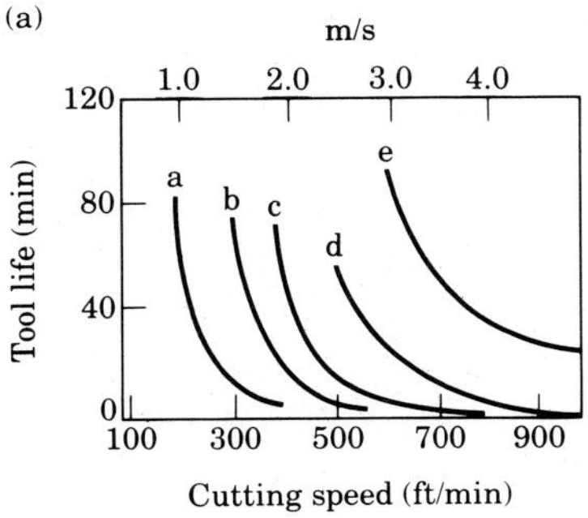

	Hardness (HB)	Ferrite	Pearlite
a. As cast	265	20%	80%
b. As cast	215	40	60
c. As cast	207	60	40
d. Annealed	183	97	3
e. Annealed	170	100	—

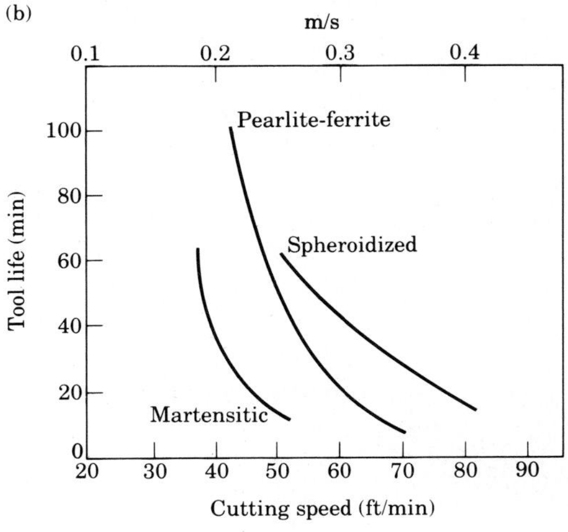

FIGURE 8.28
Effect of workpiece microstructure on tool life in turning. Tool life is given in terms of the time (in minutes) required to reach a flank wear land of a specified dimension. (a) Ductile cast iron. (b) Steels, with identical hardness. Note the rapid decrease in tool life as the cutting speed increases.

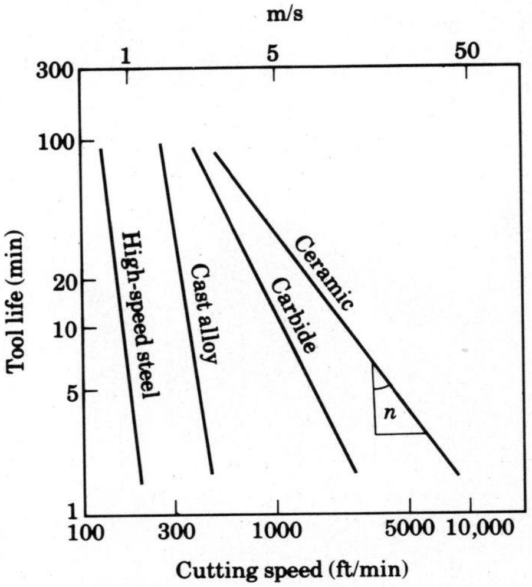

FIGURE 8.29

Tool-life curves for a variety of cutting-tool materials. The negative inverse of the slope of these curves is the exponent n in tool-life equations.

life. Also note the large difference in tool life for different workpiece microstructures. Heat treatment is important largely because of increasing workpiece hardness. For example, ferrite has a hardness of about 100 *HB*, pearlite 200 *HB*, and martensite 300–500 HB. Impurities and hard constituents in the material are also important considerations because they reduce tool life by their abrasive action.

We usually plot tool-life curves on log–log paper, from which we can easily determine the exponent n (Fig. 8.29). The range of n values that have been determined experimentally is given in Table 8.5. These curves are usually linear over a certain range of cutting speeds but are rarely so over a wide range. Moreover, the exponent n can indeed become negative at low cutting speeds. Thus tool-life curves may actually reach a maximum and then curve downward. Therefore, caution should be exercised when using tool-life equations beyond the range of cutting speeds for which they are applicable.

Because of the influence of temperature on the physical and mechanical properties of materials, we would expect that wear is strongly influenced by

TABLE 8.5

RANGE OF n VALUES FOR VARIOUS CUTTING TOOLS

High-speed steels	0.08–0.2
Cast alloys	0.1–0.15
Carbides	0.2–0.5
Ceramics	0.5–0.7

temperature. Experimental investigations have shown that there is indeed a direct relation between flank wear and temperature generated during cutting (Fig. 8.30). Although cutting speed has been found to be the most significant process variable in tool life, depth of cut and feed rate are also important. Thus Eq. (8.31) can be modified as follows:

$$VT^n d^x f^y = C, \tag{8.32}$$

where d is the depth of cut and f is the feed rate (in mm/rev or in./rev) in turning.

The exponents x and y must be determined experimentally for each cutting condition. For example, taking $n = 0.15$, $x = 0.15$, and $y = 0.6$ as typical values encountered in practice, we see that cutting speed, feed rate, and depth of cut are of decreasing order of importance.

Equation (8.32) can be rewritten as

$$T = C^{1/n} V^{-1/n} d^{-x/n} f^{-y/n} \tag{8.33}$$

or

$$T \simeq C^7 V^{-7} d^{-1} f^{-4}. \tag{8.34}$$

For a constant tool life, the following observations can be made from Eq. (8.34).

a) If the feed rate or the depth of cut is increased, the cutting speed must be decreased, and vice versa.
b) Depending on the exponents, a reduction in speed can then result in an increase in the volume of the material removed because of the increased feed rate and/or depth of cut.

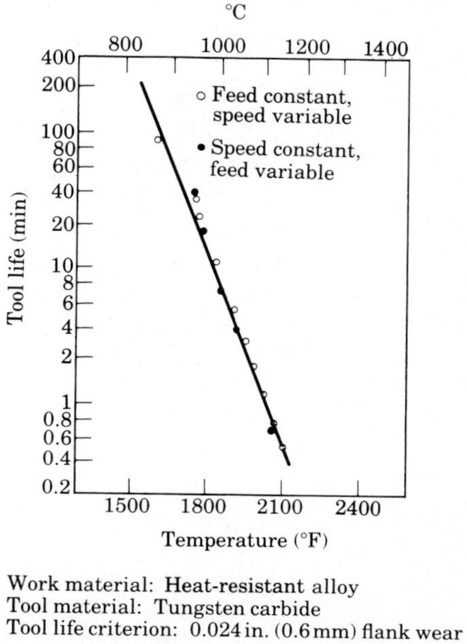

FIGURE 8.30 Relationship between measured temperature during cutting and tool life (flank wear). Note that high cutting temperatures severely reduce tool life. See also Eq. (8.30). *Source:* After H. Takeyama and Y. Murata.

TABLE 8.6
ALLOWABLE AVERAGE WEAR LAND (*VB*) FOR CUTTING TOOLS IN VARIOUS OPERATIONS

	ALLOWABLE WEAR LAND (mm)	
OPERATION	*HIGH-SPEED STEELS*	*CARBIDES*
Turning	1.5	0.4
Face milling	1.5	0.4
End milling	0.3	0.3
Drilling	0.4	0.4
Reaming	0.15	0.15

Allowable wear land. The *allowable wear land* (VB) for various conditions is given in Table 8.6. For improved dimensional accuracy and surface finish, the allowable wear land may be made smaller than the values given in Table 8.6. The recommended cutting speed for a high-speed-steel tool is generally the one that gives a tool life of 60–120 min and for carbide tools 30–60 min.

Optimum cutting speed. We have shown that as cutting speed increases, tool life is rapidly reduced. On the other hand, if cutting speeds are low, tool life is long but the rate at which material is removed is also low. Thus there is an optimum cutting speed.

• **Example 8.4: Effect of cutting speed on material removal.**

We can appreciate the effect of cutting speed on the volume of metal removed between tool resharpenings or replacements by analyzing Fig. 8.28(a). Assume that we are machining the material in the "a" condition, that is, as cast with a hardness of 265 HB. If our cutting speed is 1 m/s (200 ft/min), we note that tool life is about 40 min. Hence the tool travels a distance of (1 m/s)(60 s/min)(40 min) = 2400 m before it is resharpened or replaced. If we change the cutting speed to 2 m/s, tool life is about 5 min. Hence the tool travels (2)(60)(5) = 600 m.

Since the volume of material removed is directly proportional to the distance the tool has traveled, we see that by decreasing the cutting speed, we can remove more material between tool changes. Note, however, that the lower the cutting speed the longer is the time required to machine a part. These variables have important economic impact, and we will discuss them further in Section 8.13. •

8.3.2 Crater wear

The most significant factors in *crater* wear are temperature and the degree of chemical affinity between the tool and the workpiece. The factors affecting flank wear also influence crater wear. We have shown that the rake face of the tool is

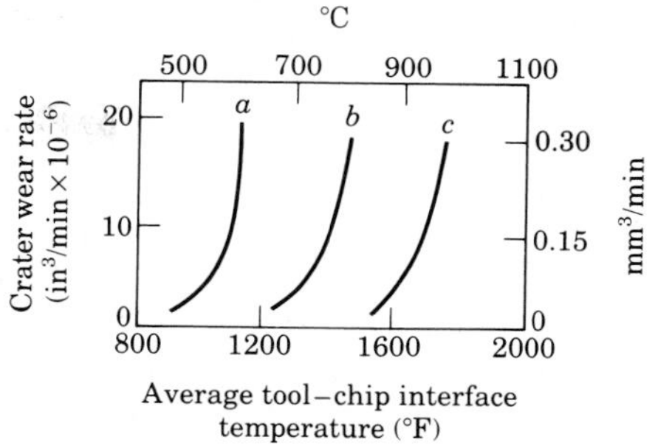

FIGURE 8.31 Relation between crater wear rate and average tool–chip interface temperature in turning: (a) high-speed steel tool; (b) C-1 carbide; (c) C-5 carbide. Note that crater wear increases rapidly within a narrow range of temperature. *Source:* After K. J. Trigger and B. T. Chao.

subjected to high levels of stress and temperature, in addition to sliding at relatively high speeds. Peak temperatures, for instance, can be on the order of 1100 °C (2000 °F) (Fig. 8.23b).

Interestingly the location of maximum crater wear generally coincides with the location of maximum temperature. Experimental evidence indicates a direct relation between crater-wear rate and tool–chip interface temperature (Fig. 8.31). Note the sharp increase in crater wear after a certain temperature range has been reached.

The cross-section of the tool–chip interface in cutting steel at high speeds is shown in Fig. 8.32. Note the location of the crater-wear pattern and the

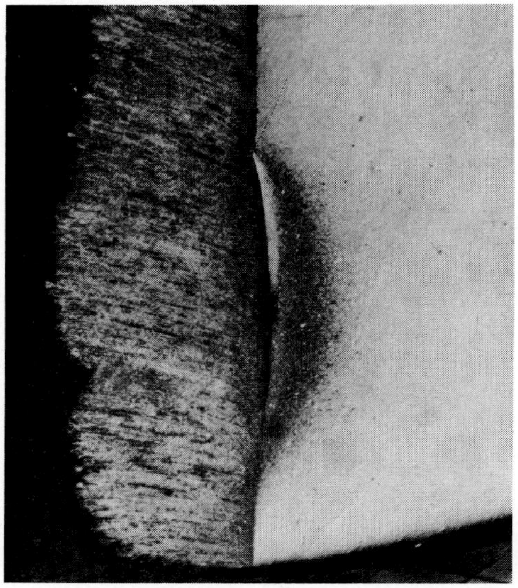

FIGURE 8.32 Interface of chip (left) and rake face of tool (right) and crater wear in cutting AISI 1004 steel at 3 m/s (585 ft/min). Discoloration of the tool indicates high temperature (loss of temper). Note how the crater wear pattern coincides with the discoloration pattern. Compare this pattern with the temperature distribution shown in Fig. 8.21. *Source:* Courtesy of P. K. Wright.

discoloration of the tool (loss of temper) as a result of high temperatures. Note also how well the discoloration profile agrees with the temperature profile shown in Fig. 8.21.

The effect of temperature on crater wear has been described in terms of a *diffusion* mechanism (the movement of atoms across the tool–chip interface). Diffusion depends on the tool–workpiece material combination and on temperature, pressure, and time. As these quantities increase, the diffusion rate increases. Unless these factors are favorable, crater wear will not take place by diffusion. Wear will then be caused by other factors, such as those outlined for flank wear. High temperatures may also cause softening and plastic deformation of the tool because of the decrease in yield strength with temperature. This type of deformation is generally observed in machining high-strength materials.

The various patterns of wear for a carbide tool in cutting austenitic stainless steel are shown in Fig. 8.33. A BUE forms at low speeds. As the cutting speed is

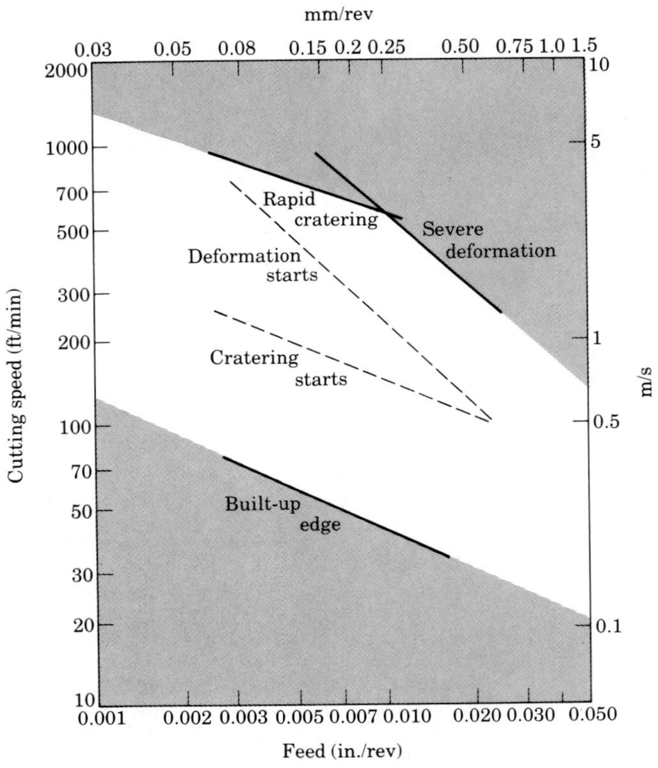

FIGURE 8.33 Plastic deformation and wear of a carbide tool in cutting austenitic stainless steel, as a function of cutting speed and feed. Note that deformation and crater wear take place at higher speeds, indicating higher temperatures. Charts like this are useful in determining the optimum range of feed and speed in cutting operations. *Source:* After P. A. Dearnley and E. M. Trent.

increased, crater wear and plastic deformation of the cutting edge of the tool take place owing to increased temperature. The complex interaction of all the parameters involved prevents generalization about the wear behavior of cutting tools. Each particular set of parameters results in a specific type of wear on the cutting tool.

8.3.3 Chipping

The term *chipping* is used to describe the breaking away of a piece from the cutting edge of the tool. The chipped pieces may be very small (microchipping or macrochipping), or they may involve relatively large fragments (gross chipping or fracture). Unlike wear, which is a more gradual process, chipping is a phenomenon that results in a sudden loss of tool material. Two main causes of chipping are mechanical shock (impact by interrupted cutting, such as in milling) and thermal fatigue. Chipping may occur in a region in the tool where a small crack or defect already exists. Thermal cracks (Fig. 8.26a) are caused by the thermal cycling of the tool in interrupted cutting. Thermal cracks are generally perpendicular to the cutting edge.

High positive rake angles can contribute to chipping because of the small included angle of the tool tip. Crater wear may also progress toward the tool tip and weaken it, thus causing chipping. Chipping or fracture can be reduced by selecting tool materials with high-impact and thermal shock resistance.

8.3.4 General observations

Because of the many factors involved, including the characteristics of the machine tool, the wear behavior of cutting tools varies significantly. In addition to the wear processes we have already described, other phenomena also occur in tool wear (Fig. 8.26). For example, as a result of the high temperatures generated during cutting, tools could soften and undergo plastic deformation because their yield strength decreases. This type of deformation generally occurs in machining high-strength metals and alloys. Thus tools must maintain their strength and hardness at the elevated temperatures encountered in cutting.

The wear groove or notch on cutting tools (Fig. 8.26) has been attributed to the fact that this region is the boundary where the chip is no longer in contact with the tool. This boundary, or *depth-of-cut line*, oscillates because of inherent variations in the cutting operation and accelerates the wear process. Furthermore, this region is in contact with the machined surface from the previous cut. Since a machined surface may develop a thin work-hardened layer, this contact could contribute to the formation of the wear groove.

Because they are hard and abrasive, scale and oxide layers on a workpiece surface increase wear. In such cases, the depth of cut should be greater than the thickness of the oxide film or the work-hardened layer. Thus the value of d in Fig. 8.22 should be greater than the thickness of the scale on the workpiece. In other words, light cuts should not be taken on rusted workpieces.

8.3.5 Techniques for measuring and monitoring tool wear

Tool-wear measuring techniques fall into two categories: direct and indirect. The *direct* method involves optical measurement of wear, such as by periodically observing changes in the tool profile. This is the most common and reliable technique and is done using a microscope (toolmakers' microscope). This procedure, however, requires that the cutting operation be stopped. Another direct method involves observing the tool-side face of the chip (see Fig. 8.6) for the presence and amount of crater wear particles using special instrumentation.

Indirect methods of measuring wear involve the correlation of wear with process variables such as forces, power, temperature rise, surface finish, and vibrations. One recent development is the *acoustic emission technique*, which utilizes a piezoelectric transducer attached to a tool holder. The transducer picks up signals that are acoustic emissions resulting from the stress waves generated during cutting. By analyzing the signals, we can monitor tool wear and chipping (Fig. 8.34). We can also record and monitor forces and vibrations during cutting, and changes in their patterns as the cut progresses, for indications of tool wear and fracture.

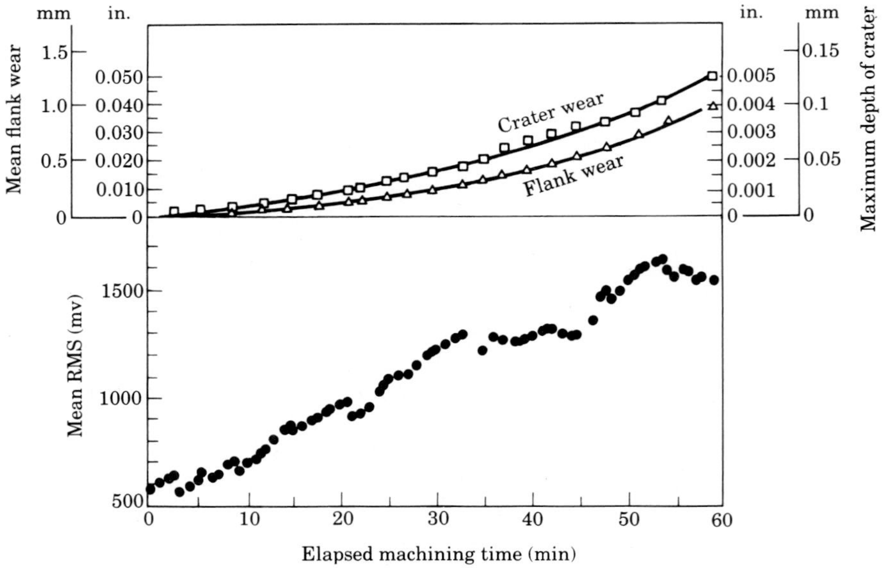

FIGURE 8.34 Relationship between mean flank wear, maximum crater wear, and acoustic emission (noise generated during cutting) as a function of machining time. This technique is being developed as a means for monitoring wear rate in various cutting processes without interrupting the operation. *Source:* After M. S. Lan and D. A. Dornfeld, *Proc. NAMRC-X*, 1982, pp. 305–311.

Because direct observation methods interrupt the steady-state nature of the cut, they influence the economics of machining operations in manufacturing plants. Implementing on-line monitoring of the rate of tool wear, that is, while the cutting operation is taking place, is therefore more desirable, particularly for computer-controlled machine tools. The indirect methods stated above, including monitoring of power, is being used for this purpose. There are, however, difficulties involved concerning reliability and calibration in the use of these techniques. Continued progress is being made in refining measurement techniques. Some instrumentation for tool-condition monitoring is now commercially available.

Considerable variability has been observed in wear tests in machining but reproducibility of wear data is difficult in view of the many factors that affect wear. Among these factors are metallurgic and other property variations in the materials involved, the machine tool used, application of cutting fluids, environmental conditions, and inaccuracies in controlling process parameters and wear-measurement techniques.

8.4 Surface Finish and Integrity

Surface finish influences not only the dimensional accuracy of machined parts, but also their properties. Whereas *surface finish* describes the geometric features of surfaces, *surface integrity* pertains to properties such as fatigue life and corrosion resistance, which are influenced strongly by the type of surface produced. Factors influencing surface integrity are temperatures generated during processing, residual stresses, metallurgical (phase) transformations, and surface plastic deformation, tearing, and cracking. Surface roughness for machining and other processes is given in Fig. 8.35.

The built-up edge, with its significant effect on tool profile, has the greatest influence on surface roughness. Figure 8.36 shows surfaces obtained in two different cutting operations. Note the considerable damage to the surfaces from BUE. Ceramic and diamond tools generally produce better surface finish than other tools, largely because of their much lower tendency to form BUE.

A tool that is not sharp has a large tip radius (see Fig. 8.26c), just like a dull pencil or knife does. If this radius, not to be confused with radius R in Fig. 8.26(b), is large in relation to the depth of cut, the tool will rub over the machined surface (see Fig. 8.15). This rubbing generates heat and induces surface residual stresses, which in turn, may cause surface damage such as tearing and cracking.

In turning, as in other cutting operations, the tool leaves a spiral profile—*feed marks*—on the machined surface as it moves across the workpiece. The higher the feed f and the smaller the radius R, the more prominent these marks are. Although

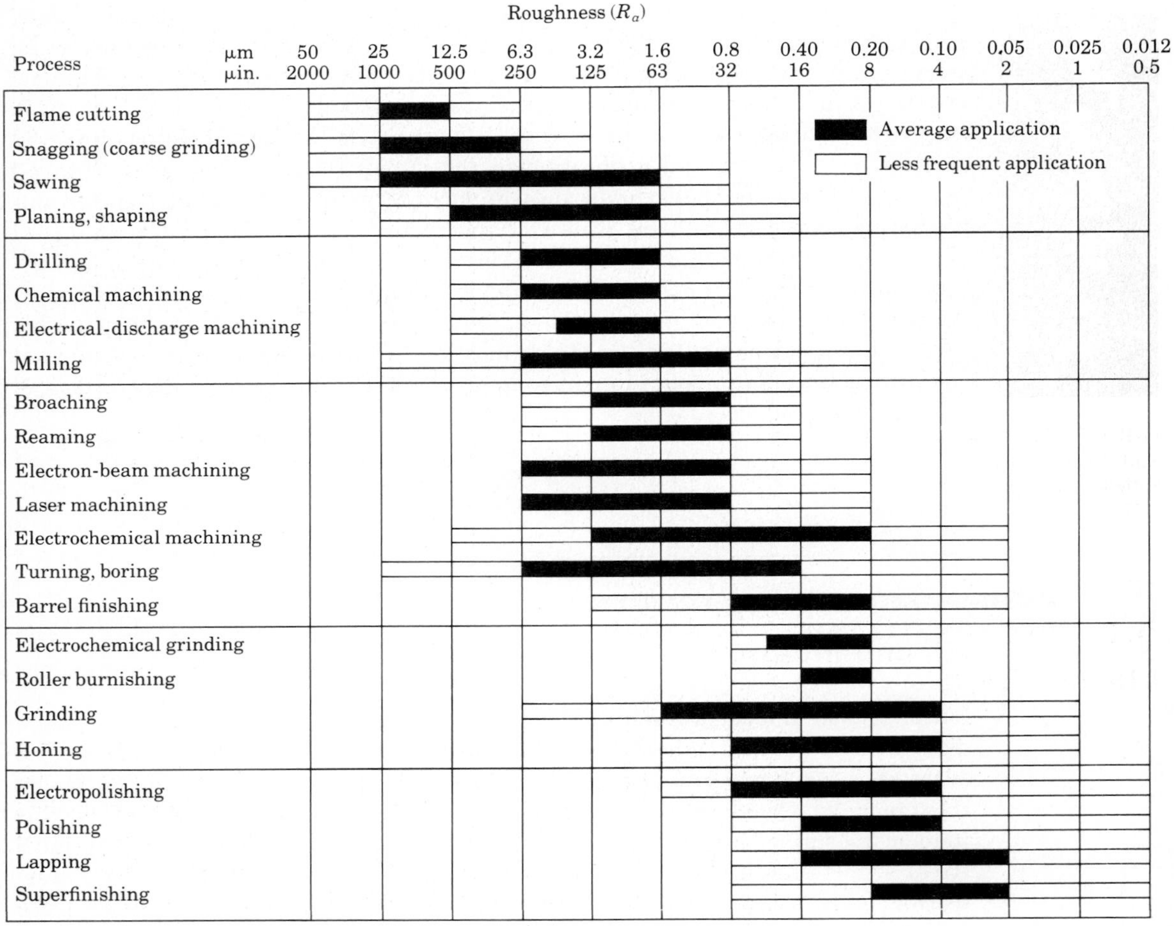

FIGURE 8.35 Range of surface roughnesses obtained in various machining processes. Note the wide range within each group. (See also Fig. 9.12.)

not significant in rough machining operations, these marks are important in finish machining.

We describe vibration and chatter in some detail in Section 8.11. For now, we should recognize that if the tool vibrates or chatters during cutting, it will adversely affect surface finish. The reason is that a vibrating tool changes the dimensions of the cut periodically. Excessive chatter can also cause chipping and premature failure of the more brittle cutting tools such as ceramics and diamond.

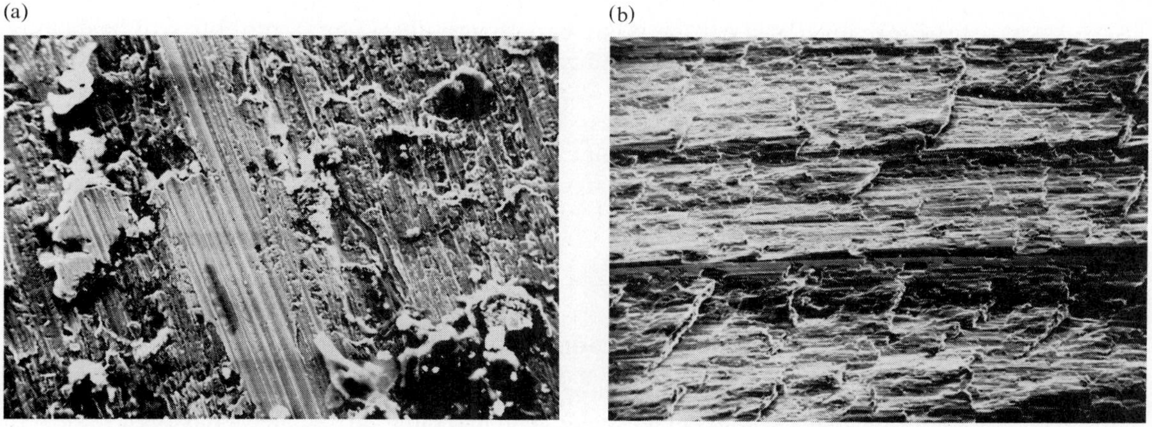

FIGURE 8.36
Surfaces produced on steel by cutting, as observed with a scanning electron microscope: (a) turned surface and (b) surface produced by shaping. *Source:* J. T. Black and S. Ramalingam.

8.5 Machinability

We usually define *machinability* of a material in terms of three factors: (1) surface finish and integrity of the machined part, (2) tool life obtained, and (3) force and power requirements. Thus good machinability indicates good surface finish and integrity, long tool life, and low force and power requirements. An additional parameter is chip curl. As we stated earlier, long, thin curled chips, if not broken up, can severely interfere with the cutting operation by becoming entangled in the cutting zone. Thus the type of chip a material produces is also a factor in its machinability.

Because of the complex nature of cutting operations, establishing relationships to define quantitatively the machinability of a material is difficult. In manufacturing plants, tool life and surface roughness are generally considered to be the most important factors in machinability. Although not used much any more, approximate *machinability ratings* are available.

• Example 8.5: Machinability ratings.

Machinability ratings are based on a tool life of $T = 60$ min. The standard is AISI 1112 steel, which is given a rating of 100. Thus for a tool life of 60 min, this steel should be machined at a cutting speed of 100 ft/min (0.5 m/s). Higher speeds will reduce tool life, and lower speeds will increase it.

For example, 3140 steel has a machinability rating of 55. This means that when it is machined at a cutting speed of 55 ft/min (0.275 m/s), tool life will be 60 min. Nickel has a rating of 200, indicating that it should be machined at 200 ft/min (1 m/s) to obtain a tool life of 60 min. Machinability ratings for various materials are as follows: free-cutting brass 300, 2011 wrought aluminum 200, pearlitic gray iron 70, Inconel 30, and precipitation-hardening 17-7 steel 20.

As the hardness of the material increases, its machinability rating decreases proportionately. However, such ratings are only approximate and should be used with caution.

8.5.1 Machinability of steels

Because steels are among the most important engineering materials, their machinability has been studied extensively. The machinability of steels has been improved mainly by adding lead and sulfur to obtain so-called *free-machining steels.*

Leaded steels. Lead is added to the molten steel and takes the form of dispersed fine lead particles. Lead is insoluble in iron, copper, and aluminum and their alloys. Thus during cutting, the lead particles are sheared and smeared over the tool–chip interface. Because of its low shear strength, the lead acts as a solid lubricant.

This behavior has been verified by the presence of high concentrations of lead on the tool-side face of chips in machining leaded steels. In addition to this effect, lead probably lowers the shear stress in the primary shear zone, thus reducing cutting forces and power consumption. Lead may be used in either nonsulfurized or resulfurized steels. If the presence of lead is objectionable, it may be replaced by bismuth. Leaded steels are identified by the letter *L* between the second and third numerals; thus, for example, 10L45. However, in stainless steels, similar use of the letter *L* in their identification means low carbon, which improves their corrosion resistance.

Resulfurized steels. Sulfur in steels forms manganese sulfide inclusions (second-phase particles), which act as stress raisers in the primary shear zone (Fig. 8.37). As a result, the chips produced are small and break up easily, thus improving machinability. The shape, orientation, distribution, and concentration of these inclusions significantly influence machinability. Elements such as tellurium and selenium (both chemically similar to sulfur) in resulfurized steels act as inclusion modifiers.

Calcium-deoxidized steels. An important development is calcium-deoxidized steels, in which oxide flakes of calcium aluminosilicate (CaO, SiO_2, and Al_2O_3) are formed. These flakes, in turn, reduce the strength of the secondary shear zone, thus decreasing tool–chip interface friction and wear, hence temperature. Consequently, these steels undergo less crater wear, especially at high cutting speeds.

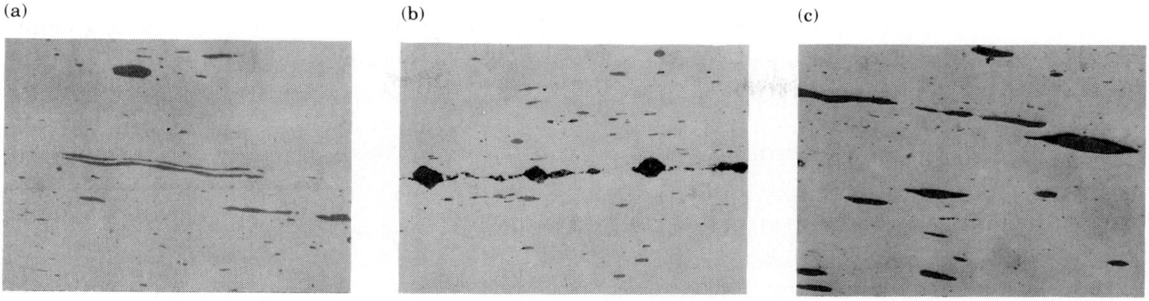

FIGURE 8.37
Photomicrographs showing various types of inclusions in low-carbon, resulfurized free-machining steels. (a) Manganese sulfide inclusions in AISI 1215 steel. (b) Manganese sulfide inclusions and glassy manganese silicate-type oxide (dark) in AISI 1215 steel. (c) Manganese sulfide with lead particle as tails in AISI 12L14 steel. *Source:* Courtesy of H. Yaguchi, Inland Steel Company.

Effects of other elements in steels on machinability. The presence of aluminum and silicon in steels is always harmful because they combine with oxygen and form aluminum oxide and silicates. These compounds are hard and abrasive, thus increasing tool wear and reducing machinability.

Carbon, manganese, and phosphorus have various effects on the machinability of steels, depending on their composition. As the carbon content increases, machinability decreases. However, plain low-carbon steels (less than 0.15% C) can produce poor surface finish by forming a built-up edge. Cast steels are more abrasive, although their machinability is similar to wrought steels. Tool and die steels are very difficult to machine and usually require annealing prior to machining. Machinability of most steels is generally improved by cold working, which reduces the tendency for built-up edge formation.

Other alloying elements, such as nickel, chromium, molybdenum, and vanadium, which improve the properties of steels, generally reduce machinability. The effect of boron is negligible. The role of gaseous elements such as oxygen, hydrogen, and nitrogen has not been clearly established. Any effect that they may have would depend on the presence and quantity of other alloying elements.

In selecting various elements to improve machinability, we should consider the possible detrimental effects of these elements on the properties and strength of the machined part in service. At elevated temperatures, for example, lead causes embrittlement of steels (hot shortness), although at room temperature it has no effect on mechanical properties. Sulfur can severely reduce hot workability of steels, because of the presence of iron sulfide, unless sufficient manganese is present to prevent the formation of iron sulfide. At room temperature, the mechanical properties of resulfurized steels depend on the orientation of the deformed manganese sulfide inclusions (anisotropy).

Stainless steels. Austenitic (300 series) steels are generally difficult to machine. Chatter could be a problem, thus requiring machine tools with high stiffness. However, ferritic stainless steels (also 300 series) have good machinability.

Martensitic (400 series) steels are abrasive and tend to form built-up edge, and require tool materials with high hot hardness and crater-wear resistance. Precipitation-hardening stainless steels are strong and abrasive, requiring hard and abrasion-resistant tool materials.

8.5.2 Machinability of various other metals

Aluminum is generally easy to machine. However, the softer grades tend to form built-up edge. High cutting speeds, high rake angles, and high relief angles are recommended. Wrought alloys with high silicon content and cast aluminum alloys may be abrasive and hence require harder tool materials. Dimensional control may be a problem in machining aluminum since it has a low elastic modulus and a relatively high thermal coefficient of expansion.

Beryllium is similar to cast irons but is more abrasive and toxic. Hence it requires machining in a controlled environment.

Gray cast irons are generally machinable but are abrasive. Free carbides in castings reduce their machinability and cause tool chipping or fracture, thus requiring tools with high toughness. Nodular and malleable irons are machinable with hard tool materials.

Cobalt-base alloys are abrasive and highly work hardening. They require sharp and abrasion-resistant tool materials and low feeds and speeds.

Wrought copper can be difficult to machine because of built-up edge formation, although cast copper alloys are easy to machine. Brasses are easy to machine, especially with the addition of lead (leaded free-machining brass). Bronzes are more difficult to machine than brass.

Magnesium is very easy to machine, has good surface finish, and prolongs tool life. However, care should be exercised because of its high rate of oxidation and the danger of fire.

Molybdenum is ductile and work hardening. Hence it can produce poor surface finish, thus requiring sharp tools.

Nickel-base alloys are work hardening, abrasive, and strong at high temperatures. Their machinability is similar to that of stainless steels.

Tantalum is very work hardening, ductile, and soft. Hence it produces a poor surface finish; tool wear is high.

The poor thermal conductivity of titanium (lowest of all engineering metals) and its alloys causes significant temperature rise and built-up edge. Thus it can be difficult to machine.

Because tungsten is brittle, strong, and very abrasive, its machinability is low. Machinability improves greatly at elevated temperatures.

Zirconium has good machinability. However, it requires a coolant-type cutting fluid because of the danger of explosion and fire.

8.5.3 Machinability of various materials

Graphite is abrasive. It requires hard, abrasion-resistant, sharp tools. Thermoplastics generally have low thermal conductivity, low elastic modulus, and low softening temperature. Consequently, machining them requires tools with positive rake to reduce cutting forces, large relief angles, small depths of cut and feed, relatively high speeds, and proper support of the workpiece. Tools should be sharp. External cooling of the cutting zone may be necessary to keep the chips from becoming "gummy" and sticking to the tools. Cooling can usually be done with a jet of air, vapor mist, or water-soluble oils. Residual stresses may develop during machining. To relieve residual stresses, machined parts can be annealed at temperatures ranging from 80 °C to 160 °C (175 °F to 315 °F) for a period of time and then cooled slowly and uniformly to room temperature.

Thermosetting plastics are brittle and sensitive to thermal gradients during cutting. Their machinability is generally similar to that of thermoplastics.

Because of the fibers present, reinforced plastics are very abrasive and are difficult to machine. Fiber tearing and pulling is a significant problem. Furthermore, composites require careful removal of machining debris to avoid contact with and inhaling of fibers.

8.5.4 Thermally assisted machining

Metals and alloys that are difficult to machine at room temperature can be machined more easily at elevated temperatures, thus lowering cutting forces and increasing tool life. In *thermally assisted machining* (*hot machining*), the source of heat is a torch, high-energy beam (such as laser or electron beam), or plasma arc, focused to an area just ahead of the cutting tool. Most applications in hot machining are in turning. Heating and maintaining a uniform temperature distribution within the workpiece may be difficult to control. Except in isolated cases, thermally assisted machining offers no significant advantage over machining at room temperature with the use of appropriate cutting tools and fluids.

8.5.5 Automachining

Automachining is a term that describes cutting of metals using single-point tools made of the workpiece material. The cutting tool is not subjected to any additional treatment or coatings. When used on a lathe, it produces continuous chips and good surface finish on materials such as 2011-T3, 2024-T3, and 301 stainless steel. The surface finish is fair in automachining of mild steel and poor in copper. The mechanism by which automachining is possible has been determined experimentally to be the formation of a BUE. The BUE tip acts as the cutting edge; the chip does not contact the original tool at its rake face during cutting. As the hardness of the BUE can be considerably higher than that of the workpiece material (see Fig. 8.7a), cutting is actually performed by the harder BUE. Although it is not likely to be of industrial significance, automachining is nevertheless an interesting method of metal cutting.

8.6 Cutting-Tool Materials

Cutting-tool materials and their proper selection are among the most important factors in machining operations, as are mold and die materials for forming and shaping processes. We have noted previously that the tool is subjected to high temperatures, contact stresses, and rubbing on the workpiece surface, as well as by the chip climbing up the rake face of the tool. Consequently, a cutting tool must have the following characteristics in order to produce good-quality and economical parts.

- *Hardness*, particularly at elevated temperatures (hot hardness), so that the hardness and strength of the tool are maintained at the temperatures encountered in cutting operations (Fig. 8.38).
- *Toughness*, so that impact forces on the tool in interrupted cutting operations, such as milling or turning a splined shaft, do not chip or fracture the tool.

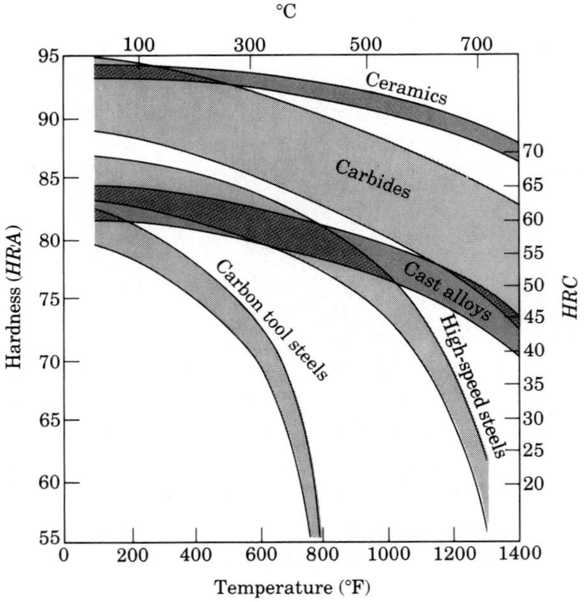

FIGURE 8.38 Hardness of various cutting-tool materials as a function of temperature (hot hardness). The wide range in each group of materials results from the variety of tool compositions and treatments available for that group.

- *Wear resistance*, so that an acceptable tool life is obtained before the tool is resharpened or replaced.
- *Chemical stability* or *inertness* with respect to the workpiece material, so that any adverse reactions contributing to tool wear are avoided.

Various cutting-tool materials having a wide range of properties are available (Tables 8.7 and 8.8). Tool materials are usually divided into the following general categories, which are listed in the approximate chronological order in which they were developed.

1. Carbon and medium-alloy steels
2. High-speed steels
3. Cast-cobalt alloys
4. Carbides
5. Coated tools
6. Ceramics
7. Cubic boron nitride
8. Silicon nitride
9. Diamond

Many of these materials are also used for dies and molds. In this section, we present the characteristics, applications, and limitations of these tool materials in machining operations. We discuss characteristics such as hot hardness, toughness, impact strength, wear resistance, thermal shock resistance, and costs, as well as the range of cutting speeds and depth of cut for optimum performance. We show that proper selection of tools is a critical factor in the quality of surfaces produced and the economics of machining.

8.6.1 Carbon and medium-alloy steels

Carbon steels are the oldest of tool materials and have been used widely for drills, taps, broaches, and reamers since the 1880s. Low-alloy and medium-alloy steels were developed later for similar applications but with longer tool life. Although inexpensive and easily shaped and sharpened, these steels do not have sufficient hot hardness and wear resistance for cutting at high speeds where, as you have seen, the temperature rises significantly. Note in Fig. 8.38, for example, how rapidly the hardness of carbon steels decreases as the temperature increases. Consequently, the use of these steels is limited to low-speed cutting operations.

8.6.2 High-speed steels

High-speed steel (HSS) tools are so named because they were developed to cut at high speeds. First produced in the early 1900s, *high-speed steels* are the most highly alloyed of the tool steels. They can be hardened to various depths, have good wear resistance, and are relatively inexpensive. Because of their high toughness and resistance to fracture, high-speed steels are especially suitable for high positive-rake-angle tools (small included angle) and for machine tools with low stiffness that are subject to vibration and chatter.

TABLE 8.7
TYPICAL PROPERTIES OF TOOL MATERIALS

PROPERTY	HIGH-SPEED STEELS	CAST ALLOYS	CARBIDES WC	CARBIDES TiC	CERAMICS	CUBIC BORON NITRIDE	DIAMOND*
Hardness	83–86 HRA	82–84 HRA 46–62 HRC	90–95 HRA 1800–2400 HK	91–93 HRA 1800–3200 HK	91–95 HRA 2000–3000 HK	4000–5000 HK	7000–8000 HK
Compressive strength							
MPa	4100–4500	1500–2300	4100–5850	3100–3850	2750–4500	6900	6900
psi × 10^3	600–650	220–335	600–850	450–560	400–650	1000	1000
Transverse rupture strength							
MPa	2400–4800	1380–2050	1050–2600	1380–1900	345–950	700	1350
psi × 10^3	350–700	200–300	150–375	200–275	50–135	105	200
Impact strength							
J	1.35–8	0.34–1.25	0.34–1.35	0.79–1.24	<0.1	<0.5	<0.2
in.-lb	12–70	3–11	3–12	7–11	<1	<5	<2
Modulus of elasticity							
GPa	200	—	520–690	310–450	310–410	850	820–1050
psi × 10^6	30	—	75–100	45–65	45–60	125	120–150
Density							
kg/m^3	8600	8000–8700	10,000–15,000	5500–5800	4000–4500	3500	3500
lb/in^3	0.31	0.29–0.31	0.36–0.54	0.2–0.22	0.14–0.16	0.13	0.13
Volume of hard phase, %	7–15	10–20	70–90	—	100	95	95
Melting or decomposition temperature							
°C	1300	—	1400	1400	2000	1300	700
°F	2370	—	2550	2550	3600	2400	1300
Thermal conductivity, W/m. K	30–50	—	42–125	17	29	13	500–2000
Coefficient of thermal expansion, × 10^{-6}/°C	12	—	4–6.5	7.5–9	6–8.5	4.8	1.5–4.8

* Single crystal. The values for polycrystalline diamond are generally lower, except impact strength, which is higher.

TABLE 8.8

GENERAL CHARACTERISTICS OF CUTTING-TOOL MATERIALS. THESE TOOL MATERIALS HAVE A WIDE RANGE OF COMPOSITIONS AND PROPERTIES. THUS OVERLAPPING CHARACTERISTICS EXIST IN MANY CATEGORIES OF TOOL MATERIALS.

	CARBON AND LOW- TO MEDIUM-ALLOY STEELS	HIGH-SPEED STEELS	CAST-COBALT ALLOYS	CEMENTED CARBIDES	COATED CARBIDES	CERAMICS	POLYCRYSTALLINE CUBIC BORON NITRIDE	DIAMOND
Hot hardness	——	——	Increasing	——	——	——	——	——→
Toughness	←——	——	Increasing	——	——	——	——	——
Impact strength	←——	——	Increasing	——	——	——	——	——
Wear resistance	——	——	Increasing	——	——	——	——	——→
Chipping resistance	←——	——	Increasing	——	——	——	——	——
Cutting speed	——	——	Increasing	——	——	——	——	——→
Depth of cut	Light to medium	Light to heavy	Light to heavy	Light to heavy	Light to heavy	Light to heavy	Light to heavy	Very light for single-crystal diamond
Finish obtainable	Rough	Rough	Rough	Good	Good	Very good	Very good	Excellent
Method of processing	Wrought	Wrought, cast, HIP* sintering	Cast and HIP sintering	Cold pressing and sintering	CVD†	Cold pressing and sintering or HIP sintering	High-pressure, high-temperature sintering	High-pressure, high-temperature sintering
Fabrication	Machining and grinding	Machining and grinding	Grinding	Grinding		Grinding	Grinding and polishing	Grinding and polishing
Thermal-shock resistance	←——	——	——	Increasing	——	——	——	——
Tool material cost	——	——	Increasing	——	——	——	——	——→

* Hot-isostatic pressing.

† Chemical-vapor deposition.

Source: Komanduri, R., *Kirk–Othmer Encyclopedia of Chemical Technology,* 3d ed. New York: Wiley, 1978.

There are two basic types of high-speed steels: *molybdenum* (M series) and *tungsten* (T series). The M series contains up to about 10 percent molybdenum, with chromium, vanadium, tungsten, and cobalt as alloying elements. The T series contains 12–18 percent tungsten, with chromium, vanadium, and cobalt as alloying elements. The M series generally has higher abrasion resistance than the T series, undergoes less distortion during heat treating, and is less expensive. Consequently, 95 percent of all high-speed steel tools produced in the United States are made of M-series steels.

High-speed steel tools are available in wrought, cast, and sintered (powder-metallurgy) forms. They can be coated for improved performance. High-speed steels account for the largest tonnage of tool materials used today, followed by various die steels and carbides. They are used in a wide variety of cutting operations requiring complex tool shapes such as drills, reamers, taps, and gear cutters.

8.6.3 Cast-cobalt alloys

Introduced in 1915, *cast-cobalt alloys* have the following ranges of composition: 38–53 percent cobalt, 30–33 percent chromium, and 10–20 percent tungsten. Because of their high hardness, typically 58–64 HRC, they have good wear resistance and maintain their hardness at elevated temperatures. Commonly known as *Stellite* tools, these alloys are cast and ground into relatively simple tool shapes. However, they are not as tough as high-speed steels and are sensitive to impact forces. Consequently, they are less suitable than high-speed steels for interrupted cutting operations. These tools are now used only for special applications that involve deep, continuous roughing operations at relatively high feeds and speeds—as much as twice the rates possible with high-speed steels.

8.6.4 Carbides

The three groups of tool materials we have just described (alloy steels, high-speed steels, and cast alloys) have the necessary toughness, impact strength, and thermal shock resistance but have important limitations, such as strength and hardness, particularly hot hardness. Consequently, they cannot be used as effectively where high cutting speeds, hence high temperatures, are involved and thus tool life can be short.

To meet the challenge of higher speeds (Fig. 8.39a) for higher production rates, *carbides* (also known as *cemented* or *sintered carbides*) were introduced in the 1930s. Because of their high hardness over a wide range of temperatures, high elastic modulus and thermal conductivity, and low thermal expansion, carbides are among the most important tool and die materials. The two basic groups of carbides used for machining operations are tungsten carbide and titanium carbide.

Tungsten carbide (WC) is generally used for cutting nonferrous abrasive materials and cast irons. It is a composite material, consisting of tungsten-carbide particles bonded together in a cobalt matrix, hence the term *cemented* carbides. These tools are manufactured by powder-metallurgy techniques (see Chapter 11), in which WC powders are crushed together with cobalt in a ball mill, with the cobalt

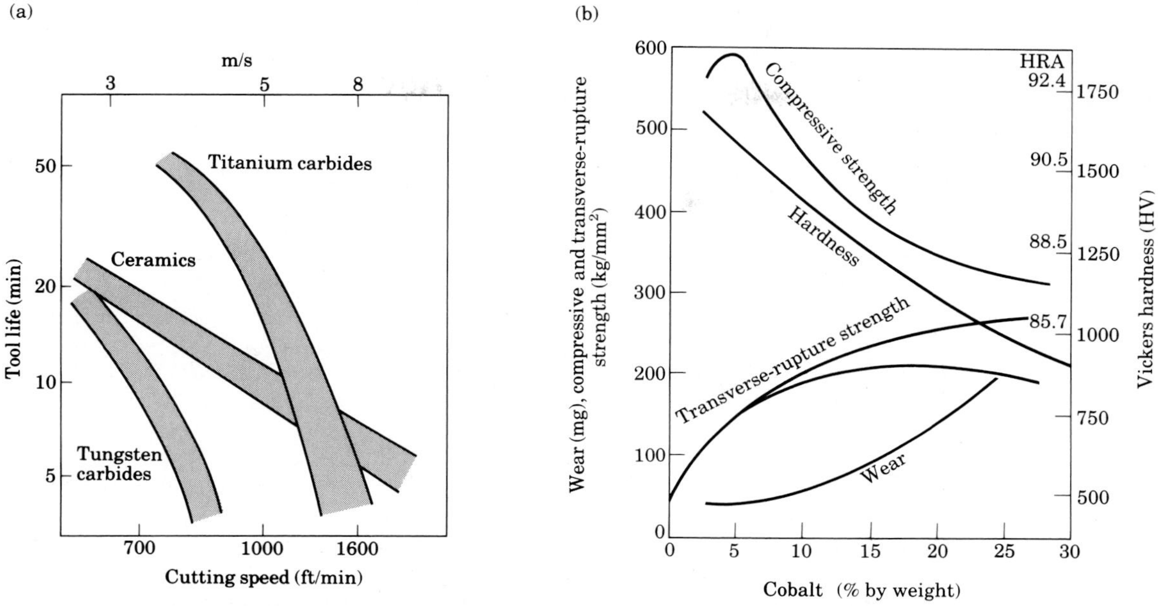

FIGURE 8.39
(a) Tool-life curves for various tool materials in medium and light turning operations as a function of cutting speed. Note how the curve for ceramics crosses over the curve for titanium carbides as speed, hence temperature, increases. (b) Effect of cobalt content in tungsten-carbide tools on mechanical properties. Note that hardness is directly related to compressive strength and hence, inversely, with wear.

coating the WC particles. These particles, which are 1–5 μm (40–200 μin.) in size, are then pressed and sintered into the desired insert shapes.

The amount of cobalt significantly affects the properties of carbide tools (Fig. 8.39b). As the cobalt content increases, the strength, hardness, and wear resistance of WC decrease, while its toughness increases because of the higher toughness of cobalt. To improve hot hardness and crater-wear resistance, WC may be compounded with carbides of titanium and tantalum. These carbides can then be used for machining steels. Table 8.9 gives typical applications for tungsten-carbide cutting tools according to the C-system used in the United States. The ISO (International Standards Organization) system of classification uses the symbols P, M, and K to identify various carbides (Table 8.10).

Titanium carbide (TiC) has higher wear resistance than tungsten carbide but is not as tough. With a nickel–molybdenum alloy as the matrix, TiC is suitable for machining hard materials, mainly steels and cast irons, and for cutting at speeds higher than those for tungsten carbide (Figs. 8.39a and 8.40).

Stiffness of the machine tool is of major importance in using carbide tools. Light feeds, low speeds, and chatter are detrimental because they tend to damage the tool's cutting edge. Cutting fluids are generally not needed, but if used to minimize

TABLE 8.9
CLASSIFICATION OF TUNGSTEN CARBIDES ACCORDING TO MACHINING APPLICATION

CLASSIFICATION NUMBER	MATERIALS TO BE MACHINED	MACHINING OPERATION	TYPE OF CARBIDE	CHARACTERISTICS OF		TYPICAL PROPERTIES	
				CUT	*CARBIDE*	*HARDNESS (HRA)*	*TRANSVERSE RUPTURE STRENGTH (MPa)*
C-1	Cast iron, nonferrous metals, and nonmetallic materials requiring abrasion resistance	Roughing cuts	Wear-resistant grades; generally straight WC–Co with varying grain sizes	Increasing cutting speed ↑	Increasing hardness and wear resistance ↑	89.0	2,400
C-2		General purpose				92.0	1,725
C-3		Finishing				92.5	1,400
C-4		Precision boring and fine finishing		↓ Increasing feed rate	↓ Increasing strength and binder content	93.5	1,200
C-5	Steels and steel alloys requiring crater and deformation resistance	Roughing cuts	Crater-resistant grades; various WC–Co compositions with TiC and/or TaC alloys	Increasing cutting speed ↑	Increasing hardness and wear resistance ↑	91.0	2,070
C-6		General purpose				92.0	1,725
C-7		Finishing				93.0	1,380
C-8		Precision boring and fine finishing		↓ Increasing feed rate	↓ Increasing strength and binder content	94.0	1,035

TABLE 8.10
ISO CLASSIFICATION OF CARBIDE CUTTING TOOLS ACCORDING TO USE

SYMBOL	WORKPIECE MATERIAL	COLOR	DESIGNATION IN INCREASING ORDER OF WEAR RESISTANCE AND DECREASING ORDER OF TOUGHNESS IN EACH CATEGORY
P	Ferrous metals with long chips	Blue	P01, P10, P20, P30, P40, P50
M	Ferrous metals with long or short chips; nonferrous metals	Yellow	M10, M20, M30, M40
K	Ferrous metals with short chips; nonferrous metals; nonmetallic materials	Red	K01, K10, K20, K30, K40

the heating and cooling of the tool in interrupted cutting operations, they should be applied continuously and in large quantities.

Inserts. The more traditional cutting tools are made of carbon steels and high-speed steels. These tools are formed in one piece and ground to various shapes. Other such tools include drills and milling cutters. After the cutting edge wears, the tool has to be removed from its holder and reground. Although a supply of sharp or resharpened tools is usually available from tool rooms, tool-changing operations are not efficient. The need for a more effective method has led to the development of *inserts*, which are individual cutting tools with a number of cutting points (Fig. 8.41).

Inserts are usually clamped on the tool *shank* with various locking mechanisms, or they may be *brazed* to the tool shank (Fig. 8.42). However, because of the

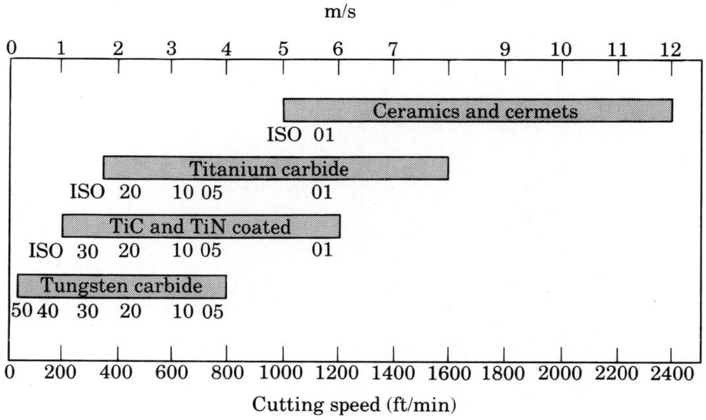

FIGURE 8.40
Approximate cutting-speed ranges for optimum use of various cutting-tool materials. Increasing the cutting speed has a significant economic impact in machining operations. The various grades of tools are indicated by the ISO numbers.

FIGURE 8.41
Typical carbide inserts with various shapes and chip-breaker features. The holes in the inserts are standardized for interchangeability. *Source:* Courtesy of GTE Valenite.

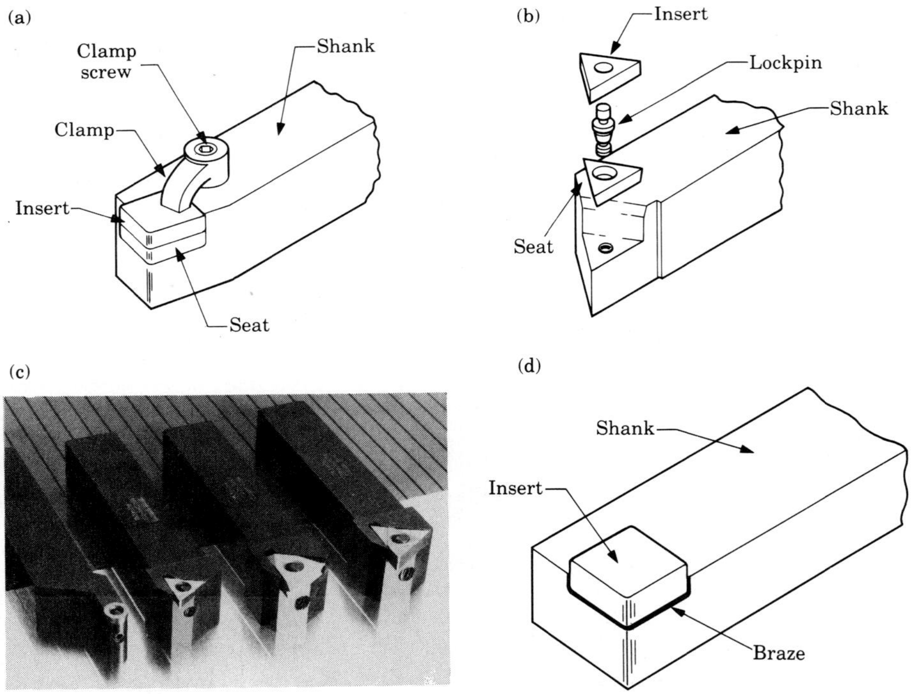

FIGURE 8.42
Methods of attaching inserts to tool shank: (a) clamping; and (b) wing lockpins. (c) Examples of inserts attached to tool shank with threadless lockpins, which are secured with side screws. *Source:* Courtesy of GTE Valenite. (d) Brazed insert on a tool shank.

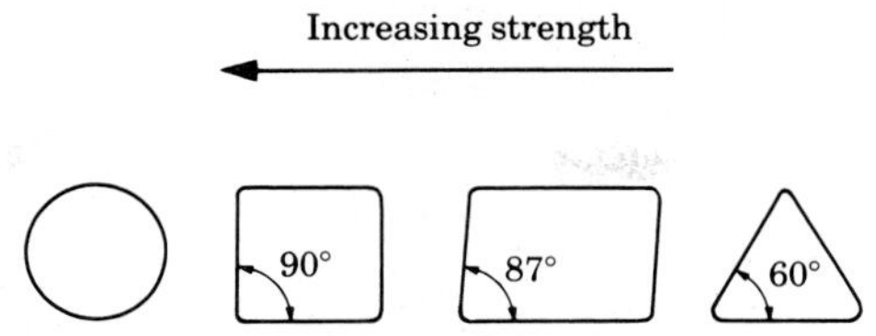

FIGURE 8.43
Relative edge strength of different shapes of inserts. Strength refers to the cutting edge shown by the angles.

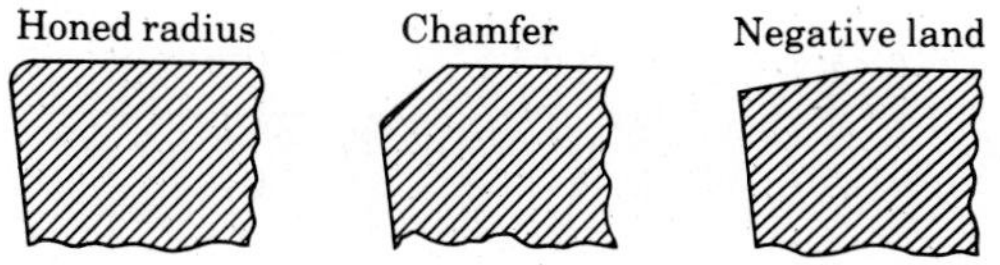

FIGURE 8.44
Edge preparation of inserts to improve edge strength and resistance to chipping and fracture.

difference in thermal expansion between the insert and the tool-shank materials, brazing must be done carefully to avoid cracking or warping. Clamping is the preferred method because each insert has a number of cutting edges, and after one edge is worn, it is *indexed* (rotated in its holder) for another cutting edge.

Carbide inserts are available in a variety of shapes, such as square, triangle, diamond, and round, with or without chip-breaker features for chip-flow control. The strength of the cutting edge of an insert depends on its shape. The smaller the angle (Fig. 8.43), the less strength the edge has. In order to further improve edge strength and prevent chipping, all insert edges are usually honed, chamfered, or produced with a negative land (Fig. 8.44). Most inserts are honed to a radius of about 0.025 mm (0.001 in.).

8.6.5 Coated tools

New alloys and engineered materials have been developed continuously, particularly since the 1960s. These materials have high strength but are generally abrasive and are highly reactive chemically with tool materials. The difficulty of machining these materials efficiently—and the need for improving the performance in machining the more common engineering materials—has led to important developments in *coated tools*. Because of their unique properties, coated tools can be used at high cutting speeds, thus reducing the time required for machining

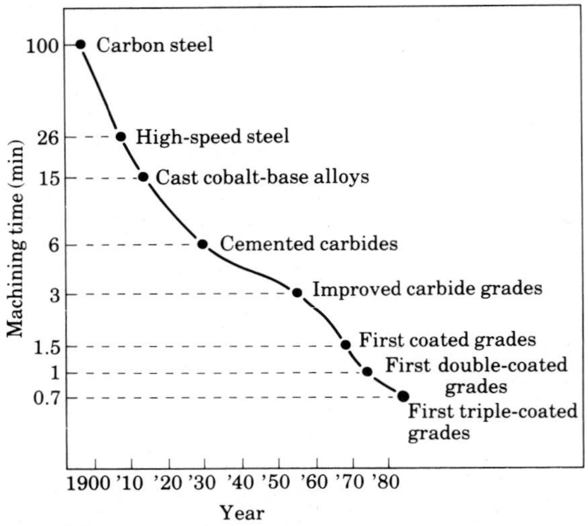

FIGURE 8.45

Relative time required to machine with various cutting-tool materials, indicating the year the tool materials were introduced. *Source:* Sandvik Coromant.

operations and costs. Figure 8.45 shows that cutting time has been reduced by more than 100 times since 1900. Since the late 1950s alone, coated tools have reduced cutting time some 4 times. Coated tools are now being used for cutting operations, with tool life as much as 10 times that of uncoated tools.

Coating materials. Coating materials commonly used are titanium nitride, titanium carbide, and ceramics. Other materials, such as hafnium nitride, are being investigated. Coatings on tools, generally 5–10 μm (200–400 μin.) in thickness, are applied by various techniques (see Section 4.5.1). Honing is an important procedure used to maintain the strength of the coating along the edges of the tool. Otherwise the coating may chip off at sharp edges. Coatings for cutting tools, as well as for dies, should have the following general characteristics:

a) High hardness at elevated temperatures.
b) Chemical stability and inertness to the workpiece material.
c) Low thermal conductivity.
d) Good bonding to the substrate to prevent flaking or spalling.
e) Little or no porosity.

The effectiveness of coatings, in turn, are enhanced by hardness, toughness, and high thermal conductivity of the substrate.

Titanium nitride. Titanium-nitride (TiN) coatings have low coefficient of friction, high hardness, resistance to high temperature, and good adhesion to the

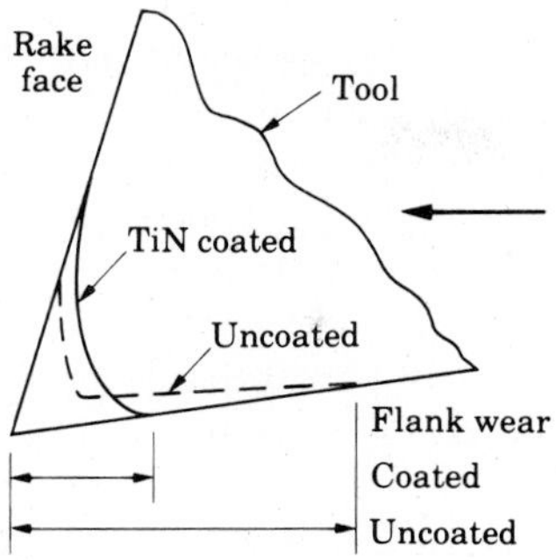

FIGURE 8.46
Wear patterns on high-speed-steel uncoated and titanium-nitride coated tools. Note that flank wear is lower for the coated tool.

substrate. Consequently, they greatly improve the life of high-speed-steel tools, as well as the lives of carbide tools, drills, and cutters. Titanium-nitride coated tools, which are gold in color, perform well at higher cutting speeds and feeds.

Flank wear is significantly lower than for uncoated tools (Fig. 8.46), and flank surfaces can be reground after use, since regrinding does not remove the coating on the rake face of the tool. However, coated tools do not perform as well at low cutting speeds because the coating can be worn off by chip adhesion. Hence the use of appropriate cutting fluids to discourage adhesion is important.

Titanium carbide. Titanium-carbide (TiC) coatings on tungsten-carbide inserts have high flank-wear resistance in machining abrasive materials.

Ceramics. Their resistance to high temperature, chemical inertness, low thermal conductivity, and resistance to flank and crater wear make ceramics suitable coatings for tools. The most commonly used ceramic coating is aluminum oxide (Al_2O_3). However, because they are very stable (not chemically reactive), oxide coatings generally bond weakly with the substrate.

Multiple coatings. The desirable properties of the coatings just described can be combined and thus optimized in multiple coatings. Tungsten-carbide tools are now available with two or three layers of such coatings and are particularly effective in machining cast irons and steels.

The first layer over the substrate is TiC, followed by Al_2O_3, and then TiN. The first layer should bond well with the substrate; the outer layer should resist wear and have low thermal conductivity; the intermediate layer should bond well and be compatible with both layers. Typical applications of multiple-coated tools are:

a) High-speed, continuous cutting: TiC/Al_2O_3.
b) Heavy-duty, continuous cutting: $TiC/Al_2O_3/TiN$.
c) Light, interrupted cutting: TiC/TiC + TiN/TiN.

8.6.6 Ceramics

These tool materials, introduced in the early 1950s, consist primarily of fine-grained, high-purity aluminum oxide. They are cold pressed under high pressure, sintered at high temperature, and called *white*, or *cold-pressed*, ceramics. Additions of titanium carbide and zirconium oxide help improve properties such as toughness and thermal-shock resistance.

Black, or *hot-pressed*, ceramics (carboxides) were introduced in the 1960s. They typically contain 70 percent aluminum oxide and 30 percent titanium carbide, and are also called *cermets* (from ceramic and metal). Other cermets contain molybdenum carbide, niobium carbide, and tantalum carbide.

Ceramic tools have very high abrasion resistance and hot hardness (Fig. 8.47). Chemically, they are more stable than high-speed steels and carbides. Thus they have less tendency to adhere to metals during cutting and hence less tendency to form a built-up edge. Consequently, good surface finish is obtained with ceramic tools in cutting cast irons and steels. However, ceramics lack toughness, resulting in premature tool failure by chipping or catastrophic failure.

Ceramic inserts are available in shapes similar to carbide inserts. They are effective in very high speed, uninterrupted cutting operations, such as finishing or semifinishing by turning. To reduce thermal shock, cutting should be performed either dry or with a copious amount of steady stream of cutting fluid. Improper or intermittent applications of fluid can cause thermal shock in the ceramic tool.

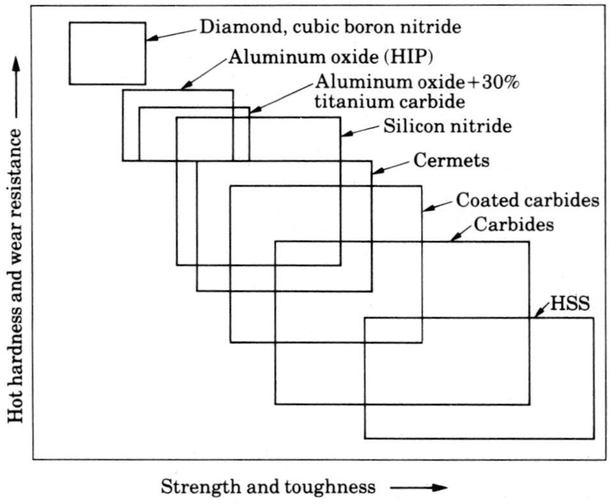

FIGURE 8.47
Ranges of properties for various groups of tool materials (see also various tables in this chapter).

Ceramic tool shape and setup are important. Negative rake angles, hence large included angles, are generally preferred in order to avoid chipping. Tool failure can be reduced by increasing the stiffness and damping capacity of machine tools and mountings, thus reducing vibration and chatter.

• Example 8.6: Tool materials and their *n* value in the Taylor tool-life equation.

In Table 8.5 we note that the n values for the cutting tools listed range from 0.08 to 0.7. To what factors can this difference be attributed?

SOLUTION. From the tool-life equation $T \propto 1/V^{1/n}$, we note that n is a measure of the sensitivity of the tool life to cutting speed. The smaller the value of n, the higher the sensitivity, and hence the shorter the tool life or the higher the wear rate. As we describe in Section 4.4, wear is related to hardness; the higher the hardness, the lower the wear. Thus, as a first approximation, tool life will increase with tool hardness. Furthermore, because temperature affects hardness, we note in Fig. 8.38 that, in decreasing order of hardness, we have ceramics, carbides, and high-speed steels.

We next consider the chemical stability or inertness of the tool materials. While ceramics have high and carbides have good chemical stability, high-speed steels can react with the workpiece material at elevated temperatures. Also, from Eq. (8.30) we note that temperature rise is higher with high-speed steels than with carbides, due partly to the higher friction and lower thermal conductivity of steels compared to carbides (see Table 8.7). Combining these observations, we can now present a general and relative rating of three tool materials in terms of their characteristics, as follows:

Material	Hot hardness	Inertness	Temperature rise
High-speed steels	Low	Low	High
Carbides	High	High	Low
Ceramics	Highest	Highest	Low

An inspection of this table indicates that the value of n for high-speed steels should be the lowest, and for ceramics the highest.

8.6.7 Cubic boron nitride

Next to diamond, *cubic boron nitride* (CBN) is the hardest material presently available. The CBN cutting tool was introduced in 1962. It is made by bonding a 0.5–1-mm (0.02–0.04-in.) layer of polycrystalline cubic boron nitride to a carbide

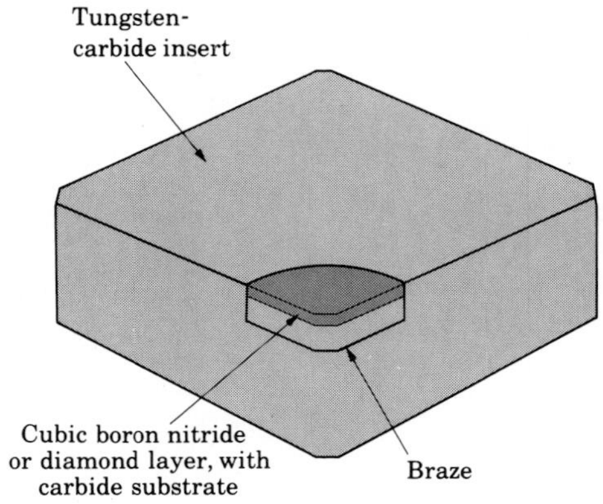

FIGURE 8.48
Construction of polycrystalline cubic boron nitride or diamond layer on a tungsten-carbide insert.

substrate by sintering under pressure (Fig. 8.48). While the carbide provides shock resistance, the CBN layer provides very high wear resistance and cutting-edge strength. Cubic boron nitride tools are also made in small sizes without a substrate. Because CBN tools are brittle, stiffness of the machine tool is important. At elevated temperatures, CBN is chemically inert to iron and nickel and its resistance to oxidation is high. It is therefore particularly suitable for cutting hardened ferrous and high-temperature alloys. Cubic boron nitride is also used as an abrasive.

8.6.8 Silicon-nitride base tools

Developed in the 1970s, silicon-nitride (SiN) base tool materials consist of silicon nitride with various additions of aluminum oxide, yttrium oxide, and titanium carbide. These tools have high toughness, hot hardness, and good thermal-shock resistance. An example of an SiN-base material is *sialon*, so called for the elements silicon, aluminum, oxygen, and nitrogen in its composition. It has higher thermal-shock resistance than silicon nitride and is recommended for machining cast irons and nickel-base superalloys at intermediate cutting speeds. Because of chemical affinity, SiN-base tools are not suitable for machining steels. New developments include composite ceramic tool materials. Consisting of silicon nitride reinforced with silicon-carbide whiskers, these materials have high fracture toughness and resistance to thermal shock.

8.6.9 Diamond

The hardest substance of all known materials is diamond. It has low friction, high wear resistance, and ability to maintain a sharp cutting edge. It is used when good surface finish and dimensional accuracy are required, particularly with soft nonferrous alloys and abrasive nonmetallic materials. *Single-crystal* diamonds of various carats are used for special applications, such as machining copper-front surface mirrors.

Because diamond is brittle, tool shape is important. Low rake angles (large included angles) are normally used to provide a strong cutting edge. Special attention should be given to proper mounting and crystal orientation in order to obtain optimum tool use. Diamond wear may occur by microchipping, caused by thermal stresses and oxidation, and transformation to carbon, caused by the heat generated during cutting.

Single-crystal diamond tools have been largely replaced by *polycrystalline-diamond tools* (*compacts*), which are also used as wire-drawing dies for fine wire. These materials consist of very small synthetic crystals, fused by a high-pressure, high-temperature process to a thickness of about 0.5–1 mm (0.02–0.04 in.) and bonded to a carbide substrate, similar to CBN tools (see Fig. 8.48). The random orientation of the diamond crystals prevents the propagation of cracks through the structure, thus improving its toughness.

Diamond tools can be used satisfactorily at almost any speed, but are suitable mostly for light, uninterrupted finishing cuts. In order to minimize tool fracture, the diamond must be resharpened as soon as it becomes dull. Because of its strong chemical affinity, diamond is not recommended for machining plain-carbon steels and titanium, nickel, and cobalt-base alloys. Diamond is also used as an abrasive in grinding and polishing operations.

8.6.10 Cutting-tool reconditioning

When tools, particularly high-speed steels, become worn, they are *reconditioned* (resharpened) for further use. They are usually ground on tool and cutter grinders in toolrooms having special fixtures. The reconditioning may be carried out either by hand, which requires considerable operator skill, or on computer-controlled tool and cutter grinders. Other methods may also be used to recondition tools and cutters. Reconditioning may also involve recoating used tools with titanium nitride.

Consistency and precision in reconditioning is important. Resharpened tools should be inspected for their shape and surface finish. As we described earlier, inserts are usually discarded after use. However, whether tools should be reconditioned or discarded depends on the relative costs involved. Skilled labor is costly, as are computer-controlled grinders. Thus an additional consideration is the possible recycling of tool materials, since many contain expensive materials of strategic importance, such as tungsten and cobalt.

8.7 Cutting Fluids

Also called lubricants and coolants, *cutting fluids* are used extensively in machining operations to:

- Reduce friction and wear, thus improving tool life and surface finish.
- Reduce forces and energy consumption.

- Cool the cutting zone, thus reducing workpiece temperature and distortion.
- Wash away the chips.
- Protect the newly machined surfaces from environmental corrosion.

A cutting fluid can interchangeably be a coolant and a lubricant. Its effectiveness in cutting operations depends on a number of factors, such as the method of application, temperature, cutting speed, and type of machining operation. As we have shown, temperature increases as cutting speed increases. Thus cooling of the cutting zone is of major importance at high cutting speeds. On the other hand, if the speed is low, such as in broaching or tapping, lubrication—not cooling—is the important factor. Lubrication reduces the tendency for built-up edge formation and thus improves surface finish.

The relative severity of various machining operations is shown qualitatively in Fig. 8.49, which also includes the relative cutting speeds employed. Severity is defined as the magnitude of temperatures and forces encountered, the tendency for built-up edge formation, and the ease with which chips are disposed of from the cutting zone. Note how important cutting-fluid effectiveness is as severity increases.

8.7.1 Cutting-fluid action

Although we describe the basic lubrication mechanisms in metalworking operations in greater detail in Section 4.4.3, we need to discuss briefly here the mechanisms by which cutting fluids influence machining operations. In view of the high contact pressures and relative sliding at the tool–chip interface, how does a cutting fluid penetrate this interface to influence the cutting process?

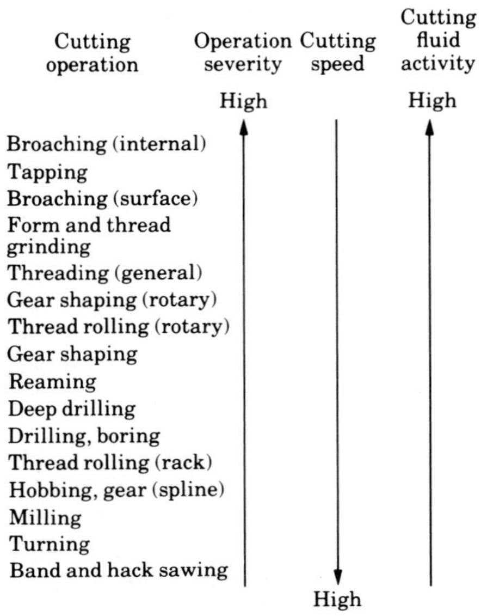

FIGURE 8.49 Relative severity of machining operations. The higher the severity, the greater is the need for an effective cutting fluid. *Source:* Mobil Oil Corporation.

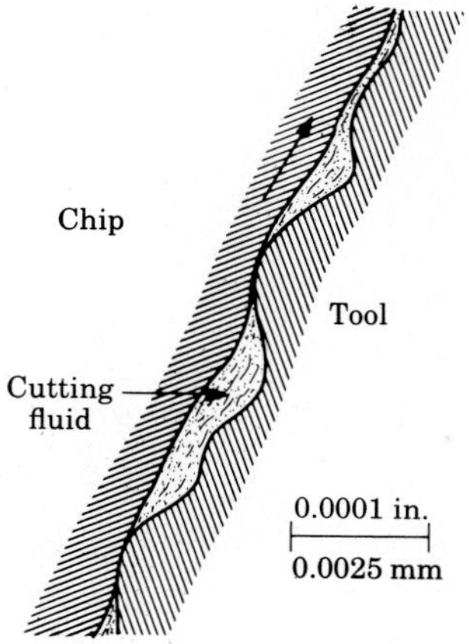

FIGURE 8.50
Schematic illustration of tool–chip interface, showing capillary passages allowing the cutting fluid to penetrate the interface and improve lubrication and cooling. *Source:* M. E. Merchant.

It appears that the fluid is drawn into the tool–chip interface by the *capillary action* of the interlocking network of surface asperities (Fig. 8.50). Studies have shown that the cutting fluid gains access to the interface by seeping from the *sides* of the chip. Because of the small size of this capillary network, the cutting fluid should have a small molecular size and proper wetting (surface tension) characteristics.

There are situations, however, in which the use of cutting fluids can be detrimental. In interrupted cutting operations, such as milling, the cooling action of the cutting fluid extends the range of temperature amplitude to which the cutter teeth are subjected. This condition can lead to thermal cracks because of thermal cycling of the tool. Cutting fluids may also cause the chip to become curlier, thus concentrating the stresses near the tool tip. These stresses, in turn, concentrate the heat closer to the tip and reduce tool life.

• Example 8.7: Effect of cutting fluids on machining.

A machining operation is being carried out with an effective cutting fluid. Explain the changes in the mechanics of the cutting operation and total energy consumption if the fluid is shut off.

SOLUTION. An effective cutting fluid is a good lubricant. Thus when the fluid is shut off, the friction at the tool–chip interface will increase. The following chain of events then takes place:

a) Fluid is shut off.
b) Friction at the tool–chip interface increases.
c) The shear angle decreases.
d) The shear strain increases.
e) The chip is thicker.
f) A built-up edge is likely to form.

As a consequence:

a) The shear energy in the primary zone increases.
b) The friction energy in the secondary zone increases.
c) Hence the total energy increases.
d) Surface finish is likely to deteriorate.
e) The temperature in the cutting zone increases; hence tool wear increases.
f) Tolerances may be difficult to maintain because of the increased temperature and expansion of the workpiece during machining.

8.7.2 Methods of application

The most common method of applying cutting fluid is *flood cooling*. Flow rates range from 10 L/min (3 gal/min) for single-point tools, to 225 L/min (60 gal/min) per cutter for multiple-tooth cutters, such as in milling. In operations such as gun drilling and end milling, fluid pressures of 700–14,000 kPa (100–2000 psi) are used to wash away the chips.

Mist cooling is another method of applying cutting fluids and is used particularly with water-base fluids. Although it requires venting (to prevent inhaling of fluid particles by the machine operator and others nearby) and has limited cooling capacity, mist cooling supplies fluid to otherwise inaccessible areas and provides better visibility of the workpiece being machined. It is particularly effective in grinding operations and at air pressures of 70–600 kPa (10–80 psi).

8.7.3 Effects of cutting fluids

The selection of a cutting fluid should also include other considerations. These are effects on workpiece material and machine tools and biological and environmental effects.

Effects on workpiece material. When selecting a cutting fluid, we should consider whether the machined component will be subjected to environmental attack and high service stresses, thus possibly leading to stress-corrosion cracking. This consideration is particularly important for cutting fluids having sulfur and chlorine additives. Additional considerations are staining of the workpiece by cutting fluids, especially on copper and aluminum. Machined parts should be cleaned and washed in order to remove any cutting-fluid residue.

Effects on machine tools. Just as a cutting fluid may adversely affect the workpiece material, it can similarly affect the machine tool and its various components, such as bearings and slideways. The choice of fluid must therefore include consideration of its compatibility with the machine-member materials.

Biological and environmental effects. Because the machine-tool operator is usually in close proximity to cutting fluids, the effects of operator contact with fluids should be of primary concern. Fumes, smoke, and odors from cutting fluids can cause severe skin reactions, as well as respiratory problems. Considerable progress has been made in ensuring the safe use of cutting fluids in manufacturing plants. Additionally, the effect on the external environment, particularly with regard to degradation of the fluid and its ultimate disposal, is important. Disposal practices must comply with federal, state, and local laws and regulations.

8.8 Cutting Processes and Machine Tools for Producing Round Shapes

In this section we describe processes that produce parts that are basically round in shape. Typical products made include parts as small as miniature screws for eyeglass-frame hinges and as large as shafts, pistons, cylinders, gun barrels, and turbines for hydroelectric power plants. These processes are usually performed by turning the workpiece on a lathe. *Turning* means that the part is rotating while it is being machined. The starting material is usually a workpiece that has been made by other processes, such as casting, shaping, forging, extrusion, and drawing. Turning processes are versatile and capable of producing a wide variety of shapes, as we outline in Fig. 8.51:

- *Turning* straight, conical, curved, or grooved workpieces, such as shafts, spindles, pins, handles, and various machine components.
- *Facing*, to produce a flat surface at the end of the part, such as parts that are attached to other components, or to produce grooves for O-ring seats.
- Producing various shapes by *form tools*, such as for functional purposes or for appearance.
- *Boring*, to enlarge a hole made by a previous process or in a tubular workpiece or to produce internal grooves.
- *Drilling*, to produce a hole, which may be followed by boring to improve its accuracy and surface finish.
- *Parting*, also called *cutting off*, to cut a piece from the end of a part, as in making slugs or blanks for additional processing into discrete products.
- *Threading*, to produce external and internal threads in workpieces.
- *Knurling*, to produce a regularly shaped roughness on cylindrical surfaces, as in making knobs.

These operations may be performed at various rotational speeds of the workpiece, depths of cut, d, and feed, f, depending on the workpiece and tool materials, the surface finish and dimensional accuracy required, and the capacity of the machine tool. Cutting speeds are usually in the range of 0.15–4 m/s

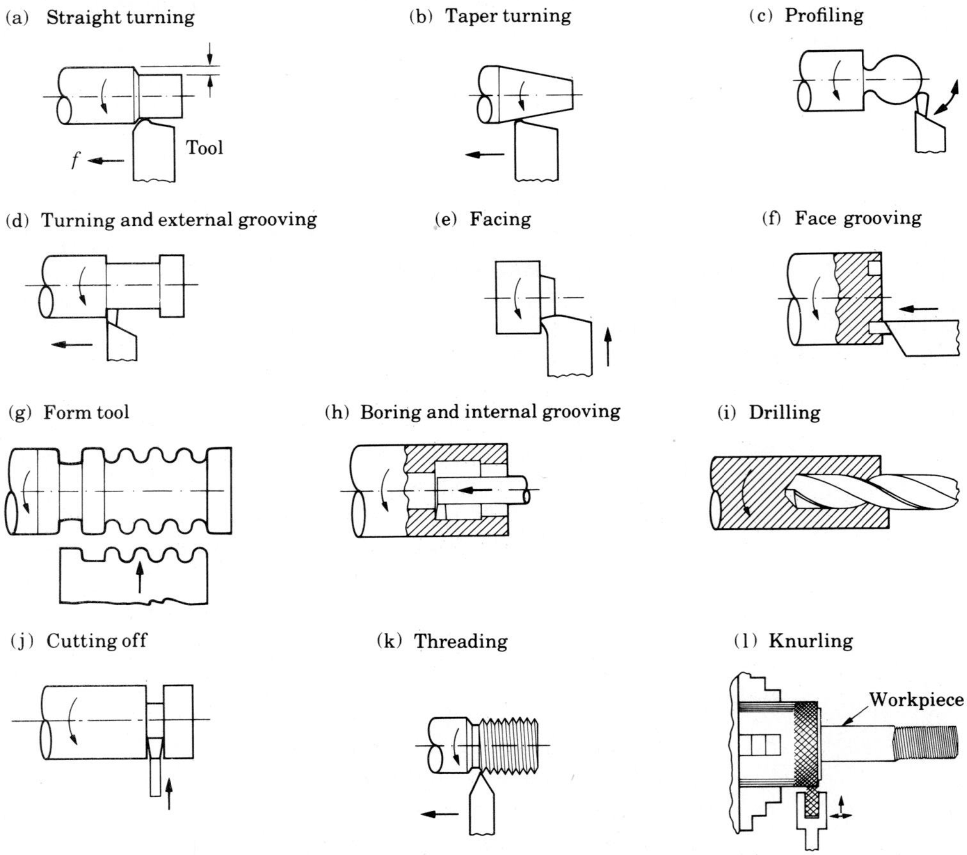

FIGURE 8.51
Various cutting operations that can be performed on a lathe.

(30–800 ft/min). *Roughing cuts*, which are performed for large-scale material removal, usually involve depths of cut greater than 0.5 mm (0.02 in.) and feeds on the order of 0.2–2 mm/rev (0.008–0.08 in./rev). *Finishing cuts* usually involve lower depths of cut and feed.

8.8.1 Turning and lathes

Lathes are generally considered to be the oldest machine tools. Although woodworking lathes were first developed during the period 1000–1 B.C., metalworking lathes with lead screws were not built until the late 1700s. The most common lathe, shown schematically in Fig. 8.52, was originally called an *engine lathe* because it was powered with overhead pulleys and belts from nearby engines. Although simple and versatile, an engine lathe requires a skilled machinist because all controls are manipulated by hand. Consequently, it is inefficient for repetitive operations and for

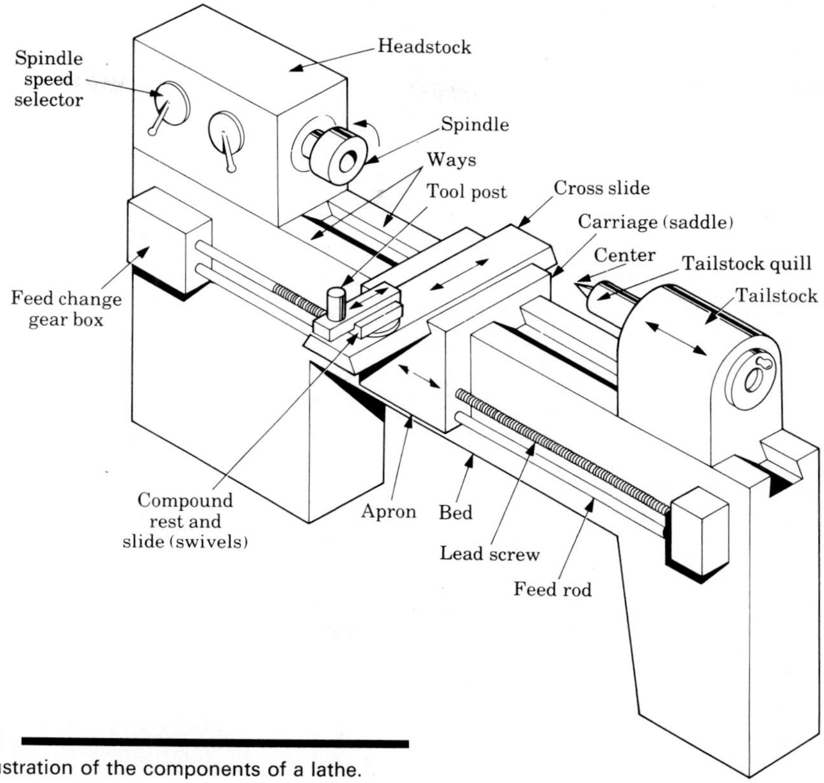

FIGURE 8.52
Schematic illustration of the components of a lathe.

large production runs. However, various types of automation can be added to improve efficiency, as we describe in the rest of this section.

Lathe components. Lathes are equipped with a variety of components and accessories. The *bed* supports all the other major components of the lathe. The *carriage* or *carriage assembly*, which slides along the ways, consists of an assembly of the cross-slide, tool post, and apron. The cutting tool is mounted on the *tool post*, usually with a *compound rest* that swivels for tool positioning and adjustments. The *headstock*, which is fixed to the bed, is equipped with motors, pulleys, and V-belts that supply power to the *spindle* at various rotational speeds. Headstocks have a hollow spindle to which workholding devices, such as *chucks* and *collets*, are attached. The *tailstock*, which can slide along the ways and be clamped at any position, supports the other end of the workpiece. The *feed rod*, which is powered by a set of gears from the headstock, rotates during operation of the lathe and provides movement to the carriage and the cross-slide. The *lead screw*, which is used for cutting threads accurately, is engaged with the carriage by closing a split nut around the lead screw.

A lathe is usually specified by its *swing*, that is, the maximum diameter of the workpiece that can be machined, by the maximum distance between the headstock and tailstock centers, and by the length of the bed. A variety of lathe designs are available, including *bench lathes*, *toolroom lathes*, *engine lathes*, *gap lathes*, and *special-purpose lathes*.

Tracer lathes are machine tools and attachments that are capable of turning parts with various contours. Also called *duplicating lathes* or *contouring lathes*, the cutting tool follows a path that duplicates the contour of a template through a hydraulic or electrical system. *Automatic lathes*, also called *chucking machines*, or *chuckers*, which are usually vertical and do not have tailstocks, are used for machining individual pieces of regular or irregular shapes. *Turret lathes* are capable of performing multiple cutting operations on the same workpiece, such as turning, boring, drilling, thread cutting, and facing. Several cutting tools are mounted on the hexagonal *main turret*. The lathe usually has a *square turret* on the cross-slide, with as many as four cutting tools mounted on it. Computer-controlled lathes (Fig. 8.53) are the most advanced, with their movement and control actuated by computer numerical controls (CNC).

Turning parameters and process capabilities. The majority of turning operations use simple single-point cutting tools. Figure 8.54 shows the geometry of a typical right-hand cutting tool for turning. Such tools are described by a standardized nomenclature.

Rake angles are important in controlling the direction of chip flow and to the strength of the tool tip. Positive angles improve the cutting operation by reducing forces and temperatures. However, positive angles produce a small included angle of the tool tip, which, depending on the toughness of the tool material, may cause

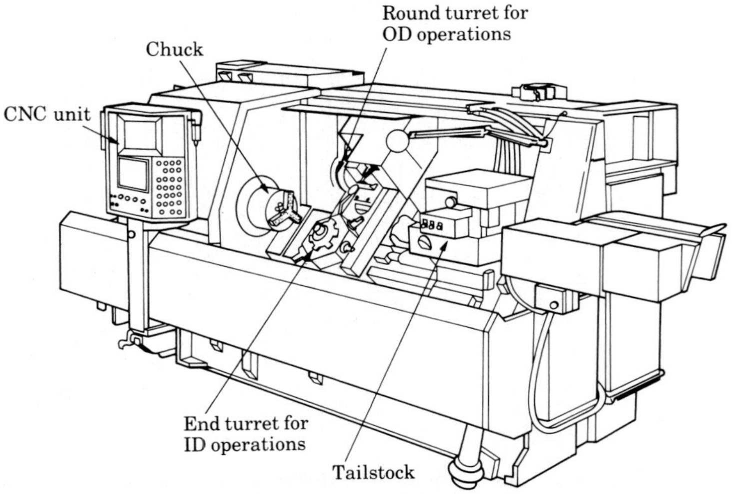

FIGURE 8.53
A computer numerical control lathe. Note the two turrets on this machine. *Source:* Jones & Lamson, Textron, Inc.

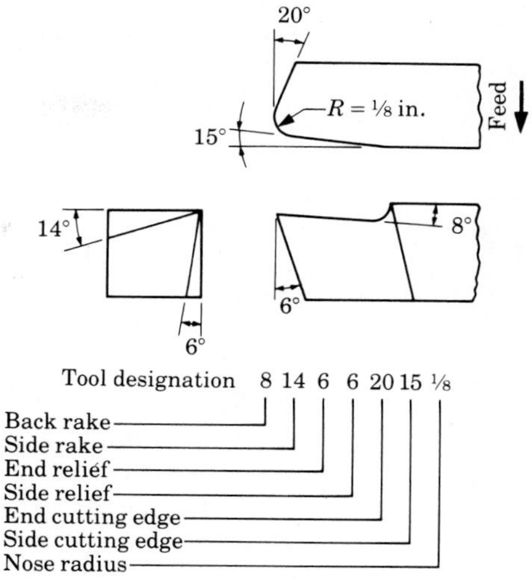

FIGURE 8.54 Designation and symbols for a right-hand cutting tool.

premature tool failure. Side rake angle is more important than back rake angle, although the latter usually controls the direction of chip flow.

Relief angles control the interference and rubbing at the tool–workpiece interface. If the relief angle is too large, the tool may chip off; if too small, flank wear may be excessive. *Cutting edge angles* affect chip formation, tool strength, and cutting forces to various degrees. *Nose radius* affects surface finish and strength of the tool tip. The sharper the radius, the rougher will be the surface finish of the workpiece and the lower will be the strength of the tool. However, large nose radii can lead to tool chatter (see Section 8.11).

The *material removal rate* (MRR) is the volume of material removed per unit time. In turning we can calculate it by referring to Fig. 8.55 and noting that $D_{avg} = (D_o + D_f)/2$. Thus the material removal rate is

$$\text{MRR} = (\pi)(D_{avg})(d)(f)(N), \tag{8.35}$$

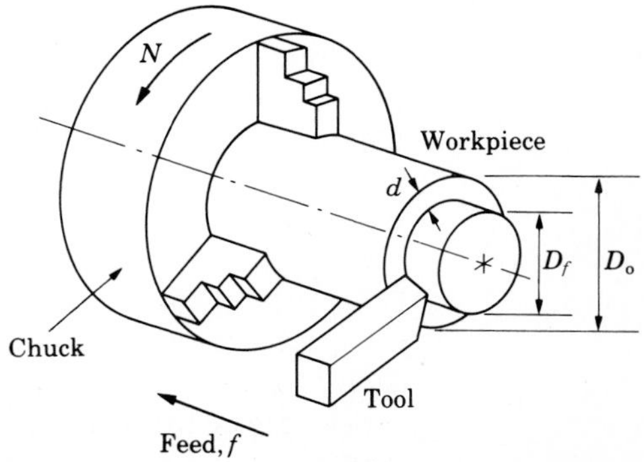

FIGURE 8.55 Schematic illustration of a turning operation showing depth of cut, *d*, and feed, *f*. Cutting speed is the surface speed of the workpiece.

where f is the feed in mm/rev (in./rev) and N is the workpiece rotational speed per unit time. For cases when $D_o \gg d$, we can replace D_{avg} by D_o.

Similarly the *cutting time*, t, for a workpiece of length L is

$$t = \frac{L}{fN}. \tag{8.36}$$

This cutting time does not include the time required for tool approach and retraction. Equations (8.35) and (8.36) are also applicable to boring operations, where D_o is the bore diameter.

Cutting screw threads. Threads are produced primarily by forming (*thread rolling*), which constitutes the largest quantity of threaded parts produced, or cutting. Casting threaded parts is also possible, although dimensional accuracy and production rate are not as high as those obtained in the other processes. Figure 8.51(k) shows that turning operations are capable of producing threads on round bar stock. When threads are produced externally or internally by cutting with a lathe-type tool, the process is called *thread cutting* or *threading*. When cut internally with a special threaded tool (*tap*), the process is called *tapping*. External threads may also be cut with a die or by milling. Although it adds considerably to cost, threads may be ground for improved accuracy and surface finish.

Automatic screw machines are designed for high-production–rate machining of screws and similar threaded parts. Because they are capable of producing other components, they are now called *automatic bar machines*. All operations on these machines are performed automatically, with tools attached to a special turret. The bar stock is fed forward automatically after each screw or part is machined to finished dimensions and cut off. These machines may be equipped with single or multiple spindles, and capacities range from 3- to 150-mm ($\frac{1}{8}$- to 6-in.) diameter bar stock.

8.8.2 Boring and boring machines

The basic boring operation, as carried out on a lathe, is shown in Fig. 8.51(h). *Boring* consists of producing circular internal profiles in hollow workpieces or on a hole made by drilling or another process and is carried out using cutting tools that are similar to those used in turning. However, because the boring bar has to reach the full length of the bore, tool deflection and hence maintaining dimensional accuracy can be a significant problem. The boring bar must be sufficiently stiff—that is, a material with high elastic modulus, such as carbides—in order to minimize deflection and avoid vibration and chatter. Boring bars have been designed with capabilities for damping vibrations.

Although boring operations on relatively small workpieces can be carried out on a lathe, *boring mills* are used for large workpieces. These machines are either vertical or horizontal, and are capable of performing operations such as turning, facing, grooving, and chamfering. A *vertical* boring machine (Fig. 8.56) is similar to a lathe but with a vertical axis of workpiece rotation. In *horizontal* boring machines

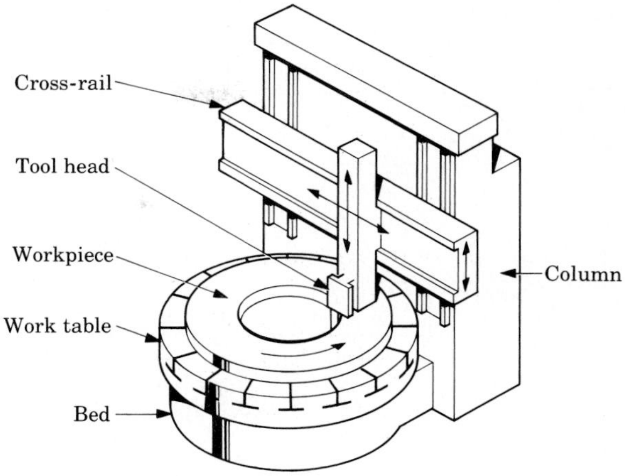

FIGURE 8.56 Schematic illustration of the components of a vertical boring mill.

the workpiece is mounted on a table that can move horizontally both axially and radially. The cutting tool is mounted on a spindle that rotates in the headstock, which is capable of both vertical and longitudinal motions. Drills, reamers, taps, and milling cutters can also be mounted on the spindle.

8.8.3 Drilling and drilling machines

One of the most common of all machining processes is drilling. *Drills* usually have a high length-to-diameter ratio (Fig. 8.57) and thus are capable of producing deep holes. However, they are somewhat flexible, depending on their diameter, and should be used with care in order to drill holes accurately and to prevent the drill from breaking. Moreover, the chips are produced within the workpiece, and the chips have to move in a direction opposite to the axial movement of the drill. Consequently, chip disposal and the effectiveness of cutting fluids can present significant problems.

The most common drill is the standard-point twist drill (Fig. 8.57). The main features of the drill point are a *point angle*, *lip-relief angle*, *chisel-edge angle*, and *helix angle*. The geometry of the drill tip is such that the normal rake angle and velocity of the cutting edge vary with the distance from the center of the drill. Other types of drills (Fig. 8.58) include the *step* drill, *core* drill, *counterboring* and *countersinking* drills, *center* drill, and *spade* drill. *Crankshaft* drills have good centering ability, and because chips tend to break up easily, these drills are suitable for drilling deep holes. *Gun drilling* requires a special drill and is used for drilling deep holes; hole depth-to-diameter ratios can be 300 or higher.

In the *trepanning* technique a cutting tool produces a hole by removing a disk-shaped piece (*core*), usually from flat plates. Thus a hole is produced without reducing all the material to chips. The process can be used to make disks up to 150 mm (6 in.) in diameter from flat sheet or plate.

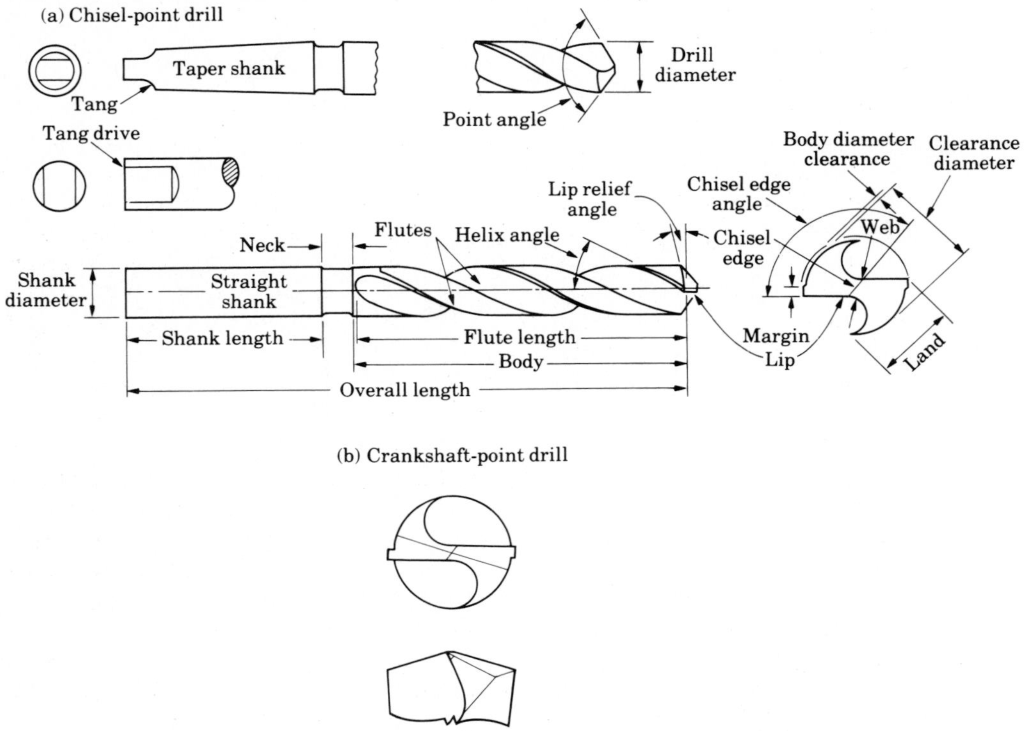

FIGURE 8.57
(a) Standard chisel-point drill indicating various features. (b) Crankshaft-point drill.

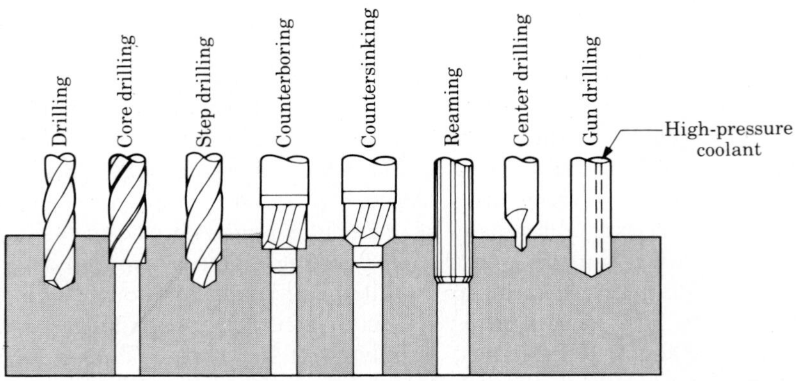

FIGURE 8.58
Various types of drills and drilling operations.

The *material removal rate* (MRR) in drilling is the ratio of volume removed to time. Thus

$$\text{MRR} = \frac{\pi D^2}{4} fN, \tag{8.37}$$

where D is the drill diameter, f is the feed, and N is the rpm of the drill.

The *thrust force* in drilling is the force that acts in the direction of the hole axis. If this force is excessive, it can cause the drill to break or bend. The thrust force depends on factors such as the strength of the workpiece material, feed, rotational speed, cutting fluids, drill diameter, and drill geometry. Although some attempts have been made, accurate calculation of the thrust force on the drill has proven to be difficult. Experimental data are available as an aid in the design and use of drills and drilling equipment. Thrust forces in drilling range from a few newtons for small drills to as high as 100 kN (22.5 klb) in drilling high-strength materials with large drills. Similarly, drill torque can range as high as 4000 N·m (3000 ft-lb).

The torque during drilling also is difficult to calculate. It can be obtained by using the data in Table 8.4 by noting that power dissipated during drilling is the product of torque and rotational speed. Thus by first calculating the MRR, we can calculate the torque on the drill. *Drill life*, as well as tap life, is usually measured by the number of holes drilled before the drill becomes dull and forces increase.

Drilling machines are used for drilling holes, tapping, reaming, and other general-purpose, small-diameter boring operations. Drilling machines are generally vertical, the most common type of which is a *drill press*. Its major components are shown in Fig. 8.59. The workpiece is placed on an adjustable table, either by

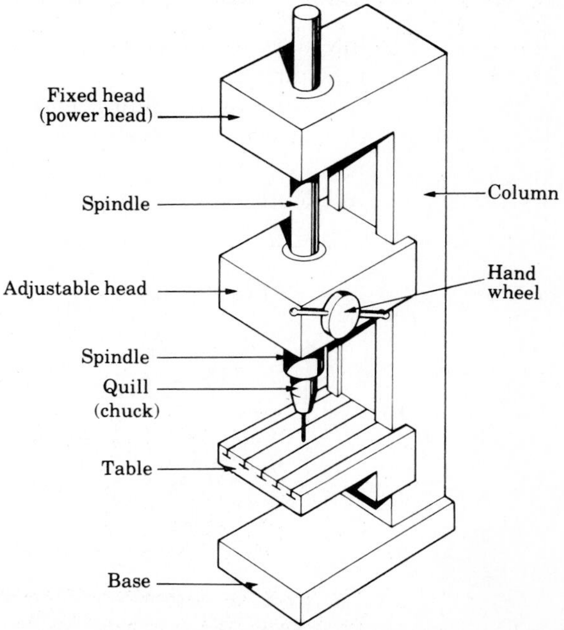

FIGURE 8.59 Schematic illustration of the components of a vertical drill press.

clamping it directly into the slots and holes on the table or by using a vise, which in turn can be clamped to the table. The workpiece should be properly clamped, for both safety and accuracy, because the drilling torque can be high enough to rotate the workpiece. The drill is lowered manually by hand wheel or by power feed at preset rates. Manual feeding requires some skill in judging the appropriate feed rate. In order to maintain proper cutting speeds at the cutting edges of drills, the spindle speed on drilling machines has to be adjustable to accommodate different sizes of drills. Adjustments are made by means of pulleys, gear boxes, or variable-speed motors. Drill presses are usually designated by the largest workpiece diameter that can be accommodated on the table. Sizes typically range from 150 mm (6 in.) to 1250 mm (50 in.).

8.8.4 Reaming and reamers

Reaming is an operation to make an existing hole dimensionally more accurate than can be obtained by drilling alone and to improve its surface finish. The most accurate holes are produced by the following sequence of operations: centering, drilling, boring, and reaming. For even better accuracy and surface finish, holes may be internally ground and honed.

A *reamer* (Fig. 8.60) is a multiple-cutting–edge tool with straight or helically fluted edges that removes very little material. The shanks may be straight or tapered, as in drills. The basic types of reamers are *hand* and *machine* (*chucking*) reamers. Other types are *rose* with cutting edges with wide margins and no relief, *fluted*, *shell*, *expansion*, and *adjustable* reamers.

8.8.5 Tapping and taps

Internal threads in workpieces can be produced by tapping. A *tap* is basically a threading tool with multiple cutting teeth (Fig. 8.61). Taps are generally available with three or four flutes. Three-fluted taps are stronger because of the larger amount of material available in the flute. *Tapered* taps are designed to reduce the torque required for tapping through holes, *bottoming* taps for tapping blind holes to their full depth, and *collapsible* taps for large diameter holes. After tapping is completed,

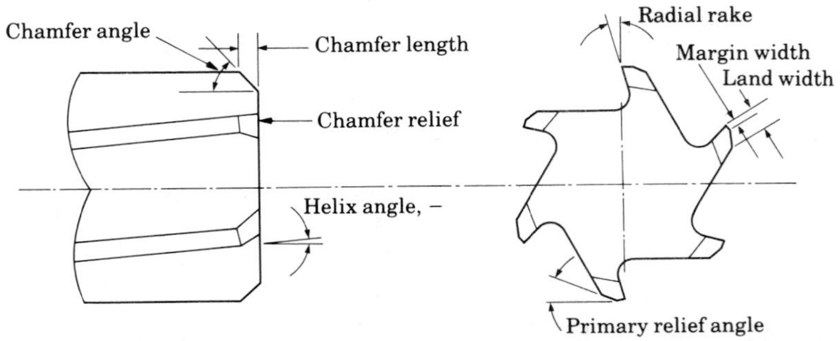

FIGURE 8.60
Terminology for a helical reamer.

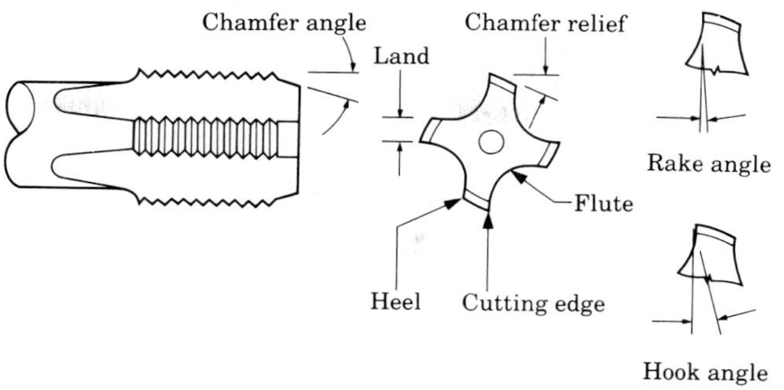

FIGURE 8.61
Terminology for a tap.

the tap is mechanically collapsed and removed from the hole, without having to rotate it. Tap sizes range up to 100 mm (4 in.).

8.9 Cutting Processes for Producing Various Shapes

Several cutting processes and machine tools (Fig. 8.62) are capable of producing complex shapes with the use of multitooth, as well as single-point, cutting tools. We

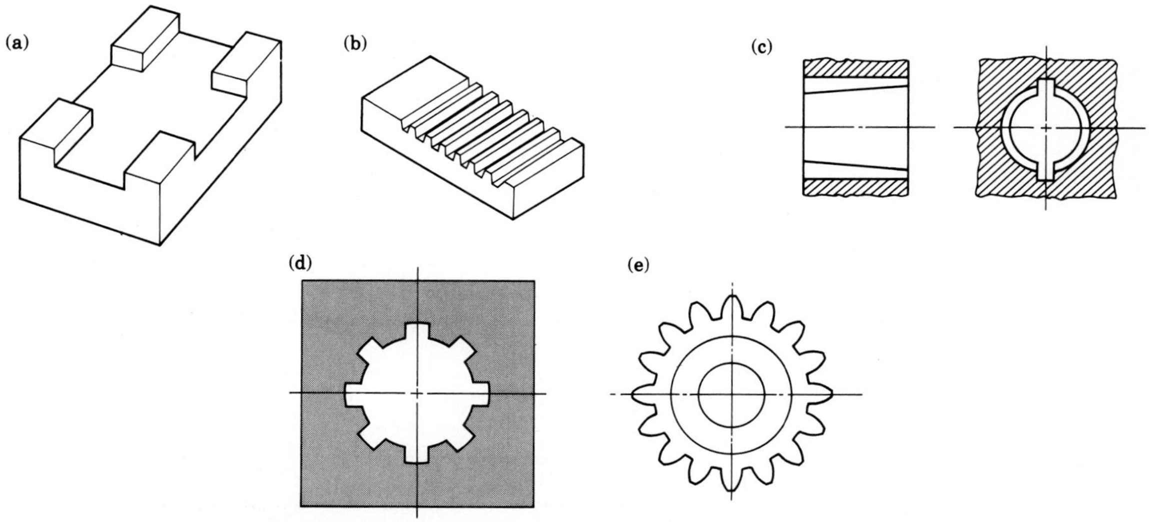

FIGURE 8.62
Typical parts and shapes produced by the cutting processes described in Section 8.9.

begin with one of the most versatile processes, called milling, in which a multitooth cutter rotates along various axes with respect to the workpiece. We then cover planing, shaping, and broaching, in which flat and shaped surfaces are produced, and the tool or workpiece travels along a straight path. Next, we describe sawing processes, which are generally used to prepare blanks, usually from rods, flat plate, and sheet, for further operations by forming, machining, and welding. We briefly cover filing, which is used to remove small amounts of material, usually from edges and corners.

8.9.1 Milling and milling machines

Milling includes a number of versatile machining operations, which are capable of producing a variety of configurations (Fig. 8.63). A *milling cutter* is a multitooth tool that produces a number of chips in one revolution.

In *slab* milling, also called *peripheral* milling, the axis of cutter rotation is parallel to the workpiece surface to be machined (Fig. 8.63a). The cutter has a number of teeth along its circumference, each tooth acting like a single-point cutting tool. The cutting speed is the peripheral speed of the cutter, or

$$V = \pi DN, \tag{8.38}$$

where D is the cutter diameter and N is the rotational speed of the cutter.

Note that the thickness of the chip in slab milling (Fig. 8.64a) varies along its length because of the relative longitudinal motion between cutter and workpiece. We can determine the approximate undeformed chip thickness t_c (*chip depth of cut*) from the equation

$$t_c = 2f\sqrt{\frac{d}{D}}, \tag{8.39}$$

where f is the feed per tooth of the cutter, measured along the workpiece surface, and d is the depth of cut. As the value of t_c becomes greater, the force on the cutter tooth increases.

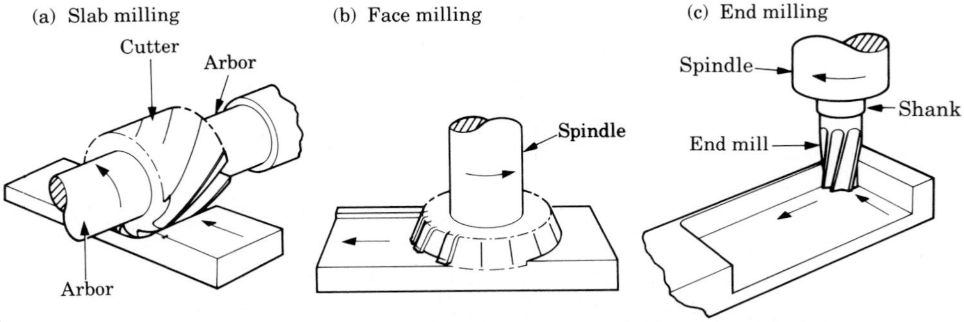

FIGURE 8.63
Basic types of milling cutters and operations.

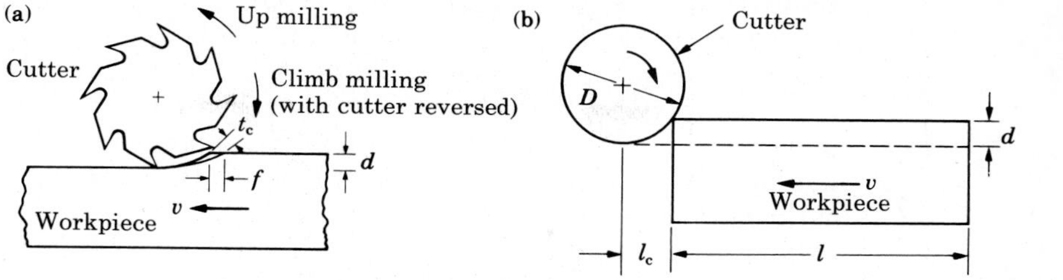

FIGURE 8.64
(a) Slab (peripheral) milling operation, showing depth of cut d, feed per tooth f, chip depth of cut t, and workpiece speed v. (b) Schematic illustration of cutter travel distance l_c to reach full depth of cut.

Cutters may have *straight* or *helical* teeth. The helical teeth on the cutter shown in Fig. 8.63(a) are preferred over straight teeth because the contact load is less, resulting in a smoother operation and reducing tool forces and chatter.

Feed per tooth is determined from the equation

$$f = v/Nn, \tag{8.40}$$

where v is the workpiece speed and n is the number of teeth on the cutter periphery. Depths of cut are usually 1–8 mm (0.04–0.3 in.). Feeds range from about 0.1 mm/tooth (0.004 in./tooth) to 0.5 mm/tooth (0.02 in./tooth). The higher the strength and the lower the machinability of the material, the lower these values become.

The cutting time, t, is given by the expression

$$t = (l + l_c)/v, \tag{8.41}$$

where l is the length of the workpiece (Fig. 8.64b) and l_c is the extent of the cutter's first contact with the workpiece. Based on the assumption that $l_c \ll l$, the material removal rate is

$$\text{MRR} = \frac{lwd}{t} = wdv, \tag{8.42}$$

where w is the width of the cut, which for a workpiece narrower than the length of the cutter, is the same as the width of the workpiece.

The direction of cutter rotation in slab milling, as well as in other milling operations, is important. In *climb* milling, also called *down* milling, cutting starts with the chip at its thickest location (Fig. 8.64a). The advantage is that the downward component of cutting forces holds the workpiece in place, particularly slender parts. Because of the resulting high impact forces, however, this operation must have a rigid setup, and backlash must be eliminated in the gear mechanisms that rotate the cutter. Climb milling is not suitable for machining workpieces having surface scale, such as hot-worked metals and castings. The scale is hard and abrasive and causes excessive wear and damage to the cutter teeth during processing. For these reasons, tool life can be short.

In *up* milling, also called *conventional* milling, the maximum thickness of the chip is at the end of the cut. Advantages are that tooth engagement is not a function of workpiece surface characteristics, and contamination or scale on the surface does not affect tool life. The cut is smoother, provided that the cutter teeth are sharp, although there is a greater tendency for tool chatter than in climb milling. In up milling the workpiece has a tendency to be pulled upward, so proper clamping is important.

In *face* milling the cutter is mounted on a spindle having an axis of rotation perpendicular to the workpiece surface (Fig. 8.63b). The cutter rotates at a speed N and the workpiece moves along a straight path at a speed v. When the cutter rotates as shown in Fig. 8.65(a), the operation is climb milling; when it rotates in the opposite direction (Fig. 8.65b), the operation is up milling. The terminology for face-milling cutters is shown in Fig. 8.66.

The cutter in *end* milling is shown in Fig. 8.63(c) and has either straight or tapered shanks for smaller and larger cutter sizes, respectively. The cutter usually rotates on an axis vertical to the workpiece, although it can be tilted to machine tapered surfaces. Flat surfaces as well as various profiles can be produced by end

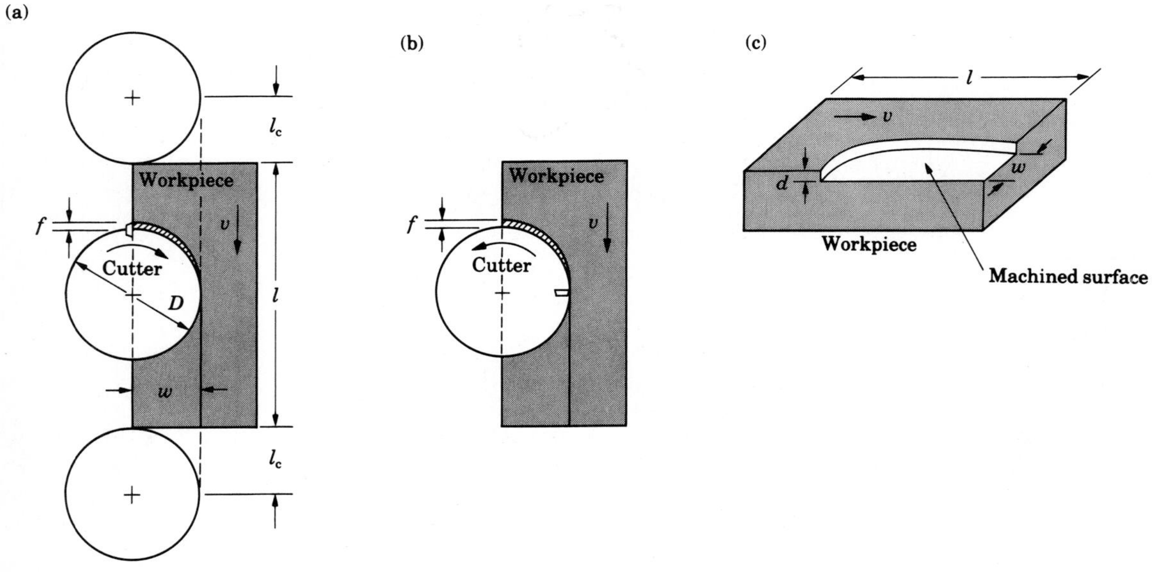

FIGURE 8.65
Face-milling operation showing (a) climb, or down, milling; (b) up, or conventional, milling; (c) dimensions in face milling. The width of cut *w* is not necessarily the same as the cutter radius. *Source:* Ingersoll Cutting Tool Company.

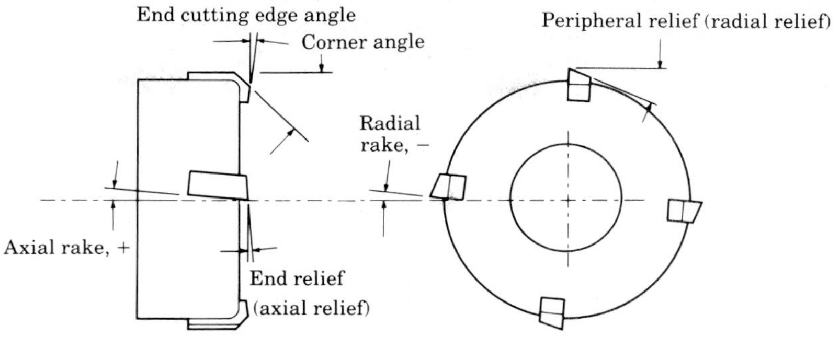

FIGURE 8.66
Terminology for a face-milling cutter.

milling. The end face of the cutter has cutting teeth, and thus it can be used as a drill to start a cavity. End mills are also available with hemispherical ends for producing curved surfaces, as in making dies. *Hollow end mills* have internal cutting teeth and are used for machining the cylindrical surface of solid round workpieces, as in preparing stock with accurate diameters for automatic screw machines.

Several other milling operations are shown in Fig. 8.67. These include straddle milling, form milling, slotting, and slitting, using various cutters.

Because of their capabilities to perform a variety of cutting operations, milling machines (first built in 1876) are among the most versatile and useful machine tools. A wide selection of milling machines with numerous features is available.

Used for general-purpose milling operations, *column-and-knee–type* machines are the most common milling machines. The spindle to which the milling cutter is

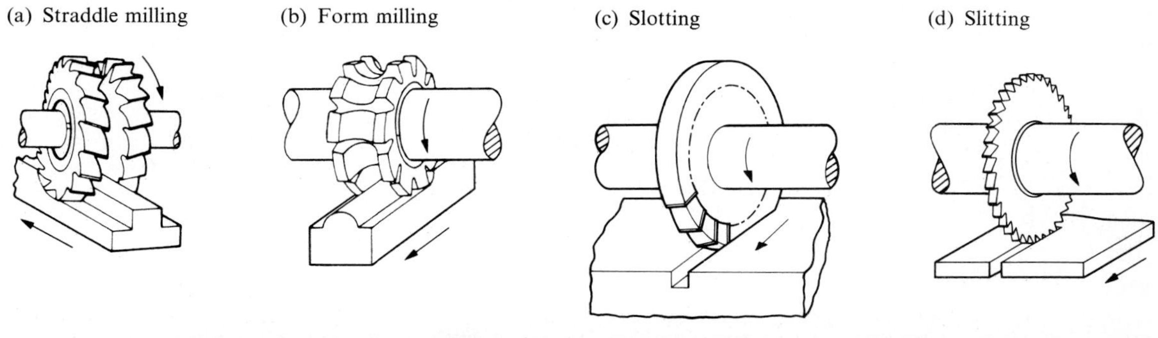

FIGURE 8.67
Cutters for (a) straddle milling, (b) form milling, (c) slotting, and (d) slitting with milling cutters.

attached may be horizontal (Fig. 8.68) for slab milling or vertical for face and end milling, boring, and drilling operations (Fig. 8.69). The various components are moved manually or by power. The basic components of these machines are:

a) *Work table*, on which the workpiece is clamped, using the T-slots. The table moves longitudinally with respect to the saddle.
b) *Saddle*, which supports the table and can move transversely.
c) *Knee*, which supports the saddle and gives the table vertical movement for adjusting the depth of cut.
d) *Overarm* in horizontal machines, which is adjustable to accommodate different arbor lengths.
e) *Head*, which contains the spindle and cutter holders. In vertical machines the head may be fixed or vertically adjustable and can be swiveled in a vertical plane on the column for milling tapered surfaces.

In *bed-type* machines, the work table is mounted directly on the bed, which replaces the knee, and can move only longitudinally (Fig. 8.70). These milling machines are not as versatile as others, but have high stiffness and are used for high-production-rate work. The spindles may be horizontal or vertical and of duplex or triplex types, that is, with two or three spindles for simultaneously machining two or three workpiece surfaces.

Several other types of milling machines are available. *Planer-type* machines, which are similar to bed-type machines, are equipped with several heads and cutters to mill various surfaces. They are used for heavy workpieces and are more efficient than planers (see Section 8.9.2) when used for similar purposes. *Rotary-table* machines are similar to vertical milling machines and are equipped with one or more heads for face-milling operations. Using tracer fingers, *duplicating* machines (copy milling machines) reproduce parts from a master model. They are used in the

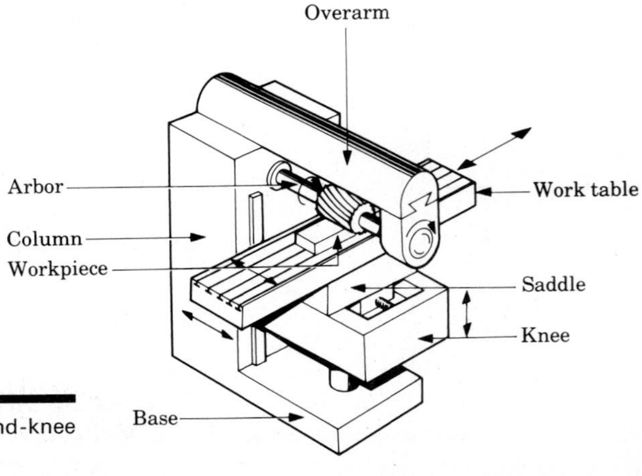

FIGURE 8.68
Schematic illustration of a horizontal-spindle column-and-knee type milling machine. *Source:* G. Boothroyd.

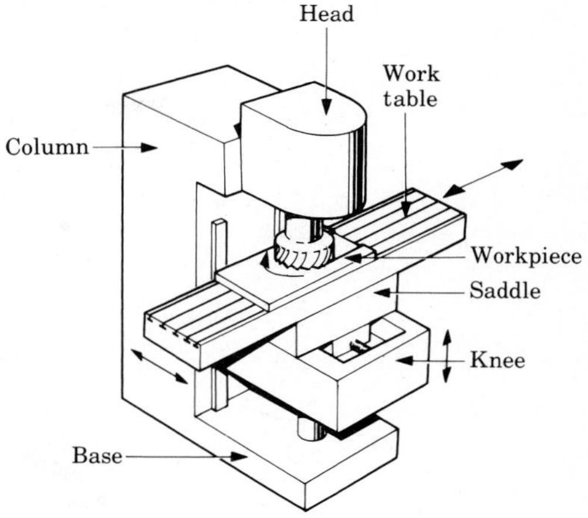

FIGURE 8.69
Schematic illustration of a vertical-spindle column-and-knee type milling machine. *Source:* G. Boothroyd.

automotive and aerospace industries for machining complex parts and dies (die sinking). Various milling-machine components are being replaced rapidly with *computer numerical control* (CNC) machines. These machine tools are versatile and capable of milling, drilling, boring, and tapping with repetitive accuracy. Other developments include *profile* milling machines, with five-axis movements.

8.9.2 Planing and planers

Planing is a relatively simple cutting process by which flat surfaces, as well as various cross-sections with grooves and notches, are produced along the length of the workpiece. Planing is usually done on large workpieces—as large as 25 m × 15 m (75 ft × 40 ft).

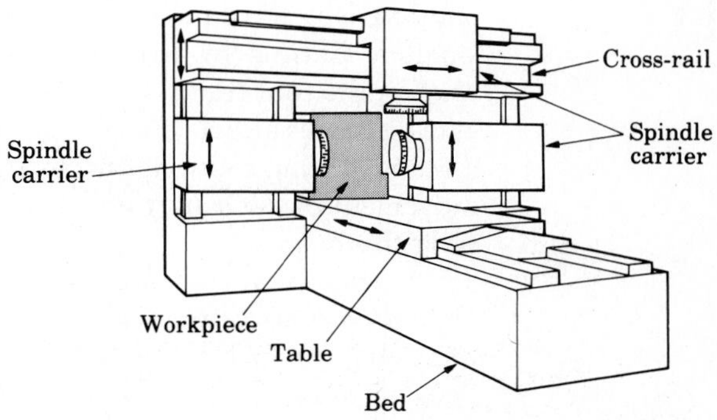

FIGURE 8.70
Schematic illustration of a triplex milling machine. Note the vertical-spindle cutter and two horizontal-spindle cutters. *Source:* ASM International.

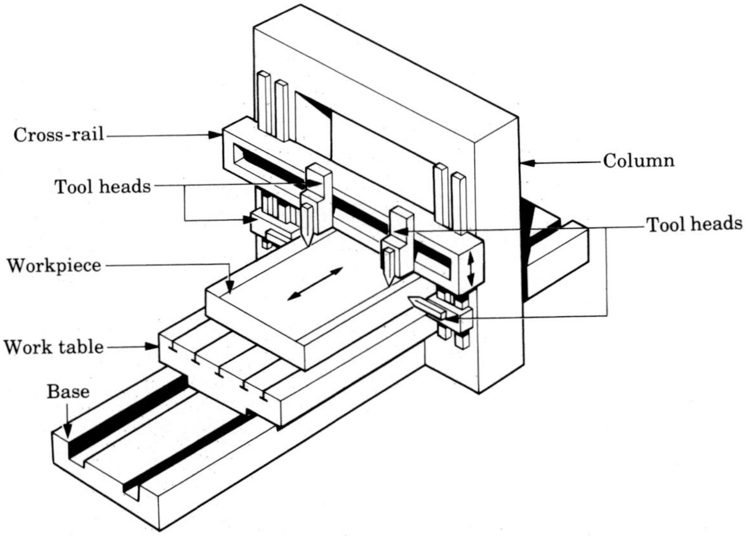

FIGURE 8.71
Schematic illustration of a planer. *Source:* G. Boothroyd.

In a *planer* (Fig. 8.71), the workpiece is mounted on a table that travels along a straight path. A horizontal cross-rail, which can be moved vertically along the ways in the column, is equipped with one or more tool heads. The cutting tools are attached to the heads, and machining is done along a straight path. Because of the reciprocating motion of the workpiece, elapsed noncutting time during the return stroke is significant both in planing and in shaping. Consequently, these operations are not efficient or economical, except for low-quantity production. Efficiency of the operation can be improved by equipping planers with tool holders and tools that cut in both directions of table travel.

8.9.3 Shaping and shapers

Shaping is used to machine parts much like planing does, except that the parts are smaller. Cutting by shaping is basically the same as in planing. In a *horizontal shaper*, the tool travels along a straight path, and the workpiece is stationary. The cutting tool is attached to the tool head, which is mounted on the ram. The ram has a reciprocating motion, and in most machines cutting is done during the forward movement of the ram (*push cut*); in others it is done during the return stroke of the ram (*draw cut*).

Vertical shapers (*slotters*) are used for machining notches, keyways, and dies. Shapers are also capable of producing complex shapes, such as cutting a helical impeller, in which the workpiece is rotated during the cut through a master cam. Because of low production rate, shapers are generally used in toolrooms, job shops, and repair work.

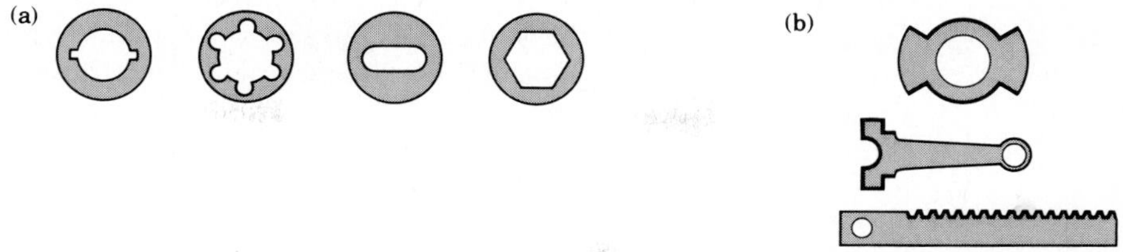

FIGURE 8.72
(a) Typical parts made by internal broaching. (b) Parts made by surface broaching. Heavy lines indicate broached surfaces. *Source:* General Broach and Engineering Company.

8.9.4 Broaching and broaching machines

The *broaching* operation is similar to shaping with multiple teeth and is used to machine internal and external surfaces, such as holes of circular, square, or irregular section, keyways, teeth of internal gears, multiple spline holes, and flat surfaces (Fig. 8.72). A *broach* is, in effect, a long multitooth cutting tool (Fig. 8.73) with successively deeper cuts. Thus the total depth of material removed in one stroke is the sum of the depths of cut of each tooth. A broach can remove material as deep as 6 mm (0.25 in.) in one stroke. Broaching can produce parts with good surface finish and dimensional accuracy; hence it competes favorably with other processes, such as boring, milling, shaping, and reaming, to produce similar shapes. Although broaches can be expensive, the cost is justified because of their use for high-quantity production runs.

The terminology for a broach is given in Fig. 8.73(b). The rake (hook) angle depends on the material cut, as in turning and other cutting operations, and usually ranges between 0° and 20°. The clearance angle is usually 1°–4°; finishing teeth have smaller angles. Too small a clearance angle causes rubbing of the teeth against the broached surface. The pitch of the teeth depends on factors such as length of the

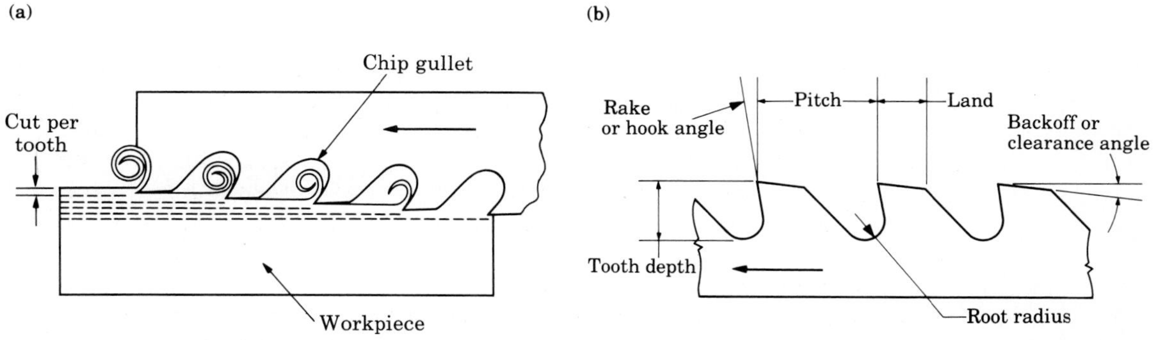

FIGURE 8.73
(a) Cutting action of a broach, showing various features. (b) Terminology for a broach.

workpiece (length of cut), tooth strength, and size and shape of chips. The tooth depth and pitch must be sufficiently large to accommodate the chips produced during broaching, particularly for long workpieces, but at least two teeth should be in contact with the workpiece at all times. The following formula may be used to obtain the pitch for a broach to cut a surface of length l:

$$\text{Pitch} = k\sqrt{l}, \tag{8.43}$$

where $k = 1.76$ for l in mm, and $k = 0.35$ for l in inches. An average pitch for small broaches is in the range of 3.2–6.4 mm (0.125–0.25 in.) and for large ones in the range of 12.7–25 mm (0.5–1 in.). The cut per tooth depends on the workpiece material and the surface finish desired. It is usually in the range of 0.025–0.075 mm (0.001–0.003 in.) for medium-size broaches and can be larger than 0.25 mm (0.01 in.) for larger broaches.

Broaches are available with various tooth profiles, including some that have chip breakers. Broaches are also made round with circular cutting teeth and are used to enlarge holes (Fig. 8.74). Note that the cutting teeth on the broach have three regions: roughing, semifinishing, and finishing. A variety of broaches are made for producing various internal and external shapes. Irregular internal shapes are usually broached by starting with a round hole in the workpiece, produced by drilling or boring.

A recent advance in broaching technology is *turn broaching* of crankshafts. The crankshaft rotates between centers and the broach, which is equipped with multiple inserts and passes tangentially across the part. The process is thus a combination of broaching and skiving. Straight as well as circular broaches have been used successfully for such applications. Machines that broach a number of crankshafts simultaneously have been built. Main bearings for engines are also broached in this way.

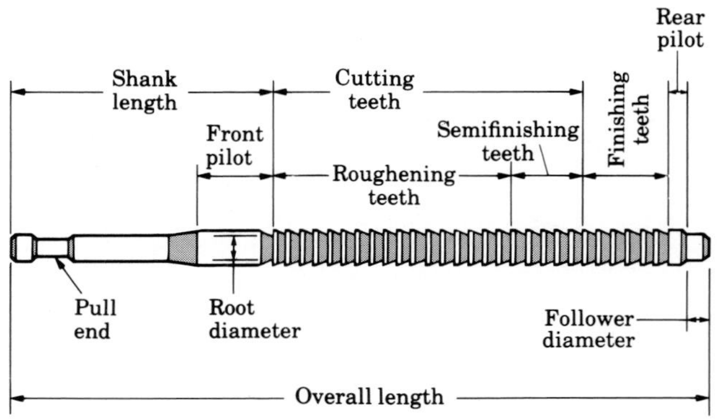

FIGURE 8.74
Terminology for a pull-type internal broach used for enlarging long holes.

Broaching machines either pull or push the broaches and are made horizontal or vertical. *Push broaches* are usually shorter, generally in the range of 150–350 mm (6–14 in.). *Pull broaches* tend to straighten the hole, whereas pushing permits the broach to follow any irregularity of the leader hole. Horizontal machines are capable of longer strokes. Broaching machines are relatively simple in construction, have only linear motions, and are usually actuated hydraulically, although some are moved by crank, screw, or rack. Many styles of broaching machines are manufactured, some with multiple heads, allowing a variety of shapes and parts to be produced, including helical splines and rifled gun barrels. Sizes range from machines for making needlelike parts to those used for broaching gun barrels. The force required to pull or push the broach depends on the strength of the workpiece material, total depth of cut, and width of cut. Tooth profile and cutting fluids also affect this force. Capacities of broaching machines are as high as 0.9 MN (100 tons) of pulling force.

8.9.5 Sawing and saws

Sawing is a cutting operation in which the cutting tool is a blade (*saw*) having a series of small teeth, with each tooth removing a small amount of material. This process is used for all metallic and nonmetallic materials that are machinable by other cutting processes and is capable of producing various shapes. The width of cut (*kerf*) in sawing is usually narrow, and thus sawing wastes little material.

Typical saw-tooth and saw-blade configurations are shown in Fig. 8.75. Tooth spacing is usually in the range of 0.08–1.25 teeth per mm (2–32 per in.). A wide variety of tooth forms and spacing and blade thicknesses, widths, and sizes are

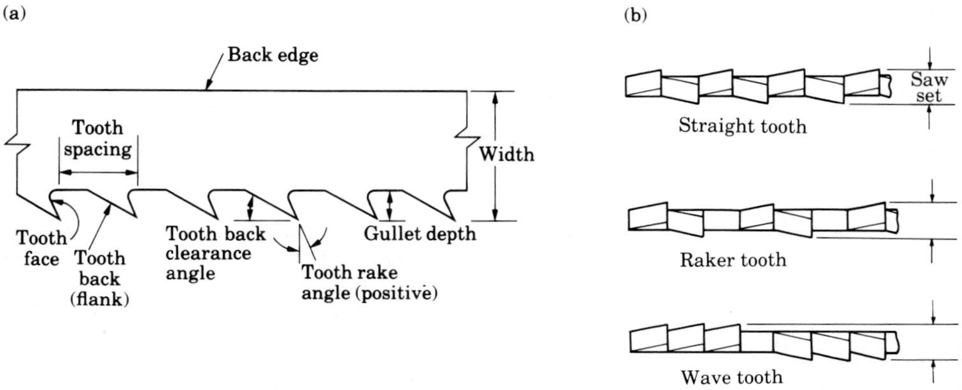

FIGURE 8.75
(a) Terminology for saw teeth. (b) Types of saw teeth, staggered to provide clearance for the saw blade to prevent binding during sawing.

available. Blades are made from carbon and high-speed steels. Carbide-steel or high-speed-steel tipped steel blades are used for sawing harder materials (Fig. 8.76).

In order to prevent the saw from binding and rubbing during cutting, the teeth are alternately set in opposite directions so that the kerf is wider than the blade (Fig. 8.75b). At least two or three teeth should always be engaged with the workpiece in order to prevent snagging (catching of the saw tooth on the workpiece). This is why sawing thin materials satisfactorily is difficult. The thinner the stock, the finer the saw teeth should be—and the greater the number of teeth per unit length. Cutting speed in sawing usually ranges up to 1.5 m/s (300 ft/min), with lower speeds for high-strength metals. Cutting fluids are generally used to improve the quality of cut and the life of the saw.

Hacksaws have straight blades and reciprocating motions; they may be manually or power operated. *Circular* saws (*cold* saws) are generally used for high-production–rate sawing of large cross-sections. *Band* saws have long, flexible, continuous blades and have continuous cutting action. Blades and high-strength wire can be coated with diamond powder (*diamond-edged* blades and *diamond* saws. They are suitable for sawing hard metallic, nonmetallic, and composite materials.

Friction sawing is a process in which a mild-steel blade or disk rubs against the workpiece at speeds of up to 125 m/s (25,000 ft/min). The frictional energy is converted into heat, which rapidly softens a narrow zone in the workpiece. The action of the blade or disk, which is sometimes provided with teeth or notches, pulls and ejects the softened metal from the cutting zone. The heat generated in the workpiece produces a heat-affected zone on the cut surfaces; thus properties can be adversely affected by the sawing process. Because only a small portion of the blade is engaged with the workpiece at any time, the blade cools rapidly as it passes through the air.

The friction-sawing process is suitable for hard ferrous metals and reinforced plastics but not for nonferrous metals because they have a tendency to stick to the blade. Disks for friction sawing as large as 1.8 m (6 ft) in diameter are used to cut off large steel sections in rolling mills. Friction sawing can also be used to remove flash from castings.

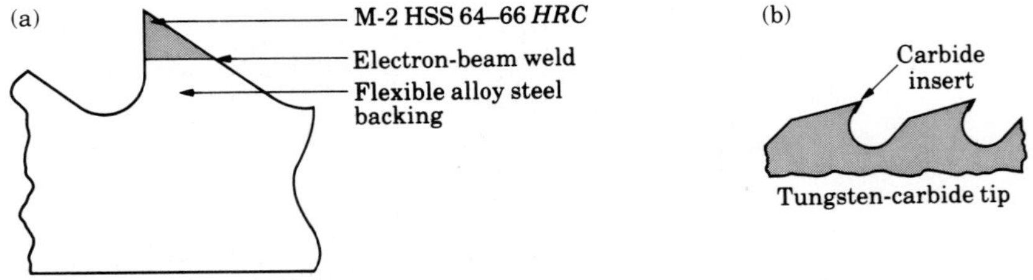

FIGURE 8.76
(a) High-speed-steel teeth welded on steel blade. (b) Carbide inserts brazed to blade teeth.

8.9.6 Filing

Filing is small-scale removal of material from a surface, corner, or hole. First developed in about 1000 B.C., files are usually made of hardened steels and are available in a variety of cross-sections, including flat, round, half round, square, and triangular. Files have many tooth forms and grades of coarseness, such as smooth-cut, second-cut, and bastard-cut. Although filing is usually done by hand, various machines with automatic features are available for high production rates, with files reciprocating at up to 500 strokes/min. *Band files* consist of file segments, each about 75 mm (3 in.) long, that are riveted to flexible steel bands and used in a manner similar to band saws. Disk type files are also available.

Rotary files and *burs* are available and are used for special applications. These cutters are usually conical, cylindrical, or spherical in shape and have various tooth profiles. Similar to that of reamers, their cutting action removes small amounts of material. The rotational speeds range from 1500 rpm for cutting steel with large burs to as high as 45,000 rpm for magnesium with small burs.

8.9.7 Gear manufacturing by cutting

Because of their capability for transmitting motion and power, gears are among the most important of all mechanical components. Gears can be manufactured by casting, forging, extrusion, drawing, thread rolling, powder metallurgy, and blanking (for making thin gears such as those used in watches and small clocks). Nonmetallic gears can be made by injection molding and casting.

In *form cutting* the cutting tool is similar to a form-milling cutter in the shape of the space between the gear teeth (Fig. 8.77). The gear-tooth shape is reproduced by cutting the gear blank around its periphery. The cutter travels axially along the length of the gear tooth at the appropriate depth to produce the gear tooth. After each tooth is cut, the cutter is withdrawn, the gear blank is rotated (indexed), and the cutter proceeds to cut another tooth. The process continues until all teeth are cut. Each cutter is designed to cut a range of tooth numbers. The precision of the form-cut tooth profile depends on the accuracy of the cutter and the machine and its stiffness.

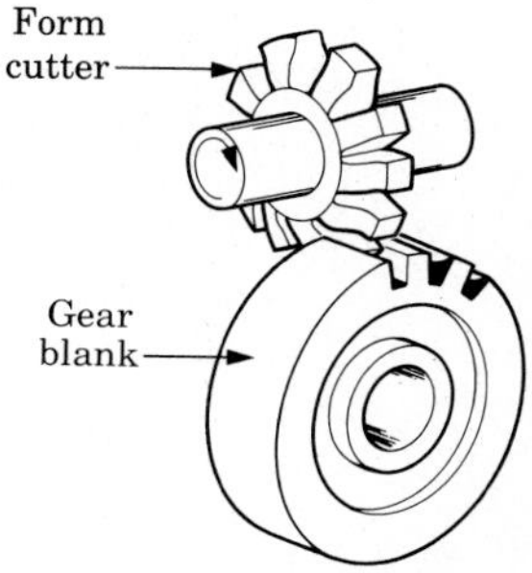

FIGURE 8.77
Producing gear teeth on a blank by form cutting.

Form cutting can be done on milling machines, with the cutter mounted on an arbor and the gear blank mounted in a dividing head. Because the cutter has a fixed geometry, form cutting can be used only to produce gear teeth that have constant width, that is, on spur or helical gears but not on bevel gears. Internal gears and gear teeth on straight surfaces, such as in rack and pinion, are form cut with a shaped cutter, using a machine similar to a shaper.

Broaching can also be used to produce gear teeth and is particularly applicable to internal teeth. The process is rapid and produces fine surface finish with high dimensional accuracy. However, because broaches are expensive—and a separate broach is required for each size of gear—this method is suitable mainly for high-quantity production. Gear teeth may be cut on special machines with a single-point cutting tool that is guided by a *template* in the shape of the gear-tooth profile. Because the template can be made much larger than the gear tooth, dimensional accuracy is improved.

In *gear generating*, the tool may be a (a) pinion-shaped cutter, (b) rack-shaped straight cutter, or (c) hob. The pinion-shaped cutter can be considered as one of the gears in a conjugate pair and the other as the gear blank (Fig. 8.78a). This type of cutter is used in gear generating on machines called *gear shapers* (Fig. 8.78b). The cutter has an axis parallel to that of the gear blank and rotates slowly with the blank at the same pitch-circle velocity with an axial reciprocating motion. A train of gears provides the required relative motion between the cutter shaft and the gear-blank shaft. Cutting may take place either at the downstroke or upstroke of the machine. Because the clearance required for cutter travel is small, gear shaping is suitable for gears that are located close to obstructing surfaces such as flanges (as in the gear blank in Fig. 8.78b). The process can be used for low-quantity as well as high-quantity production.

On a *rack shaper* the generating tool is a segment of a rack (Fig. 8.78c), which reciprocates parallel to the axis of the gear blank. Because it is not practical to have more than 6–12 teeth on a rack cutter, the cutter must be disengaged at suitable intervals and returned to the starting point, the gear blank meanwhile remaining fixed. A gear-cutting *hob* is basically a worm, or screw, that has been made into a gear-generating tool by machining a series of longitudinal slots or gashes into it to form cutting teeth. When hobbing a spur gear, the angle between the hob and gear blank axes is 90° minus the lead angle at the hob threads. All motions in hobbing are rotary, and the hob and gear blank rotate continuously as in two gears meshing until all teeth are cut.

Straight *bevel gears* are generally roughed out in one cut with a form cutter on machines that index automatically. The gear is then finished to the proper shape on a gear generator. The generating method is analogous to the rack-generating method that we described. The cutters reciprocate across the face of the bevel gear as does the tool on a shaper. The machines that are used to make spiral bevel gears operate on essentially the same principle. The spiral cutter is basically a face-milling cutter that has a number of straight-sided cutting blades protruding from its periphery.

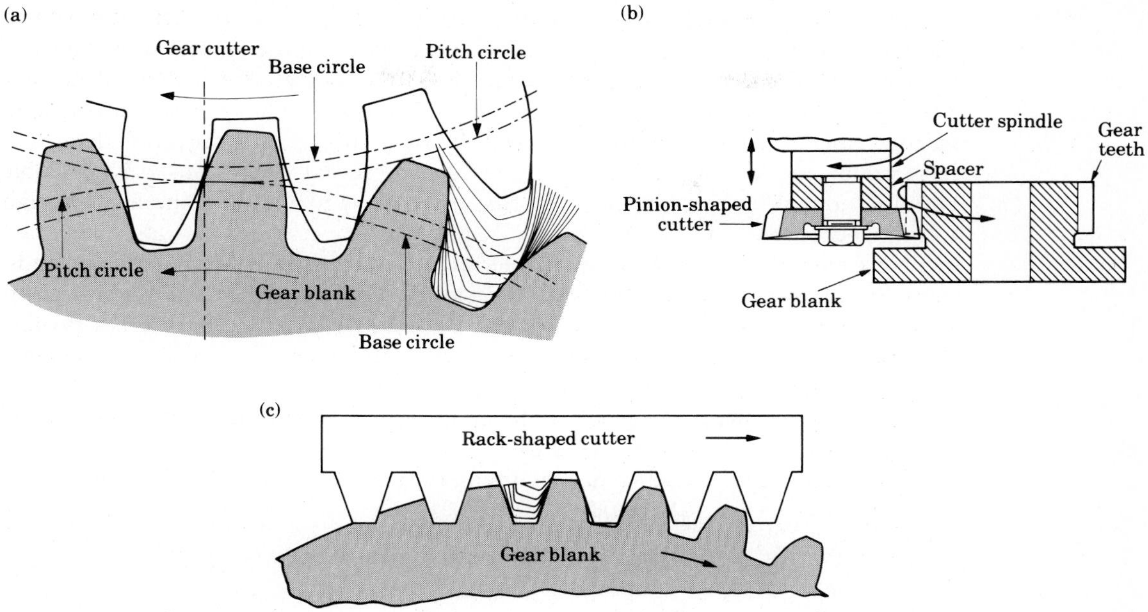

FIGURE 8.78
(a) Schematic illustration of gear generating with a pinion-shaped gear cutter. (b) Schematic illustration of gear generating in a gear shaper using a pinion-shaped cutter. Note that the cutter reciprocates vertically. (c) Gear generating with rack-shaped cutter.

As produced by any of the processes described, the surface finish and dimensional accuracy of gear teeth may not be accurate enough for certain applications. Several *finishing* operations are available, including shaving, burnishing, grinding, honing and lapping.

8.10 Machining Centers

In describing machining processes and machinery in the preceding sections, we noted that each machine tool, regardless of how well it is automated, is traditionally designed to perform basically one type of operation. We have also shown that in manufacturing operations most parts require a number of different cutting operations on their various surfaces.

An important concept developed in the late 1950s is *machining centers* (Fig. 8.79). A machining center is a computer-controlled machine tool with automatic tool-changing capability. The machining center is designed to perform a variety of cutting operations on different surfaces of the workpiece, which is placed on a *module* that can be oriented in various directions. Thus, after a particular cutting operation, say milling, has been completed, the workpiece does not have to be moved to another machine for additional operations, say drilling, reaming, and threading. In other words, the tools and the machine are brought to the workpiece.

Note in Fig. 8.79 that the work module on which the workpiece is placed is capable of moving in three principal linear directions, as well as rotating around a vertical axis. Up to 120 cutting tools can be stored in a *magazine* (tool storage). Auxiliary tool storage is available on some machining centers for up to 480 tools. After all cutting operations have been completed, the module automatically moves away with the finished workpiece, and a new module containing another workpiece to be machined moves into its position.

The major characteristics of machining centers are:

a) They are capable of handling a variety of part sizes and shapes efficiently, economically, and with repetitively high dimensional accuracy with tolerances on the order of ± 0.0025 mm (0.0001 in.).
b) The machines are versatile, having many axes of linear and angular movements and the capability of quick changeover from one type of product to another. Thus the need for a variety of machine tools and floor space is reduced significantly.
c) The time required for loading and unloading workpieces, changing tools, gaging, and troubleshooting is reduced, thus improving productivity,

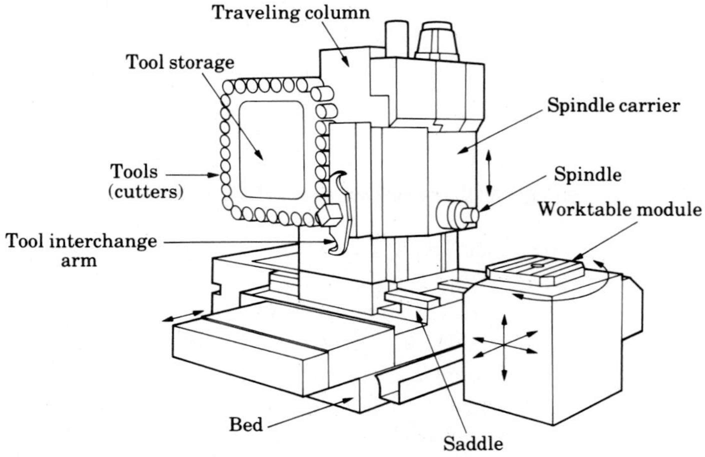

FIGURE 8.79
Schematic illustration of a horizontal-spindle machining center, equipped with an automatic tool changer. Tool magazines can store 120 cutting tools or more. *Source:* Courtesy of Cincinnati Milacron, Inc.

reducing labor (particularly skilled-labor) requirements, and minimizing machining costs.

d) Machining centers are highly automated, so that one operator can attend two or more machines at the same time.

The two major types of machining centers are vertical spindle and horizontal spindle, although many machines have the capability of using both axes. Machining centers are available in a variety of designs and sizes. Capacities range up to 75 kW (100 hp)—and even higher—and spindle speeds usually run 4000–8000 rpm—but some as high as 75,000 rpm.

Vertical-spindle machines are suitable for machining flat surfaces with deep cavities, such as in mold and die making. All operations and movements in the machine are directed and modified through the computer-control panel. Because the thrust forces in vertical machining are directed downward, these machines have high stiffness and produce parts with good dimensional accuracy. These machines are generally less expensive than horizontal-spindle machines.

Horizontal-spindle machines (see Fig. 8.79) are suitable for large, including tall, workpieces requiring machining on a number of their surfaces, with the use of a work table module (with workpieces mounted on a *pallet*) that can be rotated to various angular positions.

Universal machining centers have a variety of features and are capable of machining all surfaces of a workpiece. New developments include turret-type machines capable of performing turning operations. In-process and post-process gaging and inspection of machined workpieces are also new features of machining centers. The stiffness of machining centers, hence dimensional accuracy, is being improved continually. Many machines are being constructed on a modular basis, so that various peripheral equipment can be installed and modified as the demand for and type of products change.

Machining centers come in a wide variety of sizes. The smallest machine is designed for parts up to 100 mm (4 in.), a spindle speed of up to 12,000 rpm, and capacity of 2.2 kW (3 hp). The largest machining center built to date (Fig. 8.80) is capable of machining workpieces 11.5 m × 5.6 m × 5.4 m (38 ft × 18.5 ft × 18 ft), and weighing up to 200 tons. The structure consists of double columns. The master head in which the spindle is housed is mounted on a cross-rail and can be set to vertical and horizontal positions. Auxiliary spindle units are transported by robots.

Machining centers are equipped with *automatic tool changers*. Tools are usually stored in a drum, magazine, or chain. Tools are automatically selected for the shortest route to the spindle. The tool-interchange arm (see Fig. 8.79) is a common design: It swings around to pick up the tool and place it in the spindle. Tool changing times are on the order of a few seconds. Tools are identified by coded tags, bar coding, or memory chips applied to the toolholders. Machining centers are also equipped with a tool-checking station that feeds information to the computer-numerical control to compensate for any variations in tool settings.

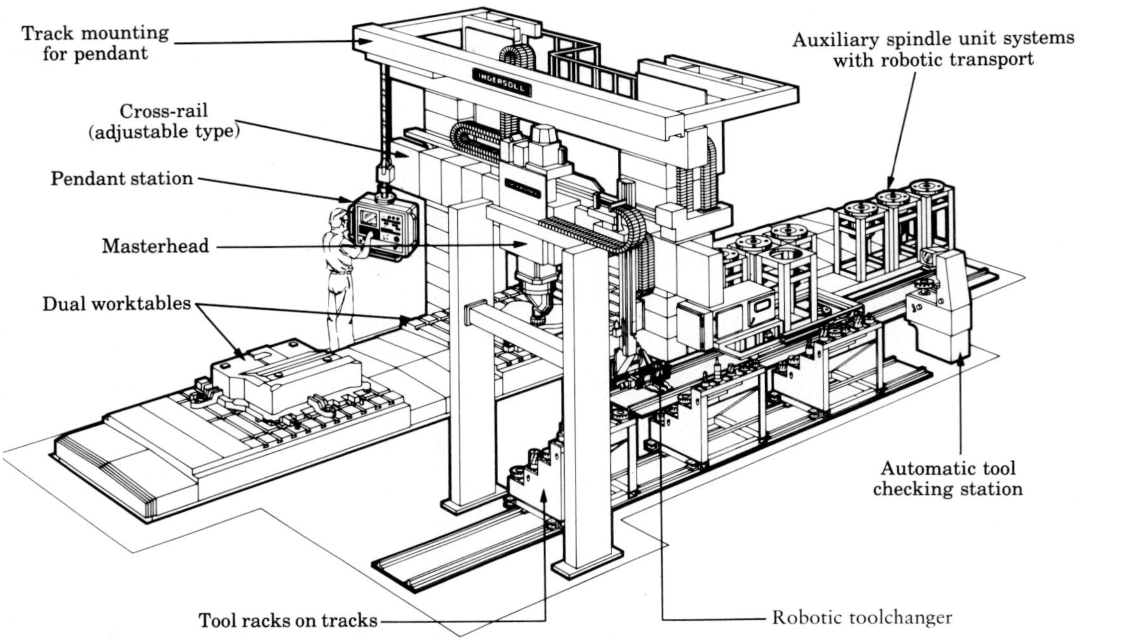

FIGURE 8.80
Schematic illustration of a large machining center, showing various components. *Source:* The Ingersoll Milling Machine Company.

8.11 Vibration and Chatter

In describing cutting processes and machine tools, we pointed out the importance of machine stiffness in controlling dimensional accuracy and surface finish of parts. In this section we describe the adverse effects of low stiffness on product quality and machining operations. Low stiffness affects the level of vibration and chatter in tools and machines. If uncontrolled, vibration and chatter can result in the following:

- Poor surface finish (right central region in Fig. 8.81).
- Loss of dimensional accuracy of the workpiece.
- Premature wear, chipping, and failure of the cutting tool, which is crucial with brittle tool materials, such as ceramics, some carbides, and diamond.
- Damage to machine-tool components from excessive vibrations.
- Objectionable noise generated, particularly if it is of high frequency, such as the squeal heard in turning brass on a lathe.

FIGURE 8.81 Chatter marks (right of center of photograph) on the surface of a turned part. *Source:* General Electric Company.

Extensive studies have shown that vibration and chatter in cutting are complex phenomena. Cutting operations cause two basic types of vibration: forced vibration and self-excited vibration.

Forced vibration is generally caused by some periodic applied force present in the machine tool, such as from gear drives, imbalance of the machine-tool components, misalignment, or motors and pumps. In processes such as milling or turning a splined shaft or a shaft with a keyway, forced vibrations are caused by the periodic engagement of the cutting tool with and exit from the workpiece surface.

The basic solution to forced vibrations is to isolate or remove the forcing element. If the forcing frequency is at or near the natural frequency of a component of the machine-tool system, one of the frequencies may be raised or lowered. The amplitude of vibration can be reduced by increasing the stiffness or damping the system (see the discussion below). Although changing the cutting-process parameters generally does not appear to greatly influence forced vibrations, changing the cutting speed and the tool geometry can be helpful.

Generally called *chatter*, *self-excited vibration* is caused by the interaction of the chip-removal process and the structure of the machine tool. Self-excited vibrations usually have a very high amplitude. Chatter typically begins with a disturbance in the cutting zone. Such disturbances include lack of homogeneity in the workpiece material or its surface condition, changes in type of chips produced, or a change in frictional conditions at the tool–chip interface as also influenced by cutting fluids and their effectiveness.

The most important type of self-excited vibration in machining is *regenerative chatter*. It is caused when a tool cuts a surface that has a roughness or disturbances from the previous cut. Because the depth of cut varies, the resulting variations in the cutting force subject the tool to vibrations and the process continues repeatedly —hence the term regenerative. You may observe this type of vibration while driving over a rough road (the so-called washboard effect).

Self-excited vibrations can generally be controlled by increasing the dynamic stiffness of the system and by damping. We define *dynamic stiffness* as the ratio of the amplitude of the force applied to the amplitude of vibration. Since a machine tool has different stiffnesses at different frequencies, changes in cutting parameters, such as cutting speed, can influence chatter.

8.11.1 Damping

We define *damping* as the rate at which vibrations decay. Damping is an important factor in controlling machine-tool vibration and chatter.

Internal damping of structural materials. Internal damping results from the energy loss in materials during vibration. Thus, for example, steel has less damping than gray cast iron, and composite materials have more damping than gray iron (Fig. 8.82). You can observe this difference in the damping capacity of materials by striking them with a gavel and listening to the sound. Try, for example, striking pieces of steel, concrete, and wood.

Joints in the machine-tool structure. Although less significant than internal damping, bolted joints in the structure of a machine tool are also a source of damping. Because friction dissipates energy, small relative movements along dry (unlubricated) joints dissipate energy and thus improve damping. In joints where oil is present, the internal friction of the oil layers also dissipates energy, thus contributing to damping. Note, for example, that stirring oil with a stick requires force; thus energy is dissipated in stirring it.

In describing the machine tools for various cutting operations, we noted that all machines consist of a number of large and small components, assembled into a structure. Consequently, this type of damping is cumulative, owing to the presence of a number of joints in a machine tool. Note in Fig. 8.83 how damping increases as the number of components on a lathe and their contact area increase. Thus the more joints, the greater is the amount of energy dissipated and the higher is the damping.

External damping. External damping is accomplished with external dampers, which are similar to shock absorbers on automobiles. Special vibration absorbers have been developed and installed on machine tools for this purpose.

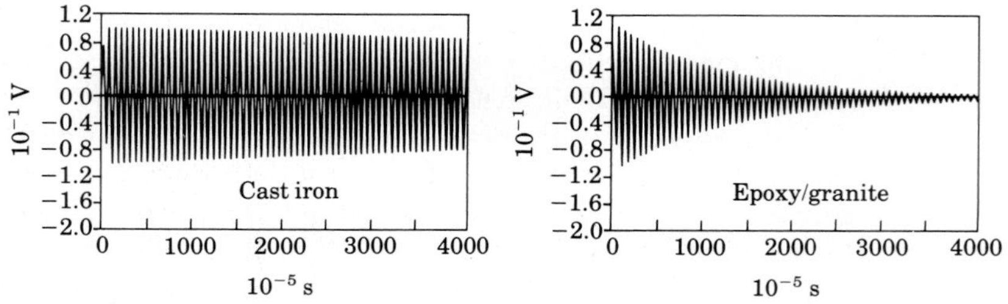

FIGURE 8.82

Relative damping capacity of gray cast iron and epoxy–granite composite material. The vertical scale is the amplitude of vibration and the horizontal scale is time. *Source:* Cincinnati Milacron, Inc.

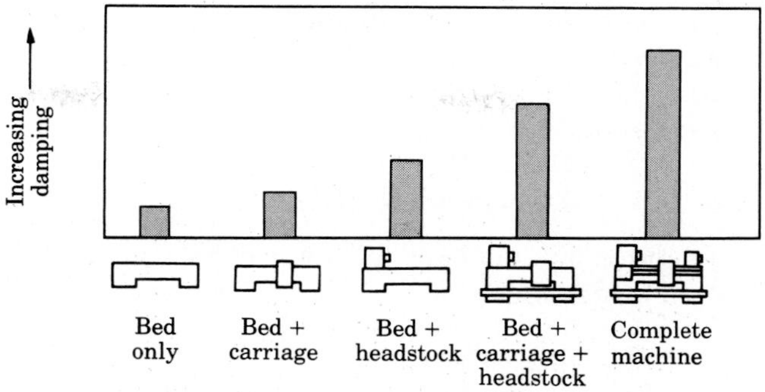

FIGURE 8.83

Damping of vibrations as a function of the number of components on a lathe. Joints dissipate energy; thus the greater the number of joints, the higher the damping will be. *Source:* J. Peters.

8.11.2 Factors influencing chatter

Studies have indicated that the tendency for a particular workpiece to chatter during cutting is proportional to the cutting forces and the depth and width of cut. Consequently, because cutting forces increase with strength (hardness), the tendency to chatter generally increases as the hardness of the workpiece material increases. Thus aluminum and magnesium alloys have less tendency to chatter than do martensitic and precipitation-hardening stainless steels, nickel alloys, and high-temperature and refractory alloys.

An important factor in chatter is the type of chip produced during cutting operations. As we have shown, continuous chips involve steady cutting forces. Consequently, they generally do not cause chatter. Discontinuous chips and serrated chips, on the other hand, may do so. These chips are produced periodically, and the resulting variations in force during cutting can cause chatter.

8.12 Machine-Tool Structures

In this section we discuss the *material* and *design* aspects of machine tools as structures that have certain desired characteristics. Today's markets have stringent requirements for high quality and precision in manufactured products, often made with difficult-to-machine materials, with precise specifications concerning dimensional accuracy and surface finish and integrity. Consequently, the design and construction of machine tools are important aspects of manufacturing engineering.

The proper design of machine-tool structures requires a thorough knowledge of the materials available for construction, their forms and properties, the dynamics of the particular machining process, and the forces involved. *Stiffness* and *damping* are important factors in machine-tool structures. Stiffness involves both the dimensions of the structural components and the elastic modulus of the materials used. Damping, as you have seen, involves the type of materials used, as well as the number and nature of the joints in the structure.

8.12.1 Materials and design

Traditionally, the base and some of the major components of machine tools have been made of gray or nodular cast iron, which has the advantages of low cost and good damping capacity, but are heavy. Lightweight designs are desirable because of ease of transportation, higher natural frequencies, and lower inertial forces of moving members. Lightweight designs and design flexibility require fabrication processes such as (a) mechanical fastening (bolts and nuts) of individual components and (b) welding. However, this approach to fabrication increases labor, as well as material, costs because of the preparations involved.

Wrought steels are the likely choice for such lightweight structures because of their low cost, availability in various section sizes and shapes (such as channels, angles, and tubes), desirable mechanical properties, and favorable characteristics (such as formability, machinability, and weldability). Tubes, for example, have high stiffness-to-weight ratios. On the other hand, the benefit of the higher damping capacity of castings and composites isn't available with steels.

In addition to stiffness, another factor that contributes to lack of precision of the machine tool is the thermal expansion of its components, causing distortion. The source of heat may be internal, such as bearings, ways, motors, and heat generated from the cutting zone, or external, such as nearby furnaces, heaters, sunlight, and fluctuations in cutting fluid and ambient temperatures.

Equally important in machine-tool precision are foundations, their mass, and how they are installed in a plant. For example, in one grinder for grinding 2.75-m (9-ft) diameter marine-propulsion gears with high precision, the concrete foundation is 6.7 m (22 ft deep). The large mass of concrete and the machine base reduces the amplitude of vibrations and its adverse effects.

8.12.2 Recent developments

Significant developments concerning the materials used for machine-tool bases and components have taken place. These developments are summarized in the following paragraphs.

Concrete. Concrete has been used for machine-tool bases since the early 1970s. Compared to cast iron, concrete is less expensive, its curing time is about three weeks (much shorter than for making castings), and it has good damping capacity. However, concrete is brittle and thus not suitable for applications involving impact loads. Concrete can also be poured into cast-iron base structures to increase their mass and improve machine-tool damping capacity. Filling the cavities of bases with loose sand is also an effective means of improving damping capacity.

Acrylic concretes. A mixture of concrete and polymer (polymethylmethacrylate), acrylic concretes can easily be cast into desired shapes for machine bases and various components. They were first introduced in the 1980s, and several new compositions are being developed. These materials can also be used for sandwich construction with cast irons, thus combining the advantages of each type of material.

Granite–epoxy composite. A castable composite has been developed, with a composition of about 93 percent crushed granite and 7 percent epoxy binder. First used in precision grinders in the early 1980s, this composite material has several favorable properties: (a) good castability, which allows design versatility in machine tools; (b) high stiffness-to-weight ratio; (c) thermal stability; (d) resistance to environmental degradation; and (e) good damping capacity (see Fig. 8.82).

8.13 Machining Economics

Besides technical considerations, the economics of metal removal processes is very important. In machining a certain part, we may want to determine the parameters that will give us either the minimum cost per part or the maximum production rate.

The *total cost* per piece is composed of four items:

$$C_p = C_m + C_s + C_\ell + C_t, \tag{8.44}$$

where C_p is the cost (say, in dollars) per piece; C_m is the machining cost; C_s is the cost of setting up for machining, such as mounting the cutter and fixtures and preparing the machine for the particular operation; C_ℓ is the cost of loading, unloading, and machine handling; and C_t is the tooling cost, which includes tool changing, regrinding, and depreciation of the cutter.

The *machining cost* is

$$C_m = T_m(L_m + B_m), \tag{8.45}$$

where T_m is the machining time per piece; L_m is the labor cost of production operator per hour; and B_m is the burden rate, or overhead charge, of the machine, including depreciation, maintenance, indirect labor, and the like.

The *setup cost* C_s is a fixed figure in dollars per piece. The C_ℓ cost is

$$C_\ell = T_\ell(L_m + B_m), \tag{8.46}$$

where T_ℓ is the time involved in loading and unloading the part, changing speeds, feed rates, and so on.

The *tooling cost* is

$$C_t = \frac{1}{N_p}\left[T_c(L_m + B_m) + T_g(L_g + B_g) + D_c\right], \tag{8.47}$$

where N_p is the number of parts machined per tool grind; T_c is the time required to change the tool; T_g is the time required to grind the tool; L_g is the labor cost of tool grinder operator per hour; B_g is the burden rate of tool grinder per hour; and D_c is the depreciation of the tool in dollars per grind.

The *time* needed to produce one part is

$$T_p = T_\ell + T_m + \frac{T_c}{N_p}, \tag{8.48}$$

where T_m has to be calculated for a particular operation. For example, let's consider a turning operation. The machining time is

$$T_m = \frac{L}{fN} = \frac{\pi LD}{fV}, \tag{8.49}$$

where L is the length of cut, f is the feed, N is the rpm of the workpiece, D is the workpiece diameter, and V is the cutting speed. Note that appropriate units must be used in all these equations.

From the tool-life equation we have

$$VT^n = C.$$

Hence

$$T = \left(\frac{C}{V}\right)^{1/n}, \tag{8.50}$$

where T is the time, in minutes, required to reach a flank wear of certain dimension, after which the tool has to be reground or changed. The number of pieces per tool grind is thus simply

$$N_p = \frac{T}{T_m}. \tag{8.51}$$

The combination of Eqs. (8.49), (8.50), and (8.51) gives

$$N_p = \frac{fC^{1/n}}{\pi LDV^{(1/n)-1}}. \tag{8.52}$$

We can now define the *cost per piece* C_p in Eq. (8.44) in terms of several variables. To find the optimum cutting speed and also the optimum tool life for *minimum cost*, we have to differentiate C_p with respect to V and set it to zero. Thus

$$\frac{\partial C_p}{\partial V} = 0. \tag{8.53}$$

We then find that the optimum cutting speed V_o is

$$V_o = \frac{C(L_m + B_m)^n}{\{[(1/n) - 1][T_c(L_m + B_m) + T_g(L_g + B_g) + D_c]\}^n}, \tag{8.54}$$

and the optimum tool life T_o is

$$T_o = [(1/n) - 1]\frac{T_c(L_m + B_m) + T_g(L_g + B_g) + D_c}{L_m + B_m}. \tag{8.55}$$

To find the optimum cutting speed and also the optimum tool life for *maximum production*, we have to differentiate T_p with respect to V and set it to zero. Thus

$$\frac{\partial T_p}{\partial V} = 0. \tag{8.56}$$

We find that the *optimum cutting speed* V_o now becomes

$$V_o = \frac{C}{\{[(1/n) - 1]T_c\}^n}, \tag{8.57}$$

and the *optimum tool life* T_o is

$$T_o = [(1/n) - 1]T_c. \tag{8.58}$$

A qualitative plot of minimum cost per piece and the minimum time per piece, that is, the maximum production rate, is given in Fig. 8.84. The cost of a machined

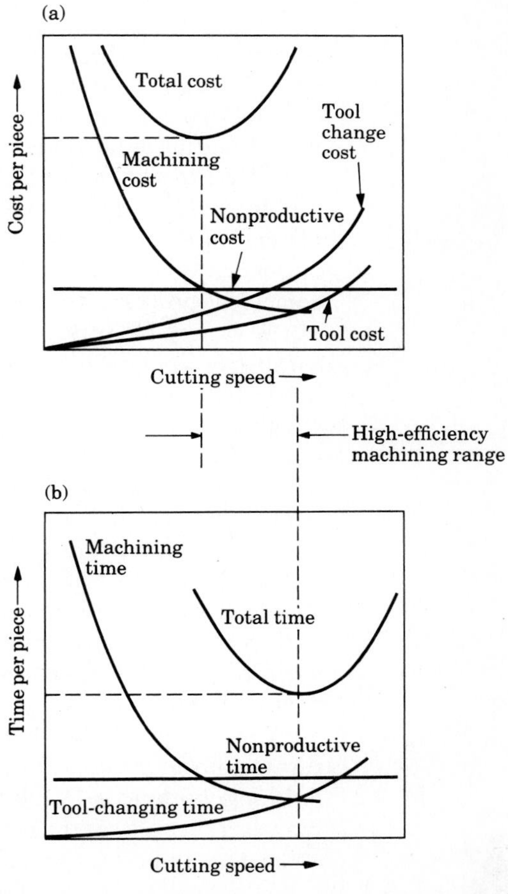

FIGURE 8.84 Graphs showing (a) cost per piece and (b) time per piece in machining. Note the optimum speeds for both cost and time. The range between the two is known as the *high-efficiency machining range*.

surface also depends on the finish required (see Section 9.16); the machining cost increases rapidly with finer surface finish.

This analysis indicates the importance of identifying all relevant parameters in a machining operation, determining various cost factors, obtaining relevant tool-life curves for the particular operation, and properly measuring the various time intervals involved in the overall operation. The importance of obtaining accurate data is clearly shown in Fig. 8.84: Small changes in cutting speed can have a significant effect on the minimum cost or time per piece.

SUMMARY

Cutting processes are among the most important of manufacturing operations. They are often necessary in order to impart the desired surface finish and dimensional accuracy to components, particularly those with complex shapes that cannot be produced economically or properly by other shaping techniques.

A large number of variables have significant influence on the mechanics of chip formation in cutting operations. Commonly observed chip types are continuous, built-up edge, discontinuous, and segmented. Among important process variables are tool shape and material, cutting conditions such as speed, feed, and depth of cut, use of cutting fluids, and the characteristics of the machine tool, as well as the characteristics of the workpiece material. Parameters influenced by these variables are forces and power consumption, tool wear, surface finish and integrity, temperature, and dimensional accuracy of the workpiece. Machinability of materials depends not only on their intrinsic properties, but also on proper selection and control of process variables.

A variety of cutting-tool materials have been developed over the past century for specific applications in machining operations. These materials have a wide range of mechanical and physical properties, such as hot hardness, toughness, chemical stability, and resistance to chipping and wear. Various coatings have been developed, resulting in major improvements in tool life. The selection of appropriate tool materials depends not only on the material to be machined, but also on process parameters and the characteristics of the machine tool.

Cutting fluids are an important factor in machining operations. Generally, slower operations with high tool pressures require a fluid with good lubricating characteristics. In high-speed operations with significant temperature rise, fluids with cooling capacity are preferred. Selection should include consideration of the various adverse effects of cutting fluids on products, machinery, personnel, and the environment.

Cutting processes that produce external and internal circular profiles are turning, boring, and drilling. Reaming, tapping, and die threading are processes for finishing workpieces. Chip formation in all these processes is essentially the same. However, because of the three-dimensional nature of the cut, chip movement and its control are important considerations, since otherwise they interfere with the cutting operation. Chip removal can be a significant problem especially in drilling and tapping and can lead to tool breakage. Each process should be studied in order to understand the interrelationships of design parameters, such as dimensional accuracy, surface finish, and integrity, and process parameters, such as speed, feed, depth of cut, tool material and shape, and cutting fluids.

Some of the most versatile machining processes are milling, planing, shaping, broaching, and sawing. Milling is one of the most useful processes because of its capability to produce a variety of shapes from workpieces. Although there are similarities with processes such as turning, drilling, and boring, most of these processes utilize multitooth tools and cutters at various axes with respect to the workpiece. The machine tools employed have various features, attachments, and considerable flexibility in operation.

Because of their versatility and capability of performing a variety of cutting operations on small and large workpieces, machining centers have become one of the most important developments in machine tools. Various types of machining centers are available, and their selection depends on factors such as part complexity, the number and type of cutting operations involved, dimensional accuracy, and production rate required.

Vibration and chatter in machining are important considerations for workpiece dimensional accuracy and surface finish, as well as tool life. Stiffness and damping capacity of machine tools are important factors in controlling vibration and chatter. New materials are being developed and used for constructing machine-tool structures.

The economics of machining processes depend on nonproductive, machining, tool-change, and tool costs. Optimum cutting speeds can be obtained for minimum machining time per piece and minimum cost per piece, respectively.

BIBLIOGRAPHY

Fundamentals

Armarego, E.J.A., and R.H. Brown, *The Machining of Metals*. Englewood Cliffs, N.J.: Prentice-Hall, 1969.

Arshinov, V., and G. Alekseev, *Metal Cutting Theory and Cutting Tool Design*. Moscow: Mir Publishers, 1976.

Bhattacharyya, A., and I. Ham, *Design of Cutting Tools*. Dearborn, Mich.: Society of Manufacturing Engineers, 1969.

Boothroyd, G., and W.A. Knight, *Fundamentals of Machining and Machine Tools*, 2d ed. New York: Marcel Dekker, 1989.

Kaczmarek, J., *Principles of Machining*. Stevenage, England: Peter Peregrinus, 1976.

Kobayashi, A., *Machining of Plastics*. New York: McGraw-Hill, 1967.

Kronenberg, M., *Machining Science and Application*. Oxford: Pergamon, 1966.

Machinability Testing and Utilization of Machining Data. Metals Park, Ohio: American Society for Metals, 1979.

Mills, B., and A.H. Redford, *Machinability of Engineering Materials.* London: Applied Science Publishers, 1983.

Oxley, P.L.B., *Mechanics of Machining: An Analytical Approach to Assessing Machinability.* New York: Wiley, 1987.

Shaw, M.C., *Metal Cutting Principles.* New York: Oxford, 1984.

Tool and Manufacturing Engineers Handbook, 4th ed., *Vol. 1: Machining.* Dearborn, Mich.: Society of Manufacturing Engineers, 1983.

Trent, E.M., *Metal Cutting*, 2d ed. London: Butterworths, 1984.

Venkatesh, V.C., and H. Chandrasekaran, *Experimental Techniques in Metal Cutting*, rev. ed. New Delhi: Prentice-Hall, 1987.

Zorev, N.N., *Metal Cutting Mechanics.* New York: Pergamon, 1966.

Cutting Tools

Cutting Tool Materials. Metals Park, Ohio: American Society for Metals, 1981.

Kalpakjian, S. (ed.), *New Developments in Tool Materials and Applications.* Chicago: Illinois Institute of Technology, 1977.

King, A.G., and W.M. Wheildon, *Ceramics in Machining Processes.* New York: Academic Press, 1966.

Komanduri, R. (ed.), *Advances in Hard Material Tool Technology.* Pittsburgh: Carnegie Press, 1976.

Komanduri, R., "Tool Materials," in *Kirk–Othmer Encyclopedia of Chemical Technology*, 3d ed., v. 23. New York: Wiley, 1978.

Machinery's Handbook. New York: Industrial Press, revised periodically.

Machining Data Handbook, 3d ed., 2 vols. Cincinnati: Machinability Data Center, 1980.

Metals Handbook, 9th ed., *Vol. 3: Properties and Selection: Stainless Steels, Tool Materials and Special Purpose Metals.* Metals Park, Ohio: American Society for Metals, 1980.

Metals Handbook, 9th ed., *Vol. 16: Machining.* Metals Park, Ohio: ASM International, 1989.

Roberts, G.A., and R.A. Cary, *Tool Steels*, 4th ed. Metals Park, Ohio: American Society for Metals, 1980.

Trent, E.M., *Metal Cutting*, 2d ed. Stoneham, Mass.: Butterworths, 1984.

Tool and Manufacturing Engineers Handbook, 4th ed., *Vol. 1: Machining.* Dearborn, Mich.: Society of Manufacturing Engineers, 1983.

Cutting Fluids

Cutting and Grinding Fluids. Dearborn, Mich.: Society of Manufacturing Engineers, 1967.

Nachtman, E.S., and S. Kalpakjian, *Lubricants and Lubrication in Metalworking Operations.* New York: Marcel Dekker, 1985.

Olds, N.J., *Lubricants, Cutting Fluids and Coolants.* Boston: Cahners, 1973.

Schey, J.A., *Tribology in Metalworking—Friction, Wear and Lubrication.* Metals Park, Ohio: American Society for Metals, 1983.

Zintac, B. (ed.), *Improving Production with Coolants and Lubricants.* Dearborn, Mich.: Society of Manufacturing Engineers, 1982.

Machining Processes

Gear Processing and Manufacturing, 2d ed. Dearborn, Mich.: Society of Manufacturing Engineers, 1984.

Lambert, B. (ed.), *Milling: Methods and Machines*. Dearborn, Mich.: Society of Manufacturing Engineers, 1980.

Machining Data Handbook, 2 vols., 3d ed. Cincinnati: Machinability Data Center, 1980.

Metals Handbook, 9th ed., *Vol. 16, Machining*. Metals Park, Ohio: ASM International, 1989.

Milling Handbook of High-Efficiency Metal Cutting. Detroit, Mich.: Carboloy Systems Department, General Electric Company, 1980.

Tool and Manufacturing Engineering Handbook, 4th ed., *Vol. 1, Machining*. Dearborn, Mich.: Society of Manufacturing Engineers, 1983.

Turning Handbook of High-Efficiency Cutting. Detroit: Carboloy Systems Department, General Electric Company, 1980.

Weck, M., *Handbook of Machine Tools*, 4 vols. New York: Wiley, 1984.

Machine Tools

Juneja, B.L., *Fundamentals of Metal Cutting and Machine Tools*. New York: Wiley, 1987.

Koenigsberger, F., *Design Principles of Metal-Cutting Machine Tools*. New York: Pergamon, 1964.

Koenigsberger, F., and J. Tlusty, *Machine Tool Structures*. New York: Pergamon, 1970.

Metals Handbook, 9th ed., *Vol. 16, Machining*. Metals Park, Ohio: ASM International, 1989.

Schlesinger, G., F. Koenigsberger, and M. Burdekin, *Testing Machine Tools*, 8th ed. Oxford: Pergamon, 1978.

Tool and Manufacturing Engineering Handbook, 4th ed., *Vol. 1, Machining*. Dearborn, Mich.: Society of Manufacturing Engineers, 1983.

Weck, M., *Handbook of Machine Tools*, 4 vols. New York: Wiley, 1984.

QUESTIONS

8.1 Researchers have observed that the actual shear strains obtained in metal cutting are higher than those calculated from the properties of the material. To what factors would you attribute this difference? Explain.

8.2 Describe the effects of material properties and process variables on serrated chip formation.

8.3 Explain why the cutting force F_c increases with increasing depth of cut and decreasing rake angle.

8.4 How does a dull tool tip affect a cutting operation?

8.5 Describe the trends that you observe in Table 8.1.

8.6 To what factors would you attribute the large difference in the specific energies within each group of materials in Table 8.4?

8.7 Describe the effects of cutting fluids on chip formation. Explain why and how they influence the cutting operation.

8.8 Under what conditions would you discourage the use of cutting fluids?

8.9 Describe the trends that you observe in Tables 8.2 and 8.3.

8.10 Explain why the maximum temperature in cutting is located at about the middle of the tool–chip interface. Remember that there are two principal sources of heat: the shear plane and the tool–chip interface.

8.11 State whether each of the following statements is correct, explaining your reasons:

a) For the same shear angle, there are two rake angles that give the same cutting ratio.
b) For the same depth of cut and rake angle, the type of cutting fluid used has no influence on chip thickness.
c) If the cutting speed, shear angle, and rake angle are known, the chip velocity can be calculated.
d) The chip becomes thinner as the rake angle increases.
e) The function of a chip breaker is to decrease the curvature of the chip.

8.12 Allowing temperatures to rise too high in cutting operations is generally undesirable. Explain why.
8.13 Why may the same tool life be obtained at two different cutting speeds?
8.14 Inspect Table 8.8 and identify tool materials that would not be particularly suitable for interrupted cutting operations. Explain why they would not.
8.15 Explain the disadvantages in a cutting operation in which the type of chip produced is discontinuous.
8.16 We stated that tool life can be almost infinite at low cutting speeds. Would you recommend that all machining be done at low speeds? Explain.
8.17 How would you explain the effect of cobalt content on the properties of carbides as shown in Fig. 8.39(b)?
8.18 Explain why studying the types of chips produced is important in understanding machining operations.
8.19 How would you expect the cutting force to vary in the case of serrated chip formation? Explain.
8.20 Wood is a highly anisotropic material (it is orthotropic). Explain the effects of cutting wood at different angles to the grain direction on chip formation.
8.21 Describe the advantages of oblique cutting.
8.22 Explain why removing more material between tool resharpenings by lowering the cutting speed is possible.
8.23 Explain the significance of Eq. (8.8).
8.24 Explain why there are so many different types of cutting-tool materials.
8.25 How would you go about measuring the hot hardness of cutting tools?
8.26 Describe the reasons for making cutting tools with multiple coatings of different materials.
8.27 Explain the advantage and limitations of inserts. Why were they developed?
8.28 Make a list of alloying elements used in high-speed steels. Explain why they are such effective cutting-tool materials.
8.29 What are the purposes of chamfers on cutting tools?
8.30 Why does temperature have such an important effect on cutting tool performance?
8.31 Ceramic and cermet cutting tools have certain advantages over carbide tools. Why then are they not replacing carbide tools to a greater extent?
8.32 Why are chemical stability and inertness important in cutting tools?
8.33 What precautions would you take in cutting with brittle tool materials?
8.34 Why do cutting fluids have different effects at different cutting speeds? Is the control of cutting-fluid temperature important? If so, why?
8.35 Which of the two materials, diamond or cubic boron nitride, is more suitable for cutting steels? Why?
8.36 Describe the properties that the substrate for multicoated tools should have.
8.37 List and explain the considerations involved in whether a cutting tool should be reconditioned or discarded.

8.38 List the parameters that influence temperature in metal cutting and explain why and how they do so.

8.39 List and explain factors that contribute to poor surface finish.

8.40 Explain the functions of different angles on a single-point lathe cutting tool. How does the chip thickness vary as the side cutting edge angle is increased?

8.41 The helix angle for drills is different for different groups of materials. Why?

8.42 A turning operation is being carried out on a long round bar at a constant depth of cut. Explain what changes, if any, may occur in the machined diameter from one end of the bar to the other. Give reasons for any changes that may occur.

8.43 Describe the relative characteristics of climb milling and up milling.

8.44 In Fig. 8.76(a) high-speed steel cutting teeth are welded to a steel blade. Would you recommend that the entire blade be made of high-speed steel? Explain your reasons.

8.45 Explain the technical requirements that led to the development of machining centers. What are the distinctive features of these centers? Why do their spindle speeds vary widely? Are there cutting operations that cannot be performed on these centers? Explain.

8.46 Describe the adverse effects of vibrations and chatter in machining.

8.47 Make a list of components of machine tools that could be made of ceramics and explain why ceramics would be suitable.

8.48 In Fig. 8.14, why do the forces start at a finite value when the feed is zero?

8.49 Is temperature rise in cutting related to the hardness of the workpiece? Explain.

8.50 Describe the effects of tool wear on the workpiece and on the machining operation in general.

8.51 Refer to Fig. 8.33 and explain why the curves go downward as feed increases.

8.52 Explain whether having a high or low n value in the Taylor tool-life equation is desirable.

8.53 Explain why the cutting force is a function of cutting speed, feed, and rake angle.

8.54 Devise a method whereby you can perform an orthogonal cutting operation with a round workpiece on a lathe.

8.55 Cutting tools are sometimes designed so that the tool–chip contact length is controlled. (See, for example, B.T. Chao and K.J. Trigger, *Trans. ASME*, *J. Eng. Ind.*, vol. 81, 1959, pp. 139–151.) Discuss the effect of this approach on the mechanics of orthogonal cutting.

8.56 It has been stated that the thermal conductivity of the tool material is relatively unimportant in the mechanics of cutting compared to that of the workpiece. Can you offer an explanation? (See, for example, E.G. Loewen and M.C. Shaw, *Trans. ASME*, vol. 76, 1954, pp. 217–231.)

8.57 Assume that you are asked to estimate the cutting force in slab milling with a straight-tooth cutter, without running a test. Describe the procedure that you would follow.

8.58 Explain the possible reasons why a knife cuts better when it is moved back and forth. Consider factors such as the material cut, friction, and the dimensions of the knife.

PROBLEMS

8.1 Assume that in orthogonal cutting the rake angle is 10° and the coefficient of friction is 0.5. Determine the percentage change in chip thickness when the friction is doubled.

8.2 Derive Eq. (8.11).

8.3 Taking carbide as an example and using Eq. (8.30), determine how much the feed should be changed in order to keep the mean temperature constant when the cutting speed is doubled.

8.4 With appropriate diagrams show how the use of a cutting fluid can change the magnitude of the thrust force F_t in orthogonal cutting.

8.5 An 8-in. diameter stainless steel bar is being turned on a lathe at 500 rpm and a depth of cut $d = 0.1$ in. If the power of the motor is 2 hp, what is the maximum feed that you can have before the motor stalls?

8.6 Using the Taylor equation for tool wear and letting $n = 0.5$ and $C = 400$, calculate the percentage increase in tool life if the cutting speed is reduced by (a) 20% and (b) 50%.

8.7 Determine the n and C values for the four tool materials shown in Fig. 8.29.

8.8 Estimate the machining time required in rough turning a 1-m long, annealed aluminum-alloy round bar, 100 mm in diameter, using a high-speed steel tool. Estimate the time for a carbide tool.

8.9 A 6-in. long, $\frac{1}{2}$-in. diameter 304 stainless steel rod is being reduced in diameter to 0.480 in. by turning on a lathe. The spindle rotates at 400 rpm, and the tool is traveling at an axial speed of 8 in./min. Calculate the cutting speed, material removal rate, time of cut, power required, and cutting force.

8.10 A 0.5-in. diameter drill is used on a drill press operating at 200 rpm. If the feed is 0.005 in./rev, what is the material removal rate? What is the MRR if the drill diameter is tripled?

8.11 A hole is being drilled in a block of magnesium alloy with a 10-mm drill at a feed of 0.2 mm/rev. The spindle is running at 800 rpm. Calculate the material removal rate and estimate the torque on the drill.

8.12 Show that the distance l_c in slab milling is approximately equal to $\sqrt{Dd}$ for situations where $D \gg d$.

8.13 A slab-milling operation is being carried out on a 12-in. long, 4-in. wide annealed mild-steel block at a feed of 0.01 in./tooth and depth of cut of $\frac{1}{8}$ in. The cutter is 2 in. in diameter, has 20 straight cutting teeth, and rotates at 100 rpm. Calculate the material removal rate and cutting time, and estimate the power required.

8.14 Refer to Fig. 8.65 and assume that $D = 150$ mm, $w = 60$ mm, $l = 500$ mm, $d = 3$ mm, $v = 0.01$ mm/s, and $N = 100$ rpm. The cutter has 10 inserts and the workpiece material is a high-strength aluminum alloy. Calculate the material removal rate, cutting time, and feed per tooth, and estimate the power required.

8.15 Estimate the time required for face milling an 8-in. long, 2-in. wide brass block with an 8-in. diameter cutter with 10 high-speed steel teeth.

8.16 A 10-in. long, 1-in. thick plate is being cut on a band saw at 100 ft/min. The saw has 12 teeth per in. If the feed per tooth is 0.003 in., how long will it take to saw the plate along its length?

8.17 A single-thread hob is used to cut 40 teeth on a spur gear. The cutting speed is 100 ft/min and the hob is 4 in. in diameter. Calculate the rotational speed of the spur gear.

8.18 In deriving Eq. (8.20), we assumed that the friction angle β was independent of the shear angle ϕ. Is this a valid assumption? Explain.

8.19 An orthogonal cutting operation is being carried out under the following conditions: Depth of cut = 0.1 mm, width of cut = 5 mm, chip thickness = 0.2 mm, cutting speed = 2 m/s, rake angle = 10°, cutting force = 500 N, and thrust force = 200 N. Calculate the percentage of the total energy that is dissipated in the shear plane during cutting.

8.20 An orthogonal cutting operation is being carried out under the following conditions: Depth of cut = 0.010 in., width of cut = 0.1 in., cutting ratio = 0.3, cutting speed = 400 ft/min, rake angle = 0°, cutting force = 200 lb, thrust force = 150 lb, workpiece density = 0.26 lb/in^3, workpiece specific heat = 0.12 BTU/lb.°F. Assume that (a) the sources of heat are the shear plane and the tool–chip interface, (b) the thermal conductivity of the tool is zero and there is no heat loss to the environment, and (c) the temperature of the chip is uniform throughout. If the temperature rise in the chip is 668 °F, calculate the percentage of the energy dissipated in the shear plane that goes into the workpiece.

8.21 With a simple analytical expression prove the validity of the last sentence in Example 8.2.

8.22 The angle ψ between the shear plane and the direction of maximum grain elongation (see Fig. 8.5a) is given by the expression (see *J. App. Phys.*, vol. 18, 1947, p. 489)

$$\psi = 0.5 \cot^{-1}(\gamma/2),$$

where γ is the shear strain, as given by Eq. (8.3). Assume that you are given a piece of the chip obtained from orthogonal cutting of an annealed metal. The rake angle and cutting speed are also given, but you have not seen the setup on which the chip was produced. Outline the procedure that you would follow to estimate the power consumed in producing this chip. Further assume that you have access to a fully equipped laboratory and a technical library.

9

Material Removal Processes: Abrasive, Chemical, Electrical, and High-Energy Beams

9.1 Introduction

In all the cutting processes described in Chapter 8, the tool is made of a certain material and has a clearly defined geometry. Furthermore, the cutting process is carried out by chip removal, the mechanics of which are reasonably well understood. There are many situations in manufacturing, however, where the workpiece material is either too *hard* or too *brittle*, or its *shape* is difficult to produce with sufficient accuracy by any of the cutting methods described.

One of the best methods for producing such parts is to use *abrasives*. An abrasive is a small, hard particle having sharp edges and an irregular shape, unlike the cutting tools we described earlier. Abrasives are capable of removing small amounts of material from a surface by a cutting process that produces tiny chips.

Abrasive processes are generally among the last operations performed on manufactured products. These processes, however, are not necessarily confined to fine or small-scale material removal. They are also used for large-scale removal operations and can indeed compete economically with some machining processes, such as milling and turning.

Because they are hard, abrasives are also used in finishing very hard or heat-treated parts; shaping hard nonmetallic materials, such as ceramics and glasses; removing unwanted weld beads; cutting off lengths of bars, structural shapes, masonry, and concrete; and cleaning surfaces with jets of air or water containing abrasive particles.

There are situations, however, where none of these processes is satisfactory or economical for the following reasons.

a) The hardness and strength of the material is very high, typically above 400 HB.
b) The workpiece is too flexible or slender to support the cutting or grinding forces, or parts are difficult to clamp in workholding devices.
c) The shape of the part is complex, such as internal and external profiles, or small-diameter holes.
d) Surface finish and tolerances better than those obtainable by other processes are required.
e) Temperature rise or residual stresses in the workpiece are undesirable or unacceptable.

These requirements led to the development of chemical, electrical, and other means of material removal in the 1940s. These methods are called *nontraditional* or *unconventional machining*. When selected and applied properly these processes offer significant economic and technical advantages over the traditional machining methods described in the preceding chapter.

9.2 Abrasives

Abrasives commonly used in material removal processes in manufacturing operations are (a) aluminum oxide (Al_2O_3), (b) silicon carbide (SiC), (c) cubic boron nitride (CBN), and (d) diamond. These abrasives are considerably harder than cutting-tool materials (Table 9.1). Because of their extreme hardness, diamond and CBN are generally called *superabrasives*. As used in manufacturing processes, abrasives are generally very small in size as compared to cutting tools or inserts, and have sharp edges, allowing the removal of very small quantities of material from the workpiece surface. Consequently, very fine surface finish and dimensional accuracy can be obtained.

The size of an *abrasive grain* (or grit) is identified by a number. The smaller the grain size, the larger the number. For example, grain size 10 is regarded as very

TABLE 9.1
KNOOP HARDNESS FOR VARIOUS MATERIALS AND ABRASIVES

Common glass	300–500	Zirconium carbide	2100
Flint, quartz	800–1100	Titanium nitride	2000
Zirconium oxide	1000	Titanium carbide	1800–3200
Hardened steels	700–1300	Silicon carbide	2100–3000
Emery, garnet, topaz	1350	Boron carbide	2800
Tungsten carbide	1800–2400	Cubic boron nitride	4000–5000
Aluminum oxide	2000–3000	Diamond	7000–8000

coarse, 100 as fine, and 500 as very fine. Sandpaper or emery cloth, for example, are also identified in this manner, with the grain size printed on the back of the paper or cloth.

Abrasives found in nature are emery, corundum (alumina, aluminum oxide), quartz, garnet, and diamond. However, natural abrasives contain unknown amounts of impurities and possess nonuniform properties. Consequently, their performance is unreliable. As a result, the aluminum oxide and silicon carbide used are now almost totally synthetic.

9.3 Bonded Abrasives

Because each abrasive grain usually removes only a very small amount of material at a time, high rates of material removal can be obtained only if a large number of these grains act together. This is done by using *bonded abrasives*, typically in the form of a *grinding wheel*. A simple grinding wheel is shown schematically in Fig. 9.1. The abrasive grains are spaced at some distance from each other and are held

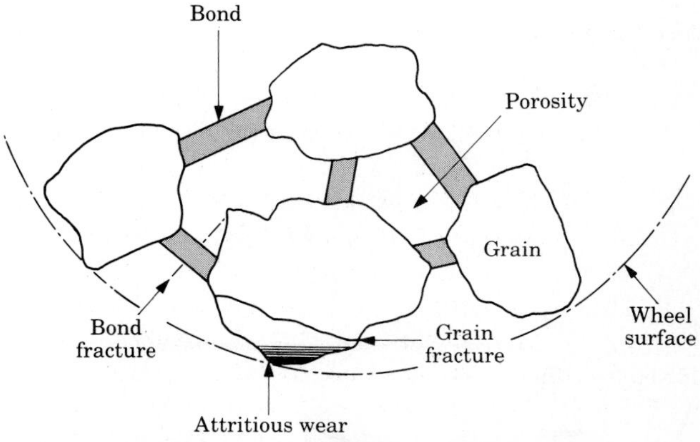

FIGURE 9.1
Schematic illustration of a physical model of a grinding wheel, showing its structure and wear and fracture patterns.

together by bonds, which act as supporting posts or braces between the grains. Some of the commonly used types of grinding wheel are shown in Fig. 9.2, with their grinding surfaces indicated by arrows. An estimated 250,000 different types and sizes of abrasive wheels are made today.

Bonded abrasives are marked with a standardized system of letters and numbers, indicating the type of abrasive, grain size, grade, structure, and bond type. Figure 9.3 shows the marking system for aluminum-oxide and silicon-carbide bonded abrasives. Figure 9.4 shows the marking system for diamond and cubic boron nitride bonded abrasives.

9.3.1 Bond types

Vitrified. Essentially a glass, *vitrified bond* is also called a ceramic bond, particularly outside the United States. It is the most common and widely used bond. The raw materials consist of feldspar (a crystalline mineral) and clays. They are mixed with the abrasives, moistened, and molded under pressure into the shape of grinding wheels. These "green" products, which are similar to powder-metallurgy parts, are then fired slowly, up to a temperature of about 1250 °C (2300 °F), to fuse the glass and develop structural strength. The wheels are then cooled slowly to avoid thermal cracking, finished to size, inspected for quality and dimensional accuracy, and tested for defects.

Vitrified bonds produce wheels that are strong, stiff, porous, and resistant to oil, acids, and water. Because they are brittle, they lack resistance to mechanical and

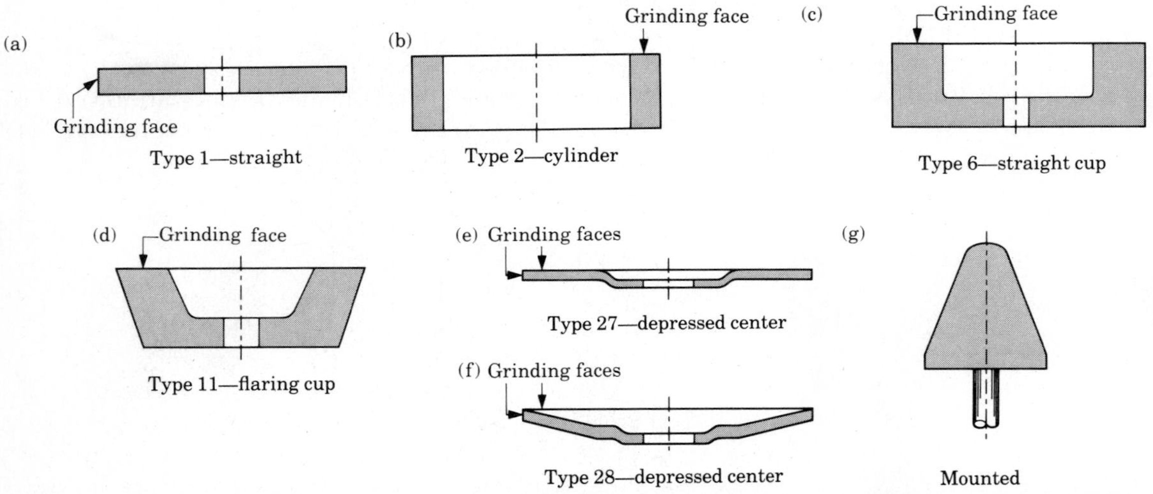

FIGURE 9.2 Some common types of grinding wheels. Note that each wheel has a specific grinding face. Grinding on other surfaces is improper and unsafe.

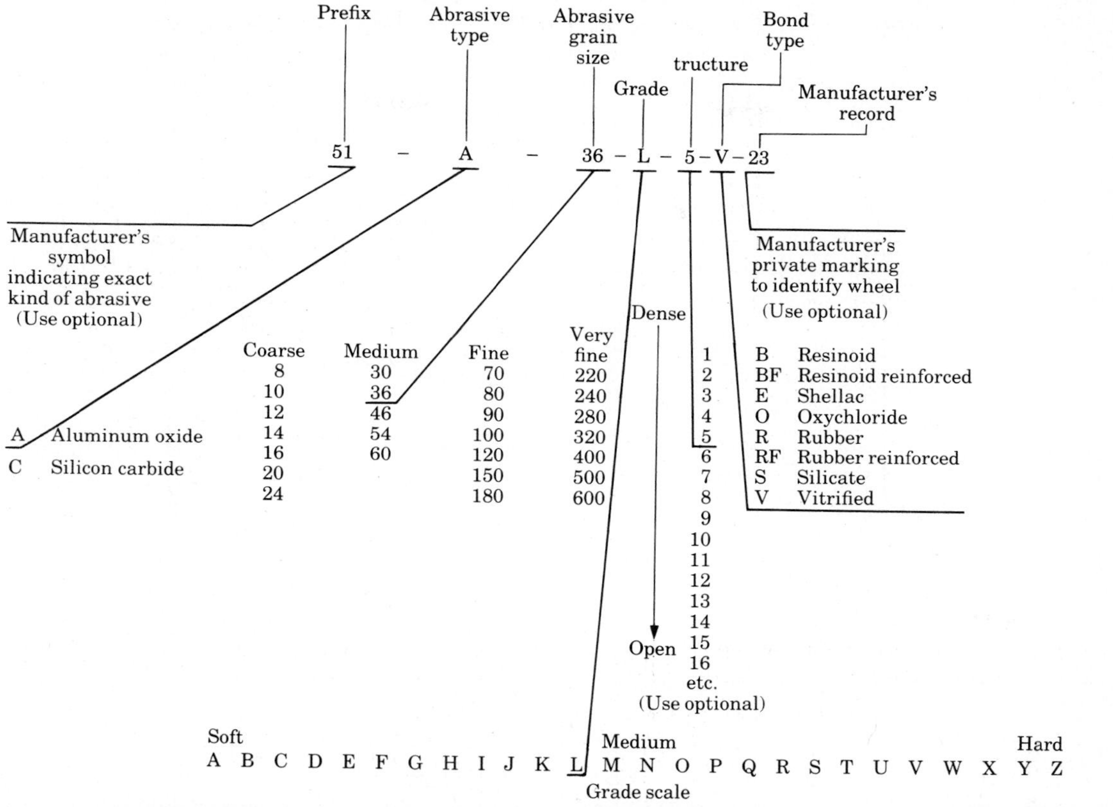

FIGURE 9.3
Standard marking system for aluminum-oxide and silicon-carbide bonded abrasives.

thermal shock. However, vitrified wheels are also available with steel backing plates or cups for better structural support during their use.

Resinoid. Resinoid bonding materials are thermosetting resins, and are available in a wide range of compositions and properties. Because the bond is an organic compound, wheels with resinoid bonds are also called *organic wheels*. The basic manufacturing technique consists of mixing the abrasive with liquid or powdered phenolic resins and additives, pressing the mixture into the shape of a grinding wheel, and curing it at temperatures of about 175 °C (350 °F).

Because the elastic modulus of thermosetting resins is lower than that of glasses, resinoid wheels are more flexible than vitrified wheels. *Reinforced wheels* are available. One or more layers of fiberglass mats of various mesh sizes provide the reinforcement. Its purpose is to retard the disintegration of the wheel should it break for some reason, rather than to improve its strength. Large-diameter resinoid

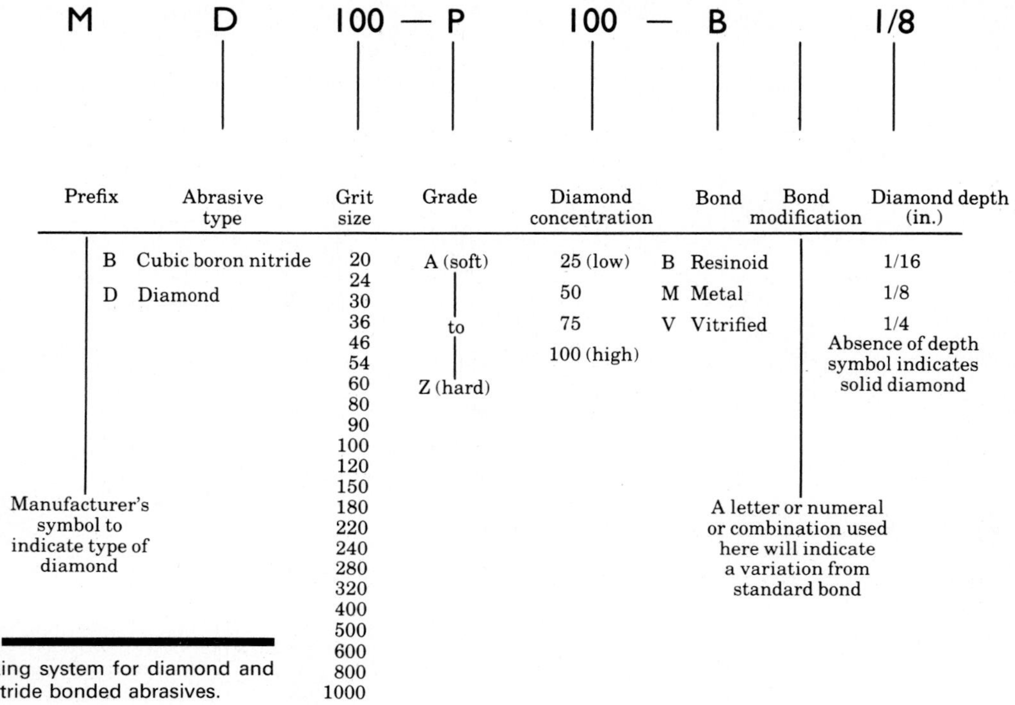

FIGURE 9.4 Standard marking system for diamond and cubic boron nitride bonded abrasives.

wheels can be supported additionally with one or more internal steel rings, which are inserted during molding of the wheel.

Rubber. The most flexible bond used in abrasive wheels is rubber. The manufacturing process consists of mixing crude rubber, sulfur, and the abrasive grains together, rolling the mixture into sheets, cutting out circles, and heating them under pressure to vulcanize the rubber. Thin wheels can be made in this manner and are used like saws for cutting-off operations.

Metal bonds. Using powder-metallurgy techniques, the abrasive grains, which are usually diamond or cubic boron nitride, are bonded to the periphery of a metal wheel, to depths of 6 mm (0.25 in.) or less. Bonding is carried out under high pressure and temperature, without the use of bonding materials. The wheel itself may be made of steel or bronze, although plastics can also be used.

Other bonds. In addition to those we described above, other bonds include silicate, shellac, and oxychloride bonds. However, they have limited uses, and we won't discuss them further here. A new development is the use of polyimide as a substitute for the phenolic in resinoid wheels. It is tough and has resistance to high temperatures.

9.3.2 Grade and structure

The *grade* of a bonded abrasive is a measure of the bond's strength. Thus it includes both the type and the amount of bond in the wheel. Because strength and hardness are directly related, the grade is also referred to as the *hardness* of a bonded abrasive. Thus a hard wheel has a stronger bond and/or a larger amount of bonding material between the grains than a soft wheel.

The *structure* is a measure of the *porosity* (spacing between the grains in Fig. 9.1) of the bonded abrasive. Some porosity is essential to provide clearance for the grinding chips; otherwise they would interfere with the grinding process. The structure of bonded abrasives ranges from dense to open.

9.4 Mechanics of Grinding

Grinding is basically a chip-removal process in which the cutting tool is an individual abrasive grain. The following are major differences between the action of a single grain and a single-point tool.

a) The individual grain has an irregular geometry and is spaced randomly along the periphery of the wheel (Fig. 9.5).
b) The radial positions of the grains vary.

FIGURE 9.5 The grinding surface of an abrasive wheel (A46-J8V) showing grains, porosity, wear flats on grains (see also Fig. 9.8), and metal chips from the workpiece adhering to the grains. Note the random distribution and shape of the abrasive grains. Magnification: 50 ×.

TABLE 9.2
TYPICAL RANGE OF SPEEDS AND FEEDS FOR ABRASIVE PROCESSES

PROCESS VARIABLE	CONVENTIONAL GRINDING	CREEP-FEED GRINDING	BUFFING	POLISHING
Wheel speed (m/s)	25–50	25–50	30–60	25–40
Work speed (m/s)	0.2–1	0.002–0.02	—	—
Feed (mm/pass)	0.01–0.05	1–6	—	—

c) The average rake angle of the grains is highly negative: $-60°$ or even lower. Consequently, the shear angles are very low.
d) The cutting speeds are very high: typically 30 m/s (6000 ft/min) (Table 9.2).

An example of chip formation by an abrasive grain is shown in Fig. 9.6. Note the negative rake angle, the low shear angle, and the small size of the chip. A variety of metal chips can be observed in grinding. (Chips are easily collected on a piece of adhesive tape held against the sparks of a grinding wheel.)

The mechanics of grinding and the variables involved can best be studied by analyzing the surface grinding operation (Fig. 9.7). A grinding wheel of diameter D is removing a layer of metal at a depth d, known as the *wheel depth of cut*. An individual grain on the periphery of the wheel is moving at a tangential velocity V (*up grinding*, as shown in Fig. 9.7), and the workpiece is moving at a velocity v. The grain is removing a chip whose undeformed thickness (*grain depth of cut*) is t and the undeformed length is ℓ.

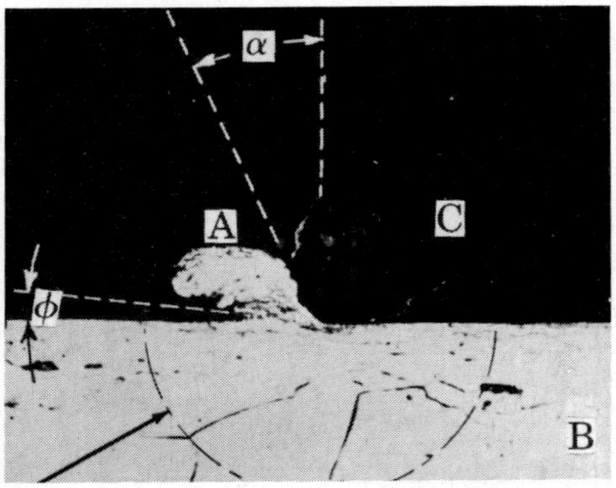

FIGURE 9.6
Grinding chip being produced by a single abrasive grain: (A) chip, (B) workpiece, (C) abrasive grain. Note the large negative rake angle of the grain. The inscribed circle is 0.065 mm (0.0025 in.) in diameter. *Source:* M. E. Merchant.

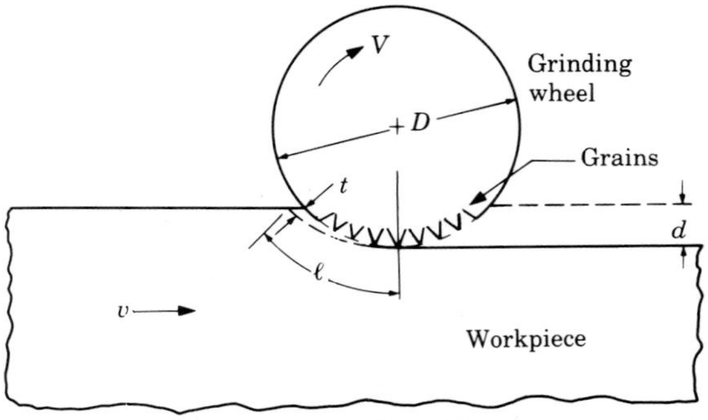

FIGURE 9.7
Variables in surface grinding. In actual grinding, the wheel depth of cut d and contact length l are much smaller than the wheel diameter D. The dimension t is called the grain depth of cut.

For $v \ll V$, the *undeformed chip length* ℓ is approximately

$$\ell \simeq \sqrt{Dd}, \tag{9.1}$$

where D is the diameter of the wheel and d is the depth of cut.

For external (cylindrical) grinding (see Section 9.6),

$$\ell = \sqrt{\frac{Dd}{1 + D/D_w}}, \tag{9.2}$$

and for internal grinding,

$$\ell = \sqrt{\frac{Dd}{1 - D/D_w}}, \tag{9.3}$$

where D_w is the diameter of the workpiece.

We can derive the relationship between t and other process variables as follows. Let C be the number of cutting points per unit area of wheel surface; v and V are the surface speeds of the workpiece and the wheel, respectively (Fig. 9.7). If we let the width of the workpiece be unity, the number of chips produced per unit time is VC, and the volume of material removed per unit time is vd.

If we let r be the ratio of the chip width w to the average chip thickness, the volume of a chip with a rectangular cross-sectional area and constant width is

$$\text{Vol}_{\text{chip}} = \frac{wt\ell}{2} = \frac{rt^2\ell}{4}. \tag{9.4}$$

The volume of material removed per unit time, then, is the product of the number of chips produced per unit time and the volume of each chip, or

$$VC\frac{rt^2\ell}{4} = vd,$$

and because $\ell = \sqrt{Dd}$,

$$t = \sqrt{\frac{4v}{VCr}\sqrt{\frac{d}{D}}}. \tag{9.5}$$

Experimental observations indicate the value of C to be roughly on the order of 0.1 to 10 per square mm (10^2 to 10^3 per square inch); the finer the grain size of the wheel, the larger this number. The magnitude of r is between 10 and 20 for most grinding operations. If we substitute typical values for a grinding operation into Eqs. (9.1)–(9.5), we obtain very small quantities for ℓ and t. For example, typical values for t are in the range of 0.3–0.4 μm (12–160 μin.).

• Example 9.1: Chip dimensions.

Estimate the undeformed chip length and undeformed chip thickness for a typical surface grinding operation.

SOLUTION. The formulas for undeformed length and thickness, respectively, are

$$\ell = \sqrt{Dd} \quad \text{and} \quad t = \sqrt{\frac{4v}{VCr}\sqrt{\frac{d}{D}}}.$$

From Table 9.2, we select the following values:

$$v = 0.5 \text{ m/s} \quad \text{and} \quad V = 30 \text{ m/s}.$$

We also assume that

$$d = 0.05 \text{ mm} \quad \text{and} \quad D = 200 \text{ mm},$$

and let

$$C = 2 \text{ per mm}^2 \quad \text{and} \quad r = 15.$$

Then

$$\ell = \sqrt{(200)(0.05)} = 3.2 \text{ mm} = 0.126 \text{ in.}$$

and

$$t = \sqrt{\frac{(4)(0.5)}{(30)(2)(15)}\sqrt{\frac{0.05}{200}}} = 0.006 \text{ mm} = 2.3 \times 10^{-4} \text{ in.}$$

Because of plastic deformation, the actual length of the chip will be shorter and the thickness greater than these values (see Fig. 9.6).

If we assume that the *force* on the grain (see cutting force F_c in Section 8.2.3) is proportional to the cross-sectional area of the undeformed chip, we can show that the relative grain force is

$$\text{Relative grain force} \propto \frac{v}{VC}\sqrt{\frac{d}{D}}. \tag{9.6}$$

The actual force is the product of the relative grain force and the strength of the metal being ground.

The *specific energy* consumed in producing a grinding chip consists of three components:

$$u = u_{\text{chip}} + u_{\text{ploughing}} + u_{\text{sliding}}, \tag{9.7}$$

where u_{chip} is the specific energy required for chip formation by plastic deformation and $u_{\text{ploughing}}$ is the specific energy required for ploughing, which is plastic deformation without chip removal (Fig. 9.8). The last term, u_{sliding}, can best be understood by observing the grain in Fig. 9.9. The grain develops a *wear flat* as a result of the grinding operation (similar to flank wear in cutting tools). The wear flat slides along the surface being ground and, because of friction, requires energy for sliding. The larger the wear flat, the higher the grinding force is.

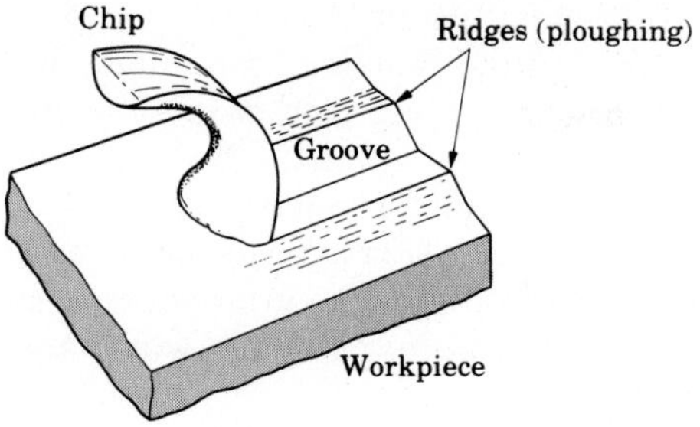

FIGURE 9.8
Chip formation and ploughing of the workpiece surface by an abrasive grain.

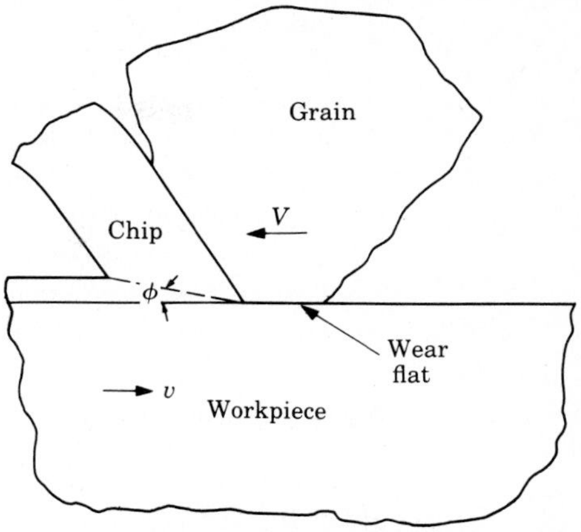

FIGURE 9.9
Schematic illustration of chip formation by an abrasive grain. Note the negative rake angle, the small shear angle, and the wear flat on the grain.

Typical specific energy requirements in grinding are given in Table 9.3. Note that these energy levels are much higher than those in cutting operations with single-point tools (see Table 8.4). This difference has been attributed to the following factors.

a) *Size effect*: As previously stated, the size of grinding chips is quite small compared to chips from other cutting operations, or by about two orders of magnitude. As described in Section 2.10.3, the smaller the size of a piece of metal, the greater is its strength. Thus grinding involves higher specific energy than cutting. Recent studies have indicated that extremely high dislocation densities occur in the shear zone during chip formation, thus influencing the grinding energies involved.

TABLE 9.3
APPROXIMATE UNIT POWER REQUIREMENTS FOR SURFACE GRINDING

WORKPIECE MATERIAL	HARDNESS	UNIT POWER* (W·s/mm³)
Aluminum	150 HB	6.8–27
Cast iron (class 40)	215 HB	12–60
Low-carbon steel (1020)	110 HB	13.7–68
Titanium alloy	300 HB	16.4–55
Tool steel (T15)	67 HRC	17.7–82

* Divide by 2.73 to obtain hp·min/in³.

b) *Wear flat*: Because a wear flat requires frictional energy for sliding, this energy can contribute substantially to the total energy consumed. The size of the wear flat in grinding is much larger than the grinding chip, unlike in metal cutting by a single-point tool where flank wear land is small compared to the size of the chip.

c) *Chip morphology*: Because the average rake angle of a grain is highly negative (Fig. 9.6), the shear strains are very large. This result indicates that the energy required for plastic deformation to produce a grinding chip is higher than in other cutting processes. Furthermore, ploughing consumes energy without contributing to chip formation.

9.4.1 Temperature

Temperature rise in grinding is an important consideration because it can adversely affect the surface properties and cause residual stresses on the workpiece. Furthermore, temperature gradients in the workpiece cause distortions by differential thermal expansion and contraction. When a portion of the heat generated is conducted into the workpiece, it expands the part being ground, thus making it difficult to control dimensional accuracy.

The work expended in grinding is mainly converted into heat. The surface temperature rise ΔT has been found to be a function of the ratio of the total energy input to the surface area ground. Thus in surface grinding if w is the width and L is the length of the surface area ground,

$$\Delta T \propto \frac{uwLd}{wL} \propto ud. \tag{9.8}$$

If we introduce size effect and assume that u varies inversely with the undeformed chip thickness t, then the temperature rise is

$$\Delta T \propto \frac{d}{t} \propto d^{3/4} \sqrt{\frac{VC}{v} \sqrt{D}}. \tag{9.9}$$

The peak temperatures in chip generation during grinding may be as high as 1650 °C (3000 °F). However, the time involved in producing a chip is extremely short—on the order of microseconds—hence melting may or may not occur. Because the chips carry away much of the heat generated (as in metal cutting), only a fraction of the heat generated is conducted to the workpiece. Experiments indicate that in grinding, as much as one half the energy is conducted to the workpiece. (This is higher than in metal cutting.) The heat generated by sliding and ploughing is conducted mostly into the workpiece.

Sparks. The sparks observed in metal grinding are actually glowing chips. They glow because of the exothermic reaction of the hot chips with oxygen in the atmosphere. Sparks have not been observed with any metal ground in an oxygen-free environment. The color, intensity, and shape of the sparks depend on the composition of the metal being ground.

If the heat generated by exothermic reaction is sufficiently high, the chip may melt and, because of surface tension, acquire a round shape and solidify as a shiny spherical particle. Observation of these particles under scanning electron microscopy has revealed that they are hollow and have a fine dendritic structure, indicating that they were once molten (by exothermic oxidation of hot chips in air) and resolidified rapidly. (It has been suggested that some of the spherical particulars may also be produced by plastic deformation and rolling of chips at the grit–workpiece interface.)

9.4.2 Effects of temperature

Temperature rise in grinding can significantly affect surface properties and residual stresses. Temperature gradients distort the part because of differential thermal expansion and contraction. Furthermore, if the heat generated is allowed to be conducted into the workpiece, it expands the part being ground and hence controlling tolerances is difficult.

Tempering. Excessive temperature rise caused by grinding can temper and soften the surfaces of steel components, which are often ground in the hardened state. Grinding process parameters must therefore be chosen carefully to avoid excessive temperature rise. The use of grinding fluids can effectively control temperatures.

Burning. If the temperature is excessive the surface may burn. Burning produces a bluish color on steels, which indicates oxidation at high temperatures. A burn may not be objectionable in itself. However, the surface layers may undergo metallurgical transformations, with martensite formation in high-carbon steels from reaustenization followed by rapid cooling. This is known as *metallurgical burn*, which also is a serious problem with nickel-base alloys.

High temperatures in grinding may also lead to thermal cracking of the surface of the workpiece, known as *heat checking*. Cracks are usually perpendicular to the grinding direction; however, under severe grinding conditions, parallel cracks may also develop.

Residual stresses. Temperature change and gradients within the workpiece are mainly responsible for residual stresses in grinding. Other contributing factors are the physical interactions of the abrasive grain in chip formation and the sliding of the wear flat along the workpiece surface, causing plastic deformation of the

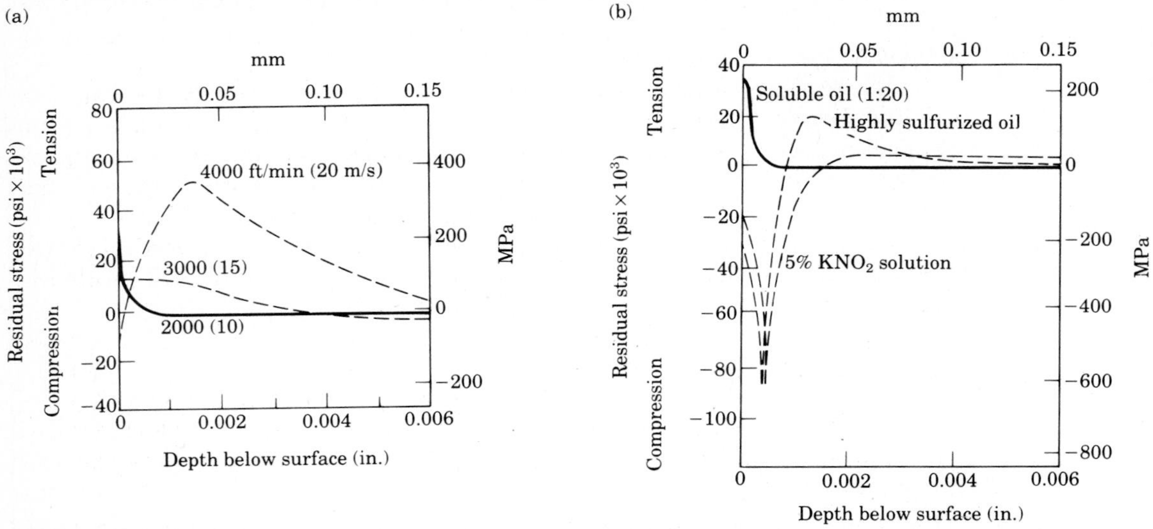

FIGURE 9.10
Residual stresses developed on the workpiece surface in grinding tungsten: (a) effect of wheel speed and (b) effect of grinding fluid. Tensile residual stresses on a surface are detrimental to the fatigue life of ground components. The variables in grinding can be controlled to minimize residual stresses. This is known as low-stress grinding. *Source:* After N. Zlatin, et al., 1963.

surface. Two examples of residual stresses in grinding are shown in Fig. 9.10, demonstrating the effects of wheel speed and the type of grinding fluid used. The method and direction of the application of grinding fluid can also have a significant effect on residual stresses.

Because of the deleterious effect of tensile residual stresses on fatigue strength, process parameters should be chosen carefully. Residual stresses can usually be lowered (*low-stress* or *gentle* grinding) by using softer grade wheels (free cutting wheels), lower wheel speeds, and higher work speeds.

9.5 Wear of Bonded Abrasives

The overall wear of a bonded abrasive is caused by three distinct mechanisms: attritious wear, fracture of the grain, and fracture of the bond. The sharp cutting edges of an abrasive grain become dull by *attrition*, developing the wear flat shown in Fig. 9.9. This type of wear is caused by the interaction of the grain with the workpiece material. Complex physical and chemical reactions take place between the two materials. These reactions include oxidation of the grain surface resulting from high temperatures, diffusion, chemical degradation or decomposition, fracture at a microscopic scale, melting, and plastic flow.

Attritious wear is low when the two materials are chemically inert with respect to each other, thus lowering the tendency for reaction and adhesion. For example, the rate of attritious wear of aluminum oxide on steel is much lower than that for silicon carbide. The reason is that aluminum oxide is relatively inert with respect to iron. On the other hand, silicon carbide can be dissolved in molten iron. Thus the selection of the type of abrasive for low attritious wear is based on the reactivity of the grain and the workpiece and mechanical properties, such as the relative hardness and toughness. The environment and the type of grinding fluid used also have an effect on grain–workpiece interactions.

Abrasive grains are brittle and, hence, their *fracture* characteristics in grinding are important. If the wear flat from attritious wear is excessive, the grain becomes dull and cutting is inefficient. Preferably the grain should fracture or fragment at a moderate rate so that new sharp cutting edges are produced. (This action is equivalent to breaking a dull piece of chalk in order to be able to draw fine lines on the board.)

Friability describes the fracture behavior of abrasives. High friability indicates low strength or low fracture resistance of the grain. Thus a highly friable grain fragments more easily under the dynamic forces in a grinding operation than one with low friability. The shape and size of the grain also affect friability. Block-shaped particles are less friable than those that are platelike, and smaller particles are stronger and hence less friable than larger ones.

Aluminum oxide has lower friability than silicon carbide. This characteristic indicates that aluminum oxide has less tendency to fragment (or fail by microchipping) during grinding. Selection of a material for a particular application should also include consideration of the attritious wear rate. A grain–workpiece combination with high attritious wear and low friability indicates a dull grain with a large wear flat; cutting is inefficient and surface damage is likely to occur. The following combinations are generally recommended.

a) Aluminum oxide: with steels, ferrous alloys, and alloy steels.
b) Silicon carbide: with cast iron, nonferrous metals, and hard and brittle materials, such as carbides, ceramics, marble, and glass.
c) Diamond: with cemented carbides and some hardened steels.
d) Cubic boron nitride: with steels and cast irons at 50 HRC or above, and high-temperature superalloys.

The strength of the bond (*grade*) is also a significant parameter in grinding. If the bond is too strong, dull grains cannot be dislodged and cutting becomes inefficient. However, if the bond is too weak, the wear rate of the wheel is too high, tolerances may not be held, and the operation becomes uneconomical. In general, softer bonds are recommended for materials that are difficult to grind and for reducing residual stresses and thermal damage. Hard-grade wheels are used for softer metals and to remove large amounts of material at high rates.

Dressing is the process of conditioning worn grains on the surface of a bonded abrasive in order to produce sharp new grains. Wheels are dressed when they become dull from excessive attritious wear (*glazing*, or the shiny appearance of the wheel surface), or when the wheel becomes *loaded*. Loading is a condition in which the porosities on the surface of the wheel become filled or clogged with chips or other materials during grinding. A loaded wheel cuts very inefficiently, generates frictional heat, and causes surface damage.

Dressing is done by various methods. In one method, a specially shaped diamond, or diamond cluster, is moved across the width of the grinding face of a rotating wheel, removing a layer of abrasives from the wheel surface. In another very coarse method, a set of star-shaped steel disks is pressed against the wheel and removes material from the surface by crushing the grains. Dressing can also be done with abrasive sticks or with other abrasive wheels. Dressing techniques and the rate at which the surface of the wheel is dressed are significant factors in grinding forces and surface finish. A finely dressed wheel produces a fine surface finish. Dressing is also done to generate a certain shape or form on a grinding wheel for the purpose of grinding profiles on workpieces. One example is thread grinding. Dressing is also a means of *truing* the wheel (making the wheel into a true circle).

9.5.1 Grinding ratio

The *grinding ratio* G is defined as

$$G = \frac{\text{Volume of metal removed}}{\text{Volume of wheel wear}}. \tag{9.10}$$

In a particular grinding operation this ratio depends on many factors, including the type of wheel, the workpiece material, the grinding fluid, and process parameters such as depth of cut and speeds. In practice, G ratios vary over a wide range, or from about 2 to 200 and higher.

A particular grinding wheel may *act soft* or *hard*, regardless of its grade, just as a pencil acts soft when we use it to write on rough paper (wear rate is high) and acts hard on smooth paper. In grinding, this behavior is the result of the force acting on the grain. Thus the greater the force the greater is the tendency for grain fragmentation or bond fracture to occur, hence the softer the wheel acts.

From Eq. (9.6) it is evident that this force increases with work speed v and wheel depth of cut d and decreases with increasing wheel speed V, number of cutting points C (more dense structure), and wheel diameter D. Thus a wheel in surface grinding acts soft when v and d increase or when V, C, and D decrease. This characteristic, in turn, affects the grinding ratio G. Trying to obtain a high G ratio with a longer wheel life is not always desirable, as this result could indicate dulling of the grains and possible surface damage. A lower G value may be quite acceptable when an overall economic analysis justifies it.

Example 9.2: Wheel behavior.

A surface-grinding operation is being carried out with the wheel running at a constant speed. Does the wheel act soft or hard as the wheel wears down over a period of time?

SOLUTION. Referring to Eq. (9.6),

$$\text{Relative grain force} \propto \frac{v}{VC}\sqrt{\frac{d}{D}},$$

we note that the only parameter that changes over time in this operation is the wheel diameter D (assuming that d remains constant and the wheel is dressed periodically). As D becomes smaller, the relative grain force increases, and the wheel acts softer. Some grinding machines are equipped with variable-speed spindle motors to accommodate these changes and also wheels of different diameter.

9.6 Grinding Operations and Machines

Grinding operations, which account for about 20 percent of all machining in the United States, are carried out with a variety of wheel–workpiece configurations. The selection of a grinding process for a particular application depends on part shape, part size, ease of fixturing, and the production rate required.

The basic types of grinding operations are surface, cylindrical, internal, and centerless grinding (Fig. 9.11). The relative movement of the wheel may be along the surface of the workpiece (*traverse* grinding, *through feed* grinding, *cross-feeding*), or it may be radially into the workpiece (*plunge* grinding). Surface grinders comprise the largest percentage of grinders in use in industry, followed by bench grinders (usually with two wheels), cylindrical grinders, and tool and cutter grinders. The least used are internal grinders. Grinding machines are available for various workpiece geometries and sizes. Special machines have been built with features for automatic workpiece loading, clamping, cycling, and gaging and wheel dressing. The range of surface roughness and tolerances obtained in grinding and other processes described in this chapter are shown in Fig. 9.12.

9.6.1 Surface grinding

Surface grinding is performed with a horizontal-spindle (Fig. 9.13) or a vertical-spindle wheel on which a number of parts can be ground at the same time. The workpiece is secured on a magnetic chuck, which is attached to the work table. A

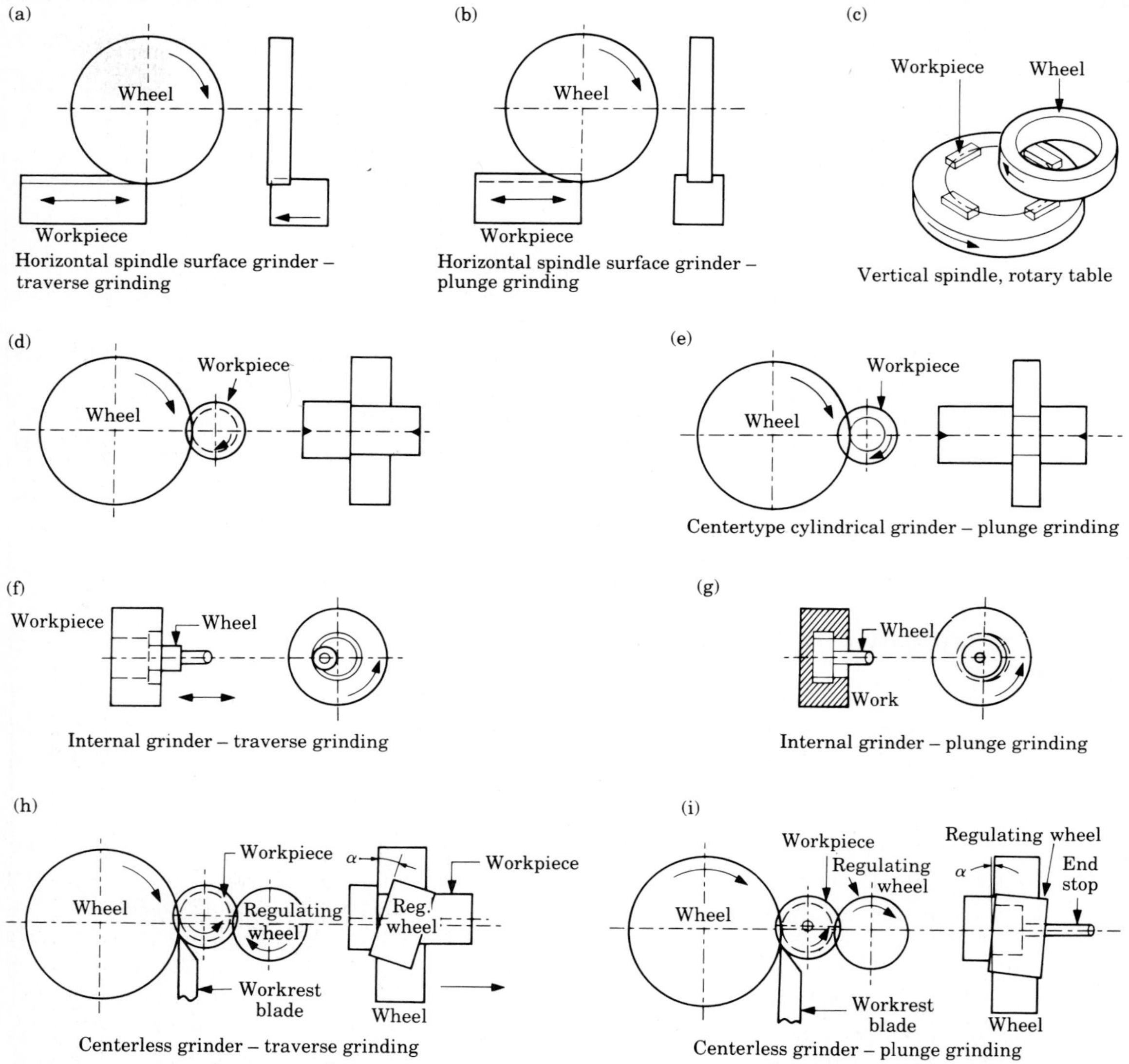

FIGURE 9.11 Schematic illustration of various grinding operations. Grinding is a versatile and important finishing operation, although it can also be used for large-scale removal operations. *Source:* Adapted from *Machinability Data Handbook*, 3d ed., 1980, Metcut Research Associates Inc.

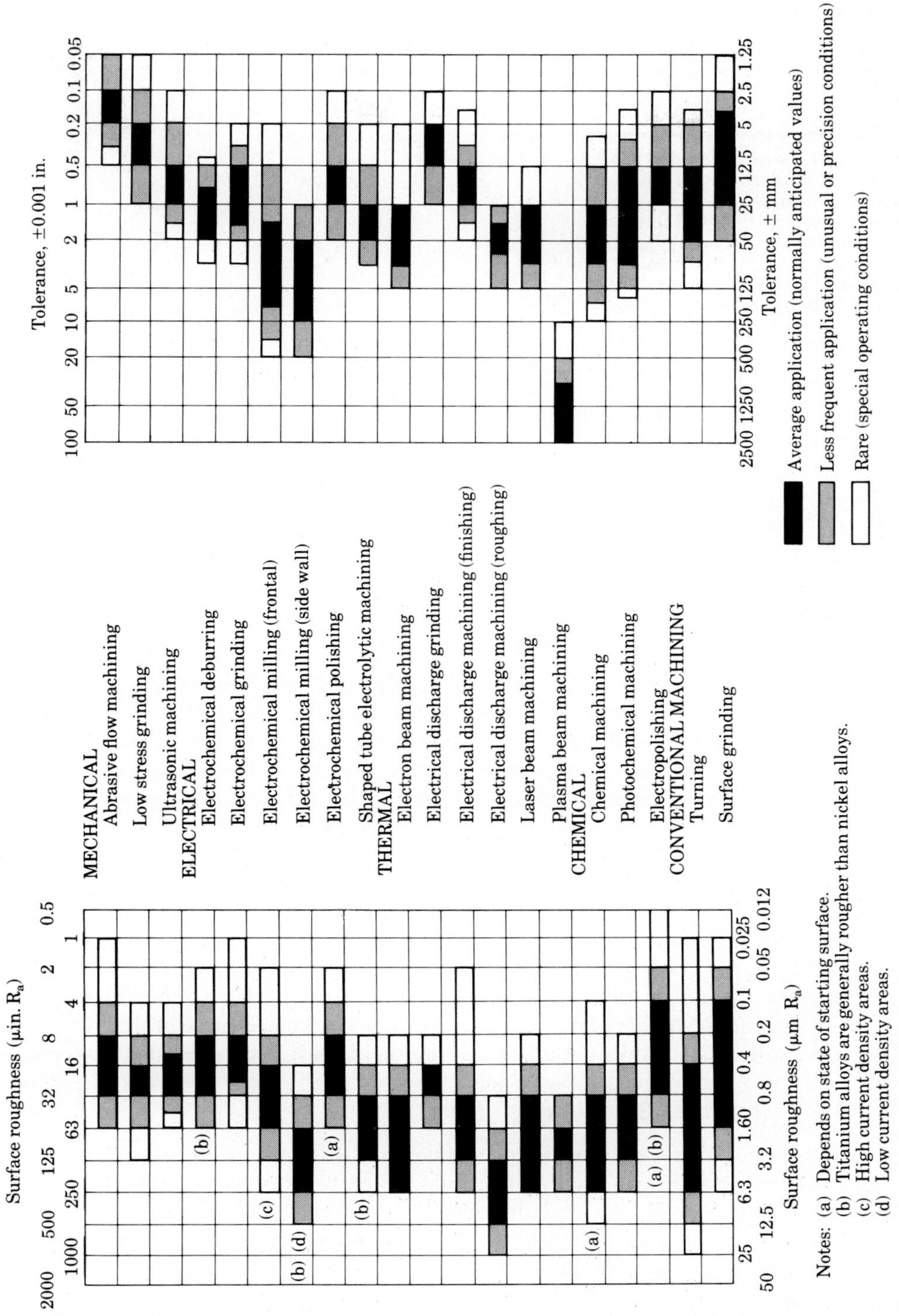

FIGURE 9.12

Surface roughness and tolerances obtained in various material removal processes. Note the wide range of roughness within each machining process. See also Fig. 4.22. *Source*: Compiled from data in *Machinability Data Handbook*, 3d ed., by permission of the Machinability Data Center. © 1980 by Metcut Research Associates Inc.

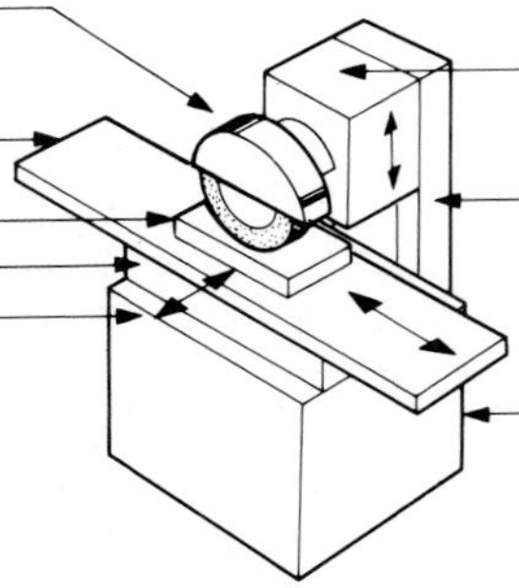

FIGURE 9.13 Schematic illustration of a surface grinder. Workpieces are usually held in place with a magnetic chuck. Note the wheel guard to protect the operator in case of wheel breakage. These machines may also be equipped with fixtures to apply a grinding fluid during the operation. The fluid should be turned on after the wheel is turned on, and it should be turned off before the wheel is stopped.

straight wheel (Type 1, Fig. 9.2) is mounted on a horizontal spindle. Grinding is done with the table reciprocating in the longitudinal direction and feeding laterally after each stroke (cross feed).

The size of a surface grinder is identified by the dimensions of the surface that can be ground on that machine. In addition to the design shown in Fig. 9.13, other designs include grinders with vertical spindles and rotary tables for grinding a number of pieces in one operation (see Fig. 9.11c).

9.6.2 Cylindrical grinding

In this operation (see Fig. 9.11d and e), the external cylindrical surface and the shoulders of a workpiece are ground. Examples include crankshafts, axles, spindles, and rolls for rolling mills. Threads and workpieces with two or more diameters (*plunge grinding*) are also ground on these machines. In cylindrical grinding the workpiece reciprocates along its axis, although for large and long workpieces the grinding wheel reciprocates. The latter type is called a *roll grinder*.

Cylindrical grinders are identified by the maximum diameter and length of the workpiece that can be ground. In universal grinders, both the workpiece and the wheel axis can be swiveled around a horizontal plane, permitting the grinding of tapered shafts and similar parts. *Thread grinding* is done on cylindrical grinders, as well as centerless grinders, with specially dressed wheels matching the shape of the threads. Threads produced by grinding are the most accurate of any manufacturing process and have a very fine surface finish. The workpiece and wheel movements are synchronized to produce the pitch of the thread, usually in about six passes.

9.6.3 Internal Grinding

In this operation a small wheel grinds the inside diameter of the part (see Fig. 9.11f and g), such as bearing races or bushings. The workpiece is held inside a rotating chuck in the headstock and the wheel rotates at 30,000 rpm or higher. Internal grinders also have features whereby the headstock can be swiveled on a horizontal plane to grind tapered holes.

9.6.4 Centerless grinding

Centerless grinding is a process for continuously grinding cylindrical surfaces in which the workpiece is supported not by centers (hence the term centerless) or chucks but by a blade (see Fig. 9.11h and i). Typical parts made by centerless grinding are roller bearings, piston pins, shafts, and similar components. This continuous production process requires little operator skill. Parts with diameters as small as 0.1 mm (0.004 in.) can be ground.

In *through-feed grinding* the workpiece is supported on a workrest blade and is ground between two wheels. Grinding is done by the larger wheel, while the smaller wheel regulates the axial movement of the workpiece. The *regulating wheel*, which is rubber bonded, is tilted and runs at speeds of only about 1/20 those of the grinding wheel.

Parts with variable diameters, such as bolts, valve tappets, and distributor shafts, can be ground by centerless grinding. Called *infeed*, or plunge, grinding, the process is similar to plunge or form grinding with cylindrical grinders. Tapered pieces are centerless ground by *end-feed* grinding. High production rate thread grinding can be done with centerless grinders using specially dressed wheels. In *internal centerless* grinding, the workpiece is supported between three rolls and is internally ground. Typical applications are sleeve-shaped parts and rings.

9.6.5 Creep-feed grinding

Grinding has traditionally been associated with small rates of material removal (Table 9.2) and fine finishing operations. However, grinding can also be used for large-scale metal removal operations similar to milling, shaping, and planing. In *creep-feed* grinding, developed in the late 1950s, the wheel depth of cut d is as much as 6 mm (0.25 in.), and the workpiece speed is low (Fig. 9.14). The wheels are mostly softer grade resin bonded with open structure to keep temperatures low and

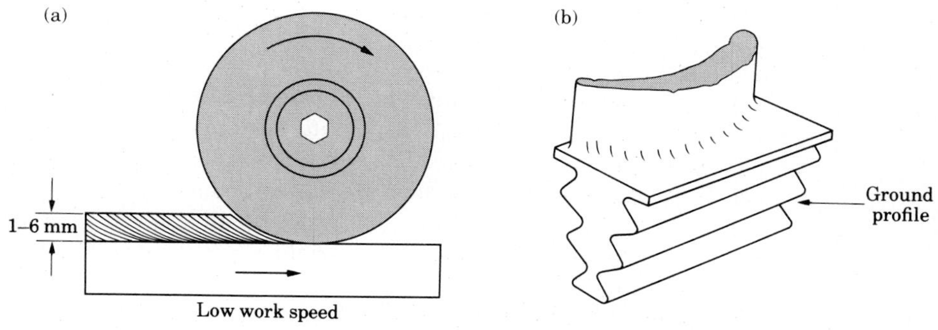

FIGURE 9.14
(a) Schematic illustration of the creep-feed grinding process. Note the large wheel depth of cut. (b) The fir-tree root of this turbine blade can be produced by creep-feed grinding with shaped wheels. This operation can also be performed by some of the processes described in this chapter.

improve surface finish. The wheels are rarely dressed during operation because surface finish is of secondary importance. The machines used for creep-feed grinding have special features, such as high power—up to 225 kW (300 hp)—high stiffness and damping capacity, variable and well-controlled spindle and work-table speeds, and ample capacity for grinding fluids.

Its overall economics and competitive position with other material-removal processes indicate that creep-feed grinding can be economical for specific applications, such as in grinding cavities, key seats, twist-drill flutes, and the roots of turbine blades (Fig. 9.14b). The wheel is dressed to the shape of the workpiece to be produced. Consequently, the workpiece does not have to be previously milled, shaped, or broached.

9.6.6 Other grinding operations

A variety of special-purpose grinders are available. *Bench grinders* are used for routine offhand grinding of tools and small parts. They are usually equipped with two wheels mounted on the two ends of the shaft of an electric motor. *Pedestal grinders* are placed on the floor and used similarly to bench grinders.

Universal tool and cutter grinders are used for grinding single-point or multipoint tools and cutters. They are equipped with special workholding devices for accurate positioning of the tools to be ground. *Tool-post grinders* are self-contained units and are usually attached to the tool post of a lathe. The workpiece is mounted on the headstock and is ground by moving the tool post. These grinders are versatile, but the lathe should be protected from abrasive debris.

Swing-frame grinders are used in foundries for grinding large castings. Rough grinding of castings is called *snagging*, and is usually done on floorstand grinders using wheels as large as 0.9 m (36 in.) in diameter. *Portable grinders*, either air or electrically driven, or with a flexible shaft connected to an electric motor or gasoline engine are available for operations such as grinding off weld beads and cutting-off operations, usually on large workpieces or structures.

9.6.7 Grinding chatter

Chatter is particularly bothersome in grinding because it adversely affects surface finish and wheel performance. Vibrations during grinding may be caused by bearings, spindles, and unbalanced wheels, as well as external sources, such as from nearby machinery. The grinding process can itself cause regenerative chatter. The analysis of chatter in grinding is similar to that for machining operations (see Section 8.11). Thus the important variables are stiffness of the machine tool and damping. Additional factors that are unique to grinding chatter are nonuniformities in the grinding wheel, dressing techniques used, and uneven wheel wear.

Because these variables produce characteristic chatter marks on ground surfaces, careful study of these marks can often lead to the source of the problem. General guidelines have been established to reduce the tendency for chatter in grinding, such as using soft-grade wheels, dressing the wheel frequently, changing dressing techniques, reducing the material-removal rate, and supporting the workpiece rigidly.

9.7 Ultrasonic Machining

In ultrasonic machining, material is removed from a surface by microchipping or erosion with abrasive particles. The tip of the tool (Fig. 9.15a) vibrates at low amplitude [0.05 to 0.125 mm (0.002 to 0.005 in.)] and at high frequency (20 kHz). This vibration, in turn, transmits a high velocity to fine abrasive grains between the tool and the surface of the workpiece.

The grains are usually boron carbide, but aluminum oxide and silicon carbide are also used. Grain size ranges from 100 (for roughing) to 1000 (for finishing). The grains are in a water slurry with concentrations ranging from 20 percent to 60 percent by volume. The slurry also carries away the debris from the cutting area.

Ultrasonic machining is best suited for hard, brittle materials, such as ceramics, carbides, glass, precious stones, and hardened steels. The tip of the tool is usually made of low-carbon steel and undergoes wear. It is attached to a transducer through the toolholder. With fine abrasives, tolerances of 0.0125 mm (0.0005 in.) or better can be held in this process. Figures 9.15(b) and (c) show two applications of ultrasonic machining.

Microchipping in ultrasonic machining is possible because of the high stresses produced by particles striking a surface. The time of contact between the particle and the surface is very short (10 to 100 μs) and the area of contact is very small.

The time of contact, t_o, can be expressed as

$$t_o \simeq \frac{5r}{c_o}\left(\frac{c_o}{v}\right)^{1/5}, \tag{9.11}$$

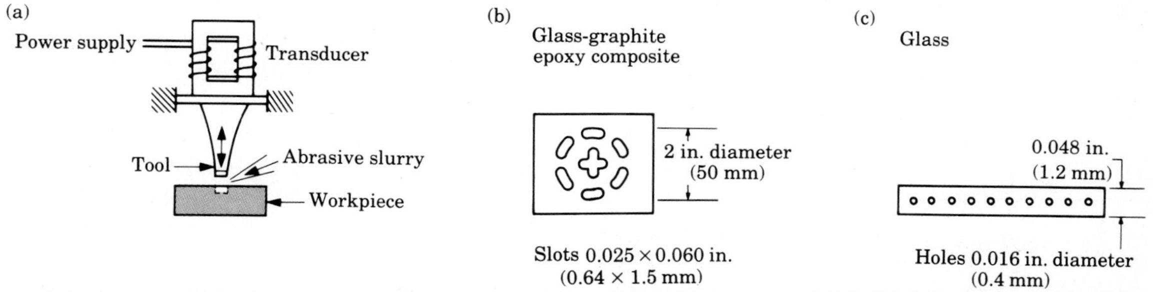

FIGURE 9.15
(a) Schematic illustration of the ultrasonic machining process by which material is removed by microchipping and erosion. (b) and (c) Typical examples of holes produced by ultrasonic machining. Note the dimensions of cut and the type of workpiece materials.

where r is the radius of a spherical particle, c_o is the elastic wave velocity in the workpiece ($c_o = \sqrt{E/\rho}$), and v is the velocity with which the particle strikes the surface. The force F of the particle on the surface is obtained from the rate of change of momentum. That is,

$$F = \frac{d(mv)}{dt}, \tag{9.12}$$

where m is the mass of the particle.

The *average force* $\bar{F}$ of a particle striking the surface and rebounding is

$$\bar{F} = \frac{2mv}{t_o}. \tag{9.13}$$

Substitution of numerical values into Eq. (9.13) indicates that even small particles can exert significant forces and, because of the very small contact area, produce very high stresses. In brittle materials, these stresses are sufficiently high to cause microchipping and surface erosion.

• Example 9.3: Effect of temperature on impact force.

Explain what change, if any, takes place in the magnitude of the impact force of a particle in ultrasonic machining as the temperature of the workpiece is increased.

SOLUTION. The force of a particle is given by Eq. (9.13). For this problem, m and v are constant. For the contact time t_o in Eq. (9.11), we can now write

$$t_o \propto \frac{1}{c_o^{4/5}} \propto \frac{1}{E^{2/5}}.$$

With increasing temperature, the modulus of elasticity E decreases, and thus t_o increases. Therefore the impact force decreases according to Eq. (9.13). •

In *abrasive-jet machining*, a jet of air or carbon dioxide containing abrasive particles is aimed at the workpiece surface under controlled conditions. The impact of the particles is capable of cutting holes or slots in very hard metallic and nonmetallic materials. Because the flow of free abrasives tends to round off corners, designs should avoid sharp corners.

9.8 Finishing Operations

In addition to those described thus far, several processes are generally used on workpieces as the final finishing operation. These processes mainly utilize abrasive grains. Commonly used finishing operations are described in this section in the order of improved surface finish produced. Finishing operations can contribute significantly to production time and product cost. Thus they should be specified with due consideration to their costs and benefits.

Typical examples of *coated abrasives* are sandpaper and emery cloth. The grains used in coated abrasives are more pointed than those used for grinding wheels. The grains are electrostatically deposited on flexible backing materials, such as paper or cloth (Fig. 9.16), with their long axes perpendicular to the plane of the backing. The matrix (coating) is made of resins.

Coated abrasives are available as sheets and belts and usually have a much more open structure than the abrasives on grinding wheels. Coated abrasives are used extensively in finishing flat or curved surfaces on metallic and nonmetallic parts, of metallographic specimens, and in woodworking. The precision of surface finish obtained depends primarily on grain size.

Coated abrasives are also used as belts for high-rate material removal. *Abrasive belts* have become an important production process, in some cases replacing conventional grinding operations. Belt speeds range between 12 m/s and 30 m/s (2500 ft/min and 6000 ft/min). Machines for abrasive-belt operations require proper belt support and rigid construction to minimize vibrations.

Honing is an operation used primarily to give holes a fine surface finish. The honing tool consists of a set of aluminum-oxide or silicon-carbide sticks. They are mounted on a mandrel that rotates in the hole, applying a radial force with a reciprocating axial motion, thus producing a cross-hatched pattern. The sticks can be adjusted radially for different hole sizes. The fineness of surface finish can be controlled by the type and size of abrasive used, the speed of rotation, and the pressure applied. A fluid is used to remove chips and to keep temperatures low. Honing is also done on external cylindrical and flat surfaces. *Superfinishing* is similar to honing, but the pressure applied is very light and the motion of the hone has a short stroke. The process is controlled so that the grains do not travel along the same path along the surface of the workpiece.

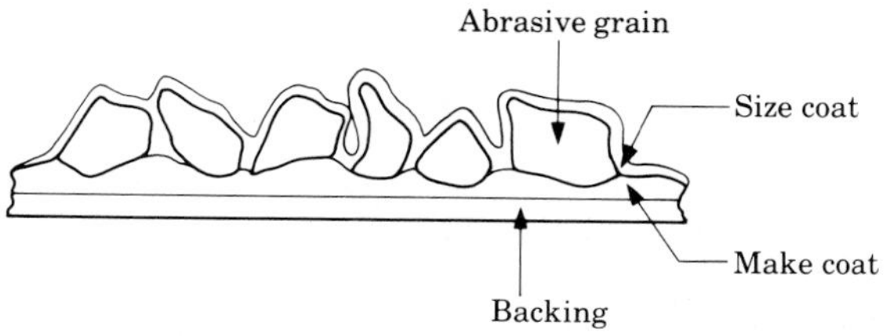

FIGURE 9.16

Schematic illustration of the structure of a coated abrasive. Sandpaper, developed in the sixteenth century, and emery cloth are common examples of coated abrasives.

Lapping is a finishing operation used on flat or cylindrical surfaces. The lap (Fig. 9.17) is usually made of cast iron, copper, leather, or cloth. The abrasive particles are embedded in the lap, or they may be carried through a slurry. Tolerances on the order of ± 0.0004 mm (0.000015 in.) can be obtained with the use of fine abrasives—up to size 900. Surface finish can be as smooth as 0.025–0.1 μm (1–4 μin.).

Production lapping on flat or cylindrical pieces is done on machines such as that shown in Fig. 9.17(b). Lapping is also done on curved surfaces, such as spherical objects and glass lenses, using specially shaped laps. Running-in of mating gears can be done by lapping. Depending on the hardness of the workpiece, lapping pressures range from 7 to 140 kPa (1 to 20 psi).

Polishing is a process that produces a smooth, lustrous surface finish. Two basic mechanisms are involved in the polishing process: (a) fine-scale abrasive removal, and (b) softening and smearing of surface layers by frictional heating during polishing. The shiny appearance of polished surfaces results from the smearing action. Polishing is done with disks or belts of fabric, leather, or felt and coated with fine powders of aluminum oxide or diamond. Parts with irregular shapes, sharp corners, deep recesses, and sharp projections are difficult to polish.

Buffing is similar to polishing, with the exception that very fine abrasives are used on soft disks made of cloth or hide. The abrasive is supplied externally from a stick of abrasive compound. Polished parts may be buffed to obtain an even finer surface finish.

Mirror-like finishes can be obtained on metal surfaces by *electropolishing*, a process that is the reverse of electroplating. Because there is no mechanical contact with the workpiece, this process is particularly suitable for polishing irregular shapes. The electrolyte attacks projections and peaks on the workpiece surface at a higher rate than the rest, thus producing a smooth surface. Electropolishing is also used for deburring operations.

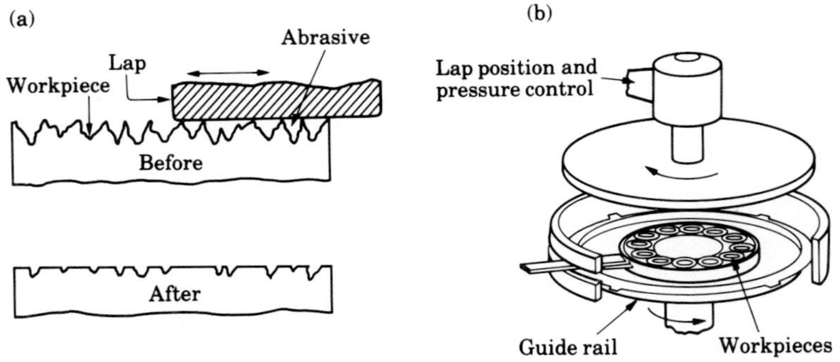

FIGURE 9.17
(a) Schematic illustration of the lapping process. (b) Production lapping on flat surfaces, cylindrical surfaces.

9.8.1 Deburring

Burrs are thin ridges, usually triangular in shape, that develop along the edges of a workpiece from shearing sheet materials, trimming forgings and castings, and machining. Burrs may interfere with the assembly of parts and can cause jamming of parts, misalignment, and short circuits in electrical components. Furthermore, burrs may reduce the fatigue life of components. Because they are usually sharp, they can be a safety hazard to personnel. On the other hand, burrs on thin drilled or tapped components, such as tiny parts in watches, can provide extra thickness and, thus, improve the holding torque of screws.

Several *deburring* processes are available. Burrs may be removed manually with files or mechanically by cutting, wire brushing, sanding, tumbling, vibratory finishing, abrasive flow, ultrasonics, and abrasive jets or water jets. The need for deburring may be reduced by adding chamfers to sharp edges on parts.

Vibratory and *barrel finishing* processes are used to improve the surface finish and remove burrs from large numbers of relatively small workpieces. In this batch-type operation, specially shaped abrasive pellets are placed in a container along with the parts to be deburred. The container is either vibrated or tumbled. The impact of individual abrasives and metal particles removes sharp edges and burrs from the parts.

In *shot blasting* (also called *grit blasting*), abrasive particles (usually sand) are propelled by a high-velocity jet of air, or by a rotating wheel, onto the surface of the workpiece. Shot blasting is particularly useful in deburring metallic and nonmetallic materials and stripping, cleaning, and removing surface oxides. The surface produced has a matte finish. Small-scale polishing and etching can also be done by this process on bench-type units (*microabrasive blasting*).

In *abrasive-flow machining*, abrasive grains, such as silicon carbide or diamond, are mixed in a viscous matrix, which is then forced back and forth through the openings and passageways in the workpiece. The movement of the abrasive matrix under pressure erodes away burrs and sharp corners and polishes the part. The process is particularly suitable for workpieces with internal cavities that are inaccessible by other means. Pressures applied range from 0.7 MPa to 11 MPa (100 psi to 1600 psi).

9.9 Grinding Fluids

The functions of grinding fluids are similar to those we described for cutting fluids. Although grinding and other abrasive-removal processes can be performed dry, the use of a fluid is important. It prevents temperature rise in the workpiece and improves the part's surface finish and dimensional accuracy. Fluids also improve the efficiency of the operation by reducing wheel loading and wear and lowering power consumption.

TABLE 9.4
GENERAL RECOMMENDATIONS FOR GRINDING FLUIDS

MATERIAL	GRINDING FLUID
Aluminum	E, EP
Beryllium	D, E, CSN
Copper	CSN, E, MO + FO
Magnesium	D, MO
Nickel	CSN, EP
Refractory metals	EP
Steels	CSN, E
Titanium	CSN, E

D: dry; E: emulsion; EP: extreme pressure; CSN: chemicals and synthetics; MO: mineral oil; FO: fatty oil.

Grinding fluids are typically water-base emulsions for general grinding and oils for thread grinding (Table 9.4). They may be applied as a stream (flood) or as mist, which is a mixture of fluid and air. Because of the high surface speeds involved, an airstream or air blanket around the periphery of the wheel usually prevents the fluid from reaching the cutting zone. Special nozzles have been designed in which the grinding fluid is applied under high pressure and effectively.

9.10 Chemical Machining

We know that certain chemicals attack metals and etch them, thereby removing small amounts of material from the surface. Thus *chemical machining* (CM) was developed, whereby material is removed from a surface by chemical dissolution, using chemical *reagents*, or *etchants*, such as acids and alkaline solutions. Parts may also be deburred by chemical means.

9.10.1 Chemical milling

In *chemical milling*, shallow cavities are produced on plates, sheets, forgings, and extrusions for overall reduction of weight. Chemical milling has been used on a wide variety of metals, with depths of removal to as much as 12 mm (0.5 in.). Selective attack by the chemical reagent on different areas of the workpiece surfaces is controlled by removable *masking* (Fig. 9.18), or by partial immersion in the reagent. The material-removal rate is a maximum at about 0.1 mm/min.

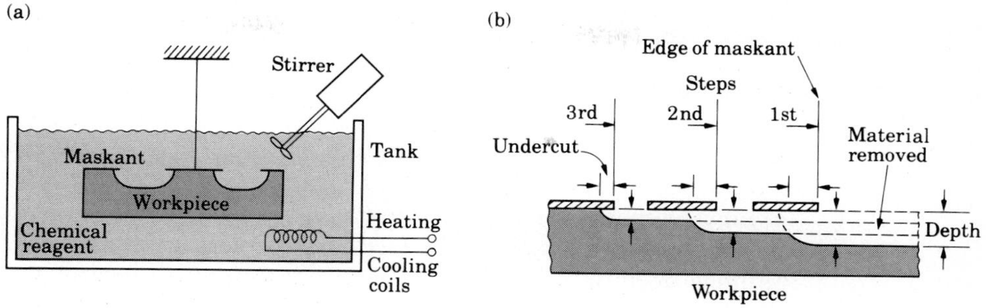

FIGURE 9.18
(a) Schematic illustration of the chemical machining process. Note that no forces or machine tools are involved in this process. (b) Stages in producing a profiled cavity by chemical machining.

This process is used in the aerospace industry for removing shallow layers of material from large aircraft, missile skin panels, and extruded parts for airframes. Tank capacities for reagents are as large as 3.7 m × 15 m (12 ft × 50 ft). The process is also used to fabricate microelectronic devices. The range of surface finish and tolerances obtained by chemical machining and other machining processes is shown in Fig. 9.12.

Some surface damage may result from chemical milling because of preferential etching and intergranular attack, which adversely affect surface properties. Chemical milling of welded and brazed structures may produce uneven material removal. Chemical milling of castings may result in uneven surfaces caused by porosity and nonuniformity of structure.

9.10.2 Chemical blanking

Chemical blanking is similar to blanking of sheet metal, with the exception that material is removed by chemical dissolution rather than by shearing. Typical applications for chemical blanking are burr-free etching of printed circuit boards, decorative panels, and thin sheet-metal stampings.

9.10.3 Photochemical blanking

Also called photoetching, *photochemical blanking* is a modification of chemical milling. Material is removed, usually from flat thin sheet, by photographic techniques. Complex burr-free shapes can be blanked on metals as thin as 0.0025 mm (0.0001 in.). Typical applications for photochemical blanking are fine screens, printed-circuit cards, electric-motor laminations, flat springs, and masks for color television. Although skilled labor is required, tooling costs are low, and the process can be automated.

9.11 Electrochemical Machining

Electrochemical machining (ECM) is basically the reverse of electroplating. Electrolytes dissolve the reaction products formed on the workpiece (*anode*) by electrochemical action, thus removing material from the surface and producing a cavity. Modifications of this process are used for turning, facing, slotting, trepanning, and profiling operations in which the electrode becomes the cutting tool.

The shaped tool (*cathode*) is generally made of brass, copper, or bronze. The electrolyte is usually either sodium chloride mixed in water or sodium nitrate. It is pumped at a high rate through the passages in the tool (Fig. 9.19). A dc power supply in the range of 5–25 V maintains current densities, which for most applications are 1.5–8 A/mm^2 (1000–5000 A/in^2) of active machined surface. Machines having current capacities as high as 40,000 A and as small as 5 A are available. The penetration rate of the tool is proportional to the current density, and the tool does not undergo any wear. The material-removal rate is in the range of 2.5 to 12 mm/min.

Electrochemical machining is generally used for machining complex cavities in high-strength materials, particularly in the aerospace industry for mass production of turbine blades, jet-engine parts, and nozzles. It is also used for machining forging-die cavities and producing small holes.

The ECM process leaves a burr-free surface; in fact, it can also be used as a deburring process. It does not cause any thermal damage to the part, and the lack of

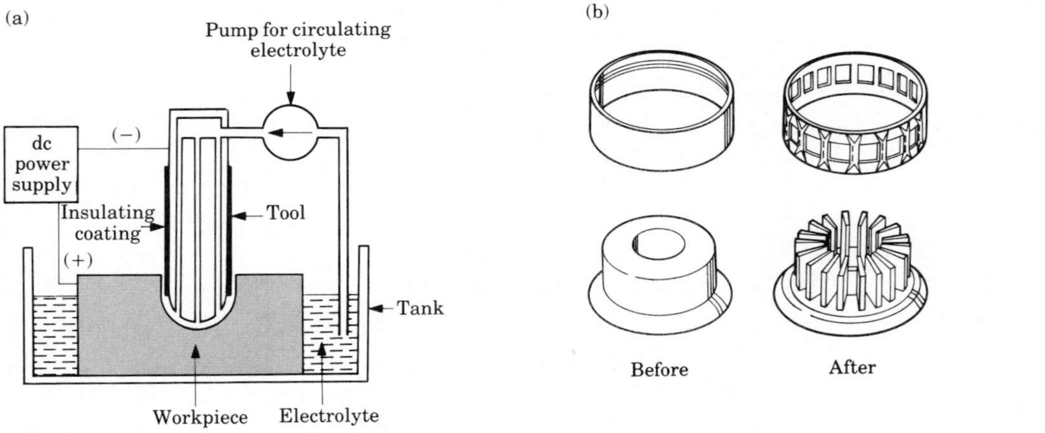

FIGURE 9.19 (a) Schematic illustration of the electrochemical-machining process. ECM machine. *Source:* Courtesy of Anocut, Inc. (b) Typical parts made by electrochemical machining.

tool forces prevents distortion of the part. However, the mechanical properties of components made by ECM should be compared carefully to those of other material-removal methods, as they typically have lower fatigue strength compared to that obtained for ground and polished surfaces.

9.12 Electrochemical Grinding

Electrochemical grinding (ECG) combines electrochemical machining with conventional grinding. The equipment used in electrochemical grinding is similar to a grinder, except that the wheel is a rotating cathode with abrasive particles (Fig. 9.20a). The wheel is metal-bonded with diamond or aluminum-oxide abrasives, and rotates at a surface speed of 20–35 m/s (4000–7000 ft/min). The abrasives serve as insulators between the wheel and the workpiece and mechanically remove electrolytic products from the working area. A flow of electrolyte, usually sodium nitrate, is provided for the electrochemical machining phase of the operation.

The majority of metal removal in ECG is by electrolytic action. Thus wheel wear is very low, and current densities ranging from 1–3 A/mm^2 (500–2000 A/in^2). Finishing cuts are usually made by the grinding action but only to produce a surface with good finish and dimensional accuracy. The material-removal rate is typically 1.5 cm^3/min per 1000 A. This process is suitable for applications similar to those for milling, grinding, and sawing (Fig. 9.20b). It is not adaptable to cavity-sinking operations, such as die making. The ECG process has been successfully applied to carbides and high-strength alloys.

Electrochemical honing combines the fine abrasive action of honing with electrochemical action. Although the equipment is costly, the process is as much as five times faster than conventional honing. It is used primarily for finishing internal cylindrical surfaces.

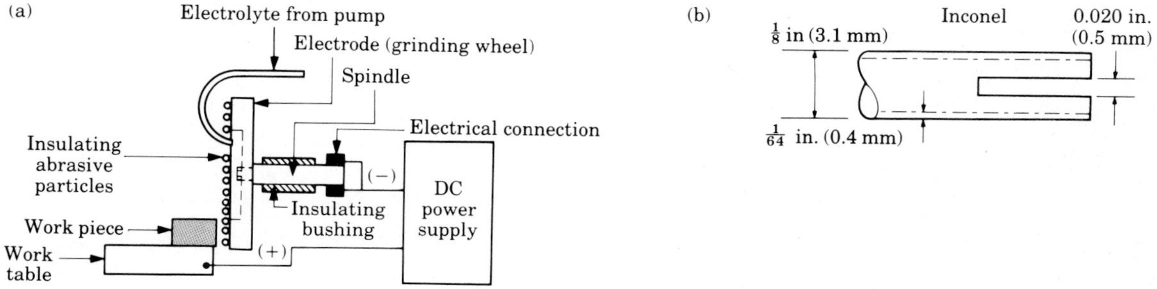

FIGURE 9.20
(a) Schematic illustration of the electrochemical-grinding process. (b) Thin slot produced on a round nickel-alloy tube by this process.

9.13 Electrical-Discharge Machining

The principle of *electrical-discharge machining* (EDM), also called *electrodischarge* or *spark-erosion* machining, is based on erosion of metals by spark discharges. We know that when two current-conducting wires are allowed to touch each other, an arc is produced. If we look closely at the point of contact between the two wires, we note that a small portion of the metal has been eroded away, leaving a small crater. In other words, we have removed a small amount of material from the surface of the wire. Although this phenomenon has been known since the discovery of electricity, it was not until the 1940s that a machining process based on this principle was developed.

The EDM system consists of a shaped tool (*electrode*) and the workpiece, connected to a dc power supply and placed in a *dielectric fluid* (Fig. 9.21). When the potential difference between the tool and the workpiece is sufficiently high, a transient spark discharges through the fluid, removing a very small amount of metal from the workpiece surface. The discharge is repeated at rates of between 50 kHz and 500 kHz, with voltages usually ranging between 50 V and 300 V, and currents from 0.1 A to 500 A. A wide variety of EDM machines, many with computer controls, are available.

The dielectric fluid (a) acts as an insulator until the potential is sufficiently high, (b) carries away the debris in the gap, and (c) provides a cooling medium. The gap between the tool and the workpiece is critical; hence the downward feed of the tool is controlled by a servomechanism, which automatically maintains a constant gap. The most common dielectric fluids are mineral oils, although kerosene and distilled and deionized water may be used in specialized applications.

Electrical-discharge machining has become an important manufacturing process. It has numerous applications, such as producing die cavities for large automotive-body components, small-diameter deep holes using tungsten wire as the electrode, narrow slots, turbine blades, and various intricate shapes (Fig. 9.22).

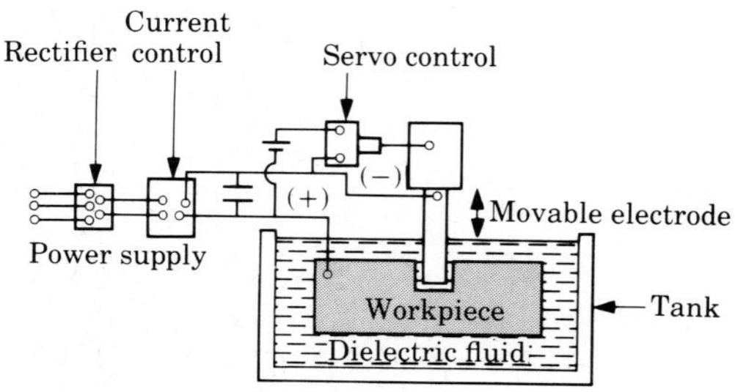

FIGURE 9.21 Schematic illustration of the electrical-discharge machining process.

(a) (b)

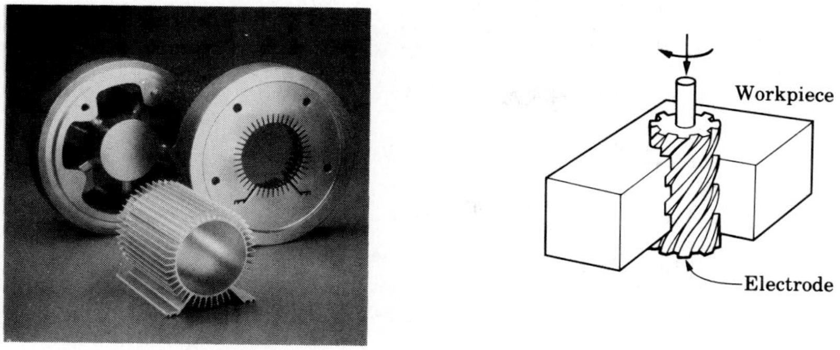

FIGURE 9.22
(a) Examples of cavities produced by the electrical-discharge machining process, using shaped electrodes. Two round parts (rear) are the set of dies for extruding the aluminum piece shown in front. *Source:* Courtesy of AGIE USA Ltd. (b) A spiral cavity produced by a rotating electrode. *Source: American Machinist.*

Stepped cavities can be produced by controlling the relative movements of the workpiece in relation to the electrode. In another setup, internal cavities are produced by a rotating electrode with a movable tip. The electrode is rotated mechanically during machining.

The EDM process can be used on any material that is an electrical conductor. The melting point and latent heat of melting are important physical properties that determine the volume of metal removed per discharge. As these values increase, the rate of material removal slows. The volume of material removed per discharge is typically in the range of 10^{-6}–10^{-4} mm^3 (10^{-10}–10^{-8} in^3). Since the process doesn't involve mechanical energy, hardness, strength, and toughness of the workpiece material don't necessarily influence the removal rate.

The frequency of discharge or the energy per discharge is usually varied to control the removal rate. The rate and surface roughness increase with increasing current density and decreasing frequency of sparks. Metal-removal rates usually range from 0.1 to 25 cm^3/h (0.005 to 1.5 in^3/h). Higher rates are possible but they produce a very rough finish, having a molten and resolidified (recast) structure with poor surface integrity and low fatigue properties. Thus finishing cuts are made at low removal rates, or the recast layer is removed later by finishing operations.

Electrodes for EDM are usually made of graphite, although brass, copper, or copper–tungsten alloy may be used. The tools are shaped by forming, casting, powder metallurgy, or machining. Tool wear is an important factor since it affects dimensional accuracy and the shape produced. Tool wear in EDM is related to the melting points of the materials involved: the lower the melting point, the higher the wear rate. Consequently, graphite electrodes have the highest wear resistance. Tool wear can be minimized by reversing the polarity and using copper tools, a process called *no-wear EDM*.

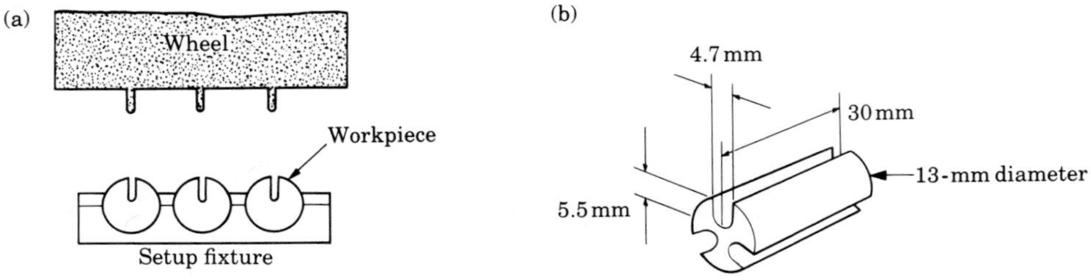

FIGURE 9.23
(a) Longitudinal grooves produced in a round rod by the electrical–discharge grinding process. (b) Sample dimensions. Note that the workpieces are indexed by 120° to produce each set of grooves.

9.13.1 Electrical-discharge grinding

The grinding wheel in *electrical-discharge grinding* (EDG) is made of graphite or brass and contains no abrasives. Material is removed from the surface of the workpiece by repetitive spark discharges between the rotating wheel and the workpiece. An application of electrical-discharge grinding is shown in Fig. 9.23, which shows longitudinal grooves being produced on the workpiece using a shaped wheel.

The EDG process can be combined with electrochemical grinding. The process is then called *electrochemical-discharge grinding* (ECDG). Material is removed by chemical action, with the electrical discharges from the graphite wheel breaking up the oxide film, and is washed away by the electrolyte flow. The process is used primarily for grinding carbide tools and dies but can also be used for fragile parts, such as surgical needles, thin-walled tubes, and honeycomb structures. The ECDG process is faster than EDG, but power consumption is higher.

In *sawing* with EDM, a setup similar to a band or circular saw (but without any teeth) is used with the same electrical circuit as in EDM. Narrow cuts can be made at high rates of metal removal. Because cutting forces are negligible, the process can be used on slender components.

9.13.2 Traveling-wire EDM

A variation of EDM is *traveling-wire EDM*, as shown in Fig. 9.24, or *electrical-discharge wire cutting*. In this process, which is similar to contour cutting with a band saw, a slowly moving wire travels along a prescribed path, cutting the workpiece, with the discharge sparks acting like cutting teeth. This process is used to cut plates as thick as 150 mm (6 in.), and for making punches, tools, and dies from hard metals. Machines are equipped with computer controls to control the cutting path of the wire.

The wire is usually made of brass, copper, or tungsten and is typically about 0.25 mm (0.01 in.) in diameter, making narrow cuts possible. The wire is usually used only once, as it is relatively inexpensive. It travels at sufficiently high and constant velocity, 2.5–150 mm/s (0.1–6 in./s), and a constant gap (kerf) is maintained during the cut.

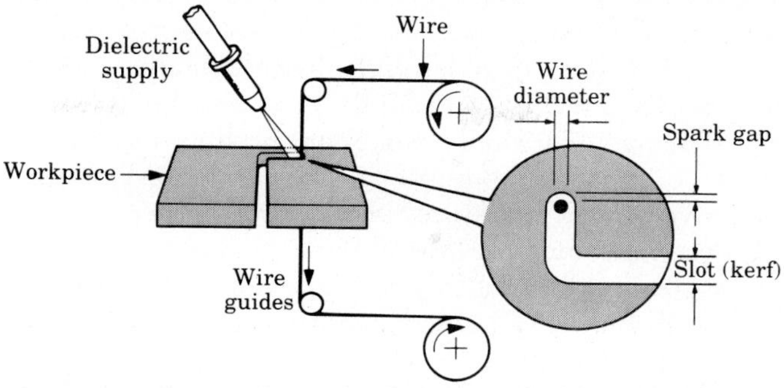

FIGURE 9.24
Schematic illustration of the traveling-wire EDM process (also called electrical-discharge wire cutting). This operation is similar to cutting on a band saw, with spark discharges acting like cutting teeth. As much as 50 hours of accurate cutting can be performed with one reel of wire, which is discarded after use.

9.14
High-Energy Beam Machining

In *laser-beam machining* (LBM), the source of energy is a laser (an acronym for *L*ight *A*mplification by *S*timulated *E*mission of *R*adiation), which focuses optical energy on the surface of the workpiece (Fig. 9.25). The highly focused, high-density energy melts and evaporates portions of the workpiece in a controlled manner.

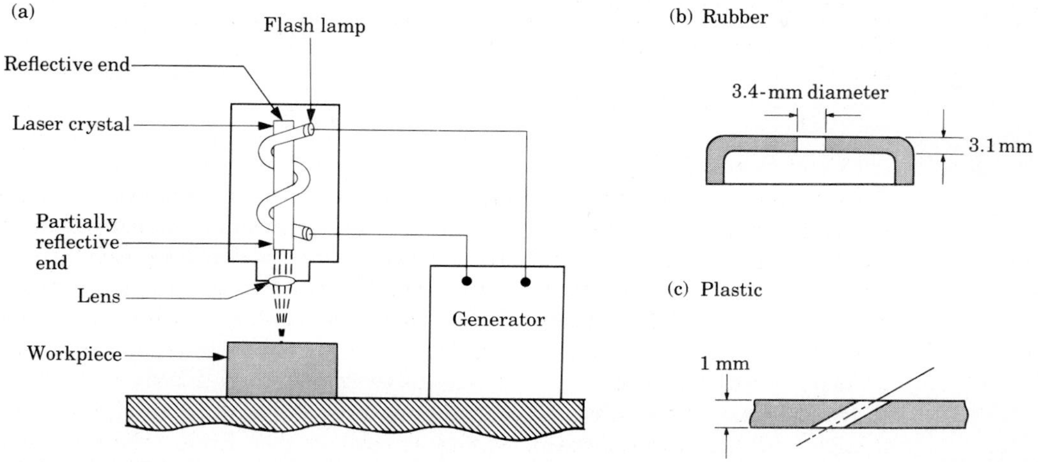

FIGURE 9.25
(a) Schematic illustration of the laser-beam machining process. (b) and (c) Examples of holes produced in nonmetallic parts by LBM.

This process, which does not require a vacuum, is used to machine a variety of metallic and nonmetallic materials. It is also used for small-scale cutting operations, such as slitting, and drilling holes as small as 0.005 mm (0.0002 in.), with hole depth to diameter ratios of 50 to 1. The cooling passages in the first-stage vanes of the Boeing 747 turbine engines are produced by this process.

Important physical parameters in LBM are the reflectivity and thermal conductivity of the workpiece surface and its specific heat and latent heats of melting and evaporation. The lower these quantities, the more efficient the process is. The surface produced by LBM is usually rough and has a heat-affected zone which, in critical applications, may have to be removed or heat treated.

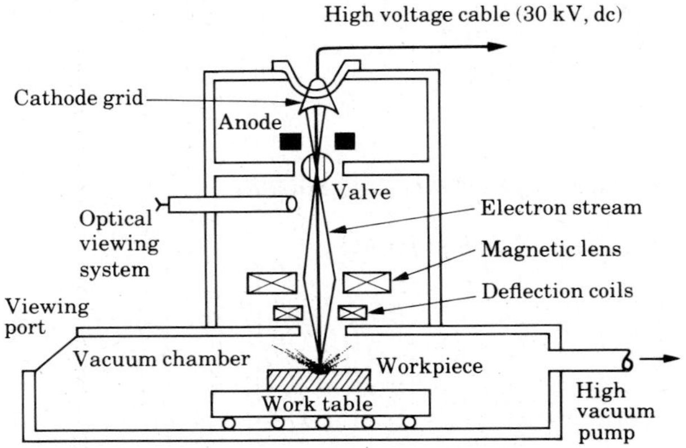

FIGURE 9.26
Schematic illustration of the electron-beam machining process. Unlike LBM, this process requires a vacuum, hence workpiece size is limited.

Laser beams may be used in combination with a gas stream, such as oxygen, nitrogen, or argon (*laser-beam torch*), for cutting thin sheet materials. Laser-beam machining is used widely in drilling and cutting composite materials, especially for the electronics industry. The abrasive nature of composites and the cleanliness of the operation make laser-beam machining an attractive alternative to traditional machining methods. Laser beams are also used for small-scale heat-treating and welding operations.

The source of energy in *electron-beam machining* (EBM) is high-velocity electrons, which strike the surface of the workpiece (Fig. 9.26). Its applications are similar to those of laser-beam machining, except that EBM requires a vacuum. *Plasma* (ionized gas) *beams* are also used to cut sheet and plate at high speeds. Material-removal rates in this process are much higher than in the EDM and LBM processes, and parts can be machined with good reproducibility.

9.15 Hydrodynamic Machining

We know that when we put our hand across a jet of water or air, we feel a considerable force acting on it. This force results from the momentum change of the stream—and in fact is the principle on which the operation of water or gas turbines is based. In *hydrodynamic machining* (HDM), also called *water-jet machining* (Fig. 9.27), this force is utilized in cutting and deburring operations.

The water jet acts like a saw and cuts a narrow groove in the material. Although pressures as high as 1400 MPa (200 ksi) can be generated, a pressure level of about 400 MPa (60 ksi) is generally used for efficient operation. Jet-nozzle diameters usually range between 0.1 mm and 0.3 mm (0.004 in. and 0.012 in.).

A variety of materials can be cut with this technique, including plastics, fabrics, rubber, wood products, paper, leather, insulating materials, brick, and composite materials. Thicknesses range up to 25 mm (1 in.) and higher. The advantages of this process are that cuts can be started at any location without the need for predrilled holes, no heat is produced, no deflection of the rest of the workpiece takes place (hence the process is suitable for flexible materials), little wetting of the workpiece takes place, and the burr produced is minimal.

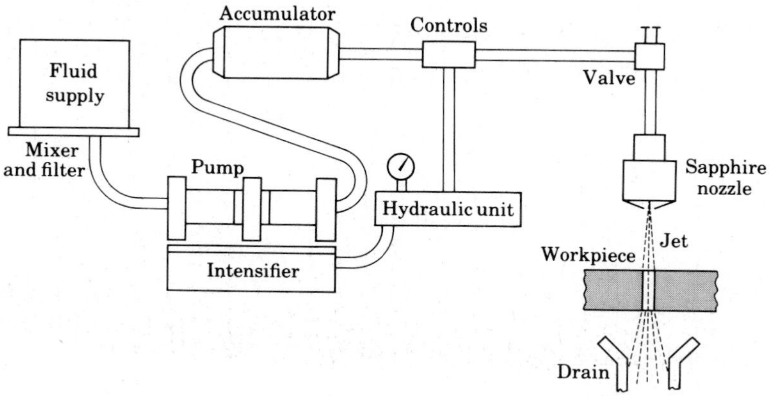

FIGURE 9.27 Schematic illustration of the hydrodynamic machining process, also called water-jet machining.

9.16 Process Economics

We have shown that grinding may be used both as a finishing operation and as a large-scale removal operation (as in creep-feed grinding). The use of grinding as a finishing operation is often necessary because forming and machining processes alone usually cannot produce parts with the desired dimensional accuracy and surface finish. However, because it is an additional operation, grinding contributes significantly to product cost. Creep-feed grinding, on the other hand, has proved to be an economical alternative to machining operations such as milling, even though wheel wear is high.

All finishing operations contribute to product cost. On the basis of the discussion thus far, you can see that as the surface finish improves, more operations are required, and hence the cost increases. Note in Fig. 9.28 how rapidly cost increases as surface finish is improved by processes such as grinding and honing.

Much progress has been made in automating the equipment involved in finishing operations, including computer controls. Consequently, labor costs and production times have been reduced, even though such machinery may require significant capital investment. If finishing is likely to be an important factor in manufacturing a particular product, the conceptual and design stages should involve an analysis of the degree of surface finish and dimensional accuracy required.

Furthermore, all processes that precede finishing operations should be analyzed for their capability to produce a more acceptable surface finish and dimensional

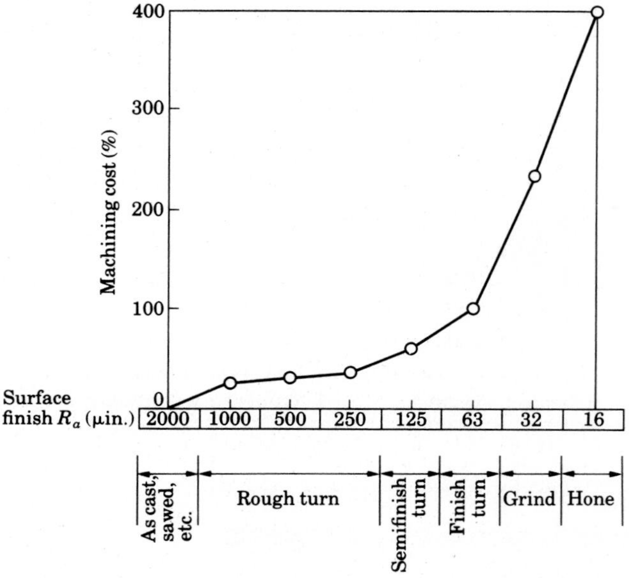

FIGURE 9.28
Increase in the cost of machining and finishing a part as a function of the surface finish required.

accuracy. This can be accomplished through proper selection of tools and process parameters and the characteristics of the machine tools involved.

We have also shown the unique applications of nontraditional processes, particularly for difficult-to-machine materials and complex internal and external profiles. The economic production run for a particular process depends on the cost of tooling and equipment, the material-removal rate, operating costs, and the level of operator skill required, as well as secondary and finishing operations that may be necessary. In chemical machining, the costs of reagents, maskants, and disposal, together with the cost of cleaning the parts, are important factors. In electrical-discharge machining, the cost of electrodes and the need to replace them periodically are significant.

The rate of material removal, hence production rate, can vary significantly in these processes. The cost of tooling and equipment also varies significantly, as does the operator skill required. The high capital investment for machines such as electrical and high-energy beam machining should be justified in terms of the production runs and the feasibility of manufacturing the same part by other means, if at all possible.

SUMMARY

Grinding and various abrasive-removal processes are capable of producing the finest accuracy and surface finish in manufactured products. The majority of abrasive processes are basically finishing operations that are usually performed on machined or cold-worked parts. However, abrasives are also used for large-scale material-removal processes, such as creep-feed grinding and snagging in foundries.

A variety of abrasive processes and machinery are available for surface, external, and internal grinding. The selection of abrasives and process variables in these operations must be controlled in order to obtain the desired surface finish and dimensional accuracy. Otherwise, damage to surfaces such as burning, heat checking, and harmful residual stresses may develop. Several finishing operations are available for deburring. Because they contribute significantly to product cost, proper selection and implementation of finishing operations are important.

Machining processes involve not only single-point or multipoint tools but also other methods using chemical, electrical, and high-energy-beam sources of energy. The mechanical properties of the workpiece material are not significant because these processes rely on mechanisms that do not involve the strength, hardness, ductility, or toughness of the material. Rather, they involve physical, chemical, and electrical properties.

Chemical and electrical methods of machining are particularly suitable for hard materials and complex shapes. They do not produce forces (hence can be used for slender and flexible workpieces), significant temperatures, or residual stresses. However, the effects of these processes on surface integrity must be understood, as they can damage surfaces considerably, thus reducing fatigue life.

BIBLIOGRAPHY

Abrasive Processes

Andrew, C., T.D. Howes, and T.R.A. Pearce, *Creep Feed Grinding*. New York: Industrial Press, 1985.

Drozda, T. (ed.), *Manufacturing Engineering Explores Grinding Technology*. Dearborn, Mich.: Society of Manufacturing Engineers, 1982.

Farago, F.T., *Abrasive Methods Engineering*, Vol. 1, 1976; Vol. 2, 1980. New York: Industrial Press.

Gillespie, L.K., *Deburring Technology for Improved Manufacturing*. Dearborn, Mich.: Society of Manufacturing Engineers, 1981.

King, R.I., and R.S. Hahn, *Handbook of Modern Grinding Technology*. New York: Chapman and Hall/Methuen, 1987.

Lewis, K.B., and W.F. Schleicher, *The Grinding Wheel: A Textbook of Modern Grinding Practice*, 3d ed. Cleveland: The Grinding Wheel Institute, 1976.

Machinery's Handbook, revised periodically. New York: Industrial Press.

Machining Data Handbook, 3d ed., 2 vols. Cincinnati: Machinability Data Center, 1980.

Malkin, S., *Grinding Technology: Theory and Applications of Machining with Abrasives*. New York: Wiley, 1989.

McKee, R.L., *Machining with Abrasives*. New York: Van Nostrand Reinhold, 1982.

Metals Handbook, 9th ed., *Vol. 16: Machining*. Metals Park, Ohio: ASM International, 1989.

Metzger, J.L., *Superabrasive Grinding*. Stoneham, Mass.: Butterworths, 1986.

Tool and Manufacturing Engineers Handbook, 4th ed., *Vol. 1: Machining*. Dearborn, Mich.: Society of Manufacturing Engineers, 1983.

Nontraditional Processes

Benedict, G.F., *Nontraditional Manufacturing Processes*. New York: Marcel Dekker, 1987.

Faust, C.L. (ed.), *Fundamentals of Electrochemical Machining*. Princeton, N.J.: Electrochemical Society, 1971.

Harris, W.T., *Chemical Milling: The Technology of Cutting Materials by Etching*. New York: Oxford, 1976.

Lange, K. (ed.), *Handbook of Metal Forming* (Chapter 32, Die Manufacture). New York: McGraw-Hill, 1985.

Machinery's Handbook, revised periodically. New York: Industrial Press.

Machining Data Handbook, 3d ed., 2 Vols. Cincinnati: Machinability Data Center, 1980.

McGeough, J.A., *Advanced Methods of Machining*. London: Chapman and Hall, 1988.

McGeough, J.A., *Principles of Electrochemical Machining*. London: Chapman and Hall, 1974.

Metals Handbook, 9th ed., *Vol. 16: Machining*. Metals Park, Ohio: ASM International, 1989.

Ready, J.F., *Industrial Applications of Lasers*. New York: Academic Press, 1978.

Source Book on Applications of the Laser in Metalworking. Metals Park, Ohio: American Society for Metals, 1979.

Wilson, J.F., *Practice and Theory of Electrochemical Machining*. New York: Wiley, 1971.

Tool and Manufacturing Engineers Handbook, 4th ed., *Vol. 1: Machining*. Dearborn, Mich.: Society of Manufacturing Engineers, 1983.

QUESTIONS

9.1 Why are grinding operations necessary for parts that have been machined by other processes?

9.2 Explain why there are so many different types and sizes of abrasive wheels.

9.3 Explain the large difference between the specific energies involved in grinding (Table 9.3) and machining (Table 8.3).

9.4 Give examples of applications for the grinding wheels shown in Table 9.2.

9.5 Explain why the same grinding wheel may act soft or hard.

9.6 Explain the factors involved in selecting the appropriate type of abrasive for a particular grinding operation.

9.7 What are the effects of wear flat on the grinding process?

9.8 The grinding ratio *G* depends on the (a) type of grinding wheel, (b) workpiece hardness, (c) wheel depth of cut, (d) wheel and workpiece speeds, and (e) type of grinding fluid. Explain why.

9.9 List and explain the precautions you would take when grinding with high precision. Comment on the machine, process parameters, the grinding wheel, and grinding fluids.

9.10 Describe the methods you would use to determine the number of active cutting points per unit surface area on the periphery of a straight (Type 1; see Fig. 9.2a) grinding wheel. What is the significance of this number?

9.11 Describe the difficulties involved in grinding thermoplastics and explain why.

9.12 Explain why ultrasonic machining is not suitable for soft, ductile metals.

9.13 Why should a soft grade grinding wheel be used for hardened steels?

9.14 Explain why the processes described in this chapter can adversely affect the fatigue strength of materials.

9.15 Describe the factors that may cause chatter in grinding operations and explain why.

9.16 Outline the methods that are generally available for deburring parts.

9.17 In which of the processes described in this chapter are physical properties of the workpiece material important. Why?

9.18 Give the technical and economic reasons why material-removal processes described in this chapter may be preferred—or may even be necessary—over those described in Chapter 8.

9.19 What processes would you recommend for die sinking in a die block? Explain.

9.20 In Fig. 9.2 the proper grinding surfaces for each type of wheel are shown with an arrow. Explain why the other surfaces of the wheels should not be used for grinding.

9.21 Why is preshaping or premachining of parts sometimes desirable in the nontraditional machining processes described in this chapter?

9.22 Why has the traveling-wire EDM process become so widely accepted in industry?

9.23 Make a list of material-removal processes that may be suitable for the following workpiece materials: (a) ceramics, (b) cast iron, (c) thermoplastics, (d) thermosets, (e) diamond, and (f) annealed copper.

9.24 Explain why producing sharp corners and profiles using some of the processes described in this chapter is difficult.

9.25 How do specific energy *u* and grinding ratio *G* vary with respect to wheel depth of cut and hardness of the workpiece material?

9.27 The shaft of a Type 1 grinding wheel is attached to a flywheel only, which is rotating at a certain initial rpm. With this setup a surface grinding operation is being carried out on a long workpiece at a constant workpiece speed. Discuss the factors involved in estimating the distance ground before the wheel comes to a stop.

PROBLEMS

9.1 For a surface grinding operation, calculate the chip dimensions for the following process variables: $D = 8$ in., $d = 0.001$ in., $v = 100$ ft/min, $V = 6000$ ft/min, $C = 500$ per in^2, and $r = 20$.

9.2 If the workpiece strength in grinding is doubled, what should be the percentage decrease in the wheel depth of cut d in order to maintain the same grain force, all other variables being the same?

9.3 Derive a formula for the material removal rate (MRR) in surface grinding in terms of process parameters. Use the same terminology as in the text.

9.4 Assume that a surface grinding operation is being carried out under the following conditions: $D = 200$ mm, $d = 0.1$ mm, $v = 0.4$ m/s, and $V = 30$ m/s. These conditions are then changed to: $D = 150$ mm, $d = 0.1$ mm, $v = 0.3$ m/s, and $V = 25$ m/s. How much does the temperature rise differ from the initial condition?

9.5 For a surface grinding operation derive an expression for the power dissipated in imparting kinetic energy to the chips and comment on the magnitude of this energy. Use the same terminology as in the text.

9.6 Calculate the average impact force on a steel plate by a spherical aluminum oxide abrasive particle of grit size 200, dropped from a height of 1 ft.

9.7 A 40-mm deep hole, 20 mm in diameter, is being produced by electrochemical machining. High production rate is more important than machined surface quality. Estimate the maximum current and the time required to perform this operation.

9.8 If the operation in Problem 9.7 were performed on an electrical-discharge machine, what would be the estimated machining time?

9.9 A cutting-off operation is being performed with a laser beam. The workpiece being cut is $\frac{1}{2}$ in. thick and 2 in. wide. If the kerf is $\frac{1}{16}$ in. wide, estimate the time required to perform this operation.

10

Processing of Polymers and Reinforced Plastics

10.1 Introduction

Because of their many unique and diverse properties, *polymers* (*plastics*) have increasingly replaced metallic components in applications such as automobiles, civilian and military aircraft, sporting goods, and office equipment. These substitutions reflect the advantages of plastics in terms of high strength-to-weight ratio, design possibilities, wide choice of colors and transparencies, ease of manufacturing, and relatively low cost.

Compared to metals, plastics are generally characterized by low density, low strength and stiffness (Table 10.1), low electrical and thermal conductivity, good resistance to chemicals, and high coefficient of thermal expansion. However, the useful temperature range for most plastics is generally low—up to about 300 °C

TABLE 10.1
RANGE OF MECHANICAL PROPERTIES FOR VARIOUS ENGINEERING PLASTICS AT ROOM TEMPERATURE

MATERIAL	UTS (MPa)	E (GPa)	ELONGATION (%)	POISSON'S RATIO (ν)
ABS	28–55	1.4–2.8	75–5	—
ABS, reinforced	100	7.5	—	0.35
Acetal	55–70	1.4–3.5	75–25	—
Acetal, reinforced	135	10	—	0.35–0.40
Acrylic	40–75	1.4–3.5	50–5	—
Cellulosic	10–48	0.4–1.4	100–5	—
Epoxy	35–140	3.5–17	10–1	—
Epoxy, reinforced	70–1400	21–52	4–2	—
Fluorocarbon	7–48	0.7–2	300–100	0.46–0.48
Nylon	55–83	1.4–2.8	200–60	0.32–0.40
Nylon, reinforced	70–210	2–10	10–1	—
Phenolic	28–70	2.8–21	2–0	—
Polycarbonate	55–70	2.5–3	125–10	0.38
Polycarbonate, reinforced	110	6	6–4	—
Polyester	55	2	300–5	0.38
Polyester, reinforced	110–160	8.3–12	3–1	—
Polyethylene	7–40	0.1–1.4	1000–15	0.46
Polypropylene	20–35	0.7–1.2	500–10	—
Polypropylene, reinforced	40–100	3.5–6	4–2	—
Polystyrene	14–83	1.4–4	60–1	0.35
Polyvinyl chloride	7–55	0.0–4	450–40	—

(600 °F)—and they are not as dimensionally stable in service, over a period of time, as metals are.

The word plastics is from the Greek word *plastikos*, meaning it can be molded and shaped. Plastics can be machined, cast, formed, and joined into many shapes with relative ease. Little or no additional surface-finishing operations are required, which is an important advantage over metals. Plastics are commercially available as sheet, plate, film, rods, and tubing of various cross-sections.

The earliest polymers were made of *natural organic materials* from animal and vegetable products, cellulose being the most common example. With various chemical reactions, cellulose is modified into cellulose acetate, used in making photographic films (celluloid), sheets for packaging, textile fibers, as well as cellulose nitrate for plastics, explosives, rayon (a cellulose textile fiber), and varnishes. The earliest *synthetic* (manmade) polymer was a phenol-formaldehyde, a thermoset developed in 1906 and called Bakelite (a trade name, after L. H. Baekeland, 1863–1944).

The development of modern plastics technology began in the 1920s with raw materials extracted from coal and petroleum products. Ethylene was the first example of the building block for polyethylene. Ethylene is the product of the reaction between acetylene and hydrogen, and acetylene is the product of the reaction between coke and methane. These materials are known as *synthetic organic polymers*. Although in polyethylene only carbon and hydrogen atoms are involved, other compounds can be obtained with chlorine, fluorine, sulfur, silicon, nitrogen, and oxygen. As a result, an extremely wide range of polymers with an equally wide range of properties have been developed.

Among the most important developments are *reinforced plastics* (*composite materials*). These materials are a combination of two or more chemically distinct and insoluble phases having properties and structural performance superior to those of the constituents acting independently. Typical applications are in aircraft, sporting good, boat, and ladder manufacturing.

10.2 The Structure of Plastics

Plastics are composed of polymer molecules and various additives. Polymers are *long-chain molecules*, also called *macromolecules* or *giant molecules*, which are formed by polymerization, that is, by linking and cross-linking of different monomers. A *monomer* is the basic building block of polymers. The word *mer*, from the Greek *meros*, meaning part, indicates the smallest repetitive unit, similar to the term unit cell used in connection with crystal structures of metals.

The term *polymer* means many mers or units, generally repeated hundreds or thousands of times in a chainlike structure. Most monomers are organic materials in which carbon atoms are joined in *covalent* bonds (electron sharing) with other atoms, such as hydrogen, oxygen, nitrogen, fluorine, chlorine, silicon, and sulfur. A simple monomer is the ethylene molecule, which consists of carbon and hydrogen atoms (Fig. 10.1).

10.2.1 Polymerization

Molecules can be linked in repeating units to make longer and larger molecules by polymerization processes, which are chemical reactions. In these reactions, the double bonds between the carbon atoms open, and the molecules arrange themselves end to end linearly. Thus ethylene molecules, for example, link to become the polymer known as polyethylene (Fig. 10.1b). Polymerization processes are complex and we can describe them only briefly here. Although there are many variations, two basic polymerization processes are condensation and addition polymerization.

In *condensation polymerization*, the bonds between the molecules are formed by the application of heat and pressure, with the aid of a catalyst, or initiator. The reaction byproducts, such as water, are condensed; hence the term condensation. This process is also known as *step-growth* or *step-reaction* polymerization because the polymer molecule grows step by step.

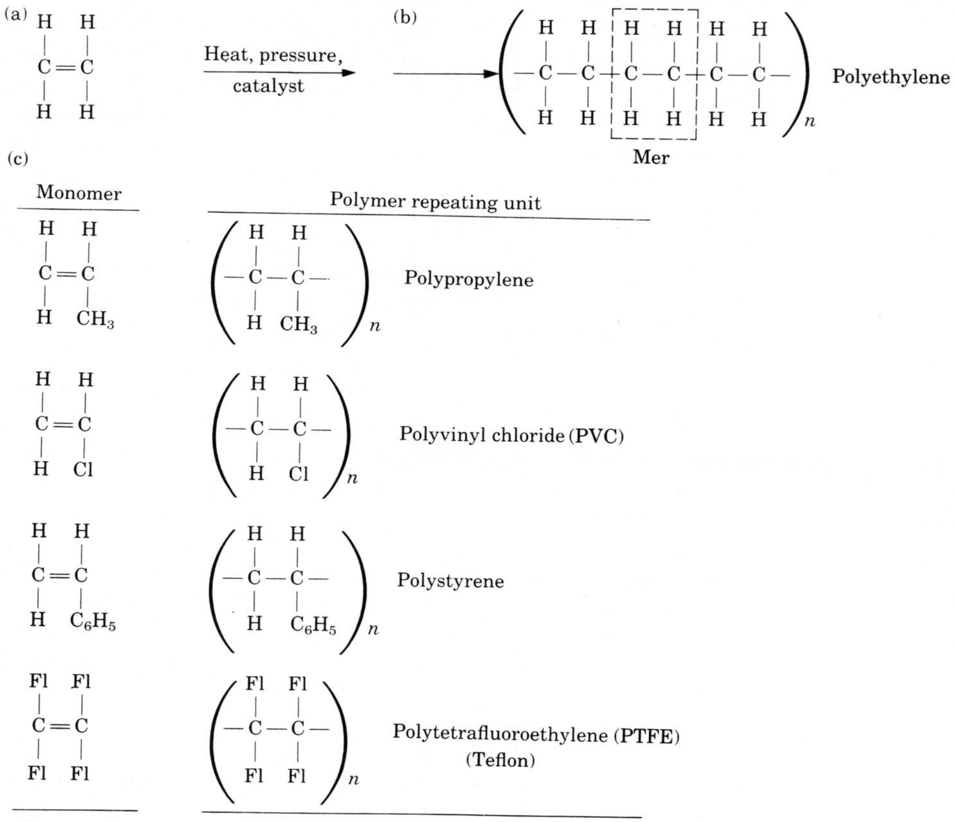

FIGURE 10.1 Basic structure of polymer molecules: (a) ethylene molecule; (b) polyethylene, a linear chain of many ethylene molecules; (c) molecular structure of various polymers. These are examples of the basic building blocks for plastics.

In *addition polymerization*, also known as *chain-growth* or *chain-reaction* polymerization, bonding takes place without reaction byproducts. It is called chain-reaction because of the high rate at which long molecules form simultaneously, usually within a few seconds. This rate is much higher than that for condensation polymerization.

Molecular weight. The sum of the molecular weights of the mers in the polymer chain is the *molecular weight* of the polymer. These chains are of different lengths and their arrangement is *amorphous*, that is, without any long-range order. The amorphous arrangement of polymer molecules is often described as a bowl of spaghetti, or worms in a bucket, all intertwined with each other. Because the lengths of the various chains differ, we determine and express the *average* molecular weight of a polymer on a statistical basis by averaging. For a polymer to have sufficient strength and toughness, the chains must be at least a certain length.

Degree of polymerization. The *degree of polymerization* is defined as the ratio of the molecular weight of the polymer to the molecular weight of the mer, or the number of mers per average molecule. This number can range from a few hundred to millions. The higher the number, the higher is the polymer's resistance to flow, a behavior similar to that obtained by increasing fluid viscosity. This behavior, in turn, affects the processing and shaping of the polymer into useful products.

Bonding. Within each molecular chain are covalent bonds, also known as *primary bonds* because of their strength. The bonds between different chains, and between the overlapping portions of the same chain, are known as *secondary bonds*, such as van der Waals bonds, hydrogen bonds, and ionic bonds. Secondary bonds are much weaker than the covalent bonds within the chain, the difference in strength being from one to two orders of magnitude.

Linear polymers. The chainlike polymers shown in Fig. 10.1 are called *linear polymers* because of their linear structure (Fig. 10.2a). A linear molecule is not straight in shape. In addition to those shown, other linear polymers are polyamides (nylon 6,6) and polyvinyl fluoride. Generally, a polymer consists of more than one type of structure. Thus a linear polymer may contain some branched and cross-linked chains.

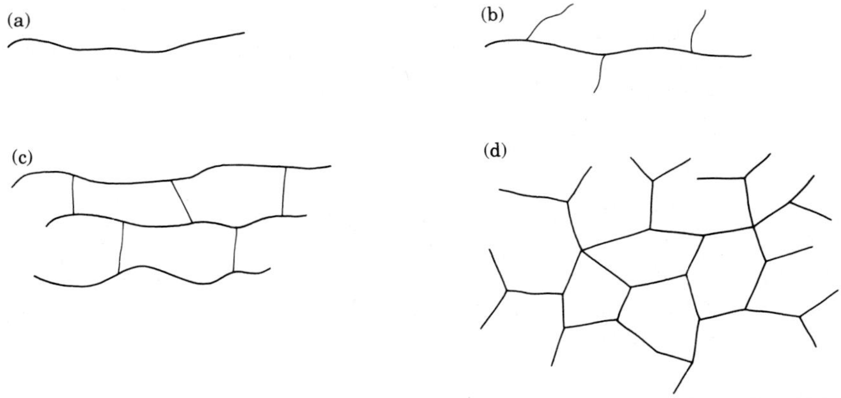

FIGURE 10.2

Schematic illustration of polymer chains. (a) Linear structure. Thermoplastics such as acrylics, nylons, polyethylene, and polyvinyl chloride have linear structures. (b) Branched structure, such as in polyethylene. (c) Cross-linked structure. Many rubbers or elastomers have this structure. Vulcanization of rubber produces this structure. (d) Network structure, which is basically highly cross-linked. Examples are thermosetting plastics, such as epoxies and phenolics.

Copolymers and terpolymers. If the repeating units in a polymer chain are all of the same type, we call the molecule a *homopolymer*. However, as with solid-solution metal alloys, two or three different types of monomers can be combined to impart certain special properties and characteristics to the polymer. *Copolymers* contain two types of polymers, such as styrene-butadiene, used widely for automobile tires. *Terpolymers* contain three types, such as ABS (acrylonitrile-butadiene-styrene) used for helmets, telephones, and refrigerator liners.

Branched polymers. The properties of a polymer depend not only on the type of monomers, but also on their arrangement in the molecular structure. In *branched polymers* (Fig. 10.2b), side-branch chains are attached to the main chain during the synthesis of the polymer. Branching interferes with the relative movement of the molecular chains. It affects the resistance to deformation of the polymer by making it stronger. The density of branched polymers is lower than that of linear-chain polymers. The behavior of branched polymers can be compared to that of linear-chain polymers by making an analogy with a pile of tree branches (branched polymers) and a bundle of straight logs (linear). You will note that it is more difficult to move a branch within the pile of branches than to move a log in its bundle. The three-dimensional entanglements of branches make movements more difficult, a phenomenon akin to increased strength.

Cross-linked polymers. Generally three-dimensional in structure, *cross-linked polymers* have adjacent chains linked by covalent bonds (Fig. 10.2c). Polymers with cross-linked chain structure are called *thermosets*, or *thermosetting* plastics, such as epoxies, phenolics, and silicones. Cross-linking has a major influence on the properties of polymers (generally imparting hardness, strength, stiffness, brittleness, and better dimensional stability), as well as in the vulcanization of rubber. *Network polymers* consist of spatial (three-dimensional) networks of three active covalent bonds (Fig. 10.2d). A highly cross-linked polymer is also considered a network polymer. Thermoplastic polymers that are already formed or shaped can be cross linked to obtain greater strength by subjecting them to high-energy radiation, such as ultraviolet light, x-rays, and electron beams. However, excessive radiation can cause degradation of the polymer.

• Example 10.1: Degree of polymerization in polyvinyl chloride (PVC).

Determine the molecular weight of a polyvinyl chloride mer. If a PVC polymer has an average molecular weight of 50,000, what is the degree of polymerization?

SOLUTION. From Fig. 10.1(c), we note that each PVC mer has 3 hydrogen atoms, 2 carbon atoms, and 1 chlorine atom. Since the atomic number of each element is 1, 12, and 35.5, respectively, we have the weight of a PVC mer as $(3)(1) + (2)(12) + (1)(35.5) = 62.5$. Hence the degree of polymerization is $50{,}000/62.5 = 800$. •

10.2.2 Crystallinity

In the preceding section, we described the arrangement of the long-chain molecules as amorphous and randomly intertwined. However, it is possible to impart some crystallinity to polymers and thereby modify their characteristics. This may be done either during the synthesis of the polymer or by deformation during its subsequent processing.

The crystalline regions in polymers are called *crystallites* (Fig. 10.3). These crystals are formed when the long molecules fold over themselves in an orderly manner, similar to folding a fire hose in a cabinet or facial tissues in a box. Thus we can regard a partially crystalline polymer as a two-phase material, one phase being crystalline and the other amorphous.

By controlling the rate of solidification during cooling and chain configuration, it is possible to impart different degrees of crystallinity to polymers, although never completely 100 percent. Crystallinity ranges from an almost complete crystal, up to about 95 percent by volume in the case of polyethylene, to slightly crystallized but mostly amorphous polymers.

Effects of crystallinity. The mechanical and physical properties of polymers are greatly influenced by the *degree of crystallinity.* As it increases, polymers become stiffer, harder, less ductile, more dense, less rubbery, and more resistant to solvents and heat. The increase in density with increasing crystallinity is caused by crystallization shrinkage and a more efficient packing of the molecules in the crystal lattice. For example, the highly crystalline form of polyethylene, known as high-density polyethylene (HDPE), has a specific gravity of 0.97 and is stronger, stiffer,

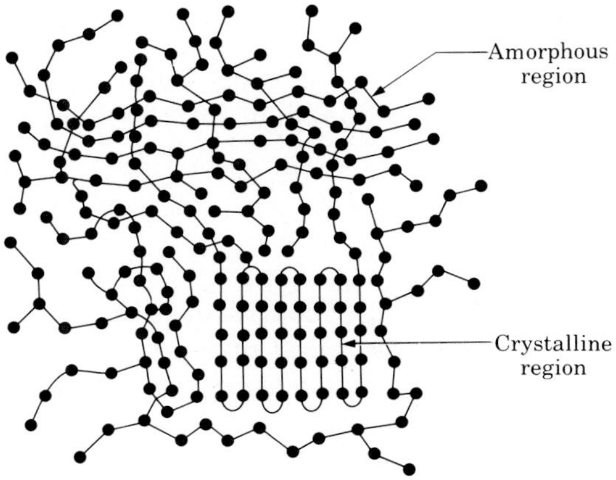

FIGURE 10.3
Amorphous and crystalline regions in a polymer. The crystalline region (crystallite) has an orderly arrangement of molecules. The higher the crystallinity, the harder, stiffer, and less ductile is the polymer.

tougher, and less ductile than low-density polyethylene (LDPE), which is about 60 percent crystalline and has a specific gravity of 0.915.

Optical properties are also affected by the degree of crystallinity. Because the index of refraction is proportional to density, the greater the density difference between the amorphous and crystalline phases, the greater is the opaqueness of the polymer (transmits less light). Polymers that are completely amorphous can be transparent, such as polycarbonate and acrylics.

10.2.3 Glass-transition temperature

Amorphous polymers do not have a specific melting point, but they undergo a distinct change in their mechanical behavior across a narrow range of temperature. At low temperatures they are hard, rigid, brittle, and glassy and at high temperatures are rubbery or leathery. The temperature at which this transition occurs is called the *glass-transition temperature*, T_g, and is also called the *glass point* or *glass temperature*. The term glass is included in this definition because glasses, which are amorphous solids, behave in the same manner (as you can see by holding a glass rod over a flame and observing its behavior).

Although most amorphous polymers exhibit this behavior, there are some exceptions, such as polycarbonate, which is not rigid or brittle below its glass-transition temperature. Polycarbonate is tough at ambient temperature and is thus used for safety helmets and shields.

To determine T_g, we measure the specific volume of the polymer and plot it against temperature to find the sharp change in the slope of the curve (Fig. 10.4). However, in the case of highly cross-linked polymers, the slope of the curve changes gradually near T_g, making it difficult to determine T_g for these polymers. The glass-transition temperature varies with different polymers (Table 10.2). For example, room temperature is above T_g for some polymers and below it for others. Unlike amorphous polymers, partly crystalline polymers have a distinct melting point, T_m (Fig. 10.4; see also Table 10.2). Because of the structural changes occurring, the specific volume of the polymer drops suddenly as its temperature is reduced.

10.2.4 Blends

To improve the brittle behavior of amorphous polymers below their glass-transition temperature, we can blend them (mix them with another polymer), usually with small quantities of an *elastomer*. These tiny particles are dispersed throughout the amorphous polymer, enhancing its toughness and impact strength by improving its resistance to crack propagation. Such polymers are known as *rubber modified*. More recent trends in blending involve several components, or *polyblends*, that utilize the favorable properties of different polymers. Some advances have been made in *miscible blends* (mixing without separation of two phases), a process similar to alloying of metals, enabling polymer blends to become more ductile.

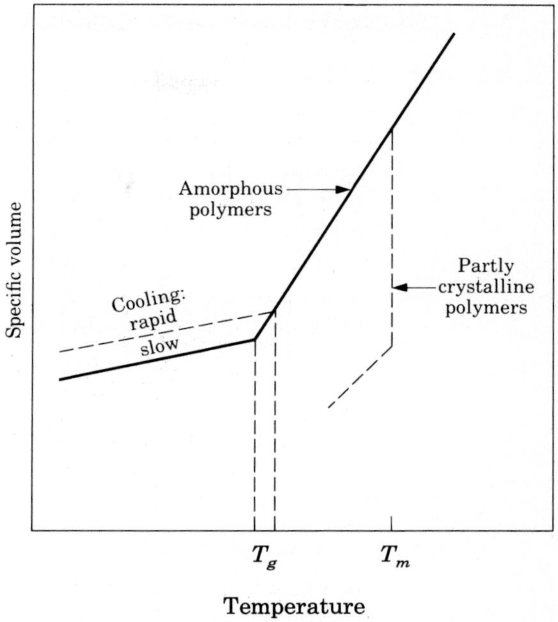

FIGURE 10.4
Specific volume of polymers as a function of temperature. Amorphous polymers, such as acrylic and polycarbonate, have a glass-transition temperature, T_g, but do not have a specific melting point, T_m. Partly crystalline polymers, such as polyethylene and nylons, contract sharply at their melting points during cooling.

TABLE 10.2
GLASS-TRANSITION AND MELTING TEMPERATURES OF SOME POLYMERS

MATERIAL	T_g(°C)	T_m(°C)
Nylon 6,6	57	265
Polycarbonate	150	265
Polyester	73	265
Polyethylene		
High density	−90	137
Low density	−110	115
Polymethylmethacrylate	105	—
Polypropylene	−14	176
Polystyrene	100	239
Polytetrafluoroethylene	−90	327
Polyvinyl chloride	87	212
Rubber	−73	—

10.3

Behavior of Thermoplastics

In addition to the structure and composition of thermoplastics, their behavior depends on numerous variables. Among the most important are temperature and rate of deformation. Below the glass-transition temperature, T_g, amorphous polymers are glassy (described as rigid, brittle, or hard), and they behave like an elastic solid. If the load exceeds a certain critical value the polymer fractures, just as a piece of glass does at room temperature. In the glassy region, the relationship between stress and strain is linear, or

$$\sigma = E\varepsilon. \tag{10.1}$$

If we test the polymer in torsion, we have

$$\tau = G\gamma. \tag{10.2}$$

The glassy behavior can be represented by a spring having a stiffness equivalent to the modulus of elasticity of the polymer (Fig. 10.5a). Note that the strain is completely recovered when the load is removed at time t_1.

If we raise the temperature of the thermoplastic to near T_g and then a little above, the polymer retains some of its elasticity while also becoming viscous. This is known as *viscoelastic* behavior and can be shown by the spring and dashpot models in Fig. 10.5(c) and (d), known as the *Maxwell* and *Kelvin* (or *Voigt*) models, respectively. When a constant load is applied, the polymer first stretches at a high strain rate, and then continues to elongate over a period of time because of its viscous behavior. Note in these models that the elastic portion of the elongation is reversible (elastic recovery), but the viscous portion is not.

The viscous behavior is expressed by

$$\tau = \eta(dv/dy) = \eta\dot{\gamma} \tag{10.3}$$

where η is the viscosity, and dv/dy is the shear strain rate $\dot{\gamma}$, as shown in Fig. 10.6. When the shear stress τ is directly proportional to the shear strain rate, the behavior is known as *Newtonian*. Note from Eq. (10.3) that the viscous behavior is similar to the strain rate sensitivity of metals, described in Section 2.2.7, and given by the expression

$$\sigma = C\dot{\varepsilon}^m \tag{10.4}$$

where we note that for Newtonian behavior, $m = 1$. Thermoplastics have high m values, indicating that they can undergo large uniform deformations in tension before fracture. Note how, unlike ordinary metals, the necked region elongates

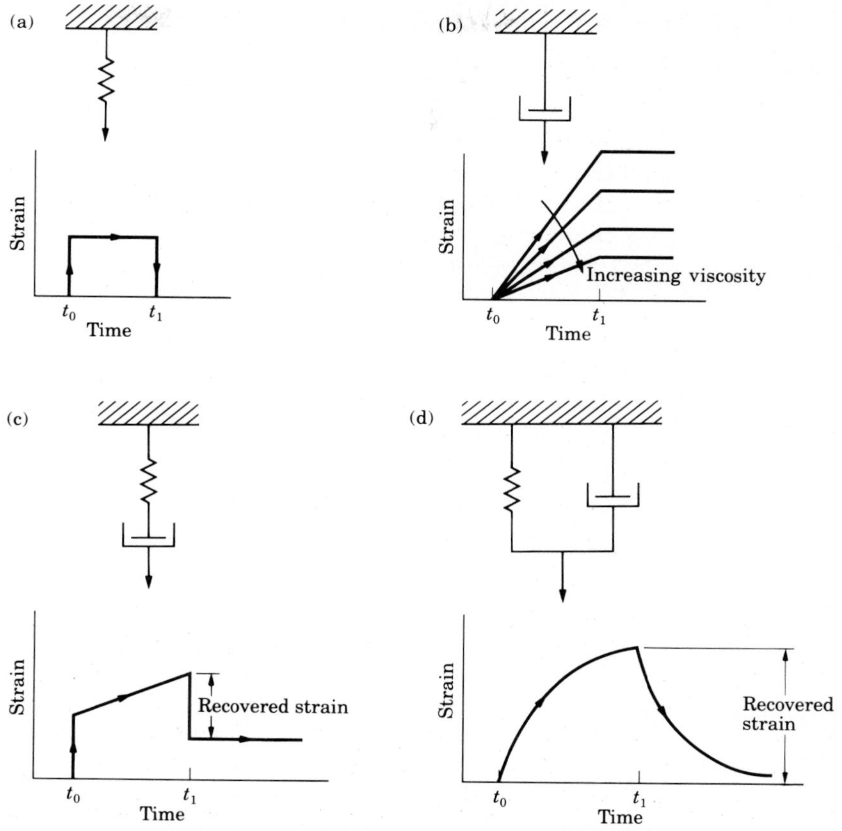

FIGURE 10.5
Various deformation modes for polymers: (a) elastic; (b) viscous; (c) viscoelastic (Maxwell model); and (d) viscoelastic (Voigt or Kelvin model).

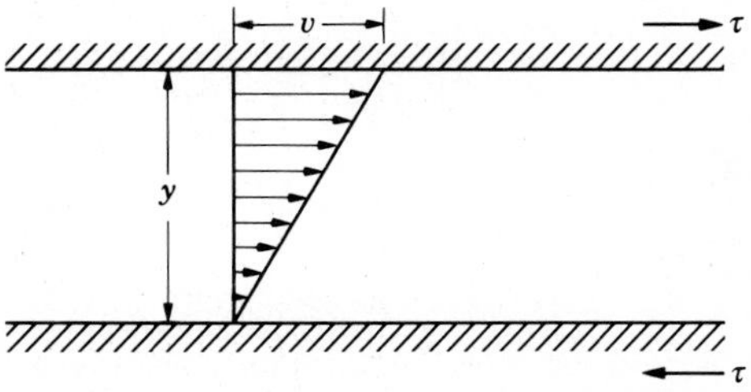

FIGURE 10.6
Parameters used to describe viscosity; see Eq. (10.3).

considerably (Fig. 10.7). This behavior can easily be demonstrated by stretching a piece of the plastic (polypropylene) holder for 6-pack beverage cans. This characteristic, which is the same as in superplastic metals, enables thermoplastics to be formed into complex shapes, such as bottles for soft drinks, cookie and meat trays, and lighted signs, as described in Section 10.9. Note also that the effect of increasing the rate of load application is to increase the strength of the polymer.

Between T_g and T_m, thermoplastics also exhibit leathery and rubbery behavior, depending on their structure and degree of crystallinity, as shown in Fig. 10.8. A term combining the strains caused by elastic behavior (e_e) and viscous flow (e_v) is called the *viscoelastic modulus* E_r, and is expressed as

$$E_r = \frac{\sigma}{(e_e + e_v)}. \tag{10.5}$$

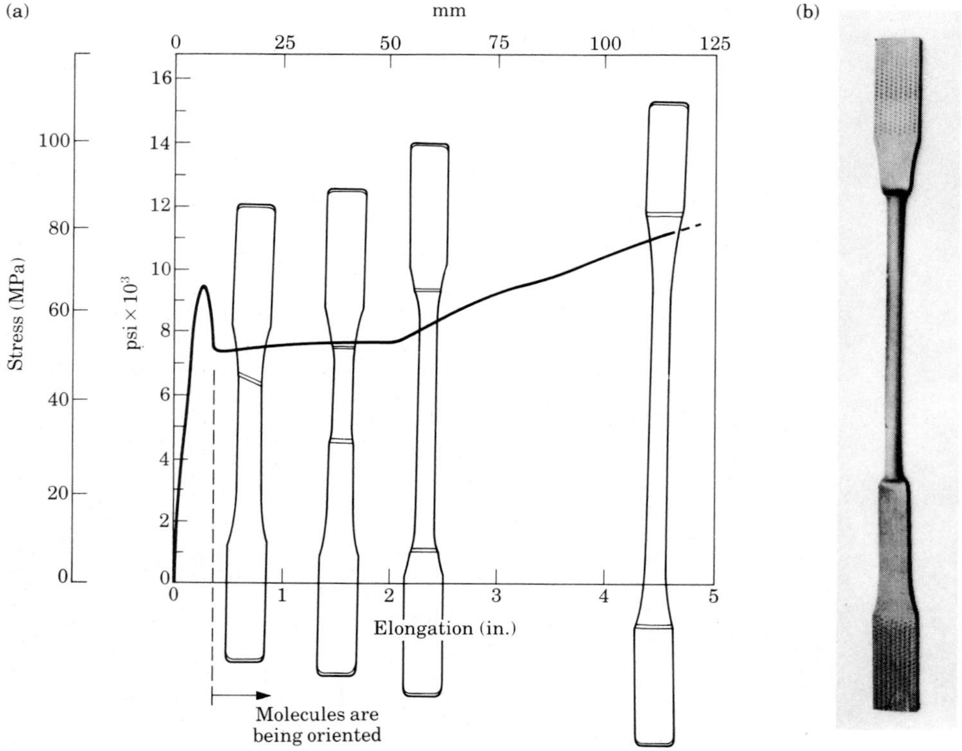

FIGURE 10.7
(a) Load-elongation curve for polycarbonate, a thermoplastic. *Source:* R. P. Kambour and R. E. Robertson. (b) High-density polyethylene tensile-test specimen, showing uniform elongation (the long, narrow region in the specimen).

This modulus, shown as the ordinate in Fig. 10.8, essentially represents a time-dependent elastic modulus.

The viscosity η of polymers—a measure of the resistance of their molecules in sliding along each other—depends on temperature and on their structure, molecular weight, and pressure. The effect of temperature can be represented by

$$\eta = \eta_0 e^{E/kT} \tag{10.6}$$

where η_0 is a material constant, E is the activation energy (energy required to initiate a reaction), k is Boltzmann's constant (thermal energy constant, or 13.8×10^{-24} J/K), and T is the temperature (K). Thus as the temperature of the polymer is increased, η decreases because of the higher mobility of the molecules. Note that viscosity may be regarded as the inverse of fluidity of molten metals in casting. Increasing the molecular weight (hence increasing chain length) increases η because of the greater number of secondary bonds present. As the molecular weight distribution widens, there are more shorter chains and η decreases. The effect of increasing pressure is to increase viscosity because of reduced *free volume* or *free space*, defined as the volume in excess of the true volume of the crystal in the crystalline regions of the polymer.

Based on experimental observations that, at the glass-transition temperature T_g, polymers have a viscosity η of about 10^{12} Pa·s, an empirical relationship between viscosity and temperature has been developed for linear thermoplastics:

$$\log \eta = 12 - \frac{17.5\,\Delta T}{52 + \Delta T}, \tag{10.7}$$

where $\Delta T = T - T_g$, in K or °C. Thus we can estimate the viscosity of the polymer at any temperature.

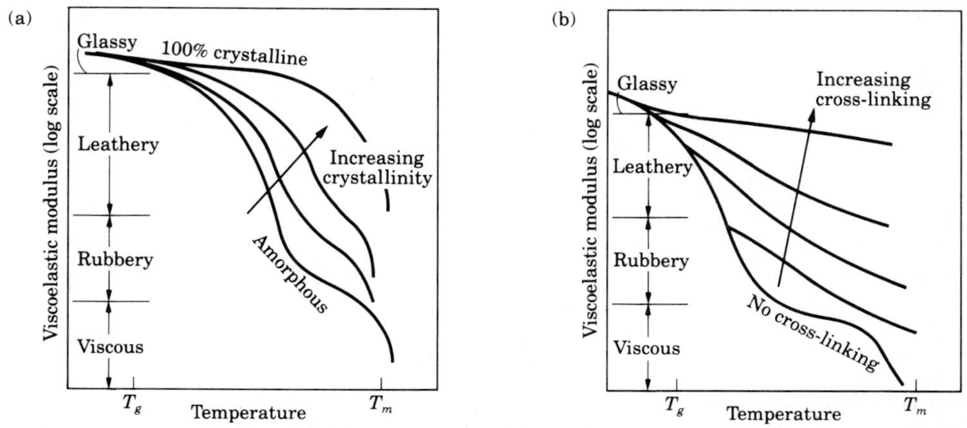

FIGURE 10.8
Viscoelastic modulus and behavior of polymers as a function of temperature and (a) degree of crystallinity and (b) cross-linking.

• Example 10.2: Lowering the viscosity of a polymer.

In processing a batch of polycarbonate at 170 °C to make a part, we find that its viscosity is twice that desired. We want to determine the temperature at which this polymer should be processed.

SOLUTION. From Table 10.2, we find that T_g for polycarbonate is 150 °C. We find its viscosity at 170 °C by using Eq. (10.7):

$$\log \eta = 12 - \frac{17.5(20)}{52 + 20} = 7.14.$$

Hence $\eta = 13.8$ MPa·s. Because this magnitude is twice what we want, the new viscosity should be 6.9 MPa·s. Therefore

$$\log (6.9 \times 10^6) = 12 - \frac{17.5(\Delta T)}{52 + \Delta T},$$

and $\Delta T = 21.7$, or the new temperature is $150 + 21.7 = 171.7$ °C. Note that we rounded the numbers for temperature and that viscosity is very sensitive to temperature.

10.3.1 Creep and stress relaxation

We defined the terms *creep* and *stress relaxation* in Section 2.8. Because of their viscoelastic behavior, thermoplastics are particularly susceptible to these phenomena. Referring back to the dashpot models in Fig. 10.5(b), (c), and (d), we note that under a constant load, the polymer undergoes further strain—thus it creeps. The recovered strain depends on the stiffness of the spring, hence the modulus of elasticity.

Stress relaxation in polymers occurs over a period of time, according to

$$\tau = \tau_0 e^{-t/\lambda}, \tag{10.8}$$

where τ_0 is the shear stress at time zero, τ is the stress at time t, and λ is the *relaxation time*. This expression may also be given in terms of the normal stresses σ and σ_0. The relaxation time is defined as

$$\lambda = \eta/G, \tag{10.9}$$

where G is the shear modulus of the polymer and $\gamma = G/\tau$. Thus the relaxation time is a function of the material and its temperature. As both η and G depend on temperature to varying degrees, the relaxation time also depends on temperature.

• Example 10.3: Stress relaxation in a thermoplastic member under tension.

A piece of thermoplastic is stretched between two rigid supports at a stress level of 5 MPa. After 30 days, the stress level has decayed to half the original level. How long will it take for the stress level to reach one-tenth of the original value?

SOLUTION. Substituting these data into Eq. (10.8) as modified for normal stress, we find that

$$2.5 = 5e^{-30/\lambda}.$$

Hence

$$\ln\left(\frac{2.5}{5}\right) = \frac{-30}{\lambda}, \quad \text{or} \quad \lambda = 43.3 \text{ days}.$$

Therefore

$$\ln\left(\frac{0.5}{5}\right) = \frac{-t}{43.3}, \quad \text{or} \quad t = 99.7 \text{ days}.$$

10.3.2 Mechanical properties

The mechanical properties of several polymers listed in Table 10.1 indicate that thermoplastics are about two orders of magnitude less stiff than metals. Their ultimate tensile strength is about one order of magnitude lower than that of metals (see Tables 2.1 and 2.2). Typical stress–strain curves for some thermoplastics and thermosets at room temperature are shown in Fig. 10.9. Note that these plastics

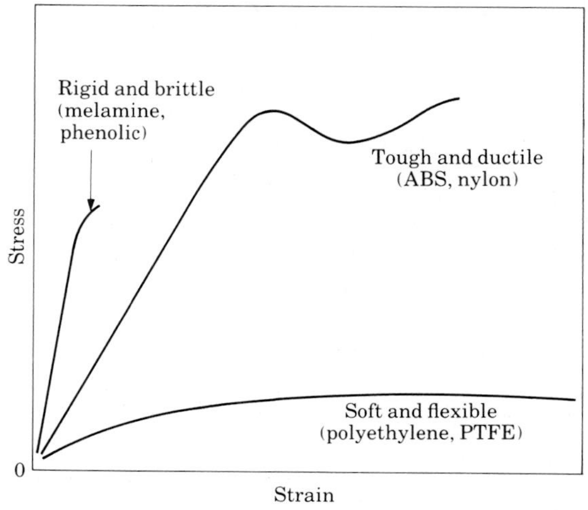

FIGURE 10.9
General terminology describing the behavior of three types of plastics. PTFE (polytetrafluoroethylene) is Teflon, a trade name. *Source:* R. L. E. Brown.

exhibit different behaviors, which we describe as rigid, soft, brittle, flexible, and so on. Plastics undergo fatigue and creep phenomena, just as metals do.

The deformation of thermoplastics at room temperature not only increases their strength (Fig. 10.10) but also results in anisotropy (*orientation* of the molecules), much like that observed in cold working of metals. The increased strength and toughness of polymers, with cold working, makes these materials attractive for applications in which improved mechanical properties are required. Note in Fig. 10.10 that with greater cold-rolling reduction, the material becomes more and more strain hardening; that is, the curve has an increasing, positive slope. Thus, as described in Section 2.2.4, the uniform elongation increases with greater cold-rolling reduction, and hence the material can be stretched further during processing.

The typical effects of temperature on the strength and elastic modulus of thermoplastics is similar to those for metals. Thus with increasing temperature, the strength and modulus of elasticity decrease, and toughness increases (Figs. 10.11 and 10.12). The effect of temperature on impact strength is shown

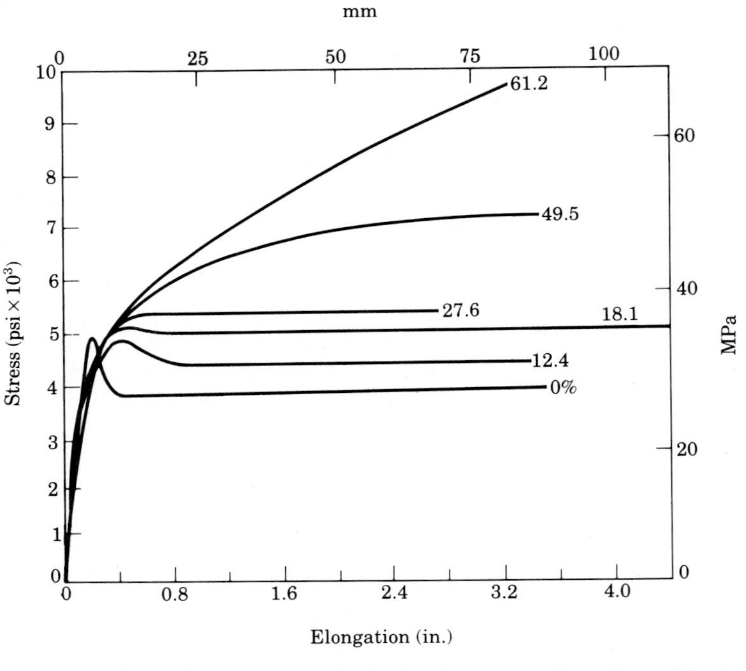

FIGURE 10.10

Stress-elongation curves for ABS (acrylonitrile-butadiene-styrene) as a function of thickness reduction by cold-rolling. Tested in the direction of rolling. *Source:* L. J. Broutman and S. Kalpakjian, *SPE J*, vol. 25, no. 10, 1969, pp. 46–52.

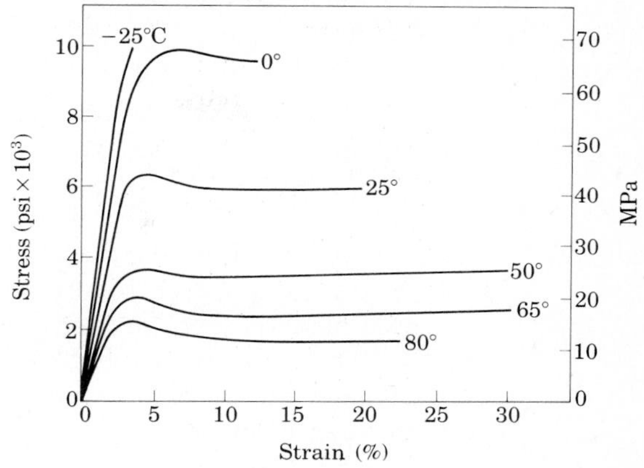

FIGURE 10.11

Effect of temperature on the stress–strain curve for cellulose acetate, a thermoplastic. Note the large drop in strength and increase in ductility with a relatively small increase in temperature. *Source:* After T. S. Carswell and H. K. Nason.

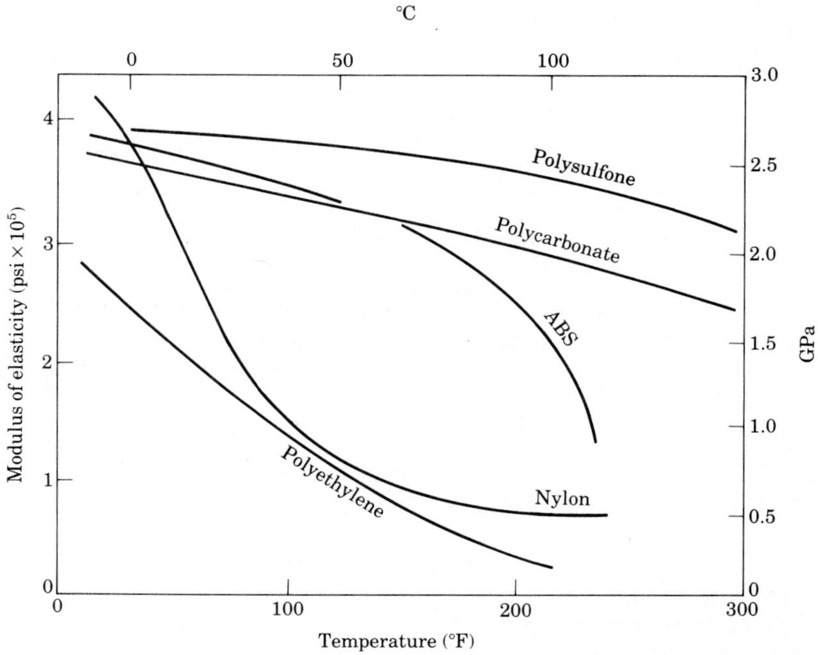

FIGURE 10.12

Effect of temperature on the modulus of elasticity for several plastics. Note that some plastics are more sensitive to temperature than others.

in Fig. 10.13. Note the large difference in the impact behavior of various polymers.

Crazing. Some thermoplastics, such as polystyrene and polymethylmethacrylate, develop localized, wedge-shaped, narrow regions of highly deformed material when subjected to tensile stresses or to bending. This phenomenon is called *crazing*. Although they may appear to be like cracks, crazes are spongy material, typically containing about 50 percent voids. With increasing tensile load on the specimen, these voids coalesce and eventually lead to fracture of the polymer. Crazing has been observed both in transparent glassy polymers and in other types.

The environment and the presence of solvents, lubricants, and water vapor enhance the formation of crazes (*environmental stress cracking* and *solvent crazing*). Residual stresses in the material contribute to crazing and cracking of the polymer. Radiation and visible light can adversely affect the strength of polymers.

A related phenomenon is *stress whitening*. When subjected to tensile stresses, such as by folding or bending, the plastic becomes lighter in color. This phenomenon is usually attributed to the formation of microvoids in the material. As a result, the material becomes less translucent (transmits less light), or more opaque. You can easily demonstrate this result by bending plastic components commonly found in household products and toys.

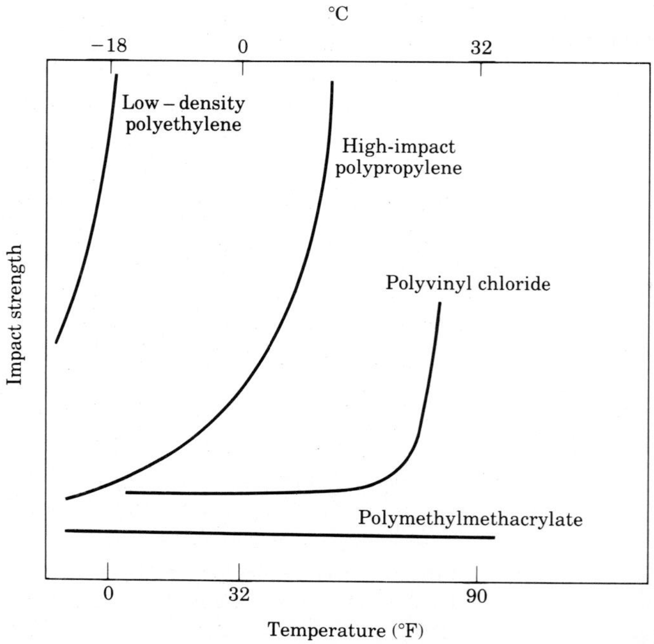

FIGURE 10.13 Effect of temperature on the impact strength for various plastics. Note that small changes in temperature have a significant effect on the impact strength of some plastics. *Source:* After P. C. Powell.

Water absorption. An important limitation of some polymers, such as nylons, is their ability to absorb water. Water acts as a plasticizing agent; that is, it makes the polymer more plastic. Thus in a sense, it lubricates the chains in the amorphous region. Typically, with increasing moisture absorption, the glass-transition temperature, the yield stress, and the elastic modulus of the polymer are lowered severely. Dimensional changes also occur because of water absorption, such as in a humid environment.

Thermal and electrical properties. Compared to metals, plastics are generally characterized by low thermal and electrical conductivity, low specific gravity (ranging from 0.90 to 2.2), and relatively high coefficient of thermal expansion, which is about an order of magnitude higher.

Most polymers are electrical insulators and hence are used for insulators and as packaging material for electronic components. However, the electrical conductivity of some polymers can be increased by doping (introducing certain impurities in the polymer, such as metal powder). One of the earliest conducting polymers developed was polyacetylene. The electrical conductivity of polymers increases with moisture absorption.

10.4 Thermosets

When the long-chain molecules in a polymer are cross-linked in a three-dimensional arrangement, the structure in effect becomes one giant molecule with strong covalent bonds. As we previously stated, these polymers are called thermosetting plastics, or thermosets, because during polymerization the network is completed and the shape of the part is permanently set. This *curing* reaction, unlike that in thermoplastics, is irreversible. We can liken the response of a thermosetting plastic to temperature to baking a cake or boiling an egg. Once the cake is baked and cooled, or the egg boiled and cooled, reheating it will not change its shape.

Some thermosets, such as epoxy, polyester, and urethane, cure at room temperature. Although curing takes place at ambient temperature, the heat of the reaction cures the plastic. Thermosetting polymers do not have a sharply defined glass-transition temperature. The polymerization process for thermosets generally takes place in two stages. The first is at the chemical plant where the molecules are partially polymerized into linear chains. The second stage is at the parts-producing plant where cross-linking is completed under heat and pressure during the molding and shaping of the part.

Because of the nature of their bonds, the strength and hardness of thermosets, unlike thermoplastics, are not affected by temperature or rate of deformation. A typical thermoset is phenolic, which is a product of the reaction between phenol and formaldehyde. Common products made from this polymer are the handles and knobs on cooking pots and pans and components of light switches and outlets. Thermosetting plastics generally possess better mechanical, thermal, and chemical properties, electrical resistance, and dimensional stability than do thermoplastics. However, if the temperature is increased sufficiently, the thermosetting polymer begins to burn up, degrade, and char.

10.5 Additives

In order to impart certain specific properties, polymers are usually compounded with *additives*. These additives modify and improve certain characteristics of polymers, such as their stiffness, strength, color, weatherability, flammability, arc resistance for electrical applications, and ease of subsequent processing.

Fillers used are generally wood flour (fine sawdust), silica flour (fine silica powder), clay, powdered mica, and short fibers of cellulose, glass, and asbestos. Because of their low cost, fillers are important in reducing the overall cost of polymers. Depending on their type, fillers improve the strength, hardness, toughness, abrasion resistance, dimensional stability, and/or stiffness of plastics. These properties are greatest at various percentages of different types of polymer–filler combinations. As in reinforced plastics, a filler's effectiveness depends on the nature of the bond between the filler material and the polymer chains.

Plasticizers are added to some polymers to impart flexibility and softness by lowering their glass-transition temperature. Plasticizers are low molecular weight solvents with high boiling points (nonvolatile). They reduce the strength of the secondary bonds between the long-chain molecules, thus making the polymer soft and flexible. The most common use of plasticizers is in polyvinyl chloride (PVC), which remains flexible during its many uses. Other applications of plasticizers are in thin sheet, film, tubing, shower curtains, and clothing materials.

Most polymers are adversely affected by ultraviolet radiation (sunlight) and oxygen, which weaken and break the primary bonds, resulting in the scission (splitting) of the long-chain molecules. The polymer then degrades and becomes brittle and stiff. On the other hand, degradation may be beneficial, as in the disposal of plastic objects by subjecting them to environmental attack.

A typical example of protection against ultraviolet radiation is the compounding of rubber with *carbon black* (soot). The carbon black absorbs a high percentage of the ultraviolet radiation. Protection against degradation by oxidation, particularly at elevated temperatures, is done by adding antioxidants to the polymer. Various coatings are another means of protection of polymers.

The wide variety of colors available in plastics is obtained by adding *colorants*. They are either organic (dyes) or inorganic (pigments) materials. The selection of a colorant depends on service temperature and exposure to light. Pigments, which are dispersed particles, generally have greater resistance than dyes to temperature and light.

If the temperature is sufficiently high, most polymers will ignite. The flammability (ability to support combustion) of polymers varies considerably, depending on their composition (such as the chlorine and fluorine content). Polymethylmethacrylate, for example, continues to burn when ignited, whereas polycarbonate self-extinguishes. The flammability of polymers can be reduced either by making them from less flammable raw materials, or by the addition of *flame retardants*, such as compounds of chlorine, bromine, and phosphorus.

Lubricants may be added to polymers to reduce friction during their subsequent processing into useful products and to prevent parts from sticking to the molds. Lubrication is also important in preventing thin polymer films from sticking to each other.

10.6 General Properties and Applications of Thermoplastics

In this section we outline the general characteristics and typical applications of major thermoplastics, particularly as they relate to manufacturing plastic products and their service life. General recommendations for various plastics applications are given in Table 10.3.

TABLE 10.3
GENERAL RECOMMENDATIONS FOR PLASTIC PRODUCTS

DESIGN REQUIREMENT	APPLICATIONS	PLASTICS
Mechanical strength	Gears, cams, rollers, valves, fan blades, impellers, pistons	Acetal, nylon, phenolic, polycarbonate
Functional and decorative	Handles, knobs, camera and battery cases, trim moldings, pipe fittings	ABS, acrylic, cellulosic, phenolic, polyethylene, polypropylene, polystyrene, polyvinyl chloride
Housings and hollow shapes	Power tools, pumps, housings, sport helmets, telephone cases	ABS, cellulosic, phenolic, polycarbonate, polyethylene, polypropylene, polystyrene
Functional and transparent	Lenses, goggles, safety glazing, signs, food-processing equipment, laboratory hardware	Acrylic, polycarbonate, polystyrene, polysulfone
Wear resistance	Gears, wear strips and liners, bearings, bushings, roller-skate wheels	Acetal, nylon, phenolic, polyimide, polyurethane, ultrahigh molecular weight polyethylene

Acetals (from acetic and alcohol) have good strength, stiffness, and resistance to creep, abrasion, moisture, heat, and chemicals. Typical applications are mechanical parts and components where high performance is required over a long period: bearings, cams, gears, bushings, rollers, impellers, wear surfaces, pipes, valves, shower heads, and housings. Their common trade name is Delrin.

Acrylics (polymethylmethacrylate, PMMA) possess moderate strength, good optical properties, and weather resistance. They are transparent but can be made opaque, are generally resistant to chemicals, and have good electrical resistance. Typical applications are lenses, lighted signs, displays, window glazing, skylights, bubble tops, automotive lenses, windshields, lighting fixtures, and furniture. Their common trade names are Plexiglas and Lucite.

Acrylonitrile-butadiene-styrene (ABS) is dimensionally stable and rigid and has good impact, abrasion, and chemical resistance, strength and toughness, low-temperature properties, and electrical resistance. Typical applications are pipes, fittings, chrome-plated plumbing supplies, helmets, tool handles, automotive components, boat hulls, telephones, luggage, housing, appliances, refrigerator liners, and decorative panels.

Cellulosics have a wide range of mechanical properties, depending on composition. They can be made rigid, strong, and tough. However, they weather poorly and are affected by heat and chemicals. Typical applications are tool handles, pens, knobs, frames for eyeglasses, safety goggles, machine guards, helmets, tubing and pipes, lighting fixtures, rigid containers, steering wheels, packaging film, signs, billiard balls, toys, and decorative parts.

Fluorocarbons possess good resistance to temperature, chemicals, weather, and electricity. They also have unique nonadhesive properties and low friction. Typical applications are linings for chemical process equipment, nonstick coatings for cookware, electrical insulation for high-temperature wire and cable, gaskets, low friction surfaces, bearings, and seals. Their common trade name is Teflon.

Polyamides (from the words poly, amine, and carboxyl acid) are available in two main types: nylons and aramids. *Nylons* (a coined word) have good mechanical properties and abrasion resistance. They are self-lubricating and resistant to most chemicals. All nylons are hygroscopic (absorb water). Moisture absorption reduces mechanical properties and increases part dimensions. Typical applications are gears, bearings, bushings, rollers, fasteners, zippers, electrical parts, combs, tubing, wear-resistant surfaces, guides, and surgical equipment. *Aramids* (aromatic polyamides) have very high tensile strength and stiffness. Typical applications include fibers for reinforced plastics (composite materials), bulletproof vests, cables, and radial tires. Their common trade name is Kevlar.

Polycarbonates are versatile and have good mechanical and electrical properties. They also have high impact resistance, and can be made resistant to chemicals. Typical applications are safety helmets, optical lenses, bullet-resistant window glazing, signs, bottles, food-processing equipment, windshields, load-bearing electrical components, electrical insulators, medical apparatus, business machine components, guards for machinery, and parts requiring dimensional stability. Their common trade name is Lexan.

Polyesters (thermoplastic; see also Section 10.7) have good mechanical, electrical, and chemical properties, good abrasion resistance, and low friction. Typical applications are gears, cams, rollers, load-bearing members, pumps, and electromechanical components. Their common trade names are Dacron, Mylar, and Kodel.

Polyethylenes possess good electrical and chemical properties. Their mechanical properties depend on composition and structure. Three major classes are low density (LDPE), high density (HDPE), and ultrahigh molecular weight (UHMWPE). Typical applications for LDPE are housewares, bottles, garbage cans, ducts, bumpers, luggage, toys, tubing, bottles, and packaging material; for HDPE, machinery parts, belts and straps, wear-resistant surfaces, sleds, canoes, and camper tops; and for UHMWPE, parts requiring high-impact toughness and abrasive wear resistance.

Polyimides have the structure of a thermoplastic but the nonmelting characteristic of a thermoset (see Section 10.7).

Polypropylenes have good mechanical, electrical, and chemical properties and good resistance to tearing. Typical applications are automotive trim and components, medical devices, appliance parts, wire insulation, TV cabinets, pipes, fittings, drinking cups, dairy-product and juice containers, luggage, ropes, and weather stripping.

Polystyrenes have properties that depend on composition. They are inexpensive, they have generally average properties, and they are somewhat brittle. Typical applications include disposable containers, packaging, trays for meats, cookies, and candy, foam insulation, appliances, automotive and radio/TV components, housewares, and toys and furniture parts (as a wood substitute).

Polysulfones have excellent resistance to heat, water, and steam, are highly resistant to some chemicals, but are attacked by organic solvents. Typical applications are steam irons, coffeemakers, hotwater containers, medical equipment that requires sterilization, power-tool and appliance housings, aircraft cabin interiors, and electrical insulators.

Polyvinyl chloride (PVC) has a wide range of properties, is inexpensive and water resistant, and can be made rigid or flexible. It is not suitable for applications requiring strength and heat resistance. Rigid PVC is tough and hard. Typical applications are rigid PVC for the construction industry, such as pipes and conduits, automobile windshields, and signs; flexible PVC for wire and cable coatings, low-pressure flexible tubing and hose, footware, imitation leather, upholstery, records, gaskets, seals, trim, film, sheet, and coatings. Their common trade names are Saran and Tygon.

Vinyls. See polyvinyl chloride.

10.7

General Properties and Applications of Thermosetting Plastics

In this section we outline the general characteristics and typical applications of major thermosetting plastics.

Alkyds (from alkyl, meaning alcohol, and acid) possess good electrical insulating properties, impact resistance, and dimensional stability and have low water absorption. Typical applications are electrical and electronic components.

Aminos (urea and melamine) have properties that depend on composition. Generally, aminos are hard and rigid and are resistant to abrasion, creep, and electrical arcing. Typical applications are small appliance housings, countertops, circuit breakers, handles, and distributor caps. Urea is used for electrical and electronic components, melamine for dinnerware.

Epoxies have excellent mechanical and electrical properties, dimensional stability, strong adhesive properties, and good resistance to heat and chemicals. Typical applications are electrical components requiring mechanical strength and high insulation, tools and dies, and adhesives. Fiber-reinforced epoxies have excellent mechanical properties and are used in pressure vessels, rocket motor casings, tanks, and similar structural components.

Phenolics, although brittle, are rigid and dimensionally stable, and have high resistance to heat, water, electricity, and chemicals. Typical applications are knobs, handles, laminated panels, telephones, bond material to hold abrasive grains together in grinding wheels, and electrical components, such as wiring devices, connectors, and insulators.

Polyesters (thermosetting; see also Section 10.6) have good mechanical, chemical, and electrical properties. Polyesters are generally reinforced with glass or other fibers. Typical applications are boats, luggage, chairs, automotive bodies, swimming pools, material for impregnating cloth, paper, and decorations. They also are available as casting resins.

Polyimides possess good mechanical, physical, and electrical properties at elevated temperatures. They also have creep resistance and low friction and wear characteristics. Polyimides have the nonmelting characteristics of a thermoset but the structure of a thermoplastic. Typical applications are pump components (bearings, seals, valve seats, retainer rings, and piston rings), electrical connectors for high-temperature use, aerospace parts, high-strength impact-resistant structures, sports equipment, and safety vests.

Silicones have properties that depend on composition. Generally, they possess excellent electrical properties over a wide range of humidity and temperature, weather well, and resist chemicals and heat (see also Section 10.8). Typical applications are electrical components requiring strength at elevated temperatures, oven gaskets, heat seals, and waterproof materials.

10.8 Elastomers (Rubbers)

Elastomers comprise a large family of amorphous polymers having a low glass-transition temperature. They have the characteristic ability to undergo large elastic deformations without rupture; they are soft and have a low elastic modulus. The term elastomer is derived from the words *elastic* and *mer*. These polymers are highly kinked (tightly twisted or curled). They stretch but then return to their original shape after the load is removed. Cross-linking also takes place, the best example being the elevated temperature *vulcanization* of rubber with sulfur, discovered by Charles Goodyear in 1839 and named for Vulcan, the Roman god of fire. Once the elastomer is cross-linked, it cannot be reshaped. For example, an automobile tire, which is one giant molecule, cannot be reshaped.

The terms rubber and elastomer are often used interchangeably. Generally, an *elastomer* is defined as being capable of recovering substantially in shape and size after the load has been removed. *Rubber* is defined as being capable of recovering from large deformations quickly.

The hardness of elastomers, which is measured with a durometer, increases with increasing cross-linking of the molecular chains. A variety of additives can be blended with elastomers to impart specific properties, as with plastics. Elastomers have a wide range of applications, such as high-friction and nonskid surfaces, protection against corrosion and abrasion, electrical insulation, and shock and vibration insulation. Examples include tires, hoses, weather stripping, footwear, linings, gaskets, seals, printing rolls, and flooring.

A characteristic of elastomers is their hysteresis loss in stretching or compression (Fig. 10.14). The clockwise loop indicates energy loss, whereby mechanical energy is converted into heat. This property is desirable for absorbing vibrational energy (damping) and sound deadening.

The base for *natural rubber* is *latex*, a milklike sap obtained from the inner bark of a tropical tree. It has good resistance to abrasion and fatigue and high frictional

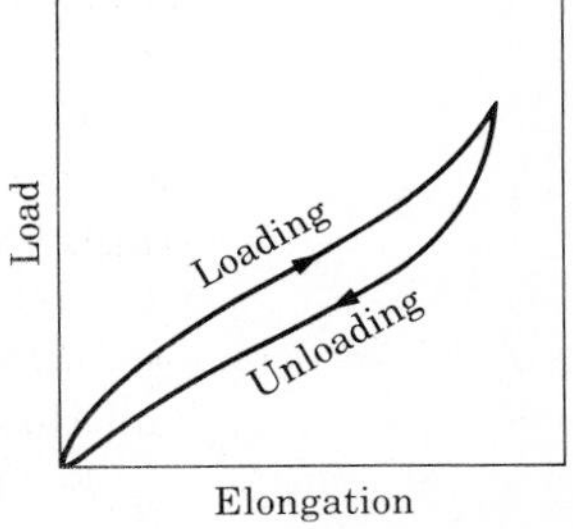

FIGURE 10.14 Typical load-elongation curve for rubbers. The clockwise loop, indicating loading and unloading paths, is the hysteresis loss. Hysteresis gives rubbers the capacity to dissipate energy, damp vibration, and absorb shock loading, as in automobile tires and vibration dampeners placed under machinery.

properties, but it has low resistance to oil, heat, ozone, and sunlight. Typical applications are tires, seals, shoe heels, coupling, and engine mounts.

Further developed natural rubbers are the *synthetic rubbers*. Examples are synthetic natural rubber, butyl, styrene butadiene, polybutadiene, and ethylene propylene. Compared to natural rubbers, they have improved resistance to heat, gasoline, and chemicals and higher useful temperature range. Examples of synthetic rubbers that are resistant to oil are neoprene, nitrile, urethane, and silicone. Typical applications of synthetic rubbers are tires, shock absorbers, seals, and belts.

Silicones (see also Section 10.7) have the highest useful temperature range, up to 315 °C (600 °F), but their other properties—such as strength and resistance to wear and oils—are generally inferior to other elastomers. Typical applications are seals, gaskets, thermal insulation, high-temperature electrical switches, and electronic apparatus.

Polyurethane has very good overall properties of high strength, stiffness, and hardness and exceptional resistance to abrasion, cutting, and tearing. Typical applications are seals, gaskets, cushioning, and diaphragms for rubber forming of sheet metals.

10.9 Forming and Shaping Polymers

The processing of polymers involves operations similar to those used to form and shape metals, as we described in the preceding chapters. Plastics can be molded, cast, and formed, as well as machined and joined, into many shapes with relative ease and with little or no additional operations required. Polymers melt or cure at relatively low temperatures and hence, unlike metals, are easy to handle and require less energy to process.

Plastics are usually shipped to manufacturing plants as pellets or powders and are melted just before the shaping process. Plastics are also available as sheet, plate, rod, and tubing, which may be formed into a variety of products. Liquid plastics are used especially in making reinforced plastic parts.

10.9.1 Extrusion

In extrusion, raw materials in the form of thermoplastic pellets, granules, or powder are placed into a hopper and fed into the extruder barrel (Fig. 10.15). The barrel is equipped with a *screw* that blends and conveys the pellets down the barrel. The internal friction from the mechanical action of the screw, along with heaters around the extruder's barrel, heats the pellets and liquefies them. The screw action also builds up pressure in the barrel.

Screws have three distinct sections: (1) a feed section that conveys the material from the hopper area into the central region of the barrel; (2) a melt, or transition,

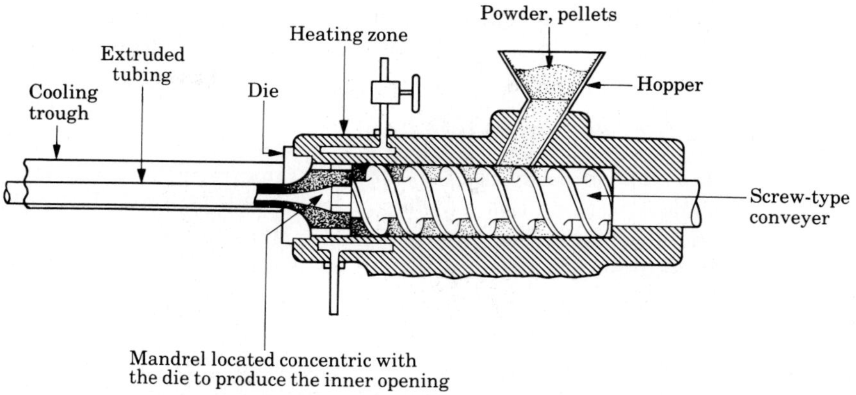

FIGURE 10.15
Schematic illustration of extrusion of plastic tubing in a screw-type extruder. Various solid cross-sections are also extruded by this process.

section where the heat generated from shearing of the plastic causes melting to begin; and (3) a pumping section where additional shearing and melting occurs, with pressure buildup at the die. The lengths of these sections can be changed to accommodate the melting characteristics of different plastics.

The molten plastic is forced through a die—similar to extruding metals. The extruded product is then cooled, either by air or by passing it through a water-filled channel. Controlling the rate and uniformity of cooling is important to minimize product shrinkage and distortion. The extruded product can also be drawn (sized) by a puller after it has cooled. The extruded product is then coiled or cut into desired lengths. Complex shapes with constant cross-section can be extruded with relatively inexpensive tooling. This process is also used to extrude elastomers.

Because this operation is continuous, long products with cross-sections of solid rods, channels, pipe, window frames, and architectural components, as well as sheet, are extruded through dies of various geometries. Plastic-coated wire, cable, or strips for electrical applications are also extruded and coated by this process. The wire is fed into the die opening at a controlled rate with the extruded plastic.

Pellets, which are used for other plastics-processing methods described in this chapter, are also made by extrusion. Here the extruded product is a small-diameter rod and is chopped into short lengths, or *pellets*, as it is extruded. With some modifications, extruders can also be used as simple melters for other shaping processes, such as injection molding and blow molding.

Process parameters such as extruder screw speed, barrel-wall temperatures, die design, and cooling and drawing speeds should be controlled carefully in order to extrude products having uniform dimensional accuracy. To filter out unmelted or congealed resin, a metal screen is usually placed just before the die and is replaced periodically.

Extruders are generally rated by the diameter of the barrel and by the length-to-diameter (L/D) ratio of the barrel. Typical commercial units are 25–200 mm (1–8 in.) in diameter, with L/D ratios of 5 to 30.

10.9.2 Injection molding

Injection molding is essentially the same process as hot-chamber die casting. The pellets or granules are fed into a heated cylinder, where they are melted. The melt is then forced into a split-die chamber (Fig. 10.16a), either by a hydraulic plunger or by the rotating screw system of an extruder. Newer equipment is of the *reciprocating screw* type (Fig. 10.16b). As the pressure builds up at the die entrance, the rotating screw begins to move backward under pressure to a predetermined distance, thus controlling the volume of material to be injected. The screw stops rotating and is pushed forward hydraulically, forcing the molten plastic into the die cavity. The pressure in the mold cavity is usually in the range of 35–140 MPa (5–20 ksi).

Typical injection-molded products are cups, containers, housings, tool handles, knobs, electrical and communication components (such as telephone receivers), toys, and plumbing fittings. Although for thermoplastics the dies are relatively cool, thermosets are molded in heated dies where *polymerization* and *cross-linking* take place. In either case, after the part is sufficiently cooled (for thermoplastics) or set or cured (for thermosets), the dies are opened and the part is ejected. The dies are then

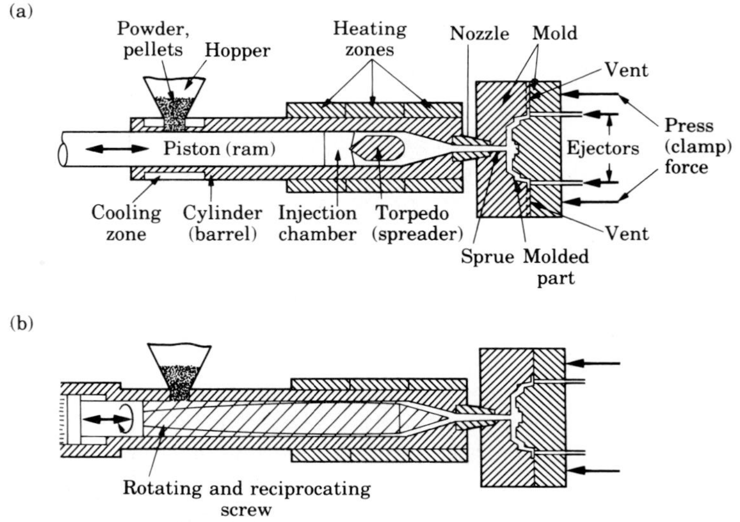

FIGURE 10.16
Injection molding with (a) plunger and (b) reciprocating rotating screw. Telephone receivers, plumbing fittings, tool handles, and housings are made by injection molding.

closed and the process is repeated automatically. Elastomers also are injection molded by these processes.

Because the material is molten when injected into the mold, complex shapes and good dimensional accuracy can be achieved, as in die casting. Dies with moving and unscrewing mandrels are not unusual, and allow the molding of parts with multiple cavities and internal and external threads.

Injection molding is a high-rate production process, with good dimensional control. Typical cycle times range from 5 to 60 seconds, but can be several minutes for thermosetting materials. The dies, generally made of tool steels or beryllium–copper, may have multiple cavities so that more than one part can be made in one cycle. Proper die design and control of material flow in the die cavities are important factors in the quality of the product. Other factors affecting quality are injection pressure, temperature, and condition of the resin.

Injection-molded parts are generally molded to final desired dimensions, and no subsequent finishing operations are required. However, the plastic in the channels (sprue and runners, as in metal-casting molds) that connect the mold cavity to the end of the barrel should be removed. These are normally trimmed off but can also be removed by tumbling. When thermoplastic resins are molded, the sprues and runners can be chopped and recycled. Newer and more expensive molds have heated sprues and runners, which eliminate the need for trimming.

Injection-molding machines are generally horizontal and are rated according to the capacity of the mold and the clamping force on the dies. Although in most machines this force generally ranges from 0.9 MN to 2.2 MN (100 tons to 250 tons), the largest machine in operation has a capacity of 45 MN (5000 tons), and can produce parts weighing 25 kg (55 lb). However, parts typically weigh 100–600 g (3–20 oz).

• Example 10.4: Injection molding of parts.

A 250-ton injection molding machine is to be used to make 4.5-in. diameter spur gears, 0.5 in. thick. The gears have fine teeth profile. How many gears can be injection molded in one set of molds? Does the thickness of the gears influence the answer?

SOLUTION. Because of the fine detail involved, the pressures required in the mold cavity will probably be on the order of 100 MPa (15 ksi). The cross-sectional (projected) area of the gear is $\pi(4.5)^2/4 = 15.9\ \text{in}^2$. If we assume that the parting plane of the two halves of the mold is in the middle of the gear, the force required is $(15.9)(15{,}000) = 238{,}500$ lb. The capacity of the machine is 200 tons, so we have $(250)(2000) = 500{,}000$ lb of clamping force available. Therefore the mold can accommodate two cavities, thus producing two gears per cycle. Because it does not influence the cross-sectional area of the gear, the thickness of the gear does not directly influence the pressures involved, and the answer is the same.

•

Reaction-injection molding. In the *reaction-injection molding* (RIM) process, a mixture of two or more reactive fluids is forced under high pressure into the mold cavity (Fig. 10.17). Chemical reactions take place rapidly in the mold and the polymer solidifies, producing a thermoset part. Major applications are automotive bumpers and fenders, thermal insulation for refrigerators and freezers, and stiffeners for structural components. Various reinforcing fibers, such as glass or graphite, may also be used to improve the product's strength and stiffness.

Structural foam molding. The *structural foam molding* process is used to make plastic products that have a solid skin and a cellular inner structure. Typical products are furniture components, TV cabinets, business-machine housings, and storage-battery cases. Although there are several foam molding processes, they are basically similar to injection molding or extrusion. Both thermoplastics and thermosets can be used for foam molding.

In *injection foam molding*, thermoplastics are mixed with a blowing agent (usually an inert gas such as nitrogen), which expands the material. The core of the part is cellular and the skin is rigid. The thickness of the skin can be as much as 2 mm (0.08 in.), and part densities are as low as 40 percent of the density of the solid plastic. Thus parts have a high stiffness-to-weight ratio.

10.9.3 Blow molding

Blow molding is a modified extrusion and injection molding process, wherein a tube is extruded (usually turned so that it is vertical), clamped into a mold with a cavity

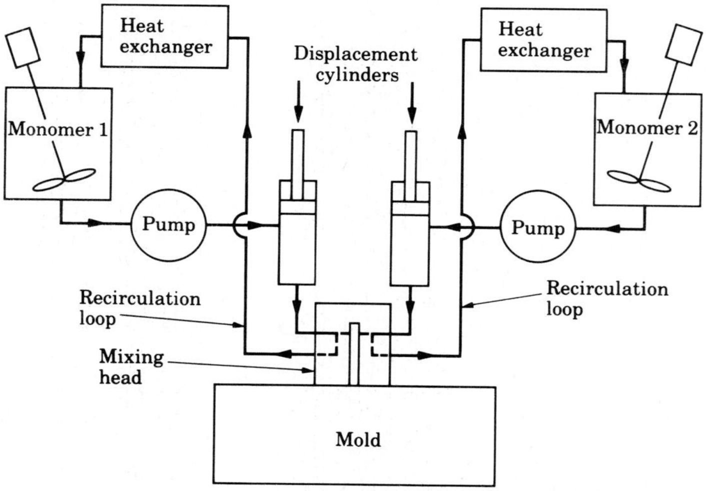

FIGURE 10.17 Schematic illustration of the reaction-injection molding process. *Source:* Modern Plastics Encyclopedia.

much larger than the tube diameter, and then blown outward to fill the mold (Fig. 10.18a). Thus blow molding is similar to blowing up a balloon inside a bottle. Blowing is usually done with a hot-air blast, at a pressure of 350–700 kPa (50–100 psi).

In some operations, the extrusion is continuous, and the molds move with the tubing. The molds close around the tubing, close off both ends (thereby breaking the tube into sections), and then move away as air is injected. The part is then cooled and ejected. In other operations, extrusion is reciprocating as in injection molding, and a short tube, or *parison*, is injected into the oversize mold. The far end of the parison is pinched off and the air is injected into the other end. Typical products made by these processes are plastic beverage bottles and hollow containers.

Blown film is a further modification of continuous blow molding. A thin-walled tube is extruded vertically and expanded into a balloon shape by blowing air through the center of the extrusion die until the desired film thickness is reached (Fig. 10.18b). The extruded tube is expanded approximately 1.5–2.5 times the diameter of the extrusion die (blowup ratio). The balloon is usually cooled by air from a cooling ring around it, which can also act as a barrier to further expansion of the balloon. Blown film is sold as wrapping film (after slitting the cooled bubble) or

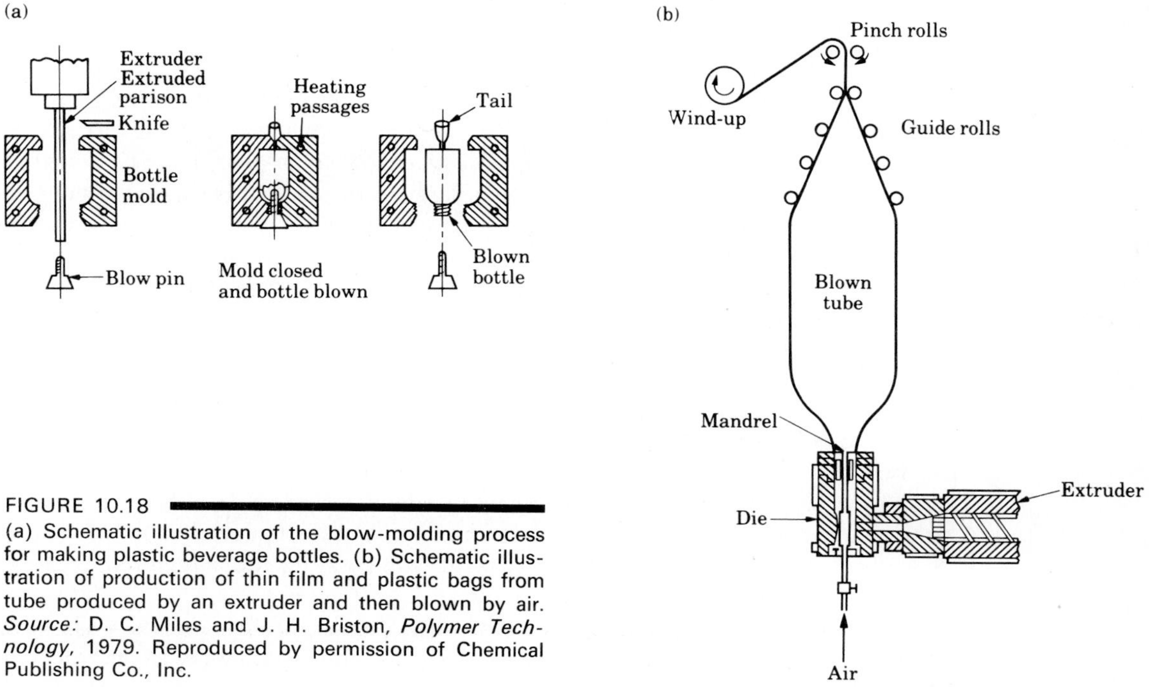

FIGURE 10.18 (a) Schematic illustration of the blow-molding process for making plastic beverage bottles. (b) Schematic illustration of production of thin film and plastic bags from tube produced by an extruder and then blown by air. *Source:* D. C. Miles and J. H. Briston, *Polymer Technology*, 1979. Reproduced by permission of Chemical Publishing Co., Inc.

as bags (where the bubble is pinched and cut off). Film is also produced by shaving solid round billets of plastics, especially polytetrafluoroethylene (PTFE), by skiving, that is, shaving with specially designed knives.

• Example 10.5: Blow molding of film.

Assume that a typical plastic shopping bag, made by blown film, has a lateral (width) dimension of 400 mm. (a) What should be the extrusion die diameter? (b) These bags are relatively strong, so how is this strength achieved?

SOLUTION. (a) The perimeter of the bag is (2)(400) = 800 mm. As the original cross-section of the film was round, we find that the blown diameter should be $\pi D = 800$, or $D = 255$ mm. Recall that in this process a tube is expanded 1.5–2.5 times the extrusion die diameter. Taking the maximum value of 2.5, we calculate the die diameter as 255/2.5 = 100 mm.

(b) Note in Fig. 10.18(b) that after being extruded, the bubble is being pulled upward by the pinch rolls. Thus in addition to diametral stretching and the attendant molecular orientation, the film is stretched and oriented in the longitudinal direction. The biaxial orientation of the polymer molecules significantly improves the strength and toughness of the blown film.

10.9.4 Rotational molding

Most thermoplastics and some thermosets can be formed into large hollow parts by *rotational molding*. The thin-walled metal mold is made of two pieces (split female mold) and is designed to be rotated about two perpendicular axes (Fig. 10.19). A premeasured quantity of finely ground plastic material is placed inside a warm mold. The mold is then heated, usually in a large oven, while it is rotated about the two axes. This action tumbles the powder against the mold where heating fuses the powder without melting it. In some parts, a chemical cross-linking agent is added to the powder, and cross-linking occurs after the part is formed in the mold by continued heating.

Typical parts made by rotational molding are tanks of various sizes, trash cans, boat hulls, buckets, housings, toys, carrying cases, and footballs. Various metallic or plastic inserts may also be molded into the parts made by this process.

Liquid polymers, called *plastisols* (vinyl plastisols being the most common), can also be used in a process called *slush molding*. The mold is heated and rotated simultaneously. The particles of plastic material are forced against the inside walls of the heated mold by centrifugal forces. Upon contact, the material melts and coats the walls of the mold. The part is cooled while still rotating and is then removed by opening the mold.

Rotational molding can produce parts with complex hollow shapes, with wall thicknesses as small as 0.4 mm (0.016 in.). Parts as large as 1.8 m × 1.8 m × 3.6 m (6 ft × 6 ft × 12 ft) have been formed. The outer surface finish of the part is a

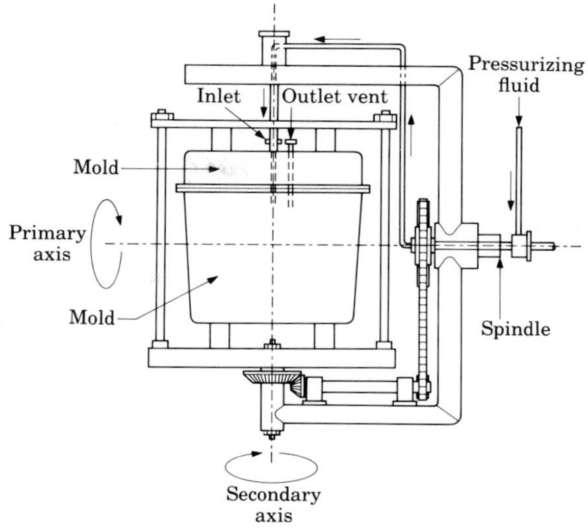

FIGURE 10.19
The rotational molding (rotomolding) process. Trash cans, buckets, and plastic footballs can be made by this process.

replica of the surface finish of the mold walls. Cycle times are longer than in other processes, but equipment costs are low. Quality control considerations usually involve proper weight of powder placed in the mold, proper rotation of the mold, and the temperature–time relationship during the oven cycle.

10.9.5 Thermoforming

Thermoforming is a series of processes for forming thermoplastic sheet or film over a mold with the application of heat and pressure differentials (Fig. 10.20). In this

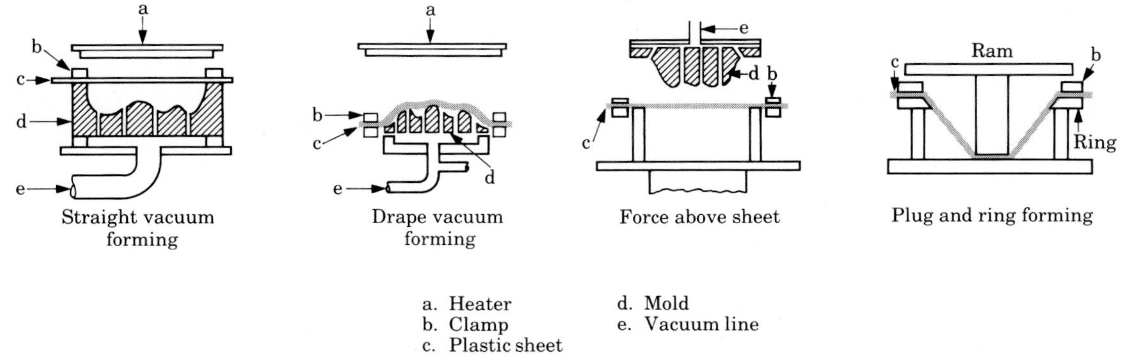

FIGURE 10.20
Various thermoforming processes for thermoplastic sheet. These processes are commonly used in making advertising signs, cookie and candy trays, panels for shower stalls, and packaging.

process, a sheet is heated in an oven to the sag (softening) point—but not to the melting point. The sheet is then removed from the oven and placed over a mold and through the application of a vacuum is pulled against the mold. Since the mold is usually at room temperature, the shape of the plastic is set upon contacting the mold. Because of the low strength of the materials formed, the pressure differential caused by the vacuum is usually sufficient for forming, although air pressure or mechanical means is also applied for some parts.

Typical parts made this way are advertising signs, refrigerator liners, packaging, appliance housings, and panels for shower stalls. Parts with openings or holes cannot be formed because the pressure differential cannot be maintained during forming. Because thermoforming is a drawing and stretching operation, much like sheet-metal forming, the material should exhibit high uniform elongation—otherwise, it will neck and fail. Thermoplastics have a high capacity for uniform elongation by virtue of their high strain-rate sensitivity exponent, m. The sheets used in thermoforming are made by the calendering process (Fig. 10.21), or by sheet extrusion. In *calendering*, a warm plastic mass is fed through a series of heated rolls (*masticated*) and is then stripped off in the form of a sheet.

Molds for thermoforming are usually made of aluminum since high strength is not a requirement. The holes in the molds are usually less than 0.5 mm (0.02 in.) in order not to leave any marks on the formed sheets. Tooling is inexpensive and production rates are high. Quality considerations include tears, nonuniform wall thickness, improperly filled molds, and poor part definition (surface details).

10.9.6 Compression molding

In *compression molding*, a preshaped part, a premeasured volume of powder, or a viscous mixture of liquid resin and filler material is placed directly in a heated mold cavity. Forming is done under pressure with a plug or the upper half of the die (Fig. 10.22). Compression molding is similar to a forging operation and has the same problem of flash formation and the need to remove it by trimming or some other means. Typical parts made are dishes, handles, container caps, fittings, electrical and electronic components, washing-machine agitators, and housings. Fiber-reinforced plastics may also be formed by this process.

Compression molding is used mainly with thermosetting plastics, with the original material in a partially polymerized state. Cross-linking is completed in the

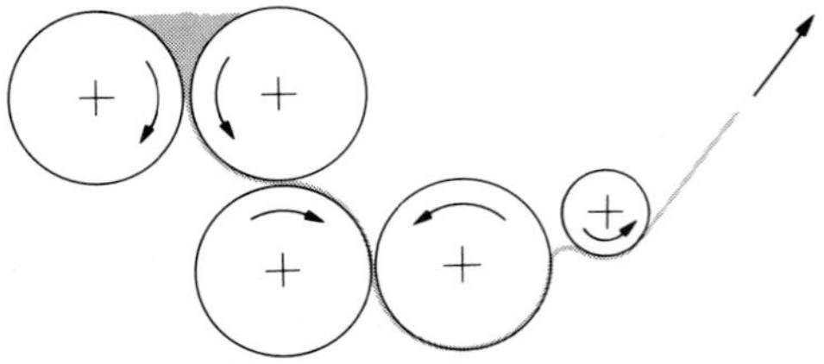

FIGURE 10.21 Schematic illustration of calendering. Sheets produced by this process are subsequently used in thermoforming.

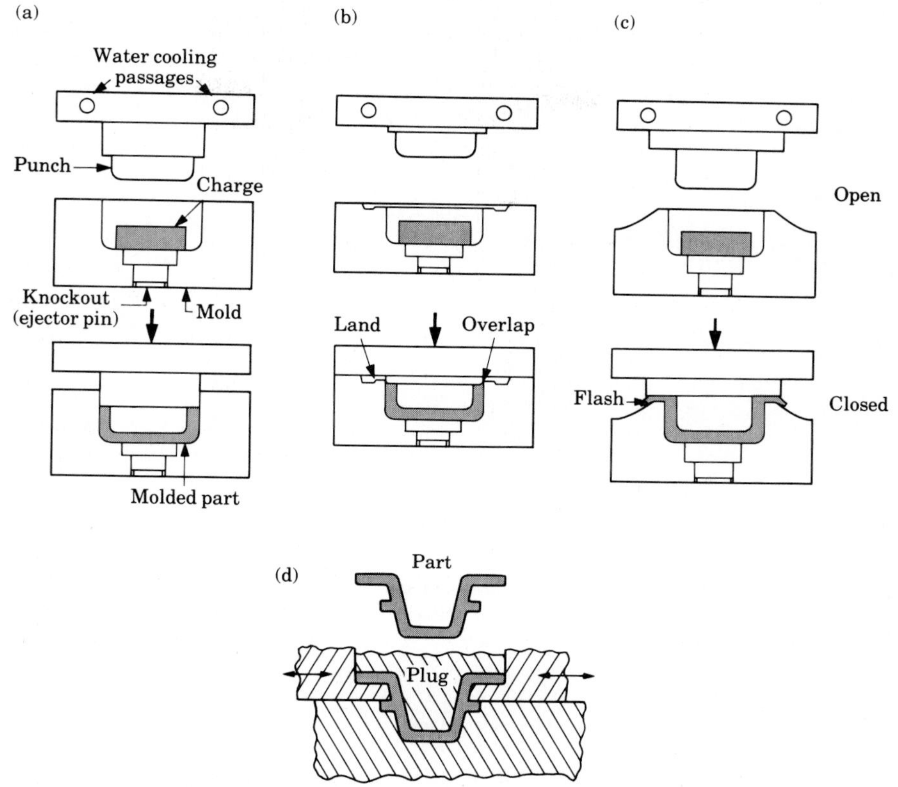

FIGURE 10.22
Types of compression molding, a process similar to forging: (a) positive, (b) semipositive, and (c) flash. The flash in part (c) has to be trimmed off. (d) Die design for making a compression-molded part with undercuts.

heated die, with curing times ranging from 0.5 min to 5 min, depending on the material and part geometry and its thickness. The thicker the material, the longer it will take to cure. Elastomers also are shaped by compression molding.

Because of their relative simplicity, die costs in compression molding are generally lower than in injection molding. Three types of compression molds are available: (1) *flash-type* for shallow or flat parts, (2) *positive* for high density, and (3) *semipositive* for quality production. Undercuts in parts are not recommended; however, dies can be designed to open sideways (Fig. 10.22d) to allow removal of the part. In general, the complexity and dimensional control of parts made by compression molding are less than with injection molding. Product quality is determined by die design, temperature, pressure, and cycle time.

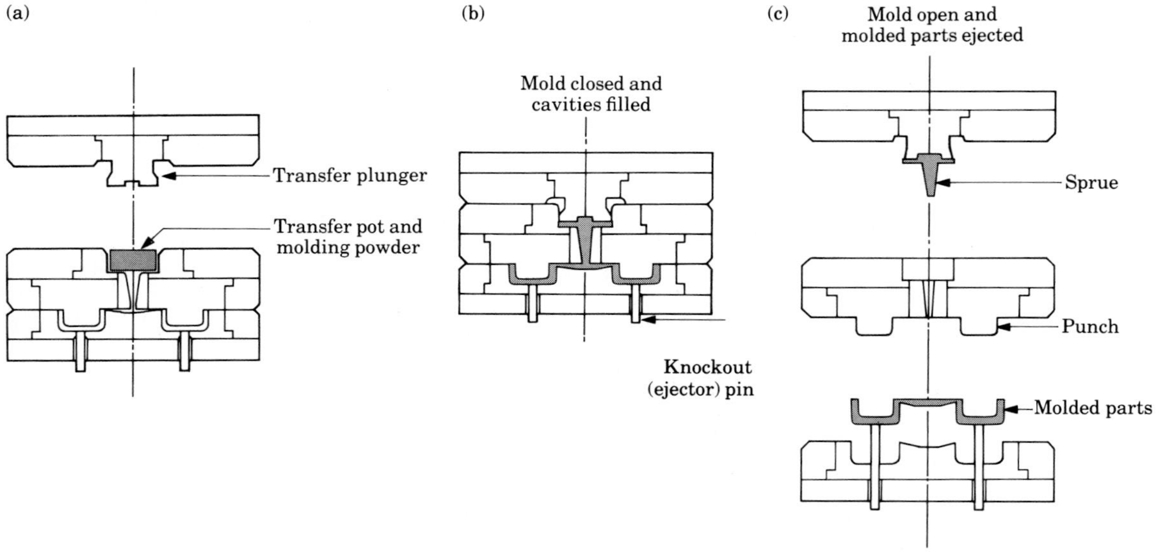

FIGURE 10.23 Sequence of operations in transfer molding for thermosetting plastics. This process is particularly suitable for intricate parts with varying wall thickness.

10.9.7 Transfer molding

Transfer molding represents a further development of compression molding. The uncured thermosetting material is placed in a heated transfer pot or chamber (Fig. 10.23). After the material is heated, it is injected into heated, closed molds. Depending on the type of machine used, a ram, plunger, or rotating screw-feeder forces the material to flow through the narrow channels into the mold cavity. This flow generates considerable heat, which raises the temperature of the material and homogenizes it. Curing takes place by cross-linking. Because the resin is molten as it enters the molds, the complexity of the part and dimensional control approach those for injection molding.

Typical parts made by transfer molding are electrical and electronic components and rubber and silicone parts. The process is particularly suitable for intricate shapes having varying wall thicknesses. Usually made of tool steel, molds tend to be more expensive than those for compression molding, and material is wasted in the channels of the mold during filling.

10.9.8 Casting

Some thermoplastics, such as nylons and acrylics, and thermosetting plastics, such as epoxies, phenolics, polyurethanes, and polyester, can be cast in rigid or flexible molds into a variety of shapes (Fig. 10.24a). Typical parts cast are large gears,

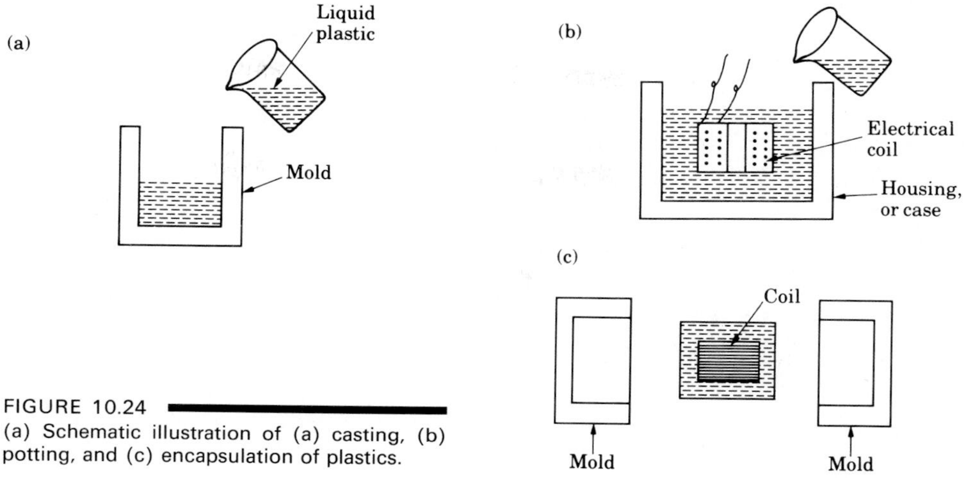

FIGURE 10.24
(a) Schematic illustration of (a) casting, (b) potting, and (c) encapsulation of plastics.

bearings, wheels, thick sheets, and components requiring resistance to abrasive wear.

In casting thermoplastics, a mixture of monomer, catalyst, and various additives is heated and poured into the mold. The part is formed after polymerization takes place at ambient pressure. Intricate shapes can be formed with flexible molds, which are then peeled off. Centrifugal casting is also used with plastics, including reinforced plastics with short fibers. Thermosets are cast in a similar manner. Typical parts produced are similar to those made by thermoplastic castings. Degassing may be necessary for product integrity.

A variation of casting that is important to the electrical and electronics industry is potting and encapsulation. This involves casting the plastic around an electrical component, thus embedding it in the plastic. *Potting* (Fig. 10.24b) is done in a housing or case, which is an integral part of the product. In *encapsulation* (Fig. 10.24c), the component is covered with a layer of the solidified plastic. In both applications the plastic serves as a dielectric (nonconductor). Structural members, such as hooks and studs, may also be partly encapsulated.

10.9.9 Cold forming and solid-phase forming

Processes that have been used in cold working of metals can also be used to form many thermoplastics at room temperature, such as rolling, deep drawing, extrusion, closed-die forging, coining, and rubber forming. Typical materials formed are polypropylene, polycarbonate, ABS, and rigid PVC. The important considerations are that (a) the material be sufficiently ductile at room temperature (hence polystyrenes, acrylics, and thermosets cannot be formed); and (b) the material's deformation must be nonrecoverable (to minimize springback and creep).

The advantages of cold forming of plastics over other methods of shaping are:

- Strength, toughness, and uniform elongation are increased.
- Plastics with high molecular weight can be used to make parts with superior properties.
- Forming speeds are not affected by part thickness since there is no heating or cooling involved. Typical cycle times are shorter than molding processes.

Solid-phase forming is carried out at a temperature about 10–20 °C (20–40 °F) below the melt temperature of the plastic, if a crystalline polymer, and formed while still in a solid state. The advantages over cold forming are that forming forces and springback are lower. These processes are not as widely used as hot-processing methods and are generally restricted to special applications.

10.10 Reinforced Plastics (Composites)

We have shown that plastics possess mechanical properties (particularly strength, stiffness, and creep resistance) that are generally inferior to those for metals and alloys. These properties can be improved by embedding reinforcements of various types, such as glass or graphite fibers, to produce *reinforced plastics* (Fig. 10.25). Table 10.1 shows that reinforcements improve the strength, stiffness, and creep resistance of plastics—and their strength-to-weight and stiffness-to-weight ratios.

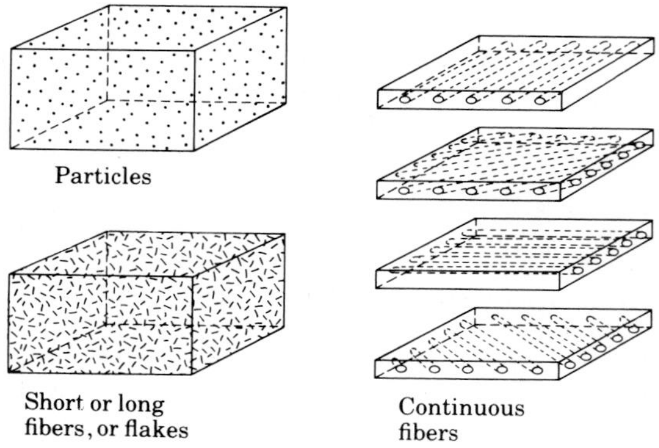

FIGURE 10.25
Schematic illustration of several methods of reinforcing plastics. The plastic is the matrix in these structures. The four layers in the illustration of continuous fibers are assembled into a laminate, similar to plywood.

Composite materials have found increasingly wider applications in aircraft, automobiles, boats, ladders, and sporting goods. Metals and ceramics also can be embedded with fibers or particles to improve their properties.

The oldest example of composites is the addition of straw to clay for making mud huts and bricks for structural use, dating back to 4000 B.C. In that application the straws are the reinforcing fibers, and the clay is the matrix. Another example of a composite material is the reinforcing of masonry and concrete with iron rods, begun in the 1800s. In fact, concrete itself is a composite material, consisting of cement, sand, and gravel. In reinforced concrete, steel rods impart the necessary tensile strength to the composite, since concrete is brittle and generally has little or no useful tensile strength.

10.10.1 Structure of reinforced plastics

Reinforced plastics consist of *fibers* (the discontinuous or dispersed phase) in a plastic *matrix* (the continuous phase). Commonly used fibers are glass, graphite, aramids, and boron. These fibers are strong and stiff (Table 10.4) and have high specific strength (strength-to-weight ratio) and specific modulus (stiffness-to-weight ratio), as shown in Fig. 10.26. However, they are generally brittle and abrasive and lack toughness. Thus fibers, by themselves, have little structural value. The plastic matrix is less strong and less stiff but tougher than the fibers. Thus reinforced plastics combine the advantages of each of the two constituents. When more than one type of fiber is used in a reinforced plastic, the composite is called a *hybrid*, which generally has better properties yet.

In addition to high specific strength and specific modulus, reinforced plastic structures have improved fatigue resistance, greater toughness, and higher creep resistance than unreinforced plastics. These structures are relatively easy to design, fabricate, and repair.

TABLE 10.4
PROPERTIES OF REINFORCING FIBERS

TYPE	TENSILE STRENGTH (MPa)	ELASTIC MODULUS (GPa)	DENSITY (kg/m^3)	RELATIVE COST
Boron	3500	380	2600	Highest
Carbon				
High strength	3000	275	1900	Low
High modulus	2000	415	1900	Low
Glass				
E type	3500	73	2480	Lowest
S type	4600	85	2540	Lowest
Kevlar				
29	2800	62	1440	High
49	2800	117	1440	High

Note: These properties vary significantly depending on the material and method of preparation.

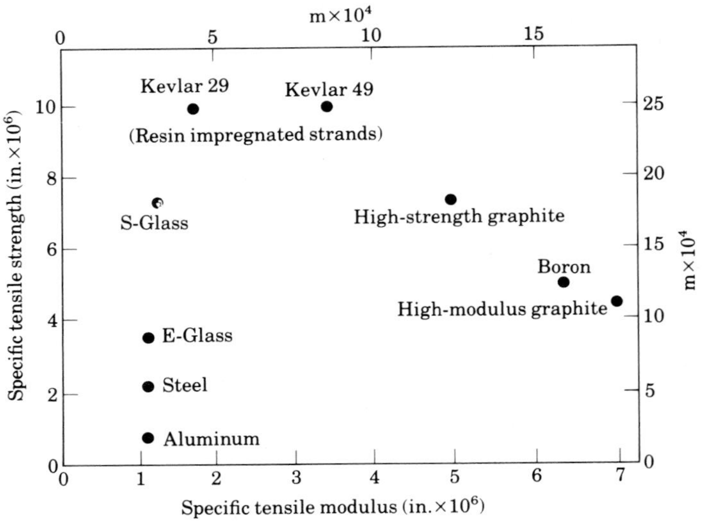

FIGURE 10.26
Specific tensile strength (tensile strength-to-density ratio) and specific tensile modulus (modulus of elasticity-to-density ratio) for various fibers used in reinforced plastics. Note the wide range of specific strengths and stiffnesses available.

The percentage of fibers (by volume) in reinforced plastics usually ranges between 10 percent and 60 percent. Practically, the percentage of fiber in a matrix is limited by the average distance between adjacent fibers or particles. The highest practical fiber content is 65 percent; higher percentages generally result in diminished structural properties.

Reinforcing fibers. Glass fibers are the most widely used and least expensive of all fibers. The composite material is called *glass-fiber reinforced plastic* (GFRP) and may contain between 30 percent and 60 percent glass fibers by volume. Glass fibers are made by drawing molten glass through small openings in a platinum die. There are two principal types of glass fibers: (1) the E type, a borosilicate glass, which is used most; and (2) the S type, a magnesia-alumina-silicate glass, which has higher strength and stiffness and is more expensive.

Although they are more expensive than glass fibers, graphite fibers (Fig. 10.27a) have a combination of low density, high strength, and high stiffness. The product is called *carbon-fiber reinforced plastic* (CFRP). All graphite fibers are made by pyrolysis of organic *precursors*, commonly polyacrylonitrile (PAN) because of its lower cost. Rayon and pitch (the residue from catalytic crackers in petroleum refining) can also be used as precursors. *Pyrolysis* is the term for inducing chemical changes by heat, such as burning a length of yarn, which becomes carbon and black in color. The temperatures for carbonizing range up to about 3000 °C (5400 °F).

The difference between carbon and graphite, although the words are often used

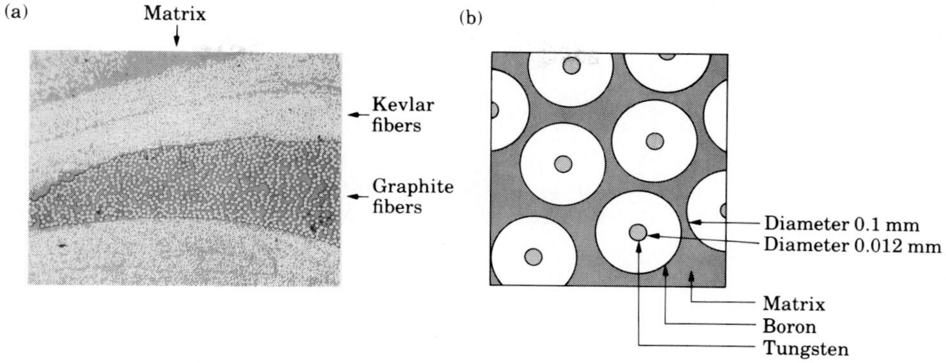

FIGURE 10.27 (a) Cross-section of a tennis racket, showing graphite and aramid (Kevlar) reinforcing fibers. *Source:* J. Dvorak, Mercury Marine Corporation, and F. Garrett, Wilson Sporting Goods Co. (b) Cross-section of boron-fiber reinforced composite material.

interchangeably, depends on the temperature of pyrolysis and the purity of the material. Carbon fibers are generally 93–95 percent carbon, and graphite fibers are usually more than 99 percent carbon.

Marketed under the trade name Kevlar, *aramids* are the toughest fibers available and have the highest specific strength of any fiber (see Fig. 10.26). They can undergo some plastic deformation before fracture and thus have higher toughness than brittle fibers. However, aramids absorb moisture, which reduces their properties and complicates their application, as hygrothermal stresses must be considered.

Boron fibers consist of boron deposited (by chemical vapor-deposition techniques) on tungsten fibers (Fig. 10.27b), although boron can also be deposited on carbon fibers. These fibers have favorable properties, such as high strength and stiffness in tension and compression and resistance to high temperatures. However, because of the use of tungsten, they have high density and are expensive, thus increasing the cost and weight of the reinforced plastic component.

Other fibers that are being used are nylon, silicon carbide, silicon nitride, aluminum oxide, sapphire, steel, tungsten, molybdenum, boron carbide, boron nitride, and tantalum carbide. *Whiskers* are also used as reinforcing fibers. They are tiny needlelike single crystals that grow to 1 μm to 10 μm (40 μin. to 400 μin.) in diameter and have aspect ratios (length to diameter) ranging from 100 to 15,000. Because of their small size, either they are free of imperfections or the imperfections they contain do not significantly affect their strength, which approaches the theoretical strength of the material.

Fiber size and length. The mean diameter of fibers used in reinforced plastics is usually about 0.01 mm (0.0004 in.). The fibers are very strong and rigid in tension. The reason is that the molecules in the fibers are oriented in the longitudinal direction, and their cross-sections are so small that the probability is low that any defects exist in the fiber. Glass fibers, for example, can have tensile strengths as high as 4600 MPa (650 ksi), whereas the strength of glass in bulk form is much lower. Thus glass fibers are stronger than steel.

Fibers are classified as short or long, both also called *chopped fibers*. Short fibers generally have an aspect ratio between 20 and 60 and long fibers between 200 and 500. In addition to the discrete fibers that we have described, reinforcements in composites may be in the form of continuous *roving* (slightly twisted strand of fibers), *yarn* (twisted strand), *woven* fabric (similar to cloth), and *mats* of various combinations. Reinforcing elements may also be in the form of particles and flakes.

Matrix materials. The matrix in reinforced plastics has three functions:

1. Support and transfer the stresses to the fibers, which carry most of the load.
2. Protect the fibers against physical damage and the environment.
3. Prevent propagation of cracks in the composite by virtue of the ductility and toughness of the plastic matrix.

Matrix materials are usually epoxy, polyester, phenolic, fluorocarbon, polyethersulfone, and silicon. The most commonly used are epoxies (80 percent of all reinforced plastics) and polyesters, which are less expensive than epoxies. Polyimides, which resist exposure to temperatures in excess of 300 °C (575 °F), are being developed for use with graphite fibers. Some thermoplastics, such as polyetheretherketone, are also being developed as matrix materials. They generally have higher toughness than thermosets, but their resistance to temperature is lower.

10.10.2 Properties of reinforced plastics

The properties of reinforced plastics depend on the kind, shape, and orientation of the reinforcing material, the length of the fibers, and the volume fraction (percentage) of the reinforcing material. Short fibers are less effective than long fibers (Fig. 10.28), and their properties are strongly influenced by time and temperature. Long fibers transmit the load through the matrix better and thus are commonly used in critical applications, particularly at elevated temperatures. Fiber reinforcement also affects the physical and other properties of composites (Fig. 10.29).

A critical factor in reinforced plastics is the strength of the bond between the fiber and the polymer matrix, since the load is transmitted through the fiber–matrix interface. Weak bonding causes *fiber pullout* and *delamination* of the structure, particularly under adverse environmental conditions. Poor bonding in composites is analogous to a brick structure with poor bonding between the bricks and the mortar.

Bonding can be improved by special surface treatments for better adhesion at the interface, such as coatings and the use of coupling agents. Glass fibers, for example, are treated with a chemical called silane (SiH_4) for improved wetting and bonding between the fiber and the matrix. You can appreciate the importance of

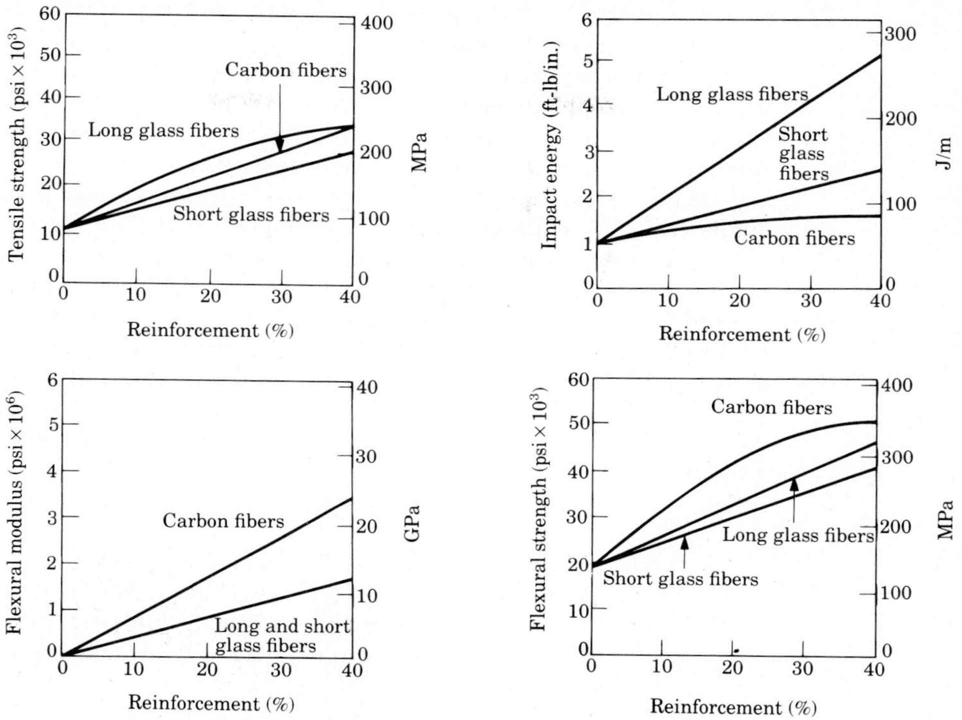

FIGURE 10.28

Effect of the amount of reinforcing fibers and fiber length on the mechanical properties of reinforced nylon. Note the significant improvement with increasing percentage of fiber reinforcement. *Source:* Courtesy of Wilson Fiberfill International.

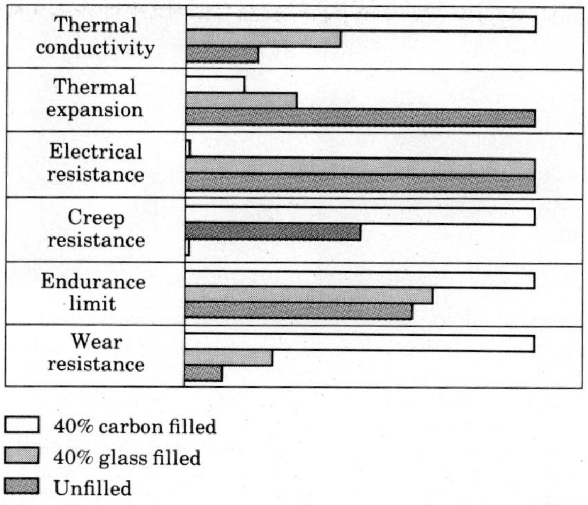

FIGURE 10.29

The effect of type of fibers on various properties of fiber-reinforced nylon 6,6. *Source:* NASA.

proper bonding by looking at Figs. 10.30(a) and (b), which show the fracture surfaces of reinforced plastics.

Generally, the greatest stiffness and strength in reinforced plastics is obtained when the fibers are aligned in the direction of the tension force. This composite, of course, is highly anisotropic (Fig. 10.31). As a result, other properties of the composite, such as stiffness, creep resistance, thermal and electrical conductivity, and thermal expansion, are also anisotropic. The transverse properties of such a unidirectionally reinforced structure are much lower than the longitudinal. Note, for example, how easily you can split a fiber-reinforced packaging tape, yet how strong it is when you pull on it (tension).

For a specific service condition, we can give a reinforced plastic part an optimum configuration. For example, if the reinforced plastic part is to be subjected to forces in different directions (such as thin-walled, pressurized vessels), the fibers are crisscrossed in the matrix. Reinforced plastics may also be made with various other materials and shapes of the polymer matrix in order to impart specific properties, such as permeability (ability to diffuse through) and dimensional stability, as well as making processing easier and reducing costs.

Strength and elastic modulus of reinforced plastics. The strength of a reinforced plastic with longitudinal fibers can be determined in terms of the strength of the fibers and matrix, respectively, and the volume fraction of fibers in the composite. In the following equations, c refers to the composite, f to the fiber, and m to the matrix. The total strength P_c of the composite is

$$P_c = P_f + P_m, \tag{10.10}$$

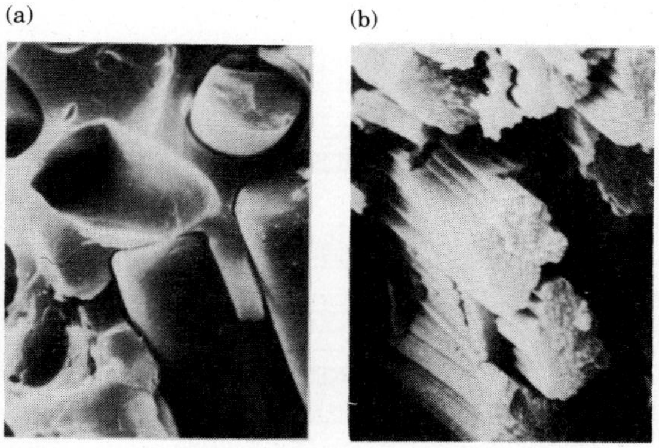

FIGURE 10.30
(a) Fracture surface of glass-fiber reinforced epoxy composite. The fibers are 10 μm (400 μin.) in diameter and have random orientation. (b) Fracture surface of a graphite-fiber reinforced epoxy composite. The fibers, 9–11 μm in diameter, are in bundles and are all aligned in the same direction. *Source:* L. J. Broutman.

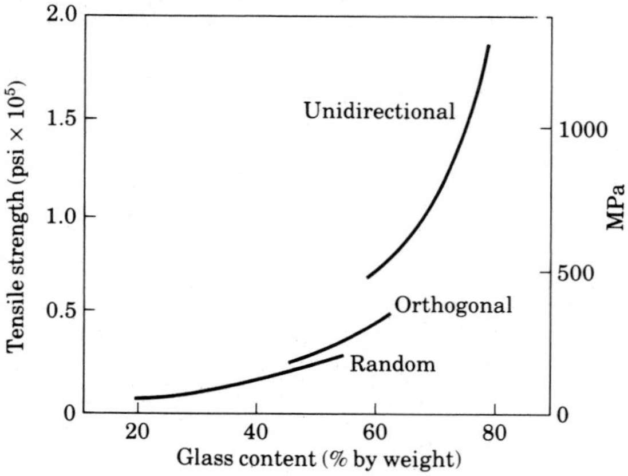

FIGURE 10.31
The tensile strength of glass-reinforced polyester as a function of fiber content and fiber direction in the matrix. *Source:* R. M. Ogorkiewicz, *The Engineering Properties of Plastics*. Oxford: Oxford University Press, 1977.

which we can write as

$$\sigma_c A_c = \sigma_f A_f + \sigma_m A_m. \qquad (10.11)$$

However, we know that $A_c = A_f + A_m$. Let's now denote x as the area fraction of the fibers in the composite. (Note that x also represents the volume fraction, because the fibers are uniformly longitudinal in the matrix.) We may now rewrite Eq. (10.11) as follows:

$$\sigma_c = x\sigma_f + (1 - x)\sigma_m. \qquad (10.12)$$

We can now calculate the fraction of the total load carried by the fibers. First, we note that, in the composite under a tension load, the strains sustained by the fibers and the matrix are the same, and then recall from Section 2.2 that

$$e = \frac{\sigma}{E} = \frac{P}{AE}.$$

Consequently,

$$\frac{P_f}{P_m} = \frac{A_f E_f}{A_m E_m}. \qquad (10.13)$$

As we know the relevant quantities for a specific situation, and using Eq. (10.10), we can determine the fraction P_f/P_c. Then, using the foregoing relationships, we can also calculate the elastic modulus E_c of the composite by replacing σ in Eq. (10.12) with E. Thus

$$E_c = xE_f + (1 - x)E_m. \qquad (10.14)$$

• Example 10.6: Strength of reinforced plastics.

A graphite–epoxy reinforced plastic with longitudinal fibers contains 20 percent graphite fibers with a strength of 2500 MPa and an elastic modulus of 300 GPa. The strength of the epoxy matrix is 120 MPa, with an elastic modulus of 100 GPa. Calculate the elastic modulus of the composite and fraction of the load supported by the fibers.

SOLUTION. The data given are $x = 0.2$, $E_f = 300$ GPa, $E_m = 100$ GPa, $\sigma_f = 2500$ MPa, and $\sigma_m = 120$ MPa. Using Eq. (10.14), we find that

$$E_c = 0.2(300) + (1 - 0.2)100 = 60 + 80 = 140 \text{ GPa}.$$

We obtain the load fraction P_f/P_m from Eq. (10.13):

$$\frac{P_f}{P_m} = \frac{0.2(300)}{0.8(100)} = \frac{60}{80} = 0.75.$$

As

$$P_c = P_f + P_m \quad \text{and} \quad P_m = \frac{P_f}{0.75},$$

we find that

$$P_c = P_f + \frac{P_f}{0.75} = 2.33P_f \quad \text{or} \quad P_f = 0.43P_c.$$

Thus the fibers support 43 percent of the load, even though they occupy only 20 percent of the cross-sectional area (hence volume) of the composite.

10.10.3 Applications

The first application of reinforced plastics (in 1907) was for an acid-resistant tank, made of a phenolic resin with asbestos fibers. Formica, commonly used for counter tops, was developed in the 1920s. Epoxies were first used as a matrix in the 1930s and beginning in the 1940s, boats were made with fiberglass, and reinforced plastics were used for aircraft, electrical equipment, and sporting goods. Major developments in composites began in the 1970s, and these materials are now called *advanced composites.*

Reinforced plastics are typically used in military and commercial aircraft and rocket components, helicopter blades, helmets, automotive bodies, leaf springs, drive shafts, pipes, ladders, pressure vessels, sporting goods, boat hulls, and various other structures. Applications of reinforced plastics include components in DC-10, L-1011, and Boeing 727, 757, and 767 commercial aircraft. By virtue of the resulting weight savings (Table 10.5), reinforced plastics have reduced fuel consumption by about 2 percent. Substituting aluminum in large commercial aircraft with graphite–epoxy reinforced plastics could reduce both weight and production costs by 30 percent, with improved fatigue and corrosion resistance. The structure of the Lear Fan 2100 passenger aircraft is almost totally made of graphite–epoxy reinforced plastic. Nearly 90 percent of the structure of the lightweight Voyager aircraft, which circled the earth without refueling, was made of carbon-reinforced plastic. Boron-fiber reinforced composites are used in military aircraft, golf club shafts, tennis rackets, and fishing rods.

TABLE 10.5
APPROXIMATE WEIGHT SAVINGS IN THE SUBSTITUTION OF COMMERCIAL AIRCRAFT COMPONENTS WITH GRAPHITE/EPOXY REINFORCED PLASTICS

COMPONENT	WEIGHT SAVING (%)
McDonnell–Douglas DC-10 vertical tail	20
Lockheed L-1011 aileron	26
Boeing 727 elevator	26
McDonnell–Douglas DC-10 rudder	27
Boeing 737 horizontal tail	27
Lockheed L-1011 vertical tail	28

Careful inspection and testing of reinforced plastics is essential in critical applications, in order to ensure that good bonding between the reinforcing fiber and the matrix has been obtained throughout. In some instances, the cost of inspection can be as high as one quarter of the total cost of the composite product.

10.11 Processing Reinforced Plastics

The unique structure of reinforced plastics requires special methods to shape them into useful products. They can usually be fabricated by the methods described in this chapter, with some provision for the presence of more than one type of material in the composite. The reinforcement may be loose fibers, woven fabric or mat, roving (slightly twisted fiber), or continuous lengths of fiber. In order to obtain good bonding between the reinforcing fibers and the polymer matrix, it is necessary to impregnate and coat the reinforcement with the polymer.

When the impregnation is done as a separate step, the resulting partially dry sheets are called *prepregs*, *bulk-molding compound* (BMC), or *sheet-molding compound* (SMC), depending on their form. They should be stored at a sufficiently low temperature to delay curing. Alternatively, the resin and the fibers can be mixed together at the time they are placed in the mold.

Commercial reinforced plastics are available as sheet-molding compounds. Continuous strands of reinforcing fiber are chopped into short fibers (Fig. 10.32) and deposited over a layer of resin paste, usually a polyester mixture, carried on a polymer film such as polyethylene. A second layer of resin paste is deposited on top, and the sheet is pressed between rollers. It is then ready for molding into the desired shapes.

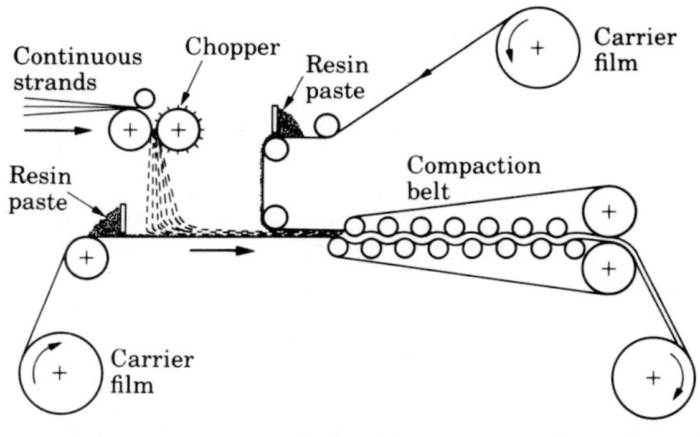

FIGURE 10.32 Manufacturing process for producing reinforced plastic sheets. The sheet is still viscous at this stage and can later be shaped into various products. *Source:* T.-W. Chou, R. L. McCullough, and R. B. Pipes.

A typical procedure for making reinforced plastic prepregs is shown in Fig. 10.33(a). The continuous fibers are aligned and subjected to surface treatment to enhance adhesion to the polymer matrix. They are then coated by dipping them in a resin bath and made into a sheet or tape (Fig. 10.33b). Individual pieces of the sheet are then assembled into laminated structures. Typical products are flat or corrugated architectural paneling, panels for construction and electric insulation, and structural components of aircraft.

10.11.1 Molding

In compression molding, the material is placed between two molds and pressure is applied. Depending on the material, the molds may be either at room temperature or heated to accelerate hardening. The material may be in bulk form (bulk-molding compound), which is a viscous, sticky mixture of polymers, fibers, and additives. It is generally shaped into a log which is cut into the desired mass. Fiber lengths generally range from 3 mm to 50 mm (0.125 in. to 2 in.), although longer fibers (75 mm; 3 in.) may alsc be used.

Sheet-molding compounds can also be used in molding. Sheet-molding compound is similar to BMC, except that the resin–fiber mixture is laid between plastic sheets to make a sandwich that can be easily handled. The sheets are removed when the SMC is placed in the mold.

In *vacuum-bag molding* (Fig. 10.34), prepregs are laid in a mold to form the desired shape. In this case, the pressure required to form the shape and good bonding is obtained by covering the lay-up with a plastic bag and creating a vacuum. If additional heat and pressure are desired, the entire assembly is put in an autoclave. Care should be exercised to maintain fiber orientation, if specific fiber

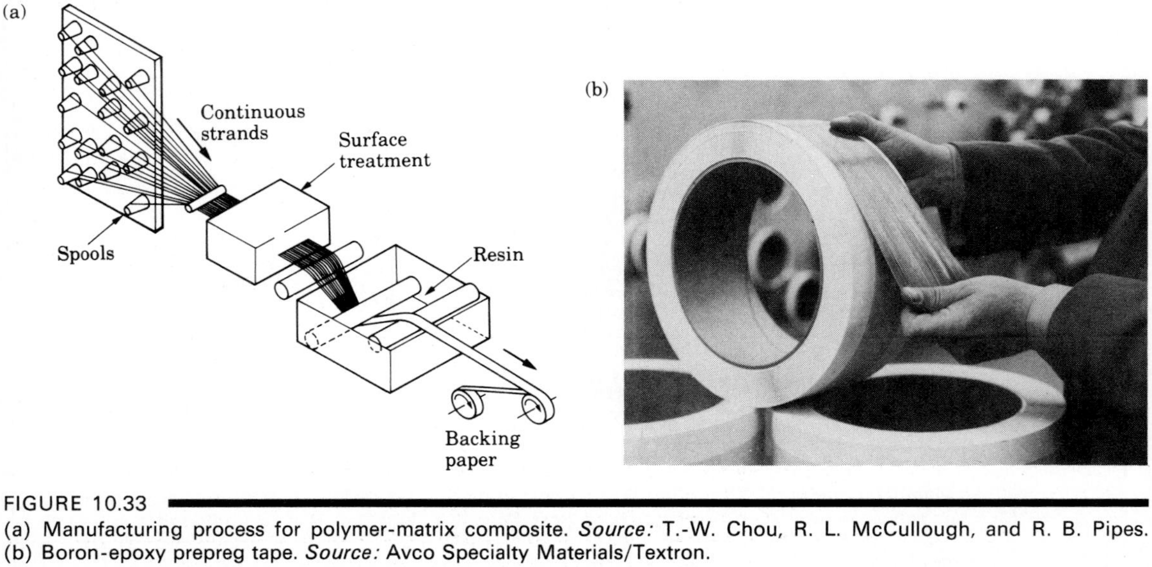

FIGURE 10.33
(a) Manufacturing process for polymer-matrix composite. *Source:* T.-W. Chou, R. L. McCullough, and R. B. Pipes. (b) Boron-epoxy prepreg tape. *Source:* Avco Specialty Materials/Textron.

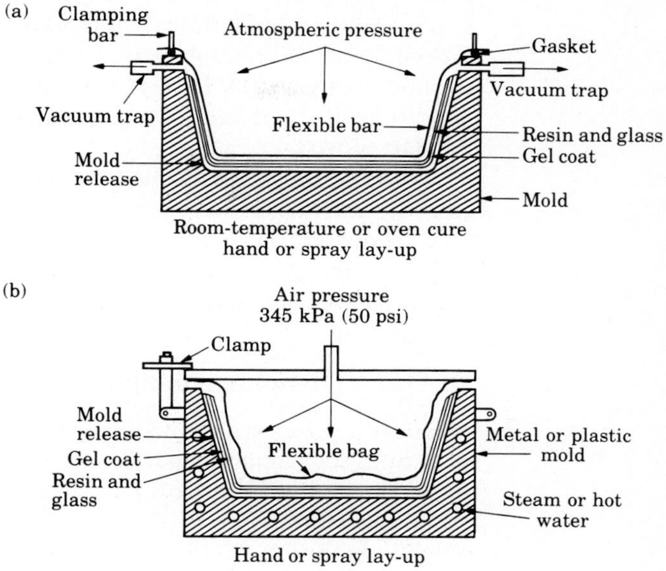

FIGURE 10.34
(a) Vacuum-bag forming. (b) Pressure-bag forming. *Source:* T. H. Meister.

orientations are desired. In chopped-fibers materials, no specific orientation is intended.

In order to prevent the resin from sticking to the vacuum bag and to facilitate removal of excess resin, several sheets of various materials (*release cloth*, *bleeder cloth*) are placed on top of the prepreg sheets. The molds can be made of metal, usually aluminum, but more often are made from the same resin (with reinforcement) as the material to be cured. This eliminates any problem with differential thermal expansion between the mold and the part.

Contact molding processes use a single male or female mold (Fig. 10.35) made of materials such as reinforced plastics, wood, or plaster. Contact molding is used in

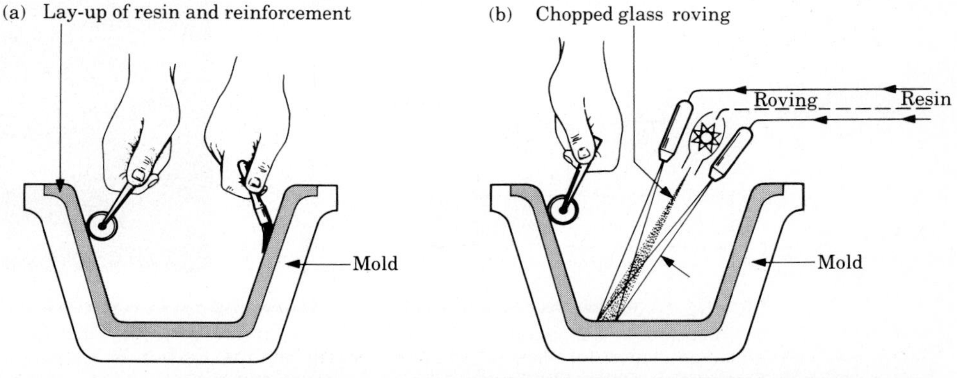

FIGURE 10.35
Manual methods of processing reinforced plastics: (a) hand lay-up and (b) spray-up.

making products with high surface area-to-thickness ratios, such as swimming pools, boats, tub and shower units, and housings. This is a "wet" method, in which the reinforcement is impregnated with the resin at the time of molding. The simplest method is called *hand lay-up*. The materials are placed and formed in the mold by hand (Fig. 10.35a), and the squeezing action expels any trapped air and compacts the part.

Molding may also be done by spraying (*spray-up*; Fig. 10.35b). Although spraying can be automated, these processes are relatively slow and labor costs are high. However, they are simple and the tooling is inexpensive. Only the mold-side surface of the part is smooth and the choice of materials is limited. Many types of boats, as well as buckets for powerline servicing equipment, are made by this process.

10.11.2 Filament winding and pultrusion

Filament winding is a process whereby the resin and fibers are combined at the time of curing. Axisymmetric parts, such as pipes and storage tanks, are produced on a rotating mandrel. The reinforcing filament, tape, or roving is wrapped continuously around the form. The reinforcements are impregnated by passing them through a polymer bath (Fig. 10.36a). The process can be modified by wrapping the mandrel with prepreg material.

The products made by filament winding are very strong because of their highly reinforced structure. Filament winding has also been used for strengthening cylindrical or spherical pressure vessels (Fig. 10.36b) made of materials such as aluminum and titanium. The presence of a metal inner lining makes the part impermeable. Filament winding can be used directly over solid-rocket propellant forms.

Pultrusion. Long shapes with various constant profiles, such as rods, profiles, or tubing (similar to extruded metal products), are made by the *pultrusion* process.

(a)

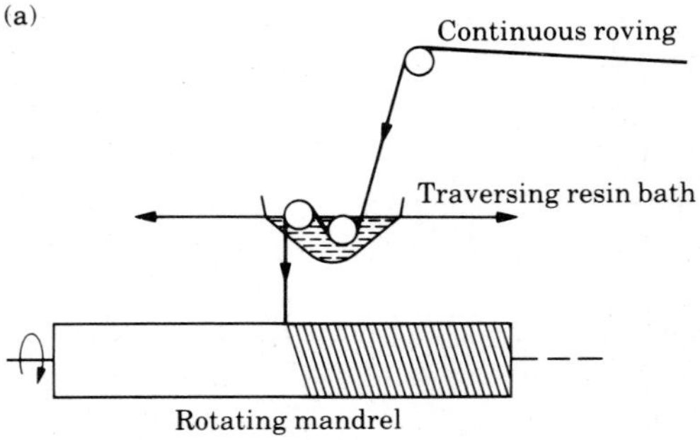

(b)

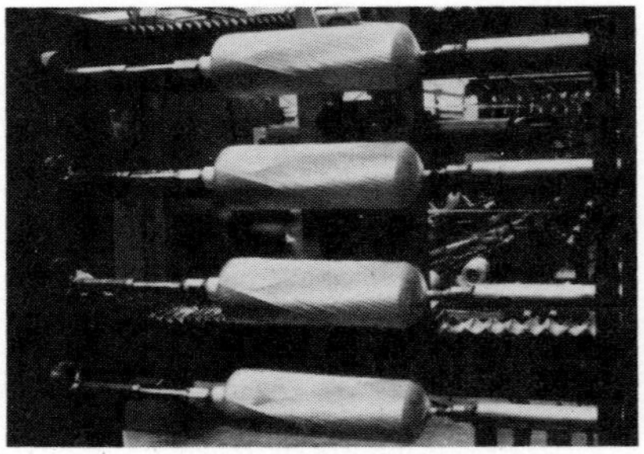

FIGURE 10.36

(a) Schematic illustration of the filament-winding process. (b) Fiberglass being wound over aluminum liners for slide-raft inflation vessels for the Boeing 767 aircraft. *Source:* Brunswick Corporation, Defense Division.

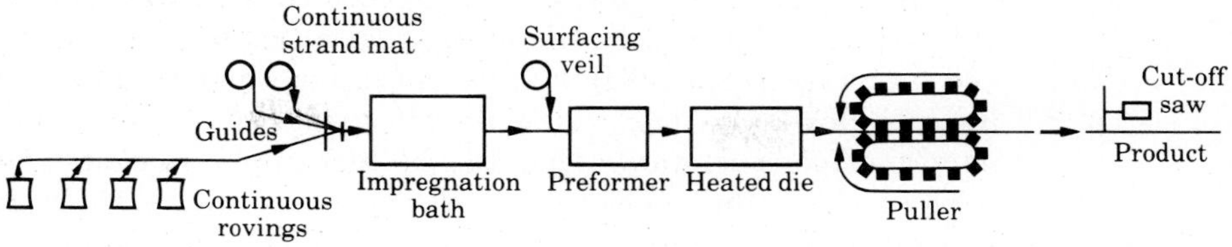

FIGURE 10.37
Schematic illustration of the pultrusion process.

Typical products are golf clubs, drive shafts, and structural members such as ladders, walkways, and handrails. In this process the continuous reinforcement (roving or fabric) is pulled through a thermosetting polymer bath, and then through a long heated steel die (Fig. 10.37). The product is cured during its travel through the die. The most common material used in pultrusion is polyester with glass reinforcements.

10.12 Design Considerations

Design considerations for forming and shaping plastics are somewhat similar to those for processing of metals. However, the mechanical and physical properties of plastics should be carefully considered during design and material and process selection.

Selection of an appropriate material from an extensive list requires consideration of service requirements and possible long-range effects on properties and behavior, such as dimensional stability and wear. Compared to metals, plastics have lower strength and stiffness, although the strength-to-weight and stiffness-to-weight ratio for reinforced plastics is higher than for many metals. Thus section sizes should be selected accordingly, with a view to maintaining a sufficiently high section modulus for improved stiffness. Reinforcement with fibers and particles can also be highly effective in achieving this objective, as can be designing sections with a high ratio of moment of inertia to cross-sectional area.

One of the major design advantages of reinforced plastics is the directional nature of the strength of the material. Forces applied to the material are transferred by the resin matrix to the fibers, which are much stronger and stiffer than the matrix. When fibers are all oriented in one direction, the resulting material is exceptionally strong in the fiber direction. This property is often utilized in designing reinforced plastic structures. For strength in two principal directions, the unidirectional materials are often laid at different angles to each other. If strength in the third (thickness) direction is desired, a different type of material (commonly an aluminum honeycomb layer) is used to form a sandwich structure.

Physical properties, especially high coefficient of thermal expansion, and hence contraction, are important. Improper part design or assembly can lead to warping and shrinking (Fig. 10.38a). Plastics can easily be molded around metallic parts and inserts. However, their compatibility with metals when so assembled is an important consideration. Through prudent application and orientation of fiber reinforcement, the coefficient of thermal expansion can be modified. In fact, structures with zero or even negative coefficients can be produced.

The overall part geometry often determines the particular forming or molding process. Table 10.6 is a guide to making this selection. Even after a particular process is selected, the design of the part and die should be such that it will not cause problems concerning shape generation (Fig. 10.38b), dimensional control, and surface finish. As in casting metals and alloys, material flow in the mold cavities should be controlled properly.

Large variations in section sizes (Fig. 10.38c) and abrupt changes in geometry should be avoided for better product quality and increased mold life. Furthermore, contraction in large cross-sections tends to cause porosity in plastic parts. Conversely, because of a lack of stiffness, removing thin sections from molds after shaping may be difficult. The low elastic modulus of plastics further requires that shapes be selected properly for improved stiffness of the component (Fig. 10.38d), particularly when saving materials is important. These considerations are similar to those in designing metal castings and forgings.

The properties of the final product depend on the original material and its processing history. Cold working of polymers improves their strength and toughness. On the other hand, because of the nonuniformity of deformation (even in simple rolling), residual stresses develop in polymers, as they do in metals. Residual stresses can also be generated by thermal cycling of the part. Also, as in metals, the magnitude and direction of residual stresses, however produced, are important

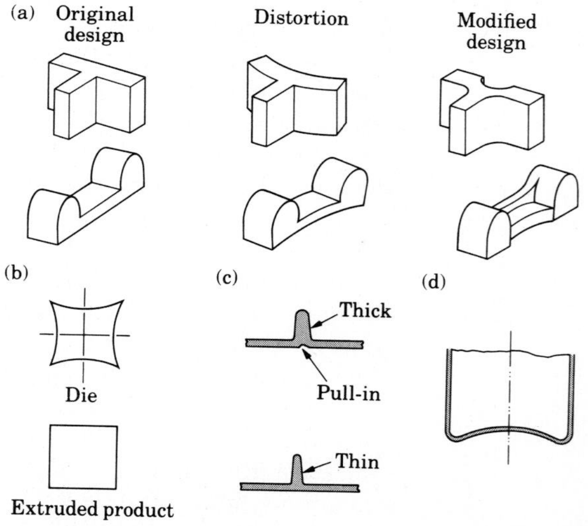

FIGURE 10.38 Examples of design modifications to eliminate or minimize distortion of plastic parts. (a) Suggested design changes to minimize distortion. *Source:* F. Strasser. (b) Die design (exaggerated) for extrusion of square sections. Without this design, product cross-sections swell because of the recovery of the material, which is known as die swell. (c) Design change in a rib to minimize pull-in caused by shrinkage during cooling. (d) Stiffening of bottom of thin plastic containers by doming, which is similar to the process used to make the bottoms of aluminum beverage cans.

TABLE 10.6

CHARACTERISTICS OF VARIOUS MOLDING AND FORMING PROCESSES FOR PLASTICS

	SHAPE LIMITATIONS	INTRICATE, COMPLICATED SHAPES	CONTROLLED WALL THICKNESS	OPEN, HOLLOW SHAPES	ENCLOSED, HOLLOW SHAPES	LARGE ENCLOSED VOLUME	VERY SMALL ITEMS	PLAN AREA, 1 m^2 (10 ft^2)	FACTOR LIMITING MAXIMUM SIZE	INSERTS	MOLDED-IN HOLES	THREADS
NONREINFORCED MATERIALS												
Compression molding	Moldable	Yes	Yes	Yes					Press	Yes	Yes	Yes
Transfer molding	Moldable	Yes	Yes	Yes			Yes		Press	Yes	Yes	Yes
Injection molding	Moldable	Yes	Yes	Yes			Yes		Press	Yes	Yes	Yes
Extrusion	Constant cross-section	Yes	Yes						Die	Yes		
Rotational molding	Hollow			Yes	Yes	Yes		Yes	Available machine	Yes	Yes	Yes
Blow molding	Hollow, thin-wall			Yes	Yes	Yes		Yes	Mold			Yes
Thermoforming	Thin-wall			Yes				Yes	Available machine			
Casting	Moldable	Yes	Yes						Mold	Yes	Yes	
Forging	Moldable	Yes	Yes						Die			
Foam molding	Moldable	Yes	Yes	Yes				Yes	Press		Yes	
FIBER-REINFORCED MATERIALS												
Injection molding	Moldable	Yes	Yes	Yes					Press	Yes	Yes	Yes
Hand lay-up and spray-up	Large, thin-wall	Yes	Yes	Yes		Yes, by joining		Yes	Mold, or transport of parts	Yes	Yes	
Compression-type molding	Moldable	Yes	Yes	Yes					Press	Yes	Yes	Yes
Preform molding	Moldable	Yes	Yes	Yes				Yes	Press			
Cold-press molding	Moldable	Yes	Yes	Yes				Yes	Press			
Filament winding	Surface of revolution		Yes						Available machine			
Pultrusion	Constant cross-section	Yes	Yes						Die			

Source: After R. L. E. Brown, *Design and Manufacture of Plastic Parts.*

factors. These stresses can relax over a period of time and cause distortion of the part during its service life.

10.13 The Economics of Forming and Shaping Plastics

We have described a number of processes used to form and shape plastics and composite materials. As in all other processes, design and manufacturing decisions

TABLE 10.7
COMMON SHAPING PROCESSES FOR THERMOPLASTICS

	COMPRESSION MOLDING	TRANSFER MOLDING	INJECTION MOLDING	EXTRUSION	ROTATIONAL MOLDING	BLOW MOLDING	THERMOFORMING	REACTION INJECTION MOLDING	CASTING	FORGING	FOAM MOLDING	REINFORCED PLASTIC MOLDING	VACUUM MOLDING	PULTRUSION	CALENDERING
Acetal			•	•	•	•	•				•	•			•
ABS			•	•	•	•	•			•					•
Acrylic	•		•	•		•	•		•						•
Cellulose acetate	•		•	•			•								•
Nylon			•	•	•	•		•	•	•		•			•
Polyimide			•												
Polycarbonate			•	•		•	•								•
Polyethylene			•	•	•	•	•			•	•	•			•
Polypropylene	•		•	•	•	•	•		•	•		•			•
Polystyrene			•	•	•	•	•				•	•			•
Polysulfone			•			•	•					•		•	
Polyurethane			•	•	•						•	•	•		•
PVC	•	•	•	•	•	•	•				•	•	•		•
Polyvinyl acetate	•	•	•	•	•	•	•				•				•
Tetrafluoroethylene	•	•	•	•											•

are ultimately based on performance and cost, including the costs of equipment, tooling, and production. The final selection of a process depends greatly on production volume. High equipment and tooling costs can be acceptable if the production run is large, as is the case in casting and forging. Various types of equipment are used in plastics forming and shaping processes. The most expensive are injection-molding machines, with cost directly proportional to clamping force.

The optimum number of cavities in the die for making the product in one cycle is an important consideration, as in die casting. For small parts, a number of cavities can be made in a die, with runners to each cavity. If the part is large, then only one cavity may be accommodated. As the number of cavities increases, so does the cost of the die. Larger dies may be considered for larger numbers of cavities, increasing die cost even further. On the other hand, more parts will be produced per machine cycle, thus increasing the production rate. Hence a detailed analysis has to be made to determine the optimum number of cavities, die size, and machine capacity. Similar considerations apply to other plastics processing methods. Tables 10.7–10.9 are general guides to the selection of processes and economical processing of plastics and composite materials. Note, for example, the high capital costs for molding plastics and the wide range of production rates. For composite materials, equipment and tooling costs for most molding operations are generally high. Production rates and economic production quantities vary widely.

TABLE 10.8
COMMON SHAPING PROCESSES FOR THERMOSETS

	COMPRESSION MOLDING	TRANSFER MOLDING	INJECTION MOLDING	ROTATIONAL MOLDING	THERMOFORMING	REACTION INJECTION MOLDING	CASTING	FOAM MOLDING	REINFORCED PLASTICS MOLDING	LAMINATING
Alkyd	•	•	•				•		•	
Allyl					•		•		•	•
Epoxy				•		•	•	•	•	•
Melamine	•	•	•					•	•	•
Phenolic	•	•	•				•	•	•	•
Polyester	•					•	•	•	•	
Polyurethane						•				
Silicone							•	•	•	
Urea	•	•	•					•		

TABLE 10.9
ECONOMIC PRODUCTION QUANTITIES FOR VARIOUS MOLDING METHODS*

MOLDING METHOD	RELATIVE INVESTMENT REQUIRED		RELATIVE PRODUCTION RATE	ECONOMIC PRODUCTION QUANTITY
	EQUIPMENT	*TOOLING*		
Hand lay-up	VL	L	L	VL
Spray-up	L	L	L	L
Casting	M	L	L	L
Vacuum-bag molding	M	L	VL	VL
Compression-molded BMC	H	VH	H	H
SMC and preform	H	VH	H	H
Pressure-bag molding	H	H	L	L
Centrifugal casting	H	H	M	M
Filament winding	H	H	L	L
Pultrusion	H	H	H	H
Rotational molding	H	H	L	M
Injection molding	VH	VH	VH	VH

* VL, very low; L, low; M, medium; H, high; VH, very high.
Source: After J. G. Bralla (ed.), *Handbook of Product Design for Manufacturing*. New York: McGraw-Hill, 1986.

SUMMARY

Polymers are an important class of materials because they possess a very wide range of mechanical, physical, and chemical properties. Compared to metals, plastics are generally characterized by lower density, strength, elastic modulus, and thermal and electrical conductivity and a higher coefficient of thermal expansion.

Plastics are composed of polymer molecules and various additives. The smallest repetitive unit in a polymer chain is called a mer. Monomers are linked by polymerization processes to form larger molecules. Two major classes of polymers are thermoplastics and thermosets. Thermoplastics become soft and are easy to form at elevated temperatures; they return to their original properties when cooled. Thermosets, which are obtained by cross-linking polymer chains, do not become soft to any significant extent with increasing temperature. Polymer structures can be modified by various means to impart a wide range of properties to plastics.

Elastomers have the characteristic ability to undergo large elastic deformations and return to their original shapes when unloaded. Consequently, they have important applications as tires, seals, footware, hose, belts, and shock absorbers.

Additives in polymers have various functions, such as improving strength, hardness, abrasion resistance, flame retardation, and lubrication. Other functions are to impart flexibility, softness, color, and stability against ultraviolet radiation and oxygen.

Reinforced plastics are an important class of materials that have superior mechanical properties and are lightweight. The reinforcing fibers are usually glass, graphite, aramids, and boron; epoxies commonly serve as a matrix material. Reinforced plastics are being developed rapidly, with a wide variety of present and future applications in aircraft, transportation, structural components, containers, and sporting goods.

Plastics can be formed and shaped by a variety of processes, such as extrusion, molding, casting, and thermoforming. The starting material is usually in the form of pellets and powders. Thermosets are generally molded and cast, and thermoplastics are formed by these processes as well as by thermoforming and techniques used for metalworking. The high strain-rate sensitivity of thermoplastics allows extensive stretching in forming operations; thus complex and deep shapes can be produced. The design of plastic parts should include considerations of their low strength and stiffness, as well as physical properties such as high thermal expansion and low resistance to temperature.

Reinforced plastics are shaped into important structural components using liquid plastics, prepregs, and bulk- and sheet-molding compounds. Fabricating techniques include various molding methods, filament winding, and pultrusion. Important factors in fabricating reinforced-plastic components are the type and orientation of the fibers, and the strength of the bond between fibers and matrix and between different layers of materials. Inspection techniques are available to check the integrity of these products.

BIBLIOGRAPHY

Polymers

Billmeyer, F.W., Jr., *Textbook of Polymer Science*, 3d ed. New York: Wiley, 1984.
Brydson, J.A., *Plastics Materials*, 4th ed. London: Butterworths, 1982.
Crawford, R.J., *Plastics Engineering*, 2d ed. Oxford: Pergamon, 1987.
DuBois, J.H., and F.W. John, *Plastics*, 6th ed. New York: Van Nostrand Reinhold, 1981.
Hall, C., *Polymer Materials*. New York: Macmillan, 1981.
Kaufman, H.S., *Introduction to Polymer Science and Technology*. New York: Wiley, 1986.
MacDermott, C.P., *Selecting Thermoplastics for Engineering Applications*. New York: Marcel Dekker, 1984.
Margolis, J.M., *Engineering Thermoplastics: Properties and Applications*. New York: Marcel Dekker, 1985.
McCrum, N.G., C.P. Buckley, and C.B. Bucknall, *Principles of Polymer Engineering*. Oxford: Oxford University Press, 1988.
Moore, G.R., and D.E. Kline, *Properties and Processing of Polymers for Engineers*. Englewood Cliffs, N.J.: Prentice-Hall, 1984.
Rosen, S.L., *Fundamental Principles of Polymeric Materials*, 2d ed. New York: Wiley, 1982.
Rudin, A., *The Elements of Polymer Science and Engineering*. New York: Academic Press, 1982.

Seymour, R.B., *Polymers for Engineering Applications*. Metals Park, Ohio: ASM International, 1987.

Ward, I.M., *Mechanical Properties of Solid Polymers*, 2d ed. New York: Wiley, 1983.

General References

Ash, M., and I. Ash, *Encyclopedia of Plastics, Polymers and Resins*, 3 vols. New York: Chemical Publishing Co., 1980–81.

Brandrup, J., and E.H. Immergut (eds.), *Polymer Handbook*, 3d ed. New York: Wiley, 1989.

Chanda, M., and S.K. Roy, *Plastics Technology Handbook*. New York: Marcel Dekker, 1987.

Engineered Materials Handbook, Vol. 2: Engineering Plastics. Metals Park, Ohio: ASM International, 1988.

Frados, J. (ed.), *Plastics Engineering Handbook*, 4th ed. New York: Van Nostrand Reinhold, 1976.

Goodman, S.H., *Handbook of Thermoset Plastics*. Park Ridge, N.J.: Noyes Publications, 1986.

Harper, C.A., *Handbook of Plastics and Elastomers*. New York: McGraw-Hill, 1975.

Modern Plastics Encyclopedia. New York: McGraw-Hill; published annually.

Saechtling, H., *International Plastics Handbook*. New York: Macmillan, 1983.

Skotheim, T.A., *Handbook of Conducting Polymers*, 2 vols. New York: Marcel Dekker, 1986.

Wallace, B.M. (ed.), *Handbook of Thermoplastic Elastomers*. New York: Van Nostrand Reinhold, 1979.

Reinforced Plastics

Agarwal, B.D., and L.J. Broutman, *Analysis and Performance of Fibre Composites*. New York: Wiley, 1980.

Brandrup, J., and E.H. Immergut (eds.), *Polymer Handbook*, 3d ed. New York: Wiley, 1989.

Clegg, D.W., *Mechanical Properties of Reinforced Thermoplastics*. New York: Elsevier, 1985.

Delmonte, J., *Technology of Carbon and Graphite Fiber Composites*. New York: Van Nostrand Reinhold, 1981.

Hull, D., *An Introduction to Composite Materials*. Cambridge, England: Cambridge University Press, 1981.

Kaelble, D.H., *Computer-Aided Design of Polymers and Composites*. New York: Marcel Dekker, 1985.

Kowata, K., and T. Akasaka (eds.), *Composite Materials: Mechanical Properties and Fabrication*. London: Applied Science Publishers, 1982.

Sheldon, R.P., *Composite Polymeric Materials*. London: Applied Science Publishers, 1982.

Shook, G. (ed.), *Reinforced Plastics for Commercial Composites: Source Book*. Metals Park, Ohio: American Society for Metals, 1986.

General References

Engineered Materials Handbook, Vol. 1: Composites. Metals Park, Ohio: ASM International, 1987.

Grayson, M. (ed.), *Encyclopedia of Composite Materials and Components*. New York: Wiley, 1983.

Lubin, G. (ed.), *Handbook of Composites*. New York: Van Nostrand Reinhold, 1982.
Schwartz, M., *Composite Materials Handbook*. New York: McGraw-Hill, 1984.
Weeton, J.W. (ed.), *Engineers' Guide to Composite Materials*. Metals Park, Ohio: American Society for Metals, 1986.

Processing and Design

Ash, M., and I. Ash, *Encyclopedia of Plastics, Polymers and Resins*, 3 vols. New York: Chemical Publishing Co., 1980–1981.
Astarita, G., and L. Nicolais, *Polymer Processing and Properties*. New York: Plenum, 1985.
Beck, R.D., *Plastic Product Design*, 2d ed. New York: Van Nostrand, 1980.
Becker, W.E. (ed.), *Reaction Injection Molding*. New York: Van Nostrand Reinhold, 1979.
Benjamin, B.S., *Structural Design with Plastics*, 2d ed. New York: Van Nostrand Reinhold, 1982.
Brown, R.L.E., *Design and Manufacture of Plastic Parts*. New York: Wiley, 1980.
Chanda, M., and S.K. Roy, *Plastics Technology Handbook*. New York: Marcel Dekker, 1987.
Dym, J.B., *Product Design with Plastics: A Practical Manual*. New York: Industrial Press, 1982.
Engineered Materials Handbook, Vol. 1: Composites. Metals Park, Ohio: ASM International, 1987.
Fenner, R.T., *Principles of Polymer Processing*. New York: Macmillan, 1979.
Florian, J., *Practical Thermoforming: Principles and Applications*. New York: Marcel Dekker, 1987.
Frados, J. (ed.), *Plastics Engineering Handbook*, 4th ed. New York: Van Nostrand Reinhold, 1976.
Grayson, M. (ed.), *Encyclopedia of Composite Materials and Components*. New York: Wiley, 1983.
Kelly, A., and S.T. Mileiko (eds.), *Fabrication of Composites*. Amsterdam: Elsevier, 1983.
Kowata, K., and T. Akasaka (eds.), *Composite Materials: Mechanical Properties and Fabrication*. London: Applied Science Publishers, 1982.
Levy, S., *Plastics Extrusion Technology Handbook*. New York: Industrial Press, 1981.
Levy, S., and J.H. DuBois, *Plastics Product Design Engineering Handbook*, 2d ed. New York: Van Nostrand Reinhold, 1985.
Meyer, R.W., *Handbook of Pultrusion Technology*. New York: Methuen, 1985.
Middleman, S., *Fundamentals of Polymer Processing*. New York: McGraw-Hill, 1977.
Miles, D.C., and J.H. Briston, *Polymer Technology*. New York: Chemical Publishing Company, 1979.
Miller, E. (ed.), *Plastics Products Design Handbook. Part A: Materials and Components*, 1981; *Part B: Processes and Design for Processes*. 1983. New York: Marcel Dekker.
Modern Plastics Encyclopedia. New York: McGraw-Hill, annual.
Schwartz, M. (ed.), *Composite Materials Handbook*. New York: McGraw-Hill, 1984.
Schwartz, M. (ed.), *Fabrication of Composite Materials: Source Book*. Metals Park, Ohio: American Society for Metals, 1985.
Stoeckert, K. (ed.), *Mold Making Handbook*. New York: Macmillan, 1983.
Sweeney, F.M., *Reaction Injection Molding Machinery and Processes*. New York: Marcel Dekker, 1987.
Tadmor, Z., and C.G. Gogos, *Principles of Polymer Processing*. New York: Wiley, 1979.
Throne, J.L., *Plastics Process Engineering*. New York: Marcel Dekker, 1979.
Wendle, B.C., *Structural Foam*. New York: Marcel Dekker, 1985.

QUESTIONS

10.1 What are the major differences between polymers and metals?

10.2 Describe the basic differences between thermoplastics and thermosetting plastics. Why are thermosets generally brittle?

10.3 How do the water absorption characteristics of some polymers affect their engineering applications? Give specific examples.

10.4 Why would we want to synthesize a polymer to a high degree of crystallinity?

10.5 Inspect various plastic components in an automobile and state whether they are made of thermoplastics or thermosets.

10.6 Explain the significance of orientation in polymers.

10.7 What is the significance of the glass-transition temperature, T_g?

10.8 What properties do elastomers have that thermoplastics in general do not have?

10.9 Observe the behavior of the specimen shown in Fig. 10.7 and state whether the material has a high or low m (strain-rate sensitivity exponent). Explain.

10.10 Why does cross-linking improve the strength of polymers?

10.11 Describe the differences between the significance of covalent, ionic, hydrogen, and van der Vaals bonds. Identify the primary and secondary bonds.

10.12 Can polymers be made to conduct electricity? If so, how?

10.13 Explain why elastomers were developed.

10.14 Make a list of products or parts that are not currently made of plastics and give the possible reasons why they are not.

10.15 Is the substitution of plastics for metals in products traditionally made of metal viewed negatively by the public? Why or why not?

10.16 Assume that you are manufacturing a product in which all the gears are made of metal. A visiting salesperson asks you to consider replacing some of these metal gears with plastic ones. Make a list of questions that you would raise before making a decision.

10.17 Describe the design considerations involved in replacing a metal container for a beverage with a plastic container.

10.18 Review the three curves in Fig. 10.9 and name applications for each type of material. Explain your choices.

10.19 Repeat Problem 10.18 for the curves in Fig. 10.13.

10.20 Is it possible for a material to have a hysteresis behavior that is the opposite of that shown in Fig. 10.14, whereby the arrows are counterclockwise? Explain.

10.21 Describe the mechanism that allows thermoplastics to be stretched to a far greater extent than metals.

10.22 Explain the advantages and limitations of reinforced plastics.

10.23 Distinguish between composites and alloys.

10.24 What fundamental differences are there between the properties of reinforcing fibers and matrix materials?

10.25 List and explain the factors that influence the strength of reinforced plastics.

10.26 Identify metals and alloys that have strengths comparable to those of reinforced plastics.

10.27 Figure 10.28 shows the influence of fiber length on the properties of reinforced plastics. Why does fiber length have an influence?

10.28 What advantages do hybrid composites have over other composites?

10.29 Would a composite material with a strong and stiff matrix and soft and flexible reinforcement have any practical uses? Explain.

10.30 Make a list of products for which the use of composite materials could be advantageous because of their anisotropic properties.
10.31 Name applications in which both specific strength and specific stiffness are important.
10.32 Why are fibers capable of supporting a major portion of the load in composite materials?
10.33 Other than those described in this chapter, what materials can be regarded as composite materials?
10.34 What are the forms of materials for processing plastics into products?
10.35 Why is injection molding capable of producing parts with complex shapes and fine detail?
10.36 How is thin plastic film produced?
10.37 What similarities and differences are there between compression molding of plastics and impression-die forging of metals?
10.38 Describe the advantages of cold forming of plastics over other processing methods.
10.39 What are the characteristics of filament-wound products?
10.40 Explain the major design considerations in forming and shaping reinforced plastics.
10.41 Describe the advantages of applying traditional metalworking techniques to forming plastics.
10.42 Explain why some forming processes are more suitable for certain plastics than for others.
10.43 Inspect various plastic components and identify the processes that could be used in making them.
10.44 Would you use thermosetting plastics for injection molding? Explain.
10.45 Outline the precautions that you would take in shaping reinforced plastics.
10.46 An injection-molded nylon gear is found to contain small pores. Explain why drying the material before molding it will solve this problem.
10.47 Explain why operations such as blow molding and film-bag making are done vertically and why buildings housing these processes have ceilings 10–15 m (35–50 ft) high.
10.48 Some plastic products have lids with integral hinges; that is, no other material or part is used at the junction of the two parts. Describe a method for making this product.
10.49 Make a list of the processing methods used for reinforced plastics. Identify each with the following fiber arrangement capabilities: (a) uniaxial, (b) cross-ply, (c) in-plane random, and (d) three-dimensional random.

PROBLEMS

10.1 Calculate the areas under the stress–strain curve (toughness) for the material in Fig. 10.11, plot them as a function of temperature, and describe your observations.
10.2 Calculate the average increase in the properties of plastics in Tables 10.1 as a result of their reinforcement and describe your observations.
10.3 Take an injection-molded product and estimate the clamping force that would be required in making it.
10.4 In Example 10.5, what would be the percentage of the load supported by the fibers if their strength is 1250 MPa and the matrix strength is 240 MPa?

11

Processing of Powder Metals and Ceramics

11.1 Introduction

In the manufacturing processes we described in the preceding chapters, the raw materials used are either in a molten state or in solid form. In this chapter we describe how metal parts can be made by compacting metal powders in suitable dies and sintering them (heating without melting). This process is called *powder metallurgy* (P/M). One of its first uses was in the early 1900s to make the tungsten filaments for incandescent light bulbs. The availability of a wide range of powder compositions, the capability to produce parts to net dimensions (net-shape forming), and the economics of the overall operation make this process attractive for many applications.

Typical products made by powder metallurgy techniques are gears, cams, bushings, cutting tools, porous products such as filters and oil-impregnated bearings, and automotive components, such as piston rings, valve guides, connecting rods, and hydraulic pistons (Fig. 11.1). Advances in technology now permit structural parts of aircraft—such as landing gear, engine-mount supports, engine disks, impellers, and engine nacelle frames—to be made by P/M.

Pure metals, alloys, or mixtures of metallic and nonmetallic materials can be used in powder metallurgy. The most commonly used metals are iron, copper, aluminum, tin, nickel, titanium, and refractory metals. For parts made of brass, bronze, steels, and stainless steels, prealloyed powders are used, where each powder particle itself is an alloy.

Powder metallurgy has become competitive with processes such as casting, forging, and machining, particularly for relatively complex parts made of high-strength and hard alloys. Parts made by this process have good dimensional accuracy, and their sizes range from tiny balls for ball-point pens to parts weighing about 50 kg (100 lb), although most parts weigh less than 2.5 kg (5 lb).

(a) (b)

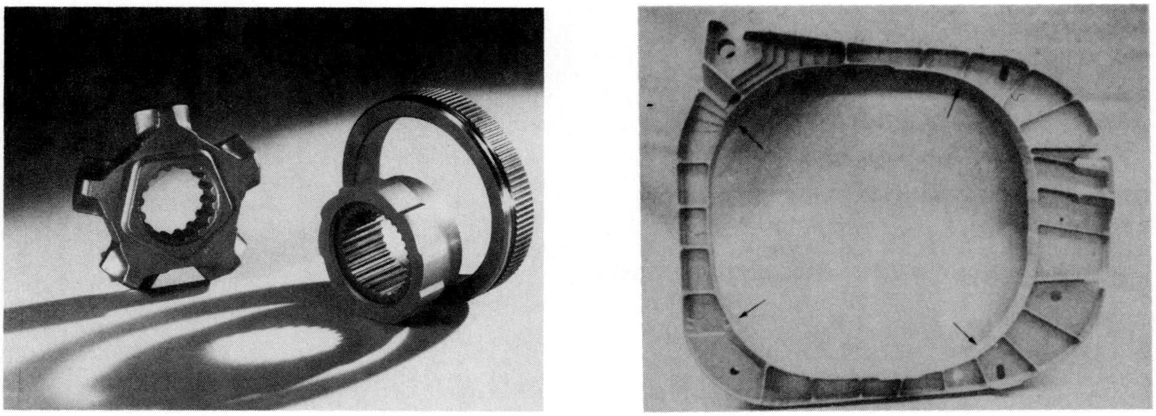

FIGURE 11.1

Examples of parts made by powder-metallurgy processes. (a) Left, steel chain-drive sprocket; right, race and cam for automatic transmission. *Source:* Metal Powder Industries Federation. (b) Titanium-alloy engine nacelle frame for F-14 fighter aircraft made of four pieces and welded (at arrows) by the electron-beam process. *Source:* Colt-Crucible.

11.2

Production of Metal Powders

Basically, the powder metallurgy process consists of the following operations:

a) Powder production
b) Blending
c) Compaction
d) Sintering
e) Finishing operations.

For improved quality and dimensional accuracy, or for special applications, additional processing such as coining, sizing, forging, machining, infiltration, and resintering may be carried out.

11.2.1 Methods of powder production

There are several methods of producing metal powder (Table 11.1). Metal sources are generally bulk metals and alloys, ores, salts, and other compounds. Most metal

TABLE 11.1
METHODS OF PRODUCING METAL POWDERS

POWDER METAL	ATOMIZATION	REDUCTION	ELECTROLYTIC DEPOSITION	THERMAL DECOMPOSITION	COMMINUTION	PRECIPITATION FROM LIQUID OR GAS
Aluminum	X					
Aluminum alloys	X					
Beryllium			X		X	
Cobalt		X				
Copper	X	X	X			X
Copper alloys	X					
Iron	X	X	X	X	X	
Iron alloys (low-alloy steel, stainless steel, tool steel)	X					
Molybdenum		X				
Nickel		X		X		X
Nickel alloys	X				X	
Silver	X		X			X
Tantalum		X	X			
Tin	X					
Titanium	X	X		X		
Tungsten		X				
Zirconium				X		

powders can be produced by more than one method, the choice depending on the requirements of the end product. Particle sizes range from 0.1 to 1000 μm (4 μin. to 0.04 in.). The shape, size distribution, porosity, chemical purity, and bulk and surface characteristics of the particles depend on the particular process used (Fig. 11.2). These characteristics are important because they significantly affect permeability and flow characteristics during compaction and in subsequent sintering operations.

Atomization. *Atomization* produces a liquid-metal stream by injecting molten metal through a small orifice (Fig. 11.3a). The stream is broken up by jets of inert gas, air, or water. The size of particles formed depends on the temperature of the metal, the rate of flow, nozzle size, and jet characteristics. In one variation of this method, a consumable electrode is rotated rapidly in a helium-filled chamber (Fig. 11.3b). The centrifugal force breaks up the molten tip of the electrode, producing metal particles.

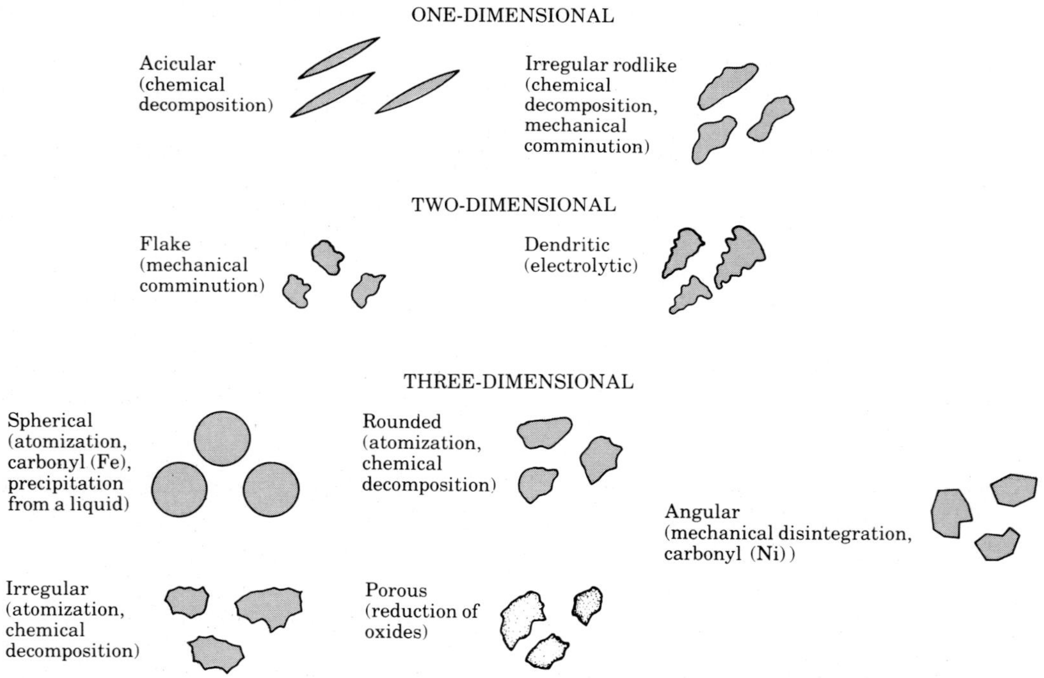

FIGURE 11.2

Particle shapes in metal powders and the processes by which they are produced. Iron powders are produced by many of these processes. *Source:* After P. K. Johnson.

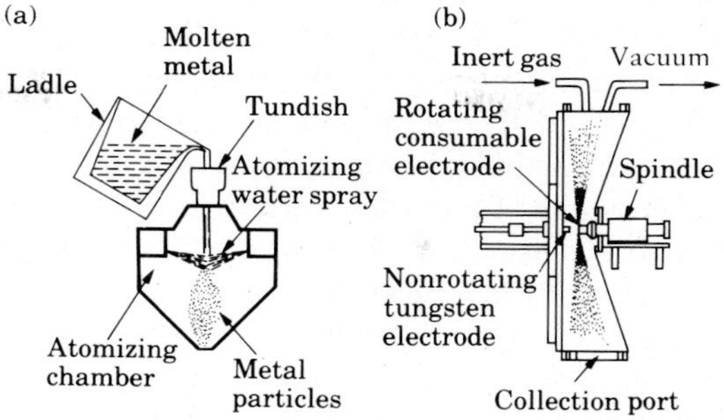

FIGURE 11.3
Methods of metal powder production by atomization: (a) melt atomization; and (b) atomization with a rotating consumable electrode.

Reduction. *Reduction* of metal oxides (removal of oxygen) uses gases, such as hydrogen and carbon monoxide, as reducing agents. Thus very fine metallic oxides are reduced to the metallic state. The powders produced by this method are spongy and porous and have uniformly sized spherical or angular shapes.

Electrolytic deposition. *Electrolytic deposition* utilizes either aqueous solutions or fused salts. The powders produced are among the purest.

Carbonyls. Metal *carbonyls*, such as iron carbonyl, $Fe(CO)_5$, and nickel carbonyl, $Ni(CO)_4$, are formed by letting iron or nickel react with carbon monoxide. The reaction products are then decomposed to iron and nickel, producing small, dense, and uniform spherical particles of high purity.

Comminution. Mechanical *comminution* (*pulverization*) involves crushing, milling in a *ball mill*, or grinding brittle or less ductile metals into small particles. With brittle materials, the powder particles have angular shapes, whereas with ductile metals they are flaky and are not particularly suitable for powder metallurgy applications.

Mechanical alloying. In *mechanical alloying*, developed in the 1960s, powders of two or more pure metals are mixed in a ball mill. Under the impact of the hard balls, the powders fracture and weld together by diffusion, forming alloy powders.

Other methods. Other less commonly used methods are *precipitation* from a chemical solution, production of fine metal chips by *machining*, and *vapor condensation*.

Particle *size* is determined by screening, that is, passing the metal powder through screens of various mesh sizes. Screen analysis is by means of a vertical stack of screens with increasing mesh size as the powder flows downward through the screens. The larger the mesh size is, the smaller the opening in the screen. For example, a mesh size of 30 has an opening of 600 μm, size 100 has 150 μm, and size 400 has 38 μm. (Note that this method is similar to numbering of abrasive grits; the larger the number, the smaller the size of the abrasive particle.)

The *shape* of particles has a major influence on their processing characteristics. The shape is usually described in terms of aspect ratio or shape index. *Aspect ratio* is the ratio of the largest dimension to the smallest dimension of the particle. This ratio ranges from unity for a spherical particle, to about 10 for flakelike or needlelike particles. *Shape index* or *factor* (SF) is a measure of the surface area to the volume of the particle with reference to a spherical particle of equivalent diameter.

• Example 11.1: Particle shape-factor determination.

Determine the shape factor for (a) a spherical particle, (b) a cubic particle, and (c) a cylinder with a length-to-diameter ratio of 2. Let the smallest dimension be unity.

SOLUTION.

a) The surface area A of a sphere of diameter D is $A = \pi D^2$ and its volume is $V = \pi D^3/6$, so the shape factor will be SF $= A/V = 6D = 6$. Note that $D = (6V/\pi)^{1/3}$.

b) The surface area of a cube with lateral dimensions of unity is 6 and its volume is 1. Thus $A/V = 6$. The equivalent diameter for a sphere is $D = (6V/\pi)^{1/3} = (6/\pi)^{1/3} = 1.24$. Hence SF $= (1.24)(6) = 7.44$.

c) The surface area of this cylinder is $A = 2\pi D^2/4 + 2(\pi D) = 2.5\pi$. Its volume is $2\pi D^2/4 = \pi/2$. Thus $A/V = 5$. The equivalent diameter for a sphere is $D = [(6\pi/2)(1/\pi)]^{1/3} = 1.44$. Hence SF $= (1.44)(6) = 7.21$.

11.2.2 Blending metal powders

Blending (mixing) powders is the second step in powder metallurgy processing and is carried out for the following purposes.

- Because the powders made by various processes may have different sizes and shapes, they have to be mixed to obtain uniformity. The ideal mix is one in which all the particles of each material are distributed uniformly.
- Powders of different metallic and other materials may be mixed in order to impart special physical and mechanical properties and characteristics to the P/M product.

- Lubricants may be mixed with the powders to improve their flow characteristics. The results are reduced friction between the metal particles, improved flow of the powder metals into the dies, and longer die life. Lubricants typically are stearic acid or zinc stearate, in proportions of 0.25–5 percent by weight.

Powder mixing must be carried out under controlled conditions to avoid contamination or deterioration. Deterioration is caused by excessive mixing, which may alter the shape of the particles and work-harden them, thus making the subsequent compacting operation more difficult. Powders can be mixed in air or in inert atmospheres (to avoid oxidation) and in liquids, which act as lubricants and make the mix more uniform. Several different types of blending equipment are available. These operations are being increasingly controlled by microprocessors to improve and maintain quality.

Hazards. Because of their high surface area-to-volume ratio, metal powders are explosive, particularly aluminum, magnesium, titanium, zirconium, and thorium. Great care must be exercised both during blending and during storage and handling. Precautions include grounding equipment and avoiding sparks (by using nonsparking tools and avoiding friction as a source of heat), dust clouds, open flames, and chemical reactions.

• Example 11.2: Density of metal powder–lubricant mix.

We have stated that zinc stearate is a lubricant that commonly is mixed with metal powders prior to compaction, in proportions up to 5 percent by weight. Calculate the theoretical and apparent densities of an iron powder–zinc stearate mix, assuming that (a) 1000 g of iron powder is mixed with 20 g of lubricant, (b) the density of the lubricant is 1.10 g/cm^3, (c) the theoretical density of the iron powder is 7.86 g/cm^3 (from Table 3.2), and (d) the apparent density of the iron powder is 2.75 g/cm^3 (from Fig. 11.5a).

SOLUTION. The volume of the mixture is

$$V = \left(\frac{1000}{7.86}\right) + \left(\frac{20}{1.10}\right) = 145.41 \text{ cm}^3.$$

The combined weight of the mix is 1020 g, so its theoretical density is 1020/145.41 = 7.01 g/cm^3. The apparent density of the iron powder is 2.75 g/cm^3, so it is (2.75/7.86)100 = 35% of the theoretical density. Assuming a similar percentage for the zinc stearate, we can estimate the apparent density of the mix as (0.35)(7.01) = 2.45 g/cm^3.

11.3 Compaction of Metal Powders

Compaction is the step in which the blended powders are pressed into shapes in dies (Figs. 11.4a and b), using presses that are either hydraulically or mechanically activated. The purposes of compaction are to obtain the required shape, density, and particle-to-particle contact and to make the part strong enough to be processed further. The pressed powder is known as a *green compact*. The powder must flow easily to properly feed into the die cavity. Pressing is generally carried out at room temperature, although it can be done at elevated temperatures.

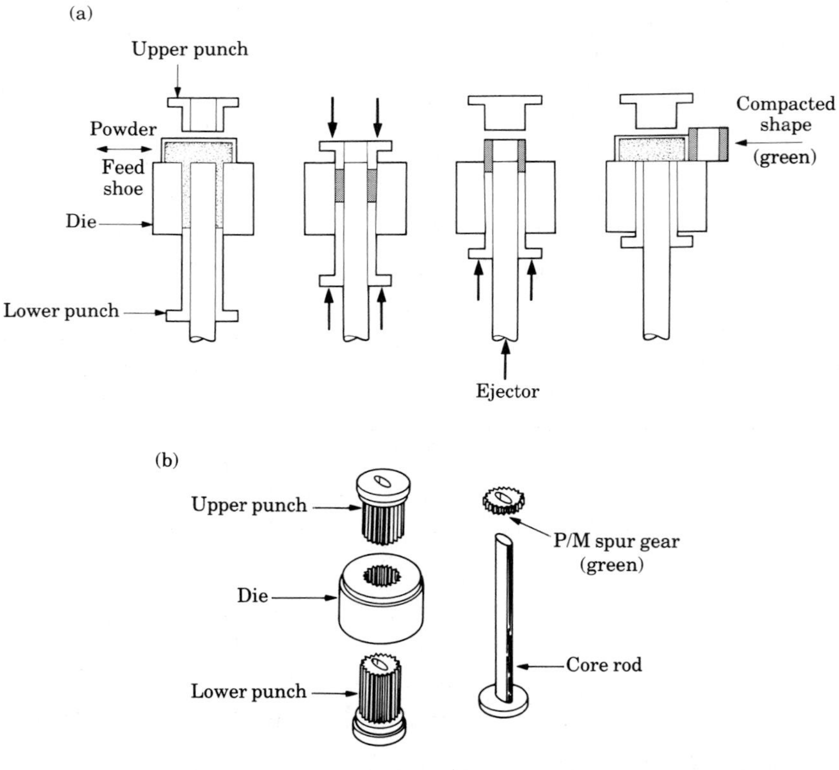

FIGURE 11.4
(a) Compaction of metal powder to form a bushing. The pressed powder part is called green compact. (b) Typical tool and die set for compacting a spur gear. *Source:* Metal Powder Industries Federation.

The density of the green compact depends on the pressure applied (Fig. 11.5a). As the compacting pressure is increased, the density approaches that of the theoretical density of the metal in its bulk form. Another important factor is the size distribution of the particles. If all the particles are the same size, there will always be some porosity when they are packed together—theoretically, at least 24 percent by volume. Imagine, for example, a box filled with tennis balls; there are always open spaces between the balls. However, introducing smaller particles will fill the spaces between the larger particles and thus result in a higher density of the compact.

The higher the density, the higher will be the strength and elastic modulus of the part as well as its electrical conductivity (Fig. 11.5b). The reason is that the higher the density, the higher will be the amount of solid metal in the same volume—hence the greater its resistance to external forces. Because of friction between the metal particles in the powder, and the friction between the punches and the die walls, the density can vary considerably within the part. This variation can be minimized by proper punch and die design and friction control. For example, it may be necessary

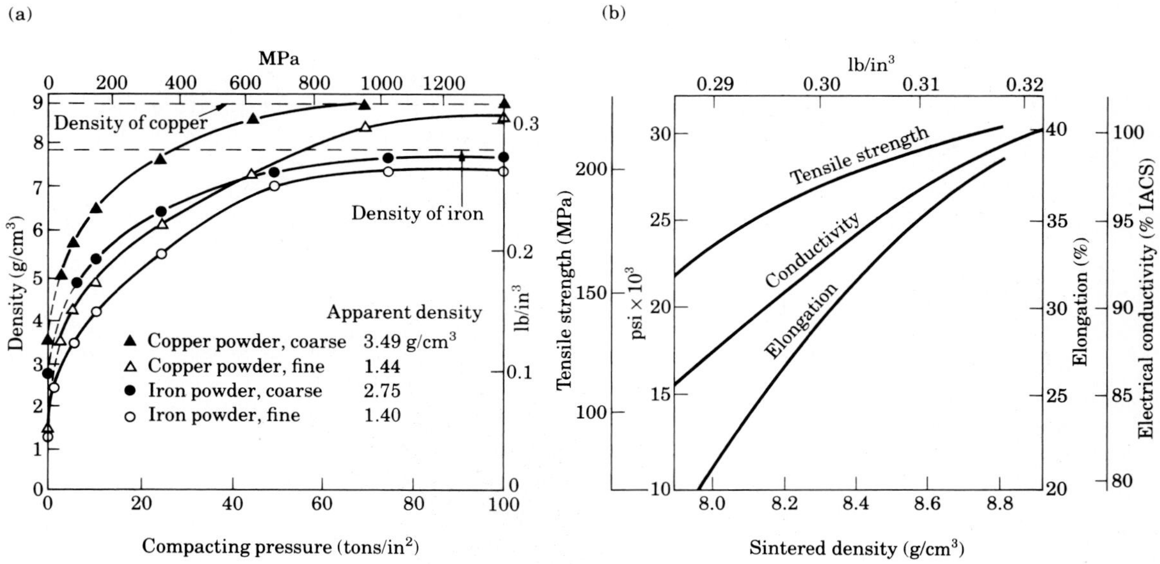

FIGURE 11.5 (a) Density of copper and iron powder compacts as a function of compacting pressure. Density greatly influences the mechanical and physical properties of P/M parts. *Source:* F. V. Lenel, *Powder Metallurgy: Principles and Applications.* Princeton, N.J.: Metal Powder Industries Federation, 1980. (b) Effect of density on tensile strength, elongation, and electrical conductivity of copper powder. IACS means International Annealed Copper Standard for electrical conductivity. *Source:* After J. L. Everhart.

to use multiple punches, with separate movements in order to ensure that the density is more uniform throughout the part (Fig. 11.6).

11.3.1 Equipment

The pressure required for pressing metal powders ranges from 70 MPa (10 ksi) for aluminum to 800 MPa (120 ksi) for high-density iron parts (Table 11.2). The compacting pressure required depends on the characteristics and shape of the particles, method of blending, and lubrication.

Press capacities are on the order of 1.8–2.7 MN (200–300 tons), although presses with much higher capacities are used for special applications. Most applications require less than 100 tons. For small tonnage, crank or eccentric type mechanical presses are used; for higher capacities, toggle or knucklejoint presses are employed. Hydraulic presses can be used with capacities as high as 45 MN (5000 tons) for large parts.

The selection of the press depends on part size and configuration, density requirements, and production rate. An important consideration with regard to pressing speed is the entrapment of air in the die cavity. The presence of air will prevent proper compaction and the higher the speed, the greater is the tendency for the press to trap air.

11.3.2 Pressure distribution in powder compaction

Figure 11.6(e) shows that, in single-action pressing, the pressure decays rapidly toward the bottom of the compact. We can determine the pressure distribution along the length of the compact by using the slab method of analysis of deformation

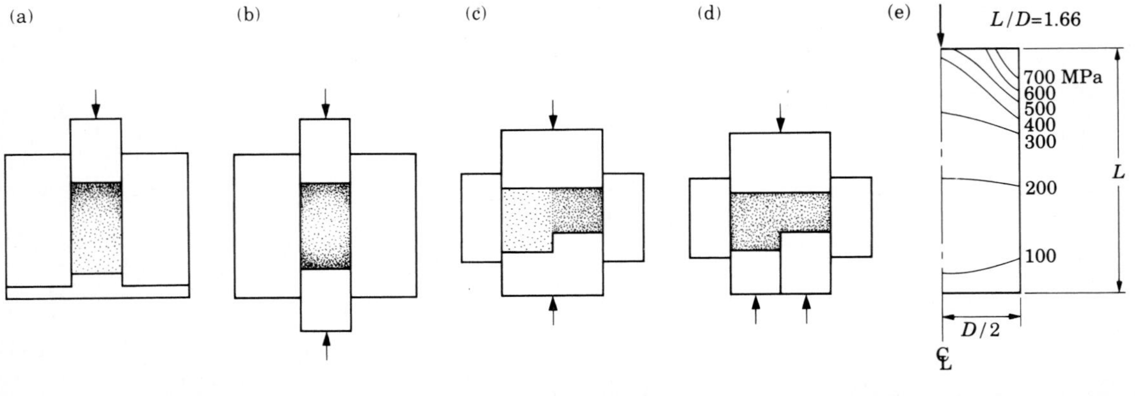

FIGURE 11.6

Density variation in compacting metal powders in different dies: (a) and (c) single-action press; (b) and (d) double-action press. Note in (d) the greater uniformity of density in pressing with two punches with separate movements compared with (c). There are situations in which density variation, hence property variation, within a part may be desirable. (e) Pressure contours in compacted copper powder in a single-action press. *Source:* P. Duwez and L. Zwell.

TABLE 11.2
COMPACTING PRESSURES FOR VARIOUS METAL POWDERS

METAL	PRESSURE (MPa)
Aluminum	70–275
Brass	400–700
Bronze	200–275
Iron	350–800
Tantalum	70–140
Tungsten	70–140
OTHER MATERIALS	
Aluminum oxide	110–140
Carbon	140–165
Cemented carbides	140–400
Ferrites	110–165

processes described in Section 6.2.2. As we did in Fig. 6.4, we first describe the operation in terms of its coordinate system, as shown in Fig. 11.7: D is the diameter of the compact, L is its length, and p_0 is the pressure applied by the punch. We take an element dx thick and place on it all the relevant stresses, namely, the compacting

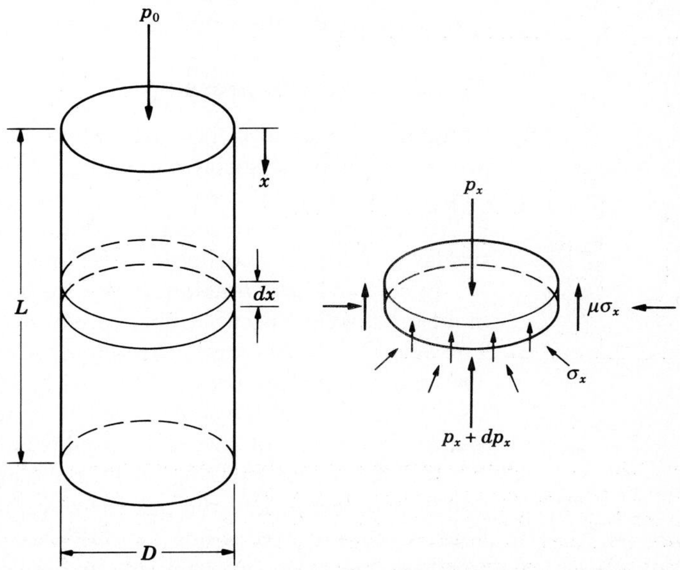

FIGURE 11.7
Coordinate system and stresses acting on an element in compaction of powders. The pressure is assumed to be uniform across the cross-section (see also Fig. 6.4).

pressure p_x, die-wall pressure σ_x, and frictional stress $\mu\sigma_x$. Note that the frictional stresses act upward on the element, because the punch movement is downward.

Balancing the vertical forces acting on this element, we have

$$\left(\frac{\pi D^2}{4}\right)p_x - \left(\frac{\pi D^2}{4}\right)(p_x + dp_x) - (\pi D)(\mu\sigma_x)\,dx = 0,$$

which we can simplify to

$$D\,dp_x + 4\mu\sigma_x\,dx = 0.$$

We have one equation but two unknowns (p_x and σ_x). Let's now introduce a factor k:

$$\sigma_x = kp_x,$$

which is a measure of the interparticle friction during compaction. Thus if there is no friction between the particles, $k = 1$, the powder behaves like a fluid, and we have $\sigma_x = p_x$, signifying a state of hydrostatic pressure. We now have the expression

$$dp_x + \frac{4\mu k p_x\,dx}{D} = 0, \quad \text{or} \quad \frac{dp_x}{p_x} = \frac{-4\mu k\,dx}{D}.$$

This expression is similar to that given in Section 6.2.2 for upsetting. In the same manner as there, we integrate this expression, noting that the boundary condition in this case is $p_x = p_0$ when $x = 0$:

$$p_x = p_0 e^{-4\mu k x/D}. \qquad (11.1)$$

Thus the pressure within the compact decays as the coefficient of friction, the parameter k, and the length-to-diameter ratio increase.

11.3.3 Isostatic pressing

Compaction can also be carried out or improved by a number of additional processes, such as isostatic pressing, rolling, and forging. Because the density of compacted powders can vary significantly, green compacts are subjected to *hydrostatic pressure* in order to achieve more uniform compaction. This process is similar to cupping your hands when making snow balls.

In *cold isostatic pressing* (CIP), the metal powder is placed in a flexible rubber mold made of neoprene rubber, urethane, polyvinyl chloride, or other elastomers. The assembly is then pressurized hydrostatically in a chamber, usually with water. The most common pressure is 400 MPa (60 ksi), although pressures of up to 1000 MPa (150 ksi) have been used. The applications of CIP and other compacting methods, in terms of size and complexity of part, are shown in Fig. 11.8.

In *hot isostatic pressing* (HIP), the container is usually made of a high-melting-point sheet metal, and the pressurizing medium is inert gas or vitreous (glasslike) fluid. Common conditions for HIP are 100 MPa (15 ksi) at 1100 °C (2000 °F), although the trend is toward higher pressures and temperatures. The main

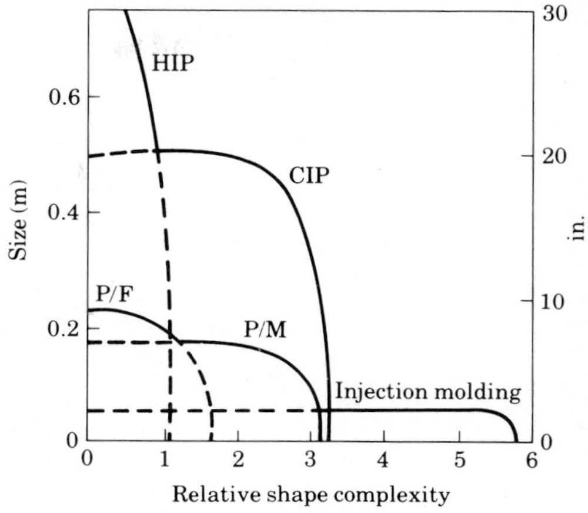

FIGURE 11.8 Capabilities of part size and shape complexity according to various P/M operations. P/F means powder forging. *Source:* Metal Powder Industries Federation.

advantage of HIP is its ability to produce compacts with essentially 100 percent density, good metallurgical bond among the particles, and good mechanical properties.

The HIP process is relatively expensive and is used mainly in making superalloy components for the aerospace industry. It is routinely used as a final densification step for tungsten-carbide cutting tools and P/M tool steels. The process is also used to close internal porosity and improve properties in superalloy and titanium-alloy castings for the aerospace industry.

The main advantage of isostatic pressing is that, because of uniformity of pressure from all directions and the absence of die-wall friction, it produces compacts of practically uniform grain structure and density, irrespective of shape. Parts with high length-to-diameter ratios have been produced, with very uniform density, strength, and toughness and good surface details.

11.3.4 Other compacting and shaping processes

Rolling. In *powder rolling*, also called *roll compaction*, the powder is fed to the roll gap in a two-high rolling mill, and is compacted into a continuous strip at speeds of up to 0.5 m/s (100 ft/min). The process can be carried out at room or elevated temperatures. Sheet metal for electrical and electronic components and for coins can be made by powder rolling.

Extrusion. Powders can be compacted by *extrusion*; the powder is encased in a metal container and extruded. After sintering, preformed P/M parts may be reheated and forged in a closed die to their final shape.

Injection molding. In *injection molding*—a relatively new process—the metal powders are blended with a polymer. The mixture then undergoes a process similar to die casting. The molded greens are placed in a low-temperature oven to burn off the plastic and then sintered in a furnace. The major advantage of injection molding over conventional compaction is that relatively complex shapes, with wall thicknesses as small as 5 mm (0.2 in.), can be molded and removed easily from the dies.

Pressureless compaction. In *pressureless compaction*, the die is filled with metal powder by gravity, and the powder is sintered directly in the die. Because of the resulting low density, pressureless compaction is used principally for porous parts, such as filters.

Ceramic molds. Molds for shaping metal powders are made by the technique used in investment casting. After the ceramic mold is made, it is filled with metal powder and placed in a steel container. The space between the mold and the container is filled with particulate material. The container is then evacuated, sealed, and subjected to hot isostatic pressing. Titanium-alloy compressor rotors for missile engines have been made by this process.

11.3.5 Punch and die materials

The selection of punch and die materials for P/M depends on the abrasiveness of the powder metal and the number of parts to be made. Most common die materials are air- or oil-hardening tool steels, such as D2 or D3, with a hardness range of 60–64 HRC. Because of their greater hardness and wear resistance, tungsten-carbide dies are used for more severe applications. Punches are generally made of similar materials.

Close control of die and punch dimensions is essential for proper compaction and die life. Too large a clearance between the punch and the die will allow the metal powder to enter the gap, interfere with the operation, and result in eccentric parts. Diametral clearances are generally less than 25 μm (0.001 in.). Die and punch surfaces must be lapped or polished—and in the direction of tool movements for improved die life and overall performance.

• Example 11.3: Pressure decay in compaction.

Assume that a powder mix has $k = 0.5$ and $\mu = 0.3$. At what depth will the pressure in a straight cylindrical compact 10 mm in diameter become (a) zero and (b) one-half the pressure at the punch?

SOLUTION. For case (a) from Eq. (11.1), $p_x = 0$. Consequently, we have the expression

$$0 = p_0 e^{-(4)(0.3)(0.5)x/10}, \quad \text{or} \quad e^{-0.06x} = 0.$$

The value of x must be ∞ for the pressure to decay to 0.

For case (b), we have $p_x/p_0 = 0.5$. Therefore

$$e^{-0.06x} = 0.5, \quad \text{or} \quad x = 11.55 \text{ mm}.$$

•

11.4 Sintering

Sintering is the process whereby compressed metal powder is heated in a controlled-atmosphere furnace to a temperature below its melting point, but sufficiently high to allow bonding (fusion) of the individual particles. Prior to sintering, the compact is brittle and its strength, known as *green strength*, is low. The nature and strength of the bond between the particles, and hence of the sintered compact, depend on the mechanisms of diffusion, plastic flow, evaporation of volatile materials in the compact, recrystallization, grain growth, and pore shrinkage.

The principal governing variables in sintering are temperature, time, and the atmosphere in the sintering furnace. Sintering temperatures (Table 11.3) are generally within 70–90 percent of the melting point of the metal or alloy. Sintering times range from a minimum of about 10 minutes for iron and copper alloys to as much as 8 hours for tungsten and tantalum. Continuous sintering furnaces are used for most production today. These furnaces have three chambers: (1) a burn-off chamber to volatilize the lubricants in the green compact in order to improve bond strength and prevent cracking; (2) a high-temperature chamber for sintering; and (3) a cooling chamber.

Proper control of the furnace atmosphere is essential for successful sintering and to obtain optimum properties. An oxygen-free atmosphere is essential to control the carburization and decarburization of iron and iron-base compacts and to prevent oxidation of powders. A vacuum is generally used for sintering refractory metal alloys and stainless steels. The gases most commonly used for sintering a variety of other metals are hydrogen, dissociated or burned ammonia, partially combusted hydrocarbon gases, and nitrogen.

TABLE 11.3
SINTERING TEMPERATURE AND TIME FOR VARIOUS METALS

MATERIAL	TEMPERATURE (°C)	TIME (MIN)
Copper, brass, and bronze	760–900	10–45
Iron and iron-graphite	1000–1150	8–45
Nickel	1000–1150	30–45
Stainless steels	1100–1290	30–60
Alnico alloys (for permanent magnets)	1200–1300	120–150
Tungsten carbide	1430–1500	20–30
Molybdenum	2050	120
Tungsten	2350	480

Sintering mechanisms are complex and depend on the composition of metal particles as well as processing parameters (Fig. 11.9). As temperature increases, two adjacent particles begin to form a bond by diffusion (*solid-state bonding*). As a result, the strength, density, ductility, and thermal and electrical conductivities of the compact increase. At the same time, however, the compact shrinks; hence allowances should be made for *shrinkage*, as in casting.

If two adjacent particles are of different metals, *alloying* can take place at the interface of the two particles. One of the particles may be a lower melting-point metal than the other. In that case, one particle may melt and, because of surface tension, surround the particle that has not melted (*liquid-phase sintering*). An example is cobalt in tungsten carbide tools and dies. Stronger and denser parts can be obtained in this way.

Depending on temperature, time, and processing history, different structures and porosities can be obtained in a sintered compact. However, porosity cannot be completely eliminated because voids remain after compaction and gases evolve during sintering. Porosities can consist of either a network of interconnected pores or closed holes. Their presence is an important consideration in making P/M filters and bearings.

Another method, which is still at an experimental stage, is *spark sintering*. In this process, loose metal powders are placed in a graphite mold, heated by an electric current, subjected to a high-energy discharge, and compacted, all in one step. The rapid discharge strips any oxide coating, such as found on aluminum, or contaminants from the surfaces of the particles, and thus encourages good bonding during compaction at elevated temperatures.

Typical mechanical properties and applications for several sintered P/M alloys are given in Table 11.4. Note the effect of heat treating on the properties of

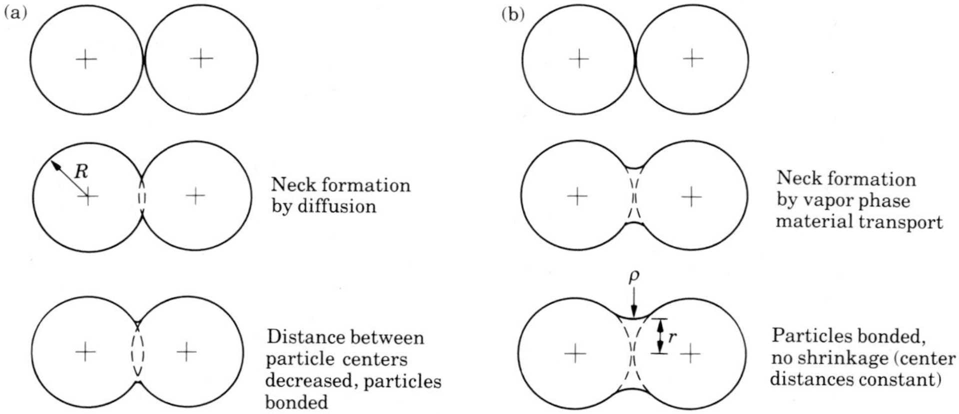

FIGURE 11.9

Schematic illustration of two mechanisms for sintering metal powders: (a) solid-state material transport; and (b) liquid-phase material transport. R = particle radius, r = neck radius, and ρ = neck profile radius.

TABLE 11.4
PROPERTIES AND TYPICAL APPLICATIONS OF POWDER METALLURGY PARTS

MATERIAL	MPIF TYPE	CONDITION	DENSITY (kg/m^3)	ULTIMATE TENSILE STRENGTH (MPa)	HARDNESS	ELASTIC MODULUS (GPa)	ELONGATION IN 25 mm (%)	TYPICAL APPLICATIONS
P/M brass (leaded) 77.0–80.0 Cu, 1.0–2.0 Pb, 0.3 Fe max, 0.1 Sn max, bal Zn	T	* Sintered	7400	165	55 HRH	83	13	Mechanical components resistant to atmospheric corrosion.
	U	Sintered	7800	195	68 HRH	90	19	Ordnance components, builders' hardware, lock parts, housings, nuts, gears.
	W	Sintered	8200	220	75 HRH	95	23	
601AB (Alcoa) 0.25 Cu, 0.6 Si, 1.0 Mg, 1.5 lubricant, bal Al		Sintered	2550	145	65–70 HRH	—	6	Similar to wrought 6061; strength, ductility, corrosion resistance.
		Heat treated	2550	240	80–85 HRE	—	2	
P/M iron 0.3 C max	N	Sintered	5800	110	10 HRH	72	2	Structural (lightly loaded gears); magnetic (motor pole pieces); self-lubricating bearings; structural, wear resisting (small levers and cams) as carbonitrided.
	P	Sintered	6200	130	70 HRH	90	2.5	
	R	Sintered	6600	165	80 HRH	110	5	
	S	Sintered	7000	205	15 HRB	130	9	
	T	Sintered	7400	275	30 HRB	160	15	
P/M austenitic stainless steels 303	P	Sintered	6200	240	—	—	1	Type 303, mechanical components; type 410, structural, corrosion resisting components.
303	R	Sintered	6600	360	—	—	2	
410	N	Sintered	5800	290	—	—	<1	
410	P	Sintered	6200	380	—	—	<1	

Source: Metal Powder Industries Federation (MPIF).

aluminum. To evaluate the differences between the properties of P/M, wrought, and cast metals and alloys, compare this table with tables in Chapters 2, 3, and 5.

• Example 11.4: Shrinkage in sintering.

In solid-state bonding during sintering of a powder-metal green compact, the linear shrinkage is 4%. If the desired sintered density is 95% of the theoretical density of the metal, what should be the density of the green compact?

SOLUTION. We define linear shrinkage as $\Delta L/L_o$, where L_o is the original length. We can then express the volume shrinkage during sintering as

$$V_{\text{sint}} = V_{\text{green}}\left(1 - \frac{\Delta L}{L_o}\right)^3. \quad (11.2)$$

The volume of the green compact has to be larger than that for the sintered part. However, the mass does not change during sintering, so we can rewrite this expression in terms of the density ρ as

$$\rho_{\text{green}} = \rho_{\text{sint}}\left(1 - \frac{\Delta L}{L_o}\right)^3. \quad (11.3)$$

Thus

$$\rho_{\text{green}} = 0.95(1 - 0.04)^3 = 0.84, \quad \text{or} \quad 84\%.$$

11.5 Secondary and Finishing Operations

In order to further improve the properties of sintered P/M products or to give them special characteristics, several additional operations may be carried out after sintering. *Coining* and *sizing* are additional compacting operations, performed under high pressure in presses. The purposes of these operations are to impart dimensional accuracy to the sintered part and improve its strength and surface finish by additional densification.

An important recent development is the use of *preformed* and sintered alloy powder compacts, which are subsequently cold or hot forged to the desired final shapes. These products have good surface finish and tolerances, with uniform fine grain size. The superior properties obtained make this technology particularly suitable for applications such as highly stressed automotive and jet-engine components.

The inherent porosity of powder metallurgy components can be utilized by *impregnating* them with a fluid. A typical application is to impregnate the sintered part with oil, which is usually done by immersing the part in heated oil. Bearings and bushings that are internally lubricated, with up to 30 percent oil by volume, are made by this method. Internally lubricated components have a continuous supply of lubricant during their service lives. Universal joints are now being made with grease-impregnated P/M techniques, no longer requiring grease fittings.

Infiltration is a process whereby a slug of lower melting-point metal is placed against the sintered part, and the assembly is heated to a temperature sufficient to melt the slug. The molten metal infiltrates the pores by capillary action, resulting in a relatively pore-free part with good density and strength. The most common application is the infiltration of iron-base compacts with copper. The advantages are that hardness and tensile strength are improved and the pores are filled, thus preventing moisture penetration, which could cause corrosion. Infiltration may also be done with lead whereby, because of the low shear strength of lead, the infiltrated part has lower frictional characteristics than the uninfiltrated one.

Powder-metal parts may be subjected to other finishing operations, including:

a) Heat treating, for improved hardness and strength.
b) Machining by milling, drilling, and tapping to produce threaded holes.
c) Grinding, for improved dimensional accuracy and surface finish.
d) Plating, for wear resistance, improved appearance, and corrosion resistance.

• Example 11.5: Powder metallurgy gears for a garden tractor.

Components such as gears, bushings, and some structural parts of garden tractors have been made by P/M techniques, replacing the traditional casting or forging methods. Gears have been manufactured competitively using high-quality powders with high compressibility and requiring low compacting pressures. These parts range from medium to high density and are suitable for severe applications with high loads and for high-wear surfaces.

In one application, a reduction gear for a garden tractor was made from iron powder and infiltrated with copper. Although its strength was acceptable, the wear rate under high loads was very high. This resulted in loss of tooth profile, side loading on the bearings, and a high noise level. To improve wear resistance and strength, a new powder was selected containing 2.0 percent nickel, 0.5 percent graphite, 0.5 percent molybdenum, and the balance atomized iron powder. Because of the size of the part, the pressing loads were very high. The part was redesigned and the tooling was made with three punches to compact the part with three different densities.

By this method high density was obtained in those sections of the part requiring high strength and wear resistance. The high density also permitted carburization for hardness improvement. The presintered part was re-pressed to a density of 7.3–7.5 g/cm^3 in the tooth area for improved strength, while the hub of the gear remained at its pressed density of 6.4–6.6 g/cm^3. The entire part required a compacting load of 4 MN (450 tons). The weight of the gear was 1 kg (2.25 lb).

11.6 Design Considerations for Powder Metallurgy

Because of the unique properties of metal powders, their flow characteristics in the die, and the brittleness of green compacts, certain design principles should be followed (Fig. 11.10):

- The shape of the compact must be kept as simple and uniform as possible. Sharp changes in contour, thin sections, variations in thickness, and high length-to-diameter ratios should be avoided.

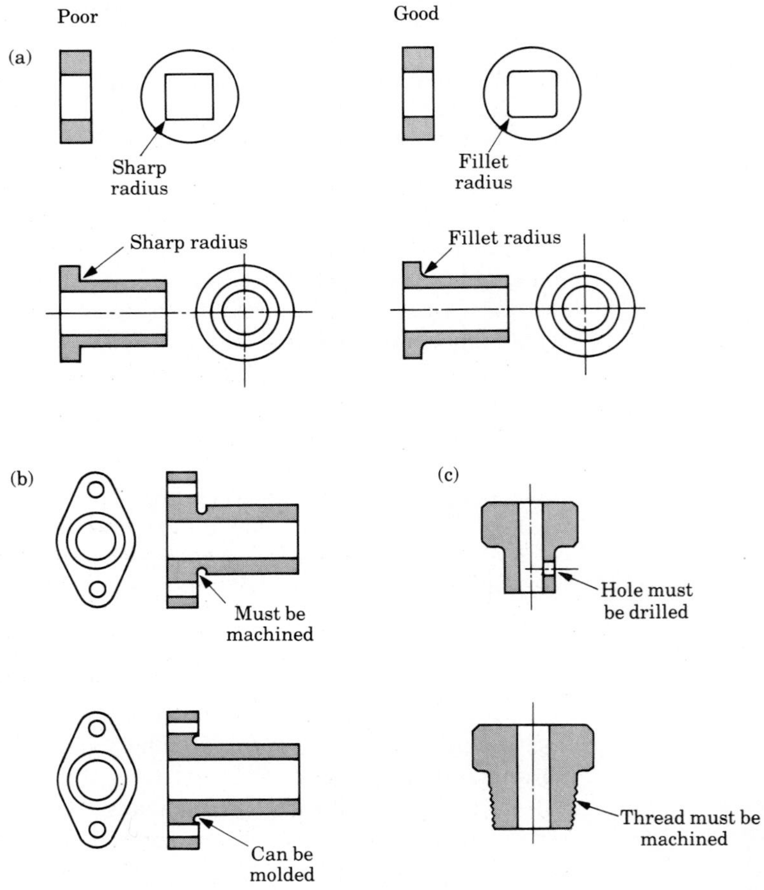

FIGURE 11.10 Examples of P/M parts, showing poor and good designs. Note that sharp radii and reentry corners should be avoided, and that threads and transverse holes have to be produced separately by additional machining operations.

- Provision must be made for ejection of the green compact from the die without damaging the compact. Thus holes or recesses should be parallel to the axis of punch travel. Chamfers should also be provided.
- As with most other processes, P/M parts should be made with the widest tolerances, consistent with their intended applications, in order to increase tool and die life and reduce production costs.

Tolerances of sintered P/M parts are usually on the order of ± 0.05–0.1 mm (± 0.002–0.004 in.). Tolerances improve significantly with additional operations such as sizing, machining, and grinding.

11.7 Economics of Powder Metallurgy

Because P/M can produce parts at or near net shape, thus eliminating many secondary manufacturing and assembly operations, it has become increasingly competitive with casting, forging, and machining. However, because of the high initial cost of punches, dies, and equipment for P/M processing, production volume must be high enough to warrant this expenditure. Although there are exceptions, the process is generally economical for quantities above 10,000 pieces.

The near net-shape capability of P/M reduces or eliminates scrap. Weight comparisons of aircraft components produced by forging and P/M processes are shown in Table 11.5. Note that these P/M parts are subjected to material-removal processes; thus the final parts weigh less than those made by either of the two processes.

TABLE 11.5
FORGED AND P/M TITANIUM PARTS AND WEIGHT COMPARISONS

	WEIGHT (kg)		
PART	*FORGED BILLET*	*HIP*	*FINAL PART*
Boeing 747 walking beam	25	14	9.5
General Dynamics F-16 pivot shaft	67	24	14.5
McDonnell Douglas F-15 drop-out link	52	29	6.8
Northrop F-18 arrestor hook	82	25	12.7
Pratt & Whitney F-100 fan disk	54	29	12.2
Williams International F-107 compressor rotor	15	2.8	1.6

11.8 Ceramics and Their Structure

Ceramics are compounds of metallic and nonmetallic elements. The term *ceramics* refers both to the material and to the ceramic product itself. In Greek the word *keramos* means potter's clay and *keramikos* means clay products. Because of the large number of possible combinations of elements, a great variety of ceramics is now available for widely different consumer and industrial applications.

The earliest use of ceramics was in pottery and bricks, dating back to before 4000 B.C. Ceramics have been used for many years in automotive spark plugs as an electrical insulator and for high-temperature strength. They are becoming increasingly important in heat engines and various other applications (Table 11.6), as well as tool and die materials. More recent applications of ceramics are in automotive components, such as exhaust-port liners, coated pistons, and cylinder liners, with the desirable properties of strength and corrosion resistance at high operating temperatures.

The structure of ceramic crystals is among the most complex of all materials, containing various elements of different sizes. The bonding between these atoms is generally covalent (electron sharing, hence strong bonds) and ionic (primary bonding between oppositely charged ions, thus strong bonds). These bonds are much stronger than metallic bonds. Consequently, the properties of ceramics are

TABLE 11.6
CATEGORIES AND USES OF CERAMICS

	TRADITIONAL CERAMICS
Abrasive products	Abrasive wheels, emery cloth and sand paper, nozzles for sandblasting, ball milling
Clay products	Brick, pottery, sewer pipe
Construction	Brick, concrete, tile, plaster, glass
Glass	Bottles, laboratory ware, glazing
Refractories	Brick, crucibles, molds, cement
Whitewares	Dishes, tiles, plumbing, enamels
	ENGINEERING CERAMICS
Automotive and aerospace	Turbine components, heat shields and exchangers, reentry components, seals
Electronics	Semiconductors, insulators, transducers, lasers, dielectrics, heating elements
High temperature	Refractories, brazing fixtures, kilns
Manufacturing	Cutting tools, wear and corrosion resistant components, glass ceramics, magnets, fiber optics, bearings
Medical	Laboratory ware, controls, prosthetics, dental

significantly higher than those for metals, particularly their hardness and thermal and electrical resistance.

Ceramics are available as a single crystal or in polycrystalline form, consisting of many grains. Grain size has a major influence on the strength and properties of ceramics. The finer the grain size, the higher are the strength and toughness—hence the term *fine ceramics*.

11.8.1 Raw materials

Among the oldest raw materials for ceramics is *clay*, a fine-grained sheetlike structure, the most common example being *kaolinite* (from Kao-ling, a hill in China). It is a white clay, consisting of silicate of aluminum with alternating weakly bonded layers of silicon and aluminum ions (Fig. 11.11). When added to kaolinite, water attaches itself to the layers (adsorption), makes them slippery, and gives wet clay its well-known softness and plastic properties (*hydroplasticity*) that make it formable.

Other major raw materials for ceramics that are found in nature are *flint* (rock of very fine grained silica, SiO_2) and *feldspar* (a group of crystalline minerals consisting of aluminum silicates, potassium, calcium, or sodium). In their natural state, these raw materials generally contain impurities of various kinds, which have to be removed prior to further processing of the materials into useful products with reliable performance. Highly refined raw materials produce ceramics with improved properties.

11.8.2 Oxide ceramics

Alumina. Also called *corundum* or *emery*, *alumina* (aluminum oxide, Al_2O_3) is the most widely used *oxide ceramic*, either in pure form or as a raw material to be mixed with other oxides. It has high hardness and moderate strength. Although alumina exists in nature, it contains unknown amounts of impurities and possesses nonuniform properties. As a result, its behavior is unreliable. Aluminum oxide, as well as silicon carbide and many other ceramics, are now almost totally manufactured synthetically so that we can control their quality.

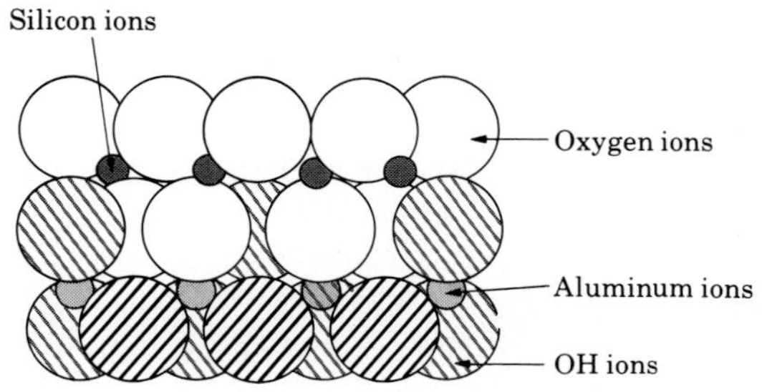

FIGURE 11.11 The crystal structure of kaolinite, commonly known as clay.

First made in 1893, synthetic aluminum oxide is obtained by the fusion of molten bauxite (an aluminum oxide ore that is the principal source of aluminum), iron filings, and coke in electric furnaces. It is then crushed and graded by size by passing the particles through standard screens. Parts made of aluminum oxide are cold pressed and sintered (*white ceramics*). Their properties are improved by minor additions of other ceramics, such as titanium oxide and titanium carbide. Structures containing various alumina and other oxides are known as *mullite* and *spinel* and are used as refractory materials for high-temperature applications. The mechanical and physical properties of alumina are particularly suitable for applications such as electrical and thermal insulation and as cutting tools and abrasives.

Zirconia and partially stabilized zirconia. *Zirconia* (zirconium oxide, ZrO_2, white in color) has good toughness, resistance to thermal shock, wear, and corrosion, low thermal conductivity, and low friction coefficient. A more recent development is *partially stabilized zirconia* (PSZ), which has high strength and toughness and better reliability in performance than zirconia. It is obtained by doping the zirconia with oxides of calcium, yttrium, or magnesium. This process forms a material with fine particles of tetragonal zirconia in a cubic lattice.

Another important characteristic of PSZ is the fact that its coefficient of thermal expansion is only about 20 percent lower than that of cast iron, and its thermal conductivity is about one-third that of other ceramics. Consequently, it is very suitable for heat-engine components, such as cylinder liners and valve bushings, to keep the cast-iron engine assembly intact. New developments to further improve the properties of PSZ include *transformation-toughened zirconia* (TTZ) which has higher toughness because of dispersed tough phases in the ceramic matrix.

11.8.3 Other ceramics

Carbides. Typical examples of *carbides* are those of tungsten (WC) and titanium (TiC), used as cutting tools and die materials, and silicon carbide (SiC), used as abrasives, as in grinding wheels. *Tungsten carbide* consists of tungsten-carbide particles with cobalt as a binder. The amount of binder has a major influence on the material's properties. Toughness increases with cobalt content, whereas hardness, strength, and wear resistance decrease. *Titanium carbide* has nickel and molybdenum as the binder and is not as tough as tungsten carbide.

Silicon carbide (SiC) has good wear, thermal shock, and corrosion resistance. It has a low friction coefficient and retains strength at elevated temperatures. It is suitable for high-temperature components in heat engines and is also used as an abrasive. First produced in 1891, synthetic silicon carbide is made from silica sand, coke, and small amounts of sodium chloride and sawdust. The process is similar to making synthetic aluminum oxide.

Nitrides. Another important class of ceramics are the *nitrides*, particularly cubic boron nitride (CBN), titanium nitride (TiN), and silicon nitride (Si_3N_4).

Cubic boron nitride, the second hardest known substance, after diamond, has special applications, such as abrasives in grinding wheels and as cutting tools. It does not exist in nature and was first made synthetically in the 1970s, with techniques similar to those used in making synthetic diamond.

Titanium nitride is used widely as coatings on cutting tools. It improves tool life by virtue of its low frictional characteristics.

Silicon nitride has high resistance to creep at elevated temperatures, low thermal expansion, and high thermal conductivity, and hence resists thermal shock. It is suitable for high-temperature structural applications, such as in automotive engine and gas-turbine components.

Sialon. *Sialon* consists of silicon nitride, with various additions of aluminum oxide, yttrium oxide, and titanium carbide (see Section 8.6.8).

Cermets. *Cermets* are combinations of ceramics bonded with a metallic phase. Introduced in the 1960s, they combine the high-temperature oxidation resistance of ceramics and the toughness, thermal-shock resistance, and ductility of metals. An application of cermets is cutting tools, a typical composition being 70 percent aluminum oxide and 30 percent titanium carbide. Other cermets contain various oxides, carbides, and nitrides. They have been developed for high-temperature applications such as nozzles for jet engines and aircraft brakes. Cermets can be regarded as composite materials and can be used in various combinations of ceramics and metals bonded by powder-metallurgy techniques.

11.8.4 Silica

Abundant in nature, *silica* is a polymorphic material; that is, it can have different crystal structures. The cubic structure is found in refractory bricks used for high-temperature furnace applications. Most glasses contain more than 50 percent silica. The most common form of silica is *quartz*, which is a hard, abrasive hexagonal crystal. It is used extensively as oscillating crystals of fixed frequency in communications applications, since it exhibits the piezoelectric effect.

Silicates are products of the reaction of silica with oxides of aluminum, magnesium, calcium, potassium, sodium, and iron. Examples are clay, asbestos, mica, and silicate glasses. *Lithium aluminum silicate* has very low thermal expansion and thermal conductivity and good thermal-shock resistance. However, it has very low strength and fatigue life. Thus it is suitable only for nonstructural applications, such as catalytic converters, regenerators, and heat-exchanger components.

11.9 General Properties and Applications of Ceramics

Compared to metals, ceramics have the following relative characteristics: brittle, high strength and hardness at elevated temperatures, high elastic modulus, low toughness, low density, low thermal expansion, and low thermal and electrical conductivity. However, because of the wide variety of ceramic material composition and grain size, the mechanical and physical properties of ceramics vary significantly. For example, the electrical conductivity of ceramics can be modified from poor to good, which is the principle behind semiconductors. Because of their sensitivity to flaws, defects, and cracks (surface or internal), the presence of different types and levels of impurities, and different methods of manufacturing, ceramics can have a wide range of properties.

11.9.1 Mechanical properties

The mechanical properties of several engineering ceramics are presented in Table 11.7. Note that their strength in tension (transverse rupture strength) is approximately one order of magnitude lower than their compressive strength. The reason is their sensitivity to cracks, impurities, and porosity. Such defects lead to the initiation and propagation of cracks under tensile stresses, severely reducing tensile strength. Thus reproducibility and reliability (acceptable performance over a specified period of time) is an important aspect in the service life of ceramic components. Tensile strength of polycrystalline ceramic parts increases with decreasing grain size.

TABLE 11.7
PROPERTIES OF VARIOUS CERAMICS AT ROOM TEMPERATURE

MATERIAL	SYMBOL	TRANSVERSE RUPTURE STRENGTH (MPa)	COMPRESSIVE STRENGTH (MPa)	ELASTIC MODULUS (GPa)	HARDNESS (HK)	POISSON'S RATIO (ν)	DENSITY (kg/m³)
Aluminum oxide	Al_2O_3	140–240	1000–2900	310–410	2000–3000	0.26	4000–4500
Cubic boron nitride	CBN	725	7000	850	4000–5000	—	3480
Diamond	—	1400	7000	830–1000	7000–8000	—	3500
Silica, fused	SiO_2	—	1300	70	550	0.25	—
Silicon carbide	SiC	100–750	700–3500	240–480	2100–3000	0.14	3100
Silicon nitride	Si_3N_4	480–600	—	300–310	2000–2500	0.24	3300
Titanium carbide	TiC	1400–1900	3100–3850	310–410	1800–3200	—	5500–5800
Tungsten carbide	WC	1030–2600	4100–5900	520–700	1800–2400	—	10,000–15,000
Partially stabilized zirconia	PSZ	620	—	200	1100	0.30	5800

Note: These properties vary widely depending on the condition of the material.

Tensile strength is empirically related to porosity as follows:

$$\text{UTS} \simeq \text{UTS}_0 e^{-nP}, \qquad (11.4)$$

where P is the volume fraction of pores in the solid, UTS_0 is the tensile strength at zero porosity, and the exponent n ranges between 4 and 7.

The modulus of elasticity is likewise affected by porosity, as given by

$$E \simeq E_0(1 - 1.9P + 0.9P^2), \qquad (11.5)$$

where E_0 is the modulus at zero porosity. Equation (11.4) is valid up to 50 percent porosity. Common earthenware has a porosity ranging between 10 percent and 15 percent, whereas hard porcelain is about 3 percent.

Although there are exceptions and unlike most metals and thermoplastics, ceramics generally lack impact toughness and thermal-shock resistance because of their inherent lack of ductility. Once initiated, a crack propagates rapidly. In addition to undergoing fatigue failure under cyclic loading, ceramics (and particularly glasses) exhibit a phenomenon called *static fatigue*. When subjected to a static tensile load over a period of time, these materials may suddenly fail. This phenomenon occurs in environments where water vapor is present. Static fatigue, which does not occur in a vacuum or dry air, has been attributed to a mechanism similar to stress-corrosion cracking of metals.

Ceramic components that are to be subjected to tensile stresses may be prestressed, much like prestressed concrete. Prestressing shaped ceramic components subjects them to compressive stresses. Methods used include (a) heat treatment and chemical tempering, (b) laser treatment of surfaces, (c) coating with ceramics with different thermal expansion, and (d) surface-finishing operations, such as grinding, in which compressive residual stresses are induced on the surfaces.

Significant advances are being made in improving the toughness and other properties of ceramics. Among these are proper selection and processing of raw materials, control of purity and structure, and use of reinforcements and with particular emphasis during design on advanced methods of stress analysis in ceramic components.

11.9.2 Physical properties

Most ceramics have relatively low specific gravity, ranging from about 3 to 5.8 for oxide ceramics, compared to 7.86 for iron. They have very high melting or decomposition temperature. Thermal conductivity of ceramics varies by as much as three orders of magnitude, depending on their composition, whereas metals vary by one order. Thermal conductivity of ceramics, as well as other materials, decreases with increasing temperature and porosity because air is a poor thermal conductor.

The thermal conductivity k is related to porosity by

$$k = k_0(1 - P), \qquad (11.6)$$

where k_0 is the thermal conductivity at zero porosity.

The thermal expansion characteristics of ceramics are shown in Fig. 11.12. Thermal expansion and thermal conductivity induce thermal stresses that can lead to thermal shock or thermal fatigue. The tendency for *thermal cracking* (called *spalling* when a piece or a layer from the surface breaks off) is lower with low thermal expansion and high thermal conductivity. For example, fused silica has high thermal shock resistance because of its virtually zero thermal expansion.

A familiar example that illustrates the importance of low thermal expansion is the heat-resistant ceramics for cookware and stove tops. They can sustain high thermal gradients, from hot to cold, and vice versa. Moreover, the relative thermal expansion of ceramics and metals is an important reason for the use of ceramic components in heat engines. The fact that the thermal conductivity of partially stabilized zirconia components is close to that of the cast iron in engine blocks is an additional advantage in the use of PSZ in heat engines.

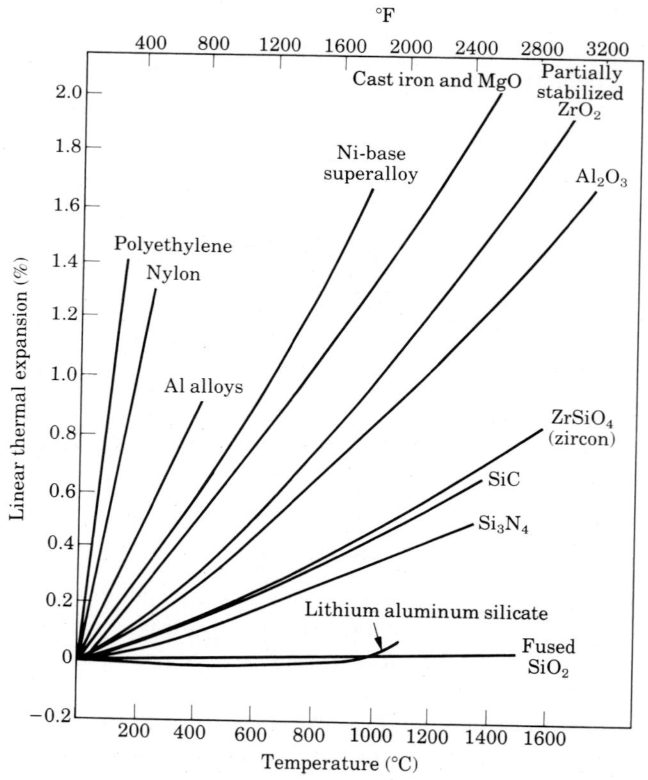

FIGURE 11.12 Effect of temperature on thermal expansion for several ceramics, metals, and plastics. Note that the expansion for cast iron and for partially stabilized zirconia (PSZ) are within about 20 percent. This makes the two materials compatible for internal combustion engine applications.

TABLE 11.8
COEFFICIENTS OF THERMAL EXPANSION FOR SOME ANISOTROPIC CERAMICS

MATERIAL	COEFFICIENT OF THERMAL EXPANSION ($\times 10^6$/°C)	
	*NORMAL TO c-AXIS**	*PARALLEL TO c-AXIS*
Graphite	1	27
Al_2O_3 (alumina)	8.3	9
$3Al_2O_3$, 2 SiO_2 (mullite)	4.5	5.7
TiO_2	6.8	8.3
$ZrSiO_4$	3.7	6.2
SiO_2 (quartz)	14	9

* See Fig. 3.2.

An additional characteristic is the *anisotropy of thermal expansion* exhibited by oxide ceramics, whereby thermal expansion varies in different directions of the ceramic (Table 11.8). This behavior causes thermal stresses that can lead to cracking of the ceramic component.

The optical properties of ceramics can be controlled by various formulations and control of structure, imparting different degrees of transparency and colors. Single-crystal sapphire, for example, is completely transparent, zirconia is white, and fine-grained polycrystalline aluminum oxide is a translucent gray. Porosity influences the optical properties of ceramics, much like trapped air in ice cubes, which makes the ice less transparent and gives it a white appearance.

• Example 11.6: Effect of porosity on properties.

If a fully dense ceramic has the properties of $UTS_0 = 100$ MPa, $E_0 = 400$ GPa, and $k_0 = 0.5$ W/m·K, what are these properties at 10% porosity? Let $n = 0.5$ and $P = 0.1$.

SOLUTION. Using Eqs. (11.4)–(11.6), we have

$$UTS = 100e^{-(5)(0.1)} = 61 \text{ MPa},$$

$$E = 400[1 - (1.9)(0.1) + (0.9)(0.1)^2] = 328 \text{ GPa},$$

and

$$k = 0.5(1 - 0.1) = 0.45 \text{ W/m} \cdot \text{K}.$$

•

11.9.3 Applications

As shown in Table 11.6, ceramics have numerous consumer and industrial applications. Several types of ceramics are used in the electrical and electronics industry because of their high electrical resistivity, dielectric strength (voltage required for electrical breakdown per unit thickness), and magnetic properties suitable for applications such as magnets for speakers. An example is *porcelain*, which is a white ceramic composed of kaolin, quartz, and feldspar. Certain ceramics also have good piezoelectric properties.

The capability of ceramics to maintain their strength and stiffness at elevated temperatures (Figs. 11.13 and 11.14) makes them very attractive for high-temperature applications. Their high resistance to wear makes them suitable for applications such as cylinder liners, bushings, seals, and bearings. The higher operating temperatures made possible by the use of ceramic components means more efficient fuel burning and reduced emissions. Currently, internal combustion engines are only about 30 percent efficient, but with the use of ceramic components the operating performance can be improved by at least 30 percent. Ceramics being used successfully, especially in gasoline and diesel engine components and as rotors, are silicon nitride, silicon carbide, and partially stabilized zirconia. Coating metal with ceramics is another application, which may be done to reduce wear, prevent corrosion, and provide a thermal barrier.

Other attractive properties of ceramics are their low density and high elastic modulus. Thus engine weight can be reduced and, in other applications, the inertial forces generated by moving parts are lower. High-speed components for machine tools, for example, are candidates for ceramics. Furthermore, the higher elastic modulus of ceramics makes them attractive for improving the stiffness, while reducing the weight, of machines. Ceramics are also used as ball bearings and

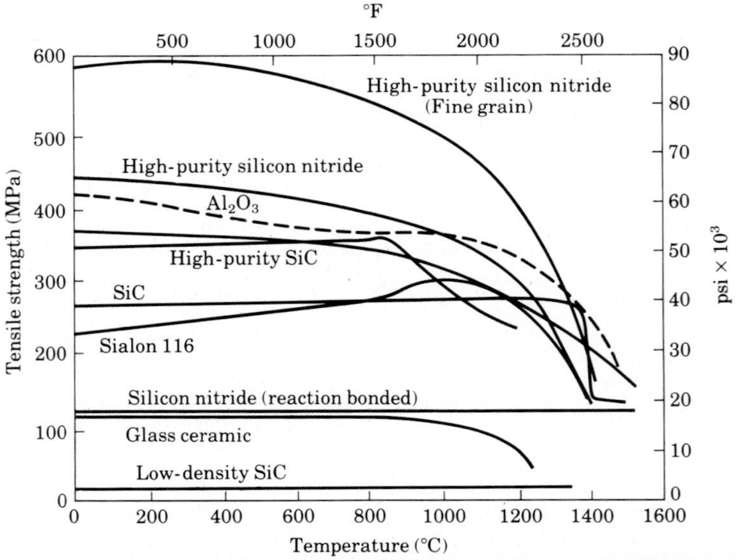

FIGURE 11.13

Effect of temperature on the strength of various engineering ceramics. Note that much of the strength is maintained at high temperatures.

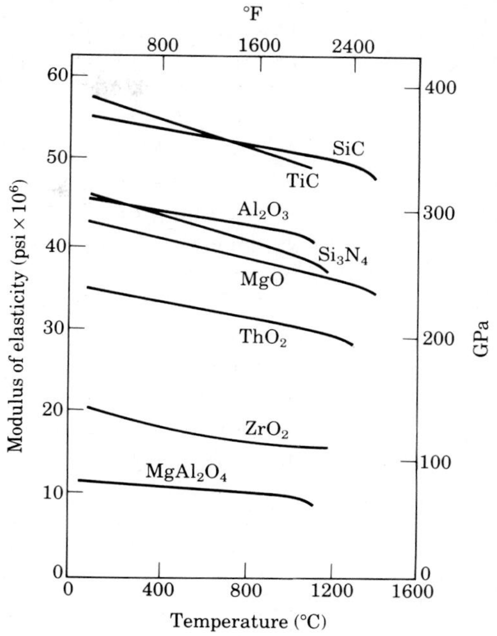

FIGURE 11.14
Effect of temperature on the modulus of elasticity for several ceramics. *Source:* D. W. Richerson, *Modern Ceramic Engineering*. New York: Marcel Dekker, Inc., 1982.

rollers. Because of their strength and inertness, ceramics are used as biomaterials to replace joints in the human body, as prosthetic devices, and for dental work.

11.10 Shaping Ceramics

Several techniques have been developed for processing ceramics into useful products. Generally, the procedure involves the following steps: crushing or grinding the raw materials into very fine particles, mixing them with additives to impart certain desirable characteristics, and shaping, drying, and firing the material.

The first step in processing ceramics is *crushing* (also called comminution or milling) of the raw materials. Crushing is generally done in a ball mill (Fig. 11.15),

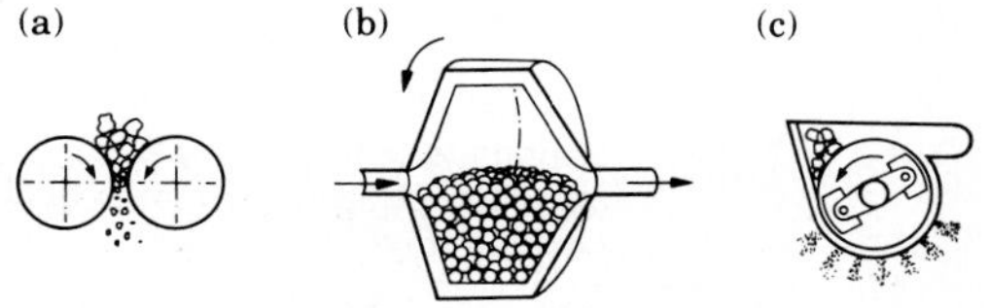

FIGURE 11.15
Methods of crushing ceramics to obtain very fine particles: (a) roll crushing, (b) ball mill, and (c) hammer milling.

either dry or wet. Wet crushing is more effective because it keeps the particles together and prevents the suspension of fine particles in air. The ground particles are then mixed with *additives*, the functions of which are one or more of the following:

a) Binder for the ceramic particles.
b) Lubricant for mold release and to reduce internal friction between particles during molding.
c) Wetting agent to improve mixing.
d) Plasticizer to make the mix more plastic and formable.
e) Deflocculent to make the ceramic–water suspension. Deflocculation changes the electrical charges on the particles of clay so that they repel instead of attract each other. Water is added to make the mixture more pourable and less viscous. Typical deflocculants are Na_2CO_3 and Na_2SiO_3 in amounts of less than 1 percent.
f) Various agents to control foaming and sintering.

The three basic shaping processes for ceramics are casting, plastic forming, and pressing.

11.10.1 Casting

The most common casting process is *slip casting*, also called *drain casting* (Fig. 11.16). A *slip* is a suspension of ceramic particles in a liquid, generally water. In this process, the slip is poured into a porous mold made of plaster of paris. The slip must have sufficient fluidity and low viscosity to flow easily into the mold, much like the fluidity of molten metals. After the mold has absorbed some of the water from the outer layers of the suspension, it is inverted and the remaining suspension is poured

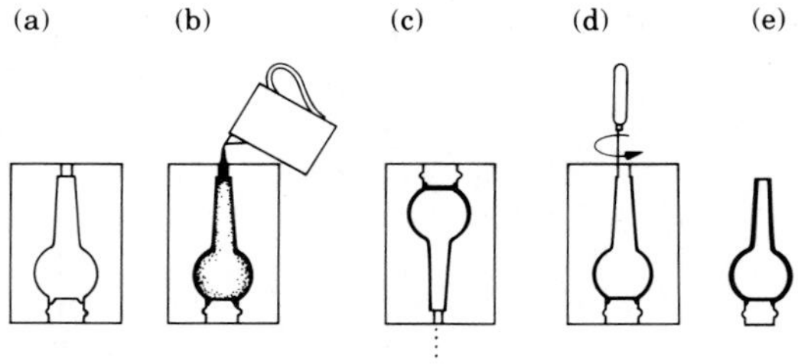

FIGURE 11.16
Sequence of operations in slip casting a ceramic part. After the slip has been poured, the part is dried and fired in an oven to give it strength and hardness. *Source:* F. H. Norton, *Elements of Ceramics.* Copyright © 1974, Addison-Wesley Publishing Company, Inc.

out (for making hollow objects, as in slush casting of metals). The top of the part is then trimmed, the mold is opened, and the part is removed.

Large and complex parts, such as plumbing ware, art objects, and dinnerware, can be made by slip casting. Although dimensional control is limited and the production rate is low, mold and equipment costs are low. In some applications, components of the product (such as handles for cups and pitchers) are made separately and then joined, using the slip as an adhesive.

Thin sheets of ceramics, less than 1.5-mm (0.06-in.) thick, can be made by a casting technique called the *doctor-blade process*. The slip is cast over a moving plastic belt and its thickness is controlled by a blade. Other processes include *rolling* the slip between pairs of rolls and casting the slip over a paper tape, which is then burned off during firing.

For solid ceramic parts, the slip is supplied continuously into the mold to replenish the absorbed water; the suspension is not drained from the mold. At this stage the part is a soft solid or semirigid. The higher the concentration of solids in the slip, the less water has to be removed. The part, called *green*, as in powder metallurgy, is then fired.

11.10.2 Plastic forming

Plastic forming (also called *soft*, *wet*, or *hydroplastic forming*) can be done by various methods, such as extrusion, injection molding, or molding and jiggering (as done on a potter's wheel). Plastic forming tends to orient the layered structure of clays along the direction of material flow. This leads to anisotropic behavior of the material, both in subsequent processing and in the final properties of the ceramic product.

In *extrusion*, the clay mixture, containing 20–30 percent water, is forced through a die opening by screw-type equipment. The cross-section of the extruded product is constant, but there are limitations to wall thickness for hollow extrusions. Tooling costs are low and production rates are high. The extruded products may be subjected to additional shaping operations.

11.10.3 Pressing

Dry pressing. Similar to powder-metal compaction, *dry pressing* is used for relatively simple shapes. Typical parts are whiteware, refractories, and abrasive products. The process has the same high production rates and close control of tolerances as in P/M. The moisture content of the mixture is generally below 4 percent but may be as high as 12 percent. Organic and inorganic binders, such as stearic acid, wax, starch, and polyvinyl alcohol, are usually added to the mixture and also act as lubricants. The pressure for pressing is between 35 MPa and 200 MPa (5 ksi and 30 ksi). Modern presses used for dry pressing are highly automated. Dies, usually made of carbides or hardened steel, must have high wear resistance to withstand the abrasive ceramic particles and can be expensive.

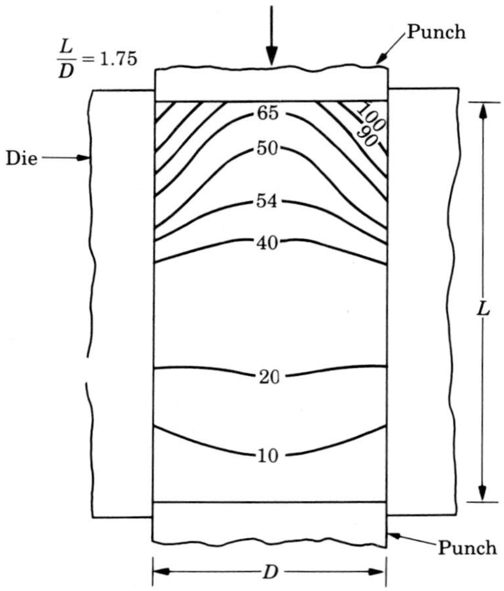

FIGURE 11.17 Density variation in pressed compacts in a single-action press. The variation increases with increasing L/D ratio. See also Fig. 11.7(e). *Source:* Adapted from W. D. Kingery, et al., *Ceramic Fabrication Processes,* MIT Press, Cambridge, Mass., 1963.

Density can vary greatly in dry-pressed ceramics (Fig. 11.17) because of friction between particles and at the mold walls, as in P/M compaction. Density variations cause warping during firing. Warping is particularly severe for parts having high length-to-diameter ratios; the recommended maximum ratio is 2:1. Several methods may be used to minimize density variations. Design of tooling is important. Vibratory pressing and impact forming are used, particularly for nuclear-reactor fuel elements. Isostatic pressing also reduces density variations.

Wet pressing. In *wet pressing* the part is formed in a mold while under high pressure in a hydraulic or mechanical press. This process is generally used to make intricate shapes. Moisture content usually ranges from 10 percent to 15 percent. Production rates are high but part size is limited, dimensional control is difficult because of shrinkage during drying, and tooling costs can be high.

Isostatic pressing. Used extensively in powder metallurgy, as you have seen, *isostatic pressing* is also used for ceramics in order to obtain uniform density distribution throughout the part. Automotive spark-plug insulators are made by this method. Silicon-nitride vanes for high-temperature use are also made by hot isostatic pressing.

Jiggering. A combination of processes is used to make ceramic plates. Clay slugs are first extruded, then formed into a *bat* over a plaster mold, and finally jiggered on a rotating mold (Fig. 11.18). *Jiggering* is a motion in which the clay bat

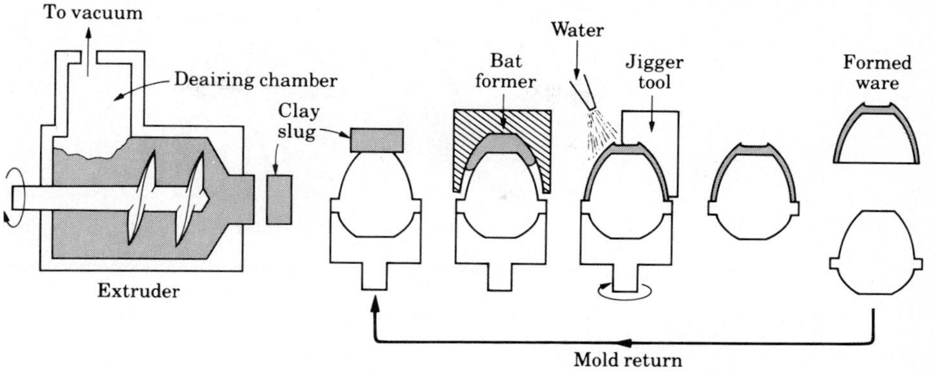

FIGURE 11.18
Extruding and jiggering operations. *Source:* R. F. Stoops.

is formed with templates or rollers. The part is then dried and fired. The process is limited to axisymmetric parts and has limited dimensional accuracy, but the operation can be automated.

Injection molding. We have previously described the advantages of *injection molding* of plastics and powder metals. This process is now being used extensively for precision forming of ceramics for high-technology applications, as in rocket engine components. The raw material is mixed with a thermoplastic polymer and injection molded.

Hot pressing. In *hot pressing*, also called *pressure sintering*, pressure and temperature are applied simultaneously. This method reduces porosity, making the part denser and stronger. Hot isostatic pressing may also be used in this operation, particularly to improve the quality of high-technology ceramics. Because of the presence of both pressure and temperature, die life in hot pressing can be short. Protective atmospheres are usually employed, and graphite is a commonly used punch and die material.

11.10.4 Drying and firing

After the ceramic has been shaped by any of the methods described, the next step is to dry and fire the part to give it the proper strength. *Drying* is a critical stage because of the tendency for the part to warp or crack from variations in moisture content and thickness within the part and the complexity of its shape. Control of atmospheric humidity and temperature is important in order to reduce warping and cracking. Loss of moisture results in shrinkage of the part by as much as 15–20

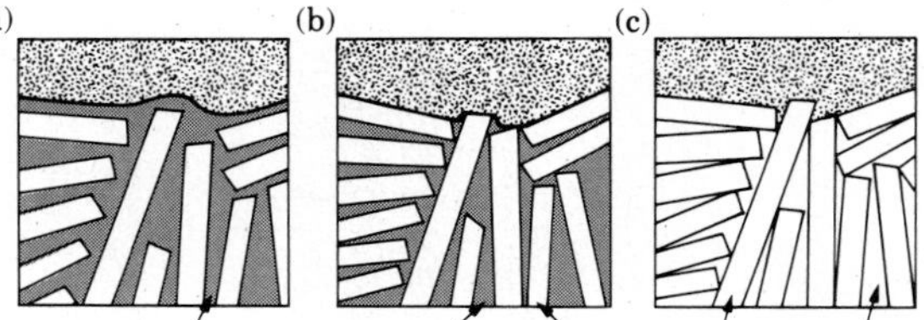

FIGURE 11.19
Shrinkage of wet clay caused by removal of water during drying. Shrinkage may be as much as 20 percent by volume. *Source:* F. H. Norton, *Elements of Ceramics.* Copyright © 1974, Addison-Wesley Publishing Company, Inc.

percent of the original moist size (Fig. 11.19). In a humid environment the evaporation rate is low, and consequently the moisture gradient across the thickness of the part is lower than that in a dry environment. The low moisture gradient, in turn, prevents a large, uneven gradient in shrinkage from the surface to the interior during drying.

Firing (sintering) involves heating the part to an elevated temperature in a controlled environment, similar to sintering in powder metallurgy. Some shrinkage occurs during firing. Firing gives the ceramic part its strength and hardness. The improvement in properties results from (a) development of a strong bond between the complex oxide particles in the ceramic and (b) reduced porosity.

11.10.5 Finishing operations

After firing, additional operations may be performed to give the part its final shape, remove surface flaws, and improve surface finish and tolerances. The processes used can be grinding, lapping, and ultrasonic, chemical, and electrical-discharge machining. The choice of the process is important in view of the brittle nature of most ceramics and the additional costs involved in these processes. The effect of the finishing operation on properties of the product must also be considered. Because of notch sensitivity, the finer the finish, the higher the part's strength will be. To improve appearance and strength, and to make them impermeable, ceramic products are often coated with a *glaze* material, which forms a glassy coating after firing.

• Example 11.7: Dimensional changes during shaping of ceramic components.

A solid cylindrical ceramic part is to be made whose final length must be $L = 20$ mm. It has been established that for this material, linear shrinkages during drying and firing are 7 percent and 6 percent, respectively, based on the dried dimension L_d. Calculate (a) the initial length L_o of the part and (b) the dried porosity P_d if the porosity of the fired part, P_f, is 3 percent.

SOLUTION.

(a) On the basis of the information given and remembering that firing is preceded by drying, we can write

$$(L_d - L)/L_d = 0.06$$

or

$$L = (1 - 0.06)L_d;$$

hence

$$L_d = 20/0.94 = 21.28 \text{ mm}$$

and

$$L_o = (1 + 0.07)L_d = (1.07)(21.28) = 22.77 \text{ mm}.$$

(b) Since the final porosity is 3 percent, the actual volume V_a of the ceramic material is

$$V = (1 - 0.03)V_f = 0.97V_f,$$

where V_f is the fired volume of the part. Since the linear shrinkage during firing is 6 percent, we can determine the dried volume V_d of the part as

$$V_d = \frac{V_f}{(1 - 0.06)^3} = 1.2V_f.$$

Hence

$$\frac{V_a}{V_d} = \frac{0.97}{1.2} = 0.81 = 81 \text{ percent}.$$

Therefore, the porosity P_d of the dried part is 19 percent.

11.11 Glasses

Glass is an amorphous solid with the structure of a liquid. In other words, it has been *supercooled*, that is, cooled at a rate too high for crystals to form. Generally, we define glass as an inorganic product of fusion that has cooled to a rigid condition without crystallizing. Glass has no distinct melting or freezing point; thus its behavior is similar to amorphous polymers.

Glass beads were produced in about 2000 B.C., followed by glass blowing in about 200 B.C. Silica was used for all glass products until the late 1600s. Rapid developments in glasses began in the early 1900s. Presently there are some 750 different types of commercially available glasses. The uses of glass range from window glass, bottles, and cookware to glasses with special mechanical, electrical, high-temperature, chemical, corrosion, and optical characteristics. Special glasses are used in fiber optics for communication by light with little loss in signal power and in glass fibers with very high strength for reinforced plastics.

All glasses contain at least 50 percent silica, which is known as a *glass former*. The composition and properties of glasses, except strength, can be modified greatly by the addition of oxides of aluminum, sodium, calcium, barium, boron, magnesium, titanium, lithium, lead, and potassium. Depending on their function, these oxides are known as *intermediates*, or *modifiers*. Glasses are generally resistant to chemical attack, and are ranked by their resistance to acid, alkali, or water corrosion.

11.11.1 Types of glasses

Almost all commercial glasses are categorized by type (Tables 11.9 and 11.10):

1. Soda–lime glass (the most common).
2. Lead–alkali glass.
3. Borosilicate glass.
4. Aluminosilicate glass.
5. 96 percent silica glass.
6. Fused silica.

Glasses are also classified as colored, opaque (white and translucent), multiform (variety of shapes), optical, photochromatic (darkens when exposed to light, as in sunglasses), photosensitive (changing from clear to opal), fibrous (drawn into long fibers, as in fiberglass), and foam or cellular glass (containing bubbles, thus a good thermal insulator).

TABLE 11.9
VARIOUS PROPERTIES OF GLASSES

	SODA–LIME GLASS	LEAD GLASS	BOROSILICATE GLASS	96 PERCENT SILICA	FUSED SILICA
Density	High	Highest	Medium	Low	Lowest
Strength	Low	Low	Moderate	High	Highest
Resistance to thermal shock	Low	Low	Good	Better	Best
Electrical resistivity	Moderate	Best	Good	Good	Good
Hot workability	Good	Best	Fair	Poor	Poorest
Heat treatability	Good	Good	Poor	None	None
Chemical resistance	Poor	Fair	Good	Better	Best
Impact-abrasion resistance	Fair	Poor	Good	Good	Best
Ultraviolet-light transmission	Poor	Poor	Fair	Good	Good
Relative cost	Lowest	Low	Medium	High	Highest

TABLE 11.10
COMPOSITION AND PROPERTIES OF SOME GLASSES

TYPE	MAJOR COMPONENTS (%) SiO_2	Al_2O_3	CaO	Na_2O	B_2O_3	MgO	PbO	Other	TYPICAL USES
Fused silica	99.5 +								High-frequency electrical insulation
96% silica (Vycor)	96.3	0.4		<0.2	2.9			K_2O <2	Chemical ware, home appliances, sun lamps
Borosilicate (Pyrex)	81	2		4	13			K_2O 0.4	Oven ware, laboratory glasses, industrial glass piping, gauge glasses
Soda-lime (Plate glass)	71–73	1	10–12	12–14		1–4			Windows, containers, ash trays, glass blocks, electric bulbs
Fiber (E-glass)	54	14	16–22	1.5	10	4			Fiber reinforcement
Lead glass	67			6	6		17	K_2O 10	Table ware, optical lenses, crystal glassware, neon tubes, capacitors
Glass ceramics	40–70	10–35				10–30		TiO_2–7–15	Cookware, heat exchangers
Aluminosilicate	57	20.5	5.5	1	4	12			Resistors, stove top ware

Glasses are referred to as hard and soft, usually in the sense of a thermal property rather than mechanical, as in hardness. Thus a soft glass softens at a lower temperature than does a hard glass. Soda–lime and lead–alkali glasses are considered soft and the rest as hard.

11.11.2 Mechanical properties

For all practical purposes, we regard the behavior of glass, as for most ceramics, as perfectly elastic and brittle. The modulus of elasticity range for most commercial glasses is 55–90 GPa (8–13 million psi), and their Poisson's ratio 0.16–0.28. Hardness of glasses, as a measure of resistance to scratching, range from 5 to 7 on the Mohs scale, equivalent to a range of approximately 350–500 HK.

Glass in bulk form has a strength of less than 140 MPa (20 ksi). The relatively low strength of bulk glass is attributed to the presence of small flaws and microcracks on its surface, some or all of which may be introduced during normal handling of the glass by inadvertent abrading. These defects reduce the strength of glass by two to three orders of magnitude, compared to its ideal (defect free) strength. Glasses can be strengthened by thermal or chemical treatments to obtain high strength and toughness.

The strength of glass can theoretically reach as high as 35 GPa (5 million psi). When molten glass is freshly drawn into fibers (fiberglass), its tensile strength ranges from 0.2 GPa to 7 GPa (30 ksi to 1000 ksi), with an average value of about 2 GPa (300 ksi). Thus glass fibers are stronger than steel and are used to reinforce plastics in applications such as boats, automobile bodies, furniture, and sports equipment.

The strength of glass is usually measured by bending it. The surface of the glass is first thoroughly abraded (roughened) to ensure that the test gives a reliable strength level in actual service under adverse conditions. The phenomenon of static fatigue observed in ceramics is also exhibited by glasses. If a glass item must withstand a load for 1000 hours or longer, the maximum stress that can be applied to it is approximately one-third the maximum stress that the same item can withstand during the first second of loading.

11.11.3 Physical properties

Glasses have low thermal conductivity and high electrical resistivity and dielectric strength. Their thermal expansion coefficient is lower than those for metals and plastics, and may even approach zero. Titanium silicate glass (a clear synthetic high-silica glass), for example, has a near-zero coefficient of expansion. Fused silica, a clear synthetic amorphous silicon dioxide of very high purity, also has a near-zero coefficient of expansion (see Fig. 11.12). Optical properties of glasses, such as reflection, absorption, transmission, and refraction, can be modified by varying their composition and treatment.

11.11.4 Glass ceramics

Although glasses are amorphous, *glass ceramics* (such as Pyroceram, a trade name) have a high crystalline component to their microstructure. Glass ceramics contain large proportions of several oxides, and thus their properties are a combination of those for glass and ceramics. Most glass ceramics are stronger than glass. These products are first shaped and then heat treated, with *devitrification* (recrystallization) of the glass occurring. Unlike most glasses, which are clear, glass ceramics are generally white or gray in color.

The hardness of glass ceramics ranges approximately from 520 HK to 650 HK. They have a near-zero coefficient of thermal expansion; hence they have good thermal shock resistance and are strong because of the absence of porosity usually found in conventional ceramics. The properties of glass ceramics can be improved by modifying their composition and by heat-treatment techniques. First developed in 1957, glass ceramics are suitable for cookware, heat exchangers for gas-turbine engines, radomes (housings for radar antenna), and electrical and electronics applications.

• Example 11.8: Ovenware.

Common household ovenware carries the following instruction: For oven and microwave, no stovetop or broiler. Explain the reasons for this type of instruction for this product.

SOLUTION. This product is made of borosilicate, which has the property of low thermal expansion; hence thermal stresses are low when the product is subjected to thermal gradients. Under extreme temperature conditions, however, thermal stresses can be high enough to fracture this material. The contact temperatures on a stovetop, or the radiation temperatures in a broiler, are very high compared to those in an oven. Furthermore, this product is subjected to a rapid temperature rise on its exposed surface on a stovetop or in a broiler. Thus temperature gradients are much more severe compared to a preheated oven, where heat is transmitted mainly through convection at temperatures generally below 230 °C (450 °F).

11.12 Forming and Shaping Glass

Glass products can generally be categorized as:

- Flat sheet or plate, ranging in thickness from about 0.8 mm to 10 mm (0.03 in. to 0.4 in.), such as window glass, glass doors, and table tops.
- Rods and tubing used for chemicals, neon lights, and decorative artifacts.
- Discrete products, such as bottles, vases, headlights, and television tubes.
- Glass fibers to reinforce composite materials and for fiber optics.

All glass forming and shaping processes begin with molten glass, which has the appearance of red-hot viscous syrup, supplied from a melting furnace or tank.

Flat sheet glass can be made by drawing or rolling from the molten state, or by a floating method, all of which are continuous processes. Figure 11.20(a) shows the *drawing* process for making flat sheet or plate by a machine in which the molten glass passes through a pair of rolls, similar to an old-fashioned clothes wringer. The solidifying glass is squeezed between these rolls, forming a sheet, which is then moved forward over a set of smaller rolls. In the *rolling* process (Fig. 11.20b), the

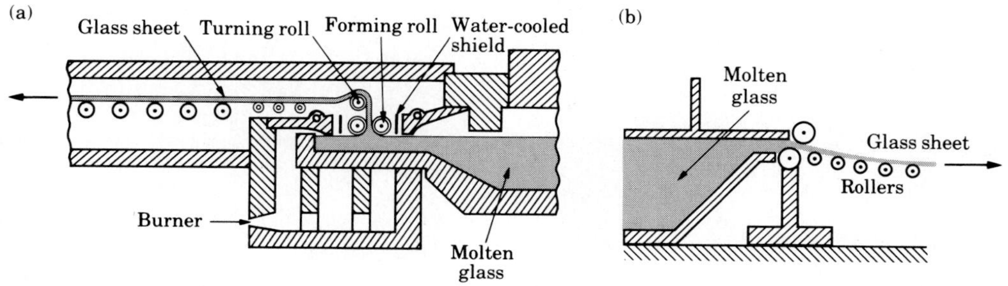

FIGURE 11.20
(a) Continuous process for drawing sheet glass from a molten bath. *Source:* W. D. Kingery, *Introduction to Ceramics.* New York: John Wiley & Sons, Inc., 1976. (b) Rolling glass to produce flat sheet.

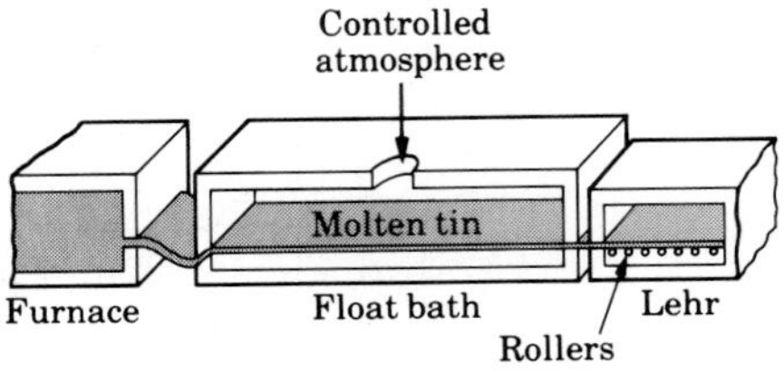

FIGURE 11.21 The float method of forming sheet glass. *Source:* Corning Glass Works.

molten glass is squeezed between rollers, forming a sheet. The surfaces of the glass can be embossed with a pattern by shaping the roller surfaces accordingly. Glass sheet produced by drawing and rolling has a rough surface appearance. In making plate glass, both surfaces have to be ground parallel and polished.

In the *float* method of production (Fig. 11.21), molten glass from the furnace is fed into a bath in which the glass, under controlled atmosphere, floats on a bath of molten tin. The glass then moves over rollers into another chamber (called a *lehr*) and solidifies. *Float glass* has a smooth (fire-polished) surface and needs no further grinding or polishing.

Glass tubing is manufactured by the process shown in Fig. 11.22. Molten glass is wrapped around a rotating hollow cylindrical or cone-shaped mandrel and is drawn out by a set of rolls. Air is blown through the mandrel to keep the glass tube from collapsing. These machines may be horizontal, vertical, or slanted downward. Glass rods are made in a similar manner, but air is not blown through the mandrel; thus the drawn product becomes a solid rod.

Continuous fibers are drawn through multiple (200–400) orifices in heated platinum plates, at speeds as high as 500 m/s (1700 ft/s). Fibers as small as 2 μm (80 μin.) in diameter can be produced by this method. In order to protect their surfaces, the fibers are subsequently coated with chemicals. Short glass fibers, used as thermal insulating material (*glass wool*) or for acoustic insulation, are made by a *centrifugal spraying process* in which molten glass is fed into a rotating head.

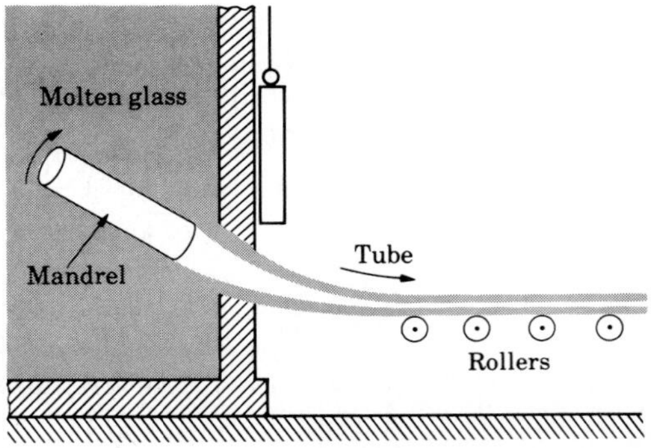

FIGURE 11.22 Manufacturing process for glass tubing. Air is blown through the mandrel to keep the tube from collapsing. *Source:* Corning Glass Works.

11.12.1 Manufacturing discrete glass products

Several processes are used in making discrete glass objects. These processes include blowing, pressing, centrifugal casting, and sagging.

Blowing. The *blowing* process is used to make hollow thin-walled glass items, such as bottles and flasks, and is similar to blow molding of thermoplastics. The steps involved in the production of an ordinary glass bottle by the blowing process are shown in Fig. 11.23. Blown air expands a hollow gob of heated glass against the

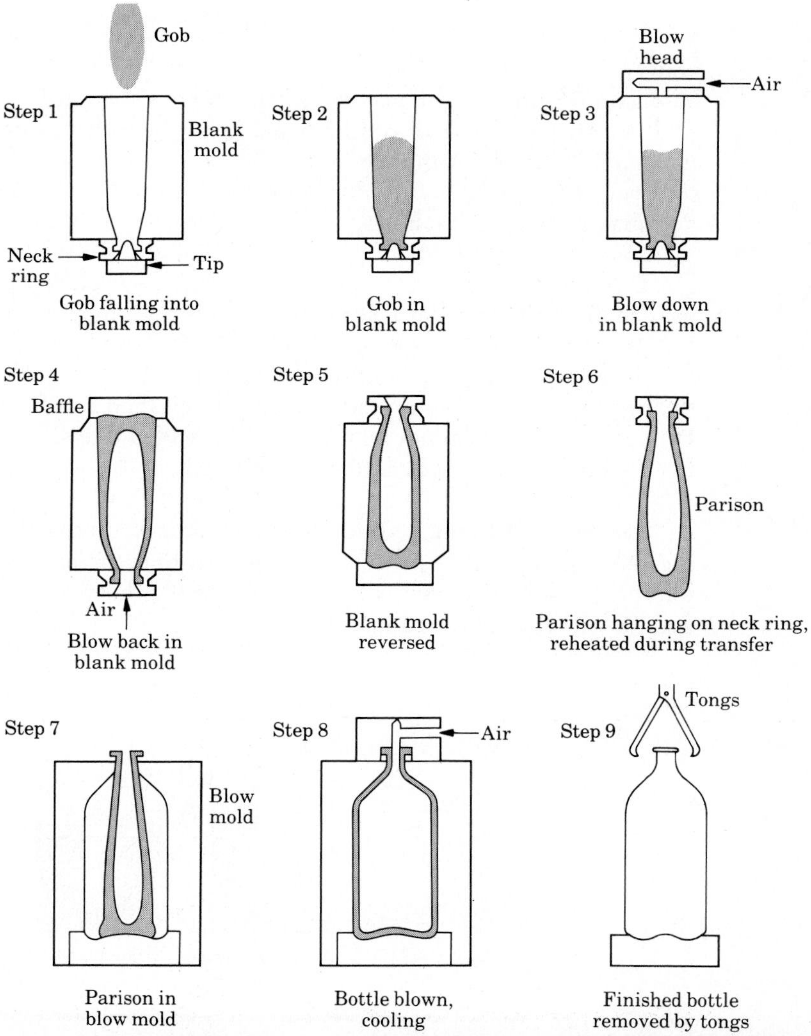

FIGURE 11.23

Stages in manufacturing an ordinary glass bottle. *Source*: F. H. Norton, *Elements of Ceramics*. Copyright © 1974, Addison-Wesley Publishing Company, Inc.

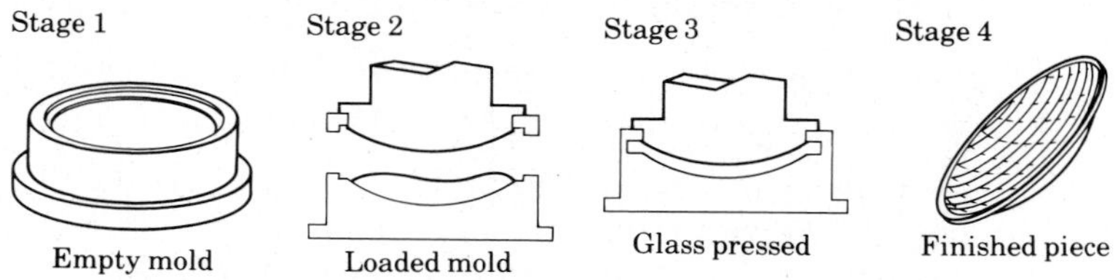

FIGURE 11.24
Manufacturing a glass item by pressing in a mold. *Source:* Corning Glass Works.

walls of the mold. The molds are usually coated with a parting agent, such as oil or emulsion, to prevent the part from sticking to the mold.

The surface finish of products made by the blowing process is acceptable for most applications. Although it is difficult to control the wall thickness of the product, the process is used for high rates of production. Light bulbs are made in automatic blowing machines, at a rate of over 1000 bulbs per minute.

Pressing. In *pressing*, a gob of molten glass is placed in a mold and is pressed into shape with the use of a plunger. The mold may be made in one piece (Fig. 11.24), or it may be a split mold (Fig. 11.25). After pressing, the solidifying glass acquires the shape of the mold-plunger cavity. Because of the confined environment, the product has greater dimensional accuracy than can be obtained with blowing. However, pressing cannot be used on thin-walled items, or for parts such as bottles from which the plunger cannot be retracted.

Centrifugal casting. Also known as *spinning* in the glass industry (Fig. 11.26), the *centrifugal casting* process is similar to that for metals. The centrifugal force pushes the molten glass against the mold wall where it solidifies. Typical products are TV picture tubes and missile nose cones.

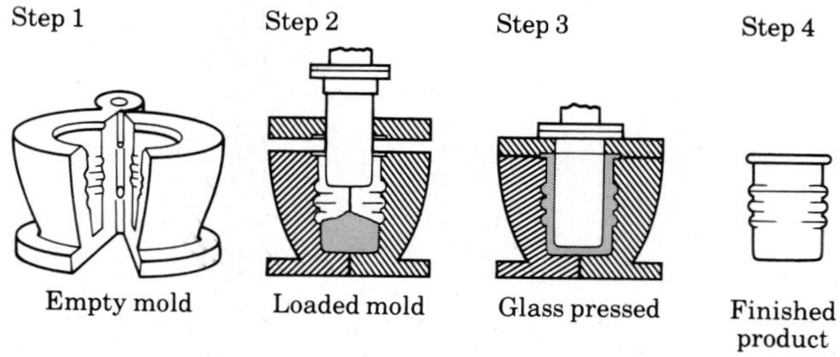

FIGURE 11.25
Pressing glass in a split mold. *Source:* E. B. Shand, *Glass Engineering Handbook.* New York: McGraw-Hill, 1958. Reprinted with permission.

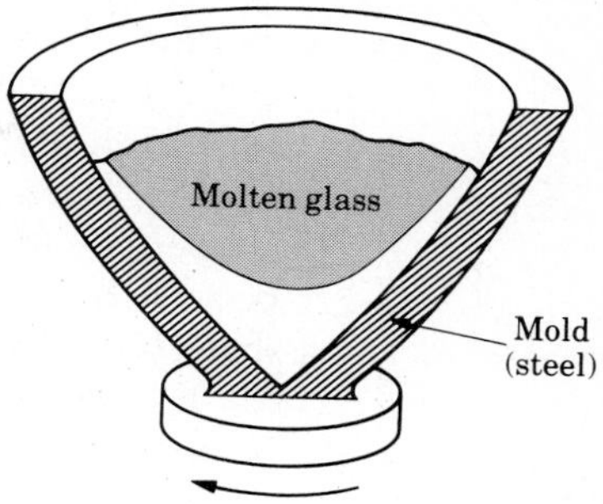

FIGURE 11.26
Centrifugal casting of glass. Television-tube funnels are made by this process. *Source:* Corning Glass Works.

Sagging. Shallow dish-shaped or lightly embossed glass parts can be made by the *sagging* process. A sheet of glass is placed over the mold and is heated. The glass sags by its own weight and takes the shape of the mold. The process is similar to thermoforming with thermoplastics but without pressure or a vacuum. Typical applications are dishes, sunglass lenses, mirrors for telescopes, and lighting panels.

11.12.2 Techniques for treating glass

As produced by the methods we described, glass can be strengthened by thermal tempering, chemical tempering, and laminating. Glass products may also undergo annealing and other finishing operations.

In *thermal tempering* (also called *physical tempering* or *chill tempering*), the surfaces of the hot glass are cooled rapidly (Fig. 11.27). As a result, the surfaces shrink, and because the bulk is still hot, tensile stresses develop on the surfaces. As the bulk of the glass begins to cool, it contracts. The solidified surfaces are now

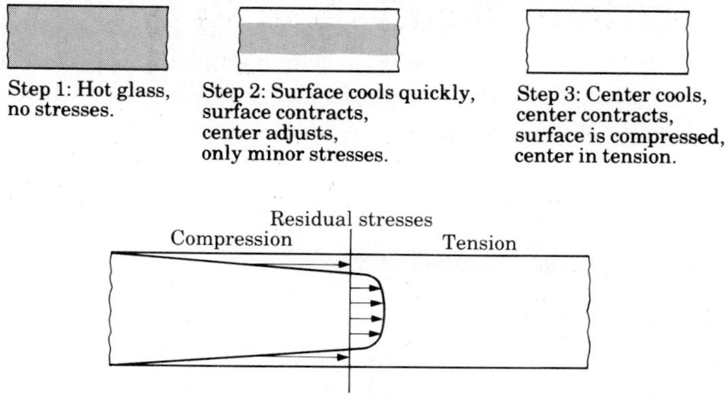

FIGURE 11.27
Residual stresses in tempered glass plate and stages involved in inducing compressive surface residual stresses for improved strength.

forced to contract, thus developing residual compressive surface stresses and interior tensile stresses. Compressive surface stresses improve the strength of the glass, as they do in other materials. Note that the higher the coefficient of thermal expansion of the glass and the lower its thermal conductivity, the higher the level of residual stresses developed and hence the stronger the glass becomes. Thermal tempering takes a relatively short time (minutes) and can be applied to most glasses. Because of the large amount of energy stored from residual stresses, *tempered glass* shatters into a large number of pieces when broken.

In *chemical tempering*, the glass is heated in a bath of molten KNO_3, K_2SO_4, or $NaNO_3$, depending on the type of glass. Ion exchange takes place, with larger atoms replacing the smaller atoms on the surface of the glass. As a result, residual compressive stresses are developed on the surface. This condition is similar to that created by forcing a wedge between two bricks in a wall. The time required for chemical tempering is longer (about one hour) than for thermal tempering. Chemical tempering may be performed at various temperatures. At low temperatures, part distortion is minimal and complex shapes can be treated. At elevated temperatures, there may be some distortion of the part, but the product can be used at higher temperatures without loss of strength.

Another strengthening method is to make *laminated glass*, which is two pieces of flat glass with a thin sheet of tough plastic between them. The process is also called *laminate strengthening*. When laminated glass is broken, its pieces are held together by the plastic sheet. You probably have seen this phenomenon in a shattered automobile windshield.

As in metal products, residual stresses can develop in glass products if they are not cooled slowly enough. In order to ensure that the product is free from these stresses, it is *annealed* by a process similar to stress-relief annealing of metals. The glass is heated to a certain temperature and cooled gradually. Depending on the size, thickness, and type of glass, annealing times may range from a few minutes, to as long as 10 months, as in the case of a 600-mm (24-in.) mirror for a telescope.

In addition to annealing, glass products may be subjected to further operations, such as cutting, drilling, grinding, and polishing. Sharp edges and corners can be smoothened by grinding, as in glass tops for desks and shelves, or by holding a torch against the edges (*fire polishing*), which rounds them by localized softening and surface tension.

11.13 Design Considerations

Ceramic and glass products require careful selection of composition, processing methods, finishing operations, and methods of assembly into other components. Limitations such as general lack of tensile strength, sensitivity to defects, and low impact toughness are important. These limitations have to be balanced against desirable characteristics, such as their hardness, scratch resistance, compressive strength at room and elevated temperatures, and diverse physical properties.

Control of processing parameters and the quality and level of impurities in the raw materials used are important. As in all design decisions, there are priorities, limitations, and various factors that should be considered, including the number of parts needed and the costs of tooling, equipment, and labor.

Dimensional changes and warping and cracking possibilities during processing are significant factors in selecting methods for shaping these materials. When a ceramic or glass component is part of a larger assembly, compatibility with other components is another important consideration. Particularly important are thermal expansion (as in seals) and the type of loading. The potential consequences of part failure are always a significant factor in designing ceramic products.

11.14 Graphite

Graphite is a crystalline form of carbon having a *layered structure* of basal planes or sheets of close-packed carbon atoms. Consequently, graphite is weak when sheared along the layers. This characteristic, in turn, gives graphite its low frictional properties as a solid lubricant. However, frictional properties are low only in an environment of air or moisture; graphite is abrasive and a poor lubricant in vacuum.

Although brittle, graphite has high electrical and thermal conductivity and resistance to thermal shock and high temperature (although it begins to oxidize at 500 °C (930 °F)). It is therefore an important material for applications such as electrodes, heating elements, brushes for motors, high-temperature fixtures and furnace parts, mold materials such as crucibles for melting and casting of metals, and seals (because of low friction and wear). Unlike other materials, the strength and stiffness of graphite increases with temperature.

An important use of graphite is as fibers in composite materials and reinforced plastics (see Chapter 10). A characteristic of graphite is its resistance to chemicals; hence it is used as filters for corrosive fluids. Also, its low thermal neutron absorption cross-section and high scattering cross-section make graphite suitable for nuclear applications. Ordinary pencil "lead" is a graphite and clay mixture.

Graphite is generally graded in decreasing order of grain size: industrial, fine grain, and micrograin. As in ceramics, the mechanical properties of graphite improve with decreasing grain size. Micrograin graphite can be impregnated with copper and is used as electrodes for electrical discharge machining and for furnace fixtures. Amorphous graphite is known as lampblack (black soot) and is used as a pigment. Graphite is usually processed by molding or forming, oven baking, and then machining to the final shape. It is available commercially in square, rectangular, or round shapes of various sizes.

11.15 Diamond

The second principal form of carbon is *diamond*, which has a covalently bonded structure. It is the hardest substance known (7000–8000 HK). This characteristic makes diamond an important cutting-tool material, as a single crystal or in polycrystalline form, as an abrasive in grinding wheels for grinding hard materials, and for dressing of grinding wheels (sharpening of abrasive grains). Diamond is also used as a die material for drawing thin wire, of less than 0.06 mm (0.0025 in.) in diameter. Diamond is brittle and it begins to decompose in air at about 700 °C (1300 °F). In nonoxidizing environments it resists higher temperatures.

Synthetic, or *industrial diamond*, was first made in 1955 and is used extensively for industrial applications. One method of manufacturing it is to subject graphite to a hydrostatic pressure of 14 GPa (2 million psi) and a temperature of 3000 °C (5400 °F). Synthetic diamond is identical to natural diamond for industrial applications, and has superior properties because of lack of impurities. It is available in various sizes and shapes, the most common abrasive grain size being 0.01 mm (0.004 in.) in diameter.

11.16 Composites

New developments in composite materials are continually taking place, with a wide range and form of polymeric, metallic, and ceramic materials being used both as fibers and as matrix materials. Research and development activities in this area are concerned with improving strength, toughness, stiffness, resistance to high-temperatures, and reliability in service.

11.16.1 Metal-matrix composites

The advantage of a metal matrix over a polymer matrix is its higher resistance to elevated temperatures and higher ductility and toughness. The limitations are higher density and greater difficulty in processing components. Matrix materials in these composites are usually aluminum, aluminum-lithium, magnesium, and titanium, although other metals are also being investigated. Fiber materials are graphite, aluminum oxide, silicon carbide, and boron, with beryllium and tungsten as other possibilities.

Because of their high specific stiffness, light weight, and high thermal conductivity, boron fibers in aluminum matrix have been used for structural tubular supports in the space shuttle orbiter. Other applications are bicycle frames and sporting goods. Studies of techniques for optimum bonding of fibers to the metal matrix are in progress. Metal-matrix composite applications are presently in gas turbines, electrical components, and various structural components (Table 11.11).

11.16.2 Ceramic-matrix composites

Ceramic-matrix composites are another important recent development in engineered materials. As we described earlier, ceramics are strong and stiff, resist high temperatures, but generally lack toughness. New matrix materials that retain their strength to 1700 °C (3100 °F) are silicon carbide, silicon nitride, aluminum oxide, and mullite (a compound of aluminum, silicon, and oxygen). Also under development are carbon–carbon matrix composites that retain much of their strength up to 2500 °C (4500 °F), although they lack oxidation resistance at high temperatures. Present applications for ceramic-matrix composites are in jet and automotive engines, deep-sea mining equipment, pressure vessels, and various structural components.

11.16.3 Other composites

Composites may also consist of coatings of various kinds on base metals or substrates. Examples are plating of aluminum and other metals over plastics for decorative purposes and enamels, dating to before 1000 B.C., or similar vitreous (glasslike) coatings on metal surfaces for various functional or ornamental purposes.

Composites are made into cutting tools and dies, such as cemented carbides, usually tungsten carbide and titanium carbide, with cobalt and nickel, respectively, as a binder. Other composites are grinding wheels made of aluminum oxide, silicon carbide, diamond, or cubic boron nitride abrasive particles, held together with

TABLE 11.11
METAL-MATRIX COMPOSITE MATERIALS AND APPLICATIONS

FIBER	MATRIX	APPLICATIONS
Graphite	Aluminum	Satellite, missile, and helicopter structures
	Magnesium	Space and satellite structures
	Lead	Storage-battery plates
	Copper	Electrical contacts and bearings
Boron	Aluminum	Compressor blades and structural supports
	Magnesium	Antenna structures
	Titanium	Jet-engine fan blades
Alumina	Aluminum	Superconductor restraints in fusion power reactors
	Lead	Storage-battery plates
	Magnesium	Helicopter transmission structures
Silicon carbide	Aluminum, titanium	High-temperature structures
	Superalloy (cobalt-base)	High-temperature engine components
Molybdenum, tungsten	Superalloy	High-temperature engine components

various organic, inorganic, or metallic binders. Another category of composites is cermets.

A composite developed relatively recently is granite particles in an epoxy matrix. It has high strength, good vibration damping capacity (better than gray cast iron), and good frictional characteristics. It is used as machine-tool beds for some precision grinders.

SUMMARY

The powder-metallurgy process consists of preparing metal powders, compacting them into shapes, and sintering them to impart strength, hardness, and toughness. Although size and weight are limited, the process is capable of producing relatively complex parts economically, in net-shape form to close tolerances, from a wide variety of metal and alloy powders. Control of powder shape and quality, process variables, and sintering atmospheres is an important consideration in product quality. Parts may be subjected to additional metalworking, machining, and finishing operations to impart certain geometric features and to improve properties and dimensional accuracy.

Density and mechanical and physical properties can be controlled by tooling design and compacting pressure. Some critical parts, such as jet-engine components, are now being made by P/M techniques. By controlling porosity, products such as filters and oil-impregnated bearings can be made. The P/M process is suitable for medium- to high-volume production runs, and has competitive advantages over other methods of production, such as casting, forging, and machining.

Ceramics, which are compounds of metallic and nonmetallic materials, are generally characterized by high hardness and compressive strength, high-temperature resistance, and chemical inertness. These properties make ceramics particularly attractive for heat-engine components, cutting tools, and components requiring wear and corrosion resistance. Ceramic products are shaped by various casting, plastic forming, or pressing techniques. They are then dried and fired to impart strength and hardness. Because of their inherent brittleness, ceramics are processed with due consideration of distortion and cracking. Control of raw-material quality and processing parameters are important factors.

Glasses are supercooled liquids; that is, the rate of cooling is so high that they do not have time to solidify into a crystalline structure. They do not have a clearly defined melting point. Glasses are available in a wide variety of compositions and mechanical, physical, and optical properties. Their strength can be improved by thermal or chemical treatments. Glass ceramics are predominantly crystalline in structure and have better properties than glasses. Glass products are made by a variety of shaping processes, which are similar to those used for plastics and ceramics.

Graphite has high-temperature and electrical applications and is also used as fibers to reinforce plastics. Diamond, the hardest substance known, is used as cutting tools, dies for wire drawing, and abrasives for grinding wheels.

New developments concern metal-matrix and ceramic-matrix composites and honeycomb structures. These materials have several important aircraft and aerospace applications, as well as high-temperature or lightweight industrial applications.

BIBLIOGRAPHY

Powder Metallurgy

Ashbrook, R.L. (ed.), *Rapid Solidification Technology: Source Book*. Metals Park, Ohio: American Society for Metals, 1983.

Bradbury, S. (ed.), *Powder Metallurgy Equipment Manual*. Princeton, N.J.: Powder Metallurgy Equipment Association, 1986.

Bradbury, S. (ed.), *Source Book on Powder Metallurgy*. Metals Park, Ohio: American Society for Metals, 1979.

German, R.M., *Powder Metallurgy Science*. Princeton, N.J.: Metal Powder Industries Federation, 1984.

Gessinger, G.H., *Powder Metallurgy of Superalloys*. London: Butterworths, 1984.

Hanes, E.D., D.A. Seifert, and C.R. Watts, *Hot Isostatic Processing*. New York: Springer-Verlag, 1980.

Hausner, H.H., and M.K. Mal, *Handbook of Powder Metallurgy*. New York: Chemical Publishing Company, 1982.

James, P.J. (ed.), *Isostatic Pressing Technology*. London: Applied Science Publishers, 1983.

Klar, E. (ed.), *Powder Metallurgy: Applications, Advantages, and Limitations*. Metals Park, Ohio: American Society for Metals, 1983.

Lenel, F.V., *Powder Metallurgy: Principles and Applications*. New York: American Powder Metallurgy Institute, 1980.

Metals Handbook, 9th ed., Vol. 7: *Powder Metallurgy*. Metals Park, Ohio: American Society for Metals, 1984.

Powder Metallurgy Design Guidebook. New York: American Powder Metallurgy Institute, revised periodically.

Ceramics

Doremus, R.H., *Glass Science*. New York: Wiley, 1973.

Encyclopedia of Glass, Ceramics, Clay and Cement. New York: Wiley, 1984.

Kingery, W.D., H.K. Bowen, and D.R. Uhlmann, *Introduction to Ceramics*, 2d ed. New York: Wiley, 1976.

Kirchner, H.P., *Strengthening of Ceramics: Treatments, Tests, and Design Applications*. New York: Marcel Dekker, 1979.

Lewis, M.H., *Glasses and Glass-Ceramics*. London: Chapman & Hall, 1987.

McColm, I.J., *Ceramic Science for Materials Technologists*. Glasgow: Leonard Hill (Chapman and Hall), 1983.

McLellen, G.W., and E.B. Shand, *Glass Engineering Handbook*. New York: McGraw-Hill, 1984.

Norton, F.H., *Elements of Ceramics*, 2d ed. Reading, Mass.: Addison-Wesley, 1974.

Richerson, D.W., *Modern Ceramic Engineering*. New York: Marcel Dekker, 1982.

Samsonov, C.V., and J.M. Vinitsku, *Handbook of Refractory Compounds*. New York: Plenum, 1980.
Schwartz, M.M. (ed.), *Engineering Applications of Ceramic Materials, Source Book*. Metals Park, Ohio: American Society for Metals, 1985.

Ceramics Processing

Ford, R.W., *Ceramics Drying*. Elmsford, New York: Pergamon, 1986.
Hlavac, J., *The Technology of Glass and Ceramics*. Amsterdam: Elsevier, 1983.
Kingery, W.D., H.K. Bowen, and D.R. Uhlmann, *Introduction to Ceramics*, 2d ed. New York: Wiley, 1976.
Norton, F.H., *Elements of Ceramics*, 2d ed. Reading, Mass.: Addison-Wesley, 1974.
Onoda, G.Y., and L.L. Hench (ed.), *Ceramic Processing Before Firing*. New York: Wiley-Interscience, 1978.
Reed, J.S., *Introduction to the Principles of Ceramic Processing*. New York: Wiley, 1988.
Richerson, D.W., *Modern Ceramic Engineering*. New York: Marcel Dekker, 1982.
Scholes, S.R., and C.H. Greene, *Modern Glass Practice*, 7th ed. Boston: Cahners Books, 1975.
Schwartz, M.M. (ed.), *Engineering Applications of Ceramic Materials—Source Book*. Metals Park, Ohio: American Society for Metals, 1985.
Wang, F.F.Y. (ed.), *Ceramic Fabrication Processes*. New York: Academic Press, 1984.
Zschommler, W., *Precision Optical Glassworking*. New York: Macmillan, 1984.

QUESTIONS

Powder Metallurgy

11.1 Explain why metal powders are blended.
11.2 Is green strength important in powder metals processing? Explain.
11.3 Explain why density differences may be desirable within some P/M parts. Give specific examples.
11.4 Give the reasons for injection molding of metal powders becoming an important process.
11.5 Describe what happens during sintering.
11.6 What is mechanical alloying and what are its advantages over conventional alloying of metals?
11.7 How are design considerations for powder-metallurgy parts different from those for casting and forging?
11.8 Describe the effects of different shapes and sizes of metal powders in P/M processing, including comments on the shape factor (SF) of the particles.
11.9 Are there applications in which you would not use a P/M product? Explain.
11.10 Explain the reasons for the shapes of the curves and their relative positions shown in Fig. 11.5(a).
11.11 Should green compacts be brought up to the sintering temperature slowly or rapidly? Explain the advantages and limitations of each method.
11.12 Explain the effects of using fine powders and coarse powders, respectively, in making P/M parts.

11.13 Are the requirements for punch and die materials in powder metallurgy different from those for forging and extrusion? Explain.

11.14 Describe the relative advantages and limitations of cold and hot isostatic pressing, respectively.

11.15 Why do mechanical and physical properties depend on the density of P/M parts? Explain with appropriate diagrams.

11.16 What type of press is required to compact parts by the set of punches shown in Fig. 11.6(d)?

11.17 Tool steels are now being made by P/M techniques. Explain the advantages of this method over traditional methods, such as casting and subsequent metalworking techniques.

Ceramics and Others

11.18 Compare the major differences between the properties of ceramics and those of metals and plastics.

11.19 Explain why ceramics are weaker in tension than in compression.

11.20 Explain why the mechanical and physical properties of ceramics decrease with increasing porosity.

11.21 What engineering applications could benefit from the fact that ceramics maintain their modulus of elasticity at elevated temperatures?

11.22 Explain why the mechanical property data in Table 11.7 have such a broad range. What is its significance in engineering practice?

11.23 List the factors that you would take into account when replacing a metal component with a ceramic component.

11.24 How are ceramics made tougher?

11.25 Describe applications in which static fatigue can be important.

11.26 Explain the difficulties involved in making large ceramic components.

11.27 Explain why ceramics are effective cutting-tool materials. Would ceramics also be suitable for die materials for metal forming? Explain.

11.28 Describe applications in which a ceramic material with a zero coefficient of thermal expansion is desirable.

11.29 Give reasons for the development of ceramic-matrix components. Name some possible applications.

11.30 List the factors that are important in drying ceramic components and explain why they are important.

11.31 The higher the coefficient of thermal expansion of the glass and the lower its thermal conductivity, the higher the level of residual stresses developed during processing. Explain why.

PROBLEMS

11.1 Estimate the number of particles in a 500-g sample of iron powder, if the particle size is 100 μm.

11.2 Assume that the surface of a copper particle is covered with an oxide layer 0.1 μm in thickness. What is the volume occupied by this layer if the copper particle itself is 50 μm in diameter? What would be the role of this oxide layer on subsequent processing of the powder metal?

11.3 Determine the shape factor (SF) of (a) a flakelike particle with a surface area to thickness ratio of 15 by 15 by 1 and (b) an ellipsoid with an axial ratio of 5 by 2 by 1.

11.4 We have seen in Section 2.10.2 and Example 2.6 that the energy in brittle fracture is dissipated as surface energy. We have also noted that the comminution process of powder preparation generally involves brittle fracture. What are the relative energies involved in making spherical powders of diameters 1, 10, and 100 μm, respectively?

11.5 Referring to Fig. 11.5(a), what should be the volume of loose, fine iron powder in order to make a solid cylindrical compact 20 mm in diameter and 10 mm high?

11.6 In Fig. 11.6(e), we note that the pressure is not uniform across the diameter of the compact. What is the reason for it?

11.7 Plot the family of pressure ratio p_x/p_o curves as a function of x for the following ranges of process parameters: $\mu = 0$ to 1, $k = 0$ to 1, $D = 5$ mm to 50 mm.

11.8 Derive an expression similar to Eq. (11.1) for compaction in a square die with dimensions a by a.

11.9 In Example 11.7, calculate (a) the porosity of the dried part if the porosity of the fired part is to be 9 percent and (b) the initial length L_o of the part if the linear shrinkages during drying and firing are 8 and 7 percent, respectively.

11.10 Plot the UTS, E, and k values for ceramics as a function of porosity P, and describe and explain the trends that you observe in their behavior.

12

Joining and Fastening Processes

12.1 Introduction

Joining is an all inclusive term, which covers processes such as welding, brazing, soldering, adhesive bonding, and mechanical joining. These processes are an important and necessary aspect of manufacturing operations for the following reasons:

- The product is impossible to manufacture as a single piece.
- The product—such as a cooking pot with a handle—is easier and more economical to manufacture as individual components, which are then assembled.

- Products such as automobile engines, hair dryers, printers, and soldered-joint electronic devices may have to be taken apart for repair or maintenance during their service lives.
- Different properties may be desirable for functional purposes of the product. Surfaces subjected to friction and wear, or corrosion and environmental attack, generally require characteristics different from those of the component's bulk. Examples are carbide cutting tips brazed to the shank of a drill and brake shoes bonded to a metal backing.
- Transporting the product in individual components and assembling them at home or at the customer's plant may be easier and less costly. Some bicycles and toys, most machine tools, and mechanical or hydraulic presses are assembled after the components have been transported to the appropriate site.

Joining processes may be classified according to the state of the materials being joined, the extent to which external heating and/or pressure is utilized, and the use of filler metals in the joint. *Liquid-state* joining processes, such as oxyfuel, arc, and resistance welding, involve fusion—that is, partial melting of the metals to be joined. These processes require application of heat and/or pressure by some suitable means to permanently fuse or bond two pieces. The source of heat may be chemical, electrical, or optical (such as lasers). *Solid-state* processes, such as ultrasonic, explosion, and friction welding, do not utilize fillers or bonding materials. The joint is developed through external pressure and a heat source, supplied externally or developed internally, such as through friction.

Several joining processes, such as brazing, soldering, and most adhesive bonding processes, are *liquid–solid state* in nature. In these processes, the braze metal, solder, or an adhesive is applied to the joint in a liquid state while the materials to be joined remain in a solid state; that is, they do not melt. Mechanical *fasteners*, such as bolts, nuts, and screws, as well as seaming, crimping, and stitching are other methods of joining. Some of these joints are nonpermanent and are used on products and machines that have to be taken apart for maintenance or repair.

12.2 Oxyfuel Gas Welding

Oxyfuel gas welding, as well as the arc welding and resistance welding processes that we describe in subsequent sections, involve *melting* and *fusion* of the joint between two members. Fusion is the melting together and coalescing of materials by means of heat. The thermal energy required for these welding operations is usually supplied by chemical and electrical means. *Filler metals*, which are metals added to the weld area during welding of the joint, may or may not be used. Fusion welds made without the addition of filler metals are known as *autogenous* welds.

Oxyfuel gas welding (OFW) is a general term used to describe any welding process that uses a *fuel gas* combined with oxygen to produce a flame. This flame is used as the source of heat to melt the metals at the joint. The most common gas welding process uses *acetylene fuel*, and is known as *oxyacetylene welding*. Developed in the early 1900s, this process utilizes the heat generated by the combustion of acetylene gas (C_2H_2) in a mixture with oxygen in a *torch*.

The heat is generated in accordance with the following chemical reactions. The primary combustion process, which occurs in the inner core of the flame (Fig. 12.1), is

$$C_2H_2 + O_2 \rightarrow 2CO + H_2 + \text{heat}. \quad (12.1)$$

This reaction dissociates the acetylene into carbon monoxide and hydrogen and produces about one-third of the total heat generated in the flame. The second reaction is

$$2CO + H_2 + 1.5O_2 \rightarrow 2CO_2 + H_2O + \text{heat}, \quad (12.2)$$

which results in burning of the hydrogen and combustion of the carbon monoxide, producing about two-thirds of the total heat. The temperatures developed in the flame as a result of these reactions can reach 3300 °C (6000 °F). The reaction of hydrogen with oxygen produces water vapor.

12.2.1 Types of flames

The proportions of acetylene and oxygen in the gas mixture are an important factor in oxyfuel gas welding. At a ratio of 1 : 1, that is, when there is no excess oxygen, it

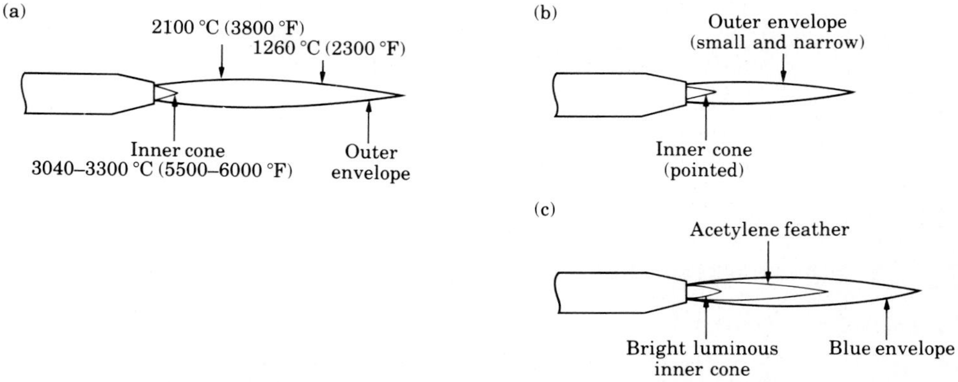

FIGURE 12.1
Three types of oxyacetylene flames used in oxyfuel gas welding and cutting operations: (a) neutral flame; (b) oxidizing flame; and (c) carburizing, or reducing, flame. The gas mixture is basically equal volumes of oxygen and acetylene.

is considered to be a *neutral flame*. With a greater oxygen supply, it becomes an *oxidizing flame*. This flame is harmful, especially for steels, because it oxidizes the steel. Only in copper and copper-base alloys is an oxidizing flame desirable because a thin protective layer of slag forms over the molten metal. If the supply of oxygen is lowered, it becomes a *reducing* or *carburizing flame*. The temperature of a reducing, or excess-acetylene, flame is lower. Hence it is suitable for applications requiring low heat, such as brazing, soldering, and flame hardening. Other fuel gases such as hydrogen and methylacetylene propadiene can be used in oxyfuel gas welding. However, the temperatures developed are low, and hence they are used for welding metals with low melting points, such as lead, and parts that are thin and small.

12.2.2 Filler metals

Filler metals are used to supply additional material to the weld zone during welding. They are available as rods or wire, and are made of metals similar to those to be welded. These consumable *filler rods* may be bare, or they may be coated with *flux*. The purpose of the flux is to retard oxidation of the surfaces of the parts being welded, by generating a gaseous shield around the weld zone. The flux also helps dissolve and remove oxides and other substances from the workpiece, resulting in a stronger joint. The slag developed protects the molten puddle of metal against oxidation as it cools.

12.2.3 Welding practice and equipment

Oxyfuel gas welding can be used with most ferrous and nonferrous metals for any thickness of workpiece, but the relatively low heat input limits the process economically to less than 6 mm (0.25 in.). A variety of joints can be produced by this method. Small joints may consist of a single weld bead. Deep V-groove joints are made in multiple passes. Cleaning the surface of each weld bead prior to depositing a second layer is important for joint strength and avoiding defects. Hand or power wire brushes may be used for this purpose. The equipment for oxyfuel gas welding basically consists of a welding torch, which is available in various sizes and shapes, connected by hoses to high-pressure gas cylinders and equipped with pressure gauges and regulators.

12.3 Thermit Welding

Thermit welding (TW) gets its name from *thermite*, which is based on the word *therm*, meaning heat; the word Thermit is a registered trademark. The process involves exothermic (heat producing) reactions between metal oxides and metallic reducing agents. The heat of the reaction is then utilized in welding.

The most common mixture of materials used in welding steel and cast iron is finely divided particles of iron oxide (Fe_3O_4), aluminum oxide (Al_2O_3), iron, and aluminum. The basic reactions are

$$\tfrac{3}{4}Fe_3O_4 + 2Al \rightarrow \tfrac{9}{4}Fe + Al_2O_3 + \text{heat}; \qquad (12.3)$$

$$3FeO + 2Al \rightarrow 3Fe + Al_2O_3 + \text{heat}; \qquad (12.4)$$

$$Fe_2O_3 + 2Al \rightarrow 2Fe + Al_2O_3 + \text{heat}. \qquad (12.5)$$

This nonexplosive mixture produces a maximum theoretical temperature of 3200 °C (5800 °F) within less than a minute. In practice, however, this temperature is only about 2200–2400 °C (4000–4350 °F). The mixture may also contain other materials to impart special properties to the weld. The reaction is started by applying a magnesium fuse to special compounds of peroxides, chlorates, or chromates, known as oxidizing agents, with an ignition temperature of about 1200 °C (2200 °F).

Welding copper, brasses, and bronzes, and copper alloys to steels, involves the following reactions:

$$3CuO + 2Al \rightarrow 3Cu + Al_2O_3 + \text{heat}; \qquad (12.6)$$

$$3Cu_2O + 2Al \rightarrow 6Cu + Al_2O_3 + \text{heat}. \qquad (12.7)$$

Oxides of copper, nickel, chromium, and manganese are also used in Thermit welding, resulting in temperatures ranging up to 5000 °C (9000 °F).

12.4 Arc-Welding: Consumable Electrode

In *arc welding*, developed in the mid-1800s, the heat required is obtained through electrical energy. Using either a *consumable* or *nonconsumable electrode* (rod or wire), an arc is produced between the tip of the electrode and the workpiece to be welded, using ac or dc power supplies. This arc produces temperatures in the range of 5000–30,000 °C (9000–54,000 °F), which are much higher than those developed in oxyfuel gas welding. The arc also produces radiation, which may dissipate as much as 20 percent of the total energy. Arc welding includes various welding processes, which we describe below.

12.4.1 Shielded metal-arc welding

Shielded metal-arc welding (SMAW) is one of the oldest, simplest, and most versatile joining processes. Currently, about 50 percent of all industrial welding is performed by this process. The electric arc is generated by touching the tip of a coated electrode against the workpiece and then withdrawing it quickly to a distance sufficient to maintain the arc (Fig. 12.2). The electrodes are in the shape of thin, long sticks, so this process is also known as stick welding.

The heat generated melts a portion of the tip of the electrode, its coating, and the base metal in the immediate area of the arc. A weld forms after the molten metal—a mixture of the base metal (workpiece), electrode metal, and substances from the coating on the electrode—solidifies in the weld area.

A bare section at the end of the electrode is clamped in an electrode holder. The holder is connected to one terminal of the power source, while the other terminal is connected to the workpiece being welded. The current usually ranges between 50 A and 300 A, with power requirements generally less than 10 kW. The current may be ac or dc, and the polarity of the electrode may be positive (reverse polarity) or negative (straight polarity). The choice depends on the type of electrode, type of metals to be welded, arc atmosphere, and the depth of the heated zone. Too low a current causes incomplete fusion, and too high a current can damage the electrode coating and reduce its effectiveness.

The SMAW process has the advantage of being relatively simple and versatile, requiring a relatively small variety of electrodes. The equipment consists of a power supply, power cables, and electrode holder. This process is commonly used in general construction, shipbuilding, and pipelines, as well as for maintenance work, since the equipment is portable and can be easily maintained. It is especially useful for work in remote areas where portable fuel-powered generators can be used as the power supply. The SMAW process is best suited for workpiece thicknesses of 3–19 mm (0.12–0.75 in.), although this range can be easily extended using special techniques and highly skilled operators.

12.4.2 Submerged arc welding

In *submerged arc welding* (SAW), the weld arc is shielded by granular flux, consisting of lime, silica, manganese oxide, calcium fluoride, and other elements.

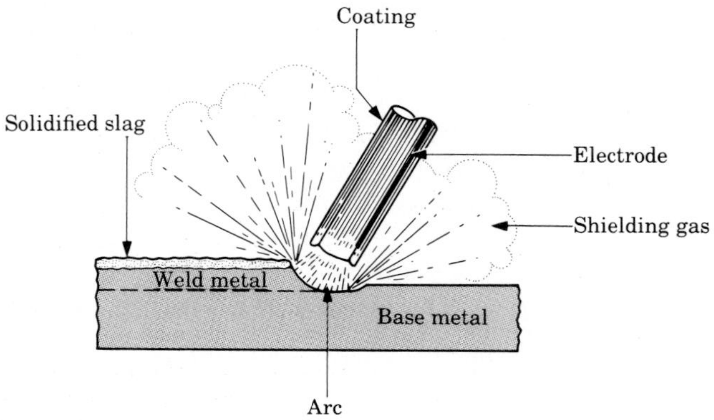

FIGURE 12.2 Schematic illustration of the shielded metal-arc welding process. About 50 percent of all large-scale industrial welding operations use this process.

The flux is fed into the weld zone by gravity flow through a nozzle (Fig. 12.3). The thick layer of flux completely covers the molten metal and prevents spatter and sparks—and without the intense radiation and fumes of the SMAW process. The flux also acts as a thermal insulator, allowing deep penetration of heat into the workpiece. The welder must wear gloves, but other than tinted safety glasses, face shields generally are unnecessary.

The consumable electrode is a coil of bare round wire 1.5–10 mm ($\frac{1}{16}$–$\frac{3}{8}$ in.) in diameter, and is fed automatically through a tube (welding gun). Electric currents usually range between 600 A and 2000 A, from either ac or dc power sources, at up to 440 V. Because the flux is fed by gravity, the SAW process is somewhat limited to welds in a flat or horizontal position with backup piece. Circular welds can be made on pipes, provided that they are rotated during welding. The unfused flux can be recovered, treated, and reused. The quality of the weld is very high, with good toughness, ductility, and uniformity of properties. The SAW process provides very high welding productivity, depositing 4–10 times the amount of weld metal per hour as the SMAW process.

12.4.3 Gas metal-arc welding

In *gas metal-arc welding* (GMAW), the weld area is shielded by an external source of inert gas, such as argon, helium, carbon dioxide, or various other gas mixtures

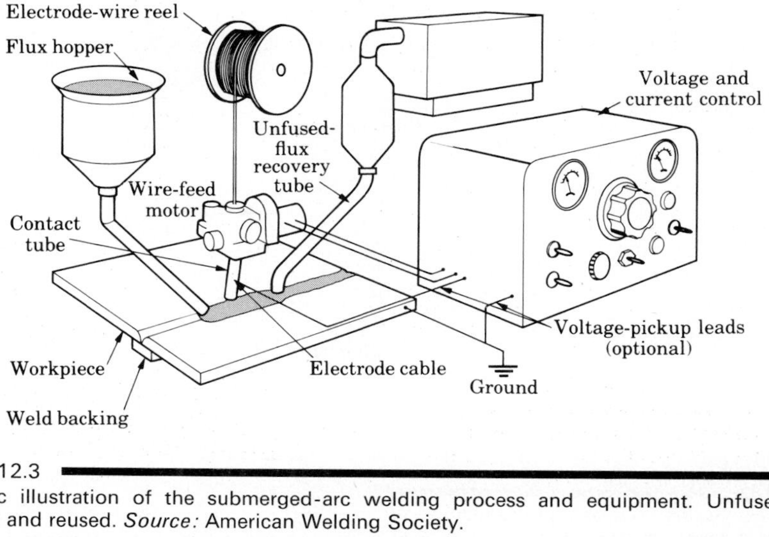

FIGURE 12.3 Schematic illustration of the submerged-arc welding process and equipment. Unfused flux is recovered and reused. *Source:* American Welding Society.

(Fig. 12.4). The consumable bare wire is fed automatically through a nozzle into the weld arc. In addition to the use of inert shielding gases, deoxidizers are usually present in the electrode metal itself, in order to prevent oxidation of the molten weld puddle. The welds made by this process are thus relatively free of slag, and hence multiple weld layers can be deposited at the joint without the necessity for intermediate cleaning of slag.

Metal can be transferred three ways in the GMAW process: spray, globular, and short circuiting. In *spray transfer*, small droplets of molten metal from the electrode are transferred to the weld area at rates of several hundred droplets per second. In *globular transfer*, carbon-dioxide rich gases are utilized, and globules propelled by the forces of the electric arc transfer the metal, resulting in considerable spatter. In *short circuiting*, the metal is transferred in individual droplets, at rates of more than 50 per second, as the electrode tip touches the molten weld arc and short circuits.

The GMAW process was developed in the 1950s and was formerly called *metal inert-gas* (MIG) welding. It is suitable for welding a variety of ferrous and nonferrous metals and is used extensively in the metal-fabrication industry. Because of the relatively simple nature of the process, training operators is easy. This process is rapid, versatile, and economical; welding productivity is double that of the SMAW process. The GMAW process can easily be automated and lends itself readily to flexible manufacturing systems and robotics. It has virtually replaced the SMAW process in present-day welding applications in manufacturing plants.

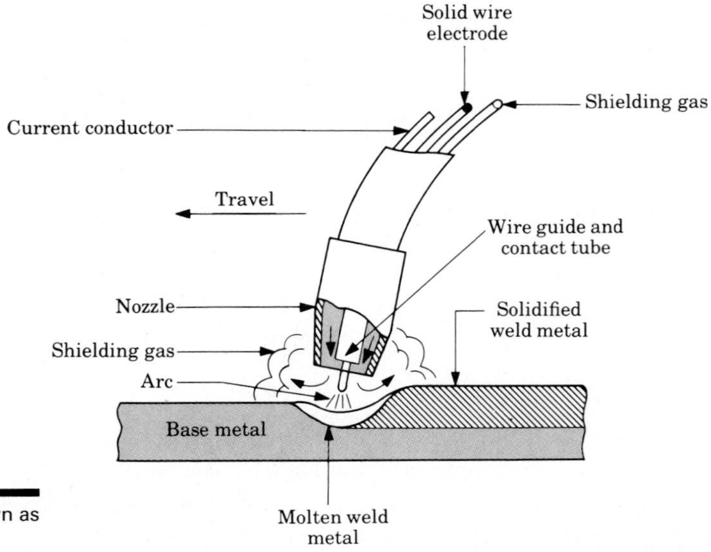

FIGURE 12.4
Gas metal-arc welding process, formerly known as MIG (for metal inert gas).

12.4.4 Flux-cored arc welding

The *flux-cored arc welding process* (FCAW, Fig. 12.5) is similar to gas metal-arc welding, with the exception that the electrode is tubular in shape and is filled with flux (hence the term flux cored). Cored electrodes produce a more stable arc, improve weld contour, and improve the mechanical properties of the weld metal. The flux in these electrodes is much more flexible than the brittle coating used on SMAW electrodes. Thus the tubular electrode can be provided in long coiled lengths. The electrodes are usually 1.5–2.5 mm ($\frac{1}{16}$–$\frac{3}{32}$ in.) in diameter. The power required is about 20 kW. Self-shielded cored electrodes are also available. These electrodes do not require external shielding because they contain emissive fluxes that shield the weld area against the surrounding atmosphere.

The flux-cored arc welding process combines the versatility of SMAW with the continuous and automatic electrode-feeding feature of GMAW. It is economical and is used for welding a variety of joints, mainly with steels and stainless steels. The higher weld metal deposition rate of the FCAW process, over that of GMAW, has led to its use primarily in joining sections 25 mm (1 in.) and thicker. Recent development of tubular electrodes with very small diameters has extended the use of this process to smaller workpiece section sizes. A major advantage of FCAW is the

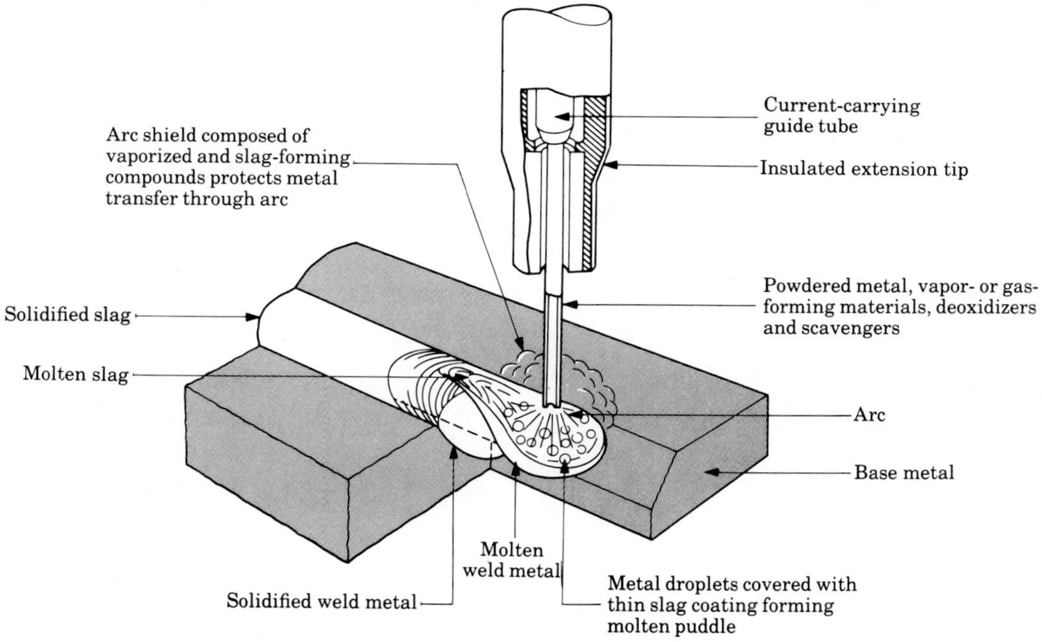

FIGURE 12.5
Schematic illustration of the flux-cored arc welding process. This operation is similar to gas metal-arc welding shown in Fig. 12.4.

ease with which specific weld metal chemistries can be developed. By adding alloys to the flux core, virtually any alloy composition can be developed. This process is easy to automate and is readily adaptable to flexible manufacturing systems and robotics.

12.4.5 Electrogas welding

Electrogas welding (EGW) is used primarily for welding the edges of sections vertically in one pass with the pieces placed edge to edge (butt) (Fig. 12.6). It is classified as a machine-welding process because it requires special equipment. The weld metal is deposited into a weld cavity between the two pieces to be joined. The space is enclosed by two water-cooled copper dams (shoes) to prevent the molten slag from running off. Mechanical drives move the shoes upward. Circumferential welds are also possible, with the workpiece rotating.

A single electrode is fed through a guide tube and a continuous arc is maintained, using flux-cored electrodes at up to 750 A, or solid electrodes at 400 A. Power requirements are about 20 kW. Shielding is by inert gas, such as carbon dioxide, argon, or helium, depending on the type of material being welded. The gas may be provided from an external source, or it may be produced from a flux-cored electrode, or both. The equipment for electrogas welding is reliable, and training

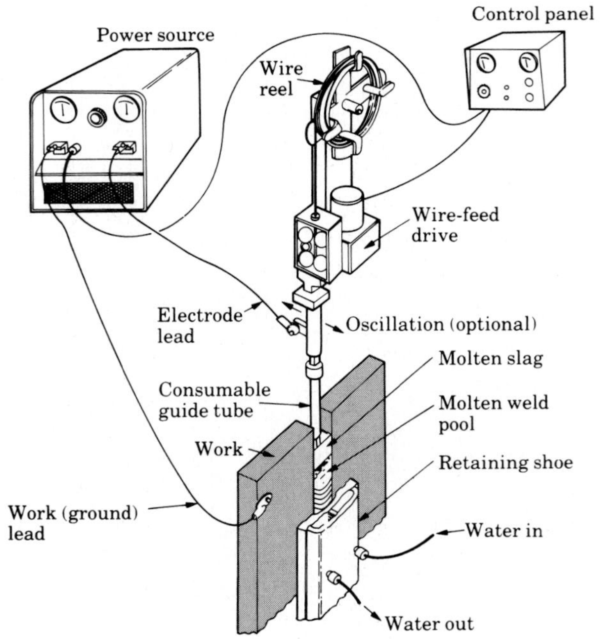

FIGURE 12.6
Equipment used for electrogas welding operations. *Source:* American Welding Society.

operators is relatively simple. Weld thickness ranges from 12 mm to 75 mm (0.5 in. to 3 in.) on steels, titanium, and aluminum alloys, and the welds are of high quality.

12.4.6 Electroslag welding

Developed in the 1950s, *electroslag welding* (ESW) and its applications are similar to electrogas welding. The main difference is that the arc is started between the electrode tip and the bottom of the part to be welded. Flux is added and melted by the heat of the arc. After the molten slag reaches the tip of the electrode, the arc is extinguished. Energy is supplied continuously through the electrical resistance of the molten slag. Thus because the arc is extinguished, ESW is not strictly an arc welding process. Single or multiple solid as well as flux-cored electrodes may be used. The guide may be nonconsumable (conventional method) or consumable.

Electroslag welding is capable of welding plates with thicknesses ranging from 50 mm to more than 900 mm (2 in. to more than 36 in.). Welding is done in one pass. The current required is about 600 A at 40–50 V, although higher currents are used for thick plates. Travel speed of the weld is 0.2–0.6 mm/s (0.5–1.5 in./min). The weld quality is good and the process is used for heavy structural steel sections, such as heavy machinery and nuclear-reactor vessels.

12.4.7 Electrodes

Electrodes for the consumable arc welding processes described are classified according to the strength of the deposited weld metal, current (ac or $\pm$ dc), and the type of coating. Electrodes are identified by numbers and letters, or by color code, particularly if they are too small to imprint with identification. Typical coated electrode dimensions are 150–460 mm (6–18 in.) long and 1.5–8 mm ($\frac{1}{16}$–$\frac{5}{16}$ in.) in diameter. The thinner the sections to be welded and the lower the current required, the smaller the diameter of the electrode should be.

The specifications for electrodes and filler metals, including tolerances, quality control procedures, and processes, are stated by the American National Standards Institute (ANSI) and in the Aerospace Materials Specifications (AMS) by the Society of Automotive Engineers (SAE). Among others, the specifications require that the wire diameter not vary more than 0.05 mm (0.002 in.) from nominal size, and that the coatings be concentric with the wire. Electrodes are sold by weight and are available in a wide variety of sizes and specifications.

Electrodes are coated with claylike materials that include silicate binders and powdered materials such as oxides, carbonates, fluorides, metal alloys, and cellulose (cotton cellulose and wood flour). The coating, which is brittle and has complex interactions during welding, functions to (a) stabilize the arc; (b) generate gases (carbon dioxide and water vapor, and carbon monoxide and hydrogen in small amounts) to act as a shield against the surrounding atmosphere; (c) control the rate at which the electrode melts; (d) act as a flux to protect the weld against formation of inclusions, and protect the molten weld pool; and (e) add alloying elements to the weld zone to enhance the properties of the weld, including deoxidizers to prevent the weld from becoming brittle.

12.5 Arc-Welding: Nonconsumable Electrode

Unlike the arc-welding processes that use consumable electrodes, nonconsumable-electrode processes typically use a tungsten electrode. As one pole of the arc, it generates the heat required for welding. A shielding gas is supplied from an external source. We describe below the advantages, limitations, and typical applications of these processes.

12.5.1 Gas tungsten-arc welding

In *gas tungsten-arc welding* (GTAW), formerly known as TIG welding (for tungsten inert gas), the filler metal is supplied from a filler wire (Fig. 12.7). Because the tungsten electrode is not consumed in this operation, a constant and stable arc gap is maintained at a constant current level. The filler metals are similar to the metals to be welded, and flux is not used. The shielding gas is usually argon or helium, or a mixture of the two. Welding with GTAW may be done without filler metals, as in welding close-fit joints.

The GTAW process is used for a wide variety of metals, particularly aluminum, magnesium, titanium, and refractory metals. It is especially suitable for thin metals. The cost of the inert gas makes this process more expensive than SMAW, but it provides welds with very high quality and surface finish. It is used in a variety of critical applications with a wide range of workpiece thicknesses and shapes.

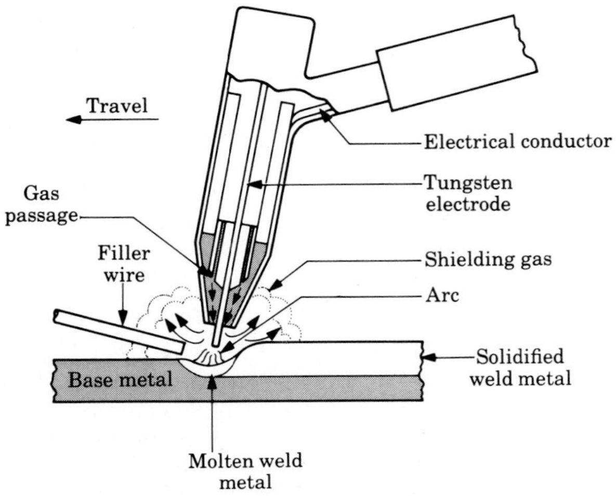

FIGURE 12.7 Gas tungsten-arc welding process, formerly known as TIG (for tungsten inert gas).

12.5.2 Atomic hydrogen welding

Atomic hydrogen welding (AHW) uses an arc in a shielding atmosphere of hydrogen. The arc is between two tungsten or carbon electrodes. Thus the workpiece is not part of the electrical circuit, as it is in GTAW. The hydrogen gas also cools the electrodes.

12.5.3 Plasma-arc welding

In *plasma-arc welding* (PAW), developed in the 1960s, a concentrated plasma arc is produced and aimed at the weld area. The arc is stable and reaches temperatures as high as 33,000 °C (60,000 °F). A *plasma* is ionized hot gas, composed of nearly equal numbers of electrons and ions. The plasma is initiated between the tungsten electrode and the orifice, using a low-current pilot arc. Unlike other processes, the plasma arc is concentrated because it is forced through a relatively small orifice. Operating currents are usually below 100 A, but they can be higher for special applications. When a filler metal is used, it is fed into the arc, as in GTAW. Arc and weld-zone shielding is supplied through an outer shielding ring by gases such as argon, helium, or mixtures.

There are two methods of plasma-arc welding. In the *transferred-arc* method (Fig. 12.8a), the workpiece being welded is part of the electrical circuit. The arc thus transfers from the electrode to the workpiece—hence the term transferred. In the *nontransferred* method (Fig. 12.8b), the arc is between the electrode and the nozzle, and the heat is carried to the workpiece by the plasma gas.

Compared to other arc-welding processes, plasma-arc welding has greater energy concentration (hence deeper and narrower welds can be made), better arc stability, and higher welding speeds, such as 2–16 mm/s (5–40 in./min). A variety of metals can be welded, with part thicknesses generally less than 6 mm (0.25 in.). The

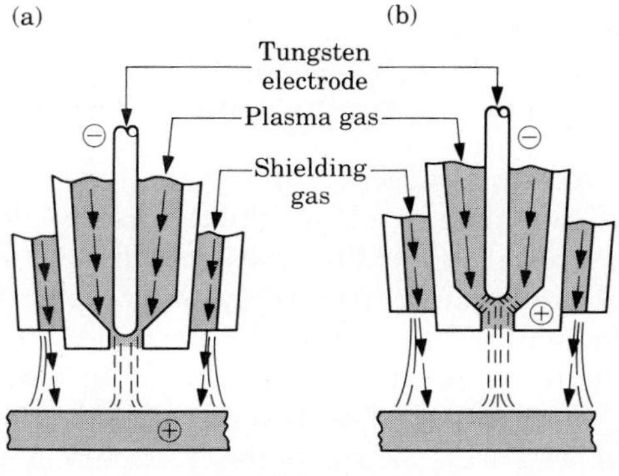

FIGURE 12.8
Two types of plasma-arc welding processes: (a) transferred and (b) nontransferred. Deep and narrow welds are made by this process at high welding speeds.

high heat concentration can penetrate completely through the joint (*keyhole* technique), with thicknesses as much as 20 mm (0.75 in.) for some titanium and aluminum alloys. Plasma-arc welding, rather than the GTAW process, is often used for butt and lap joints because of higher energy concentration, better arc stability, and higher welding speeds.

12.6 Resistance Welding

Resistance welding (RW) covers a number of processes in which the heat required for welding is produced by means of the electrical resistance between the two members to be joined. These processes have major advantages, such as not requiring consumable electrodes, shielding gases, or flux.

The heat generated in resistance welding is given by the general expression

$$H = I^2Rt, \tag{12.8}$$

where

H = heat generated, in joules (watt-seconds);
I = current, in amperes;
R = resistance, in ohms; and
t = time of current flow, in seconds.

The total resistance in these processes, such as in the resistance spot welding shown in Fig. 12.9, is the sum of the following:

a) Resistance of the electrodes.
b) Electrode–workpiece contact resistances.
c) Resistances of the individual parts to be welded.
d) Workpiece–workpiece contact resistances (*faying surfaces*).

For high heat generation at the junction, workpiece–workpiece contact resistances should be kept high, and the rest should be kept low. However, because these resistances are generally low (typically 100 $\mu\Omega$), the current required is high. The actual temperature rise at the joint depends on the specific heat and thermal conductivity of the metals to be joined. Thus because they have high thermal conductivity, metals such as aluminum and copper require high heat concentrations. Similar as well as dissimilar metals can be joined by resistance welding. The workpieces are generally in the secondary circuit of a transformer. The transformer converts high-voltage, low-current power to low-voltage, high-current power. The magnitude of the current in resistance welding operations may be as high as 100,000 A, although the voltage is typically only 0.5–10 V.

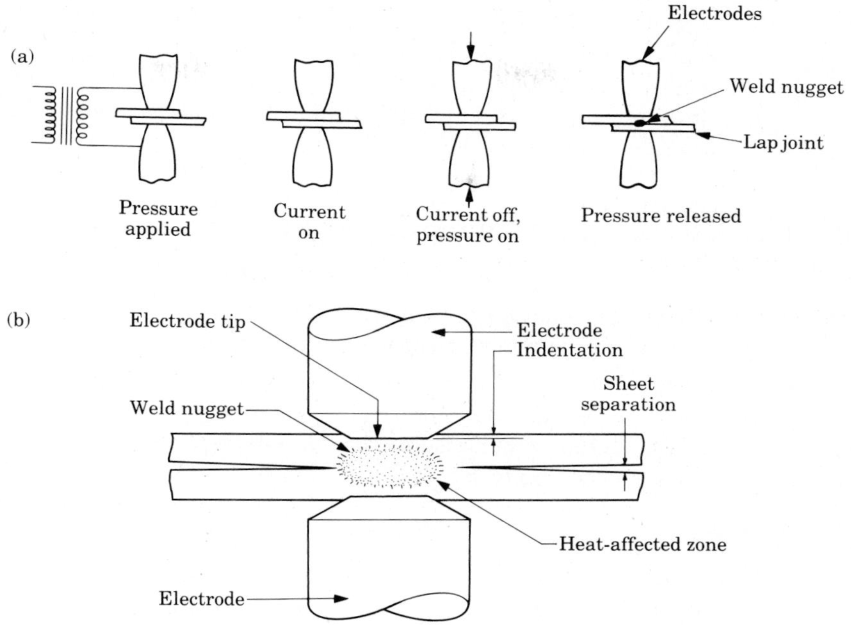

FIGURE 12.9

(a) Sequence in the resistance spot welding process. (b) Cross-section of a spot weld, showing weld nugget and light indentation by the electrode on sheet surfaces. This is one of the most common processes used in sheet-metal fabrication and automotive-body assembly.

12.6.1 Resistance spot welding

In *resistance spot welding* (RSW), the tips of two opposing solid cylindrical electrodes contact the lap joint of two sheet metals, and resistance heating produces a spot weld (Fig. 12.9a). In order to obtain a good bond in the *weld nugget*, pressure is also applied until the current is turned off. Accurate control and timing of the electric current and pressure are essential in resistance welding. The strength of the bond depends on surface roughness and the cleanliness of the mating surfaces. Thus oil, paint, or thick oxide layers should be removed before welding. The presence of uniform, thin oxide layers and other contaminants is not critical.

The *weld nugget* (Fig. 12.9b) is generally 6–10 mm (0.25–0.375 in.) in diameter. The surface of the weld spot has a slightly discolored indentation. Currents range from 3000 A to 40,000 A, depending on the materials being welded and their thickness. Spot welding is the simplest and most commonly used resistance welding process. Welding may be performed by means of single (most common) or multiple electrodes, and the required pressure is supplied through mechanical or pneumatic means.

Spot welding is widely used for fabricating sheet metal. Examples range from attaching handles to stainless-steel cookware, to rapid spot welding of automobile bodies with multiple electrodes. An automobile may have as many as 10,000 spot welds. Modern equipment used for spot welding is computer controlled for optimum timing of current and pressure, and the spot-welding guns are manipulated by programmable robots.

• Example 12.1: Heat generated in resistance spot welding.

Assume that two 1-mm (0.04-in.) thick steel sheets are being spot welded at a current of 5000 A and current flow time $t = 0.1$ s, and using electrodes 5 mm (0.2 in.) in diameter. Estimate the heat generated and its distribution in the weld zone.

SOLUTION. Let's assume that the effective resistance in this operation is 200 $\mu\Omega$. Then, according to Eq. (12.8),

$$\text{Heat} = (5000)^2(0.0002)(0.1) = 500 \text{ J}.$$

From the information given, we estimate the weld nugget volume to be 30 mm^3 (0.0018 in^3). If we assume the density for steel to be 8000 kg/m^3 (0.008 g/mm^3), the weld nugget has a mass of 0.24 g. Since the heat required to melt 1 g of steel is about 1400 J, the heat required to melt the weld nugget is (1400)(0.24) = 336 J. Consequently, the remaining heat (164 J) is dissipated into the metal surrounding the nugget.

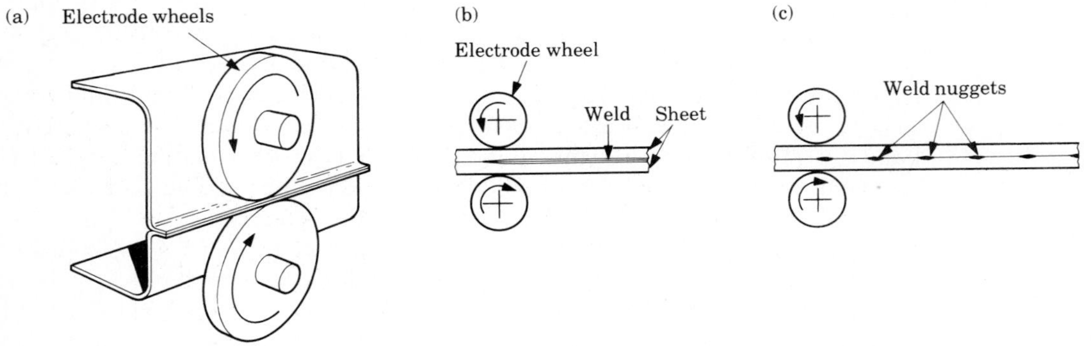

FIGURE 12.10
(a) Seam welding process, with rolls acting as electrodes. (b) Overlapping spots in a seam weld. (c) Roll spot welds.

12.6.2 Resistance seam welding

Resistance seam welding (RSEW) is a modification of resistance spot welding, wherein the electrodes are replaced by rotating wheels or rollers (Fig. 12.10). With continuous ac power supply, the electrically conducting rollers produce continuous spot welds whenever the current reaches a sufficiently high level in the ac cycle. These are actually overlapping spot welds and produce a joint that is liquid tight and gas tight (Fig. 12.10b). With intermittent application of current to the rollers, a series of spot welds at various intervals can be made along the length of the seam (Fig. 12.10c). This procedure is called roll spot welding. The RSEW process is used to make the longitudinal (side) seam of cans for household products, mufflers, gasoline tanks, and other containers. The typical welding speed is 25 mm/s (60 in./min) for thin sheet.

12.6.3 High-frequency resistance welding

High-frequency resistance welding (HFRW) is similar to seam welding, except that high-frequency current (up to 450 kHz) is employed. A typical application is making butt-welded tubing. In one method, the current is conducted through two sliding contacts (Fig. 12.11a) to the edges of roll-formed tubes. The heated edges are then pressed together by passing the tube through a pair of squeeze rolls. In another method, the roll-formed tube is subjected to high-frequency induction heating (Fig. 12.11b). Spiral pipe and tubing, and finned tubes (for heat exchangers), may be made by these techniques.

12.6.4 Resistance projection welding

In *resistance projection welding* (RPW), high electrical resistance at the joint is developed by embossing one or more projections (dimples) on one of the surfaces to

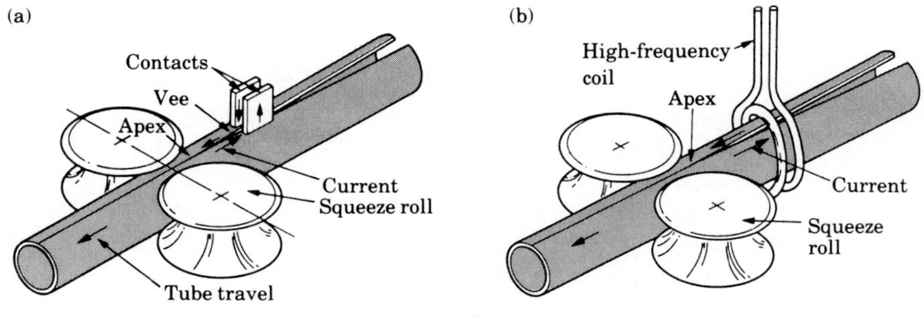

FIGURE 12.11
Methods of high-frequency butt welding of tubes.

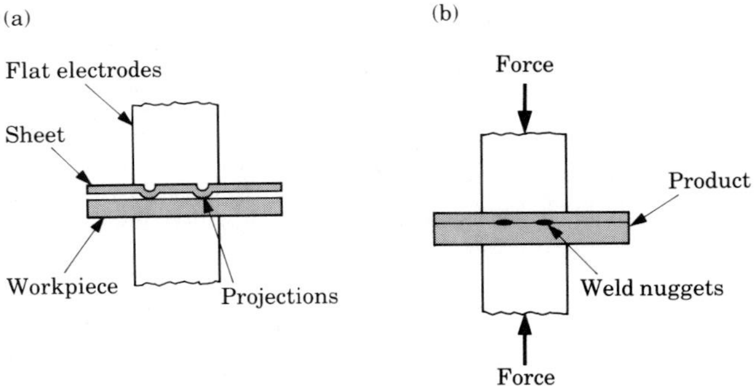

FIGURE 12.12 Schematic illustration of resistance projection welding: (a) before and (b) after. The projections are produced by embossing operations.

be welded (Fig. 12.12). High localized temperatures are generated at the projections, which are in contact with the flat mating part. The electrodes—made of copper-base alloys and water cooled to keep their temperature low—are large and flat. Weld nuggets, similar to those in spot welding, are formed as the electrodes exert pressure to compress the projections. The projections may be round or oval for design or strength purposes.

Spot welding equipment can be used for RPW by modifying the electrodes. Although embossing workpieces is an added expense, this process produces a number of welds in one stroke, extends electrode life, and is capable of welding metals of different thicknesses. Nuts and bolts are also welded to sheet and plate by this process, with projections that are produced by machining or forging. Joining a network of wires such as metal baskets, grills, oven racks, and shopping carts is also considered resistance projection welding because of the small contact area between crossing wires (grids).

12.6.5 Flash welding

In *flash welding* (FW), also called flash butt welding, heat is generated from the arc as the ends of the two members begin to make contact, developing an electrical resistance at the joint (Fig. 12.13). Because of the arc's presence, this process is also classified as arc welding. After the proper temperature is reached and the interface begins to melt, an axial force is applied at a controlled rate, and a weld is formed by plastic deformation (upsetting) of the joint. Some metal is expelled from the joint as a shower of sparks during the flashing process.

Because impurities and contaminants are squeezed out during this operation, the quality of the weld is good, although a significant amount of material may be burned off during the welding process. The joint may later be machined to improve its appearance. The machines for FW are usually automated and large, with a variety of power supplies ranging from 10 to 1500 kVA.

The flash welding process is suitable for end-to-end or edge-to-edge joining of similar or dissimilar metals 1–75 mm (0.05–3 in.) in diameter and sheet and bars

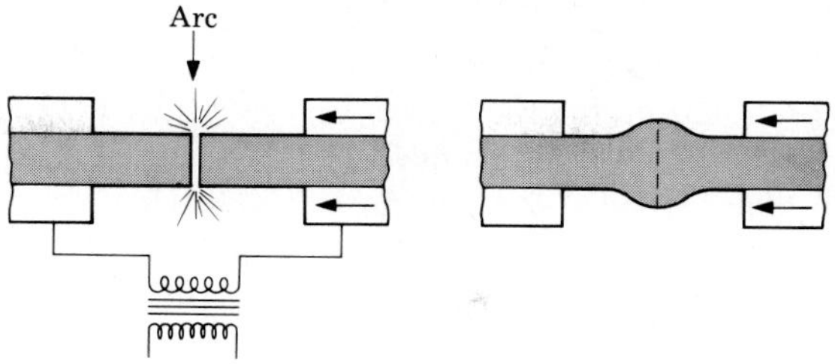

FIGURE 12.13
Flash welding process for end-to-end welding of solid rods or tubular parts.

0.02–25 mm (0.01–1 in.) thick. Thinner sections have a tendency to buckle under the axial force applied during welding. This process is also used to repair broken band-saw blades, with fixtures that are attached to the band-saw frame. This process can be automated for reproducible welding operations. Typical FW applications are joining pipe and tubular shapes for metal furniture and windows. It is also used for welding the ends of coils of sheet or wire in continuously operating rolling mills and wire-drawing equipment.

12.6.6 Stud welding

Stud welding (SW) is similar to flash welding and is also called stud arc welding. The stud, such as a small part or a threaded rod or hanger, serves as one of the electrodes while being joined to another member, which is usually a flat plate (Fig. 12.14). In order to concentrate the heat generated, prevent oxidation, and retain the molten metal in the weld zone, a disposable ceramic ring (ferrule) is placed around the joint. The equipment for stud welding can be automated, with various controls for arcing and applying pressure. Portable stud welding equipment is also available.

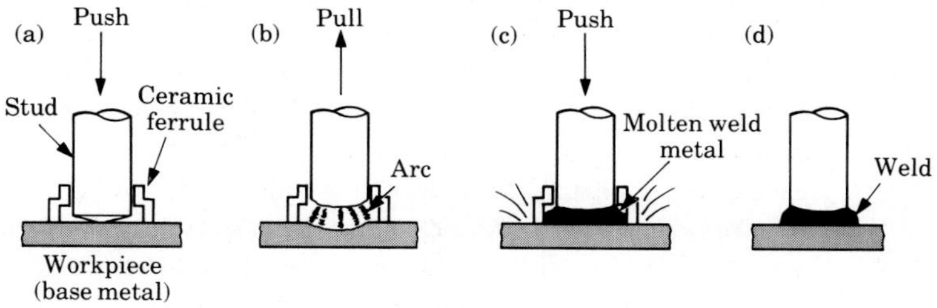

FIGURE 12.14
Sequence of operations in stud welding, which is used for welding bars, threaded rods, and various fasteners on metal plates.

12.6.7 Percussion welding

The resistance welding processes we have described require a transformer to meet power requirements. However, the electrical energy for welding may be stored in a condenser. *Percussion welding* (PEW) utilizes this technique, in which the power is discharged in a very short time (1–10 ms), developing high localized heat at the joint. This process is useful where heating of the components adjacent to the joint is to be avoided, such as in electronic components.

12.7 Solid-State Welding

In this section we describe processes in which joining takes place without fusion of the workpieces. Unlike the oxyfuel gas, arc, and resistance welding processes, no liquid (molten) phase is present in the joint. We also describe welding processes that use energy sources other than those discussed thus far, including laser and electron beams.

We can best demonstrate the principle of *solid-state bonding* with the following example. If two clean surfaces are brought into atomic contact with each other under sufficient pressure—and in the absence of oxide films and other contaminants—they form bonds and produce a strong joint. Heat and some movement of the mating surfaces by plastic deformation (as in forge and cold welding and roll bonding) may be employed to improve the strength of the joint.

Applying external heat improves the bond by diffusion (as in diffusion bonding). Small interfacial movements on the faying surfaces disturb the surfaces, breaking up oxide films and generating new and clean surfaces, thus improving the strength of the bond (as in ultrasonic welding). Heat may also be generated by friction, which is utilized in friction welding. Very high contact pressures are also utilized in joining processes, as in explosion welding.

12.7.1 Forge welding

In *forge welding* (FOW), both elevated temperature and pressure are applied to obtain a strong bond between the members being joined. The components are heated and pressed or hammered together with suitable tools, dies, or rollers. Local plastic deformation at the interface breaks up the oxide films and other contaminants, thus improving bond strength. However, compared to other welding processes, the resulting joint does not have particularly high load-bearing strength.

Forge welding is an ancient process, dating back to the period 1000–1 B.C. It has been practiced widely by blacksmiths using anvils and hammers, with charcoal as the heat source, making iron and steel chain links and medieval swords. Because of the difficulties involved in controlling the process—and the considerable skill and labor involved—this process has been largely replaced by various other joining methods.

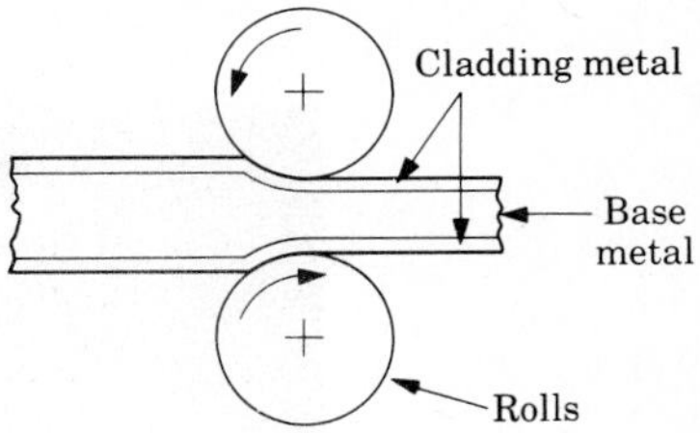

FIGURE 12.15
Schematic illustration of the roll bonding, or cladding, process.

12.7.2 Cold welding

In *cold welding* (CW), pressure is applied to the workpieces, either through dies or rolls. Because of the plastic deformation involved, it is necessary that at least one, but preferably both, of the mating parts be ductile. The interface is usually cleaned by wire brushing prior to welding.

However, in bonding two dissimilar metals that are mutually soluble, brittle intermetallic compounds may form, resulting in a weak and brittle joint. An example is the bonding of aluminum and steel, where a brittle intermetallic compound is formed at the interface. The best bond strength and ductility is obtained with two similar materials. Cold welding can be used to join small workpieces made of soft, ductile metals. Applications include electrical connections and welding wire stock.

Roll bonding. The pressure required for cold welding can be applied through a pair of rolls (Fig. 12.15). Hence the process is called *roll bonding*. Developed in the 1960s, roll bonding is used for manufacturing certain U.S. coins. The process can be carried out at elevated temperatures (hot roll bonding). Typical examples are cladding pure aluminum over aluminum-alloy sheet and stainless steel over mild steel for corrosion resistance.

• Example 12.2: Roll bonding of the U.S. quarter.

The technique used for manufacturing composite U.S. quarters is roll bonding of two outer layers of 75 percent copper–25 percent nickel (cupronickel), each 1.2 mm (0.048 in.) thick, with an inner layer of pure copper 5.1 mm (0.20 in.) thick. To obtain good bond strength, the faying surfaces are chemically cleaned and wire brushed. The strips are first rolled to a thickness of 2.29 mm (0.090 in.). A second rolling operation reduces the final thickness to 1.36 mm (0.0535 in.). The strips thus undergo a total reduction in thickness of 82 percent. Since volume constancy is maintained in plastic deformation, there is a major increase in the surface area between the layers, thus generating clean interfacial surfaces. This extension in surface area under the high pressure of the rolls, combined with the solid solubility of nickel in copper, produces a strong bond.

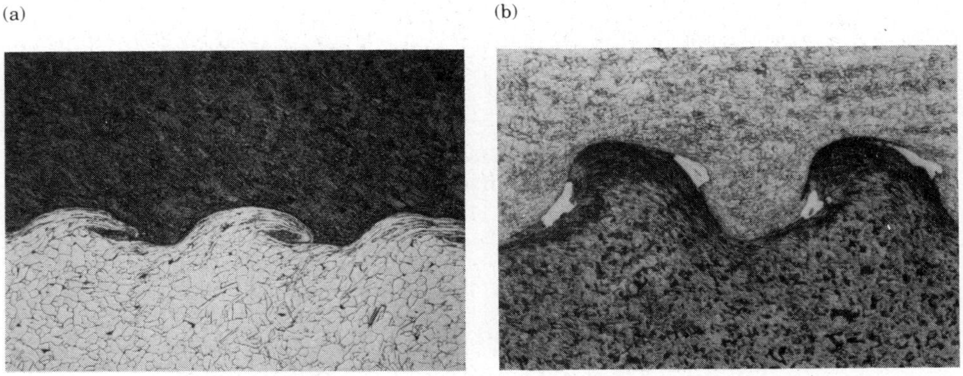

FIGURE 12.16
Cross-sections of explosion-welded joints: (a) titanium (top) on low-carbon steel (bottom) and (b) Incoloy 800 (iron–nickel base alloy) on low-carbon steel. *Source:* Courtesy of E. I. du Pont de Nemours & Co.

12.7.3 Explosion welding

In *explosion welding* (EXW), pressure is applied by detonating a layer of explosive that has been placed over one of the members being joined, called the *flyer plate*. The contact pressures developed are extremely high, and the plate's kinetic energy striking the mating member produces a wavy interface. This impact mechanically interlocks the two surfaces (Fig. 12.16). Cold pressure welding by plastic deformation also takes place. The flyer plate is placed at an angle, and any oxide films present at the interface are broken up and propelled from the interface. As a result, bond strength in explosion welding is very high.

This process was developed in the 1950s. It is particularly suitable for cladding plates and slab with dissimilar metals. Plates as large as 6 m × 2 m (20 ft × 7 ft) have been explosively clad. These plates may then be rolled into thinner sections. Tube and pipe are often joined to the holes in head plates of boilers and heat exchangers by placing the explosive inside the tube; explosion expands the tube. Explosion welding is inherently dangerous and requires safe handling by well-trained and experienced personnel.

12.7.4 Ultrasonic welding

In *ultrasonic welding* (USW), the faying surfaces of the two members are subjected to a static normal force and oscillating shearing (tangential) stresses. The shearing stresses are applied by the tip of a transducer (Fig. 12.17), similar to that used for ultrasonic machining. The frequency of oscillation generally ranges from 10 kHz to 75 kHz, although both lower and higher frequencies than these can be employed. The energy required increases with the thickness and hardness of the materials being joined. Proper coupling between the transducer and the tip (called *sonotrode*, from the word *sonic*, as contrasted to electrode) is important for efficient operation. These units are now operated with solid-state frequency converters.

The shearing stresses cause lateral movement and plastic deformation at the workpiece interfaces, breaking up oxide films and contaminants and thus allowing good contact and producing a strong bond. Temperatures generated in the weld zone are usually in the range of one-third to one-half the melting point (absolute scale) of the metals joined. Therefore no melting and fusion take place. In certain situations, however, temperatures can be sufficiently high to cause metallurgical changes in the weld zone.

The ultrasonic welding process is reliable and versatile. It can be used with a wide variety of metallic and nonmetallic materials, including dissimilar metals (bimetallic strips). It is used extensively in the electronics industry for lap welding of sheet, foil, and thin wire and in packaging with foils. The welding tip can be replaced with rotating disks for seam welding of structures.

12.7.5 Friction welding

In the joining processes that we have described so far, the energy required for welding, such as chemical, electrical, and ultrasonic, is supplied externally. In *friction welding* (FRW), the required heat for welding is, as the name implies, generated through friction at the interface of the two members being joined. Thus the source of energy is mechanical. You can demonstrate the significant rise in temperature from friction by rubbing your hands together fast or sliding down a rope rapidly.

In friction welding, one of the members remains stationary while the other is placed in a chuck or collet and rotated at a high constant speed. The two members

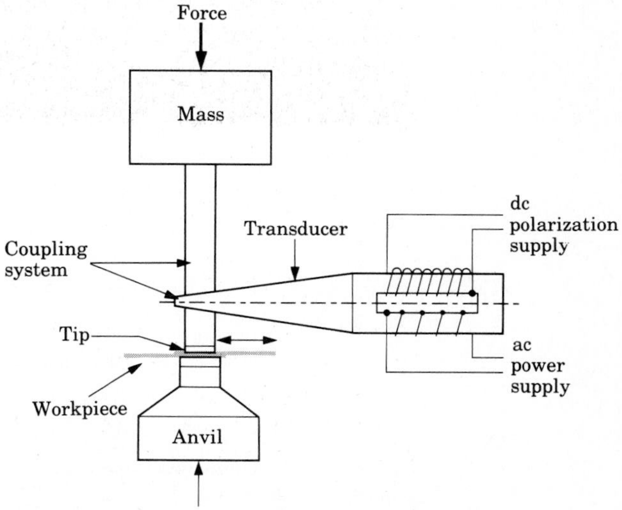

FIGURE 12.17 Components of an ultrasonic welding machine for lap welds. The lateral vibrations of the tool tip cause plastic deformation and bonding at the interface of the workpieces.

to be joined are then brought into contact under an axial force (Fig. 12.18). After sufficient contact is established, the rotating member is brought to a quick stop, so that the weld is not destroyed by shearing, while the axial force is increased. The rotating member must be clamped securely to the chuck or collet to resist both torque and axial forces without slipping.

The pressure at the interface and the resulting friction produce sufficient heat for melting and fusion to take place. The weld zone is usually confined to a narrow region, depending on (a) the amount of heat generated, (b) the thermal conductivity of the materials, and (c) the materials' mechanical properties at elevated temperatures. The shape of the welded joint depends on the rotational speed and the axial pressure applied. These factors must be controlled to obtain a uniformly strong joint. Oxides and other contaminants at the interface are removed by the radial movement of the hot metal at the interface (Fig. 12.18d).

Developed in the 1940s, friction welding can be used to join a wide variety of materials, provided that one of the components has some rotational symmetry. Solid, as well as tubular parts, can be joined by this method, with good joint strength. Solid steel bars up to 100 mm (4 in.) in diameter and pipes up to 250 mm (10 in.) outside diameter have been welded successfully by friction. The surface speed of the rotating member may be as high as 15 m/s (3000 ft/min). Because of the combined heat and pressure, the interface in FRW develops a flash by plastic deformation (upsetting) of the heated zone. This flash, if objectionable, can easily be removed by machining.

Inertia friction welding is a modification of FRW. The energy required for frictional heating in inertia friction welding is supplied through the kinetic energy of a flywheel. The flywheel is accelerated to the proper speed, the two members are

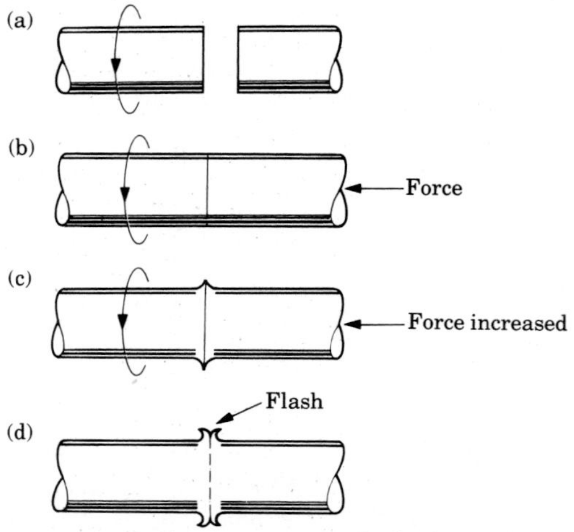

FIGURE 12.18
Sequence of operations in the friction welding process. (a) Left part is rotated at high speed. (b) Right part is brought into contact under an axial force. (c) Axial force is increased; flash begins to form. (d) Left part stops rotating. Weld is completed. Flash can be removed by machining.

brought into contact, and an axial force is applied. As friction at the interface slows the flywheel, the axial force is increased. The weld is completed when the flywheel comes to a stop. The timing of this sequence is important. If the timing is not properly controlled, weld quality will be poor. The rotating mass of inertia friction welding machines can be adjusted for applications requiring different levels of energy, which depend on workpiece size and properties.

12.7.6 Diffusion bonding

Diffusion bonding, or diffusion welding (DFW), is a solid-state joining process in which the strength of the joint results primarily from diffusion and, to a lesser extent, small plastic deformation of the faying surfaces. This process requires temperatures of about $0.5T_m$ (where T_m is the melting point of the metal on the absolute scale) in order to have a sufficiently high diffusion rate between the parts being joined. The bonded interface in DFW has essentially the same physical and mechanical properties as the base metal. Its strength depends on pressure, temperature, time of contact, and the cleanliness of the faying surfaces. These requirements can be lowered by using filler metal at the interfaces.

Although developed in the 1970s as a modern welding technology, the principle of diffusion bonding actually dates back centuries, when goldsmiths bonded gold over copper. Called filled gold, a thin layer of gold foil is obtained by hammering; the gold is then placed over copper, and a weight is placed on top of it. The assembly is then placed in a furnace and left until a good bond is obtained. Pressure may be applied by dead weights or a press, as well as by using differential gas pressure or utilizing relative thermal expansion of the parts to be joined. The parts are usually heated in a furnace or by electrical resistance. High-pressure autoclaves are also used for bonding complex parts.

Diffusion bonding is generally most suitable for dissimilar metal pairs. However, it is also used for reactive metals, such as titanium, beryllium, zirconium, and refractory metal alloys, and for composite materials. Because diffusion involves migration of the atoms across the joint, this process is slower than other welding processes. Although DFW is used for fabricating complex parts in low demand for the aerospace, nuclear, and electronics industries, it has been automated to make it suitable and economical for moderate-volume production. In Section 7.10 we described fabrication of sheet-metal structures by combining diffusion bonding with superplastic forming.

12.8 Electron-Beam Welding

In *electron-beam welding* (EBW), heat is generated by high-velocity narrow-beam electrons. The kinetic energy of the electrons is converted into heat as they strike the workpiece. This process requires special equipment to focus the beam on the

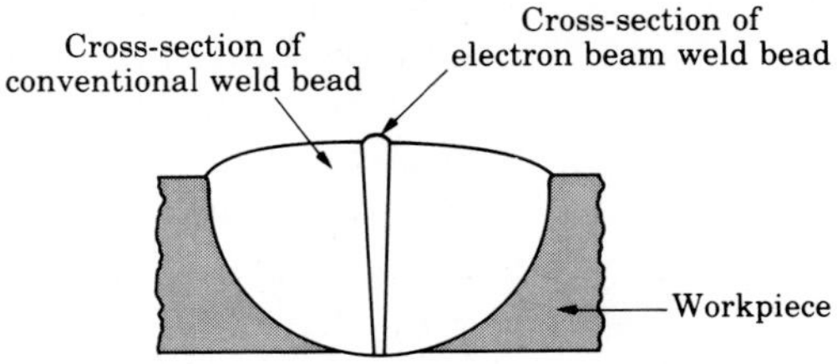

FIGURE 12.19 Comparison of the size of weld beads in conventional (tungsten-arc) and electron-beam welding.

workpiece in a vacuum. The higher the vacuum, the more the beam penetrates and the greater the depth-to-width ratio is. Almost any metal can be welded by EBW, with workpiece thicknesses ranging from foil to plate. The intense energy is also capable of producing holes in the workpiece (keyhole). Generally, no shielding gas, flux, or filler metal is required. Capacities of electron beam guns range up to 175 kV and 1000 milliamps.

Developed in the 1960s, EBW has the capability to make high-quality welds that are almost parallel sided, are deep and narrow, and have small heat-affected zones. Depth-to-width ratios range between 10:1 and 30:1. The size of welds made by EBW and conventional processes are compared in Fig. 12.19. Using servo controls, parameters can be controlled accurately, with welding speeds as high as 200 mm/s (40 ft/min). The electron beam can be projected a distance of several meters, thus making welding at otherwise inaccessible locations possible. Almost any metal can be butt or lap welded with this process, with thicknesses to as much as 150 mm (6 in.).

Distortion and shrinkage in the weld area is minimal, although the weld has a propensity to crack along its centerline. Precisely controlling the welding parameters can usually eliminate this situation. Electron-beam welding equipment generates x-rays, and hence proper monitoring and periodic maintenance are important.

12.9 Laser-Beam Welding

Laser-beam welding (LBW) utilizes a focused high-power coherent monochromatic (one wavelength) light beam as the source of heat. Because the beam has high energy density, it has deep penetrating power. It can be directed, shaped, and focused precisely on the workpiece. Consequently, this process is particularly suitable for welding narrow and deep joints.

Laser-beam welding can be used successfully on a variety of materials with thicknesses of up to 25 mm (1 in.) and is particularly effective on thin workpieces. Welding speeds range from 40 mm/s (8 ft/min), to as high as 1.3 m/s (250 ft/min) for thin metals. Because of the nature of the process, welding can be done in otherwise inaccessible locations. Laser-beam welding produces welds of good quality, with minimum shrinkage and distortion, and with depth-to-width ratios as high as 30:1. However, as in EBW, this high ratio can cause centerline cracking in the weld.

The major advantages of LBW over EBW are:

- The beam can be transmitted through air, so a vacuum is not required.
- Because laser beams can be shaped, directed, and focused optically, the process can easily be automated.
- The beams do not generate x-rays.
- The quality of the weld is better, with less tendency for incomplete fusion, spatter, and porosity.

12.10 The Welded Joint

A typical fusion weld joint is shown in Fig. 12.20, where three distinct zones can be identified: (a) the *base metal*, that is, the metal to be welded; (b) the *heat-affected zone* (HAZ); and (c) the *weld metal*, that is, the region that has melted during welding.

The metallurgy and properties of the second and third zones depend strongly on the metals joined, the welding process, filler metals used, if any, and process

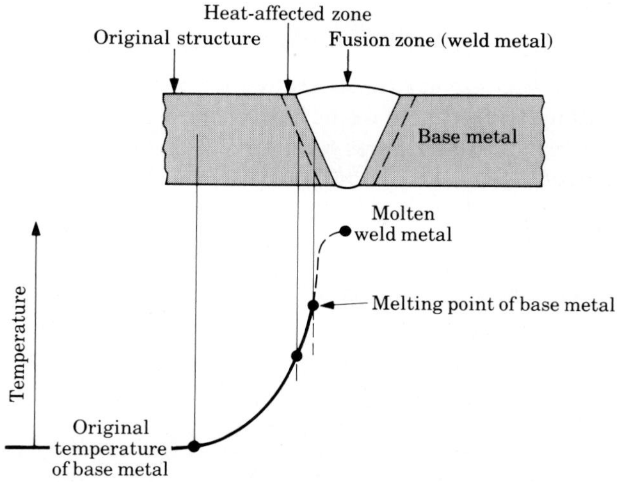

FIGURE 12.20
Characteristics of a typical fusion weld zone in oxyfuel gas and arc welding.

variables. A joint produced without a filler metal is called *autogenous*, and the weld zone is composed of the molten and resolidified base metal. A joint made with a filler metal has a central zone called the weld metal and is composed of a mixture of the base and weld metals. The study of a weld joint requires an understanding of metal and alloy solidification and phase diagrams. This subject is complex and we merely introduce it here.

12.10.1 Solidification of the weld metal

After applying heat and introducing filler metal, if any, into the weld area, the molten weld joint is allowed to cool naturally to ambient temperature. The solidification process is similar to that in casting and begins with the formation of columnar (dendritic) grains. These grains are relatively long and form parallel to the heat flow. Because metals are much better heat conductors than the surrounding air, the grains lie parallel to the plane of the two plates or sheets being welded (Fig. 12.21a). The grains in a shallow weld are shown in Figs. 12.21(b) and 12.22. Grain structure and size depend on the specific alloy, the welding process, and the filler metal used.

The weld metal is basically a cast structure and, because it has cooled slowly, it generally has coarse grains. Consequently, this structure has generally low strength, toughness, and ductility. However, the proper selection of filler-metal composition or heat treatments following welding can improve the joint's mechanical properties. The results depend on the particular alloy, its composition, and the thermal cycling to which the joint is subjected. Cooling rates may, for example, be controlled and reduced by preheating the general weld area prior to welding. Preheating is particularly important for metals with high thermal conductivity, such as aluminum and copper; otherwise, the heat during welding dissipates rapidly.

12.10.2 Heat-affected zone

The *heat-affected zone* is within the base metal itself. It has a microstructure different from that of the base metal before welding, because it has been subjected to elevated temperatures for a period of time during welding. The portions of the base metal that are far enough away from the heat source do not undergo any structural changes during welding.

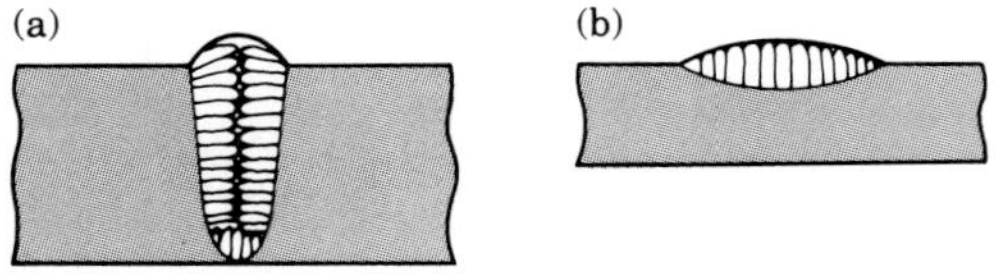

FIGURE 12.21
Grain structure in (a) a deep and (b) a shallow weld. Note that the grains in the solidified weld metal are perpendicular to the surface of the base metal.

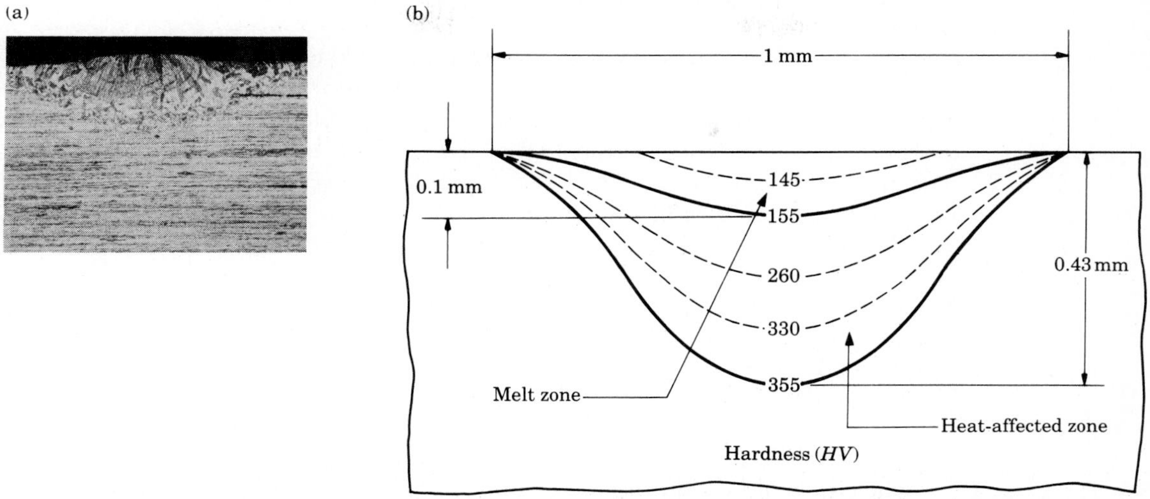

FIGURE 12.22
(a) Weld bead on a cold-rolled nickel strip produced by a laser beam. (b) Microhardness profile across the weld bead. Note the softer condition of the weld bead compared to the base metal. *Source:* IIT Research Institute.

The properties and microstructure of HAZ depend on (a) the rate of heat input (usually expressed per unit length of weld) and cooling; and (b) the temperature to which this zone was raised. The HAZ and the corresponding phase diagram for 0.3 percent carbon steel are shown in Fig. 12.23. In addition to metallurgical factors (such as original grain size, grain orientation, and degree of prior cold work), the

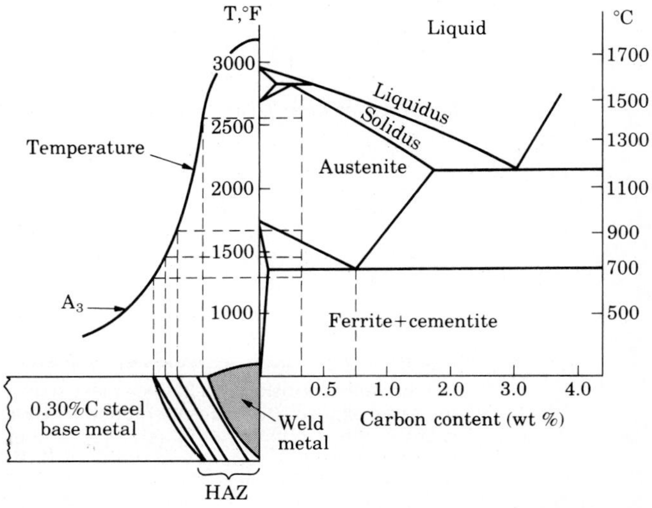

FIGURE 12.23
Schematic illustration of various regions in a fusion weld zone and the corresponding phase diagram for 0.30 percent carbon steel. *Source:* American Welding Society.

specific heat and thermal conductivity of the metals influence the HAZ's size and characteristics.

The strength and hardness of the heat-affected zone depend partly on how the original strength and hardness of the particular alloy were developed prior to welding. They may have been developed by cold working, solid-solution strengthening, or precipitation hardening or by various heat treatments. Of these strengthening methods, the simplest to analyze is base metal that has been cold worked, say, by cold rolling or forging.

The heat applied during welding recrystallizes the elongated grains (preferred orientation) of the cold-worked base metal. Grains that are away from the weld metal will recrystallize into fine equiaxed grains. However, grains close to the weld metal, having been subjected to elevated temperatures for a longer period of time, will grow. This growth will result in a region that is softer and has less strength. Such a joint will be weakest in its heat-affected zone. The grain structure of such a weld—exposed to corrosion by chemical reaction—is shown in Fig. 12.24. The center vertical line is where the two workpieces meet.

The effects of heat during welding on the HAZ for joints made with dissimilar metals, and for alloys strengthened by other methods, are complex and beyond the scope of this text. Details can be found in the more specialized texts listed in the bibliography at the end of this chapter.

12.10.3 Weld quality

Because of its history of thermal cycling and attendant microstructural changes, a welded joint may develop certain imperfections and discontinuities.

Porosity in welds is caused by trapped gases released during solidification of the weld area, chemical reactions during welding, or contaminants. Most welded joints contain some porosity, which is generally spherical in shape or in the form of elongated pockets. The distribution of porosity in the weld zone may be random, or it may be concentrated in a certain region.

Porosity in welds can be reduced by the following methods:

a) Proper selection of electrodes and filler metals.

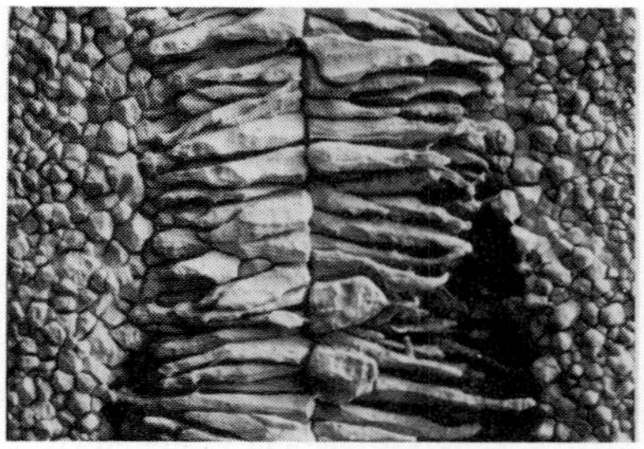

FIGURE 12.24
Intergranular corrosion of a 310 stainless steel welded tube after exposure to a caustic solution. The weld line is at the center of the photograph. Scanning electron micrograph at 20×. *Source:* Courtesy of B. R. Jack, Allegheny Ludlum Steel Corp.

b) Improving welding techniques, such as preheating the weld area or increasing the rate of heat input.
c) Proper cleaning and preventing contaminants from entering the weld zone.

Slag inclusions are compounds such as oxides, fluxes, and electrode-coating materials that are trapped in the weld zone. If shielding gases are not effective during welding, contamination from the environment may also contribute to such inclusions. Welding conditions are important, and with proper techniques the molten slag will float to the surface of the molten weld metal and not be entrapped. Slag inclusions may be prevented by:

a) Cleaning the weld-bead surface before the next layer is deposited by using a hand or power wire brush.
b) Providing adequate shielding gas.
c) Changing the type of electrode.

Incomplete fusion (or lack of fusion) produces poor weld beads, such as those shown in Fig. 12.25. A better weld can be obtained by:

a) Raising the temperature of the base metal.
b) Cleaning the weld area prior to welding.
c) Changing the joint design and type of electrode.
d) Providing adequate shielding gas.

Incomplete penetration occurs when the depth of the welded joint is insufficient. Penetration can be improved by:

a) Increasing the heat input.
b) Lowering travel speed during welding.
c) Changing the joint design.

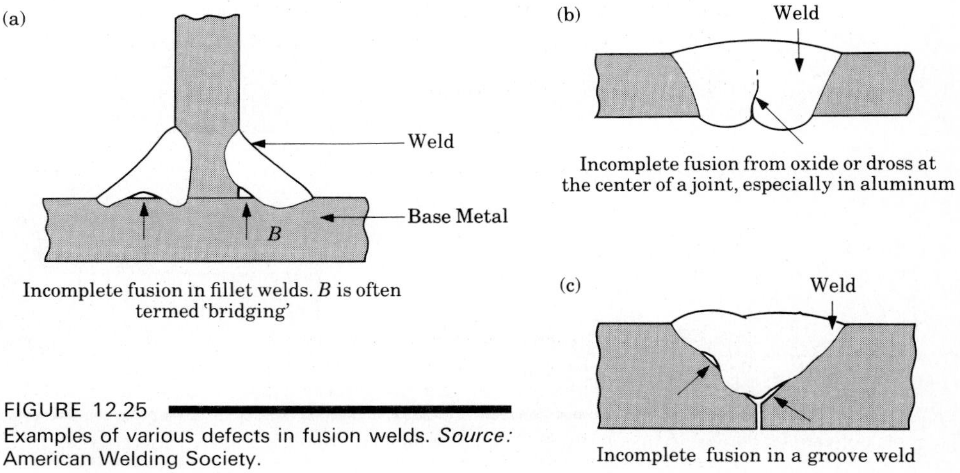

FIGURE 12.25 Examples of various defects in fusion welds. *Source:* American Welding Society.

Weld profile is important not only because of its effects on the strength and appearance of the weld, but also because it can indicate incomplete fusion or the presence of slag inclusions in multiple-layer welds (Fig. 12.26). *Underfilling* results when the joint is not filled with the proper amount of weld metal. *Undercutting* results from melting away the base metal and subsequently generating a groove in the shape of a sharp recess or notch. Unless it is not deep or sharp, an undercut can act as a stress raiser and reduce the fatigue strength of the joint—and may lead to premature failure. *Overlap* is a surface discontinuity generally caused by poor welding practice and selection of the wrong materials. A proper weld is shown in Fig. 12.26(c).

Cracks may occur in various locations and directions in the weld area. The types of cracks are typically longitudinal, transverse, crater, underbead, and toe cracks (Fig. 12.27). These cracks generally result from a combination of the following factors:

a) Temperature gradients that cause thermal stresses in the weld zone.
b) Variations in the composition of the weld zone that cause different contractions.
c) Embrittlement of grain boundaries by segregation of elements, such as sulfur, to the grain boundaries as the solid–liquid boundary moves when the weld metal begins to solidify.
d) Hydrogen embrittlement.
e) Inability of the weld metal to contract during cooling—a situation similar to hot tears that develop in castings.

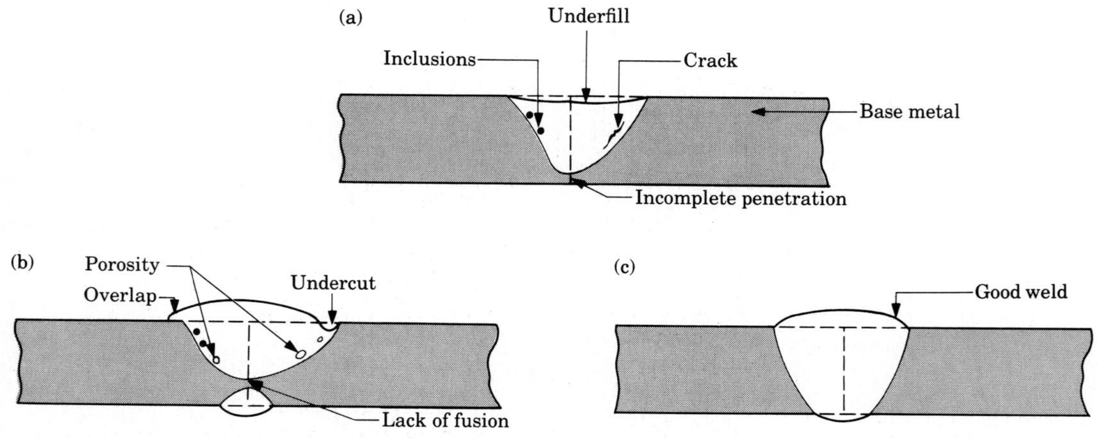

FIGURE 12.26
Examples of various defects in fusion welds. *Source:* American Welding Society.

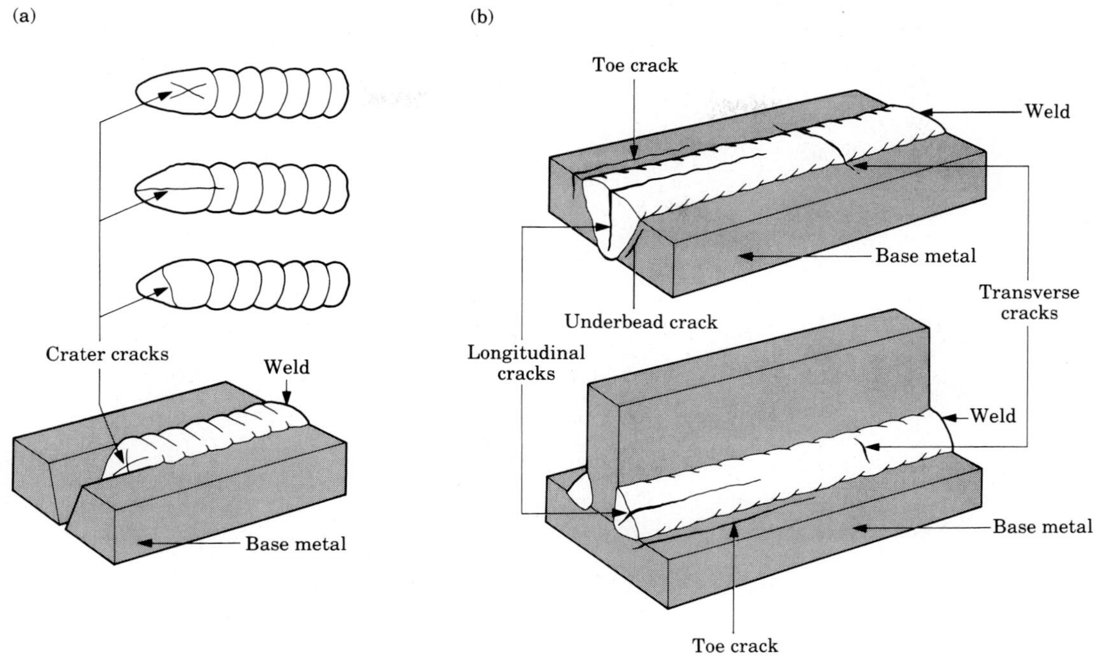

FIGURE 12.27
Types of cracks in welded joints caused by thermal stresses that develop during solidification and contraction of the weld bead and the welded structure. (a) Crater cracks. (b) Various types of cracks in butt and T joints.

Cracks are classified as *hot* or *cold cracks*. Hot cracks occur while the joint is still at elevated temperatures. Cold cracks develop after the weld metal has solidified. Some crack-prevention measures are:

a) Change the weld design to minimize stresses from shrinkage during cooling.
b) Change welding-process parameters, procedures, and sequence.
c) Preheat components being welded.
d) Avoid rapid cooling of the components after welding.

In describing the anisotropy of plastically deformed metals, we stated that because of the alignment of nonmetallic impurities and inclusions (stringers) the workpiece is weaker when tested in its thickness direction. This condition is particularly evident in rolled plates and structural shapes. In welding such components, *lamellar tears* may develop because of shrinkage of the restrained

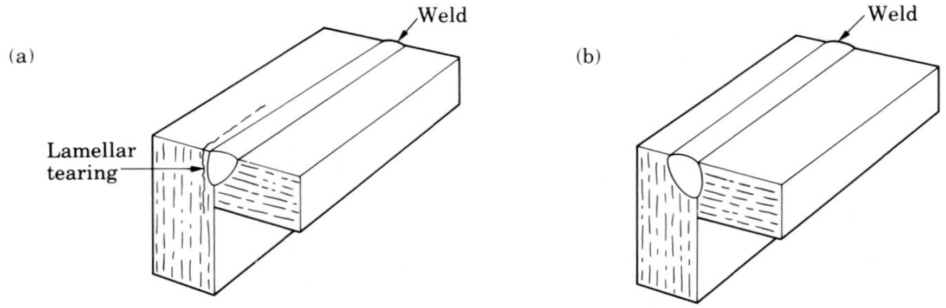

FIGURE 12.28
(a) Lamellar tears in heavy welded construction, showing lack of ductility of the vertical plate in its thickness direction, resulting from the presence of stringers (mechanical fibering). (b) Avoiding tears by changing the shape of the weld groove, with deeper penetration into the vertical plate.

members in the structure during cooling (Fig. 12.28). Such tears can be avoided by providing for shrinkage of the members or by changing the joint design to make the weld bead penetrate the weaker member farther.

Surface damage and irregularities. During welding, some of the metal may spatter and be deposited as small droplets on adjacent surfaces. In arc welding processes, the electrode may inadvertently touch the parts being welded at places not in the weld zone (arc strikes). Such surface defects may be objectionable for reasons of appearance or subsequent use of the welded part. If severe, these defects may adversely affect the properties of the welded structure, particularly for notch-sensitive metals. Using proper welding techniques and procedures is important in avoiding surface damage.

Residual stresses. Because of localized heating and cooling during welding, expansion and contraction of the weld area causes residual stresses in the workpiece. Residual stresses can cause (a) distortion, warping, and buckling of the welded parts (Fig. 12.29); (b) stress-corrosion cracking; and (c) further distortion if a portion of the welded structure is subsequently removed, say, by machining or sawing.

We can best describe the type and distribution of residual stresses in welds by reference to Fig. 12.30(a). When two plates are being welded, a long narrow region is subjected to elevated temperatures, whereas the plates as a whole are essentially at ambient temperature. As the weld is completed and time elapses, the heat from the weld area dissipates laterally to the plates as the weld area cools. The plates thus begin to expand longitudinally while the welded length begins to contract.

These two opposing effects cause residual stresses that are typically distributed as shown in Fig. 12.30(b). Note that the magnitude of compressive residual stresses in the plates diminishes to zero at a point away from the weld area. Because no external forces are acting on the welded plates, the tensile and compressive forces represented by these residual stresses must balance each other.

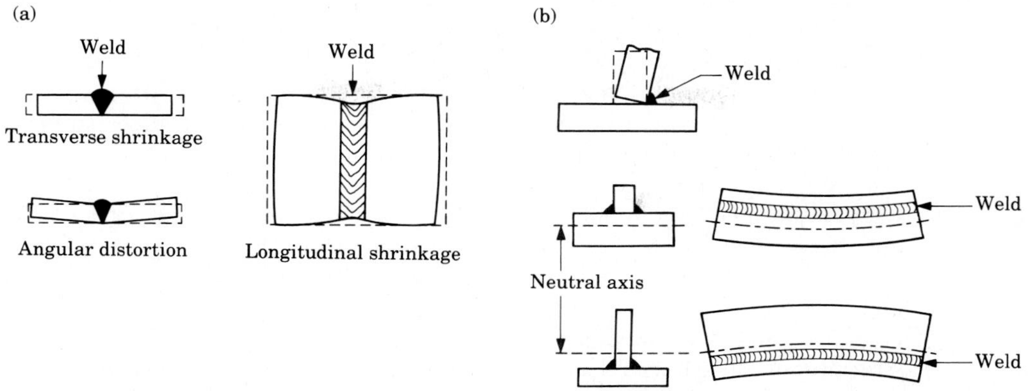

FIGURE 12.29 Distortion of parts after welding: (a) butt joints and (b) fillet welds. Distortion is caused by differential thermal expansion and contraction of different parts of the welded assembly. Warping can be reduced or eliminated by proper fixturing of the parts prior to welding.

In complex welded structures, residual stress distributions are three dimensional and difficult to analyze. The preceding example involves two plates that are not restrained from movement. In other words, the plates are not an integral part of a larger structure. If they are restrained, reaction stresses will be generated because the plates are not free to expand or contract. This situation arises particularly in structures with high stiffness.

Stress relieving of welds. The problems caused by residual stresses, such as distortion, buckling, or cracking, can be avoided by preheating the base metal or the parts to be welded. Preheating improves weldability by reducing the cooling rate

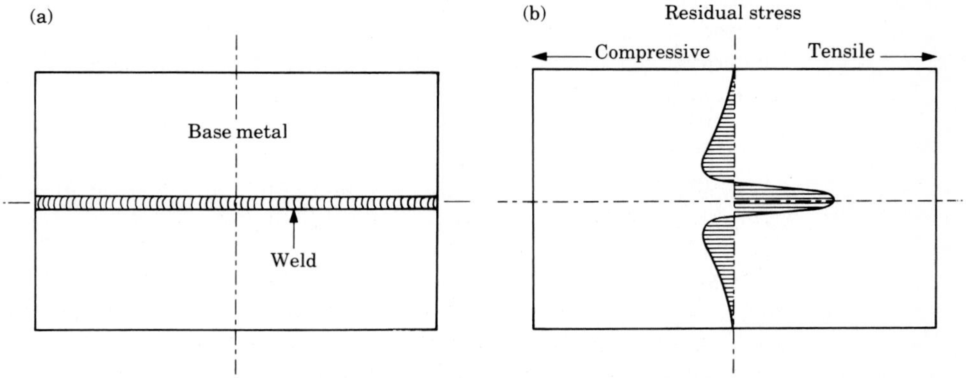

FIGURE 12.30 Residual stresses developed in a straight butt joint.

and the level of thermal stresses (by reducing the elastic modulus). This technique also reduces shrinkage and possible cracking of the joint. The workpieces may be heated in a furnace or electrically or inductively, and for thin sections, by radiant lamps or hot-air blast. For optimum results, preheating temperatures and cooling rates must be controlled carefully in order to maintain acceptable strength and toughness in the welded structure.

Residual stresses can be reduced by *stress relieving* the welded structure. The temperature and time required for stress relieving depend on the type of material and magnitude of the residual stresses developed. These parameters must be selected carefully in order to retain the base metal's original properties of strength and toughness. For structures that are too large to be stress relieved in a furnace, the technique of in-situ (meaning "in the original shape") stress relieving is utilized. The structure (such as a vessel) is completely insulated, and the entire structure is used as a furnace to bring it up to the required temperature. The technique of partial stress relief may also be employed. In this case, only the immediate areas surrounding the welds are subjected to the stress-relief cycle.

Other methods of stress relieving include peening, hammering, or surface rolling the weld bead area. These processes induce compressive residual stresses, thus reducing or eliminating tensile residual stresses in the weld. For multilayer welds, the first and last layers should not be peened in order to protect them against possible peening damage.

Residual stresses can also be relieved, or reduced, by plastically deforming of the structure by a small amount. This technique can be used in some welded structures, such as pressure vessels, by pressurizing the vessels internally (proof-stressing). In order to reduce the possibility of sudden fracture under high internal pressure, the weld must be made properly and be free from notches and discontinuities, which could act as points of stress concentration.

Vibratory means of stress relieving in which the welded part is vibrated at one of its resonant frequencies may also be used. This technique is relatively new, and its results are controversial. The metallurgical structure and hardness of the weld area is not affected by this process. In addition to stress relieving, welds may also be heat treated by various techniques in order to modify their properties. These techniques include annealing, normalizing, or quenching and tempering of steels and solution treatment and aging of various alloys.

12.10.4 Testing welded joints

As in all manufacturing processes, the quality of a welded joint is established by testing. Several standardized tests and test procedures have been established and are available from organizations such as ASTM, AWS, ANSI, ASME, ASCE, and federal agencies. Welded joints may be tested either destructively or nondestructively. Each technique has certain capabilities, sensitivity, limitations, reliability, and need for special equipment and operator skill.

Destructive techniques. Five methods of destructively testing welded joints are commonly used. Longitudinal and transverse *tension tests* are performed on specimens removed from actual welded joints and from the weld metal area.

Specimens in the *tension-shear test* are specially prepared to simulate actual welded joints and procedures. The specimens are subjected to tension, and the shear strength of the weld metal and the location of fracture are determined.

Several *bend tests* have been developed to determine the ductility and strength of welded joints. In one test, the welded specimen is bent around a fixture (wrap-around bend test). In another, the specimens are tested in three-point transverse bending. These tests help establish the relative ductility and strength of welded joints. *Fracture toughness tests* commonly utilize impact testing techniques. Charpy V-notch specimens are prepared and tested for toughness. Other toughness tests include the drop-weight test in which the energy is supplied by a falling weight.

In addition to mechanical tests, welded joints may be subjected to *corrosion and creep resistance tests*. Because of the difference in the composition and microstructure of the materials in the weld zone, preferential corrosion may take place there. *Spot welded joint tests* may be used to determine weld-nugget strength using the (a) tension-shear, (b) cross-tension, (c) twist, and (d) peel tests. Because they are easy to perform and inexpensive, tension-shear tests are commonly used in fabricating facilities.

Nondestructive techniques. Welded structures often have to be tested non-destructively, particularly for critical applications where weld failure can be catastrophic, such as in pressure vessels, load-bearing structural members, and power plants. Nondestructive testing techniques usually consist of visual, radiographic, magnetic particle, liquid penetrant, and ultrasonic methods.

12.11 Weldability

We may define *weldability* of a metal as its capacity to be welded into a specific structure that has certain properties and characteristics and that will satisfactorily meet its service requirements. Weldability involves a large number of variables, making generalizations difficult. Material characteristics—such as alloying elements, impurities, inclusions, grain structure, and processing history—of the base metal and filler metal are important.

Because of the melting, solidification, and microstructural changes involved, a thorough knowledge of the phase diagram and the response of the metal or alloy to elevated temperatures over a period of time is essential. Also influencing weldability are the mechanical and physical properties of strength, toughness, ductility, notch sensitivity, elastic modulus, specific heat, melting point, thermal expansion, surface-tension characteristics of the molten metal, and corrosion.

Preparation of surfaces for welding is important, as are the nature and properties of surface oxide films and adsorbed gases. The welding process employed significantly affects the temperatures developed and their distribution in the weld zone. Other factors are shielding gases, fluxes, moisture content of the coating on electrodes, welding speed, welding position, cooling rate, preheating, and post-welding techniques (such as stress relieving and heat treating).

The following list states generally the weldability of specific metals, which can vary if special welding techniques are used.

a) Plain-carbon steels: Excellent for low-carbon steels; fair to good for medium-carbon steels; poor for high-carbon steels.
b) Low-alloy steels: Depends greatly on composition.
c) High-alloy steels: Generally good under well-controlled conditions.
d) Stainless steels: Weldable by various processes.
e) Aluminum alloys: Weldable at a high rate of heat input.
f) Copper alloys: Similar to that of aluminum alloys.
g) Magnesium alloys: Weldable with the use of protective shielding gas and fluxes.
h) Nickel alloys: Similar to that of stainless steels.
i) Titanium alloys: Weldable with the proper use of shielding gases.
j) Tantalum: Similar to that of titanium.
k) Tungsten: Weldable under well-controlled conditions.
l) Molybdenum: Similar to that of tungsten.
m) Columbium: Good.
n) Beryllium: Weldable under well-controlled conditions.

12.12 Welding Design and Process Selection

Besides the material characteristics that we have described thus far, selection of a joint and a welding process (Tables 12.1 and 12.2) involves the following considerations:

a) Configuration of the parts or structure to be welded and their thickness and size.
b) The methods used to manufacture component parts.
c) Service requirements, such as the type of loading and stresses generated.
d) Location, accessibility, and ease of welding.
e) Effects of distortion and discoloration.
f) Appearance.
g) Costs involved in edge preparation, welding, and post-processing of the weld, including machining and finishing operations.

As in all manufacturing processes, the optimum choice is the one that meets all design and service requirements at minimum cost.

Standardized symbols used in engineering drawings to describe the type of weld and its characteristics are shown in Fig. 12.31. These symbols identify the type of

TABLE 12.1
OVERVIEW OF COMMERCIAL JOINING PROCESSES*

MATERIAL	THICKNESS	JOINING PROCESS															B							
		SMAW	SAW	GMAW	FCAW	GTAW	PAW	ESW	EGW	RW	FW	OFW	DFW	FRW	EBW	LBW	TB	FB	IB	RB	DB	IRB	DFB	S
Carbon steel	S	X	X	X		X				X	X	X			X	X	X	X	X	X	X	X	X	X
	I	X	X	X	X	X				X	X	X		X	X	X	X	X	X	X	X		X	X
	M	X	X	X	X					X	X	X		X	X	X	X	X	X				X	
	T	X	X	X	X			X	X		X	X		X	X			X					X	
Low-alloy steel	S	X	X	X		X				X	X	X	X		X	X	X	X	X	X	X	X	X	X
	I	X	X	X	X	X				X	X		X	X	X	X	X	X	X				X	X
	M	X	X	X	X						X		X	X	X	X	X	X	X				X	
	T	X	X	X	X			X		X	X		X	X	X		X					X		
Stainless steel	S	X	X	X		X	X			X	X	X	X		X	X	X	X	X	X	X	X	X	X
	I	X	X	X	X	X	X			X	X		X	X	X	X	X	X	X				X	X
	M	X	X	X	X		X				X		X	X	X	X	X	X	X				X	
	T	X	X	X	X			X			X		X	X	X			X					X	
Cast iron	I	X										X					X	X	X				X	X
	M	X	X	X	X							X					X	X	X				X	X
	T	X	X	X	X							X						X					X	
Nickel and alloys	S	X		X		X	X			X	X	X			X	X	X	X	X	X	X	X	X	X
	I	X	X	X		X	X			X	X			X	X	X	X	X	X				X	X
	M	X	X	X			X				X			X	X	X	X	X					X	
	T	X		X				X			X			X	X			X					X	
Aluminum and alloys	S	X		X		X	X			X	X	X	X	X	X	X	X	X	X	X	X	X	X	X
	I	X		X		X				X	X		X	X	X	X	X	X			X		X	X
	M	X		X		X					X			X	X		X	X			X		X	
	T	X		X				X	X		X				X			X					X	
Titanium and alloys	S			X		X	X			X	X		X		X	X		X	X			X	X	
	I			X		X	X				X		X	X	X	X		X					X	
	M			X		X	X				X		X	X	X	X		X					X	
	T			X							X		X		X		X	X					X	

(continued)

TABLE 12.1
OVERVIEW OF COMMERCIAL JOINING PROCESSES* *(continued)*

MATERIAL	THICKNESS	JOINING PROCESS																						
																	B							
		SMAW	SAW	GMAW	FCAW	GTAW	PAW	ESW	EGW	RW	FW	OFW	DFW	FRW	EBW	LBW	TB	FB	IB	RB	DB	IRB	DFB	S
Copper and alloys	S			X		X	X				X				X		X	X	X	X			X	X
	I			X			X				X			X	X		X	X		X			X	X
	M			X							X			X	X		X	X					X	
	T			X							X				X			X					X	
Magnesium and alloys	S			X		X				X					X	X	X	X			X		X	
	I			X		X				X	X			X	X	X	X	X			X		X	
	M			X							X			X	X	X		X					X	
	T			X							X				X									
Refractory alloys	S			X		X	X			X	X				X		X	X	X	X		X	X	
	I			X			X				X				X		X	X					X	
	M									X	X													
	T																							

* This table presented as a general survey only. In selecting processes to be used with specific alloys, the reader should refer to other appropriate sources of information. *Source:* Courtesy of the American Welding Society.

LEGEND

PROCESS CODE

SMAW—Shielded Metal-Arc Welding
SAW—Submerged Arc Welding
GMAW—Gas Metal-Arc Welding
FCAW—Flux-Cored Arc Welding
GTAW—Gas Tungsten-Arc Welding
PAW—Plasma Arc Welding
ESW—Electroslag Welding
EGW—Electrogas Welding
RW—Resistance Welding
FW—Flash Welding
OFW—Oxyfuel Gas Welding
DFW—Diffusion Welding
FRW—Friction Welding
EBW—Electron Beam Welding
LBW—Laser Beam Welding
B—Brazing
- TB—Torch Brazing
- FB—Furnace Brazing
- IB—Induction Brazing
- RB—Resistance Brazing
- DB—Dip Brazing
- IRB—Infrared Brazing
- DFB—Diffusion Brazing

S—Soldering

THICKNESS

S—Sheet: up to 3 mm ($\frac{1}{8}$ in.)
I—Intermediate: 3 to 6 mm ($\frac{1}{8}$ to $\frac{3}{4}$ in.)
M—Medium: 6 to 19 mm ($\frac{1}{4}$ to $\frac{3}{4}$ in.)
T—Thick: 19 mm ($\frac{3}{4}$ in.) and up

TABLE 12.2

GENERAL CHARACTERISTICS OF JOINING PROCESSES

PROCESS	OPERATION	ADVANTAGE	SKILL LEVEL REQUIRED	WELDING POSITION	CURRENT TYPE	DISTOR-TION*	COST OF EQUIP-MENT
SMAW	Manual	Portable and flexible	High	All	ac, dc	1 to 2	Low
SAW	Automatic	High deposition	Medium to low	Flat and horizontal	ac, dc	1 to 2	Medium
GMAW	Semiautomatic or automatic	Most metals	High to low	All	dc	2 to 3	Medium to high
GTAW	Manual or automatic	Most metals	High to low	All	ac, dc	2 to 3	Medium
FCAW	Semiautomatic or automatic	High deposition	High to low	Flat and horizontal	dc	1 to 3	Medium
OFW	Manual	Portable and flexible	High	All	—	2 to 4	Low
EBW	Semiautomatic or automatic	Most metals	Medium to high	All	—	3 to 5	High

*1, highest; 5, lowest.

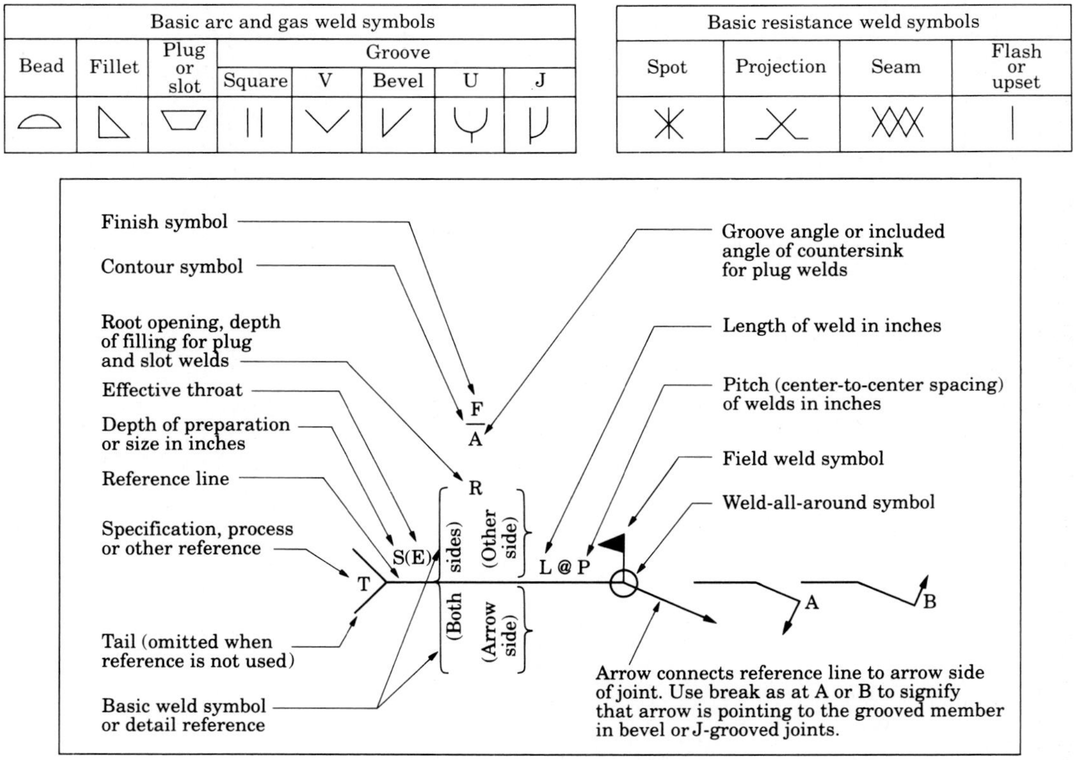

FIGURE 12.31

Standard identification and symbols for welds.

weld, groove design, weld size and length, welding process, sequence of operations, and various other information. Further information about these symbols can be found in the references cited at the end of this chapter.

• Example 12.3: Weld design selection.

The figures below show three different types of weld design. In example (a), the two vertical joints can be welded externally or internally. Full-length external welding takes considerable time and requires more weld material than the alternative design, which consists of intermittent internal welds. Moreover, in the alternative method, the appearance of the structure is improved and distortion is reduced. In example (b), although both designs require the same amount of weld material and welding time, it can be shown that the design on the right can carry three times the moment M of the one on the left.

In example (c), the weld on the left requires about twice the amount of weld material than the design on the right. Note also that because more material must be machined, the design on the left will require more time for edge preparation and more base metal will be wasted.

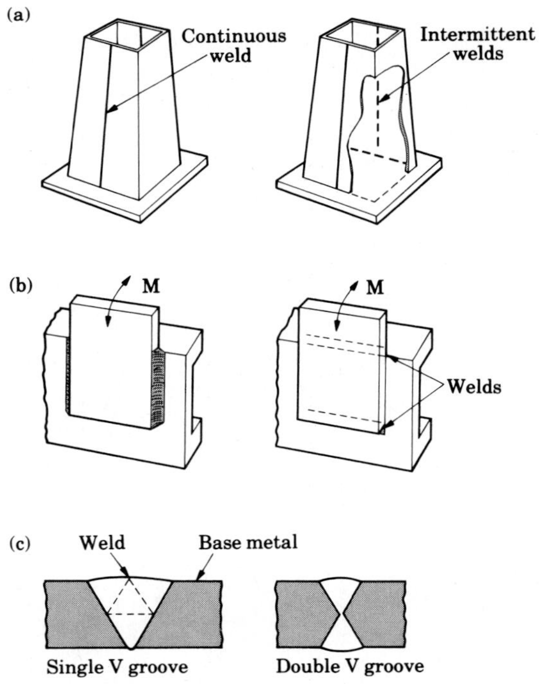

12.13 Liquid-Solid State Joining

12.13.1 Brazing

Brazing is a joining process in which a filler metal is placed at or between the faying surfaces to be joined, and the temperature is raised to melt the filler metal but not the workpieces (Fig. 12.32a). The molten metal fills the closely fitting space by capillary action. Upon cooling and solidification of the filler metal, a strong joint is obtained. Brazing was first used as far back as 3000–2000 B.C. Actually, there are two types of brazing processes: (a) that which we have already described, and (b) braze welding (Fig. 12.32b), in which the filler metal is deposited at the joint with a technique similar to oxyfuel gas welding.

Filler metals used for brazing melt above 450 °C (840 °F). The temperatures employed in brazing are below the melting point (solidus temperature) of the metals to be joined. Thus this process is unlike liquid-state welding processes in which the workpieces must melt in the weld area for fusion to occur. Problems associated with heat-affected zones, warping, and residual stresses are therefore reduced in brazing.

The strength of the brazed joint depends on (a) joint design and (b) the adhesion at the interfaces of the workpiece and filler metal. Consequently, the surfaces to be brazed should be chemically or mechanically cleaned to ensure full capillary action; hence the use of a flux is important. Typical joint clearance in brazing ranges from 0.025 mm to 0.2 mm (0.001 in. to 0.008 in.). The clearances must fit within a very small tolerance range because larger clearances reduce the strength of the brazed joint. A variety of special fixtures may be used to hold the parts together, some with provision for thermal expansion and contraction, during brazing.

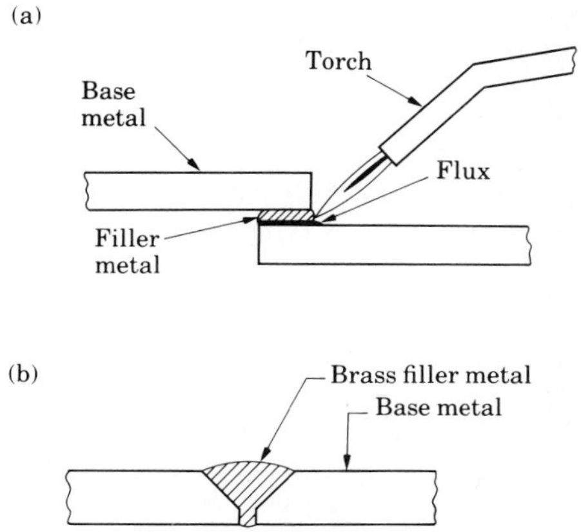

FIGURE 12.32
(a) Brazing and (b) braze welding operations.

TABLE 12.3
TYPICAL FILLER METALS FOR BRAZING VARIOUS METALS AND ALLOYS

BASE METAL	FILLER METAL	BRAZING TEMPERATURE, (°C)
Aluminum and its alloys	Aluminum–silicon	570–620
Magnesium alloys	Magnesium–aluminum	580–625
Copper and its alloys	Copper–phosphorus	700–925
Ferrous and nonferrous (except aluminum and magnesium)	Silver and copper alloys, copper–phosphorus	620–1150
Iron-, nickel-, and cobalt-base alloys	Gold	900–1100
Stainless steels, nickel- and cobalt-base alloys	Nickel–silver	925–1200

Several *filler metals* (*braze metals*) are available and have a range of brazing temperatures (Table 12.3). They come in a variety of shapes, such as wire, rings, shims, and filings. The choice of filler metal and its composition are important in order to avoid embrittlement of the joint (by grain boundary penetration of liquid metal), formation of brittle intermetallic compounds at the joint, and galvanic corrosion in the joint. Thus studying the relevant phase diagrams prior to the final selection of a filler metal for a particular application is essential. Note that filler metals for brazing, unlike other welding operations, generally have significantly different compositions than the metals to be joined.

Because of diffusion between the filler metal and the base metal, mechanical and metallurgical properties of joints can change in subsequent processing or during the service life of brazed components. For example, when titanium is brazed with pure tin filler metal, it is possible for the tin to completely diffuse into the titanium base metal by subsequent aging or heat treatment. When that happens, the joint no longer exists.

Use of a *flux* is essential in brazing in order to prevent oxidation and to remove oxide films from workpiece surfaces. Brazing fluxes are generally made of borax, boric acid, borates, fluorides, and chlorides. Wetting agents may also be added to improve both the wetting characteristics of the molten filler metal and capillary action. Surfaces to be brazed must be clean and free from rust, oil, and other contaminants. Clean surfaces are essential to obtain the proper wetting and spreading characteristics of the molten filler metal in the joint and maximum bond strength. Sand blasting may also be used to improve surface finish of faying surfaces. Because they are corrosive, fluxes should be removed after brazing has been completed. Fluxes can usually be removed by washing with hot water. The heating methods used also identify the various brazing processes.

Torch brazing. The heat source in *torch brazing* (TB) is oxyfuel gas with a carburizing flame. Brazing is performed by first heating the joint with the torch, then depositing the brazing rod or wire in the joint. Suitable part thicknesses are usually in the range of 0.25–6 mm (0.01–0.25 in.).

Furnace brazing. As the name suggests, *furnace brazing* (FB) is carried out in a furnace. The parts are precleaned and preloaded with brazing metal in appropriate configurations before being placed in the furnace (Fig. 12.33). Furnaces may be batch type for complex shapes or continuous type for high production runs, especially for small parts with simple joint designs.

Induction brazing. The source of heat in *induction brazing* (IB) is induction heating by high-frequency ac current. Parts are preloaded with filler metal and are placed near the induction coils for rapid heating. Unless a protective atmosphere is utilized, fluxes are usually used. Part thicknesses are usually less than 3 mm (0.125 in.).

Resistance brazing. In *resistance brazing* (RB), the source of heat is through electrical resistance of the components to be brazed. Electrodes are utilized for this purpose, as in resistance welding. Parts are either preloaded with filler metal, or it is supplied externally during brazing. Parts that are commonly brazed by this process have thicknesses of 0.1–12 mm (0.004–0.5 in.).

Dip brazing. *Dip brazing* (DB) is carried out by dipping the assemblies to be brazed into either a molten filler-metal bath or a molten salt bath (at a temperature just above the melting point of the filler metal), which serves as the heat source. All workpiece surfaces are thus coated with the filler metal. Consequently, dip brazing in metal baths is used only for small parts, such as sheet, wire, and fittings, usually of less than 5 mm (0.2 in.) in thickness or diameter.

Infrared brazing. The heat source in *infrared brazing* (IRB) is a high-intensity quartz lamp. This process is particularly suitable for brazing very thin components, usually less than 1 mm (0.04 in.) thick, including honeycomb structures. The radiant energy is focused on the joint, and the process can be carried out in a vacuum.

Diffusion brazing. *Diffusion brazing* (DFB) is carried out in a furnace where—with proper control of temperature and time—the filler metal diffuses into the faying surfaces of the components to be joined. The brazing time required may

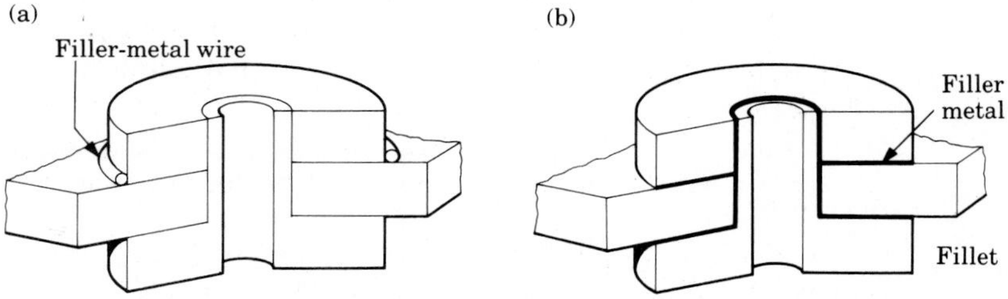

FIGURE 12.33
An example of brazing: (a) before and (b) after. Note that the filler metal is a shaped wire.

range from 30 minutes to 24 hours. Diffusion brazing is used for strong lap or butt joints and for difficult joining operations. Because the rate of diffusion at the interface does not depend on the thickness of the components, part thicknesses may range from foil to as much as 50 mm (2 in.).

Braze welding. The joint in *braze welding* is prepared as in fusion welding. Using an oxyacetylene torch with an oxidizing flame, filler metal is deposited at the joint (see Fig. 12.32b), rather than by capillary action as in brazing. Thus much more filler metal is used, compared to brazing. However, temperatures in braze welding are generally lower than in fusion welding, and part distortion is minimal. The use of a flux is essential in this process. The principal use of braze welding is to maintain and repair parts, such as ferrous castings and steel components.

12.13.2 Soldering

In *soldering*, the filler metal, called *solder*, melts below 450 °C (840 °F). As in brazing, the solder fills the joint by capillary action between closely fitting or closely placed components. Heat sources for soldering are usually soldering irons, torches, or ovens. Soldering with copper–gold and tin–lead alloys was first practiced as far back as 4000–3000 B.C.

There are several soldering methods, which are similar to brazing methods:

a) Torch soldering (TS).
b) Furnace soldering (FS).
c) Iron soldering (INS), using a soldering iron.
d) Induction soldering (IS).
e) Resistance soldering (RS).
f) Dip soldering (DS).
g) Infrared soldering (IRS).
h) Wave soldering (WS), used for automated soldering of printed circuit boards.
i) Ultrasonic soldering, in which a transducer subjects the molten solder to ultrasonic cavitation, which removes the oxide films from the surfaces to be joined. The need for a flux is thus eliminated.

Solders are usually tin–lead alloys in various proportions. For better joint strength and special applications, other solder compositions that can be used are tin–zinc, lead–silver, cadmium–silver, and zinc–aluminum alloys (Table 12.4). In soldering, fluxes are used as in welding and brazing and for the same purposes. Fluxes are generally of two types: (a) inorganic acids or salts, such as zinc

TABLE 12.4
TYPICAL SOLDERS AND THEIR APPLICATIONS

Tin–lead	General purpose
Tin–zinc	Aluminum
Lead–silver	Strength at higher than room temperature
Cadmium–silver	Strength at high temperatures
Zinc–aluminum	Aluminum; corrosion resistance

ammonium chloride solutions, which clean the surface rapidly, but after soldering, the flux residues should be removed by washing thoroughly with water to avoid corrosion; and (b) noncorrosive resin-based fluxes, used in electrical applications.

Soldering is used extensively in the electronics industry and in making containers for liquid- or air-tight joints. Unlike brazing, soldering temperatures are relatively low. Thus a soldered joint has very limited use at elevated temperatures. Moreover, because solders do not generally have much strength, they are not used for load-bearing structural members. Because of the small faying surfaces, butt joints are rarely made with solders. In other situations, joint strength is improved by mechanical interlocking of the joint.

Copper and precious metals, such as silver and gold, are easy to solder, as is tinplate for food containers. Aluminum and stainless steels are difficult to solder because of their strong, thin oxide film. However, these and other metals can be soldered using special fluxes that modify surfaces. Soldering can be used to join various metals and thicknesses. Although manual operations require skill and are time-consuming, soldering speeds can be high with automated equipment.

12.13.3 Adhesive bonding

Numerous components and products can be joined and assembled using an *adhesive*, rather than by any of the joining methods we have described thus far. For a long time, *adhesive bonding* has been a common method of joining and assembly for applications such as labeling, packaging, bookbinding, home furnishings, and footware. Plywood, developed in 1905, is a typical example of adhesive bonding of several layers of wood with glue. Adhesive bonding has been gaining increased acceptance in manufacturing ever since its first use on a large scale in assembling load-bearing components in aircraft during World War II (1939–1945).

Many types of adhesives are available—and continue to be developed—which provide adequate joint strength, including fatigue strength. The three basic types of adhesives are:

- *Natural adhesives*, such as starch, dextrin (a gummy substance obtained from starch), soya flour, and animal products.
- *Inorganic adhesives*, such as sodium silicate and magnesium oxychloride.
- *Synthetic organic adhesives*, which may be thermoplastics (used for non-structural and some structural bonding) or thermosetting polymers (used primarily for structural bonding).

Because of their strength, synthetic organic adhesives are the most important in manufacturing processes, particularly for load-bearing applications. They are classified as:

a) Chemically reactive.
b) Pressure sensitive.
c) Hot melt.
d) Evaporative.
e) Film adhesives.

Adhesives are available in various forms, such as liquids, pastes, solutions, emulsions, powder, tape, and film. When applied, adhesives generally are about 0.1 mm (0.004 in.) thick. Depending on the particular application, an adhesive must have one or more of the following properties:

a) Strength.
b) Toughness.
c) Resistance to various fluids and chemicals.
d) Resistance to environmental degradation, including moisture.
e) Capability to wet the interface to be bonded.

Adhesive joints are designed to withstand shear (Fig. 12.34), compressive, and tensile forces, but they should not be subjected to peeling forces (Fig. 12.35). Note, for example, how easily you can peel adhesive tape from a surface. During peeling the behavior of an adhesive may be brittle, or it may be ductile and tough, requiring large forces to peel it.

A wide variety of similar and dissimilar metallic and nonmetallic materials and components with different shapes, sizes, and thicknesses can be bonded to each other by adhesives. Adhesive bonding can be combined with mechanical joining methods to further improve the strength of the bond.

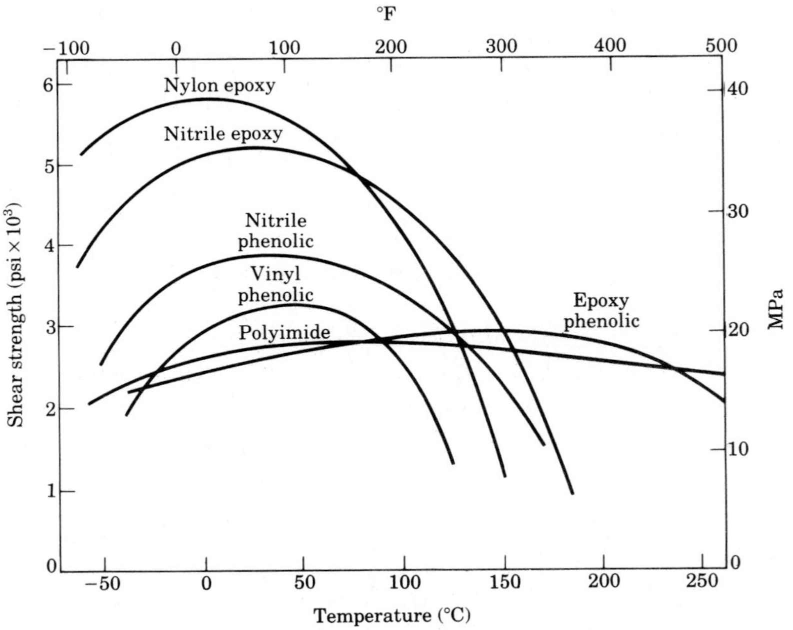

FIGURE 12.34
Shear strength of several adhesives as a function of temperature.

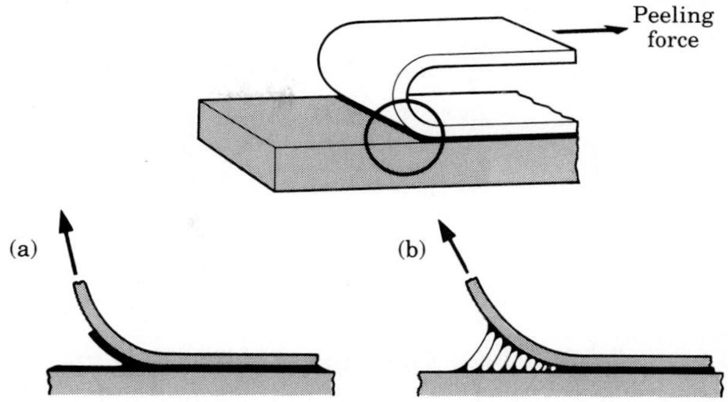

FIGURE 12.35
Characteristic behavior of (a) brittle and (b) tough adhesives in a peeling test. This test is similar to peeling adhesive tape from a solid surface. Adhesive joints should not be subjected to this type of loading in service.

An important consideration in the use of adhesives in production is curing time, which can range from a few seconds at high temperatures to many hours at room temperature, particularly for thermosetting adhesives. (In using various glues, you undoubtedly have read instructions on the labels to this effect). Thus production rates may be low. Furthermore, adhesive bonds for structural applications are rarely suitable for service above 250 °C (500 °F). Joint design and bonding methods require care and skill. Special equipment, such as fixtures, presses, tooling, and autoclaves and ovens for curing, is usually needed.

Nondestructive inspection of the quality and strength of adhesively bonded components can be difficult. Some of the techniques that we described in Section 4.8, such as acoustic impact (tapping), holography, infrared detection, and ultrasonic testing, are effective nondestructive testing methods.

Major industries that use adhesive bonding extensively are aerospace, automotive, appliances, and building products. Applications include attaching rear-view mirrors to windshields, automotive brake-lining assemblies, laminated windshield glass, appliances, helicopter blades, honeycomb structures, and aircraft bodies and control surfaces.

The major advantages of adhesive bonding are:

a) It provides a bond at the interface either for structural strength or for nonstructural applications such as sealing, insulating, preventing electrochemical corrosion between dissimilar metals, and reducing vibration and noise through internal damping at the joints.
b) It distributes the load at an interface, thus eliminating localized stresses that result from joining the components with welds or mechanical fasteners, such as bolts and screws. Moreover, structural integrity of the sections is maintained, since no holes are required, and the appearance of the components is generally improved.
c) Very thin and fragile components can be bonded with adhesives without contributing significantly to weight.

d) Porous materials and materials of very different properties and sizes can be joined.
e) Because it is usually carried out at a temperature between room temperature and about 200 °C (400 °F), there is no significant distortion of the components or change in their original properties. This is particularly important for materials that are heat sensitive.

The major limitations of adhesive bonding are:

a) Limited service temperatures.
b) Possibly long bonding time.
c) Need for great care in surface preparation.
d) Difficulty in testing bonded joints nondestructively, particularly for large structures.
e) Reliability of adhesively bonded structures during their service life.

Designs for adhesive bonding should ensure that joints are subjected to compressive, tensile, and shear forces, but not peeling or cleavage. They vary considerably in strength; hence selection of the appropriate design is important and should include considerations such as type of loading and the environment. Butt joints require large bonding surfaces. Simple lap joints tend to distort under tension because of the force couple at the joint. The coefficients of expansion of the components to be bonded should preferably be close in order to avoid internal stresses during adhesive bonding.

Surface preparation is essential in adhesive bonding. Joint strength depends greatly on absence of dirt, dust, oil, and other contaminants. Thick, weak, or loose oxide films on workpieces are detrimental to adhesive bonding. On the other hand, a porous or thin, strong oxide film may be desirable, particularly one with some surface roughness to improve adhesion. Various compounds and primers are available to modify surfaces to increase adhesive bond strength. Liquid adhesives may be applied by brushing, spraying, and rollers.

12.14 Mechanical Joining

Two or more components may have to be joined or fastened in such a way that they can be taken apart sometime during the product's service life. Numerous objects, including mechanical pencils, caps and lids on containers, mechanical watches, engines, and bicycles, have components that are fastened mechanically. *Mechanical joining* may be preferred for the following reasons:

a) Ease of manufacturing.
b) Ease of assembly and transportation.
c) Ease of parts replacement, maintenance, or repair.

d) Designs requiring movable joints, such as hinges, sliding mechanisms for drawers and doors, and adjustable components and fixtures.
e) Lower overall cost of manufacturing the product.

The most common method of mechanical joining is by fastening, using bolts, nuts, screws, pins, and a variety of other *fasteners*. These processes are also known as mechanical assembly.

Mechanical joining generally requires that the components have holes through which fasteners are inserted. These joints may be subjected to both shear and tensile stresses and should be designed to resist these forces.

Hole preparation is an important aspect of mechanical joining. A hole in a solid body can be produced by punching, drilling, chemical, and electrical means, and by high-energy beams, depending on the type of material, its properties, and thickness. Recall from previous discussions that holes may be produced as an integral part of the product during casting, forging, extrusion, and powder metallurgy. For improved accuracy and surface finish, many of these holemaking operations may be followed by finishing operations, such as shaving, deburring, reaming, and honing.

Because of their fundamental differences, each holemaking operation produces a hole with different surface finish and properties and dimensional characteristics. The most significant influence of a hole in a solid body is its tendency to reduce the component's fatigue life (stress concentration). Fatigue life can best be improved by inducing compressive residual stresses on the cylindrical surface of the hole. These stresses are usually developed by pushing a round rod (*drift pin*) through the hole and expanding it by a very small amount. This process plastically deforms the surface layers of the hole, in a manner similar to shot peening or roller burnishing.

12.14.1 Threaded fasteners

Bolts, screws, and nuts are among the most commonly used threaded fasteners. Texts on machine design describe numerous standards and specifications, including thread dimensions, tolerances, pitch, strength, and the quality of materials used to make these fasteners.

Bolts and screws may be secured with nuts (machine screws), or they may be self-tapping, whereby the screw either cuts or forms the thread into the part to be fastened. This method is particularly effective and economical in plastic products. In this way, fastening does not require a tapped hole and the need for—and cost of—a nut.

If the joint is to be subjected to vibration, such as in aircraft and various types of engines and high-speed machinery, several specially designed nuts and lock washers are available. They increase the frictional resistance in the torsional direction, thus preventing vibrational loosening of the fasteners.

12.14.2 Rivets

The most common method of permanent or semipermanent mechanical joining is by *riveting*. Hundreds of thousands of rivets may be used in the construction and assembly of large commercial aircraft. Holes for the rivets are required in the components to be joined. Installing a rivet consists of placing the rivet in the hole and deforming the end of its shank by upsetting (heading). Riveting may be done either at room temperature or hot, using special tools or explosives in the rivet cavity.

12.14.3 Other methods of fastening

Many types of fasteners are used in numerous joining and assembly applications. We describe the most common types below.

Metal stitching or stapling. The process of *metal stitching* or *stapling* (Fig. 12.36) is much like that of ordinary stapling of papers. This operation is fast and particularly suitable for joining thin metallic and nonmetallic materials, and it does not require holes in the components. A common example is the stapling of cardboard containers for appliances and other consumer products.

Seaming. *Seaming* is based on the simple principle of folding two thin pieces of material together. Seaming is much like joining two pieces of paper, in the absence of a paper clip, by folding them at the corner. The most common examples of seaming are the tops of beverage cans and containers for food and household products. In seaming, the materials should be able to undergo bending and folding at very small radii. Otherwise, they will crack and the seams will not be airtight or watertight. The performance of seams may be improved with adhesives, coatings, seals, or by soldering.

Crimping. *Crimping* is a method of joining without using fasteners. It can be done with beads or dimples (Fig. 12.37), which can be produced by shrinking or swaging operations. Crimping can be used both on tubular and flat parts, provided that the materials are thin and ductile enough to withstand the large localized deformations. Caps on glass bottles are attached by crimping, as are connectors for electrical wiring.

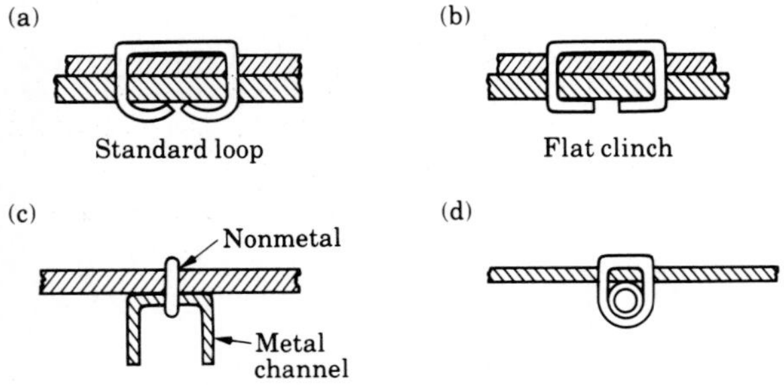

FIGURE 12.36
Examples of metal stitching.

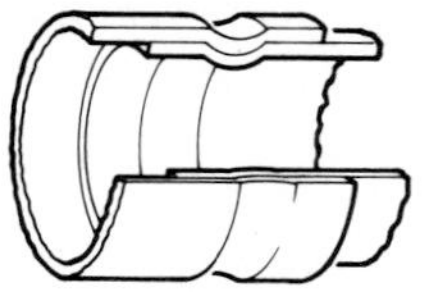

FIGURE 12.37
An example of mechanical joining by crimping.

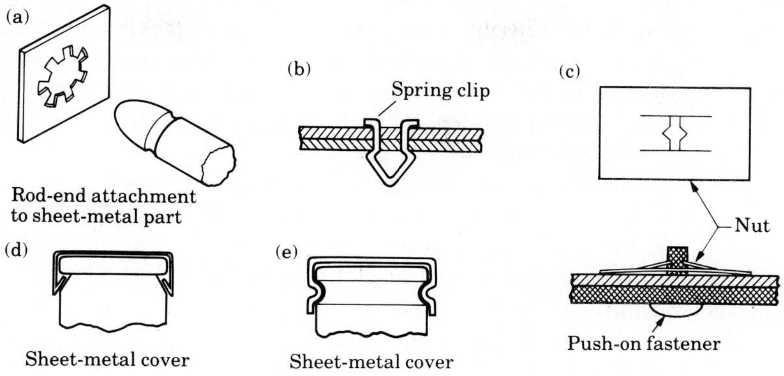

FIGURE 12.38
Examples of spring and snap-in fasteners to facilitate assembly.

Snap-in fasteners. Various spring and snap-in fasteners are shown in Fig. 12.38. Such fasteners are widely used in automotive bodies and household appliances. They are economical and permit easy and rapid component assembly.

Shrink and press fits. Components may also be assembled by shrink fitting and press fitting. *Shrink fitting* is based on the principle of differential thermal expansion and contraction of two components. Typical applications are the assembly of die components and mounting gears and cams on shafts. In *press fitting*, one component is forced over another, resulting in high joint strength.

12.15 Joining Plastics

Plastics can be joined by many of the methods we have described for joining metals and nonmetallic materials, including (a) fusion welding with heat, (b) ultrasonic welding, (c) friction welding, (d) adhesive bonding, (e) bonding with solvents, and (f) mechanical joining, particularly with self-tapping screws. Mechanical joining is particularly effective with most plastics because of their inherent toughness and resilience.

12.15.1 Thermoplastics

Because thermoplastics soften and melt as temperature is increased, they can be joined by the fusion-welding techniques—but at much lower temperatures. This method is particularly effective with plastics that cannot be bonded easily with adhesives. Plastics such as polyvinyl chloride, polyethylene, polypropylene, acrylics, and acrylonitrile butadiene styrene (ABS) can be joined in this manner. For example, specially designed portable fusion sealing systems have been developed to allow in-field joining of pipe, which is usually made of polyethylene, used for natural-gas delivery.

The heat source in fusion welding of thermoplastics is usually hot air or other gases. The heat melts the joint and, with the application of pressure to ensure a good bond, allows fusion to take place at the interface. Filler materials of the same type of polymer may be used. Other heating methods may also be utilized in joining thermoplastics: (1) heated tools and dies (for joining thin materials, since plastics are poor thermal conductors), (2) electrical-resistance wire, and (3) high-frequency heating.

Coextruded multiple food wrappings consist of different types of films, which are bonded by heat during extrusion. Each film has a different function, such as to keep out moisture, keep out oxygen, and facilitate heat sealing during packaging. Some wrappings have as many as seven layers, all bonded together during production of the film.

Oxidation can be a problem in joining some polymers, such as polyethylene, causing degradation. In these cases an inert shielding gas such as nitrogen is used to prevent oxidation. Because of low thermal conductivity of thermoplastics, the heat source may burn or char the surfaces of the components if applied at a high rate, possibly causing difficulties in obtaining sufficiently deep fusion.

Adhesive bonding of plastics is best illustrated in joining sections of polyvinyl chloride (PVC) pipe, used extensively in home irrigation systems, and ABS pipe, used in drain, waste, and vent systems. The adhesive is applied to the connecting sleeve and pipe surfaces, using a primer to improve adhesion (much like using primers in painting), and the pieces are pushed together. Adhesive bonding of polyethylene, polypropylene, and polytetrafluoroethylene (Teflon) can be difficult because adhesives do not stick readily to their surfaces. Their surfaces usually have to be treated chemically to improve bonding. The use of adhesive primers and double-sided adhesive tapes are also effective.

12.15.2 Thermosets

Because they do not soften or melt with temperature, thermosetting plastics, such as epoxy and phenolics, are usually joined by using (1) threaded or other molded-in inserts, (2) mechanical fasteners, and (3) solvent bonding.

The basic process of thermoset bonding with solvents involves:

1. Roughening the surfaces with an abrasive.
2. Wiping the surfaces with a solvent.
3. Pressing the surfaces together and holding until sufficient bond strength is developed.

12.16 Cutting

A piece of metal can be separated into two or more pieces, or into various contours, not only by mechanical means such as sawing, but also by a source of heat that removes a narrow zone in the workpiece. The sources of heat that we have described thus far, namely, torches and electric arcs, can be used for this purpose.

Oxyfuel gas cutting (OFC) is similar to oxyfuel welding, but the heat source is now used to remove a narrow zone from a metal plate or sheet (Fig. 12.39). This process is particularly suitable for steels. The basic reactions with steel are:

$$Fe + O \rightarrow FeO + \text{heat}; \tag{12.9}$$

$$3Fe + 2O_2 \rightarrow Fe_3O_4 + \text{heat}; \tag{12.10}$$

$$4Fe + 3O_2 \rightarrow 2Fe_2O_3 + \text{heat}. \tag{12.11}$$

The highest heat is generated in the second reaction, resulting in a temperature rise of about 870 °C (1600 °F). This temperature is not sufficiently high to cut steels, so the workpiece is preheated with fuel gas, and oxygen is introduced later (see nozzle cross-section in Fig. 12.39). The higher the carbon content of the steel, the higher the preheating temperature must be. Cutting occurs mainly by oxidation and burning of the steel, with some melting taking place. Cast irons and steel castings can also be cut by this method. The process generates a kerf, similar to that produced by sawing with a saw blade.

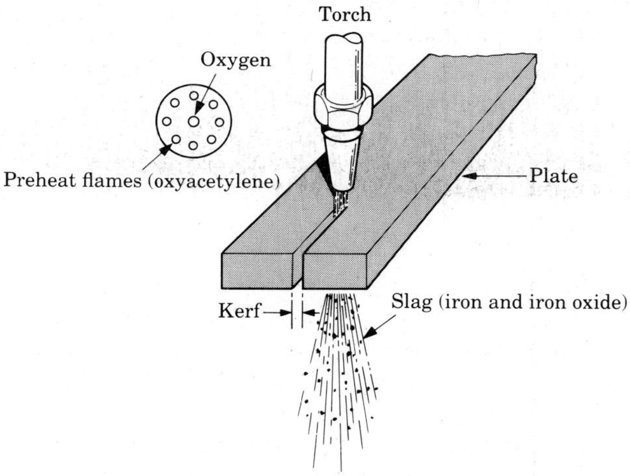

FIGURE 12.39
Flame cutting of steel plate with oxyacetylene torch and cross-section of torch nozzle.

For metals and alloys that do not oxidize as readily or have high thermal conductivity, such as stainless steels and copper, iron powders or fluxes are introduced into the flame. These elements increase the local temperature and help remove the oxides in the cutting zone, which otherwise would obstruct the heat transfer from the torch to the metal.

The maximum thickness that can be cut by OFC depends mainly on the gases used. With oxyacetylene gas, the maximum thickness is about 300–350 mm (12–14 in.); with oxyhydrogen, about 600 mm (24 in.). Kerf widths range from about 1.5 mm to 10 mm (0.06 to 0.4 in.), with reasonably good control of tolerances. The flame leaves *drag lines* on the cut surface, which is rougher than surfaces produced by sawing, blanking, or other operations using cutting tools. Distortion caused by uneven temperature distribution can be a problem in OFC.

Although long used for salvage and repair work, oxyfuel gas cutting has become an important manufacturing process. Torches may be guided along various paths manually, mechanically, or by automatic machines using programmable controllers and robots. Also, two or more layers of flat sheet can be cut with OFC (stack cutting), thus improving productivity and reducing cutting cost. Underwater cutting is done with specially designed torches that produce a blanket of compressed air between the flame and the surrounding water.

Arc cutting processes are based on the same principles as arc-welding processes. A variety of materials can be cut at high speeds by arc cutting.

In *air carbon-arc cutting* (AAC), a carbon electrode is used, and the molten metal is blown away by a high-velocity air jet. Thus the metal being cut doesn't have to oxidize. This process is used especially for gouging and scarfing (removal of metal from a surface). The AAC process is noisy, and the molten metal can be blown substantial distances, causing safety hazards.

Plasma-arc cutting (PAC) produces the highest temperatures. It is used for rapid cutting of nonferrous and stainless-steel plates. The cutting productivity of this process is higher than that of oxyfuel gas methods. It produces good surface finish and narrow kerfs. It also is the most popular cutting process utilizing programmable controls that is used in manufacturing today.

Lasers and *electron beams* are used for very accurately cutting a wide variety of metals. Surface finish is better and kerf width is narrower than that for other thermal cutting processes. Proper safety precautions are important.

SUMMARY

Oxyfuel gas, arc, and resistance welding are among the most commonly used joining operations. Gas welding uses chemical energy, whereas arc and resistance welding use electrical energy to supply the necessary heat for welding. In all these processes, heat is used to bring the joint being welded to a liquid state. Shielding gases are used to protect the molten weld pool and weld area against oxidation. Filler rods may or may not be used in oxyfuel gas and arc welding to fill the weld area. Resistance welding operations do not require filler rods.

A number of other joining processes that are based on producing a strong joint under pressure and/or heat are available. Surface preparation and cleanliness are important in some of these processes. Pressure is applied mechanically or by explosives. Heat may be supplied externally, including high-energy beams, or generated internally, as in friction welding. Among important developments is the combining of the diffusion bonding and superplastic forming processes. Productivity is improved, as is the capability to make complex parts economically.

Joining processes that do not rely on fusion or pressure at the interfaces include brazing and soldering. Instead, these processes utilize filler material that requires some temperature rise in the joint. They can be used to join dissimilar metals of intricate shapes and various thicknesses.

Adhesive bonding has gained increased acceptance in major industries, such as the aerospace and automotive. In addition to good bond strength, adhesives have other favorable characteristics, such as sealing, insulating, preventing electrochemical corrosion between dissimilar metals, and reducing vibration and noise through internal damping in the bond. Surface preparation and joint design are important factors in adhesive bonding.

Mechanical joining is one of the oldest and most common joining methods. Bolts, screws, and nuts are common fasteners for machine components and structures, which are likely to be taken apart for maintenance, ease of transportation, and various other reasons. Rivets are semipermanent or permanent fasteners used in buildings, bridges, and transportation equipment. A wide variety of other fasteners and fastening techniques is available for numerous permanent or semipermanent applications.

The metallurgy of the welded joint is an important aspect of all welding processes because it determines the strength and toughness of the joint. The welded joint consists of solidified metal and a heat-affected zone, with a wide variation in microstructure and properties, depending on the metals joined and the filler metals. Because of severe thermal gradients in the weld zone, distortion, residual stresses, and cracking can be a significant problem.

BIBLIOGRAPHY

General

Arata, Y. (ed.), *Plasma, Electron, and Laser Beam Technology*. Metals Park, Ohio: American Society for Metals, 1986.

Burgess, N.T. (ed.), *Quality Assurance of Welded Construction*. London: Applied Science Publishers, 1983.

Cary, H.B., *Modern Welding Technology*. Englewood Cliffs, N.J.: Prentice-Hall, 1979.

Crossland, B., *Explosive Welding of Metals and its Applications*. Oxford: Clarendon Press, 1982.

Davies, A.C., *The Science and Practice of Welding*, 7th ed. Cambridge, England: Cambridge University Press, 1977.

Esterling, K.E., *Introduction to the Physical Metallurgy of Welding*. Stoneham, Mass.: Butterworths, 1983.
Galyen, J., G. Sear, and C. Tuttle, *Welding: Fundamentals and Procedures*. New York: Wiley, 1985.
Gray, T.G.F., J. Spence, and T.H. North, *Rational Welding Design*. New York: Butterworths, 1975.
Houldcroft, P.T., *Welding and Cutting: A Guide to Fusion Welding and Associated Cutting Processes*. New York: Industrial Press, 1988.
Koellhoffer, L., *Welding Processes and Practices*. New York: Wiley/Chichester, 1987.
Lancaster, J.F., *Metallurgy of Welding*, 3d ed. London: George Allen and Unwin, 1980.
Lindberg, R.A., and N.R. Braton, *Welding and Other Joining Processes*. Boston: Allyn and Bacon, 1976.
Masubushi, K., *Analysis of Welded Structures—Residual Stresses and Distortion and Their Consequences*. New York: Pergamon, 1980.
Metals Handbook, 9th ed., *Vol. 6: Welding, Brazing, and Soldering*. Metals Park, Ohio: American Society for Metals, 1983.
Mohler, R., *Practical Welding Technology*. New York: Industrial Press, 1983.
Principles of Industrial Welding. Cleveland, Ohio: The James F. Lincoln Arc Welding Foundation, 1978.
Romans, D., and E.N. Simons, *Welding Processes and Technology*. London: Pitman, 1974.
Saperstein, Z.P. (ed.), *Control of Distortion and Residual Stress in Weldments*. Metals Park, Ohio: American Society for Metals, 1977.
Schwartz, M.M. (ed.), *Source Book on Innovative Welding Processes*. Metals Park, Ohio: American Society for Metals, 1981.
Tool and Manufacturing Engineers Handbook, *Vol. 4, Quality Control and Assembly*. Dearborn, Mich.: Society of Manufacturing Engineers, 1986.
Welding Handbook, 7th ed., 5 vols. Miami: American Welding Society, 1976.
Welding Handbook, 8th ed., 3 vols. Miami: American Welding Society, 1987.

Brazing and Soldering

Allen, B.M., *Soldering Handbook*. London: Iliffe, 1969.
Metals Handbook, 9th ed., *Vol 6: Welding, Brazing, and Soldering*. Metals Park, Ohio: American Society for Metals, 1983.
Principles of Industrial Welding. Cleveland, Ohio: The James F. Lincoln Arc Welding Foundation, 1978.
Soldering Manual. Miami: American Welding Society, 1978.
Solders and Soldering. New York: Lead Industries Association, 1982.
Source Book on Brazing and Brazing Technology. Metals Park, Ohio: American Society for Metals, 1980.
Schwartz, M.M., *Brazing*. Metals Park, Ohio: ASM International, 1987.
Thwaites, C.J., *Capillary Joining—Brazing and Soft-Soldering*. New York: Wiley, 1982.

Adhesive Bonding

DeFryne, G. (ed.), *High Performance Adhesive Bonding*. Dearborn, Mich.: Society of Manufacturing Engineers, 1982.
Haviland, G.S., *Machinery Adhesives for Locking, Retaining, and Sealing*. New York: Marcel Dekker, 1986.

Landrock, A.H., *Adhesives Technology Handbook*. Park Ridge, N.J.: Noyes, 1985.
Parmley, R.O. (ed.), *Standard Handbook of Fastening and Joining*, 2d ed. New York: McGraw-Hill, 1989.
Schneberger, G.L. (ed.), *Adhesives in Manufacturing*. New York: Marcel Dekker, 1983.
Shields, J., *Adhesives Handbook*, 3d ed. London: Butterworths, 1984.
Skeist, I. (ed.), *Handbook of Adhesives*, 2d ed. New York: Van Nostrand Reinhold, 1977.
Tool and Manufacturing Engineers Handbook, Vol. 4: Quality Control and Assembly. Dearborn, Mich.: Society of Manufacturing Engineers, 1986.

Mechanical Joining

Blake, A., *What Every Engineer Should Know about Threaded Fasteners*. New York: Marcel Dekker, 1986.
Lincoln, B., K.J. Gomes, and J.F. Braden, *Mechanical Fastening of Plastics*. New York: Marcel Dekker, 1983.
Parmley, R.O. (ed.), *Standard Handbook of Fastening and Joining*, 2d ed. New York: McGraw-Hill, 1989.

QUESTIONS

12.1 Explain why so many different welding processes have been developed.

12.2 Why do dendrites form in the particular directions shown in Fig. 12.21?

12.3 Explain fusion as it relates to welding operations.

12.4 Why is an oxidizing flame desirable in welding copper alloys?

12.5 Why is the quality of submerged-arc welding very good?

12.6 Explain why the electroslag welding process is suitable for thick plates and heavy structural sections.

12.7 What are the similarities and differences between consumable and nonconsumable electrodes?

12.8 What advantages do resistance welding processes have over others?

12.9 What does the strength of a weld nugget depend on?

12.10 Discuss your observations concerning Table 12.1.

12.11 What is the effect of thermal conductivity of the workpiece on kerf width in oxyfuel cutting?

12.12 Could plasma-arc cutting be used for nonmetallic materials? Explain.

12.13 Explain the significance of the pressure applied through the electrodes during resistance welding operations.

12.14 What factors influence the shape of the upset joint in flash welding, as shown in Fig. 12.13?

12.15 Why is diffusion bonding when combined with superplastic forming of sheet metals an attractive fabrication process? Does it have any limitations? Explain.

12.16 Can roll bonding be applied to a variety of part configurations? Explain with appropriate sketches.

12.17 What factors influence the size of the weld beads shown in Fig. 12.19?

12.18 Discuss the factors that influence the strength of (a) a diffusion-bonded and (b) a cold-welded component.

12.19 Describe part designs that cannot be joined by friction welding processes.

12.20 Describe the difficulties involved in making deep narrow welds.

12.21 Explain how you would fabricate the structures shown in Fig. 7.50 with methods other than diffusion bonding/superplastic forming.

12.22 Discuss factors that contribute to the shape of grains in a weld zone.

12.23 What are the characteristics of the heat-affected zone? How important are they?

12.24 Discuss factors that influence weld quality.

12.25 Explain why aluminum and cast iron can be difficult to weld.

12.26 How does the weldability of steel change as its carbon content increases? Why?

12.27 Explain why some joints have to be preheated prior to joining.

12.28 What are the similarities and differences between casting of metals and fusion welds?

12.29 Are there common factors in the weldability, castability, formability, and machinability of metals? Explain with appropriate examples.

12.30 Explain the effect of the stiffness of various components to be welded on weld defects.

12.31 What are the relative advantages of braze welding and fusion welding?

12.32 What type of materials and part configurations would be difficult to join by brazing?

12.33 How different is adhesive bonding from other joining processes? What limitations does it have?

12.34 Why have mechanical joining methods been developed? Give specific examples of their applications.

12.35 What precautions should be taken in mechanical joining of dissimilar metals? Do these precautions also apply to nonmetallic materials?

12.36 Name several products that have been assembled by (a) seaming, (b) stitching, (c) soldering, (d) brazing, (e) spot welding, and (f) riveting.

12.37 Soldering is generally applied to thin sections. Why?

12.38 Is the strength of an adhesively bonded structure as high as that obtained by diffusion bonding? Explain.

12.39 Suggest methods of attaching a round thermosetting plastic bar perpendicular to a flat metal plate.

12.40 What joining methods would be suitable to assemble a thermoplastic cover over a metal frame? Assume that the cover is to be removed periodically.

12.41 How would you explain the large difference in the size of the weld beads shown in Fig. 12.19?

12.42 List and explain the rules that must be followed to avoid cracks in welded joints.

12.43 Describe the differences between torch cutting of ferrous and nonferrous metals.

12.44 In resistance projection welding (see Fig. 12.12), describe the factors that influence flattening of the interface after welding.

PROBLEMS

12.1 Two 1.5-mm thick copper plates are being spot welded using a current of 8000 A and a current flow time of $t = 2$ s. The electrodes are 4 mm in diameter. Estimate the heat generated in the weld zone.

12.2 Figure P12.1(a) shows a metal sheave consisting of two matching pieces of hot-rolled low-carbon steel sheets. These two pieces can be joined either by spot welding or by V-groove welding. Discuss the advantages and limitations of each process for this application.

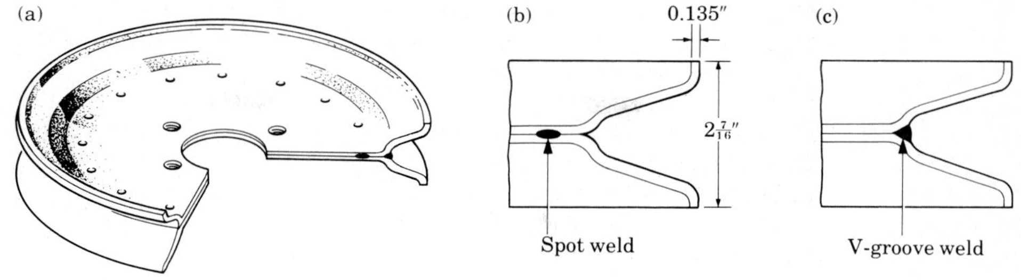

FIGURE P12.1

12.3 Inspect a variety of sheet-metal containers and cans used for household products and beverages. List them and discuss your observations about the joining or fastening processes used.

12.4 Figure P12.2 shows various joints that can be used for brazing two pieces of metal. Inspect them and identify those that are desirable and those that are poor designs. Explain your reasons.

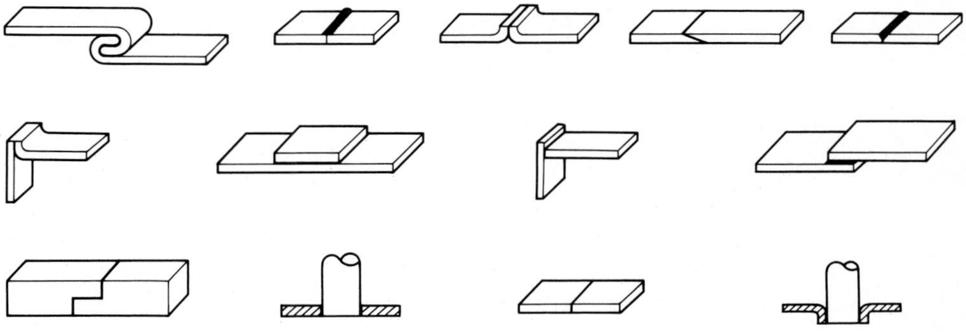

FIGURE P12.2

12.5 Figure P12.3 shows various joint designs for adhesive bonding. Inspect them and comment on their desirability. Explain your reasons. Suggest additional designs of your own.

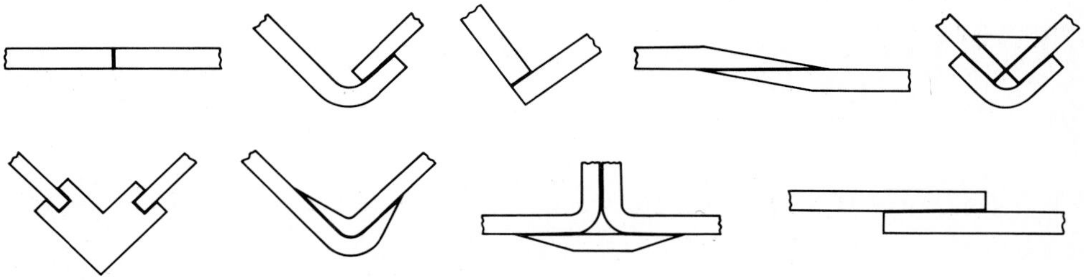

FIGURE P12.3

13

Manufacturing Automation

13.1 Introduction

Until about four decades ago, most manufacturing operations were carried out on traditional machinery, such as lathes, milling machines, and presses, which lacked flexibility and required considerable skilled labor. Each time a different product was manufactured, the machinery had to be retooled, and the movement of materials had to be rearranged. The development of new products and parts with complex shapes required numerous trial-and-error attempts by the operator to set the proper processing parameters on the machine. Furthermore, because of human involvement, making parts that were exactly alike was difficult.

These circumstances meant that processing methods were generally inefficient and that labor costs were a significant portion of overall production costs. The need for reducing the labor share of product cost gradually became apparent, as was the need to improve the efficiency and flexibility of manufacturing operations. This need was particularly significant in terms of increased competition, both nationally and from other industrialized countries.

Productivity also became a major concern. Defined as the optimum use of all resources—materials, energy, capital, labor, and technology—or as output per employee per hour, *productivity* basically measures operating efficiency. With rapid advances in the science and technology of manufacturing and their gradual implementation, the efficiency of manufacturing operations began to improve and the percentage of total cost represented by labor costs declined.

How can productivity be improved? Mechanization of machinery and operations had, by and large, reached its peak by the 1940s. *Mechanization* runs a process or operation with the use of various mechanical, hydraulic, pneumatic, or electrical devices. Take, for example, the use of a simple can opener. Opening a thousand cans by hand would take a long time, would require much physical effort, and would be tedious. You would soon lose interest, and your efficiency would drop off. On the other hand, using an electric can opener—which is a mechanical device—takes less time and effort, but the job is still tedious, and your efficiency is still likely to drop after a while. Note that, in mechanized systems, the operator still directly controls the process, and must check each step of the machine's performance. If a tool breaks during machining, if parts overheat during heat treatment, if surface finish begins to deteriorate during grinding, and if dimensional tolerances become too large in metal forming, the operator has to intervene and change one or more process parameters.

The next step in improving the efficiency of manufacturing operations was *automation*, from the Greek word *automatos*, meaning self-acting. The word automation was coined in the mid-1940s by the U.S. automobile industry to indicate automatic handling of parts between production machines, together with their continuous processing at the machines. During the past three decades, major advances and breakthroughs in the types and extent of automation have occurred. These important developments were made possible largely through rapid advances in the capacity and sophistication of *control systems* and *computers*.

In this chapter, we first review the history and principles of automation and how they help us *integrate* various operations and activities in a manufacturing plant to improve productivity. We then introduce the important concept of control of machines and systems through *numerical control* and *adaptive control* techniques. An essential element in manufacturing is the movement of raw materials and parts in various stages of completion throughout the plant. We present the basic concepts of *material handling*, and how they have been developed into various systems, including the use of *industrial robots*, to improve their efficiency. We then introduce the subject of *sensor technology*, which is an essential element in the control and optimization of machinery, processes, and systems. Finally, we describe the concept and applications of *flexible fixturing* and *assembly operations*. These methods enable us to take full advantage of advanced manufacturing technologies, particularly those of flexible manufacturing systems.

13.2 Automation

Automation can generally be defined as the process of following a predetermined sequence of operations with little or no human labor, using specialized equipment and devices that perform and control manufacturing operations. The meaning and concept of automation has been variously interpreted as follows:

a) Semiautomatic or automatic material handling, workpiece loading and unloading in machines and workholding fixtures, and use of other labor-saving devices.
b) Automatic cycle control of machines and equipment, including use of mechanical devices, numerical control of machines, and use of computers.
c) Complete computer-based control of all aspects of manufacturing operations from raw materials to the finished product.

Automation can be fully achieved with various devices, sensors, actuators, techniques, and equipment that are capable of observing the manufacturing process, making decisions concerning the changes that should be made in the operation, and controlling all aspects of the operation. Automation is and will continue to be an *evolutionary*, rather than a revolutionary, concept. All of us are familiar with the evolution of automation, beginning with hand tools and hand-operated simple machines, continuing to mechanized processes and machines, and finally moving to higher levels of automation. In the rest of this chapter and in Chapter 14, we give several examples of the more recent stages in this important evolution.

Automation in manufacturing plants has been implemented successfully in the following basic areas of activity:

- *Manufacturing processes*. Machining, forging, cold extrusion, and grinding operations are examples of processes that have been automated extensively.
- *Material handling*. Materials and parts in various stages of completion are moved throughout a plant by computer-controlled equipment without human guidance.
- *Inspection*. Parts are automatically inspected for quality, dimensional accuracy, and surface finish, either at the time of manufacturing (in-process inspection), or after they are made (postprocess inspection).
- *Assembly*. Individually manufactured parts are automatically assembled into a product.
- *Packaging*. Products are packaged automatically.

Although metalworking processes were developed as early as 4000 B.C., it was not until the beginning of the Industrial Revolution in the 1750s that automation began to be introduced in the production of goods. As Table 1.1 in the General Introduction shows, machine tools, such as turret lathes, automatic screw machines, and automatic bottle-making equipment, were developed in the late 1890s and early 1900s. Mass-production techniques and transfer machines were developed in the 1920s. These machines had fixed automatic mechanisms and were designed to produce specific products. These developments were best represented by the automobile industry, which produced passenger cars at a high production rate and low cost.

The major breakthrough in automation began with numerical control (NC) of machine tools in the late 1940s. Since this historic development, rapid progress has been made in automating many aspects of manufacturing (Table 13.1). These involve the introduction of computers into automation, computerized numerical

TABLE 13.1
DEVELOPMENTS IN THE HISTORY OF AUTOMATION OF MANUFACTURING PROCESSES

DATE	DEVELOPMENT
1500–1600	Water power for metalworking; rolling mills for coinage strips.
1600–1700	Hand lathe for wood; mechanical calculator.
1700–1800	Boring, turning, and screw cutting lathe; drill press.
1800–1900	Copying lathe; turret lathe; universal milling machine; advanced mechanical calculators.
1808	Sheet-metal cards with punched holes for automatic control of weaving patterns in looms.
1863	Automatic piano player (Pianola).
1900–1920	Geared lathe; automatic screw machine; automatic bottlemaking machine.
1920	First use of the word *robot.*
1920–1940	Transfer machines; mass production.
1940	First electronic computing machine.
1943	First digital electronic computer.
1945	First use of the word *automation*.
1948	Invention of the transistor.
1952	First prototype numerical-control machine tool.
1954	Development of the symbolic language APT (Automatically Programmed Tool); adaptive control.
1957	Commercially available NC machine tools.
1959	Integrated circuits; first use of the term *group technology*.
1960s	Industrial robots.
1965	Large-scale integrated circuits.
1969	Programmable controllers.
1970	First integrated manufacturing system; spot welding of automobile bodies with robots.
1970s	Microprocessors; minicomputer controlled robot; flexible manufacturing systems; group technology; integrated manufacturing systems.
1980s	Artificial intelligence; intelligent robots; smart sensors; untended manufacturing cells.

control (CNC), adaptive control, industrial robots, and computer-integrated manufacturing (CIM) systems, including computer-aided design and computer-aided manufacturing (CAD/CAM).

13.2.1 Goals and applications of automation

Automation has several primary goals:

- Integrate various aspects of manufacturing operations, so as to improve product quality and uniformity, minimize cycle times and effort, and thus reduce labor costs.
- Improve productivity by reducing manufacturing costs through better control of production. Parts are loaded, fed, and unloaded on machines more efficiently. Machines are used more effectively and production is organized more efficiently.
- Reduce human involvement, boredom, and possibilities of human error.
- Reduce workpiece damage caused by manual handling of parts.
- Raise the level of safety for personnel, especially under hazardous working conditions.
- Economize on floor space in the manufacturing plant by arranging machines, material movement, and related equipment more efficiently.

Automation and production quantity. Production quantity is crucial in determining the type of machinery and equipment—and the level of automation—required to produce parts economically. Before we proceed further, let's define some basic production terms. *Total production quantity* is defined as the total number of parts to be made. This quantity can be produced in individual batches of various *lot sizes*. Lot size greatly influences the economics of production, as we describe in Chapter 14. *Production rate* is defined as the number of parts produced per unit time, such as per day, month, or year. The approximate and generally accepted ranges of production volume are shown in Table 13.2 for some typical applications. As you might expect, *experimental* or *prototype* products represent the lowest volume.

TABLE 13.2
APPROXIMATE ANNUAL VOLUME OF PRODUCTION

TYPE OF PRODUCTION	NUMBER PRODUCED	TYPICAL PRODUCTS
Experimental or prototype	1–10	—
Piece or small batch	10–5000	Aircraft, special machinery, dies
Batch or high volume	5000–100,000	Trucks, agricultural machinery, jet engines, diesel engines
Mass production	100,000 and over	Automobiles, appliances, fasteners

Small quantities per year (Fig. 13.1) can be manufactured in *job shops* using various standard general-purpose machine tools (*stand alone* machines) or machining centers. These operations have high part variety, meaning that different parts can be produced in a short time without extensive changes in production operations and tooling. However, in job shops machinery generally requires skilled labor to operate, production quantity and rate are low, and hence cost per part can be high (Fig. 13.2). Production of parts involving a large labor component is known as *labor intensive*.

Piece-part production usually involves very small quantities and is suitable for job shops. The majority of piece-part production is in lot sizes of 50 or less. *Small-batch* production quantities usually range from 10 to 100, and general-purpose machines and machining centers with various computer controls are used. *Batch* production usually involves lot sizes between 100 and 5000 and utilizes machinery similar to that used for small-batch production but with specially designed fixtures for higher production rates.

Mass production generally involves quantities of 100,000 and over and requires special-purpose machinery, called *dedicated machines*, and automated equipment for transferring materials and parts. Although the machinery, equipment, and specialized tooling are expensive, labor skills required and labor costs are relatively low because of the high level of automation. However, these production systems are organized for a specific product and, consequently, lack flexibility. Thus, as Fig.

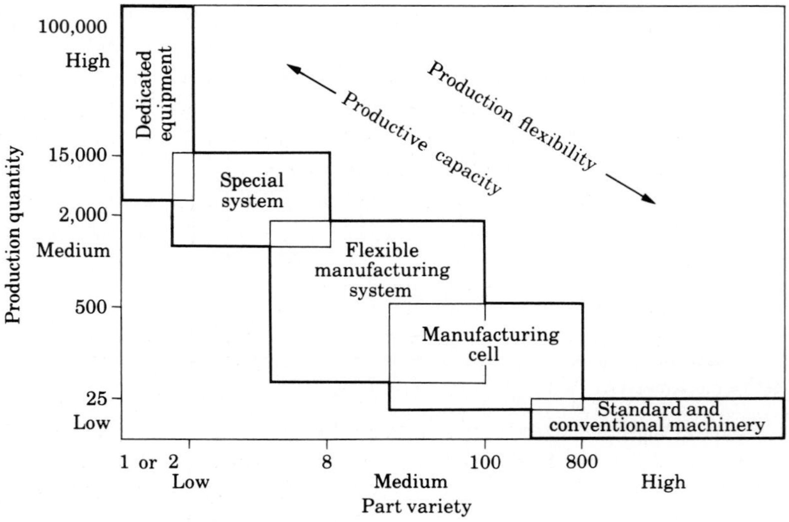

FIGURE 13.1
Capabilities of production systems and machines for various production quantities and part variety. See also Chapter 14.

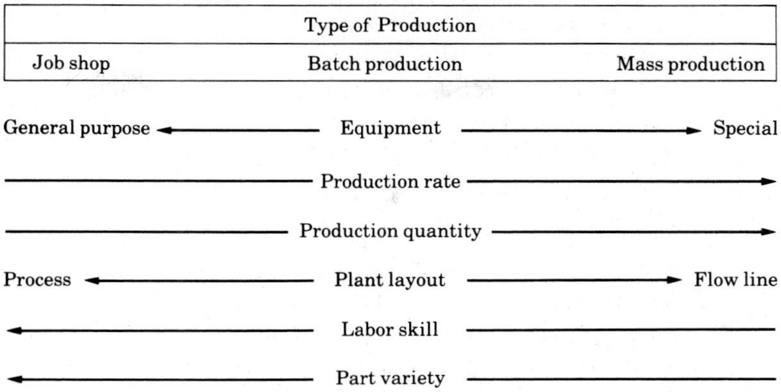

FIGURE 13.2
General characteristics of three types of production methods.

13.1 shows, production flexibility and productive capacity are inversely related. Most manufacturing facilities operate with a variety of machines in combination.

Applications of automation. Automation can be applied to manufacturing all types of goods, from raw materials to finished products, and in all types of production from job shops to large manufacturing facilities. The decision to automate a new or existing production facility requires the following additional considerations:

a) Type of product manufactured.
b) Quantity and rate of production required.
c) The particular phase of manufacturing operation to be automated.
d) High initial cost of equipment.
e) Reliability and maintenance problems associated with automated systems.
f) Economic benefit.

13.2.2 Hard automation

In *hard*, or *fixed-position*, *automation*, the production machines are designed to basically produce a standardized product, such as engine blocks, valves, gears, and spindles. Although product size and processing parameters (such as speed, feed, and depth of cut) can be changed, these machines are specialized. They lack flexibility and cannot be modified to any significant extent to accommodate products that have widely different shapes and dimensions (see Group Technology, Section 14.7). Because these machines are expensive to design and construct, their economic use requires mass production of parts in very large quantities.

Machines used in hard-automation applications are usually built on the *building-block*, or *modular*, principle. They are generally called *transfer machines*, and consist of the following two major components: powerhead production units and transfer mechanisms.

Powerhead production units. Consisting of a frame or bed, electric driving motors, gearboxes, and tool spindles (Fig. 13.3), *powerhead production units* are self-contained. The components are commercially available in various standard sizes and capacities. They can easily be regrouped for producing a different part and thus have certain adaptability and flexibility. Transfer machines consist of two or more powerhead units, which can be arranged on the shop floor in linear, circular, or U patterns. The weight and shape of parts influence the arrangement selected. The arrangement is also important for continuity of operation in the event of tool failure or machine breakdown in one or more of the units. Buffer storage features are incorporated in these machines to permit continued operation.

Transfer mechanisms. *Transfer mechanisms* are used to move the workpiece from one station to another in the machine—or from one transfer machine to another—to enable various operations to be performed on the part. Workpieces are transferred by several methods: (1) rails along which the parts, usually placed on pallets, are pushed or pulled by various mechanisms (Fig. 13.3a); (2) rotary indexing tables (Fig. 13.3b); and (3) overhead conveyors. Transfer of parts from station to station is usually controlled by sensors and other devices. Tools on transfer machines can be changed easily using toolholders with quick-change features. These machines may be equipped with various automatic gaging and inspection systems. These systems are utilized between operations to ensure that the dimensions of a part produced in one station are within acceptable tolerances before that part is transferred to the next station. Transfer machines are also used extensively in automatic assembly.

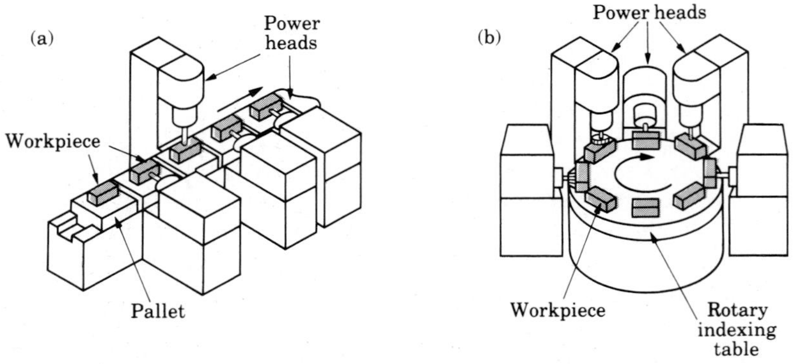

FIGURE 13.3
Arrangement of powerhead units in (a) straight and (b) circular patterns.

13.2.3 Soft automation

We stated that hard automation generally involves mass-production machines that lack flexibility. In *soft*, or *flexible* or *programmable*, *automation*, greater flexibility is achieved through numerical control of the machine and its various functions, using various programs. Soft automation is an important development, because the machine can be easily and readily reprogrammed to produce a part that has a different shape or dimensions than the one just produced. Because of this capability, soft automation can produce parts with complex shapes. Further advances in flexible automation, with extensive use of modern computers, has led to the development of *flexible manufacturing systems* (see Section 14.9), with high levels of efficiency and productivity.

13.3 Numerical Control

Numerical control (NC) is a method of controlling the movements of machine components by directly inserting coded instructions in the form of numerical data (numbers and letters) into the system. The system automatically interprets these data and converts them to output signals. These signals, in turn, control various machine components, such as turning spindles on and off, changing tools, moving the workpiece or the tools along specific paths, and turning cutting fluids on and off.

In order to appreciate the importance of numerical control of machines, let's briefly review how a process such as machining has been carried out traditionally. After studying the working drawings of a part, the operator sets up the appropriate process parameters (such as cutting speed, feed, depth of cut, cutting fluid, and so on), determines the sequence of operations to be performed, clamps the workpiece in a workholding device such as a chuck or collet, and proceeds to make the part. Depending on part shape and the dimensional accuracy specified, this approach usually requires skilled operators. Furthermore, the machining procedure followed may depend on the operator's judgment, and because of the possibilities of human error, the parts produced by the same operator may not all be identical. Part quality may thus depend on the particular operator or even the same operator on different days or different hours of the day. Because of our increased concern with product quality and reducing manufacturing costs, such variability and its effects on product quality are no longer acceptable. This situation can be eliminated by numerical control of the machining operation.

We can illustrate the importance of numerical control by the following example. Assume that holes have to be drilled on a part in the positions shown in

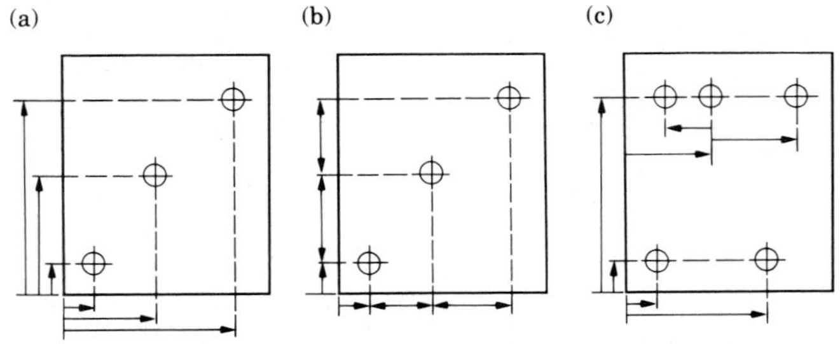

FIGURE 13.4
Positions of drilled holes in a workpiece. Three methods of measurements are shown: (a) absolute dimensioning, referenced from one point at the lower left of the part; (b) incremental dimensioning, made sequentially from one hole to another; and (c) mixed dimensioning, a combination of both methods.

Fig. 13.4. In the traditional manual method of machining this part, the operator positions the drill with respect to the workpiece, using as reference points any of the three methods shown. The operator then proceeds to drill these holes. Let's assume that 100 parts, having exactly the same shape and dimensional accuracy, have to be drilled. Obviously, this operation is going to be tedious because the operator has to go through the same motions again and again. Moreover, the probability is high that, for various reasons, some of the parts machined will be different from others. Let's further assume that during this production run, the order for these parts is changed, so that 10 of the parts now require holes in different positions. The machinist now has to reset the machine, which will be time consuming and subject to error. Such operations can be performed easily by numerical control machines that are capable of producing parts repeatedly and accurately and of handling different parts by simply loading different part programs.

In numerical control, data concerning all aspects of the machining operation, such as locations, speeds, feeds, and cutting fluid, are stored on magnetic tape, cassette, floppy or hard disks, or paper or plastic (Mylar, which is a thermoplastic polyester) tape. Data are stored on punched 25-mm (1-in.) wide paper or plastic tape, as originally developed and still used. The concept of NC control is that holes in the tape represent specific information in the form of alphanumeric codes. The presence (on) or absence (off) of these holes is read by sensing devices in the control panel, which then actuate relays and other devices (called *hard-wired controls*). These devices control various mechanical and electrical systems in the machine. This method eliminates manual setting of machine positions and tool paths or the use of templates and other mechanical guides and devices. Complex operations, such as turning a part having various contours and die sinking in a milling machine, can be carried out.

Numerical control machines are now used extensively in small- and medium-quantity production (typically 500 parts or less) of a wide variety of parts in small shops and large manufacturing facilities. Older machines can be retrofitted with numerical control.

The basic concept of numerical control probably was implemented in the early 1800s, when punched holes in sheet metal cards were used to automatically control weaving machines. Needles were activated by sensing the presence or absence of a hole in the card. This invention was followed by automatic piano players (Pianola), in which the keys were activated by air flowing through holes punched in a perforated roll of paper.

The principle of numerically controlling the movements of machine tools was conceived in the 1940s by J. Parsons in his attempt to machine complex helicopter blades. The first prototype NC machine was built in 1952 at the Massachusetts Institute of Technology. It was a vertical-spindle, two-axis copy milling machine, retrofitted with servomechanisms, and the machining operations performed consisted of end milling and face milling on an aluminum plate. The numerical data to be punched into the paper tapes were generated by a digital computer, which was being developed at the same time at MIT. The experiments successfully machined parts accurately and repeatedly without operator intervention. On the basis of this success, the machine-tool industry began building and marketing NC machine tools, the latest developments in which are machining centers.

Numerical control has the following advantages over conventional methods of machine control:

- Flexibility of operation and ability to produce complex shapes with good dimensional accuracy, repeatability, reduced scrap loss, and high production rates, productivity, and product quality.
- Tooling costs are reduced, since templates and other fixtures are not required.
- Machine adjustments are easy to make with minicomputers and digital readouts.
- More operations can be performed with each setup, and less lead time for setup and machining is required compared to conventional methods. Design changes are facilitated, and inventory is reduced.
- Programs can be prepared rapidly and can be recalled at any time utilizing microprocessors. Less paperwork is involved.
- Faster prototype production is possible.
- Required operator skill is less, and the operator has more time to attend to other tasks in the work area.

The major limitations of NC are the relatively high initial cost of the equipment and the need for programming and special maintenance, requiring trained personnel. Because NC machines are complex systems, breakdowns can be very costly, so preventive maintenance is essential. However, these limitations are often easily outweighed by the overall economic advantages of NC.

In the next step in the development of numerical control, the control hardware mounted on the NC machine was converted to local computer control with software. Two types of computerized systems were developed: direct numerical control and computer numerical control.

In *direct numerical control* (DNC), as originally conceived and developed in the 1960s, several machines are directly controlled step by step by a central main frame computer. In this system, the operator has access to the central computer through a remote terminal. Thus handling tapes and the need for computers on each machine are eliminated. With DNC, the status of all machines in a manufacturing facility could be monitored and assessed from the central computer. However, DNC had the crucial disadvantage that if the computer went down, all the machines became inoperative.

A more recent definition of DNC (now meaning *distributed* numerical control) includes the use of a central computer serving as the control system over a number of individual computer numerical control machines with onboard minicomputers. This system provides large memory and computational capabilities, thus offering flexibility, while overcoming the previous disadvantage of DNC.

Computer numerical control (CNC) is a system in which a minicomputer or microprocessor is an integral part of the control panel of a machine or equipment (onboard computer). The part program may be prepared at a remote site by the programmer. However, the machine operator can now easily and manually program onboard computers. The operator can modify the programs directly, prepare programs for different parts, and store the programs. Because of the availability of small computers with large memory, microprocessors, and program editing capabilities, CNC systems are widely used today. We cannot overemphasize the importance of the availability of low-cost, programmable controllers in the successful implementation of CNC in manufacturing plants.

The advantages of CNC over conventional NC systems are:

- Increased flexibility. The machine can produce a certain part, followed by other parts with different shapes and at reduced cost.
- Greater accuracy.
- More versatility. Editing and debugging programs, reprogramming, and plotting and printing part shape are simpler.

13.3.1 Principles of NC machines

The basic elements and operation of a typical NC machine are outlined in Fig. 13.5. The functional elements in numerical control and the components involved are:

a) *Data input*. Numerical information is read and stored in the tape or in computer memory.
b) *Data processing*. The programs are read into the machine control unit for processing.
c) *Data output*. This information is translated into commands, typically pulsed commands to the servomotor (Fig. 13.6). The servomotor moves the table on which the workpiece is placed to specific positions, through linear or rotary movements, by means of stepping motors, lead screws, and other devices.

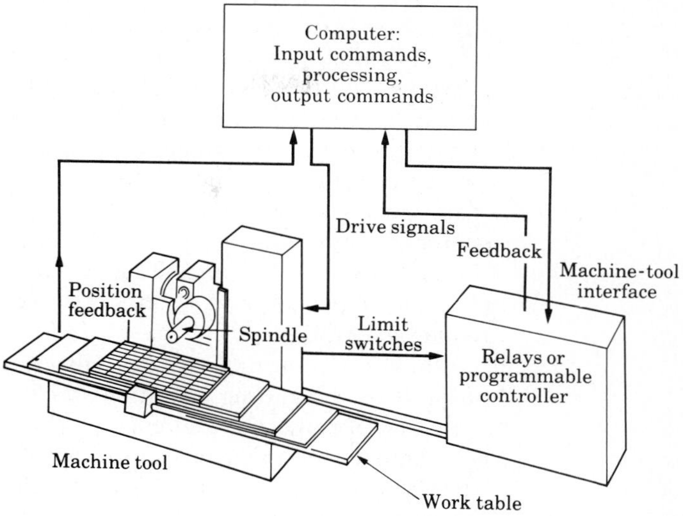

FIGURE 13.5 Schematic illustration of the major components of a numerical control machine tool. *Source: Flexible Automation* 87/88, *The International CNC Reference Book.*

Types of control circuits. An NC machine can be controlled through two types of circuits: open loop and closed loop. In the *open-loop* system (Fig. 13.6a), the signals are given to the servomotor by the processor, but the movements and final destinations of the work table are not checked for accuracy. The *closed-loop* system (Fig. 13.6b) is equipped with various transducers, sensors, and counters that measure the position of the table accurately. Through *feedback* control, the position of the table is compared against the signal. Table movements terminate when the proper coordinates are reached. The closed-loop system is more complicated and more expensive than the open-loop system.

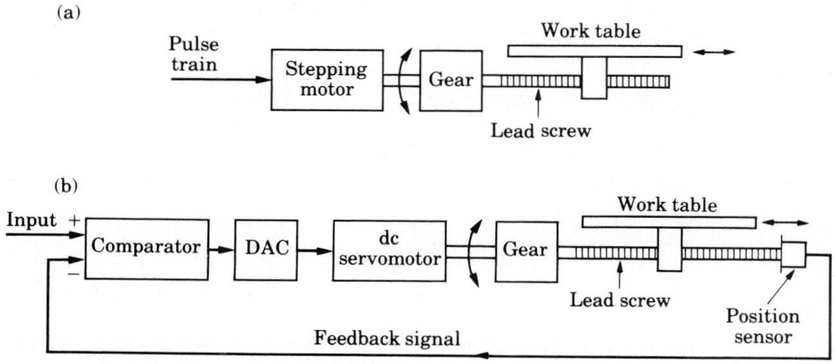

FIGURE 13.6 Schematic illustration of the components of (a) open-loop and (b) closed-loop control systems for a numerical control machine. DAC means digital-to-analog converter.

Types of control systems. There are two basic types of control systems in numerical control: point-to-point and contouring. In the *point-to-point* system, also called *positioning*, each axis of the machine is driven separately by lead screws and, depending on the type of operation, at different velocities. The machine moves initially at maximum velocity in order to reduce nonproductive time but decelerates as the tool reaches its numerically defined position. Thus in an operation such as drilling or punching, the positioning and cutting take place sequentially (Fig. 13.7a). After the hole is drilled or punched, the tool retracts, moves rapidly to another position, and repeats the operation. The path followed from one position to another is important in only one respect: The time required should be minimized for efficiency. Point-to-point systems are used mainly in drilling, punching, and straight milling operations.

In the *contouring* system, also known as the *continuous path* system, positioning and cutting operations are both along controlled paths but at different velocities. Because the tool cuts as it travels along a prescribed path (Fig. 13.7b), accurate control and synchronization of velocities and movements are important. The contouring system is used on lathes, milling machines, grinders, welding machinery, and machining centers.

Movement along the path, or *interpolation*, occurs incrementally, by one of several basic methods (Fig. 13.8). Examples of actual paths in drilling, boring, and milling operations are shown in Fig. 13.9. In all interpolations, the path controlled is that of the center of rotation of the tool. Compensation for different tools, different diameter tools, or tool wear during machining, can be made in the NC program.

In *linear* interpolation, the tool moves in a straight line from start to end in two or three axes. Theoretically, all types of profiles can be produced by this method by making the increments between the points small. However, a large amount of data has to be processed in order to do so. In *circular* interpolation, the input required

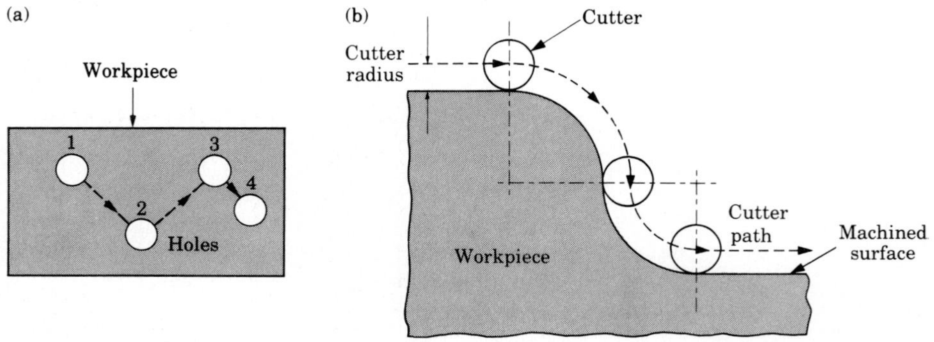

FIGURE 13.7 Movement of tools in numerical-control machining. (a) Point-to-point, in which the drill bit drills a hole at position 1, is retracted and moved to position 2, and so on. (b) Continuous path by a milling cutter. Note that the cutter path is compensated for by the cutter radius. This path can also be compensated for cutter wear.

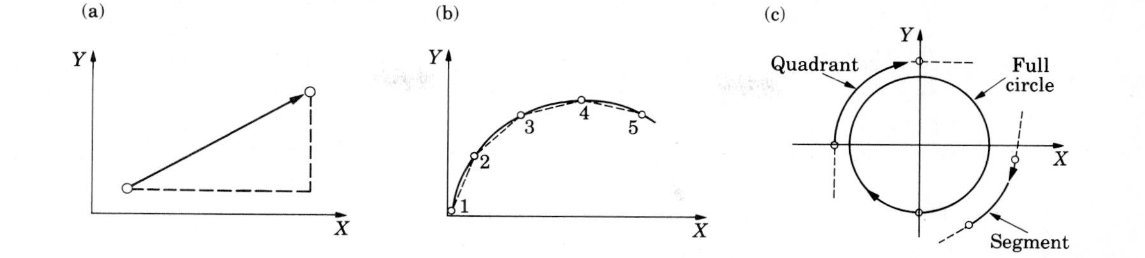

FIGURE 13.8
Types of interpolation: (a) linear, (b) continuous path approximated by incremental straight lines, and (c) circular.

for the path is the coordinates of the end points, the coordinates of the center of the circle and its radius, and the direction of the tool along the arc. In *parabolic* and *cubic* interpolation, the path is approximated by curves using higher order mathematical equations. This method is effective in 4-axis or 5-axis machines and is particularly useful in die sinking operations for automotive bodies. These interpolations are also used for industrial robot movements.

Accuracy in numerical control. Positioning accuracy in NC machines is defined by how accurately the machine can be positioned to a certain coordinate system. An NC machine usually has a positioning accuracy of at least $\pm 3\ \mu m$ (0.0001 in.). *Repeatability*, defined as the closeness of agreement of repeated position movements under the same operating conditions of the machine, is usually around $\pm 8\ \mu m$ (0.0003 in.). *Resolution*, defined as the smallest increment of motion of the machine components, is usually about 2.5 μm (0.0001 in.).

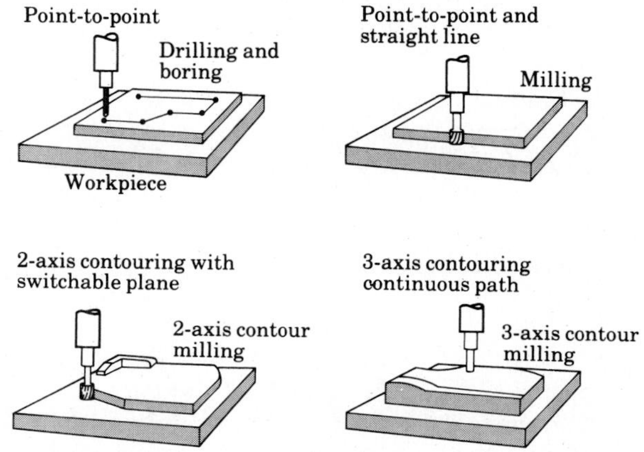

FIGURE 13.9
Schematic illustration of drilling, boring, and milling with various paths.

The *stiffness* of the machine tool and *backlash* in its gear drives and lead screws are important for accuracy. Backlash can be eliminated with special backlash take-up circuits, whereby the tool always approaches a particular position on the workpiece from the same direction, as should be done in traditional machining operations. Rapid response to command signals requires that friction and inertia be minimized, say, by reducing the mass of moving components of the machine.

13.3.2 Programming for numerical control

A program for numerical control consists of a sequence of directions that causes an NC machine to carry out a certain operation, machining being the most commonly used process. *Programming for NC* may be done by an internal programming department, on the shop floor, or purchased from an outside source. Also, programming may be done manually or with computer assistance.

The program contains instructions and commands. Geometric instructions pertain to relative movements between the tool and the workpiece. Processing instructions pertain to spindle speeds, feeds, tools, and so on. Travel instructions pertain to the type of interpolation and slow or rapid movements of the tool or work table. Switching commands pertain to on/off position for coolant supplies, spindle rotation, direction of spindle rotation, tool changes, workpiece feeding, clamping, and so on.

Manual programming. *Manual part programming* consists of first calculating dimensional relationships of the tool, workpiece, and work table, based on the engineering drawings of the part, and manufacturing operations to be performed and their sequence. A program sheet is then prepared, which consists of the necessary information to carry out the operation, such as cutting tools, spindle speeds, feeds, depth of cut, cutting fluids, power, and tool or workpiece relative positions and movements. Based on this information, the part program is prepared. Usually a paper tape is first prepared for trying out and debugging the program. Depending on how often it is to be used, the tape may be made of more durable Mylar.

Manual programming can be done by someone knowledgeable about the particular process and able to understand, read, and change part programs. Because they are familiar with machine tools and process capabilities, skilled machinists can do manual programming with some training in programming. However, the work is tedious, time consuming, and uneconomical—and is used mostly in simple point-to-point applications.

Computer-aided programming. *Computer-aided part programming* involves special symbolic programming languages that determine the coordinate points of corners, edges, and surfaces of the part. *Programming language* is the means of communicating with the computer and involves the use of symbolic characters. The programmer describes the component to be processed in this language, and the computer converts it to commands for the NC machine. Several languages having various features and applications are commercially available. The first language that used English-like statements was developed in the late 1950s and is called APT (for Automatically Programmed Tools). This language, in its various expanded forms, is still the most widely used for both point-to-point and continuous-path programming.

Computer-aided part programming has the following significant advantages over manual methods:

- Use of symbolic language that is relatively easy to use. Several programs have been developed: ADAPT, IFAPT, MINIAPT, UNIAPT, AUTOSPOT, AUTOMAP, AUTOPROMPT, CAMPI, SPLIT, and COMPACT II.
- Reduced programming time. Programming is capable of accommodating a large amount of data concerning machine characteristics and process variables, such as power, speeds, feed, tool shape, compensation for tool shape changes, tool wear, deflections, and coolant use.
- Reduced possibility of human error, which can occur in manual programming.
- Capability of simple changeover of machining sequence or from machine to machine.
- Lower cost because less time is required for programming.

Selection of a particular NC programming language depends on the following factors:

a) Level of expertise of the personnel in the manufacturing facility.
b) Complexity of the part.
c) Type of equipment and computers available.
d) Time and costs involved in programming.

Because numerical control involves the insertion of data concerning workpiece materials and processing parameters, programming must be done by operators or programmers who are knowledgeable about the relevant aspects of the manufacturing processes being used. Before production begins, programs should be verified, either by viewing a simulation of the process on a CRT screen or by making the part from an inexpensive material, such as aluminum, wood, or plastic, rather than the material specified for the finished part.

13.4 Adaptive Control

Adaptive control (AC) is defined as automatic on-line adjustment of process parameters in manufacturing operations. The purposes of adaptive control are to:

- Optimize production rate.
- Optimize product quality.
- Minimize cost.

Although adaptive control has been used widely in continuous processing in the chemical industry and oil refineries for some time, its successful application to machining, grinding, forming, and other manufacturing processes is relatively recent. Application of AC in manufacturing operations is particularly important in situations where workpiece dimensions and quality is not uniform, such as a poor casting or an improperly heat-treated part.

Developed in the late 1950s, adaptive control is a logical extension of computer numerical control systems. The part programmer sets the processing parameters, on the basis of existing knowledge of the material and data on the particular manufacturing process. In CNC machines, these parameters are held constant during a particular process cycle. In AC, on the other hand, the system is capable of automatic adjustments during processing through closed-loop feedback control (Fig. 13.10).

13.4.1 Principles and applications of adaptive control

The basic functions common to adaptive control systems are to:

a) Identify unknown parameters or measure performance, using sensors that detect parameters such as force, torque, vibration, and temperature.
b) Decide on a control strategy. Based on preset upper and lower thresholds, the system determines whether a particular measurement is exceeding the index of performance.
c) Modify the process parameters through input to the controller.

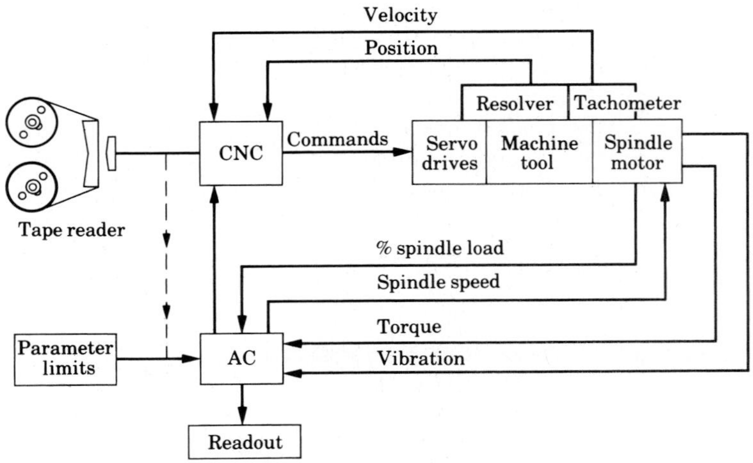

FIGURE 13.10
Schematic illustration of the application of adaptive control (AC) for a turning operation. The system monitors parameters such as cutting force, torque, and vibrations, and if excessive, modifies process variables such as feed and depth of cut to bring them to acceptable levels. *Source: Flexible Automation* 87/88, *The International CNC Reference Book*.

You may recognize that human reactions to occurrences in everyday life already contain adaptive control. When you drive a car on a rough road, you know how to steer to avoid potholes by visually and continuously observing the condition of the road, or your body feels the car's movements and vibrations. You then react by changing direction and speed to minimize the effects of the rough road on your car and to increase the comfort of the ride.

In an operation such as turning on a lathe, the adaptive control system senses real-time cutting forces, torque, temperature, tool wear rate, tool chipping or fracture, and surface finish of the workpiece. The system converts this information into commands that modify the process parameters on the machine tool to hold them constant (or within certain limits) or to optimize the cutting operation.

Those systems that place a constraint on a process variable (such as forces, torque, or temperature) are called *adaptive control constraint* (ACC) systems. Thus if the thrust force, the cutting force, and hence the torque increase excessively—because of a hard region in a cast workpiece, say—the adaptive control system changes the speed and/or feed to lower the cutting forces to acceptable levels (Fig. 13.11). Without adaptive control or direct operator intervention as in traditional machining operations, high forces may cause tools to chip or break and the workpiece to deflect excessively, thus losing dimensional accuracy. Systems that optimize an operation are called *adaptive control optimization* (ACO) systems. Optimization may involve maximizing material removal rate between tool changes or resharpening or improving surface finish. Most systems are currently based on ACC because development and implementation of ACO is complex.

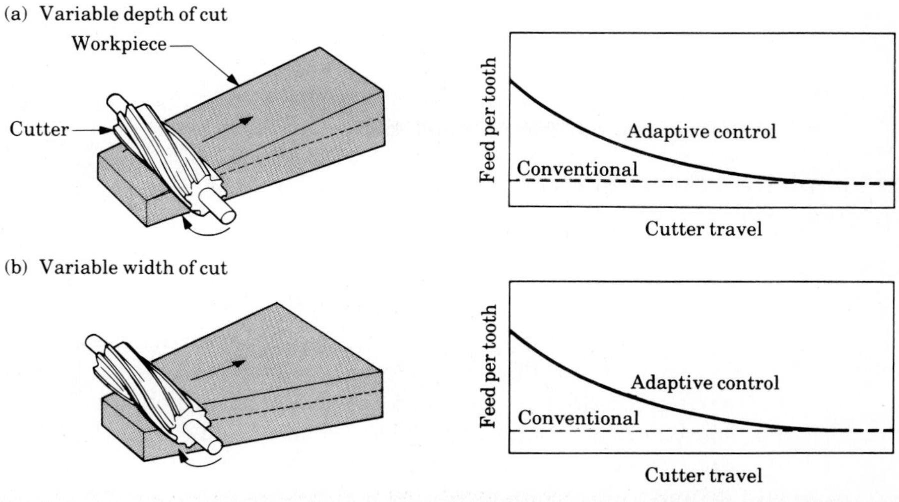

FIGURE 13.11
An example of adaptive control in milling. As the depth or width of cut increases, the cutting forces and torque increase. The system senses this increase and automatically reduces the feed to avoid excessive forces or tool breakage and to maintain cutting efficiency. *Source:* Y. Koren.

Response time must be short for AC to be effective, particularly in high-speed operations. Assume, for example, that a turning operation is being performed on a lathe at a spindle speed of 1000 rpm, and the tool suddenly breaks, adversely affecting the surface finish and dimensional accuracy of the part. In order for the AC system to be effective, the sensing system must respond within a very short time; otherwise the damage to the workpiece will be extensive.

For adaptive control to be effective in manufacturing operations, quantitative relationships must be known and stored in the computer software as mathematical models. For example, if tool wear rate in a machining operation is excessive, the computer must be able to know how much of a change (and whether increase or decrease) in speed and/or feed is necessary to reduce the wear rate to an acceptable level. The system should also be able to compensate for dimensional changes on the workpiece caused by tool wear and temperature rise.

If the operation is grinding, for example, the computer software must contain quantitative relationships among process variables (wheel and work speeds, feed, type of wheel) and parameters such as wheel wear, dulling of abrasive grains, grinding forces, temperature, surface finish, and part deflections. Similarly, for bending of a sheet in a V-die, data on the dependence of springback on punch travel and other material and process variables must be stored in the computer software. Because of the many factors involved, mathematical equations for such quantitative relationships in manufacturing processes are difficult to establish. Compared to the other parameters involved, forces and torque have been found to be the easiest to monitor by AC. Various solid-state power controls are available commercially, in which power is displayed or interfaced with data acquisition systems. Coupled with CNC, adaptive control is potentially a powerful tool in optimizing manufacturing operations.

13.5 Material Handling and Movement

During a typical manufacturing operation, raw materials and parts are moved from storage to machines, from machine to machine, from inspection to assembly and inventory, and finally to shipment. Workpieces are loaded on machines (a forging is mounted on a milling machine bed, or sheet metal is fed into a press for stamping), parts are removed from one machine and loaded on another (a machined forging is subsequently ground), and parts are inspected for flaws and dimensional accuracy prior to being assembled into a finished product. Similarly, tools, molds, dies, and various other equipment and fixtures are also moved in manufacturing plants. Cutting tools are mounted on lathes, dies are placed in presses or hammers, grinding wheels are mounted on spindles, and parts are mounted on special fixtures for dimensional measurement and inspection.

These materials must be moved either manually or by some mechanical means, and time is required to transport them from one location to another. Thus we may define *material handling* as the functions and systems associated with the transportation, storage, and control of materials and parts in the total manufacturing cycle of a product. The total time required for manufacturing depends on part size and shape and the set of operations required. Idle time and the time required for transporting materials can constitute the majority of the time consumed (Fig. 13.12). Because handling materials adds cost but not value to a product, it should be reduced as much as possible.

Plant layout is an important aspect of the flow of materials and components throughout the manufacturing cycle. The arrangement of production machinery and material-handling equipment should be orderly and efficient. The time and distances required for moving raw materials and parts should be minimized, and storage areas and service centers should be organized accordingly. For parts requiring multiple operations, equipment should be grouped around the operator or the industrial robot (see Cellular Manufacturing, Section 14.8).

Material handling should therefore be an integral part of planning, implementing, and controlling manufacturing operations. Furthermore, material handling should be repeatable and predictable. Here we define repeatability as the closeness of agreement of repeated positions under the same conditions to the same location. Consider, for example, what happens if a part or workpiece is loaded improperly in a forging die or in the chuck of a lathe. The consequences of such action may well be broken dies and tools, improperly made parts, or parts that are out of tolerance. This action can also present safety hazards and possibly cause injury to the operator, as well as to other nearby personnel.

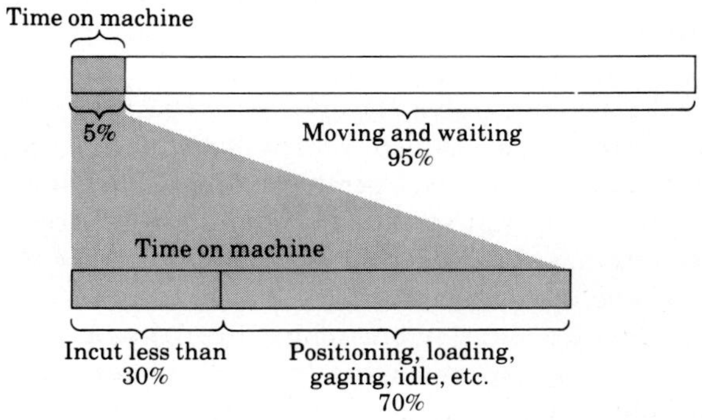

FIGURE 13.12

Time spent on various phases in a typical machining operation. Note that very little time is spent on actual machining. *Source:* Cincinnati Milacron, Inc.

13.5.1 Methods of material handling

Several factors have to be considered in choosing a suitable material-handling method for a particular manufacturing operation:

a) Shape, weight, and characteristics of parts.
b) Types of movement and distances involved and the position and orientation of parts during movement and at their final destination.
c) Conditions of the path along which parts are to be transported.
d) Degree of automation and control desired and integration with other systems and equipment.
e) Operator skill required.
f) Economic considerations.

For small-batch manufacturing operations, raw materials and parts can be handled and transported by hand, but this method is generally costly. Moreover, because it involves human beings, this practice can be unpredictable and unreliable and can be unsafe to the operator, depending on the weight and shape of the parts to be moved and environmental factors, such as heat and smoke in foundries and forging plants. In automated manufacturing plants, computer controlled material and parts flow is being rapidly implemented. These changes have improved repeatability and lowered labor costs.

13.5.2 Equipment

Various types of equipment can be used to move materials, such as conveyors, rollers, self-powered monorails, carts, forklift trucks, and various mechanical, electrical, magnetic, pneumatic, and hydraulic devices and manipulators. *Manipulators* are designed to be controlled directly by the operator, or they are automated for repeated operations, such as loading and unloading parts from machine tools, presses, and furnaces. Manipulators are capable of gripping and moving heavy parts and orienting them as required between manufacturing and assembly operations. Industrial robots, specially designed pallets, and *automated guided vehicles* (AGVs) are used extensively in flexible manufacturing systems to move parts and orient them as required (Fig. 13.13). Thus flexible material handling and movement with real-time control has become an integral part of modern manufacturing.

Automated guided vehicles, which are the latest development in material movement in plants, operate automatically along pathways with in-floor wiring or tapes for optical scanning and without any operator intervention. This transport system has great flexibility and is capable of random delivery to different workstations. It optimizes the movement of materials and parts in cases of congestion around workstations, machine breakdown (downtime), or failure of part of the system. The movements of AGVs are planned so that they interface with *automated storage/retrieval systems* (AS/RS). The latter utilizes warehouse spaces efficiently and reduces manpower.

Coding systems have been developed to locate and identify parts throughout the manufacturing system and to transfer them to their appropriate stations. *Bar codes*

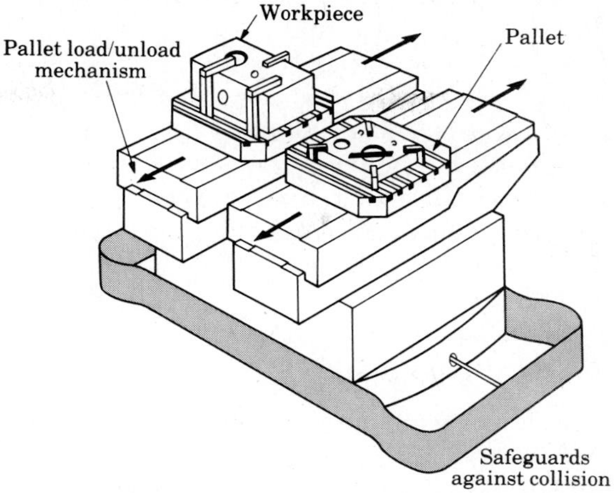

FIGURE 13.13
Schematic illustration of an automated guided vehicle (AGV), showing two different workpieces on pallets. *Source:* P. Ranky.

are the most widely used and the least costly. They are printed on labels, attached to the parts themselves, and read by fixed or portable code readers using light pens. Other identification systems are based on acoustic waves, optical character recognition, and machine vision (see Section 13.7).

13.6 Industrial Robots

The word *robot* was coined in 1920 by the Czech author K. Čapek in his play R.U.R. (Rossum's Universal Robots), and is derived from the word *robota*, meaning work. An *industrial robot* is a reprogrammable multifunctional manipulator, designed to move materials, parts, tools, or other specialized devices by means of variable programmed motions and to perform a variety of other tasks. The term robot also includes manipulators that are activated directly by an operator.

More generally, an industrial robot has been described by the International Standards Organization (ISO) as follows: A machine formed by a mechanism including several degrees of freedom, often having the appearance of one or several arms ending in a wrist capable of holding a tool, a workpiece, or an inspection device. In particular, its control unit must use a memorizing device and it may sometimes use sensing or adaptation appliances to take into account environment and circumstances. These multipurpose machines are generally designed to carry out a repetitive function and can be adapted to other functions.

Introduced in the early 1960s, the first industrial robots were used in hazardous operations, such as handling toxic and radioactive materials and loading and unloading hot workpieces from furnaces and handling them in foundries. Some rule-of-thumb applications for robots are the three D's (dull, dirty, and dangerous, including demeaning but necessary tasks), and the three H's (hot, heavy, and hazardous). From their early uses for worker protection and safety in manufacturing plants, industrial robots have been further developed to improve productivity, increase product quality, and reduce labor costs. Computer-controlled robots were commercialized in the early 1970s, with the first robot controlled by a minicomputer appearing in 1974.

13.6.1 Components

While learning about robot components and capabilities, you might simultaneously observe the flexibility and capability of diverse movements of your arm, wrist, hand, and fingers in reaching for and grabbing an object from a shelf or in operating your car. Figure 13.14 shows the basic components of an industrial robot.

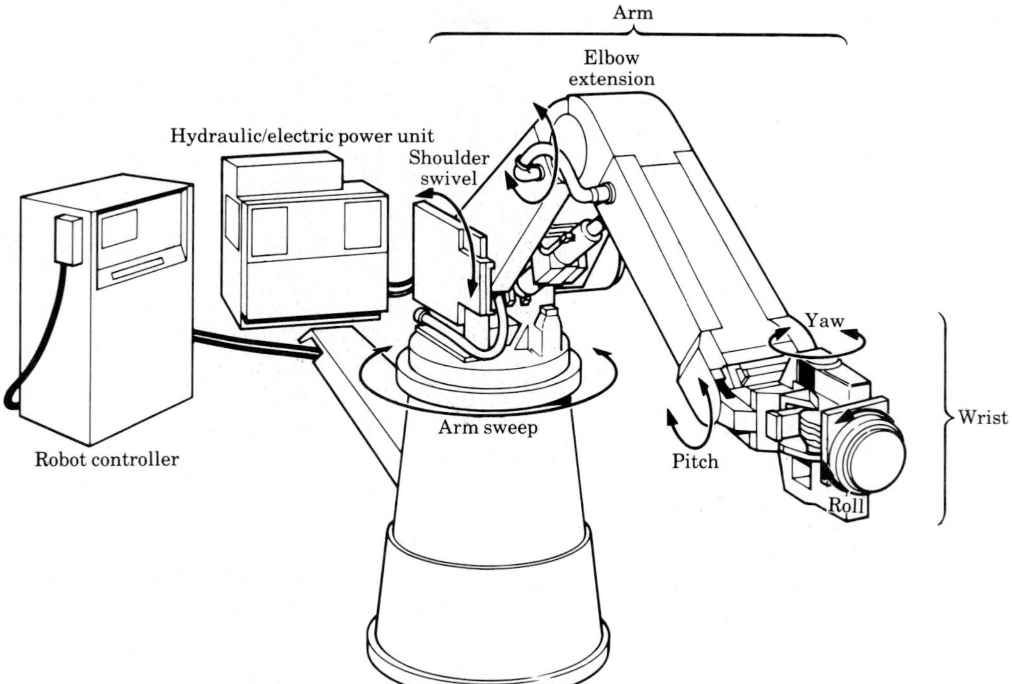

FIGURE 13.14
Components of a typical industrial robot. *Source:* Cincinnati Milacron, Inc.

Manipulator. Also called *arm and wrist*, the *manipulator* is a mechanical unit that provides motions (trajectories) similar to that of a human arm and hand. The end of the wrist can reach a point in space with a specific orientation. There are three degrees of freedom each in linear and rotational movements, respectively. Manipulation is carried out using mechanical devices, such as linkages, gears, and various joints.

End effector. The end of the wrist in a robot is equipped with an *end-effector*, also called *end-of-arm tooling*. Depending on the type of operation, conventional end effectors are equipped with:

a) Grippers, hooks, scoops, electromagnets, vacuum cups, and adhesive fingers, for materials handling.
b) Spray gun for painting.
c) Attachments for spot and arc welding and arc cutting.
d) Power tools, such as drills, nut drivers, and burs.
e) Measuring instruments, such as dial indicators.

End effectors are generally custom made to meet special handling requirements. Mechanical grippers are the most commonly used and are equipped with either two or more fingers. The selection of an appropriate end effector for a specific application depends on factors such as the payload, environment, reliability, and cost.

Power supply. Each motion of the manipulator, in linear and rotational axes, is controlled and regulated by independent actuators, using electric, pneumatic, or hydraulic power supplies. Each source of energy and the motors involved have their own characteristics, advantages, and limitations.

Control system. Also known as the *controller*, the *control system* is the communications and information processing system that gives commands for the movements of the robot. It is the brain of the robot and stores data to initiate and terminate movements of the manipulator. It interfaces with computers and other equipment, such as manufacturing cells or assembly systems.

Feedback devices, such as transducers, are an important part of the control system. They transmit information to the control system on the position of various robot joints and linkages. In closed-loop control, the system automatically measures the degree to which the robot's movements conform to the desired response. It then utilizes this feedback to drive the system into conformance. Hence, the system has a self-correcting capability.

Robots with a fixed set of motions have open-loop control. In this system commands are given and the robot arm goes through its motions, but—unlike feedback in closed-loop systems—accuracy of the movements is not monitored. The system does not have a self-correcting capability.

As in numerical control machines, the types of control in industrial robots are point-to-point and continuous-path. Depending on the particular task, the positioning repeatability required may be as small as 0.050 mm (0.002 in.), as in assembly operations for electronic printed circuitry. Specialized robots can reach such accuracy, although most robots are unable to do so.

13.6.2 Classification

Robots may be classified by basic type: (a) cartesian or rectilinear, (b) cylindrical, (c) spherical or polar, and (d) articulated, or revolute, jointed, or anthropomorphic (Fig. 13.15). Robots may be attached permanently to the floor of a manufacturing plant, may move along overhead rails (*gantry robot*), or may be equipped with wheels to move along the factory floor (*mobile robot*). However, a broader classification of robots currently in use is most helpful for our purposes here. We use it to describe below various types of robots.

Fixed- and variable-sequence robots. The *fixed-sequence robot*, also called a *pick-and-place robot*, is programmed for a specific sequence of operations. Its movements are from point to point, and the cycle is repeated continuously. These robots are simple and relatively inexpensive. The *variable-sequence robot* can be programmed for a specific sequence of operations but can be reprogrammed to perform another sequence of operations.

Playback robot. An operator leads, or walks, the *playback robot* and its end effector through the desired path. In other words, the operator teaches the robot by showing it what to do. The robot memorizes and records the path and sequence of motions and can repeat them continuously without any further action or guidance by the operator. Another method, the *teach pendant*, utilizes hand-held button boxes that are connected to the control panel and used to control and guide the robot and its tooling through the work to be performed. These movements are then registered in the memory of the controller and automatically reenacted by the robot.

Numerically controlled robot. The *numerically controlled robot* is programmed and operated much like a numerically controlled machine. The robot is

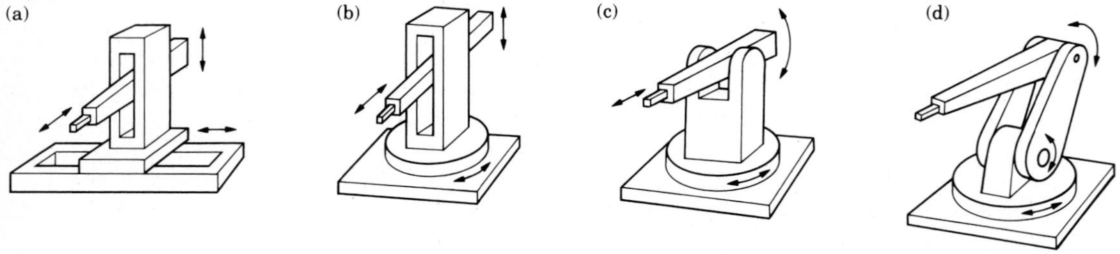

FIGURE 13.15
Types of industrial robots: (a) cartesian (rectilinear), (b) cylindrical, (c) spherical (polar), and (d) articulated (revolute, jointed, or anthropomorphic).

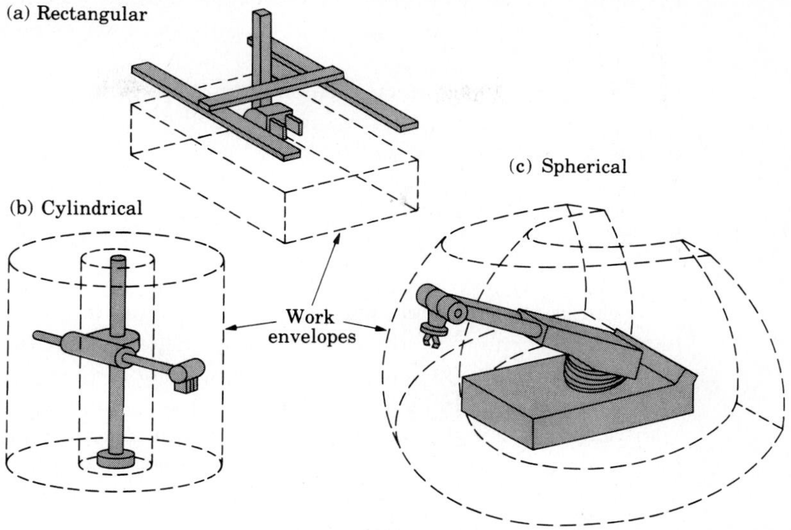

FIGURE 13.16
Work envelopes for three types of robots. The choice depends on the particular application.

servocontrolled by digital data, and its sequence of movements can be changed with relative ease. As in NC machines, there are two basic types of controls: point-to-point and continuous path. Point-to-point robots are easy to program and have higher load-carrying capacity and *work envelope* (the maximum extent or reach of the robot hand or working tool in all directions, also called working envelope; Fig. 13.16). Continuous path robots have greater precision than point-to-point robots, but have lower load-carrying capacity. Some advanced robots have a complex system of path control, enabling high speed movements with great accuracy.

Intelligent (sensory) robot. The *intelligent*, or *sensory, robot*, is capable of sensory perception, either tactile (touching) or nontactile (that is, nontouching, as by vision systems; Section 13.7). Much like human beings, the robot observes and evaluates the immediate environment by perception and pattern recognition, makes appropriate decisions for the next movement, and proceeds with them. The functioning of a sensory robot requires a visual or tactile capability to judge proximity to objects. Because its operation is so complex, powerful computers are required to control this type of robot.

13.6.3 Applications and selection of robots

Major applications of industrial robots include the following.

a) Material handling, loading, unloading, and transferring workpieces in manufacturing operations. Examples are casting and molding.

b) Spot welding automobile and truck bodies, producing welds of good quality. Robots perform other, similar operations, such as arc welding, arc cutting, and riveting.
c) Machining operations, such as deburring, grinding, and polishing, with appropriate tools attached to their end effectors.
d) Applying adhesives and sealants.
e) Spray painting, particularly of complex shapes, and cleaning operations.
f) Automated assembly (Fig. 13.17).
g) Inspection and gaging in various stages of manufacture.

Factors that influence the selection of robots in manufacturing plants are: (a) load-carrying capacity, (b) speed of movement, (c) reliability, (d) repeatability, (e) arm configuration, (f) degrees of freedom, (g) control system, (h) program memory, and (i) work envelope. In addition to these technical factors, cost and benefit considerations are significant aspects of robot selection and their use. The increasing availability, reliability, and reduced costs of sophisticated intelligent robots are having a major economic impact on manufacturing operations and gradually replacing human labor. Whereas hourly wages are steadily rising, particularly in industrial nations, the cost of robot operation per hour has increased more slowly.

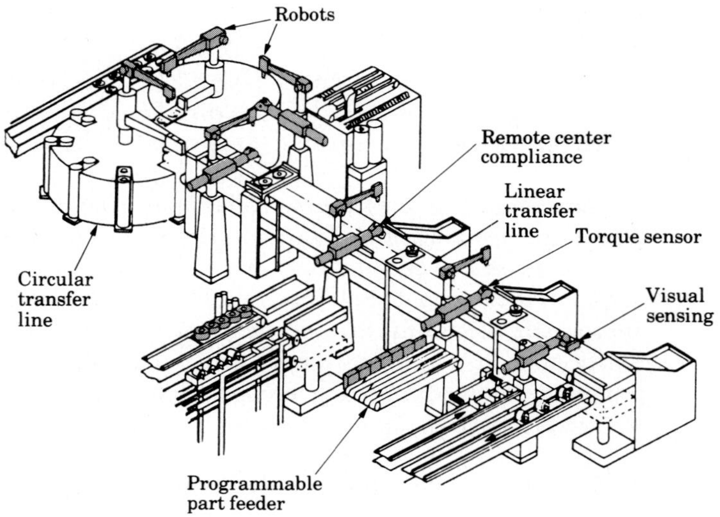

FIGURE 13.17

Automated assembly operations using industrial robots and circular and linear transfer lines. *Source: Computers in America,* January 1984.

13.7 Sensor Technology

A *sensor* is a device that produces a signal for purposes of detecting or measuring a property, such as position, force, torque, pressure, temperature, humidity, speed, acceleration, and vibration. Traditionally, sensors, actuators, and switches have been used to set limits on the performance of machines. Familiar examples are stops on machine tools to restrict work-table movements, pressure and temperature gages with automatic shutoff features, and governors on engines to prevent excessive speed of operation. Sensor technology is essential to data acquisition, monitoring, communication, and computer control of machines and systems.

Because they convert one quantity to another, sensors are also often referred to as *transducers*, meaning to transfer. *Analog* sensors produce a signal, such as voltage, that is proportional to the measured quantity. *Digital* sensors have numeric or digital outputs that can be directly transferred to computers. Analog-to-digital converters (ADC) are available to interface analog sensors with computers.

13.7.1 Classification

Sensors that are of interest in manufacturing may be classified generally as:

a) *Mechanical*—for measuring quantities such as position, shape, velocity, force, torque, pressure, vibration, strain, and mass.
b) *Electrical*—for measuring voltage, current, charge, and conductivity.
c) *Magnetic*—for measuring magnetic field, flux, and permeability.
d) *Thermal*—for measuring temperature, flux, conductivity, and specific heat.
e) *Others*—such as acoustic, ultrasonic, chemical, optical, and radiation.

Depending on its application, a sensor may consist of metallic, nonmetallic, organic, or inorganic materials and fluids, gases, plasmas, or semiconductors. Using the special characteristics of these materials, sensors convert the quantity or property measured to analog or digital output. An ordinary mercury thermometer, for example, is based on the principle of thermal expansion of mercury. Similarly, a machine part or a physical obstruction or barrier in a space can be detected by breaking the beam of light sensed by a photoelectric cell. A proximity sensor, which senses and measures the distance between it and an object or a moving member of a machine, can be based on acoustics, magnetics, capacitance, or optics. Other actuators physically contact the object and, usually by electromechanical means, take appropriate action. Sensors are essential to the control of advanced and intelligent robots. Sensors are being developed with capabilities that resemble those of human beings, as in the two principal methods of sensing: tactile and visual.

Tactile sensing. *Tactile sensing* is the continuous sensing of variable contact forces, commonly by an array of sensors. Such a system is capable of performing within an arbitrary three-dimensional space. Fragile parts, such as glass bottles and electronic devices, can be handled by robots with *compliant*, or *smart*, end effectors. These effectors can sense the force applied to the object being handled, using piezoelectric devices, strain gages, magnetic induction, ultrasonics, and optical systems of fiber optics and light-emitting diodes. Tactile sensors capable of measuring and controlling gripping forces and moments in three axes are available commercially (Fig. 13.18).

The force sensed is monitored and controlled through closed-loop feedback devices. Compliant grippers with force feedback and sensory perception can be complicated, require powerful computers, and hence are costly. Anthropomorphic end effectors are being designed to simulate the human hand and fingers, with the capability of sensing touch, force, movement, and pattern. The ideal tactile sensor must also sense slip, a capability of human fingers and hand that we tend to take for granted, yet is so important in the use of robots.

Visual sensing (machine vision). In *visual sensing*, cameras optically sense the presence and shape of the object (Fig. 13.19). There are two basic systems of machine vision: linear array and matrix array. In *linear array*, only one dimension is sensed, such as the presence of an object or some feature on its surface. *Matrix arrays* sense up to three dimensions and are, for example, capable of detecting a properly inserted component in a printed circuit or a properly made solder joint. When used in automated inspection systems, they can also detect cracks and flaws.

In visual sensing, a microprocessor processes the image, usually in less than one second. The image is measured, and the measurements are digitized (*image*

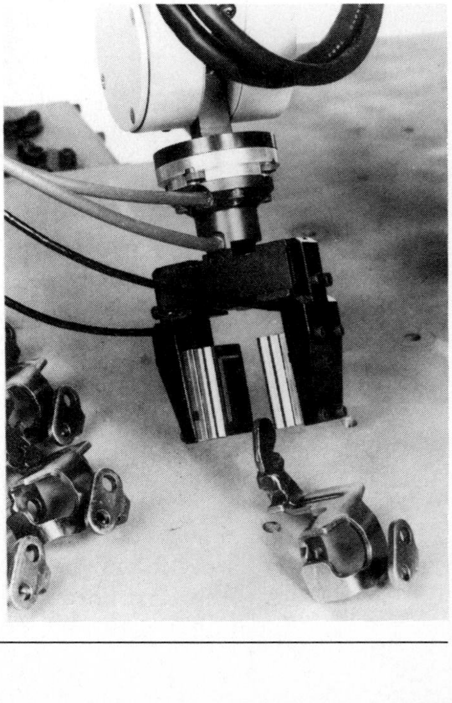

FIGURE 13.18
A robot gripper with tactile sensors. *Source:* Courtesy of Lord Corporation.

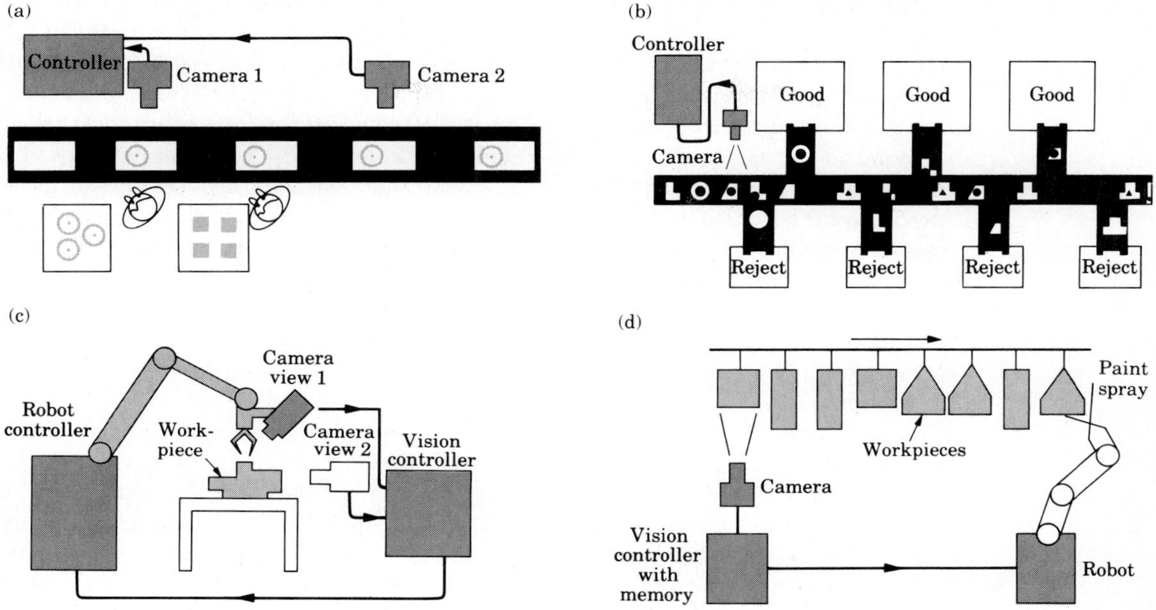

FIGURE 13.19

Examples of machine-vision application. (a) Online inspection of parts. (b) Identification of parts with various shapes, inspection, and rejection of defective parts. (c) Use of cameras to provide positional input to a robot relative to the workpiece. (d) Painting parts with different shapes with input from a camera. The system's memory allows the robot to correctly identify the particular shape to be painted and to proceed with the correct movements of a paint spray attached to the end effector. *Source: Manufacturing Engineering,* November 1982.

recognition). Machine vision is capable of on-line identification and inspection of parts—and rejection of defective parts. Several applications of machine vision in manufacturing are shown in Fig. 13.19. With visual sensing capabilities, end effectors are able to pick up parts and grip them in proper orientation and location. However, picking parts from a bin has proven to be a difficult task, because of the random orientation of parts in close proximity to each other.

The selection of a sensor for a particular application depends on factors such as the quantity to be measured or sensed, the environment, the sensor's interaction with other components in the system, expected service life, level of sophistication, difficulties associated with the sensor's use, power source, and cost.

13.8 Flexible Fixturing

In previous chapters, we have described several workholding devices, such as chucks, collets, mandrels, and various fixtures, many of which are operated manually. Others are designed and operated at various levels of mechanization and automation, such as power chucks driven by mechanical, hydraulic, or electrical means. Workholding devices have certain ranges of capacity: Collets can accommodate round bars within a certain range of diameters; four-jaw chucks can accommodate square or prismatic workpieces having certain dimensions; and other devices and fixtures are made for specific workpiece shapes and dimensions. In manufacturing operations, the words clamp, jig, and fixture are often used interchangeably and sometimes in pairs, as in jigs and fixtures. *Clamps* are simple multifunctional devices, whereas *fixtures* are generally designed for specific purposes, with on and off features and shapes that are usually replicas of workpieces. *Jigs* have various reference surfaces and points for accurate alignment of parts and tools and are widely used in mass production.

The emergence of flexible manufacturing systems has necessitated the use of workholding devices and fixtures that have a certain built-in flexibility. Generally called *flexible fixturing*, these devices are capable of quickly accommodating a range of part shapes and dimensions, without the necessity of changing or extensively adjusting the fixtures or requiring operator intervention, both of which would adversely affect productivity. A schematic illustration of a flexible fixturing system is shown in Fig. 13.20. The strain gage attached to the clamp senses the magnitude of the clamping force, and the system adjusts this force to keep the workpiece securely clamped on the work table.

Proper design of flexible workholding devices and fixtures is essential to the operation of advanced manufacturing systems. These devices should position the workpiece automatically and accurately and maintain its location precisely and with sufficient clamping force during the manufacturing operation. Fixtures should accommodate parts repeatedly in the same position and must have sufficient stiffness to resist without excessive deflection the forces developed. Fixtures and

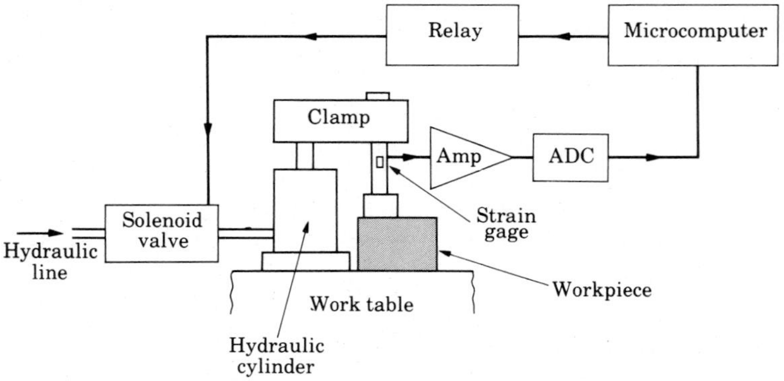

FIGURE 13.20
Schematic illustration of a flexible fixturing setup. The clamping force is sensed by the strain gage and the system automatically adjusts this force. *Source:* P. K. Wright.

clamps should have low profiles so as to avoid colliding with tools and dies. Collision avoidance is also an important factor in programming tool paths.

13.9 Automated Assembly

The individual parts and components made by various manufacturing processes are *assembled* into finished products by various methods. Some products are simple, having only two or three components to assemble, which can be done with relative ease, such as an ordinary pencil with an eraser, a frying pan with a wooden handle, or a beverage can. Most products, however, consist of many parts and their assembly requires considerable care and planning.

Traditionally, assembly has involved much manual work, contributing significantly to cost. The total assembly operation is usually broken into individual assembly operations, with an operator assigned to carry out each operation. Assembly costs are typically 25–50 percent of the total cost of manufacturing, with the percentage of workers involved in assembly operations ranging from 20 to 60 percent. In the electronics industries, some 40–60 percent of total wages are paid to assembly workers.

As production costs and quantities of products to be assembled increased, the necessity for automated assembly became obvious. Beginning with the hand assembly of muskets in the late 1700s and early 1800s with interchangeable parts, assembly methods have been vastly improved over the years. The first instance of large-scale modern assembly was the assembly of flywheel magnetos for the Model T Ford automobile, which eventually led to mass production of the automobile.

The choice of an assembly method and system depends on the required production rate and total quantities, product market life, labor availability, and most important, cost. *Automated assembly* can effectively minimize overall product cost. Although it requires a significant capital investment, automated machinery may be adapted with relative ease to permit assembly of a new product, thus becoming economical in the long run.

13.9.1 Assembly methods

You have seen that parts are manufactured with certain tolerances. Taking roller bearings as an example, you know that although they have the same nominal dimensions, some rollers in a lot are smaller than others by a small amount. Likewise, some bearing races are smaller than others in the lot.

There are two methods of assembly for such high-volume products: random assembly and selective assembly. In *random assembly*, parts are put together by selecting them randomly from the lots. In *selective assembly*, the rollers and races are segregated by groups of sizes, from smallest to largest. The parts are then selected to mate properly. Thus the smallest diameter rollers are mated with inner races having the largest outside diameter and with outer races having the smallest inside diameters.

13.9.2 Assembly systems

There are three basic types of assembly systems: synchronous, nonsynchronous, and continuous. In *synchronous systems*, also called *indexing*, individual parts and components are supplied and assembled at a constant rate at fixed individual stations. The rate of movement is based on the station that takes the longest time to complete its portion of the assembly. This system is used primarily for high-volume, high-speed assembly of small components. Transfer systems move the partially assembled parts from workstation to workstation by various mechanical means, as well as robots. Two typical transfer systems (*rotary* indexing and *in-line* indexing) are shown in Fig. 13.21. These systems can operate in either a fully automatic or a semiautomatic mode. However, a breakdown of one station can shut down the whole assembly operation. The part feeders supply the individual parts to be assembled and place them on other components, which are secured on work carriers or fixtures. The feeders move the individual parts by vibratory or other means through delivery chutes and ensure their proper orientation by various ingenious means (Fig. 13.22). Properly orienting parts and avoiding jamming are essential in all automated assembly operations.

In *nonsynchronous systems*, each station operates independently, and any imbalance is accommodated in storage (buffer) between stations. Thus if there are sufficient parts in the buffer, that particular station need not operate. Furthermore, if one station becomes inoperative for some reason, the assembly line continues to operate until all the parts in the buffer have been used. Nonsynchronous systems are

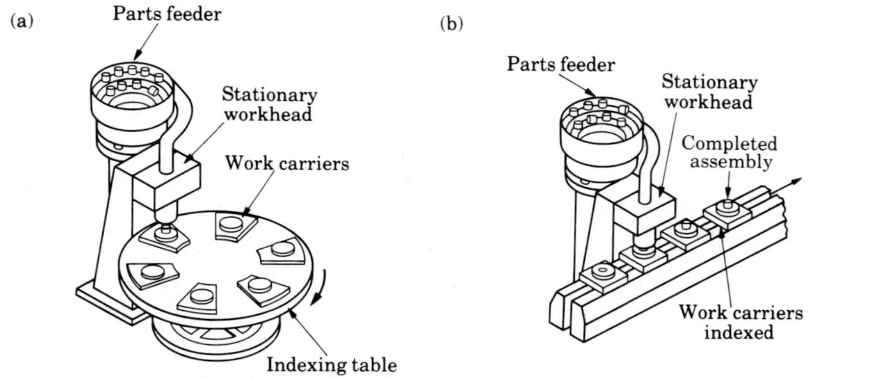

FIGURE 13.21

Transfer systems for automated assembly: (a) rotary indexing machine and (b) in-line indexing machine. *Source:* Reprinted from G. Boothroyd, courtesy of Marcel Dekker, Inc.

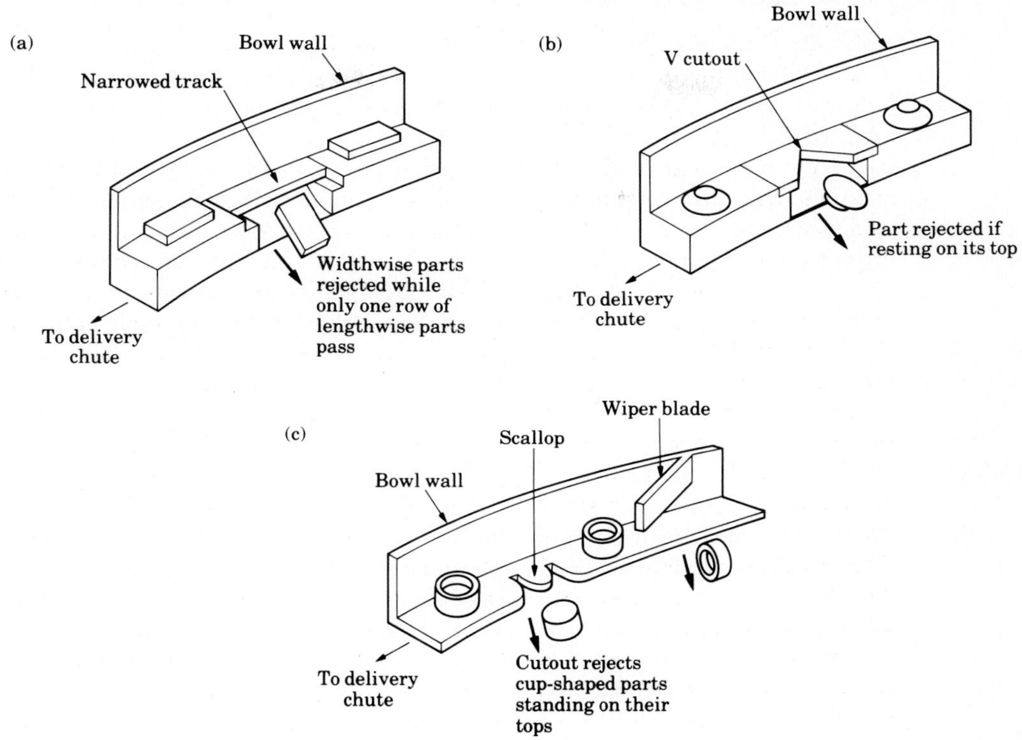

FIGURE 13.22
Various guides for proper orientation of parts for automated assembly. *Source:* Reprinted from G. Boothroyd, courtesy of Marcel Dekker, Inc.

suitable for large components with many parts to be assembled and for types of assembly in which the times required for individual assembly operations vary widely. The speed of operation in this system is slower than the synchronous system.

In *continuous systems*, the product is assembled while moving at a constant speed on pallets or similar work carriers. The parts to be assembled are brought to the product on various workheads, and their movements are synchronized with the constant movement of the product. Typical applications of this system are in bottling and packaging plants and mass-production lines for automobiles and appliances.

Although these systems are set up for a certain product line, they can be modified for increased flexibility in order to assemble product lines that change with market demand. Such *flexible assembly systems* (FAS) utilize computer controls, interchangeable and programmable workheads and feeding devices, coded pallets, and automated guiding devices.

13.9.3 Design for assembly

Although the functions of a product and its design for manufacturing have been matters of considerable interest for some time, only recently has *design for assembly* (DFA) attracted special attention, particularly design for automated assembly. The need for reducing labor costs has necessitated greater use of automated assembly. In spite of the use of sophisticated mechanisms, robots, and computer controls, aligning and placing a simple square peg in a square hole with small clearances can be difficult in automated assembly. Because of the dexterity of the human hand and fingers—and their capability for feedback through various senses—assembly workers can manually assemble even complex parts without much difficulty. However, in hand assembly, the worker uses both hands, thus greatly increasing the flexibility of the operation, whereas automated assembly usually involves just one workhead.

SUMMARY

There are several levels of automation, from simple automation of machines to unmanned manufacturing cells to—ultimately—the factory of the future. Automation has been implemented successfully in manufacturing processes, material handling, inspection, assembly, and packaging. Production quantity and rate are important factors in determining the economical levels of automation. True automation began with the numerical control of machines, which has the capability of flexibility of operation, lower cost, and ease of making different parts with lower operator skill. Manufacturing operations are further optimized, both in quality and cost, by adaptive control techniques, which continuously monitor the operation and make necessary adjustments in process parameters.

Great advances have been made in material handling, particularly with the implementation of industrial robots and automated guided vehicles. The role of sensors is crucial in the implementation of these technologies, and a wide variety of sensors based on various principles have been developed and installed. Other advances include flexible fixturing and automated assembly techniques that reduce the need for worker intervention and that lower manufacturing costs. The efficient and economic implementation of these techniques requires that design for assembly be recognized as an important factor in the total design and manufacturing process.

BIBLIOGRAPHY

Numerical Control

Groover, M.P., *Automation, Production Systems, and Computer-Integrated Manufacturing.* Englewood Cliffs, N.J.: Prentice-Hall, 1987.

Harrington, J., Jr., *Understanding the Manufacturing Process.* New York: Marcel Dekker, 1984.

Koren, Y., *Computer Control of Manufacturing Systems.* New York: McGraw-Hill, 1983.

Rapello, R.G., *Essentials of Numerical Control*. Englewood Cliffs, N.J.: Prentice-Hall, 1986.
Tool and Manufacturing Engineers Handbook, 4th ed., *Vol. 1: Machining*. Dearborn, Mich.: Society of Manufacturing Engineers, 1983.

Material Movement and Robots

Ballard, D.H., and C.M. Brown, *Computer Vision*. Englewood Cliffs, N.J.: Prentice-Hall, 1982.
Basics of Material Handling. Pittsburgh: The Material Handling Institute, 1981.
Craig, J.J., *Introduction to Robotics: Mechanics and Control*. Reading, Mass.: Addison-Wesley, 1985.
Critchlow, A.J., *Introduction to Robotics*. New York: Macmillan, 1985.
Gevarter, W.B., *Intelligent Machines: An Introductory Perspective of Artificial Intelligence and Robotics*. Englewood Cliffs, N.J.: Prentice-Hall, 1985.
Groover, M.P., M. Weiss, R.N. Nagel, and N.G. Odrey, *Industrial Robotics: Technology, Programming, and Applications*. New York: McGraw-Hill, 1986.
Holzbock, W.G., *Robotic Technology: Principles and Practice*. New York: Van Nostrand Reinhold, 1986.
Koren, Y., *Robotics for Engineers*. New York: McGraw-Hill, 1985.
McDonald, A.C., *Robot Technology: Theory, Design, and Applications*. Englewood Cliffs, N.J.: Prentice-Hall, 1986.
Ranky, P.G., and C.Y. Ho, *Robot Modelling, Control and Applications with Software*. New York: Springer-Verlag, 1985.
Snyder, W.E., *Industrial Robots: Computer Interfacing and Control*. Englewood Cliffs, N.J.: Prentice-Hall, 1985.
Villers, P., and J.M. Brady, *The Robotic Handbook*. Menlo Park, California: Benjamin-Cummings, 1987.
Zeldman, M., *What Every Engineer Should Know About Robots*. New York: Marcel Dekker, 1984.

Assembly and Fixtures

Andreasen, M.M., S. Kahler, and T. Lund (eds.), *Design for Assembly*. Bedford, England: IFS (Publications) Ltd., 1983.
Boothroyd, G., C. Poli, and L.E. Murch, *Automatic Assembly*. New York: Marcel Dekker, 1982.
Heginbotham, W. (ed.), *Programmable Assembly*. Bedford, England: IFS (Publications) Ltd., 1984.
Owen, T., *Assembly with Robots*. Englewood Cliffs, N.J.: Prentice-Hall, 1985.
Rathmill, K. (ed.), *Robotic Assembly*. Bedford, England: IFS (Publications) Ltd., 1985.
Tool and Manufacturing Engineers Handbook, 4th ed., *Vol. 4: Assembly, Testing, and Quality Control*. Dearborn, Mich.: Society of Manufacturing Engineers, 1986.
Wilson, F.W. (ed.), *Handbook of Fixture Design*. New York: McGraw-Hill, 1984.

QUESTIONS AND PROBLEMS

13.1 Discuss your observations concerning Fig. 13.2.

13.2 What are the relative advantages and limitations of the two arrangements for power heads shown in Fig. 13.3?

13.3 Discuss methods of on-line gaging of workpiece diameters in turning operations. Explain the relative advantages and limitations of each.

13.4 Comment on the accuracies of hole positions in the three different methods shown in Fig. 13.4.

13.5 What is the function of the digital-to-analog converter (DAC) in Fig. 13.6(b)?

13.6 Is drilling the only application for the point-to-point system shown in Fig. 13.7(a)? Explain.

13.7 Give an example of a forming operation that is suitable for adaptive control similar to that shown in Fig. 13.11.

13.8 What are the implications of Fig. 13.12?

13.9 What should be the characteristics of the guidance system for automated guided vehicles?

13.10 Design end effectors for the applications listed in Section 13.6.

13.11 Describe possible applications for industrial robots not discussed in this chapter.

13.12 What determines the number of robots in an automated assembly line such as that shown in Fig. 13.17?

13.13 Describe situations in which the shape and size of the work envelope of a robot (Fig. 13.16) can be critical.

13.14 Give some applications for the systems shown in Figs. 13.19(a) and (c).

13.15 Based on a system similar to that shown in Fig. 13.20, design a flexible fixturing setup for a lathe chuck.

13.16 Select some common parts and devise means for proper orientation, other than those shown in Fig. 13.22.

13.17 Give examples of when tactile sensors would not be suitable. Explain why.

13.18 Give examples of products that are suitable for the types of production shown in Fig. 13.2.

14

Integrated Manufacturing Systems

14.1 Introduction

In describing manufacturing processes and operations in this text, we discussed each of them primarily as an individual activity with its own capabilities, applications, advantages, and limitations. On several occasions, we described the implementation and benefits of mechanization, automation, and computer control of various stages of manufacturing operations. In this chapter we focus our attention on the *integration* of manufacturing activities. Integration means that manufacturing processes, operations, and their management are treated as a *system*, allowing complete control of the manufacturing facility. In this way, productivity, product quality, reliability, and cost can be optimized.

Beginning with the 1960s, certain trends developed that had a major impact on manufacturing:

- International competition increased rapidly.
- Market conditions fluctuated widely.

- Customers demanded high-quality, low-cost products.
- Product variety increased substantially.
- Product life cycles became shorter.
- Labor costs increased significantly.
- Computer-controlled equipment became available.

These trends led to recognition of the need for an integrated approach to manufacturing. In *integrated manufacturing* the traditionally separate functions of research and development, design, production, assembly, inspection, and quality control are all linked. Consequently, integration requires that quantitative relationships among product design, materials, manufacturing process and equipment capabilities, and related activities be well understood. Thus changes in material requirements, product types, and market demand can be accommodated. The goal: to find the optimum method of producing high-quality, low-cost products efficiently.

In Chapter 13 we described the principles of automation, its benefits, and how it has been gradually introduced into manufacturing facilities, progressing from mechanization to numerical control machines, adaptive control, automated material handling, industrial robots, and automated assembly. We discussed the role and importance of sensors and various devices in automation. In spite of their benefits, however, these advances cannot by themselves be responsive to changing market demand. Why? Because the elements of automation exist in isolation; consequently, their performance can be optimized relative only to their individual tasks. However, within the context of a *system*, all operations become interdependent elements.

Therefore, machines, tooling, and manufacturing operations must have a certain built-in flexibility to respond to change and ensure *on-time delivery* of products to the customer. You can understand the significance of on-time product delivery by noting your own dissatisfaction when you do not receive something on its promised date. In industry, failure of on-time delivery can upset production plans and schedules and, consequently, have a significant economic impact. In a highly competitive environment, failure of on-time product delivery can cost a company its *competitive edge* because the customer will simply change suppliers.

In the following sections, we describe major advances in integrating all aspects of manufacturing that have led to *computer-integrated manufacturing systems.* Among these advances are:

- Computer-aided design.
- Computer-aided manufacturing.
- Computer-aided process planning.
- Group technology.
- Cellular manufacturing.
- Flexible manufacturing systems.
- Just-in-time production.
- Artificial intelligence.

We present several examples to illustrate the major impact of these activities on the technology and economics of manufacturing. Because some of these advances are still in the development stage, some controversy exists among engineers and corporate management regarding their merits and economic justification. Thus understanding the characteristics, applications, potentials, advantages, and limitations of these technological developments and sound evaluations of their economic benefits are essential. Manufacturing engineers should be ready to implement new technologies as they become viable economic alternatives.

We conclude this chapter with a discussion and overview of the *factory of the future*. The prediction is that manufacturing operations in the future will be performed with little human labor; hence these facilities are also being called *untended factories*. Because of its potential impact on the workforce, the social implications of this trend are also discussed.

14.2 Manufacturing Systems

Although the word *system* is derived from the Greek *systema*, meaning to combine, it has come to mean an arrangement of physical entities characterized by identifiable and quantifiable interacting parameters. Because manufacturing entails a large number of interdependent activities consisting of distinct entities such as materials, tools, machines, power, and human beings, it should properly be regarded as a system. It is, in fact, a complex system because it is comprised of many diverse physical and human elements (see Fig. 1.2), some of which are difficult to predict and control, such as raw-material prices, market changes, and human behavior and performance.

Ideally, we should be able to represent a system by *mathematical* and *physical models*, which show us the nature and extent of interdependence. In a manufacturing system, a change or disturbance anywhere in the system requires that the system adjust itself in order to continue functioning efficiently. For example, if the supply of a particular raw material is reduced and hence its cost is increased because of, say, geopolitical reasons, alternative materials must be selected. This selection should be made only after careful consideration of the effect of this change on product quality, production rate, and costs.

Similarly, if demand for a product is such that its shape, size, or capacity fluctuates randomly and rapidly, the system must be able to produce the modified product on short lead time and without the need for major capital investment in machinery and tooling. Although the labor share of product cost has been decreasing steadily over the years (in some cases it is now less than 10 percent of total product cost), the system must also be capable of absorbing some of the cost if worker pay escalates.

Modeling such a complex system can be difficult because we lack reliable data on many of the variables and cannot easily predict and control some of these variables. For example, machine characteristics, performance, and response to external disturbances cannot be modeled precisely, raw-material costs are difficult to predict accurately, and human behavior and performance are difficult to model.

In spite of these difficulties, considerable progress has been made in modeling and simulating manufacturing systems. In this way the effects of disturbances, such as product demand, materials changes, and machine performance, can be predicted with reasonable accuracy.

14.3 Computer-Integrated Manufacturing

The various levels of automation in manufacturing operations we described in Chapter 13 can be extended further by including information processing functions utilizing an extensive network of interactive computers. The result is *computer-integrated manufacturing* (CIM), which is a broad term describing the computerized integration of *all* aspects of design, planning, manufacturing, distribution, and management.

Computer-integrated manufacturing is a *methodology* and a *goal*, rather than an assemblage of equipment and computers. The technology for implementing CIM is well understood and available. However, because CIM ideally involves the total operation of a company, it must be comprehensive and have an extensive database. Consequently, if implemented all at once, CIM can be prohibitively expensive, particularly for small and medium-size companies.

Considerable experience and training is required in different areas of CIM and at all levels within a company if CIM is to operate effectively. Implementation of CIM in existing plants may begin with modules in various phases of the operation. For new manufacturing plants, comprehensive, long-range strategic planning covering all phases of the operation is necessary to take full advantage of CIM. Thus such plans should include availability of resources; organizational mission, goals, and culture; existing and possible future technology; and degree of integration required.

Computer-integrated manufacturing systems consist of *subsystems* that are integrated into a whole. These subsystems consist of business planning and support, product design, manufacturing process planning, process control, shop floor monitoring systems, and process automation. The subsystems are designed, developed, and applied in such a manner that the output of one subsystem serves as the input of another subsystem. Organizationally, these subsystems are usually divided into business planning and business execution functions, respectively. *Planning* functions include activities such as forecasting, scheduling, material-requirements planning, invoicing, and accounting. *Execution* functions include production and process control, material handling, testing, and inspection.

The effectiveness of CIM depends greatly on the presence of a large-scale, *integrated communications system* involving computers, machines, and their controls. Major problems have arisen in factory communication because of the difficulty of interfacing different types of computers purchased from different vendors by the company at various times. The trend is strongly toward standardization to make communications equipment compatible. The benefits of CIM include:

- Responsive to shorter product life cycles and changing market demand.
- Emphasis on product quality and its uniformity through better process control.
- Better use of materials, machinery, and personnel and reduction of inventory, thus improving productivity.
- Better control of production and management of the total manufacturing operation, resulting in lower product cost.

An efficient computer-integrated manufacturing system requires a single database that is shared by the entire manufacturing organization. A *database*, or *databank*, consists of current, detailed, and accurate data relating to products, designs, machines, processes, materials, production, finances, purchasing, sales, marketing, and inventory. This vast array of data is stored in computer memory and recalled or modified as necessary, either by individuals in the organization or by the CIM system itself in controlling various aspects of design and production.

A database generally consists of the following items, some of which are classified as technical and others as nontechnical:

a) Product data, such as part shape, dimensions, and specifications.
b) Production data, such as the manufacturing processes involved in making parts and products.
c) Operational data, such as scheduling, lot sizes, and assembly requirements.
d) Resources data, such as capital, machines, equipment, tooling, and personnel, and their capabilities.

Databases are built by individuals and various sensors in the machinery and equipment used in production. Data from the latter are collected automatically by a *data acquisition system* (DAS), such as number of parts being produced per unit of time, their dimensional accuracy, surface finish, weight, and so on, at specified rates of sampling. The components of DAS include minicomputers, microprocessors, transducers, and analog-to-digital converters (ADCs). Data acquisition systems are also capable of analyzing data and transferring them to other computers for purposes such as statistical analysis, data presentation, and forecasting product demand.

Several factors are important in the use and implementation of databases. They should be timely, accurate, easily accessible, easily shared, and user friendly. In the event that something goes wrong with the data, correct data should be recovered and restored. Because it is used for many purposes and by many people, the database must be flexible and responsive to the needs of different users. Companies must protect data against tampering or unauthorized use. CIM systems can be accessed by designers, manufacturing engineers, process planners, financial officers, and the management of the company by using appropriate access codes.

14.4 Computer-Aided Design

In traditional engineering design practice, a draftsperson typically uses a drafting board, drawing equipment, templates, and various handbooks and reference sources—a slow, often tedious, and not always accurate process. Activities such as generating and revising engineering drawings, modifying designs, and analyzing and optimizing designs are done with limited technological assistance. Furthermore, the inherently iterative, trial-and-error nature of the traditional design process usually leads to a lengthy product design phase.

Consider, for example, the design of a lawnmower in which the capacity of the motor is to be increased. The new motor has larger dimensions, thus requiring changes in the mower's design and method of mounting the motor. Such a change can involve significant effort, as the assembly and detail drawings have to be modified, new calculations have to be made, and the design changes have to be checked. Then revised drawings and manufacturing plans have to be prepared.

Computer-aided design (CAD) involves the use of computers to create design drawings. Computer-aided design is usually associated with *interactive computer graphics*, known as a *CAD system*. Computer-aided design systems are powerful tools and are used in mechanical design and geometric modeling of products and components. They simplify finite-element analysis of stresses, strains, deflections, and temperature distribution in structures and load-bearing members and generation, storage, and retrieval of NC data. These systems are also used extensively in the design of integrated circuits and other electronic devices.

When using a CAD system, the designer can visualize the object to be designed more easily on the graphics screen and consider alternative designs or modify a particular design quickly to meet the necessary design requirements or changes. The designer can then subject the design to a variety of engineering analyses and identify potential problems, such as excessive loads, deflection, and temperature. The speed and accuracy of such analyses far surpasses traditional methods. Some CAD systems are capable of simulating additional designs and optimum product performance and analyzing requirements for materials, tools, dies, and manufacturing processes.

The CAD system generates working drawings for the product and its components quickly and accurately. These drawings generally are of higher quality and consistency than those produced by traditional manual drafting. They can be reproduced any number of times and at different levels of reduction and enlargement. In addition to the design's geometric and dimensional features, other information, such as a list of materials, specifications, and manufacturing instructions, are stored in the CAD database. Using such information, the designer can analyze the economics of various designs.

All data in the databases can readily be transmitted electronically to any part of the manufacturing organization. Designers, manufacturing engineers, and management have access to the data, thus greatly enhancing communication among different activities.

14.4.1 CAD system elements

The design process in a CAD system consists of four stages. We describe each stage briefly.

Geometric modeling. In *geometric modeling*, a physical object or any of its parts is described mathematically or analytically. The designer first constructs a geometric model by giving commands that create or modify lines, surfaces, solids, dimensions, and text that together are an accurate and complete two- or three-dimensional representation of the object. The results of these commands are displayed and can be moved around on the screen, and any section desired can be magnified to view details. These data are digital and are stored in the database contained in computer memory. They can be further used to analyze and optimize design, and for manufacturing.

The models can be presented in three different ways. In *line* representation (*wire frame*; Fig. 14.1) all edges are visible as solid lines. This image can be ambiguous, particularly for complex shapes. However, various colors are generally used for different parts of the object, thus making the object easier to visualize. In the *surface* model, all visible surfaces are shown in the model, and in the *solid* model, all surfaces are shown but the data describe the interior volume. Solid

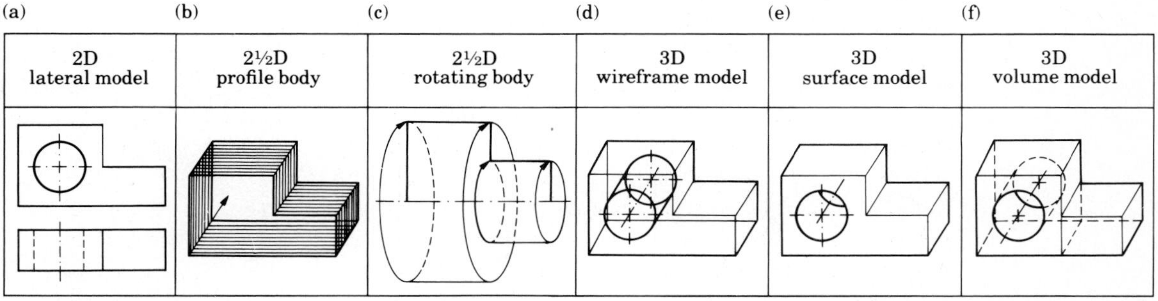

FIGURE 14.1
Types of modeling for CAD. *Source: Flexible Automation, 87/88, The International CNC Reference Book.*

models can be constructed from "swept volumes" (Figs. 14.1b and c) or a combination of shapes called primitives of solids (Fig. 14.2).

The three types of wire-frame representations are 2-D, $2\frac{1}{2}$D, and 3-D. The 2-D image shows the profile of the object or part. The $2\frac{1}{2}$-D image can be obtained by translational sweep, that is, moving the 2-D object along the z axis. For round objects, a $2\frac{1}{2}$-D model can be generated by simply rotating a 2-D model around its axis.

Design analysis and optimization. After the design's geometric features have been determined, the design is subjected to an engineering analysis. This phase may consist of analyzing stresses, strains, deflections, vibrations, heat transfer, temperature distribution, casting, solidification, or tolerance. Various sophisticated software packages having capabilities to compute these quantities accurately and rapidly are available. Because of the relative ease with which such analyses can now be made, designers are increasingly willing to thoroughly analyze a design before it moves on to production. Experiments and field measurements may be necessary to determine the effects of loads, temperature, and other variables on the designed components.

Design review and evaluation. An important design stage is review and evaluation to check for any interference between various components in order to avoid difficulties during assembly or use of the part, and whether moving members,

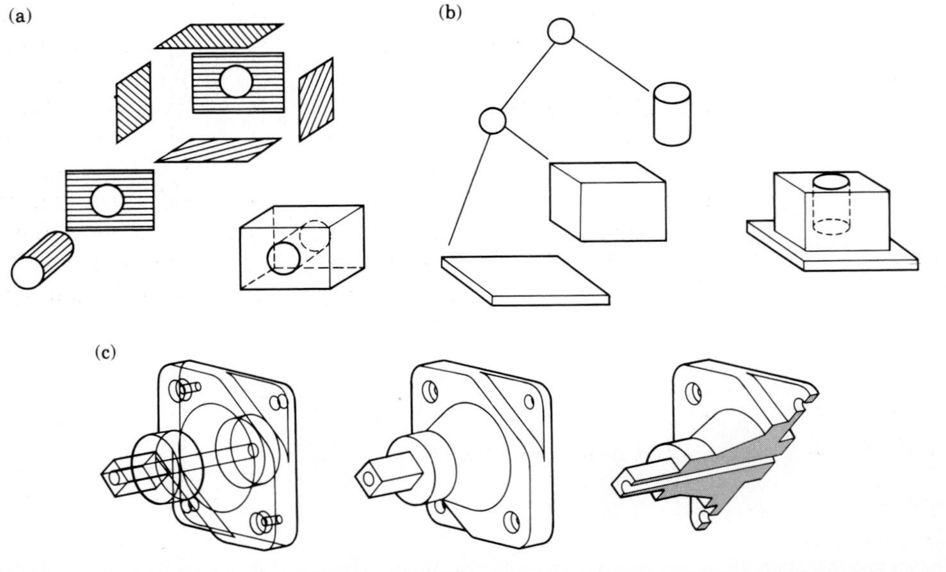

FIGURE 14.2
(a) Boundary representation of solids, showing the enclosing surfaces of the solid model and the generated solid model. (b) A solid model represented as compositions of solid primitives. (c) Three representations of the same part by CAD. *Source:* P. Ranky.

such as linkages, are going to operate as intended. Software having animation capabilities to identify potential problems with moving members and other dynamic situations is available. During the design review and evaluation stage, the part is precisely dimensioned and toleranced, as required for manufacturing it.

Documentation and drafting. After the preceding stages have been completed, the design is reproduced by automated drafting machines for documentation and reference. Detail and working drawings are also developed and printed. The CAD system is also capable of developing and drafting sectional views of the part, scaling the drawings, and performing transformations to present various views of the part. Although in CAD systems much of the design process is carried out on workstations connected to a main frame computer, smaller systems are being developed for use on personal computers (PCs).

14.5 Computer-Aided Manufacturing

Computer-aided manufacturing (CAM) is the use of computers and computer technology to assist in all phases of manufacturing a product, including process and production planning, machining, scheduling, management, and quality control. Computer-aided manufacturing encompasses many of the technologies that we have described in Chapter 13 and this chapter. Because of the benefits, computer-aided design and computer-aided manufacturing are sometimes combined into CAD/CAM systems. This combination allows the transfer of information from design into the planning for manufacture of a product without the need to manually reenter the part geometry. The database developed in CAD is stored and processed further by CAM into the necessary data and instructions for operating and controlling production machinery, material-handling equipment, and automated testing and inspection for product quality. The emergence of CAD/CAM has had a major impact on manufacturing by standardizing product development and reducing design effort, tryout, and prototype work, thus resulting in significantly reduced costs and improved productivity.

14.6 Computer-Aided Process Planning

In order for a manufacturing operation to be efficient, all of its diverse activities must be planned. *Process planning* is concerned with selecting methods of production, tooling, fixtures and machinery, sequence of operations, and assembly. It is traditionally done by process planners. The sequence of processes and operations to be performed, machines to be used, standard time for each operation, and similar information are documented on a *routing sheet*. This task is highly labor intensive and time consuming and relies heavily on the experience of the process planner.

Computer-aided process planning (CAPP) accomplishes this complex task by viewing the total operation as an integrated system, so that the individual operations and steps involved in making each part are coordinated with others and are performed efficiently and reliably. Computer-aided process planning is an important link between CAD and CAM. Although CAPP requires extensive software and good coordination with CAD/CAM, as well as other aspects of integrated manufacturing systems discussed in the rest of this chapter, it is a powerful tool for efficiently planning and scheduling manufacturing operations. It is particularly effective in small-volume, high-variety parts production involving machining, forming, and assembly operations.

14.6.1 CAPP system advantages

The advantages of CAPP systems over traditional process planning methods include standardization of process plans, thus improving the productivity of process planners, reducing lead times, reducing planning costs, and improving the consistency of product quality and reliability. Process plans can be prepared for parts having similar shapes and features, and they can be retrieved easily to produce new parts. Process plans can be modified to suit specific needs. Route sheets can be prepared more quickly. Compared to traditionally handwritten route sheets, computer printouts are neater and much more legible. Other functions, such as cost estimating and work standards, can be incorporated into CAPP.

14.6.2 CAPP system elements

The software for CAPP contains an extensive database on all aspects of manufacturing, such as properties and dimensions of tools and dies, characteristics and capabilities of processes and machinery, and the effects of process variables on dimensions, surface finish, and product quality. The software also includes the necessary information on the particular part to be manufactured, such as its dimensions, shape, tolerances, material, and surface finish desired. Based on this input, CAPP automatically generates an appropriate process plan to manufacture that part. It specifies parameters such as cutting speeds, feeds, power requirements, and time for machining. The system is capable of modifications should the type of product or some aspect of it be changed.

There are two types of CAPP systems. In the *retrieval system*, the computer files contain a standard process plan for the part to be manufactured. A search is made for the part's code number, which is based on its shape and manufacturing characteristics. The plan is then retrieved, displayed for review, and printed as a route sheet. Minor modifications can also be made to an existing process planning system, called a *variant CAPP system*. If a particular part is not in the computer files, a similar part with a similar code number is accessed, and an existing route sheet is retrieved. If a route sheet does not exist, one is made for the new part and stored in the computer files.

The second system is called a *generative CAPP system*, because it generates a process plan based on the same logical procedures that would be followed by a traditional process planner in making a particular part. Such a system is complex because it must contain knowledge of a part's shape, dimensions, process capabilities, manufacturing methods and machinery required, and the sequence of operations to be performed. Computer process-planning capabilities can be integrated with production-system planning and control. These activities are a subsystem of CIM. Several functions, such as capacity planning for plants to meet production schedules, inventory control, purchasing, and production scheduling, can be performed.

14.6.3 Material-requirements planning and manufacturing resource planning

Computer-based systems for managing inventories and delivery schedules of raw materials and tools are called *material-requirements planning* (MRP). This activity, sometimes regarded as a method of inventory control, involves keeping complete records of inventories of materials, supplies, parts in various stages of production, orders, purchasing, and scheduling. Several files of data are usually involved in a master production schedule.

A further development is *manufacturing resource planning* (MRP-II), which controls all aspects of manufacturing planning through feedback. Although the system is complex, MRP-II is capable of final production scheduling, monitoring actual performance and output, and comparing them against the master production schedule.

14.7 Group Technology

Group technology (GT) is a manufacturing concept that seeks to take advantage of the *design* and *processing similarities* among parts. This concept was first developed in Europe at about the turn of the century. It involved categorizing parts and recording them manually in card files or catalogs; designs were retrieved manually, as needed. This concept began to evolve further in the 1950s, and the term *group technology* was first used in 1959. However, not until the use of interactive minicomputers became widespread in the 1970s did the use of GT grow significantly.

In large manufacturing facilities, tens of thousands of different parts are made and placed in inventory. Although many of these parts have certain similarities in shape and method of manufacture, each part was traditionally viewed as a separate entity and produced in batches, involving considerable duplication and redundancy. The characteristic of similar parts (Fig. 14.3) suggests that benefits can be obtained by *classifying* and *coding* these parts into families. Results of surveys in manufacturing plants have repeatedly shown the prevalence of similar parts. One survey found that 90 percent of 3000 parts made by a company fell into five major families of parts. Such surveys consist of breaking each product into its components and identifying similar parts. For example, a pump can be broken into the basic components of motor, housing, shaft, seals, and flanges. In spite of the variety of pumps manufactured, each of these components is basically the same in terms of design and manufacturing methods. Consequently, all shafts can be placed in one family, unless there are major differences in shape, materials, and production method.

This approach becomes more attractive in view of consumer demand for an ever-greater variety of products in smaller quantities. Under these conditions, maintaining high efficiency in batch operations is difficult. Thus overall manufacturing efficiency is hurt because nearly 75 percent of manufacturing today is batch production. Manufacturing surveys have also shown that designs are not necessarily optimized, either for product manufacture or assembly. Thus questions have been raised, for example, about why a particular part should have so many different sizes of fastener.

The traditional product flow in a batch manufacturing operation is shown in Fig. 14.4. Note that machines of the same type are arranged in groups, that is, groups of lathes, of milling machines, of drill presses, and of grinders. In such a layout (*functional layout*), there is usually considerable random movement, as shown by the arrows that indicate movement of materials and parts. Such an arrangement is not efficient because it wastes time and effort. The machines in

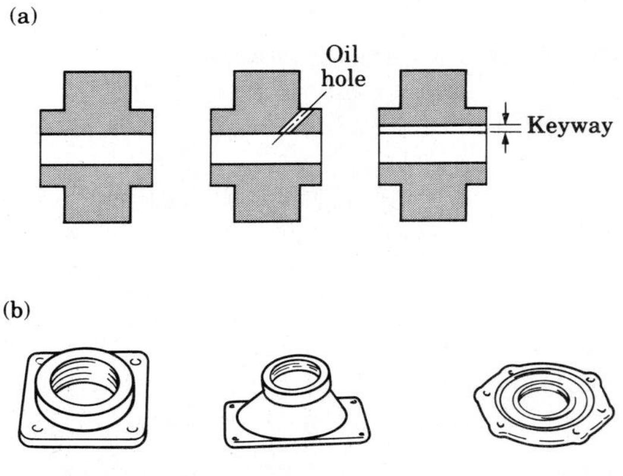

FIGURE 14.3
(a) Grouping parts according to their geometric similarities. (b) Grouping parts according to their manufacturing similarities. *Source:* Organization for Industrial Research.

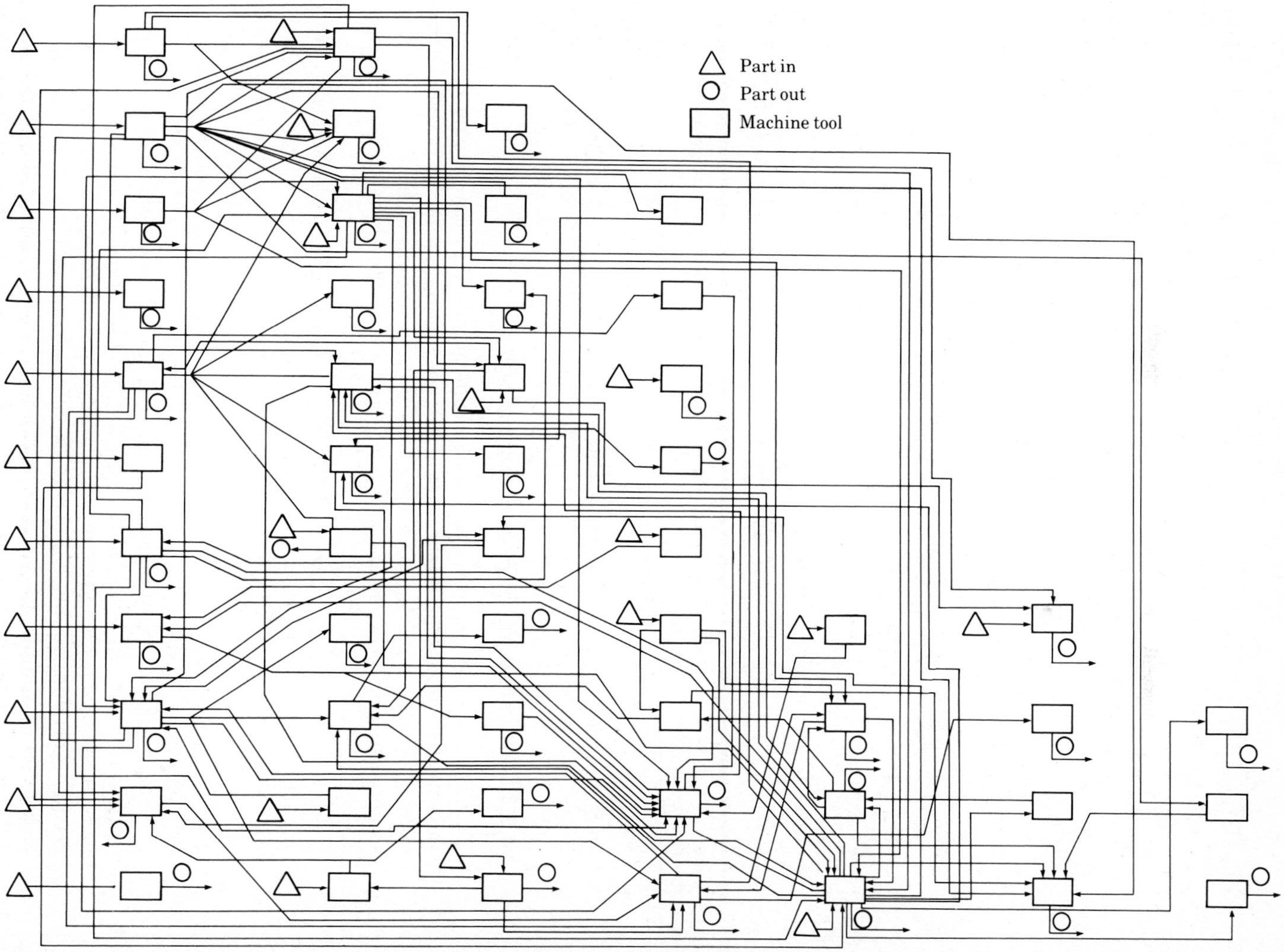

FIGURE 14.4

Schematic illustration of actual flow lines through a plant, showing routings for 150 very similar parts through 51 different machine tools. Eighty-seven different process plans were used in this production. *Source:* T. Baer, *Mechanical Engineering,* November 1985.

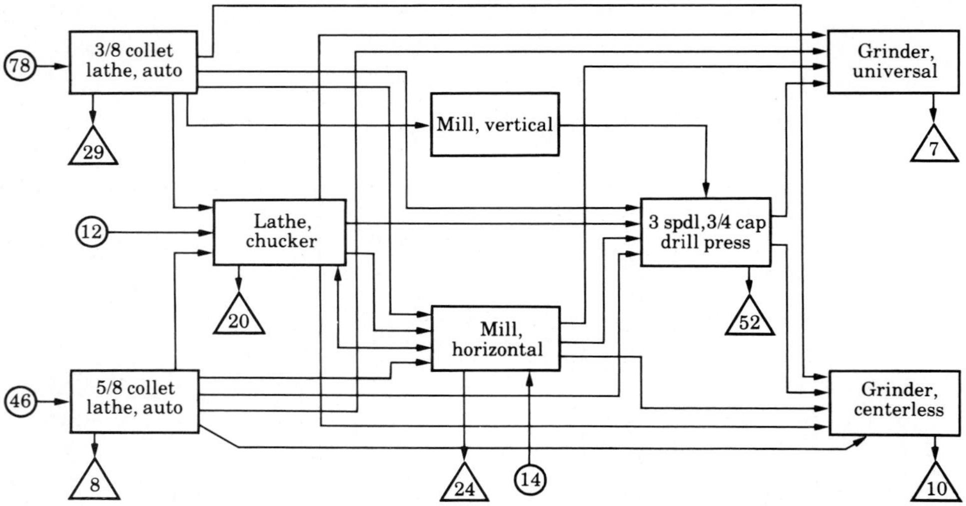

FIGURE 14.5
Flow lines for the parts in Fig. 14.4, showing standardized routings according to group technology analysis. *Source:* T. Baer, *Mechanical Engineering,* November 1985.

cellular manufacturing are arranged in a more efficient product flow line (*group layout*; Fig. 14.5).

14.7.1 Advantages of group technology

The advantages of group technology include standardization of part design and minimization of design duplication. New part designs can be developed using similar previous designs, thus saving a significant amount of time and effort. The designer can quickly determine whether a similar part already exists in the computer files.

Data reflecting the experience of the designer and manufacturing process planner are stored in the database. Thus a new and less experienced engineer can quickly benefit from that experience by retrieving any of the previous designs and process plans. Costs can be estimated more easily, and relevant statistics on materials, processes, number of parts produced, and other factors can be more easily obtained.

Process plans are standardized and scheduled more efficiently, orders are grouped for more efficient production, and machine utilization is improved. Setup times are reduced, and parts are produced more efficiently with better and more consistent product quality. Similar tools, fixtures, and machinery are shared in the production of a family of parts. Programming for NC is more automatic.

With the implementation of CAD/CAM, cellular manufacturing, and CIM, group technology is capable of improving productivity and reducing costs in batch production to approach those of mass production. Potential savings in each of the various design and manufacturing phases can range from 5 percent to 75 percent.

14.7.2 Classification and coding of parts

Parts are identified and grouped into families by *classification and coding* (C/C) *systems*. Classification and coding of parts is a complex and critical first step in GT. It may be done according to the part's design attributes or manufacturing attributes (see Fig. 14.3).

Design attributes pertain to similarities in geometric features and consist of the following:

a) External and internal shapes and dimensions.
b) Aspect ratio (length-to-width or length-to-diameter).
c) Tolerances specified.
d) Surface finish specified.
e) Part function.

Manufacturing attributes pertain to similarities in the methods and sequence of manufacturing operations performed on the part. As we have stated several times, selection of a manufacturing process or processes depends on many factors, among which are the shape, dimensions, and other geometric features of the part. Consequently, manufacturing and design attributes are interrelated. Manufacturing attributes consist of the following:

a) Primary processes.
b) Secondary and finishing processes.
c) Tolerances and surface finish.
d) Sequence of operations performed.
e) Tools, dies, fixtures, and machinery.
f) Production quantity and production rate.

From these lists, you can see that coding can be time consuming and requires considerable experience in designing and manufacturing products in general. In its simplest form, coding can be done by viewing the shape of the parts in a generic way and classifying them. Examples are parts with rotational symmetry, parts that are rectilinear in shape, and parts that have large surface-to-thickness ratios. Parts being reviewed and classified should be representative of the company's product lines. A more thorough method is to review all of the data and drawings concerning the design *and* production of parts.

Parts may also be classified by studying the production flow of parts during their manufacturing cycle (called *production flow analysis*, or PFA). One drawback to PFA is that a particular route sheet does not necessarily indicate that the total operation is optimized. In fact, depending on the experience of the process planner, route sheets for manufacturing the same part can be quite different. We have previously noted the benefits of computer-aided process planning in avoiding such problems.

14.7.3 Coding systems

Coding can be based on a company's own system or one of several classification and coding systems that are available commercially. Because of widely varying product lines and organizational needs, none of the C/C systems has been universally adopted. Whether the system was developed in-house or was purchased, it should be compatible with the company's other systems, such as NC machinery and CAPP systems.

The code structure for part families consists of numbers, letters, or a combination of both. Each specific part or component of a product is assigned a code, which may pertain to design attributes only (generally less than 12 digits) or manufacturing attributes only, although most advanced systems include both, using as many as 30 digits.

The three basic levels of coding vary in degree of complexity. They are:

- *Hierarchical coding.* In this code, also called *monocode*, the interpretation of each succeeding digit depends on the value of the preceding digit. Each symbol amplifies the information contained in the preceding digit, so a digit in the code cannot be interpreted alone. The advantage of this system is that a short code can contain a large amount of information. However, this method is difficult to apply in a computerized system.
- *Polycodes.* Each digit in this code, also known as *chain type*, has its own interpretation, which does not depend on the preceding digit. This structure tends to be relatively long, but allows identification of specific part attributes and is well suited to computer implementation.
- *Decision-tree coding.* This system, also called *hybrid codes*, is the most advanced and combines both design and manufacturing attributes.

Two major industrial coding systems are the Opitz and the MultiClass. The Opitz system, developed in the 1960s in West Germany, was the first comprehensive coding system presented. The basic code consists of nine digits (12345 6789) representing design and manufacturing data (Fig. 14.6). Four additional codes (ABCD) may be used to identify the type and sequence of production operations. This system has two drawbacks: (1) it is possible to have different codes for parts that have similar manufacturing attributes, and (2) a number of parts with different shapes can have the same code.

The MultiClass system was originally developed under the name MICLASS (for Metal Institute Classification System) by the Netherlands Organization for Applied Scientific Research and marketed in the United States by The Organization for Industrial Research and is shown in Fig. 14.7. This system was developed to help automate and standardize several design, production, and management functions. MultiClass, which uses up to 30 digits, is used interactively with a computer, which asks the user a number of questions. On the basis of the answers, the computer automatically assigns a code number to the part. The software is available in modules that can be linked.

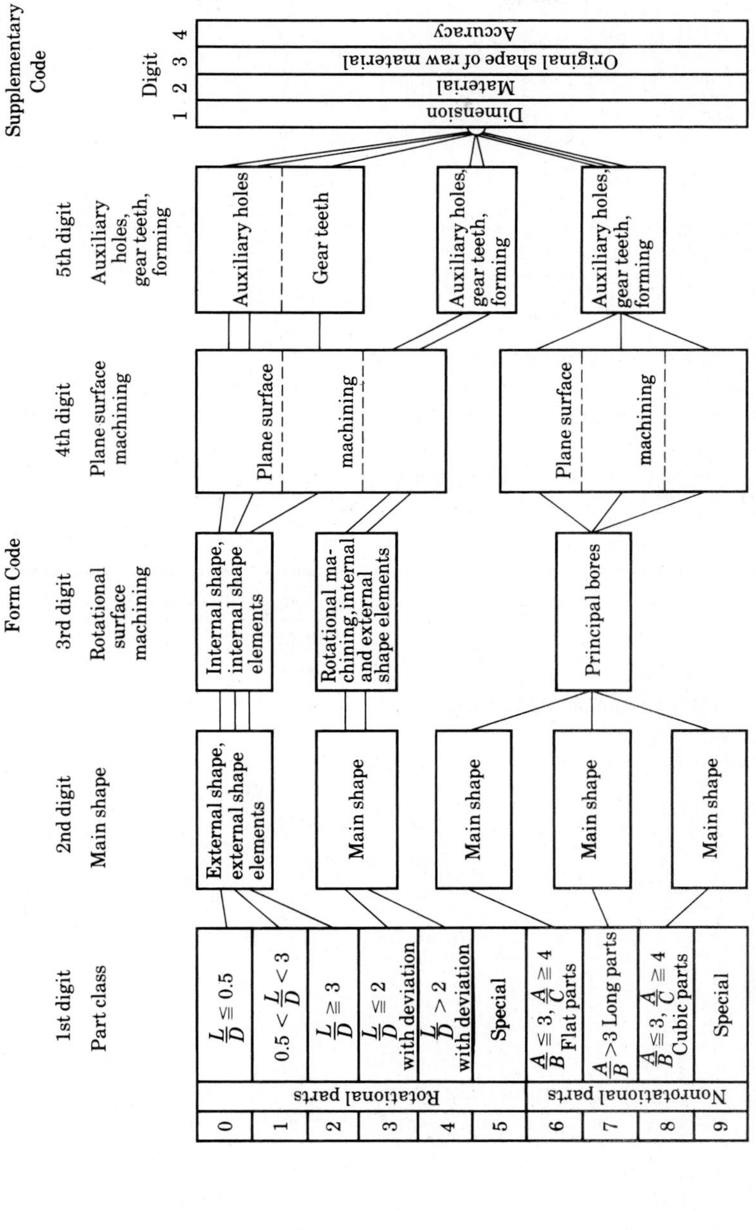

FIGURE 14.6
Classification and coding system according to Opitz, consisting of 5 digits and a supplementary code of 4 digits.

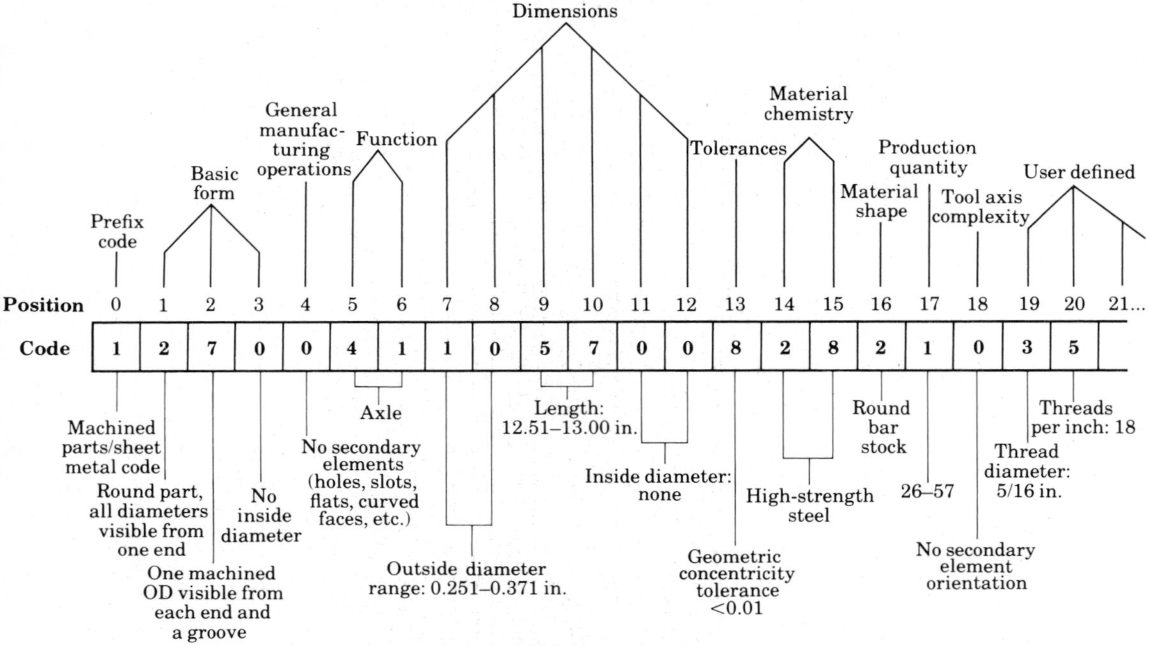

FIGURE 14.7
Typical MultiClass code for a machined part. *Source:* Organization for Industrial Research.

14.8 Cellular Manufacturing

The concept of group technology can be implemented effectively in *cellular manufacturing*, consisting of one or more manufacturing cells. A *manufacturing cell* is a small unit of one to several workstations within a manufacturing system. The workstations usually contain one (single machine cell) or more machines (group machine cell), each performing a different operation on the part. The machines can be modified, retooled, and regrouped for different product lines within the same family of parts. Manufacturing cells are particularly effective in producing families of parts that have relatively constant demand.

Cellular manufacturing thus far has been utilized primarily in machining and sheet-metal forming operations. The machine tools commonly used in manufacturing cells are lathes, machining centers, milling machines, drills, and grinders. For sheet forming, equipment consists of shearing, punching, bending, and other forming machines. This equipment may be special-purpose machines or CNC machines. Cellular manufacturing has some degree of automatic control for:

a) Loading and unloading raw materials and workpieces at workstations.
b) Changing tools at workstations.

c) Transferring workpieces and tools between workstations.
d) Scheduling and controlling the total operation in the cell.

Central to these activities is a material-handling system to transfer materials and parts among workstations. In attended (manned) machining cells, materials can be moved and transferred manually by the operator, unless the parts are too heavy or the movements are hazardous, or by an industrial robot located centrally in the cell. Automated inspection and testing equipment can also be a part of this cell.

Manned manufacturing cells are typically attended by one operator, who oversees all the machines in a cell. Because of the variety of machines and processes involved, the operator becomes multifunctional and is not subjected to the tedium involved with working on the same machine. Consequently, productivity is improved, which is an important feature of manned cells.

Because of the unique features of manufacturing cells, their design and implementation in traditional plants require plant reorganization and rearrangement of existing product flow lines. The machines may be arranged along a line, in a U-shape, in an L-shape, or in a loop. For a group-machine cell where materials are handled by the operator, the U-shaped arrangement is convenient and efficient because the operator can easily reach the various machines. For mechanized material handling, the linear arrangement and loop layout are more efficient. Selecting the best machine and material-handling equipment arrangement also depends on factors such as production rate and type of product and its shape, size, and weight.

We have stressed that, in view of rapid changes in market demand and the need for more product variety in smaller quantity, flexibility of manufacturing operations is highly desirable. Manufacturing cells can be made flexible by using CNC machines and machining centers, with industrial robots or other mechanized systems for handling materials. An example of *flexible manufacturing cells* (FMC) for machining operations is shown in Fig. 14.8. In another flexible manufacturing

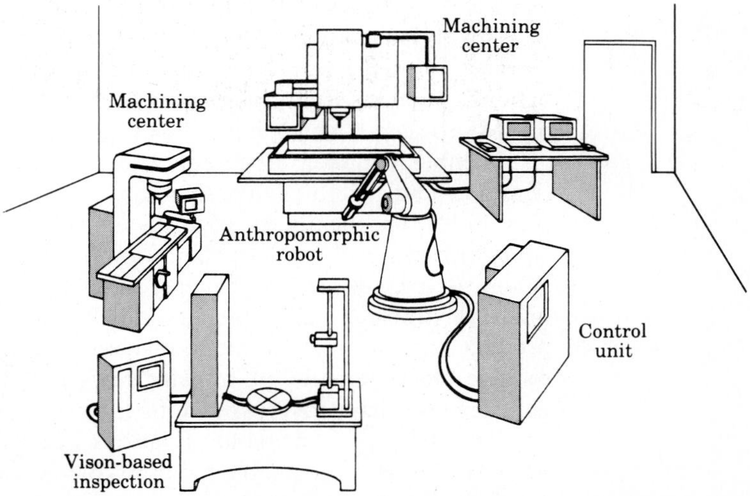

FIGURE 14.8 Schematic view of a flexible manufacturing cell, showing two machine tools, an automated part inspection system, and a central robot serving these machines. *Source:* P. K. Wright.

cell, turbine blades are made by closed-die forging. A robot places cylindrical billets into a furnace, and then transfers them to the forging hammer for forming the blade. A second robot transfers the forged blades to a cropper to trim the flash. The blades are then inspected by machine vision for dimensional accuracy.

Flexible manufacturing cells are usually untended, and their design and operation are more exacting than for other cells. The selection of machines and robots, including the types and capacities of end effectors and their control systems, are important to the proper functioning of the FMC. Significant changes in demand for part families should be considered during design to ensure that the equipment involved has the proper capacity and flexibility. As in other flexible manufacturing systems, the cost of flexible cells is high. Proper maintenance of tools and machinery is essential, as are two or three shift operations of the cells for economic reasons.

14.9 Flexible Manufacturing Systems

A *flexible manufacturing system* (FMS) integrates all major elements of manufacturing that we have described into a highly automated system. First utilized in the late 1960s, FMS consists of a number of manufacturing cells, each containing an industrial robot serving several CNC machines and an automated material-handling system, all interfaced with a central computer. Different computer instructions for the manufacturing process can be downloaded for each successive part passing through the workstation. This system is highly automated and is capable of optimizing each step of the total manufacturing operation. These steps may involve one or more processes and operations, such as machining, grinding, cutting, forming, powder metallurgy, heat treating, and finishing, as well as handling raw materials, inspection, and assembly. The most common applications of FMS to date have been in machining and assembly operations.

Flexible manufacturing systems represent the highest level of efficiency, sophistication, and productivity that has been achieved in manufacturing plants (Fig. 14.9). The flexibility of FMS is such that it can handle a *variety of part configurations* and produce them *in any order*. Hence FMS can be regarded as a system that combines the benefits of two systems: (1) the highly productive but inflexible transfer lines; and (2) job-shop production, which can produce large product variety on stand-alone machines but is inefficient. Table 14.1 shows the relative characteristics of transfer lines and FMS. Note that in FMS the time required for changeover to a different part is very short. The quick response to product and market-demand variations is a major attribute of FMS. A variety of FMS technology is available from machine-tool manufacturers.

(a)

(b)

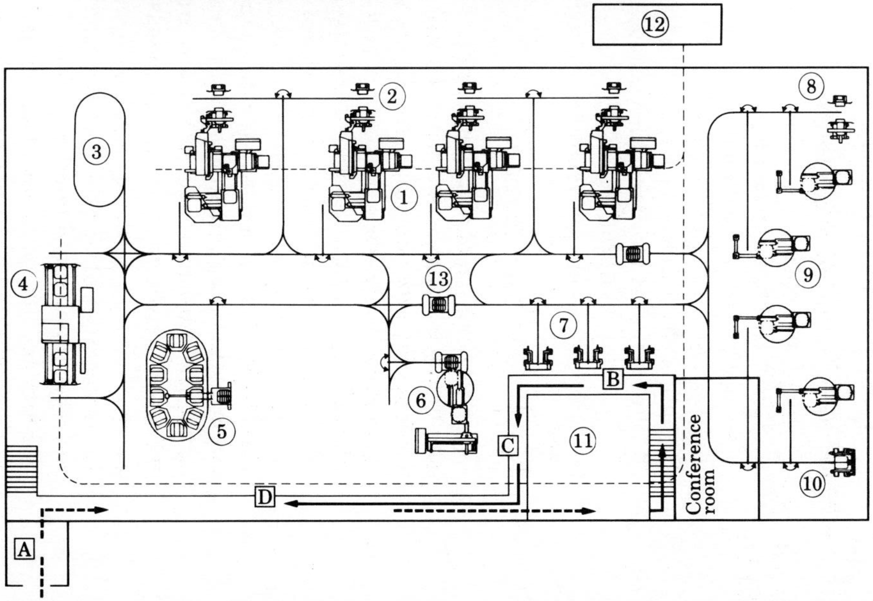

KEY:

1 Four CNC machining centers

2 Four tool-interchange stations, one per machine, for tool storage chain delivery via computer-controlled cart

3 Cart maintenance station; coolant monitoring and maintenance area

4 Parts wash station, automatic handling

5 Automatic workchanger (10 pallets) for online pallet queue

6 One inspection module—horizontal type coordinate measuring machine

7 Three queue stations for tool delivery chains

8 Tool delivery chain load/unload station

9 Four part load/unload stations

10 Pallet/fixture build station

11 Control center, computer room (elevated)

12 Centralized chip/coolant collection/recovery system (- - - - flume path)

13 Three computer-controlled carts, with wire-guided path

↷ Cart turnaround station (up to 360° around its own axis)

A Vestibule: construction schedule; general system layout in model form; production capabilities.

B Catwalk: fixture buildup station; part load/unload stations; CNC machining centers with 2-pallet automatic workchangers; tool delivery chain load/unload stations; queue stations for tool delivery chains.

C Control center: FMS control (DEC PDP 11/44 with peripherals); material-handling control (PDP 11/24); and computerized coordinate measuring machine (control PDP 11/24).

D Automatic workchanger pallet queue; automatic parts wash station; cart maintenance area; automatic coolant maintenance area.

FIGURE 14.9

(a) A general view of a flexible manufacturing system, showing several machine tools and an automated guided vehicle carrying a workpiece on a pallet. (b) Identification of components and machinery in the flexible manufacturing system shown in (a). *Source:* Courtesy of Cincinnati Milacron, Inc.

TABLE 14.1
COMPARISON OF THE CHARACTERISTICS OF TRANSFER LINES AND FLEXIBLE-MANUFACTURING SYSTEMS

CHARACTERISTIC	TRANSFER LINE	FMS
Types of parts made	Generally few	Infinite
Lot size	>100	1–50
Part changing time	$\frac{1}{2}$ to 8 hr	1 min
Tool change	Manual	Automatic
Adaptive control	Difficult	Available
Inventory	High	Low
Production during breakdown	None	Partial
Efficiency	60–70%	85%
Justification for capital expenditure	Simple	Difficult

Compared to conventional systems, the benefits of FMS are:

a) Parts can be produced randomly and in batch sizes as small as one and at lower unit cost.
b) Direct labor and inventories are reduced, with savings estimated at 80–90 percent over conventional systems.
c) Machine utilization is as high as 90 percent, thus improving productivity.
d) Shorter lead times are required for product changes.
e) Production is more reliable because the system is self-correcting and product quality is uniform.
f) Work-in-progress inventories are reduced.

14.9.1 FMS elements

The basic elements of a flexible manufacturing system are workstations, automated handling and transport of materials and parts, and control systems. The types of machines in workstations depend on the type of production. For machining operations, they usually consist of a variety of three- to five-axis machining centers, CNC lathes, milling machines, drill presses, and grinders. Also included are various other pieces of equipment for automated inspection (including coordinate-measuring machines), assembly, and cleaning. Other types of operations include sheet metal forming, punching and shearing, and forging, which include heating furnaces, forging machines, trimming presses, heat-treating facilities, and cleaning equipment. The workstations in FMS are arranged to yield the greatest efficiency in production, with an orderly flow of materials, parts, and products through the system.

Because of FMS flexibility, the material-handling, storage, and retrieval systems are very important. Material handling is controlled by a central computer and performed by industrial robots, conveyors, cranes, and automated guided vehicles. The system should be capable of transporting raw materials, blanks, and parts in various stages of completion to any machine, in random order, at any time. Prismatic parts are usually moved on specially designed *pallets*. Parts with rotational symmetry, such as those for turning operations, are usually moved by robots and mechanical devices.

The computer control system of FMS is its brains and includes various software and hardware. It controls the machinery and equipment in workstations and for transporting raw materials, blanks, and parts in various stages of completion from machine to machine. It also stores data and provides communication terminals that display the data visually.

14.9.2 Scheduling

Because FMS involves a major capital investment, efficient machine utilization is essential; that is, machines should not stand idle. Consequently, proper scheduling and process planning are crucial. Scheduling for FMS is *dynamic*, unlike that in job shops where a relatively rigid schedule is followed to perform a set of operations during a certain period of time. The scheduling system for FMS clearly specifies the types of operation to be performed on each part and identifies the machines or manufacturing cells to be used. Dynamic scheduling is capable of responding to quick changes in product type and hence is responsive to real-time decisions. Because of the flexibility in FMS, no setup time is wasted in switching manufacturing operations since the system is capable of performing different operations in different orders on different machines. However, the characteristics, performance, and reliability of each unit in the system must be checked to ensure that as parts move from workstation to workstation they are of acceptable quality and dimensional accuracy.

14.9.3 Economic justification

FMS installations are very costly, typically starting at well over $1 million. Consequently, a thorough cost–benefit analysis must be conducted before a final decision is made. This analysis should include factors such as costs of capital, energy, materials, and labor, expected markets for the products to be manufactured, and anticipated fluctuations in market demand and product type. An additional factor is the time and effort required for installing and debugging the system. Typically, an FMS system can take 2–5 years to install and at least 6 months to debug. Although FMS requires few, if any, machine operators, the personnel involved with the total operation must be trained and highly skilled. These personnel include manufacturing engineers, maintenance engineers, and computer programmers.

As we indicated in Fig. 13.1, the most effective FMS applications have been in medium-volume batch production. High-volume, low-variety parts production is best obtained from transfer machines (dedicated equipment). Low-volume, high-variety parts production can best be done on conventional standard machinery, with or without numerical control, or by machining centers. When a variety of parts is to be produced, FMS is suitable for production volumes of 15,000–35,000 aggregate parts per year. For individual parts of the same configuration, production may reach 100,000 units per year.

14.10 Just-in-Time Production

The *just-in-time* (JIT) *production* concept was developed in Japan to eliminate waste of materials, machines, capital, manpower, and inventory throughout the manufacturing system. The JIT concept has the following goals:

- Purchase supplies just in time to be used.
- Produce parts just in time to be made into subassemblies.
- Produce subassemblies just in time to be assembled into finished products.
- Produce and deliver finished products just in time to be sold.

In traditional manufacturing, parts are made in batches, placed in inventory, and used whenever necessary. This approach is known as a *push system*, meaning that parts are made according to a schedule and are in inventory to be used if and when they are needed. Just-in-time is a *pull system*, meaning that parts are produced to order, and production is matched with demand for final assembly of products. There are no stockpiles, with the ideal production quantity being one (*zero inventory*, *stockless production*, *demand scheduling*). Moreover, parts are inspected by the worker as they are manufactured and are used within a short period of time. In this way, the worker maintains continuous production control, immediately identifying and correcting defective parts. The worker takes pride in product quality. Unnecessary motions in stockpiling and retrieving parts from storage are eliminated.

Implementation of the JIT concept requires that all aspects of manufacturing operations be carefully reviewed and monitored so that all nonvalue-adding operations and use of resources are eliminated. This approach emphasizes pride and dedication in producing high-quality products, eliminating idle resources, and joint problem solving by teamwork among workers, engineers, and management to quickly solve any problems that occur during production or assembly.

The ability to detect production problems in producing parts just in time has been likened to the level of water (representing the inventory levels) in a lake covering a bed of boulders. The boulders represent production problems. When the water level is high (high inventories), the boulders are not exposed. On the other hand, when the level is low (low inventories), the boulders are exposed; they can be found, observed, and removed. This analogy indicates that high inventory levels can mask quality and production problems involving parts that are stockpiled. Delayed inspection, especially, causes faulty parts production.

The just-in-time concept also includes delivery of supplies and parts from outside sources, thus reducing in-plant inventory. Suppliers are expected to deliver pre-inspected goods as they are needed for production, often on a daily basis. This requires reliable suppliers and close cooperation and trust between the company and its vendors. The JIT concept of purchasing and delivery is a significant departure from traditional purchasing of supplies from one or more vendors, whereby deliveries are made, with lead times of weeks or months, in larger quantities and stored in inventory. Although a buffer is built into the traditional system, it tends to create large inventory levels and to make quality control of incoming supplies difficult.

The concept was first demonstrated on a large scale in 1953 at the Toyota Motor Company, under the name *kanban*, meaning visible record. These records usually consist of two types of cards (kanbans). One is the *production card*, which authorizes the production of one container or cart of identical, specified parts at a workstation. The other card is the *conveyance* or *move card*, which authorizes the transfer of one container or cart of parts from that workstation to the workstation where the parts will be used. The cards contain information on the type of part, place of issue, part number, and the number of items in the container or cart. Recently, these cards have been replaced by bar-coded plastic tags and other devices. The number of containers in circulation at any time is completely controlled and can be scheduled as desired for maximum efficiency.

The advantages of JIT are:

- Low inventory carrying costs.
- Fast detection of defects in production or in delivery of supplies and hence low scrap loss.
- Reduced inspection and rework of parts.
- High-quality parts at low cost.

Implementation of just-in-time production has resulted in reductions of 20–40 percent in product cost, 60–80 percent in inventory, up to 90 percent in rejection rates, 90 percent in lead times, 50 percent in scrap, rework, and warranty costs, and increases of 30–50 percent in direct labor productivity and 60 percent in indirect labor productivity.

14.11 Manufacturing Automation Protocol

We have pointed out that, because of its many benefits, total integration of diverse manufacturing activities is a major goal of modern manufacturing. In order to maintain a high level of coordination and efficiency of operation in integrated manufacturing, an extensive, high-speed, and interactive *communications network* is required.

A significant advance in communications technology is the *local area network* (LAN). In this hardware and software system, logically related groups of machines and equipment communicate with each other. Recall the importance and efficiency of grouping related machines and activities by such means as group technology and manufacturing cells using robots and various other material-handling equipment. A LAN ties these groups to each other, bringing different phases of manufacturing into a unified operation. A local area network can be quite large and complex, linking hundreds of machines and devices in several buildings. Various network layouts (Fig. 14.10) of copper cables or fiber optics are used over distances ranging from a few meters to as much as 30 km (20 mi).

Typically, a manufacturing cell is built with machines and equipment bought from one vendor, another cell with machines from another vendor, and yet another from another vendor. Thus a variety of programmable devices are involved, driven

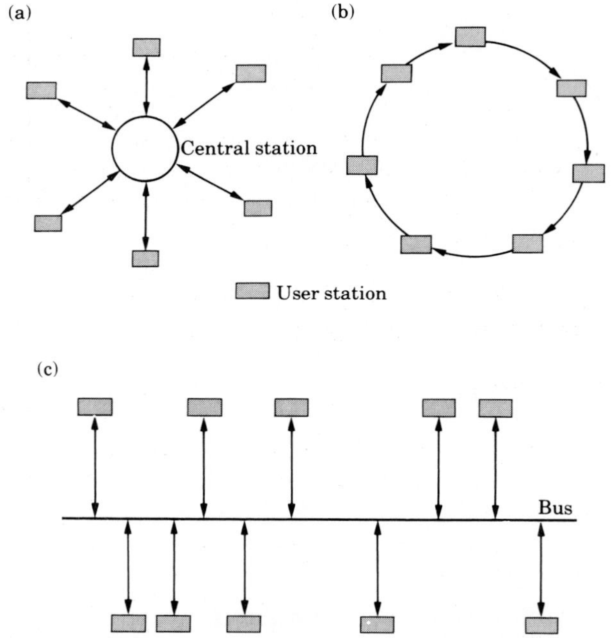

FIGURE 14.10 Three basic types of topology for local area network (LAN). (a) The *star* topology is suitable for situations that are not subject to frequent configuration changes. All messages pass through a central station. Telephone systems in office buildings usually have this type of topology. (b) In the *ring* topology all individual user stations are connected in a continuous ring. The message is forwarded from one station to the next until it reaches its assigned destination. Although the wiring is relatively simple, the failure of one station shuts down the entire network. (c) In the *bus* topology all stations have independent access to the bus. This system is reliable and is easier to service. Because its arrangement is similar to the layout of machines in a factory, its installation is relatively easy and can be rearranged when machines are rearranged.

by several computers and microprocessors purchased at various times from different vendors and having various capacities and levels of sophistication. Each cell's computers have their own specifications and proprietary standards and cannot communicate beyond the cell with others, unless equipped with custom-built interfaces. This situation has created "islands of automation," and in some cases up to 50 percent of the cost of automation was related to difficulties in communication between manufacturing cells and other parts of the organization.

The existence of automated cells that could only function independently from each other—and without a common base for information transfer—led to the need for a *communications standard* to improve communications and the efficiency of computer-integrated manufacturing. The first step toward standardization began with formation of a task force at the General Motors Corporation in 1980. Its charter was to "identify communications standards which provide for multi-vendor data communications in the factory environment." After considerable effort, and based on existing national and international standards, a set of communications standards known as *manufacturing automation protocol* (MAP) was developed.

Because of the diverse interests involved, implementation of MAP is a gradual process. However, its capabilities and effectiveness were demonstrated in 1984 by the successful interconnection of devices from a number of vendors. As a result, the importance of a worldwide communications standard is now recognized, and vendors are designing their products in compliance with these standards.

14.12 Artificial Intelligence

Artificial intelligence (AI), in its broadest sense, is the use of computers to replace human intelligence. *Intelligence* may be defined as the mental capacity to apprehend, retain, and extend facts and be able to reason about them and make appropriate decisions based on them. In human beings, senses and complex brain processes perform these functions.

Since the late 1940s when work on artificial intelligence began, its nature, capabilities, and potential benefits have been argued—heatedly at times. One basic view is that computers are approaching intelligent behavior with their capabilities of speech recognition, vision, and extensive information storage, retrieval, and processing. Another basic view of artificial intelligence is that the computer will eventually be able to think like human beings, including the ability to learn for itself, reason, deduct, explain, and plan, as well as control manufacturing operations. Predicting the future accurately is difficult. However, AI is likely to affect profoundly automation and the overall economics of manufacturing operations. Because of advances in memory expansion—VLSI chip designs and their decreasing costs (Fig. 14.11)—AI programs that can now be run on personal computers have been developed. Thus AI has become accessible to office desk and shop floors.

14.12.1 AI elements

In general, artificial intelligence encompasses the following activities: expert systems, natural language, and machine vision. We describe each of these elements briefly.

Expert systems. An *expert system*, also called a *knowledge-based system*, may be defined as an intelligent computer program that has the capability to solve difficult problems using *knowledge* and *inference* procedures (Fig. 14.12). The knowledge, or body of information, is expressed in computer codes, usually in the form of *if–then rules*. The codes model the collective experience and knowledge of human experts and are based on facts, data, assumptions, and definitions. Ideally the system would be capable of making decisions using a *heuristic* approach, that is, making good judgments by discovery and revelation, and making good guesses, just as an expert would.

Because of the difficulty involved in accurately modeling the many years of experience of an expert, and the complex inductive reasoning and decision-making capabilities of humans, developing knowledge-based systems requires considerable time and effort. The problem is particularly difficult because human beings rely largely on *intuition* and *feel* in making decisions.

Expert systems work on a real-time basis, and their short reaction times provide rapid responses to problems. The most common AI programming languages are LISP and PROLOG. A recent development is expert-system software *shells*, also called *framework systems*. These software packages are essentially expert-system outlines that allow a person to write specific applications to suit the special needs of a company. Writing these programs requires considerable experience and time.

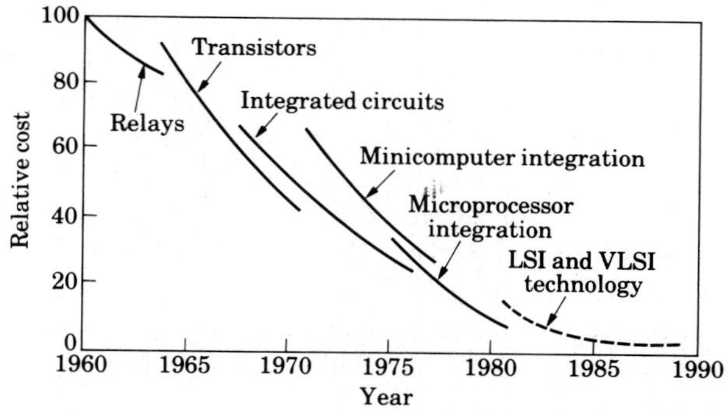

FIGURE 14.11

Relative cost of electronic and computer components. *Source: Flexible Automation,* 87/88, *The International CNC Reference Book.*

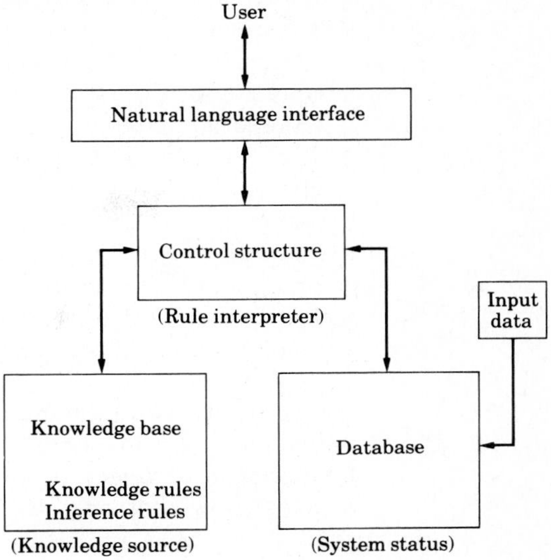

FIGURE 14.12 Basic structure of an expert system. The knowledge base consists of knowledge rules (general information about the problem) and inference rules (the way conclusions are reached). The results are communicated to the user through the natural language interface. *Source:* K. W. Goff, *Mechanical Engineering,* October 1985.

Natural language processing. Traditionally, obtaining information from a database in computer memory has required utilization of computer programs that translate natural language to machine language. Natural language interfaces with database systems are now in various stages of development. These systems allow a user to obtain information by entering English language commands in the form of simple, typed queries.

Software shells are available and are used in applications such as scheduling material flow in manufacturing and analyzing information in databases. Significant progress is being made on computers that will have speech synthesis and recognition capabilities, thus eliminating the need to type commands on keyboards.

Machine vision. We described the basic features of machine vision in Section 13.7. Computers and software for artificial intelligence can be combined with cameras and other optical sensors. These machines can then perform operations such as inspecting, identifying, and sorting parts and guiding robots (intelligent robots), which would otherwise require human intervention.

14.13 Factory of the Future

On the basis of the advances made to date in all aspects of manufacturing technology and computer controls, we may envisage the *factory of the future* as a fully automated facility in which human beings would not be directly involved with production on the shop floor (hence the term *untended factories*). All manufacturing, material handling, assembly, and inspection would be done by automated and computer-controlled machinery and equipment.

Similarly, activities such as processing incoming orders, production planning and scheduling, cost accounting, and various decision-making processes (usually performed by management) would also be done automatically by computers. The role of human beings would be confined to activities such as supervising, maintaining (especially preventive maintenance), and upgrading machines and equipment; shipping and receiving supplies and finished products; providing security for the plant facilities; and programming, upgrading, and monitoring computer programs, and monitoring, maintaining, and upgrading hardware.

Industries such as some food, petroleum, and chemical already operate automatically with little human intervention. These are continuous processes and, unlike piece part manufacturing, are easier to automate fully. Even so, the direct involvement of fewer people in manufacturing products is already apparent: Surveys show that only 10–15 percent of the workforce is directly involved in production. Most of the workforce is involved in gathering and processing information.

Virtually untended manufacturing cells already make products such as engine blocks, axles, and housings for clutches and air compressors. For large-scale, flexible manufacturing systems, however, highly trained and skilled personnel will always be needed to plan, maintain, and oversee operations.

The reliability of machines, control systems, and power supply is crucial to full factory automation. A local or general breakdown in machinery, computers, power, or communications networks will, without rapid human intervention, cripple production. The computer-integrated factory of the future should be capable of automatically rerouting materials and production flows to other machines and to the control of other computers in case of such emergencies.

Important considerations are the nature and extent of the fully automated factory's impact on employment, with all its ramifications. The fear of unemployment persists despite strong evidence that advances in technology create more jobs—or at least maintain the same number—rather than eliminating them. Although projections indicate that there will be a decline in the number of machine tool operators and tool and die workers, there will be major increases in the number of computer service technicians and maintenance electricians.

Thus the generally low-skilled, direct labor force engaged in traditional manufacturing will shift to an indirect labor force, with special training or retraining required in areas such as computer programming, information processing, CAD/CAM, and other high-technology tasks that are essential parts of computer-integrated manufacturing. Development of more user-friendly computer software is making retraining of the work force much easier.

Opinions about the impact of untended factories diverge widely. Consequently, predicting the nature of future manufacturing strategies with any certainty is difficult. Although economic considerations and trade-offs are crucial, companies that do not install computer-integrated operations are in jeopardy. It is now widely recognized that, in a highly competitive marketplace, rapid adaptability is essential to the survival of a manufacturing organization.

SUMMARY

Computers are having a major impact on all aspects of manufacturing processes and operations. Integrated manufacturing systems are being implemented to various degrees to optimize operations, reduce costs, and improve product quality and productivity. Computer-integrated manufacturing has become the important means of improving productivity, responding to changing market demand, and better controlling manufacturing and management functions. With extensive use of computers and rapid developments in sophisticated software, designs and their analysis and simulation are more detailed and thorough.

New developments in manufacturing operations, such as group technology, cellular manufacturing, and flexible manufacturing systems, are contributing significantly to improved productivity. Artificial intelligence is likely to open new opportunities in all aspects of manufacturing engineering. The factory of the future will extend these developments further, while their benefits and economic justification continue to be debated at all levels of management and the workforce.

BIBLIOGRAPHY

Aleksander, I., *Designing Intelligent Systems.* New York: UNIPUB, 1984.

Barr, P., R. Krimper, M. Lazear, and C. Stammen, *CAD: Principles and Applications.* Englewood Cliffs, N.J.: Prentice-Hall, 1985.

Begg, V., *Developing Expert CAD Systems.* New York: UNIPUB, 1984.

Besant, C.B. *Computer-Aided Design and Manufacture*, 2d ed. New York: Wiley, 1983.

Chang, T.-C., and R.A. Wysk, *An Introduction to Automated Process Planning Systems.* Englewood Cliffs, N.J.: Prentice-Hall, 1985.

Gallagher, C.C., and W.A. Knight, *Group Technology Production Methods in Manufacture.* New York: Wiley, 1986.

Greenwood, N.R., *Implementing Flexible Manufacturing Systems.* New York: Wiley, 1988.

Groover, M.P., *Automation, Production Systems, and Computer-Integrated Manufacturing.* Englewood Cliffs, N.J.: Prentice-Hall, 1987.

Groover, M.P., and E.W. Zimmers, Jr., *CAD/CAM: Computer-Aided Design and Manufacturing.* Englewood Cliffs, N.J.: Prentice-Hall, 1984.

Hales, L.E., *Computer-Aided Facilities Planning.* New York: Marcel Dekker, 1984.

Ham, I., K. Hitomi, and T. Yoshida, *Group Technology: Applications to Production Management.* Boston: Kluwer-Nijhoff, 1985.

Harmon, P., and D. King, *Expert Systems.* New York: Wiley, 1985.

Harrington, J., Jr., *Understanding the Manufacturing Process: Key to Successful CAD/CAM Implementation.* New York: Marcel Dekker, 1984.

Hitomi, K., *Manufacturing Systems Engineering*. London: Taylor & Francis, Ltd., 1979.
Hyer, N.L. (ed.), *Group Technology at Work*. Dearborn, Mich.: Society of Manufacturing Engineers, 1984.
Koren, Y., *Computer Control of Manufacturing Systems*. New York: McGraw-Hill, 1983.
LeMaistre, C., and A. El-Sawy, *Computer Integrated Manufacturing: A Systems Approach*. New York: UNIPUB/Kraus International Publications, 1987.
Medland, A.J., and P. Burnett, *CAD/CAM in Practice*. New York: Wiley, 1986.
Menipaz, E., *Essentials of Production Operations Management*. Englewood Cliffs, N.J.: Prentice-Hall, 1986.
Milner, D.A., and V. Vasiliou, *Computer-Aided Engineering for Manufacture*. New York: McGraw-Hill, 1987.
Pao, Y.C., *Elements of Computer-Aided Design and Manufacturing CAD/CAM*. New York: Wiley, 1985.
Proceedings of Group Technology Conference. London: Institution of Production Engineers (annual since 1971).
Proceedings of International Conference on Flexible Manufacturing Systems. Bedford, England: IFS (Publications), Ltd. (annual since 1982).
Ranky, P.G., *Computer Integrated Manufacturing: An Introduction with Case Studies*. Englewood Cliffs, N.J.: Prentice-Hall, 1986.
Ranky, P.G., *The Design and Operation of FMS*. New York: North-Holland, 1983.
Rembold, U., C. Blume, and R. Dillmann, *Computer-Integrated Manufacturing Technology and Systems*. New York: Marcel Dekker, 1985.
Shingo, S., *A Revolution in Manufacturing: The S.M.E.D. System*. Cambridge, Mass.: Productivity Press, 1986.
Suzaki, K., *The New Manufacturing Challenge*. New York: Free Press, 1987.
Teicholz, E. (ed.), *CAD/CAM Handbook*. New York: McGraw-Hill, 1985.
Teicholz, E., and J.N. Orr (eds.), *Computer-Integrated Manufacturing Handbook*. New York: McGraw-Hill, 1987.
Tool and Manufacturing Engineers Handbook, 4th ed., *Vol. 5: Manufacturing Management*. Dearborn, Mich.: Society of Manufacturing Engineers, 1987.
Voss, C.A., *Just-In-Time Manufacturing*. Bedford, England: IFS (Publications), Ltd., 1987.
Weatherall, A., *Computer-integrated Manufacturing*. Stoneham, Mass.: Butterworths, 1988.
Wright, P.K., and D.A. Bourne, *Manufacturing Intelligence*. Reading, Mass.: Addison-Wesley, 1988.

QUESTIONS AND PROBLEMS

14.1 Give several examples of mathematical and physical models relevant to manufacturing engineering. Describe the relative advantages and limitations of each type of model.

14.2 Describe the various subsystems of a CIM system.

14.3 Give examples of primitives of solids other than those shown in this chapter.

14.4 Think of a part to be manufactured and prepare a routing sheet for it. Explain why another person might develop a different routing sheet for the same part.

14.5 Why was group technology developed?

14.6 Review various parts and products discussed in this text and group them similarly to those shown in Fig. 14.3.

14.7 Explain the logic behind the arrangements shown in Fig. 14.5.

14.8 Why were manufacturing cells developed?

14.9 Think of two different products and design a manufacturing cell for making each of them, describing the features of the machines and equipment involved. Consider alternative designs and justify your final selection.

14.10 Why is a flexible manufacturing system capable of producing a wide range of lot sizes? What would be the minimum lot size? Why?

14.11 Discuss your observations concerning the information given in Table 14.1.

14.12 Discuss your observations concerning Fig. 14.9.

14.13 Why is just-in-time production called a pull system?

14.14 Give examples of manufacturing engineering in which artificial intelligence could be effective. Give other examples in which it would not be effective and explain why.

14.15 Why is machine vision a part of artificial intelligence?

14.16 Give examples of situations in which machine vision cannot be applied properly and reliably. Explain why.

14.17 Comment on the advantages and limitations of implementing flexible manufacturing systems in small versus large companies.

14.18 Design a manufacturing system that implements machine vision. What would be its advantages over other methods?

14.19 Discuss your own thoughts about the factory of the future.

15

Competitive Aspects and Economics of Manufacturing

15.1 Introduction

Manufacturing high-quality products at the lowest possible cost requires an understanding of the often complex relationships among many factors. We have shown that product design, selection of materials, and manufacturing processes are interrelated. Designs are often modified to improve product performance, take advantage of the characteristics of new materials, and make manufacturing and assembly easier, thus reducing costs. Because of the wide variety of materials and processes available today, the task of producing a high-quality product by selecting the best materials and processes while minimizing costs has become a major challenge. Meeting this challenge requires not only a thorough knowledge of the characteristics of materials and processes, but also innovative and creative approaches to design and manufacturing technology.

The cost of a product often determines its marketability and customer satisfaction. Consequently, each production step must be analyzed with a view toward ensuring an optimal economic outcome (Fig. 15.1). In this chapter we consider the following approaches to minimize costs while fulfilling design, quality, and service requirements:

a) Selection of materials.
b) Product design and quantity of materials.
c) Substitution of materials.
d) Selection of manufacturing processes.
e) Process capabilities.
f) Manufacturing costs.
g) Cost reduction.

Although we have described the economics of various manufacturing processes, we take a broader view in this chapter and summarize the important overall manufacturing cost factors. We look at cost reduction methods, particularly value

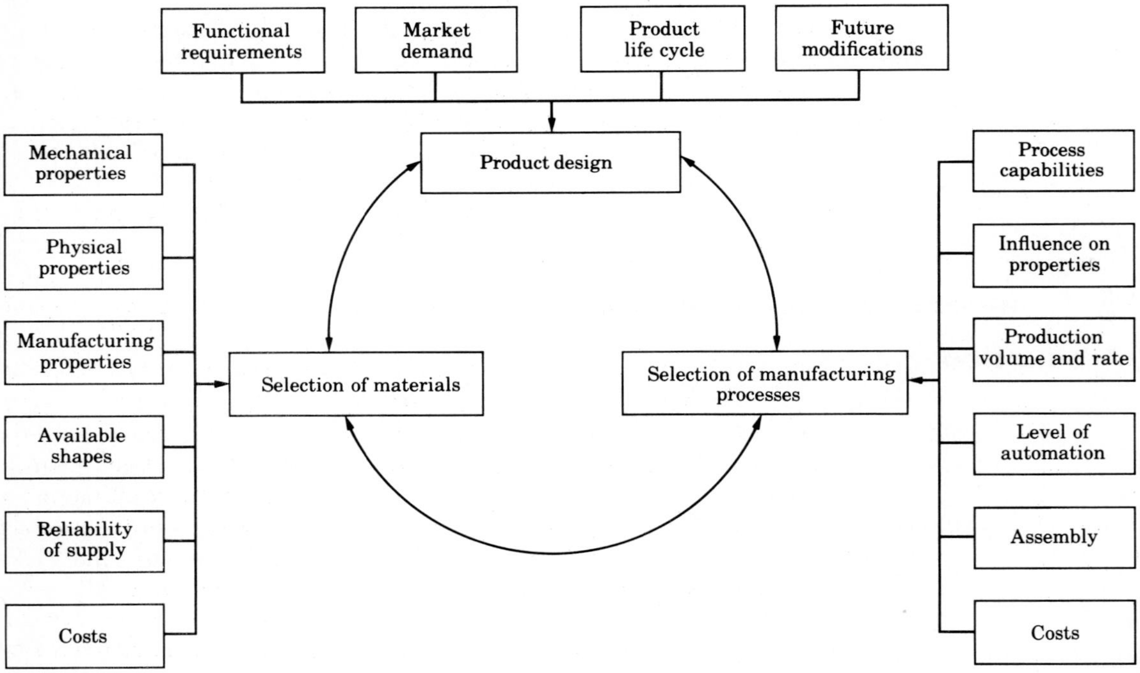

FIGURE 15.1
Interrelationships among various factors influencing product cost.

engineering, which is a powerful tool that can be used to evaluate the cost of each manufacturing step in terms of its share of the value added to the product. Finally, we list a number of questions that manufacturing engineers and other members of a manufacturing organization must ask concerning the design and manufacture of a product.

15.2 Selection of Materials

In selecting materials for a product, we must clearly understand the *functional* requirements for each of its individual components. We described the general criteria for selecting materials in Chapter 1 and now discuss selecting materials in further detail.

15.2.1 Material properties and availability

Mechanical properties are strength, toughness, ductility, stiffness, hardness, and resistance to fatigue, creep, and impact. Physical properties include density, melting point, specific heat, thermal and electrical conductivity, thermal expansion, and magnetic properties. Chemical properties that we are primarily concerned with are oxidation and corrosion resistance. We described these properties and their relevance to product design and manufacturing in various chapters.

Manufacturing properties are typically castability, formability, machinability, weldability, and hardenability by heat treatment. Because materials have to be formed, shaped, machined, ground, fabricated, and heat treated into individual parts and components of specific shapes and dimensions, these properties are crucial to proper selection of materials.

Recall that the quality of a material can greatly influence its manufacturing properties. A rod or bar with a longitudinal seam will develop cracks during upsetting and heading operations. Bars with internal defects and inclusions will crack during seamless-tube production. Porous cast workpieces will produce poor surface finish when machined. Blanks that are heat treated nonuniformly or bars that are not stress relieved will distort during finishing operations, such as grinding.

Similarly, incoming stock that has variations in composition and microstructure cannot be heat treated or machined properly. Sheet-metal stock with variations in its cold-worked conditions will spring back differently during bending and other forming operations. If prelubricated sheet-metal blanks are supplied with nonuniform lubricant thickness, their formability, surface finish, and overall quality will be adversely affected.

When selecting materials, we need to know the shapes and sizes of commercially available materials (Table 15.1) in order to avoid additional processing, unless it is economically justified. Materials are available in various forms: casting, extrusion, forging, bar, plate, sheet, foil, rod, wire, or powder. Purchasing materials in shapes that generally require the least additional processing means that we also have to consider characteristics such as surface quality, tolerances, and straightness. The better and more consistent these characteristics are, the less additional processing is required.

In Chapter 1, we pointed out geopolitical factors that can affect the supply of strategic materials. Other factors such as strikes, shortages, and reluctance of suppliers to produce materials in a particular shape, quality, or quantity also affect reliability of supply. Even though availability of materials may not be a problem for a country as a whole, it can be a problem for certain industries or because of the location of a manufacturing plant.

15.2.2 Costs of materials and processing

Because of its processing history, the *unit cost* of a material (cost per unit weight or volume) depends not only on the material itself, but also on its shape, size, and condition. Because more operations are involved in its production, the unit cost of thin wire, for example, is higher than that of a round bar made of the same material. The unit cost of metal foil is higher than that of metal plate. Similarly, powder metals are more expensive than bulk metals. The cost of materials generally

TABLE 15.1
COMMERCIALLY AVAILABLE FORMS OF MATERIALS

MATERIAL	AVAILABLE AS
Aluminum	P, F, B, T, W, S, I
Copper and brass	P, f, B, T, W, s, I
Magnesium	P, B, T, w, S, I
Steels and stainless steels	P, B, T, W, S, I
Precious metals	P, F, B, t, W, I
Zinc	P, F, B, W, I
Plastics	P, f, B, T, w
Elastomers	P, b, T
Ceramics (alumina)	p, B, T, s
Glass	P, B, T, W, s
Graphite	P, B, T, W, s

Note: P, plate or sheet; F, foil; B, bar; T, tubing; W, wire; S, structural shapes; I, ingots for casting. Lowercase letter indicates limited availability.

TABLE 15.2
MATERIALS PRICE INDEX (1982 = 100)

MATERIAL	JANUARY 1987	JANUARY 1988
Steel	101.7	107.3
Aluminum	100.1	118.2
Copper	104.8	168.4
Zinc	110.6	114.1
Magnesium	114.2	114.2
Titanium and its alloys	73.5	71.2
Nickel and Ni-Cu alloys	81.8	95.3
Plastics	102.6	119.8

decreases as the purchase quantity increases, much like packaged food products you purchase at supermarkets: the larger the quantity per package, the lower the cost per unit weight.

The cost of a particular material is subject to fluctuations (Table 15.2), which may be caused by factors as simple as supply and demand or as complex as geopolitics. If a product is no longer cost competitive in the marketplace, alternative and less costly materials should be selected. For example, the copper shortage in the 1940s caused the U.S. government to mint pennies from zinc-plated steel. Similarly, when the price of copper increased substantially during the 1960s, electrical wiring for residential use was, for a time, made from aluminum.

If scrap is produced during manufacturing, as in sheet-metal fabricating, forging, and machining, the value of the scrap is deducted from the material's cost to obtain net material cost. The value of the scrap depends on the type of metal and the demand for it. The value is usually 10–40 percent of the original cost of the material. Typical percentages of scrap produced in selected manufacturing processes are given in Table 15.3.

TABLE 15.3
APPROXIMATE AMOUNT OF SCRAP PRODUCED IN VARIOUS MANUFACTURING PROCESSES

PROCESS	SCRAP (%)
Machining	10–60
Hot closed-die forging	20–25
Sheet-metal forming	10–25
Cold or hot extrusion, forging	15
Permanent-mold casting	10
Powder metallurgy	5

15.2.3 Product design and quantity of materials

With high production rates and less labor, the cost of materials becomes a significant portion of product cost. Although the cost of materials cannot be reduced below a certain level, efforts can be made to reduce the amount of material involved in components that are to be mass produced. Since the overall shape of the part is usually optimized during the design and prototype stages, further reductions in the amount of material used can be obtained only by reducing the part's thickness. To do so requires selection of materials having high strength-to-weight or stiffness-to-weight ratios.

Many engineering designs have traditionally been done intuitively, usually based on experience and empirical information. Although the components in some products may be underdesigned, parts are generally overdesigned (called *irrational design*), presumably with safety in mind. New techniques such as finite-element analysis, minimum-weight design, design optimization, and computer-aided design and manufacturing have greatly facilitated design analysis and optimization and product simulation.

Implementing new design techniques and minimizing the amount of materials utilized can, however, present significant problems in manufacturing. Producing parts with thin cross-sections can be difficult, such as welding, casting, forging and sheet-stretching operations on thin sections. In forging, thin parts require high forces. In welding, they tend to distort. In casting, thin sections present difficulties in mold-cavity filling and in maintaining dimensional accuracy and good surface finish. In sheet-forming operations, formability may be reduced.

Conversely, utilizing parts with thick sections can slow production in processes such as casting and injection molding because of the length of time required for cooling and removing the part from the mold. Abrupt changes in part cross-section, such as from thick to thin, contribute to problems and may lead to defective parts.

15.3 Substitution of Materials

There is hardly a product on the market today for which substitution of materials has not played a major role in helping companies maintain their domestic and international competitive positions. Automobile and aircraft manufacturing are examples of major industries where substitution of materials is an important ongoing activity. A similar trend is evident in sporting goods and other consumer-product areas.

Although new products continually appear on the market, the majority of both design and manufacturing activities is concerned with improving existing products. Product improvements are resulting from substitution of materials, as well as implementation of new or improved processing techniques, better control of processing parameters, and plant automation.

There are a number of reasons for substituting materials in existing products, including:

a) Reduction in costs of materials and processing.
b) Ease of manufacturing.
c) Improvements in performance, such as reduction in weight and better resistance to wear, fatigue, corrosion, and other characteristics.
d) Increase in stiffness-to-weight and strength-to-weight ratios.
e) Simplification of assembly and installation and for conversion to automated parts assembly.
f) Ease of maintenance and repair.
g) Unreliable domestic and overseas supply of materials.
h) Legislation and regulations prohibiting the use of certain materials in products for environmental control.

• Example 15.1: Aluminum versus steel beverage cans.

The market for beverage cans in the United States is estimated at $4 billion a year and offers an example of how material substitution can have a major economic impact. For many years, beverage cans were made of steel in three pieces: the top, the bottom, and the seam-welded cylindrical body. Since the development of the aluminum can in 1958, however, aluminum has gained an increasingly larger share of the market, so that 95 percent of the 78 billion beverage cans produced today are made of aluminum. The two-piece aluminum cans (top and body) have such advantages as lightness, recyclability, and the lack of seams that were unsightly and rust-prone in steel cans. Because of the value of aluminum, it is cost effective to recycle aluminum cans—another attractive feature. There are now about 10,000 can collection points, and more than 50 percent of the cans produced are recycled.

The steel industry is now making great strides to capture a larger share of this market. Two-piece steel cans (tin plate) have now been developed, although the top of the can is still made of aluminum because of the lower effort involved in opening the tops. Because the cost of aluminum has increased at a higher rate than steel, steel cans have the potential of becoming competitive. Furthermore, steel cans can be magnetically separated from other garbage and can be recycled, although the economics of recycling is still a matter of dispute. The steel industry is planning collection centers for cans in large cities for recycling. Some brewing companies are now showing interest in switching to steel cans. It is estimated that the cost saving with steel cans is on the order of $5 to $7 per thousand cans, or an estimated $500 million a year.

• Example 15.2: Material substitution in a vacuum cleaner.

The materials used in a household (Hoover) vacuum cleaner are shown in the accompanying figure. As shown in the table on the following page, some of these materials were different in older models of this product. The table indicates the reasons for the changes made. *Source: Metal Progress.*

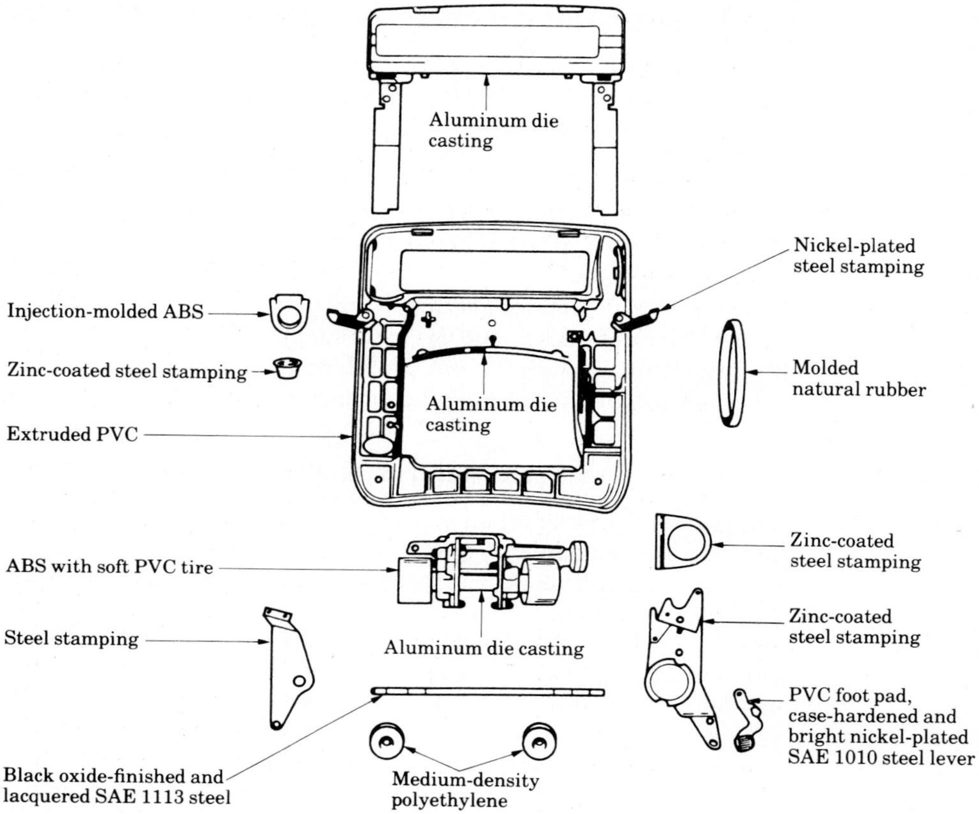

PART	FORMER MATERIAL	NEW MATERIAL	REASON FOR CHANGE
Bottom plate	Steel stampings	Die-cast aluminum	Easier servicing
Wheels	Phenolic resin	Medium-density polyethylene	Less noise
Wheel mounting	Screw-machine parts	Preassembled with a cold-headed steel shaft	Lower cost, simpler to replace
Switch toggle	Phenolic resin	ABS	Tougher
Handle tube	Lock-seam steel tubing	Electric seam-welded tubing	Lower cost, better dimensional control
Motor hood	Cellulose acetate	ABS	Better heat and moisture resistance, no warpage
Rug nozzle	ABS	High-impact styrene	Lower cost
Hose	PVC-coated wire with single-ply PVC covering	PVC-coated wire with double-ply covering separated by a nylon reinforcement	More durable, lower cost
Bellows, cord insulation, bumper strips	Rubber	PVC	Lower cost, better aging, better choice of colors, less marking

15.3.1 Substitution of materials in the automobile industry

The automobile is a good example of effective substitution of materials in order to achieve one or more of the objectives just listed. Several examples are:

a) All or part of the metal body replaced by plastic or reinforced plastic parts.
b) Metal bumpers, gears, pumps, fuel tanks, housings, covers, clamps, and many other similar components replaced by plastic substitutes.
c) Metal engine components replaced by ceramic and reinforced plastic parts.
d) Composite-material driveshafts used in place of all-metal driveshafts.
e) Various other substitutions, such as cast-iron to cast-aluminum engine blocks, forged to cast crankshafts, and forged to cast, powder-metallurgy, or composite-material connecting rods.

The automobile industry is a major consumer of both metallic and nonmetallic materials. Consequently, there is constant competition among suppliers, particularly in the steel, aluminum, and plastics industries. Automotive industry engineers and management are constantly investigating the advantages and limitations of these principal materials with regard to their applications and costs.

15.3.2 Substitution of materials in the aircraft industry

In the aircraft and aerospace industries, conventional aluminum alloys (2000 and 7000 series) are being replaced by aluminum–lithium alloys and titanium alloys because of their higher strength-to-weight ratios. Forged parts are being replaced with powder-metallurgy parts that are manufactured with better control of impurities and microstructure, which enhance the overall economics of the substitutions. Furthermore, the P/M parts require less machining and produce less scrap of expensive materials. Advanced composite materials and honeycomb structures are replacing traditional aluminum airframe components (Figs. 15.2 and 15.3), and metal-matrix composites are replacing some of the aluminum and titanium structural components.

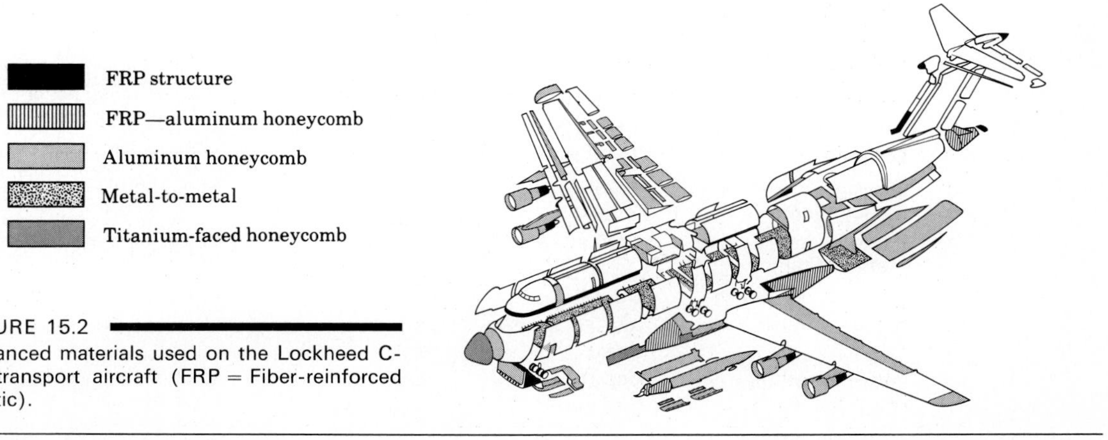

FIGURE 15.2 Advanced materials used on the Lockheed C-5A transport aircraft (FRP = Fiber-reinforced plastic).

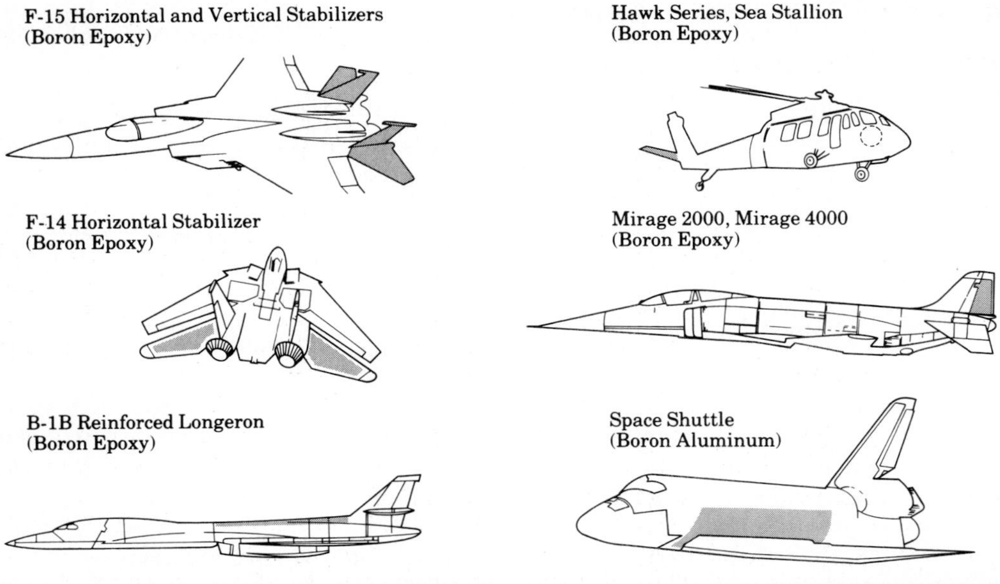

FIGURE 15.3
Boron epoxy and boron aluminum composite materials in aerospace applications. *Source:* AVCO Specialty Materials Division.

• Example 15.3: Material changes from C-5A to C-5B military cargo aircraft.

The accompanying table shows the changes made in materials for various components of the new aircraft and the reasons for the changes. *Source:* H. B. Allison, Lockheed-Georgia.

ITEM	C-5A MATERIAL	C-5B MATERIAL	REASON FOR CHANGE
Wing panels	7075-T6511	7175-T73511	Durability
Main frame			
Forgings	7075-F	7049-01	Stress corrosion resistance
Machined frames	7075-T6	7049-T73	
Frame straps	7075-T6 plate	7050-T7651 plate	
Fuselage skin	7079-T6	7475-T61	Material availability
Fuselage underfloor end fittings	7075-T6 forging	7049-T73 forging	Stress corrosion resistance
Wing/pylon attach fitting	4340 alloy steel	PH13-8Mo	Corrosion prevention
Aft ramp lock hooks	D6-AC	PH13-8Mo	Corrosion prevention
Hydraulic lines	AM350 stainless steel	21-6-9 stainless steel	Improved field repair
Fuselage failsafe straps	6Al-4V titanium	7475-T61 aluminum	Titanium strap disbond

•

15.4 Selection of Manufacturing Processes

In this section we describe the importance of proper manufacturing process and machinery selection, and how the selection process relates to characteristics of materials, tolerances, surface finishes obtained, and cost. As we have pointed out, many traditional manufacturing processes have been automated and are being computer controlled to optimize the processes. Computerization is also effectively increasing product reliability and quality and reducing labor costs.

The choice of a manufacturing process is dictated by various considerations, such as:

a) Characteristics and properties of the workpiece material.
b) Shape, size, and thickness of the part.
c) Tolerances and surface finish requirements.
d) Functional requirements for the expected service life of the component.
e) Production volume (quantity).
f) Level of automation required to meet production volume and rate.
g) Costs involved in individual and combined aspects of the manufacturing operation.

We have shown that some materials can be processed at room temperature, but others require elevated temperatures, which means additional furnaces and appropriate tooling. Some materials are soft and ductile, whereas others are hard, brittle, and abrasive, thus requiring special processing techniques and tool and die materials. Different materials have different manufacturing characteristics, such as castability, forgeability, workability, machinability, and weldability. Few materials have the same favorable characteristics in all categories. For example, a material that is castable or forgeable may later present problems in machining, grinding, or finishing operations that may be required in order to produce a product with acceptable surface finish and dimensional accuracy.

• Example 15.4: Process selection in making a part.

Assume that you are asked to make the part shown in the accompanying figure. You should first determine the part's function, the type of loads and environment to which it is to be subjected, the tolerances and surface finish required, and so on. For the sake of discussion, let's assume that the part is round, 125 mm (5 in.) long and that the large and small diameters are 38 mm and 25 mm (1.5 in. and 1.0 in.), respectively. Let's further assume that, for functional requirements such as stiffness, hardness, and resistance to elevated temperatures, this part should be made of metal.

Which manufacturing process would you choose, and how would you organize the production facilities to manufacture a cost-competitive, high-quality product?

We have stated several times that parts should be produced at or near their final shape (net or near-net-shape manufacturing). This approach eliminates much secondary processing, such as machining, grinding, and other finishing operations, reduces total manufacturing time, and thus lowers the cost.

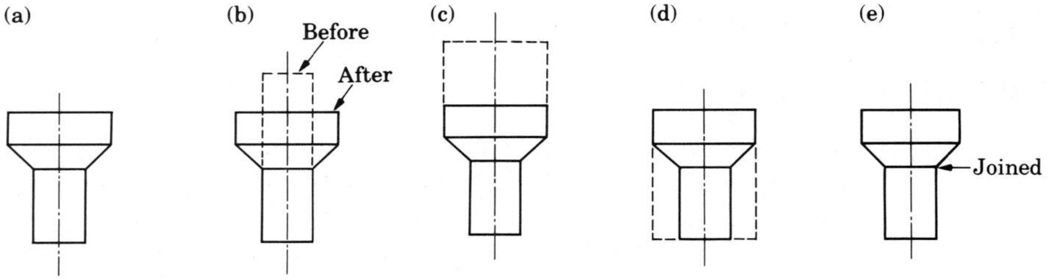

This part is relatively simple and could be suitably manufactured by different methods. You can readily see that the part can be made by (a) casting and powder metallurgy, (b) upsetting, (c) extrusion, (d) machining, or (e) joining two pieces together. For net-shape processing, the two logical processes are casting and powder metallurgy, each process having its own characteristics and various tooling and labor costs. This part can also be made by cold, warm, or hot forming. One method is by upsetting a 25-mm (1-in.) diameter round bar in a suitable die to form the larger end. The partial extrusion of a 38-mm (1.5-in.) diameter bar to reduce the diameter to 25 mm is another possibility. Each of these processes shapes the material without any waste.

This part can also be made by machining a 38-mm diameter bar stock to obtain the 25-mm diameter section. However, machining will take much longer than forming, and some material will be wasted as metal chips. On the other hand, unlike net-shape processes, which require special dies, machining does not require special tooling, and the operation can easily be carried out on a lathe. Note also that you can make this part in two pieces and then join them by welding, brazing, or adhesive bonding.

Each process has its own characteristics, production rate, surface finish, dimensional accuracy, and final product properties. Note that all these processes require materials in different shapes, ranging from metal powders to round bars. Because of the various operations involved in producing the materials, costs depend not only on the type of material (ingot, powder, drawn rod, extrusion), but also on its size and shape. Per unit weight, square bars are more expensive than round bars, and cold-rolled plate and sheet are more expensive than hot-rolled plate and sheet. Also, hot-rolled bars are much less expensive than powders of the same metal.

By now, you should realize that even a simple part such as this requires considerable thought before you can finally select a process and initiate a manufacturing plan. In addition to technical requirements, process selection also depends on factors such as required production quantity and rate, as we describe in the next section. Summarizing our discussion, we can say that if only a few parts are needed, machining this part is the economical method. However, as the production quantity increases, producing this part by a heading operation or by cold extrusion would be the proper choice. Method (e) would be the most appropriate choice if the top and bottom portions of this part were made of different metals.

15.5 Process Capabilities

We have shown that all manufacturing processes have certain advantages and limitations. Casting and injection molding, for example, can generally produce more complex shapes than can forging and powder metallurgy. The reason? Because the molten metal or plastic is capable of filling complex mold cavities. On the other hand, forgings generally can be made into complex shapes by additional machining and finishing operations, and have toughnesses that are generally superior to castings and powder metallurgy products.

Recall that the shape of a product may be such that it should be fabricated from several parts, then joined with fasteners or by brazing, welding, or adhesive bonding techniques. The reverse may be true for other products, whereby manufacturing them in one piece might be more economical because of significant assembly-operation costs.

Other factors that you have to consider in process selection are the minimum section size or dimensions that can be satisfactorily produced (Fig. 15.4). For

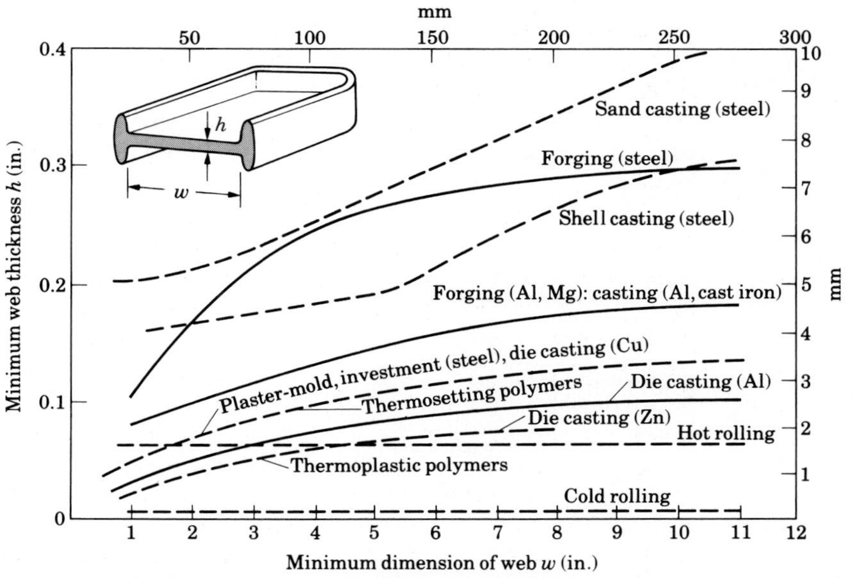

FIGURE 15.4

Process capabilities for minimum part dimensions. *Source:* J. A. Schey, *Introduction to Manufacturing Processes,* 2d ed. New York: McGraw-Hill, 1987.

example, very thin sections can be obtained by cold rolling, but processes like sand casting or forging prohibit forming such thin sections.

15.5.1 Tolerances and surface finish

Tolerances and surface finishes produced are important aspects of manufacturing. They have a particular bearing on subsequent assembly operations, proper operation of various machines and instruments, and consistency of overall appearance. The ranges of surface finishes and tolerances obtained in manufacturing processes are given in various chapters and summarized in Figs. 15.5 and 15.6. We have shown that in order to obtain finer surface finishes and closer tolerances, additional finishing operations, better control of processing parameters, and the use of higher quality equipment may be required.

The closer the tolerance required, the higher the cost of manufacturing will be (Fig. 15.7); the finer the surface finish required, the longer manufacturing will take, increasing the cost (Fig. 15.8). In machining aircraft structural members made of titanium alloys, for example, as much as 60 percent of the cost of machining the part is consumed in the final machining pass in order to hold proper tolerances and surface finishes. Unless specifically required with proper technical and economic

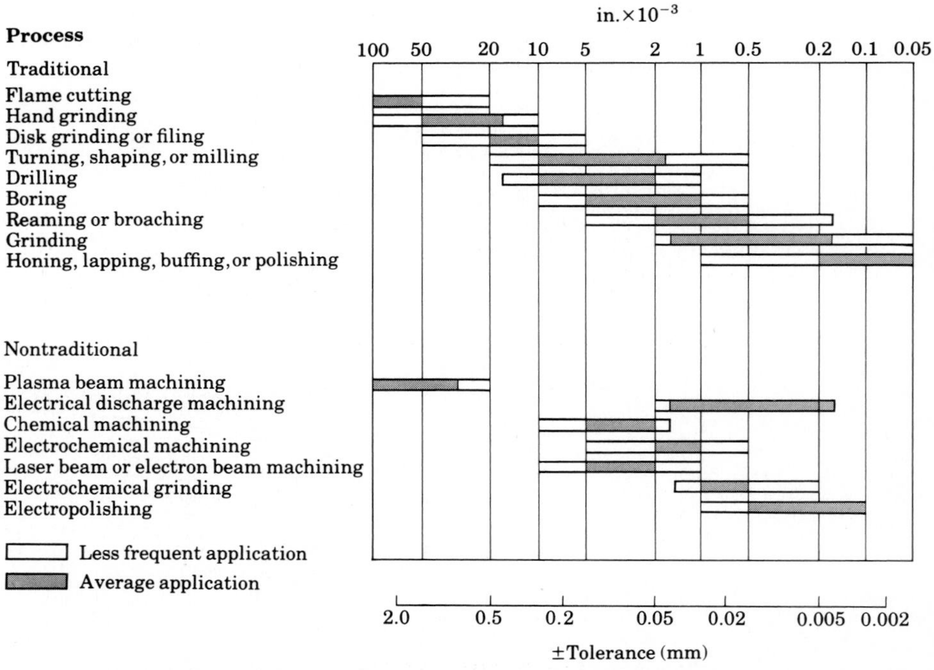

FIGURE 15.5
Tolerances produced by various processes.

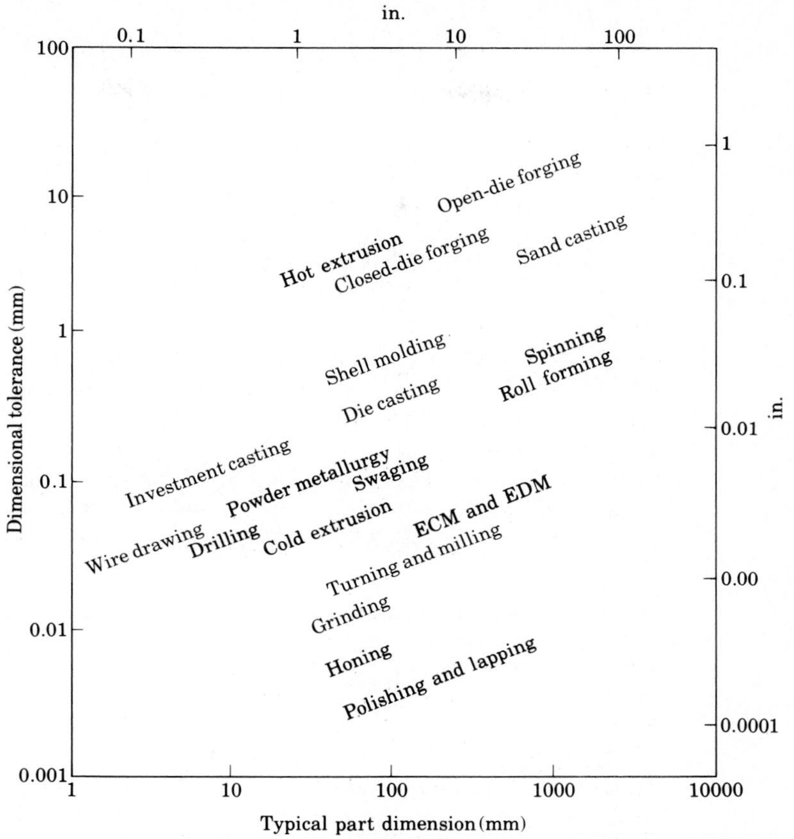

FIGURE 15.6

Tolerances as a function of component size for various manufacturing processes. Note that, because many factors are involved, there is a broad range for tolerances.

FIGURE 15.7

Relationship between relative cost and tolerance.

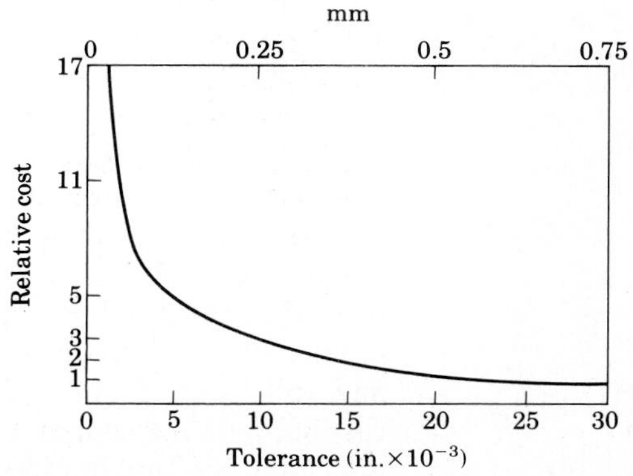

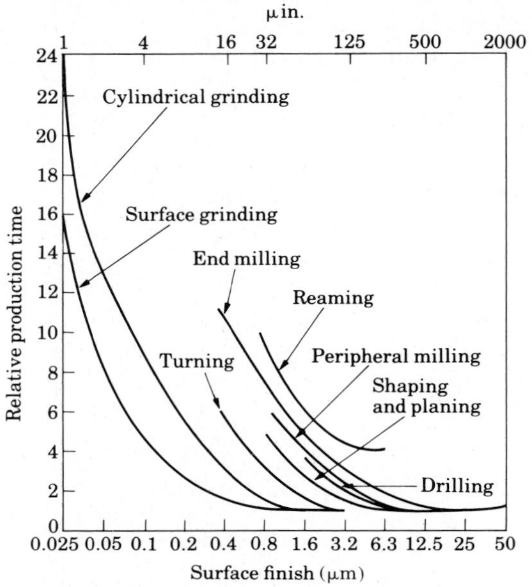

FIGURE 15.8 Relative production time as a function of surface finish produced by various manufacturing methods. *Source: American Machinist.*

justification, parts should be made with as coarse a surface finish and with as wide a tolerance as functionally and aesthetically acceptable. Thus the importance of interaction and continuous communication between designer and manufacturing engineer is obvious.

15.5.2 Production volume and rate

Depending on the type of product, the *production volume*, or *quantity* (lot size), can vary greatly. For example, paper clips, bolts, washers, spark plugs, bearings, and ball-point pens are produced in very large quantities. On the other hand, jet engines for large commercial aircraft, diesel engines for locomotives, and propellers for ocean liners are manufactured in limited quantities. Production quantity plays a significant role in process and equipment selection. In fact, an entire manufacturing discipline is devoted to determining mathematically the optimum production quantity for a specific process, called the *economic order quantity*.

A significant factor in manufacturing process selection is the *production rate*, defined as the number of pieces to be produced per unit of time, such as per hour, month, or year. Processes such as powder metallurgy, die casting, deep drawing, and roll forming are high-production-rate operations. On the other hand, sand casting, conventional and electrochemical machining, spinning, superplastic forming, adhesive and diffusion bonding, and processing of reinforced plastics are relatively slow operations. These rates can be increased by automation or the use of multiple machines. However, a slower production rate does not necessarily mean that the manufacturing process is inherently uneconomical.

• Example 15.5: Economical quantities for different production methods.

The accompanying figure shows three different methods for making a type 347 stainless steel threaded cap. Although the basic shape of the cap was the same, each method required a different design. The original order for this part was 100 caps. After a brief analysis, it was obvious that the most economical and quickest method of production was machining the cap from a solid round bar stock, as shown in (a). When the order was increased to 1000 caps, it was determined that a two-piece assembly might be more economical. The threaded part was machined from a smaller bar stock and the other piece was made of 0.0625-in. thick sheet, stamped and pressed on the machined part, as shown in (b). It was found, however, that the cost per piece was not significantly different from that of method (a).

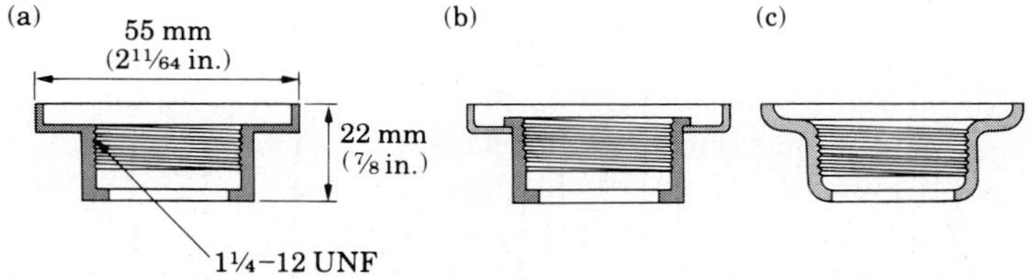

The order was increased to 5000 caps, which called for a further review of the cap design. It was determined that an economical means of producing this cap in larger quantities was to make it in one piece, by first press forming (drawing) a 0.093-in. thick sheet and then machining the threads. The sheet had to be thicker than in method (b) because of the required threading. Because of the smaller amount of material removed by machining and the speed with which press forming could be done, both material and labor costs in this method were lower than those for the other two methods. As a result, the manufacturing cost per cap was reduced by nearly 50 percent. *Source:* ASM International.

15.5.3 Lead time

The choice of a process is greatly influenced by the time required to start production, or *lead time*. Processes such as forging, extrusion, die casting, roll forming, and sheet-metal forming may require extensive—and expensive—dies and tooling. In contrast, most machining and grinding processes have an inherent flexibility, and utilize tooling that can be adapted to many requirements in a relatively short time. Recall also our discussion of machining centers, flexible manufacturing cells, and flexible manufacturing systems, which are capable of responding quickly and effectively to product changes both in type and quantity.

15.6 Manufacturing Costs

We have shown in various chapters that economics is one of the most important aspects of any manufacturing operation. In this section we describe the various costs involved in manufacturing a product in order to identify those factors that can help minimize them, while maintaining quality.

Designing and manufacturing a product according to certain specifications and to meet service requirements is only one aspect of production. In order for the product to be successfully marketed, its cost must be competitive with similar products in the marketplace. Although different methods are used to account for product cost, the total cost consists of (a) materials costs, (b) tooling costs, (c) labor costs, (d) fixed costs, and (e) capital costs. We described *material costs* sufficiently in Section 15.2.

15.6.1 Tooling costs

Tooling costs are the costs involved in making the tools, dies, molds, patterns, and special jigs and fixtures necessary for manufacturing a product or component. The tooling cost is greatly influenced by the production process selected. For example, if a part is to be made by casting, the tooling cost for die casting is higher than that for sand casting. Similarly, the tooling cost in machining or grinding is much lower than that for powder metallurgy, forging, or extrusion. In machining operations, the choice of cutting tool materials can be significant. For example, carbide tools are more expensive than high-speed steel tools, but tool life is longer.

If a part is to be manufactured by spinning, the tooling cost for conventional spinning is much lower than that for power spinning. Tooling for rubber forming processes is less expensive than that for male-and-female die sets used for drawing and stamping of sheet metals. High tooling costs, on the other hand, can be justified for high-volume production of a single item. As we have stated previously, the expected life of tools and dies, and their obsolescence because of product changes, are also important considerations.

15.6.2 Labor costs

Labor costs are generally divided into direct and indirect costs. The *direct labor cost* is for the labor directly involved in manufacturing the part (productive labor). This cost includes all labor from the time materials are first handled to the time the product is finished. This time is generally referred to as *floor-to-floor time*. For example, a machine operator picks up a round bar from a bin, machines it to the shape of a threaded rod, and returns it to another bin. The direct labor cost is calculated by multiplying the labor rate (hourly wage, including benefits) by the time that the worker spends producing the part.

The time required for producing a part depends not only on required part size, shape, and dimensional accuracy, but also on the workpiece material. For example, with regard to cutting speeds we pointed out that for high-temperature alloys they are lower than those for aluminum or plain-carbon steels. Consequently, the cost of machining certain aerospace materials is much higher than that for the more common alloys.

Indirect labor costs are those that are involved in servicing the total manufacturing operation. This cost is comprised of activities such as supervision, repair, maintenance, quality control, engineering, research, sales, and also generally includes the cost of office staff. Because these costs do not contribute directly to the production of finished parts or are not chargeable to a specific product, they are referred to as *overhead*, or the *burden rate*, which is charged proportionally to all products. The personnel involved are categorized as nonproductive labor.

15.6.3 Fixed costs

Fixed costs include the costs of power, fuel, taxes on real estate, rent, and insurance, and capital, including depreciation and interest. The company would have to pay these costs regardless of whether it made a particular product. Thus fixed costs are not sensitive to production volume. *Capital costs* represent the capital investment in land, buildings, machinery, and equipment and represent major expenses for most manufacturing operations.

Let's assume that a company has decided to begin manufacturing a variety of valves. A new plant has to be built, or an old plant remodeled, with all the necessary machinery, support equipment, and facilities. To cast and machine the valve bodies, melting furnaces, casting equipment, machine tools, quality control equipment, and other related equipment and machinery of various sizes and types have to be purchased. If valve designs are to be changed often, or if product lines vary greatly, the manufacturing equipment must have sufficient flexibility to accommodate these requirements. Machining centers and flexible manufacturing systems are especially suitable for this purpose. The equipment and machinery we have just listed are capital cost items and, as you can appreciate, require major investment.

In view of generally high equipment costs, particularly those involving flexible manufacturing systems, high production rates are often required to justify large expenditures and hold product unit cost at a competitive level. Lower unit costs can be achieved by continuous production, involving round-the-clock operation, so long as demand warrants. Proper equipment maintenance is essential to ensure high productivity. Any breakdown of machinery (*downtime*) can be very expensive, ranging from a few hundred dollars to thousands of dollars per hour.

15.6.4 Relative costs

The costs we have described are interrelated, with *relative costs* depending on many factors. Consequently, the unit cost of the product can vary widely. For example, some parts may be made from expensive materials but require very little processing, such as minting gold coins. Thus the cost of materials relative to direct labor costs is high. On the other hand, some products may require several production steps to process relatively inexpensive materials, such as carbon steels. An electric motor, for example, is made of relatively inexpensive materials, yet many different manufacturing operations are involved in making the housing, rotor, bearings, brushes, and other components of the motor. In such cases, assembly operations can become a significant portion of the overall cost.

Undoubtedly you have noted that, over a period of time, the costs of products such as calculators, computers, and digital watches have decreased, whereas the costs of other products such as automobiles, aircraft, houses, and books have gone up. These differences result from the rates of change in various costs over time, including labor, machinery, materials, market demands, domestic and international competition, and worldwide economic trends (demand, exchange rates, and tariffs), and the inevitable impact of computers on all aspects of manufacturing.

• Example 15.6: Cost comparison in making gears by machining and powder metallurgy.

The oil pump gear for a truck is made of steel and requires a tensile strength of 700 MPa (100,000 psi) and hardness of 260–302 HB or equivalent. The gear is 40 mm (1.56 in.) in diameter and 50 mm (2 in.) long, and has a semicircular keyway.

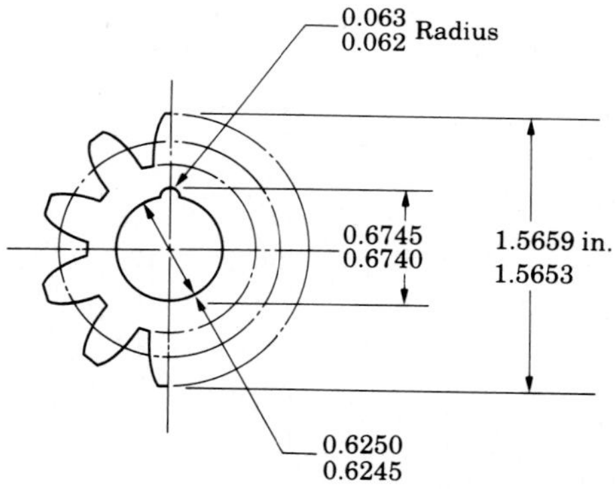

Full dimensions and tolerances are shown in the accompanying figure. This gear can be produced by machining or by powder-metallurgy techniques. The following table shows cost comparisons between the two processes, using a cost base of 100 for a hobbed gear. The analysis indicates that it is much more economical to produce this gear by P/M techniques. Note that the hobbing operation constitutes the largest portion of the total cost of making the gear by machining. In the P/M method, the largest cost is the cost of the metal powder. *Source:* S. S. McGee and F. K. Burgess, *International Journal of Powder Metallurgy and Powder Technology*, October 1976.

HOBBED GEAR	PERCENT OF TOTAL	P/M GEAR	PERCENT OF HOBBED TOTAL
Material:		Material:	
SAE 1045, including 2% setup scrap and 46% chips	15.10	MPIF FC-0208-S (7.0 g/cm^3 density) 5% scrap	9.97
Operations:		Operations:	
Bar chuck, cut off and bore	8.49	Compact (100-ton press)	2.37
Broach keyway	3.17	Sinter	2.56
Hob teeth	47.50	Harden	1.92
Harden	1.92	Grind ends perpendicular to pitch diameter	5.93
Grind ends perpendicular to pitch diameter	5.93	Deburr	0.53
Deburr	0.53	Inspect	0.26
Inspect	0.26	Perishable tools and gages, per piece	8.19
Perishable tools and gages, per piece	17.10		31.73
	100.00		

15.6.5 Manufacturing costs and production volume

One of the most significant factors in manufacturing costs is production volume. Large production volumes require high production rates. High production rates require the use of mass-production techniques, involving special machinery and proportionately less labor and plants operating on two or three shifts per day. On the other hand, small production volumes usually mean larger direct labor involvement.

Figure 15.9 shows a downward trend in the unit-cost curves as production volume increases. This result reflects proportionately lower tooling and capital costs per piece. Once the tooling is made, a greater number of produced parts will lower the cost of tooling per part. Note in Fig. 15.10 that at low production volumes, material costs are low relative to direct labor costs, whereas at high volumes the opposite is true.

As we described in Section 13.2.1, *small-batch production* is usually done on general-purpose machines, such as lathes, milling machines, and hydraulic presses. The equipment is versatile, and parts with different shapes and sizes can be produced by appropriate changes in the tooling. However, direct labor costs are high because these machines are usually operated by skilled labor.

For larger quantities (*medium-batch production*), these same general-purpose machines can be equipped with various jigs and fixtures, or they can be computer

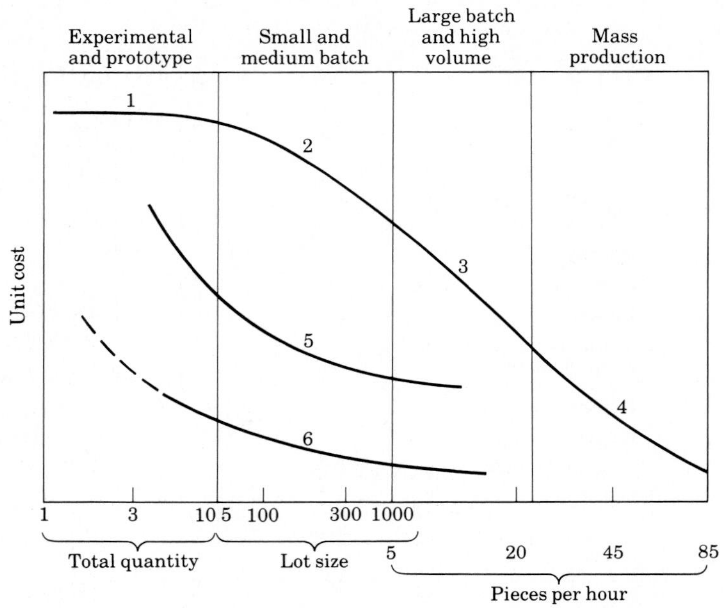

FIGURE 15.9 Unit cost of product as a function of production method and quantity. 1 = tool-room machinery; 2 = general-purpose machine tools; 3 = special-purpose machines; 4 = automatic transfer lines; 5 = NC and CNC machine tools; and 6 = computer-integrated manufacturing systems. *Source:* Cincinnati Milacron, Inc.

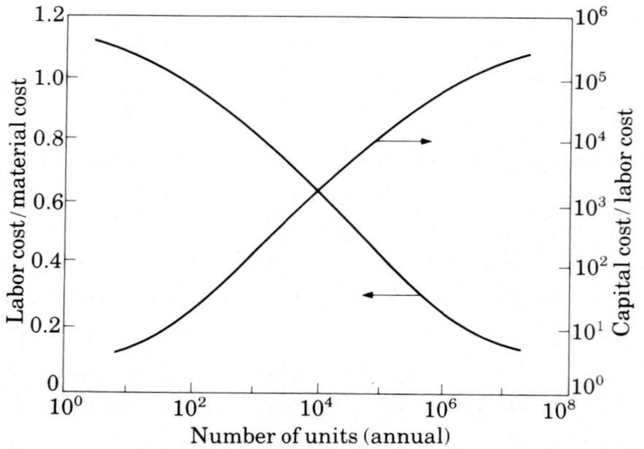

FIGURE 15.10 Relative cost of labor, materials, and capital as a function of annual production volume. Note the difference in the scales for the two ordinates in the figure. *Source:* After N. P. Suh.

controlled. To reduce labor costs further, machining centers and flexible manufacturing systems have been developed. For quantities of 100,000 and higher, the machines involved are designed for specific purposes (*dedicated machines*) and perform a variety of specific operations with very little labor involved.

15.7 Value Engineering

We have identified several areas of activity in manufacturing where cost reduction is possible. As we stated in Chapter 1, manufacturing adds value to materials as they become discrete products and are marketed. Because this value is added in individual steps during creation of the product, utilization of value engineering (value analysis, value control, and value management) is important.

Value engineering is a system that evaluates each step in design, materials, processes, and operations so as to manufacture a product that performs its intended functions and has the lowest possible cost. A monetary value is established for each of two product attributes: (a) use value, reflecting the functions of the product; and (b) esteem or prestige value, reflecting the attractiveness of the product that makes its ownership desirable. The *value* of a product is then defined as the ratio of product function and performance to the cost of the product. Thus the goal of value engineering is to obtain maximum performance per unit cost.

Value engineering is an important and all-encompassing, interdisciplinary activity. It is usually coordinated by a value engineer and conducted jointly by designers, engineers, quality control, purchasing, and marketing personnel, and managers. In order for value engineering to be effective, it requires the full support of the company's top management. Implementation of value engineering in manufacturing facilities has resulted in benefits such as significant cost savings, reduced lead times, better product quality and performance, reduced product weight and size, and shorter manufacturing times. To properly assess the value of each step in manufacturing a product, several groups of questions have to be asked.

Product design

- Can the design be simplified without adversely affecting its functions? Have all alternative designs been investigated? Can unnecessary features or some components be eliminated or combined with others? Can the design be made lighter and smaller?
- Are the dimensional tolerances and surface finish specified necessary? Can they be relaxed?
- Will the product be difficult to assemble and take apart for maintenance? Is the use of fasteners minimized?
- Does each part have to be manufactured in the plant? Are some parts commercially available as standard products from outside sources?

Materials

- Do the materials selected have properties that are much above minimum requirements and specifications?
- Can some materials be replaced by others that are cheaper?
- Do the materials selected have the proper manufacturing characteristics?
- Are the materials to be ordered available in standard sizes, dimensions, surface finish, and tolerances?
- Is material supply reliable? Are there likely to be significant price fluctuations?

Manufacturing processes

- Have all alternative manufacturing processes been investigated?
- Are the methods chosen economical for the type of material, shape to be produced, and desired production rate? Can requirements for tolerances, surface finish, and product quality be met consistently?
- Can the part be formed and shaped to final dimensions rather than made by material-removal processes? Are machining, secondary processes, and finishing operations required? Are they necessary?
- Is tooling available in the plant? Can it be purchased as standard items?
- Is scrap produced? If so, what is the value of the scrap?
- Are processing parameters optimized? Have all the automation and computer-control possibilities been explored for all phases of the manufacturing operation? Can group technology be implemented for parts with similar geometric and manufacturing attributes?
- Are inspection techniques and quality control being implemented properly?

SUMMARY

Competitive aspects of production and costs are among the most significant considerations in manufacturing. Regardless of how well a product meets design specifications and quality standards, it must also meet economic criteria in order to be competitive in the domestic and international marketplace. The total cost of a product includes several costs, such as costs of materials, tooling, capital, labor, and overhead. Materials costs can be reduced through careful selection so that the least costly material can be identified and selected, while maintaining design and service requirements, functions, and specifications for good product quality.

Substitution of materials, modification of designs, and relaxing of tolerance and surface finish requirements are important methods of cost reduction. Certain guidelines for designing products for economic production have been established. Although labor costs are becoming a small percentage of product costs, they can be reduced further through the use of automated and computer-controlled machinery Automation requires significant capital expenditures; however, a well-planned production facility can significantly reduce labor costs and improve productivity.

BIBLIOGRAPHY

Bralla, J.G. (ed.), *Handbook of Product Design for Manufacturing—A Practical Guide for Low-Cost Production.* New York: McGraw-Hill, 1986.

Charles, J.A., and F.A.A. Crane, *Selection and Use of Engineering Materials.* Stoneham, Mass.: Butterworths, 1984.

Chow, W.W., *Cost Reduction in Product Design.* New York: Van Nostrand Reinhold, 1978.

Dieter, G.E., *Engineering Design.* New York: McGraw-Hill, 1983.

Enrick, N.L., and H.E. Mottley, *Manufacturing Analysis for Productivity and Quality/Cost Enhancement*, 2d ed. New York: Industrial Press, 1983.

Malstrom, E.M., *What Every Engineer Should Know About Manufacturing Cost Estimating.* New York: Marcel Dekker, 1981.

Malstrom, E.M. (ed.), *Manufacturing Cost Engineering Handbook.* New York: Marcel Dekker, 1984.

Manufacturing Engineering Management, 4th ed., *Vol. 5: Tool and Manufacturing Engineers Handbook.* Dearborn, Mich.: Society of Manufacturing Engineers, 1988.

Michaels, J.V., *Design to Cost.* New York: Wiley, 1989.

Ostwald, P.F., *Cost Estimating*, 2d ed. Englewood Cliffs, N.J.: Prentice-Hall, 1984.

Smolik, D.P., *Material Requirements of Manufacturing.* New York: Van Nostrand Reinhold, 1983.

Suh, N.P., *The Principles of Design.* New York: Oxford University Press, 1990.

Trucks, H.E., and G. Lewis (eds.), *Designing for Economical Production*, 2d ed. Dearborn, Mich.: Society of Manufacturing Engineers, 1987.

QUESTIONS

15.1 Why is a knowledge of available shapes of materials important? Give five specific examples.

15.2 Is it always desirable to purchase stock that is close to the final dimensions of a part to be manufactured? Explain, giving some examples.

15.3 Describe the potential problems involved in reducing the quantity of materials in products.

15.4 Why has material substitution been crucial in automotive and aerospace industries?

15.5 Discuss your thoughts concerning the replacement of aluminum beverage cans with steel cans.

15.6 What is meant by process capabilities? Select four different and specific manufacturing processes and describe their capabilities.

15.7 Why is production volume significant in process selection?

15.8 Explain why the larger the quantity per package of food products, the lower the cost per unit weight is.

15.9 Explain why the value of the scrap produced in a manufacturing process depends on the type of material.

15.10 Comment on the magnitude and range of scrap shown in Table 15.3.

15.11 Describe the difference between the cost and the price of a product. Are customers likely to pay more for a product that has good styling and design, even though these features do not add significantly to the function and performance of the product? Explain with appropriate examples.

15.12 Explain how the high cost of some of the metalworking machinery described in this text can be justified.

15.13 Other than size, what factors are involved in the range of machine prices?

15.14 Explain the reasons for the relative positions of the curves shown in Fig. 15.4.

15.15 What factors are involved in the shape of the curve shown in Fig. 15.7?

15.16 How do you explain the trends in Fig. 15.9?

15.17 Why does the abscissa in Fig. 15.9 change from quantity or lot size to production rate?

15.18 Discuss your observations concerning Fig. 15.10.

15.19 Discuss the trade-offs involved in selecting one of the two materials for each application listed. Also discuss typical conditions to which these products are subjected in their normal use.

a) Steel versus plastic paper clips.
b) Forged versus cast crankshafts.
c) Forged versus powder-metallurgy connecting rods.
d) Plastic versus sheet metal light-switch plates.
e) Glass versus metal water pitchers.
f) Sheet metal versus cast hub caps.
g) Steel versus copper nails.
h) Wood versus metal handles for hammers.
i) Sheet metal versus reinforced plastic chairs.

15.20 Discuss the manufacturing process or processes that are suitable for making the products listed in Problem 15.19. Explain whether they would need additional operations such as coating, plating, heat treating, and finishing. If so, make recommendations and give the reasons for them.

15.21 Assuming that the starting material (stock) is a round rod and only one specimen is needed, choose a process and the machinery by which a tensile-test specimen can be made. Discuss their relative advantages and limitations. Describe how the process you selected may be changed for economical production as the number of specimens to be made increases.

15.22 Discuss fully the factors that influence the choice between the following pairs of processes.

a) Sand casting versus die casting a fractional electric-motor housing.
b) Machining versus forming a large gear.
c) Forged versus powder-metallurgy gear.
d) Casting versus stamped sheet-metal frying pan.
e) Aluminum tubing versus cast iron outdoor furniture.
f) Welded versus cast machine tool structures.
g) Thread rolled versus machined bolt for high strength application.
h) Thermoformed plastic versus molded thermoset fan blade for household use.

15.23 Comment on the differences, if any, between the designs, materials, and processing and assembly methods used for making products such as hand tools and ladders for professional versus consumer use.

15.24 Structural sections such as I-beams are usually made by shape rolling as shown in Fig. 6.47. Comment on the influence of the size of the cross-section of the beam and the quantity required on the production method selected.

PROBLEMS

15.1 Make suggestions as to how to reduce the dependence of production time on surface finish, as shown in Fig. 15.8.

15.2 Select three different products and make a survey of the change in their prices over the past 10 years. Discuss the reasons why their prices have changed.

15.3 In Table 15.1 we have listed several materials and their commercially available shapes. By contacting suppliers of materials, complete this list to include the following materials: (a) titanium, (b) superalloys, (c) lead, (d) tungsten, and (e) amorphous metals.

15.4 Select three products commonly found in your home. State your opinions as to (a) what materials were used in the product and why and (b) how they are made and why those particular processes were employed.

15.5 Inspect the components under the hood of your automobile. Identify three parts each that have been produced to (a) net-shape and (b) near-net-shape condition. Comment on the design and production aspects of these parts and how the manufacturer achieved the net-shape condition.

15.6 The double-flare V-groove weld is used to weld two solid cylindrical bars along their length. Assuming that the material is carbon steel, comment on whether you would still use this process if the diameter of the bars are 2 mm (0.080 in.) each. If not, which processes could be used to make it? If the material is changed to copper, would your answer be different? Explain.

15.7 Several methods can be used to make the metal part shown. List these methods, and for each method, give (a) the details concerning the equipment needed, (b) estimated production time per part, and (c) amount of scrap produced.

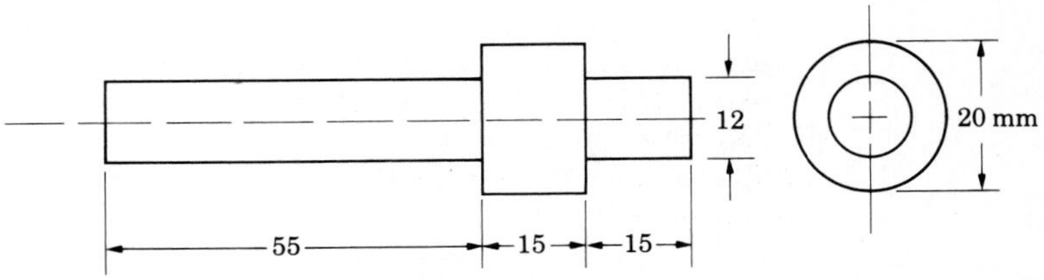

15.8 Review the manufacturing processes for the part in Problem 15.7 and explain which process would produce a part that (a) may not have sufficient fatigue strength, (b) may not resist high temperatures, (c) would require additional finishing operations, and (d) would require the largest equipment.

15.9 Assume that the part shown in Problem 15.7 is made of (a) thermoplastic and (b) thermosetting plastic. Explain which processes would be suitable for each of these materials.

15.10 The sheet-metal part shown is made of steel. Consider either press-brake forming or contour roll forming as the alternative processes to form this part. Discuss how this part can be formed by either of the processes, the design and materials for tooling for

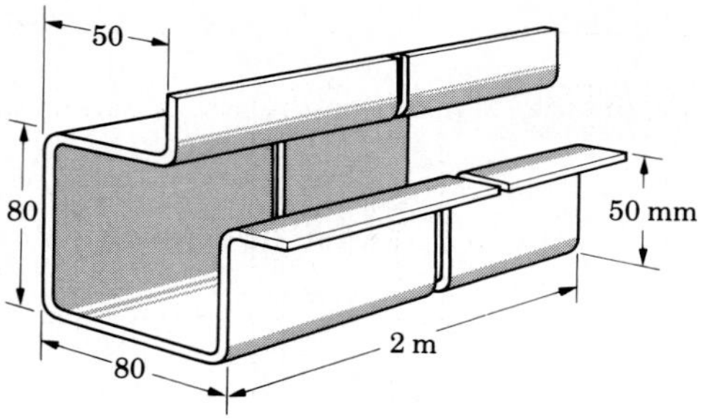

each process, and how your selection of a process may be changed as (a) the length of the part increases and (b) the number of parts required increases.

15.11 The part shown is a carbon-steel segment gear. The small hole is for clamping the part on a spindle with a bolt and nut. When at first the demand was low, this part was made by machining from bar stock. As the demand increased, the process was changed to extrusion. (a) Describe the procedure for machining this part from bar stock and the processes and machines you would recommend. (b) Explain how this part can be extruded. (c) Would you produce the slot at the bottom during extrusion by proper die design, or would you prefer to slit it after the part is extruded? How would you produce the slot on a production basis? Explain the relative advantages and limitations.

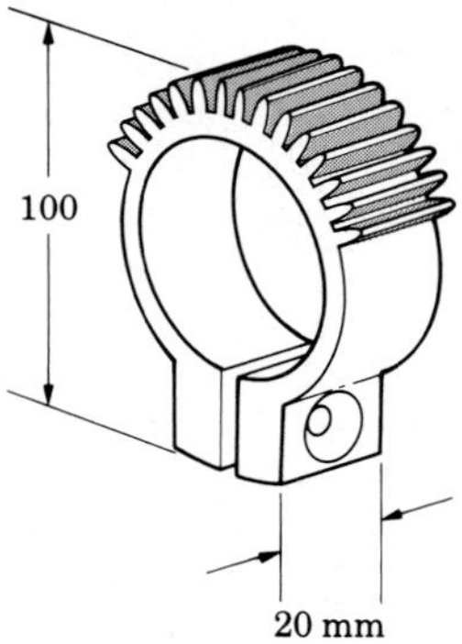

15.12 The part shown is a cable terminal made of 304 stainless steel. The cable is slipped into the hole on the left, and the cylindrical end of the part is swaged over the cable, securing the cable. Assuming that the manufacturing facility has several types of equipment and that the quantity needed is 500, suggest processes by which you can manufacture this part. Discuss details of each process recommended. Is swaging the only method by which you can attach the cable? Discuss other alternatives.

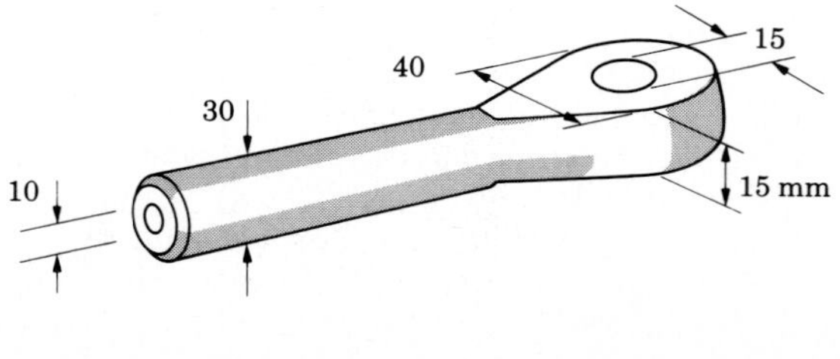

Index